Atomic Numbers and Atomic Masses
of the Elements

Based on $^{12}_{6}C$. Numbers in parentheses are the mass numbers of the most stable isotopes of radioactive elements.

Element	Symbol	Atomic Number	Atomic Mass	Element	Symbol	Atomic Number	Atomic Mass
Actinium	Ac	89	(227)	Mendelevium	Md	101	(260)
Aluminum	Al	13	26.98	Mercury	Hg	80	200.59
Americium	Am	95	(243)	Molybdenum	Mo	42	95.94
Antimony	Sb	51	121.76	Neodymium	Nd	60	144.24
Argon	Ar	18	39.95	Neon	Ne	10	20.18
Arsenic	As	33	74.92	Neptunium	Np	93	(237)
Astatine	At	85	(210)	Nickel	Ni	28	58.69
Barium	Ba	56	137.33	Niobium	Nb	41	92.91
Berkelium	Bk	97	(247)	Nitrogen	N	7	14.01
Beryllium	Be	4	9.01	Nobelium	No	102	(259)
Bismuth	Bi	83	208.98	Osmium	Os	76	190.23
Bohrium	Bh	107	(264)	Oxygen	O	8	16.00
Boron	B	5	10.81	Palladium	Pd	46	106.42
Bromine	Br	35	79.90	Phosphorus	P	15	30.97
Cadmium	Cd	48	112.41	Platinum	Pt	78	195.08
Calcium	Ca	20	40.08	Plutonium	Pu	94	(242)
Californium	Cf	98	(251)	Polonium	Po	84	(209)
Carbon	C	6	12.01	Potassium	K	19	39.10
Cerium	Ce	58	140.12	Praseodymium	Pr	59	140.91
Cesium	Cs	55	132.91	Promethium	Pm	61	(145)
Chlorine	Cl	17	35.45	Protactinium	Pa	91	(231)
Chromium	Cr	24	52.00	Radium	Ra	88	(226)
Cobalt	Co	27	58.93	Radon	Rn	86	(222)
Copper	Cu	29	63.55	Rhenium	Re	75	186.21
Curium	Cm	96	(248)	Rhodium	Rh	45	102.91
Dubnium	Db	105	(262)	Rubidium	Rb	37	85.47
Dysprosium	Dy	66	162.50	Ruthenium	Ru	44	101.07
Einsteinium	Es	99	(252)	Rutherfordium	Rf	104	(261)
Erbium	Er	68	167.26	Samarium	Sm	62	150.36
Europium	Eu	63	151.96	Scandium	Sc	21	44.96
Fermium	Fm	100	(257)	Seaborgium	Sg	106	(266)
Fluorine	F	9	19.00	Selenium	Se	34	78.96
Francium	Fr	87	(223)	Silicon	Si	14	28.09
Gadolinium	Gd	64	157.25	Silver	Ag	47	107.87
Gallium	Ga	31	69.72	Sodium	Na	11	22.99
Germanium	Ge	32	72.59	Strontium	Sr	38	87.62
Gold	Au	79	196.97	Sulfur	S	16	32.07
Hafnium	Hf	72	178.49	Tantalum	Ta	73	180.95
Hassium	Hs	108	(269)	Technetium	Tc	43	(98)
Helium	He	2	4.00	Tellurium	Te	52	127.60
Holmium	Ho	67	164.93	Terbium	Tb	65	158.93
Hydrogen	H	1	1.01	Thallium	Tl	81	204.38
Indium	In	49	114.82	Thorium	Th	90	(232)
Iodine	I	53	126.90	Thulium	Tm	69	168.93
Iridium	Ir	77	192.22	Tin	Sn	50	118.71
Iron	Fe	26	55.85	Titanium	Ti	22	47.87
Krypton	Kr	36	83.80	Tungsten	W	74	183.84
Lanthanum	La	57	138.91	Uranium	U	92	(238)
Lawrencium	Lr	103	(262)	Vanadium	V	23	50.94
Lead	Pb	82	207.19	Xenon	Xe	54	131.29
Lithium	Li	3	6.94	Ytterbium	Yb	70	173.04
Lutetium	Lu	71	174.97	Yttrium	Y	39	88.91
Magnesium	Mg	12	24.30	Zinc	Zn	30	65.38
Manganese	Mn	25	54.94	Zirconium	Zr	40	91.22
Meitnerium	Mt	109	(268)				

General, Organic, and Biological Chemistry

THIRD EDITION

H. STEPHEN STOKER

Weber State University

Houghton Mifflin Company   Boston   New York

Publisher: Charles Hartford
Executive Editor: Richard Stratton
Development Editor: Rita Lombard
Editorial Assistant: Rosemary Mack
Project Editor: Merrill Peterson
Senior Production/Design Coordinator: Jill Haber
Manufacturing Manager: Florence Cadran
Senior Marketing Manager: Katherine Greig
Marketing Associate: Alexandra Shaw

Cover photograph © 2002 Stuart Hughs/Getty Images

Printed in the U.S.A.

Library of Congress Control Number: 2003100105

ISBN: 0-618-26597-X

456789-DOW-07 06 05 04

General, Organic, and Biological Chemistry

Brief Contents

Contents

PART III ■ BIOLOGICAL CHEMISTRY

Preface

Writing an introductory college text, particularly one that encompasses as wide a range of topics as *General, Organic, and Biological Chemistry* does, is a huge undertaking. When the first edition of the text was published six years ago, my hopes were high. Thus, the positive responses of instructors and students who used the first and second editions of this text have been gratifying—and have led to the new edition you hold in your hands. This third edition represents a renewed commitment to the goals I initially set out to meet.

I wrote the text to fit the needs of the many students in the fields of nursing, allied health, biological sciences, agricultural sciences, food sciences, and public health, who are required to take such a course. The students who will use this text often have little or no background in chemistry and, hence approach the course with a good deal of trepidation. Thus, this text's development of chemical topics always starts at *ground level*. Though clearly some chemical principles cannot be divorced entirely from mathematics, the amount and level of mathematics used is purposefully minimized. Early chapters of the text focus on fundamental chemical principles. The later chapters, built on the foundation of these principles, develop the concepts and applications central to the fields of organic chemistry and biochemistry.

Focus on Biochemistry: Most students taking this course have a greater interest in the biochemistry portion of the course than the preceding two parts. But biochemistry, of course, cannot be understood without a knowledge of the fundamentals of organic chemistry, and understanding organic chemistry in turn depends on knowing the key concepts of general chemistry. Thus, in writing this text, I essentially started from the back and worked forward. I began by determining what topics would be considered in the biochemistry chapters and then tailored the organic and then general sections to support that presentation. Users of the two previous editions confirm that this approach ensures an efficient but thorough coverage of the principles needed to understand biochemistry.

Emphasis on Visual Support: I believe strongly in visual reinforcement of key concepts in a textbook; thus, this book uses art and photos wherever possible to teach key concepts. Artwork is used to make connections and highlight what is important for the student to know. The use of color in reaction equations to emphasize the portions of a molecule that undergo change has been greatly expanded in this new edition. Colors are likewise assigned to things like valence shells and classes of compounds to help students follow trends. Computer-generated, three-dimensional molecular models now accompany many discussions in the organic and biochemistry sections of the text. Color photographs show applications of chemistry to help make concepts real and more readily remembered. Seventy-four pieces of new art have been added to this edition.

Visual summary features, called ***Chemistry at a Glance***, pull together material from several sections of a chapter to help students see the larger picture. For example, the *Chemistry at a Glance* on page 171 summarizes intermolecular forces; the one on page 257 presents buffer solutions; page 317 displays the physical and chemical properties of alkanes, the simplest type of organic compound; the one on page 535 shows the bonding features present among structural units in di- and polysaccharides; and page 667 summarizes DNA replication. The *Chemistry at a Glance* features serve both as overviews for the student reading the material for the first time and as review tools for the student preparing for exams. Given the popularity of the *Chemistry at a Glance* summaries in the previous editions, this feature has been expanded to all chapters; there are nine new *Chemistry at a Glance* features and numerous existent ones have been expanded.

Commitment to Student Learning: In addition to the study help *Chemistry at a Glance* offers, the text is built on a strong foundation of learning aids designed to help students master the course material.

- **Problem-solving Pedagogy:** Because problem solving is often difficult for students in this course to master, I have taken special care to provide support to help students build their skills. Within the chapters, worked *Examples* follow the explanation of many concepts. These examples walk students through the thought processes involved in problem solving, carefully outlining all the steps involved. Each is immediately followed by a *Practice Exercise*, to reinforce the information just presented. A dozen new *Examples* have been added to this edition.

- **Chemical Connections:** In every chapter *Chemical Connections* show chemistry as it appears in everyday life. These boxes focus on topics that are relevant to students' future careers in the health and environmental fields and on those that are important for informed citizens to understand. Many of the health-related *Chemical Connections* have been updated to include the latest research findings, and a number of new boxes on environmental and social issues have been added.

- **Margin Notes:** Liberally distributed throughout the text, *margin notes* provide tips for remembering and distinguishing between concepts, highlight links across chapters, and describe interesting historical background information.

- **Defined Terms:** All definitions are highlighted in the text when they are first presented, using boldface and italic type. Each defined term appears as a complete sentence; students are never forced to deduce a definition from context. In addition, the definitions of all terms appear in a separate *Glossary* found at the end of the text. A major emphasis in this new edition has been "refinements" in the defined terms arena. All defined terms were reexamined to see if they could be stated with greater clarity. The result was a "rewording" of many defined terms. In addition, the number of defined terms has been expanded with 75 new definitions having been added to the text.

- **Review Aids:** Several review aids appear at the ends of the chapters. *Concepts to Remember* and *Key Reactions and Equations* provide concise review of the material presented in the chapter. A *Key Terms Review* lists all the key terms in the chapter alphabetically and cross-references the section of the chapter in which they appear. These aids help students prepare for exams.

- **End-of-chapter problems:** An extensive set of end-of-chapter problems complements the worked examples within the chapters. Each end-of-chapter problem set is divided into two sections: *Exercises and Problems*, and *Additional Problems*. The *Exercises and Problems* are organized by topic and paired, with each pair testing similar material and the answer to the odd-numbered member of the pair at the back of the book. These problems always involve only a single concept. The *Additional Problems* involve more than one concept and are more difficult than the Exercises and Problems.

Content Changes: Coverage of a number of topics has been expanded in this edition. The two driving forces in expanded coverage considerations were (1) the requests of users and reviewers of the previous editions, and (2) the author's desire to incorporate new research findings, particularly in the area of biochemistry, into the text. Topics with expanded coverage include

- Solubility of solutes in water
- Chemical effects of radiation
- Oxidation and reduction as it applies to organic molecules
- Dietary considerations relative to carbohydrates
- Dietary considerations relative to fats and oils
- Classification of lipids in terms of function
- Classification of proteins in terms of structural characteristics
- Biochemical functions for vitamins

- Human genome project results and protein synthesis
- Genetic engineering
- Oxidative phosphorylation and ATP production

Three margin note features have been incorporated into all organic chemistry chapters. The first summarizes physical state information at room temperature (solid, liquid, gas) for the simplest members of each family of organic compounds. The second contrasts common and IUPAC naming systems for the simple members of each family of compounds. The third introduces line-angle drawing representations of the structures of the simplest members of each family of organic compounds to prepare the student for the further use of such line-angle notation in the structures of biochemical molecules.

The Package

Alternate Edition: For instructors who prefer to use only the organic and biochemistry portions of a text, an alternate edition of this book, *Organic and Biological Chemistry,* is available.

■ Study Help for Students

Student Website (accessible through http://www.hmco.com/college/chemistry/) Available free of charge, this dedicated website offers a wealth of resources to help students succeed, including:

- Chapter overviews
- Self-quizzing using Houghton Mifflin's ACE system
- Glossary terms, key reactions, and key concepts in flashcard format
- Additional Chemical Connections with links to related sites
- Additional real-life applications with relevant links
- Career preparation information, including practice questions for the NCLEX exam, descriptions of common nursing specialties, frequently asked questions about entering the nursing field (and their answers), and web links to nursing-related job information

Study Guide with Answers to Selected Problems by Danny V. White of American River College and Joanne A. White includes, for each chapter, a brief overview, activities and practice problems to reinforce skills, and a practice test. The answers section includes answers for all odd-numbered end of chapter exercises.

Experiments for General, Organic, and Biological Chemistry by Michael S. Matta of Southern Illinois University is a printed lab manual containing 45 experiments. Suggestions for demonstrations and structural studies are also included.

Math Review CD-ROM. This CD offers brief tutorials on basic mathematical concepts that appear in the text, including solving simple algebraic equations, scientific notation, conversions, reading a graph, and ratio and proportion.

■ Course Support for Instructors

Instructor Website (accessible through http://www.hmco.com/college/chemistry/) allows access to all student website resources (above) as well as additional instructor course/classroom resources such as downloadable PowerPoint slides and useful links.

Instructor's Resource Manual with Test Bank by H. Stephen Stoker includes answers to all end-of-chapter exercises and a printed test bank of over 1,500 multiple choice and matching problems.

HMClass Prep with HMTesting Version 6.0 CD-ROM Package (ISBN 0-618-340386) This package includes both HMClass Prep and HMTesting on one CD-ROM. It allows an instructor to access both lecture aids and testing software in one place. These components cannot be ordered separately.

- HMClass Prep includes everything an instructor will need to develop lectures–PowerPoint slides with all text figures, tables, concept checks, and virtually all photos, as well as an Instructor's Resource Guide and MS Word files of the Printed Test Bank.

- HM Testing Version 6.0 combines a flexible test-editing program with a comprehensive gradebook function for easy administration and tracking. It enables instructors to administer tests via network server or the web. The HM Testing database contains a wealth of algorithmically generated questions and can produce multiple-choice, true/false, fill-in-the-blank, and essay tests. Questions can be customized based on the chapter being covered, the question format, level of difficulty, and specific topics. HM Testing provides for the utmost security in accessing both test questions and grades.

Instructor's Guide for Experiments in General, Organic, and Biological Chemistry offers reagent/solvent lists, equipment lists, waste disposal instructions, mixing instructions, preparatory notes, planning notes, teaching notes, quiz questions, expected results, and answers to questions and problems for each experiment in the Bibliobase customizable lab manuals.

Overhead Transparencies provide over 140 full-color acetates of important figures, tables, and images from the text.

Acknowledgements

I would like to gratefully acknowledge reviewers of earlier editions, whose influence continues to be felt.

■ Reviewers of the First Edition

Hugh Akers, *Lamar University*; Steven Albrecht, *Oregon State University*; Margaret Asirvatham, *University of Colorado*; George Bandik, *University of Pittsburgh*; Gerald Berkowitz, *Erie Community College*; Robert Bogess, *Radford University*; Christine Brzezowski, *University of Utah*; Harry Conley, *Murray State*; Karen Eichstadt, *Ohio University*; William Euler, *University of Rhode Island*; Arthur Glasfeld, *Reed College*; Fabian Fang, *California State University—Long Beach*; John Fulkrod, *University of Minnesota—Duluth*; Marvin Hackert, *University of Texas at Austin*; Henry Harris, *Armstrong State College*; Leland Harris, *University of Arizona*; Larry Jackson, *Montana State University*; James Jacob, *University of Rhode Island*; James Johnson, *Sinclair Community College*; Eugene Klein, *Tennessee Technological University*; Norman Kulevsky, *University of North Dakota*; James W. Long, *University of Oregon*; Ralph Martinez, *Humboldt State University*; Scott Mohr, *Boston University*; Melvyn Mosher, *Missouri Southern University*; Elva Mae Nicholson, *Eastern Michigan University*; Frasier Nyasulu, *University of Washington*; John Ohlsson, *University of Colorado*; Roger Penn, *Sinclair Community College*; Helen Place, *Washington State University*; John Reasoner, *Western Kentucky University*; Norman Rose, *Portland State University*; Michael Ryan, *Marquette University*; John Searle, *College of San Mateo*; Dan Sullivan, *University of Nebraska at Omaha*; Emanuel Terezakis, *Community College of Rhode Island*; Ruiess Van Fossen Bravo, *Indiana University of Pennsylvania*; Donald Williams, *University of Louisville*; Les Wynston, *California State University Long Beach*.

■ Reviewers of the Second Edition

Vicky L.H. Bevilacqua, *Kennesaw State University*; David R. Bjorkman, *East Carolina University*; Frank D. Bay, *North Shore Community College*; Tim Champion, *Johnson C. Smith University*; Alison J. Dobson, *Georgia Southern University*; Naomi Eliezer, *Oakland University*; Wes Fritz, *College of DuPage*; Caroline Gil, *Lexington Community College*; Robert Gooden, *Southern University—Baton Rouge*; Ellen Kime-Hunt, *Riverside Community College*; Peter Krieger, *Palm Beach Community College*; Cathy MacGowan, *Armstrong Atlantic State University*; Lawrence L. Mack, *Bloomsburg University*; Charmaine B. Mamantov, *University of Tennessee, Knoxville*; Joann S. Monko, *Kutztown University*; Elva Mae Nicholson, *Eastern Michigan University*; Michael Shanklin, *Palo Alto College*; Hugh Akers, *Lamar University*; Eric Holmberg, *University of Alaska*; Marvin Jaffe, *Borough of Manhattan Community College*; David Johnson, *Biola University*; Fred Johnson, *Brevard Community College*; Daniel Jones, *University of North Carolina—Charlotte*; Peter Krieger, *Palm Beach Community College*; Da-hong Lu, *Fitchburg State College*; Cynthia Martin, *Des Moines Area Community College*; Elva Mae Nicholson, *Eastern Michigan University*; Mary Palaszek, *Grand Valley State University*; Diane Payne, *Villa Julie College*; Janet Rogers, *Edinboro University of Pennsylvania*; Jackie Scholars, *Bellevue University*; Michelle Sulikowski, *Texas A&M University*; Joanne Tscherne, *Bergen Community College*; James Yuan, *Old Dominion University*

I also want to thank the following reviewers for their valuable comments and suggestions which helped to guide my revision efforts for this edition:

Diane Payne, *Villa Julie College*

Kristan Lenning, *Lexington Community College*

Sidney Alozie, *Bronx Community College*

Barbara Keller, *Lake Superior State University*

Naomi Eliezer, *Oakland University*

Josh Smith, *Humboldt State University*

Garon Smith, *The University of Montana*

Mundiyath Venugopalan, *Western Illinois University*

Renee Rosentreter, *Idaho State University*

Laura Kibler-Herzog, *Georgia State University*

Peter Krieger, *Palm Beach Community College*

Peter Olds, *Laney College*

Marcia Miller, *University of Wisconsin—Eau Claire*

Sara Hein, *Winona State University*

Special thanks go to Stephen Z. Goldberg, *Adelphi University*, for his help in ensuring this book's accuracy by reviewing manuscript, proofs, and artwork.

I also give special thanks to the people at Houghton Mifflin who guided the revision through various stages of development and production: Richard Stratton, Executive Editor, Chemistry; Rita Lombard, Development Editor; Charline Lake, Production Services Manager; Jill Haber, Senior Production/Design Coordinator; Katherine Greig, Senior Marketing Manager, Science; and Alexandra Shaw, Marketing Associate, made especially significant suggestions for refining the features and text. I would also like to thank Rosemary Mack, Editorial Assistant, for her skillful management of the ancillary package; Charlotte Miller, Art Editor, for her thoughtful and creative contributions to the illustration program; Naomi Kornhauser, Photo Editor; and Jessyca Broekman, Photo Editor; for helping to enrich the text with photographs and molecular models; and Jean Hammond, Designer, for the complementary design; Merrill Peterson, Project Editor, and Michele Ostovar, Assistant Project Editor, for making the production process for this text a smooth one.

H. Stephen Stoker, Weber State University

Exciting Photo Program
Throughout the text, **photos** help students see the everyday applications of the chemistry they are learning

Chapter Outlines give students a road map for where they are going.

6 Chemical Calculations: Formula Masses, Moles, and Chemical Equations

The energy associated with a lightning discharge causes many different chemical reactions to occur within the atmosphere. In this chapter, we learn how to write chemical equations to describe such chemical reactions.

In this chapter we discuss "chemical arithmetic," the quantitative relationships between elements and compounds. Anyone who deals with chemical processes needs to understand at least the simpler aspects of this topic. All chemical processes, regardless of where they occur—in the human body, at a steel mill, on top of the kitchen stove, or in a clinical laboratory setting—are governed by the same mathematical rules.

We have already presented some information about chemical formulas (Section 1.10). In this chapter we discuss formulas again, and here we look beyond describing the composition of compounds in terms of constituent atoms. A new unit, the mole, will be introduced and its usefulness discussed. Chemical equations will be considered for the first time. We will learn how to represent chemical reactions by using chemical equations and quantitative relationships from these equations.

6.1 Formula Masses

Our entry into the realm of "chemical arithmetic" is a discussion of the quantity called formula mass. A **formula mass** *is the sum of the atomic masses of all the atoms in the chemical formula of a substance.* Formula masses, like the atomic masses from which they are calculated, are relative masses based on the $^{12}_{6}C$ relative-mass scale (Section 3.3). Example 6.1 illustrates how formula masses are calculated.

127

CHEMICAL CONNECTIONS

Combustion Reactions, Carbon Dioxide, and Global Warming

Most fuels used in our society, including coal, petroleum, and natural gas, are carbon-containing substances. When such fuels are burned (combustion; Section 9.1), carbon dioxide is one of the combustion products. Nearly all such combustion-generated CO_2 enters the atmosphere.

Significant amounts of the CO_2 entering the atmosphere are absorbed into the oceans because of this compound's solubility in water, and plants also remove CO_2 from the atmosphere via the process of photosynthesis. However, these removal mechanisms are not sufficient to remove all the combustion-generated CO_2; it is being generated faster than it can be removed. Consequently, atmospheric concentrations of CO_2 are slowly increasing, as the following graph shows.

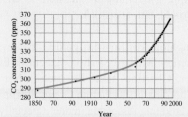

The concentration unit on the vertical axis is parts per million (ppm)—the number of molecules of CO_2 per million molecules present in air. Data for periods before 1958 were derived from analysis of air trapped in bubbles in glacial ice.

Increasing atmospheric CO_2 levels pose an environmental concern because within the atmosphere, CO_2 acts as a heat-trapping agent. During the day, Earth receives energy from the sun, mostly in the form of *visible light*. At night, as Earth cools, it re-radiates the energy it received during the day in the form of *infrared light* (heat energy). Carbon dioxide does not absorb

visible light, but it has the ability to absorb infrared light. The CO_2 thus traps some of the heat energy re-radiated by the surface of the Earth as it cools, preventing this energy from escaping to outer space. Because this action is similar to that of glass in a greenhouse, CO_2 is called a *greenhouse gas*. The warming caused by CO_2 as it prevents heat loss from Earth is called the *greenhouse effect* or *global warming*.

Some scientists believe that the presence of increased concentrations of CO_2 (and other greenhouse gases present in the atmosphere in lower concentrations than CO_2) is beginning to cause a change in our climate as the result of a small increase in the average temperature of Earth's surface. Some computer models predict an average global temperature increase of 1 to 3°C toward the end of the twenty-first century if atmospheric CO_2 concentrations continue to increase at their current rate. Because numerous other factors are also involved in determining climate, however, predictions cannot be made with certainty. Much research concerning this situation is in progress, and many governments around the world are now trying to reduce the amount of combustion-generated CO_2 that enters the atmosphere.

The other greenhouse gases besides CO_2 include CH_4, N_2O, and CFCs (chlorofluorocarbons). Atmospheric concentrations of these other greenhouse gases are lower than that of CO_2. However, because they are more effective absorbers of infrared radiation than CO_2 is, they make an appreciable contribution to the overall greenhouse effect. Further information about CFCs is given in the Chemical Connections feature "Chlorofluorocarbons and the Ozone Layer" in Chapter 12. Estimated contributions of various greenhouse gases to global warming are as follows:

Chemical Connections boxes show chemistry as it appears in everyday life. Topics are relevant to students' future careers in the health and environmental fields and are important for informed citizens to understand. See page 209 for an example.

3. *The oxidation numbers of Groups IA and IIA metals are always +1, and +2, respectively.*
4. *The oxidation number of hydrogen is +1 in most hydrogen-containing compounds.*

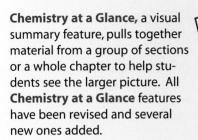

Chemistry at a Glance, a visual summary feature, pulls together material from a group of sections or a whole chapter to help students see the larger picture. All **Chemistry at a Glance** features have been revised and several new ones added.

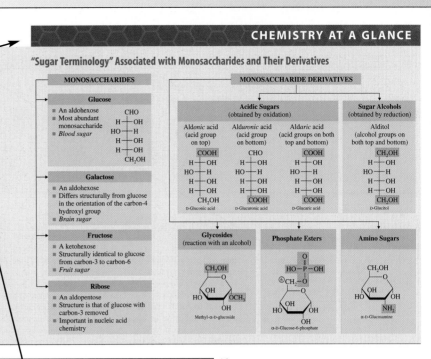

CHEMISTRY AT A GLANCE

"Sugar Terminology" Associated with Monosaccharides and Their Derivatives

MONOSACCHARIDES

Glucose
- An aldohexose
- Most abundant monosaccharide
- *Blood sugar*

Galactose
- An aldohexose
- Differs structurally from glucose in the orientation of the carbon-4 hydroxyl group
- *Brain sugar*

Fructose
- A ketohexose
- Structurally identical to glucose from carbon-3 to carbon-6
- *Fruit sugar*

Ribose
- An aldopentose
- Structure is that of glucose with carbon-3 removed
- Important in nucleic acid chemistry

MONOSACCHARIDE DERIVATIVES

Acidic Sugars (obtained by oxidation)

Ald*onic* acid (acid group on top) — D-Gluconic acid

Ald*uronic* acid (acid group on bottom) — D-Glucuronic acid

Ald*aric* acid (acid groups on both top and bottom) — D-Glucaric acid

Sugar Alcohols (obtained by reduction)

Alditol (alcohol groups on both top and bottom) — D-Glucitol

Glycosides (reaction with an alcohol) — Methyl-α-D-glucoside

Phosphate Esters — α-D-Glucose-6-phosphate

Amino Sugars — α-D-Glucosamine

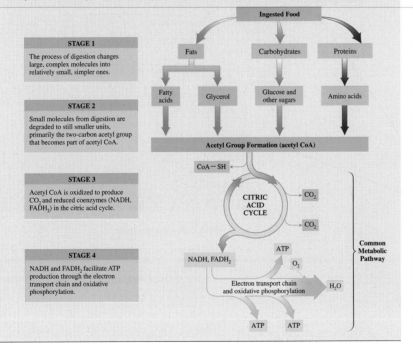

CHEMISTRY AT A GLANCE

Simplified Summary of the Four Stages of Biochemical Energy Production

Ingested Food

STAGE 1
The process of digestion changes large, complex molecules into relatively small, simpler ones.

STAGE 2
Small molecules from digestion are degraded to still smaller units, primarily the two-carbon acetyl group that becomes part of acetyl CoA.

STAGE 3
Acetyl CoA is oxidized to produce CO_2 and reduced coenzymes (NADH, $FADH_2$) in the citric acid cycle.

STAGE 4
NADH and $FADH_2$ facilitate ATP production through the electron transport chain and oxidative phosphorylation.

Fats → Fatty acids, Glycerol
Carbohydrates → Glucose and other sugars
Proteins → Amino acids

Acetyl Group Formation (acetyl CoA)

CoA—SH

CITRIC ACID CYCLE → CO_2, CO_2

NADH, $FADH_2$

ATP

O_2

Electron transport chain and oxidative phosphorylation → H_2O

ATP ATP

Common Metabolic Pathway

...that has cyclic forms (hemiacetal forms) can react with an alcohol to ...cetal), as we noted in Section 18.12. This same type of reaction can ...a *disaccharide,* a carbohydrate in which two monosaccharides are ...ection 18.3). In disaccharide formation, one of the monosaccharide ...as a hemiacetal, and the other functions as an alcohol.

...de + monosaccharide ⟶ disaccharide + H_2O
    (Functioning as an alcohol)   (Glycoside)

Glycosidic linkage

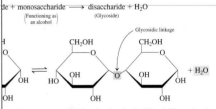

$+ H_2O$

...nks the two monosaccharides of a disaccharide (glycoside) together is ...nkage. A **glycosidic linkage** *is the bond in a disaccharide resulting from ...the hemiacetal carbon atom —OH group of one monosaccharide and ...e other monosaccharide.* It is always a carbon–oxygen–carbon bond in

---

The formation of citryl CoA is a condensation reaction (Section 14.7) because a new carbon–carbon bond is formed.

with the four-carbon keto dicarboxylate species oxaloacetate. This results in the transfer of the acetyl group from coenzyme A to oxaloacetate, producing the C_6 citrate species and free coenzyme A.

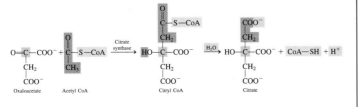

FIGURE 23.2 A schematic representation of a eukaryotic cell with selected internal components identified.

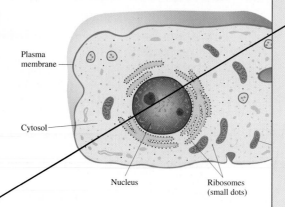

Plasma membrane

Cytosol

Nucleus

Ribosomes (small dots)

Margin Notes, such as those on page 695, summarize key information, give tips for remembering or distinguishing between similar ideas, and provide additional details and links between concepts.

▶ A protective mechanism exists to prevent lysosome enzymes from destroying the cell in which they are found if they should be accidently released (via membrane rupture or leakage). The optimum pH (Section 21.6) for lysosome enzyme activity is 4.8. The cytoplasmic pH of 7.0–7.3 renders them inactive.

▶ *Mitochondria,* pronounced my-toe-KON-dree-ah, is plural. The singular form of the term is *mito-chondrion.* The threadlike shape of mitochondria is responsible for this organelle's name; *mitos* is Greek for "thread," and *chondrion* is Greek for "granule."

FIGURE 23.3 (a) A schematic representation of a mitochondrion, showing key features of its internal structure. (b) An electron micrograph of a single mitochondrial crista, showing the ATP synthase knobs extending into the matrix.

"trapped" by the body's immune system (Section 20.17) are degra enzymes from lysosomes.

A **mitochondrion** *is an organelle that is responsible for the ger energy for a cell.* Much of the discussion of this chapter deals with chemical reactions that occur within mitochondria. Further deta structure will help us understand more about how these reactions c

Mitochondria are sausage-shaped organelles containing both ar a *multifolded inner membrane* (see Figure 23.3). The outer mem 50% lipid and 50% protein, is freely permeable to small molecules. which i nonper compar outer m protrud

Th interior small sp implies membra brane is

EXAMPLE 8.1

Predicting Solute Solubility Using Solubility Rules

■ With the help of Table 8.2, predict the solubility of each of the following solutes in the solvent indicated.

a. CH_4 (a nonpolar gas) in water
b. Ethyl alcohol (a polar liquid) in chloroform (a polar liquid)
c. AgCl (an ionic solid) in water
d. Na_2SO_4 (an ionic solid) in water
e. $AgNO_3$ (an ionic solid) in water

Solution

a. Insoluble. They are of unlike polarity because water is polar.
b. Soluble. Both substances are polar, so they should be relatively soluble in one another—like dissolves like.
c. Insoluble. Table 8.2 indicates that all chlorides except those of silver, lead, and mercury(I) are soluble. Thus AgCl is one of the exceptions.
d. Soluble. Table 8.2 indicates that all ionic sodium-containing compounds are soluble.
e. Soluble. Table 8.2 indicates that all compounds containing the nitrate ion (NO_3^-) are soluble.

Practice Exercise 8.1

With the help of Table 8.2, predict the solubility of each of the following solutes in the solvent indicated.

a. NO_2 (a polar gas) in water
b. CCl_4 (a nonpolar liquid) in benzene (a nonpolar liquid)
c. NaBr (an ionic solid) in water
d. $MgCO_3$ (an ionic solid) in water
e. $(NH_4)_3PO_4$ (an ionic solid) in water

8.5 Solution Concentration Units

Because solutions are mixtures (Section 8.1), they have a variable composition. Specifying what the composition of a solution is involves specifying solute concentrations. In general, **concentration** *is the amount of solute present in a specified amount of solution.* Many methods of expressing concentration exist, and certain methods are better suited for some purposes than others. In this section we consider two methods: *percent concentration* and *molarity.*

CHEMICAL CONNECTIONS

Solubility of Vitamins

Polarity plays an important role in the solubility of many substances in the fluids and tissues of the human body. For example, consider vitamin solubilities. The 13 known vitamins fall naturally into two classes: fat-soluble and water-soluble. The fat-soluble vitamins are A, D, E, and K. Water-soluble vitamins are vitamin C and the eight B vitamins (thiamine, riboflavin, niacin, vitamin B_6, folic acid, vitamin B_{12}, pantothenic acid, and biotin). Water-soluble vitamins have polar molecular structures, as does water. By contrast, fat-soluble vitamins have nonpolar molecular structures that are compatible with the nonpolar nature of fats.

Vitamin C is water-soluble. Because of this, vitamin C is not stored in the body and must be ingested in our daily diet. Unused vitamin C is eliminated rapidly from the body via body fluids. Vitamin A, on the other hand, is fat-soluble. It can be,

and is, stored by the body in fat tissue for later use. If vitamin A is consumed in excess quantities (from excessive vitamin supplements), illness can result. Because of its limited water solubility, vitamin A cannot be rapidly eliminated from the body by body fluids.

The water-soluble vitamins can be easily leached out of foods as they are prepared. As a rule of thumb, you should eat foods every day that are rich in the water-soluble vitamins. Taking megadose vitamin supplements of water-soluble vitamins is seldom effective. The extra amounts of these vitamins are usually picked up by the extracellular fluids, carried away by blood, and excreted in the urine. As one person aptly noted, "If you take supplements of water-soluble vitamins, you may have the most expensive urine in town."

Within the chapters, worked-out **Examples** follow the explanation of many concepts. These examples walk students through the thought process involved in problem solving, carefully outlining all the steps involved. They are immediately followed by a **Practice Exercise** to reinforce the information just presented. See, for example, page 183.

CONCEPTS TO REMEMBER

Triacylglycerol digestion and absorption. Triacylglycerols are digested (hydrolyzed) in the intestine and then reassembled after passage into the intestinal wall. Chylomicrons transport the reassembled triacylglycerols from intestinal cells to the bloodstream.

Triacylglycerol storage and mobilization. Triacylglycerols are stored as fat droplets in adipose tissue. When they are needed for energy, enzyme-controlled hydrolysis reactions liberate the fatty acids, which then enter the bloodstream and travel to tissues where they are utilized.

Glycerol metabolism. Glycerol is first phosphorylated and then oxidized to dihydroxyacetone phosphate, a glycolysis pathway intermediate. Through glycolysis and the common metabolic pathway, the glycerol can be converted to CO_2 and H_2O.

Fatty acid degradation. Fatty acid degradation is accomplished through the fatty acid (β oxidation) spiral. The degradation process involves removal of carbon atoms, two at a time, from the carboxyl end of the fatty acid. There are four repeating reactions that accom-

pany the removal of each two-carbon unit. A turn of the cycle a produces one molecule each of acetyl CoA, NADH, and FADH

Ketone bodies. Acetoacetate, β-hydroxybutyrate, and acetone a known as ketone bodies. They are synthesized in the liver from acetyl CoA as a result of excessive fatty acid degradation. Durin starvation and in unchecked diabetes, the level of ketone bodies the blood becomes very high.

Fatty acid biosynthesis. Fatty acid biosynthesis, lipogenesis, oc through the addition of two-carbon units to a growing acyl chai The added two-carbon units come from malonyl CoA. A multie zyme complex, an acyl carrier protein (ACP), and NADPH are portant parts of the biosynthetic process.

Biosynthesis of cholesterol. Cholesterol is biosynthesized from acetyl CoA in a complex series of reactions in which isoprene are key intermediates. Cholesterol is the precursor for the vario classes of steroid hormones.

KEY REACTIONS AND EQUATIONS

1. Digestion of triacylglycerols (Section 25.1)

Triacylglycerol + H_2O $\xrightarrow{\text{Lipase}}$

fatty acids + glycerol + monoacylglycerols

2. Mobilization of triacylglycerols (Section 25.2)

Triacylglycerol + $3H_2O$ $\xrightarrow{\text{Lipase}}$ 3 fatty acids + glycerol

3. Glycerol metabolism (Section 25.3)

Glycerol + ATP + NAD$^+$ $\xrightarrow{\text{Two}}$

> **Concepts to Remember, Key Reactions and Equations,** and **Key Terms** provide concise review of the material presented in the chapter, helping students prepare for exams.

KEY TERMS

Adipocyte (25.2)
Adipose tissue (25.2)
Chylomicron (25.1)

Fatty acid micelle (25.1)
Fatty acid spiral (25.4)
Ketogenesis (25.6)

Ketone body (25.6)
Lipogenesis (25.7)
Triacylglycerol mobilizat

> Extensive and varied **Exercises and Problems** at the end of each chapter are organized by topic and paired, with answers to the odd-numbered problems at the back of the book. These problems always involve only a single concept.

EXERCISES AND PROBLEMS

The members of each pair of problems in this section test similar material.

■ **Digestion and Absorption of Lipids (Section 25.1)**

25.1 What percent of dietary lipids are triacylglycerols?

25.2 What are the solubility characteristics of triacylglycerols?

25.3 What effect do salivary enzymes have on triacylglycerols?

25.4 What effect do stomach fluids have on triacylglycerols?

25.5 Why does ingestion of lipids make one feel "full" for a long time?

25.6 The process of lipid digestion occurs primarily at two sites within the human body.
  a. What are the identities of these two sites?
  b. What is the relative amount of TAG digestion that occurs at each site?
  c. What type of digestive enzyme functions at each site?
  .... lipid digestion?

■ **Glycerol Metabolism (Section 25.3)**

25.19 In what order are the compounds glycerol 3 dihydroxyacetone phosphate encountered in glycerol?

25.20 How many reactions are needed to convert glycolysis intermediate?

25.21 How many ATP molecules are expended in glycerol to a glycolysis intermediate?

25.22 What are the two fates of glycerol after it h to a glycolysis intermediate?

■ **Oxidation of Fatty Acids (Section 25.4)**

25.23 Where in a cell does fatty acid activation ta

25.24 What is the chemical form for an activated

ADDITIONAL PROBLEMS

16.81 With the help of Figure 16.6 and IUPAC naming rules, specify the number of carbon atoms present and the number of carboxyl groups present in each of the following carboxylic acids.
  a. Oxalic acid   b. Heptanoic acid
  c. *cis*-3-Heptenoic acid d. Citric acid
  e. Pyruvic acid   f. Dichloroethanoic acid

16.82 Malonic, maleic, and malic acids are dicarboxylic acids with similar-sounding names. How do the structures of these acids differ from each other?

16.83 The general molecular formula for an alkane is C_nH_{2n+2}. What is the general molecular formula for an unsaturated unsubstituted monocarboxylic acid containing one carbon–carbon double bond?

16.84 Draw structural formulas and give IUPAC names for all possible saturated unsubstituted monocarboxylic acids that contain six carbon atoms. There are eight isomers.

16.85 Monocarboxylic acids and esters with the same number of carbon atoms and the same degree of unsaturation are isomeric. Give IUPAC names for all possible esters that are isomeric with butanoic acid. There are four isomers.

16.86 Assign IUPAC names to the following compounds.

16.87 A sample of ethyl alcohol is divided into A is added to an aqueous solution of a str and allowed to react. The organic product mixed with portion B of the ethyl alcoho added and the solution is heated. What is final product of this reaction scheme?

16.88 For each of the following reactions, draw the organic product(s).
a.
$$CH_3-CH_2-\overset{\overset{\displaystyle O}{\|}}{C}-O-CH_3 + NaOH$$
b.
$$CH_3-CH_2-\overset{\overset{\displaystyle O}{\|}}{C}-OH + CH_3-SH$$
c.
$$CH_3-\overset{\overset{\displaystyle O}{\|}}{C}-OH + NaOH \longrightarrow$$
d.

> **Additional Problems** involve more than one concept and are more difficult than the Exercises and Problems.

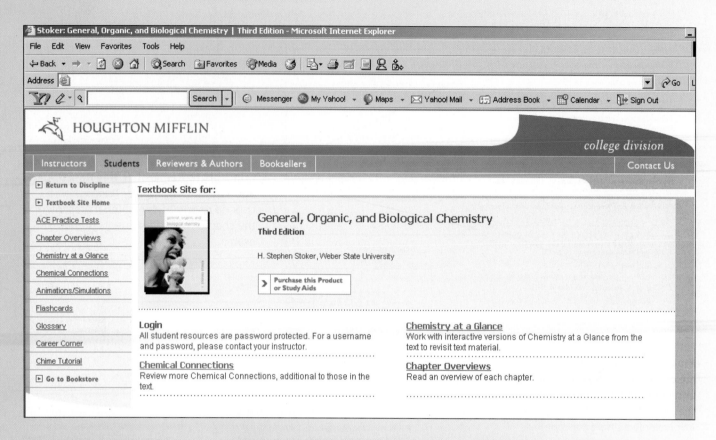

A password protected **Student Website** is accessible through http://college.hmco.com (select "chemistry".) It includes a wealth of resources to help students in the course, including:

* Chapter overviews
* Self-quizzing using Houghton Mifflin's ACE system
* Electronic flashcards of key terms, reactions, and concepts
* Additional *Chemical Connections* features
* Additional real-life applications with relevant links
* Career-related information

The **Instructor Website**, accessible through the address above, allows access to all student resources, plus instructor/classroom resources such as dowloadable PowerPoint slides and useful links.

1 Basic Concepts About Matter

Heat and light generation, as well as numerous changes in matter, occur as wood is burned in a bonfire.

In this chapter we address the question "What exactly is chemistry about?" In addition, we consider common terminology associated with the field of chemistry. Much of this terminology is introduced in the context of the ways in which matter is classified. Like all other sciences, chemistry has its own specific language. It is necessary to restrict the meanings of some words so that all chemists (and those who study chemistry) can understand a given description of a chemical phenomenon in the same way.

1.1 Chemistry: The Study of Matter

Chemistry *is the field of study concerned with the characteristics, composition, and transformations of matter.* What is matter? **Matter** *is anything that has mass and occupies space.* The term *mass* refers to the amount of matter present in a sample.

Matter includes all things—both living and nonliving—that can be seen (such as plants, soil, and rocks), as well as things that cannot be seen (such as air and bacteria). Not considered to be matter are the various forms of energy, such as heat, light, and electricity. However, chemists must be concerned with energy as well as with matter, because nearly all changes that matter undergoes involve the release or absorption of energy.

The scope of chemistry is extremely broad, and it touches every aspect of our lives. An iron gate rusting, a chocolate cake baking, the diagnosis and treatment of a heart attack, the propulsion of a jet airliner, and the digesting of food all fall within the realm

▶ The universe is composed entirely of matter and energy.

▶ The term *chemistry* is derived from the word *alchemy,* which denotes practices carried out during the Middle Ages in an attempt to transform something common into something precious (in particular, lead into gold). Alchemy originated in Alexandrian Egypt, and the term *alchemy* is derived from the Greek *al* ("the") and *khemia* (a native name for Egypt).

FIGURE 1.1 Everyday activities such as making a batch of muffins involve chemistry.

of chemistry (see Figure 1.1). The key to understanding such diverse processes is an understanding of the fundamental nature of matter, which is what we now consider.

1.2 Physical States of Matter

Three physical states exist for matter: solid, liquid, and gas. The classification of a given matter sample in terms of physical state is based on whether its shape and volume are definite or indefinite.

> The *volume* of a sample of matter is a measure of the amount of space occupied by the sample.

    Solid *is the physical state characterized by a definite shape and a definite volume.* A dollar coin has the same shape and volume whether it is placed in a large container or on a table top (Figure 1.2a). For solids in powdered or granulated forms, such as sugar or salt, a quantity of the solid takes the shape of the portion of the container it occupies, but each individual particle has a definite shape and definite volume. **Liquid** *is the physical state characterized by an indefinite shape and a definite volume.* A liquid always takes the shape of its container to the extent that it fills the container (Figure 1.2b). **Gas** *is the physical state characterized by an indefinite shape and an indefinite volume.* A gas always completely fills its container, adopting both the container's volume and its shape (Figure 1.2c).

    The state of matter observed for a particular substance depends on its temperature, the surrounding pressure, and the strength of the forces holding its structural particles together. At the temperatures and pressures normally encountered on Earth, water is one of the few substances found in all three of its physical states: solid ice, liquid water, and gaseous steam (Figure 1.3). Under laboratory conditions, states other than those com-

FIGURE 1.2 (a) A solid has a definite shape and a definite volume. (b) A liquid has an indefinite shape—it takes the shape of its container—and a definite volume. (c) A gas has an indefinite shape and an indefinite volume—it assumes the shape and volume of its container.

        (a)   (b)   (c)

FIGURE 1.3 Water can be found in the solid, liquid, and vapor (gaseous) forms simultaneously, as shown here at Yellowstone National Park.

monly observed can be attained for almost all substances. Oxygen, which is nearly always thought of as a gas, becomes a liquid at $-183°C$ and a solid at $-218°C$. The metal iron is a gas at extremely high temperatures (above 3000°C).

1.3 Properties of Matter

Various kinds of matter are distinguished from each other by their properties. A **property** *is a distinguishing characteristic of a substance that is used in its identification and description.* Each substance has a unique set of properties that distinguishes it from all other substances. Properties of matter are of two general types: physical and chemical.

A **physical property** *is a characteristic of a substance that can be observed without changing the basic identity of the substance.* Common physical properties include color, odor, physical state (solid, liquid, or gas), melting point, boiling point, and hardness.

During the process of determining a physical property, the physical appearance of a substance may change, but the substance's identity does not. For example, it is impossible to measure the melting point of a solid without changing the solid into a liquid. Although the liquid's appearance is much different from that of the solid, the substance is still the same; its chemical identity has not changed. Hence melting point is a physical property.

A **chemical property** *is a characteristic of a substance that describes the way the substance undergoes or resists change to form a new substance.* For example, copper objects turn green when exposed to moist air for long periods of time (Figure 1.4); this is a chemical property of copper. The green coating formed on the copper is a new substance that results from the copper's reaction with oxygen, carbon dioxide, and water present in air. The properties of this new substance (the green coating) are very different from those of metallic copper. On the other hand, gold objects resist change when exposed to air for long periods of time. The lack of reactivity of gold with air is a chemical property of gold.

Most often the changes associated with chemical properties result from the interaction (reaction) of a substance with one or more other substances. However, the presence of a second substance is not an absolute requirement. Sometimes the presence of energy (usually heat or light) can trigger the change called decomposition. That hydrogen peroxide, in the presence of either heat or light, decomposes into the substances water and oxygen is a chemical property of hydrogen peroxide.

FIGURE 1.4 The green color of the Statue of Liberty (present before it was restored) results from the reaction of the copper skin of the statue with the components of air. That copper will react with the components of air is a chemical property of copper.

▶ Chemical properties describe the ability of a substance to form new substances, either by reaction with other substances or by decomposition. Physical properties are properties associated with a substance's physical existence. They can be determined without reference to any other substance, and determining them causes no change in the identity of the substance.

CHEMICAL CONNECTIONS

"Good" versus "Bad" Properties for a Chemical Substance

It is important not to judge the significance or usefulness of a chemical substance on the basis of just one or two of the many chemical and physical properties it exhibits. Possession of a "bad" property, such as toxicity or a strong noxious odor, does not mean that a chemical substance has nothing to contribute to the betterment of human society.

A case in point is the substance carbon monoxide. Everyone knows that it is a gaseous air pollutant present in automobile exhaust and cigarette smoke and that it is toxic to human beings. For this reason, some people automatically label carbon monoxide a "bad" substance, a substance we do not need or want.

Indeed, carbon monoxide is toxic to human beings. It impairs human health by reducing the oxygen-carrying capacity of the blood. Carbon monoxide does this by interacting with the hemoglobin in red blood cells in a way that prevents the hemoglobin from distributing oxygen throughout the body. Someone who dies from carbon monoxide poisoning actually dies from lack of oxygen.

The fact that carbon monoxide is colorless, odorless, and tasteless is very significant. Because of these properties, carbon monoxide gives no warning of its initial presence. There are several other common air pollutants that are more toxic than carbon monoxide. However, they have properties that give warning of their presence and hence they are not considered as "dangerous" as carbon monoxide.

Despite its toxicity, carbon monoxide plays an important role in the maintenance of the high standard of living we now enjoy. Its contribution lies in the field of iron metallurgy and the production of steel. The isolation of iron from iron ores, necessary for the production of steel, involves a series of high-temperature reactions, carried out in a blast furnace, in which the iron content of molten iron ores reacts with carbon monoxide. These reactions release the iron from its ores. The carbon monoxide needed in steel making is obtained by reacting coke (a product derived by heating coal to a high temperature without air being present) with oxygen.

The industrial consumption of the metal iron, both in the United States and worldwide, is approximately ten times greater than that of all other metals combined. Steel production accounts for nearly all of this demand for iron. Without steel, our standard of living would drop dramatically, and carbon monoxide is necessary for the production of steel.

Is carbon monoxide a "good" or a "bad" chemical substance? The answer to this question depends on the context in which the carbon monoxide is encountered. In terms of air pollution, it is a "bad" substance. In terms of steel making, it is a "good" substance. A similar "good–bad" dichotomy exists for almost every chemical substance.

FIGURE 1.5 The melting of ice cream is a physical change involving a change of state; solid turns to liquid.

▶ Physical changes need not involve a change of state. Pulverizing an aspirin tablet into a powder and cutting a piece of adhesive tape into small pieces are physical changes that involve only the solid state.

When we specify chemical properties, we usually give conditions such as temperature and pressure because they influence the interactions between substances. For example, the gases oxygen and hydrogen are unreactive toward each other at room temperature, but they interact explosively at a temperature of several hundred degrees.

1.4 Changes in Matter

Changes in matter are common and familiar occurrences. Changes take place when food is digested, paper is burned, and a pencil is sharpened. Like properties of matter, changes in matter are classified into two categories: physical and chemical.

A **physical change** *is a process in which a substance changes its physical appearance but not its chemical composition.* A new substance is never formed as a result of a physical change.

A change in physical state is the most common type of physical change. Melting, freezing, evaporation, and condensation are all changes of state (see Figure 1.5). In any of these processes, the composition of the substance undergoing change remains the same even though its physical state and appearance change. The melting of ice does not produce a new substance; the substance is water both before and after the change. Similarly, the steam produced from boiling water is still water.

A **chemical change** *is a process in which a substance undergoes a change in chemical composition.* Chemical changes always involve conversion of the material or materials under consideration into one or more new substances, each of which has properties and composition distinctly different from those of the original materials. Consider, for example, the rusting of iron objects left exposed to moist air (Figure 1.6). The reddish brown substance (the rust) that forms is a new substance with chemical properties that are obviously different from those of the original iron.

EXAMPLE 1.1

Correct Use of the Terms
Physical and *Chemical* in
Describing Changes

■ Complete each of the following statements about changes in matter by placing the word *physical* or *chemical* in the blank.

a. The fashioning of a piece of wood into a round table leg involves a _____ change.

b. The vigorous reaction of potassium metal with water to produce hydrogen gas is a _____ change.

c. Straightening a bent piece of iron with a hammer is an example of a _____ change.

d. The ignition and burning of a match involve a _____ change.

Solution

a. *Physical.* The table leg is still wood. No new substances have been formed.

b. *Chemical.* A new substance, hydrogen, is produced.

c. *Physical.* The piece of iron is still a piece of iron.

d. *Chemical.* New gaseous substances, as well as heat and light, are produced as the match burns.

Practice Exercise 1.1

Complete each of the following statements about changes in matter by placing the word *physical* or *chemical* in the blank.

a. The destruction of a newspaper through burning involves a _____ change.

b. The grating of a piece of cheese is a _____ change.

c. The heating of a blue powdered material to produce a white glassy substance and a gas is a _____ change.

d. The crushing of ice cubes to make ice chips is a _____ change.

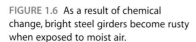

Answers to Practice Exercises are given at the end of the chapter.

Chemists study the nature of changes in matter to learn how to bring about favorable changes and prevent undesirable ones. The control of chemical change has been a major factor in attainment of the modern standard of living now enjoyed by most people in the developed world. The many plastics, synthetic fibers, and prescription drugs now in common use are derived via controlled chemical change.

The Chemistry at a Glance feature on page 6 reviews the ways in which the terms *physical* and *chemical* are used to describe the properties of substances and the changes that substances undergo. Note that the term *physical*, used as a modifier, always conveys the idea that the composition (chemical identity) of a substance did not change, and that the term *chemical*, used as a modifier, always conveys the idea that the composition of a substance did change.

1.5 Pure Substances and Mixtures

In addition to its classification by physical state (Section 1.2), matter can also be classified in terms of its chemical composition as a pure substance or as a mixture. A **pure substance** *is a single kind of matter that cannot be separated into other kinds of matter by any physical means.* All samples of a pure substance contain only that substance and nothing else. Pure water is water and nothing else. Pure sucrose (table sugar) contains only that substance and nothing else.

A pure substance always has a definite and constant composition. This invariant composition dictates that the properties of a pure substance are always the same under a given set of conditions. Collectively, these definite and constant physical and chemical properties constitute the means by which we identify the pure substance.

A **mixture** *is a physical combination of two or more pure substances in which each substance retains its own chemical identity.* Components of a mixture retain their identity because they are physically mixed rather than chemically combined. Consider a mixture of small rock salt crystals and ordinary sand. Mixing these two substances changes neither the salt nor the sand in any way. The larger, colorless salt particles are easily distinguished from the smaller, light-gray sand granules.

FIGURE 1.6 As a result of chemical change, bright steel girders become rusty when exposed to moist air.

Substance is a general term used to denote any variety of matter. *Pure substance* is a specific term that is applied to matter that contains only a single substance.

All samples of a pure substance, no matter what their source, have the same properties under the same conditions.

Use of the Terms *Physical* and *Chemical*

PHYSICAL
This term conveys the idea that the composition (chemical identity) of a substance DOES NOT CHANGE.

CHEMICAL
This term conveys the idea that the composition (chemical identity) of a substance DOES CHANGE.

Physical Properties

Properties observable without changing composition

- Color and shape
- Solid, liquid, or gas
- Boiling point, melting point

Physical Changes

Changes observable without changing composition

- Change in physical state (melting, boiling, freezing, etc.)
- Change in state of subdivision with no change in physical state (pulverizing a solid)

Chemical Properties

Properties that describe how a substance changes (or resists change) to form a new substance

- Flammability (or nonflammability)
- Decomposition at a high temperature (or lack of decomposition)
- Reaction with chlorine (or lack of reaction with chlorine)

Chemical Changes

Changes in which one or more new substances are formed

- Decomposition
- Reaction with another substance

Most naturally occurring samples of matter are mixtures. Gold and diamond are two of the few naturally occurring pure substances. Despite their scarcity in nature, numerous pure substances are known. They are obtained from natural mixtures by using various types of separation techniques or are synthesized in the laboratory from naturally occurring materials.

One characteristic of any mixture is that its components can be separated by using physical means. In our salt–sand mixture, the larger salt crystals could be—though very tediously—"picked out" from the sand. A somewhat easier separation method would be to dissolve the salt in water, which would leave the undissolved sand behind. The salt could then be recovered by evaporation of the water. Figure 1.7 shows a heterogeneous mixture of potassium dichromate (orange crystals) and iron filings. A magnet can be used to separate the components of this mixture.

Another characteristic of a mixture is variable composition. Numerous different salt–sand mixtures, with compositions ranging from a slightly salty sand mixture to a slightly sandy salt mixture, could be made by varying the amounts of the two components.

FIGURE 1.7 (a) A magnet (on the left) and a mixture consisting of potassium dichromate (the orange crystals) and iron filings. (b) The magnet can be used to separate the iron filings from the potassium dichromate.

FIGURE 1.8 Matter falls into two basic classes: pure substances and mixtures. Mixtures, in turn, may be homogeneous or heterogeneous.

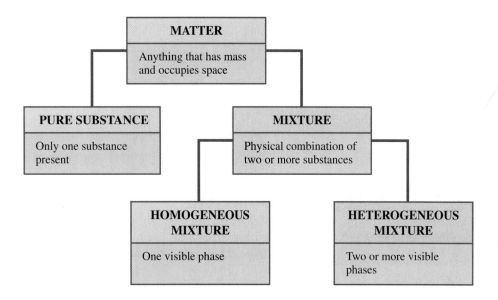

Mixtures are subclassified as heterogeneous or homogeneous. This subclassification is based on visual recognition of the mixture's components. A **heterogeneous mixture** *is a mixture that contains visibly different phases (parts), each of which has different properties.* A nonuniform appearance is a characteristic of all heterogeneous mixtures. Examples include chocolate chip cookies and blueberry muffins. Naturally occurring heterogeneous mixtures include rocks, soils, and wood.

A **homogeneous mixture** *is a mixture that contains only one visibly distinct phase (part), which has uniform properties throughout.* The components present in a homogeneous mixture cannot be visually distinguished. A sugar–water mixture in which all of the sugar has dissolved has an appearance similar to that of pure water. Air is a homogeneous mixture of gases; motor oil and gasoline are multicomponent homogeneous mixtures of liquids; and metal alloys such as 14-karat gold (a mixture of copper and gold) are examples of homogeneous mixtures of solids.

Figure 1.8 summarizes what we have learned thus far about various classifications of matter.

1.6 Elements and Compounds

Chemists have isolated and characterized an estimated 8.5 million pure substances. A very small number of these pure substances, 113 to be exact, are different from all of the others. They are elements. All of the rest, the remaining millions, are compounds. What distinguishes an element from a compound?

An **element** *is a pure substance that cannot be broken down into simpler pure substances by ordinary chemical means such as a reaction, an electric current, heat, or a beam of light.* The metals gold, silver, and copper are all elements.

A **compound** *is a pure substance that can be broken down into two or more simpler pure substances by chemical means.* Water is a compound. By means of an electric current, water can be broken down into the gases hydrogen and oxygen, both of which are elements. The ultimate breakdown products for any compound are elements. A compound's properties are always different from those of its component elements, because the elements are chemically rather than physically combined in the compound (Figure 1.9).

Even though two or more elements are obtained from decomposition of compounds, compounds are not mixtures. Why is this so? Remember, substances can be combined either physically or chemically. Physical combination of substances produces

Both elements and compounds are pure substances.

The definition for the term *element* that is given here will do for now. After considering the concept of atomic number (Section 3.2), we will give a more precise definition.

Every known compound is made up of some combination of 2 or more of the 113 known elements. In any given compound, the elements are combined chemically in fixed proportions by mass.

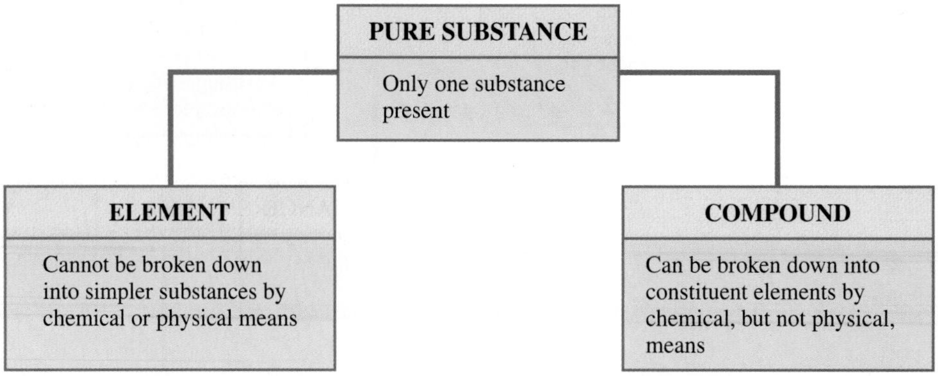

FIGURE 1.9   A pure substance can be
either an element or a compound.

PURE SUBSTANCE

Only one substance
present

ELEMENT

Cannot be broken down
into simpler substances by
chemical or physical means

COMPOUND

Can be broken down into
constituent elements by
chemical, but not physical,
means

▶ There are three major property
distinctions between compounds
and mixtures.
1. Compounds have properties
   distinctly different from those
   of the substances that combined
   to form the compound. The
   components of mixtures retain their
   individual properties.
2. Compounds have a definite
   composition. Mixtures have a variable composition.
3. Physical methods are sufficient to
   separate the components of a
   mixture. The components of a compound cannot be separated by physical methods; chemical methods are
   required.

a mixture. Chemical combination of substances produces a compound, a substance in
which combining entities are *bound* together. No such binding occurs during physical
combination.

The Chemistry at a Glance feature on page 9 summarizes what we have learned thus
far about matter, including pure substances, elements, compounds, and mixtures.

Figure 1.10 summarizes the thought processes that a chemist goes through in classifying a sample of matter as a heterogeneous mixture, a homogeneous mixture, an element,
or a compound. This figure is based on the following three questions about a sample of
matter:

1. Does the sample of matter have the same properties throughout?
2. Are two or more different substances present?
3. Can the pure substance be broken down into simpler substances?

FIGURE 1.10   Questions used in classifying
matter into various categories.

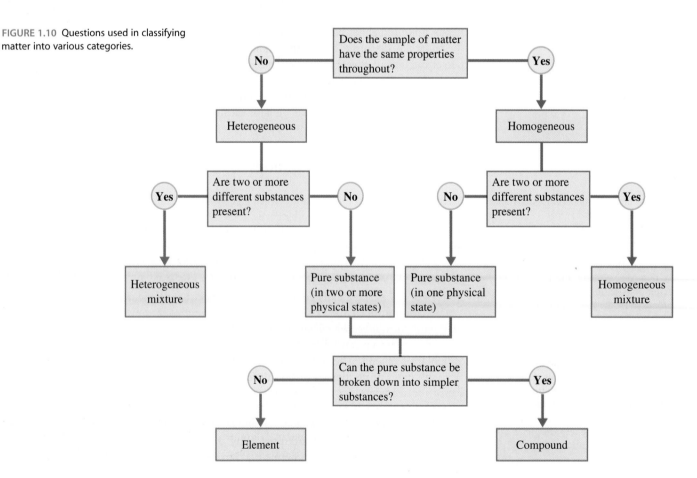

Classes of Matter

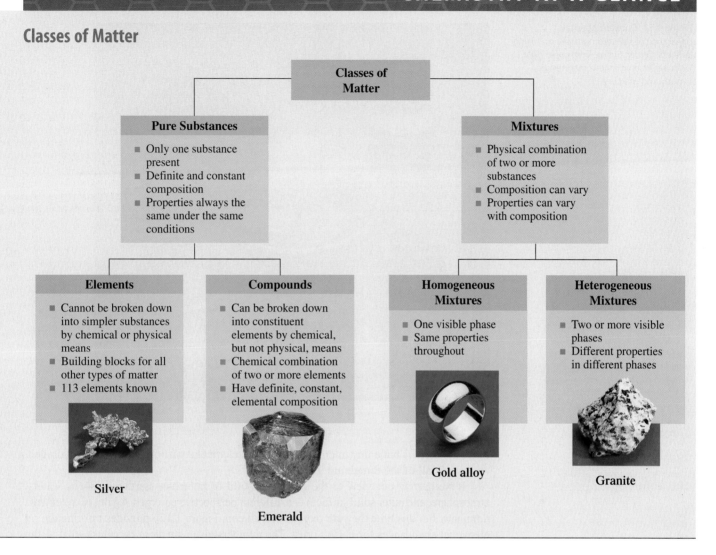

Classes of Matter

Pure Substances
- Only one substance present
- Definite and constant composition
- Properties always the same under the same conditions

Mixtures
- Physical combination of two or more substances
- Composition can vary
- Properties can vary with composition

Elements
- Cannot be broken down into simpler substances by chemical or physical means
- Building blocks for all other types of matter
- 113 elements known

Silver

Compounds
- Can be broken down into constituent elements by chemical, but not physical, means
- Chemical combination of two or more elements
- Have definite, constant, elemental composition

Emerald

Homogeneous Mixtures
- One visible phase
- Same properties throughout

Gold alloy

Heterogeneous Mixtures
- Two or more visible phases
- Different properties in different phases

Granite

1.7 Discovery and Abundance of the Elements

The discovery and isolation of the 113 known elements, the building blocks for all matter, have taken place over a period of several centuries. Most of the discoveries have occurred since 1700, the 1800s being the most active period.

Eighty-eight of the 113 elements occur naturally, and 25 have been synthesized in the laboratory by bombarding samples of naturally occurring elements with small particles. Figure 1.11 shows samples of selected naturally occurring elements. The synthetic (laboratory-produced) elements are all unstable (radioactive) and usually revert quickly to the naturally occurring elements (see Section 11.5).

The naturally occurring elements are not evenly distributed on Earth and in the universe. What is startling is the nonuniformity of the distribution. A small number of elements account for the majority of elemental particles (atoms). (An atom is the smallest particle of an element that can exist; see Section 1.9.)

Studies of the radiation emitted by stars enable scientists to estimate the elemental composition of the universe (Figure 1.12a). Results indicate that two elements, hydrogen and helium, are absolutely dominant. All other elements are mere "impurities" when their abundances are compared with those of these two dominant elements. In this big picture,

A student who attended a university in the year 1700 would have been taught that 13 elements existed. In 1750 he or she would have learned about 16 elements, in 1800 about 34, in 1850 about 59, in 1900 about 82, and in 1950 about 98. Today's total of 113 elements was reached in 1999.

Any increase in the number of known elements from 113 will result from the production of additional synthetic elements. Current chemical theory strongly suggests that all naturally occurring elements have been identified. The isolation of the last of the known naturally occurring elements, rhenium, occurred in 1925.

FIGURE 1.11 Outward physical appearance of selected naturally occurring elements. *Center:* Sulfur. *From upper right, clockwise:* Arsenic, iodine, magnesium, bismuth, and mercury.

in which Earth is but a tiny microdot, 91% of all elemental particles (atoms) are hydrogen, and nearly all of the remaining 9% are helium.

If we narrow our view to the chemical world of humans—Earth's crust (its waters, atmosphere, and outer solid surface)—a different perspective emerges. Again, two elements dominate, but this time they are oxygen and silicon. Figure 1.12b provides information on elemental abundances for Earth's crust. The numbers given are atom percents—that is, the percentage of total atoms that are of a given type. Note that the eight elements listed (the only elements with atom percents greater than 1%) account for over 98% of total atoms in Earth's crust. Note also the dominance of oxygen and silicon; these two elements account for 80% of the atoms that make up the chemical world of humans.

FIGURE 1.12 Abundance of elements in the universe and in Earth's crust (in atom percent).

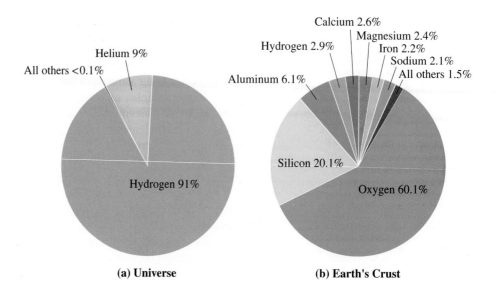

(a) Universe   **(b) Earth's Crust**

CHEMICAL CONNECTIONS

Elements Necessary for Human Life

The distribution of elements in the human body and other living systems is very different from that found in Earth's crust. This distribution is the result of living systems *selectively* taking up matter from their external environment rather than simply accumulating matter representative of their surroundings.

Eleven elements are found in the human body in atom percent levels of 0.01 or greater, as shown in the accompanying table.

Element	Percent of Total Number of Atoms in the Human Body
hydrogen	63
oxygen	25.5
carbon	9.5
nitrogen	1.4
calcium	0.31
phosphorus	0.22
potassium	0.06
sulfur	0.05
chlorine	0.03
sodium	0.03
magnesium	0.01

The high abundances of hydrogen and oxygen in the body reflect its high water content. Hydrogen is over twice as abundant as oxygen, largely because water contains hydrogen and oxygen in a 2-to-1 atom ratio.

The compositions of the four major classes of compounds present in the human body (carbohydrates, fats, proteins, and nucleic acids) involve the elements carbon, nitrogen, sulfur, and

phosphorus, in addition to hydrogen and oxygen. All four of these types of compounds contain hydrogen, oxygen, and carbon. Proteins also contain nitrogen and sulfur, and nucleic acids (DNA, RNA) also contain nitrogen and phosphorus.

The following table offers additional information about the functions of the 11 most abundant elements in the human body. Several other elements, including iron, copper, and iodine, are needed in trace amounts.

Elements Found in Major Types of Biochemical Molecules

hydrogen	present in water, carbohydrates, fats, proteins, and nucleic acids
oxygen	present in water, carbohydrates, fats, proteins, and nucleic acids
carbon	present in carbohydrates, fats, proteins, and nucleic acids
nitrogen	present in proteins and nucleic acids
sulfur	present in proteins
phosphorus	present in nucleic acids

Additional Elements Needed by the Human Body

calcium	present in bones and teeth; needed for proper blood clotting
potassium	helps regulate movement of body fluids
chlorine	helps regulate movement of body fluids
sodium	helps regulate movement of body fluids
magnesium	enhances nerve action and muscle tissue action

1.8 Names and Chemical Symbols of the Elements

Each element has a unique name that, in most cases, was selected by its discoverer. A wide variety of rationales for choosing a name have been applied. Some elements bear geographical names; germanium is named after the native country of its German discoverer, and the elements francium and polonium are named after France and Poland. The elements mercury, uranium, neptunium, and plutonium are all named for planets. Helium gets its name from the Greek word *helios,* for "sun," because it was first observed spectroscopically in the sun's corona during an eclipse. Some elements carry names that reflect specific properties of the element or of the compounds that contain it. Chlorine's name is derived from the Greek *chloros,* denoting "greenish-yellow," the color of chlorine gas. Iridium gets its name from the Greek *iris,* meaning "rainbow"; this alludes to the varying colors of the compounds from which it was isolated.

Abbreviations called chemical symbols also exist for the names of the elements. A **chemical symbol** *is a one- or two-letter designation for an element derived from the element's name.* These symbols are used more frequently than the elements' names. Chemical symbols can be written more quickly than the names, and they occupy less space. A complete list of the known elements and their chemical symbols is given in Table 1.1. The symbols and names of the more frequently encountered elements are shown in color in this table.

Note that the first letter of a chemical symbol is always capitalized and that the second is not. Two-letter symbols are often, but not always, the first two letters of the element's name.

▶ Learning the symbols of the more common elements is an important key to success in studying chemistry. Knowledge of chemical symbols is essential for writing chemical formulas (Section 1.10) and chemical equations (Section 6.6).

TABLE 1.1
The Chemical Symbols for the Elements. The names and symbols of the more frequently encountered elements are shown in color.

Ac	actinium	Ge	germanium	Pr	praseodymium
Ag	silver*	H	hydrogen	Pt	platinum
Al	aluminum	He	helium	Pu	plutonium
Am	americium	Hf	hafnium	Ra	radium
Ar	argon	Hg	mercury*	Rb	rubidium
As	arsenic	Ho	holmium	Re	rhenium
At	astatine	Hs	hassium	Rf	rutherfordium
Au	gold*	I	iodine	Rh	rhodium
B	boron	In	indium	Rn	radon
Ba	barium	Ir	iridium	Ru	ruthenium
Be	beryllium	K	potassium*	S	sulfur
Bh	bohrium	Kr	krypton	Sb	antimony*
Bi	bismuth	La	lanthanum	Sc	scandium
Bk	berkelium	Li	lithium	Se	selenium
Br	bromine	Lu	lutetium	Sg	seaborgium
C	carbon	Lr	lawrencium	Si	silicon
Ca	calcium	Md	mendelevium	Sm	samarium
Cd	cadmium	Mg	magnesium	Sn	tin*
Ce	cerium	Mn	manganese	Sr	strontium
Cf	californium	Mo	molybdenum	Ta	tantalum
Cl	chlorine	Mt	meitnerium	Tb	terbium
Cm	curium	N	nitrogen	Tc	technetium
Co	cobalt	Na	sodium*	Te	tellurium
Cr	chromium	Nb	niobium	Th	thorium
Cs	cesium	Nd	neodymium	Ti	titanium
Cu	copper*	Ne	neon	Tl	thallium
Db	dubnium	Ni	nickel	Tm	thulium
Dy	dysprosium	No	nobelium	U	uranium
Er	erbium	Np	neptunium	V	vanadium
Es	einsteinium	O	oxygen	W	tungsten*
Eu	europium	Os	osmium	Xe	xenon
F	fluorine	P	phosphorus	Y	yttrium
Fe	iron*	Pa	protactinium	Yb	ytterbium
Fm	fermium	Pb	lead*	Zn	zinc
Fr	francium	Pd	palladium	Zr	zirconium
Ga	gallium	Pm	promethium		
Gd	gadolinium	Po	polonium		

Only 109 elements are listed in this table. Elements 110–112 and 114, discovered (synthesized) in the period 1994–1999, are yet to be named.
*These elements have symbols that were derived from non-English names.

Eleven elements have symbols that bear no relationship to the element's English-language name. In ten of these cases, the symbol is derived from the Latin name of the element; in the case of the element tungsten, a German name is the symbol's source. Most of these elements have been known for hundreds of years and date back to the time when Latin was the language of scientists. Elements whose symbols are derived from non-English names are marked with an asterisk in Table 1.1.

1.9 Atoms and Molecules

Consider the process of subdividing a sample of the element gold (or any other element) into smaller and smaller pieces. It seems reasonable that eventually a "smallest possible piece" of gold would be reached that could not be divided further and still be the element

FIGURE 1.13 A computer reconstruction of the surface of a sample of graphite (carbon) as observed with a scanning tunneling microscope. The image reveals the regular pattern of individual carbon atoms. The color was added to the image by computer.

▶ Reasons for the tendency of atoms to assemble into molecules and information on the binding forces involved are considered in Chapter 4.

▶ The Latin word *mole* means "a mass." The word *molecule* denotes "a little mass."

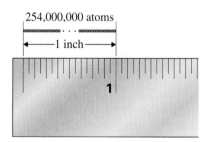

FIGURE 1.14 254 million atoms arranged in a straight line would extend a distance of approximately 1 inch.

FIGURE 1.15 Molecular structure of (a) chlorine, (b) phosphorus, and (c) sulfur.

gold. This smallest possible unit of gold is called a gold atom. An **atom** *is the smallest particle of an element that can exist and still have the properties of the element.*

A sample of any element is composed of atoms of a single type, those of that element. In contrast, a compound must have two or more types of atoms present, because by definition at least two elements must be present (Section 1.6).

No one ever has seen or ever will see an atom with the naked eye; they are simply too small for such observation. However, sophisticated electron microscopes, with magnification factors in the millions, have made it possible to photograph "images" of individual atoms (Figure 1.13).

Atoms are incredibly small particles. Atomic dimensions, although not directly measurable, can be calculated from measurements made on large-size samples of elements. The diameter of an atom is approximately four-billionths of an inch. If atoms of such diameter were arranged in a straight line, it would take 254 million of them to extend a distance of 1 inch (see Figure 1.14).

Free atoms are rarely encountered in nature. Instead, under normal conditions of temperature and pressure, atoms are almost always found together in aggregates or clusters ranging in size from two atoms to numbers too large to count. When the group or cluster of atoms is relatively small and bound together tightly, the resulting entity is called a molecule. A **molecule** *is a group of two or more atoms that functions as a unit because the atoms are tightly bound together.* This resultant "package" of atoms behaves in many ways as a single, distinct particle would.

A **diatomic molecule** *is a molecule that contains two atoms.* It is the simplest type of molecule that can exist. Next is complexity are *triatomic* molecules. A **triatomic molecule** *is a molecule that contains three atoms.* Continuing on numerically, we have *tetratomic* molecules, *pentatomic* molecules, and so on.

The atoms present in a molecule may all be of the same kind, or two or more kinds may be present. On the basis of this observation, molecules are classified into two categories: *homoatomic* and *heteroatomic*. A **homoatomic molecule** *is a molecule in which all atoms present are of the same kind.* A substance containing homoatomic molecules must be an element. The fact that homoatomic molecules exist indicates that *individual* atoms are not always the preferred structural unit for an element. The gaseous elements hydrogen, oxygen, nitrogen, and chlorine exist in the form of diatomic molecules. There are four atoms present in a gaseous phosphorus molecule and eight atoms present in a gaseous sulfur molecule (see Figure 1.15).

A **heteroatomic molecule** *is a molecule in which two or more kinds of atoms are present.* Substances that contain heteroatomic molecules must be compounds, because the presence of two or more kinds of atoms reflects the presence of two or more kinds of elements. The number of atoms present in the heteroatomic molecules associated with compounds varies over a wide range. A water molecule contains 3 atoms: 2 hydrogen atoms and 1 oxygen atom. The compound sucrose (table sugar) has a much larger molecule: 45 atoms are present, of which 12 are carbon atoms, 22 are hydrogen atoms, and 11 are oxygen atoms. Figure 1.16 shows general models for four simple types of heteroatomic molecules. Comparison of parts (c) and (d) of this figure shows that molecules with the same number of atoms need not have the same arrangement of atoms.

(a)
Chlorine molecule

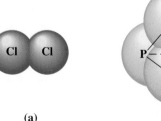

(b)
Phosphorus molecule

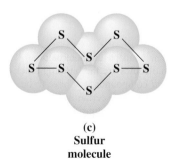

(c)
Sulfur molecule

FIGURE 1.16 Depictions of various simple heteroatomic molecules using models. Spheres of different sizes and colors represent different kinds of atoms.

(a) A diatomic molecule containing one atom of A and one atom of B

(b) A triatomic molecule containing two atoms of A and one atom of B

(c) A tetraatomic molecule containing two atoms of A and two atoms of B

(d) A tetraatomic molecule containing three atoms of A and one atom of B

EXAMPLE 1.2

Classifying Molecules on the Basis of Numbers of and Types of Atoms

■ Classify each of the following molecules as (1) *diatomic, triatomic,* etc., (2) *homoatomic* or *heteroatomic,* and (3) representing an *element* or a *compound.*

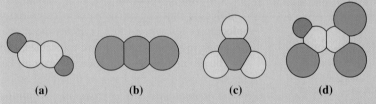

Solution

a. Tetraatomic (four atoms); heteroatomic (two kinds of atoms); a compound (two kinds of atoms)

b. Triatomic (three atoms); homoatomic (only one kind of atom); an element (one kind of atom)

c. Tetraatomic (four atoms); heteroatomic (two kinds of atoms); a compound (two kinds of atoms)

d. Hexatomic (six atoms); heteroatomic (three kinds of atoms); a compound (three kinds of atoms)

Practice Exercise 1.2

Classify each of the following molecules as (1) *diatomic, triatomic,* etc., (2) *homoatomic* or *heteroatomic,* and (3) representing an *element* or a *compound.*

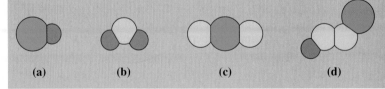

▶ The concept that heteroatomic molecules are the building blocks for *all* compounds will have to be modified when certain solids, called ionic solids, are considered in Section 4.6.

A molecule is the smallest particle of a compound capable of a stable independent existence. Continued subdivision of a quantity of table sugar to yield smaller and smaller amounts would ultimately lead to the isolation of one single "unit" of table sugar: a molecule of table sugar. This table sugar molecule could not be broken down any further and still exhibit the physical and chemical properties of table sugar. The table sugar molecule could be broken down further by chemical (not physical) means to produce atoms, but if that occurred, we would no longer have table sugar. The *molecule* is the limit of *physical* subdivision. The *atom* is the limit of *chemical* subdivision.

1.10 Chemical Formulas

Information about compound composition can be presented in a concise way by using a chemical formula. A **chemical formula** *is a notation made up of the symbols of the elements present in a compound and numerical subscripts (located to the right of each symbol) that indicate the number of atoms of each element present in a molecule of the compound.*

The chemical formula for the compound aspirin is $C_9H_8O_4$. This chemical formula conveys the information that an aspirin molecule contains three different elements—carbon (C), hydrogen (H), and oxygen (O)—and 21 atoms—9 carbon atoms, 8 hydrogen atoms, and 4 oxygen atoms.

When only one atom of a particular element is present in a molecule of a compound, that element's symbol is written without a numerical subscript in the formula for the compound. The formula for rubbing alcohol, C_3H_6O, reflects this practice for the element oxygen.

In order to write formulas correctly, one must follow the capitalization rules for elemental symbols (Section 1.8). Making the error of capitalizing the second letter of an element's symbol can dramatically alter the meaning of a chemical formula. The formulas $CoCl_2$ and $COCl_2$ illustrate this point; the symbol Co stands for the element cobalt, whereas CO stands for one atom of carbon and one atom of oxygen.

Sometimes chemical formulas contain parentheses; an example is $Al_2(SO_4)_3$. The interpretation of this formula is straightforward; in a formula unit, there are present 2 aluminum (Al) atoms and 3 SO_4 groups. The subscript following the parentheses always indicates the number of units in the formula of the polyatomic entity inside the parentheses. In terms of atoms, the formula $Al_2(SO_4)_3$ denotes 2 aluminum (Al) atoms, $3 \times 1 = 3$ sulfur (S) atoms, and $3 \times 4 = 12$ oxygen (O) atoms. Example 1.3 contains further comments about chemical formulas that contain parentheses.

▶ Further information about the use of parentheses in formulas (when and why) will be presented in Section 4.11. The important concern now is being able to interpret formulas that contain parentheses in terms of total atoms present.

EXAMPLE 1.3

Interpreting Chemical Formulas

■ For each of the following chemical formulas, determine how many atoms of each element are present in one molecule of the substance.

a. HCN—hydrogen cyanide, a poisonous gas
b. $C_{18}H_{21}NO_3$—codeine, a pain-killing drug
c. $Ca_{10}(PO_4)_6(OH)_2$—hydroxyapatite, present in tooth enamel

Solution

a. One atom each of the elements hydrogen, carbon, and nitrogen is present. Remember that the subscript 1 is implied when no subscript is written.
b. This formula indicates that 18 carbon atoms, 21 hydrogen atoms, 1 nitrogen atom, and 3 oxygen atoms are present in one molecule of the compound.
c. There are 10 calcium atoms. The amounts of phosphorus, hydrogen, and oxygen are affected by the subscripts outside the parentheses. There are 6 phosphorus atoms and 2 hydrogen atoms present. Oxygen atoms are present in two locations in the formula. There are a total of 26 oxygen atoms: 24 from the PO_4 subunits (6×4) and 2 from the OH subunits (2×1).

Practice Exercise 1.3

For each of the following chemical formulas, determine how many atoms of each element are present in one molecule of the substance.

a. H_2SO_4—sulfuric acid, an industrial acid
b. $C_{17}H_{20}N_4O_6$—riboflavin, a B vitamin
c. $Ca(NO_3)_2$—calcium nitrate, used in fireworks to give a reddish color

Chemistry. Chemistry is the field of study that is concerned with the characterization, composition, and transformations of matter.

Matter. Matter, the substances of the physical universe, is anything that has mass and occupies space. Matter exists in three physical states: solid, liquid, and gas.

Properties of matter. Properties, the distinguishing characteristics of a substance that are used in its identification and description, are of two types: physical and chemical. Physical properties are properties that we can observe without changing a substance into another substance. Chemical properties are properties that matter exhibits as it undergoes or resists changes in chemical composition. The failure of a substance to undergo change in the presence of another substance is considered a chemical property.

Changes in matter. Changes that can occur in matter are classified into two types: physical and chemical. A physical change is a process that does not alter the basic nature (chemical composition) of the substance under consideration. No new substances are ever formed as a result of a physical change. A chemical change is a process that involves a change in the basic nature (chemical composition) of the substance. Such changes always involve conversion of the material or materials under consideration into one or more new substances that have properties and composition distinctly different from those of the original materials.

Pure substances and mixtures. All specimens of matter are either pure substances or mixtures. A pure substance is a form of matter that always has a definite and constant composition. A mixture is a physical combination of two or more pure substances in which the pure substances retain their identity.

Types of mixtures. Mixtures can be classified as heterogeneous or homogeneous on the basis of the visual recognizability of the components present. A heterogeneous mixture contains visibly different parts or phases, each of which has different properties. A homogeneous mixture contains only one phase, which has uniform properties throughout it.

Types of pure substances. A pure substance can be classified as either an element or a compound on the basis of whether it can be broken down into two or more simpler substances by ordinary chemical means. Elements cannot be broken down into simpler substances. Compounds yield two or more simpler substances when broken down. There are 113 pure substances that qualify as elements. There are millions of compounds.

Atoms and molecules. An atom is the smallest particle of an element that can exist and still have the properties of the element. Free isolated atoms are rarely encountered in nature. Instead, atoms are almost always found together in aggregates or clusters. A molecule is a group of two or more atoms that functions as a unit because the atoms are tightly bound together.

Types of molecules. Molecules are of two types: homoatomic and heteroatomic. Homoatomic molecules are molecules in which all atoms present are of the same kind. A pure substance containing homoatomic molecules is an element. Heteroatomic molecules are molecules in which two or more different kinds of atoms are present. Pure substances that contain heteroatomic molecules must be compounds.

Chemical symbols. Chemical symbols are a shorthand notation for the names of the elements. Most consist of two letters; a few involve a single letter. The first letter of a chemical symbol is always capitalized, and the second letter is always lower-case.

Chemical formulas. Chemical formulas are used to specify compound composition in a concise manner. They consist of the symbols of the elements present in the compound and numerical subscripts (located to the right of each symbol) that indicate the number of atoms of each element present in a molecule of the compound.

Atom (1.9)
Chemical change (1.4)
Chemical formula (1.10)
Chemical property (1.3)
Chemical symbol (1.8)
Chemistry (1.1)
Compound (1.6)
Diatomic molecule (1.9)

Element (1.6)
Gas (1.2)
Heteroatomic molecule (1.9)
Heterogeneous mixture (1.5)
Homoatomic molecule (1.9)
Homogeneous mixture (1.5)
Liquid (1.2)
Matter (1.1)

Mixture (1.5)
Molecule (1.9)
Physical change (1.4)
Physical property (1.3)
Property (1.3)
Pure substance (1.5)
Solid (1.2)
Triatomic molecule (1.9)

The members of each pair of problems in this section test similar material.

■ **Chemistry: The Study of Matter (Section 1.1)**

1.1 Classify each of the following as matter or energy (nonmatter).
   a. Air   b. Pizza   c. Sound
   d. Light   e. Gold   f. Virus

1.2 Classify each of the following as matter or energy (nonmatter).
   a. Electricity   b. Bacteria   c. Silver
   d. Cake   e. Water   f. Magnetism

■ **Physical States of Matter (Section 1.2)**

1.3 Give a characteristic that distinguishes
   a. liquids from solids   b. gases from liquids

1.4 Give a characteristic that is the same for
   a. liquids and solids   b. gases and liquids

1.5 Indicate whether each of the following substances would take the shape of its container and also have a definite volume.

a. Copper wire   b. Oxygen gas
c. Granulated sugar   d. Liquid water

1.6 Indicate whether each of the following substances would take the shape of its container and also have an indefinite volume.
a. Aluminum powder   b. Carbon dioxide gas
c. Clean air   d. Gasoline

■ Properties of Matter (Section 1.3)

1.7 The following are properties of the substance magnesium. Classify each property as physical or chemical.
a. Solid at room temperature
b. Ignites upon heating in air
c. Hydrogen gas is produced when it is dissolved in acids
d. Has a density of 1.738 g/cm^3 at 20°C

1.8 The following are properties of the substance magnesium. Classify each property as physical or chemical.
a. Silvery-white in color
b. Does not react with cold water
c. Melts at 651°C
d. Finely divided form burns in oxygen with a dazzling white flame

1.9 Indicate whether each of the following statements describes a physical or a chemical property.
a. Silver salts discolor the skin by reacting with skin protein.
b. Hemoglobin molecules have a red color.
c. Beryllium metal vapor is extremely toxic to humans.
d. Aspirin tablets can be pulverized with a hammer.

1.10 Indicate whether each of the following statements describes a physical or a chemical property.
a. Diamonds are very hard substances.
b. Gold metal does not react with nitric acid.
c. Lithium metal is light enough to float on water.
d. Mercury is a liquid at room temperature.

■ Changes in Matter (Section 1.4)

1.11 Does each of the following statements describe a physical change or a chemical change?
a. An Alka-Seltzer tablet is dropped into water.
b. A table leg is fashioned from a piece of wood.
c. Water is made into ice cubes.
d. Leaves turn red in autumn.

1.12 Does each of the following statements describe a physical change or a chemical change?
a. A board is sawed into two pieces.
b. Frozen orange juice is reconstituted by adding water to it.
c. A container of grape juice left for an extended period of time becomes fermented.
d. Ski goggles get fogged up.

1.13 Correctly complete each of the following sentences by placing the word *chemical* or *physical* in the blank.
a. The freezing over of a pond's surface is a _____ process.
b. The crushing of some ice to make ice chips is a _____ procedure.
c. The destruction of a newspaper through burning it is a _____ process.
d. Pulverizing a hard sugar cube using a mallet is a _____ procedure.

1.14 Correctly complete each of the following sentences by placing the word *chemical* or *physical* in the blank.
a. The reflection of light by a shiny metallic object is a _____ process.

b. The heating of a blue powdered material to produce a white glassy-type substance and a gas is a _____ procedure.
c. A burning candle produces light by _____ means.
d. The grating of a piece of cheese is a _____ technique.

■ Pure Substances and Mixtures (Section 1.5)

1.15 Classify each of the following statements as true or false.
a. All heterogeneous mixtures must contain three or more substances.
b. Pure substances cannot have a variable composition.
c. Substances maintain their identity in a heterogeneous mixture but not in a homogeneous mixture.
d. Pure substances are seldom encountered in the "everyday" world.

1.16 Classify each of the following statements as true or false.
a. All homogeneous mixtures must contain at least two substances.
b. Heterogeneous mixtures, but not homogeneous mixtures, can have a variable composition.
c. Pure substances cannot be separated into other kinds of matter by physical means.
d. The number of known pure substances is less than 100,000.

1.17 Assign each of the following descriptions of matter to one of the following categories: *heterogeneous mixture, homogeneous mixture,* or *pure substance.*
a. Two substances present, two phases present
b. Two substances present, one phase present
c. One substance present, two phases present
d. Three substances present, three phases present

1.18 Assign each of the following descriptions of matter to one of the following categories: *heterogeneous mixture, homogeneous mixture,* or *pure substance.*
a. Three substances present, one phase present
b. One substance present, three phases present
c. One substance present, one phase present
d. Two substances present, three phases present

1.19 Classify each of the following as a *heterogeneous mixture,* a *homogeneous mixture,* or a *pure substance.* Also indicate how many phases are present, assuming all components are present in the same container.
a. Water and dissolved salt
b. Water and sand
c. Water, ice, and oil
d. Carbonated water (soda water) and ice

1.20 Classify each of the following as a *heterogeneous mixture,* a *homogeneous mixture,* or a *pure substance.* Also indicate how many phases are present, assuming all components are present in the same container.
a. Water and dissolved sugar
b. Water and oil
c. Water, wax, and pieces of copper metal
d. Salt water and sugar water

■ Elements and Compounds (Section 1.6)

1.21 From the information given, classify each of the pure substances A through D as elements or compounds, or indicate that no such classification is possible because of insufficient information.
a. Analysis with an elaborate instrument indicates that substance A contains two elements.
b. Substance B decomposes upon heating.
c. Heating substance C to 1000°C causes no change in it.
d. Heating substance D to 500°C causes it to change from a solid to a liquid.

1.22 From the information given, classify each of the pure substances A through D as elements or compounds, or indicate that no such classification is possible because of insufficient information.
 a. Substance A cannot be broken down into simpler substances by chemical means.
 b. Substance B cannot be broken down into simpler substances by physical means.
 c. Substance C readily dissolves in water.
 d. Substance D readily reacts with the element chlorine.

1.23 From the information given in the following equations, classify each of the pure substances A through G as elements or compounds, or indicate that no such classification is possible because of insufficient information.
 a. A + B → C   b. D → E + F + G

1.24 From the information given in the following equations, classify each of the pure substances A through G as elements or compounds, or indicate that no such classification is possible because of insufficient information.
 a. A → B + C   b. D + E → F + G

1.25 Indicate whether each of the following statements is true or false.
 a. Both elements and compounds are pure substances.
 b. A compound results from the physical combination of two or more elements.
 c. In order for matter to be heterogeneous, at least two compounds must be present.
 d. Compounds, but not elements, can have a variable composition.

1.26 Indicate whether each of the following statements is true or false.
 a. Compounds can be separated into their constituent elements by chemical means.
 b. Elements can be separated into their constituent compounds by physical means.
 c. A compound must contain at least two elements.
 d. A compound is a physical mixture of different elements.

■ **Discovery and Abundance of the Elements (Section 1.7)**

1.27 Indicate whether each of the following statements about elements is true or false.
 a. All except two of the elements are naturally occurring.
 b. New elements have been identified within the last 10 years.
 c. Oxygen is the most abundant element in the universe as a whole.
 d. Two elements account for over 75% of the atoms in Earth's crust.

1.28 Indicate whether each of the following statements about elements is true or false.
 a. The majority of the known elements have been discovered since 1900.
 b. At present, 108 elements are known.
 c. Silicon is the second most abundant element in Earth's crust.
 d. Elements that do not occur in nature can be produced in a laboratory setting.

■ **Names and Chemical Symbols of the Elements (Section 1.8)**

1.29 Name the element that each of the following symbols represents.
 a. Ag   b. Au   c. Ca
 d. Na   e. P   f. S

1.30 Name the element that each of the following symbols represents.
 a. Cl   b. Fe   c. He
 d. Hg   e. Ni   f. Pb

1.31 What are the chemical symbols of the following elements?
 a. Tin   b. Copper   c. Aluminum
 d. Boron   e. Barium   f. Argon

1.32 What are the chemical symbols of the following elements?
 a. Zinc   b. Silicon   c. Beryllium
 d. Magnesium   e. Cobalt   f. Fluorine

1.33 In which of the following sequences of elements do all of the elements have two-letter symbols?
 a. Magnesium, nitrogen, phosphorus
 b. Bromine, iron, calcium
 c. Aluminum, copper, chlorine
 d. Boron, barium, beryllium

1.34 In which of the following sequences of elements do all of the elements have symbols that start with a letter that is *not* the first letter of the element's English name?
 a. Silver, gold, mercury   b. Copper, helium, neon
 c. Cobalt, chromium, sodium   d. Potassium, iron, lead

■ **Atoms and Molecules (Section 1.9)**

1.35 Indicate whether each of the following statements is true or false. If a statement is false, change it to make it true. (Such a rewriting should involve more than merely converting the statement to the negative of itself.)
 a. The atom is the limit of chemical subdivision for both elements and compounds.
 b. Triatomic molecules must contain at least two kinds of atoms.
 c. A molecule of a compound must be heteroatomic.
 d. Only heteroatomic molecules may contain three or more atoms.

1.36 Indicate whether each of the following statements is true or false. If a statement is false, change it to make it true. (Such a rewriting should involve more than merely converting the statement to the negative of itself.)
 a. A molecule of an element may be homoatomic or heteroatomic, depending on which element is involved.
 b. The limit of chemical subdivision for a compound is a molecule.
 c. Heteroatomic molecules do not maintain the properties of their constituent elements.
 d. Only one kind of atom may be present in a homoatomic molecule.

1.37 Which of the terms *heteroatomic, homoatomic, diatomic, triatomic, element,* and *compound* apply to each of the following molecules? More than one term will apply in each case.
 a. Q—X   b. Q—Z—X   c. X—X   d. X—Q—X

1.38 Which of the terms *heteroatomic, homoatomic, diatomic, triatomic, element,* and *compound* apply to each of the following molecules? More than one term will apply in each case.
 a. Q—Q   b. Q—Z—Z   c. X—X—Q   d. Z—Q—X

1.39 Draw a diagram of each of the following molecules, using circular symbols of your choice to represent atoms.
 a. A diatomic molecule of a compound
 b. A molecule that is triatomic and homoatomic
 c. A molecule that is tetraatomic and contains three kinds of atoms
 d. A molecule that is triatomic, is symmetrical, and contains two elements

1.40 Draw a diagram of each of the following molecules, using circular symbols of your choice to represent atoms.
   a. A triatomic molecule of an element
   b. A molecule that is diatomic and heteroatomic
   c. A molecule that is triatomic and contains three elements
   d. A molecule that is triatomic, is not symmetrical, and contains two kinds of atoms

■ **Chemical Formulas (Section 1.10)**

1.41 What is the chemical formula for each of the following molecules?
   a. (Q—X)   b. (Q—Z—X)   c. (X—X)   d. (X—Q—X)

1.42 What is the chemical formula for each of the following molecules?
   a. (Q—Q)   b. (Z—Z—X)
   c. (X—X—X)   d. (X—Q—Q—X)

1.43 On the basis of its formula, classify each of the following substances as an element or a compound.
   a. $LiClO_3$ b. CO   c. Co   d. $CoCl_2$
   e. $COCl_2$ f. BN   g. S_8 h. Sn

1.44 On the basis of its formula, classify each of the following substances as an element or a compound.
   a. BaO   b. Ba   c. BaO_2 d. OF_2
   e. O_3 f. O_2 g. Pm   h. $NaNO_3$

1.45 Write a formula for each of the following substances by using the information given about a molecule of the substance.
   a. A molecule of caffeine contains 8 atoms of carbon, 10 atoms of hydrogen, 4 atoms of nitrogen, and 2 atoms of oxygen.
   b. A molecule of table sugar contains 12 atoms of carbon, 22 atoms of hydrogen, and 11 atoms of oxygen.
   c. A molecule of hydrogen cyanide is triatomic and contains the elements hydrogen, carbon, and nitrogen.
   d. A molecule of sulfuric acid contains 2 atoms of hydrogen, 1 atom of sulfur, and the element oxygen and is heptaatomic.

1.46 Write a formula for each of the following substances by using the information given about a molecule of the substance.
   a. A molecule of nicotine contains 10 atoms of carbon, 14 atoms of hydrogen, and 2 atoms of nitrogen.
   b. A molecule of vitamin C contains 6 atoms of carbon, 8 atoms of hydrogen, and 6 atoms of oxygen.
   c. A molecule of nitrous oxide contains twice as many atoms of nitrogen as of oxygen and is triatomic.
   d. A molecule of nitric acid contains 3 atoms of oxygen and the elements hydrogen and nitrogen and is pentaatomic.

1.47 Determine the number of elements and the number of each type of atom present in molecules represented by the following formulas.
   a. H_2CO_3 b. NH_4ClO_4 c. $CaSO_4$ d. C_4H_{10}

1.48 Determine the number of elements and the number of each type of atom present in molecules represented by the following formulas.
   a. $KHCO_3$ b. NaSCN   c. $Na_3P_3O_{10}$ d. NH_4ClO

ADDITIONAL PROBLEMS

1.49 A hard sugar cube is pulverized, and the resulting granules are heated in air until they discolor and then finally burst into flame and burn.
   a. List all *physical* changes to substances mentioned in the preceding narrative.
   b. List all *chemical* changes to substances mentioned in the preceding narrative.

1.50 Assign each of the following descriptions of matter to one of the following categories: *element, compound,* or *mixture.*
   a. One substance present, one phase present, substance cannot be decomposed by chemical means
   b. One substance present, three elements present
   c. Two substances present, two phases present
   d. Two elements present, composition is definite and constant

1.51 Indicate whether each of the following samples of matter is a *heterogeneous mixture,* a *homogeneous mixture,* a *compound,* or an *element.*
   a. A colorless gas, only part of which reacts with hot iron
   b. A "cloudy" liquid that separates into two layers upon standing for 2 hours
   c. A green solid, all of which melts at the same temperature to produce a liquid that decomposes upon further heating
   d. A colorless gas that cannot be separated into simpler substances using physical means and that reacts with copper to produce both a copper–nitrogen compound and a copper–oxygen compound

1.52 Assign each of the following descriptions of matter to one of the following categories: *element, compound,* or *mixture.*
   a. One substance present, one phase present, one kind of homoatomic molecule present

   b. Two substances present, two phases present, all molecules are heteroatomic
   c. One phase present, two kinds of homoatomic molecules present
   d. One phase present, all molecules are triatomic, all molecules are heteroatomic, all molecules are identical

1.53 Certain words can be viewed whimsically as sequential combinations of symbols of elements. For example, the given name Stephen is made up of the following sequence of chemical symbols: S-Te-P-He-N. Analyze each of the following given names in a similar manner.
   a. Barbara
   b. Eugene
   c. Heather
   d. Allan

1.54 Classify each of the following pairs of substances as (1) two elements, (2) two compounds, (3) an element and a compound, or (4) a single pure substance.
   a. (Q—X) and (Q—Q)   b. (Q—X) and (X)
   c. (Q) and (X)   d. (Q—X) and (Q—X)

1.55 In each of the following pairs of formulas, indicate whether the first formula listed denotes *more total atoms, the same number of total atoms,* or *fewer total atoms* than the second formula listed.
   a. HN_3 and NH_3
   b. $CaSO_4$ and $Mg(OH)_2$
   c. $NaClO_3$ and $Be(CN)_2$
   d. $Be_3(PO_4)_2$ and $Mg(C_2H_3O_2)_2$

1.56 On the basis of the given information, determine the numerical value of the subscript x in each of the following chemical formulas.
   a. BaS_2O_x; formula unit contains 6 atoms
   b. $Al_2(SO_x)_3$; formula unit contains 17 atoms
   c. SO_xCl_x; formula unit contains 5 atoms
   d. $C_xH_{2x}Cl_x$; formula unit contains 8 atoms

1.57 A mixture contains the following five pure substances: N_2, N_2H_4, NH_3, CH_4, and CH_3Cl.

   a. How many different kinds of molecules that contain four or fewer atoms are present in the mixture?
   b. How many different kinds of atoms are present in the mixture?
   c. How many total atoms are present in a mixture sample containing five molecules of each component?
   d. How many total hydrogen atoms are present in a mixture sample containing four molecules of each component?

ANSWERS TO PRACTICE EXERCISES

1.1 a. chemical   b. physical   c. chemical   d. physical

1.2 a. diatomic (two atoms); heteroatomic (two kinds of atoms); compound (two kinds of atoms)
   b. triatomic (three atoms); heteroatomic (two kinds of atoms); compound (two kinds of atoms)
   c. triatomic (three atoms); heteroatomic (two kinds of atoms); compound (two kinds of atoms)

   d. tetratomic (four atoms); heteroatomic (three kinds of atoms); compound (three kinds of atoms)

1.3 a. 2 hydrogen atoms, 1 sulfur atom, and 4 oxygen atoms
   b. 17 carbon atoms, 20 hydrogen atoms, and 6 oxygen atoms
   c. 1 calcium atom, 2 nitrogen atoms, and 6 oxygen atoms

2 Measurements in Chemistry

Measurements can never be exact; there is always some degree of uncertainty.

It would be extremely difficult for a carpenter to build cabinets without being able to use hammers, saws, and drills. They are the tools of a carpenter's trade. Chemists also have "tools of the trade." The tool they use most is called *measurement*. Understanding measurement is indispensable in the study of chemistry. Questions such as "How much . . . ?," "How long . . . ?," and "How many . . . ?" simply cannot be answered without resorting to measurements. This chapter will help you learn what you need to know to deal properly with measurement. Much of the material in the chapter is mathematical. This is necessary; measurements require the use of numbers.

2.1 Measurement Systems

We all make measurements on a routine basis. For example, measurements are involved in following a recipe for making brownies, in determining our height and weight, and in fueling a car with gasoline. **Measurement** *is the determination of the dimensions, capacity, quantity, or extent of something*. In chemical laboratories, the most common types of measurements are those of mass, volume, length, time, temperature, pressure, and concentration.

Two systems of measurement are in use in the United States at present: (1) the English system of units and (2) the metric system of units. Common measurements of commerce, such as those used in a grocery store, are made in the *English system*. The units of this system include the inch, foot, pound, quart, and gallon. The *metric system* is used in scientific work. The units of this system include the gram, meter, and liter.

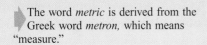 The word *metric* is derived from the Greek word *metron,* which means "measure."

FIGURE 2.1 Metric system units are becoming increasingly evident on highway signs.

▶ The modern version of the metric system is called the International System, or SI (the abbreviation is taken from the French name, *le Système International*).

The United States is in the process of voluntary conversion to the metric system for measurements of commerce. Metric system units now appear on numerous consumer products. Soft drinks now come in 1-, 2-, and 3-liter containers. Road signs in some states display distances in both miles and kilometers (Figure 2.1). Canned and packaged goods such as cereals and mixes on grocery store shelves now have the masses of their contents listed in grams as well as in pounds and ounces.

Interrelationships between units of the same type, such as volume or length, are less complicated in the metric system than in the English system. Within the metric system, conversion from one unit size to another can be accomplished simply by multiplying or dividing by units of 10, because the metric system is a decimal unit system—that is, it is based on multiples of 10. The metric system is simply more convenient to use.

2.2 Metric System Units

In the metric system, there is one base unit for each type of measurement (length, mass, volume, and so on). The names of fractional parts of the base unit and multiples of the base unit are constructed by adding prefixes to the base unit. These prefixes indicate the size of the unit relative to the base unit. Table 2.1 lists common metric system prefixes, along with their symbols or abbreviations and mathematical meanings. The prefixes in color are the ones most frequently used.

The meaning of a metric system prefix is independent of the base unit it modifies and always remains constant. For example, the prefix *kilo-* always means 1000; a *kilo*second is 1000 seconds, a *kilo*watt is 1000 watts, and a *kilo*calorie is 1000 calories. Similarly, the prefix *nano-* always means one-billionth; a *nano*meter is one-billionth of a meter, a *nano*gram is one-billionth of a gram, and a *nano*liter is one-billionth of a liter.

EXAMPLE 2.1

Recognizing the Mathematical Meanings of Metric System Prefixes

▶ The use of numerical prefixes should not be new to you. Consider the use of the prefix *tri-* in the words *tri*angle, *tri*cycle, *tri*o, *tri*nity, and *tri*ple. Each of these words conveys the idea of three of something. The metric system prefixes are used in the same way.

■ Write the name of the metric system prefix associated with the listed power of 10 or the power of 10 associated with the listed metric system prefix.

a. nano- **b.** micro- **c.** deci- **d.** 10^3 **e.** 10^6 **f.** 10^9

Solution

a. The prefix nano- denotes 10^{-9} (one-billionth).
b. The prefix micro- denotes 10^{-6} (one-millionth).
c. The prefix deci- denotes 10^{-1} (one-tenth).
d. 10^3 (one thousand) is denoted by the prefix *kilo-*.
e. 10^6 (one million) is denoted by the prefix *mega-*.
f. 10^9 (one billion) is denoted by the prefix *giga-*.

TABLE 2.1

Common Metric System Prefixes with Their Symbols and Mathematical Meanings

	Prefix[a]	Symbol	Mathematical meaning[b]
Multiples	giga-	G	1,000,000,000 (10^9, billion)
	mega-	M	1,000,000 (10^6, million)
	kilo-	k	1000 (10^3, thousand)
Fractional	deci-	d	0.1 (10^{-1}, one-tenth)
parts	centi-	c	0.01 (10^{-2}, one-hundredth)
	milli-	m	0.001 (10^{-3}, one-thousandth)
	micro-	μ (Greek mu)	0.000001 (10^{-6}, one-millionth)
	nano-	n	0.000000001 (10^{-9}, one-billionth)
	pico-	p	0.000000000001 (10^{-12}, one-trillionth)

[a]Other prefixes also are available but are less commonly used.
[b]The power-of-10 notation for denoting numbers is considered in Section 2.6.

■ Metric Length Units

The **meter** (m) *is the base unit of length in the metric system*. It is about the same size as the English yard; 1 meter equals 1.09 yards (Figure 2.2a). The prefixes listed in Table 2.1 enable us to derive other units of length from the meter. The kilometer (km) is 1000 times larger than the meter; the centimeter (cm) and millimeter (mm) are, respectively, one-hundredth and one-thousandth of a meter. Most laboratory length measurements are made in centimeters rather than meters because of the meter's relatively large size.

> *Length* is measured by determining the distance between two points.

■ Metric Mass Units

The **gram** (g) *is the base unit of mass in the metric system*. It is a very small unit compared with the English ounce and pound (Figure 2.2b). It takes approximately 28 grams to equal 1 ounce and nearly 454 grams to equal 1 pound. Both grams and milligrams (mg) are commonly used in the laboratory, where the kilogram (kg) is generally too large.

> *Mass* is measured by determining the amount of matter in an object.

The terms *mass* and *weight* are often used interchangeably in measurement discussions; technically, however, they have different meanings. **Mass** is a *measure of the total quantity of matter in an object*. **Weight** is a *measure of the force exerted on an object by the pull of gravity*.

The mass of a substance is a constant; the weight of an object varies with the object's geographical location. For example, matter weighs less at the equator than it would at the North Pole because the pull of gravity is less at the equator. Because Earth is not a perfect sphere, but bulges at the equator, the magnitude of gravitational attraction is less at the equator. An object would weigh less on the moon than on Earth because of the smaller size of the moon and the correspondingly lower gravitational attraction. Quantitatively, a 22.0-lb mass weighing 22.0 lb at Earth's North Pole would

> Students often erroneously think that the terms *mass* and *weight* have the same meaning. *Mass* is a measure of the amount of material present in a sample. *Weight* is a measure of the force exerted on an object by the pull of gravity.

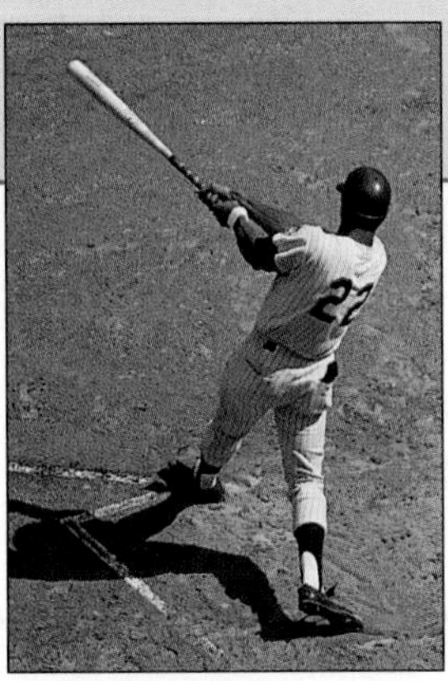

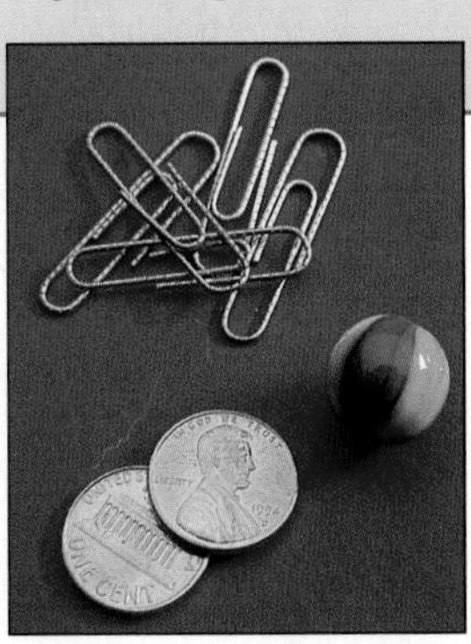

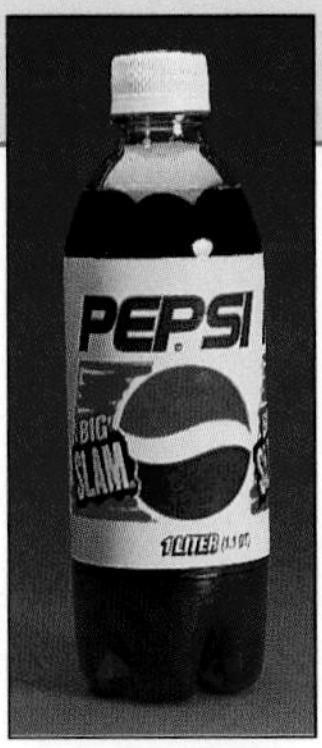

FIGURE 2.2 Comparisons of the base metric system units of length (meter), mass (gram), and volume (liter) with common objects.

(a) Length	**(b) Mass**	**(c) Volume**
A meter is slightly larger than a yard.	A gram is a small unit compared to a pound.	A liter is slightly larger than a quart.
1 meter = 1.09 yards.	1 gram = 1/454 pound.	1 liter = 1.06 quarts.
A baseball bat is about 1 meter long.	Two pennies, five paperclips, and a marble have masses of about 5, 2, and 5 grams, respectively.	Most beverages are now sold by the liter rather than by the quart.

weigh 21.9 lb at Earth's equator and only 3.7 lb on the moon. In outer space, an astronaut may be weightless but never massless. In fact, he or she has the same mass in space as on Earth.

■ Metric Volume Units

The **liter** (**L**) *is the base unit of volume in the metric system.* The abbreviation for liter is a capital L rather than a lower-case l because a lower-case l is easily confused with the number 1. A liter is a volume equal to that occupied by a cube that is 10 centimeters on each side. Because the volume of a cube is calculated by multiplying length times width times height (which are all the same for a cube), we have

$$1 \text{ liter} = \text{volume of a cube with 10 cm edges}$$
$$= 10 \text{ cm} \times 10 \text{ cm} \times 10 \text{ cm}$$
$$= 1000 \text{ cm}^3$$

A liter is also equal to 1000 milliliters; the prefix *milli-* means one-thousandth. Therefore,

$$1000 \text{ mL} = 1000 \text{ cm}^3$$

Dividing both sides of this equation by 1000 reveals that

$$1 \text{ mL} = 1 \text{ cm}^3$$

Consequently, the units milliliter and cubic centimeter are the same. In practice, mL is used for volumes of liquids and gases, and cm³ for volumes of solids. Figure 2.3 shows the relationship between 1 mL (1 cm³) and its parent unit, the liter, in terms of cubic measurements.

A liter and a quart have approximately the same volume; 1 liter equals 1.06 quarts (Figure 2.2c). The milliliter and deciliter (dL) are commonly used in the laboratory. Deciliter units are routinely encountered in clinical laboratory reports detailing the composition of body fluids (Figure 2.4). A deciliter is equal to 100 mL (0.100 L).

> *Volume* is measured by determining the amount of space occupied by a three-dimensional object.

Total volume of large cube = 1000 cm³ = 1 L

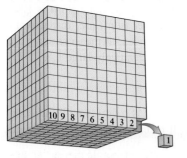

$$1 \text{ cm}^3 = 1 \text{ mL}$$

FIGURE 2.3 A cube 10 cm on a side has a volume of 1000 cm³, which is equal to 1 L. A cube 1 cm on a side has a volume of 1 cm³, which is equal to 1 mL.

> Another abbreviation for the unit cubic centimeter, used in medical situations, is cc.

$$1 \text{ cm}^3 = 1 \text{ cc}$$

FIGURE 2.4 The use of the concentration unit milligrams per deciliter (mg/dL) is common in clinical laboratory reports dealing with the composition of human body fluids.

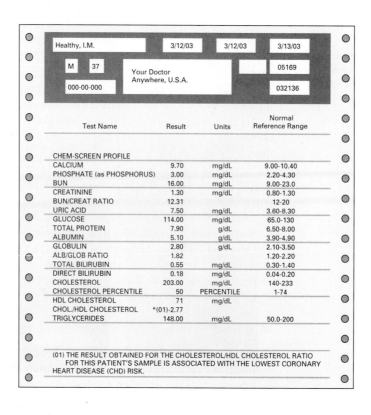

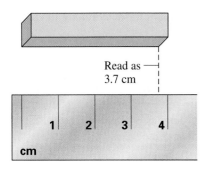

Ruler A

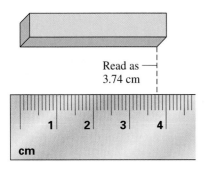

Ruler B

FIGURE 2.5 The scale on a measuring device determines the magnitude of the uncertainty for the recorded measurement. Measurements made with ruler A will have greater uncertainty than those made with ruler B.

2.3 Exact and Inexact Numbers

In scientific work, numbers are grouped in two categories: *exact numbers and inexact numbers*. An **exact number** *is a number whose value has no uncertainty associated with it—that is, it is known exactly.* Exact numbers occur in definitions (for example, there are exactly 12 objects in a dozen, not 12.01 or 12.02); in counting (for example, there can be 7 people in a room, but never 6.99 or 7.03); and in simple fractions (for example, 1/3, 3/5, and 5/9).

An **inexact number** *is a number whose value has a degree of uncertainty associated with it.* Inexact numbers result any time a measurement is made. It is impossible to make an *exact* measurement; some uncertainty will always be present. Flaws in measuring device construction, improper calibration of an instrument, and the skills (or lack of skills) possessed by a person using a measuring device all contribute to error (uncertainty). Section 2.4 considers further the origins of the uncertainty associated with measurements and also the methods used to "keep track" of such uncertainty.

2.4 Uncertainty in Measurement and Significant Figures

As noted in the previous section, because of the limitations of the measuring device and the limited powers of observation of the individual making the measurement, every measurement carries a degree of uncertainty or error. Even when very elaborate measuring devices are used, some degree of uncertainty is always present.

■ Origin of Measurement Uncertainty

To illustrate how measurement uncertainty arises, let us consider how two different rulers, shown in Figure 2.5, are used to measure a given length. Using ruler A, we can say with certainty that the length of the rod is between 3 and 4 centimeters. We can further say that the actual length is closer to 4 centimeters and estimate it to be 3.7 centimeters. Ruler B has more subdivisions on its scale than ruler A. It is marked off in tenths of a centimeter instead of in centimeters. Using ruler B, we can definitely say that the length of the rod is between 3.7 and 3.8 centimeters and can estimate it to be 3.74 centimeters.

Note how both length measurements (ruler A and ruler B) contain some digits (all those except the last one) that are exactly known and one digit (the last one) that is estimated. It is this last digit, the estimated one, that produces uncertainty in a measurement. Note also that the uncertainty in the second length measurement is less than that in the first one—an uncertainty in the hundredths place compared with an uncertainty in the tenths place. We say that the second measurement is more *precise* than the first one; that is, it has less uncertainty than the first measurement.

Only one estimated digit is ever recorded as part of a measurement. It would be incorrect for a scientist to report that the length of the metal rod in Figure 2.5 is 3.745 centimeters as read by using ruler B. The value 3.745 contains two estimated digits, the 4 and the 5, and indicates a measurement with less uncertainty than what is actually obtainable with that particular measuring device. Again, only one estimated digit is ever recorded as part of a measurement.

Because measurements are never exact, two types of information must be conveyed whenever a numerical value for a measurement is recorded: (1) the magnitude of the measurement and (2) the uncertainty of the measurement. The magnitude is indicated by the digit values. Uncertainty is indicated by the number of significant figures recorded. **Significant figures** *are the digits in a measurement that are known with certainty plus one digit that has uncertainty.*

■ Guidelines for Determining Significant Figures

Recognizing the number of significant figures in a measured quantity is easy for measurements we make ourselves, because we know the type of instrument we are using and its limitations. However, when someone else makes the measurement, such information is often not available. In such cases, we follow a set of guidelines for determining the number of significant figures in a measured quantity.

The term *significant figures* is often verbalized in shortened form as "sig figs."

1. In any measurement, all nonzero digits are significant.
2. *Zeros* may or may not be significant because zeros can be used in two ways: (1) to position a decimal point and (2) to indicate a measured value. Zeros that perform the first function are not significant, and zeros that perform the second function are significant. When zeros are present in a measured number, we follow these rules:
   a. *Leading zeros,* those at the beginning of a number, are never significant.

   0.0141 has three significant figures.

   0.0000000048 has two significant figures.

   b. *Confined zeros,* those between nonzero digits, are always significant.

   3.063 has four significant figures.

   0.001004 has four significant figures.

   c. *Trailing zeros,* those at the end of a number, are significant if a decimal point is present in the number.

   56.00 has four significant figures.

   0.05050 has four significant figures.

   d. *Trailing zeros,* those at the end of a number, are not significant if the number lacks an explicitly shown decimal point.

   59,000,000 has two significant figures.

   6010 has three significant figures.

It is important to remember what is "significant" about significant figures. The number of significant figures in a measurement conveys information about the uncertainty associated with the measurement. The "location" of the last significant digit in the numerical value of a measurement specifies the measurement's uncertainty; is this last significant digit located in the hundredths, tenths, ones, or tens position, etc. Consider the following measurement values (with the last significant digit "boxed" for emphasis.

4620.0 (five significant figures) has an uncertainty of tenths.
4620 (three significant figures) has an uncertainty in the tens place.
462,000 (three significant figures) has an uncertainty in the thousands place.

EXAMPLE 2.2

Determining the Number of Significant Figures in a Measurement and the Uncertainty Associated with the Measurement

■ For each of the following measurements, give the number of significant figures present and the uncertainty associated with the measurement.
a. 5623.00   **b.** 0.0031   **c.** 97,200   **d.** 637

Solution

a. Six significant figures are present because trailing zeros are significant when a decimal point is present. The uncertainty is in the hundredths place (± 0.01), the location of the last significant digit.
b. Two significant figures are present because leading zeros are never significant. The uncertainty is in the ten-thousandths place (± 0.0001), the location of the last significant digit.
c. Three significant figures are present because the trailing zeros are not significant (no explicit decimal point is shown). The uncertainty is in the hundreds place (± 100).
d. Three significant figures are present and the uncertainty is in the ones place (± 1).

Practice Exercise 2.2

For each of the following measurements, give the number of significant figures present and the uncertainty associated with the measurement.
a. 727.23   **b.** 0.1031   **c.** 47,230   **d.** 637,000,000

The Chemistry at a Glance feature on page 27 reviews the rules that govern which digits in a measurement are significant.

Significant Figures

SIGNIFICANT FIGURES

The digits in a measurement known with certainty plus one estimated digit.

Nonzero Digits

The digits 1, 2, 3, 4, 5, 6, 7, 8, and 9 are always significant.

The Digit Zero

Zeros may or may not be significant, depending on whether they mark the decimal point or indicate a measured value.

Leading Zeros

Zeros located at the beginning of a number are NEVER significant.

Confined Zeros

Zeros located between nonzero digits are ALWAYS significant.

Trailing Zeros

Zeros located at the end of a number are significant only if the number has an explicitly shown decimal point.

0.0070002000

Not significant   Significant   Significant because of decimal point

2.5 Significant Figures and Mathematical Operations

When measurements are added, subtracted, multiplied, or divided, consideration must be given to the number of significant figures in the computed result. Mathematical operations should not increase (or decrease) the uncertainty of experimental measurements.

Hand-held electronic calculators generally "complicate" uncertainty considerations because they are not programmed to take significant figures into account. Consequently, the digital readouts display more digits than are warranted (Figure 2.6). It is a mistake to record these extra digits, because they are not significant figures and hence are meaningless.

■ Rounding Off Numbers

When we obtain calculator answers that contain too many digits, it is necessary to drop the nonsignificant digits, a process that is called rounding off. **Rounding off** *is the process of deleting unwanted (nonsignificant) digits from calculated numbers.* There are two rules for rounding off numbers.

1. *If the first digit to be deleted is 4 or less, simply drop it and all the following digits.* For example, the number 3.724567 becomes 3.72 when rounded to three significant figures.
2. *If the first digit to be deleted is 5 or greater, that digit and all that follow are dropped and the last retained digit is increased by one.* The number 5.00673 becomes 5.01 when rounded to three significant figures.

FIGURE 2.6 The digital readout on an electronic calculator usually shows more digits than are needed—and more than are acceptable. Calculators are not programmed to account for significant figures.

■ Operational Rules

Significant-figure considerations in mathematical operations that involve measured numbers are governed by two rules, one for multiplication and division and one for addition and subtraction.

1. *In multiplication and division, the number of significant figures in the answer is the same as the number of significant figures in the measurement that contains the fewest significant figures.* For example,

$$\underset{\underset{\text{figures}}{\text{Four significant}}}{6.038} \times \underset{\underset{\text{figures}}{\text{Three significant}}}{2.57} = 15.51766 \quad \text{(calculator answer)}$$

$$= \underset{\underset{\text{figures}}{\text{Three significant}}}{15.5} \quad \text{(correct answer)}$$

The calculator answer is rounded to three significant figures because the measurement with the fewest significant figures (2.57) contains only three significant figures.

2. *In addition and subtraction, the answer has no more digits to the right of the decimal point than are found in the measurement with the fewest digits to the right of the decimal point.* For example,

$$
\begin{array}{ll}
9.333 & \leftarrow \text{Uncertain digit (thousandths)} \\
+\ 1.4 & \leftarrow \text{Uncertain digit (tenths)} \\
\hline
10.733 & \text{(calculator answer)} \\
10.7 & \text{(correct answer)} \\
\end{array}
$$
Uncertain digit (tenths)

The calculator answer is rounded to the tenths place because the uncertainty in the number 1.4 is in the tenths place.

> Concisely stated, the significant-figure operational rules are
>
> × or ÷: Keep smallest number of significant figures in answer.
> + or −: Keep smallest number of decimal places in answer.

Note the contrast between the rule for multiplication and division and the rule for addition and subtraction. In multiplication and division, significant figures are counted; in addition and subtraction, decimal places are counted. It is possible to gain or lose significant figures during addition or subtraction, but *never* during multiplication or division. In our previous sample addition problem, one of the input numbers (1.4) has two significant figures, and the correct answer (10.7) has three significant figures. This is allowable in addition (and subtraction) because we are counting decimal places, not significant figures.

EXAMPLE 2.3

Expressing Answers to the Proper Number of Significant Figures

■ Perform the following computations, expressing your answers to the proper number of significant figures.

a. 6.7321×0.0021 **b.** $\dfrac{16{,}340}{23.42}$ **c.** 6.000×4.000

d. $8.3 + 1.2 + 1.7$ **e.** $3.07 \times (17.6 - 13.73)$

Solution

a. The calculator answer to this problem is

$$6.7321 \times 0.0021 = 0.01413741$$

The input number with the least number of significant figures is 0.0021.

$$6.7321 \times 0.0021$$

Five significant   Two significant
figures   figures

Thus the calculator answer must be rounded to two significant figures.

0.01413741   becomes   0.014
Calculator answer   Correct answer

b. The calculator answer to this problem is

$$\frac{16{,}340}{23.42} = 697.69427$$

Both input numbers contain four significant figures. Thus the correct answer will also contain four significant figures.

697.69427   becomes   697.7
Calculator answer   Correct answer

c. The calculator answer to this problem is

$$6.000 \times 4.000 = 24$$

Both input numbers contain four significant figures. Thus the correct answer must also contain four significant figures, and

24   becomes   24.00
(calculator   (correct
answer)   answer)

Note here how the calculator answer had too few significant figures. Most calculators cut off zeros after the decimal point even if those zeros are significant. Using too few significant figures in an answer is just as wrong as using too many.

d. The calculator answer to this problem is

$$8.3 + 1.2 + 1.7 = 11.2$$

All three input numbers have uncertainty in the tenths place. Thus the last retained digit in the correct answer will be that of tenths. (In this particular problem, the calculator answer and the correct answer are the same, a situation that does not occur very often.)

e. This problem involves the use of both multiplication and subtraction significant-figure rules. We do the subtraction first.

17.6 − 13.73 = 3.87   (calculator answer)

= 3.9   (correct answer)

This answer must be rounded to tenths because the input number 17.6 involves only tenths. We now do the multiplication.

3.07 × 3.9 = 11.973   (calculator answer)

= 12   (correct answer)

The number 3.9 limits the answer to two significant figures.

> Calculators are not programmed to take significant figures into account, which means that students must always adjust their calculator answer to the correct number of significant figures. Sometimes this involves deleting a number of digits through rounding, and other times it involves adding zeros to increase the number of significant figures.

Practice Exercise 2.3

Perform the following computations, expressing your answers to the proper number of significant figures. Assume that all numbers are measured numbers.

a. 5.4430×1.203

b. $\dfrac{17.4}{0.0031}$

c. $7.4 + 20.74 + 3.03$

d. $4.73 \times (2.2 + 8.9)$

Some numbers used in computations may be *exact numbers* rather than measured numbers. Because exact numbers (Section 2.3) have no uncertainty associated with them, they possess an unlimited number of significant figures. Therefore, such numbers never limit the number of significant figures in a computational answer.

2.6 Scientific Notation

Up to this point in the chapter, we have expressed all numbers in decimal notation, the everyday method for expressing numbers. Such notation becomes cumbersome for very large and very small numbers (which occur frequently in scientific work). For example, in one drop of blood, which is 92% water by mass, there are approximately

$$1,600,000,000,000,000,000,000 \text{ molecules}$$

of water, each of which has a mass of

$$0.00000000000000000000030 \text{ gram}$$

Recording such large and small numbers is not only time-consuming but also open to error; often, too many or too few zeros are recorded. Also, it is impossible to multiply or divide such numbers with most calculators because they can't accept that many digits. (Most calculators accept either 8 or 10 digits.)

A method called *scientific notation* exists for expressing in compact form multidigit numbers that involve many zeros. **Scientific notation** *is a system in which a decimal number is expressed as the product of a number between 1 and 10 and 10 raised to a power.* The ordinary decimal number is called a *coefficient* and is written first. The number 10 raised to a power is called an *exponential term*. The coefficient is always multiplied by the exponential term.

> Scientific notation is also called exponential notation.

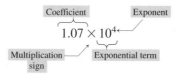

The two previously cited numbers that deal with molecules of water are expressed in scientific notation as

$$1.6 \times 10^{21} \text{ molecules}$$

and

$$3.0 \times 10^{-22} \text{ gram}$$

Obviously, scientific notation is a much more concise way of expressing numbers. Such scientific notation is compatible with most calculators.

■ Converting from Decimal to Scientific Notation

The procedure for converting a number from decimal notation to scientific notation has two parts.

1. *The decimal point in the decimal number is moved to the position behind the first nonzero digit.*
2. *The exponent for the exponential term is equal to the number of places the decimal point has been moved.* The exponent is positive if the original decimal number is 10 or greater and is negative if the original decimal number is less than 1. For numbers between 1 and 10, the exponent is zero.

The following two examples illustrate the use of these procedures:

$$93,000,000 = 9.3 \times 10^7$$

Decimal point is
moved 7 places

$$0.0000037 = 3.7 \times 10^{-6}$$

Decimal point is
moved 6 places

■ Significant Figures and Scientific Notation

How do significant-figure considerations affect scientific notation? The answer is simple. *Only significant figures become part of the coefficient.* The numbers 63, 63.0, and 63.00, which respectively have two, three, and four significant figures, when converted to scientific notation become, respectively,

> The decimal and scientific notation forms of a number *always* contain the same number of significant figures.

$$6.3 \times 10^1 \quad \text{(two significant figures)}$$
$$6.30 \times 10^1 \quad \text{(three significant figures)}$$
$$6.300 \times 10^1 \quad \text{(four significant figures)}$$

Multiplication and division of numbers expressed in scientific notation are common procedures. For these two types of operations, the coefficients, which are decimal numbers, are combined in the usual way. The rules for handling the exponential terms are

1. To multiply exponential terms, *add* the exponents.
2. To divide exponential terms, *subtract* the exponents.

EXAMPLE 2.4

Multiplication and Division in Scientific Notation

■ Carry out the following mathematical operations involving numbers that are expressed in scientific notation.

a. $(2.33 \times 10^3) \times (1.55 \times 10^4)$

b. $\dfrac{8.42 \times 10^6}{3.02 \times 10^4}$

Solution

a. Multiplying the two coefficients gives

$$2.33 \times 1.55 = 3.6115 \quad \text{(calculator answer)}$$
$$= 3.61 \quad \text{(correct answer)}$$

Remember that the coefficient obtained by multiplication can have only three significant figures in this case, the same number as in both input numbers for the multiplication.

Multiplication of the two powers of 10 to give the exponential term requires that we add the exponents.

$$10^3 \times 10^4 = 10^{3+4} = 10^7$$

Combining the new coefficient with the new exponential term gives the answer.

$$3.61 \times 10^7$$

b. Performing the indicated division of the coefficients gives

$$\frac{8.42}{3.02} = 2.7880794 \quad \text{(calculator answer)}$$
$$= 2.79 \quad \text{(correct answer)}$$

Because both input numbers have three significant figures, the answer also has three significant figures.

The division of exponential terms requires that we subtract the exponents.

$$\frac{10^6}{10^4} = 10^{(+6)-(+4)} = 10^2$$

(continued)

Combining the coefficient and the exponential term gives

$$2.79 \times 10^2$$

Practice Exercise 2.4

Carry out the following mathematical operations involving numbers that are expressed in scientific notation.

a. $(4.057 \times 10^3) \times (2.001 \times 10^7)$

b. $\dfrac{4.1 \times 10^{-10}}{3.112 \times 10^{-7}}$

2.7 Conversion Factors and Dimensional Analysis

With both the English unit and metric unit systems in common use in the United States, we often must change measurements from one system to their equivalent in the other system. The mathematical tool we use to accomplish this task is a general method of problem solving called *dimensional analysis.* Central to the use of dimensional analysis is the concept of conversion factors. A **conversion factor** is *a ratio that specifies how one unit of measurement is related to another.*

Conversion factors are derived from equations (equalities) that relate units. Consider the quantities "1 minute" and "60 seconds," both of which describe the same amount of time. We may write an equation describing this fact.

$$1 \text{ min} = 60 \text{ sec}$$

This fixed relationship is the basis for the construction of a pair of conversion factors that relate seconds and minutes.

$$\frac{1 \text{ min}}{60 \text{ sec}} \quad \text{and} \quad \frac{60 \text{ sec}}{1 \text{ min}}$$

These two quantities are the same.

Note that conversion factors always come in pairs, one member of the pair being the reciprocal of the other. Also note that the numerator and the denominator of a conversion factor always describe the same amount of whatever we are considering. One minute and 60 seconds denote the same amount of time.

■ Conversion Factors Within a System of Units

Most students are familiar with and have memorized numerous conversion factors within the English system of measurement (English-to-English conversion factors). Some of these factors, with only one member of a conversion factor pair being listed, are

$$\frac{12 \text{ in.}}{1 \text{ ft}} \qquad \frac{3 \text{ ft}}{1 \text{ yd}} \qquad \frac{4 \text{ qt}}{1 \text{ gal}} \qquad \frac{16 \text{ oz}}{1 \text{ lb}}$$

> In order to avoid confusion with the word *in*, the abbreviation for inches, in., includes a period. This is the only unit abbreviation in which a period appears.

Such conversion factors contain an unlimited number of significant figures because the numbers within them arise from definitions.

Metric-to-metric conversion factors are similar to English-to-English conversion factors in that they arise from definitions. Individual conversion factors are derived from the meanings of the metric system prefixes (Section 2.2). For example, the set of conversion factors involving kilometer and meter come from the equality

> In order to obtain metric-to-metric conversion factors, you need to know the meaning of the metric system prefixes in terms of powers of 10 (see Table 2.1).

$$1 \text{ kilometer} = 10^3 \text{ meters}$$

and those relating microgram and gram come from the equality

$$1 \text{ microgram} = 10^{-6} \text{ gram}$$

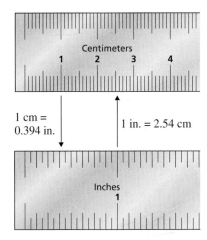

1 cm = 0.394 in.

1 in. = 2.54 cm

FIGURE 2.7 It is experimentally determined that 1 inch equals 2.54 centimeters, or 1 centimeter equals 0.394 inch.

The two pairs of conversion factors are

$$\frac{10^3 \text{ m}}{1 \text{ km}} \quad \text{and} \quad \frac{1 \text{ km}}{10^3 \text{ m}} \qquad \frac{1 \ \mu g}{10^{-6} \text{ g}} \quad \text{and} \quad \frac{10^{-6} \text{ g}}{1 \ \mu g}$$

Note that the numerical equivalent of the prefix is always associated with the base (unprefixed) unit in a metric-to-metric conversion factor.

The number 1 always goes with the *prefixed* unit. ⟶

$$\frac{1 \text{ mL}}{10^{-3} \text{ L}}$$

The power of 10 always ↗ goes with the *unprefixed* unit.

■ Conversion Factors Between Systems of Units

Conversion factors that relate metric units to English units and vice versa are not exact defined quantities because they involve two different systems of measurement. The numbers associated with these conversion factors must be determined experimentally (see Figure 2.7). Table 2.2 lists commonly encountered relationships between metric system and English system units. These few conversion factors are sufficient to solve most of the problems that we will encounter.

Metric-to-English conversion factors can be specified to differing numbers of significant figures. For example,

$$1.00 \text{ lb} = 454 \text{ g}$$
$$1.000 \text{ lb} = 453.6 \text{ g}$$
$$1.0000 \text{ lb} = 453.59 \text{ g}$$

TABLE 2.2
Equalities and Conversion Factors That Relate the English and Metric Systems of Measurement

	Metric to English	English to Metric
Length		
1.00 inch = 2.54 centimeters	$\dfrac{1.00 \text{ in.}}{2.54 \text{ cm}}$	$\dfrac{2.54 \text{ cm}}{1.00 \text{ in.}}$
1.00 meter = 39.4 inches	$\dfrac{39.4 \text{ in.}}{1.00 \text{ m}}$	$\dfrac{1.00 \text{ m}}{39.4 \text{ in.}}$
1.00 kilometer = 0.621 mile	$\dfrac{0.621 \text{ mi}}{1.00 \text{ km}}$	$\dfrac{1.00 \text{ km}}{0.621 \text{ mi}}$
Mass		
1.00 pound = 454 grams	$\dfrac{1.00 \text{ lb}}{454 \text{ g}}$	$\dfrac{454 \text{ g}}{1.00 \text{ lb}}$
1.00 kilogram = 2.20 pounds	$\dfrac{2.20 \text{ lb}}{1.00 \text{ kg}}$	$\dfrac{1.00 \text{ kg}}{2.20 \text{ lb}}$
1.00 ounce = 28.3 grams	$\dfrac{1.00 \text{ oz}}{28.3 \text{ g}}$	$\dfrac{28.3 \text{ g}}{1.00 \text{ oz}}$
Volume		
1.00 quart = 0.946 liter	$\dfrac{1.00 \text{ qt}}{0.946 \text{ L}}$	$\dfrac{0.946 \text{ L}}{1.00 \text{ qt}}$
1.00 liter = 0.265 gallon	$\dfrac{0.265 \text{ gal}}{1.00 \text{ L}}$	$\dfrac{1.00 \text{ L}}{0.265 \text{ gal}}$
1.00 milliliter = 0.034 fluid ounce	$\dfrac{0.034 \text{ fl oz}}{1.00 \text{ mL}}$	$\dfrac{1.00 \text{ mL}}{0.034 \text{ fl oz}}$

In a problem-solving context, which "version" of a conversion factor is used depends on how many significant figures there are in the other numbers of the problem. Conversion factors should never limit the number of significant figures in the answer to a problem. The conversion factors in Table 2.2 are given to three significant figures, which is sufficient for the applications we will make of them.

■ Dimensional Analysis

Dimensional analysis *is a general problem-solving method in which the units associated with numbers are used as a guide in setting up calculations.* In this method, units are treated in the same way as numbers; that is, they can be multiplied, divided, or canceled. For example, just as

$$5 \times 5 = 5^2 \qquad \text{(5 squared)}$$

we have

$$cm \times cm = cm^2 \qquad \text{(cm squared)}$$

Also, just as the 3s cancel in the expression

$$\frac{\cancel{3} \times 5 \times 7}{\cancel{3} \times 2}$$

the centimeters cancel in the expression

$$\frac{(\cancel{cm}) \times (in.)}{(\cancel{cm})}$$

"Like units" found in the numerator and denominator of a fraction will always cancel, just as like numbers do.

The following steps show how to set up a problem using dimensional analysis.

Step 1: *Identify the known or given quantity (both numerical value and units) and the units of the new quantity to be determined.*

This information will always be found in the statement of the problem. Write an equation with the given quantity on the left and the units of the desired quantity on the right.

Step 2: *Multiply the given quantity by one or more conversion factors in such a manner that the unwanted (original) units are canceled, leaving only the desired units.*

The general format for the multiplication is

$$(\text{Information given}) \times (\text{conversion factors}) = (\text{information sought})$$

The number of conversion factors used depends on the individual problem.

Step 3: *Perform the mathematical operations indicated by the conversion factor setup.*

When performing the calculation, double-check to make sure that all units except the desired set have canceled.

EXAMPLE 2.5

Unit Conversions Within the Metric System

■ A standard aspirin tablet contains 324 mg of aspirin. How many grams of aspirin are in a standard aspirin tablet?

Solution

Step 1: The given quantity is 324 mg, the mass of aspirin in the tablet. The unit of the desired quantity is grams.

$$324 \text{ mg} = ? \text{ g}$$

Step 2: Only one conversion factor will be needed to convert from milligrams to grams, one that relates milligrams to grams. The two forms of this conversion factor are

$$\frac{1 \text{ mg}}{10^{-3} \text{ g}} \quad \text{and} \quad \frac{10^{-3} \text{ g}}{1 \text{ mg}}$$

The second factor is the one needed because it allows for cancellation of the milligram units, leaving us with grams as the new units.

$$324 \text{ mg} \times \left(\frac{10^{-3} \text{ g}}{1 \text{ mg}} \right) = ? \text{ g}$$

Step 3: Combining numerical terms as indicated generates the final answer.

$$\left(324 \times \frac{10^{-3}}{1} \right) \text{ g} = 0.324 \text{ g}$$

Number from first factor   Numbers from second factor

The answer is given to three significant figures because the given quantity in the problem, 324 mg, has three significant figures. The conversion factor used arises from a definition and thus does not limit significant figures in any way.

Practice Exercise 2.5

Analysis shows the presence of 203 μg of cholesterol in a sample of blood. How many grams of cholesterol are present in this blood sample?

EXAMPLE 2.6

Unit Conversions Between the Metric and English Systems

■ Capillaries, the microscopic vessels that carry blood from small arteries to small veins, are on the average only 1 mm long. What is the average length of a capillary in inches?

Solution

Step 1: The given quantity is 1 mm, and the units of the desired quantity are inches.

$$1 \text{ mm} = ? \text{ in.}$$

Step 2: The conversion factor needed for a one-step solution, millimeters to inches, is not given in Table 2.2. However, a related conversion factor, meters to inches, is given. Therefore, we first convert millimeters to meters and then use the meters-to-inches conversion factor in Table 2.2.

$$\text{mm} \longrightarrow \text{m} \longrightarrow \text{in.}$$

The correct conversion factor setup is

$$1 \text{ mm} \times \left(\frac{10^{-3} \text{ m}}{1 \text{ mm}} \right) \times \left(\frac{39.4 \text{ in.}}{1.00 \text{ m}} \right) = ? \text{ in.}$$

All of the units except for inches cancel, which is what is needed. The information for the first conversion factor was obtained from the meaning of the prefix *milli-*.

This setup illustrates the fact that sometimes the given units must be changed to intermediate units before common conversion factors, such as those found in Table 2.2, are applicable.

Step 3: Collecting the numerical factors and performing the indicated math gives

$$\left(\frac{1 \times 10^{-3} \times 39.4}{1 \times 1.00} \right) \text{ in.} = 0.0394 \text{ in.} \quad \text{(calculator answer)}$$

$$= 0.04 \text{ in.} \quad \text{(correct answer)}$$

The calculator answer must be rounded to one significant figure because 1 mm, the given quantity, contains only one significant figure.

(continued)

Conversion Factors

Characteristics of Conversion Factors

- Ratios that specify how units are related to each other
- Derived from equations that relate units

    1 minute = 60 seconds

- Come in pairs, one member of the pair being the reciprocal of the other

    $$\frac{1 \text{ min}}{60 \text{ sec}} \quad \text{and} \quad \frac{60 \text{ sec}}{1 \text{ min}}$$

- Conversion factors originate from two types of relationships:

    (1) defined relationships
    (2) measured relationships

Conversion Factors from DEFINED Relationships

- All English-to-English and metric-to-metric conversion factors
- Such conversion factors have an unlimited number of significant figures

    12 inches = 1 foot (exactly)
    4 quarts = 1 gallon (exactly)
    1 kilogram = 10^3 grams (exactly)

- Metric-to-metric conversion factors are derived using the meaning of the metric system prefixes

Conversion Factors from MEASURED Relationships

- All English-to-metric and metric-to-English conversion factors
- Such conversion factors have a specific number of significant figures, depending on the precision of the defining relationship

    1.00 lb = 454 g (three sig figs)
    1.000 lb = 453.6 g (four sig figs)
    1.0000 lb = 453.59 g (five sig figs)

Prefixes That INCREASE Base Unit Size

kilo- 10^3
mega- 10^6
giga- 10^9

Prefixes That DECREASE Base Unit Size

deci- 10^{-1}
centi- 10^{-2}
milli- 10^{-3}
micro- 10^{-6}
nano- 10^{-9}

Practice Exercise 2.6

Blood analysis reports often give the amounts of various substances present in the blood in terms of milligrams per deciliter. What is the measure, in quarts, of 1.00 deciliter?

The Chemistry at a Glance feature above summarizes what we have learned about conversion factors in this section.

FIGURE 2.8 (a) The penny is less dense than the mercury it floats on. (b) Liquids that do not dissolve in one another and that have different densities float on one another, forming layers. The top layer is gasoline, with a density of about 0.8 g/mL. Next is water (plus food coloring), with a density of 1.0 g/mL. The next layer is carbon tetrachloride, with a density of 1.6 g/mL. The bottom layer is mercury, with a density of 13.6 g/mL.

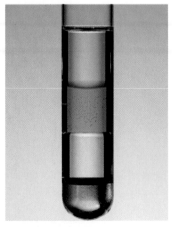

(a)   (b)

2.8 Density

Density *is the ratio of the mass of an object to the volume occupied by that object.*

$$\text{Density} = \frac{\text{mass}}{\text{volume}}$$

People often speak of a substance as being heavier or lighter than another substance. What they actually mean is that the two substances have different densities; a specific volume of one substance is heavier or lighter than the same volume of the second substance (Figure 2.8). Equal masses of substances with different densities occupy different volumes; the contrast in volume is often very striking (see Figure 2.9).

A correct density expression includes a number, a mass unit, and a volume unit. Although any mass and volume units can be used, densities are generally expressed in grams per cubic centimeter (g/cm^3) for solids, grams per milliliter (g/mL) for liquids, and grams per liter (g/L) for gases. Table 2.3 gives density values for a number of substances. Note that temperature must be specified with density values, because substances expand and contract with changes in temperature. For the same reason, the pressure of gases is also given with their density values.

FIGURE 2.9 Both of these items have a mass of 23 grams, but they have very different volumes; therefore, their densities are different as well.

▶ Density is a *physical property* peculiar to a given substance or mixture under fixed conditions.

E X A M P L E   2 . 7

Calculating Density Given a Mass and a Volume

■ A student determines that the mass of a 20.0-mL Sample of olive oil is 18.4 g. What is the density of the olive oil in grams per milliliter?

Solution

To calculate density, we substitute the given mass and volume values into the defining formula for density.

$$\text{Density} = \frac{\text{mass}}{\text{volume}} = \frac{18.4 \text{ g}}{20.0 \text{ mL}} = 0.92 \frac{\text{g}}{\text{mL}} \quad \text{(calculator answer)}$$

$$= 0.920 \frac{\text{g}}{\text{mL}} \quad \text{(correct answer)}$$

Because both input numbers contain three significant figures, the density is specified to three significant figures.

Practice Exercise 2.7

A sample of table sugar (sucrose) with a mass of 2.500 g occupies a volume of 1.575 cm^3. What is the density, in grams per cubic centimeter, of this sample of table sugar?

TABLE 2.3
Densities of Selected Substances

Solids (25°C)			
gold	19.3 g/cm³	table salt	2.16 g/cm³
lead	11.3 g/cm³	bone	1.7–2.0 g/cm³
copper	8.93 g/cm³	table sugar	1.59 g/cm³
aluminum	2.70 g/cm³	wood (pine)	0.30–0.50 g/cm³
Liquids (25°C)			
mercury	13.55 g/mL	water	0.997 g/mL
milk	1.028–1.035 g/mL	olive oil	0.92 g/mL
blood plasma	1.027 g/mL	ethyl alcohol	0.79 g/mL
urine	1.003–1.030 g/mL	gasoline	0.56 g/mL
Gases (25°C and 1 atmosphere pressure)			
chlorine	3.17 g/L	nitrogen	1.25 g/L
carbon dioxide	1.96 g/L	methane	0.66 g/L
oxygen	1.42 g/L	hydrogen	0.08 g/L
air (dry)	1.29 g/L		

CHEMICAL CONNECTIONS

Body Density and Percent Body Fat

More than half the adult population of the United States is overweight. But what does "overweight" mean? In years past, people were considered overweight if they weighed more for their height than called for in standard height/mass charts. Such charts are now considered outdated. Today, we realize that body composition is more important than total body mass. The proportion of fat to total body mass—that is, the percent of body fat—is the key to defining *overweight*. A very muscular person, for example, can be overweight according to height/mass charts although he or she has very little body fat. Some athletes fall into this category. Body composition ratings, tied to percent body fat, are listed here.

Body composition rating	Percent body fat	
	Men	**Women**[a]
excellent	less than 13	less than 18
good	13–17	18–22
average	18–21	23–26
fair	22–30	27–35
poor	greater than 30	greater than 35

[a]Women are genetically predisposed to maintain a higher percentage of body fat.

The percentage of fat in a person's body can be determined by hydrostatic (underwater) weighing. Fat cells, unlike most other human body cells and fluids, are less dense than water. Consequently, a person with a high percentage of body fat is buoyed up by water more than is a lean person. The hydrostatic-weighing technique for determining body fat is based on this difference in density. A person is first weighed in air and then weighed again submerged in water. The difference between these two masses (with a correction for residual air in the lungs

and for the temperature of the water) is used to calculate body density. The higher the density of the body, the lower the percent of body fat. Sample values relating body density and percent body fat are given here.

Body density (g/mL)	Percent body fat
1.070	12.22
1.062	15.25
1.052	19.29
1.036	25.35
1.027	29.39

▶ Density may be used as a conversion factor to convert from mass to volume or vice versa.

Density can be used as a conversion factor that relates the volume of a substance to its mass. This use of density enables us to calculate the volume of a substance if we know its mass. Conversely, the mass can be calculated if the volume is known.

Density conversion factors, like all other conversion factors, have two reciprocal forms. For a density of 1.03 g/mL, the two conversion factor forms are

$$\frac{1.03 \text{ g}}{1 \text{ mL}} \quad \text{and} \quad \frac{1 \text{ mL}}{1.03 \text{ g}}$$

EXAMPLE 2.8

Converting from Mass to Volume by Using Density as a Conversion Factor

■ Blood plasma has a density of 1.027 g/mL at 25°C. What volume, in milliliters, does 125 g of plasma occupy?

Solution

Step 1: The given quantity is 125 g of blood plasma. The units of the desired quantity are milliliters. Thus our starting point is

$$125 \text{ g} = ? \text{ mL}$$

Step 2: The conversion from grams to milliliters can be accomplished in one step because the given density, used as a conversion factor, directly relates grams to milliliters. Of the two conversion factor forms

$$\frac{1.027 \text{ g}}{1 \text{ mL}} \quad \text{and} \quad \frac{1 \text{ mL}}{1.027 \text{ g}}$$

we will use the latter, because it allows for cancellation of gram units, leaving milliliters.

$$125 \text{ g} \times \left(\frac{1 \text{ mL}}{1.027 \text{ g}} \right) = ? \text{ mL}$$

Step 3: Doing the necessary arithmetic gives us our answer:

$$\left(\frac{125 \times 1}{1.027} \right) \text{ mL} = 121.71372 \text{ mL} \quad \text{(calculator answer)}$$

$$= 122 \text{ mL} \quad \text{(correct answer)}$$

Even though the given density contained four significant figures, the correct answer is limited to three significant figures. This is because the other given number, the mass of blood plasma, had only three significant figures.

Practice Exercise 2.8

If your blood has a density of 1.05 g/mL at 25°C, how many grams of blood would you lose if you made a blood bank donation of 1.00 pint (473 mL) of blood?

2.9 Temperature Scales and Heat Energy

Heat is a form of energy. Temperature is an indicator of the tendency of heat energy to be transferred. Heat energy flows from objects of higher temperature to objects of lower temperature.

Three different temperature scales are in common use: Celsius, Kelvin, and Fahrenheit (Figure 2.10). Both the Celsius and the Kelvin scales are part of the metric measurement system; the Fahrenheit scale belongs to the English measurement system. Degrees of different size and different reference points are what produce the various temperature scales.

The *Celsius scale* is the scale most commonly encountered in scientific work. The normal boiling and freezing points of water serve as reference points on this scale, the former having a value of 100° and the latter 0°. Thus there are 100 "degree intervals" between the two reference points.

FIGURE 2.10 The relationships among the Celsius, Kelvin, and Fahrenheit temperature scales are determined by the degree sizes and the reference point values. The reference point values are not drawn to scale.

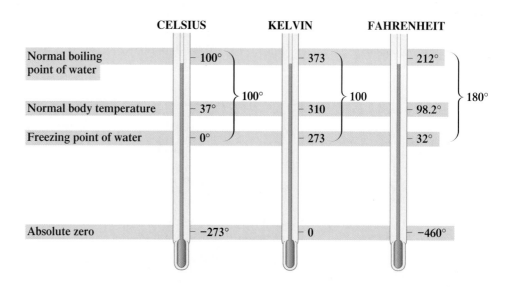

Zero on the Kelvin scale is known as *absolute zero*. It corresponds to the lowest temperature allowed by nature. How fast particles (molecules) move depends on temperature. The colder it gets, the more slowly they move. At absolute zero, movement stops. Scientists in laboratories have been able to attain temperatures as low as 0.0001 K, but a temperature of 0 K is impossible.

The *Kelvin scale* is a close relative of the Celsius scale. Both have the same size degree, and the number of degrees between the freezing and boiling points of water is the same. The two scales differ only in the numbers assigned to the reference points. On the Kelvin scale, the boiling point of water is 373 kelvins (K) and the freezing point of water is 273 K. The choice of these reference points makes all temperature readings on the Kelvin scale positive values. Note that the degree sign (°) is not used with the Kelvin scale. For example, we say that an object has a temperature of 350 K (*not* 350°K).

The *Fahrenheit scale* has a smaller degree size than the other two temperature scales. On this scale, there are 180 degrees between the freezing and boiling points of water as contrasted to 100 degrees on the other two scales. Thus the Celsius (and Kelvin) degree size is almost two times ($\frac{9}{5}$) larger than the Fahrenheit degree. Reference points on the Fahrenheit scale are 32° for the freezing point of water and 212° for the normal boiling point of water.

■ **Conversions Between Temperature Scales**

Because the size of the degree is the same, the relationship between the Kelvin and Celsius scales is very simple. No conversion factors are needed; all that is required is an adjustment for the differing numerical scale values. The adjustment factor is 273, the number of degrees by which the two scales are offset from one another.

$$K = °C + 273$$
$$°C = K - 273$$

The relationship between the Fahrenheit and Celsius scales can also be stated in an equation format.

$$°F = \frac{9}{5}(°C) + 32 \quad \text{or} \quad °C = \frac{5}{9}(°F - 32)$$

EXAMPLE 2.9

Converting from One Temperature Scale to Another

■ Body temperature for a person with a high fever is found to be 104°F. To what is this temperature equivalent on the following scales?

a. Celsius scale   **b.** Kelvin scale

Solution

a. We substitute 104 for °F in the equation

$$°C = \frac{5}{9}(°F - 32)$$

Then solving for °C gives

$$°C = \frac{5}{9}(104 - 32) = \frac{5}{9}(72) = 40°$$

b. Using the answer from part **a** and the equation

$$K = °C + 273$$

we get, by substitution,

$$K = 40 + 273 = 313$$

Practice Exercise 2.9

In the human body, heat stroke occurs at a temperature of 41°C. To what is this temperature equivalent on the following scales?

a. Fahrenheit scale   **b.** Kelvin scale

TABLE 2.4

Specific Heats of Selected Common Substances

Substance	Specific heat (cal/g · °C)ᵃ
water, liquid	1.00
ethyl alcohol	0.58
olive oil	0.47
wood	0.42
aluminum	0.21
glass	0.12
silver	0.057
gold	0.031

ᵃThe unit notation cal/g · °C means calories per gram per degree Celsius.

> In discussions involving nutrition, the energy content of foods, and dietary tables, the term *Calorie* (spelled with a capital C) is used. The dietetic Calorie is actually 1 kilocalorie (1000 calories). The statement that an oatmeal raisin cookie contains 60 Calories means that 60 kcal (60,000 cal) of energy is released when the cookie is metabolized (undergoes chemical change) within the body.

> Water has the highest specific heat of all common substances.

■ Heat Energy and Specific Heat

The form of energy most often required for or released by chemical reactions and physical changes is *heat energy*. A commonly used unit for the measurement of heat energy is the calorie. A **calorie** (cal) *is the amount of heat energy needed to raise the temperature of 1 gram of water by 1 degree Celsius.* For large amounts of heat energy, the measurement is usually expressed in kilocalories.

$$1 \text{ kilocalorie} = 1000 \text{ calories}$$

Another unit for heat energy that is used with increasing frequency is the joule (J). The relationship between the joule (which rhymes with *pool*) and the calorie is

$$1 \text{ calorie} = 4.184 \text{ joules}$$

Heat energy values in calories can be converted to joules by using the conversion factor

$$\frac{4.184 \text{ J}}{1 \text{ cal}}$$

Specific heat *is the quantity of heat energy, in calories, necessary to raise the temperature of 1 gram of a substance by 1 degree Celsius.* Specific heats for a number of substances in various states are given in Table 2.4.

The higher the specific heat of a substance, the less its temperature will change as it absorbs a given amount of heat. Water has a relatively high specific heat; it is thus a very effective coolant. The moderate climates of geographical areas near large bodies of water—the Hawaiian Islands, for example—are related to the ability of water to absorb large amounts of heat without undergoing drastic temperature changes. Desert areas (obviously lacking in water) experience low- and high-temperature extremes (see Figure 2.11).

Specific heat is an important quantity because it can be used to calculate the number of calories required to heat a known mass of a substance from one temperature to another. It can also be used to calculate how much the temperature of a substance increases

FIGURE 2.11 Water's high specific heat causes it to have a moderating effect in climates like that of Hawaii. The desert, lacking water, experiences extreme temperatures.

Normal Human Body Temperature

Studies show that "normal" human body temperature varies from individual to individual. For oral temperature measurements, this individual variance spans the range from 96°F to 101°F.

Furthermore, individual body temperatures vary with exercise and with the temperature of the surroundings. When excessive heat is produced in the body by strenuous exercise, oral temperature can rise as high as 103°F. On the other hand, when the body is exposed to cold, oral temperature can fall to values considerably below 96°F. A rapid fall in temperature of 2°F to 3°F produces uncontrollable shivering.

Each individual also has a characteristic pattern of temperature variation during the day, with differences of as much as 1°F to 3°F between high and low points. Body temperature is typically lowest in the very early morning, after several hours of sleep, when one is inactive and not digesting food. During the day, body temperature rises to a peak and begins to fall again. "Morning people"—people who are most productive early in the day—have a body temperature peak at midmorning or midday. "Night people"—people who feel as though they are just getting started as evening approaches and who work best late at night—have a body temperature peak in the evening.

What, then, is the average (normal) human body temperature? Reference books list the value 98.6°F (37.0°C) as the answer to this question. The source for this value is a study involving over 1 million human body temperature readings that was published in 1868, over 130 years ago.

A 1992 study, published in the *Journal of the American Medical Association,* questions the validity of this average value (98.6°F). This new study notes that the 1868 study was carried out using thermometers that were more difficult to get accurate readings from than modern thermometers. The 1992 study is based on oral temperature readings obtained using electronic thermometers. Findings of this new study include the following:

1. The range of temperatures was 96.0°F to 100.8°F.
2. The mean (average) temperature was 98.2°F (36.8°C).
3. At 6 A.M., the temperature 98.9°F is the upper limit of the normal temperature range.
4. In late afternoon (4 P.M.), the temperature 99.9°F is the upper limit of the normal temperature range.
5. Women have a slightly higher average temperature than men (98.4°F versus 98.1°F).
6. Over the temperature range 96°F to 101°F, there is an average increase in heart rate of 2.44 beats per minute for each 1°F rise in temperature.

As a result of this study, future reference books will probably use 98.2°F rather than 98.6°F as the value for average (normal) human body temperature.

▶ The property of specific heat varies *slightly* with temperature and pressure. We will ignore such variations in this text.

when it absorbs a known number of calories of heat. The equation used for such calculations is

$$\text{Heat absorbed} = \text{specific heat} \times \text{mass} \times \text{temperature change}$$

If any three of the four quantities in this equation are known, the fourth quantity can be calculated. If the units for specific heat are cal/g · °C, the units for mass are grams, and the units for temperature change are °C, then the heat absorbed has units of calories.

EXAMPLE 2.10

Calculating the Amount of Heat Released as the Result of a Temperature Decrease

■ If a hot-water bottle contains 1200 g of water at 65°C, how much heat, in calories, will it have supplied to a person's "aching back" by the time it has cooled to 37°C (assuming all of the heat energy goes into the person's back)?

Solution

We will substitute known quantities into the equation

$$\text{Heat released} = \text{specific heat} \times \text{mass} \times \text{temperature change}$$

Table 2.4 shows that the specific heat of liquid water is 1.00 cal/g · °C. The mass of the water is given as 1200 g. The temperature change in going from 65°C to 37°C is 28°C. Substituting these values into the preceding equation gives

$$\text{Heat released} = \left(\frac{1.00 \text{ cal}}{\text{g} \cdot °\text{C}} \right) \times (1200 \text{ g}) \times (28°\text{C})$$

$$= 33{,}600 \text{ cal} \quad \text{(calculator answer)}$$

$$= 34{,}000 \text{ cal} \quad \text{(correct answer)}$$

The given quantity of 1200 g and the temperature difference of 28°C, both of which have only two significant figures, limit the answer to two significant figures.

Practice Exercise 2.10

How much heat energy, in calories, must be absorbed by 125.0 g of water to raise its temperature by 12°C?

CONCEPTS TO REMEMBER

The metric system. The metric system, the measurement system preferred by scientists, is a decimal system in which larger and smaller units of a quantity are related by factors of 10. Prefixes are used to designate relationships between the basic unit and larger or smaller units of a quantity. Units in the metric system include the gram (mass), liter (volume), and meter (length).

Exact and inexact numbers. Numbers are of two kinds: exact and inexact. An exact number has a value that has no uncertainty associated with it. Exact numbers occur in definitions, in counting, and in simple fractions. An inexact number has a value that has a degree of uncertainty associated with it. Inexact numbers are generated any time a measurement is made.

Significant figures. Significant figures in a measurement are those digits that are certain, plus a last digit that has been estimated. The maximum number of significant figures possible in a measurement is determined by the design of the measuring device.

Calculations and significant figures. Calculations should never improve (or decrease) the precision of experimental measurements. In multiplication and division, the number of significant figures in the answer is the same as that in the measurement containing the fewest significant figures. In addition and subtraction, the answer has no more digits to the right of the decimal point than are found in the measurement with the fewest digits to the right of the decimal point.

Scientific notation. Scientific notation is a system for writing decimal numbers in a more compact form that greatly simplifies the mathematical operations of multiplication and division. In this system, numbers are expressed as the product of a number between 1 and 10 and 10 raised to a power.

Dimensional analysis. Dimensional analysis is a general problem-solving method in which the units associated with numbers are used as a guide in setting up calculations. A given quantity is multiplied by one or more conversion factors in such a manner that the unwanted (original) units are canceled, leaving only the desired units.

Density. Density is the ratio of the mass of an object to the volume occupied by that object. A correct density expression includes a number, a mass unit, and a volume unit.

Temperature scales. The three major temperature scales are the Celsius, Kelvin, and Fahrenheit scales. The size of the degree for the Celsius and Kelvin scales is the same; they differ only in the numerical values assigned to the reference points. The Fahrenheit scale has a smaller degree size than the other two temperature scales.

Heat energy and specific heat. The most commonly used unit of measurement for heat energy is the calorie. A calorie is the amount of heat energy needed to raise the temperature of 1 gram of water by 1 degree Celsius. The specific heat of a substance is the quantity of heat energy, in calories, that is necessary to raise the temperature of 1 gram of the substance by 1 degree Celsius.

KEY REACTIONS AND EQUATIONS

1. Density of a substance (Section 2.8)

$$\text{Density} = \frac{\text{mass}}{\text{volume}}$$

2. Conversion of temperature readings from one scale to another (Section 2.9)

$$K = °C + 273 \qquad\qquad °C = K - 273$$

$$°F = \frac{9}{5}(°C) + 32 \qquad °C = \frac{5}{9}(°F - 32)$$

3. Heat energy absorbed by a substance (Section 2.9)

$$\frac{\text{Heat energy}}{\text{absorbed}} = \frac{\text{specific}}{\text{heat}} \times \text{mass} \times \frac{\text{temperature}}{\text{change}}$$

KEY TERMS

Calorie (2.9)
Conversion factor (2.7)
Density (2.8)
Dimensional analysis (2.7)
Exact number (2.3)
Gram (2.1)

Inexact number (2.3)
Liter (2.2)
Mass (2.2)
Measurement (2.1)
Meter (2.2)
Rounding off (2.5)

Scientific notation (2.6)
Significant figures (2.4)
Specific heat (2.9)
Weight (2.2)

EXERCISES AND PROBLEMS

The members of each pair of problems in this section test similar material.

■ **Metric System Units (Section 2.2)**

2.1 Write the name of the metric system prefix associated with each of the following mathematical meanings.
   a. 10^3 b. 10^{-3} c. 10^{-6} d. 1/10

2.2 Write the name of the metric system prefix associated with each of the following mathematical meanings.
   a. 10^{-2} b. 10^{-9} c. 10^6 d. 1/1000

2.3 Write out the names of the metric system units that have the following abbreviations.
   a. cm   b. kL   c. μL   d. ng

2.4 Write out the names of the metric system units that have the following abbreviations.
   a. mg   b. pg   c. Mm   d. dL

2.5 Arrange each of the following from smallest to largest.
   a. milligram, centigram, nanogram
   b. gigameter, megameter, kilometer
   c. microliter, deciliter, picoliter
   d. milligram, kilogram, microgram

2.6 Arrange each of the following from smallest to largest.
   a. milliliter, gigaliter, microliter
   b. centigram, megagram, decigram
   c. micrometer, picometer, kilometer
   d. nanoliter, milliliter, centiliter

■ **Exact and Inexact Numbers (Section 2.3)**

2.7 Indicate whether the number in each of the following statements is an *exact* or an *inexact* number.
   a. A classroom contains 32 chairs.
   b. There are 60 seconds in a minute.
   c. A bowl of cherries weighs 3.2 pounds.
   d. A newspaper article contains 323 words.

2.8 Indicate whether the number in each of the following statements is an *exact* or an *inexact* number.
   a. A classroom contains 63 students.
   b. The car is traveling at a speed of 56 miles per hour.
   c. The temperature on the back porch is $-3°F$.
   d. There are 3 feet in a yard.

2.9 Indicate whether each of the following quantities would involve an *exact* number or an *inexact* number.
   a. The length of a swimming pool
   b. The number of gummi bears in a bag
   c. The number of quarts in a gallon
   d. The surface area of a living room rug

2.10 Indicate whether each of the following quantities would involve an *exact* number or an *inexact* number.
   a. The number of pages in a chemistry textbook
   b. The number of teeth in a bear's mouth
   c. The distance from Earth to the sun
   d. The temperature of a heated oven

■ **Uncertainty in Measurement and Significant Figures (Section 2.4)**

2.11 Indicate to what decimal position readings should be recorded (nearest 0.1, 0.01, etc.) for measurements made with the following devices.
   a. A thermometer with a smallest scale marking of 1°C
   b. A graduated cylinder with a smallest scale marking of 0.1 mL
   c. A volumetric device with a smallest scale marking of 10 mL
   d. A ruler with a smallest scale marking of 1 mm

2.12 Indicate to what decimal position readings should be recorded (nearest 0.1, 0.01, etc.) for measurements made with the following devices.
   a. A ruler with a smallest scale marking of 1 cm
   b. A device for measuring angles with a smallest scale marking of 1°
   c. A thermometer with a smallest scale marking of 0.1°F
   d. A graduated cylinder with a smallest scale marking of 10 mL

2.13 Determine the number of significant figures in each of the following measured values.
   a. 6.000   b. 0.0032   c. 0.01001
   d. 65,400   e. 76.010   f. 0.03050

2.14 Determine the number of significant figures in each of the following measured values.
   a. 23,009   b. 0.00231   c. 0.3330
   d. 73,000   e. 73.000   f. 0.40040

2.15 In which of the following pairs of numbers do both members of the pair contain the same number of significant figures?
   a. 11.01 and 11.00   b. 2002 and 2020
   c. 0.000066 and 660,000   d. 0.05700 and 0.05070

2.16 In which of the following pairs of numbers do both members of the pair contain the same number of significant figures?
   a. 345,000 and 340,500   b. 2302 and 2320
   c. 0.6600 and 0.66   d. 936 and 936,000

2.17 Identify the *estimated digit* in each of the measured values in Problem 2.13.

2.18 Identify the *estimated digit* in each of the measured values in Problem 2.14.

2.19 What is the uncertainty associated with each of the measured values in Problem 2.13?

2.20 What is the uncertainty associated with each of the measured values in Problem 2.14?

■ **Significant Figures and Mathematical Operations (Section 2.5)**

2.21 Round off each of the following numbers to the number of significant figures indicated in parentheses.
   a. 0.350763 (three)   b. 653,899 (four)
   c. 22.55555 (five)   d. 0.277654 (four)

2.22 Round off each of the following numbers to the number of significant figures indicated in parentheses.
   a. 3883 (two)   b. 0.0003011 (two)
   c. 4.4050 (three)   d. 2.1000 (three)

2.23 Without actually solving, indicate the number of significant figures that should be present in the answers to the following multiplication and division problems.
   a. $10.300 \times 0.30 \times 0.300$ b. $3300 \times 3330 \times 333.0$
   c. $\dfrac{6.0}{33.0}$ d. $\dfrac{6.000}{33}$

2.24 Without actually solving, indicate the number of significant figures that should be present in the answers to the following multiplication and division problems.
   a. $3.00 \times 0.0003 \times 30.00$ b. $0.3 \times 0.30 \times 3.0$

c. $\dfrac{6.00}{33,000}$ d. $\dfrac{6.00000}{3}$

2.25 Carry out the following multiplications and divisions, expressing your answer to the correct number of significant figures. Assume that all numbers are measured numbers.

a. $2.0000 \times 2.00 \times 0.0020$ b. 4.1567×0.00345

c. $0.0037 \times 3700 \times 1.001$ d. $\dfrac{6.00}{33.0}$

e. $\dfrac{530,000}{465,300}$ f. $\dfrac{4.670 \times 3.00}{2.450}$

2.26 Carry out the following multiplications and divisions, expressing your answer to the correct number of significant figures. Assume that all numbers are measured numbers.

a. $2.000 \times 0.200 \times 0.20$ b. 3.6750×0.04503

c. $0.0030 \times 0.400 \times 4.00$ d. $\dfrac{6.0000}{33.00}$

e. $\dfrac{45,000}{1.2345}$ f. $\dfrac{3.000 \times 6.53}{13.567}$

2.27 Carry out the following additions and subtractions, expressing your answer to the correct number of significant figures. Assume that all numbers are measured numbers.

a. $12 + 23 + 127$ b. $3.111 + 3.11 + 3.1$
c. $1237.6 + 23 + 0.12$ d. $43.65 - 23.7$

2.28 Carry out the following additions and subtractions, expressing your answer to the correct number of significant figures. Assume that all numbers are measured numbers.

a. $237 + 37.0 + 7.0$ b. $4.000 + 4.002 + 4.20$
c. $235.45 + 37 + 36.4$ d. $3.111 - 2.07$

■ **Scientific Notation (Section 2.6)**

2.29 Express the following numbers in scientific notation.

a. 120.7 b. 0.0034 c. 231.00
d. $23,000$ e. 0.200 f. 0.1011

2.30 Express the following numbers in scientific notation.

a. 37.06 b. 0.00571 c. 437.0
d. 4370 e. 0.20340 f. $230,000$

2.31 Which number in each pair of numbers is the larger of the two numbers?

a. 1.0×10^{-3} or 1.0×10^{-6}
b. 1.0×10^{3} or 1.0×10^{-2}
c. 6.3×10^{4} or 2.3×10^{4}
d. 6.3×10^{-4} or 1.2×10^{-4}

2.32 Which number in each pair of numbers is the larger of the two numbers?

a. 2.0×10^{2} or 2.0×10^{-2}
b. 1.0×10^{6} or 3.0×10^{6}
c. 4.4×10^{-4} or 4.4×10^{-5}
d. 9.7×10^{3} or 8.3×10^{2}

2.33 How many significant figures are present in each of the following measured numbers?

a. 1.0×10^{2} b. 5.34×10^{6}
c. 5.34×10^{-4} d. 6.000×10^{3}

2.34 How many significant figures are present in each of the following measured numbers?

a. 1.01×10^{2} b. 1.00×10^{2}
c. 6.6700×10^{8} d. 6.050×10^{-3}

2.35 Carry out the following multiplications and divisions, expressing your answer in scientific notation to the correct number of significant figures.

a. $(3.20 \times 10^{7}) \times (1.720 \times 10^{5})$
b. $(3.71 \times 10^{-4}) \times (1.117 \times 10^{2})$
c. $(1.00 \times 10^{3}) \times (5.00 \times 10^{3}) \times (3.0 \times 10^{-3})$
d. $\dfrac{3.0 \times 10^{-5}}{1.5 \times 10^{2}}$ e. $\dfrac{4.56 \times 10^{7}}{3.0 \times 10^{-4}}$
f. $\dfrac{(2.2 \times 10^{6}) \times (2.3 \times 10^{-6})}{(1.2 \times 10^{-3}) \times (3.5 \times 10^{-3})}$

2.36 Carry out the following multiplications and divisions, expressing your answer in scientific notation to the correct number of significant figures.

a. $(4.0 \times 10^{4}) \times (1.32 \times 10^{8})$
b. $(2.23 \times 10^{-6}) \times (1.230 \times 10^{-2})$
c. $(3.200 \times 10^{7}) \times (1.10 \times 10^{-2}) \times (2.3 \times 10^{-7})$
d. $\dfrac{6.0 \times 10^{-5}}{3.0 \times 10^{3}}$ e. $\dfrac{5.132 \times 10^{7}}{1.12 \times 10^{3}}$
f. $\dfrac{(3.2 \times 10^{2}) \times (3.31 \times 10^{6})}{(4.00 \times 10^{-3}) \times (2.0 \times 10^{6})}$

■ **Conversion Factors and Dimensional Analysis (Section 2.7)**

2.37 Give both forms of the conversion factor that you would use to relate the following sets of units to each other.

a. gram and kilogram   b. meter and nanometer
c. liter and milliliter   d. gram and pound
e. kilometer and mile   f. gallon and liter

2.38 Give both forms of the conversion factor that you would use to relate the following sets of units to each other.

a. liter and milliliter   b. meter and micrometer
c. gram and centigram   d. inch and meter
e. kilogram and pound   f. liter and quart

2.39 Convert each of the following measurements to meters.

a. 1.6×10^{3} dm   b. 24 nm
c. 0.003 km   d. 3.0×10^{8} mm

2.40 Convert each of the following measurements to meters.

a. 2.7×10^{3} mm   b. 24 μm
c. 0.003 pm   d. 4.0×10^{5} cm

2.41 The human stomach produces approximately 2500 mL of gastric juice per day. What is the volume, in liters, of gastric juice produced?

2.42 A typical normal loss of water through sweating per day for a human is 450 mL. What is the volume, in liters, of sweat produced per day?

2.43 The mass of premature babies is customarily determined in grams. If a premature baby weighs 1550 g, what is its mass in pounds?

2.44 The smallest bone in the human body, which is in the ear, has a mass of 0.0030 g. What is the mass of this bone in pounds?

2.45 What volume of water, in gallons, would be required to fill a 25-mL container?

2.46 What volume of gasoline, in milliliters, would be required to fill a 17.0-gal gasoline tank?

2.47 An individual weighs 83.2 kg and is 1.92 m tall. What are the person's equivalent measurements in pounds and feet?

2.48 An individual weighs 135 lb and is 5 ft 4 in. tall. What are the person's equivalent measurements in kilograms and meters?

■ **Density (Section 2.8)**

2.49 A sample of mercury is found to have a mass of 524.5 g and to have a volume of 38.72 cm^3. What is its density in grams per cubic centimeter?

2.50 A sample of sand is found to have a mass of 12.0 g and to have a volume of 2.69 cm^3. What is its density in grams per cubic centimeter?

2.51 Acetone, the solvent in nail polish remover, has a density of 0.791 g/mL. What is the volume, in milliliters, of 20.0 g of acetone?

2.52 Silver metal has a density of 10.40 g/cm^3. What is the volume, in cubic centimeters, of a 100.0-g bar of silver metal?

2.53 The density of homogenized milk is 1.03 g/mL. How much does 1 cup (236 mL) of homogenized milk weigh in grams?

2.54 Nickel metal has a density of 8.90 g/cm^3. How much does 15 cm^3 of nickel metal weigh in grams?

■ **Temperature Scales and Heat Energy (Section 2.9)**

2.55 An oven for baking pizza operates at approximately 525°F. What is this temperature in degrees Celsius?

2.56 A comfortable temperature for bathtub water is 95°F. What temperature is this in degrees Celsius?

2.57 Mercury freezes at −38.9°C. What is the coldest temperature, in degrees Fahrenheit, that can be measured using a mercury thermometer?

2.58 The body temperature for a hypothermia victim is found to have dropped to 29.1°C. What is this temperature in degrees Fahrenheit?

2.59 Which is the higher temperature, −10°C or 10°F?

2.60 Which is the higher temperature, −15°C or 4°F?

2.61 A substance has a specific heat of 0.63 cal/g · °C. What is its specific heat in J/g · °C?

2.62 A substance has a specific heat of 0.24 cal/g · °C. What is its specific heat in J/g · °C?

2.63 If it takes 18.6 cal of heat to raise the temperature of 12.0 g of a substance by 10.0°C, what is the specific heat of the substance?

2.64 If it takes 35.0 cal of heat to raise the temperature of 25.0 g of a substance by 12.0°C, what is the specific heat of the substance?

2.65 How many calories of heat energy are required to raise the temperature of 42.0 g of each of the following substances from 20.0°C to 40.0°C?
a. Silver   b. Liquid water   c. Aluminum

2.66 How many calories of heat energy are necessary to raise the temperature of 20.0 g of each of the following substances from 25°C to 55°C?
a. Gold   b. Ethyl alcohol   c. Olive oil

ADDITIONAL PROBLEMS

2.67 A person is told that there are 12 inches in a foot and also that a piece of rope is 12 inches long. What is the fundamental difference between the value of 12 in these two pieces of information?

2.68 Round off the number 4.7205059 to the indicated number of significant figures.
a. six   b. five   c. four   d. two

2.69 Write each of the following numbers in scientific notation to the number of significant figures indicated in parentheses.
a. 0.00300300 (three)   b. 936,000 (two)
c. 23.5003 (three)   d. 450,000,001 (six)

2.70 For each of the pairs of units listed, indicate whether the first unit is larger or smaller than the second unit, and then indicate how many times larger or smaller it is.
a. milliliter, liter   b. kiloliter, microliter
c. nanoliter, deciliter   d. centiliter, megaliter

2.71 Indicate how each of the following conversion factors should be interpreted in terms of significant figures present.

a. $\dfrac{2.540 \text{ cm}}{1.000 \text{ in.}}$ b. $\dfrac{453.6 \text{ g}}{1.000 \text{ lb}}$ c. $\dfrac{2.113 \text{ pt}}{1.00 \text{ L}}$ d. $\dfrac{10^{-9} \text{ m}}{1 \text{nm}}$

2.72 A one-gram sample of a powdery white solid is found to have a volume of two cubic centimeters. Calculate the solid's density using the following uncertainty specifications and express your answers in scientific notation.

a. 1.0 g and 2.0 cm^3
b. 1.000 g and 2.00 cm^3
c. 1.0000 g and 2.0000 cm^3
d. 1.000 g and 2.0000 cm^3

2.73 Calculate the volume, in *milliliters*, for each of the following.
a. 75.0 g of gasoline (density = 0.56 g/mL)
b. 75.0 g of sodium metal (density = 0.93 g/cm^3)
c. 75.0 g of ammonia gas (density = 0.759 g/L)
d. 75.0 g of mercury (density = 13.6 g/mL)

2.74 Which quantity of heat energy in each of the following pairs of heat energy values is the larger?
a. 2.0 joules or 2.0 calories
b. 1.0 kilocalories or 92 calories
c. 100 Calories or 100 calories
d. 2.3 Calories or 1000 kilocalories

2.75 The concentration of salt in a salt solution is found to be 4.5 mg/mL. What is the salt concentration in each of the following units?
a. mg/L   b. pg/mL   c. g/L   d. kg/m^3

2.76 If one U.S. dollar is equal to 5.94 French francs and one franc is equal to 0.102 British pound, then what is the value in dollars of seven pounds?

ANSWERS TO PRACTICE EXERCISES

2.1 a. 10^{-3} b. 10^{-12} c. 10^6 d. micro-
e. centi-   f. deci-

2.2 a. five, ±0.01   b. four, ±0.0001
c. four, ±10   d. three, ±1,000,000

2.3 a. 6.548   b. 5600   c. 31.2   d. 52.5

2.4 a. 8.118×10^{10} b. 1.3×10^{-3}

2.5 2.03×10^{-4} g

2.6 0.106 qt

2.7 1.587 g/cm^3

2.8 497 g

2.9 a. 106°F   b. 314 K

2.10 1500 cal

3

Atomic Structure and the Periodic Table

Music consists of a series of tones that build octave after octave. Similarly, elements have properties that recur period after period.

I n Chapter 1 we learned that all matter is made up of small particles called atoms and that 113 different types of atoms are known, each type of atom corresponding to a different element. Furthermore, we found that compounds result from the chemical combination of different types of atoms in various ratios and arrangements.

Until the last two decades of the nineteenth century, scientists believed that atoms were solid, indivisible spheres without an internal structure. Today, this model of the atom is known to be incorrect. Evidence from a variety of sources indicates that atoms are made up of even smaller particles called *subatomic particles*. In this chapter we consider the fundamental types of subatomic particles, how they arrange themselves within an atom, and the relationship between an atom's subatomic makeup and its chemical identity.

3.1 Internal Structure of an Atom

▶ Atoms of all 113 elements contain the same three types of subatomic particles. Different elements differ only in the numbers of the various subatomic particles they contain.

Atoms possess internal structure; that is, they are made up of even smaller particles, which are called subatomic particles. A **subatomic particle** *is a very small particle that is a building block for atoms.* Three types of subatomic particles are found within atoms: electrons, protons, and neutrons. Key properties of these three types of particles are summarized in Table 3.1. An **electron** *is a subatomic particle that possesses a negative (−) electrical charge.* It is the smallest, in terms of mass, of the three types of subatomic particles. A **proton** *is a subatomic particle that possesses a positive (+) electrical charge.* Protons and electrons carry the *same amount* of charge; the charges, however, are opposite (positive versus negative). A **neutron** *is a subatomic particle that has no charge*

TABLE 3.1
Charge and Mass Characteristics of Electrons, Protons, and Neutrons

	Electron	Proton	Neutron
Charge	-1	$+1$	0
Actual mass (g)	9.109×10^{-28}	1.673×10^{-24}	1.675×10^{-24}
Relative mass (based on the electron being 1 unit)	1	1837	1839

associated with it; that is, it is neutral. Both protons and neutrons are massive particles compared to electrons; they are almost 2000 times heavier.

■ Arrangement of Subatomic Particles Within an Atom

The arrangement of subatomic particles within an atom is not haphazard. *All* protons and *all* neutrons present are found at the center of an atom in a very tiny volume called the *nucleus* (Figure 3.1). The **nucleus** *is the very small, dense, positively charged center of an atom.* A nucleus is always positively charged because it contains positively charged protons. Because the nucleus houses the heavy subatomic particles (protons and neutrons), almost all (over 99.9%) of the mass of an atom is concentrated in its nucleus. The small size of the nucleus, coupled with its large amount of mass, causes nuclear material to be extremely dense.

> The radius of a nucleus is approximately 10,000 times smaller than the radius of an entire atom.

The outer (extranuclear) region of an atom contains all of the electrons. In this region, which accounts for most of the volume of an atom, the electrons move rapidly about the nucleus. The electrons are attracted to the positively charged protons of the nucleus by the forces that exist between particles of opposite charge. The motion of the electrons in the extranuclear region determines the volume (size) of the atom in the same way that the blade of a fan determines a volume by its circular motion. The volume occupied by the electrons is sometimes referred to as the *electron cloud.* Because electrons are negatively charged, the electron cloud is also negatively charged. Figure 3.1 illustrates the nuclear and extranuclear regions of an atom.

Closely resembling the term *nucleus* is the term *nucleon.* A **nucleon** *is any subatomic particle found in the nucleus of an atom.* Thus both protons and neutrons are nucleons, and the nucleus can be regarded as containing a collection of nucleons (protons and neutrons).

■ Charge Neutrality of an Atom

An atom as a whole is electrically neutral; that is, it has no *net* electrical charge. For this to be the case, the same amount of positive and negative charge must be present in the atom. Equal numbers of positive and negative charges cancel one another. Thus equal numbers of protons and electrons are present in an atom.

Number of protons = number of electrons

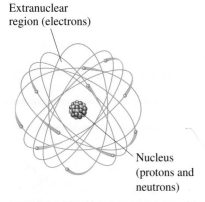

Extranuclear region (electrons)

Nucleus (protons and neutrons)

FIGURE 3.1 The protons and neutrons of an atom are found in the central nuclear region, or nucleus, and the electrons are found in an electron cloud outside the nucleus. Note that this figure is not drawn to scale; the correct scale would be comparable to a penny (the nucleus) in the center of a baseball field (the atom).

■ Size Relationships Within an Atom

To help you visualize the size relationships among the parts of an atom, imagine enlarging (magnifying) the nucleus until it is the size of a baseball (about 2.9 inches in diameter). If the nucleus were this large, the whole atom would have a diameter of approximately 2.5 miles. The electrons would still be smaller than the periods used to end sentences in this text, and they would move about at random within that 2.5-mile region.

The concentration of nearly all of the mass of an atom in the nucleus can also be illustrated by using our imagination. If a coin the same size as a copper penny contained copper nuclei (copper atoms stripped of their electrons) rather than copper atoms (which are mostly empty space), the coin would weigh 190,000,000 tons! Nuclei are indeed very dense matter.

Despite the existence of subatomic particles, we will continue to refer to atoms as the fundamental building blocks for all types of matter. Subatomic particles do not lead an independent existence for any appreciable length of time; they gain stability by joining together to form atoms.

3.2 Atomic Number and Mass Number

> Atomic number and mass number are always *whole* numbers because they are obtained by counting whole objects (protons, neutrons, and electrons).

An **atomic number** *is the number of protons in the nucleus of an atom.* Because an atom has the same number of electrons as protons (Section 3.1), the atomic number also specifies the number of electrons present.

Atomic number = number of protons = number of electrons

The symbol Z is used as a general designation for atomic number.

A **mass number** *is the sum of the number of protons and the number of neutrons in the nucleus of an atom.* Thus the mass number gives the number of subatomic particles present in the nucleus.

Mass number = number of protons + number of neutrons

The mass of an atom is almost totally accounted for by the protons and neutrons present—hence the term *mass number.* The symbol A is used as a general designation for mass number.

The number and identity of subatomic particles present in an atom can be calculated from its atomic and mass numbers in the following manner.

> The *sum* of the mass number and the atomic number for an atom $(A + Z)$ corresponds to the total number of subatomic particles present in the atom (protons, neutrons, and electrons).

Number of protons = atomic number = Z
Number of electrons = atomic number = Z
Number of neutrons = mass number − atomic number = $A - Z$

Note that neutron count is obtained by subtracting atomic number from mass number.

EXAMPLE 3.1

Determining the Subatomic Particle Makeup of an Atom Given Its Atomic Number and Mass Number

■ An atom has an atomic number of 9 and a mass number of 19.
a. Determine the number of protons present.
b. Determine the number of neutrons present.
c. Determine the number of electrons present.

Solution

a. There are 9 protons because the atomic number is always equal to the number of protons present.
b. There are 10 neutrons because the number of neutrons is always obtained by subtracting the atomic number from the mass number.

$$\underbrace{(\text{Protons} + \text{neutrons})}_{\text{Mass number}} - \underbrace{\text{protons}}_{\substack{\text{Atomic} \\ \text{number}}} = \text{neutrons}$$

c. There are 9 electrons because the number of protons and the number of electrons are always the same in an atom.

Practice Exercise 3.1

An atom has an atomic number of 11 and a mass number of 23.

a. Determine the number of protons present.
b. Determine the number of neutrons present.
c. Determine the number of electrons present.

■ Electrons and Chemical Properties

The chemical properties of an atom, which are the basis for its identification, are determined by the number and arrangement of the electrons about the nucleus. When two atoms interact, the outer part (electrons) of one interacts with the outer part (electrons) of the other. The small nuclear centers never come in contact with each other in a chemical reaction. The number of electrons about a nucleus may be considered to be determined by the number of protons in the nucleus; charge balance requires an equal number of the two (Section 3.1). Hence the number of protons (which is the atomic number) characterizes an atom. All atoms with the same atomic number have the same chemical properties and are atoms of the same element.

In Section 1.6, an element was defined as a pure substance that cannot be broken down into simpler substances by ordinary chemical means. Although this is a good historical definition for an element, we can now give a more rigorous definition by using the concept of atomic number. An **element** *is a pure substance in which all atoms present have the same atomic number.*

An alphabetical listing of the 113 known elements, with their atomic numbers as well as other information, is found on the inside front cover of this text. If you check the atomic number column in this tabulation, you will find an entry for each of the numbers in the sequence 1 to 112 plus 114. The highest-atomic-numbered element that occurs naturally is uranium (element 92); elements 93 to 112 and 114 have been made in the laboratory but are not found in nature (Section 1.7). The fact that there are no gaps in the numerical sequence 1 to 92 is interpreted by scientists to mean that there are no "missing elements" yet to be discovered in nature.

3.3 Isotopes and Atomic Masses

Charge neutrality (Section 3.1) requires the presence in an atom of an equal number of protons and electrons. However, because neutrons have no electrical charge, their numbers in atoms do not have to be the same as the number of protons or electrons. Most atoms contain more neutrons than either protons or electrons.

Studies of atoms of various elements also show that the number of neutrons present in atoms of an element is not constant; it varies over a small range. This means that not all atoms of an element have to be identical. They must have the same number of protons and electrons, but they can differ in the number of neutrons.

■ Isotopes

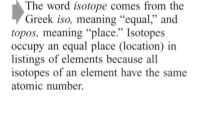
The word *isotope* comes from the Greek *iso,* meaning "equal," and *topos,* meaning "place." Isotopes occupy an equal place (location) in listings of elements because all isotopes of an element have the same atomic number.

Atoms of an element that differ in neutron count are called isotopes. **Isotopes** *are atoms of an element that have the same number of protons and the same number of electrons but different numbers of neutrons.* Different isotopes always have the same atomic number and different mass numbers.

Most elements found in nature exist in isotopic forms, with the number of naturally occurring isotopes ranging from two to ten. For example, all silicon atoms have 14 protons and 14 electrons. Most silicon atoms also contain 14 neutrons. However, some silicon atoms contain 15 neutrons and others contain 16 neutrons. Thus three different kinds of silicon atoms exist.

Isotopes of an element have the same chemical properties, but their physical properties are often slightly different. Isotopes of an element have the same chemical properties because they have the same number of electrons. They have slightly different physical properties because they have different numbers of neutrons and therefore different masses.

When it is necessary to distinguish between isotopes of an element, the following notation is used:

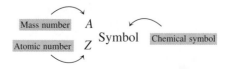

There are a few elements for which all naturally occurring atoms have the same number of neutrons—that is, for which all atoms are identical. They include the elements Be, F, Na, Al, P, and Au.

A mass number, in contrast to an atomic number, lacks uniqueness. Atoms of different elements can have the same mass number. For example, carbon-14 and nitrogen-14 have the same mass numbers. Atoms of different elements, however, cannot have the same atomic number.

An analogy involving isotopes and identical twins may be helpful: Identical twins need not weigh the same, even though they have identical "gene packages." Likewise, isotopes, even though they have different masses, have the same number of protons.

The atomic number is written as a *subscript* to the left of the elemental symbol for the atom. The mass number is written as a *superscript* to the left of the elemental symbol. Thus the three silicon isotopes are designated, respectively, as

$$^{28}_{14}\text{Si}, \qquad ^{29}_{14}\text{Si}, \qquad \text{and} \qquad ^{30}_{14}\text{Si}$$

Names for isotopes include the mass number. $^{28}_{14}\text{Si}$ is called silicon-28, and $^{29}_{14}\text{Si}$ is called silicon-29. The atomic number is not included in the name because it is the same for all isotopes of an element.

The various isotopes of a given element are of varying abundance; usually one isotope is predominant. Silicon is typical of this situation. The percentage abundances for its three isotopes are 92.21% ($^{28}_{14}\text{Si}$), 4.70% ($^{29}_{14}\text{Si}$), and 3.09% ($^{30}_{14}\text{Si}$). Percentage abundances are number percentages (numbers of atoms) rather than mass percentages. A sample of 10,000 silicon atoms contains 9221 $^{28}_{14}\text{Si}$ atoms, 470 $^{29}_{14}\text{Si}$ atoms, and 309 $^{30}_{14}\text{Si}$ atoms.

There are 286 isotopes that occur naturally. In addition, over 2000 more have been synthesized in the laboratory via nuclear rather than chemical reactions (Section 11.5). All these synthetic isotopes are unstable (radioactive). Despite their instability, many are used in chemical research, as well as in medicine.

■ Atomic Masses

The existence of isotopes means that atoms of an element can have several different masses. For example, silicon atoms can have any one of three masses because there are

CHEMICAL CONNECTIONS

Protium, Deuterium, and Tritium: The Three Isotopes of Hydrogen

Measurable differences in physical properties are found among isotopes for elements with low atomic numbers. This results from differences in mass among isotopes being relatively large compared to the masses of the isotopes themselves. The situation is greatest for the element hydrogen, the element with the lowest atomic number.

Three isotopes of hydrogen exist: ^1H, ^2H, and ^3H. With a single proton and no neutrons in its nucleus, hydrogen-1 is by far the most abundant isotope (99.985%). Hydrogen-2, with a neutron in addition to a proton in its nucleus, has an abundance of 0.015%. The presence of the additional neutron in ^2H doubles its mass compared to that of ^1H. Hydrogen-3 has two neutrons and a proton in its nucleus and has a mass triple that of ^1H. Only minute amounts of ^3H, which is radioactive (unstable: see Section 11.1), occur naturally.

In discussions involving hydrogen isotopes, special names and symbols are given to the isotopes—something that does not occur for any other element. Hydrogen-1 is usually called hydrogen but is occasionally called *protium*. Hydrogen-2 has the name *deuterium* (symbol D), and hydrogen-3 is called *tritium* (symbol T). The following table contrasts the properties of H_2 and D_2.

Water in which both hydrogen atoms are deuterium (D_2O) is called "heavy water." The properties of heavy water are measurably different from those of "ordinary" (H_2O).

Compound	Melting point	Boiling point	Density (at 0°C and 1 atmosphere pressure)
H_2O	0.0°C	100.0°C	0.99987 g/mL
D_2O	3.82°C	101.4°C	1.1047 g/mL

Heavy water (D_2O) can be obtained from natural water by distilling a sample of natural water, because the D_2O has a slightly higher boiling point than H_2O. Pure deuterium (D_2) is produced by decomposing the D_2O. Heavy water is used in the operation of nuclear power plants (to slow down free neutrons present in the reactor core).

Tritium, the heaviest hydrogen isotope, is used in nuclear weapons. Because of the minute amount of naturally occurring tritium, it must be synthesized in the laboratory using bombardment reactions (Section 11.5).

Isotope	Melting point	Boiling point	Density (at 0°C and 1 atmosphere pressure)
H_2	−259°C	−253°C	0.090 g/L
D_2	−253°C	−250°C	0.18 g/L

three silicon isotopes. Which of these three silicon isotopic masses is used in situations in which the mass of the element silicon needs to be specified? The answer is none of them. Instead we use a *weighted-average mass* that takes into account the existence of isotopes and their relative abundances.

The *weighted-average mass* of the isotopes of an element is known as the element's atomic mass. An **atomic mass** *is the calculated average mass for the isotopes of an element, expressed on a scale where* ^{12}C *serves as the reference point.* What we need to calculate an atomic mass are the masses of the various isotopes on the $^{12}_{6}C$ reference scale and the percentage abundance of each isotope.

The $^{12}_{6}C$ reference scale mentioned in the definition of *atomic mass* is a scale scientists have set up for comparing the masses of atoms. On this scale, the mass of a $^{12}_{6}C$ atom is defined to be exactly 12 atomic mass units (amu). The masses of all other atoms are then determined relative to that of $^{12}_{6}C$. For example, if an atom is twice as heavy as $^{12}_{6}C$, its mass is 24 amu, and if an atom weighs half as much as an atom of $^{12}_{6}C$, its mass is 6 amu.

Example 3.2 shows how an atomic mass is calculated by using the amu ($^{12}_{6}C$) scale, the percentage abundances of isotopes, and the number of isotopes of an element.

> The terms *atomic mass* and *atomic weight* are often used interchangeably. *Atomic mass,* however, is the correct term.

EXAMPLE 3.2

Calculation of an Element's Atomic Mass

■ Naturally occurring chlorine exists in two isotopic forms, $^{35}_{17}Cl$ and $^{37}_{17}Cl$. The relative mass of $^{35}_{17}Cl$ is 34.97 amu, and its abundance is 75.53%; the relative mass of $^{37}_{17}Cl$ is 36.97 amu, and its abundance is 24.47%. What is the atomic mass of chlorine?

Solution

An element's atomic mass is calculated by multiplying the relative mass of each isotope by its fractional abundance and then totaling the products. The fractional abundance for an isotope is its percentage abundance converted to decimal form (divided by 100).

$^{35}_{17}Cl$: $\left(\dfrac{75.53}{100}\right) \times 34.97 \text{ amu} = (0.7553) \times 34.97 \text{ amu} = 26.41 \text{ amu}$

$^{37}_{17}Cl$: $\left(\dfrac{24.47}{100}\right) \times 36.97 \text{ amu} = (0.2447) \times 36.97 \text{ amu} = 9.047 \text{ amu}$

$$\text{Atomic mass of Cl} = (26.41 + 9.047) \text{ amu}$$
$$= 35.46 \text{ amu}$$

This calculation involved an element containing just two isotopes. A similar calculation for an element having three isotopes would be carried out the same way, but it would have three terms in the final sum; an element possessing four isotopes would have four terms in the final sum.

Practice Exercise 3.2

Naturally occurring copper exists in two isotopic forms, $^{63}_{29}Cu$ and $^{65}_{29}Cu$. The relative mass of $^{63}_{29}Cu$ is 62.93 amu, and its abundance is 69.09%; the relative mass of $^{65}_{29}Cu$ is 64.93 amu, and its abundance is 30.91%. What is the atomic mass of copper?

The alphabetical listing of the known elements printed inside the front cover of this text gives the calculated atomic mass for each of the elements; it is the last column of numbers. Table 3.2 gives isotopic data for the elements with atomic numbers 1 through 12.

The Chemistry at a Glance feature on page 53 summarizes what has been said about atoms in Sections 3.1 through 3.3.

Atomic Structure

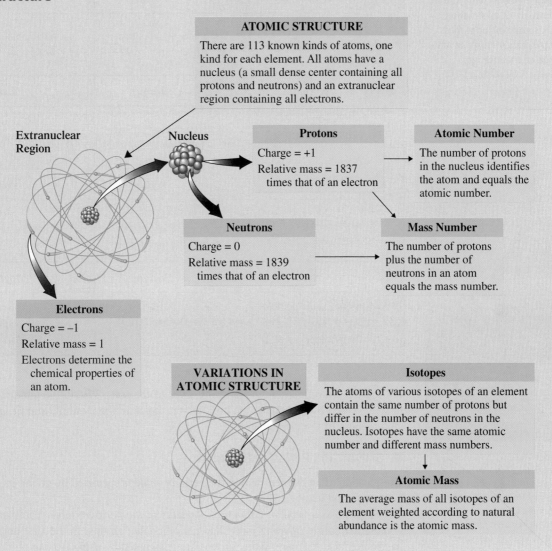

ATOMIC STRUCTURE

There are 113 known kinds of atoms, one kind for each element. All atoms have a nucleus (a small dense center containing all protons and neutrons) and an extranuclear region containing all electrons.

Extranuclear Region

Nucleus

Protons

Charge = +1
Relative mass = 1837 times that of an electron

Atomic Number

The number of protons in the nucleus identifies the atom and equals the atomic number.

Neutrons

Charge = 0
Relative mass = 1839 times that of an electron

Mass Number

The number of protons plus the number of neutrons in an atom equals the mass number.

Electrons

Charge = −1
Relative mass = 1
Electrons determine the chemical properties of an atom.

VARIATIONS IN ATOMIC STRUCTURE

Isotopes

The atoms of various isotopes of an element contain the same number of protons but differ in the number of neutrons in the nucleus. Isotopes have the same atomic number and different mass numbers.

Atomic Mass

The average mass of all isotopes of an element weighted according to natural abundance is the atomic mass.

3.4 The Periodic Law and the Periodic Table

During the early part of the nineteenth century, scientists began to look for order in the increasing amount of chemical information that had become available. They knew that certain elements had properties that were very similar to those of other elements, and they sought reasons for these similarities in the hope that these similarities would suggest a method for arranging or classifying the elements.

In 1869, these efforts culminated in the discovery of what is now called the *periodic law,* proposed independently by the Russian chemist Dmitri Mendeleev (Figure 3.2) and the German chemist Julius Lothar Meyer. Given in its modern form, the **periodic law** *states that when elements are arranged in order of increasing atomic number, elements with similar properties occur at periodic (regularly recurring) intervals.*

A periodic table represents a compact graphical method for representing the behavior described by the periodic law. A **periodic table** *is a graphical display of the elements in order of increasing atomic number in which elements with similar properties fall in the same column of the display.* The most commonly used form of the periodic table is shown

TABLE 3.2

Isotopic Data for Elements with Atomic Numbers 1 Through 12. Information given for each isotope includes mass number, isotopic mass in terms of amu, and percentage abundance.

1	Hydrogen	2	Helium	3	Lithium
$^1_1$H 1.008 amu 99.985% $^2_1$H 2.014 amu 0.015% $^3_1$H 3.016 amu trace		$^3_2$He 3.016 amu trace $^4_2$He 4.003 amu 100%		$^6_3$Li 6.015 amu 7.42% $^7_3$Li 7.016 amu 92.58%	
4	**Beryllium**	**5**	**Boron**	**6**	**Carbon**
$^9_4$Be 9.012 amu 100%		$^{10}_5$B 10.013 amu 19.6% $^{11}_5$B 11.009 amu 80.4%		$^{12}_6$C 12.000 amu 98.89% $^{13}_6$C 13.003 amu 1.11% $^{14}_6$C 14.003 amu trace	
7	**Nitrogen**	**8**	**Oxygen**	**9**	**Fluorine**
$^{14}_7$N 14.003 amu 99.63% $^{15}_7$N 15.000 amu 0.37%		$^{16}_8$O 15.995 amu 99.759% $^{17}_8$O 16.999 amu 0.037% $^{18}_8$O 17.999 amu 0.204%		$^{19}_9$F 18.998 amu 100%	
10	**Neon**	**11**	**Sodium**	**12**	**Magnesium**
$^{20}_{10}$Ne 19.992 amu 90.92% $^{21}_{10}$Ne 20.994 amu 0.26% $^{22}_{10}$Ne 21.991 amu 8.82%		$^{23}_{11}$Na 22.990 amu 100%		$^{24}_{12}$Mg 23.985 amu 78.70% $^{25}_{12}$Mg 24.986 amu 10.13% $^{26}_{12}$Mg 25.983 amu 11.17%	

FIGURE 3.2 Dmitri Ivanovich Mendeleev (1834–1907). Mendeleev constructed a periodic table as part of his effort to systematize chemistry. He received many international honors for his work, but his reception at home in czarist Russia was mixed. Element 101 carries his name.

▶ Using the information on a periodic table, you can quickly determine the number of protons and electrons for atoms of an element. However, no information concerning neutrons is available from a periodic table; mass numbers are not part of the information given, because they are not unique to an element.

▶ The elements within a given periodic-table group show numerous similarities in properties, the degree of similarity varying from group to group. In no case are the group members "clones" of one another. Each element has some individual characteristics not found in other elements of the group. By analogy, the members of a human family often bear many resemblances to each other, but each member also has some (and often much) individuality.

in Figure 3.3 (see also the inside front cover of the text). Within the table, each element is represented by a rectangular box, which contains the symbol, atomic number, and atomic mass of the element.

■ Groups and Periods of Elements

The location of an element within the periodic table is specified by giving its group number and period number.

A **group** *is a vertical column of elements in the periodic table.* There are two notations in use for designating individual periodic-table groups. In the first notation, which has been in use for many years, groups are designated by using Roman numerals and the letters A and B. In the second notation, which an international scientific commission has recommended, the Arabic numbers 1 through 18 are used. Note that in Figure 3.3 both group notations are given at the top of each group. The elements with atomic numbers 8, 16, 34, 52, and 84 (O, S, Se, Te, and Po) constitute Group VIA (old notation) or Group 16 (new notation).

Several groups of elements have common (non-numerical) names that are used so frequently that they should be learned. The Group IA elements, except for hydrogen, are called the *alkali metals,* and the Group IIA elements the *alkaline earth metals.* On the opposite side of the periodic table from the IA and IIA elements are found the *halogens* (the Group VIIA elements) and the *noble gases* (Group VIIIA).

A **period** *is a horizontal row of elements in the periodic table.* For identification purposes, the periods are numbered sequentially with Arabic numbers, starting at the top of the periodic table. In Figure 3.3, period numbers are found on the left side of the table. The elements Na, Mg, Al, Si, P, S, Cl, and Ar are all members of Period 3, the third row of elements. Period 4 is the fourth row of elements, and so on. There are only two elements in Period 1, H and He.

The location of any element in the periodic table is specified by giving its group number and its period number. The element gold, with an atomic number of 79, belongs to Group IB (or 11) and is in Period 6. The element nitrogen, with an atomic number of 7, belongs to Group VA (or 15) and is in Period 2.

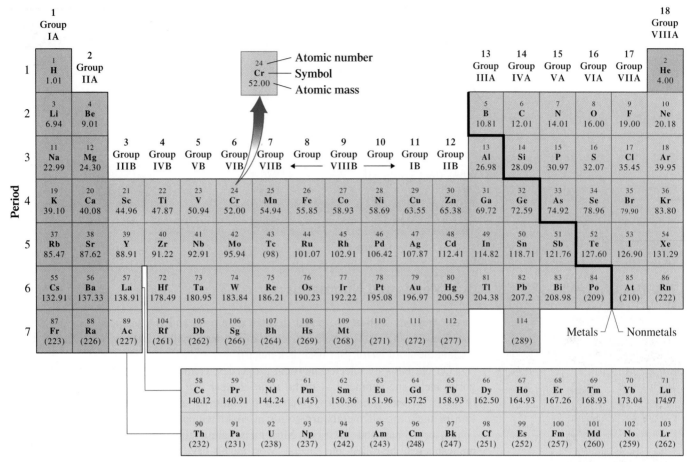

FIGURE 3.3 The periodic table of the elements is a graphical way to show relationships among the elements. Elements with similar chemical properties fall in the same vertical column.

> When the phrase "the first ten elements" is used, it means the first ten elements in the periodic table, the elements with atomic numbers 1 through 10.

■ The Shape of the Periodic Table

Within the periodic table of Figure 3.3, the practice of arranging the elements according to increasing atomic number is violated in Groups IIIB and IVB. Element 72 follows element 57, and element 104 follows element 89. The missing elements, elements 58 through 71 and 90 through 103 are located in two rows at the bottom of the periodic table. Technically, the elements at the bottom of the table should be included in the body of the table, as shown in Figure 3.4. However, in order to have a more compact table, we place them at the bottom of the table as shown in Figure 3.3.

FIGURE 3.4 In this periodic table, elements 58 through 71 and 90 through 103 (in color) are shown in their proper positions.

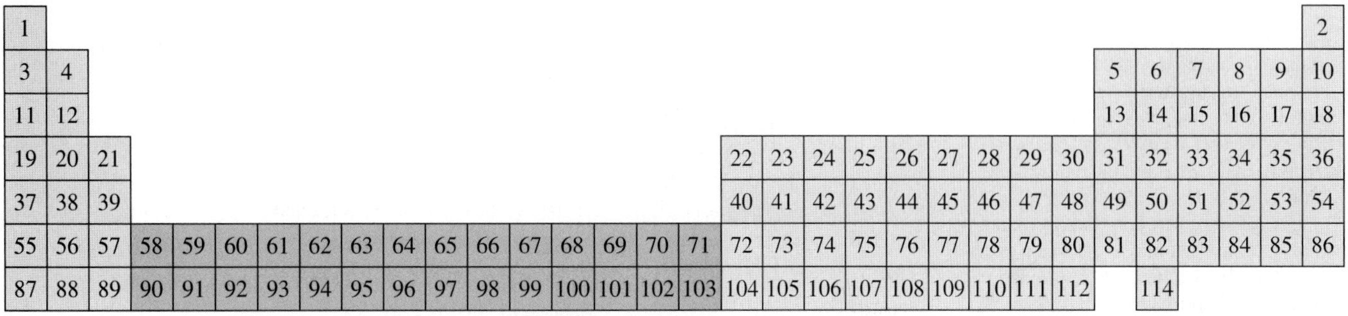

FIGURE 3.5 (a) Some familiar metals are aluminum, lead, tin, and zinc. (b) Some familiar nonmetals are sulfur (yellow), phosphorus (dark red), and bromine (reddish-brown).

(a)

(b)

3.5 Metals and Nonmetals

In the previous section, we noted that the Group IA and IIA elements are known, respectively, as the alkali metals and the alkaline earth metals. Both of these designations contain the word *metal*. But what is a metal?

On the basis of selected physical properties, elements are classified into the categories metal and nonmetal. A **metal** *is an element that has the characteristic properties of luster, thermal conductivity, electrical conductivity, and malleability.* With the exception of mercury, all metals are solids at room temperature (25°C). Metals are good conductors of heat and electricity. Most metals are ductile (can be drawn into wires) and malleable (can be rolled into sheets). Most metals have high luster (shine), high density, and high melting points. Among the more familiar metals are the elements iron, aluminum, copper, silver, gold, lead, tin, and zinc (see Figure 3.5a).

A **nonmetal** *is an element characterized by the absence of the properties of luster, thermal conductivity, electrical conductivity, and malleability.* Many of the nonmetals, such as hydrogen, oxygen, nitrogen, and the noble gases, are gases. The only nonmetal found as a liquid at room temperature is bromine. Solid nonmetals include carbon, iodine, sulfur, and phosphorus (Figure 3.5b). In general, the nonmetals have lower densities and lower melting points than metals. Table 3.3 contrasts selected physical properties of metals and nonmetals.

▶ *Metals* generally are malleable, ductile, and lustrous and are good thermal and electrical conductors. *Nonmetals* tend to lack these properties. In many ways, the general properties of metals and nonmetals are opposites.

■ Periodic Table Locations for Metals and Nonmetals

The majority of the elements are metals. Only 22 elements are nonmetals. It is not necessary to memorize which elements are nonmetals and which are metals; this information is obtainable from a periodic table (Figure 3.6). The steplike heavy line that runs through

TABLE 3.3
Selected Physical Properties of Metals and Nonmetals

Metals	Nonmetals
1. High electrical conductivity that decreases with increasing temperature	1. Poor electrical conductivity (except carbon in the form of graphite)
2. High thermal conductivity	2. Good heat insulators (except carbon in the form of diamond)
3. Metallic gray or silver luster[a]	3. No metallic luster
4. Almost all are solids[b]	4. Solids, liquids, or gases
5. Malleable (can be hammered into sheets)	5. Brittle in solid state
6. Ductile (can be drawn into wires)	6. Nonductile

[a]Except copper and gold.
[b]Except mercury; cesium and gallium melt on a hot summer day (85°F) or when held in a person's hand.

FIGURE 3.6 This portion of the periodic table shows the dividing line between metals and nonmetals. All elements that are not shown are metals.

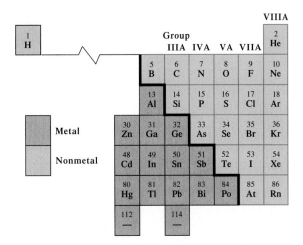

the right third of the periodic table separates the metals on the left from the nonmetals on the right. Note also that the element hydrogen is a nonmetal.

The fact that the vast majority of elements are metals in no way indicates that metals are more important than nonmetals. Most nonmetals are relatively common and are found in many important compounds. For example, water (H_2O) is a compound involving two nonmetals.

An analysis of the abundance of the elements in Earth's crust (Figure 1.12) in terms of metals and nonmetals shows that the two most abundant elements, which account for 80.2% of all atoms, are nonmetals—oxygen and silicon. The four most abundant elements in the human body (see the Chemical Connections feature on page 11 in Chapter 1), which comprise over 99% of all atoms in the body, are nonmetals—hydrogen, oxygen, carbon, and nitrogen. Besides these abundant elements, trace elements are also important in the functioning of the human body (see the Chemical Connections feature on page 58).

3.6 Electron Arrangements Within Atoms

As electrons move about an atom's nucleus, they are restricted to specific regions within the extranuclear portion of the atom. Such restrictions are determined by the amount of energy the electrons possess. Furthermore, electron energies are limited to certain values and a specific "behavior" is associated with each allowed energy value.

The space in which electrons move rapidly about a nucleus is divided into subspaces called *shells, subshells,* and *orbitals.*

■ Electron Shells

Electrons within an atom are grouped into main energy levels called electron shells. An **electron shell** *is a region of space about a nucleus that contains electrons that have approximately the same energy and that spend most of their time approximately the same distance from the nucleus.*

▶ Electrons that occupy the first electron shell are closer to the nucleus and have a lower energy than electrons in the second electron shell.

Electron shells are numbered 1, 2, 3, and so on, outward from the nucleus. Electron energy increases as the distance of the electron shell from the nucleus increases. An electron in shell 1 has the minimum amount of energy that an electron can have.

The maximum number of electrons that an electron shell can accommodate varies; the higher the shell number (n), the more electrons that can be present. In higher-energy shells, the electrons are farther from the nucleus and a greater volume of space is available for them; hence more electrons can be accommodated. (Conceptually, electron shells may be considered to be nested one inside another, somewhat like the layers of flavors inside a jawbreaker or similar type of candy.)

The lowest-energy shell ($n = 1$) accommodates a maximum of 2 electrons. In the second, third, and fourth shells, 8, 18, and 32 electrons, respectively, are allowed. The relationship among these numbers is given by the formula $2n^2$, where n is the shell number. For example, when $n = 4$, the quantity $2n^2 = 2(4)^2 = 32$.

CHEMICAL CONNECTIONS

Importance of Metallic and Nonmetallic Trace Elements for Human Health

Within the past three decades, biochemists have developed techniques that can detect substances in smaller and smaller quantities in living cells. As a result, we now know that living organisms need minute amounts of certain elements—called trace elements—to function properly. These trace elements now number 15, and more may be discovered. Ten of the 15 known *trace elements* are metals, and 5 are nonmetals. The identity and functions of these trace elements are listed in the accompanying tables.

The trace elements are present in milligram quantities in the human body. If you could collect them all together, you would not have enough material to fill a teaspoon. Each, however, plays a vital role in the operation of living cells, and each is responsible for a function for which there is no substitute.

Knowledge about the functions of trace elements is difficult to obtain, because it is so difficult to provide an experimental diet that lacks the one element under study. Infinitesimal amounts of some of the elements are all that is needed to invalidate the experiment. Thus research in this area consists primarily of animal studies involving highly refined, purified diets in environments that are free of all contamination.

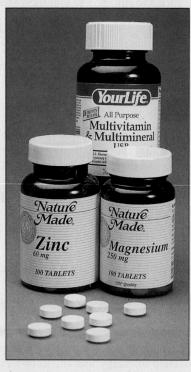

Many food supplements incorporate trace elements required by humans.

Need in Humans Established and Quantified

Metals
Iron: forms part of hemoglobin (the oxygen-carrying protein of red blood cells) and myoglobin (the oxygen-holding protein in muscle cells)
Zinc: occurs in more than 70 enzymes that perform specific tasks in the eyes, liver, kidneys, muscles, skin, bones, and male reproductive organs

Nonmetals
Iodine: occurs in three thyroid gland hormones that regulate metabolic rate
Selenium: part of an enzyme that acts as an antioxidant for polyunsaturated fatty acids

---

Need in Humans Established But Not Completely Quantified

Metals
Copper: necessary for the absorption and use of iron in the formation of hemoglobin; also a factor in the formation of the protective covering of nerves
Manganese: facilitator, with enzymes, of many different metabolic processes
Cobalt: part of vitamin B_{12}; necessary for nerve cell function and blood formation
Molybdenum: facilitator, with enzymes, of numerous cell processes
Chromium: associated with insulin and required for the release of energy from glucose

Nonmetals
Fluorine: involved in the formation of bones and teeth; helps make teeth resistant to decay

---

Need Established in Animals But Not Yet in Humans

Metals
Nickel: deficiencies harm the liver and other organs
Tin: necessary for growth
Vanadium: necessary for growth, bone development, and normal reproduction

Nonmetals
Silicon: involved in bone calcification
Boron: involved in bone development and minimization of demineralization in osteoporosis

■ Electron Subshells

Within each electron shell, electrons are further grouped into energy sublevels called electron subshells. An **electron subshell** *is a region of space within an electron shell that contains electrons that have the same energy.* We can draw an analogy between the relationship of shells and subshells and the physical layout of a high-rise apartment complex. The shells are analogous to the floors of the apartment complex, and the subshells are the counterparts of the various apartments on each floor.

The number of subshells within a shell is the same as the shell number. Shell 1 contains one subshell, shell 2 contains two subshells, shell 3 contains three subshells, and so on.

Subshells within a shell differ in size (that is, the maximum number of electrons they can accommodate) and energy. The higher the energy of the contained electrons, the larger the subshell.

Subshell size (type) is designated using the letters *s, p, d,* and *f.* Listed in this order, these letters denote subshells of increasing energy and size. The lowest-energy subshell within a shell is always the *s* subshell, the next highest is the *p* subshell, then the *d* subshell, and finally the *f* subshell. An *s* subshell can accommodate 2 electrons, a *p* subshell 6 electrons, a *d* subshell 10 electrons, and an *f* subshell 14 electrons.

Both a number and a letter are used in identifying subshells. The number gives the shell within which the subshell is located, and the letter gives the type of subshell. Shell 1 has only one subshell—the 1*s.* Shell 2 has two subshells—the 2*s* and 2*p.* Shell 3 has three subshells—the 3*s,* 3*p,* and 3*d,* and so on. Figure 3.7 summarizes the relationships between electron shells and electron subshells for the first four shells.

> The letters used to label the different types of subshells come from old spectroscopic terminology associated with the lines in the spectrum of the element hydrogen. These lines were denoted as *s*harp, *p*rincipal, *d*iffuse, and *f*undamental. Relationships exist between such lines and the arrangement of electrons in an atom.

■ Electron Orbitals

Electron subshells have within them a certain, definite number of locations (regions of space), called electron orbitals, where electrons may be found. In our apartment complex analogy, if shells are the counterparts of floor levels and subshells are the apartments, then

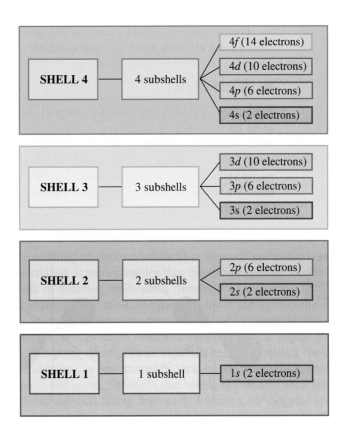

FIGURE 3.7 The number of subshells within a shell is equal to the shell number, as shown here for the first four shells. Each individual subshell is denoted with both a number (its shell) and a letter (the type of subshell it is in).

SHELL 4 — 4 subshells — 4*f* (14 electrons) / 4*d* (10 electrons) / 4*p* (6 electrons) / 4s (2 electrons)

SHELL 3 — 3 subshells — 3*d* (10 electrons) / 3*p* (6 electrons) / 3s (2 electrons)

SHELL 2 — 2 subshells — 2*p* (6 electrons) / 2*s* (2 electrons)

SHELL 1 — 1 subshell — 1*s* (2 electrons)

CHEMICAL CONNECTIONS

Electrons in Excited States

When an atom has its electrons positioned in the lowest-energy orbitals available, it is said to be in its *ground state*. An atom's ground state is the normal (most stable) state for the atom.

It is possible to elevate electrons in an atom to higher-energy unoccupied orbitals by subjecting the atom to a beam of light energy, an electrical discharge, or an influx of heat energy. With electrons in higher, normally unoccupied orbitals, the atom is said to be in an *excited state*. An excited state is an unstable state that has a short life span. Quickly, the excited electrons drop back down to their previous positions (the ground state). Accompanying the transition from excited state to ground state is a release of energy. Often this release of energy is in the form of visible light.

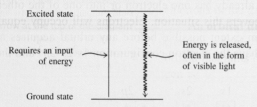

The principle of electron excitation through input of energy has been found to have useful applications in many areas.

1. *"Neon" Advertising Signs.* In such signs, gaseous atoms are excited by an electric discharge to produce a variety of colors; the color depends on the identity of the gas. Neon gas produces an orange-red light, argon gas a blue-purple light, and krypton gas a white light.
2. *Street and Highway Lights.* Such lights involve energy emitted by electrically excited metal atoms. Mercury vapor lamps produce a yellow light that can penetrate fog farther than does light from a sodium vapor lamp. On the other hand, sodium vapor lamps are more energy-efficient in their operation.
3. *Fireworks.* Metal atoms excited by heat are responsible for the color of fireworks. Strontium (red color), barium (green color), copper (blue color), and aluminum (white color) are some of the metals involved. The metals are present in the fireworks in the form of metal-containing compounds rather than as pure metals.

4. *Identification of Elements.* Each of the known elements, when electronically excited in the gaseous state, produces a unique pattern of wavelengths of emitted light (radiation). This emission pattern, which is called an atomic spectrum, serves as a "fingerprint" for the element and can be used to distinguish the element from any other.
5. *Analysis of Human Body Fluids.* Instruments called atomic spectrometers are now used to analyze body fluids for the presence of particular elements (in free or combined form). It is possible to determine concentrations of species with such instruments, which are now found in almost all clinical chemistry laboratories. The concentration of sodium and potassium in a particular fluid can be obtained from the intensity of the light emitted by excited atoms of these elements. Atomic spectroscopy makes it possible to measure the amount of lead in a patient's blood or urine (in cases of lead poisoning) by using a sample as small as 0.01 cm^3.

The different colors of fireworks result when heat excites the electrons of different kinds of metal atoms present.

EXAMPLE 3.3

Writing an Electron Configuration

■ Write the electron configurations for the following elements.

a. Strontium (atomic number = 38)   **b.** Lead (atomic number = 82)

Solution

a. The number of electrons in a strontium atom is 38. Remember that the atomic number gives the number of electrons (Section 3.2). We will need to fill subshells, in order of increasing energy, until 38 electrons have been accommodated.

The $1s$, $2s$, and $2p$ subshells fill first, accommodating a total of 10 electrons among them.

$$1s^2 2s^2 2p^6 \ldots$$

Next, according to Figures 3.11 and 3.12, the $3s$ subshell fills and then the $3p$ subshell.

$$1s^2 2s^2 2p^6 \left(3s^2 3p^6\right) \ldots$$

We have accommodated 18 electrons at this point. We still need to add 20 more electrons to get our desired number of 38.

The $4s$ subshell fills next, followed by the $3d$ subshell, giving us 30 electrons at this point.

$$1s^2 2s^2 2p^6 3s^2 3p^6 \left(4s^2 3d^{10}\right) \ldots$$

Note that the maximum electron population for d subshells is 10 electrons.

Eight more electrons are needed, which are added to the next two higher subshells, the $4p$ and the $5s$. The $4p$ subshell can accommodate 6 electrons, and the $5s$ can accommodate 2 electrons.

$$1s^2 2s^2 2p^6 3s^2 3p^6 4s^2 3d^{10} \left(4p^6 5s^2\right)$$

To double-check that we have the correct number of electrons, 38, we add the superscripts in our final electron configuration.

$$2 + 2 + 6 + 2 + 6 + 2 + 10 + 6 + 2 = 38$$

The sum of the superscripts in any electron configuration should add up to the atomic number if the configuration is for a neutral atom.

b. To write this configuration, we continue along the same lines as in part **a**, remembering that the maximum electron subshell populations are $s = 2$, $p = 6$, $d = 10$, and $f = 14$.

Lead, with an atomic number of 82, contains 82 electrons, which are added to subshells in the following order. (The line of numbers beneath the electron configuration is a running total of added electrons and is obtained by adding the superscripts up to that point. We stop when we have 82 electrons.)

$$1s^2 2s^2 2p^6 3s^2 3p^6 4s^2 3d^{10} 4p^6 5s^2 4d^{10} 5p^6 6s^2 4f^{14} 5d^{10} 6p^2$$

$$2 \quad 4 \quad 10 \quad 12 \quad 18 \quad 20 \quad 30 \quad 36 \quad 38 \quad 48 \quad 54 \quad 56 \quad 70 \quad 80 \quad 82$$

Running total of electrons added

Note in this electron configuration that the $6p$ subshell contains only 2 electrons, even though it can hold a maximum of 6. We put only 2 electrons in this subshell because that is sufficient to give 82 total electrons. If we had completely filled this subshell, we would have had 86 total electrons, which is too many.

Practice Exercise 3.3

Write the electron configurations for the following elements.

a. Manganese (atomic number = 25)   **b.** Xenon (atomic number = 54)

3.8 The Electronic Basis for the Periodic Law and the Periodic Table

For many years, there was no explanation available for either the periodic law or why the periodic table has the shape that it has. We now know that the theoretical basis for both the periodic law and the periodic table is found in electronic theory. As we saw earlier in the chapter (Section 3.2), when two atoms interact, it is their electrons that interact.

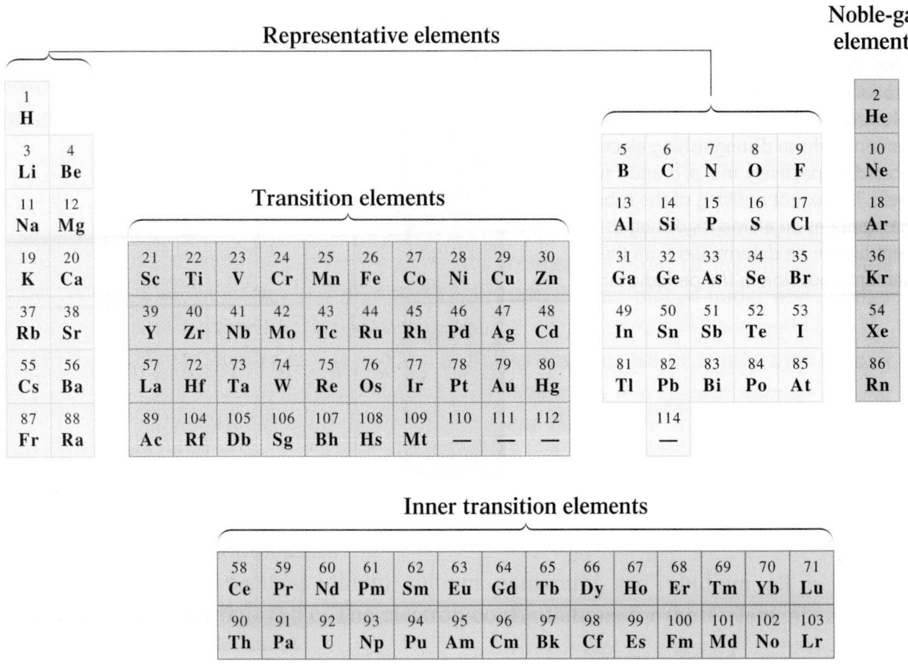

FIGURE 3.14 A classification scheme for the elements based on their electron configurations. Representative elements occupy the s area and most of the p area shown in Figure 3.13. The noble-gas elements occupy the last column of the p area. The transition elements are found in the d area, and the inner transition elements are found in the f area.

Representative elements

Noble-gas elements

1 H																	2 He
3 Li	4 Be											5 B	6 C	7 N	8 O	9 F	10 Ne
11 Na	12 Mg											13 Al	14 Si	15 P	16 S	17 Cl	18 Ar
19 K	20 Ca	21 Sc	22 Ti	23 V	24 Cr	25 Mn	26 Fe	27 Co	28 Ni	29 Cu	30 Zn	31 Ga	32 Ge	33 As	34 Se	35 Br	36 Kr
37 Rb	38 Sr	39 Y	40 Zr	41 Nb	42 Mo	43 Tc	44 Ru	45 Rh	46 Pd	47 Ag	48 Cd	49 In	50 Sn	51 Sb	52 Te	53 I	54 Xe
55 Cs	56 Ba	57 La	72 Hf	73 Ta	74 W	75 Re	76 Os	77 Ir	78 Pt	79 Au	80 Hg	81 Tl	82 Pb	83 Bi	84 Po	85 At	86 Rn
87 Fr	88 Ra	89 Ac	104 Rf	105 Db	106 Sg	107 Bh	108 Hs	109 Mt	110 —	111 —	112 —	114 —					

Transition elements (d area)

Inner transition elements

58 Ce	59 Pr	60 Nd	61 Pm	62 Sm	63 Eu	64 Gd	65 Tb	66 Dy	67 Ho	68 Er	69 Tm	70 Yb	71 Lu
90 Th	91 Pa	92 U	93 Np	94 Pu	95 Am	96 Cm	97 Bk	98 Cf	99 Es	100 Fm	101 Md	102 No	103 Lr

3.9 Classification of the Elements

The elements can be classified in several ways. The two most common classification systems are

1. A system based on selected physical properties of the elements, in which they are described as metals or nonmetals. This classification scheme was discussed in Section 3.5.
2. A system based on the electron configurations of the elements, in which elements are described as *noble-gas, representative, transition,* or *inner transition elements.*

The classification scheme based on electron configurations of the elements is depicted in Figure 3.14.

A **noble-gas element** *is an element located in the far right column of the periodic table.* These elements are all gases at room temperature, and they have little tendency to form chemical compounds. With one exception, the distinguishing electron for a noble gas completes the *p* subshell; therefore, noble gases have electron configurations ending in p^6. The exception is helium, in which the distinguishing electron completes the first shell—a shell that has only two electrons. Helium's electron configuration is $1s^2$.

A **representative element** *is an element located in the* s *area or the first five columns of the* p *area of the periodic table.* The distinguishing electron in these elements partially or completely fills an *s* subshell or partially fills a *p* subshell. The representative elements include most of the more common elements.

A **transition element** *is an element located in the* d *area of the periodic table.* Each has its distinguishing electron in a *d* subshell.

An **inner transition element** *is an element located in the* f *area of the periodic table.* Each has its distinguishing electron in an *f* subshell. There is very little variance in the properties of either the 4*f* or the 5*f* series of inner transition elements.

The Chemistry at a Glance feature on page 69 contrasts the three element classification schemes that have been considered so far in this chapter: by physical properties (Section 3.5), by electron configuration (Section 3.9), and by non-numerical periodic table group names (Section 3.4).

▶ The electron configurations of the noble gases will be an important focal point when we consider chemical bonding theory in Chapters 4 and 5.

Element Classification Schemes and the Periodic Table

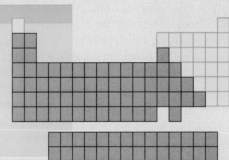

CLASSIFICATION BY PHYSICAL PROPERTIES

Nonmetals
- No metallic luster
- Poor electrical conductivity
- Good heat insulators
- Brittle and nonmalleable

Metals
- Metallic gray or silver luster
- High electrical and thermal conductivity
- Malleable and ductile

CLASSIFICATION BY ELECTRONIC PROPERTIES

Representative elements
- Found in s area and first five columns of the p area
- Some are metals, some nonmetals

Noble-gas elements
- Found in last column of p area plus He (s area)
- All are nonmetals

Transition elements
- Found in d area
- All are metals

Inner transition elements
- Found in f area
- All are metals

PERIODIC TABLE GROUPS WITH SPECIAL NAMES

Alkali metals
- Group IA elements (except for H, a nonmetal)
- Electron configurations end in s^1

Alkaline earth metals
- Group IIA elements
- Electron configurations end in s^2

Halogens
- Group VIIA
- Electron configurations end in p^5

Noble gases
- Group VIIIA elements
- Electron configurations end in p^6, except for He, which ends in s^2

3.10 Experimental Evidence for the Existence of Subatomic Particles

A significant body of evidence is consistent with and supports the existence, nature, and arrangement of subatomic particles within an atom as described in this chapter. Two historically important types of experiments illustrate some of the sources of this evidence. *Discharge tube experiments* resulted in the original concept that the atom contained negatively and positively charged particles. *Metal foil experiments* provided evidence for the existence of a nucleus within an atom.

■ Discharge Tube Experiments

Experiments involving gas discharge tubes provided the first evidence that *electrons* and *protons* are present within an atom. A simplified diagram of a gas discharge tube is shown in Figure 3.15. The apparatus consists of a sealed glass tube containing two metal disks

Negative electrode (cathode) Glass tube Positive electrode (anode)

−   +

To vacuum pump

FIGURE 3.15 A simplified version of a gas discharge tube.

Neon signs, fluorescent lights, television tubes, and computer monitors are all basic components of our modern technological society. The forerunner for all four of these developments was the gas discharge tube.

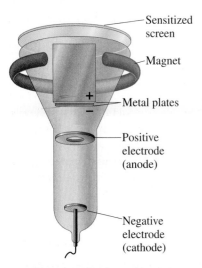

FIGURE 3.16 A cathode ray tube with perpendicular magnetic and electric fields was used in characterizing cathode rays (electrons).

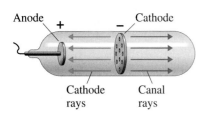

FIGURE 3.17 A gas discharge tube modified to detect canal rays.

called *electrodes*. The glass tube also has a side arm for attachment to a vacuum pump. During operation, the electrodes are connected to a source of electrical power. (The electrode attached to the positive side of the electrical power source is called the *anode;* the one attached to the negative side is known as the *cathode.*) Use of the vacuum pump allows the amount of gas within the tube to be varied. The smaller the amount of gas present, the lower the pressure within the tube.

Early studies with gas discharge tubes, conducted in the mid-1800s, showed that when the tube was almost evacuated (low pressure), electricity flowed from one electrode to the other, and the residual gas became luminous (it glowed). Different gases in the tube gave different colors to the glow. After the pressure in the tube was reduced to still lower levels (very little gas remaining), it was found that the luminosity disappeared but the electrical conductance continued, as shown by a greenish glow given off by the tube's glass walls. This glow was the initial indication of what became known as *cathode rays.* Their discovery marked the beginning of nearly 40 years of discharge tube experimentation that ultimately led to the characterization of both the electron and the proton.

The term *cathode rays* comes from the observation that when an obstacle is placed between the negative electrode (cathode) and the opposite glass wall, a sharp shadow the shape of the obstacle is cast on that wall. This indicates that the rays are coming from the cathode.

Further studies showed that these cathode rays caused certain minerals such as sphalerite (zinc sulfide) to glow. Glass plates were coated with sphalerite and observed under high magnification while being bombarded with cathode rays. The light emitted by the sphalerite coating consisted of many pinpoint flashes. This observation suggested that cathode rays were in reality a stream of extremely small particles.

Joseph John Thomson (1856–1940), an English physicist, discovered a great deal about the nature of cathode rays. Using a variety of materials as cathodes, he showed that cathode ray production is a general property of matter. By using a specially designed cathode ray tube (see Figure 3.16), he also found that cathode rays can be deflected by charged plates or a magnetic field. The rays were repelled by the north pole or negative plate and attracted to the south pole or positive plate, which indicated that they were negatively charged. In 1897, Thomson concluded that cathode rays were streams of negatively charged particles, which today we call *electrons.* Further experiments by others proved that his conclusions were correct.

In 1886, the German physicist Eugene Goldstein (1850–1930) showed that positive particles were also present in discharge tubes. He used a discharge tube in which the cathode was a metal plate with a large number of holes drilled in it. The usual cathode rays were observed to stream from cathode to anode. In addition, rays of light appeared to stream from each of the holes in the cathode in a direction opposite to that of the cathode rays (see Figure 3.17). Because these rays were observed streaming through the holes or channels in the cathode, Goldstein called them *canal rays.*

Further research showed that canal rays are of many different types, in contrast to cathode rays, which are of only one type, and that the particles making up canal rays were much heavier than those of cathode rays. The type of canal ray produced depended on the gas in the tube. The simplest canal rays were eventually identified as the particles now called *protons.*

Canal rays are now known to be gas atoms that have lost one or more electrons. Their origin and behavior in a discharge tube can be understood as follows: Electrons (cathode rays) emitted from the cathode collide with residual gas molecules (air) on the way to the anode. Some of these electrons have enough energy to knock electrons away from the gas molecules, leaving behind a positive particle (the remainder of the gas molecule). These positive particles are attracted to the cathode, and some of them pass through the holes or channels. The fact that atoms can lose electrons under certain conditions will be discussed further in Chapter 4.

On the basis of discharge tube experiments, Thomson proposed in 1898 that the atom was composed of a sphere of positive electricity, which contained most of the mass, and that small negative electrons were attached to the surface of the positive sphere. He postulated that a high voltage could pull off surface electrons to produce cathode rays. Thomson's model of the atom, sometimes referred to as the "raisin muffin" or "plum pudding" model—with the electrons as the raisins—is known to be incorrect. Its significance is that it set the stage for an experiment, commonly called the gold foil experiment, that led to the currently accepted concept of the arrangement of protons and electrons in the atom.

■ Metal Foil Experiments

In 1911, Ernest Rutherford (1871–1937) designed an experiment to test the Thomson model of the atom. In this experiment, thin sheets of metal were bombarded with alpha particles from a radioactive source. Alpha particles, which are positively charged, are ejected at high speeds from some radioactive materials (see Section 11.3). The phenomenon of radioactivity had been discovered in 1896 and gave further evidence that electrical charges exist in atoms. Gold was chosen as the target metal because it is easily hammered into very thin sheets. The experimental setup for Rutherford's experiment is shown in Figure 3.18. Alpha particles do not appreciably penetrate lead, so a lead plate with a slit was used to produce a narrow alpha particle beam.

Rutherford expected that because they were so energetic, all the alpha particles would pass straight through the thin gold foil, hit the fluorescent screen, and produce a flash of light. His reasoning was based on the Thomson model, in which the mass and positive charge of the gold atoms were distributed uniformly through each atom. Rutherford assumed that as each positive alpha particle neared the foil, it would be confronted by a uniform positive charge. All particles would be affected the same way (no deflection), which would support the Thomson model of the atom.

The results from the experiment were very surprising. Most of the particles—more than 99%—went straight through as expected. A few, however, were appreciably deflected by something that had to be much heavier than the alpha particles themselves. A very few particles were deflected almost directly back toward the alpha particle source. Similar results were obtained when elements other than gold were used as targets.

Extensive study of his results led Rutherford to propose the following explanation.

1. A very dense, small nucleus exists in the center of the atom. This nucleus contains most of the mass of the atom and all of the positive charge.
2. Electrons occupy most of the total volume of the atoms and are located outside the nucleus.
3. When an alpha particle scores a direct hit on a nucleus, it is deflected back along the incoming path.
4. A near miss of a nucleus by an alpha particle results in repulsion and deflection.
5. Most of the alpha particles pass through without any interference, because most of the atomic volume is empty space.
6. Electrons have so little mass that they do not deflect the much larger alpha particles (an alpha particle is almost 8000 times heavier than an electron).

Many other experiments have since verified Rutherford's conclusions that at the center of an atom there is a nucleus that is very small and very dense.

> At a later date, Rutherford described the unexpected results of his gold foil experiment as follows: "It is about as credible as if you had fired a 15-inch shell at a piece of tissue paper and it came back and hit you."

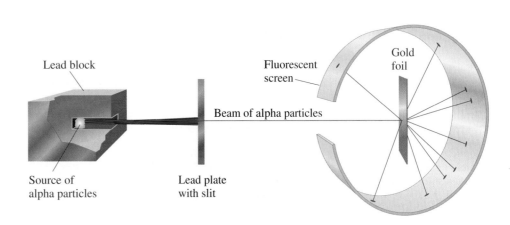

FIGURE 3.18 Rutherford's gold foil–alpha particle experiment. Most of the alpha particles went straight through the foil, but a few were deflected at large angles.

Subatomic particles. Subatomic particles, the very small building blocks from which atoms are made, are of three major types: electrons, protons, and neutrons. Electrons are negatively charged, protons are positively charged, and neutrons have no charge. All neutrons and protons are found at the center of the atom in the nucleus. The electrons occupy the region about the nucleus. Protons and neutrons have much larger masses than the electron.

Atomic number and mass number. Each atom has a characteristic atomic number and mass number. The atomic number is equal to the number of protons in the nucleus of the atom. The mass number is equal to the total number of protons and neutrons in the nucleus.

Isotopes. Isotopes are atoms that have the same number of protons and electrons but have different numbers of neutrons. The isotopes of an element always have the same atomic number and different mass numbers. Isotopes of an element have the same chemical properties.

Atomic mass. The atomic mass of an element is a calculated average mass. It depends on the percentage abundances and masses of the naturally occurring isotopes of the element.

Periodic law and periodic table. The periodic law states that when elements are arranged in order of increasing atomic number, elements with similar chemical properties occur at periodic (regularly recurring) intervals. The periodic table is a graphical representation of the behavior described by the periodic law. In a modern periodic table, vertical columns contain elements with similar chemical properties. A group in the periodic table is a vertical column of elements. A period in the periodic table is a horizontal row of elements.

Metals and nonmetals. Metals exhibit luster, thermal conductivity, electrical conductivity, and malleability. Nonmetals are characterized by the absence of the properties associated with metals. The majority of the elements are metals. The steplike heavy line that runs through the right third of the periodic table separates the metals on the left from the nonmetals on the right.

Electron shell. A shell contains electrons that have approximately the same energy and spend most of their time approximately the same distance from the nucleus.

Electron subshell. A subshell contains electrons that all have the same energy. The number of subshells in a particular shell is equal to the shell number. Each subshell can hold a specific maximum number of electrons. These values are 2, 6, 10, and 14 for s, p, d, and f subshells, respectively.

Electron orbital. An orbital is a region of space about a nucleus where an electron with a specific energy is most likely to be found. Each subshell consists of one or more orbitals. For s, p, d, and f subshells there are 1, 3, 5, and 7 orbitals, respectively. No more than two electrons may occupy any orbital.

Electron configuration. An electron configuration is a statement of how many electrons an atom has in each of its subshells. The principle that electrons normally occupy the lowest-energy subshell available is used to write electron configurations.

Orbital diagram. An orbital diagram is a statement of how many electrons an atom has in each of its orbitals. Electrons occupy the orbitals of a subshell such that each orbital within the subshell acquires one electron before any orbital acquires a second electron. All electrons in such singly occupied orbitals must have the same spin.

Electron configurations and the periodic law. Chemical properties repeat themselves in a regular manner among the elements because electron configurations repeat themselves in a regular manner among the elements.

Electron configurations and the periodic table. The groups of the periodic table consist of elements with similar electron configurations. Thus the location of an element in the periodic table can be used to obtain information about its electron configuration.

Classification system for the elements. On the basis of electron configuration, elements can be classified into four categories: noble gases (far right column of the periodic table); representative elements (s and p areas of the periodic table, with the exception of the noble gases); transition elements (d area of the periodic table); and inner transition elements (f area of the periodic table).

1. Relationships involving atomic number and mass number for a neutral atom (Section 3.2)

   Atomic number = number of protons = number of electrons

   Mass number = number of protons + number of neutrons

   Mass number = total number of subatomic particles in the nucleus

   Mass number − atomic number = number of neutrons

   Mass number + atomic number = total number of subatomic particles

2. Relationships involving electron shells, electron subshells, and electron orbitals (Section 3.6)

   Number of subshells in a shell = shell number

   Maximum number of electrons in an s subshell = 2

   Maximum number of electrons in a p subshell = 6

   Maximum number of electrons in a d subshell = 10

   Maximum number of electrons in an f subshell = 14

   Maximum number of electrons in an orbital = 2

3. Order of filling of subshells in terms of increasing energy (Section 3.7)

   $1s$, $2s$, $2p$, $3s$, $3p$, $4s$, $3d$, $4p$, $5s$, $4d$, $5p$, $6s$, $4f$, $5d$, $6p$, $7s$, $5f$, $6d$, $7p$

Atomic mass (3.3)
Atomic number (3.2)
Distinguishing electron (3.8)
Electron (3.1)

Electron configuration (3.7)
Electron orbital (3.6)
Electron shell (3.6)
Electron subshell (3.6)

Element (3.2)
Group (3.4)
Inner transition element (3.9)
Isotopes (3.3)

Mass number (3.2)
Metal (3.5)
Neutron (3.1)
Noble-gas element (3.9)
Nonmetal (3.5)

Nucleon (3.1)
Nucleus (3.1)
Orbital diagram (3.7)
Period (3.4)
Periodic law (3.4)

Periodic table (3.4)
Proton (3.1)
Representative element (3.9)
Subatomic particle (3.1)
Transition element (3.9)

EXERCISES AND PROBLEMS

The members of each pair of problems in this section test similar material.

■ Internal Structure of the Atom (Section 3.1)

3.1 Indicate which subatomic particle (proton, neutron, or electron) correctly matches each of the following phrases. More than one particle can be used as an answer.
  a. Possesses a negative charge
  b. Has no charge
  c. Has a mass slightly less than that of a neutron
  d. Has a charge equal to, but opposite in sign from, that of an electron

3.2 Indicate which subatomic particle (proton, neutron, or electron) correctly matches each of the following phrases. More than one particle can be used as an answer.
  a. Is not found in the nucleus
  b. Has a positive charge
  c. Can be called a nucleon
  d. Has a relative mass of 1837 if the relative mass of an electron is 1

3.3 Indicate whether each of the following statements about the nucleus of an atom is true or false.
  a. The nucleus of an atom is neutral.
  b. The nucleus of an atom contains only neutrons.
  c. The number of nucleons present in the nucleus is equal to the number of electrons present outside the nucleus.
  d. The nucleus accounts for almost all the mass of an atom.

3.4 Indicate whether each of the following statements about the nucleus of an atom is true or false.
  a. The nucleus of an atom contains all of the "heavy" subatomic particles.
  b. The nucleus of an atom accounts for almost all of the volume of the atom.
  c. The nucleus of an atom has an extremely low density compared to that of the atom as a whole.
  d. The nucleus of an atom can be positively or negatively charged, depending on the identity of the atom.

■ Atomic Number and Mass Number (Section 3.2)

3.5 Determine the atomic number and mass number for atoms with the following subatomic makeups.
  a. 2 protons, 2 neutrons, and 2 electrons
  b. 4 protons, 5 neutrons, and 4 electrons
  c. 5 protons, 4 neutrons, and 5 electrons
  d. 28 protons, 30 neutrons, and 28 electrons

3.6 Determine the atomic number and mass number for atoms with the following subatomic makeups.
  a. 1 proton, 1 neutron, and 1 electron
  b. 10 protons, 12 neutrons, and 10 electrons
  c. 12 protons, 10 neutrons, and 12 electrons
  d. 50 protons, 69 neutrons, and 50 electrons

3.7 Determine the number of protons, neutrons, and electrons present in atoms with the following characteristics.

  a. Atomic number = 8 and mass number = 16
  b. Mass number = 18 and $Z = 8$
  c. Atomic number = 20 and $A = 44$
  d. $A = 257$ and $Z = 100$

3.8 Determine the number of protons, neutrons, and electrons present in atoms with the following characteristics.
  a. Atomic number = 10 and mass number = 20
  b. Mass number = 110 and $Z = 48$
  c. $A = 11$ and atomic number = 5
  d. $Z = 92$ and $A = 238$

3.9 Indicate whether the *atomic number*, the *mass number*, or *both the atomic number and the mass number* are needed to determine the following.
  a. Number of protons in an atom
  b. Number of neutrons in an atom
  c. Number of nucleons in an atom
  d. Total number of subatomic particles in an atom

3.10 What information about the subatomic particles present in an atom is obtained from each of the following?
  a. Atomic number
  b. Mass number
  c. Mass number − atomic number
  d. Mass number + atomic number

3.11 Arrange the following atoms in the orders specified.
$$^{32}_{16}\text{S} \quad ^{40}_{18}\text{Ar} \quad ^{35}_{17}\text{Cl} \quad ^{37}_{19}\text{K}$$
  a. Order of increasing atomic number
  b. Order of decreasing mass number
  c. Order of increasing number of electrons
  d. Order of increasing number of neutrons

3.12 Arrange the following atoms in the orders specified.
$$^{14}_{6}\text{C} \quad ^{17}_{8}\text{O} \quad ^{13}_{7}\text{N} \quad ^{19}_{9}\text{F}$$
  a. Order of decreasing atomic number
  b. Order of increasing mass number
  c. Order of decreasing number of neutrons
  d. Order of increasing number of nucleons

3.13 Determine the number of protons, neutrons, electrons, nucleons, and total subatomic particles for each of the following atoms.
  a. $^{53}_{24}\text{Cr}$ b. $^{256}_{101}\text{Md}$ c. $^{67}_{30}\text{Zn}$ d. $^{40}_{20}\text{Ca}$

3.14 Determine the number of protons, neutrons, electrons, nucleons, and total subatomic particles for each of the following atoms.
  a. $^{103}_{44}\text{Ru}$ b. $^{34}_{16}\text{S}$ c. $^{9}_{4}\text{Be}$ d. $^{4}_{2}\text{He}$

■ Isotopes and Atomic Masses (Section 3.3)

3.15 With the help of the information listed on the inside front cover, write complete symbols for the five naturally occurring isotopes of zirconium, given that the heaviest isotope has a mass number of 96 and that the other isotopes have, respectively, 2, 4, 5, and 6 fewer neutrons.

3.16 With the help of the information listed on the inside front cover, write complete symbols for the four naturally occurring isotopes of strontium, given that the lightest isotope has a mass number of 84 and that the other isotopes have, respectively, 2, 4, and 5 more neutrons.

3.17 Indicate whether each of the following statements about sodium isotopes is true or false.
a. $^{23}_{11}Na$ has one more electron than $^{24}_{11}Na$.
b. $^{23}_{11}Na$ and $^{24}_{11}Na$ contain the same number of neutrons.
c. $^{23}_{11}Na$ has one less subatomic particle than $^{24}_{11}Na$.
d. $^{23}_{11}Na$ and $^{24}_{11}Na$ have the same atomic number.

3.18 Indicate whether each of the following statements about magnesium isotopes is true or false.
a. $^{24}_{12}Mg$ has one more proton than $^{25}_{12}Mg$.
b. $^{24}_{12}Mg$ and $^{25}_{12}Mg$ contain the same number of subatomic particles.
c. $^{24}_{12}Mg$ has one less neutron than $^{25}_{12}Mg$.
d. $^{24}_{12}Mg$ and $^{25}_{12}Mg$ have different mass numbers.

3.19 The following are selected properties for the most abundant isotope of a particular element. Which of these properties would also be the same for the second-most-abundant isotope of the element?
a. Mass number is 70
b. 31 electrons are present
c. Isotopic mass is 69.92 amu
d. Isotope reacts with chlorine to give a green compound

3.20 The following are selected properties for the most abundant isotope of a particular element. Which of these properties would also be the same for the second-most-abundant isotope of the element?
a. Atomic number is 31
b. Does not react with the element gold
c. 40 neutrons are present
d. Density is 1.03 g/mL

3.21 A certain isotope of silver is 8.91 times heavier than $^{12}_{6}C$. What is the mass of this silver isotope on the amu scale?

3.22 A certain isotope of silicon is 2.42 times heavier than $^{12}_{6}C$. What is the mass of this silicon isotope on the amu scale?

3.23 Calculate the atomic mass of each of the following elements using the given data for the percentage abundance and mass of each isotope.
a. Lithium: 7.42% ^6Li (6.01 amu) and 92.58% ^7Li (7.02 amu)
b. Magnesium: 78.99% ^{24}Mg (23.99 amu), 10.00% ^{25}Mg (24.99 amu), and 11.01% ^{26}Mg (25.98 amu)

3.24 Calculate the atomic mass of each of the following elements using the given data for the percentage abundance and mass of each isotope.
a. Silver: 51.82% ^{105}Ag (106.9 amu) and 48.18% ^{107}Ag (108.9 amu)
b. Silicon: 92.21% ^{28}Si (27.98 amu), 4.70% ^{29}Si (28.98 amu), and 3.09% ^{30}Si (29.97 amu)

3.25 The arbitrary standard for the atomic mass scale is the exact number 12 for the mass of $^{12}_{6}C$. Why, then, is the atomic mass of carbon listed as 12.011?

3.26 The atomic mass of fluorine is 19.00 amu, and the atomic mass of iron is 55.85 amu. All fluorine atoms have a mass of 19.00 amu, and not a single iron atom has a mass of 55.85 amu. Explain.

■ **The Periodic Law and the Periodic Table (Section 3.4)**

3.27 Give the symbol of the element that occupies each of the following positions in the periodic table.

a. Period 4, Group IIA   b. Period 5, Group VIB
c. Group IA, Period 2   d. Group IVA, Period 5

3.28 Give the symbol of the element that occupies each of the following positions in the periodic table.
a. Period 1, Group IA   b. Period 6, Group IB
c. Group IIIB, Period 4   d. Group VIIA, Period 3

3.29 Using the periodic table, determine the following.
a. The atomic number of the element carbon
b. The atomic mass of the element silicon
c. The atomic number of the element with an atomic mass of 88.91 amu
d. The atomic mass of the element located in Period 2 and Group IIA

3.30 Using the periodic table, determine the following.
a. The atomic number of the element magnesium
b. The atomic mass of the element nitrogen
c. The atomic mass of the element with an atomic number of 10
d. The atomic number of the element located in Group IIIA and Period 3

3.31 For each of the following sets of elements, choose the two that would be expected to have similar chemical properties.
a. $_{19}K$, $_{29}Cu$, $_{37}Rb$, $_{41}Nb$ b. $_{13}Al$, $_{14}Si$, $_{15}P$, $_{33}As$
c. $_9F$, $_{40}Zr$, $_{50}Sn$, $_{53}I$ d. $_{11}Na$, $_{12}Mg$, $_{54}Xe$, $_{55}Cs$

3.32 For each of the following sets of elements, choose the two that would be expected to have similar chemical properties.
a. $_{11}Na$, $_{14}Si$, $_{23}V$, $_{55}Cs$ b. $_{13}Al$, $_{19}K$, $_{32}Ge$, $_{50}Sn$
c. $_{37}Rb$, $_{38}Sr$, $_{54}Xe$, $_{56}Ba$ d. $_2He$, $_6C$, $_8O$, $_{10}Ne$

3.33 The following statements either define or are closely related to the terms *periodic law, period,* and *group.* Match each statement with the appropriate term.
a. This is a vertical arrangement of elements in the periodic table.
b. The properties of the elements repeat in a regular way as atomic numbers increase.
c. The chemical properties of elements 12, 20, and 38 demonstrate this principle.
d. Carbon is the first member of this arrangement.

3.34 The following statements either define or are closely related to the terms *periodic law, period,* and *group.* Match each statement with the appropriate term.
a. This is a horizontal arrangement of elements in the periodic table.
b. Element 19 begins this arrangement in the periodic table.
c. Elements 24 and 33 belong to this arrangement.
d. Elements 10, 18, and 36 belong to this arrangement.

3.35 Classify each of the following elements as a *halogen, noble gas, alkali metal,* or *alkaline earth metal.*
a. Cs   b. K   c. Ne   d. Mg
e. F   f. Ar   g. Ca   h. I

3.36 Classify each of the following elements as a *halogen, noble gas, alkali metal,* or *alkaline earth metal.*
a. Br   b. Na   c. Li   d. Xe
e. Be   f. Ba   g. Kr   h. Cl

■ **Metals and Nonmetals (Section 3.5)**

3.37 In which of the following pairs of elements are both members of the pair metals?
a. $_{17}Cl$ and $_{35}Br$ b. $_{13}Al$ and $_{14}Si$
c. $_{29}Cu$ and $_{42}Mo$ d. $_{30}Zn$ and $_{83}Bi$

3.38 In which of the following pairs of elements are both members of the pair metals?

a. $_7$N and $_{34}$Se   b. $_{16}$S and $_{48}$Cd
c. $_3$Li and $_{26}$Fe   d. $_{50}$Sn and $_{53}$I

3.39 Identify the nonmetal in each of the following sets of elements.
a. S, Na, K   b. Cu, Li, P
c. Be, I, Ca   d. Fe, Cl, Ga

3.40 Identify the nonmetal in each of the following sets of elements.
a. Al, H, Mg   b. C, Sn, Sb
c. Ti, V, F   d. Sr, Se, Sm

3.41 Classify each of the following general physical properties as a property of metallic elements or of nonmetallic elements.
a. Ductile
b. Low electrical conductivity
c. High thermal conductivity
d. Good heat insulator

3.42 Classify each of the following general physical properties as a property of metallic elements or of nonmetallic elements.
a. Nonmalleable
b. High luster
c. Low thermal conductivity
d. Brittle

■ Electron Arrangements Within Atoms (Section 3.6)

3.43 The following statements define or are closely related to the terms *shell, subshell,* and *orbital*. Match each statement with the appropriate term.
a. In terms of electron capacity, this unit is the smallest of the three.
b. This unit can contain a maximum of two electrons.
c. This unit is designated just by a number.
d. The term *energy level* is closely associated with this unit.

3.44 The following statements define or are closely related to the terms *shell, subshell,* and *orbital*. Match each statement with the appropriate term.
a. This unit can contain as many electrons as, or more electrons than, either of the other two.
b. The term *energy sublevel* is closely associated with this unit.
c. Electrons that occupy this unit do not need to have identical energies.
d. The unit is designated in the same way as the orbitals contained within it.

3.45 Indicate whether each of the following statements is true or false.
a. An orbital has a definite size and shape, which are related to the energy of the electrons it could contain.
b. All the orbitals in a subshell have the same energy.
c. All subshells accommodate the same number of electrons.
d. A $2p$ subshell and a $3p$ subshell contain the same number of orbitals.

3.46 Indicate whether each of the following statements is true or false.
a. All the subshells in a shell have the same energy.
b. An s orbital has a shape that resembles a four-leaf clover.
c. The third shell can accommodate a maximum number of 18 electrons.
d. All orbitals accommodate the same number of electrons.

3.47 Give the maximum number of electrons that can occupy each of the following electron-accommodating units.
a. One of the orbitals in the $2p$ subshell
b. One of the orbitals in the $3d$ subshell
c. The $4p$ subshell
d. The third shell

3.48 Give the maximum number of electrons that can occupy each of the following electron-accommodating units.
a. One of the orbitals in the $4d$ subshell
b. One of the orbitals in the $5f$ subshell
c. The $3d$ subshell
d. The second shell

■ Electron Configurations and Orbital Diagrams (Section 3.7)

3.49 Write complete electron configurations for atoms of each of the following elements.
a. $_6$C   b. $_{11}$Na   c. $_{16}$S   d. $_{18}$Ar

3.50 Write complete electron configurations for atoms of each of the following elements.
a. $_{10}$Ne   b. $_{13}$Al   c. $_{19}$K   d. $_{22}$Ti

3.51 On the basis of the total number of electrons present, identify the elements whose electron configurations are
a. $1s^2 2s^2 2p^4$ b. $1s^2 2s^2 2p^6$
c. $1s^2 2s^2 2p^6 3s^2 3p^1$ d. $1s^2 2s^2 2p^6 3s^2 3p^6 4s^2$

3.52 On the basis of the total number of electrons present, identify the elements whose electrons configurations are
a. $1s^2 2s^2 2p^2$ b. $1s^2 2s^2 2p^6 3s^1$
c. $1s^2 2s^2 2p^6 3s^2 3p^5$ d. $1s^2 2s^2 2p^6 3s^2 3p^6 4s^2 3d^{10} 4p^3$

3.53 Write *complete* electron configurations for atoms whose electron configurations *end* as follows.
a. $3p^5$ b. $4d^7$ c. $4s^2$ d. $3d^1$

3.54 Write *complete* electron configurations for atoms whose electron configurations *end* as follows.
a. $4p^2$ b. $3d^{10}$ c. $5s^1$ d. $4p^6$

3.55 Draw the orbital diagram associated with each of the following electron configurations.
a. $1s^2 2s^2 2p^2$ b. $1s^2 2s^2 2p^6 3s^2$
c. $1s^2 2s^2 2p^6 3s^2 3p^3$ d. $1s^2 2s^2 2p^6 3s^2 3p^6 4s^2 3d^7$

3.56 Draw the orbital diagram associated with each of the following electron configurations.
a. $1s^2 2s^2 2p^5$ b. $1s^2 2s^2 2p^6 3s^1$
c. $1s^2 2s^2 2p^6 3s^2 3p^1$ d. $1s^2 2s^2 2p^6 3s^2 3p^6 4s^2 3d^5$

3.57 How many unpaired electrons are present in the orbital diagram for each of the following elements?
a. $_7$N   b. $_{12}$Mg   c. $_{17}$Cl   d. $_{25}$Mn

3.58 How many unpaired electrons are present in the orbital diagram for each of the following elements?
a. $_9$F   b. $_{16}$S   c. $_{20}$Ca   d. $_{30}$Zn

■ Electron Configurations and the Periodic Law (Section 3.8)

3.59 Indicate whether the elements represented by the given pairs of electron configurations have similar chemical properties.
a. $1s^2 2s^1$ and $1s^2 2s^2$
b. $1s^2 2s^2 2p^6$ and $1s^2 2s^2 2p^6 3s^2 3p^6$
c. $1s^2 2s^2 2p^3$ and $1s^2 2s^2 2p^6 3s^2 3p^6 4s^2 3d^3$
d. $1s^2 2s^2 2p^6 3s^2 3p^4$ and $1s^2 2s^2 2p^6 3s^2 3p^6 4s^2 3d^{10} 4p^4$

3.60 Indicate whether the elements represented by the given pairs of electron configurations have similar chemical properties.
a. $1s^2 2s^2 2p^4$ and $1s^2 2s^2 2p^5$
b. $1s^2 2s^2$ and $1s^2 2s^2 2p^2$
c. $1s^2 2s^1$ and $1s^2 2s^2 2p^6 3s^2 3p^6 4s^1$
d. $1s^2 2s^2 2p^6$ and $1s^2 2s^2 2p^6 3s^2 3p^6 4s^2 3d^6$

■ Electron Configurations and the Periodic Table (Section 3.8)

3.61 For each of the following elements, specify the extent to which the subshell containing the distinguishing electron is filled (s^2, p^3, p^5, d^4, etc.).
a. $_{13}$Al   b. $_{23}$V   c. $_{20}$Ca   d. $_{36}$Kr

3.62 For each of the following elements, specify the extent to which the subshell containing the distinguishing electron is filled (s^2, p^3, p^5, d^4, etc.).
a. $_{10}Ne$ b. $_{19}K$ c. $_{33}As$ d. $_{30}Zn$

■ **Classification of the Elements (Section 3.9)**

3.63 Classify each of the following elements as a noble gas, representative element, transition element, or inner transition element.
a. $_{15}P$ b. $_{18}Ar$ c. $_{79}Au$ d. $_{92}U$

3.64 Classify each of the following elements as a noble gas, representative element, transition element, or inner transition element.
a. $_1H$ b. $_{44}Ru$ c. $_{51}Sb$ d. $_{86}Rn$

3.65 Classify the element with each of the following electron configurations as a representative element, transition element, noble gas, or inner transition element.
a. $1s^2 2s^2 2p^6$ b. $1s^2 2s^2 2p^6 3s^2 3p^4$
c. $1s^2 2s^2 2p^6 3s^2 3p^6 4s^2 3d^1$ d. $1s^2 2s^2 2p^6 3s^2 3p^6 4s^2$

3.66 Classify the element with each of the following electron configurations as a representative element, transition element, noble gas, or inner transition element.
a. $1s^2 2s^2 2p^6 3s^1$ b. $1s^2 2s^2 2p^6 3s^2 3p^6$
c. $1s^2 2s^2 2p^6 3s^2 3p^6 4s^2 3d^7$ d. $1s^2 2s^2 2p^6 3s^2 3p^6 4s^2 3d^{10} 4p^5$

■ **Experimental Evidence for the Existence of Subatomic Particles (Section 3.10)**

3.67 Indicate whether each of the following statements about discharge tube experiments and metal foil experiments is true or false.
a. Information obtained from discharge tube experiments led to the characterization of both protons and electrons.
b. Canal rays are negatively charged particles produced in a discharge tube.
c. Metal foil experiments led to the discovery of neutrons.
d. Many different types of cathode rays are known.

3.68 Indicate whether each of the following statements about discharge tube experiments and metal foil experiments is true or false.
a. In metal foil experiments, nearly all of the bombarding particles were stopped by the metal foil.
b. Metal foil experiments led to the concept that an atom has a nucleus.
c. Cathode rays and canal rays move in opposite directions in a discharge tube.
d. Many different types of canal rays have been observed, but only one type of cathode ray is known.

ADDITIONAL PROBLEMS

3.69 Write complete symbols ($_Z^A E$), with the help of a periodic table, for atoms with the following characteristics.
a. Contains 20 electrons and 24 neutrons
b. Radon atom with a mass number of 211
c. Silver atom that contains 157 subatomic particles
d. Beryllium atom that contains 9 nucleons

3.70 Characterize each of the following pairs of atoms as containing (1) the same number of neutrons, (2) the same number of electrons, or (3) the same total number of subatomic particles.
a. $_6^{13}C$ and $_7^{14}N$ b. $_8^{18}O$ and $_9^{19}F$
c. $_{17}^{37}Cl$ and $_{18}^{36}Ar$ d. $_{17}^{35}Cl$ and $_{17}^{37}Cl$

3.71 Indicate whether each of the following numbers are the same or different for two isotopes of an element.
a. Atomic number
b. Mass number
c. Number of neutrons
d. Number of electrons

3.72 Write the complete symbol ($_Z^A E$) for the isotope of chromium with each of the following characteristics.
a. Two more neutrons than $_{24}^{55}Cr$
b. Two fewer subatomic particles than $_{24}^{52}Cr$
c. The same number of neutrons as $_{29}^{60}Cu$
d. The same number of subatomic particles as $_{29}^{60}Cu$

3.73 How many electrons are present in nine molecules of the compound $C_{12}H_{22}O_{11}$ (table sugar)?

3.74 Which of the six elements nitrogen, beryllium, argon, aluminum, silver, and gold belong(s) in each of the following classifications?
a. Period and Roman numeral group numbers are numerically equal
b. Readily conducts electricity and heat
c. Has an atomic mass greater than its atomic number
d. All atoms have a nuclear charge greater than $+20$

3.75 The electron configuration of the isotope $_8^{16}O$ is $1s^2 2s^2 2p^4$. What is the electron configuration for the isotope $_8^{18}O$?

3.76 Write electron configurations for the following elements.
a. The Group IIIA element in the same period as $_4Be$
b. The Period 3 element in the same group as $_5B$
c. The lowest-atomic-numbered metal in Group IA
d. The Period 3 element that has three unpaired electrons

3.77 Referring only to the periodic table, determine the element of lowest atomic number whose electron configuration contains each of the following.
a. Three completely filled orbitals
b. Three completely filled subshells
c. Three completely filled shells
d. Three completely filled s subshells

ANSWERS TO PRACTICE EXERCISES

3.1 a. 11   b. 12   c. 11

3.2 63.55 amu

3.3 a. $1s^2 2s^2 2p^6 3s^2 3p^6 4s^2 3d^5$
b. $1s^2 2s^2 2p^6 3s^2 3p^6 4s^2 3d^{10} 4p^6 5s^2 4d^{10} 5p^6$

4 | Chemical Bonding: The Ionic Bond Model

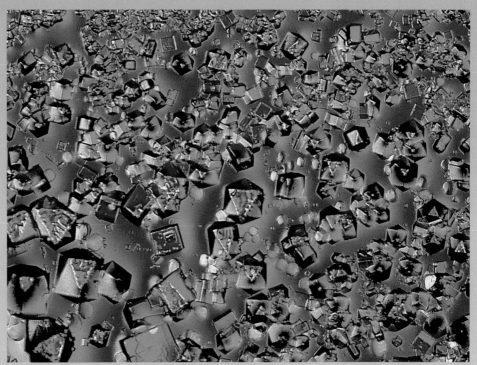

Magnification of crystals of sodium chloride (table salt), one of the most commonly encountered ionic compounds. Color has been added to the image by computer.

As scientists study living organisms and the world in which we live, they rarely encounter free isolated atoms. Instead, under normal conditions of temperature and pressure, they nearly always find atoms associated in aggregates or clusters ranging in size from two atoms to numbers too large to count. In this chapter, we will explain why atoms tend to join together in larger units, and we will discuss the binding forces (chemical bonds) that hold them together.

As we examine the nature of attractive forces between atoms, we will discover that both the tendency and the capacity of an atom to be attracted to other atoms are dictated by its electron configuration.

4.1 Chemical Bonds

Chemical compounds are conveniently divided into two broad classes called *ionic compounds* and *molecular compounds*. Ionic and molecular compounds can be distinguished from each other on the basis of general physical properties. Ionic compounds tend to have high melting points (500°C–2000°C) and are good conductors of electricity when they are in a molten (liquid) state or in solution. Molecular compounds, on the other hand, generally have much lower melting points and tend to be gases, liquids, or low-melting solids. They do not conduct electricity in the molten state. Ionic compounds, unlike molecular compounds, do not have molecules as their basic structural unit. Instead, an extended array of positively and negatively charged particles called *ions* is present (Section 4.8).

Some combinations of elements produce ionic compounds, whereas other combinations of elements form molecular compounds. What determines whether the interaction of two elements produces ions (an ionic compound) or molecules (a molecular compound)? To answer this question, we need to learn about chemical bonds. A **chemical bond** *is the attractive force that holds two atoms together in a more complex unit.* Chemical bonds form as a result of interactions between electrons found in the combining atoms. Thus the nature of chemical bonds is closely linked to electron configurations (Section 3.7).

Corresponding to the two broad categories of chemical compounds are two types of chemical attractive forces (chemical bonds): ionic bonds and covalent bonds. An **ionic bond** *is a chemical bond formed through the transfer of one or more electrons from one atom or group of atoms to another atom or group of atoms.* As its name suggests, the ionic bond model (electron transfer) is used in describing the attractive forces in ionic compounds. An **ionic compound** *is a compound in which ionic bonds are present.*

A **covalent bond** *is a chemical bond formed through the sharing of one or more pairs of electrons between two atoms.* The covalent bond model (electron sharing) is used in describing the attractions between atoms in molecular compounds. A **molecular compound** *is a compound in which atoms are joined through covalent bonds.*

Even before we consider the details of these two bond models, it is important to emphasize that the concepts of ionic and covalent bonds are actually "convenience concepts." Most bonds are not 100% ionic or 100% covalent. Instead, most bonds have some degree of both ionic and covalent character—that is, some degree of both the transfer and the sharing of electrons. However, it is easiest to understand these intermediate bonds (the real bonds) by relating them to the pure or ideal bond types called ionic and covalent.

Two fundamental concepts are necessary for understanding both the ionic and the covalent bonding models.

1. Not all electrons in an atom participate in bonding. Those that are available are called *valence electrons.*
2. Certain arrangements of electrons are more stable than others, as is explained by the *octet rule.*

Section 4.2 addresses the concept of valence electrons, and Section 4.3 discusses the octet rule.

4.2 Valence Electrons and Lewis Symbols

Certain electrons, called *valence* electrons, are particularly important in determining the bonding characteristics of a given atom. A **valence electron** *is an electron in the outermost electron shell of a representative element or noble-gas element.* Note the restriction on the use of this definition; it applies only to representative elements and noble-gas elements. For such elements, valence electrons are always found in either *s* or *p* subshells. (We will not consider in this text the more complicated valence electron definitions for transition elements or inner transition elements; here, the presence of incompletely filled *inner d* or *f* subshells is a complicating factor.)

The number of valence electrons in an atom of a representative element can be determined from the atom's electron configuration, as is illustrated in Example 4.1.

Scientists have developed a shorthand system for designating the number of valence electrons present in atoms of an element. This system involves the use of Lewis symbols. A **Lewis symbol** *is the chemical symbol of an element surrounded by dots equal in number to the number of valence electrons present in atoms of the element.* Figure 4.1 gives the Lewis symbols for the representative elements and noble gases in the first four periods of the periodic table. Lewis symbols, named in honor of the American chemist Gilbert N. Lewis (Figure 4.2), who first introduced them, are also frequently called *electron-dot structures.*

▸ Another designation for *molecular compound* is *covalent compound.* The two designations are used interchangeably. The modifier *molecular* draws attention to the basic structural unit present (the molecule), and the modifier *covalent* focuses on the mode of bond formation (electron sharing).

▸ *Purely* ionic bonds involve a complete transfer of electrons from one atom to another. *Purely* covalent bonds involve equal sharing of electrons. Experimentally, it is found that most actual bonds have some degree of both ionic and covalent character. The exceptions are bonds between identical atoms; here, the bonding is purely covalent.

▸ The term *valence* is derived from the Latin word *valentia,* which means "capacity" (to form bonds).

▸ In a Lewis symbol, the chemical symbol represents the nucleus and all of the nonvalence electrons. The valence electrons are then shown as "dots."

FIGURE 4.1 Lewis structures for selected representative and noble-gas elements.

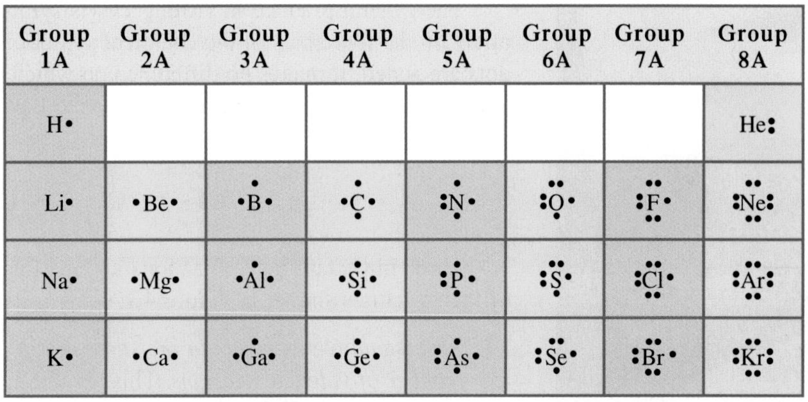

Group 1A	Group 2A	Group 3A	Group 4A	Group 5A	Group 6A	Group 7A	Group 8A
H•							He:
Li•	•Be•	•Ḃ•	•Ċ•	:Ṅ•	:Ö•	:Ḟ:	:N̈e:
Na•	•Mg•	•Ȧl•	•Ṡi•	:Ṗ•	:Ṡ•	:C̈l•	:Ȧr:
K•	•Ca•	•Ġa•	•Ġe•	:Ȧs•	:Ṡe•	:B̈r•	:K̈r:

EXAMPLE 4.1

Determining the Number of Valence Electrons in an Atom

■ Determine the number of valence electrons in atoms of each of the following elements.

a. $_{12}Mg$ **b.** $_{14}Si$ **c.** $_{33}As$

Solution

a. Atoms of the element magnesium have two valence electrons, as can be seen by examining magnesium's electron configuration.

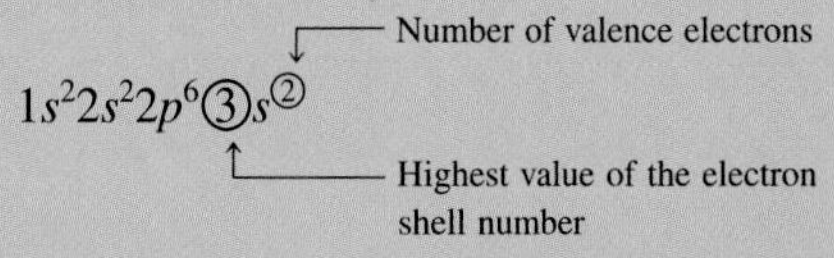

$$1s^2 2s^2 2p^6 \,\textcircled{3}s^{\textcircled{2}}$$

Number of valence electrons

Highest value of the electron shell number

The highest value of the electron shell number is $n = 3$. Only two electrons are found in shell 3: the two electrons in the $3s$ subshell.

b. Atoms of the element silicon have four valence electrons

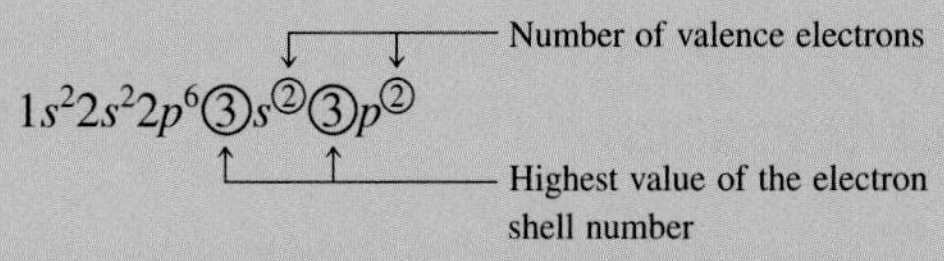

$$1s^2 2s^2 2p^6 \,\textcircled{3}s^{\textcircled{2}}\textcircled{3}p^{\textcircled{2}}$$

Number of valence electrons

Highest value of the electron shell number

Electrons in two different subshells can simultaneously be valence electrons. The highest shell number is 3, and both the $3s$ and the $3p$ subshells belong to this shell. Hence all of the electrons in both of these subshells are valence electrons.

c. Atoms of the element arsenic have five valence electrons.

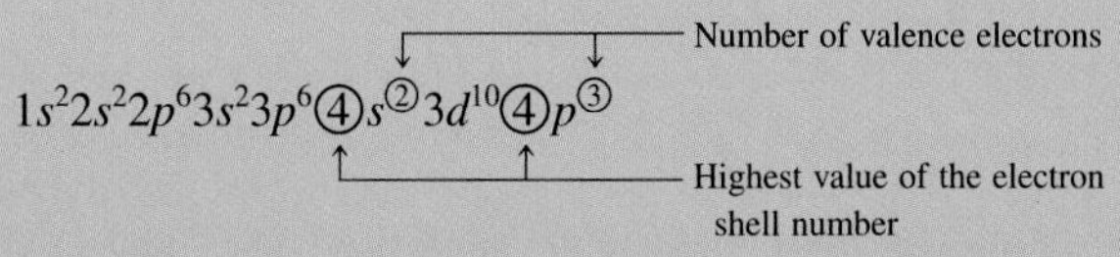

$$1s^2 2s^2 2p^6 3s^2 3p^6 \,\textcircled{4}s^{\textcircled{2}} 3d^{10} \,\textcircled{4}p^{\textcircled{3}}$$

Number of valence electrons

Highest value of the electron shell number

The $3d$ electrons are not counted as valence electrons because the $3d$ subshell is in shell 3, and this shell does not have maximum n value. Shell 4 is the outermost shell and has maximum n value.

Practice Exercise 4.1

Determine the number of valence electrons in atoms of each of the following elements.

a. $_{11}Na$ **b.** $_{16}S$ **c.** $_{35}Br$

FIGURE 4.2 Gilbert Newton Lewis (1875–1946), one of the foremost chemists of the twentieth century, made significant contributions in other areas of chemistry besides his pioneering work in describing chemical bonding. He formulated a generalized theory for describing acids and bases and was the first to isolate deuterium (heavy hydrogen).

The general practice in writing Lewis symbols is to place the first four "dots" separately on the four sides of the chemical symbol and then begin pairing the dots as further dots are added. It makes no difference on which side of the symbol the process of adding dots begins. The following notations for the Lewis symbol of the element calcium are all equivalent.

$$\text{Ca}\cdot \qquad \text{Ca}\cdot \qquad \cdot\text{Ca} \qquad \cdot\text{Ca}\cdot$$

Three important generalizations about valence electrons can be drawn from a study of the structures shown in Figure 4.1.

1. *Representative elements in the same group of the periodic table have the same number of valence electrons.* This should not be surprising. Elements in the same group in the periodic table have similar chemical properties as a result of their similar outer-shell electron configurations (Section 3.8). The electrons in the outermost shell are the valence electrons.
2. *The number of valence electrons for representative elements is the same as the Roman numeral periodic-table group number.* For example, the Lewis structures for oxygen and sulfur, which are both members of Group VIA, show six dots. Similarly, the Lewis structures of hydrogen, lithium, sodium, and potassium, which are all members of Group IA, show one dot.
3. *The maximum number of valence electrons for any element is eight.* Only the noble gases (Section 3.9), beginning with neon, have the maximum number of eight electrons. Helium, which has only two valence electrons, is the exception in the noble-gas family. Obviously, an element with a total of two electrons cannot have eight valence electrons. Although electron shells with n greater than 2 are capable of holding more than eight electrons, they do so only when they are no longer the outermost shell and thus are not the valence shell. For example, arsenic has 18 electrons in its third shell; however, shell 4 is the valence shell for arsenic.

EXAMPLE 4.2

Writing Lewis Symbols for Elements

■ Write Lewis symbols for the following elements.

a. O, S, and Se   **b.** B, C, and N

Solution

a. These elements are all Group VIA elements and thus possess six valence electrons. (The number of valence electrons and the period-table group number will always match for representative elements.) The Lewis symbols, which all have six "dots," are

$$\cdot\ddot{\underset{\cdot\cdot}{\text{O}}}{:} \qquad \cdot\ddot{\underset{\cdot}{\text{S}}}{:} \qquad \cdot\ddot{\underset{\cdot}{\text{Se}}}{:}$$

b. These elements are sequential elements in Period 2 of the periodic table; B is in Group IIIA (three valence electrons), C is in period IVA (four valence electrons), and N is in group VA (five valence electrons). The Lewis symbols for these elements are

$$\cdot\underset{\cdot}{\text{B}} \qquad \cdot\underset{\cdot}{\dot{\text{C}}} \qquad {:}\underset{\cdot}{\dot{\text{N}}}\cdot$$

Practice Exercise 4.2

Write Lewis structures for the following elements.

a. Be, Mg, and Ca   **b.** P, S, and Cl

4.3 The Octet Rule

A key concept in elementary bonding theory is that certain arrangements of valence electrons are more stable than others. The term *stable* as used here refers to the idea that a system, which in this case is an arrangement of electrons, does not easily undergo spontaneous change.

The valence electron configurations of the noble gases (helium, neon, argon, krypton, xenon, and radon) are considered the *most stable of all valence electron configurations*. All of the noble gases except helium possess eight valence electrons, which is the maximum number possible. Helium's valence electron configuration is $1s^2$. All of the other noble gases possess ns^2np^6 valence electron configurations, where n has the maximum value found in the atom.

He: $\left(1s^2\right)$
Ne: $1s^2\left(2s^22p^6\right)$
Ar: $1s^22s^22p^6\left(3s^23p^6\right)$
Kr: $1s^22s^22p^63s^23p^6\left(4s^2\right)3d^{10}\left(4p^6\right)$
Xe: $1s^22s^22p^63s^23p^64s^23d^{10}4p^6\left(5s^2\right)4d^{10}\left(5p^6\right)$
Rn: $1s^22s^22p^63s^23p^64s^23d^{10}4p^65s^24d^{10}5p^6\left(6s^2\right)4f^{14}5d^{10}\left(6p^6\right)$

<blockquote>The outermost electron shell of an atom is also called the valence electron shell.</blockquote>

Except for helium, all the noble-gas valence electron configurations have the outermost s and p subshells *completely filled*.

The conclusion that an ns^2np^6 configuration ($1s^2$ for helium) is the most stable of all valence electron configurations is based on the chemical properties of the noble gases. The noble gases are the *most unreactive* of all the elements. They are the only elemental gases found in nature in the form of individual uncombined atoms. There are no known compounds of helium, neon, and argon, and only a very few compounds of krypton, xenon, and radon are known. The noble gases have little or no tendency to form bonds to other atoms.

<blockquote>Some compounds exist whose formulation is not consistent with the octet rule, but the vast majority of simple compounds have formulas that are consistent with its precepts.</blockquote>

Atoms of many elements that lack the very stable noble-gas valence electron configuration tend to acquire it through chemical reactions that result in compound formation. This observation is known as the **octet rule:** *In compound formation, atoms of elements lose, gain, or share electrons in such a way that their electron configurations contain eight valence electrons.*

4.4 The Ionic Bond Model

Electron transfer between two or more atoms is central to the ionic bond model. This electron transfer process produces charged particles called ions. An **ion** *is an atom (or group of atoms) that is electrically charged as a result of the loss or gain of electrons.* An atom is neutral when the number of protons (positive charges) is equal to the number of electrons (negative charges). Loss or gain of electrons destroys this proton–electron balance and leaves a net charge on the atom.

<blockquote>The word ion is pronounced "eye-on."</blockquote>

If an atom *gains* one or more electrons, it becomes a *negatively* charged ion; excess negative charge is present because electrons outnumber protons. If an atom *loses* one or more electrons, it becomes a *positively* charged ion; more protons are present than electrons. There is excess positive charge (Figure 4.3). Note that the excess positive charge associated with a positive ion is never caused by proton gain but always by electron loss. If the number of protons remains constant and the number of electrons decreases, the result is net positive charge. The number of protons, which determines the identity of an element, never changes during ion formation.

<blockquote>An atom's nucleus never changes during the process of ion formation. The number of neutrons and protons remains constant.</blockquote>

The charge on an ion depends on the number of electrons that are lost or gained. Loss of one, two, or three electrons gives ions with $+1$, $+2$, or $+3$ charges, respectively. A gain of one, two, or three electrons gives ions with -1, -2, or -3 charges,

<blockquote>A loss of electrons by an atom always produces a positive ion. A gain of electrons by an atom always produces a negative ion.</blockquote>

FIGURE 4.3 Loss of an electron from a sodium atom leaves it with one more proton than electrons, so it has a net electrical charge of +1. When chlorine gains an electron, it has one more electron than protons, so it has a net electrical charge of −1.

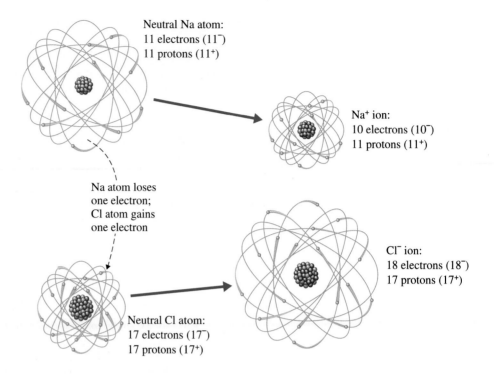

Neutral Na atom:
11 electrons (11⁻)
11 protons (11⁺)

Na⁺ ion:
10 electrons (10⁻)
11 protons (11⁺)

Na atom loses one electron; Cl atom gains one electron

Cl⁻ ion:
18 electrons (18⁻)
17 protons (17⁺)

Neutral Cl atom:
17 electrons (17⁻)
17 protons (17⁺)

respectively. (Ions that have lost or gained more than three electrons are very seldom encountered.)

The notation for charges on ions is a superscript placed to the right of the elemental symbol. Some examples of ion symbols are

Positive ions: Na^+, K^+, Ca^{2+}, Mg^{2+}, Al^{3+}

Negative ions: Cl^-, Br^-, O^{2-}, S^{2-}, N^{3-}

Note that we use a single plus or minus sign to denote a charge of 1, instead of using the notation $^{1+}$ or $^{1-}$. Also note that in multicharged ions, the number precedes the charge sign; that is, the notation for a charge of plus two is $^{2+}$ rather than $^{+2}$.

EXAMPLE 4.3

Writing Symbols for Ions

■ Give the symbol for each of the following ions.
a. The ion formed when a barium atom loses two electrons.
b. The ion formed when a phosphorus atom gains three electrons.

Solution

a. A neutral barium atom contains 56 protons and 56 electrons because barium has an atomic number of 56. The barium ion formed by the loss of 2 electrons would still contain 56 protons but would have only 54 electrons because 2 electrons were lost.

$$\frac{56\ \text{protons} = 56\ +\ \text{charges}}{54\ \text{electrons} = 54\ -\ \text{charges}}$$
$$\text{Net charge} = 2+$$

The symbol of the barium ion is thus Ba^{2+}.

b. The atomic number of phosphorus is 15. Thus 15 protons and 15 electrons are present in a neutral phosphorus atom. A gain of 3 electrons raises the electron count to 18.

$$\frac{15\ \text{protons} = 15\ +\ \text{charges}}{18\ \text{electrons} = 18\ -\ \text{charges}}$$
$$\text{Net charge} = 3-$$

The symbol for the ion is P^{3-}.

Practice Exercise 4.3

Give the symbol for each of the following ions.

a. The ion formed when cesium loses one electron.
b. The ion formed when selenium gains two electrons.

The chemical properties of a particle (atom or ion) depend on the particle's electron arrangement. Because an ion has a different electron configuration (fewer or more electrons) from the atom from which it was formed, it has different chemical properties as well. For example, the drug many people call lithium, which is used to treat mental illness (manic-depressive symptoms), does not involve lithium (Li, the element) but rather lithium ions (Li^+). The element lithium, if ingested, would be poisonous and possibly fatal. The lithium ion, ingested in the form of lithium carbonate, has entirely different effects on the human body.

4.5 Determination of Ionic Charge Sign and Magnitude

The octet rule provides a very simple and straightforward explanation for the charge magnitude associated with ions of the representative elements. *Atoms tend to gain or lose electrons until they have obtained an electron configuration that is the same as that of a noble gas.* The element sodium has the electron configuration

$$1s^2 2s^2 2p^6 3s^1$$

One valence electron is present. Sodium can attain a noble-gas electron configuration by losing this valence electron (to give it the electron configuration of neon) or by gaining seven electrons (to give it the electron configuration of argon).

$$Na\ (1s^2 2s^2 2p^6 3s^1) \quad \overset{\text{Loss of 1 } e^-}{\underset{\text{Gain of 7 } e^-}{\rightleftarrows}}$$

$Na^+ \quad (1s^2 2s^2 2p^6)$
Electron configuration of neon

$Na^{7-} \quad (1s^2 2s^2 2p^6 3s^2 3p^6)$
Electron configuration of argon

The electron loss or gain that involves the fewest electrons will always be the more favorable process, from an energy standpoint, and will be the process that occurs. Thus for sodium, the loss of one electron to form the Na^+ ion is the process that occurs.

The element chlorine has the electron configuration

$$1s^2 2s^2 2p^6 3s^2 3p^5$$

Seven valence electrons are present. Chlorine can attain a noble-gas electron configuration by losing seven electrons (to give it the electron configuration of neon) or by gaining one electron (to give it the electron configuration of argon). The latter occurs for the reason we previously cited.

$$Cl\ (1s^2 2s^2 2p^6 3s^2 3p^5) \quad \overset{\text{Loss of 7 } e^-}{\underset{\text{Gain of 1 } e^-}{\rightleftarrows}}$$

$Cl^{7+} \quad (1s^2 2s^2 2p^6)$
Electron configuration of neon

$Cl^- \quad (1s^2 2s^2 2p^6 3s^2 3p^6)$
Electron configuration of argon

The considerations we have just applied to sodium and chlorine lead to the following generalizations:

1. Metal atoms containing one, two, or three valence electrons (the metals in Groups IA, IIA, and IIIA of the periodic table) tend to lose electrons to acquire a noble-gas

CHEMICAL CONNECTIONS

Fresh Water, Seawater, Hard Water, and Soft Water: A Matter of Ions

Water is the most abundant compound on the face of Earth. We encounter it everywhere we go: as water vapor in the air, as a liquid in rivers, lakes, and oceans, and as a solid (ice and snow) both on land and in the oceans.

All water as it occurs in nature is impure in a chemical sense. The impurities present include suspended matter, microbiological organisms, dissolved gases, and dissolved minerals. Minerals dissolved in water produce ions. For example, rock salt (NaCl) dissolves in water to produce Na^+ and Cl^- ions.

The major distinction between *fresh water* and *seawater* (salt water) is the number of ions present. On a relative scale, where the total concentration of ions in fresh water is assigned a value of 1, seawater has a value of approximately 500; that is, seawater has a concentration of dissolved ions 500 times greater than that of fresh water.

The dominant ions in fresh water and seawater are not the same. In seawater, Na^+ ion is the dominant positive ion and Cl^- ion is the dominant negative ion. This contrasts with fresh water, where Ca^{2+} and Mg^{2+} ions are the most abundant positive ions and HCO_3^- (a polyatomic ion; Section 4.10) is the most abundant negative ion.

When fresh water is purified for drinking purposes, suspended particles, disease-causing agents, and objectionable odors are removed. Dissolved ions are not removed. At the concentrations at which they are normally present in fresh water, dissolved ions are not harmful to health. Indeed, some of the taste of water is caused by the ions present; water without any ions present would taste "unpleasant" to most people.

Hard water is water that contains Ca^{2+}, Mg^{2+}, and Fe^{2+} ions. The presence of these ions does not affect the drinkability of water, but it does affect other uses for the water. The hard-water ions form insoluble compounds with soap (producing scum) and lead to the production of deposits of scale in steam boilers, tea kettles, and hot water pipes.

The most popular method for obtaining *soft water* from hard water involves the process of "ion exchange." In this process, the offending hard-water ions are exchanged for Na^+ ions. Sodium ions do not form insoluble soap compounds or scale. People with high blood pressure or kidney problems are often advised to avoid drinking soft water because of its high sodium content.

electron configuration. The noble gas involved is the one preceding the metal in the periodic table.

Group IA metals form 1^+ ions.
Group IIA metals form 2^+ ions.
Group IIIA metals form 3^+ ions.

2. Nonmetal atoms containing five, six, or seven valence electrons (the nonmetals in Groups VA, VIA, and VIIA of the periodic table) tend to gain electrons to acquire a noble-gas electron configuration. The noble gas involved is the one following the nonmetal in the periodic table.

Group VIIA nonmetals form 1^- ions.
Group VIA nonmetals form 2^- ions.
Group VA nonmetals form 3^- ions.

▶ The positive charge on metal ions from Groups IA, IIA, and IIIA has a magnitude equal to the metal's periodic-table group number.

▶ Nonmetals from Groups VA, VIA, and VIIA form negative ions whose charge is equal to the group number minus 8. For example, S, in Group VIA, forms S^{2-} ions ($6 - 8 = -2$).

Elements in Group IVA occupy unique positions relative to the noble gases. They would have to gain or lose four electrons to attain a noble-gas structure. Theoretically, ions with charges of +4 or −4 could be formed by elements in this group, but in most cases, these elements form covalent bonds instead, which are discussed in Chapter 5.

An ion formed in the preceding manner with an electron configuration the same as that of a noble gas is said to be *isoelectronic* with the noble gas. **Isoelectronic species** *are an atom and ion, or two ions, that have the same number and configuration of electrons.* An atom and an ion or two ions may be isoelectronic. The following is a list of ions that are isoelectronic with the element Ne; all, like Ne, have the electron configuration $1s^2 2s^2 2p^6$.

$$N^{3-} \quad O^{2-} \quad F^- \quad Na^+ \quad Mg^{2+} \quad Al^{3+}$$

It should be emphasized that an ion that is isoelectronic with a noble gas does not have the properties of the noble gas. It has not been converted into the noble gas. The number of protons in the nucleus of the isoelectronic ion is different from that in the noble gas. This point is emphasized by the comparison in Table 4.1 between Mg^{2+} ion and Ne, the noble gas with which Mg^{2+} is isoelectronic.

TABLE 4.1

Comparison of the Characteristics of the Isoelectronic Species Mg^{2+} and Ne

	Ne atom	Mg^{2+} ion
Protons (in the nucleus)	10	12
Electrons (around the nucleus)	10	10
Atomic number	10	12
Charge	0	+2

4.6 Ionic Compound Formation

Ion formation, through the loss or gain of electrons by atoms, is not an isolated, singular process. In reality, electron loss and electron gain are always partner processes; if one occurs, the other also occurs. Ion formation requires the presence of two elements: a metal that can donate electrons and a nonmetal that can accept electrons. The electrons lost by the metal are the same ones gained by the nonmetal. The positive and negative ions simultaneously formed from such *electron transfer* attract one another. The result is the formation of an ionic compound.

> No atom can lose electrons unless another atom is available to accept them.

Lewis structures are helpful in visualizing the formation of simple ionic compounds. A **Lewis structure** *is a combination of Lewis symbols that represents either the transfer or the sharing of electrons in chemical bonds.* Lewis *symbols* involve individual elements. Lewis *structures* involve compounds. The reaction between the element sodium (with one valence electron) and chlorine (with seven valence electrons) is represented as follows with a Lewis structure:

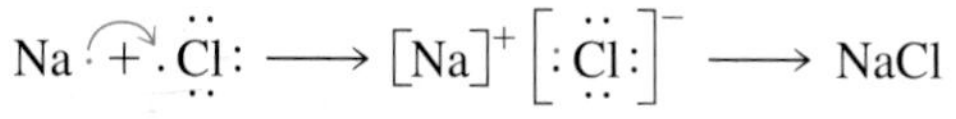

$$\text{Na} \, + \, \ddot{\underset{\cdot\cdot}{\text{Cl}}} : \longrightarrow \left[\text{Na}\right]^{+} \left[: \ddot{\underset{\cdot\cdot}{\text{Cl}}} : \right]^{-} \longrightarrow \text{NaCl}$$

The loss of an electron by sodium empties its valence shell. The next inner shell, which contains eight electrons (a noble-gas configuration), then becomes the valence shell. After the valence shell of chlorine gains one electron, it has the needed eight valence electrons.

When sodium, which has one valence electron, combines with oxygen, which has six valence electrons, the oxygen atom requires the presence of two sodium atoms to acquire two additional electrons.

$$\begin{array}{c} \text{Na} \cdot \\ \\ + \, \cdot \ddot{\underset{\cdot\cdot}{\text{O}}} : \longrightarrow \begin{matrix} \left[\text{Na}\right]^{+} \\ \\ \left[\text{Na}\right]^{+} \end{matrix} \left[: \ddot{\underset{\cdot\cdot}{\text{O}}} : \right]^{2-} \longrightarrow \text{Na}_2\text{O} \\ \\ \text{Na} \cdot \end{array}$$

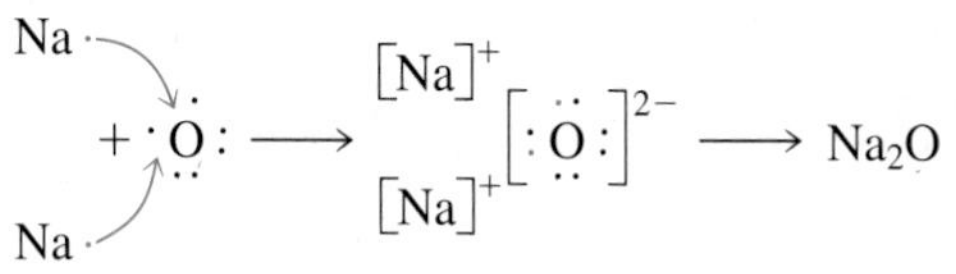

Note that because oxygen has room for two additional electrons, two sodium atoms are required per oxygen atom—hence the formula Na_2O.

An opposite situation occurs in the reaction between calcium, which has two valence electrons, and chlorine, which has seven valence electrons. Here, two chlorine atoms are necessary to accommodate electrons transferred from one calcium atom, because a chlorine atom can accept only one electron. (It has seven valence electrons and needs only one more.)

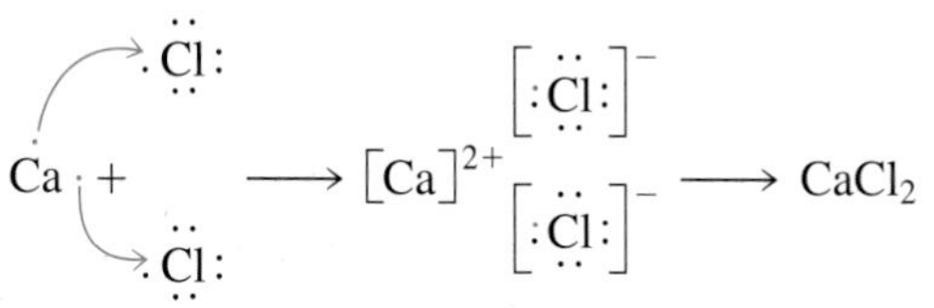

$$\begin{array}{c} \cdot \ddot{\underset{\cdot\cdot}{\text{Cl}}} : \\ \\ \text{Ca} \, + \quad \longrightarrow \left[\text{Ca}\right]^{2+} \begin{matrix} \left[: \ddot{\underset{\cdot\cdot}{\text{Cl}}} : \right]^{-} \\ \\ \left[: \ddot{\underset{\cdot\cdot}{\text{Cl}}} : \right]^{-} \end{matrix} \longrightarrow \text{CaCl}_2 \\ \\ \cdot \ddot{\underset{\cdot\cdot}{\text{Cl}}} : \end{array}$$

EXAMPLE 4.4

Using Lewis Structures to Depict Ionic Compound Formation

■ Show the formation of the following ionic compounds using Lewis structures.

a. Na_3N **b.** MgO   **c.** Al_2S_3

Solution

a. Sodium (a Group IA element) has one valence electron, which it would "like" to lose. Nitrogen (a Group VA element) has five valence electrons and would thus "like" to acquire three more. Three sodium atoms are needed to supply enough electrons for one nitrogen atom.

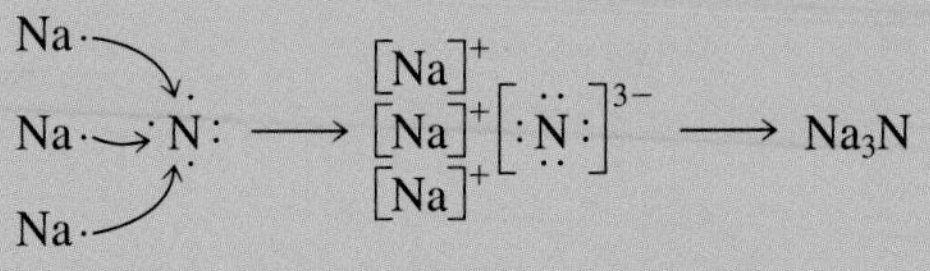

$$Na\cdot \atop Na\cdot \to N: \atop Na\cdot \qquad \longrightarrow \qquad \left[Na\right]^+ \atop \left[Na\right]^+ \left[:\ddot{N}:\right]^{3-} \atop \left[Na\right]^+ \qquad \longrightarrow \qquad Na_3N$$

b. Magnesium (a Group IIA element) has two valence electrons, and oxygen (a Group VIA element) has six valence electrons. The transfer of the two magnesium valence electrons to an oxygen atom results in each atom having a noble-gas electron configuration. Thus these two elements combine in a one-to-one ratio.

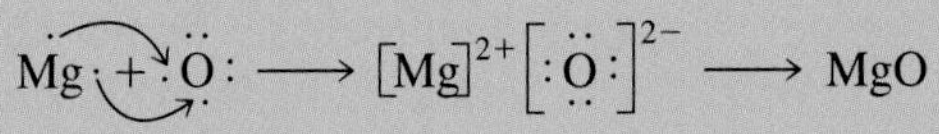

$$Mg\cdot + \cdot\ddot{O}: \longrightarrow \left[Mg\right]^{2+}\left[:\ddot{O}:\right]^{2-} \longrightarrow MgO$$

c. Aluminum (a Group IIIA element) has three valence electrons, all of which need to be lost through electron transfer. Sulfur (a Group VIA element) has six valence electrons and thus needs to acquire two more. Three sulfur atoms are needed to accommodate the electrons given up by two aluminum atoms.

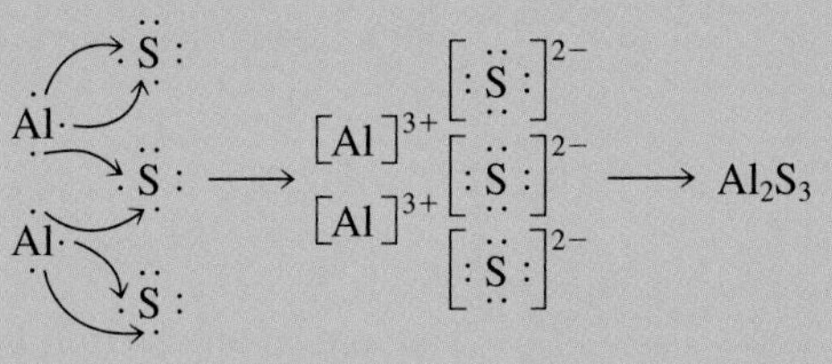

$$\begin{matrix} \cdot\ddot{S}: \\ Al\cdot \\ \cdot\ddot{S}: \\ Al\cdot \\ \cdot\ddot{S}: \end{matrix} \longrightarrow \begin{matrix} \left[:\ddot{S}:\right]^{2-} \\ \left[Al\right]^{3+} \left[:\ddot{S}:\right]^{2-} \\ \left[Al\right]^{3+} \left[:\ddot{S}:\right]^{2-} \end{matrix} \longrightarrow Al_2S_3$$

Practice Exercise 4.4

Show the formation of the following ionic compounds using Lewis structures.

a. KF   **b.** Li_2O **c.** Ca_3P_2

4.7 Formulas for Ionic Compounds

Electron loss always equals electron gain in an electron transfer process. Consequently, ionic compounds are always neutral; no net charge is present. The total positive charge present on the ions that have lost electrons always is exactly counterbalanced by the total negative charge on the ions that have gained electrons. Thus *the ratio in which positive and negative ions combine is the ratio that achieves charge neutrality for the resulting compound.* This generalization can be used, instead of Lewis structures, to determine ionic compound formulas. Ions are combined in the ratio that causes the positive and negative charges to add to zero.

The correct combining ratio when K^+ ions and S^{2-} ions combine is two to one. Two K^+ ions (each of $+1$ charge) are required to balance the charge on a single S^{2-} ion.

$$\begin{aligned} 2(K^+): \quad &(2 \text{ ions}) \times (\text{charge of } +1) = +2 \\ S^{2-}: \quad &\underline{(1 \text{ ion}) \times (\text{charge of } -2) = -2} \\ &\text{Net charge} = 0 \end{aligned}$$

The formula of the compound formed is thus K_2S.

There are three rules to remember when writing formulas for ionic compounds.

1. The symbol for the positive ions is always written first.
2. The charges on the ions that are present are *not* shown in the formula. You need to know the charges to determine the formula; however, the charges are not explicitly shown in the formula.
3. The numbers in the formula (the subscripts) give the combining ratio for the ions.

EXAMPLE 4.5

Using Ionic Charges to Determine the Formula of an Ionic Compound

■ Determine the formula for the compound that is formed when each of the following pairs of ions interact.

a. Na^+ and P^{3-} **b.** Be^{2+} and P^{3-}

Solution

a. The Na^+ and P^{3-} ions combine in a three-to-one ratio because this combination causes the charges to add to zero. Three Na^+ ions give a total positive charge of 3. One P^{3-} ion results in a total negative charge of 3. Thus the formula for the compound is Na_3P.
b. The numbers in the charges for these ions are 2 and 3. The lowest common multiple of 2 and 3 is 6 ($2 \times 3 = 6$). Thus we need 6 units of positive charge and 6 units of negative charge. Three Be^{2+} ions are needed to give the 6 units of positive charge, and two P^{3-} ions are needed to give the 6 units of negative charge. The combining ratio of ions is three to two, and the formula is Be_3P_2.

The strategy of finding the lowest common multiple of the numbers in the charges of the ions always works, and it saves you the inconvenience of drawing the Lewis structures.

Practice Exercise 4.5

Determine the formula for the compound that is formed when each of the following pairs of ions interact.

a. Ca^{2+} and F^- **b.** Al^{3+} and O^{2-}

4.8 The Structure of Ionic Compounds

An ionic compound, in the solid state, consists of positive and negative ions arranged in such a way that each ion is surrounded by nearest neighbors of the opposite charge. Any given ion is bonded by electrostatic (positive–negative) attractions to all the other ions of opposite charge immediately surrounding it. Figure 4.4 shows a two-dimensional cross section and a three-dimensional view of the arrangement of ions in the ionic compound sodium chloride (NaCl). Note in these structural representations that no given ion has a

FIGURE 4.4 (a, b) A two-dimensional cross section and a three-dimensional view of sodium chloride (NaCl), an ionic solid. Both views show an alternating array of positive and negative ions. (c) Sodium chloride crystals.

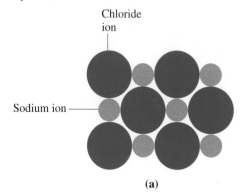

Chloride ion

Sodium ion

(a)

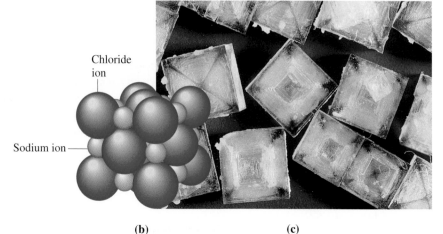

Chloride ion

Sodium ion

(b)   **(c)**

FIGURE 4.5 Cross section of the structure of the ionic solid NaCl. No molecule can be distinguished in this structure. Instead, a basic *formula unit* is present that is repeated indefinitely.

One formula unit

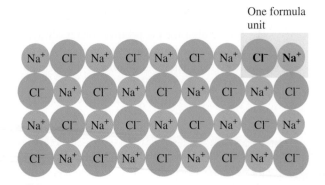

single partner. A given sodium ion has six immediate neighbors (chloride ions) that are equidistant from it. A chloride ion in turn has six immediate sodium ion neighbors.

The alternating array of positive and negative ions present in an ionic compound means that discrete molecules do not exist in such compounds. Therefore, the formulas of ionic compounds cannot represent the composition of molecules of these substances. Instead, such formulas represent the simplest combining ratio for the ions present. The formula for sodium chloride, NaCl, indicates that sodium and chloride ions are present in a one-to-one ratio in this compound. Chemists use the term *formula unit,* rather than *molecule,* to refer to the smallest unit of an ionic compound. A **formula unit** *is the smallest whole-number repeating ratio of ions present in an ionic compound that results in charge neutrality.* A formula unit is "hypothetic," because it does not exist as a separate entity; it is only "a part" of the extended array of ions that constitute an ionic solid (see Figure 4.5).

Although the formulas for ionic compounds represent only ratios, they are used in equations and chemical calculation in the same way as are the formulas for molecular species. Remember, however, that they cannot be interpreted as indicating that molecules exist for these substances; they merely represent the simplest ratio of ions present.

The ions present in an ionic solid adopt an arrangement that maximizes attractions between ions of opposite charge and minimizes repulsions between ions of like charge. The specific arrangement that is adopted depends on ion sizes and on the ratio between positive and negative ions. Arrangements are usually very symmetrical and result in crystalline solids—that is, solids with highly regular shapes. Crystalline solids usually have flat surfaces or faces that make definite angles with one another. Figure 4.6 shows crystals of various minerals.

▶ In Section 1.9 the molecule was described as the smallest unit of a pure substance that is capable of a stable, independent existence. Ionic compounds, with their formula units, are exceptions to this generalization.

FIGURE 4.6 Ionic compounds usually have crystalline forms in the solid state, such as those associated with (a) emerald, (b) fluorite, (c) topaz, and (d) ruby.

(a)

(b)

(c)

(d)

CHEMICAL CONNECTIONS

Sodium Chloride: The World's Most Common Food Additive

The *ionic compound* sodium chloride (table salt), which contains Na^+ and Cl^- ions, is a white crystalline solid that occurs naturally in large underground deposits (rock salt). It also can be obtained by solar evaporation of seawater.

Sodium chloride is the world's most common food additive. Most people find its taste innately appealing. Salt use tends to enhance other flavors, probably by suppressing the bitter flavors. In general, processed foods contain the most sodium chloride, and unprocessed foods, such as fresh fruits and vegetables, contain the least. Studies on the sodium content of foods show that as much as 75% of it is added during processing and manufacturing, 15% comes from salt added during cooking and at the table, and only 10% is naturally present in the food.

Physiologically, sodium chloride is an extremely important substance. It is the major source for both Na^+ and Cl^- ions, the two most abundant ions in blood plasma and in interstitial fluid (the fluid outside cells). Sodium ions play a major role in determining interstitial fluid volume and are essential to nerve transmission and muscle contraction. Chloride ions are essential to acid–base balance within the body and help regulate fluid flow.

A moderate daily intake of sodium chloride is needed to maintain Na^+ ion and Cl^- ion concentrations in body fluids at appropriate levels. The daily requirement is about 5 g of NaCl. An intake greater than this can be detrimental to health. Most people are aware that numerous studies on health and diet have shown a relationship between high sodium (sodium chloride) intake and hypertension (high blood pressure). Too much sodium commonly aggravates hypertension.

Many factors, including genetics, contribute to the regulation of blood pressure in the human body. Recent research on the role of sodium ions in hypertension indicates that their contribution to hypertension may be more complex than was originally thought. Sodium chloride has a greater effect on blood pressure than either sodium or chloride ions alone or in combination with other ions. The question now is whether it is the sodium ion (from any source) or just salt (sodium chloride)

that is responsible for the effects seen. Much of the early research that implicated sodium ion as a cause of hypertension may actually have reflected the effect of sodium's "silent partner," the chloride ion. Research continues.

How sodium ion (or sodium chloride) contributes to high blood pressure is not completely known, but there are indications that elevated sodium ion levels in body fluids lead to a contraction of small arteries, increasing resistance to blood flow.

Many processed foods, including pretzels, are high in sodium; that is, they contain compounds, such as sodium chloride (table salt), in which Na^+ ions are present.

The Chemistry at a Glance feature on page 90 reviews the general concepts we have considered so far about ionic compounds.

4.9 Recognizing and Naming Binary Ionic Compounds

The term *binary* means "two." A **binary compound** *is a compound in which only two elements are present.* In all binary *ionic* compounds, one element is a metal and the other element is a nonmetal. The metal is always present as the positive ion, and the nonmetal is always present as the negative ion. The joint presence of a metal and a nonmetal in a binary compound is the "recognition key" that the compound is an ionic compound.

Ionic Bonds and Ionic Compounds

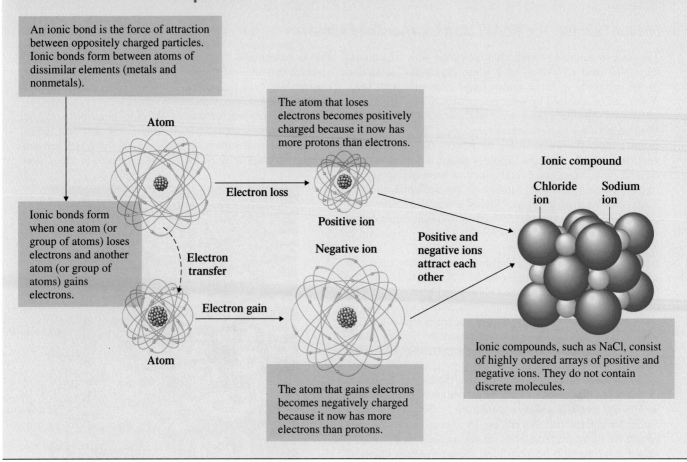

An ionic bond is the force of attraction between oppositely charged particles. Ionic bonds form between atoms of dissimilar elements (metals and nonmetals).

Ionic bonds form when one atom (or group of atoms) loses electrons and another atom (or group of atoms) gains electrons.

The atom that loses electrons becomes positively charged because it now has more protons than electrons.

Atom

Electron loss

Positive ion

Electron transfer

Electron gain

Negative ion

Atom

Positive and negative ions attract each other

The atom that gains electrons becomes negatively charged because it now has more electrons than protons.

Ionic compound

Chloride ion   Sodium ion

Ionic compounds, such as NaCl, consist of highly ordered arrays of positive and negative ions. They do not contain discrete molecules.

EXAMPLE 4.6

Recognizing a Binary Ionic Compound on the Basis of Its Chemical Formula

■ Which of the following binary compounds would be expected to be an ionic compound?

a. Al_2S_3 **b.** H_2O **c.** KF   **d.** NH_3

Solution

a. ionic; a metal (Al) and a nonmetal (S) are present
b. not ionic; two nonmetals are present
c. ionic; a metal (K) and a nonmetal (F) are present
d. not ionic; two nonmetals are present

The two compounds that are not ionic are *molecular* compounds (Section 4.1). Chapter 5 includes an extended discussion of molecular compounds. In general, molecular compounds contain just nonmetals.

Practice Exercise 4.6

Which of the following binary compounds would be expected to be an ionic compound?

a. CO_2 **b.** $MgCl_2$ **c.** Fe_2O_3 **d.** PF_3

Binary ionic compounds are named using the following rule: *The full name of the metallic element is given first, followed by a separate word containing the stem of the nonmetallic element name and the suffix -ide.* Thus, in order to name the compound NaF, we start with the name of the metal (sodium), follow it with the stem of the name of the nonmetal (fluor-), and then add the suffix *-ide*. The name becomes *sodium fluoride.*

The stem of the name of the nonmetal is the name of the nonmetal with its ending chopped off. Table 4.2 gives the stem part of the name for each of the most common nonmetallic elements. The name of the metal ion is always exactly the same as the name of the metal itself; the metal's name is never shortened. Example 4.7 illustrates the use of the rule for naming binary ionic compounds.

EXAMPLE 4.7

Naming Binary Ionic Compounds

■ Name the following binary ionic compounds.

a. MgO **b.** Al_2S_3 **c.** K_3N **d.** $CaCl_2$

Solution

The general pattern for naming binary ionic compounds is

Name of metal + stem of name of nonmetal + -ide

a. The metal is magnesium and the nonmetal is oxygen. Thus the compound's name is *magnesium oxide.*
b. The metal is aluminum and the nonmetal is sulfur; the compound's name is *aluminum sulfide.* Note that no mention is made of the subscripts present in the formula—the 2 and the 3. The name of an ionic compound never contains any reference to formula subscript numbers. There is only one ratio in which aluminum and sulfur atoms combine. Thus, just telling the names of the elements present in the compound is adequate nomenclature.
c. Potassium (K) and nitrogen (N) are present in the compound, and its name is *potassium nitride.*
d. The compound's name is *calcium chloride.*

Practice Exercise 4.7

Name the following binary ionic compounds.

a. Na_2S **b.** BeO **c.** Li_3P **d.** BaI_2

▶ All the inner transition elements (*f* area of the periodic table), most of the transition elements (*d* area), and a few representative metals (*p* area) exhibit variable ionic charge behavior.

▶ An older method for indicating the charge on metal ions uses the suffixes *-ic* and *-ous* rather than the Roman numeral system. It is mentioned here because it is still sometimes encountered. In this system, when a metal has two common ionic charges, the suffix *-ous* is used for the ion of lower charge and the suffix *-ic* for the ion of higher charge. The metal's Latin name is also used. In this older system, iron(II) ion is called ferrous ion, and iron(III) ion is called ferric ion.

Thus far in our discussion of ionic compounds, it has been assumed that the only behavior allowable for an element is that predicted by the octet rule. This is a good assumption for nonmetals and for most representative element metals. However, there are many other metals that exhibit a less predictable behavior because they are able to form more than one type of ion. For example, iron forms both Fe^{2+} ions and Fe^{3+} ions, depending on chemical circumstances.

When we name compounds that contain metals with variable ionic charges, the charge on the metal ion must be incorporated into the name. This is done by using Roman numerals. For example, the chlorides of Fe^{2+} and Fe^{3+} ($FeCl_2$ and $FeCl_3$, respectively) are named iron(II) chloride and iron(III) chloride (Figure 4.7). Likewise, CuO is named copper(II) oxide. If you are uncertain about the charge on the metal ion in an ionic compound, use the charge on the nonmetal ion (which does not vary) to calculate it. For example, in order to determine the charge on the copper ion in CuO, you can note that

TABLE 4.2

Names of Selected Common Nonmetallic Ions

Element	Stem	Name of ion	Formula of ion
bromine	brom-	bromide	Br^-
carbon	carb-	carbide	C^{4-}
chlorine	chlor-	chloride	Cl^-
fluorine	fluor-	fluoride	F^-
hydrogen	hydr-	hydride	H^-
iodine	iod-	iodide	I^-
nitrogen	nitr-	nitride	N^{3-}
oxygen	ox-	oxide	O^{2-}
phosphorus	phosph-	phosphide	P^{3-}
sulfur	sulf-	sulfide	S^{2-}

FIGURE 4.7 Copper(II) oxide (CuO) is black, whereas copper(I) oxide (Cu_2O) is reddish brown. Iron(II) chloride ($FeCl_2$) is green, whereas iron(III) chloride ($FeCl_3$) is bright yellow.

Copper (I) Oxide

Iron (III) Chloride

Copper (II) Oxide

Iron (II) Chloride

the oxide ion carries a -2 charge because oxygen is in Group VIA. This means that the copper ion must have a $+2$ charge to counterbalance the -2 charge.

EXAMPLE 4.8

Using Roman Numerals in the Naming of Binary Ionic Compounds

■ Name the following binary ionic compounds, each of which contains a metal whose ionic charge can vary.

a. AuCl   **b.** Fe_2O_3

Solution

We will need to indicate the magnitude of the charge on the metal ion in the name of each of these compounds by means of a Roman numeral.

a. To calculate the metal ion charge, use the fact that total ionic charge (both positive and negative) must add to zero.

$$(\text{Gold charge}) + (\text{chlorine charge}) = 0$$

The chloride ion has a -1 charge (Section 4.5). Therefore,

$$(\text{Gold charge}) + (-1) = 0$$

Thus,

$$\text{Gold charge} = +1$$

Therefore, the gold ion present is Au^+, and the name of the compound is *gold(I) chloride*.

b. For charge balance in this compound we have the equation

$$2(\text{iron charge}) + 3(\text{oxygen charge}) = 0$$

Note that we have to take into account the number of each kind of ion present (2 and 3 in this case). Oxide ions carry a -2 charge (Section 4.5). Therefore,

$$2(\text{iron charge}) + 3(-2) = 0$$

$$2(\text{iron charge}) = +6$$

$$\text{Iron charge} = +3$$

Here, we are interested in the charge on a single iron ion ($+3$) and not in the total positive charge present ($+6$). The compound is named *iron(III) oxide* because Fe^{3+} ions are present. As is the case for all ionic compounds, the name does not contain any reference to the numerical subscripts in the compound's formula.

Practice Exercise 4.8

Name the following binary ionic compounds, each of which contains a metal whose ionic charge can vary.

a. PbO_2 **b.** Cu_2O

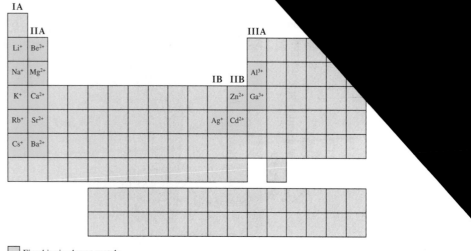

Fixed ionic charge metals

> The fixed-charge metals are those in Group IA (+1 ionic charge), those in Group IIA (+2 ionic charge), and five others (Al^{3+}, Ga^{3+}, Zn^{2+}, Cd^{2+}, and Ag^+).

In order to know when to use Roman numerals in binary ionic compound names, you must know which metals exhibit variable ionic charge and which have a fixed ionic charge. There are many more of the former (Roman numeral required) than of the latter (no Roman numeral required). Thus you should learn the identity of the metals that have a fixed ionic charge (the short list); any metal not on the short list must exhibit variable charge. Figure 4.8 shows the metals that always form a single type of ion in ionic compound formation. Ionic compounds that contain these metals are the only ones without Roman numerals in their names.

4.10 Polyatomic Ions

There are two categories of ions: monatomic and polyatomic. A **monatomic ion** *is an ion formed from a single atom through loss or gain of electrons*. All of the ions we have discussed so far have been monatomic (Cl^-, Na^+, Ca^{2+}, N^{3-}, and so on).

A **polyatomic ion** *is an ion formed from a group of atoms (held together by covalent bonds) through loss or gain of electrons*. An example of a polyatomic ion is the sulfate ion, SO_4^{2-} (Figure 4.9). This ion contains four oxygen atoms and one sulfur atom, and the whole group of five atoms has acquired a −2 charge. The whole sulfate group is the ion rather than any one atom within the group. Covalent bonding, discussed in Chapter 5, holds the sulfur and oxygen atoms together.

FIGURE 4.9 Models of polyatomic ions: (a) a sulfate ion (SO_4^{2-}) and (b) a nitrate ion (NO_3^-).

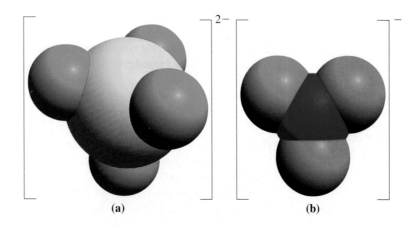

(a)   (b)

	Formula	Name of ion
	NO_3^-	nitrate
	NO_2^-	nitrite
	NH_4^+	ammonium
	N_3^-	azide
	SO_4^{2-}	sulfate
	HSO_4^-	bisulfate or hydrogen sulfate
	SO_3^{2-}	sulfite
	HSO_3^-	bisulfite or hydrogen sulfite
	$S_2O_3^{2-}$	thiosulfate
	PO_4^{3-}	phosphate
	HPO_4^{2-}	hydrogen phosphate
	$H_2PO_4^-$	dihydrogen phosphate
	PO_3^{3-}	phosphite
carbon	CO_3^{2-}	carbonate
	HCO_3^-	bicarbonate or hydrogen carbonate
	$C_2O_4^{2-}$	oxalate
	$C_2H_3O_2^-$	acetate
	CN^-	cyanide
chlorine	ClO_4^-	perchlorate
	ClO_3^-	chlorate
	ClO_2^-	chlorite
	ClO^-	hypochlorite
hydrogen	H_3O^+	hydronium
	OH^-	hydroxide
metals	MnO_4^-	permanganate
	CrO_4^{2-}	chromate
	$Cr_2O_7^{2-}$	dichromate

There are numerous ionic compounds in which the positive or negative ion (sometimes both) is polyatomic. Polyatomic ions are very stable and generally maintain their identity during chemical reactions.

Note that polyatomic ions are not molecules. They never occur alone as molecules do. Instead, they are always found associated with ions of opposite charge. Polyatomic ions are *charged pieces* of compounds, not compounds. Ionic compounds require the presence of both positive and negative ions and are neutral overall.

Table 4.3 lists the names and formulas of some of the more common polyatomic ions. The following generalizations concerning polyatomic ion names and charges emerge from consideration of the ions listed in Table 4.3.

1. Most of the polyatomic ions have a negative charge, which can vary from -1 to -3. Only two positive ions are listed in the table: NH_4^+ (ammonium) and H_3O^+ (hydronium).
2. Two of the negatively charged polyatomic ions, OH^- (hydroxide) and CN^- (cyanide), have names ending in *-ide,* and the rest of them have names ending in either *-ate* or *-ite.*
3. A number of *-ate, -ite* pairs of ions exist, as in SO_4^{2-} (sulfate) and SO_3^{2-} (sulfite). The *-ate* ion always has one more oxygen atom than the *-ite* ion. Both the *-ate* and *-ite* ions of a pair carry the same charge.
4. A number of pairs of ions exist wherein one member of the pair differs from the other by having a hydrogen atom present, as in CO_3^{2-} (carbonate) and HCO_3^- (hydrogen carbonate or bicarbonate). In such pairs, the charge on the ion that contains hydrogen is always 1 less than that on the other ion.

Learning the names of the common polyatomic ions is a memorization project. There is no shortcut. The charges and formulas for the various polyatomic ions *cannot* be easily related to the periodic table, as was the case for many of the monatomic ions.

4.11 Formulas and Names for Ionic Compounds Containing Polyatomic Ions

Formulas for ionic compounds that contain polyatomic ions are determined in the same way as those for ionic compounds that contain monatomic ions (Section 4.7). The positive and negative charges present must add to zero.

Two conventions not encountered previously in formula writing often arise when we write formulas containing polyatomic ions.

1. When more than one polyatomic ion of a given kind is required in a formula, the polyatomic ion is enclosed in parentheses, and a subscript, placed outside the parentheses, is used to indicate the number of polyatomic ions needed. An example is $Fe(OH)_3$.
2. So that the identity of polyatomic ions is preserved, the same elemental symbol may be used more than once in a formula. An example is the formula NH_4NO_3, where N appears in two locations.

Example 4.9 illustrates the use of both of these new conventions.

EXAMPLE 4.9

Writing Formulas for Ionic Compounds Containing Polyatomic Ions

■ Determine the formulas for the ionic compounds that contain these pairs of ions.

a. Na^+ and SO_4^{2-} **b.** Mg^{2+} and NO_3^- **c.** NH_4^+ and CN^-

Solution

a. In order to equalize the total positive and negative charge, we need two sodium ions ($+1$ charge) for each sulfate ion (-2 charge). We indicate the presence of two Na^+ ions with the subscript 2 following the symbol of this ion. The formula of the compound is Na_2SO_4. The convention that the positive ion is always written first in the formula still holds when polyatomic ions are present.

b. Two nitrate ions (-1 charge) are required to balance the charge on one magnesium ion ($+2$ charge). Because more than one polyatomic ion is needed, the formula contains parentheses, $Mg(NO_3)_2$. The subscript 2 outside the parentheses indicates two of what is inside the parentheses. If parentheses were not used, the formula would appear to be $MgNO_{32}$, which is not intended and conveys false information.

c. In this compound, both ions are polyatomic, which is a perfectly legal situation. Because the ions have equal but opposite charges, they combine in a one-to-one ratio. Thus the formula is NH_4CN. No parentheses are necessary because we need only one polyatomic ion of each type in a formula unit. The appearance of the symbol for the element nitrogen (N) at two locations in the formula could be prevented by combining the two nitrogens, resulting in N_2H_4C. But the formula N_2H_4C does not convey the message that NH_4^+ and CN^- ions are present. Thus, when writing formulas that contain polyatomic ions, we always maintain the identities of these ions, even if it means having the same elemental symbol at more than one location in the formula.

Practice Exercise 4.9

Determine the formulas for the ionic compounds that contain the following pairs of ions.

a. K^+ and CO_3^{2-} **b.** Ca^{2+} and OH^- **c.** NH_4^+ and HPO_4^{2-}

The names of ionic compounds containing polyatomic ions are derived in a manner similar to that for binary ionic compounds (Section 4.9). The rule for naming binary ionic compounds is as follows: Give the name of the metallic element first (including, when needed, a Roman numeral indicating ion charge), and then give a separate word containing the stem of the nonmetallic name and the suffix *-ide*.

For our present situation, *if the polyatomic ion is positive, its name is substituted for that of the metal. If the polyatomic ion is negative, its name is substituted for the non-metal stem plus* -ide. Where both positive and negative ions are polyatomic, dual substitution occurs, and the resulting name includes just the names of the polyatomic ions.

CHEMISTRY AT A GLANCE

Nomenclature of Ionic Compounds

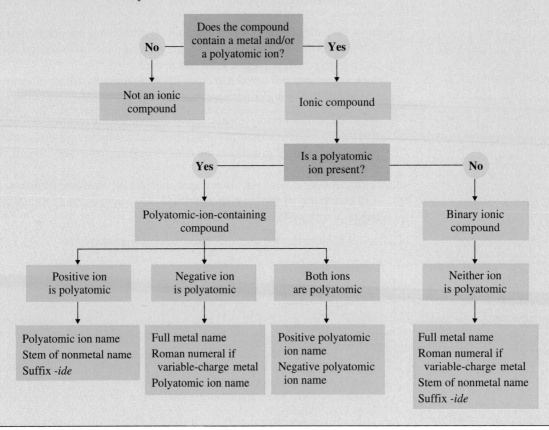

EXAMPLE 4.10

Naming Ionic Compounds in Which Polyatomic Ions Are Present

■ Name the following compounds, which contain one or more polyatomic ions.

a. $Ca_3(PO_4)_2$ **b.** $Fe_2(SO_4)_3$ **c.** $(NH_4)_2CO_3$

Solution

a. The positive ion present is the calcium ion (Ca^{2+}). We will not need a Roman numeral to specify the charge on a Ca^{2+} ion because it is always +2. The negative ion is the polyatomic phosphate ion (PO_4^{3-}). The name of the compound is *calcium phosphate*. As in naming binary ionic compounds, subscripts in the formula are not incorporated into the name.

b. The positive ion present is iron(III). The negative ion is the polyatomic sulfate ion (SO_4^{2-}). The name of the compound is *iron(III) sulfate*. The determination that iron is present as iron(III) involves the following calculation dealing with charge balance:

$$2(\text{iron charge}) + 3(\text{sulfate charge}) = 0$$

The sulfate charge is −2. (You had to memorize that.) Therefore,

$$2(\text{iron charge}) + 3(-2) = 0$$
$$2(\text{iron charge}) = +6$$
$$\text{Iron charge} = +3$$

c. Both the positive and the negative ions in this compound are polyatomic—the ammonium ion (NH_4^+) and the carbonate ion (CO_3^{2-}). The name of the compound is simply the combination of the names of the two polyatomic ions: *ammonium carbonate*.

Practice Exercise 4.10

Name the following compounds, which contain one or more polyatomic ions.

a. $Ba(NO_3)_2$ **b.** Cu_3PO_4 **c.** $(NH_4)_2SO_4$

The Chemistry at a Glance feature on page 96 summarizes the "thought processes" involved in naming ionic compounds, both those with monatomic ions and those with polyatomic ions.

CONCEPTS TO REMEMBER

Chemical bonds. Chemical bonds are the attractive forces that hold atoms together in more complex units. Chemical bonds result from the transfer of valence electrons between atoms (ionic bond) or from the sharing of electrons between atoms (covalent bond).

Valence electrons. Valence electrons, for representative elements, are the electrons in the outermost electron shell, which is the shell with the highest shell number. These electrons are particularly important in determining the bonding characteristics of a given atom.

Octet rule. In compound formation, atoms of representative elements lose, gain, or share electrons in such a way that their electron configurations become identical to those of the noble gas nearest them in the periodic table.

Ionic compounds. Ionic compounds usually involve a metal atom and a nonmetal atom. Metal atoms lose one or more electrons, producing positive ions. Nonmetal atoms acquire the electrons lost by the metal atoms, producing negative ions. The oppositely charged ions attract one another, creating ionic bonds.

Charge magnitude for ions. Metal atoms containing one, two, or three valence electrons tend to lose such electrons, producing ions of

+1, +2, or +3 charge, respectively. Nonmetal atoms containing five, six, or seven valence electrons tend to gain electrons, producing ions of −3, −2, or −1 charge, respectively.

Formulas for ionic compounds. The ratio in which positive and negative ions combine is the ratio that causes the total amount of positive and negative charges to add up to zero.

Structure of ionic compounds. Ionic solids consist of positive and negative ions arranged in such a way that each ion is surrounded by ions of the opposite charge.

Binary ionic compound nomenclature. Binary ionic compounds are named by giving the full name of the metallic element first, followed by a separate word containing the stem of the nonmetallic element name and the suffix -ide. A Roman numeral specifying ionic charge is appended to the name of the metallic element if it is a metal that exhibits variable ionic charge.

Polyatomic ions. A polyatomic ion is a group of covalently bonded atoms that has acquired a charge through the loss or gain of electrons. Polyatomic ions are very stable entities that generally maintain their identity during chemical reactions.

KEY REACTIONS AND EQUATIONS

1. Number of valence electrons for representative elements (Section 4.2)

   Number of valence electrons = periodic-table group number

2. Charges on metallic monatomic ions (Section 4.5)

   Group IA metals form 1^+ ions.
   Group IIA metals form 2^+ ions.
   Group IIIA metals form 3^+ ions.

3. Charges on nonmetallic monatomic ions (Section 4.5)

   Group VA nonmetals form 3^- ions.
   Group VIA nonmetals form 2^- ions.
   Group VIIA nonmetals form 1^- ions.

KEY TERMS

Binary compound (4.9)
Chemical bond (4.1)
Covalent bond (4.1)
Formula unit (4.8)
Ion (4.4)

Ionic bond (4.1)
Ionic compound (4.1)
Isoelectronic species (4.5)
Lewis structure (4.6)
Lewis symbol (4.2)

Molecular compound (4.1)
Monatomic ion (4.10)
Octet rule (4.3)
Polyatomic ion (4.10)
Valence electron (4.2)

EXERCISES AND PROBLEMS

The members of each pair of problems in this section test similar material.

■ Valence Electrons (Section 4.2)

4.1 How many valence electrons do atoms with the following electron configurations have?
   a. $1s^2 2s^2$
   b. $1s^2 2s^2 2p^6 3s^2$
   c. $1s^2 2s^2 2p^6 3s^2 3p^1$
   d. $1s^2 2s^2 2p^6 3s^2 3p^6 4s^2 3d^{10} 4p^2$

4.2 How many valence electrons do atoms with the following electron configurations have?
   a. $1s^2 2s^2 2p^6$
   b. $1s^2 2s^2 2p^6 3s^2 3p^1$
   c. $1s^2 2s^2 2p^6 3s^1$
   d. $1s^2 2s^2 2p^6 3s^2 3p^6 4s^2 3d^{10} 4p^5$

4.3 Give the periodic-table group number, and the number of valence electrons present, for each of the following representative elements.
   a. $_3$Li   b. $_{10}$Ne   c. $_{20}$Ca   d. $_{53}$I

4.4 Give the periodic-table group number, and the number of valence electrons present, for each of the following representative elements.
   a. $_{12}$Mg   b. $_{19}$K   c. $_{15}$P   d. $_{35}$Br

4.5 Write the complete electron configuration for each of the following representative elements.
   a. Period 2 element with four valence electrons
   b. Period 2 element with seven valence electrons
   c. Period 3 element with two valence electrons
   d. Period 3 element with five valence electrons

4.6 Write the complete electron configuration for each of the following representative elements.
   a. Period 2 element with one valence electron
   b. Period 2 element with six valence electrons
   c. Period 3 element with seven valence electrons
   d. Period 3 element with three valence electrons

■ Lewis Symbols for Atoms (Section 4.2)

4.7 Draw Lewis symbols for atoms of each of the following elements.
   a. $_{12}$Mg   b. $_{19}$K   c. $_{15}$P   d. $_{36}$Kr

4.8 Draw Lewis symbols for atoms of each of the following elements.
   a. $_{13}$Al   b. $_{20}$Ca   c. $_{17}$Cl   d. $_4$Be

4.9 Each of the following Lewis symbols represents a Period 2 element. Determine each element's identity.
   a. X·   b. :X: c. X·   d. ·X·

4.10 Each of the following Lewis symbols represents a Period 3 element. Determine each element's identity
   a. ·X·   b. ·X·   c. ·X: d. :X:

■ Notation for Ions (Section 4.4)

4.11 Give the chemical symbol for each of the following ions.
   a. An oxygen atom that has gained two electrons
   b. A magnesium atom that has lost two electrons
   c. A fluorine atom that has gained one electron
   d. An aluminum atom that has lost three electrons

4.12 Give the chemical symbol for each of the following ions.
   a. A chlorine atom that has gained one electron
   b. A sulfur atom that has gained two electrons
   c. A potassium atom that has lost one electron
   d. A beryllium atom that has lost two electrons

4.13 What would be the chemical symbol for an ion with each of the following numbers of protons and electrons?
   a. 20 protons and 18 electrons
   b. 8 protons and 10 electrons
   c. 11 protons and 10 electrons
   d. 13 protons and 10 electrons

4.14 What would be the chemical symbol for an ion with each of the following numbers of protons and electrons?
   a. 15 protons and 18 electrons
   b. 17 protons and 18 electrons
   c. 12 protons and 10 electrons
   d. 19 protons and 18 electrons

4.15 Calculate the number of protons and electrons in each of the following ions.
   a. P^{3-} b. N^{3-} c. Mg^{2+} d. Li^+

4.16 Calculate the number of protons and electrons in each of the following ions.
   a. S^{2-} b. F^- c. K^+ d. H^+

■ Ionic Charge Sign and Magnitude (Section 4.5)

4.17 What is the charge on the monatomic ion formed by each of the following elements?
   a. $_{12}$Mg   b. $_7$N   c. $_{19}$K   d. $_9$F

4.18 What is the charge on the monatomic ion formed by each of the following elements?
   a. $_3$Li   b. $_{15}$P   c. $_{16}$S   d. $_{13}$Al

4.19 Indicate the number of electrons lost or gained when each of the following atoms forms an ion.
   a. $_4$Be   b. $_{35}$Br   c. $_{38}$Sr   d. $_{34}$Se

4.20 Indicate the number of electrons lost or gained when each of the following atoms forms an ion.
   a. $_{37}$Rb   b. $_{53}$I   c. $_8$O   d. $_{11}$Na

4.21 What noble-gas element is isoelectronic with each of the following ions?
   a. O^{2-} b. P^{3-} c. Ca^{2+} d. K^+

4.22 What noble-gas element is isoelectronic with each of the following ions?
   a. F^- b. Al^{3+} c. Si^{4+} d. C^{4-}

4.23 In what group in the periodic table would representative elements that form ions with the following charges most likely be found?
   a. +2   b. −2   c. −3   d. +1

4.24 In what group in the periodic table would representative elements that form ions with the following charges most likely be found?
   a. +3   b. −4   c. +4   d. −1

4.25 Write the electron configuration of the following.
   a. An aluminum atom
   b. An aluminum ion

4.26 Write the electron configuration of the following.
   a. An oxygen atom
   b. An oxygen ion

■ Ionic Compound Formation (Section 4.6)

4.27 Using Lewis structures, show how ionic compounds are formed by atoms of
   a. Be and O   b. Mg and S
   c. K and N   d. F and Ca

4.28 Using Lewis structures, show how ionic compounds are formed by atoms of
a. Na and F   b. Li and S
c. Be and S   d. P and K

■ **Formulas for Ionic Compounds (Section 4.7)**

4.29 Write the formula for an ionic compound formed from Ba^{2+} ions and each of the following ions.
a. Cl^- b. Br^- c. N^{3-} d. O^{2-}

4.30 Write the formula for an ionic compound formed from K^+ ions and each of the following ions.
a. Cl^- b. Br^- c. N^{3-} d. O^{2-}

4.31 Write the formula for an ionic compound formed from F^- ions and each of the following ions.
a. Mg^{2+} b. Be^{2+} c. Li^+ d. Al^{3+}

4.32 Write the formula for an ionic compound formed from S^{2-} ions and each of the following ions.
a. Mg^{2+} b. Be^{2+} c. Li^+ d. Al^{3+}

4.33 Write the formulas for the ionic compounds formed from the following ions.
a. Na^+ and S^{2-} b. Ca^{2+} and I^-
c. Li^+ and N^{3-} d. Al^{3+} and Br^-

4.34 Write the formulas for the ionic compounds formed from the following ions.
a. Li^+ and O^{2-} b. Al^{3+} and N^{3-}
c. K^+ and Cl^- d. Mg^{2+} and I^-

■ **Binary Ionic Compound Nomenclature (Section 4.9)**

4.35 Which of the following pairs of elements would be expected to form a binary ionic compound?
a. Sodium and oxygen   b. Magnesium and sulfur
c. Nitrogen and chlorine   d. Copper and fluorine

4.36 Which of the following pairs of elements would be expected to form a binary ionic compound?
a. Potassium and sulfur   b. Calcium and nitrogen
c. Carbon and chlorine   d. Iron and iodine

4.37 Which of the following binary compounds would be expected to be an ionic compound?
a. Al_2O_3 b. H_2O_2 c. K_2S d. N_2H_4

4.38 Which of the following binary compounds would be expected to be an ionic compound?
a. Cu_2O b. CO c. $NaBr$ d. Be_3P_2

4.39 Name the following binary ionic compounds, each of which contains a fixed-charge metal.
a. KI b. BeO c. AlF_3 d. Na_3P

4.40 Name the following binary ionic compounds, each of which contains a fixed-charge metal.
a. $CaCl_2$ b. Ca_2C c. Be_3N_2 d. K_2S

4.41 Calculate the charge on the metal ion in the following binary ionic compounds, each of which contains a variable-charge metal.
a. Au_2O b. CuO c. SnO_2 d. SnO

4.42 Calculate the charge on the metal ion in the following binary ionic compounds, each of which contains a variable-charge metal.
a. Fe_2O_3 b. FeO c. $SnCl_4$ d. Cu_2S

4.43 Name the following binary ionic compounds, each of which contains a variable-charge metal.
a. FeO b. Au_2O_3 c. CuS d. $CoBr_2$

4.44 Name the following binary ionic compounds, each of which contains a variable-charge metal.
a. PbO b. $FeCl_3$ c. SnO_2 d. NiI_2

4.45 Name each of the following binary ionic compounds.
a. $AuCl$ b. KCl c. $AgCl$ d. $CuCl_2$

4.46 Name each of the following binary ionic compounds.
a. NiO b. FeN c. AlN d. BeO

4.47 Write formulas for the following binary ionic compounds.
a. Potassium bromide   b. Silver oxide
c. Beryllium fluoride   d. Barium phosphide

4.48 Write formulas for the following binary ionic compounds.
a. Gallium nitride   b. Zinc chloride
c. Magnesium sulfide   d. Aluminum nitride

4.49 Write formulas for the following binary ionic compounds.
a. Cobalt(II) sulfide   b. Cobalt(III) sulfide
c. Tin(IV) iodide   d. Lead(II) nitride

4.50 Write formulas for the following binary ionic compounds.
a. Iron(III) oxide   b. Iron(II) oxide
c. Nickel(III) sulfide   d. Copper(I) bromide

■ **Compounds Containing Polyatomic Ions (Sections 4.10 and 4.11)**

4.51 With the help of Table 4.3, write formulas (including charge) for each of the following polyatomic ions.
a. Sulfate   b. Chlorate
c. Hydroxide   d. Cyanide

4.52 With the help of Table 4.3, write formulas (including charge) for each of the following polyatomic ions.
a. Ammonium   b. Nitrate
c. Perchlorate   d. Phosphate

4.53 With the help of Table 4.3, write formulas (including charge) for each of the following pairs of polyatomic ions.
a. Phosphate and hydrogen phosphate
b. Nitrate and nitrite
c. Hydronium and hydroxide
d. Chromate and dichromate

4.54 With the help of Table 4.3, write formulas (including charge) for each of the following pairs of polyatomic ions.
a. Chlorate and perchlorate
b. Hydrogen phosphate and dihydrogen phosphate
c. Carbonate and bicarbonate
d. Sulfate and hydrogen sulfate

4.55 Write formulas for the compounds formed between the following positive and negative ions.
a. Na^+ and ClO_4^- b. Fe^{3+} and OH^-
c. Ba^{2+} and NO_3^- d. Al^{3+} and CO_3^{2-}

4.56 Write formulas for the compounds formed between the following positive and negative ions.
a. K^+ and CN^- b. NH_4^+ and SO_4^{2-}
c. Co^{2+} and $H_2PO_4^-$ d. Ca^{2+} and PO_4^{3-}

4.57 Name the following compounds, all of which contain polyatomic ions and fixed-charge metals.
a. $MgCO_3$ b. $ZnSO_4$
c. $Be(NO_3)_2$ d. Ag_3PO_4

4.58 Name the following compounds, all of which contain polyatomic ions and fixed-charge metals.
a. $LiOH$ b. $Al(CN)_3$
c. $Ba(ClO_3)_2$ d. $NaNO_3$

4.59 Name the following compounds, all of which contain polyatomic ions and variable-charge metals.
a. $Fe(OH)_2$ b. $CuCO_3$
c. $AuCN$ d. $Mn_3(PO_4)_2$

valence electrons and one vacancy, and hydrogen has one valence electron and one vacancy. All covalent bonds formed by these elements are single covalent bonds.

Double bonding becomes possible for elements that need two electrons to complete their octet, and triple bonding becomes possible when three or more electrons are needed to complete an octet. Note that the word *possible* was used twice in the previous sentence. Multiple bonding does not have to occur when an element has two, three, or four vacancies in its octet; single covalent bonds can be formed instead. When more than one behavior is possible, the "bonding behavior" of an element is determined by the element or elements to which it is bonded.

Let us consider the possible "bonding behaviors" for O (six valence electrons, two octet vacancies), N (five valence electrons, three octet vacancies), and C (four valence electrons, four octet vacancies).

To complete its octet by electron sharing, an oxygen atom can form either two single bonds or one double bond.

$$:\overset{\displaystyle |}{\underset{\displaystyle \cdot\cdot}{O}}\!\!— \qquad :\overset{}{\underset{\displaystyle \cdot\cdot}{O}}\!\!=$$

Two single bonds　　One double bond

Nitrogen is a very versatile element with respect to bonding. It can form single, double, or triple covalent bonds as dictated by the other atoms present in a molecule.

$$—\overset{\displaystyle \cdot\cdot}{\underset{\displaystyle |}{N}}\!\!— \qquad —\overset{\displaystyle \cdot\cdot}{N}\!\!= \qquad :N\!\!\equiv$$

Three single bonds　　One single and　　One triple bond
　　　　　　　　　　　one double bond

Note that the nitrogen atom forms three bonds in each of these bonding situations. A double bond counts as two bonds, a triple bond as three. Because nitrogen has only five valence electrons, it must form three covalent bonds to complete its octet.

Carbon is an even more versatile element than nitrogen with respect to variety of types of bonding, as illustrated by the following possibilities. In each case, carbon forms four bonds.

$$—\overset{\displaystyle |}{\underset{\displaystyle |}{C}}\!\!— \qquad —\overset{\displaystyle |}{C}\!\!= \qquad =C\!\!= \qquad —C\!\!\equiv$$

Four single bonds　　Two single bonds and　　Two double bonds　　One single bond and
　　　　　　　　　　　one double bond　　　　　　　　　　　　one triple bond

> There is a strong tendency for atoms of nonmetallic elements to form a specific number of covalent bonds. The number of bonds formed is equal to the number of electrons the nonmetallic atom must share to obtain an octet of electrons.

5.5 Coordinate Covalent Bonds

In the covalent bonds we have considered so far (single, double, and triple), all of the participating atoms in the bond contributed the same number of electrons to the bond. There is another, *less common* way in which a covalent bond can form. It is possible for one atom to supply two electrons and the other atom none to a shared electron pair. A **coordinate covalent bond** *is a covalent bond in which both electrons of a shared pair come from one of the two atoms involved in the bond.* Coordinate covalent bonding enables an atom that has two or more vacancies in its valence shell to share a pair of nonbonding electrons that are located on another atom.

The element oxygen, with two vacancies in its valence octet, quite often forms coordinate covalent bonds. Consider the Lewis structures of the molecules HOCl (hypochlorous acid) and $HClO_2$ (chlorous acid).

> An "ordinary" covalent bond can be thought of as a "Dutch-treat" bond; each atom "pays" its part of the bill. A coordinate covalent bond can be thought of as a "you-treat" bond; one atom pays the whole bill.

$$H:\overset{\displaystyle \cdot\cdot}{\underset{\displaystyle \cdot\cdot}{O}}:\overset{\displaystyle \cdot\cdot}{\underset{\displaystyle \cdot\cdot}{Cl}}: \qquad H:\overset{\displaystyle \cdot\cdot}{\underset{\displaystyle \cdot\cdot}{O}}:\overset{\displaystyle \cdot\cdot}{\underset{\displaystyle \cdot\cdot}{Cl}}:\overset{\displaystyle xx}{\underset{\displaystyle xx}{O}}\!\!{}_x^x$$

Hypochlorous acid　　　　Chlorous acid

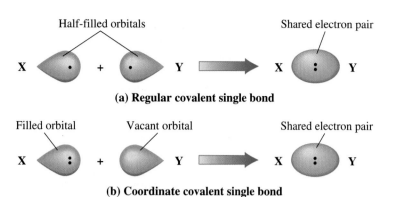

(a) Regular covalent single bond

(b) Coordinate covalent single bond

FIGURE 5.3 (a) A "regular" covalent single bond is the result of overlap of two half-filled orbitals. (b) A coordinate covalent single bond is the result of overlap of a filled and a vacant orbital.

In hypochlorous acid, all the bonds are "ordinary" covalent bonds. In chlorous acid, which differs from hypochlorous acid in that a second oxygen atom is present, the "new" chlorine–oxygen bond is a coordinate covalent bond. The second oxygen atom with six valence electrons (denoted by x's) needs two more for an octet. It shares one of the nonbonding electron pairs present on the chlorine atom. (The chlorine atom does not need any of the oxygen's electrons because it already has an octet.)

▶ Atoms participating in coordinate covalent bonds generally do not form their normal number of covalent bonds.

Atoms participating in coordinate covalent bonds generally deviate from the common bonding pattern (Section 5.5) expected for that type of atom. For example, oxygen normally forms two bonds; yet in the molecules N_2O and CO, which contain coordinate covalent bonds, oxygen forms one and three bonds, respectively.

$$: N \equiv N - \ddot{\underset{..}{O}} : \qquad : C \equiv O :$$

▶ Once a coordinate covalent bond forms, it is indistinguishable from other covalent bonds in a molecule.

Once a coordinate covalent bond is formed, there is no way to distinguish it from any of the other covalent bonds in a molecule; all electrons are identical regardless of their source. The main use of the concept of coordinate covalency is to help rationalize the existence of certain molecules and polyatomic ions whose electron-bonding arrangement would otherwise present problems. Figure 5.3 contrasts the formation of a "regular" covalent bond with that of a coordinate covalent bond.

5.6 Systematic Procedures for Drawing Lewis Structures

Could you generate the Lewis structures of HOCl, $HClO_2$, N_2O, and CO given in the preceding section without any help? Drawing Lewis structures for diatomic molecules is usually straightforward and uncomplicated. However, with triatomic and even larger molecules, students often have trouble. Here is a stepwise procedure for distributing valence electrons as bonding and nonbonding pairs within a Lewis structure.

Let us apply this stepwise procedure to the molecule SO_2, a molecule in which two oxygen atoms are bonded to a central sulfur atom (see Figure 5.4).

Step 1: *Calculate the total number of valence electrons available in the molecule by adding together the valence electron counts for all atoms in the molecule.* The periodic table is a useful guide for determining this number.

An SO_2 molecule has 18 valence electrons available for bonding. Sulfur (Group VIA) has 6 valence electrons, and each oxygen (also Group VIA) has 6 valence electrons. The total number is therefore $6 + 2(6) = 18$.

Step 2: *Write the symbols of the atoms in the molecule in the order in which they are bonded to one another, and then place a single covalent bond, involving two electrons, between each pair of bonded atoms.* For SO_2, the S atom is the central atom. Thus we have

FIGURE 5.4 The sulfur dioxide (SO_2) molecule. A computer-generated model.

$$O : S : O$$

Determining which atom is the *central atom*—that is, which atom has the most other atoms bonded to it—is the key to determining the arrangement of atoms in a molecule or polyatomic ion. Most other atoms present will be bonded to the central atom. For common binary molecular compounds, the molecular formula can help us determine the identity of the central atom. The central atom is the atom that appears only once in the formula; for example, S is the central atom in SO_3, O is the central atom in H_2O, and P is the central atom in PF_3. In molecular compounds containing hydrogen, oxygen, and an additional element, that additional element is the central atom; for example, N is the central atom in HNO_3, and S is the central atom in H_2SO_4. In compounds of this type, the oxygen atoms are bonded to the central atom, and the hydrogen atoms are bonded to the oxygens. Carbon is the central atom in nearly all carbon-containing compounds. Neither hydrogen nor fluorine is ever the central atom.

Step 3: *Add nonbonding electron pairs to the structure such that each atom bonded to the central atom has an octet of electrons. Remember that for hydrogen, an "octet" is only 2 electrons.*

For SO_2, addition of the nonbonding electrons gives

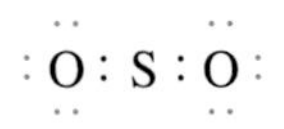

$$\ddot{\underset{..}{\text{O}}} : \text{S} : \ddot{\underset{..}{\text{O}}}$$

At this point, 16 of the 18 available electrons have been used.

Step 4: *Place any remaining electrons on the central atom of the structure.*
Placing the two remaining electrons on the S atom gives

$$: \ddot{\underset{..}{\text{O}}} : \ddot{\text{S}} : \ddot{\underset{..}{\text{O}}} :$$

Step 5: *If there are not enough electrons to give the central atom an octet, then use one or more pairs of nonbonding electrons on the atoms bonded to the central atom to form double or triple bonds.*

The S atom has only 6 electrons. Thus a nonbonding electron pair from an O atom is used to form a sulfur–oxygen double bond.

$$: \ddot{\underset{..}{\text{O}}} : \ddot{\text{S}} : \ddot{\underset{..}{\text{O}}} : \longrightarrow : \ddot{\underset{..}{\text{O}}} : \ddot{\text{S}} :: \ddot{\underset{..}{\text{O}}} :$$

This structure now obeys the octet rule.

Step 6: *Count the total number of electrons in the completed Lewis structure to make sure it is equal to the total number of valence electrons available for bonding, as calculated in Step 1. This step serves as a "double-check" on the correctness of the Lewis structure.*

For SO_2, there are 18 valence electrons in the Lewis structure of Step 5, the same number we calculated in Step 1.

EXAMPLE 5.2

Drawing a Lewis Structure Using Systematic Procedures

■ Draw Lewis structures for the following molecules.

a. PF_3, a molecule in which P is the central atom and all F atoms are bonded to it (see Figure 5.5).

b. HCN, a molecule in which C is the central atom (see Figure 5.6).

Solution

Part a.

Step 1: Phosphorus (Group VA) has 5 valence electrons, and each of the fluorine atoms (Group VIIA) has 7 valence electrons. The total electron count is $5 + 3(7) = 26$.

FIGURE 5.5 The phosphorus trifluoride (PF₃) molecule. A computer-generated model.

Step 2: Drawing the molecular skeleton with single covalent bonds (two electrons) placed between all bonded atoms gives

$$F : P : F$$
$$\overset{\cdot\cdot}{F}$$

Step 3: Adding nonbonding electrons to the structure to complete the octets of all atoms bonded to the central atom gives

$$: \overset{\cdot\cdot}{F} : P : \overset{\cdot\cdot}{F} :$$
$$: \overset{\cdot\cdot}{F} :$$

At this point, we have used 24 of the 26 available electrons.

Step 4: The central P atom has only 6 electrons; it needs 2 more. The 2 remaining available electrons are placed on the P atom, completing its octet. All atoms now have an octet of electrons.

$$: \overset{\cdot\cdot}{F} : \overset{\cdot\cdot}{P} : \overset{\cdot\cdot}{F} :$$
$$: \overset{\cdot\cdot}{F} :$$

Step 5: This step is not needed; the central atom already has an octet of electrons.

Step 6: There are 26 electrons in the Lewis structure, the same number of electrons we calculated in Step 1.

Part b.

Step 1: Hydrogen (Group IA) has 1 valence electron, carbon (Group IVA) has 4 valence electrons, and nitrogen (Group VA) has 5 valence electrons. The total number of electrons is 10.

Step 2: Drawing the molecular skeleton with single covalent bonds between bonded atoms gives

$$H : C : N$$

Step 3: Adding nonbonding electron pairs to the structure such that the atoms bonded to the central atom have "octets" gives

$$H : C : \overset{\cdot\cdot}{N} :$$

Remember that hydrogen needs only 2 electrons.

Step 4: The structure in Step 3 has 10 valence electrons, the total number available. Thus there are no additional electrons available to place on the carbon atom to give it an octet of electrons.

Step 5: To give the central carbon atom its octet, 2 nonbonding electron pairs on the nitrogen atom are used to form a carbon–nitrogen triple bond.

$$H : C : \overset{\curvearrowleft}{N} : \longrightarrow H : C ::: N :$$

Step 6: The Lewis structure has 10 electrons, as calculated in Step 1.

FIGURE 5.6 The hydrogen cyanide (HCN) molecule. A computer-generated model.

Practice Exercise 5.2

Draw Lewis structures for the following molecules.

a. SiCl₄, a molecule in which Si is the central atom and all Cl atoms are bonded to it
b. H₂CO, a molecule in which C is the central atom and to which the other three atoms are bonded.

Nitric Oxide: A Molecule Whose Bonding Does Not Follow "The Rules"

The bonding in most, *but not all,* simple molecules is easily explained using the systematic procedures for drawing Lewis structures described in Section 5.6. The molecule NO (nitric oxide) is an example of a simple molecule whose bonding does not conform to the standard rules for bonding. The presence of an odd number (11) of valence electrons in nitric oxide (5 from nitrogen and 6 from oxygen) makes it impossible to write a Lewis structure in which all electrons are paired as required by the octet rule. Thus an unpaired electron is present in the Lewis structure of NO.

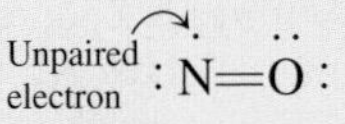

Unpaired electron : N=O :

Despite the "nonconforming" nature of the bonding in nitric oxide, it is an abundant and important molecule within our environment. This colorless, odorless, nonflammable gas is generated by numerous natural and human-caused processes, including (1) lightning passing through air, which causes the N_2 and O_2 of air to react (to a small extent) with each other to produce NO, (2) automobile engines, within which the hot walls of the cylinders again cause N_2 and O_2 of air to become slightly reactive toward each other, and (3) a burning cigarette.

The fact that NO is produced in the preceding ways has been known for many years. The environmental effects of such NO have also been well documented. The NO serves as a precursor for the formation of both acid rain and smog.

During the early 1990s, it was found that NO is also an important biochemical that is naturally present in the human body. The body generates its own NO, usually from amino acids, and once formed, the NO has a life of 10 seconds or less. Its biochemical functions within the human body include (1) helping maintain blood pressure by dilating blood vessels, (2) helping kill foreign invading molecules as part of the body's immune system response, and (3) serving as a biochemical messenger in the brain for processes associated with long-term memory.

5.7 Bonding in Compounds with Polyatomic Ions Present

Ionic compounds containing polyatomic ions (Section 4.10) present an interesting combination of both ionic and covalent bonds: covalent bonding *within* the polyatomic ion and ionic bonding *between* it and ions of opposite charge.

Polyatomic ion Lewis structures, which show the covalent bonding within such ions, are drawn using the same procedures as for molecular compounds (Section 5.6), with the accommodation that the total number of electrons used in the structure must be adjusted (increased or decreased) to take into account ion charge. The number of electrons is increased in the case of negatively charged ions and decreased in the case of positively charged ions.

In the Lewis structure for an *ionic compound* that contains a polyatomic ion, the positive and negative ions are treated separately to show that they are individual ions not linked by covalent bonds. The Lewis structure of potassium sulfate, K_2SO_4, is written as

Students often erroneously assume that the charge associated with a polyatomic ion is assigned to a particular atom within the ion. Polyatomic ion charge is not localized on a particular atom but rather is associated with the ion as a whole.

When we write the Lewis structure of an ion (monatomic or polyatomic), it is customary to use brackets and to show ionic charge outside the brackets.

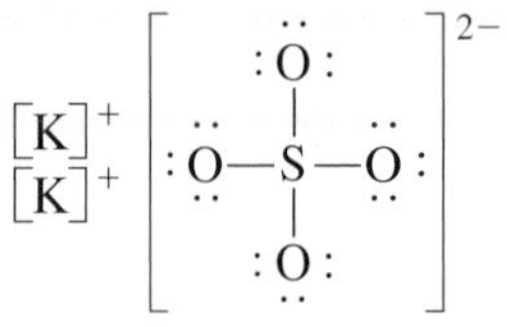

Correct structure   Incorrect structure

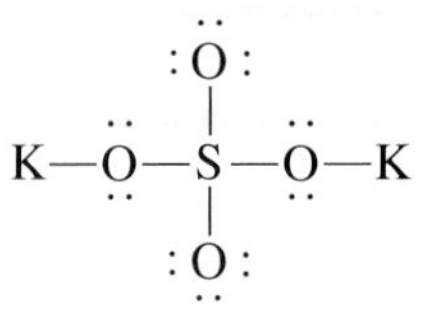

EXAMPLE 5.3

Drawing Lewis Structures for Polyatomic Ions

■ Draw a Lewis structure for SO_4^{2-}, a polyatomic ion in which S is the central atom and all O atoms are bonded to the S atom (see Figure 5.7).

Solution

Step 1: Both S and O are Group VIA elements. Thus each of the atoms has 6 valence electrons. Two extra electrons are also present, which accounts for the -2 charge on the ion. The total electron count is $6 + 4(6) + 2 = 32$.

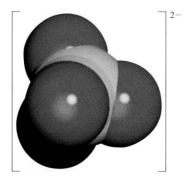

FIGURE 5.7 The sulfate ion (SO_4^{2-}). A computer-generated model.

Step 2: Drawing the molecular skeleton with single covalent bonds between bonded atoms gives

$$\left[\begin{array}{c} O \\ O:S:O \\ O \end{array}\right]^{2-}$$

Step 3: Adding nonbonding electron pairs to give each oxygen atom an octet of electrons yields

$$\left[\begin{array}{c} :\overset{\cdot\cdot}{O}: \\ :\overset{\cdot\cdot}{O}:S:\overset{\cdot\cdot}{O}: \\ :\overset{\cdot\cdot}{O}: \end{array}\right]^{2-}$$

Step 4: The Step 3 structure has 32 electrons, the total number available. No more electrons can be added to the structure, and indeed, none need to be added because the central S atom has an octet of electrons. There is no need to proceed to Step 5.

Practice Exercise 5.3

Draw a Lewis structure for BrO_3^-, a polyatomic ion in which Br is the central atom and all O atoms are bonded to it.

5.8 Molecular Geometry

Lewis structures show the numbers and types of bonds present in molecules. They do not, however, convey any information about molecular geometry—that is, molecular shape. **Molecular geometry** *is the three-dimensional arrangement of atoms within a molecule.* Indeed, Lewis structures falsely imply that all molecules have flat, two-dimensional shapes. This is not the case, as can be seen from the previously presented computer-generated models for the molecules SO_2, PF_3, and HCN (Figures 5.4 through 5.6).

Molecular geometry is an important factor in determining the physical and chemical properties of a substance. Dramatic relationships between geometry and properties are often observed in research associated with the development of prescription drugs. A small change in overall molecular geometry, caused by the addition or removal of atoms, can enhance drug effectiveness and/or decrease drug side effects. Studies also show that the human senses of taste and smell depend in part on the geometries of molecules.

For molecules that contain only a few atoms, molecular geometry can be predicted by using the information present in a molecule's Lewis structure and a procedure called valence shell electron pair repulsion (VSEPR) theory. **VSEPR theory** *is a set of procedures for predicting the three-dimensional geometry of a molecule using the information contained in the molecule's Lewis structure.*

The central concept of VSEPR theory is that electron pairs in the valence shell of an atom adopt an arrangement in space that minimizes the repulsions between the like-charged (all negative) electron pairs. The specific arrangement adopted by the electron pairs depends on the number of electron pairs present. The electron pair arrangements about a *central atom* in the cases of two, three, and four electron pairs are as follows:

1. Two electron pairs, to be as far apart as possible from one another, are found on opposite sides of a nucleus—that is, at 180° angles to one another (Figure 5.8a). Such an electron pair arrangement is said to be *linear.*
2. Three electron pairs are as far apart as possible when they are found at the corners of an equilateral triangle. In such an arrangement, they are separated by 120° angles, giving a *trigonal planar* arrangement of electron pairs (Figure 5.8b).

180°

Central atom

(a) Linear

120°

(b) Trigonal planar

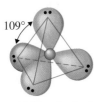

109°

(c) Tetrahedral

FIGURE 5.8 Arrangements of valence electron pairs about a central atom that minimize repulsions between the pairs.

> The preferred arrangement of a given number of valence electron pairs about a central atom is the one that maximizes the separation among them. Such an arrangement minimizes repulsions between electron pairs.

3. A *tetrahedral* arrangement of electron pairs minimizes repulsions among four sets of electron pairs (Figure 5.8c). A tetrahedron is a four-sided solid in which all four sides are identical equilateral triangles. The angle between any two electron pairs is 109°.

■ Electron Groups

Before we use VSEPR theory to predict molecular geometry, an expansion of the concept of an "electron pair" to that of an "electron group" is needed. This will enable us to extend VSEPR theory to molecules in which double and triple bonds are present. A **VSEPR electron group** *is a collection of valence electrons present in a localized region about the central atom in a molecule.* A VSEPR electron group may contain two electrons (a single covalent bond), four electrons (a double covalent bond) or six electrons (a triple covalent bond). VSEPR electron groups that contain four and six electrons repel other VSEPR electron groups in the same way electron pairs do. This makes sense. The four electrons in a double bond or the six electrons in a triple bond are localized in the region between two bonded atoms in a manner similar to the two electrons of a single bond.

> The acronym VSEPR is pronounced "vesper."

Let us now apply VSEPR theory to molecules in which two, three, and four VSEPR electron groups are present about a central atom. Our operational rules will be

1. Draw a Lewis structure for the molecule and identify the specific atom for which geometrical information is desired. (This atom will usually be the central atom in the molecule.)
2. Determine the number of VSEPR electron groups present about the central atom. The following conventions govern this determination:
   a. No distinction is made between bonding and nonbonding electron groups. Both are counted.
   b. Single, double, and triple bonds are all counted equally as "one electron group" because each takes up only one region of space about a central atom.
3. Predict the VSEPR electron group arrangement about the atom by assuming that the electron groups orient themselves in a manner that minimizes repulsions (see Figure 5.8).

■ Molecules with Two VSEPR Electron Groups

All molecules with two VSEPR electron groups are *linear.* Two common molecules with two VSEPR electron groups are carbon dioxide (CO_2) and hydrogen cyanide (HCN), whose Lewis structures are

$$\ddot{O}=C=\ddot{O}\!: \qquad H-C\equiv N\!:$$

In CO_2, the central carbon atom's two VSEPR electron groups are the two double bonds. In HCN, the central carbon atom's two VSEPR electron groups are a single bond and a triple bond. In both molecules, the VSEPR electron groups arrange themselves on opposite sides of the carbon atom, which produces a linear molecule.

■ Molecules with Three VSEPR Electron Groups

Molecules with three VSEPR electron groups have two possible molecular structures: *trigonal planar* and *angular.* The former occurs when all three VSEPR electron groups are bonding and the latter when one of the three VSEPR electron groups is nonbonding. The molecules H_2CO (formaldehyde) and SO_2 (sulfur dioxide) illustrate these two possibilities. Their Lewis structures are

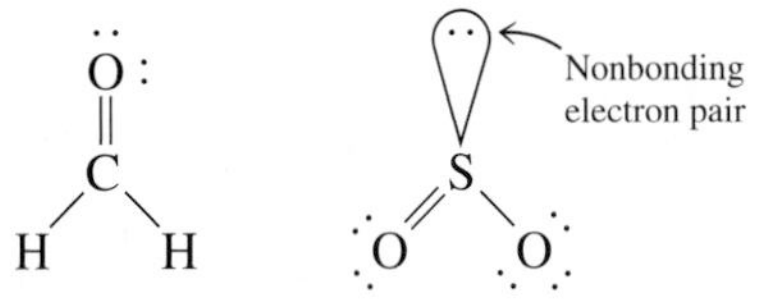

Trigonal planar   Angular

► VSEPR electron group arrangement and molecular geometry are not the same when a central atom possesses nonbonding electron pairs. The word used to describe the shape in such cases does not include the positions of the nonbonding electron groups.

► "Dotted line" and "wedge" bonds can be used to indicate the directionality of bonds, as shown below.

Bond behind page — Bonds in the plane of the page

Bond in front of page

FIGURE 5.9 Computer-generated models of (a) C_2H_2, (b) H_2O_2, and (c) HN_3.

(a)

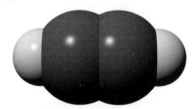

The acetylene (C_2H_2) molecule.

(b)

The hydrogen peroxide (H_2O_2) molecule.

(c)

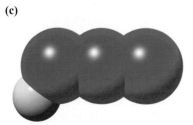

The hydrogen azide (HN_3) molecule.

In both molecules, the VSEPR electron groups are found at the corners of an equilateral triangle.

The shape of the SO_2 molecule is described as *angular* rather than *trigonal planar,* because molecular geometry describes only *atom positions.* The positions of nonbonding electron groups are not taken into account in describing molecular geometry. Do not interpret this to mean that nonbonding electron groups are unimportant in molecular geometry determinations; indeed, in the case of SO_2, it is the presence of the nonbonding electron groups that makes the molecule angular rather than linear.

■ **Molecules with Four VSEPR Electron Groups**

Molecules with four VSEPR electron groups have three possible molecular geometries: *tetrahedral* (no nonbonding electron groups present), *trigonal pyramidal* (one nonbonding electron group present), and *angular* (two nonbonding electron groups present). The molecules CH_4 (methane), NH_3 (ammonia), and H_2O (water) illustrate this sequence of molecular geometries.

Tetrahedral   Trigonal pyramidal   Angular

Nonbonding electron group

Nonbonding electron groups

In all three molecules, the VSEPR electron groups arrange themselves at the corners of a tetrahedron. Again, note that the word used to describe the geometry of the molecule does not take into account the positioning of nonbonding electron groups.

■ **Molecules with More Than One Central Atom**

The molecular shape of molecules that contain more than one central atom can be obtained by considering each central atom separately and then combining the results. Let us apply this principle to the molecules C_2H_2 (acetylene), H_2O_2 (hydrogen peroxide), and HN_3 (hydrogen azide), all of which have a four-atom "chain" structure. Their Lewis structures and VSEPR electron group counts are as follows:

Acetylene

H—C≡C—H

2 VSEPR electron groups — Linear C center

2 VSEPR electron groups — Linear C center

Hydrogen peroxide

H—O—O—H

4 VSEPR electron groups — Angular O center

4 VSEPR electron groups — Angular O center

Hydrogen azide

H—N=N=N:

3 VSEPR electron groups — Angular N center

2 VSEPR electron groups — Linear N center

These three molecules thus have, respectively, zero bends, two bends, and one bend in their four-atom chain.

H—C≡C—H

Zero bends in the chain

O—O / H, H

Two bends in the chain

N=N=N: / H

One bend in the chain

Computer-generated three-dimensional models for these three molecules are given in Figure 5.9.

The Chemistry at a Glance feature on page 114 summarizes the key concepts involved in using VSEPR theory to predict molecular geometry.

The Geometry of Molecules

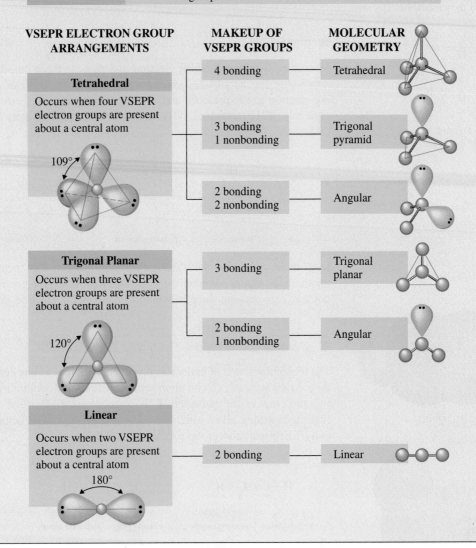

PREDICTING MOLECULAR GEOMETRY USING VSEPR THEORY	Operational rules
	1. Draw a Lewis structure for the molecule.
	2. Count the number of VSEPR electron groups about the central atom in the Lewis structure.
	3. Assign a geometry based on minimizing repulsions between electron groups.

VSEPR ELECTRON GROUP ARRANGEMENTS — **MAKEUP OF VSEPR GROUPS** — **MOLECULAR GEOMETRY**

Tetrahedral

Occurs when four VSEPR electron groups are present about a central atom

109°

- 4 bonding — Tetrahedral
- 3 bonding 1 nonbonding — Trigonal pyramid
- 2 bonding 2 nonbonding — Angular

Trigonal Planar

Occurs when three VSEPR electron groups are present about a central atom

120°

- 3 bonding — Trigonal planar
- 2 bonding 1 nonbonding — Angular

Linear

Occurs when two VSEPR electron groups are present about a central atom

180°

- 2 bonding — Linear

5.9 Electronegativity

The ionic and covalent bonding models seem to represent two very distinct forms of bonding. Actually, the two models are closely related; they are the extremes of a broad continuum of bonding patterns. The close relationship between the two bonding models becomes apparent when the concepts of *electronegativity* (discussed in this section) and *bond polarity* (discussed in the next section) are considered.

The electronegativity concept has its origins in the fact that the nuclei of various elements have differing abilities to attract shared electrons (in a bond) to themselves. Some elements are better electron attractors than other elements. **Electronegativity** *is a measure of the relative attraction that an atom has for the shared electrons in a bond.*

CHEMICAL CONNECTIONS

Molecular Geometry and Odor

Whether a substance has an odor depends on whether it can excite the olfactory nerve endings in the nose. Factors that determine whether excitation of nerve endings in the nose occurs include:

1. *Molecular volatility.* Odoriferous substances must be either gases or easily vaporized liquids or solids at room temperature; otherwise, the molecules of such substances would never reach the nose.
2. *Molecular solubility.* Olfactory nerve endings are covered with *mucus,* an aqueous solution that contains dissolved proteins and carbohydrates. Odoriferous molecules must be at least slightly soluble in this mucus.
3. *Molecular geometry and polarity.* Olfactory nerve endings have receptor sites that accommodate molecules only of particular geometries and polarity. The interaction that

causes excitation is similar to that between a key and a lock; the key must have a particular shape in order to open the lock.

One current theory for olfaction suggests that every known odor can be made from a combination of seven primary odors (see the accompanying table) in much the same way that all colors can be made from the three primary colors of red, yellow, and blue. Associated with each of the primary odors is a particular molecular polarity and/or molecular geometry.

Primary odor	Familiar substance with this odor	Molecular characteristics
ethereal	dry cleaning fluid	rodlike shape
peppermint	mint candy	wedgelike shape
musk	some perfumes	disclike shape
camphoraceous	moth balls	spherical shape
floral	rose	disc with tail (kite) shape
pungent[a]	vinegar	negative polarity interaction
putrid[a]	skunk odor	positive polarity interaction

[a]These odors are less specific than are the other five odors, which indicates that they involve different interactions.

The "taste" of a food is actually more related to a person's nose than his or her tongue; that is, "flavor" is sensed by the nose. With each breath of air, some of the incoming air goes from the back of the mouth up into the nasal passages. When food first enters the mouth, the most volatile molecules present in the food are carried via air into the nose where they are "smelled." As food is chewed, more volatile molecules are released from the food and enter the nasal passages.

Linus Pauling (Figure 5.10), whose contributions to chemical bonding theory earned him a Nobel Prize in chemistry, was the first chemist to develop a *numerical* scale of electronegativity. Figure 5.11 gives Pauling electronegativity values for the more frequently encountered representative elements. The higher the electronegativity value for an element, the greater the attraction of atoms of that element for the shared electrons in bonds. The element fluorine, whose Pauling electronegativity value is 4.0, is the most electronegative of all elements; that is, it possesses the greatest electron-attracting ability for electrons in a bond.

As Figure 5.11 shows, electronegativity values increase from left to right across periods and from bottom to top within groups of the periodic table. These two trends result in nonmetals generally having higher electronegativities than metals. This fact is consistent with our previous generalization (Section 4.5) that metals tend to lose electrons and nonmetals tend to gain electrons when an ionic bond is formed. Metals (low electronegativities, poor electron attractors) give up electrons to nonmetals (high electronegativities, good electron attractors).

Note that the electronegativity for an element is not a directly measurable quantity. Rather, electronegativity values are calculated from bond energy information and other related experimental data. Values differ from element to element because of differences in atom size, nuclear charge, and number of inner-shell (nonvalence) electrons.

5.10 Bond Polarity

When two atoms of equal electronegativity share one or more pairs of electrons, each atom exerts the same attraction for the electrons, which results in the electrons being *equally* shared. This type of bond is called a nonpolar covalent bond. A **nonpolar covalent bond** *is a covalent bond in which there is equal sharing of electrons between two atoms.*

When the two atoms involved in a covalent bond have different electronegativities, the electron-sharing situation is more complex. The atom that has the higher electronegativity attracts the electrons more strongly than the other atom, which results in an *unequal* sharing of electrons. This type of covalent bond is called a polar covalent bond. A **polar covalent bond** *is a covalent bond in which there is unequal sharing of electrons between two atoms.* Figure 5.12 pictorially contrasts a nonpolar covalent bond and a polar covalent bond using the molecules H_2 and HCl.

The significance of unequal sharing of electrons in a polar covalent bond is that it creates fractional positive and negative charges on atoms. Although both atoms involved in a polar covalent bond are initially uncharged, the unequal sharing means that the electrons spend more time near the more electronegative atom of the bond (producing a fractional negative charge) and less time near the less electronegative atom of the bond (producing a fractional positive charge). The presence of such fractional charges on atoms within a molecule often significantly affects molecular properties (Section 5.11).

The fractional charges associated with atoms involved in a polar covalent bond are always values less than 1 because complete electron transfer does not occur. Complete electron transfer, which produces an ionic bond, would produce charges of $+1$ and -1. A notation that involves the lower-case Greek letter delta (δ) is used to denote fractional charge. The symbol δ^-, meaning "fractional negative charge," is placed above the more electronegative atom of the bond, and the symbol δ^+, meaning "fractional positive charge," is placed above the less electronegative atom of the bond.

With delta notation, the direction of polarity of the bond in hydrogen chloride (HCl) is depicted as

$$\overset{\delta^+}{H}—\overset{\delta^-}{\underset{\displaystyle\cdot\cdot}{\overset{\cdot\cdot}{Cl}}}:$$

FIGURE 5.10 Linus Carl Pauling (1901–1994). Pauling received the Nobel Prize in chemistry in 1954 for his work on the nature of the chemical bond. In 1962 he received the Nobel Peace Prize in recognition of his efforts to end nuclear weapons testing.

▶ The δ^+ and δ^- symbols are pronounced "delta plus" and "delta minus." Whatever the magnitude of δ^+, it must be the same as that of δ^- because the sum of δ^+ and δ^- must be zero.

FIGURE 5.11 Abbreviated periodic table showing Pauling electronegativity values for selected representative elements.

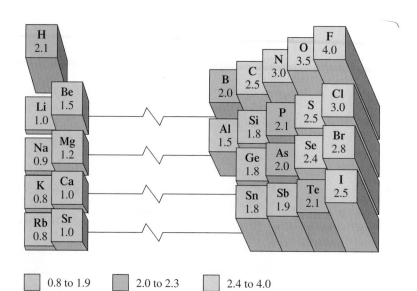

☐ 0.8 to 1.9   ☐ 2.0 to 2.3   ☐ 2.4 to 4.0

FIGURE 5.12 (a) In the nonpolar covalent bond present in H₂ (H—H), there is a symmetrical distribution of electron density between the two atoms; that is, equal sharing of electrons occurs. (b) In the polar covalent bond present in HCl (H—Cl), electron density is displaced toward the Cl atom because of its greater electronegativity; that is, unequal sharing of electrons occurs.

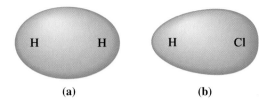

(a)   (b)

Chlorine is the more electronegative of the two elements; it dominates the electron-sharing process and draws the shared electrons closer to itself. Hence the chlorine end of the bond has the δ^- designation (the more electronegative element always has the δ^- designation).

The direction of polarity of a polar covalent bond can also be designated by using an arrow with a cross at one end ($\longmapsto$). The cross is near the end of the bond that is "positive," and the arrowhead is near the "negative" end of the bond. Using this notation, we would denote the bond in the molecule HCl as

$$\overset{\longmapsto}{H{-}\overset{..}{\underset{..}{Cl}}:}$$

An extension of the reasoning used in characterizing the covalent bond in the HCl molecule as polar leads to the generalization that most chemical bonds are not 100% covalent (equal sharing) or 100% ionic (no sharing). Instead, most bonds are somewhere in between (unequal sharing).

Bond polarity *is a measure of the degree of inequality in the sharing of electrons between two atoms in a chemical bond.* The numerical value of the electronegativity difference between two bonded atoms gives an approximate measure of the polarity of the bond. The greater the numerical difference, the greater the inequality of electron sharing and the greater the polarity of the bond. For electronegativity differences of 2.0 or greater, inequality in electron sharing is sufficiently large that the bond is considered ionic (electron transfer).

The Chemistry at a Glance feature on page 119 summarizes much that we have said about chemical bonds in this chapter.

▶ Prediction of bond type, on the basis of electronegativity differences, is as follows:

Nonpolar covalent: zero difference
Polar covalent: greater than 0 but less than 2.0
Ionic: 2.0 or greater

EXAMPLE 5.4

Using Electronegativity Difference to Predict Bond Polarity and Bond Type

■ Consider the following bonds

N—Cl   Ca—F   C—O   B—H   N—O

a. Rank the bonds in order of increasing polarity.
b. Determine the direction of polarity for each bond.
c. Classify each bond as nonpolar covalent, polar covalent, or ionic.

Solution

Let us first calculate the electronegativity difference for each of the bonds using the electronegativity values in Figure 5.11.

$$
\begin{aligned}
N{-}Cl: &\quad 3.0 - 3.0 = 0.0\\
Ca{-}F: &\quad 4.0 - 1.0 = 3.0\\
C{-}O: &\quad 3.5 - 2.5 = 1.0\\
B{-}H: &\quad 2.1 - 2.0 = 0.1\\
N{-}O: &\quad 3.5 - 3.0 = 0.5
\end{aligned}
$$

a. Bond polarity increases as electronegativity difference increases. Using the mathematical symbol <, which means "is less than," we can rank the bonds in terms of increasing bond polarity as follows:

N—Cl < B—H < N—O < C—O < Ca—F
　0.0　　　 0.1　　　 0.5　　　 1.0　　　 3.0

(continued)

b. The direction of bond polarity is from the least electronegative atom to the most electronegative atom. The more electronegative atom bears the partial negative charge (δ^-).

$$\overset{\longmapsto}{N-Cl} \qquad \overset{\longmapsto}{B-H} \qquad \overset{\longmapsto}{N-O} \qquad \overset{\longmapsto}{C-O} \qquad \overset{\longmapsto}{Ca-F}$$

c. Nonpolar covalent bonds require zero difference in electronegativity, and ionic bonds require an electronegativity difference of 2.0 or greater. The in-between region characterizes polar covalent bonds.

Nonpolar covalent:	N—Cl
Polar covalent:	B—H, N—O, and C—O
Ionic:	Ca—F

Practice Exercise 5.4

Consider the following bonds:

$$N-S \qquad H-H \qquad Na-F \qquad K-Cl \qquad F-Cl$$

a. Rank the bonds in order of increasing polarity.
b. Determine the direction of polarity for each bond.
c. Classify each bond as nonpolar covalent, polar covalent, or ionic.

5.11 Molecular Polarity

Molecules, as well as bonds (Section 5.11), can have polarity. **Molecular polarity** *is a measure of the degree of inequality in the attraction of bonding electrons to various locations within a molecule.* In terms of electron attraction, if one part of a molecule is favored over other parts, then the molecule is *polar*. A **polar molecule** *is a molecule in which there is an unsymmetrical distribution of electronic charge.* In a polar molecule, bonding electrons are more attracted to one part of the molecule than to other parts. A **nonpolar molecule** *is a molecule in which there is a symmetrical distribution of electron charge.* Attraction for bonding electrons is the same in all parts of a *nonpolar* molecule. Molecular polarity depends on two factors: (1) bond polarities and (2) molecular geometry (Section 5.7). In molecules that are symmetrical, the effects of polar bonds may cancel each other, resulting in the molecule as a whole having no polarity.

Determining the molecular polarity of a diatomic molecule is simple because only one bond is present. If that bond is nonpolar, then the molecule is nonpolar; if the bond is polar, then the molecule is polar.

Determining molecular polarity for triatomic molecules is more complicated. Two different molecular geometries are possible: linear and angular. In addition, the symmetrical nature of the molecule must be considered. Let us consider the polarities of three specific triatomic molecules: CO_2 (linear), H_2O (angular), and HCN (linear).

In the linear CO_2 molecule, both bonds are polar (oxygen is more electronegative than carbon). Despite the presence of these polar bonds, CO_2 molecules are *nonpolar*. The effects of the two polar bonds are canceled as a result of the oxygen atoms being arranged symmetrically around the carbon atom. The shift of electronic charge toward one oxygen atom is exactly compensated for by the shift of electronic charge toward the other oxygen atom. Thus one end of the molecule is not negatively charged relative to the other end (a requirement for polarity), and the molecule is nonpolar. This cancella-

> A prerequisite for determining molecular polarity is a knowledge of molecular geometry.

> Molecules in which all bonds are polar can be nonpolar if the bonds are so oriented in space that the polarity effects cancel each other.

Covalent Bonds and Molecular Compounds

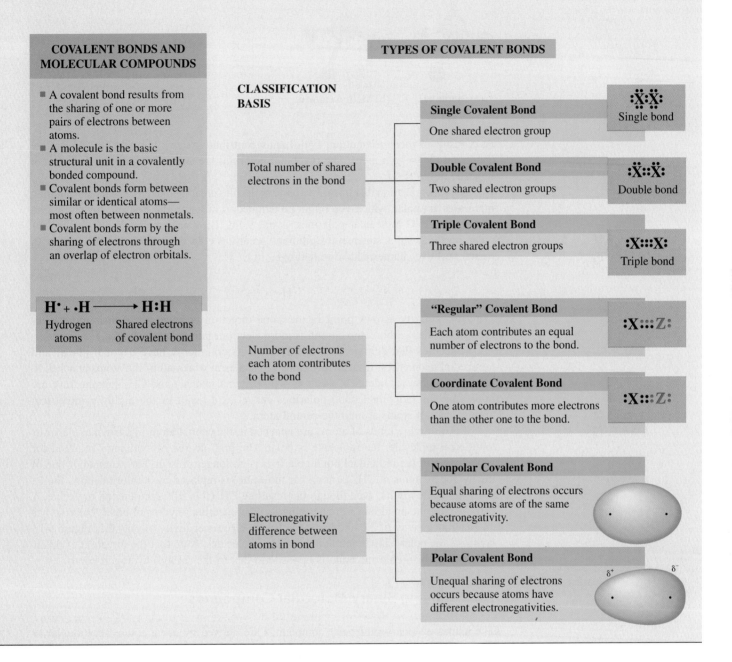

COVALENT BONDS AND MOLECULAR COMPOUNDS

- A covalent bond results from the sharing of one or more pairs of electrons between atoms.
- A molecule is the basic structural unit in a covalently bonded compound.
- Covalent bonds form between similar or identical atoms—most often between nonmetals.
- Covalent bonds form by the sharing of electrons through an overlap of electron orbitals.

$$H \cdot + \cdot H \longrightarrow H : H$$

Hydrogen atoms   Shared electrons of covalent bond

TYPES OF COVALENT BONDS

CLASSIFICATION BASIS

Total number of shared electrons in the bond

Single Covalent Bond
One shared electron group

Single bond

Double Covalent Bond
Two shared electron groups

Double bond

Triple Covalent Bond
Three shared electron groups

Triple bond

Number of electrons each atom contributes to the bond

"Regular" Covalent Bond
Each atom contributes an equal number of electrons to the bond.

Coordinate Covalent Bond
One atom contributes more electrons than the other one to the bond.

Electronegativity difference between atoms in bond

Nonpolar Covalent Bond
Equal sharing of electrons occurs because atoms are of the same electronegativity.

Polar Covalent Bond
Unequal sharing of electrons occurs because atoms have different electronegativities.

tion of individual bond polarities, with crossed arrows used to denote the polarities, is diagrammed as follows:

$$O = C = O$$

The nonlinear (angular) triatomic H_2O molecule is polar. The bond polarities associated with the two hydrogen–oxygen bonds do not cancel one another because of the nonlinearity of the molecule.

FIGURE 5.13 (a) Methane (CH_4) is a nonpolar tetrahedral molecule. (b) Methyl chloride (CH_3Cl) is a polar tetrahedral molecule. Bond polarities cancel in the first case, but not in the second.

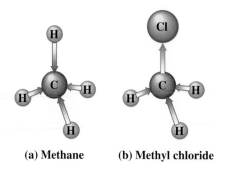

(a) Methane   **(b) Methyl chloride**

As a result of their orientation, both bonds contribute to an accumulation of negative charge on the oxygen atom. The two bond polarities are equal in magnitude but are not opposite in direction.

The generalization that linear triatomic molecules are nonpolar and nonlinear triatomic molecules are polar, which you might be tempted to make on the basis of our discussion of CO_2 and H_2O molecular polarities, is not valid. The linear molecule HCN, which is polar, invalidates this statement. Both bond polarities contribute to nitrogen's acquiring a partial negative charge relative to hydrogen in HCN.

$$\overset{\longmapsto}{H}-\overset{\longmapsto}{C}\equiv N$$

(The two polarity arrows point in the same direction because nitrogen is more electronegative than carbon, and carbon is more electronegative than hydrogen.)

Molecules that contain four and five atoms commonly have trigonal planar and tetrahedral geometries, respectively. Such molecules in which all of the atoms attached to the central atom are identical, such as SO_3 (trigonal planar) and CH_4 (tetrahedral), are *nonpolar*. The individual bond polarities cancel as a result of the highly symmetrical arrangement of atoms around the central atom.

If two or more kinds of atoms are attached to the central atom in a trigonal planar or tetrahedral molecule, the molecule is polar. The high degree of symmetry required for cancellation of the individual bond polarities is no longer present. For example, if one of the hydrogen atoms in CH_4 (a nonpolar molecule) is replaced by a chlorine atom, then a polar molecule results, even though the resulting CH_3Cl is still a tetrahedral molecule. A carbon–chlorine bond has a greater polarity than a carbon–hydrogen bond; chlorine has an electronegativity of 3.0 and hydrogen has an electronegativity of only 2.1. Figure 5.13 contrasts the polar CH_3Cl and nonpolar CH_4 molecules. Note that the direction of polarity of the carbon–chlorine bond is opposite to that of the carbon–hydrogen bonds.

5.12 Naming Binary Molecular Compounds

The names of binary molecular compounds are derived by using a rule very similar to that used for naming binary ionic compounds (Section 4.9). However, one major difference exists. Names for binary molecular compounds always contain numerical prefixes that give the number of each type of atom present in addition to the names of the elements present. This is in direct contrast to binary ionic compound nomenclature, where formula subscripts are never mentioned in the names.

> Numerical prefixes are used in naming binary molecular compounds. They are *never* used, however, in naming binary ionic compounds.

Here is the basic rule to use when constructing the name of a binary molecular compound: *The full name of the nonmetal of lower electronegativity is given first, followed by a separate word containing the stem of the name of the more electronegative nonmetal and the suffix* -ide. *Numerical prefixes, giving numbers of atoms, precede the names of both nonmetals.* Thus the compounds N_2O, N_2O_3, and N_2O_4 are dinitrogen monoxide, dinitrogen trioxide, and dinitrogen tetroxide, respectively.

Prefixes are necessary because several different compounds exist for most pairs of nonmetals. For example, all of the following nitrogen–oxygen compounds exist: NO,

TABLE 5.1
Numerical Prefixes for the Numbers 1 Through 10

Prefix	Number
mono-	1
di-	2
tri-	3
tetra-	4
penta-	5
hexa-	6
hepta-	7
octa-	8
ennea-	9
deca-	10

TABLE 5.2
Selected Binary Molecular Compounds That Have Common Names

Compound formula	Accepted common name
H_2O	water
H_2O_2	hydrogen peroxide
NH_3	ammonia
N_2H_4	hydrazine
CH_4	methane
C_2H_6	ethane
PH_3	phosphine
AsH_3	arsine

▶ When an element name begins with a vowel, an *a* or *o* at the end of the Greek prefix is dropped for phonetic reasons, as in pentoxide instead of pentaoxide.

NO_2, N_2O, N_2O_3, N_2O_4, and N_2O_5. Such diverse behavior between two elements is related to the fact that single, double, and triple covalent bonds exist. The prefixes used are the standard numerical prefixes, which are given for the numbers 1 through 10 in Table 5.1. Example 5.5 shows how these prefixes are used in nomenclature for binary covalent compounds.

EXAMPLE 5.5

Naming Binary Molecular Compounds

■ Name the following binary molecular compounds.
a. S_2Cl_2 **b.** CS_2 **c.** P_4O_{10} **d.** CBr_4

Solution

The names of each of these compounds will consist of two words. These words will have the following general formats:

$$\text{First word:} \quad (\text{prefix}) + \left(\begin{array}{c} \text{full name of least} \\ \text{electronegative nonmetal} \end{array} \right)$$

$$\text{Second word:} \quad (\text{prefix}) + \left(\begin{array}{c} \text{stem of name of more} \\ \text{electronegative nonmetal} \end{array} \right) + (\text{ide})$$

a. The elements present are sulfur and chlorine. The two portions of the name (including prefixes) are *disulfur and dichloride,* which are combined to give the name *disulfur dichloride.*
b. When only one atom of the first nonmetal is present, it is customary to omit the initial prefix *mono-.* Thus the name of this compound is *carbon disulfide.*
c. The prefix for four atoms is *tetra-* and for ten atoms is *deca-.* This compound has the name *tetraphosphorus decoxide* and the structure shown in Figure 5.14.
d. Omitting the initial *mono-* (see part **b**), we name this compound *carbon tetrabromide.*

Practice Exercise 5.5

Name the following binary molecular compounds.
a. PF_3 **b.** SO_2 **c.** P_4S_{10} **d.** $SiCl_4$

FIGURE 5.14 The tetraphosphorus decoxide (P_4O_{10}) molecule. A computer-generated molecular model.

▶ In Section 10.3, we will learn that placing hydrogen first in a formula conveys the message that the compound behaves as an acid in aqueous solution.

There is one standard exception to the use of numerical prefixes when naming binary molecular compounds. Compounds in which hydrogen is the first listed element in the formula are named without numerical prefixes. Thus the compounds H_2S and HCl are hydrogen sulfide and hydrogen chloride, respectively.

Classification of a compound as ionic or molecular determines which set of nomenclature rules is used. For *nomenclature purposes,* binary compounds in which a metal and a nonmetal are present are considered ionic, and binary compounds that contain two nonmetals are considered covalent. Electronegativity differences are *not* used in classifying a compound as ionic or molecular for nomenclature purposes.

A few binary molecular compounds have names that are completely unrelated to the rules we have been discussing. They have common names that were coined prior to the development of systematic rules. At one time, in the early history of chemistry, all compounds had common names. With the advent of systematic nomenclature, most common names were discontinued. A few, however, have persisted and are now officially accepted. The most "famous" example is the compound H_2O, which has the systematic name hydrogen oxide, a name that is never used. The compound H_2O is *water,* a name that will never change. Table 5.2 lists other compounds for which common names are used in preference to systematic names.

CONCEPTS TO REMEMBER

Molecular compounds. Molecular compounds usually involve two or more nonmetals. The covalent bonds within molecular compounds involve electron sharing between atoms. The covalent bond results from the common attraction of the two nuclei for the shared electrons.

Bonding and nonbonding electron pairs. Bonding electrons are pairs of valence electrons that are shared between atoms in a covalent bond. Nonbonding electrons are pairs of valence electrons about an atom that are not involved in electron sharing.

Types of covalent bonds. One shared pair of electrons constitutes a single covalent bond. Two or three pairs of electrons may be shared between atoms to give double and triple covalent bonds. Most often, both atoms of the bond contribute an equal number of electrons to the bond. In a few cases, however, both electrons of a shared pair come from the same atom; this is a coordinate covalent bond.

Number of covalent bonds formed. There is a strong tendency for nonmetals to form a particular number of covalent bonds. The number of valence electrons the nonmetal has and the number of covalent bonds it forms give a sum of eight.

Molecular geometry. Molecular geometry describes the way atoms in a molecule are arranged in space relative to one another. VSEPR theory is a set of procedures used to predict molecular geometry

from a compound's Lewis structure. VSEPR theory is based on the concept that valence shell electron pairs about an atom (bonding or nonbonding) orient themselves as far away from one another as possible (to minimize repulsions).

Electronegativity. Electronegativity is a measure of the relative attraction that an atom has for the shared electrons in a bond. Electronegativity values are useful in predicting the type of bond that forms (ionic or covalent).

Bond polarity. When atoms of like electronegativity participate in a bond, the bonding electrons are equally shared and the bond is nonpolar. When atoms of differing electronegativity participate in a bond, the bonding electrons are unequally shared and the bond is polar. In a polar bond, the more electronegative atom dominates the sharing process. The greater the electronegativity difference between two bonded atoms, the greater the polarity of the bond.

Molecular polarity. Molecules as a whole can have polarity. If individual bond polarities do not cancel because of the symmetrical nature of a molecule, then the molecule as a whole is polar.

Binary molecular compound nomenclature. Names for binary molecular compounds usually contain Greek numerical prefixes that give the number of each type of atom present per molecule in addition to the names of the elements.

KEY REACTIONS AND EQUATIONS

1. Molecular geometry and central atom VSEPR electron group count (Section 5.7)

$\dfrac{\text{Four VSEPR electron groups}}{\text{none of which is nonbonding}}$ = tetrahedral geometry

$\dfrac{\text{Four VSEPR electron groups}}{\text{one of which is nonbonding}}$ = trigonal pyramidal geometry

$\dfrac{\text{Four VSEPR electron groups}}{\text{two of which are nonbonding}}$ = angular geometry

$\dfrac{\text{Three VSEPR electron groups}}{\text{none of which is nonbonding}}$ = trigonal planar geometry

$\dfrac{\text{Three VSEPR electron groups}}{\text{one of which is nonbonding}}$ = angular geometry

$\dfrac{\text{Two VSEPR electron groups}}{\text{none of which is nonbonding}}$ = linear geometry

2. Bond characterization and electronegativity difference (Section 5.10)

$\dfrac{\text{Electronegativity difference}}{\text{of 2.0 or greater}}$ = ionic bond

$\dfrac{\text{Electronegativity difference greater than 0 but less than 2.0}}{}$ = polar covalent bond

$\dfrac{\text{Electronegativity difference}}{\text{of 0}}$ = nonpolar covalent bond

---

KEY TERMS

Bond polarity (5.10)
Bonding electrons (5.2)
Coordinate covalent bond (5.5)
Double covalent bond (5.3)
Electronegativity (5.9)
Molecular geometry (5.8)

Molecular polarity (5.11)
Nonbonding electrons (5.2)
Nonpolar covalent bond (5.10)
Nonpolar molecule (5.11)
Polar covalent bond (5.10)
Polar molecule (5.11)

Single covalent bond (5.3)
Triple covalent bond (5.3)
VSEPR electron group (5.8)
VSEPR theory (5.8)

EXERCISES AND PROBLEMS

The members of each pair of problems in this section test similar material.

■ **The Covalent Bond (Sections 5.1 through 5.5)**

5.1 Draw Lewis structures to illustrate the covalent bonding in the following diatomic molecules.
a. Br_2 b. HI   c. IBr   d. BrF

5.2 Draw Lewis structures to illustrate the covalent bonding in the following diatomic molecules.
a. I_2 b. HF   c. IF   d. F_2

5.3 How many nonbonding electron pairs are present in each of the following Lewis structures?

a. $:N:::N:$

b. $H:\overset{..}{\underset{..}{O}}:H$

c. $H:H$

d. $:\overset{..}{\underset{..}{O}}::\overset{..}{O}:\overset{..}{\underset{..}{O}}:$

5.4 How many nonbonding electron pairs are present in each of the following Lewis structures?
a. $:C:::O:$

b. $H:C:::N:$

c. $:\overset{..}{\underset{..}{O}}::C::\overset{..}{\underset{..}{O}}:$

d. $:\overset{..}{\underset{..}{Cl}}:\overset{..}{\underset{..}{Cl}}:$

5.5 Specify the number of single, double, and triple covalent bonds present in molecules represented by the following Lewis structures.

a. $:N:::N:$

b. $H:\overset{..}{\underset{..}{O}}:\overset{..}{\underset{..}{O}}:H$

c. $H:C:H$
$\quad\;\; :\overset{..}{\underset{..}{O}}:$

d. $H:C::C:H$
$\quad\;\;\; H \;\; H$

5.6 Specify the number of single, double, and triple covalent bonds present in molecules represented by the following Lewis structures.

a. $:C:::O:$

b. $H:\overset{}{N}:\overset{}{N}:H$
$\qquad H \;\; H$

c. $:\overset{..}{\underset{..}{O}}:\overset{}{S}:\overset{..}{\underset{..}{O}}:$
$\quad\;\; :\overset{..}{\underset{..}{O}}:$

d. $H:C:::C:H$

5.7 Convert each of the Lewis structures in Problem 5.5 into the form in which lines are used to denote shared electron pairs. Include nonbonding electron pairs in the rewritten structures.

5.8 Convert each of the Lewis structures in Problem 5.6 into the form in which lines are used to denote shared electron pairs. Include nonbonding electron pairs in the rewritten structures.

5.9 What would be the predicted formula for the simplest molecular compound formed between the following pairs of elements?
a. Nitrogen and fluorine   b. Chlorine and oxygen
c. Hydrogen and sulfur   d. Carbon and hydrogen

5.10 What would be the predicted formula for the simplest molecular compound formed between the following pairs of elements?
a. Nitrogen and hydrogen   b. Oxygen and fluorine
c. Sulfur and bromine   d. Carbon and chlorine

5.11 Identify the Period 2 nonmetal that would normally be expected to exhibit each of the following bonding capabilities.
a. Forms three single bonds
b. Forms two double bonds
c. Forms one single bond and one double bond
d. Forms two single bonds and one double bond

5.12 Identify the Period 3 nonmetal that would normally be expected to exhibit each of the following bonding capabilities.
a. Forms one triple bond
b. Forms one single bond and one triple bond
c. Forms four single bonds
d. Forms one double bond

5.13 What aspect of the following Lewis structure gives you a "hint" that the concept of coordinate covalency is needed to explain the bonding in the molecule?

$:C:::O:$

5.14 What aspect of the following Lewis structure gives you a "hint" that the concept of coordinate covalency is needed to explain the bonding in the molecule?

$:N:::N:\overset{..}{\underset{..}{O}}:$

■ **Drawing Lewis Structures (Section 5.6)**

5.15 Without actually drawing the Lewis structure, determine the total number of "dots" present in the Lewis structure of each of the following molecules. That is, determine the total number of valence electrons available for bonding in each of the molecules.
a. Cl_2O b. H_2S c. NH_3 d. SO_3

5.16 Without actually drawing the Lewis structure, determine the total number of "dots" present in the Lewis structure of each of the following molecules. That is, determine the total number of valence electrons available for bonding in each of the molecules.
a. PCl_3 b. H_2O_2 c. SF_2 d. HCl

5.17 Draw Lewis structures to illustrate the covalent bonding in the following polyatomic molecules. The first atom in each formula is the central atom to which all other atoms are bonded.
a. PH_3 b. PCl_3 c. $SiBr_4$ d. OF_2

5.18 Draw Lewis structures to illustrate the covalent bonding in the following polyatomic molecules. The first atom in each formula is the central atom to which all other atoms are bonded.
a. AsH_3 b. $AsCl_3$ c. CBr_4 d. SCl_2

5.19 Draw Lewis structures for the simplest molecular compound likely to form between these pairs of elements.
a. Sulfur and fluorine   b. Carbon and iodine
c. Nitrogen and bromine   d. Selenium and hydrogen

5.20 Draw Lewis structures for the simplest molecular compound likely to form between these pairs of elements.
a. Nitrogen and chlorine   b. Bromine and hydrogen
c. Phosphorus and fluorine   d. Selenium and bromine

5.21 Draw Lewis structures to illustrate the bonding in the following molecules. In each case, there will be at least one multiple bond present in a molecule.
a. C_3H_4: A central carbon atom has two other carbon atoms bonded to it. Each of the noncentral carbon atoms also has two hydrogen atoms bonded to it.
b. N_2F_2: The two nitrogen atoms are bonded to one another, and each nitrogen atom also has a fluorine atom bonded to it.
c. C_2H_3N: The two carbon atoms are bonded to each other. One of the carbon atoms has a nitrogen atom bonded to it, and the other carbon atom has three hydrogen atoms bonded to it.
d. C_3H_4: A central carbon atom has two other carbon atoms bonded to it. One of the noncentral carbon atoms also has one hydrogen atom bonded to it, and the other one has three hydrogen atoms bonded to it.

5.22 Draw Lewis structures to illustrate the bonding in the following molecules. In each case, there will be at least one multiple bond present in a molecule.
a. $COCl_2$: Both chlorine atoms and the oxygen atom are bonded to the carbon atom.
b. $C_2H_2Br_2$: The two carbon atoms are bonded to one another. Each carbon atom also has a bromine atom and a hydrogen atom bonded to it.
c. C_2N_2: The two carbon atoms are bonded to one another, and each carbon atom also has a nitrogen bonded to it.
d. CH_2N_2: A central carbon atom has both nitrogen atoms bonded to it. Both hydrogen atoms are bonded to one of the two nitrogen atoms.

■ **Lewis Structures for Polyatomic Ions (Section 5.7)**

5.23 Draw Lewis structures for the following polyatomic ions.
a. OH^- b. BeH_4^{2-}
c. $AlCl_4^-$ d. NO_3^-

5.24 Draw Lewis structures for the following polyatomic ions.
a. CN^- b. PF_4^+
c. BH_4^- d. ClO_3^-

5.25 Draw Lewis structures for the following compounds that contain polyatomic ions.
a. NaCN   b. K_3PO_4

5.26 Draw Lewis structures for the following compounds that contain polyatomic ions.
a. KOH   b. NH_4Br

■ **Molecular Geometry (VSEPR Theory) (Section 5.8)**

5.27 Using VSEPR theory, predict whether each of the following triatomic molecules is linear or angular (bent).

a. H : S̈ : H   b. H : Ö : C̈l :

c. : Ö : : Ö : Ö :   d. : N̈ : : N : : Ö :

5.28 Using VSEPR theory, predict whether each of the following triatomic molecules is linear or angular (bent).

a. : H : C : : : N :   b. : N̈ : : S̈ : F̈ :

c. : F̈ : S̈ : F̈ :   d. : C̈l : Ö : C̈l :

5.29 Using VSEPR theory, predict the molecular geometry of the following molecules.

a. : F̈ : N̈ : F̈ :   b. : C̈l : C : C̈l :
      : F̈ :   : Ö

c.   : Ö :   d.   H
   : C̈l : P : C̈l :   : C̈l : C : C̈l :
     : C̈l :   : C̈l :

5.30 Using VSEPR theory, predict the molecular geometry of the following molecules.

a. H : P̈ : H   b. : C̈l : C : : Ö :
     : C̈l :   H

c.   H   d.   : F̈ :
   : C̈l : C : C̈l :   H : Si : H
      H   : C̈l :

5.31 Using VSEPR theory, predict the molecular geometry of the following molecules.
a. NCl_3 b. $SiCl_4$ c. H_2Se d. SBr_2

5.32 Using VSEPR theory, predict the molecular geometry of the following molecules.
a. HOBr   b. H_2Te c. NBr_3 d. SiF_4

5.33 Using VSEPR theory, predict the molecular geometry of the following molecules.
a. H : C : : C : H   b.   H
    H   H   H : C : Ö : H
                 H

5.34 Using VSEPR theory, predict the molecular geometry of the following molecules.

a.   : Ö :   b.   H Ö :
   : Ö : : N : Ö : H   H : C : C : H
                 H

■ **Electronegativity (Section 5.9)**

5.35 Using a periodic table, but not a table of electronegativity values, arrange each of the following sets of atoms in order of increasing electronegativity.
a. Na, Al, P, Mg   b. Cl, Br, I, F
c. S, P, O, Al   d. Ca, Mg, O, C

5.36 Using a periodic table, but not a table of electronegativity values, arrange each of the following sets of atoms in order of increasing electronegativity.
a. Be, N, O, B   b. Li, C, B, K
c. S, Te, Cl, Se   d. S, Mg, K, Ca

5.37 Use the information in Figure 5.11 as a basis for answering the following questions.
   a. Which elements have electronegativity values that exceed that of the element carbon?
   b. Which elements have electronegativity values of 1.0 or less?
   c. What are the four most electronegative elements listed in Figure 5.11?
   d. By what constant amount do the electronegativity values for sequential Period 2 elements differ?

5.38 Use the information in Figure 5.11 as a basis for answering the following questions.
   a. Which elements have electronegativity values that exceed that of the element sulfur?
   b. What are the four least electronegative elements listed in Figure 5.11?
   c. Which three elements in Figure 5.11 have numerically equal electronegativities?
   d. How does the electronegativity of the element hydrogen compare to that of the Period 2 elements?

■ **Bond Polarity (Section 5.10)**

5.39 Place δ^+ above the atom that is relatively positive and δ^- above the atom that is relatively negative in each of the following bonds. Try to answer this question without referring to Figure 5.11.
   a. B—N   b. Cl—F   c. N—C   d. F—O

5.40 Place δ^+ above the atom that is relatively positive and δ^- above the atom that is relatively negative in each of the following bonds. Try to answer this question without referring to Figure 5.11.
   a. Cl—Br   b. Al—S   c. Br—S   d. O—N

5.41 Rank the following bonds in order of increasing polarity.
   a. H—Cl, H—O, H—Br   b. O—F, P—O, Al—O
   c. H—Cl, Br—Br, B—N   d. P—N, S—O, Br—F

5.42 Rank the following bonds in order of increasing polarity.
   a. H—Br, H—Cl, H—S   b. N—O, Be—N, N—F
   c. N—P, P—P, P—S   d. B—Si, Br—I, C—H

5.43 Classify each of the following bonds as nonpolar covalent, polar covalent, or ionic on the basis of electronegativity differences.
   a. C—O   b. Na—Cl   c. C—I   d. Ca—S

5.44 Classify each of the following bonds as nonpolar covalent, polar covalent, or ionic on the basis of electronegativity differences.
   a. Cl—F   b. P—H   c. C—H   d. Ca—O

■ **Molecular Polarity (Section 5.11)**

5.45 Indicate whether each of the following hypothetical triatomic molecules is polar or nonpolar. Assume that A, X, and Y have different electronegativities.
   a. A linear X—A—X molecule
   b. A linear X—X—A molecule
   c. An angular A—X—Y molecule
   d. An angular X—A—Y molecule

5.46 Indicate whether each of the following hypothetical triatomic molecules is polar or nonpolar. Assume that A, X, and Y have different electronegativities.
   a. A linear X—A—Y molecule
   b. A linear A—Y—A molecule
   c. An angular X—A—X molecule
   d. An angular X—X—X molecule

5.47 Indicate whether each of the following triatomic molecules is *polar* or *nonpolar.* The molecular geometry is given in parentheses.
   a. CS_2 (linear with C in the center position)
   b. H_2Se (angular with Se in the center position)
   c. FNO (angular with N in the center position)
   d. N_2O (linear with N in the center position)

5.48 Indicate whether each of the following triatomic molecules is *polar* or *nonpolar.* The molecular geometry is given in parentheses.
   a. SCl_2 (linear with S in the center position)
   b. OF_2 (angular with O in the center position)
   c. SO_2 (angular with S in the center position)
   d. O_3 (angular with O in the center position)

5.49 Indicate whether each of the following molecules is polar or nonpolar. The molecular geometry is given in parentheses.
   a. NCl_3 (trigonal pyramid with N at the apex)
   b. H_2Se (angular with Se in the center position)
   c. CS_2 (linear with C in the center position)
   d. $CHCl_3$ (tetrahedral with C in the center position)

5.50 Indicate whether each of the following molecules is polar or nonpolar. The molecular geometry is given in parentheses.
   a. PH_2Cl (trigonal pyramid with P at the apex)
   b. SO_3 (trigonal planar with S in the center position)
   c. CH_2Cl_2 (tetrahedral with C in the center position)
   d. CCl_4 (tetrahedral with C in the center position)

■ **Naming Binary Molecular Compounds (Section 5.12)**

5.51 Name the following binary molecular compounds.
   a. SF_4 b. P_4O_6 c. ClO_2 d. H_2S

5.52 Name the following binary molecular compounds.
   a. Cl_2O b. CO   c. PI_3 d. HI

5.53 Write formulas for the following binary molecular compounds.
   a. Iodine monochloride   b. Dinitrogen monoxide
   c. Nitrogen trichloride   d. Hydrogen bromide

5.54 Write formulas for the following binary molecular compounds.
   a. Bromine monochloride   b. Tetrasulfur dinitride
   c. Sulfur trioxide   d. Dioxygen difluoride

5.55 Write formulas for the following binary molecular compounds.
   a. Hydrogen peroxide   b. Methane
   c. Ammonia   d. Phosphine

5.56 Write formulas for the following binary molecular compounds.
   a. Ethane   b. Water
   c. Hydrazine   d. Arsine

┌──────────────────────┐
│ **ADDITIONAL PROBLEMS** │
└──────────────────────┘

5.57 How many electron dots should appear in the Lewis structure for each of the following molecules or polyatomic ions?
   a. O_2F_2 b. $C_2H_2Br_2$
   c. S_2^{2-} d. NH_4^+

5.58 In which of the following pairs of diatomic species do both members of the pair have bonds of the same multiplicity (single, double, triple)?
   a. HCl and HF   b. S_2 and Cl_2
   c. CO and NO^+ d. OH^- and HS^-

5.59 Specify the reason why each of the following Lewis structures is incorrect using the following choices: (1) not enough electron dots, (2) too many electron dots, or (3) improper placement of a correct number of electron dots.

a. $: O \equiv\kern-0.3em= O :$

b. $H : \overset{\cdot\cdot}{\underset{\cdot\cdot}{O}} : Cl$

c. $H : \overset{\cdot\cdot}{O} ::: \overset{\cdot\cdot}{O} : H$

d. $\left[: \overset{\cdot\cdot}{N} :: \overset{\cdot\cdot}{O} : \right]^{+}$

5.60 Specify both the electron pair geometry about the central atom and the molecular geometry for each of the following species.

a. SiH_4
b. NH_4^+
c. $ClNO$
d. NO_3^-

5.61 Classify each of the following molecules as polar or nonpolar, or indicate that no such classification is possible because of insufficient information.

a. a molecule in which all bonds are polar
b. a molecule in which all bonds are nonpolar
c. a molecule with two bonds, both of which are polar
d. a molecule with two bonds, one that is polar and one that is nonpolar

5.62 Indicate which molecule in each of the following pairs of molecules is *more* polar.

a. BrCl and BrI
b. CO_2 and SO_2
c. SO_3 and NF_3
d. H_3CF and Cl_3CF

5.63 Four hypothetical elements, A, B, C, and D, have electronegativities A = 3.8, B = 3.3, C = 2.8, and D = 1.3. These elements form the compounds BA, DA, DB, and CA. Arrange these compounds in order of increasing *ionic* bond character.

5.64 Successive substitution of F atoms for H atoms in the molecule CH_4 produces the molecules CH_3F, CH_2F_2, CHF_3, and CF_4.

a. Draw Lewis structures for each of the five molecules.
b. Using VSEPR theory, predict the geometry of each of the five molecules.
c. Give the polarity (polar or nonpolar) of each of the five molecules

5.65 Name each of the following binary compounds. (*Caution:* At least one of the compounds is ionic.)

a. NaCl
b. BrCl
c. K_2S
d. Cl_2O

ANSWERS TO PRACTICE EXERCISES

5.1

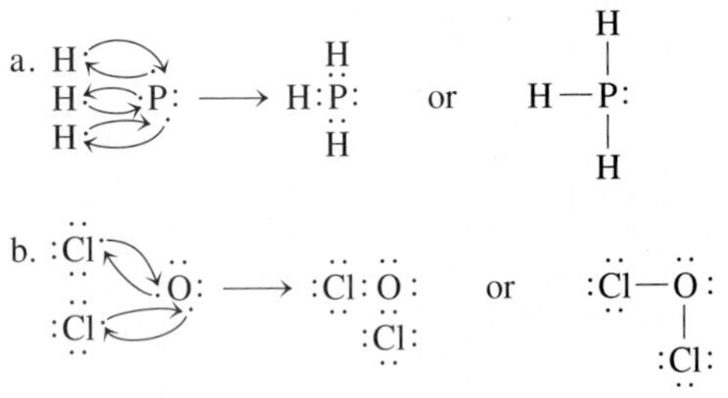

a.

b.

5.2
a. $: \overset{\cdot\cdot}{\underset{\cdot\cdot}{Cl}} : Si : \overset{\cdot\cdot}{\underset{\cdot\cdot}{Cl}} :$ with $:\overset{\cdot\cdot}{\underset{\cdot\cdot}{Cl}}:$ above and below

b. $H : \overset{\overset{\cdot\cdot}{O}}{\underset{}{C}} : H$

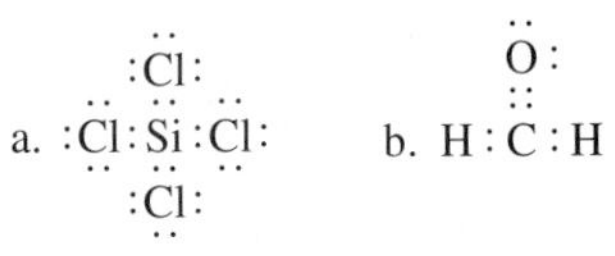

5.3
$\left[: \overset{\cdot\cdot}{O} : Br : \overset{\cdot\cdot}{O} : \atop : \overset{\cdot\cdot}{O} : \right]^{-}$

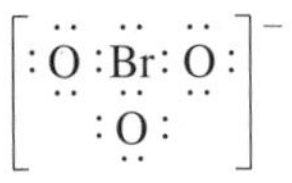

5.4
a. $H{-}H < N{-}S < F{-}Cl < K{-}Cl < Na{-}F$;
b. $H{-}H \quad N{-}S \quad F{-}Cl \quad K{-}Cl \quad Na{-}F$;
c. Nonpolar covalent: $H{-}H$
   Polar covalent: $N{-}S, F{-}Cl$
   Ionic: $K{-}Cl, Na{-}F$

5.5
a. phosphorus trifluoride
b. sulfur dioxide
c. tetraphosphorus decasulfide
d. silicon tetrachloride

6 | Chemical Calculations: Formula Masses, Moles, and Chemical Equations

The energy associated with a lightning discharge causes many different chemical reactions to occur within the atmosphere. In this chapter, we learn how to write chemical equations to describe such chemical reactions.

In this chapter we discuss "chemical arithmetic," the quantitative relationships between elements and compounds. Anyone who deals with chemical processes needs to understand at least the simpler aspects of this topic. All chemical processes, regardless of where they occur—in the human body, at a steel mill, on top of the kitchen stove, or in a clinical laboratory setting—are governed by the same mathematical rules.

We have already presented some information about chemical formulas (Section 1.10). In this chapter we discuss formulas again, and here we look beyond describing the composition of compounds in terms of constituent atoms. A new unit, the mole, will be introduced and its usefulness discussed. Chemical equations will be considered for the first time. We will learn how to represent chemical reactions by using chemical equations and how to derive quantitative relationships from these equations.

6.1 Formula Masses

▶ Many chemists use the term *molecular mass* interchangeably with *formula mass* when dealing with substances that contain discrete molecules. It is incorrect, however, to use the term *molecular mass* when dealing with ionic compounds, because such compounds do not have molecules as their basic structural unit (Section 4.8).

Our entry point into the realm of "chemical arithmetic" is a discussion of the quantity called formula mass. A **formula mass** *is the sum of the atomic masses of all the atoms represented in the chemical formula of a substance.* Formula masses, like the atomic masses from which they are calculated, are relative masses based on the $^{12}_{6}C$ relative-mass scale (Section 3.3). Example 6.1 illustrates how formula masses are calculated.

■ Calculate the formula mass of each of the following substances.

a. SnF_2 (tin(II) fluoride, a toothpaste additive)

b. $Al(OH)_3$ (aluminum hydroxide, a water purification chemical)

Solution

Formula masses are obtained simply by adding the atomic masses of the constituent elements, counting each atomic mass as many times as the symbol for the element occurs in the formula.

a. A formula unit of SnF_2 contains three atoms: one atom of Sn and two atoms of F. The formula mass, the collective mass of these three atoms, is calculated as follows:

$$1 \text{ atom Sn} \times \left(\frac{118.71 \text{ amu}}{1 \text{ atom Sn}} \right) = 118.71 \text{ amu}$$

$$2 \text{ atoms F} \times \left(\frac{19.00 \text{ amu}}{1 \text{ atom F}} \right) = \underline{38.00 \text{ amu}}$$

$$\text{Formula mass} = 156.71 \text{ amu}$$

We derive the conversion factors in the calculation from the atomic masses listed on the inside front cover of the text. Our rules for the use of conversion factors are the same as those discussed in Section 2.7.

Conversion factors are usually not explicitly shown in a formula mass calculation, as they are in the preceding calculation; the calculation is simplified as follows:

Sn:	1×118.71 amu $=$	118.71 amu
F:	2×19.00 amu $=$	$\underline{38.00 \text{ amu}}$
	Formula mass $=$	156.71 amu

b. The formula for this compound contains parentheses. Improper interpretation of parentheses (see Section 4.11) is a common error made by students doing formula mass calculations. In the formula $Al(OH)_3$, the subscript 3 outside the parentheses affects both of the symbols inside the parentheses. Thus we have

Al:	1×26.98 amu $=$	26.98 amu
O:	3×16.00 amu $=$	48.00 amu
H:	3×1.01 amu $=$	$\underline{3.03 \text{ amu}}$
	Formula mass $=$	78.01 amu

In this text, we will always use atomic masses rounded to the hundredths place, as we have done in this example. This rule allows us to use, without rounding, the atomic masses given inside the front cover of the text. A benefit of this approach is that we always use the same atomic mass for a given element and thus become familiar with the atomic masses of the common elements.

Practice Exercise 6.1

Calculate the formula mass of each of the following substances.

a. $Na_2S_2O_3$ (sodium thiosulfate, a photographic chemical)

b. $(NH_2)_2CO$ (urea, a chemical fertilizer for crops)

FIGURE 6.1 Oranges may be bought in units of mass (4-lb bag) or units of amount (3 oranges).

6.2 The Mole: A Counting Unit for Chemists

The quantity of material in a sample of a substance can be specified either in terms of units of *mass* or in terms of units of *amount*. Mass is specified in terms of units such as grams, kilograms, and pounds. The amount of a substance is specified by indicating the number of objects present—3, 17, or 437, for instance.

FIGURE 6.2 A basic process in chemical laboratory work is determining the mass of a substance.

▶ How large is the number 6.02×10^{23}? It would take an ultramodern computer that can count 100 million times a second 190 million years to count 6.02×10^{23} times. If each of the 6 billion people on Earth were made a millionaire (receiving 1 million dollar bills), we would still need 100 million other worlds, each inhabited with the same number of millionaires, in order to have 6.02×10^{23} dollar bills in circulation.

▶ Why the number 6.02×10^{23}, rather than some other number, was chosen as the counting unit of chemists is discussed in Section 6.3. A more formal definition of the mole will also be presented in that section.

We all use both units of mass and units of amount on a daily basis. For example, when buying oranges at the grocery store, we can decide on quantity in either mass units (4-lb bag or 10-lb bag) or amount units (3 oranges or 8 oranges) (Figure 6.1). In chemistry, as in everyday life, both mass and amount methods of specifying quantity are used. In laboratory work, practicality dictates working with quantities of known mass (Figure 6.2). Counting out a given number of atoms for a laboratory experiment is impossible because we cannot see individual atoms.

When we perform chemical calculations after the laboratory work has been done, it is often useful and even necessary to think of the quantities of substances present in terms of numbers of atoms or molecules instead of mass. When this is done, very large numbers are always encountered. Any macroscopic-sized sample of a chemical substance contains many trillions of atoms or molecules.

In order to cope with this large-number problem, chemists have found it convenient to use a special unit when counting atoms and molecules. Specialized counting units are used in many areas—for example, a *dozen* eggs or a *ream* (500 sheets) of paper (Figure 6.3).

The chemist's counting unit is the *mole*. What is unusual about the mole is its magnitude. A **mole** *is* 6.02×10^{23} *objects*. The extremely large size of the mole unit is necessitated by the extremely small size of atoms and molecules. To the chemist, *one mole* always means 6.02×10^{23} objects, just as *one dozen* always means 12 objects. Two moles of objects is two times 6.02×10^{23} objects, and five moles of objects is five times 6.02×10^{23} objects.

Avogadro's number *is the name given to the numerical value* 6.02×10^{23}. This designation honors Amedeo Avogadro (Figure 6.4), an Italian physicist whose pioneering work on gases later proved valuable in determining the number of particles present in given volumes of substances. When we solve problems dealing with the number of objects (atoms or molecules) present in a given number of moles of a substance, Avogadro's number becomes part of the conversion factor used to relate the number of objects present to the number of moles present.

From the definition

$$1 \text{ mole} = 6.02 \times 10^{23} \text{ objects}$$

two conversion factors can be derived:

$$\frac{6.02 \times 10^{23} \text{ objects}}{1 \text{ mole}} \quad \text{and} \quad \frac{1 \text{ mole}}{6.02 \times 10^{23} \text{ objects}}$$

Example 6.2 illustrates the use of these conversion factors in solving problems.

E X A M P L E   6 . 2

Calculating the Number of Objects in a Molar Quantity

■ How many objects are there in each of the following quantities?

a. 0.23 mole of aspirin molecules   **b.** 1.6 moles of oxygen atoms

Solution

Dimensional analysis (Section 2.7) will be used to solve each of these problems. Both of the problems are similar in that we are given a certain number of moles of substance and want to find the number of objects present in the given number of moles. We will need Avogadro's number to solve each of these moles-to-particles problems.

$$\boxed{\begin{array}{c} \text{Moles of} \\ \text{Substance} \end{array}} \xrightarrow[\text{involving Avogadro's number}]{\text{Conversion factor}} \boxed{\begin{array}{c} \text{Particles of} \\ \text{Substance} \end{array}}$$

a. The objects of concern are molecules of aspirin. The given quantity is 0.23 mole of aspirin molecules, and the desired quantity is the number of aspirin molecules.

$$0.23 \text{ mole aspirin molecules} = ? \text{ aspirin molecules}$$

(continued)

FIGURE 6.3 Everyday counting units—a dozen, a pair, and a ream.

FIGURE 6.4 Amedeo Avogadro (1776–1856) was the first scientist to distinguish between atoms and molecules. His name is associated with the number 6.02×10^{23}, the number of particles (atoms or molecules) in a mole.

▶ The mass value below each symbol in the periodic table is both an atomic mass in atomic mass units and a molar mass in grams. For example, the mass of one nitrogen atom is 14.01 amu, and the mass of 1 mole of nitrogen atoms is 14.01 g.

Applying dimensional analysis here involves the use of a single conversion factor, one that relates moles and molecules.

$$0.23 \text{ mole aspirin molecules} \times \left(\frac{6.02 \times 10^{23} \text{ aspirin molecules}}{1 \text{ mole aspirin molecules}} \right)$$
$$= 1.4 \times 10^{23} \text{ aspirin molecules}$$

b. This time we are dealing with atoms instead of molecules. This switch does not change the way we work the problem. We will need the same conversion factor.

The given quantity is 1.6 moles of oxygen atoms, and the desired quantity is the actual number of oxygen atoms present.

$$1.6 \text{ moles oxygen atoms} = ? \text{ oxygen atoms}$$

The setup is

$$1.6 \text{ moles oxygen atoms} \times \left(\frac{6.02 \times 10^{23} \text{ oxygen atoms}}{1 \text{ mole oxygen atoms}} \right) = 9.6 \times 10^{23} \text{ oxygen atoms}$$

Practice Exercise 6.2

How many objects are there in each of the following quantities?

a. 0.46 mole of vitamin C molecules   **b.** 1.27 moles of copper atoms

6.3 The Mass of a Mole

How much does a mole weigh? Are you uncertain about the answer to that question? Let us consider a similar but more familiar question first: "How much does a dozen weigh?" Your response is now immediate: "A dozen what?" The mass of a dozen identical objects obviously depends on the identity of the object. For example, the mass of a dozen elephants is greater than the mass of a dozen peanuts. The mass of a mole, like the mass of a dozen, depends on the identity of the object. Thus the mass of a mole, or *molar mass*, is not a set number; it varies and is different for each chemical substance (see Figure 6.5). This is in direct contrast to the *molar number*, Avogadro's number, which is the same for all chemical substances.

The **molar mass** *is the mass, in grams, of a substance that is numerically equal to the substance's formula mass.* For example, the formula mass (atomic mass) of the element sodium is 22.99 amu; therefore, 1 mole of sodium weighs 22.99 g. In Example 6.1, we calculated that the formula mass of tin(II) fluoride is 156.71 amu; therefore, 1 mole of tin(II) fluoride weighs 156.71 g. We can obtain the actual mass in grams of 1 mole of any substance by computing its formula mass (atomic mass for elements) and writing "grams" after it. Thus, when we add atomic masses to get the formula mass (in amu's) of a compound, we are simultaneously finding the mass of 1 mole of that compound (in grams).

It is not a coincidence that the molar mass of a substance and its formula mass or atomic mass match numerically. Avogadro's number has the value that it has in order to cause this relationship to exist. The numerical match between molar mass and atomic or formula mass makes calculating the mass of any given number of moles of a substance a very simple procedure. When you solve problems of this type, the numerical value of the molar mass becomes part of the conversion factor used to convert from moles to grams.

For example, for the compound CO_2, which has a formula mass of 44.01 amu, we can write the equality

$$44.01 \text{ g } CO_2 = 1 \text{ mole } CO_2$$

FIGURE 6.5 The mass of a mole is not a set number of grams; it depends on the substance. For the substances shown, the mass of 1 mole (clockwise from sulfur, the yellow solid) is as follows: sulfur, 32.07 g; zinc, 65.38 g; carbon, 12.0 g; magnesium, 24.30 g; lead, 207.2 g; silicon, 28.09 g; copper, 63.55 g; and, in the center, mercury, 200.6 g.

From this statement (equality), two conversion factors can be written:

$$\frac{44.01 \text{ g CO}_2}{1 \text{ mole CO}_2} \quad \text{and} \quad \frac{1 \text{ mole CO}_2}{44.01 \text{ g CO}_2}$$

Example 6.3 illustrates the use of gram-to-mole conversion factors like these in solving problems.

EXAMPLE 6.3

Calculating the Mass of a Molar Quantity of Compound

▶ Molar masses are conversion factors between grams and moles for any substance. Because the periodic table is the usual source of the atomic masses needed to calculate molar masses, the periodic table can be considered to be a useful source of conversion factors.

■ Acetaminophen, the pain-killing ingredient in Tylenol formulations, has the formula $C_8H_9O_2N$. Calculate the mass, in grams, of a 0.30-mole sample of this pain reliever.

Solution

We will use dimensional analysis to solve this problem. The relationship between molar mass and formula mass will serve as a conversion factor in the setup of this problem.

Moles of Substance	Conversion factor involving molar mass →	Grams of Substance

The given quantity is 0.30 mole of $C_8H_9O_2N$, and the desired quantity is grams of this same substance.

$$0.30 \text{ mole } C_8H_9O_2N = ? \text{ grams } C_8H_9O_2N$$

The calculated formula mass of $C_8H_9O_2N$ is 151.18 amu. Thus,

$$151.18 \text{ grams } C_8H_9O_2N = 1 \text{ mole } C_8H_9O_2N$$

With this relationship in the form of a conversion factor, the setup for the problem becomes

$$0.30 \text{ mole } C_8H_9O_2N \times \left(\frac{151.18 \text{ g } C_8H_9O_2N}{1 \text{ mole } C_8H_9O_2N} \right) = 45 \text{ g } C_8H_9O_2N$$

Practice Exercise 6.3

Carbon monoxide (CO) is an air pollutant that enters the atmosphere primarily in automobile exhaust. Calculate the mass in grams of a 2.61-mole sample of this air pollutant.

and 2 will divide into evenly. In order to obtain six atoms of chlorine on each side of the equation, we place the coefficient 3 in front of Cl_2 and the coefficient 2 in front of $FeCl_3$.

$$FeI_2 + ③Cl_2 \longrightarrow ②FeCl_3 + I_2$$

We now have six chlorine atoms on each side of the equation.

$$3Cl_2: \qquad 3 \times 2 = 6$$
$$2FeCl_3: \qquad 2 \times 3 = 6$$

Step 2: *Now pick a second element to balance.* We will balance the iron next. The number of iron atoms on the right side has already been set at 2 by the coefficient previously placed in front of $FeCl_3$. We will need two iron atoms on the reactant side of the equation instead of the one iron atom now present. This is accomplished by placing the coefficient 2 in front of FeI_2.

$$②FeI_2 + 3Cl_2 \longrightarrow 2FeCl_3 + I_2$$

It is always wise to pick, as the second element to balance, one whose amount is already set on one side of the equation by a previously determined coefficient. If we had chosen iodine as the second element to balance instead of iron, we would have run into problems. Because the coefficient for neither FeI_2 nor I_2 had been determined, we would have had no guidelines for deciding on the amount of iodine needed.

Step 3: *Now pick a third element to balance.* Only one element is left to balance—iodine. The number of iodine atoms on the left side of the equation is already set at four ($2FeI_2$). In order to obtain four iodine atoms on the right side of the equation, we place the coefficient 2 in front of I_2.

$$2FeI_2 + 3Cl_2 \longrightarrow 2FeCl_3 + ②I_2$$

The addition of the coefficient 2 in front of I_2 completes the balancing process; all the coefficients have been determined.

Step 4: *As a final check on the correctness of the balancing procedure, count atoms on each side of the equation.* The following table can be constructed from our balanced equation.

$$2FeI_2 + 3Cl_2 \longrightarrow 2FeCl_3 + 2I_2$$

Atom	Left side	Right side
Fe	$2 \times 1 = 2$	$2 \times 1 = 2$
I	$2 \times 2 = 4$	$2 \times 2 = 4$
Cl	$3 \times 2 = 6$	$2 \times 3 = 6$

All elements are in balance: two iron atoms on each side, four iodine atoms on each side, and six chlorine atoms on each side (see Figure 6.8).

FIGURE 6.8 When 16.90 g of the compound CaS (left photo) is decomposed into its constituent elements, the Ca and S produced (right photo) has an identical mass of 16.90 grams. Because atoms are neither created nor destroyed in a chemical reaction, the masses of reactants and products in a chemical reaction are always equal.

$$CaS \longrightarrow Ca + S$$

Note that the elements chlorine and iodine in the preceding equation are written in the form of diatomic molecules (Cl_2 and I_2). This is in accordance with the guideline given at the start of this section on the use of molecular formulas for elements that are gases at room temperature.

In Example 6.7 we will balance another chemical equation.

EXAMPLE 6.7 **Balancing a Chemical Equation**	■ Balance the following chemical equation. $$C_2H_6O + O_2 \longrightarrow CO_2 + H_2O$$

Solution

The element oxygen appears in four different places in this equation. This means we do not want to start the balancing process with the element oxygen. Always start the balancing process with an element that appears only once on both the reactant and product sides of the equation.

Step 1: *Balancing of H atoms.* There are six H atoms on the left and two H atoms on the right. Placing the coefficient 1 in front of C_2H_6O and the coefficient 3 in front of H_2O balances the H atoms at six on each side.

$$1C_2H_6O + O_2 \longrightarrow CO_2 + 3H_2O$$

Step 2: *Balancing of C atoms.* An effect of balancing the H atoms at six (Step 1) is the setting of the C atoms on the left side at two. Placing the coefficient 2 in front of CO_2 causes the carbon atoms to balance at two on each side of the equation.

$$1C_2H_6O + O_2 \longrightarrow 2CO_2 + 3H_2O$$

Step 3: *Balancing of O atoms.* The oxygen content of the right side of the equation is set at seven atoms: four oxygen atoms from $2CO_2$ and three oxygen atoms from $3H_2O$. To obtain seven oxygen atoms on the left side of the equation, we place the coefficient 3 in front of O_2; $3O_2$ gives six oxygen atoms, and there is an additional O in $1C_2H_6O$. The element oxygen is present in all four formulas in the equation.

$$1C_2H_6O + 3O_2 \longrightarrow 2CO_2 + 3H_2O$$

Step 4: *Final check.* The equation is balanced. There are two carbon atoms, six hydrogen atoms, and seven oxygen atoms on each side of the equation.

$$C_2H_6O + 3O_2 \longrightarrow 2CO_2 + 3H_2O$$

Practice Exercise 6.7

Balance the following chemical equation.

$$C_4H_{10}O + O_2 \longrightarrow CO_2 + H_2O$$

Some additional comments and guidelines concerning equations in general, and the process of balancing in particular, are given here.

1. The coefficients in a balanced equation are always the *smallest set of whole numbers* that will balance the equation. We mention this because more than one set of coefficients will balance an equation. Consider the following three equations:

$$2H_2 + O_2 \longrightarrow 2H_2O$$
$$4H_2 + 2O_2 \longrightarrow 4H_2O$$
$$8H_2 + 4O_2 \longrightarrow 8H_2O$$

All three of these equations are mathematically correct; there are equal numbers of hydrogen and oxygen atoms on both sides of the equation. However, the first equation is considered the correct form because the coefficients used there are the smallest set of whole numbers that will balance the equation. The coefficients in the second equation are two times those in the first equation, and the third equation has coefficients that are four times those of the first equation.

2. At this point, you are not expected to be able to write down the products for a chemical reaction when given the reactants. After learning how to balance equations, students sometimes get the mistaken idea that they ought to be able to write down equations from scratch. This is not so. You will need more chemical knowledge before attempting this task. At this stage, you should be able to balance simple equations, given *all* of the reactants and *all* of the products.

3. It is often useful to know the physical state of the substances involved in a chemical reaction. We specify physical state by using the symbols (*s*) for solid, (*l*) for liquid, (*g*) for gas, and (*aq*) for aqueous solution (a substance dissolved in water). Two examples of such symbol use in chemical equations are

$$2Fe_2O_3(s) + 3C(s) \longrightarrow 4Fe(s) + 3CO_2(g)$$
$$2HNO_3(aq) + 3H_2S(aq) \longrightarrow 2NO(g) + 3S(s) + 4H_2O(l)$$

6.7 Chemical Equations and the Mole Concept

The coefficients in a balanced chemical equation, like the subscripts in a chemical formula (Section 6.5), have two levels of interpretation—a microscopic level of meaning and a macroscopic level of meaning. The microscopic level of interpretation was used in the previous two sections. *The coefficients in a balanced equation give the numerical relationships among formula units consumed or produced in the chemical reaction.* Interpreted at the microscopic level, the equation

$$N_2 + 3H_2 \longrightarrow 2NH_3$$

conveys the information that one molecule of N_2 reacts with three molecules of H_2 to produce two molecules of NH_3.

At the macroscopic level of interpretation, chemical equations are used to relate mole-sized quantities of reactants and products to each other. At this level, *the coefficients in a balanced chemical equation give the fixed molar ratios between substances consumed or produced in the chemical reaction.* Interpreted at the macroscopic level, the equation

$$N_2 + 3H_2 \longrightarrow 2NH_3$$

conveys the information that 1 mole of N_2 reacts with 3 moles of H_2 to produce 2 moles of NH_3.

The coefficients in a balanced chemical equation can be used to generate conversion factors to be used in solving problems. Several pairs of conversion factors are obtainable from a single balanced equation. Consider the following balanced equation:

$$4Fe + 3O_2 \longrightarrow 2Fe_2O_3$$

Three mole-to-mole relationships are obtainable from this equation:

> ► Conversion factors that relate two different substances to one another are valid only for systems governed by the chemical equation from which they were obtained.

4 moles of Fe produces 2 moles of Fe_2O_3.

3 moles of O_2 produces 2 moles of Fe_2O_3.

4 moles of Fe reacts with 3 moles of O_2.

From each of these macroscopic-level relationships, two conversion factors can be written. The conversion factors for the first relationship are

$$\left(\frac{4 \text{ moles Fe}}{2 \text{ moles Fe}_2O_3} \right) \quad \text{and} \quad \left(\frac{2 \text{ moles Fe}_2O_3}{4 \text{ moles Fe}} \right)$$

All balanced chemical equations are the source of numerous conversion factors. The more reactants and products there are in the equation, the greater the number of derivable conversion factors. The next section details how conversion factors such as those in the preceding illustration are used in solving problems.

Relationships Involving the Mole Concept

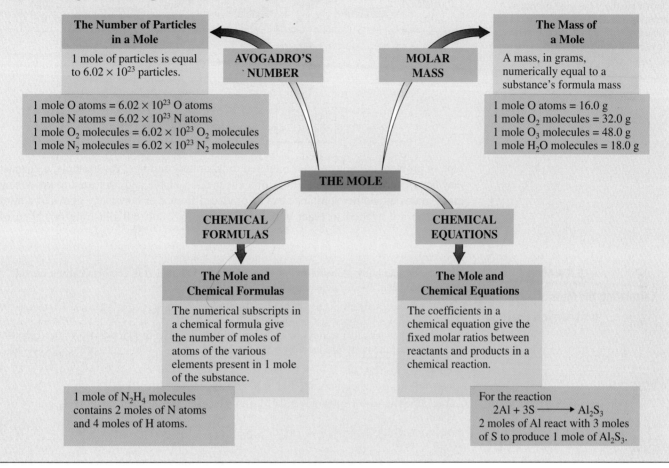

The Number of Particles in a Mole

1 mole of particles is equal to 6.02×10^{23} particles.

1 mole O atoms = 6.02×10^{23} O atoms
1 mole N atoms = 6.02×10^{23} N atoms
1 mole O_2 molecules = 6.02×10^{23} O_2 molecules
1 mole N_2 molecules = 6.02×10^{23} N_2 molecules

AVOGADRO'S NUMBER

MOLAR MASS

The Mass of a Mole

A mass, in grams, numerically equal to a substance's formula mass

1 mole O atoms = 16.0 g
1 mole O_2 molecules = 32.0 g
1 mole O_3 molecules = 48.0 g
1 mole H_2O molecules = 18.0 g

THE MOLE

CHEMICAL FORMULAS

CHEMICAL EQUATIONS

The Mole and Chemical Formulas

The numerical subscripts in a chemical formula give the number of moles of atoms of the various elements present in 1 mole of the substance.

1 mole of N_2H_4 molecules contains 2 moles of N atoms and 4 moles of H atoms.

The Mole and Chemical Equations

The coefficients in a chemical equation give the fixed molar ratios between reactants and products in a chemical reaction.

For the reaction
$$2Al + 3S \longrightarrow Al_2S_3$$
2 moles of Al react with 3 moles of S to produce 1 mole of Al_2S_3.

The Chemistry at a Glance feature on this page reviews the relationships that involve the mole.

6.8 Chemical Calculations Using Chemical Equations

When the information contained in a chemical equation is combined with the concepts of molar mass (Section 6.3) and Avogadro's number (Section 6.2), several useful types of chemical calculations can be carried out. A typical chemical-equation-based calculation gives information about one reactant or product of a reaction (number of grams, moles, or particles) and requests similar information about another reactant or product of the same reaction. The substances involved in such a calculation may both be reactants or products or may be a reactant and a product.

The conversion factor relationships needed to solve problems of this general type are given in Figure 6.9. This diagram should seem very familiar to you; it is almost identical to Figure 6.7, which you used in solving problems based on chemical formulas. There is only one difference between the two diagrams. In Figure 6.7, the subscripts in a chemical formula are listed as the basis for relating "moles of A" to "moles of B." In Figure 6.9, the same two quantities are related by using the coefficients of a balanced chemical equation.

The most common type of chemical-equation-based calculation is a "grams of A" to "grams of B" problem. In this type of problem, the mass of one substance involved in a chemical reaction (either reactant or product) is given, and information is requested about the mass of another substance involved in the reaction (either reactant or product). This

▶ The quantitative study of the relationships among reactants and products in a chemical reaction is called *chemical stoichiometry.* The word *stoichiometry,* pronounced stoy-key-om-eh-tree, is derived from the Greek *stoicheion* ("element") and *metron* ("measure"). The stoichiometry of a chemical reaction always involves the *molar relationships* between reactants and products and thus is given by the coefficients in the balanced equation for the chemical reaction.

FIGURE 6.9 In solving chemical-equation-based problems, the only "transitions" allowed are those between quantities (boxes) connected by arrows. Associated with each arrow is the concept on which the required conversion factor is based.

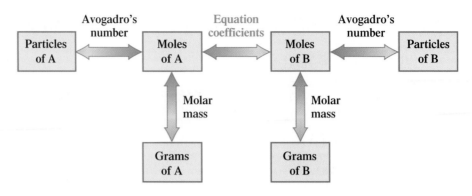

type of problem is frequently encountered in laboratory settings. For example, a chemist may have a certain number of grams of a chemical available and may want to know how many grams of another substance can be produced from it or how many grams of a third substance are needed to react with it. Examples 6.8 and 6.9 illustrate this type of problem.

EXAMPLE 6.8

Calculating the Mass of a Product in a Chemical Reaction

■ The human body converts the glucose, $C_6H_{12}O_6$, contained in foods to carbon dioxide, CO_2, and water, H_2O. The equation for the chemical reaction is

$$C_6H_{12}O_6 + 6O_2 \longrightarrow 6CO_2 + 6H_2O$$

Assume a person eats a candy bar containing 14.2 g (1/2 oz) of glucose. How many grams of water will the body produce from the ingested glucose, assuming all of the glucose undergoes reaction?

Solution

Step 1: The given quantity is 14.2 g of glucose. The desired quantity is grams of water.

$$14.2 \text{ g } C_6H_{12}O_6 = ? \text{ g } H_2O$$

In terms of Figure 6.9, this is a "grams of A" to "grams of B" problem.

Step 2: Using Figure 6.9 as a road map, we determine that the pathway for this problem is

$$\boxed{\text{Grams of A}} \xrightarrow[\text{mass}]{\text{Molar}} \boxed{\text{Moles of A}} \xrightarrow[\text{coefficients}]{\text{Equation}} \boxed{\text{Moles of B}} \xrightarrow[\text{mass}]{\text{Molar}} \boxed{\text{Grams of B}}$$

The mathematical setup for this problem is

$$14.2 \text{ g } C_6H_{12}O_6 \times \left(\frac{1 \text{ mole } C_6H_{12}O_6}{180.18 \text{ g } C_6H_{12}O_6} \right) \times \left(\frac{6 \text{ moles } H_2O}{1 \text{ mole } C_6H_{12}O_6} \right) \times \left(\frac{18.02 \text{ g } H_2O}{1 \text{ mole } H_2O} \right)$$

$$\text{g } C_6H_{12}O_6 \longrightarrow \text{moles } C_6H_{12}O_6 \longrightarrow \text{moles } H_2O \longrightarrow \text{g } H_2O$$

The 180.18 g in the first conversion factor is the molar mass of glucose, the 6 and 1 in the second conversion factor are the coefficients, respectively, of H_2O and $C_6H_{12}O_6$ in the balanced chemical equation, and the 18.02 g in the third conversion factor is the molar mass of H_2O.

Step 3: The solution to the problem, obtained by doing the arithmetic after all the numerical factors have been collected, is

$$\left(\frac{14.2 \times 1 \times 6 \times 18.02}{180.18 \times 1 \times 1} \right) \text{ g } H_2O = 8.52 \text{ g } H_2O$$

Practice Exercise 6.8

Silicon carbide, SiC, which is used as an abrasive on sandpaper, is prepared using the chemical reaction

$$SiO_2 + 3C \longrightarrow SiC + 2CO$$

How many grams of SiC can be produced from 15.0 g of C?

CHEMICAL CONNECTIONS

"Laboratory-Sized Amounts" versus "Industrial Chemistry-Sized Amounts"

The various calculations in this chapter can be considered to be "laboratory-based" calculations. Chemical substance amounts are always specified in grams, the common laboratory unit for mass. These gram-sized laboratory amounts are very small, almost "infinitesimal," when compared with industrial production figures for various "high-volume" chemicals, which are specified in terms of billions of pounds per year. About 50 of the millions of compounds known are produced in amounts exceeding 1 billion pounds per year in the United States.

The number-one chemical in the United States, in terms of production amount, is sulfuric acid (H_2SO_4), with an annual production approaching 100 billion pounds. Its production amount is almost twice that of any other chemical. So important is sulfuric acid production in the United States (and the world) that some economists use sulfuric acid production as a measure of a nation's industrial strength.

Why is so much sulfuric acid produced in the United States? What are its uses? What are its properties? Where do we encounter it in our everyday life?

Pure sulfuric acid is a colorless, corrosive, oily liquid. It is usually marketed as a concentrated (96% by mass) aqueous solution. People rarely have direct contact with this strong acid because it is seldom part of finished consumer products. The closest encounter most people have with the acid (other than in a chemical laboratory) is involvement with automobile batteries. The acid in a standard automobile battery is a 38%-by-mass aqueous solution of sulfuric acid. However, less than 1% of annual sulfuric acid production ends up in car batteries.

Approximately two-thirds of sulfuric acid production is used in the manufacture of chemical fertilizers. These fertilizer compounds are an absolute necessity if the food needs of an ever-increasing population are to be met. The connection between sulfuric acid and fertilizer revolves around the element phosphorus, which is necessary for plant growth. The starting material for phosphate fertilizer production is phosphate rock, a highly insoluble material containing calcium phosphate, $Ca_3(PO_4)_2$. The treatment of phosphate rock with H_2SO_4 results in the formation of phosphoric acid, H_3PO_4.

$$Ca_3(PO_4)_2 + 3H_2SO_4 \longrightarrow 3CaSO_4 + 2H_3PO_4$$

The phosphoric acid so produced is then used to produce soluble phosphate compounds that plants can use as a source of phosphorus. The major phosphoric acid fertilizer derivative is ammonium hydrogen phosphate—$(NH_4)_2HPO_4$.

The raw materials needed to produce sulfuric acid are simple: sulfur, air, and water. In the first step of production, elemental sulfur is burned to give sulfur dioxide gas.

$$S + O_2 \longrightarrow SO_2$$

Some SO_2 is also obtainable as a by-product of metallurgical operations associated with zinc and copper production. Next, the SO_2 gas is combined with additional O_2 (air) to produce sulfur trioxide gas.

$$2SO_2 + O_2 \longrightarrow 2SO_3$$

The SO_3 is then dissolved in water, which yields sulfuric acid as the product.

$$SO_3 + H_2O \longrightarrow H_2SO_4$$

Reactions similar to the last two steps in commercial H_2SO_4 production can also occur naturally in the atmosphere. The H_2SO_4 so produced is a major contributor to the phenomenon called acid rain (see the Chemical Connection feature on page 244 in Chapter 10).

EXAMPLE 6.9

Calculating the Mass of a Reactant Taking Part in a Chemical Reaction

■ The active ingredient in many commercial antacids is magnesium hydroxide, $Mg(OH)_2$, which reacts with stomach acid (HCl) to produce magnesium chloride ($MgCl_2$) and water. The equation for the reaction is

$$Mg(OH)_2 + 2HCl \longrightarrow MgCl_2 + 2H_2O$$

How many grams of $Mg(OH)_2$ are needed to react with 0.30 g of HCl?

(continued)

Solution

Step 1: This problem, like Example 6.8, is a "grams of A" to "grams of B" problem. It differs from the previous problem in that both the given and the desired quantities involve reactants.

$$0.30 \text{ g HCl} \longrightarrow ? \text{ g Mg(OH)}_2$$

Step 2: The pathway used to solve it will be the same as in Example 6.8.

$$\boxed{\text{Grams of A}} \xrightarrow[\text{mass}]{\text{Molar}} \boxed{\text{Moles of A}} \xrightarrow[\text{coefficients}]{\text{Equation}} \boxed{\text{Moles of B}} \xrightarrow[\text{mass}]{\text{Molar}} \boxed{\text{Grams of B}}$$

The dimensional-analysis setup is

$$0.30 \text{ g HCl} \times \left(\frac{1 \text{ mole HCl}}{36.46 \text{ g HCl}} \right) \times \left(\frac{1 \text{ mole Mg(OH)}_2}{2 \text{ moles HCl}} \right) \times \left(\frac{58.32 \text{ g Mg(OH)}_2}{1 \text{ mole Mg(OH)}_2} \right)$$

$$\text{g HCl} \longrightarrow \text{moles HCl} \longrightarrow \text{moles Mg(OH)}_2 \longrightarrow \text{g Mg(OH)}_2$$

The balanced chemical equation for the reaction is used as the bridge that enables us to go from HCl to $Mg(OH)_2$. The numbers in the second conversion factor are coefficients from this equation.

Step 3: The solution obtained by combining all of the numbers in the manner indicated in the setup is

$$\left(\frac{0.30 \times 1 \times 1 \times 58.32}{36.46 \times 2 \times 1} \right) \text{ g Mg(OH)}_2 = 0.24 \text{ g Mg(OH)}_2$$

To put our answer in perspective, we note that a common brand of antacid tablets has tablets containing 0.10 g of $Mg(OH)_2$.

Practice Exercise 6.9

The chemical equation for the photosynthesis reaction in plants is

$$6CO_2 + 6H_2O \longrightarrow C_6H_{12}O_6 + 6O_2$$

How many grams of H_2O are consumed at the same time that 20.0 g of CO_2 is consumed?

"Grams of A" to "grams of B" problems (Examples 6.8 and 6.9) are not the only type of problem for which the coefficients in a balanced equation can be used to relate the quantities of two substances. As a further example of the use of equation coefficients in problem solving, consider Example 6.10 (a "particles of A" to "moles of B" problem).

EXAMPLE 6.10

Calculating the Amount of a Substance Taking Part in a Chemical Reaction

■ Automotive airbags inflate when sodium azide, NaN_3, rapidly decomposes to its constituent elements. The equation for the chemical reaction is

$$2NaN_3(s) \longrightarrow 2Na(s) + 3N_2(g)$$

The gaseous N_2 so generated inflates the airbag (see Figure 6.10). How many moles of NaN_3 would have to decompose in order to generate 253 million (2.53×10^8) molecules of N_2?

Solution

Although a calculation of this type does not have a lot of practical significance, it tests your understanding of the problem-solving relationships discussed in this section of the text.

Step 1: The given quantity is 2.53×10^8 molecules of N_2, and the desired quantity is moles of NaN_3.

$$2.53 \times 10^8 \text{ molecules N}_2 = ? \text{ moles NaN}_3$$

In terms of Figure 6.9, this is a "particles of A" to "moles of B" problem.

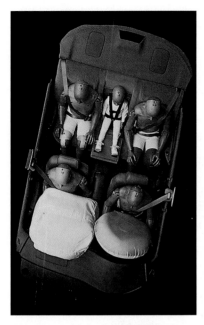

FIGURE 6.10 Testing apparatus for measuring the effects of air bag deployment.

Step 2: Using Figure 6.9 as a road map, we determine that the pathway for this problem is

$$\boxed{\text{Particles of A}} \xrightarrow[\text{number}]{\text{Avogadro's}} \boxed{\text{Moles of A}} \xrightarrow[\text{coefficients}]{\text{Equation}} \boxed{\text{Moles of B}}$$

The mathematical setup is

$$2.53 \times 10^8 \text{ molecules } N_2 \times \left(\frac{1 \text{ mole } N_2}{6.02 \times 10^{23} \text{ molecules } N_2} \right) \times \left(\frac{2 \text{ moles } NaN_3}{3 \text{ moles } N_2} \right)$$

Avogadro's number is found in the first conversion factor. The 2 and 3 in the second conversion factor are the coefficients, respectively, of NaN_3 and N_2 in the balanced chemical equation.

Step 3: The solution to the problem, obtained by doing the arithmetic after all the numerical factors have been collected, is

$$\left(\frac{2.53 \times 10^8 \times 1 \times 2}{6.02 \times 10^{23} \times 3} \right) \text{ mole } NaN_3 = 2.80 \times 10^{-16} \text{ mole } NaN_3$$

Practice Exercise 6.10

Decomposition of $KClO_3$ serves as a convenient laboratory source of small amounts of oxygen gas. The reaction is

$$2KClO_3 \longrightarrow 2KCl + 3O_2$$

How many moles of $KClO_3$ must be decomposed to produce 64 billion (6.4×10^{10}) O_2 molecules?

CONCEPTS TO REMEMBER

Formula mass. The formula mass of a substance is the sum of the atomic masses of the atoms in its chemical formula.

The mole concept. The mole is the chemist's counting unit. One mole of any substance—element or compound—consists of 6.02×10^{23} formula units of the substance. Avogadro's number is the name given to the numerical value 6.02×10^{23}.

Molar mass. The molar mass of a substance is the mass in grams that is numerically equal to the substance's formula mass. Molar mass is not a set number; it varies and is different for each chemical substance.

The mole and chemical formulas. The numerical subscripts in a chemical formula give the number of moles of atoms of the various elements present in 1 mole of the substance.

Chemical equation. A chemical equation is a written statement that uses symbols and formulas instead of words to represent how reactants undergo transformation into products in a chemical reaction.

Balanced chemical equation. A balanced chemical equation has the same number of atoms of each element involved in the reaction on each side of the equation. An unbalanced equation is brought into balance through the use of coefficients. An equation coefficient is a number that is placed to the left of the formula of a substance in an equation and that changes the amount, but not the identity, of the substance.

The mole and chemical equations. The coefficients in a balanced chemical equation give the molar ratios between substances consumed or produced in the chemical reaction described by the equation.

KEY REACTIONS AND EQUATIONS

1. Calculation of formula mass (Section 6.1)

    Formula mass = sum of atomic masses of all components
2. The mole (Section 6.3)

    1 mole = 6.02×10^{23} objects
3. Avogadro's number (Section 6.3)

    Avogadro's number = 6.02×10^{23} objects

4. Mass of a mole (Section 6.4)

    Molar mass = mass, in grams, numerically equal to a substance's formula mass
5. Balanced chemical equation (Section 6.7)

    Balanced chemical equation = same number of atoms of each kind on each side of the equation

---

KEY TERMS

EXERCISES AND PROBLEMS

The members of each pair of problems in this section test similar material.

■ **Formula Masses (Section 6.1)**

6.1 Calculate, to two decimal places, the formula mass of each of the following substances. Obtain the needed atomic masses from the inside front cover of the text.
   a. $C_{12}H_{22}O_{11}$ (sucrose, table sugar)
   b. C_7H_{16} (heptane, a component of gasoline)
   c. $C_7H_5NO_3S$ (saccharin, an artificial sweetener)
   d. $(NH_4)_2SO_4$ (ammonium sulfate, a lawn fertilizer)

6.2 Calculate, to two decimal places, the formula mass of each of the following substances. Obtain the needed atomic masses from the inside front cover of the text.
   a. $C_{20}H_{30}O$ (vitamin A)
   b. $C_{14}H_9Cl_5$ (DDT, formerly used as an insecticide)
   c. $C_8H_{10}N_4O_2$ (caffeine, a central nervous system stimulant)
   d. $Ca(NO_3)_2$ (calcium nitrate, gives fireworks their red color)

■ **The Mole as a Counting Unit (Section 6.2)**

6.3 Indicate the number of objects present in each of the following molar quantities.
   a. Number of apples in 1.00 mole of apples
   b. Number of elephants in 1.00 mole of elephants
   c. Number of atoms in 1.00 mole of Zn atoms
   d. Number of molecules in 1.00 mole of CO_2 molecules

6.4 Indicate the number of objects present in each of the following molar quantities.
   a. Number of oranges in 1.00 mole of oranges
   b. Number of camels in 1.00 mole of camels
   c. Number of atoms in 1.00 mole of Cu atoms
   d. Number of molecules in 1.00 mole of CO molecules

6.5 How many atoms are present in the following molar quantities of various elements?
   a. 1.50 moles Fe   b. 1.50 moles Ni
   c. 1.50 moles C   d. 1.50 moles Ne

6.6 How many atoms are present in the following molar quantities of various elements?
   a. 1.20 moles Au   b. 1.20 moles Ag
   c. 1.20 moles Be   d. 1.20 moles Si

6.7 Select the quantity that contains the greater number of atoms in each of the following pairs of substances.
   a. 0.100 mole C atoms of 0.200 mole Al atoms
   b. Avogadro's number of C atoms or 0.750 mole Al atoms
   c. 6.02×10^{23} C atoms or 1.50 moles Al atoms
   d. 6.50×10^{23} C atoms or Avogadro's number of Al atoms

6.8 Select the quantity that contains the greater number of atoms in each of the following pairs of substances.
   a. 0.100 mole N or 0.300 mole P atoms
   b. 6.18×10^{23} N atoms or Avogadro's number of P atoms
   c. Avogadro's number of N atoms or 1.20 moles of P atoms
   d. 6.18×10^{23} N atoms of 2.00 moles P atoms

■ **Molar Mass (Section 6.3)**

6.9 How much, in grams, does 1.00 mole of each of the following substances weigh?
   a. CO (carbon monoxide)   b. CO_2 (carbon dioxide)
   c. NaCl (table salt)   d. $C_{12}H_{22}O_{11}$ (table sugar)

6.10 How much, in grams, does 1.00 mole of each of the following substances weigh?
   a. H_2O (water)
   b. H_2O_2 (hydrogen peroxide)
   c. NaCN (sodium cyanide)
   d. KCN (potassium cyanide)

6.11 What is the mass, in grams, of each of the following quantities of matter?
   a. 0.034 mole of gold atoms
   b. 0.034 mole of silver atoms
   c. 3.00 moles of oxygen atoms
   d. 3.00 moles of oxygen molecules (O_2)

6.12 What is the mass, in grams, of each of the following quantities of matter?
   a. 0.85 mole of copper atoms
   b. 0.85 mole of nickel atoms
   c. 2.50 moles of nitrogen atoms
   d. 2.50 moles of nitrogen molecules (N_2)

6.13 How many moles are present in a sample of each of the following substances if each sample weighs 5.00 g?
   a. CO molecules   b. CO_2 molecules
   c. B_4H_{10} molecules   d. U atoms

6.14 How many moles are present in a sample of each of the following substances if each sample weighs 7.00 g?
   a. N_2O molecules   b. NO_2 molecules
   c. P_4O_{10} molecules   d. V atoms

■ **Chemical Formulas and the Mole Concept (Section 6.4)**

6.15 Write the six mole-to-mole conversion factors that can be derived from each of the following chemical formulas.
   a. H_2SO_4 b. $POCl_3$

6.16 Write the six mole-to-mole conversion factors that can be derived from each of the following chemical formulas.
   a. HNO_3 b. $C_2H_4Br_2$

6.17 How many moles of each type of atom are present in each of the following molar quantities?
   a. 2.00 moles SO_2 molecules
   b. 2.00 moles SO_3 molecules
   c. 3.00 moles NH_3 molecules
   d. 3.00 moles N_2H_4 molecules

6.18 How many moles of each type of atom are present in each of the following molar quantities?
   a. 4.00 moles NO_2 molecules
   b. 4.00 moles N_2O molecules
   c. 7.00 moles H_2O molecules
   d. 7.00 moles H_2O_2 molecules

6.19 How many *total moles* of atoms are present in each of the following molar quantities?
   a. 4.00 moles SO_3 b. 2.00 moles H_2SO_4
   c. 1.00 mole $C_{12}H_{22}O_{11}$ d. 3.00 moles $Mg(OH)_2$

6.20 How many *total moles* of atoms are present in each of the following molar quantities?
   a. 3.00 moles N_2O_4 b. 4.00 moles HNO_3
   c. 0.500 mole C_2H_6O d. 5.00 moles $(NH_4)_2S$

■ **Calculations Based on Chemical Formulas (Section 6.5)**

6.21 Determine the number of atoms in each of the following quantities of an element.
   a. 10.0 g B   b. 32.0 g Ca
   c. 2.0 g Ne   d. 7.0 g N

6.22 Determine the number of atoms in each of the following quantities of an element.
 a. 10.0 g S
 b. 39.1 g K
 c. 3.2 g U
 d. 7.0 g Be

6.23 Determine the mass, in grams, of each of the following quantities of substance.
 a. 6.02×10^{23} copper atoms
 b. 3.01×10^{23} copper atoms
 c. 557 copper atoms
 d. 1 copper atom

6.24 Determine the mass, in grams, of each of the following quantities of substance.
 a. 6.02×10^{23} silver atoms
 b. 3.01×10^{23} silver atoms
 c. 1.00×10^6 silver atoms
 d. 1 silver atom

6.25 Determine the number of moles of substance present in each of the following quantities.
 a. 10.0 g He
 b. 10.0 g N_2O
 c. 4.0×10^{10} atoms P
 d. 4.0×10^{10} atoms Be

6.26 Determine the number of moles of substance present in each of the following quantities.
 a. 25.0 g N
 b. 25.0 g Li
 c. 8.50×10^{15} atoms S
 d. 8.50×10^{15} atoms Cl

6.27 Determine the number of atoms of sulfur present in each of the following quantities.
 a. 10.0 g H_2SO_4
 b. 20.0 g SO_3
 c. 30.0 g Al_2S_3
 d. 2.00 moles S_2O

6.28 Determine the number of atoms of nitrogen present in each of the following quantities.
 a. 10.0 g N_2H_4
 b. 20.0 g HN_3
 c. 30.0 g $LiNO_3$
 d. 4.00 moles N_2O_5

6.29 Determine the number of grams of sulfur present in each of the following quantities.
 a. 3.01×10^{23} S_2O molecules
 b. 3 S_4N_4 molecules
 c. 2.00 moles SO_2 molecules
 d. 4.50 moles S_8 molecules

6.30 Determine the number of grams of oxygen present in each of the following quantities.
 a. 4.50×10^{22} SO_3 molecules
 b. 7 P_4O_{10} molecules
 c. 3.00 moles H_2SO_4 molecules
 d. 1.50 moles O_3 molecules

■ **Writing and Balancing Chemical Equations (Section 6.6)**

6.31 Indicate whether each of the following chemical equations is balanced.
 a. $SO_3 + H_2O \longrightarrow H_2SO_4$
 b. $CuO + H_2 \longrightarrow Cu + H_2O$
 c. $CS_2 + O_2 \longrightarrow CO_2 + SO_2$
 d. $AgNO_3 + KCl \longrightarrow KNO_3 + AgCl$

6.32 Indicate whether each of the following chemical equations is balanced.
 a. $H_2 + O_2 \longrightarrow H_2O$
 b. $NO + O_2 \longrightarrow NO_2$
 c. $C + O_2 \longrightarrow CO_2$
 d. $HNO_3 + NaOH \longrightarrow NaNO_3 + H_2O$

6.33 For each of the following balanced equations, indicate how many atoms of each element are present on the reactant and product sides of the chemical equation.
 a. $2N_2 + 3O_2 \longrightarrow 2N_2O_3$
 b. $4NH_3 + 6NO \longrightarrow 5N_2 + 6H_2O$
 c. $PCl_3 + 3H_2 \longrightarrow PH_3 + 3HCl$
 d. $Al_2O_3 + 6HCl \longrightarrow 2AlCl_3 + 3H_2O$

6.34 For each of the following balanced equations, indicate how many atoms of each element are present on the reactant and product sides of the chemical equation.
 a. $4Al + 3O_2 \longrightarrow 2Al_2O_3$
 b. $2Na + 2H_2O \longrightarrow 2NaOH + H_2$
 c. $2Co + 3HgCl_2 \longrightarrow 2CoCl_3 + 3Hg$
 d. $H_2SO_4 + 2NH_3 \longrightarrow (NH_4)_2SO_4$

6.35 Balance the following equations.
 a. $Na + H_2O \longrightarrow NaOH + H_2$
 b. $Na + ZnSO_4 \longrightarrow Na_2SO_4 + Zn$
 c. $NaBr + Cl_2 \longrightarrow NaCl + Br_2$
 d. $ZnS + O_2 \longrightarrow ZnO + SO_2$

6.36 Balance the following equations.
 a. $H_2S + O_2 \longrightarrow SO_2 + H_2O$
 b. $Ni + HCl \longrightarrow NiCl_2 + H_2$
 c. $IBr + NH_3 \longrightarrow NH_4Br + NI_3$
 d. $C_2H_6 + O_2 \longrightarrow CO_2 + H_2O$

6.37 Balance the following equations.
 a. $CH_4 + O_2 \longrightarrow CO_2 + H_2O$
 b. $C_6H_6 + O_2 \longrightarrow CO_2 + H_2O$
 c. $C_4H_8O_2 + O_2 \longrightarrow CO_2 + H_2O$
 d. $C_5H_{10}O + O_2 \longrightarrow CO_2 + H_2O$

6.38 Balance the following equations.
 a. $C_2H_4 + O_2 \longrightarrow CO_2 + H_2O$
 b. $C_6H_{12} + O_2 \longrightarrow CO_2 + H_2O$
 c. $C_3H_6O + O_2 \longrightarrow CO_2 + H_2O$
 d. $C_5H_{10}O_2 + O_2 \longrightarrow CO_2 + H_2O$

6.39 Balance the following equations.
 a. $PbO + NH_3 \longrightarrow Pb + N_2 + H_2O$
 b. $Fe(OH)_3 + H_2SO_4 \longrightarrow Fe_2(SO_4)_3 + H_2O$

6.40 Balance the following equations.
 a. $SO_2Cl_2 + HI \longrightarrow H_2S + H_2O + HCl + I_2$
 b. $Na_2CO_3 + Mg(NO_3)_2 \longrightarrow MgCO_3 + NaNO_3$

■ **Chemical Equations and the Mole Concept (Section 6.7)**

6.41 Write the 12 mole-to-mole conversion factors that can be derived from the following balanced equation.
$$2Ag_2CO_3 \longrightarrow 4Ag + 2CO_2 + O_2$$

6.42 Write the 12 mole-to-mole conversion factors that can be derived from the following balanced equation.
$$N_2H_4 + 2H_2O_2 \longrightarrow N_2 + 4H_2O$$

6.43 Using each of the following equations, calculate the number of moles of CO_2 that can be obtained from 2.00 moles of the first listed reactant with an excess of the other reactant.
 a. $C_7H_{16} + 11O_2 \longrightarrow 7CO_2 + 8H_2O$
 b. $2HCl + CaCO_3 \longrightarrow CaCl_2 + CO_2 + H_2O$
 c. $Na_2SO_4 + 2C \longrightarrow Na_2S + 2CO_2$
 d. $Fe_3O_4 + CO \longrightarrow 3FeO + CO_2$

6.44 Using each of the following equations, calculate the number of moles of CO_2 that can be obtained from 3.50 moles of the first listed reactant with an excess of the other reactant.
 a. $FeO + CO \longrightarrow Fe + CO_2$
 b. $3O_2 + CS_2 \longrightarrow CO_2 + 2SO_2$
 c. $2C_8H_{18} + 25O_2 \longrightarrow 16CO_2 + 18H_2O$
 d. $C_6H_{12}O_6 + 6O_2 \longrightarrow 6CO_2 + 6H_2O$

■ **Calculations Based on Chemical Equations (Section 6.8)**

6.45 How many grams of the first reactant in each of the following equations would be needed to produce 20.0 g of N_2 gas?
 a. $4NH_3 + 3O_2 \longrightarrow 2N_2 + 6H_2O$
 b. $(NH_4)_2Cr_2O_7 \longrightarrow N_2 + 4H_2O + Cr_2O_3$
 c. $N_2H_4 + 2H_2O_2 \longrightarrow N_2 + 4H_2O$
 d. $2NH_3 \longrightarrow N_2 + 3H_2$

6.46 How many grams of the first reactant in each of the following equations would be needed to produce 20.0 g of H_2O?
a. $N_2H_4 + 2H_2O_2 \longrightarrow N_2 + 4H_2O$
b. $H_2O_2 + H_2S \longrightarrow 2H_2O + S$
c. $2HNO_3 + NO \longrightarrow 3NO_2 + H_2O$
d. $3H_2 + WO_3 \longrightarrow W + 3H_2O$

6.47 The principal constituent of natural gas is methane, which burns in air according to the reaction

$$CH_4 + 2O_2 \longrightarrow CO_2 + 2H_2O$$

How many grams of O_2 are needed to produce 3.50 g of CO_2?

6.48 Tungsten (W) metal, which is used to make incandescent bulb filaments, is produced by the reaction

$$WO_3 + 3H_2 \longrightarrow 3H_2O + W$$

How many grams of H_2 are needed to produce 1.00 g of W?

6.49 The catalytic converter that is now standard equipment on American automobiles converts carbon monoxide (CO) to carbon dioxide (CO_2) by the reaction

$$2CO + O_2 \longrightarrow 2CO_2$$

What mass of O_2, in grams, is needed to react completely with 25.0 g of CO?

6.50 A mixture of hydrazine (N_2H_4) and hydrogen peroxide (H_2O_2) is used as a fuel for rocket engines. These two substances react as shown by the equation

$$N_2H_4 + 2H_2O_2 \longrightarrow N_2 + 4H_2O$$

What mass of N_2H_4, in grams, is needed to react completely with 35.0 g of H_2O_2?

6.51 Both water and sulfur dioxide are products from the reaction of sulfuric acid (H_2SO_4) with copper metal, as shown by the equation

$$2H_2SO_4 + Cu \longrightarrow SO_2 + 2H_2O + CuSO_4$$

How many grams of H_2O will be produced at the same time that 10.0 g of SO_2 is produced?

6.52 Potassium thiosulfate ($K_2S_2O_3$) is used to remove any excess chlorine from fibers and fabrics that have been bleached with that gas. The reaction is

$$K_2S_2O_3 + 4Cl_2 + 5H_2O \longrightarrow 2KHSO_4 + 8HCl$$

How many grams of HCl will be produced at the same time that 25.0 g of $KHSO_4$ is produced?

ADDITIONAL PROBLEMS

6.53 The compound 1-propanethiol, which is the eye irritant that is released when fresh onions are chopped up, has a formula mass of 76.18 amu and the formula C_3H_yS. What number does y stand for in the formula?

6.54 Select the quantity that has the greater number of atoms in each of the following pairs of quantities. Make your selection using the periodic table but without performing an actual calculation.
a. 1.00 mole S or 1.00 mole S_8
b. 28.0 g Al or 1.00 mole Al
c. 28.1 g Si or 30.0 g Mg
d. 2.00 g Na or 6.02×10^{23} atoms He

6.55 What amount or mass of each of the following substances would be needed to obtain 1.000 g of Si?
a. moles of SiH_4 b. grams of SiO_2
c. molecules of $(CH_3)_3SiCl$ d. atoms of Si

6.56 How many grams of Si would contain the same number of atoms as there are in 2.10 moles of Ar?

6.57 After the following equation was balanced, the name of one of the reactants was substituted for its formula.

$$2 \text{ butyne} + 11O_2 \longrightarrow 8CO_2 + 6H_2O$$

Using only the information found within the equation, determine the molecular formula of butyne.

6.58 Ammonium dichromate decomposes according to the following reaction.

$$(NH_4)_2Cr_2O_7 \longrightarrow N_2 + 4H_2O + Cr_2O_3$$

How many grams of each of the products can be formed from the decomposition of 75.0 g of ammonium dichromate?

6.59 Black silver sulfide can be produced from the reaction of silver metal with sulfur.

$$2Ag + S \longrightarrow Ag_2S$$

How many grams of Ag and how many grams of S are needed to produce 125 g of Ag_2S?

6.60 How many grams of beryllium (Be) are needed to react completely with 45.0 g of nitrogen (N_2) in the synthesis of Be_3N_2?

ANSWERS TO PRACTICE EXERCISES

6.1 a. 158.12 amu   b. 60.07 amu
6.2 a. 2.8×10^{23} vitamin C molecules
      b. 7.65×10^{23} copper atoms
6.3 73.1 g CO
6.4 2.28 moles C, 4.56 moles H, and 2.28 moles O
6.5 4.07×10^{21} formula units Li_2CO_3

6.6 0.026 g O
6.7 $C_4H_{10}O + 6O_2 \longrightarrow 4CO_2 + 5H_2O$
6.8 16.7 g SiC
6.9 8.19 g H_2O
6.10 7.1×10^{-14} mole $KClO_3$

Gases, Liquids, and Solids

Ice, water, and mist are simultaneously present in this winter scene in Yellowstone National Park.

In Chapters 3, 4, and 5, we considered the structure of matter from a submicroscopic point of view—in terms of molecules, atoms, protons, neutrons, and electrons. In this chapter, we are concerned with the macroscopic characteristics of matter as represented by the physical states—solid, liquid, and gas. Of particular concern are the properties exhibited by matter in the various physical states and a theory that correlates these properties with molecular behavior.

7.1 The Kinetic Molecular Theory of Matter

Solids, liquids, and gases (Section 1.2) are easily distinguished by using four common physical properties of matter (Table 7.1): (1) volume and shape, (2) density, (3) compressibility, and (4) thermal expansion. We discussed the property of density in Section 2.8. **Compressibility** *is a measure of the change in volume of a sample of matter resulting from a pressure change.* **Thermal expansion** *is a measure of the change in volume of a sample of matter resulting from a temperature change.* These distinguishing characteristics are compared in Table 7.1 for the three states of matter. The physical characteristics of the solid, liquid, and gaseous states listed in Table 7.1 can be explained by kinetic molecular theory, which is one of the fundamental theories of chemistry. The **kinetic molecular theory of matter** *is a set of five statements that are used to explain the physical behavior of the three states of matter (solids, liquids, and gases).* The basic idea of

FIGURE 7.1 The water in the lake behind the dam has potential energy as a result of its position. When the water flows over the dam, its potential energy becomes kinetic energy that can be used to turn the turbines of a hydroelectric plant.

▶ The word *kinetic* comes from the Greek *kinesis,* which means "movement." The kinetic molecular theory deals with the movement of particles.

▶ The energy released when gasoline is burned represents potential energy associated with chemical bonds.

▶ For gases, the attractions between particles (statement 3) are minimal and as a first approximation are considered to be zero (see Section 7.2).

▶ Two consequences of the elasticity of particle collisions (statement 5) are that (1) the energy of any given particle is continually changing, and (2) particle energies for a system are not all the same; a range of particle energies is always encountered.

this theory is that the particles (atoms, molecules, or ions) present in a substance, independent of the physical state of the substance, are always in motion.

The five statements of the kinetic molecular theory of matter follow.

Statement 1:   *Matter is ultimately composed of tiny particles (atoms, molecules, or ions) that have definite and characteristic sizes that do not change.*

Statement 2:   *The particles are in constant random motion and therefore possess kinetic energy.*

Kinetic energy *is energy that matter possesses because of particle motion.* An object that is in motion has the ability to transfer its kinetic energy to another object upon collision with that object.

Statement 3:   *The particles interact with one another through attractions and repulsions and therefore possess potential energy.*

Potential energy *is stored energy that matter possesses as a result of its position, condition, and/or composition* (Figure 7.1). The potential energy of greatest importance when considering the differences among the three states of matter is that which originates from electrostatic interactions between particles. An **electrostatic interaction** *is an attraction or repulsion that occurs between charged particles.* Particles of opposite charge (one positive and the other negative) attract one another, and particles of like charge (both positive or both negative) repel one another.

Statement 4:   *The velocity of the particles increases as the temperature is increased.*

The *average* velocity (kinetic energy) of all particles in a system depends on the temperature; velocity increases as temperature increases.

Statement 5:   *The particles in a system transfer energy to each other through elastic collisions.*

In an elastic collision, the total kinetic energy remains constant; no kinetic energy is lost. The difference between an *elastic* and an *inelastic* collision is illustrated by comparing the collision of two hard steel spheres with the collision of two masses of putty. The collision of spheres approximates an elastic collision (the spheres bounce off one another and continue moving, as in Figure 7.2); the putty collision has none of these characteristics (the masses "glob" together with no resulting movement).

The differences among the solid, liquid, and gaseous states of matter can be explained by the relative magnitudes of kinetic energy and potential energy (in this case, electrostatic attractions) associated with the physical state. Kinetic energy can be considered a *disruptive force* that tends to make the particles of a system increasingly independent of one another. This is because the particles tend to move away from one another as a result of the energy of motion. Potential energy of attraction can be considered a *cohesive force* that tends to cause order and stability among the particles of a system.

TABLE 7.1

Distinguishing Properties of Solids, Liquids, and Gases

Property	Solid state	Liquid state	Gaseous state
volume and shape	definite volume and definite shape	definite volume and indefinite shape; takes the shape of container to the extent that it is filled	indefinite volume and indefinite shape; takes the volume and shape of the container that it completely fills
density	high	high, but usually lower than corresponding solid	low
compressibility	small	small, but usually greater than corresponding solid	large
thermal expansion	very small: about 0.01% per °C	small: about 0.10% per °C	moderate: about 0.30% per °C

FIGURE 7.2 Upon release, the steel ball on the left transmits its kinetic energy through a series of elastic collisions to the ball on the right.

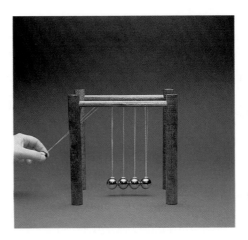

How much kinetic energy a chemical system has depends on its temperature. Kinetic energy increases as temperature increases (statement 4 of the kinetic molecular theory of matter). Thus the higher the temperature, the greater the magnitude of disruptive influences within a chemical system. Potential energy magnitude, or cohesive force magnitude, is essentially independent of temperature. The fact that one of the types of forces depends on temperature (disruptive forces) and the other does not (cohesive forces) causes temperature to be the factor that determines in which of the three physical states a given sample of matter is found. We will discuss the reasons for this in Section 7.2.

7.2 Kinetic Molecular Theory and Physical States

A **solid** *is the physical state characterized by a dominance of potential energy (cohesive forces) over kinetic energy (disruptive forces).* The particles in a solid are drawn close together in a regular pattern by the strong cohesive forces present (Figure 7.3a). Each particle occupies a fixed position, about which it vibrates because of disruptive kinetic energy. With this model, the characteristic properties of solids (Table 7.1) can be explained as follows:

1. *Definite volume and definite shape.* The strong, cohesive forces hold the particles in essentially fixed positions, resulting in definite volume and definite shape.
2. *High density.* The constituent particles of solids are located as close together as possible (touching each other). Therefore, a given volume contains large numbers of particles, resulting in a high density.
3. *Small compressibility.* Because there is very little space between particles, increased pressure cannot push the particles any closer together; therefore, it has little effect on the solid's volume.

FIGURE 7.3 (a) In a solid, the particles (atoms, molecules, or ions) are close together and vibrate about fixed sites. (b) The particles in a liquid, though still close together, freely slide over one another. (c) In a gas, the particles are in constant random motion, each particle being independent of the others present.

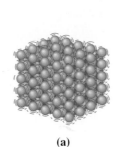

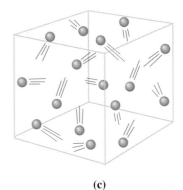

(a)   (b)   (c)

4. *Very small thermal expansion.* An increased temperature increases the kinetic energy (disruptive forces), thereby causing more vibrational motion of the particles. Each particle occupies a slightly larger volume, and the result is a slight expansion of the solid. The strong, cohesive forces prevent this effect from becoming very large.

A **liquid** *is the physical state characterized by potential energy (cohesive forces) and kinetic energy (disruptive forces) of about the same magnitude.* The liquid state consists of particles that are randomly packed but relatively near one another (Figure 7.3b). The molecules are in constant, random motion; they slide freely over one another but do not move with enough energy to separate. The fact that the particles freely slide over each other indicates the influence of disruptive forces; however, the fact that the particles do not separate indicates fairly strong cohesive forces. With this model, the characteristic properties of liquids (Table 7.1) can be explained as follows:

1. *Definite volume and indefinite shape.* The attractive forces are strong enough to restrict particles to movement within a definite volume. They are not strong enough, however, to prevent the particles from moving over each other in a random manner that is limited only by the container walls. Thus liquids have no definite shape, except that they maintain a horizontal upper surface in containers that are not completely filled.
2. *High density.* The particles in a liquid are not widely separated; they are still touching one another. Therefore, there will be a large number of particles in a given volume—a high density.
3. *Small compressibility.* Because the particles in a liquid are still touching each other, there is very little empty space. Therefore, an increase in pressure cannot squeeze the particles much closer together.
4. *Small thermal expansion.* Most of the particle movement in a liquid is vibrational because a particle can move only a short distance before colliding with a neighbor. The increased particle velocity that accompanies a temperature increase results only in increased vibrational amplitudes. The net effect is an increase in the effective volume a particle occupies, which causes a slight volume increase in the liquid.

A **gas** *is the physical state characterized by a complete dominance of kinetic energy (disruptive forces) over potential energy (cohesive forces).* As a result, the particles of a gas move essentially independently of one another, in a totally random manner (Figure 7.3c). Under ordinary pressure, the particles are relatively far apart, except when they collide with one another. In between collisions with one another or with the container walls, gas particles travel in straight lines (Figure 7.4).

The kinetic theory explanation of the properties of gases follows the same pattern that we saw earlier for solids and liquids.

1. *Indefinite volume and indefinite shape.* The attractive (cohesive) forces between particles have been overcome by high kinetic energy, and the particles are free to

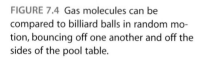

FIGURE 7.4 Gas molecules can be compared to billiard balls in random motion, bouncing off one another and off the sides of the pool table.

CHEMICAL CONNECTIONS

The Importance of Gas Densities

In the gaseous state, particles are approximately 10 times farther apart than in the solid or liquid state at a given temperature and pressure. Consequently, gases have densities much lower than those of solids and liquids. The fact that gases have low densities is a major factor in explaining many commonly encountered phenomena.

Popcorn pops because of the difference in density between liquid and gaseous water (1.0 g/mL versus 0.001 g/mL). As the corn kernels are heated, water within the kernels is converted into steam. The steam's volume, approximately 1000 times greater than that of the water from which it was generated, causes the kernels of corn to "blow up."

Changes in density that occur as a solid is converted to gases via a chemical reaction are the basis for the operation of automobile air bags and the effects of explosives. Automobile air bags are designed to inflate rapidly (in a fraction of a second) in the event of a crash and then to deflate immediately. Their activation involves mechanical shock causing a steel ball to

compress a spring that eletronically ignites a detonator cap, which in turn causes solid sodium azide (NaN_3) to decompose. The decomposition reaction is

$$2NaN_3(s) \longrightarrow 2Na(l) + 3N_2(g)$$

The nitrogen gas so generated inflates the air bag. A small amount of NaN_3 (high density) will generate over 50 L of N_2 gas at 25°C. Because the air bag is porous, it goes limp quickly as the generated N_2 gas escapes.

Millions of hours of hard manual labor are saved annually by the use of industrial explosives in quarrying rock, constructing tunnels, and mining coal and metal ores. The active ingredient in dynamite, a heavily used industrial explosive, is nitroglycerin, whose destructive power comes from the generation of large volumes of gases at high temperatures. The reaction is

$$4C_3H_5O_3(NO_2)_3(s) \longrightarrow$$
Nitroglycerin

$$12CO_2(g) + 10H_2O(g) + 6N_2(g) + O_2(g)$$

At the temperature of the explosion, about 5000°C, there is an approximately 20,000-fold increase in volume as the result of density changes. No wonder such explosives can blow materials to pieces!

The density difference associated with temperature change is the basis for the operation of hot air balloons. Hot air, which is less dense than cold air, rises.

Weather balloons and blimps are filled with helium, a gas less dense than air. Thus, such objects rise in air.

Water vapor is less dense than air. Thus, moist air is less dense than dry air. Decreasing barometric pressure (from lower-density moist air) is an indication that a storm front is approaching.

FIGURE 7.5 When a gas is compressed, the amount of empty space in the container is decreased. The size of the molecules does not change; they simply move closer together.

Gas at low pressure

Gas at higher pressure

travel in all directions. Therefore, gas particles completely fill their container, and the shape of the gas is that of the container.

2. *Low density.* The particles of a gas are widely separated. There are relatively few particles in a given volume (compared with liquids and solids), which means little mass per volume (a low density).

3. *Large compressibility.* Particles in a gas are widely separated; essentially, a gas is mostly empty space. When pressure is applied, the particles are easily pushed closer together, decreasing the amount of empty space and the volume of the gas (see Figure 7.5).

4. *Moderate thermal expansion.* An increase in temperature means an increase in particle velocity. The increased kinetic energy of the particles enables them to push back whatever barrier is confining them into a given volume, and the volume increases. You will note that the size of the particles is not changed during expansion or compression of gases, solids, or liquids; they merely move either farther apart or closer together. It is the space between the particles that changes.

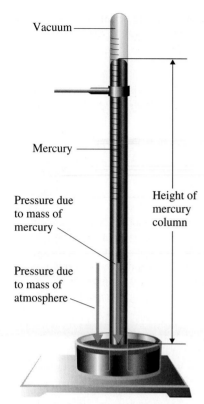

Mercury

Height of mercury column

Pressure due to mass of mercury

Pressure due to mass of atmosphere

FIGURE 7.6 The essential components of a mercury barometer are a graduated glass tube, a glass dish, and liquid mercury.

▶ "Millimeters of mercury" is the pressure unit most often encountered in clinical work in allied health fields. For example, oxygen and carbon dioxide pressures in respiration are almost always specified in millimeters of mercury.

▶ Blood pressure is measured with the aid of an apparatus known as a sphygmomanometer, which is essentially a barometer tube connected to an inflatable cuff by a hollow tube. A typical blood pressure is 120/80; this ratio means a systolic pressure of 120 mm Hg above atmospheric pressure and a diastolic pressure of 80 mm Hg above atmospheric pressure.

7.3 Gas Law Variables

The behavior of a gas can be described reasonably well by *simple* quantitative relationships called *gas laws*. A **gas law** *is a generalization that describes in mathematical terms the relationships among the amount, pressure, temperature, and volume of a gas.*

Gas laws involve four variables: amount, pressure, temperature, and volume. Three of these four variables (amount, volume, and temperature) have been previously discussed (Sections 6.2, 2.2 and 2.9, respectively). Amount is usually specified in terms of *moles* of gas present. The units *liter* and *milliliter* are generally used in specifying gas volume. Only one of the three temperature scales discussed in Section 2.9, the *Kelvin scale*, can be used in gas law calculations if the results are to be valid. We have not yet discussed pressure, the fourth gas law variable. The remainder of this section consists of a discussion of pressure. **Pressure** *is the force applied per unit area on an object—that is, the total force on a surface divided by the area of that surface.* The mathematical equation for pressure is

$$P(\text{pressure}) = \frac{F(\text{force})}{A(\text{area})}$$

For a gas, the force that creates pressure is that which is exerted by the gas molecules or atoms as they constantly collide with the walls of their container. Barometers, manometers, and gauges are the instruments most commonly used to measure gas pressures.

The air that surrounds Earth exerts pressure on every object it touches. A **barometer** *is a device used to measure atmospheric pressure.* The essential components of a simple barometer are shown in Figure 7.6. Atmospheric pressure is expressed in terms of the height of the barometer's mercury column, usually in *millimeters of mercury* (mm Hg). Another name for millimeters of mercury is *torr*, used in honor of Evangelista Torricelli, the Italian physicist who invented the barometer.

$$1 \text{ mm Hg} = 1 \text{ torr}$$

Atmospheric pressure varies with the weather and the altitude. It averages about 760 mm Hg at sea level, and it decreases by approximately 25 mm Hg for every 1000-ft increase in altitude. The pressure unit *atmosphere* (atm) is defined in terms of this average pressure at sea level. By definition,

$$1 \text{ atm} = 760 \text{ mm Hg} = 760 \text{ torr}$$

Another commonly used pressure unit is *pounds per square inch* (psi or lb/in^2). One atmosphere is equal to 14.7 psi.

$$1 \text{ atm} = 14.7 \text{ psi}$$

7.4 Boyle's Law: A Pressure–Volume Relationship

Of the several relationships that exist among gas law variables, the first to be discovered relates gas pressure to gas volume. It was formulated over 300 years ago, in 1662, by the British chemist and physicist Robert Boyle (Figure 7.7). **Boyle's law** states that *the volume of a fixed amount of a gas is* inversely proportional *to the pressure applied to the gas if the temperature is kept constant.* This means that if the pressure on the gas increases, the volume decreases proportionally; conversely, if the pressure decreases, the volume increases. Doubling the pressure cuts the volume in half; tripling the pressure reduces the volume to one-third its original value; quadrupling the pressure reduces the volume to one-fourth its original value; and so on. Figure 7.8 illustrates Boyle's law.

The mathematical equation for Boyle's law is

$$P_1 \times V_1 = P_2 \times V_2$$

where P_1 and V_1 are the pressure and volume of a gas at an initial set of conditions, and P_2 and V_2 are the pressure and volume of the same sample of gas under a new set of conditions, with the temperature and amount of gas remaining constant.

EXAMPLE 7.1

Using Boyle's Law to Calculate the New Volume of a Gas

■ A sample of O_2 gas occupies a volume of 1.50 L at a pressure of 735 mm Hg and a temperature of 25°C. What volume will it occupy, in liters, if the pressure is increased to 770 mm Hg with no change in temperature?

Solution

A suggested first step in working gas law problems that involve two sets of conditions is to analyze the given data in terms of initial and final conditions.

$$P_1 = 735 \text{ mm Hg} \qquad P_2 = 770 \text{ mm Hg}$$
$$V_1 = 1.50 \text{ L} \qquad V_2 = ? \text{ L}$$

We know three of the four variables in the Boyle's law equation, so we can calculate the fourth, V_2. We will rearrange Boyle's law to isolate V_2 (the quantity to be calculated) on one side of the equation. This is accomplished by dividing both sides of the Boyle's law equation by P_2.

$$P_1V_1 = P_2V_2 \qquad \text{(Boyle's law)}$$

$$\frac{P_1V_1}{P_2} = \frac{P_2V_2}{P_2} \qquad \begin{array}{l}\text{(Divide each side of} \\ \text{the equation by } P_2.\text{)}\end{array}$$

$$V_2 = V_1 \times \frac{P_1}{P_2}$$

Substituting the given data into the rearranged equation and doing the arithmetic give

$$V_2 = 1.50 \text{ L} \times \left(\frac{735 \text{ mm Hg}}{770 \text{ mm Hg}} \right) = 1.43 \text{ L}$$

When we know any three of the four quantities in the Boyle's law equation, we can calculate the fourth, which is usually the final pressure, P_2, or the final volume, V_2. The Boyle's law equation is valid only if the temperature and amount of the gas remains constant.

Practice Exercise 7.1

A sample of H_2 gas occupies a volume of 2.25 L at a pressure of 628 mm Hg and a temperature of 35°C. What volume will it occupy, in liters, if the pressure is decreased to 428 mm Hg with no change in temperature?

FIGURE 7.7 Robert Boyle (1627–1691), like most men of the seventeenth century who devoted themselves to science, was self-taught. It was through his efforts that the true value of experimental investigation was first recognized.

Boyle's law is consistent with kinetic molecular theory. The pressure that a gas exerts results from collisions of the gas molecules with the sides of the container. If the volume of a container holding a specific number of gas molecules is increased, the total wall area of the container will also increase, and the number of collisions in a given area (the pressure) will decrease because of the greater wall area. Conversely, if the volume of the

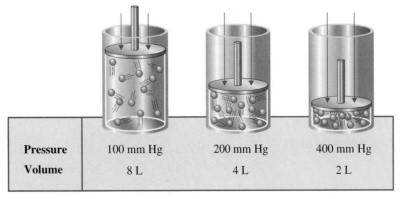

Pressure	100 mm Hg	200 mm Hg	400 mm Hg
Volume	8 L	4 L	2 L

FIGURE 7.8 Data illustrating the inverse proportionality associated with Boyle's law.

FIGURE 7.9 When the volume of a gas, at constant temperature, decreases by half (a), the average number of times a molecule hits the container walls is doubled (b).

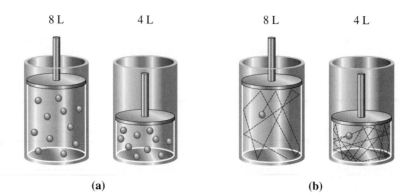

8 L   4 L   8 L   4 L

(a)   (b)

FIGURE 7.9 When the volume of a gas, at constant temperature, decreases by half (a), the average number of times a molecule hits the container walls is doubled (b).

Boyle's law explains the process of breathing. Breathing in occurs when the diaphragm flattens out (contracts). This contraction causes the volume of the thoracic cavity to increase and the pressure within the cavity to drop (Boyle's law) below atmospheric pressure. Air flows into the lungs and expands them, because the pressure is greater outside the lungs than within them. Breathing out occurs when the diaphragm relaxes (moves up), decreasing the volume of the thoracic cavity and increasing the pressure (Boyle's law) within the cavity to a value greater than the external pressure. Air flows out of the lungs. The air flow direction is always from a high-pressure region to a low-pressure region.

container is decreased, the wall area will be smaller and there will be more collisions within a given wall area. Figure 7.9 illustrates this concept.

Filling a medical syringe with a liquid demonstrates Boyle's law. As the plunger is drawn out of the syringe (see Figure 7.10), the increase in volume inside the syringe chamber results in decreased pressure there. The liquid, which is at atmospheric pressure, flows into this reduced-pressure area. This liquid is then expelled from the chamber by pushing the plunger back in. This ejection of the liquid does not involve Boyle's law; a liquid is incompressible, and mechanical force pushes it out.

7.5 Charles's Law: A Temperature–Volume Relationship

The relationship between the temperature and the volume of a gas at constant pressure is called *Charles's law* after the French scientist Jacques Charles (Figure 7.11). This law was discovered in 1787, over 100 years after the discovery of Boyle's law. **Charles's law** states that *the volume of a fixed amount of gas is* directly proportional *to its Kelvin temperature if the pressure is kept constant* (Figure 7.12). Whenever a *direct* proportion exists between two quantities, one increases when the other increases and one decreases when the other decreases. The direct-proportion relationship of Charles's law means that if the temperature increases, the volume will also increase and that if the temperature decreases, the volume will also decrease.

A balloon filled with air illustrates Charles's law. If the balloon is placed near a heat source such as a light bulb that has been on for some time, the heat will cause the balloon to increase visibly in size (volume). Putting the same balloon in the refrigerator will cause it to shrink.

Charles's law, stated mathematically, is

$$\frac{V_1}{T_1} = \frac{V_2}{T_2}$$

where V_1 is the volume of a gas at a given pressure, T_1 is the Kelvin temperature of the gas, and V_2 and T_2 are the volume and Kelvin temperature of the gas under a new set of conditions, with the pressure remaining constant.

FIGURE 7.10 Filling a syringe with a liquid is an application of Boyle's law.

EXAMPLE 7.2

Using Charles's Law to Calculate the New Volume of a Gas

■ A sample of the gaseous anesthetic cyclopropane, with a volume of 425 mL at a temperature of 27°C, is cooled at constant pressure to 20°C. What is the new volume, in milliliters, of the sample?

Solution

First, we will analyze the data in terms of initial and final conditions.

$V_1 = 425$ mL $V_2 = ?$ mL
$T_1 = 27°C + 273 = 300$ K $T_2 = 20°C + 273 = 293$ K

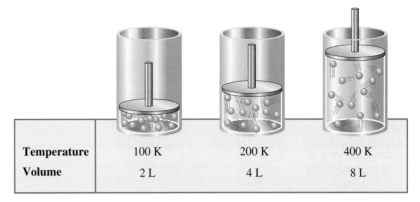

FIGURE 7.12 Data illustrating the direct proportionality associated with Charles's law.

Temperature	100 K	200 K	400 K
Volume	2 L	4 L	8 L

FIGURE 7.11 Jacques Charles (1746–1823), a French physicist, in the process of working with hot-air balloons, made the observations that ultimately led to the formulation of what is now known as Charles's law.

> When you use the mathematical form of Charles's law, the temperatures used *must be* Kelvin scale temperatures.

> Charles's law predicts that gas volume will become smaller and smaller as temperature is reduced, until eventually a temperature is reached at which gas volume becomes zero. This "zero-volume" temperature is calculated to be $-273°C$ and is known as *absolute zero* (see Section 2.8). Absolute zero is the basis for the Kelvin temperature scale. In reality, gas volume never vanishes. As temperature is lowered, at some point before absolute zero, the gas condenses to a liquid, at which point Charles's law is no longer valid.

Note that both of the given temperatures have been converted to Kelvin scale readings. This change is accomplished by simply adding 273 to the Celsius scale value (Section 2.9).

We know three of the four variables in the Charles's law equation, so we can calculate the fourth, V_2. We will rearrange Charles's law to isolate V_2 (the quantity desired) by multiplying each side of the equation by T_2.

$$\frac{V_1}{T_1} = \frac{V_2}{T_2} \qquad \text{(Charles's law)}$$

$$\frac{V_1 T_2}{T_1} = \frac{V_2 T_2}{T_2} \qquad \text{(Multiply each side by } T_2.)$$

$$V_2 = V_1 \times \frac{T_2}{T_1}$$

Substituting the given data into the equation and doing the arithmetic give

$$V_2 = 425 \text{ mL} \times \left(\frac{293 \text{ K}}{300 \text{ K}}\right) = 415 \text{ mL}$$

Practice Exercise 7.2

A sample of dry air, with a volume of 125 mL at a temperature of 53°C, is heated at constant pressure to 95°C. What is the new volume, in milliliters, of the sample?

Charles's law is consistent with kinetic molecular theory. When the temperature of a gas increases, the velocity (kinetic energy) of the gas molecules increases. The speedier particles hit the container walls harder and more often. In order for the pressure of the gas to remain constant, the container volume must increase. In a larger volume, the particles will hit the container walls less often, and the pressure can remain the same. A similar argument applies when the temperature of a gas is lowered. This time the velocity of the molecules decreases, and the wall area (volume) must also decrease in order to increase the number of collisions in a given area in a given time.

Charles's law is the principle used in the operation of a convection heater. When air comes in contact with the heating element, it expands (its density becomes less). The hot, less dense air rises, causing continuous circulation of warm air. This same principle has ramifications in closed rooms that lack effective air circulation. The warmer and less dense air stays near the top of the room. This is desirable in the summer but not in the winter.

7.6 The Combined Gas Law

Boyle's and Charles's laws can be mathematically combined to give a more versatile equation than either of the laws by themselves. The **combined gas law** states that *the product of the pressure and volume of a fixed amount of gas is inversely proportional to its Kelvin temperature*. The mathematical equation for the combined gas law is

$$\frac{P_1 V_1}{T_1} = \frac{P_2 V_2}{T_2}$$

Using this equation, we can calculate the change in pressure, temperature, or volume that is brought about by changes in the other two variables.

> ◢ Any time a gas law contains temperature terms, as is the case for both Charles's law and the combined gas law, these temperatures must be specified on the Kelvin temperature scale.

EXAMPLE 7.3

Using the Combined Gas Law to Calculate the New Volume of a Gas

■ A sample of O_2 gas occupies a volume of 1.62 L at 755 mm Hg pressure and has a temperature of 0°C. What volume, in liters, will this gas sample occupy at 725 mm Hg pressure and 50°C?

Solution

First, we analyze the data in terms of initial and final conditions.

$$P_1 = 755 \text{ mm Hg} \qquad P_2 = 725 \text{ mm Hg}$$
$$V_1 = 1.62 \text{ L} \qquad V_2 = ? \text{ L}$$
$$T_1 = 0°C + 273 = 273 \text{ K} \qquad T_2 = 50°C + 273 = 323 \text{ K}$$

We are given five of the six variables in the combined gas law, so we can calculate the sixth one, V_2. Rearranging the combined gas law to isolate the variable V_2 on a side by itself gives

$$V_2 = \frac{V_1 P_1 T_2}{P_2 T_1}$$

Substituting numerical values into this "version" of the combined gas law gives

$$V_2 = 1.62 \text{ L} \times \frac{755 \text{ mm Hg}}{725 \text{ mm Hg}} \times \frac{323 \text{ K}}{273 \text{ K}} = 2.00 \text{ L}$$

Practice Exercise 7.3

A helium-filled weather balloon, when released, has a volume of 10.0 L at 27°C and a pressure of 663 mm Hg. What volume, in liters, will the balloon occupy at an altitude where the pressure is 96 mm Hg and the temperature is −30.0°C?

7.7 The Ideal Gas Law

The **ideal gas law** *is an equation that includes the quantity of gas in a sample as well as the temperature, pressure, and volume of the sample.* Mathematically, the ideal gas law has the form

$$PV = nRT$$

In this equation, pressure, temperature, and volume are defined in the same manner as in the gas laws we have already discussed. The symbol n stands for the *number of moles* of gas present in the sample. The symbol R represents the *ideal gas constant*, the proportionality constant that makes the equation valid.

The value of the ideal gas constant (R) varies with the units chosen for pressure and volume. With pressure in atmospheres and volume in liters, R has the value

$$R = \frac{PV}{nT} = 0.0821 \frac{\text{atm} \cdot \text{L}}{\text{mole} \cdot \text{K}}$$

> ◢ The ideal gas law is used in calculations when *one* set of conditions is given with one missing variable. The combined gas law (Section 7.6) is used when *two* sets of conditions are given with one missing variable.

The value of R is the same for all gases under normally encountered conditions of temperature, pressure, and volume.

If three of the four variables in the ideal gas law equation are known, then the fourth can be calculated using the equation. Example 7.4 illustrates the use of the ideal gas law.

EXAMPLE 7.4

Using the Ideal Gas Law to Calculate the Volume of a Gas

■ The colorless, odorless, tasteless gas carbon monoxide, CO, is a by-product of incomplete combustion of any material that contains the element carbon. Calculate the volume, in liters, occupied by 1.52 moles of this gas at 0.992 atm pressure and a temperature of 65°C.

Solution

This problem deals with only one set of conditions, so the ideal gas equation is applicable. Three of the four variables in the ideal gas equation (P, n, and T) are given, and the fourth (V) is to be calculated.

$$P = 0.992 \text{ atm} \qquad n = 1.52 \text{ moles}$$
$$V = ? \text{ L} \qquad T = 65°C = 338 \text{ K}$$

Rearranging the ideal gas equation to isolate V on the left side of the equation gives

$$V = \frac{nRT}{P}$$

Because the pressure is given in atmospheres and the volume unit is liters, the R value 0.0821 is valid. Substituting known numerical values into the equation gives

$$V = \frac{(1.52 \text{ moles}) \times \left(0.0821 \dfrac{\text{atm} \cdot \text{L}}{\text{mole} \cdot \text{K}}\right)(338 \text{ K})}{0.992 \text{ atm}}$$

Note that all the parts of the ideal gas constant unit cancel except for one, the volume part.

Doing the arithmetic yields the volume of CO.

$$V = \left(\frac{1.52 \times 0.0821 \times 338}{0.992}\right) \text{ L} = 42.5 \text{ L}$$

Practice Exercise 7.4

Calculate the volume, in liters, occupied by 3.25 moles of Cl_2 gas at 1.54 atm pressure and a temperature of 213°C.

FIGURE 7.13 John Dalton (1766–1844) throughout his life had a particular interest in the study of weather. From "weather" he turned his attention to the nature of the atmosphere and then to the study of gases in general.

A sample of clean air is the most common example of a mixture of gases that do not react with one another.

7.8 Dalton's Law of Partial Pressures

In a mixture of gases that do not react with one another, each type of molecule moves around in the container as though the other kinds were not there. This type of behavior is possible because a gas is mostly empty space, and attractions between molecules in the gaseous state are negligible at most temperatures and pressures. Each gas in the mixture occupies the entire volume of the container; that is, it distributes itself uniformly throughout the container. The molecules of each type strike the walls of the container as frequently and with the same energy as though they were the only gas in the mixture. Consequently, the pressure exerted by each gas in a mixture is the same as it would be if the gas were alone in the same container under the same conditions.

The English scientist John Dalton (Figure 7.13) was the first to notice the independent behavior of gases in mixtures. In 1803, he published a summary statement concerning this behavior that is now known as Dalton's law of partial pressures. **Dalton's law of partial pressures** states that *the total pressure exerted by a mixture of gases is the sum of the partial pressures of the individual gases present.* A **partial pressure** *is the pressure that a gas in a mixture of gases would exert if it were present alone under the same conditions.*

FIGURE 7.14 A set of four identical containers can be used to illustrate Dalton's law of partial pressures. The pressure in the fourth container (the mixture of gases) is equal to the sum of the pressures in the first three containers (the individual gases).

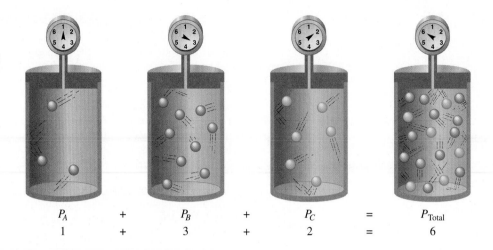

P_A	+	P_B	+	P_C	=	P_{Total}
1	+	3	+	2	=	6

Expressed mathematically, Dalton's law states that

$$P_{\text{Total}} = P_1 + P_2 + P_3 + \cdots$$

where P_{Total} is the total pressure of a gaseous mixture and P_1, P_2, P_3, and so on are the partial pressures of the individual gaseous components of the mixture.

As an illustration of Dalton's law, consider the four identical gas containers shown in Figure 7.14. Suppose we place amounts of three different gases (represented by A, B, and C) into three of the containers and measure the pressure exerted by each sample. We then place all three samples in the fourth container and measure the pressure exerted by this mixture of gases. We find that

$$P_{\text{Total}} = P_A + P_B + P_C$$

Using the actual gauge pressure values given in Figure 7.14, we see that

$$P_{\text{Total}} = 1 + 3 + 2 = 6$$

EXAMPLE 7.5

Using Dalton's Law to Calculate Partial Pressure

■ The total pressure exerted by a mixture of the three gases oxygen, nitrogen, and water vapor is 742 mm Hg. The partial pressures of the nitrogen and oxygen in the sample are 581 mm Hg and 143 mm Hg, respectively. What is the partial pressure of the water vapor present in the mixture?

Solution

Dalton's law says that

$$P_{\text{Total}} = P_{N_2} + P_{O_2} + P_{H_2O}$$

The known values for variables in this equation are

$$P_{\text{Total}} = 742 \text{ mm Hg}$$
$$P_{N_2} = 581 \text{ mm Hg}$$
$$P_{O_2} = 143 \text{ mm Hg}$$

Rearranging Dalton's law to isolate P_{H_2O} on the left side of the equation gives

$$P_{H_2O} = P_{\text{Total}} - P_{N_2} - P_{O_2}$$

Substituting the known numerical values into this equation and doing the arithmetic give

$$P_{H_2O} = 742 \text{ mm Hg} - 581 \text{ mm Hsg} - 143 \text{ mm Hg} = 18 \text{ mm Hg}$$

Practice Exercise 7.5

A gaseous mixture contains the three noble gases He, Ar, and Kr. The total pressure exerted by the mixture is 1.57 atm, and the partial pressures of the He and Ar are 0.33 atm and 0.39 atm, respectively. What is the partial pressure of the Kr in the mixture?

CHEMICAL CONNECTIONS

The Exchange of the Respiratory Gases, Oxygen and Carbon Dioxide

The concept of partial pressures helps explain the process of respiration in the human body. In general, inhaling replenishes oxygen in the blood and exhaling removes carbon dioxide from the blood. The oxygen is consumed by our cells in various cellular processes, and the carbon dioxide is the waste product from these cellular processes. Pressure gradients resulting from partial-pressure differences govern the movement of oxygen and carbon dioxide throughout the human body. Both gases always move from areas of higher partial pressure to areas of lower partial pressure. The overall exchange system is shown in the accompanying diagram.

As shown on the left side of the diagram, venous blood—blood returning to the lungs from the tissues—has been depleted of its oxygen supply (low oxygen partial pressure) and is carrying waste product carbon dioxide (high carbon dioxide partial pressure). Within the lungs (left side of diagram), carbon dioxide leaves the blood and oxygen enters the blood. The partial pressure of carbon dioxide is greater in the blood than within the lungs; for oxygen, the opposite is true.

As shown at the top of the diagram, arterial blood—blood that circulates from the lungs to the tissues—has been replenished in oxygen (high oxygen partial pressure). Oxygen from this blood diffuses into the tissues (low oxygen partial pressure). At the same time, carbon dioxide produced in the cells (high carbon dioxide partial pressure) diffuses into the bloodstream (low carbon dioxide partial pressure).

Note the actual values given in the diagram of the partial pressures for oxygen and carbon dioxide in the body. For carbon dioxide, the partial pressure varies between 40 mm Hg and 46 mm Hg; this is a small pressure gradient. For oxygen, a much larger pressure gradient exists; the gradient between arterial and venous blood is 65 mm Hg.

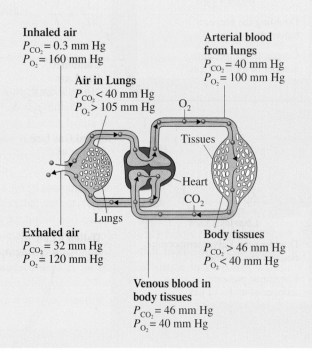

Inhaled air
$P_{CO_2} = 0.3$ mm Hg
$P_{O_2} = 160$ mm Hg

Air in Lungs
$P_{CO_2} < 40$ mm Hg
$P_{O_2} > 105$ mm Hg

Arterial blood from lungs
$P_{CO_2} = 40$ mm Hg
$P_{O_2} = 100$ mm Hg

O_2

Tissues

Heart

CO_2

Lungs

Exhaled air
$P_{CO_2} = 32$ mm Hg
$P_{O_2} = 120$ mm Hg

Body tissues
$P_{CO_2} > 46$ mm Hg
$P_{O_2} < 40$ mm Hg

Venous blood in body tissues
$P_{CO_2} = 46$ mm Hg
$P_{O_2} = 40$ mm Hg

Dalton's law of partial pressures is important when we consider the air of our atmosphere, which is a mixture of numerous gases. At higher altitudes, the total pressure of air decreases, as do the partial pressures of the individual components of air. An individual going from sea level to a higher altitude usually experiences some tiredness because his or her body is not functioning as efficiently at the higher altitude. At higher elevation, the red blood cells absorb a smaller amount of oxygen because the oxygen partial pressure at the higher altitude is lower. A person's body acclimates itself to the higher altitude after a period of time as additional red blood cells are produced by the body.

The Chemistry at a Glance feature on page 162 summarizes key concepts about the gas laws we have considered in this chapter.

7.9 Changes of State

A **change of state** *is a process in which a substance is transformed from one physical state to another physical state.* Changes of state are usually accomplished by heating or cooling a substance. Pressure change is also a factor in some systems. Changes of state are examples of physical changes—that is, changes in which chemical composition remains constant. No new substances are ever formed as a result of a change of state.

There are six possible changes of state. Figure 7.15 identifies each of these changes and gives the terminology used to describe them. Four of the six terms used in describing state changes are familiar: freezing, melting, evaporation, and condensation. The other two terms—sublimation and deposition—are not so common. *Sublimation* is the direct

FIGURE 7.23 If there were no hydrogen bonding between water molecules, the boiling point of water would be approximately −80°C; this value is obtained by extrapolation (extension of the line connecting the three heavier compounds). Because of hydrogen bonding, the actual boiling point of water, 100°C, is nearly 200°C higher than predicted. Indeed, in the absence of hydrogen bonding, water would be a gas at room temperature, and life as we know it on Earth would not be possible.

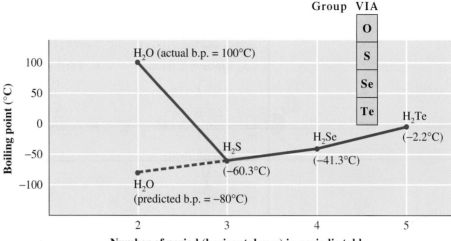

The vapor pressures (Section 7.11) of liquids that have significant hydrogen bonding are much lower than those of similar liquids wherein little or no hydrogen bonding occurs. This is because the presence of hydrogen bonds makes it more difficult for molecules to escape from the condensed state; additional energy is needed to overcome the hydrogen bonds. For this reason, boiling points are much higher for liquids in which hydrogen bonding occurs. The effect that hydrogen bonding has on boiling point can be seen by comparing water's boiling point with those of other hydrogen compounds of Group VIA elements—H_2S, H_2Se, and H_2Te (see Figure 7.23). Water is the only compound in this series where significant hydrogen bonding occurs.

■ London Forces

The third type of intermolecular force, and the weakest, is the London force, named after the German physicist Fritz London (1900–1954), who first postulated its existence. A **London force** *is a weak temporary intermolecular force that occurs between an atom or molecule (polar or nonpolar) and another atom or molecule (polar or nonpolar).* The origin of London forces is more difficult to visualize than that of dipole–dipole interactions.

London forces result from momentary (temporary) uneven electron distributions in molecules. Most of the time, the electrons in a molecule can be considered to have a predictable distribution determined by their energies and the electronegativities of the atoms present. However, there is a small statistical chance (probability) that the electrons will deviate from their normal pattern. For example, in the case of a nonpolar diatomic molecule, more electron density may temporarily be located on one side of the molecule than on the other. This condition causes the molecule to become polar for an instant. The negative side of this *instantaneously* polar molecule tends to repel electrons of adjoining molecules and causes these molecules also to become polar (*induced polarity*). The original polar molecule and all of the molecules with induced polarity are then attracted to one another. This happens many, many times per second throughout the liquid, resulting in a net attractive force. Figure 7.24 depicts the situation that prevails when London forces exist.

As an analogy for London forces, consider what happens when a bucket filled with water is moved. The water will "slosh" from side to side. This is similar to the movement of electrons. The "sloshing" from side to side is instantaneous; a given "slosh" quickly disappears. "Uneven" electron distribution is likewise a temporary situation.

The strength of London forces depends on the ease with which an electron distribution in a molecule can be distorted (polarized) by the polarity present in another molecule. In large molecules, the outermost electrons are necessarily located farther from the nucleus than are the outermost electrons in small molecules. The farther electrons are from the nucleus, the weaker the attractive forces that act on them, the more freedom they have, and the more susceptible they are to polarization. This leads to the observation that for

FIGURE 7.24 Nonpolar molecules such as H_2 can develop instantaneous dipoles and induced dipoles. The attractions between such dipoles, even though they are transitory, create London forces.

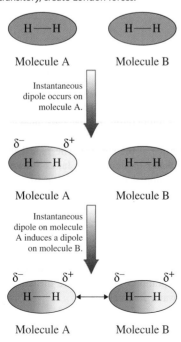

Intermolecular Forces

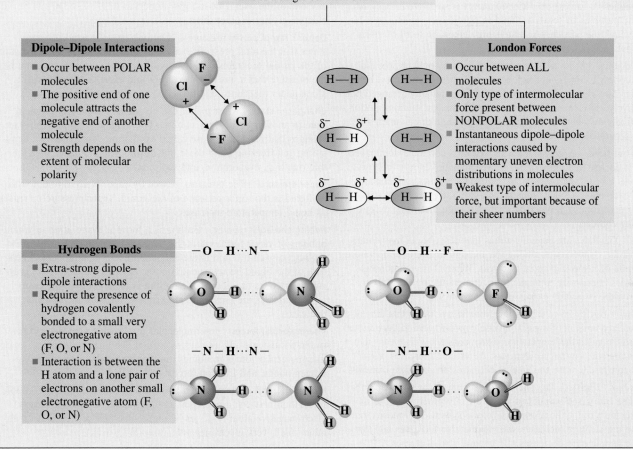

INTERMOLECULAR FORCES

- Electrostatic forces that act BETWEEN a molecule and other molecules
- Weaker than chemical bonds (intra-molecular forces)
- Strength is generally less than one-tenth that of a single covalent bond

Dipole–Dipole Interactions

- Occur between POLAR molecules
- The positive end of one molecule attracts the negative end of another molecule
- Strength depends on the extent of molecular polarity

London Forces

- Occur between ALL molecules
- Only type of intermolecular force present between NONPOLAR molecules
- Instantaneous dipole–dipole interactions caused by momentary uneven electron distributions in molecules
- Weakest type of intermolecular force, but important because of their sheer numbers

Hydrogen Bonds

- Extra-strong dipole–dipole interactions
- Require the presence of hydrogen covalently bonded to a small very electronegative atom (F, O, or N)
- Interaction is between the H atom and a lone pair of electrons on another small electronegative atom (F, O, or N)

The boiling points of substances with similar molar masses increase in this order: nonpolar molecules < polar molecules with no hydrogen bonding < polar molecules with hydrogen bonding.

related molecules, boiling points increase with molecular mass, which usually parallels size. This trend is reflected in the boiling points given in Table 7.5 for two series of related substances: the noble gases and the halogens (Group VIIA).

The Chemistry at a Glance feature on this page provides a summary of what we have discussed about intermolecular forces.

TABLE 7.5

Boiling Point Trends for Related Series of Nonpolar Molecules: (a) noble gases, (b) halogens

(a) Noble Gases (Group VIIIA Elements)			(b) Halogens (Group VIIA Elements)		
Substance	Molecular mass (amu)	Boiling point (°C)	Substance	Molecular mass (amu)	Boiling point (°C)
He	4.0	−269	F_2	39.0	−187
Ne	20.2	−246	Cl_2	70.9	−35
Ar	39.9	−186	Br_2	159.8	+59
Kr	83.8	−153			
Xe	131.3	−107			
Rn	222.0	−62			

CONCEPTS TO REMEMBER

Kinetic molecular theory. The kinetic molecular theory of matter is a set of five statements that explain the physical behavior of the three states of matter (solids, liquids, and gases). The basic idea of this theory is that the particles (atoms, molecules, or ions) present in a substance are in constant motion and are attracted or repelled by each other.

The solid state. A solid is characterized by a dominance of potential energy (cohesive forces) over kinetic energy (disruptive forces). As a result, the particles of solids are held in rigid three-dimensional lattices in which the particle's kinetic energy takes the form of vibrations about each lattice site.

The liquid state. A liquid is one characterized by neither potential energy (cohesive forces) nor kinetic energy (disruptive forces) being dominant. As a result, particles of liquids are randomly arranged but are relatively close to each other and are in constant random motion, sliding freely over each other but without enough kinetic energy to become separated.

The gaseous state. A gas is characterized by a complete dominance of kinetic energy (disruptive forces) over potential energy (cohesive forces). As a result, particles move randomly, essentially independently of each other. Under ordinary pressure, the particles of a gas are separated from each other by relatively large distances, except when they collide.

Gas laws. Gas laws are generalizations that describe, in mathematical terms, the relationships among the amount, pressure, temperature, and volume of a specific quantity of gas. When these relationships are used, it is necessary to express the temperature on the Kelvin scale. Pressure is usually expressed in atm, mm Hg, or torr.

Boyle's law. Boyle's law, the pressure–volume law, states that the volume of a fixed amount of a gas is inversely proportional to the pressure applied to the gas if the temperature is kept constant. This means that when the pressure on the gas increases, the volume decreases proportionally; conversely, when the volume decreases, the pressure increases.

Charles's law. Charles's law, the volume–temperature law, states that the volume of a fixed amount of gas is directly proportional to its Kelvin temperature if the pressure is kept constant. This means that when the temperature increases, the volume also increases and that when the temperature decreases, the volume also decreases.

The combined gas law. The combined gas law is an expression obtained by mathematically combining Boyle's and Charles's laws. A change in pressure, temperature, or volume that is brought about by changes in the other two variables can be calculated by using this law.

Ideal gas law. The ideal gas law has the form $PV = nRT$, where R is the ideal gas constant (0.0821 atm $\cdot$ L/mole $\cdot$ K). With this equation any one of the characteristic gas properties (P, V, T, or n) can be calculated, given the other three.

Dalton's law of partial pressures. Dalton's law of partial pressures states that the total pressure exerted by a mixture of gases is the sum of the partial pressures of the individual gases. A partial pressure is the pressure that a gas in a mixture would exert if it were present alone under the same conditions.

Changes of state. Most matter can be changed from one physical state to another by heating, cooling, or changing pressure. The state changes that release heat are called exothermic (condensation, deposition, and freezing), and those that absorb heat are called endothermic (melting, evaporation, and sublimation).

Vapor pressure. The pressure exerted by vapor in equilibrium with its liquid is the vapor pressure of the liquid. Vapor pressure increases as liquid temperature increases.

Boiling and boiling point. Boiling is a form of evaporation in which bubbles of vapor form within the liquid and rise to the surface. The boiling point of a liquid is the temperature at which the vapor pressure of the liquid becomes equal to the external (atmospheric) pressure exerted on the liquid. The boiling point of a liquid increases or decreases as the prevailing atmospheric pressure increases or decreases.

Intermolecular forces. Intermolecular forces are forces that act between a molecule and another molecule. The three principal types of intermolecular forces in liquids are dipole–dipole interactions, hydrogen bonds, and London forces.

Hydrogen bonds. A hydrogen bond is an extra-strong dipole–dipole interaction between a hydrogen atom covalently bonded to a very electronegative atom (F, O, or N) and a lone pair of electrons on another small, very electronegative atom (F, O, or N).

KEY REACTIONS AND EQUATIONS

1. Boyle's law (Section 7.4)
$$P_1V_1 = P_2V_2 \quad (n, T \text{ constant})$$

2. Charles's law (Section 7.5)
$$\frac{V_1}{T_1} = \frac{V_2}{T_2} \quad (n, P \text{ constant})$$

3. Combined gas law (Section 7.6)
$$\frac{P_1V_1}{T_1} = \frac{P_2V_2}{T_2} \quad (n \text{ constant})$$

4. Ideal gas law (Section 7.7)
$$PV = nRT$$

5. Ideal gas constant (Section 7.7)
$$R = 0.0821 \text{ atm} \cdot \text{L/mole} \cdot \text{K}$$

6. Dalton's law of partial pressures (Section 7.8)
$$P_{\text{Total}} = P_1 + P_2 + P_3 + \cdots$$

KEY TERMS

Barometer (7.3)
Boiling (7.12)
Boiling point (7.12)

Boyle's law (7.4)
Change of state (7.9)
Charles's Law (7.5)

Combined gas law (7.6)
Compressability (7.1)
Dalton's law of partial pressures (7.8)

Dipole–dipole interaction (7.13)
Electrostatic interaction (7.1)
Endothermic change of state (7.9)
Equilibrium (7.11)
Evaporation (7.10)
Exothermic change of state (7.9)
Gas (7.2)
Gas laws (7.3)

Hydrogen bond (7.13)
Ideal gas law (7.7)
Intermolecular force (7.13)
Kinetic energy (7.1)
Kinetic molecular theory of matter (7.1)
Liquid (7.2)
London force (7.13)
Normal boiling point (7.12)

Partial pressure (7.8)
Potential energy (7.1)
Pressure (7.3)
Solid (7.2)
Thermal expansion (7.1)
Vapor (7.10)
Vapor pressure (7.11)
Volatile substance (7.11)

EXERCISES AND PROBLEMS

The members of each pair of problems in this section test similar material.

■ **Kinetic Molecular Theory (Sections 7.1 and 7.2)**

7.1 Using kinetic molecular theory concepts, answer the following questions.
  a. What is the relationship between temperature and the average velocity with which particles move?
  b. What type of energy is related to cohesive forces?
  c. What effect does temperature have on the magnitude of disruptive forces?
  d. In which of the three states of matter are disruptive forces greater than cohesive forces?

7.2 Using kinetic molecular theory concepts, answer the following questions.
  a. How do molecules transfer energy from one to another?
  b. What type of energy is related to disruptive forces?
  c. What are the effects of cohesive forces on a system of particles?
  d. In which of the three states of matter are disruptive forces and cohesive forces of about the same magnitude?

7.3 Explain each of the following observations using kinetic molecular theory.
  a. Liquids show little change in volume with changes in temperature.
  b. Gases have a low density.

7.4 Explain each of the following observations using kinetic molecular theory.
  a. Both liquids and solids are practically incompressible.
  b. A container can be half full of a liquid but not half full of a gas.

■ **Gas Law Variables (Section 7.3)**

7.5 Carry out the following pressure unit conversions using the dimensional-analysis method of problem solving.
  a. 735 mm Hg to atmospheres
  b. 0.530 atm to millimeters of mercury
  c. 0.530 atm to torr
  d. 12.0 psi to atmospheres

7.6 Carry out the following pressure unit conversions using the dimensional-analysis method of problem solving.
  a. 73.5 mm Hg to atmospheres
  b. 1.75 atm to millimeters of mercury
  c. 735 torr to atmospheres
  d. 1.61 atm to pounds per square inch

■ **Boyle's Law (Section 7.4)**

7.7 At constant temperature, a sample of 6.0 L of O_2 at 3.0 atm pressure is compressed until the volume decreases to 2.5 L. What is the new pressure, in atmospheres?

7.8 At constant temperature, a sample of 6.0 L of N_2 at 2.0 atm pressure is allowed to expand until the volume reaches 9.5 L. What is the new pressure, in atmospheres?

7.9 A sample of ammonia (NH_3), a colorless gas with a pungent odor, occupies a volume of 3.00 L at a pressure of 655 mm Hg and a temperature of 25°C. What volume, in liters, will this NH_3 sample occupy, at the same temperature, if the pressure is increased to 725 mm Hg?

7.10 A sample of nitrogen dioxide (NO_2), a toxic gas with a reddish-brown color, occupies a volume of 4.00 L at a pressure of 725 mm Hg and a temperature of 35°C. What volume, in liters, will this NO_2 sample occupy, at the same temperature, if the pressure is decreased to 125 mm Hg?

■ **Charles's Law (Section 7.5)**

7.11 At atmospheric pressure, a sample of H_2 gas has a volume of 2.73 L at 27°C. What volume, in liters, will the H_2 gas occupy if the temperature is increased to 127°C and the pressure is held constant?

7.12 At atmospheric pressure, a sample of O_2 gas has a volume of 55 mL at 27°C. What volume, in milliliters, will the O_2 gas occupy if the temperature is decreased to 0°C and the pressure is held constant?

7.13 A sample of N_2 gas occupies a volume of 375 mL at 25°C and a pressure of 2.0 atm. Determine the temperature, in degrees Celsius, at which the volume of the gas would be 525 mL at the same pressure.

7.14 A sample of Ar gas occupies a volume of 1.2 L at 125°C and a pressure of 1.0 atm. Determine the temperature, in degrees Celsius, at which the volume of the gas would be 1.0 L at the same pressure.

■ **Combined Gas Law (Section 7.6)**

7.15 Rearrange the standard form of the combined gas law equation so that each of the following variables is by itself on one side of the equation.
  a. T_1 b. P_2 c. V_1

7.16 Rearrange the standard form of the combined gas law equation so that each of the following variables is by itself on one side of the equation.
  a. V_2 b. T_2 c. P_1

7.17 A sample of carbon dioxide (CO_2) gas has a volume of 15.2 L at a pressure of 1.35 atm and a temperature of 33°C. Determine the following for this gas sample.
  a. Volume, in liters, at $T = 35°C$ and $P = 3.50$ atm
  b. Pressure, in atmospheres, at $T = 42°C$ and $V = 10.0$ L
  c. Temperature, in degrees Celsius, at $P = 7.00$ atm and $V = 0.973$ L
  d. Volume, in milliliters, at $T = 97°C$ and $P = 6.70$ atm

7.18 A sample of carbon monoxide (CO) gas has a volume of 7.31 L at a pressure of 735 mm Hg and a temperature of 45°C. Determine the following for this gas sample.
a. Pressure, in millimeters of mercury, at $T = 357$°C and $V = 13.5$ L
b. Temperature, in degrees Celsius, at $P = 1275$ mm Hg and $V = 0.800$ L
c. Volume, in liters, at $T = 45$°C and $P = 325$ mm Hg
d. Pressure, in atmospheres, at $T = 325$°C and $V = 2.31$ L

■ **Ideal Gas Law (Section 7.7)**

7.19 What is the temperature, in degrees Celsius, of 5.23 moles of helium (He) gas confined to a volume of 5.23 L at a pressure of 5.23 atm?

7.20 What is the temperature, in degrees Celsius, of 1.50 moles of neon (Ne) gas confined to a volume of 2.50 L at a pressure of 1.00 atm?

7.21 Calculate the volume, in liters, of 0.100 mole of O_2 gas at 0°C and 2.00 atm pressure.

7.22 Calculate the pressure, in atmospheres, of 0.100 mole of O_2 gas in a 2.00-L container at a temperature of 75°C.

7.23 Determine the following for a 0.250-mole sample of CO_2 gas.
a. Volume, in liters, at 27°C and 1.50 atm
b. Pressure, in atmospheres, at 35°C in a 2.00-L container
c. Temperature, in degrees Celsius, at 1.20 atm pressure in a 3.00-L container.
d. Volume, in milliliters, at 125°C and 0.500 atm pressure

7.24 Determine the following for a 0.500-mole sample of CO gas.
a. Pressure, in atmospheres, at 35°C in a 1.00-L container
b. Temperature, in degrees Celsius, at 5.00 atm pressure in a 5.00-L container
c. Volume, in liters, at 127°C and 3.00 atm
d. Pressure, in millimeters of mercury, at 25°C in a 2.00-L container

■ **Dalton's Law of Partial Pressures (Section 7.8)**

7.25 The total pressure exerted by a mixture of O_2, N_2, and He gases is 1.50 atm. What is the partial pressure, in atmospheres, of the O_2, given that the partial pressures of the N_2 and He are 0.75 and 0.33 atm, respectively?

7.26 The total pressure exerted by a mixture of He, Ne, and Ar gases is 2.00 atm. What is the partial pressure, in atmospheres, of Ne, given that the partial pressures of the other gases are both 0.25 atm?

7.27 A gas mixture contains O_2, N_2, and Ar at partial pressures of 125, 175, and 225 mm Hg, respectively. If CO_2 gas is added to the mixture until the total pressure reaches 623 mm Hg, what is the partial pressure, in millimeters of mercury, of CO_2?

7.28 A gas mixture contains He, Ne, and H_2S at partial pressures of 125, 175, and 225 mm Hg, respectively. If all of the H_2S is removed from the mixture, what will be the partial pressure, in millimeters of mercury, of Ne?

■ **Changes of State (Section 7.9)**

7.29 Indicate whether each of the following is an exothermic or an endothermic change of state.
a. Sublimation   b. Melting   c. Condensation

7.30 Indicate whether each of the following is an exothermic or an endothermic change of state.
a. Freezing   b. Evaporation   c. Deposition

7.31 Indicate whether the liquid state is involved in each of the following changes of state.
a. Sublimation   b. Melting   c. Condensation

7.32 Indicate whether the solid state is involved in each of the following changes of state.
a. Freezing   b. Deposition   c. Evaporation

■ **Properties of Liquids (Sections 7.10 through 7.12)**

7.33 Match each of the following statements to the appropriate term: *vapor, vapor pressure, volatile, boiling,* or *boiling point.*
a. This is a temperature at which the liquid vapor pressure is equal to the external pressure on a liquid.
b. This property can be measured by allowing a liquid to evaporate in a closed container.
c. In this process, bubbles of vapor form within a liquid.
d. This temperature changes appreciably with changes in atmospheric pressure.

7.34 Match each of the following statements to the appropriate term: *vapor, vapor pressure, volatile, boiling,* or *boiling point.*
a. This state involves gaseous molecules of a substance at a temperature and pressure at which we would ordinarily think of the substance as a liquid.
b. This term describes a substance that readily evaporates at room temperature because of a high vapor pressure.
c. This process is a form of evaporation.
d. This property always increases in magnitude with increasing temperature.

7.35 Offer a clear, concise explanation for each of the following observations.
a. Liquids do not all have the same vapor pressure at a given temperature.
b. The boiling point of a liquid decreases as atmospheric pressure decreases.
c. A person emerging from an outdoor swimming pool on a breezy day gets the shivers.
d. Food will cook just as fast in boiling water with the stove set at low heat as in boiling water with the stove set at high heat.

7.36 Offer a clear, concise explanation for each of the following observations.
a. Increasing the temperature of a liquid increases its vapor pressure.
b. It takes more time to cook an egg in boiling water on a mountaintop than at sea level.
c. Food cooks faster in a pressure cooker than in an open pan.
d. Evaporation is a cooling process.

■ **Intermolecular Forces in Liquids (Section 7.13)**

7.37 Describe the molecular conditions necessary for the existence of a dipole–dipole interaction.

7.38 Describe the molecular conditions necessary for the existence of a London force.

7.39 In liquids, what is the relationship between boiling point and the strength of intermolecular forces?

7.40 In liquids, what is the relationship between vapor pressure magnitude and the strength of intermolecular forces?

7.41 For liquid-state samples of the following diatomic substances, classify the dominant intermolecular forces present as London forces, dipole–dipole interactions, or hydrogen bonds.
a. H_2 b. HF   c. CO   d. F_2

7.42 For liquid-state samples of the following diatomic substances, classify the dominant intermolecular forces present as London forces, dipole–dipole interactions, or hydrogen bonds.
a. O_2 b. HCl   c. Cl_2 d. BrCl

7.43 In which of the following substances, in the pure liquid state, would hydrogen bonding occur?

a.
```
    H       H
    |   ..  |
H — C — O — C — H
    |   ..  |
    H       H
```
b.
```
    H
    |       ..
H — C — N — H
    |   |
    H   H
```
c.
```
    H   H   H
    |   |   |
H — C — C — C — O — H
    |   |   |   ..
    H   H   H
```
d. H — I :

7.44 In which of the following substances, in the pure liquid state, would hydrogen bonding occur?

a.
```
    H   H
    |   |
H — C — C — H
    |   |
    H   H
```
b.
```
        ..
Cl — N — H
        |
        H
```
c.
```
    ..  ..
H — N — N — H
    |   |
    H   H
```
d.
```
    H   O :
    |   ||
H — C — C — H
    |
    H
```

7.45 How many hydrogen bonds can form between a single water molecule and other water molecules?

7.46 How many hydrogen bonds can form between a single ammonia molecule (NH_3) and other ammonia molecules?

ADDITIONAL PROBLEMS

7.47 A sample of NO_2 gas in a 575-mL container at a pressure of 1.25 atm and a temperature of 125°C is transferred to a new container with a volume of 825 mL.
a. What is the new pressure, in atmospheres, if no change in temperature occurs?
b. What is the new temperature, in degrees Celsius, if no change in pressure occurs?
c. What is the new temperature, in degrees Celsius, if the pressure is increased to 2.50 atm?

7.48 A sample of NO_2 gas in a nonrigid container, at a temperature of 24°C, occupies a certain volume at a certain pressure. What will be its temperature, in degrees Celsius, in each of the following situations?
a. Both pressure and volume are doubled.
b. Both pressure and volume are cut in half.
c. The pressure is doubled and the volume is cut in half.
d. The pressure is cut in half and the volume is tripled.

7.49 Match each of the listed restrictions on variables to the following gas laws: *Boyle's law, Charles's law,* and *the combined gas law.* More than one answer may be correct in a given situation.
a. The number of moles is constant.
b. The pressure is constant.
c. The temperature is constant.
d. Both the number of moles and the temperature are constant.

7.50 Suppose a helium-filled balloon, used to carry scientific instruments into the atmosphere, has a volume of 1.00×10^6 L at 25°C and a pressure of 752 mm Hg at the time it is launched. What will be the volume of the balloon, in liters, when, at a height of 37 km, it encounters a temperature of −33°C and a pressure of 75.0 mm Hg?

7.51 What is the pressure, in atmospheres, inside a 4.00-L container that contains the following amounts of O_2 gas at 40.0°C?
a. 0.72 mole
b. 4.5 moles
c. 0.72 g
d. 4.5 g

7.52 How many molecules of hydrogen sulfide (H_2S) gas are contained in 2.00 L of H_2S at 0.0°C and 1.00 atm pressure?

7.53 A 1.00-mole sample of dry ice (solid CO_2) is placed in a flexible sealed container and allowed to sublime. After complete sublimation, what will be the container volume, in liters, at 23°C and 0.983 atm pressure?

7.54 A piece of Ca metal is placed in a 1.00-L container with pure N_2. The N_2 is at a pressure of 1.12 atm and a temperature of 26°C. One hour later, the pressure has dropped to 0.924 atm and the temperature has dropped to 24°C. Calculate the number of grams of N_2 that reacted with the Ca.

7.55 Helium gas is added to an empty tank until the pressure reaches 9.0 atm. Neon gas is then added to the tank until the total pressure is 14.0 atm. Argon gas is then added to the tank until the total tank pressure reaches 29.0 atm. What is the partial pressure of each gas in the tank?

7.56 Under which of the following "pressure situations" will a liquid boil?
a. Vapor pressure and atmospheric pressure are equal.
b. Vapor pressure is less than atmospheric pressure.
c. Vapor pressure = 635 mm Hg, and atmospheric pressure = 735 mm Hg.
d. Vapor pressure = 735 torr, and atmospheric pressure = 1.00 atm.

7.57 The vapor pressure of PBr_3 reaches 400 mm Hg at 150°C. The vapor pressure of PI_3 reaches 400 mm Hg at 57°C.
a. Which substance should evaporate at the slower rate at 100°C?
b. Which substance should have the lower boiling point?
c. Which substance should have the weaker intermolecular forces?

7.58 In each of the following pairs of molecules, indicate which member of the pair would be expected to have the higher boiling point.
a. Cl_2 and Br_2
b. H_2O and H_2S
c. O_2 and CO
d. C_3H_8 and CO_2

ANSWERS TO PRACTICE EXERCISES

7.1 3.30 L H_2

7.2 141 mL air

7.3 56 L He

7.4 84.2 L Cl_2

7.5 0.85 atm for Kr

7.6 a. no; no hydrogen is present
b. yes; hydrogen attached to oxygen is present
c. no; hydrogen is present, but it is attached to carbon

8 Solutions

Ocean water is a solution in which many different substances are dissolved.

S olutions are common in nature, and they represent an abundant form of matter. Solutions carry nutrients to the cells of our bodies and carry away waste products. The ocean is a solution of water, sodium chloride, and many other substances (even gold). A large percentage of all chemical reactions take place in solution, including most of those discussed in later chapters in this text.

8.1 Characteristics of Solutions

All samples of matter are either *pure substances* or *mixtures* (Section 1.5). Pure substances are of two types: *elements* and *compounds*. Mixtures are of two types: *homogeneous* (uniform properties throughout) and *heterogeneous* (different properties in different regions).

Where do solutions fit in this classification scheme? The term *solution* is just an alternative way of saying *homogeneous mixture*. A **solution** *is a homogeneous mixture of two or more substances in which each substance retains its own chemical identity.*

It is often convenient to call one component of a solution the solvent and other components that are present solutes (Figure 8.1). A **solvent** *is the component of a solution that is present in the greatest amount.* A solvent can be thought of as the medium in which the other substances present are dissolved. A **solute** *is a component of a solution that is present in a lesser amount relative to that of the solvent.* More than one solute can be present in the same solution. For example, both sugar and salt (two solutes) can be dissolved in a container of water (solvent) to give salty sugar water.

▶ "All solutions are mixtures" is a valid statement. However, the reverse statement, "All mixtures are solutions," is not valid. Only those mixtures that are *homogeneous* are solutions.

FIGURE 8.1 The colored crystals are the solute, and the clear liquid is the solvent. Stirring produces the solution.

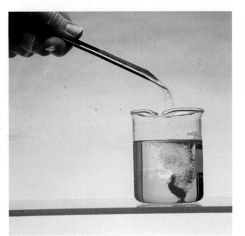

In most of the situations we will encounter, the solutes present in a solution will be of more interest to us than to the solvent. The solutes are the active ingredients in the solution. They are the substances that undergo reaction when solutions are mixed.

The general properties of a solution (homogeneous mixture) were outlined in Section 1.5. These properties, restated using the concepts of solvent and solute, are as follows:

1. A solution contains two or more components: a solvent (the substance present in the greatest amount) and one or more solutes.
2. A solution has a variable composition; that is, the ratio of solute to solvent may be varied.
3. The properties of a solution change as the ratio of solute to solvent is changed.
4. The dissolved solutes are present as individual particles (molecules, atoms, or ions). Intermingling of components at the particle level is a requirement for homogeneity.
5. The solutes remain uniformly distributed throughout the solution and will not settle out with time. Every part of a solution has exactly the same properties and composition as every other part.
6. The solute(s) generally can be separated from the solvent by physical means such as evaporation.

▶ Generally, solutions are *transparent;* that is, you can see through them. A synonym for *transparent* is *clear.* Clear solutions may be colorless or colored. A solution of potassium dichromate is a clear yellow-orange solution.

▶ Most solutes are more soluble in hot solvent than in cold solvent.

Solutions used in laboratories and clinical settings are most often liquids, and the solvent is nearly always water. However, gaseous solutions (dry air), solid solutions (metal alloys—see Figure 8.2), and liquid solutions in which water is not the solvent (gasoline, for example) are also possible and are relatively common.

8.2 Solubility

In addition to *solvent* and *solute,* several other terms are used to describe characteristics of solutions. **Solubility** *is the maximum amount of solute that will dissolve in a given amount of solvent under a particular set of conditions.* Many factors affect the numerical value of a solute's solubility in a given solvent, including the nature of the solvent itself, the temperature, and, in some cases, the pressure and presence of other solutes. Solubility is commonly expressed as grams of solute per 100 grams of solvent.

■ Effect of Temperature on Solubility

Most solids become more soluble in water with increasing temperature. The data in Table 8.1 illustrate this temperature–solubility pattern. Here, the solubilities of selected ionic solids in water are given at three different temperatures.

FIGURE 8.2 Jewelry often involves solid solutions in which one metal has been dissolved in another metal.

TABLE 8.1
Solubilities of Various Compounds in Water at 0°C, 50°C, and 100°C

Solute	Solubility (g solute/100 g H_2O)		
	0°C	50°C	100°C
lead(II) bromide ($PbBr_2$)	0.455	1.94	4.75
silver sulfate (Ag_2SO_4)	0.573	1.08	1.41
copper(II) sulfate ($CuSO_4$)	14.3	33.3	75.4
sodium chloride (NaCl)	35.7	37.0	39.8
silver nitrate ($AgNO_3$)	122	455	952
cesium chloride (CsCl)	161.4	218.5	270.5

In contrast to the solubilities of solids, gas solubilities in water decrease with increasing temperature. For example, both N_2 and O_2, the major components of air, are less soluble in hot water than in cold water. The Chemical Connections feature on page 180 considers further the topic of *temperature* and gas solubility.

■ Effect of Pressure on Solubility

Pressure has little effect on the solubility of solids and liquids in water. However, it has a major effect on the solubility of gases in water. The amount of gas that will dissolve in a liquid at a given temperature is directly related to the pressure of the gas above the liquid. In other words, as the pressure of a gas above a liquid increases, the solubility of the gas increases; conversely, as the pressure of the gas decreases, its solubility decreases. The Chemical Connections feature on page 180 considers further the topic of *pressure* and gas solubility.

■ Amounts of Solute in Solution

A **saturated solution** *is a solution that contains the maximum amount of solute that can be dissolved under the conditions at which the solution exists.* A saturated solution containing excess undissolved solute is an equilibrium situation where an amount of undissolved solute is continuously dissolving while an equal amount of dissolved solute is continuously crystallizing.

Consider the process of adding table sugar (sucrose) to a container of water. Initially, the added sugar dissolves as the solution is stirred. Finally, as we add more sugar, we reach a point where no amount of stirring will cause the added sugar to dissolve. The last-added sugar remains as a solid on the bottom of the container; the solution is saturated. Although it appears to the eye that nothing is happening once the saturation point is reached, this is not the case on the molecular level. Solid sugar from the bottom of the container is continuously dissolving in the water, and an equal amount of sugar is coming out of solution. Accordingly, the net number of sugar molecules in the liquid remains the same. The equilibrium situation in the saturated solution is somewhat similar to the evaporation of a liquid in a closed container (Section 7.10). Figure 8.3 illustrates the dynamic equilibrium process occurring in a saturated solution that contains undissolved excess solute.

Sometimes it is possible to exceed the maximum solubility of a compound, producing a *supersaturated* solution. A **supersaturated solution** *is an unstable solution that temporarily contains more dissolved solute than that present in a saturated solution.* An indirect rather than a direct procedure is needed to prepare a supersaturated solution; it involves the slow cooling, without agitation of any kind, of a high-temperature saturated solution in which no excess solid solute is present. Even though solute solubility decreases as the temperature is reduced, the excess solute often remains in solution. A supersaturated solution is an unstable situation; with time, excess solute will crystallize out, and the solution will revert to a saturated solution. A supersaturated solution will produce crystals rapidly, often in a dramatic manner, if it is slightly disturbed or if it is "seeded" with a tiny crystal of solute.

An **unsaturated solution** *is a solution which contains less than the maximum amount of solute that can be dissolved under the conditions at which the solution exists.* Most solutions we encounter fall into this category.

▶ Respiratory therapy procedures takes advantage of the fact that increased pressure increases the solubility of a gas. Patients with lung problems who are unable to get sufficient oxygen from air are given an oxygen-enriched mixture of gases to breathe. The larger oxygen partial pressure in the enriched mixture translates into increased oxygen uptake in the patient's lungs.

▶ When the amount of dissolved solute in a solution corresponds to the solute's solubility in the solvent, the solution formed is a saturated solution.

FIGURE 8.3 In a saturated solution, the dissolved solute is in dynamic equilibrium with the undissolved solute. Solute enters and leaves the solution at the same rate.

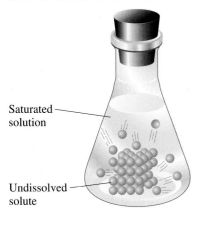

Saturated solution

Undissolved solute

CHEMICAL CONNECTIONS

Factors Affecting Gas Solubility

Both temperature and pressure affect the solubility of a gas in water. The effects are opposite. Increased temperature decreases gas solubility, and increased pressure increases gas solubility. The following table quantifies such effects for carbon dioxide (CO_2), a gas we often encounter dissolved in water.

Solubility of CO_2 (g/100 mL water)

Temperature effect (at 1 atm pressure)		Pressure effect (at 0°C)	
0°C	0.348	1 atm	0.348
20°C	0.176	2 atm	0.696
40°C	0.097	3 atm	1.044
60°C	0.058		

The effect of temperature on gas solubility has important environmental consequences because of the use of water from rivers and lakes for industrial cooling. Water used for cooling and then returned to its source at higher than ambient temperatures contains less oxygen and is less dense than when it was diverted. This lower-density "oxygen-deficient" water tends to "float" on colder water below, which blocks normal oxygen adsorption processes. This makes it more difficult for fish and

Carbon dioxide escaping from an opened bottle of a carbonated beverage.

other aquatic forms to obtain the oxygen they need to sustain life. This overall situation is known as *thermal pollution.*

Thermal pollution is sometimes unrelated to human activities. On hot summer days, the temperature of shallow water sometimes reaches the point where dissolved oxygen levels are insufficient to support some life. Under these conditions, suffocated fish may be found on the surface.

A flat taste is often associated with boiled water. This is due in part to the removal of dissolved gases during the boiling process. The removal of dissolved carbon dioxide particularly affects the taste.

The effect of pressure on gas solubility is observed every time a can or bottle of carbonated beverage is opened. The fizzing that occurs results from the escape of gaseous CO_2. The atmospheric pressure associated with an open container is much lower than the pressure used in the bottling process.

Pressure is a factor in the solubility of gases in the bloodstream. In hospitals, persons who are having difficulty obtaining oxygen are given supplementary oxygen. The result is an oxygen pressure greater than that in air. Hyperbaric medical procedures involve the use of pure oxygen. Oxygen pressure is sufficient to cause it to dissolve directly into the bloodstream, bypassing the body's normal mechanism for oxygen uptake (hemoglobin). Treatment of carbon monoxide poisoning is a situation where hyperbaric procedures are often needed.

Deep-sea divers can experience solubility-pressure problems. For every 30 feet that divers descend, the pressure increases by 1 atm. As a result, the air they breathe (particularly the N_2 component) dissolves to a greater extent in the blood. If a diver returns to the surface too quickly after a deep dive, the dissolved gases form bubbles in the blood (in the same way CO_2 does in a freshly opened can of carbonated beverage). This bubble formation may interfere with nerve impulse transmission and restrict blood flow. This painful condition, known as the *bends,* can cause paralysis or death.

Divers can avoid the bends by returning to the surface slowly and by using helium–oxygen gas mixtures, instead of air, in their breathing apparatus. Helium is less soluble in blood than N_2 and, because of its small atomic size, can escape from body tissues. Nitrogen must be removed via normal respiration.

The terms *concentrated* and *dilute* are also used to convey qualitative information about the degree of saturation of a solution. A **concentrated solution** *is a solution that contains a large amount of solute relative to the amount that could dissolve.* A concentrated solution does not have to be a saturated solution (see Figure 8.4). A **dilute solution** *is a solution that contains a small amount of solute relative to the amount that could dissolve.*

■ Aqueous and Nonaqueous Solutions

When the term *solution* is used, it is generally assumed that "aqueous solution" is meant, unless the context makes it clear that the solvent is not water.

Another set of solution terms involves the modifiers *aqueous* and *nonaqueous.* An **aqueous solution** *is a solution in which water is the solvent.* The presence of water is not a prerequisite for a solution, however. A **nonaqueous solution** *is a solution in which a substance other than water is the solvent.* Alcohol-based solutions are often encountered in a medical setting.

FIGURE 8.4 Both solutions contain the same amount of solute. A concentrated solution (left) contains a relatively large amount of solute compared with the amount that could dissolve. A dilute solution (right) contains a relatively small amount of solute compared with the amount that could dissolve.

▶ The fact that water molecules are polar is very important in the dissolving of an ionic solid in water.

8.3 Solution Formation

In a solution, solute particles are uniformly dispersed throughout the solvent. Considering what happens at the molecular level during the solution process will help us understand how this is accomplished.

In order for a solute to dissolve in a solvent, two types of interparticle attractions must be overcome: (1) attractions between solute particles (solute–solute attractions) and (2) attractions between solvent particles (solvent–solvent attractions). Only when these attractions are overcome can particles in both pure solute and pure solvent separate from one another and begin to intermingle. A new type of interaction, which does not exist prior to solution formation, arises as a result of the mixing of solute and solvent. This new interaction is the attraction between solute and solvent particles (solute–solvent attractions). These attractions are the primary driving force for solution formation.

An important type of solution process is one in which an ionic solid dissolves in water. Let us consider in detail the process of dissolving sodium chloride, a typical ionic solid, in water (Figure 8.5). The polar water molecules become oriented in such a way that the negative oxygen portion points toward positive sodium ions and the positive hydrogen portions point toward negative chloride ions. As the polar water molecules begin to surround ions on the crystal surface, they exert sufficient attraction to cause these ions to break away from the crystal surface. After leaving the crystal, an ion retains its surrounding group of water molecules; it has become a *hydrated ion.* As each hydrated ion leaves the surface, other ions are exposed to the water, and the crystal is picked apart ion by ion. Once in solution, the hydrated ions are uniformly distributed either by stirring or by random collisions with other molecules or ions.

The random motion of solute ions in solutions causes them to collide with one another, with solvent molecules, and occasionally with the surface of any undissolved solute. Ions undergoing the latter type of collision occasionally stick to the solid surface and thus leave the solution. When the number of ions in solution is low, the chances for collision with the undissolved solute are low. However, as the number of ions in solution increases, so do the chances for collisions, and more ions are recaptured by the undissolved solute. Eventually, the number of ions in solution reaches a level where ions return to the undissolved solute at the same rate at which other ions leave. At this point, the solution is saturated, and the equilibrium process discussed in the previous section is in operation.

■ Factors Affecting the Rate of Solution Formation

The rate at which a solution forms is governed by how rapidly the solute particles are distributed throughout the solvent. Three factors that affect the rate of solution formation are

FIGURE 8.5 When an ionic solid, such as sodium chloride, dissolves in water, the water molecules *hydrate* the ions. The positive ions are bound to the water molecules by their attraction for the partial negative charge on the water's oxygen atom, and the negative ions are bound to the water molecules by their attraction for the partial positive charge on the water's hydrogen atoms.

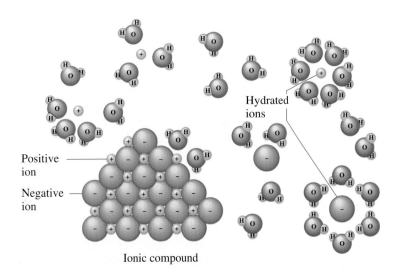

Hydrated ions

Positive ion

Negative ion

Ionic compound

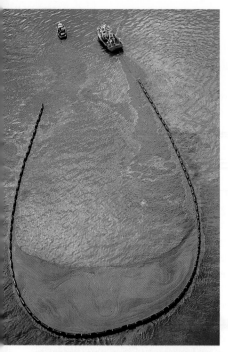

FIGURE 8.6 Oil spills can be contained to some extent by using trawlers and a boom apparatus because oil and water, having different polarities, are relatively insoluble in each other. The oil, which is of lower density, floats on top of the water.

1. *The state of subdivision of the solute.* A crushed aspirin tablet will dissolve in water more rapidly than a whole aspirin tablet. The more compact whole aspirin tablet has less surface area, and thus fewer solute molecules can interact with it at a given time.
2. *The degree of agitation during solution preparation.* Stirring solution components disperses the solute particles more rapidly, increasing the possibilities for solute–solvent interactions. Hence the rate of solution formation is increased.
3. *The temperature of the solution components.* Solution formation occurs more rapidly as the temperature is increased. At a higher temperature, both solute and solvent molecules move more rapidly (Section 7.1) so more interactions between them occur within a given time period.

8.4 Solubility Rules

In this section, we will present some rules for qualitatively predicting solute solubilities. These rules summarize in a concise form the results of thousands of experimental determinations of solute–solvent solubility.

A very useful generalization that relates polarity to solubility is that *substances of like polarity tend to be more soluble in each other than substances that differ in polarity.* This conclusion is often expressed as the simple phrase "*like dissolves like.*" Polar substances, in general, are good solvents for other polar substances but not for nonpolar substances (see Figure 8.6). Similarly, nonpolar substances exhibit greater solubility in nonpolar solvents than in polar solvents.

The generalization "like dissolves like" is a useful tool for predicting solubility behavior in many, but not all, solute–solvent situations. Results that agree with this generalization are nearly always obtained in the cases of gas-in-liquid and liquid-in-liquid solutions and for solid-in-liquid solutions in which the solute is not an ionic compound. For example, NH_3 gas (a polar gas) is much more soluble in H_2O (a polar liquid) than is O_2 gas (a nonpolar gas).

In the common case of solid-in-liquid solutions in which the solute is an ionic compound, the rule "like dissolves like" is not adequate. Their polar nature would suggest that all ionic compounds are soluble in a polar solvent such as water, but this is not the case. The failure of the generalization for ionic compounds is related to the complexity of the factors involved in determining the magnitude of the solute–solute (ion–ion) and solvent–solute (solvent–ion) interactions. Among other things, both the charge and the size of the ions in the solute must be considered. Changes in these factors affect both types of interactions, but not to the same extent.

Some guidelines concerning the solubility of ionic compounds in water, which should be used in place of "like dissolves like," are given in Table 8.2.

▶ The generalization "like dissolves like" is not adequate for predicting the solubilities of *ionic compounds* in water. More detailed solubility guidelines are needed (see Table 8.2).

TABLE 8.2
Solubility Guidelines for Ionic Compounds in Water

Ion contained in the compound	Solubility	Exceptions
Group IA (Li^+, Na^+, K^+, etc.)	soluble	
ammonium (NH_4^+)	soluble	
nitrates (NO_3^-)	soluble	
chlorides (Cl^-), bromides (Br^-), and iodides (I^-)	soluble	Ag^+, Pb^{2+}, Hg_2^{2+}
sulfates (SO_4^{2-})	soluble	Ca^{2+}, Sr^{2+}, Ba^{2+}, Pb^{2+}
carbonates (CO_3^{2-})	insoluble[a]	Group IA and NH_4^+
phosphates (PO_4^{3-})	insoluble	Group IA and NH_4^+
hydroxides (OH^-)	insoluble	Group IA, Ca^{2+}, Sr^{2+}, Ba^{2+}

[a]All ionic compounds, even the least soluble ones, dissolve to some slight extent in water. Thus the "insoluble" classification really means ionic compounds that have very limited solubility in water.

E X A M P L E 8 . 1

Predicting Solute Solubility Using Solubility Rules

■ With the help of Table 8.2, predict the solubility of each of the following solutes in the solvent indicated.

a. CH_4 (a nonpolar gas) in water
b. Ethyl alcohol (a polar liquid) in chloroform (a polar liquid)
c. AgCl (an ionic solid) in water
d. Na_2SO_4 (an ionic solid) in water
e. $AgNO_3$ (an ionic solid) in water

Solution

a. Insoluble. They are of unlike polarity because water is polar.
b. Soluble. Both substances are polar, so they should be relatively soluble in one another—like dissolves like.
c. Insoluble. Table 8.2 indicates that all chlorides except those of silver, lead, and mercury(I) are soluble. Thus AgCl is one of the exceptions.
d. Soluble. Table 8.2 indicates that all ionic sodium-containing compounds are soluble.
e. Soluble. Table 8.2 indicates that all compounds containing the nitrate ion (NO_3^-) are soluble.

Practice Exercise 8.1

With the help of Table 8.2, predict the solubility of each of the following solutes in the solvent indicated.

a. NO_2 (a polar gas) in water
b. CCl_4 (a nonpolar liquid) in benzene (a nonpolar liquid)
c. NaBr (an ionic solid) in water
d. $MgCO_3$ (an ionic solid) in water
e. $(NH_4)_3PO_4$ (an ionic solid) in water

8.5 Solution Concentration Units

Because solutions are mixtures (Section 8.1), they have a variable composition. Specifying what the composition of a solution is involves specifying solute concentrations. In general, **concentration** *is the amount of solute present in a specified amount of solution.* Many methods of expressing concentration exist, and certain methods are better suited for some purposes than others. In this section we consider two methods: *percent concentration* and *molarity.*

CHEMICAL CONNECTIONS

Solubility of Vitamins

Polarity plays an important role in the solubility of many substances in the fluids and tissues of the human body. For example, consider vitamin solubilities. The 13 known vitamins fall naturally into two classes: fat-soluble and water-soluble. The fat-soluble vitamins are A, D, E, and K. Water-soluble vitamins are vitamin C and the eight B vitamins (thiamine, riboflavin, niacin, vitamin B_6, folic acid, vitamin B_{12}, pantothenic acid, and biotin). Water-soluble vitamins have polar molecular structures, as does water. By contrast, fat-soluble vitamins have nonpolar molecular structures that are compatible with the nonpolar nature of fats.

Vitamin C is water-soluble. Because of this, vitamin C is not stored in the body and must be ingested in our daily diet. Unused vitamin C is eliminated rapidly from the body via body fluids. Vitamin A, on the other hand, is fat-soluble. It can be,

and is, stored by the body in fat tissue for later use. If vitamin A is consumed in excess quantities (from excessive vitamin supplements), illness can result. Because of its limited water solubility, vitamin A cannot be rapidly eliminated from the body by body fluids.

The water-soluble vitamins can be easily leached out of foods as they are prepared. As a rule of thumb, you should eat foods every day that are rich in the water-soluble vitamins. Taking megadose vitamin supplements of water-soluble vitamins is seldom effective. The extra amounts of these vitamins are usually picked up by the extracellular fluids, carried away by blood, and excreted in the urine. As one person aptly noted, "If you take supplements of water-soluble vitamins, you may have the most expensive urine in town."

■ Percent Concentration

There are three different ways of representing percent concentration:

1. Percent by mass (or mass–mass percent)
2. Percent by volume (or volume–volume percent)
3. Mass–volume percent

Percent by mass (or mass–mass percent) is the percentage unit most often used in chemical laboratories. **Percent by mass** *is the mass of solute in a solution divided by the total mass of solution, multiplied by 100 (to put the value in terms of percentage).*

$$\text{Percent by mass} = \frac{\text{mass of solute}}{\text{mass of solution}} \times 100$$

The solute and solution masses must be measured in the same unit, which is usually grams. The mass of the solution is equal to the mass of the solute plus the mass of the solvent.

$$\text{Mass of solution} = \text{mass of solute} + \text{mass of solvent}$$

> ▶ The concentration of butterfat in milk is expressed in terms of percent by mass. When you buy 1% milk, you are buying milk that contains 1 g of butterfat per 100 g of milk.

A solution whose mass percent concentration is 5.0% would contain 5.0 g of solute per 100.0 g of solution (5.0 g of solute and 95.0 g of solvent). Thus percent by mass directly gives the number of grams of solute in 100 g of solution. The percent-by-mass concentration unit is often abbreviated as %(m/m).

EXAMPLE 8.2

Calculating the Percent-by-Mass Concentration of a Solution

■ What is the percent-by-mass, %(m/m), concentration of sucrose (table sugar) in a solution made by dissolving 7.6 g of sucrose in 83.4 g of water?

Solution

Both the mass of solute and the mass of solvent are known. Substituting these numbers into the percent-by-mass equation

$$\%(m/m) = \frac{\text{mass of solute}}{\text{mass of solution}} \times 100$$

gives

$$\%(m/m) = \frac{7.6 \text{ g sucrose}}{7.6 \text{ g sucrose} + 83.4 \text{ g water}} \times 100$$

Remember that the denominator of the preceding equation (mass of solution) is the combined mass of the solute and the solvent.

Doing the mathematics gives

$$\%(m/m) = \frac{7.6 \text{ g}}{91.0 \text{ g}} \times 100 = 8.4\%$$

Practice Exercise 8.2

What is the percent-by-mass, %(m/m), concentration of Na_2SO_4 in a solution made by dissolving 7.6 g of Na_2SO_4 in enough water to give 87.3 g of solution?

EXAMPLE 8.3

Calculating the Mass of Solute Needed to Produce a Solution of a Given Percent-by-Mass Concentration

■ How many grams of sucrose must be added to 375 g of water to prepare a 2.75%(m/m) solution of sucrose?

Solution

Often, when a solution concentration is given as part of a problem statement, the concentration information is used in the form of a conversion factor when you solve the problem. That will be the case in this problem.

The given quantity is 375 g of H_2O (grams of solvent), and the desired quantity is grams of sucrose (grams of solute).

$$375 \text{ g } H_2O = ? \text{ g sucrose}$$

The conversion factor relating these two quantities (solvent and solute) is obtained from the given concentration. In a 2.75%-by-mass sucrose solution, there are 2.75 g of sucrose for every 97.25 g of water.

$$100.00 \text{ g solution} - 2.75 \text{ g sucrose} = 97.25 \text{ g } H_2O$$

The relationship between grams of solute and grams of solvent (2.75 to 97.25) gives us the needed conversion factor.

$$\frac{2.75 \text{ g sucrose}}{97.25 \text{ g } H_2O}$$

The problem is set up and solved, using dimensional analysis, as follows:

$$375 \text{ g } H_2O \times \left(\frac{2.75 \text{ g sucrose}}{97.25 \text{ g } H_2O} \right) = 10.6 \text{ g sucrose}$$

Practice Exercise 8.3

How many grams of $LiNO_3$ must be added to 25.0 g of water to prepare a 5.00%(m/m) solution of $LiNO_3$?

The second type of percentage unit, percent by volume (or volume–volume percent), which is abbreviated %(v/v), is used as a concentration unit in situations where the solute and solvent are both liquids or both gases. In these cases, it is more convenient to measure volumes than masses. **Percent by volume** *is the volume of solute in a solution divided by the total volume of solution, multiplied by 100.*

$$\text{Percent by volume} = \frac{\text{volume of solute}}{\text{volume of solution}} \times 100$$

Solute and solution volumes must always be expressed in the same units when you use percent by volume.

When the numerical value of a concentration is expressed as a percent by volume, it directly gives the number of milliliters of solute in 100 mL of solution. Thus a 100-mL sample of a 5.0%(v/v) alcohol-in-water solution contains 5.0 mL of alcohol dissolved in enough water to give 100 mL of solution. Note that such a 5.0%(v/v) solution could not be made by adding 5 mL of alcohol to 95 mL of water, because the volumes of two liquids are not usually additive. Differences in the way molecules are packed, as well as differences in distances between molecules, almost always result in the volume of the solution being less than the sum of the volumes of solute and solvent (see Figure 8.7). For example, the final volume resulting from the addition of 50.0 mL of ethyl alcohol to 50.0 mL of water is 96.5 mL of solution (see Figure 8.8). Working problems involving percent by volume entails using the same kinds of steps as those used for problems involving percent by mass.

The third type of percentage unit in common use is mass–volume percentage; it is abbreviated %(m/v). This unit, which is often encountered in clinical and hospital settings, is particularly convenient to use when you work with a solid solute, which is easily weighed, and a liquid solvent. Solutions of drugs for internal and external use, intravenous and intramuscular injectables, and reagent solutions for testing are usually labeled in mass–volume percent.

Mass–volume percent, *is the mass of solute in a solution (in grams) divided by the total volume of solution (in milliliters), multiplied by 100.*

$$\text{Mass–volume percent} = \frac{\text{mass of solute (g)}}{\text{volume of solution (mL)}} \times 100$$

The proof system for specifying the alcoholic content of beverages is twice the percent by volume. Hence 40 proof is 20%(v/v) alcohol; 100 proof is 50%(v/v) alcohol.

FIGURE 8.7 When volumes of two different liquids are combined, the volumes are not additive. This process is somewhat analogous to pouring marbles and golf balls together. The marbles can fill in the spaces between the golf balls. This results in the "mixed" volume being less than the sum of the "premixed" volumes.

For dilute aqueous solutions, where the density is close to 1.00 g/mL %(m/m) and %(m/v) are almost the same, because mass in grams of the solution equals the volume in milliliters of the solution.

Note that in the definition of mass–volume percent, specific mass and volume units are given. This is necessary because the units do not cancel, as was the case with mass percent and volume percent.

Mass–volume percent indicates the number of grams of solute dissolved in each 100 mL of solution. Thus a 2.3%(m/v) solution of any solute contains 2.3 g of solute in each 100 mL of solution, and a 5.4%(m/v) solution contains 5.4 g of solute in each 100 mL of solution.

EXAMPLE 8.4

Calculating the Mass of Solute Needed to Produce a Solution of a Given Mass–Volume Percent Concentration

■ Normal saline solution that is used to dissolve drugs for intravenous use is 0.92%(m/v) NaCl in water. How many grams of NaCl are required to prepare 35.0 mL of normal saline solution?

Solution

The given quantity is 35.0 mL of solution, and the desired quantity is grams of NaCl.

$$35.0 \text{ mL solution} = ? \text{ g NaCl}$$

FIGURE 8.8 Identical volumetric flasks are filled to the 50.0-mL mark with ethanol and with water. When the two liquids are poured into a 100-mL volumetric flask, the volume is seen to be less than the expected 100.0 mL; it is only 96.5 mL.

The given concentration, 0.92%(m/v), which means 0.92 g of NaCl per 100 mL of solution, is used as a conversion factor to go from milliliters of solution to grams of NaCl. The setup for the conversion is

$$35.0 \text{ mL solution} \times \left(\frac{0.92 \text{ g NaCl}}{100 \text{ mL solution}} \right)$$

Doing the arithmetic after canceling the units gives

$$\left(\frac{35.0 \times 0.92}{100} \right) \text{g NaCl} = 0.32 \text{ g NaCl}$$

Practice Exercise 8.4

How many grams of glucose ($C_6H_{12}O_6$) are needed to prepare 500.0 mL of a 4.50%(m/v) glucose–water solution?

■ Molarity

Molarity *is the moles of solute in a solution divided by the liters of solution.* The mathematical equation for molarity is

$$\text{Molarity (M)} = \frac{\text{moles of solute}}{\text{liters of solution}}$$

Note that the abbreviation for molarity is a capital M. A solution containing 1 mole of KBr in 1 L of solution has a molarity of 1 and is said to be a 1 M (1 *molar*) solution.

The molarity concentration unit is often used in laboratories where chemical reactions are being studied. Because chemical reactions occur between molecules and atoms, use of the mole—a unit that counts particles—is desirable. Equal volumes of two solutions of the same molarity contain the same number of solute molecules.

EXAMPLE 8.5

Calculating the Molarity of a Solution

■ Determine the molarities of the following solutions.

a. 4.35 moles of $KMnO_4$ are dissolved in enough water to give 750 mL of solution.
b. 20.0 g of NaOH is dissolved in enough water to give 1.50 L of solution.

Solution

a. The number of moles of solute is given in the problem statement.

$$\text{Moles of solute (KMnO}_4) = 4.35 \text{ moles}$$

The volume of the solution is also given in the problem statement, but not in the right units. Molarity requires liters for the volume units, and we are given milliliters of solution. Making the unit change yields

$$750 \text{ mL} \times \left(\frac{10^{-3} \text{ L}}{1 \text{ mL}} \right) = 0.750 \text{ L}$$

The molarity of the solution is obtained by substituting the known quantities into the equation

$$M = \frac{\text{moles of solute}}{\text{liters of solution}}$$

which gives

$$M = \frac{4.35 \text{ moles KMnO}_4}{0.750 \text{ L solution}} = 5.80 \frac{\text{moles KMnO}_4}{\text{L solution}}$$

Note that the units for molarity are always moles per liter.

(continued)

When you perform molarity concentration calculations, you need the *identity* of the solute. You cannot calculate moles of solute without knowing the chemical identity of the solute. When you perform percent concentration calculations, the *identity* of the solute is not used in the calculation; all you need is the *amount* of solute.

b. This time, the volume of solution is given in liters.

$$\text{Volume of solution} = 1.50 \text{ L}$$

The moles of solute must be calculated from the grams of solute (given) and the solute's molar mass, which is 40.00 g/mole (calculated from atomic masses).

$$20.0 \text{ g NaOH} \times \left(\frac{1 \text{ mole NaOH}}{40.00 \text{ g NaOH}}\right) = 0.500 \text{ mole NaOH}$$

Substituting the known quantities into the defining equation for molarity gives

$$M = \frac{0.500 \text{ mole NaOH}}{1.50 \text{ L solution}} = 0.333 \frac{\text{mole NaOH}}{\text{L solution}}$$

Practice Exercise 8.5

Determine the molarities of the following solutions.

a. 2.37 moles of KNO_3 are dissolved in enough water to give 650.0 mL of solution.
b. 40.0 g of KCl is dissolved in enough water to give 0.850 L of solution.

In preparing 100 mL of a solution of a specific molarity, enough solvent is added to a weighed amount of solute to give a *final* volume of 100 mL. The weighed solute is not added to a *starting* volume of 100 mL; this would produce a final volume greater than 100 mL, because the solute volume increases the total volume.

In order to find the molarity of a solution, we need to know the solution volume in liters and the number of moles of solute present. An alternative to knowing the number of moles of solute is knowing the number of grams of solute present and the solute's formula mass. The number of moles can be calculated by using these two quantities (Section 6.4).

The mass of solute present in a known volume of solution is an easily calculable quantity if the molarity of the solution is known. When we do such a calculation, molarity serves as a conversion factor that relates liters of solution to moles of solute. In a similar manner, the volume of solution needed to supply a given amount of solute can be calculated by using the solution's molarity as a conversion factor.

EXAMPLE 8.6

Calculating the Amount of Solute Present in a Given Amount of Solution

■ How many grams of sucrose (table sugar, $C_{12}H_{22}O_{11}$) are present in 185 mL of a 2.50 M sucrose solution?

Solution

The given quantity is 185 mL of solution, and the desired quantity is grams of $C_{12}H_{22}O_{11}$.

$$185 \text{ mL of solution} = ? \text{ g } C_{12}H_{22}O_{11}$$

The pathway used to solve this problem is

$$\text{mL solution} \longrightarrow \text{L solution} \longrightarrow \text{moles } C_{12}H_{22}O_{11} \longrightarrow \text{g } C_{12}H_{22}O_{11}$$

The given molarity (2.50 M) serves as the conversion factor for the second unit change; the formula mass of sucrose (which is not given and must be calculated) is used to accomplish the third unit change.

The dimensional-analysis setup for this pathway is

$$185 \text{ mL solution} \times \left(\frac{10^{-3} \text{ L solution}}{1 \text{ mL solution}}\right) \times \left(\frac{2.50 \text{ moles } C_{12}H_{22}O_{11}}{1 \text{ L solution}}\right)$$

$$\times \left(\frac{342.34 \text{ g } C_{12}H_{22}O_{11}}{1 \text{ mole } C_{12}H_{22}O_{11}}\right)$$

Canceling the units and doing the arithmetic, we find that

$$\left(\frac{185 \times 10^{-3} \times 2.50 \times 342.34}{1 \times 1 \times 1}\right) \text{g } C_{12}H_{22}O_{11} = 158 \text{ g } C_{12}H_{22}O_{11}$$

EXAMPLE 8.7

Calculating the Amount of Solution Needed to Supply a Given Amount of Solute

■ A typical dose of iron(II) sulfate ($FeSO_4$) used in the treatment of iron-deficiency anemia is 0.35 g. How many milliliters of a 0.10 M iron(II) sulfate solution would be needed to supply this dose?

Solution

The given quantity is 0.35 g of $FeSO_4$, and the desired quantity is milliliters of $FeSO_4$ solution.

$$0.35 \text{ g FeSO}_4 = ? \text{ mL FeSO}_4 \text{ solution}$$

The pathway used to solve this problem is

$$\text{g FeSO}_4 \longrightarrow \text{moles FeSO}_4 \longrightarrow \text{L FeSO}_4 \text{ solution} \longrightarrow \text{mL FeSO}_4 \text{ solution}$$

We accomplish the first unit conversion by using the formula mass of $FeSO_4$ (which must be calculated) as a conversion factor. The second unit conversion involves the use of the given molarity as a conversion factor.

$$0.35 \text{ g FeSO}_4 \times \left(\frac{1 \text{ mole FeSO}_4}{151.92 \text{ g FeSO}_4}\right) \times \left(\frac{1 \text{ L solution}}{0.10 \text{ mole FeSO}_4}\right) \times \left(\frac{1 \text{ mL solution}}{10^{-3} \text{ L solution}}\right)$$

Canceling units and doing the arithmetic, we find that

$$\left(\frac{0.35 \times 1 \times 1 \times 1}{151.92 \times 0.10 \times 10^{-3}}\right) \text{ mL solution} = 23 \text{ mL solution}$$

The Chemistry at a Glance feature on page 191 reviews the ways in which we represent solution concentrations.

■ **Dilution Calculations**

FIGURE 8.9 Frozen orange juice concentrate is diluted with water prior to drinking.

A common activity encountered when we work with solutions is that of diluting a solution of known concentration (usually called a stock solution) to a lower concentration. **Dilution** *is the process in which more solvent is added to a solution in order to lower its concentration.* The same amount of solute is present, but it is now distributed in a larger amount of solvent (the original solvent plus the added solvent).

Often, we prepare a solution of a specific concentration by adding a predetermined volume of solvent to a specific volume of stock solution (see Figure 8.9). A simple relationship exists between the volumes and concentrations of the diluted and stock solutions. This relationship is

$$\left(\begin{array}{c}\text{Concentration of}\\\text{stock solution}\end{array}\right) \times \left(\begin{array}{c}\text{volume of}\\\text{stock solution}\end{array}\right) = \left(\begin{array}{c}\text{concentration of}\\\text{diluted solution}\end{array}\right) \times \left(\begin{array}{c}\text{volume of}\\\text{diluted solution}\end{array}\right)$$

or

$$C_s \times V_s = C_d \times V_d$$

EXAMPLE 8.8

Calculating the Amount of Solvent That Must Be Added to a Stock Solution to Dilute It to a Specified Concentration

■ A nurse wants to prepare a 1.0%(m/v) silver nitrate solution from 24 mL of a 3.0%(m/v) stock solution of silver nitrate. How much water should be added to the 24 mL of stock solution?

Solution

The volume of water to be added will be equal to the difference between the final and initial volumes. The initial volume is known (24 mL). The final volume can be calculated by using the equation

$$C_s \times V_s = C_d \times V_d$$

Once the final volume is known, the difference between the two volumes can be obtained.

Substituting the known quantities into the dilution equation, which has been rearranged to isolate V_d on the left side, gives

$$V_d = \frac{C_s \times V_s}{C_d} = \frac{3.0\% \text{ (m/v)} \times 24 \text{ mL}}{1.0\% \text{ (m/v)}} = 72 \text{ mL}$$

The solvent added is

$$V_d - V_s = (72 - 24) \text{ mL} = 48 \text{ mL}$$

Practice Exercise 8.8

What is the molarity of the solution prepared by diluting 65 mL of 0.95 M Na_2SO_4 solution to a final volume of 135 mL?

CHEMICAL CONNECTIONS

Controlled-Release Drugs: Regulating Concentration, Rate, and Location of Release

In the use of both prescription and over-the-counter drugs, body concentration levels of the drug are obviously of vital importance. All drugs have an optimum concentration range where they are most effective. Below this optimum concentration range, a drug is ineffective, and above it the drug may have adverse side effects. Hence, the much-repeated warning "Take as directed."

Ordinarily, in the administration of a drug, the body's concentration level of the drug rapidly increases toward the higher end of the effective concentration range and then gradually declines and falls below the effective limit. The period of effectiveness of the drug can be extended by using the drug in a controlled-release form. This causes the drug to be released in a regulated, continuous manner over a longer period of time. The accompanying graph contrasts "ordinary-release" and "controlled-release" modes of drug action.

The use of controlled-release medication began in the early 1960s with the introduction of the decongestant Contac. Contac's controlled-release mechanism, which is now found in many drugs and used by all drug manufacturers, involves drug particles encapsulated within a slowly dissolving coating that *varies in thickness* from particle to particle. Particles of the drug with a thinner coating dissolve first. Those particles with a thicker coating dissolve more slowly, extending the period of drug release. The number of particles of various thicknesses, within a formulation, is predetermined by the manufacturer.

When drugs are taken orally, they first encounter the acidic environment of the stomach. Two problems can occur here:

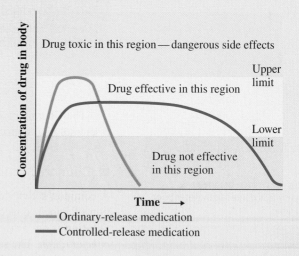

(1) The drug itself may damage the stomach lining. (2) The drug may be rendered inactive by the gastric acid present in the stomach. Controlled-release techniques are useful in overcoming these problems. Drug particle coatings are now available that are acid-resistant; that is, they do not dissolve in acidic solution. Drugs with such coatings pass from the stomach into the small intestine in undissolved form. Within the nonacidic (basic) environment of the small intestine, the dissolving process then begins.

Solutions

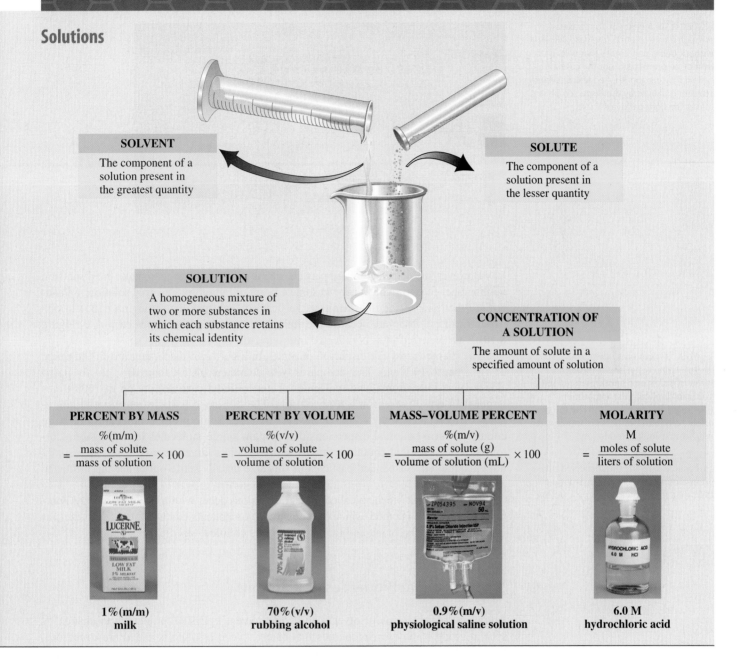

SOLVENT
The component of a solution present in the greatest quantity

SOLUTE
The component of a solution present in the lesser quantity

SOLUTION
A homogeneous mixture of two or more substances in which each substance retains its chemical identity

CONCENTRATION OF A SOLUTION
The amount of solute in a specified amount of solution

PERCENT BY MASS	PERCENT BY VOLUME	MASS–VOLUME PERCENT	MOLARITY
%(m/m)	%(v/v)	%(m/v)	M
$= \dfrac{\text{mass of solute}}{\text{mass of solution}} \times 100$	$= \dfrac{\text{volume of solute}}{\text{volume of solution}} \times 100$	$= \dfrac{\text{mass of solute (g)}}{\text{volume of solution (mL)}} \times 100$	$= \dfrac{\text{moles of solute}}{\text{liters of solution}}$
1%(m/m) milk	70%(v/v) rubbing alcohol	0.9%(m/v) physiological saline solution	6.0 M hydrochloric acid

8.6 Colloidal Dispersions

Colloidal dispersions are mixtures that have many properties similar to those of solutions, although they are not true solutions. In a broad sense, colloidal dispersions may be thought of as mixtures in which a material is *dispersed* rather than *dissolved*. A **colloidal dispersion** *is a mixture that contains dispersed particles that are intermediate in size between those of a true solution and those of an ordinary heterogeneous mixture.* The terms *solute* and *solvent* are not used to indicate the components of a colloidal dispersion. Instead, the particles dispersed in a colloidal dispersion are called the *dispersed phase,* and the material in which they are dispersed is called the *dispersing medium.*

Particles of the dispersed phase in a colloidal dispersion are so small that (1) they are not usually discernible by the naked eye, (2) they do not settle out under the influence of gravity, and (3) they cannot be filtered out using filter paper that has relatively large pores. In these respects, the dispersed phase behaves similarly to a solute in a solution.

> Some chemists use the term *colloid* instead of colloidal dispersion.

FIGURE 8.10 A beam of light travels through a true solution (the yellow liquid) without being scattered—that is, its path cannot be seen. This is not the case for a colloidal dispersion (the red liquid), where scattering of light by the dispersed phase makes the light pathway visible.

▶ Milk is a colloidal dispersion. If you shine a flashlight through a glass of salt water and a glass of milk, you can duplicate the experiment illustrated in Figure 8.10. (For the best effect, dilute the milk with some water until it just looks cloudy.)

▶ Particle size for the dispersed phase in a colloidal dispersion is larger than that for solutes in a true solution.

However, the dispersed-phase particle size is sufficiently large to make the dispersion nonhomogeneous to light. When we shine a beam of light through a true solution, we cannot see the track of the light. However, a beam of light passing through a colloidal dispersion can be observed, because the light is scattered by the dispersed phase (Figure 8.10). This scattered light is reflected into our eyes.

The diameters of the dispersed particles in a colloidal dispersion are in the range of 10^{-7} cm to 10^{-5} cm. This compares with diameters of less than 10^{-7} cm for particles such as ions, atoms, and molecules. Thus colloidal particles are up to 1000 times larger than those present in a true solution. The dispersed particles are usually aggregates of molecules, but this is not always the case. Some protein molecules are large enough to form colloidal dispersions that contain single molecules in suspension. Colloidal dispersions that contain particles with diameters larger than 10^{-5} cm are usually not encountered. Suspended particles of this size usually settle out under the influence of gravity.

Many different biochemical colloidal dispersions occur within the human body. Foremost among them is blood, which has numerous components that are colloidal in size. Fat is transported in the blood and lymph systems as colloidal-sized particles.

8.7 Colligative Properties of Solutions

Adding a solute to a pure solvent causes the solvent's physical properties to change. A special group of physical properties that change when a solute is added are called colligative properties. A **colligative property** *is a physical property of a solution that depends only on the number (concentration) of solute particles (molecules or ions) present in a given quantity of solvent and not on their chemical identities.* Examples of colligative properties include vapor-pressure lowering, boiling-point elevation, freezing-point depression, and osmotic pressure. The first three of these colligative properties are discussed in this section. The fourth, osmotic pressure, will be considered in Section 8.8.

Adding a nonvolatile solute to a solvent *lowers* the vapor pressure of the resulting solution below that of the pure solvent at the same temperature. (A nonvolatile solute is one that has a low vapor pressure and therefore a low tendency to vaporize; Section 7.11.) This lowering of vapor pressure is a direct consequence of some of the solute molecules or ions occupying positions on the surface of the liquid. Their presence decreases the probability of solvent molecules escaping; that is, the number of surface-occupying solvent molecules has been decreased. Figure 8.11 illustrates the decrease in surface concentration of solvent molecules when a solute is added. As the *number* of solute particles increases, the reduction in vapor pressure also increases; thus vapor pressure is a colligative property. What is important is not the identity of the solute molecules but the fact that they take up room on the surface of the liquid.

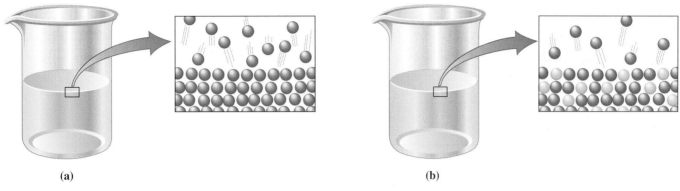

(a)   (b)

FIGURE 8.11 Close-ups of the surface of a liquid solvent (a) before and (b) after solute has been added. There are fewer solvent molecules on the surface of the liquid after solute has been added. This results in a decreased vapor pressure for the solution compared with pure solvent.

FIGURE 8.12 A water–antifreeze mixture has a higher boiling point and a lower freezing point than pure water.

▶ In the making of homemade ice cream, the function of the rock salt added to the ice is to depress the freezing point of the ice–water mixture surrounding the ice cream mix sufficiently to allow the mix (which contains sugar and other solutes and thus has a freezing point below 0°C) to freeze.

▶ The term *osmosis* comes from the Greek *osmos,* which means "push."

Adding a nonvolatile solute to a solvent *raises* the boiling point of the resulting solution above that of the pure solvent. This is logical when we remember that the vapor pressure of the solution is lower than that of pure solvent and that the boiling point is dependent on vapor pressure (Section 7.12). A higher temperature will be needed to raise the depressed vapor pressure of the solution to atmospheric pressure; this is the condition required for boiling.

A common application of the phenomenon of boiling point elevation involves automobiles. The coolant ethylene glycol (a nonvolatile solute) is added to car radiators to prevent boilover in hot weather (see Figure 8.12). The engine may not run any cooler, but the coolant–water mixture will not boil until it reaches a temperature well above the normal boiling point of water.

Adding a nonvolatile solute to a solvent *lowers* the freezing point of the resulting solution below that of the pure solvent. The presence of the solute particles within the solution interferes with the tendency of solvent molecules to line up in an organized manner, a condition necessary for the solid state. A lower temperature is necessary before the solvent molecules will form the solid.

Applications of freezing-point depression are even more numerous than those for boiling-point elevation. In climates where the temperature drops below 0°C in the winter, it is necessary to protect water-cooled automobile engines from freezing. This is done by adding antifreeze (usually ethylene glycol) to the radiator. The addition of this nonvolatile material causes the vapor pressure and freezing point of the resulting solution to be much lower than those of pure water. Also in the winter, a salt, usually NaCl or CaCl$_2$, is spread on roads and sidewalks to melt ice or prevent it from forming. The salt dissolves in the water to form a solution that will not freeze until the temperature drops much lower than 0°C, the normal freezing point of water.

8.8 Osmosis and Osmotic Pressure

The process of osmosis and the colligative property of osmotic pressure are extremely important phenomena when we consider biochemical solutions. These phenomena govern many of the processes important to a functioning human body.

■ Osmosis

Osmosis *is the passage of a solvent through a semipermeable membrane separating a dilute solution (or pure solvent) from a more concentrated solution.* The simple apparatus shown in Figure 8.13a is helpful in explaining, at the molecular level, what actually occurs during the osmotic process. The apparatus consists of a tube containing a concentrated salt–water solution that has been immersed in a dilute salt–water solution. The immersed end of the tube is covered with a semipermeable membrane. A **semipermeable membrane** *is a membrane that allows certain types of molecules to pass through it but prohibits the passage of other types of molecules.* The selectivity of a semipermeable

FIGURE 8.13 (a) Osmosis, the flow of solvent through a semipermeable membrane from a dilute to a more concentrated solution, can be observed with this apparatus. (b) The liquid level in the tube rises until equilibrium is reached. At equilibrium, the molecules move back and forth at equal rates.

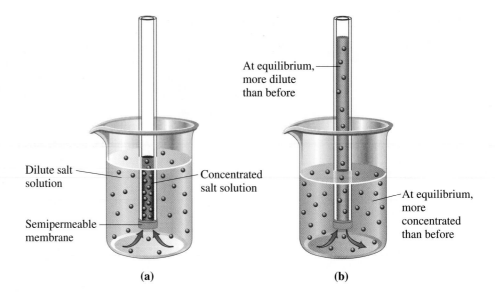

Dilute salt solution

Concentrated salt solution

Semipermeable membrane

At equilibrium, more dilute than before

At equilibrium, more concentrated than before

(a)   **(b)**

▶ An osmotic semipermeable membrane contains very small pores (holes)—too small to see—that are big enough to let small solvent molecules through but not big enough to let larger solute molecules pass through.

membrane is based on size differences between molecules. The particles that are allowed to pass through (usually just solvent molecules like water) are relatively small. Thus, the membrane functions somewhat like a sieve. Using the experimental setup of Figure 8.13a, we can observe a net flow of solvent from the dilute to the concentrated solution over the course of time. This is indicated by a rise in the level of the solution in the tube and a drop in the level of the dilute solution, as shown in Figure 8.13b.

What is actually happening on a molecular level as the process of osmosis occurs? Water is flowing in both directions through the membrane. However, the rate of flow into the concentrated solution is greater than the rate of flow in the other direction (see Figure 8.14). Why? The presence of solute molecules diminishes the ability of water molecules to cross the membrane. The solute molecules literally get in the way; they occupy some of the surface positions next to the membrane. Because there is a greater concentration of solute molecules on one side of the membrane than on the other, the flow rates differ. The flow rate is diminished to a greater extent on the side of the membrane where the greater concentration of solute is present.

The net transfer of solvent across the membrane continues until (1) the concentrations of solute particles on both sides of the membrane become equal or (2) the hydrostatic pressure on the concentrated side of the membrane (from the difference in liquid levels) becomes sufficient to counterbalance the greater escaping tendency of molecules from the dilute side. From here on, there is an equal flow of solvent in both directions across the membrane, and the volume of liquid on each side of the membrane remains constant.

▶ A process called *reverse osmosis* is used in the desalination of seawater to make drinking water. Pressure greater than the osmotic pressure is applied on the salt water side of the membrane to force solvent water across the membrane from the salt water side to the "pure" water side.

FIGURE 8.14 Enlarged views of a semipermeable membrane separating (a) pure water and a salt–water solution, and (b) a dilute salt–water solution and a concentrated salt–water solution. In both cases, water moves from the area of lower solute concentration to the area of higher solute concentration.

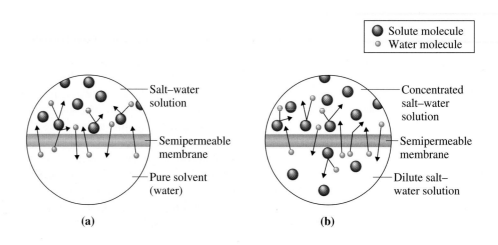

● Solute molecule
○ Water molecule

Salt–water solution

Semipermeable membrane

Pure solvent (water)

Concentrated salt–water solution

Semipermeable membrane

Dilute salt–water solution

(a)   **(b)**

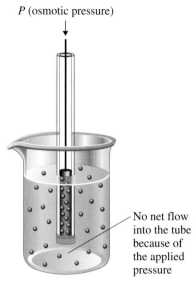

P (osmotic pressure)

No net flow into the tube because of the applied pressure

FIGURE 8.15 Osmotic pressure is the amount of pressure needed to prevent the solution in the tube from rising as a result of the process of osmosis.

▶ Osmolarity is greater for ionic solutes than for molecular solutes (solutes that do not separate into ions, such as glucose and sucrose) if the concentrations of the solutions are equal, because ionic solutes dissociate to form more than 1 mole of particles per mole of compound.

▶ The concept of osmolarity also applies to freezing-point depression and to boiling-point elevation (Section 8.7). The freezing-point depression for a 0.1 M NaCl solution ($i = 2$) is twice that for a 0.1 M glucose solution ($i = 1$).

■ Osmotic Pressure

Osmotic pressure *is the pressure that must be applied to prevent the net flow of solvent through a semipermeable membrane from a solution of lower solute concentration to a solution of higher solute concentration.* In terms of Figure 8.13, osmotic pressure is the pressure required to prevent water from rising in the tube. Figure 8.15 shows how this pressure can be measured. The greater the concentration difference between the separated solutions, the greater the magnitude of the osmotic pressure.

Cell membranes in both plants and animals are semipermeable in nature (see Figure 8.16). The selective passage of fluid materials through these membranes governs the balance of fluids in living systems. Thus osmotic-type phenomena are of prime importance for life. We say "osmotic-type phenomena" instead of "osmosis" because the semipermeable membranes found in living cells usually permit the passage of small solute molecules (nutrients and waste products) in addition to solvent. The term *osmosis* implies the passage of solvent only. The substances prohibited from passing through the membrane in osmotic-type processes are colloidal-sized molecules and insoluble suspended materials (see Section 8.9).

It is because of an osmotic-type process that plants will die if they are watered with salt water. The salt solution outside the root membranes is more concentrated than the solution in the root, so water flows out of the roots; then the plant becomes dehydrated and dies. This same principle is the reason for not drinking excessive amounts of salt water, even if you are stranded on a raft in the middle of the ocean. When salt water is taken into the stomach, water flows out of the stomach wall membranes and into the stomach; then the tissues become dehydrated. Drinking seawater will cause greater thirst because the body will lose water rather than absorb it.

■ Osmolarity

The osmotic pressure of a solution depends on the number of solute particles present. This in turn depends on the solute concentration and on whether the solute forms ions once it is in solution. Note that two factors are involved in determining osmotic pressure.

The fact that some solutes dissociate into ions in solution is of utmost importance in osmotic pressure considerations. For example, the osmotic pressure of a 1 M NaCl solution is twice that of a 1 M glucose solution, despite the fact that both solutions have equal concentrations (1 M). Sodium chloride is an ionic solute, and it dissociates in solution to give two particles (a Na^+ and a Cl^- ion) per formula unit; however, glucose is a molecular solute and does not dissociate. It is the number of particles present that determines osmotic pressure.

The concentration unit osmolarity is used to compare the osmotic pressures of solutions. **Osmolarity** *is the product of a solution's molarity and the number of particles produced per formula unit if the solute dissociates.* The equation for osmolarity is

$$\text{Osmolarity} = \text{molarity} \times i$$

where *i* is the number of particles produced from the dissociation of one formula unit of solute. The abbreviation for osmolarity is osmol.

EXAMPLE 8.9

Calculating the Osmolarity of Various Solutions

■ What is the osmolarity of each of the following solutions?

a. 2 M NaCl   **b.** 2 M CaCl₂   **c.** 2 M glucose
d. 2 M in both NaCl and glucose
e. 2 M in NaCl and 1 M in glucose

Solution

The general equation for osmolarity will be applicable in each of the parts of the problem.

$$\text{Osmolarity} = \text{molarity} \times i$$

a. Two particles per dissociation are produced when NaCl dissociates in solution.

$$NaCl \longrightarrow Na^+ + Cl^-$$

(continued)

The molarity of a 5.0%(m/v) glucose solution is 0.31 M. The molarity of a 0.92%(m/v) NaCl solution is 0.16 M. Despite the differing molarities, these two solutions have the same osmotic pressure. The concept of osmolarity explains why these solutions of different concentration can exhibit the same osmotic pressure. For glucose $i = 1$ and for NaCl $i = 2$.

FIGURE 8.16 The dissolved substances in tree sap create a more concentrated solution than the surrounding ground water. Water enters membranes in the roots and rises in the tree, creating an osmotic pressure that can exceed 20 atm in extremely tall trees.

The pickling of cucumbers and salt curing of meat are practical applications of the concept of crenation. A concentrated salt solution (brine) is used to draw water from the cells of the cucumber to produce a pickle. Salt on the surface of the meat preserves the meat by crenation of bacterial cells.

The value of i is 2, and the osmolarity is twice the molarity.

$$\text{Osmolarity} = 2 \text{ M} \times 2 = 4 \text{ osmol}$$

b. For $CaCl_2$, the value of i is 3, because three ions are produced from the dissociation of one $CaCl_2$ unit.

$$CaCl_2 \longrightarrow Ca^{2+} + 2\ Cl^-$$

The osmolarity will therefore be triple the molarity:

$$\text{Osmolarity} = 2 \text{ M} \times 3 = 6 \text{ osmol}$$

c. Glucose is a nondissociating solute. Thus the value of i is 1, and the molarity and osmolarity will be the same—two molar and two osmolar.

d. With two solutes present, we must consider the collective effects of both solutes. For NaCl, $i = 2$; and for glucose, $i = 1$. The osmolarity is calculated as follows:

$$\text{Osmolarity} = \underbrace{2 \text{ M} \times 2}_{\text{NaCl}} + \underbrace{2 \text{ M} \times 1}_{\text{glucose}} = 6 \text{ osmol}$$

e. This problem differs from the previous one in that the two solutes are not present in equal concentrations. This does not change the way we work the problem. The i values are the same as before, and the osmolarity is

$$\text{Osmolarity} = \underbrace{2 \text{ M} \times 2}_{\text{NaCl}} + \underbrace{1 \text{ M} \times 1}_{\text{glucose}} = 5 \text{ osmol}$$

Practice Exercise 8.9

What is the osmolarity of each of the following solutions?

a. 3 M $NaNO_3$ **b.** 3 M $Ca(NO_3)_2$ **c.** 3 M sucrose
d. 3 M in both $Ca(NO_3)_2$ and sucrose
e. 3 M in both $NaNO_3$ and $Ca(NO_3)_2$

Solutions of equal osmolarity have equal osmotic pressures. If the osmolarity of one solution is three times that of another, then the osmotic pressure of the first solution is three times that of the second solution. A solution with high osmotic pressure will take up more water than a solution of lower osmotic pressure; thus more pressure must be applied to prevent osmosis.

■ Isotonic, Hypertonic, and Hypotonic Solutions

The terms *isotonic solution, hypertonic solution,* and *hypotonic solution* pertain to osmotic-type phenomena that occur in the human body. A consideration of what happens to red blood cells when they are placed in three different liquids will help us understand the differences in meaning of these three terms. The liquid media are distilled water, concentrated sodium chloride solution, and physiological saline solution.

When red blood cells are placed in pure water, they swell up (enlarge in size) and finally rupture (burst); this process is called *hemolysis* (Figure 8.17a). Hemolysis is caused by an increase in the amount of water entering the cells compared with the amount of water leaving the cells. This is the result of cell fluid having a greater osmotic pressure than pure water.

When red blood cells are placed in a concentrated sodium chloride solution, a process opposite to hemolysis occurs. This time, water moves from the cells to the solution, causing the cells to shrivel (shrink in size); this process is called *crenation* (Figure 8.17b). Crenation occurs because the osmotic pressure of the concentrated salt solution surrounding the red cells is greater than that of the fluid within the cells. Water always moves in the direction of greater osmotic pressure.

Finally, when red blood cells are placed in physiological saline solution, a 0.9%(m/v) sodium chloride solution, water flow is balanced and neither hemolysis nor crenation oc-

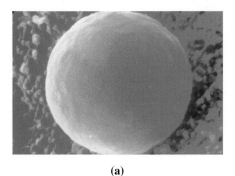
(a)

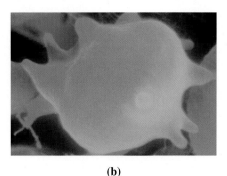
(b)

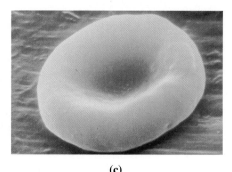
(c)

FIGURE 8.17 Effects of bathing red blood cells in various types of solutions. (a) Hemolysis in pure water (a) hypotonic solution). (b) Crenation in concentrated sodium chloride solution (a hypertonic solution). (c) Cells neither swell nor shrink in physiological saline solution (an isotonic solution).

▶ The word *tonicity* refers to the tone, or firmness, of a biological cell.

▶ The terminology "D5W," often heard in television shows involving doctors and paramedics, refers to a 5%(m/v) solution of glucose (also called dextrose, D) in water (W).

▶ The use of 5%(m/v) glucose solution for intravenous feeding has a shortcoming. A patient can accommodate only about 3 L of water in a day. Three liters of 5%(m/v) glucose water will supply only about 640 kcal of energy, an inadequate amount of energy. A resting patient requires about 1400 kcal/day.

This problem is solved by using solutions that are about 6 times as concentrated as isotonic solutions. They are administered, through a tube, directly into a large blood vessel leading to the heart (the superior vena cava) rather than through a small vein in the arm or leg. The large volume of blood flowing through this vein quickly dilutes the solution to levels that do not upset the osmotic balance in body fluids. Using this technique, patients can be given up to 5000 kcal/day of nourishment.

curs (Figure 8.17c). The osmotic pressure of physiological saline solution is the same as that of red blood cell fluid. Thus the rates of water flow into and out of the red blood cells are the same.

We will now define the terms *isotonic, hypotonic,* and *hypertonic.* An **isotonic solution** *is a solution with an osmotic pressure that is equal to that within cells.* Red blood cell fluid, physiological saline solution, and 5%(m/v) glucose water are all isotonic with respect to one another. The processes of replacing body fluids and supplying nutrients to the body intravenously require the use of isotonic solutions such as physiological saline and glucose water. If isotonic solutions were not used, the damaging effects of hemolysis or crenation would occur.

A **hypotonic solution** *is a solution with a lower osmotic pressure than that within cells.* The prefix *hypo-* means "under" or "less than normal." Distilled water is hypotonic with respect to red blood cell fluid, and these cells will hemolyze when placed in it (Figure 8.17a). A **hypertonic solution** *is a solution with a higher osmotic pressure than that within cells.* The prefix *hyper-* means "over" or "more than normal." Concentrated sodium chloride solution is hypertonic with respect to red blood cell fluid, and these cells undergo crenation when placed in it (Figure 8.17b).

It is sometimes necessary to introduce a hypertonic or hypotonic solution, under controlled conditions, into the body to correct an improper "water balance" in a patient. A hypertonic solution will cause the net transfer of water from tissues to blood; then the kidneys will remove the water. Some laxatives, such as Epsom salts, act by forming hypertonic solutions in the intestines. A hypotonic solution can be used to cause water to flow from the blood into surrounding tissue; blood pressure can be decreased in this manner. Table 8.3 summarizes the differences in meaning among the terms *isotonic, hypertonic,* and *hypotonic.*

The Chemistry at a Glance on page 198 summarizes this chapter's discussion of colligative properties of solutions.

TABLE 8.3

Characteristics of Isotonic, Hypertonic, and Hypotonic Solutions

	Type of Solution		
	Isotonic	Hypertonic	Hypotonic
osmolarity relative to body fluids	equal	greater than	less than
osmotic pressure relative to body fluids	equal	greater than	less than
osmotic effect on cells	equal water flow into and out of cells	net flow of water out of cells	net flow of water into cells

Colligative Properties of Solutions

COLLIGATIVE PROPERTIES OF SOLUTIONS

The physical properties of a solution that depend only on the concentration of solute particles in a given quantity of solute, not on the chemical identity of the particles.

VAPOR-PRESSURE LOWERING

Addition of a nonvolatile solute to a solvent makes the vapor pressure of the solution LOWER than that of the solvent alone.

BOILING-POINT ELEVATION

Addition of a nonvolatile solute to a solvent makes the boiling point of the solution HIGHER than that of the solvent alone.

FREEZING-POINT DEPRESSION

Addition of a nonvolatile solute to a solvent makes the freezing point of the solution LOWER than that of the solvent alone.

OSMOTIC PRESSURE

The pressure required to stop the net flow of water across a semipermeable membrane separating solutions of differing composition.

OSMOLARITY

Osmolarity = molarity $\times i$, where i = number of particles from the dissociation of one formula unit of solute.

HYPERTONIC SOLUTION

- Solution with an osmotic pressure HIGHER than that in cells.
- Causes cells to crenate (shrink).

ISOTONIC SOLUTION

- Solution with an osmotic pressure EQUAL to that in cells.
- Has no effect on cell size.

HYPOTONIC SOLUTION

- Solution with an osmotic pressure LOWER than that in cells.
- Causes cells to hemolyze (burst).

8.9 Dialysis

▶ In *osmosis,* only solvent passes through the membrane. In *dialysis,* both solvent and small solute particles (ions and small molecules) pass through the membrane.

Dialysis is closely related to osmosis. It is the osmotic-type process that occurs in living systems. Osmosis, you recall (Section 8.8), occurs when a solution and a solvent are separated by a semipermeable membrane that allows solvent but not solute to pass through it. There is a net transfer of solvent from the dilute solution (or pure solvent) into the more concentrated solution. **Dialysis** *is the process in which a semipermeable membrane*

FIGURE 8.18 In dialysis, there is a net movement of ions from a region of higher concentration to a region of lower concentration. (a) Before dialysis. (b) After dialysis.

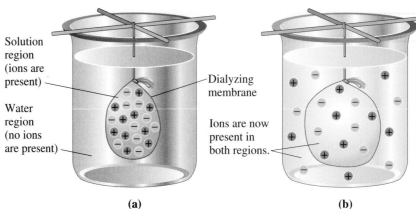

Solution region (ions are present)

Water region (no ions are present)

Dialyzing membrane

Ions are now present in both regions.

(a)   (b)

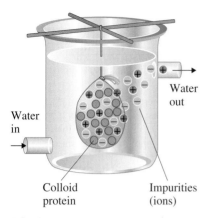

FIGURE 8.19 Impurities (ions) can be removed from a colloidal dispersion by using a dialysis procedure.

The food taken into our bodies consists primarily of molecules too large to cross cellular membranes. Digestion of food converts these large molecules into smaller molecules that can cross the membranes of cells in the intestinal wall, enter the bloodstream, and then enter cells throughout the body where they are used to produce the energy needed to "run" the body.

allows the passage of solvent, dissolved ions, and small molecules but blocks the passage of colloidal-sized particles and large molecules. Thus dialysis allows for the separation of small particles from colloids and large molecules. Many plant and animal membranes function as dialyzing membranes.

Consider the placement of an aqueous solution of sodium chloride in a dialyzing bag that is surrounded by water (Figure 8.18a). What happens? Sodium ions and chloride ions move through the dialyzing membrane into the water; that is, there is a net movement of ions from a region of high concentration to a region of low concentration. This will occur until both sides of the membrane have equal concentrations of ions (Figure 8.18b).

Dialysis can be used to purify a colloidal solution containing protein molecules and solute. The smaller solute molecules pass through the dialyzing membrane and leave the solution. The larger protein molecules remain behind. The result is a purified protein colloidal dispersion (Figure 8.19).

The human kidneys are a complex dialyzing system that is responsible for removing waste products from the blood. The removed products are then eliminated in urine. When the kidneys fail, these waste products build up and eventually poison the body.

When a person goes into shock, there is a sudden increase in the permeability of the membranes of the blood capillaries. Large colloidally dispersed molecules, such as proteins, leave the bloodstream and leak into the space between cells. This damage disrupts the normal chemistry of the blood. If a patient in shock is left untreated, death can occur.

CHEMICAL CONNECTIONS

The Artificial Kidney: A Hemodialysis Machine

Individuals who once would have died of kidney failure can now be helped through the use of artificial-kidney machines, which clean the blood of toxic waste products. In these machines, the blood is pumped through tubing made of dialyzing membrane. The tubing passes through a water bath that collects the impurities from the blood. Blood proteins and other important large molecules remain in the blood. This procedure to cleanse the blood is called *hemodialysis.*

In hemodialysis, a catheter is attached to a major artery of one arm, and the patient's blood is passed through a collection of tiny tubes with a carefully selected pore size. These tubes are immersed in a bath (dialyzing solution) that is isotonic in the normal components of blood. The isotonic solution consists of 0.6%(m/v) NaCl, 0.04%(m/v) KCl, 0.2%(m/v) NaHCO$_3$, and 1.5%(m/v) glucose. This solution does not contain urea or other wastes, which diffuse from the blood through the membrane and into the dialyzing solution.

The accompanying figure illustrates a typical hollow-fiber dialysis device.

Typically, an artificial-kidney patient must receive dialysis treatment two or three times a week, for 4 hours per treatment, in order to maintain proper health. A kidney transplant is preferable to many years on hemodialysis. However, kidney transplants are possible only when the donor kidneys are close tissue matches to the recipient.

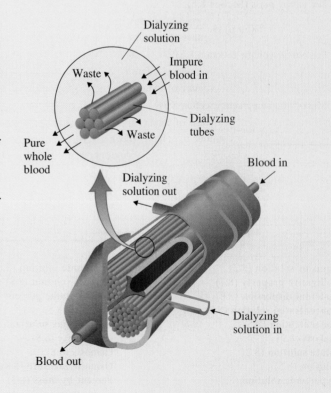

CONCEPTS TO REMEMBER

Solution components. The component of a solution that is present in the greatest amount is the *solvent.* A *solute* is a solution component that is present in a small amount relative to the solvent.

Solution characteristics. A solution is a homogeneous (uniform) mixture. Its composition and properties are dependent on the ratio of solute(s) to solvent. Dissolved solutes are present as individual particles (molecules, atoms, or ions).

Solubility. The solubility of a solute is the maximum amount of solute that will dissolve in a given amount of solvent. The extent to which a solute dissolves in a solvent depends on the structure of solute and solvent, the temperature, and the pressure. Molecular polarity is a particularly important factor in determining solubility. A saturated solution contains the maximum amount of solute that can be dissolved under the conditions at which the solution exists.

Solution concentration. Solution concentration is the amount of solute present in a specified amount of solution. Percent solute and molarity are commonly encountered concentration units. Percent concentration units include percent by mass, percent by volume, and mass–volume percent. Molarity gives the moles of solute per liter of solution.

Colloidal dispersion. A colloidal dispersion is a dispersion (suspension) of small particles of one substance in another substance. Colloidal dispersions differ from true solutions in that the dispersed particles are large enough to scatter light even though they cannot be seen with the naked eye. Many different biochemical colloidal dispersions occur within the human body.

Colligative properties of solutions. Properties of a solution that depend on the number of solute particles in solution, not on their identity, are called colligative properties. Vapor pressure lowering, boiling-point elevation, freezing-point depression, and osmotic pressure are all colligative properties.

Osmosis and osmotic pressure. Osmosis involves the passage of a solvent from a dilute solution (or pure solvent) through a semipermeable membrane into a more concentrated solution. Osmotic pressure is the amount of pressure needed to prevent the net flow of solvent across the membrane in the direction of the more concentrated solution.

Dialysis. Dialysis is the process in which a semipermeable membrane permits the passage of solvent, dissolved ions, and small molecules but blocks the passage of large molecules. Many plant and animal membranes function as dialyzing membranes.

KEY REACTIONS AND EQUATIONS

1. Percent by mass (Section 8.5)

$$\%(m/m) = \frac{\text{mass of solute}}{\text{mass of solution}} \times 100$$

2. Percent by volume (Section 8.5)

$$\%(v/v) = \frac{\text{volume of solute}}{\text{volume of solution}} \times 100$$

3. Mass–volume percent (Section 8.5)

$$\%(m/v) = \frac{\text{mass of solute (g)}}{\text{volume of solution (mL)}} \times 100$$

4. Molarity (Section 8.5)

$$M = \frac{\text{moles of solute}}{\text{liters of solution}}$$

5. Dilution of stock solution to make less-concentrated solution (Section 8.5)

$$C_s \times V_s = C_d \times V_d$$

6. Osmolarity (Section 8.8)

$$osmol = M \times i$$

KEY TERMS

Aqueous solution (8.2)
Colligative property (8.7)
Colloidal dispersion (8.6)
Concentrated solution (8.2)
Concentration (8.5)
Dialysis (8.9)
Dilute solution (8.2)
Dilution (8.5)
Hypertonic solution (8.8)

Hypotonic solution (8.8)
Isotonic solution (8.8)
Mass–volume percent (8.2)
Molarity (8.5)
Nonaqueous solution (8.2)
Osmolarity (8.8)
Osmosis (8.8)
Osmotic pressure (8.8)
Percent by mass (8.5)

Percent by volume (8.5)
Saturated solution (8.2)
Semipermeable membrane (8.8)
Solubility (8.2)
Solute (8.1)
Solution (8.1)
Solvent (8.1)
Supersaturated solution (8.2)
Unsaturated solution (8.2)

EXERCISES AND PROBLEMS

The members of each pair of problems in this section test similar material.

Solution Characteristics (Section 8.1)

8.1 Indicate whether each of the following statements about the general properties of solutions is true or false.
   a. A solution may contain more than one solute.
   b. All solutions are homogeneous mixtures.
   c. Every part of a solution has exactly the same properties as every other part.
   d. The solutes present in a solution will "settle out" with time if the solution is left undisturbed.

8.2 Indicate whether each of the following statements about the general properties of solutions is true or false.
   a. All solutions have a variable composition.
   b. For solution formation to occur, the solute and solvent must chemically react with each other.
   c. Solutes are present as individual particles (molecules, atoms, or ions) in a solution.
   d. A general characteristic of all solutions is the liquid state.

8.3 Identify the *solute* and the *solvent* in solutions composed of the following:
   a. 5.00 g of sodium chloride (table salt) and 50.0 g of water
   b. 4.00 g of sucrose (table sugar) and 1000 g of water
   c. 2.00 mL of water and 20.0 mL of ethyl alcohol
   d. 60.0 mL of methyl alcohol and 20.0 mL of ethyl alcohol

8.4 Identify the *solute* and the *solvent* in solutions composed of the following:
   a. 5.00 g of NaBr and 200.0 g of water
   b. 50.0 g of $AgNO_3$ and 1000 g of water
   c. 50.0 mL of water and 100.0 mL of methyl alcohol
   d. 50.0 mL of isopropyl alcohol and 20.0 mL of ethyl alcohol

Solubility (Section 8.2)

8.5 For each of the following pairs of solutions, select the solution for which solute solubility is greatest.
   a. Ammonia gas in water with P = 1 atm and T = 50°C
      Ammonia gas in water with P = 1 atm and T = 90°C
   b. Carbon dioxide gas in water with P = 2 atm and T = 50°C
      Carbon dioxide gas in water with P = 1 atm and T = 50°C
   c. Table salt in water with P = 1 atm and T = 60°C
      Table salt in water with P = 1 atm and T = 50°C
   d. Table sugar in water with P = 2 atm and T = 40°C
      Table sugar in water with P = 1 atm and T = 70°C

8.6 For each of the following pairs of solutions, select the solution for which solute solubility is greatest.
   a. Oxygen gas in water with P = 1 atm and T = 10°C
      Oxygen gas in water with P = 1 atm and T = 20°C
   b. Nitrogen gas in water with P = 2 atm and T = 50°C
      Nitrogen gas in water with P = 1 atm and T = 70°C
   c. Table salt in water with P = 1 atm and T = 40°C
      Table salt in water with P = 1 atm and T = 70°C
   d. Table sugar in water with P = 3 atm and T = 30°C
      Table sugar in water with P = 1 atm and T = 80°C

8.7 Use Table 8.1 to determine whether each of the following solutions is *saturated* or *unsaturated*.
   a. 1.94 g of $PbBr_2$ in 100 g of H_2O at 50°C
   b. 34.0 g of NaCl in 100 g of H_2O at 0°C
   c. 75.4 g of $CuSO_4$ in 200 g of H_2O at 100°C
   d. 0.540 g of Ag_2SO_4 in 50 g of H_2O at 50°C

8.8 Use Table 8.1 to determine whether each of the following solutions is *saturated* or *unsaturated*.
   a. 175 g of CsCl in 100 g of H_2O at 100°C
   b. 455 g of $AgNO_3$ in 100 g of H_2O at 50°C
   c. 2.16 g of Ag_2SO_4 in 200 g of H_2O at 50°C
   d. 0.97 g of $PbBr_2$ in 50 g of H_2O at 50°C

8.9 Use Table 8.1 to determine whether each of the following solutions is *dilute* or *concentrated*.
   a. 0.20 g of $CuSO_4$ in 100 g of H_2O at 100°C
   b. 1.50 g of $PbBr_2$ in 100 g of H_2O at 50°C
   c. 61 g of $AgNO_3$ in 100 g of H_2O at 50°C
   d. 0.50 g of Ag_2SO_4 in 100 g of H_2O at 0°C

8.10 Use Table 8.1 to determine whether each of the following solutions is *dilute* or *concentrated*.
   a. 255 g of $AgNO_3$ in 100 g of H_2O at 100°C
   b. 35.0 g of NaCl in 100 g of H_2O at 0°C
   c. 1.87 g of $PbBr_2$ in 100 g of H_2O at 50°C
   d. 1.87 g of $CuSO_4$ in 100 g of H_2O at 50°C

Solution Formation (Section 8.3)

8.11 Match each of the following statements about the dissolving of the ionic solid NaCl in water with the term *hydrated ion, hydrogen atom,* or *oxygen atom.*
   a. A Na^+ ion surrounded with water molecules
   b. A Cl^- ion surrounded with water molecules
   c. The portion of a water molecule that is attracted to a Na^+ ion
   d. The portion of a water molecule that is attracted to a Cl^- ion

8.12 Match each of the following statements about the dissolving of the ionic solid KBr in water with the term *hydrated ion, hydrogen atom,* or *oxygen atom.*
   a. A K^+ ion surrounded with water molecules
   b. A Br^- ion surrounded with water molecules
   c. The portion of a water molecule that is attracted to a K^+ ion
   d. The portion of a water molecule that is attracted to a Br^- ion

8.13 Indicate whether each of the following actions will *increase* or *decrease* the rate of the dissolving of a sugar cube in water.
   a. Cooling the sugar cube–water mixture
   b. Stirring the sugar cube–water mixture
   c. Breaking the sugar cube up into smaller "chunks"
   d. Crushing the sugar cube to give a granulated form of sugar

8.14 Indicate whether each of the following actions will *increase* or *decrease* the rate of the dissolving of table salt in water.
   a. Heating the table salt–water mixture
   b. Shaking the table salt–water mixture
   c. Heating the table salt prior to adding it to the water
   d. Heating the water prior to its receiving the table salt

Solubility Rules (Section 8.4)

8.15 Predict whether the following solutes are very soluble or slightly soluble in water.
   a. O_2 (a nonpolar gas)   b. CH_3OH (a polar liquid)
   c. CBr_4 (a nonpolar liquid)   d. AgCl (an ionic solid)

8.16 Predict whether the following solutes are very soluble or slightly soluble in water.
   a. NH_3 (a polar gas)   b. N_2 (a nonpolar gas)
   c. C_6H_6 (a nonpolar liquid)   d. Na_3PO_4 (an ionic solid)

8.17 Classify each of the following types of ionic compounds in the solubility categories *soluble, soluble with exceptions, insoluble,* or *insoluble with exceptions.*
a. Chlorides and sulfates
b. Nitrates and ammonium–ion containing
c. Carbonates and phosphates
d. Sodium–ion containing and potassium–ion containing

8.18 Classify each of the following types of ionic compounds in the solubility categories *soluble, soluble with exceptions, insoluble,* or *insoluble with exceptions.*
a. Nitrates and sodium–ion containing
b. Chlorides and bromides
c. Hydroxides and phosphates
d. Sulfates and iodides

8.19 In each of the following sets of ionic compounds, identify the members of the set that are soluble in water.
a. $NaCl$, Na_2SO_4, $NaNO_3$, Na_2CO_3
b. $AgNO_3$, KNO_3, $Ca(NO_3)_2$, $Cu(NO_3)_2$
c. $CaBr_2$, $Ca(OH)_2$, $CaCl_2$, $CaSO_4$
d. $NiSO_4$, $Ni_3(PO_4)_2$, $Ni(OH)_2$, $NiCO_3$

8.20 In each of the following sets of ionic compounds, identify the members of the set that are soluble in water.
a. K_2SO_4, KOH, KI, K_3PO_4
b. $NaCl$, $AgCl$, $BeCl_2$, $CuCl_2$
c. $Ba(OH)_2$, $BaSO_4$, $BaCO_3$, $Ba(NO_3)_2$
d. $CoBr_2$, $CoCl_2$, $Co(OH)_2$, $CoSO_4$

Percent Concentration Units (Section 8.5)

8.21 Calculate the mass percent of solute in the following solutions.
a. 6.50 g of $NaCl$ dissolved in 85.0 g of H_2O
b. 2.31 g of $LiBr$ dissolved in 35.0 g of H_2O
c. 12.5 g of KNO_3 dissolved in 125 g of H_2O
d. 0.0032 g of $NaOH$ dissolved in 1.2 g of H_2O

8.22 Calculate the mass percent of solute in the following solutions.
a. 2.13 g of $AgNO_3$ dissolved in 30.0 g of H_2O
b. 135 g of $CsCl$ dissolved in 455 g of H_2O
c. 10.3 g of K_2SO_4 dissolved in 93.7 g of H_2O
d. 10.3 g of KBr dissolved in 125 g of H_2O

8.23 How many grams of glucose must be added to 275 g of water in order to prepare each of the following percent-by-mass concentrations of aqueous glucose solution?
a. 1.30%   b. 5.00%   c. 20.0%   d. 31.0%

8.24 How many grams of lactose must be added to 655 g of water in order to prepare each of the following percent-by-mass concentrations of aqueous lactose solution?
a. 0.50%   b. 2.00%   c. 10.0%   d. 25.0%

8.25 Calculate the mass, in grams, of K_2SO_4 needed to prepare 32.00 g of 2.000%(m/m) K_2SO_4 solution.

8.26 Calculate the mass, in grams, of KCl needed to prepare 200.0 g of 5.000%(m/m) KCl solution.

8.27 How many grams of water must be added to 20.0 g of $NaOH$ in order to prepare a 6.75%(m/m) solution?

8.28 How many grams of water must be added to 10.0 g of $Ca(NO_3)_2$ in order to prepare a 12.0%(m/m) solution?

8.29 Calculate the volume percent of solute in each of the following solutions.
a. 20.0 mL of methyl alcohol in enough water to give 475 mL of solution

b. 4.00 mL of bromine in enough carbon tetrachloride to give 87.0 mL of solution

8.30 Calculate the volume percent of solute in each of the following solutions.
a. 60.0 mL of water in enough ethylene glycol to give 970.0 mL of solution
b. 455 mL of ethyl alcohol in enough water to give 1375 mL of solution

8.31 What is the percent by volume of isopropyl alcohol in an aqueous solution made by diluting 22 mL of pure isopropyl alcohol with water to give a volume of 125 mL of solution?

8.32 What is the percent by volume of acetone in an aqueous solution made by diluting 75 mL of pure acetone with water to give a volume of 785 mL of solution?

8.33 Calculate the mass–volume percent of $MgCl_2$ in each of the following solutions.
a. 5.0 g of $MgCl_2$ in enough water to give 250 mL of solution
b. 85 g of $MgCl_2$ in enough water to give 580 mL of solution

8.34 Calculate the mass–volume percent of $NaNO_3$ in each of the following solutions.
a. 1.00 g of $NaNO_3$ in enough water to give 75.0 mL of solution
b. 100.0 g of $NaNO_3$ in enough water to give 1250 mL of solution

8.35 How many grams of Na_2CO_3 are needed to prepare 25.0 mL of a 2.00%(m/v) Na_2CO_3 solution?

8.36 How many grams of $Na_2S_2O_3$ are needed to prepare 50.0 mL of a 5.00%(m/v) $Na_2S_2O_3$ solution?

8.37 How many grams of $NaCl$ are present in 50.0 mL of a 7.50%(m/v) $NaCl$ solution?

8.38 How many grams of glucose are present in 250.0 mL of a 10.0%(m/v) glucose solution?

Molarity (Section 8.5)

8.39 Calculate the molarity of the following solutions.
a. 3.0 moles of potassium nitrate (KNO_3) in 0.50 L of solution
b. 12.5 g of sucrose ($C_{12}H_{22}O_{11}$) in 80.0 mL of solution
c. 25.0 g of sodium chloride ($NaCl$) in 1250 mL of solution
d. 0.00125 mole of baking soda ($NaHCO_3$) in 2.50 mL of solution

8.40 Calculate the molarity of the following solutions.
a. 2.0 moles of ammonium chloride (NH_4Cl) in 2.50 L of solution
b. 14.0 g of silver nitrate ($AgNO_3$) in 1.00 L of solution
c. 0.025 mole of potassium chloride (KCl) in 50.0 mL of solution
d. 25.0 g of glucose ($C_6H_{12}O_6$) in 1.25 L of solution

8.41 Calculate the number of grams of solute in each of the following solutions.
a. 2.50 L of a 3.00 M HCl solution
b. 10.0 mL of a 0.500 M KCl solution
c. 875 mL of a 1.83 M $NaNO_3$ solution
d. 75 mL of a 12.0 M H_2SO_4 solution

8.42 Calculate the number of grams of solute in each of the following solutions.

a. 3.00 L of a 2.50 M HCl solution
b. 50.0 mL of a 12.0 M HNO_3 solution
c. 50.0 mL of a 12.0 M $AgNO_3$ solution
d. 1.20 L of a 0.032 M Na_2SO_4 solution

8.43 Calculate the volume, in milliliters, of solution required to supply each of the following.
   a. 1.00 g of sodium chloride (NaCl) from a 0.200 M sodium chloride solution
   b. 2.00 g of glucose ($C_6H_{12}O_6$) from a 4.20 M glucose solution
   c. 3.67 moles of silver nitrate ($AgNO_3$) from a 0.400 M silver nitrate solution
   d. 0.0021 mole of sucrose ($C_{12}H_{22}O_{11}$) from an 8.7 M sucrose solution

8.44 Calculate the volume, in milliliters, of solution required to supply each of the following.
   a. 4.30 g of lithium chloride (LiCl) from a 0.089 M lithium chloride solution
   b. 429 g of lithium nitrate ($LiNO_3$) from an 11.2 M lithium nitrate solution
   c. 2.25 moles of potassium sulfate (K_2SO_4) from a 0.300 M potassium sulfate solution
   d. 0.103 mole of potassium hydroxide (KOH) from an 8.00 M potassium hydroxide solution

Dilution (Section 8.5)

8.45 What is the molarity of the solution prepared by diluting 25.0 mL of 0.220 M NaCl to each of the following final volumes?
   a. 30.0 mL   b. 75.0 mL   c. 457 mL   d. 2.00 L

8.46 What is the molarity of the solution prepared by diluting 35.0 mL of 1.25 M $AgNO_3$ to each of the following final volumes?
   a. 50.0 mL   b. 95.0 mL   c. 975 mL   d. 3.60 L

8.47 For each of the following solutions, how many milliliters of water should be added to yield a solution that has a concentration of 0.100 M?
   a. 50.0 mL of 3.00 M NaCl   b. 2.00 mL of 1.00 M NaCl
   c. 1.45 L of 6.00 M NaCl   d. 75.0 mL of 0.110 M NaCl

8.48 For each of the following solutions, how many milliliters of water should be added to yield a solution that has a concentration of 0.125 M?
   a. 25.0 mL of 1.00 M $AgNO_3$
   b. 5.00 mL of 10.0 M $AgNO_3$
   c. 2.50 L of 2.50 M $AgNO_3$
   d. 75.0 mL of 0.130 M $AgNO_3$

8.49 Determine the final concentration of each of the following solutions after 20.0 mL of water has been added.
   a. 30.0 mL of 5.0 M NaCl solution
   b. 30.0 mL of 5.0 M $AgNO_3$ solution
   c. 30.0 mL of 7.5 M NaCl solution
   d. 60.0 mL of 2.0 M NaCl solution

8.50 Determine the final concentration of each of the following solutions after 30.0 mL of water has been added.
   a. 20.0 mL of 5.0 M NaCl solution
   b. 20.0 mL of 5.0 M $AgNO_3$ solution
   c. 20.0 mL of 0.50 M NaCl solution
   d. 60.0 mL of 3.0 M NaCl solution

Colligative Properties of Solutions (Section 8.7)

8.51 Why is the vapor pressure of a solution that contains a nonvolatile solute always less than that of pure solvent?

8.52 How are the boiling point and freezing point of water affected by the addition of solute?

8.53 Why does seawater evaporate more slowly than fresh water at the same temperature?

8.54 How does the freezing point of seawater compare with that of fresh water?

Osmosis and Osmotic Pressure (Section 8.8)

8.55 Indicate whether the osmotic pressure of a 0.1 M NaCl solution will be *less than, the same as,* or *greater than* that of each of the following solutions.
   a. 0.1 M NaBr   b. 0.050 M $MgCl_2$
   c. 0.1 M $MgCl_2$ d. 0.1 M glucose

8.56 Indicate whether the osmotic pressure of a 0.1 M $NaNO_3$ solution will be *less than, the same as,* or *greater than* that of each of the following solutions.
   a. 0.1 M NaCl   b. 0.1 M KNO_3
   c. 0.1 M Na_2SO_4 d. 0.1 M glucose

8.57 What is the ratio of the osmotic pressures of 0.30 M NaCl and 0.10 M $CaCl_2$?

8.58 What is the ratio of the osmotic pressures of 0.20 M NaCl and 0.30 M $CaCl_2$?

8.59 Would red blood cells *swell, remain the same size,* or *shrink* when placed in each of the following solutions?
   a. 0.9%(m/v) glucose solution
   b. 0.9%(m/v) NaCl solution
   c. 2.3%(m/v) glucose solution
   d. 5.0%(m/v) NaCl solution

8.60 Would red blood cells *swell, remain the same size,* or *shrink* when placed in each of the following solutions?
   a. distilled water
   b. 0.5%(m/v) NaCl solution
   c. 3.3%(m/v) glucose solution
   d. 5.0%(m/v) glucose solution

8.61 Will red blood cells *crenate, hemolyze,* or *remain unaffected* when placed in each of the solutions in Problem 8.59?

8.62 Will red blood cells *crenate, hemolyze,* or *remain unaffected* when placed in each of the solutions in Problem 8.60?

8.63 Classify each of the solutions in Problem 8.59 as isotonic, hypertonic, or hypotonic.

8.64 Classify each of the solutions in Problem 8.60 as isotonic, hypertonic, or hypotonic.

Dialysis (Section 8.9)

8.65 What happens in each of the following situations?
   a. A dialyzing bag containing a 1 M solution of potassium chloride is immersed in pure water.
   b. A dialyzing bag containing colloidal-sized protein, 1 M potassium chloride, and 1 M glucose is immersed in pure water.

8.66 What happens in each of the following situations?
   a. A dialyzing bag containing a 1 M solution of potassium chloride is immersed in a 1 M sodium chloride solution.
   b. A dialyzing bag containing colloidal-sized protein is immersed in a 1 M glucose solution.

ADDITIONAL PROBLEMS

8.67 With the help of Table 8.2, determine in which of the following pairs of compounds both members of the pair have like solubility in water (both soluble or both insoluble).
a. $(NH_4)_2CO_3$ and $AgNO_3$
b. $ZnCl_2$ and $Mg(OH)_2$
c. BaS and $NiCO_3$
d. $AgCl$ and $Al(OH)_3$

8.68 How many grams of solute are dissolved in the following amounts of solution?
a. 134 g of 3.00%(m/m) KNO_3 solution
b. 75.02 g of 9.735%(m/m) $NaOH$ solution
c. 1576 g of 0.800%(m/m) HI solution
d. 1.23 g of 12.0%(m/m) NH_4Cl solution

8.69 What volume of water, in quarts, is contained in 3.50 qt of a 2.00%(v/v) solution of water in acetone?

8.70 How many liters of 0.10 M solution can be prepared from 60.0 g of each of the following solutes?
a. $NaNO_3$
b. HNO_3
c. KOH
d. $LiCl$

8.71 What is the molarity of the solution prepared by concentrating, by evaporation of solvent, 2212 mL of 0.400 M potassium sulfate (K_2SO_4) solution to each of the following final volumes?
a. 1875 mL
b. 1.25 L
c. 853 mL
d. 553 mL

8.72 After all the water is evaporated from 10.0 mL of a $CsCl$ solution, 3.75 of $CsCl$ remains. Express the original concentration of the $CsCl$ solution in each of the following units.
a. mass–volume percent
b. molarity

8.73 Find the molarity of a solution obtained when 352 mL of 4.00 M sodium bromide ($NaBr$) solution is mixed with
a. 225 mL of 4.00 M $NaBr$ solution
b. 225 mL of 2.00 M $NaBr$ solution

8.74 Which of the following aqueous solutions would give rise to a greater osmotic pressure?
a. 8.00 g of $NaCl$ in 375 mL of solution or 4.00 g of $NaBr$ in 155 mL of solution
b. 6.00 g of $NaCl$ in 375 mL of solution or 6.00 g of $MgCl_2$ in 225 mL of solution

ANSWERS TO PRACTICE EXERCISES

8.1 a. soluble b. soluble
c. soluble d. insoluble
e. soluble

8.2 8.7%(m/m) Na_2SO_4

8.3 1.32 g $LiNO_3$

8.4 22.5 g $C_6H_{12}O_6$

8.5 a. 3.65 M KNO_3 b. 0.631 M KCl

8.6 95.6 g $AgNO_3$

8.7 3750 mL $NaOH$ solution

8.8 0.46 M Na_2SO_4

8.9 a. 6 osmol b. 9 osmol c. 3 osmol d. 12 osmol e. 15 osmol

9

Chemical Reactions

A double-replacement reaction involving sulfuric acid (H_2SO_4) contributes to acid-rain-caused corrosion of churches and statues. The reaction is $CaCO_3 + H_2SO_4 \longrightarrow CaSO_4 + H_2CO_3$

In the previous two chapters, we considered the properties of matter in various pure and mixed states. Nearly all of the subject matter dealt with interactions and changes of a *physical* nature. We now concern ourselves with the *chemical* changes that occur when various types of matter interact.

We first consider several types of chemical reactions and then discuss important fundamentals common to all chemical changes. Of particular concern to us will be how fast chemical changes occur (chemical reaction rates) and how far chemical changes go (chemical equilibrium).

9.1 Types of Chemical Reactions

A **chemical reaction** *is a process in which at least one new substance is produced as a result of chemical change.* An almost inconceivable number of chemical reactions are possible. The majority of chemical reactions (but not all) fall into five major categories: *combination* reactions, *decomposition* reactions, *single-replacement* reactions, *double-replacement* reactions, and *combustion* reactions.

■ Combination Reactions

A **combination reaction** *is a reaction in which a single product is produced from two (or more) reactants.* The general equation for a combination reaction involving two reactants is

$$X + Y \longrightarrow XY$$

FIGURE 9.1 When a hot nail is stuck into a pile of zinc and sulfur, a fiery combination reaction occurs and zinc sulfide forms.

$$Zn + S \longrightarrow ZnS$$

▶ In organic chemistry (Chapters 12–17), combination reactions are called *addition reactions*. One reactant, usually a small molecule, is considered to be added to a larger reactant molecule to produce a single product.

▶ In organic chemistry, decomposition reactions are often called *elimination reactions*. In many reactions, including some metabolic reactions that occur in the human body, either H_2O or CO_2 is eliminated from a molecule (a decomposition).

In such a combination reaction, two substances join together to form a more complicated product (see Figure 9.1). The reactants X and Y can be elements or compounds or an element and a compound. The product of the reaction (XY) is always a compound. Some representative combination reactions that have elements as the reactants are

$$Ca + S \longrightarrow CaS$$
$$N_2 + 3H_2 \longrightarrow 2NH_3$$
$$2Na + O_2 \longrightarrow Na_2O_2$$

Some examples of combination reactions in which compounds are involved as reactants are

$$SO_3 + H_2O \longrightarrow H_2SO_4$$
$$2NO + O_2 \longrightarrow 2NO_2$$
$$2NO_2 + H_2O_2 \longrightarrow 2HNO_3$$

■ Decomposition Reactions

A **decomposition reaction** *is a reaction in which a single reactant is converted into two (or more) simpler substances (elements or compounds).* Thus a decomposition reaction is the opposite of a combination reaction. The general equation for a decomposition reaction in which there are two products is

$$XY \longrightarrow X + Y$$

Although the products may be elements or compounds, the reactant is *always* a compound.

At sufficiently high temperatures, all compounds can be broken down (decomposed) into their constituent elements. Examples of such reactions include

$$2CuO \longrightarrow 2Cu + O_2$$
$$2H_2O \longrightarrow 2H_2 + O_2$$

At lower temperatures, compound decomposition often produces other compounds as products.

$$CaCO_3 \longrightarrow CaO + CO_2$$
$$2KClO_3 \longrightarrow 2KCl + 3O_2$$
$$4HNO_3 \longrightarrow 4NO_2 + 2H_2O + O_2$$

Decomposition reactions are easy to recognize in that they are the only type of reaction in which there is only one reactant.

■ Single-Replacement Reactions

A **single-replacement reaction** *is a reaction in which an atom or molecule replaces an atom or group of atoms from a compound.* There are always two reactants and two products in a single-replacement reaction. The general equation for a single-replacement reaction is

$$X + YZ \longrightarrow Y + XZ$$

A common type of single-replacement reaction is one in which an element and a compound are reactants and an element and a compound are products. Examples of this type of single-replacement reaction include

$$Fe + CuSO_4 \longrightarrow Cu + FeSO_4$$
$$Mg + Ni(NO_3)_2 \longrightarrow Ni + Mg(NO_3)_2$$
$$Cl_2 + NiI_2 \longrightarrow I_2 + NiCl_2$$
$$F_2 + 2NaCl \longrightarrow Cl_2 + 2NaF$$

The first two equations illustrate one metal replacing another metal from its compound. The latter two equations illustrate one nonmetal replacing another nonmetal from its compound. A more complicated example of a single-replacement reaction, in which all reactants and products are compounds, is

$$4PH_3 + Ni(CO)_4 \longrightarrow 4CO + Ni(PH_3)_4$$

■ Double-Replacement Reactions

> In organic chemistry, replacement reactions (both single and double) are often called *substitution* reactions. Substitution reactions of the single-replacement type are seldom encountered. However, double-replacement reactions are common in organic chemistry.

A **double-replacement reaction** *is a reaction in which two substances exchange parts with one another and form two different substances.* The general equation for a double-replacement reaction is

$$AX + BY \longrightarrow AY + BX$$

Such reactions can be thought of as involving "partner switching." The AX and BY partnerships are disrupted, and new AY and BX partnerships are formed in their place.

When the reactants in a double-replacement reaction are ionic compounds in solution, the parts exchanged are the positive and negative ions of the compounds present.

$$AgNO_3(aq) + NaCl(aq) \longrightarrow AgCl(s) + NaNO_3(aq)$$
$$2KI(aq) + Pb(NO_3)_2(aq) \longrightarrow 2KNO_3(aq) + PbI_2(s)$$

In most reactions of this type, one of the product compounds is in a different physical state (solid or gas) from that of the reactants (see Figure 9.2). Insoluble solids formed from such a reaction are called *precipitates;* AgCl and PbI$_2$ are precipitates in the foregoing reactions.

■ Combustion Reactions

> Hydrocarbon combustion reactions are the basis of industrial society, making possible the burning of gasoline in cars, of natural gas in homes, and of coal in factories. Gasoline, natural gas, and coal all contain hydrocarbons. Unlike most other chemical reactions, hydrocarbon combustion reactions are carried out for the energy they produce rather than for the material products.

Combustion reactions are a most common type of reaction. A **combustion reaction** *is a reaction between a substance and oxygen (usually from air) that proceeds with the evolution of heat and light (usually from a flame).* Hydrocarbons—binary compounds of carbon and hydrogen (of which many exist)—are the most common type of compound that undergoes combustion. In hydrocarbon combustion, the carbon of the hydrocarbon combines with the oxygen of air to produce carbon dioxide (CO_2). The hydrogen of the hydrocarbon also interacts with the oxygen of air to give water (H_2O) as a product. The relative amounts of CO_2 and H_2O produced depend on the composition of the hydrocarbon.

$$2C_2H_2 + 5O_2 \longrightarrow 4CO_2 + 2H_2O$$
$$C_3H_8 + 5O_2 \longrightarrow 3CO_2 + 4H_2O$$
$$C_4H_8 + 6O_2 \longrightarrow 4CO_2 + 4H_2O$$

FIGURE 9.2 A double-replacement reaction involving solutions of potassium iodide and lead(II) nitrate (both colorless solutions) produces yellow, insoluble lead(II) iodide as one of the products.

$2KI(aq) + Pb(NO_3)_2(aq) \longrightarrow$
$\qquad\qquad 2KNO_3(aq) + PbI_2(s)$

Combination reactions in which oxygen reacts with another element to form a single product are also combustion reactions. Two such reactions are

$$S + O_2 \longrightarrow SO_2$$
$$2Mg + O_2 \longrightarrow 2MgO$$

EXAMPLE 9.1

Classifying Chemical Reactions

■ Classify each of the following reactions as a *combination, decomposition, single-replacement, double-replacement,* or *combustion* reaction.

a. $2KNO_3 \longrightarrow 2KNO_2 + O_2$
b. $Zn + 2AgNO_3 \longrightarrow Zn(NO_3)_2 + 2Ag$
c. $Ni(NO_3)_2 + 2NaOH \longrightarrow Ni(OH)_2 + 2NaNO_3$
d. $3Mg + N_2 \longrightarrow Mg_3N_2$

Solution

a. Decomposition. Two substances are produced from a single substance.
b. Single-replacement. An element and a compound are reactants, and an element and a compound are products.
c. Double-replacement. Two compounds exchange parts with each other; the nickel ion (Ni^{2+}) and sodium ion (Na^+) are "swapping partners."
d. Combination. Two substances combine to form a single substance.

Practice Exercise 9.1

Classify each of the following reactions as a *combination, decomposition, single-replacement, double-replacement,* or *combustion* reaction.

a. $CH_4 + 2O_2 \longrightarrow CO_2 + 2H_2O$
b. $N_2 + 3H_2 \longrightarrow 2NH_3$
c. $Ni + Cu(NO_3)_2 \longrightarrow Cu + Ni(NO_3)_2$
d. $CuCO_3 \longrightarrow CuO + O_2$

Many, *but not all,* reactions fall into one of the five categories we have discussed in this section. Even though this classification system is not all-inclusive, it is still very useful because of the many reactions it does help correlate. The Chemistry at a Glance feature on page 210 summarizes pictorially the reaction types that we have considered in this section.

9.2 Redox and Nonredox Reactions

Chemical reactions can also be classified, in terms of whether transfer of electrons occurs, as either oxidation–reduction (redox) or nonoxidation–reduction (nonredox) reactions. An **oxidation–reduction (redox) reaction** *is a reaction in which there is a transfer of electrons from one reactant to another reactant.* A **nonoxidation–reduction (nonredox) reaction** *is a reaction in which there is no transfer of electrons from one reactant to another reactant.*

A "bookkeeping system" known as oxidation numbers is used to identify whether electron transfer occurs in a chemical reaction. An **oxidation number** *is a number that represents the charge that an atom appears to have when the electrons in each bond it is participating in are assigned to the more electronegative of the two atoms involved in the bond.*

There are several rules for determining oxidation numbers.

1. *The oxidation number of an element in its elemental state is zero.* For example, the oxidation number of copper in Cu is zero, and the oxidation number of chlorine in Cl_2 is zero.

▶ *Oxidation numbers* are also sometimes called *oxidation states.*

2. *The oxidation number of a monatomic ion is equal to the charge on the ion.* For example, the Na^+ ion has an oxidation number of $+1$, and the S^{2-} ion has an oxidation number of -2.

CHEMICAL CONNECTIONS

Combustion Reactions, Carbon Dioxide, and Global Warming

Most fuels used in our society, including coal, petroleum, and natural gas, are carbon-containing substances. When such fuels are burned (combustion; Section 9.1), carbon dioxide is one of the combustion products. Nearly all such combustion-generated CO_2 enters the atmosphere.

Significant amounts of the CO_2 entering the atmosphere are absorbed into the oceans because of this compound's solubility in water, and plants also remove CO_2 from the atmosphere via the process of photosynthesis. However, these removal mechanisms are not sufficient to remove all the combustion-generated CO_2; it is being generated faster than it can be removed. Consequently, atmospheric concentrations of CO_2 are slowly increasing, as the following graph shows.

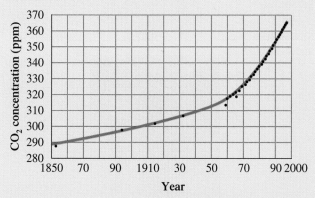

The concentration unit on the vertical axis is parts per million (ppm)—the number of molecules of CO_2 per million molecules present in air. Data for periods before 1958 were derived from analysis of air trapped in bubbles in glacial ice.

Increasing atmospheric CO_2 levels pose an environmental concern because within the atmosphere, CO_2 acts as a heat-trapping agent. During the day, Earth receives energy from the sun, mostly in the form of *visible light*. At night, as Earth cools, it re-radiates the energy it received during the day in the form of *infrared light* (heat energy). Carbon dioxide does not absorb visible light, but it has the ability to absorb infrared light. The CO_2 thus traps some of the heat energy re-radiated by the surface of the Earth as it cools, preventing this energy from escaping to outer space. Because this action is similar to that of glass in a greenhouse, CO_2 is called a *greenhouse gas*. The warming caused by CO_2 as it prevents heat loss from Earth is called the *greenhouse effect* or *global warming*.

Some scientists believe that the presence of increased concentrations of CO_2 (and other greenhouse gases present in the atmosphere in lower concentrations than CO_2) is beginning to cause a change in our climate as the result of a small increase in the average temperature of Earth's surface. Some computer models predict an average global temperature increase of 1 to 3°C toward the end of the twenty-first century if atmospheric CO_2 concentrations continue to increase at their current rate. Because numerous other factors are also involved in determining climate, however, predictions cannot be made with certainty. Much research concerning this situation is in progress, and many governments around the world are now trying to reduce the amount of combustion-generated CO_2 that enters the atmosphere.

The other greenhouse gases besides CO_2 include CH_4, N_2O, and CFCs (chlorofluorocarbons). Atmospheric concentrations of these other greenhouse gases are lower than that of CO_2. However, because they are more effective absorbers of infrared radiation than CO_2 is, they make an appreciable contribution to the overall greenhouse effect. Further information about CFCs is given in the Chemical Connections feature "Chlorofluorocarbons and the Ozone Layer" in Chapter 12. Estimated contributions of various greenhouse gases to global warming are as follows:

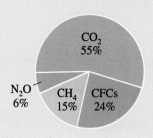

3. *The oxidation numbers of Groups IA and IIA metals are always +1, and +2, respectively.*

4. *The oxidation number of hydrogen is +1 in most hydrogen-containing compounds.*

5. *The oxidation number of oxygen is −2 in most oxygen-containing compounds.*

6. *In binary molecular compounds, the more electronegative element is assigned a negative oxidation number equal to its charge in binary ionic compounds. For example, in CCl_4 the element Cl is the more electronegative, and its oxidation number is −1 (the same as in the simple Cl^- ion).*

7. *For a compound, the sum of the individual oxidation numbers is equal to zero; for a polyatomic ion, the sum is equal to the charge on the ion.*

Example 9.2 illustrates the use of these rules.

Types of Chemical Reactions

COMBINATION REACTION

$$2Al + 3I_2 \longrightarrow 2AlI_3$$

Aluminum reacts with iodine to form aluminum iodide.

DECOMPOSITION REACTION

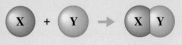

$$2HgO \longrightarrow 2Hg + O_2$$

Mercury(II) oxide decomposes to form mercury and oxygen.

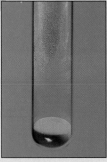

SINGLE-REPLACEMENT REACTION

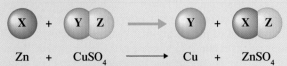

$$Zn + CuSO_4 \longrightarrow Cu + ZnSO_4$$

Zinc reacts with copper(II) sulfate to form copper and zinc sulfate.

DOUBLE-REPLACEMENT REACTION

$$AgNO_3 + NaCl \longrightarrow AgCl + NaNO_3$$

Silver nitrate reacts with sodium chloride to form silver chloride and sodium nitrate.

EXAMPLE 9.2

Assigning Oxidation Numbers to Elements in a Compound or Polyatomic Ion

■ Assign an oxidation number to each element in the following compounds or polyatomic ions.

a. P_2O_5 **b.** $KMnO_4$ **c.** NO_3^-

Solution

a. The sum of the oxidation numbers of all the atoms present must add to zero (rule 7).

$$2(\text{oxid. no. P}) + 5(\text{oxid. no. O}) = 0$$

The oxidation number of oxygen is -2 (rule 5 or rule 6). Substituting this value into the previous equation enables us to calculate the oxidation number of phosphorus.

$$2(\text{oxid. no. P}) + 5(-2) = 0$$
$$2(\text{oxid. no. P}) = +10$$
$$(\text{oxid no. P}) = +5$$

Thus the oxidation numbers for the elements involved in this compound are

$$P = +5 \quad \text{and} \quad O = -2$$

Note that the oxidation number of phosphorus is not $+10$; that is the calculated charge associated with two phosphorus atoms. Oxidation number is always specified on a *per-atom* basis.

b. The sum of the oxidation numbers of all the atoms present must add to zero (rule 7).

$$(\text{oxid. no. K}) + (\text{oxid. no. Mn}) + 4(\text{oxid. no. O}) = 0$$

The oxidation number of potassium, a Group IA element, is $+1$ (rule 3), and the oxidation number of oxygen is -2 (rule 5). Substituting these two values into the rule 7 equation enables us to calculate the oxidation number of manganese.

$$(+1) + (\text{oxid. no. Mn}) + 4(-2) = 0$$
$$(\text{oxid. no. Mn}) = 8 - 1 = +7$$

Thus the oxidation numbers for the elements involved in this compound are

$$\text{K} = +1 \qquad \text{Mn} = +7 \qquad \text{and} \qquad \text{O} = -2$$

Note that all the oxidation numbers add to zero when it is taken into account that there are four oxygen atoms.

$$(+1) + (+7) + 4(-2) = 0$$

c. The species NO_3^- is a polyatomic ion rather than a neutral compound. Thus the second part of rule 7 applies: The oxidation numbers must add to -1, the charge on the ion.

$$(\text{oxid. no. N}) + 3(\text{oxid. no. O}) = -1$$

The oxidation number of oxygen is -2 (rule 5). Substituting this value into the sum equation gives

$$(\text{oxid. no. N}) + 3(-2) = -1$$
$$(\text{oxid. no. N}) = -1 + 6 = +5$$

Thus the oxidation numbers for the elements involved in the polyatomic ion are

$$\text{N} = +5 \qquad \text{and} \qquad \text{O} = -2$$

Practice Exercise 9.2

Assign oxidation numbers to each element in the following compounds or polyatomic ions.
a. N_2O_4 **b.** $K_2Cr_2O_7$ **c.** NH_4^+

FIGURE 9.3 The burning of calcium metal in chlorine is a redox reaction. The burning calcium emits a red-orange flame.

Many elements display a range of oxidation numbers in their various compounds. For example, nitrogen exhibits oxidation numbers ranging from -3 to $+5$. Selected examples are

$$\underset{-3}{NH_3} \qquad \underset{+1}{N_2O} \qquad \underset{+2}{NO} \qquad \underset{+3}{N_2O_3} \qquad \underset{+4}{NO_2} \qquad \underset{+5}{HNO_3}$$

As shown in this listing of nitrogen-containing compounds, the oxidation number of an atom is written *underneath* the atom in the formula. This convention is used to avoid confusion with the charge on an ion.

To determine whether a reaction is a redox reaction or a nonredox reaction, we look for changes in the oxidation number of elements involved in the reaction. Changes in oxidation number are a requirement for a redox reaction. The reaction between calcium metal and chlorine gas (see Figure 9.3) is a redox reaction.

$$\underset{0}{Ca} + \underset{0}{Cl_2} \longrightarrow \underset{+2\ -1}{CaCl_2}$$

The oxidation number of Ca changes from zero to $+2$, and the oxidation number of Cl changes from zero to -1.

The decomposition of calcium carbonate is a nonredox reaction.

$$CaCO_3 \longrightarrow CaO + CO_2$$

CaCO$_3$	CaO	CO$_2$
+2 +4 −2	+2 −2	+4 −2

It is a nonredox reaction because there are no changes in oxidation number.

EXAMPLE 9.3

Using Oxidation Numbers to Determine Whether a Chemical Reaction Is a Redox Reaction

■ By using oxidation numbers, determine whether the following reaction is a redox reaction or a nonredox reaction.

$$4NH_3 + 3O_2 \longrightarrow 2N_2 + 6H_2O$$

Solution

For the reactant NH$_3$, H has an oxidation number of +1 (rule 4) and N an oxidation number of −3 (rule 7). The other reactant, O$_2$, is an element and thus has an oxidation number of zero (rule 1). The product N$_2$ also has an oxidation number of zero, because it is an element. In H$_2$O, the other product, H has an oxidation number of +1 (rule 4) and oxygen an oxidation number of −2 (rule 5).

The overall oxidation number analysis is

$$4NH_3 + 3O_2 \longrightarrow 2N_2 + 6H_2O$$

4NH$_3$	3O$_2$	2N$_2$	6H$_2$O
−3 +1	0	0	+1 −2

This reaction is a redox reaction because the oxidation numbers of both N and O change.

Practice Exercise 9.3

By using oxidation numbers, determine whether the following reaction is a redox reaction or a nonredox reaction.

$$SO_3 + H_2O \longrightarrow H_2SO_4$$

9.3 Terminology Associated with Redox Processes

Four key terms are used in describing redox processes. These terms are *oxidation, reduction, oxidizing agent,* and *reducing agent.* The definitions for these terms are closely tied to the concepts of "electron transfer" and "oxidation number change"—concepts considered in Section 9.2. It is electron transfer that links all redox processes together. Change in oxidation number is a direct consequence of electron transfer.

In a redox reaction, one reactant undergoes oxidation and another reactant undergoes reduction. **Oxidation** *is the process whereby a reactant in a chemical reaction loses one or more electrons.* **Reduction** *is the process whereby a reactant in a chemical reaction gains one or more electrons.*

Oxidation and reduction are complementary processes that always occur together. When electrons are lost by one species, they do not disappear; rather, they are always gained by another species. Thus electron transfer always involves both oxidation and reduction.

Electron loss (oxidation) always leads to an increase in oxidation number. Conversely, electron gain (reduction) always leads to a decrease in oxidation number. These generalizations are consistent with the rules for monatomic ion formation (Section 4.5); electron loss produces positive ions (increase in oxidation number), and electron gain produces

▶ Oxidation involves the *loss* of electrons, and reduction involves the *gain* of electrons. Students often have trouble remembering which is which. Two helpful mnemonic devices follow.

LEO the lion says *GER*

Loss of *Electrons*: *Oxidation.*
Gain of *Electrons*: *Reduction.*

OIL RIG

Oxidation Is Loss (of electrons).
Reduction Is Gain (of electrons).

FIGURE 9.4 An increase in oxidation number is associated with the process of oxidation, a decrease with the process of reduction.

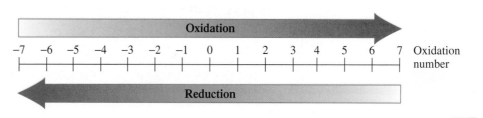

TABLE 9.1
Summary of Redox Terminology in Terms of Electron Transfer

Term	Electron transfer
oxidation	loss of electron(s)
reduction	gain of electron(s)
oxidizing agent (substance reduced)	electron(s) gained
reducing agent (substance oxidized)	electron(s) lost

negative ions (decrease in oxidation number). Figure 9.4 summarizes the relationship between change in oxidation number and the processes of oxidation and reduction.

■ Oxidizing Agents and Reducing Agents

There are two different ways of looking at the reactants in a redox reaction. First, the reactants can be viewed as being "acted on." From this perspective, one reactant is *oxidized* (the one that loses electrons) and one is *reduced* (the one that gains electrons). Second, the reactants can be looked on as "bringing about" the reaction. In this approach, the terms *oxidizing agent* and *reducing agent* are used. An **oxidizing agent** *is a reactant that causes oxidation of another reactant by accepting electrons from it.* This acceptance of electrons means that the oxidizing agent itself is reduced. Similarly, a **reducing agent** *is a reactant that causes reduction of another reactant by providing electrons for the other reactant to accept.* Thus the reducing agent and the substance oxidized are one and the same, as are the oxidizing agent and the substance reduced:

Substance oxidized = reducing agent

Substance reduced = oxidizing agent

Table 9.1 summarizes the redox terminology presented in this section in terms of electron transfer.

▶ The terms *oxidizing agent* and *reducing agent* sometimes cause confusion, because the oxidizing agent is not oxidized (it is reduced) and the reducing agent is not reduced (it is oxidized). A simple analogy is that a travel agent is not the one who takes a trip; he or she is the one who plans (causes) the trip that is taken.

EXAMPLE 9.4

Identifying the Oxidizing Agent and Reducing Agent in a Redox Reaction

■ For the redox reaction

$$FeO + CO \longrightarrow Fe + CO_2$$

identify the following.

a. The substance oxidized   b. The substance reduced
c. The oxidizing agent   d. The reducing agent

Solution

Oxidation numbers are calculated using the methods illustrated in Example 9.2.

$$FeO + CO \longrightarrow Fe + CO_2$$
$$\begin{array}{cccc} +2-2 & +2-2 & 0 & +4-2 \end{array}$$

a. Oxidation involves an increase in oxidation number. The oxidation number of C has increased from $+2$ to $+4$. Therefore, the reactant that contains C, which is CO, is the substance that has been oxidized.

b. Reduction involves a decrease in oxidation number. The oxidation number of Fe has decreased from $+2$ to zero. Therefore, the reactant that contains Fe, which is FeO, is the substance that has been reduced.

c. The oxidizing agent and the substance reduced are always one and the same. Therefore, FeO is the oxidizing agent.

d. The reducing agent and the substance oxidized are always one and the same. Therefore, CO is the reducing agent.

Practice Exercise 9.4

For the redox reaction

$$3MnO_2 + 4Al \longrightarrow 2Al_2O_3 + 3Mn$$

identify the following.

a. The substance oxidized   b. The substance reduced
c. The oxidizing agent   d. The reducing agent

FIGURE 9.7 Energy diagram graphs showing the difference between an exothermic and an endothermic reaction. (a) In an exothermic reaction, the average energy of the reactants is higher than that of the products, indicating that energy has been released in the reaction. (b) In an endothermic reaction, the average energy of the reactants is less than that of the products, indicating that energy has been absorbed in the reaction.

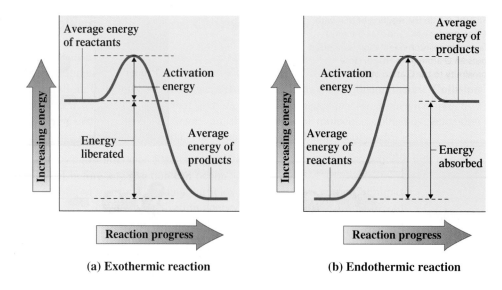

(a) Exothermic reaction **(b) Endothermic reaction**

occur. This activation energy is independent of whether a given reaction is exothermic or endothermic.

9.6 Factors That Influence Reaction Rates

A **reaction rate** *is the rate at which reactants are consumed or products produced in a given time period in a chemical reaction.* Natural processes have a wide range of reaction rates (see Figure 9.8). In this section we consider four different factors that affect reaction rate: (1) the physical nature of the reactants, (2) reactant concentrations, (3) reaction temperature, and (4) the presence of catalysts.

■ Physical Nature of Reactants

The physical nature of reactants includes not only the physical state of each reactant (solid, liquid, or gas) but also the particle size. In reactions where reactants are all in the same physical state, the reaction rate is generally faster between liquid-state reactants than between solid-state reactants and is fastest between gaseous-state reactants. Of the three states of matter, the gaseous state is the one where there is the most freedom of movement; hence, collisions between reactants are the most frequent in this state.

In the solid state, reactions occur at the boundary surface between reactants. The reaction rate increases as the amount of boundary surface area increases. Subdividing a solid into smaller particles increases surface area and thus increases reaction rate.

When the particle size of a solid is extremely small, reaction rates can be so fast that an explosion results. Although a lump of coal is difficult to ignite, the spontaneous ignition of coal dust is a real threat to underground coal-mining operations.

> For reactants in the solid state, reaction rate increases as subdivision of the solid increases.

■ Reactant Concentrations

An increase in the concentration of a reactant causes an increase in the rate of the reaction. Combustible substances burn much more rapidly in pure oxygen than in air (21% oxygen). A person with a respiratory problem such as pneumonia or emphysema is often given air enriched with oxygen, because an increased partial pressure of oxygen facilitates the absorption of oxygen in the alveoli of the lungs and thus expedites all subsequent steps in respiration (Section 7.8).

> Reaction rate increases as the concentration of reactants increases.

Increasing the concentration of a reactant means that there are more molecules of that reactant present in the reaction mixture; thus collisions between this reactant and other reactant particles are more likely. An analogy can be drawn to the game of billiards. The more billiard balls there are on the table, the greater the probability that a moving cue ball will strike one of them.

FIGURE 9.8 Natural processes occur at a wide range of reaction rates. A fire (a) is a much faster reaction than the ripening of fruit (b), which is much faster than the process of rusting (c), which is much faster than the process of aging (d).

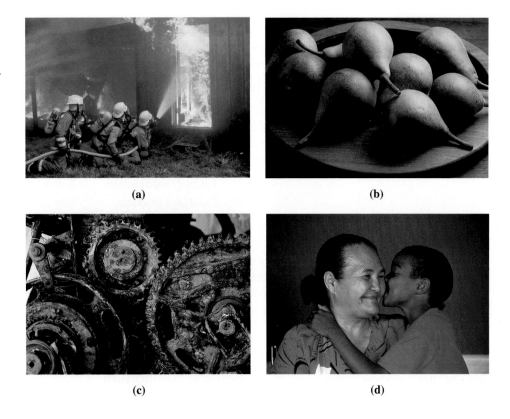

(a)

(b)

(c)

(d)

When the concentration of reactants is increased, the actual quantitative change in reaction rate is determined by the specific reaction. The rate usually increases, but not to the same extent in all cases. Sometimes the rate doubles with a doubling of concentration, but not always.

■ Reaction Temperature

The effect of temperature on reaction rates can also be explained by using the molecular-collision concept. An increase in the temperature of a system results in an increase in the average kinetic energy of the reacting molecules. The increased molecular speed causes more collisions to take place in a given time. Because the average energy of the colliding molecules is greater, a larger fraction of the collisions will result in reaction from the point of view of activation energy. As a rule of thumb, chemists have found that for the temperature ranges we normally encounter, the rate of a chemical reaction doubles for every 10°C increase in temperature (see Figure 9.9).

▶ Reaction rate increases as the temperature of the reactants increases.

FIGURE 9.9 Pictures from an "instant camera" develop more rapidly on warm days than on cold days. The chemical reactions involved in the development process occur faster at higher temperatures.

■ Presence of Catalysts

A **catalyst** *is a substance that increases a reaction rate without being consumed in the chemical reaction.* Catalysts enhance reaction rates by providing alternative reaction pathways that have lower activation energies than the original, uncatalyzed pathway. This lowering of activation energy is diagrammatically shown in Figure 9.10.

Catalysts exert their effects in varying ways. Some catalysts provide a lower-energy pathway by entering into a reaction and forming an "intermediate," which then reacts further to produce the desired products and regenerate the catalyst. The following equations, where C is the catalyst, illustrate this concept.

> Catalysts lower the activation energy for a reaction. Lowered activation energy increases the rate of a reaction.

Uncatalyzed reaction:	$X + Y \longrightarrow XY$
Catalyzed reaction:	*Step 1:* $X + C \longrightarrow XC$
(C = catalyst)	*Step 2:* $XC + Y \longrightarrow XY + C$

Solid catalysts often act by providing a surface to which reactant molecules are physically attracted and on which they are held with a particular orientation. These "held" reactants are sufficiently close to, and so favorably oriented toward, one another that the reaction takes place. The products of the reaction then leave the surface and make it available to catalyze other reactants.

The Chemistry at a Glance feature on page 219 summarizes the factors that influence reaction rates.

> Catalysts are extremely important for the proper functioning of the human body and other biochemical systems. Enzymes, which are proteins, are the catalysts within the human body (Chapter 21). They cause many reactions to take place rapidly under mild conditions and at body temperature. Without these enzymes, the reactions would proceed very slowly and then only under harsher conditions.

9.7 Chemical Equilibrium

In our discussions of chemical reactions up to this point, we have assumed that chemical reactions go to completion; that is, reactions continue until one or more of the reactants are used up. This assumption is valid as long as product concentrations are not allowed to build up in the reaction mixture. If one or more products are gases that can escape from the reaction mixture or insoluble solids that can be removed from the reaction mixture, no product buildup occurs.

When product buildup does occur, reactions do not go to completion. This is because product molecules begin to react with one another to re-form reactants. With time, a steady-state situation results wherein the rate of formation of products and the rate of re-formation of reactants are equal. At this point, the concentrations of all reactants and all products remain constant, and a state of *chemical equilibrium* is reached. **Chemical equilibrium** *is the state in which forward and reverse chemical reactions occur simultaneously at the same rate.* We discussed equilibrium situations in Sections 7.11 (vapor pressure) and 8.2 (saturated solutions), but the previous examples involved physical equilibrium rather than chemical equilibrium. Figure 9.11 illustrates an "everyday" equilibrium situation.

> A chemical reaction is in a state of chemical equilibrium when the rates of the forward and reverse reactions are equal. At this point, the concentrations of reactants and products no longer change.

FIGURE 9.10 Catalysts lower the activation energy for chemical reactions. Reactions proceed more rapidly with the lowered activation energy.

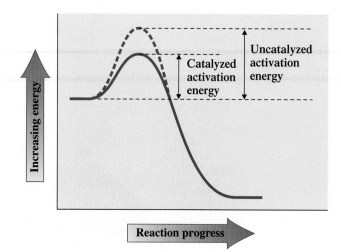

Factors That Influence Reaction Rates

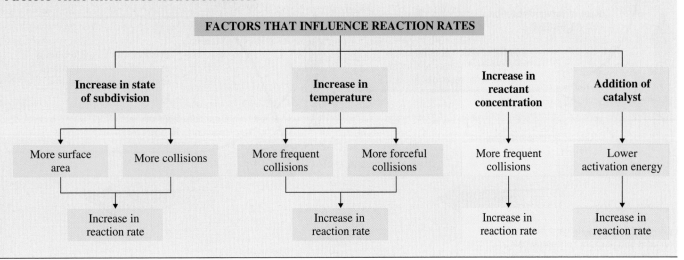

FACTORS THAT INFLUENCE REACTION RATES

Increase in state of subdivision		Increase in temperature		Increase in reactant concentration	Addition of catalyst
More surface area	More collisions	More frequent collisions	More forceful collisions	More frequent collisions	Lower activation energy
	Increase in reaction rate		Increase in reaction rate	Increase in reaction rate	Increase in reaction rate

The conditions that exist in a system in a state of chemical equilibrium can best be seen by considering an actual chemical reaction. Suppose equal molar amounts of gaseous H_2 and I_2 are mixed together in a closed container and allowed to react to produce gaseous HI.

$$H_2 + I_2 \longrightarrow 2HI$$

Initially, no HI is present, so the only reaction that can occur is that between H_2 and I_2. However, as the HI concentration increases, some HI molecules collide with one another in a way that causes a reverse reaction to occur:

$$2HI \longrightarrow H_2 + I_2$$

The initially low concentration of HI makes this reverse reaction slow at first, but as the concentration of HI increases, the reaction rate also increases. At the same time that the reverse-reaction rate is increasing, the forward-reaction rate (production of HI) is

FIGURE 9.11 These jugglers provide an illustration of equilibrium. Each throws clubs to the other at the same rate at which he receives clubs from that person. Because clubs are thrown continuously in both directions, the number of clubs moving in each direction is constant, and the number of clubs each juggler has at a given time remains constant.

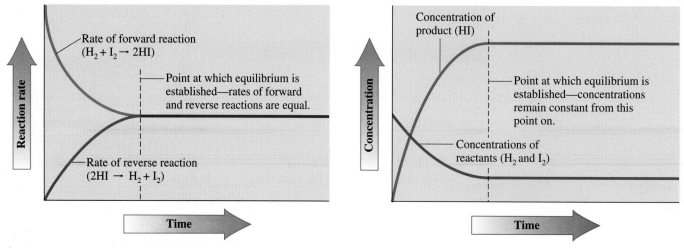

FIGURE 9.12 Graphs showing how reaction rates and reactant concentrations vary with time for the chemical system H_2–I_2–HI. (a) At equilibrium, rates of reaction are equal. (b) At equilibrium, concentrations of reactants remain constant but are not equal.

▶ At chemical equilibrium, forward and reverse reaction rates are equal. Reactant and product concentrations, although constant, do not have to be equal.

▶ Theoretically, all reactions are reversible (can go in either direction). Sometimes, the reverse reaction is so slight, however, that we say the reaction has "gone to completion" because no detectable reactants remain.

decreasing as the reactants are used up. Eventually, the concentrations of H_2, I_2, and HI in the reaction mixture reach a level at which the rates of the forward and reverse reactions become equal. At this point, a state of chemical equilibrium has been reached.

Figure 9.12a illustrates the behavior of reaction rates over time for both the forward and reverse reactions in the H_2–I_2–HI system. Figure 9.12b illustrates the important point that the reactant and product concentrations are usually not equal at the point at which equilibrium is reached.

The equilibrium involving H_2, I_2, and HI could have been established just as easily by starting with pure HI and allowing it to change into H_2 and I_2 (the reverse reaction). The final position of equilibrium does not depend on the direction from which equilibrium is approached.

It is normal procedure to represent an equilibrium by using a single equation and two half-headed arrows pointing in opposite directions. Thus the reaction between H_2 and I_2 at equilibrium is written as

$$H_2 + I_2 \rightleftharpoons 2HI$$

The half-headed arrows denote a chemical system at equilibrium.

The term *reversible* is often used to describe a reaction like the one we have just discussed. A **reversible reaction** *is a reaction in which the conversion of reactants to products (the forward reaction) and the conversion of products to reactants (the reverse reaction) occur simultaneously.* When the half-headed arrow notation is used in an equation, it means that a reaction is reversible.

9.8 Equilibrium Constants

As noted in Section 9.7, the concentrations of reactants and products are constant (not changing) in a system at chemical equilibrium. This constancy allows us to describe the extent of reaction in a given equilibrium system by a single number called an equilibrium constant. An **equilibrium constant** *is a numerical value that characterizes the relationship between the concentrations of reactants and products in a system at chemical equilibrium.*

The equilibrium constant is obtained by writing an *equilibrium expression* and then evaluating it numerically. For a hypothetical chemical reaction, where A and B are reactants, C and D are products, and w, x, y, and z are equation coefficients,

$$w\text{A} + x\text{B} \rightleftharpoons y\text{C} + z\text{D}$$

CHEMICAL CONNECTIONS

Stratospheric Ozone: An Equilibrium Situation

Ozone is oxygen that has undergone conversion from its normal diatomic form (O_2) to a triatomic form (O_3). The presence of ozone in the *lower atmosphere* is considered undesirable, because its production contributes to air pollution; it is the major "active ingredient" in smog.

Los Angeles smog.

In the *upper atmosphere* (stratosphere), ozone is a naturally occurring species whose presence is not only desirable but absolutely essential to the well-being of humans on Earth. Stratospheric ozone screens out 95% to 99% of the ultraviolet radiation that comes from the sun. It is ultraviolet light that causes sunburn and that can be a causative factor in some types of skin cancer. The upper region of the stratosphere, where ozone concentrations are greatest, is often called the ozone layer. This ozone maximization occurs at altitudes of 25 to 30 miles (see the accompanying graph).

Within the ozone layer, ozone is continually being consumed and formed through the equilibrium process

$$3O_2(g) \rightleftharpoons 2O_3(g)$$

The source for ozone is thus diatomic oxygen. It is estimated that on any given day, 300 million tons of stratospheric ozone is formed and an equal amount destroyed in this equilibrium process.

Since the mid-1970s, scientists have observed a seasonal thinning (depletion) of ozone in the stratosphere above Antarctica. This phenomenon, which is commonly called the *ozone hole,* occurs in September and October of each year, the beginning of the Antarctic spring. Up to 70% of the ozone above Antarctica is lost during these two months. (A similar, but smaller, manifestation of this same phenomenon also occurs in the north pole region.)

Winter conditions in Antarctica include extreme cold (it is the coldest location on Earth) and total darkness. When sunlight appears in the spring, it triggers the chemical reactions that lead to ozone depletion. By the end of November, weather conditions are such that the ozone-depletion reactions stop. Then the ozone hole disappears as air from nonpolar areas flows into the polar region, replenishing the depleted ozone levels.

Chlorofluorocarbons (CFCs), synthetic compounds that have been developed primarily for use as refrigerants, are considered a causative factor for this ozone hole phenomenon. How their presence in the atmosphere contributes to this situation is considered in the Chemical Connections feature "Chlorofluorocarbons and the Ozone Layer" in Chapter 12.

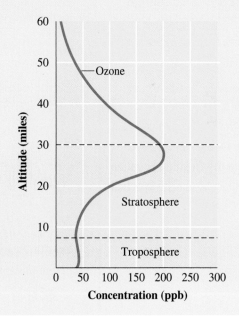

the equilibrium expression is

$$K_{eq} = \frac{[C]^y[D]^z}{[A]^w[B]^x}$$

Note the following points about this general equilibrium expression:

1. The square brackets refer to molar (moles/liter) concentrations.
2. The concentrations of the products always appear in the numerator of the equilibrium expression.

▶ In equilibrium constants, square brackets mean that concentrations are expressed in molarity units.

3. The concentrations of the reactants always appear in the denominator of the equilibrium expression.
4. When more than one concentration appears in the numerator or denominator, the terms are multiplied.
5. The coefficients in the balanced equation give us the powers to which the concentrations are raised.
6. The abbreviation K_{eq} is used to denote an equilibrium constant.

An additional convention in writing equilibrium constants, which is not apparent from the equilibrium constant definition, is that only *concentrations of gases and substances in solution* are written in an equilibrium constant expression. The reason for this convention is that other substances (pure solids and pure liquids) have constant concentrations. These constant concentrations are incorporated into the equilibrium constant itself. For example, pure water in the liquid state has a concentration of 55.5 moles/L. It does not matter whether we have 1.00, 50.0, or 750 mL of liquid water. The concentration will be the same. In the liquid state, pure water is pure water, and it has only one concentration. Similar reasoning applies to other pure liquids and pure solids. All such substances have constant concentrations.

> The concentrations of *pure liquids* and *pure solids,* which are constants, are never included in an equilibrium constant expression.

The only information we need to write an equilibrium expression is a balanced chemical equation, which includes information about physical state. Using the preceding generalizations about equilibrium expressions, for the reaction

$$4NH_3(g) + 7O_2(g) \rightleftharpoons 4NO_2(g) + 6H_2O(g)$$

we write the equilibrium expression as

$$K_{eq} = \frac{[NO_2]^4[H_2O]^6}{[NH_3]^4[O_2]^7}$$

Coefficient of NO_2 Coefficient of H_2O
Coefficient of NH_3 Coefficient of O_2

EXAMPLE 9.5

Writing the Equilibrium Constant Expression for a Chemical Reaction from the Chemical Equation for the Reaction

■ Write the equilibrium constant expression for each of the following reactions.
a. $I_2(g) + Cl_2(g) \rightleftharpoons 2ICl(g)$ **b.** $C(s) + H_2O(g) \rightleftharpoons CO(g) + H_2(g)$

Solution

a. All of the substances involved in this reaction are gases. Therefore, each reactant and product will appear in the equilibrium constant expression.

The numerator of an equilibrium expression always contains product concentrations. There is only one product, ICl. Write its concentration in the numerator and square it, because the coefficient of ICl in the equation is 2.

$$[ICl]^2$$

Next, place the concentrations of the reactants in the denominator. Their powers will be an understood (not written) 1, because the coefficient of each reactant is 1.

$$K_{eq} = \frac{[ICl]^2}{[I_2][Cl_2]}$$

The equilibrium expression is now complete.
b. The reactant carbon (C) is a solid and thus will not appear in the equilibrium expression. Therefore,

$$K_{eq} = \frac{[CO][H_2]}{[H_2O]}$$

Note that all of the powers in this expression are 1 as a result of all the coefficients in the balanced equation being equal to unity.

Practice Exercise 9.5

Write the equilibrium constant expression for each of the following reactions.

a. $2Cl_2(g) + 2H_2O(g) \rightleftharpoons 4HCl(g) + O_2(g)$
b. $NH_4Cl(s) \rightleftharpoons HCl(g) + NH_3(g)$

If the concentrations of all reactants and products are known at equilibrium, the numerical value of the equilibrium constant can be calculated by using the equilibrium expression.

EXAMPLE 9.6

Calculating the Value of an Equilibrium Constant from Equilibrium Concentrations

■ Calculate the value of the equilibrium constant for the equilibrium system

$$2NO(g) \rightleftharpoons N_2(g) + O_2(g)$$

at 1000°C, given that the equilibrium concentrations are 0.0026 M for NO, 0.024 M for N_2, and 0.024 M for O_2.

Solution

First, write the equilibrium constant expression.

$$K_{eq} = \frac{[N_2][O_2]}{[NO]^2}$$

Next, substitute the equilibrium concentrations into the equilibrium expression and solve the equation.

$$K_{eq} = \frac{[0.024][0.024]}{[0.0026]^2}$$

$$K_{eq} = 85$$

In doing the mathematics, remember that the number 0.0026 must be squared.

Practice Exercise 9.6

Calculate the value of the equilibrium constant for the equilibrium system

$$N_2(g) + 3H_2(g) \rightleftharpoons 2NH_3(g)$$

at 532°C, given that the equilibrium concentrations are 0.079 M for N_2, 0.12 M for H_2, and 0.0051 M for NH_3.

■ Temperature Dependence of Equilibrium Constants

The value of K_{eq} for a reaction depends on the reaction temperature. If the temperature changes, the value of K_{eq} also changes, and thus differing amounts of reactants and products will be present. Note that the equilibrium constant calculated in Example 9.5 is for a temperature of 1000°C. The equilibrium constant for this reaction would have a different value at a lower or a higher temperature.

Does the value of an equilibrium constant increase or decrease when reaction temperature is increased? For reactions where the forward reaction is *exothermic,* the equilibrium constant *decreases* with increasing temperature. For reactions where the forward reaction is *endothermic,* the equilibrium constant *increases* with increasing temperature (Section 9.5).

TABLE 9.2
Equilibrium Constant Values and the
Extent to Which a Chemical Reaction
Has Taken Place

Value of K_{eq}	Relative amounts of products and reactants	Description of equilibrium position
very large (10^{30})	essentially all products	far to the right
large (10^{10})	more products than reactants	to the right
near unity (between 10^3 and 10^{-3})	significant amounts of both reactants and products	neither to the right nor to the left
small (10^{-10})	more reactants than products	to the left
very small (10^{-30})	essentially all reactants	far to the left

■ Equilibrium Constant Values and Reaction Completeness

The magnitude of an equilibrium constant value conveys information about how far a reaction has proceeded toward completion. If the equilibrium constant value is large (10^3 or greater), the equilibrium system contains more products than reactants. Conversely, if the equilibrium constant value is small (10^{-3} or less), the equilibrium system contains more reactants than products. Table 9.2 further compares equilibrium constant values and the extent to which a chemical reaction has occurred.

Equilibrium position *is a qualitative indication of the relative amounts of reactants and products present at equilibrium for a chemical reaction.* As shown in the last column of Table 9.2, the terms *far to the right, to the right, neither to right nor to left, to the left,* and *far to the left* are used in describing equilibrium position. In equilibrium situations where the concentrations of products are greater than those of reactants, the equilibrium position is said to lie to the *right* because products are always listed on the right side of an equation.

Equilibrium position can also be indicated by varying the length of the arrows in the half-headed arrow notation for a reversible reaction. The longer arrow indicates the direction of the predominant reaction. For example, the arrow notation in the equation

$$CO_2 + H_2O \rightleftharpoons H_2CO_3$$

indicates that the equilibrium position lies to the right.

9.9 Altering Equilibrium Conditions: Le Châtelier's Principle

A chemical system at equilibrium is very susceptible to disruption from outside forces. A change in temperature or a change in pressure can upset the balance within the equilibrium system. Changes in the concentrations of reactants or products upset an equilibrium also.

Disturbing an equilibrium has one of two results: Either the forward reaction speeds up (to produce more products), or the reverse reaction speeds up (to produce additional reactants). Over time, the forward and reverse reactions again become equal, and a new equilibrium, different from the previous one, is established. If more products have been produced as a result of the disruption, the equilibrium is said to have *shifted to the right.* Similarly, when disruption causes more reactants to form, the equilibrium has *shifted to the left.*

An equilibrium system's response to disrupting influences can be predicted by using a principle introduced by the French chemist Henry Louis Le Châtelier (Figure 9.13). **Le Châtelier's principle** *states that if a stress (change of conditions) is applied to a system in equilibrium, the system will readjust (change the equilibrium position) in the direction that best reduces the stress imposed on the system.* We will use this principle to consider how four types of changes affect equilibrium position. The changes are (1) concentration changes, (2) temperature changes, (3) pressure changes, and (4) addition of catalysts.

▶ Products are written on the right side of a chemical equation. A *shift to the right* means more products are produced. Conversely, because reactants are written on the left side of an equation, a *shift to the left* means more reactants are produced.

▶ The surname Le Châtelier is pronounced "le-SHOT-lee-ay."

■ Concentration Changes

Adding a reactant or product to, or removing it from, a reaction mixture at equilibrium always upsets the equilibrium. If an additional amount of any reactant or product has been *added* to the system, the stress is relieved by shifting the equilibrium in the direction that *consumes* (uses up) some of the added reactant or product. Conversely, if a reactant or product is *removed* from an equilibrium system, the equilibrium shifts in a direction that *produces* more of the substance that was removed.

Let us consider the effect that concentration changes will have on the gaseous equilibrium

$$N_2(g) + 3H_2(g) \rightleftharpoons 2NH_3(g)$$

Suppose some additional H_2 is added to the equilibrium mixture. The stress of "added H_2" causes the equilibrium to shift to the right; that is, the forward reaction rate increases in order to use up some of the additional H_2.

Stress: Too much H_2
Response: Use up "extra" H_2

$$N_2(g) + 3H_2(g) \rightleftharpoons 2NH_3(g)$$
Shift to the right

$[N_2]$ decreases $[H_2]$ decreases $[NH_3]$ increases

As the H_2 reacts, the amount of N_2 also decreases (it reacts with the H_2) and the amount of NH_3 increases (it is formed as H_2 and N_2 react).

With time, the equilibrium shift to the right caused by the addition of H_2 will cease because a new equilibrium condition (not identical to the original one) has been reached. At this new equilibrium condition, most (but not all) of the added H_2 will have been converted to NH_3. Necessary accompaniments to this change are a decreased N_2 concentration (some of it reacted with the H_2) and an increased NH_3 concentration (that produced from the N_2–H_2 reaction). Figure 9.14 quantifies the changes that occur in the N_2–H_2–NH_3 equilibrium system when it is upset by the addition of H_2 for a specific set of concentrations.

Consider again the reaction between N_2 and H_2 to form NH_3.

$$N_2(g) + 3H_2(g) \rightleftharpoons 2NH_3(g)$$

Le Châtelier's principle applies in the same way to removing a reactant or product from the equilibrium mixture as it does to adding a reactant or product at equilibrium. Suppose that at equilibrium we remove some NH_3. The equilibrium position shifts to the right to replenish the NH_3.

FIGURE 9.13 Henri Louis Le Châtelier (1850–1936), although most famous for the principle that bears his name, was amazingly diverse in his interests. He worked on metallurgical processes, cements, glasses, fuels, and explosives and was also noted for his skills in industrial management.

FIGURE 9.14 Concentration changes that result when H_2 is added to an equilibrium mixture involving the system

$$N_2(g) + 3H_2(g) \rightleftharpoons 2NH_3(g)$$

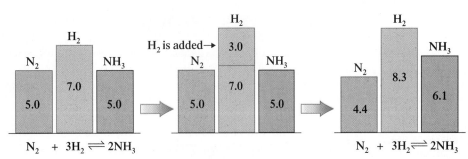

(1) Original equilibrium conditions

(2) Increase in $[H_2]$ upsets equilibrium; reaction shifts to the right as more N_2 reacts with the additional H_2.

(3) New equilibrium conditions. Compared with the original equilibrium in (1):
$[N_2]$ has decreased.
$[H_2]$ has increased because of addition. (Note that $[H_2]$ is actually decreased from conditions at (2) because some of it has reacted with N_2 to form more NH_3).
$[NH_3]$ has increased.

▶ Thousands of chemical equilibria simultaneously exist in biochemical systems. Many of them are interrelated. When the concentration of one substance changes, many equilibria are affected.

Within the human body, numerous equilibrium situations exist that shift in response to a concentration change. Consider, for example, the equilibrium between glucose in the blood and stored glucose (glycogen) in the liver:

$$\text{Glucose in blood} \rightleftharpoons \text{stored glucose} + H_2O$$

Strenuous exercise or hard work causes our blood glucose level to decrease. Our bodies respond to this stress (not enough glucose in the blood) by the liver converting glycogen into glucose. Conversely, when an excess of glucose is present in the blood (after a meal), the liver converts the excess glucose in the blood into its storage form (glycogen).

■ Temperature Changes

Le Châtelier's principle can be used to predict the influence of temperature changes on an equilibrium, provided we know whether the reaction is exothermic or endothermic. For *exothermic reactions,* heat can be treated as one of the *products;* for *endothermic reactions,* heat can be treated as one of the *reactants.*

Consider the exothermic reaction

$$H_2(g) + F_2(g) \rightleftharpoons 2HF(g) + heat$$

Heat is produced when the reaction proceeds to the right. Thus, if we add heat to an exothermic system at equilibrium (by raising the temperature), the system will shift to the left in an "attempt" to decrease the amount of heat present. When equilibrium is reestablished, the concentrations of H_2 and F_2 will be higher and the concentration of HF will have decreased. Lowering the temperature of an exothermic reaction mixture causes the reaction to shift to the right as the system acts to replace the lost heat (Figure 9.15).

The behavior, with temperature change, of an equilibrium system involving an endothermic reaction, such as

$$\text{Heat} + 2CO_2(g) \rightleftharpoons 2CO(g) + O_2(g)$$

is opposite to that of an exothermic reaction, because a shift to the left produces heat. Consequently, an increase in temperature will cause the equilibrium to shift to the right

FIGURE 9.15 Effect of temperature change on the equilibrium mixture

$$CoCl_4^{2-} + 6H_2O \rightleftharpoons$$
　　Blue

$$Co(H_2O)_6^{2+} + 4Cl^- + heat$$
　　　　　Pink

At room temperature, the equilibrium mixture is blue from $CoCl_4^{2-}$. When cooled by the ice bath, the equilibrium mixture turns pink from $Co(H_2O)_6^{2+}$. The temperature decrease causes the equilibrium position to shift to the right.

(to decrease the amount of heat present), and a decrease in temperature will produce a shift to the left (to generate more heat).

■ Pressure Changes

Pressure changes affect systems at equilibrium only when gases are involved—and then only in cases where the chemical reaction is such that a change in the total number of moles in the gaseous state occurs. This latter point can be illustrated by considering the following two gas-phase reactions:

$$\underbrace{2H_2(g) + O_2(g)}_{\text{3 moles of gas}} \longrightarrow \underbrace{2H_2O(g)}_{\text{2 moles of gas}}$$

$$\underbrace{H_2(g) + Cl_2(g)}_{\text{2 moles of gas}} \longrightarrow \underbrace{2HCl(g)}_{\text{2 moles of gas}}$$

In the first reaction, the total number of moles of gaseous reactants and products decreases as the reaction proceeds to the right. This is because 3 moles of reactants combine to give only 2 moles of products. In the second reaction, there is no change in the total number of moles of gaseous substances present as the reaction proceeds. This is because 2 moles of reactants combine to give 2 moles of products. Thus a pressure change will shift the equilibrium position in the first reaction but not in the second.

CHEMICAL CONNECTIONS

Oxygen, Hemoglobin, Equilibrium, and Le Châtelier's Principle

Hemoglobin, an iron-containing protein found in red blood cells, facilitates the transport of oxygen to body tissues. Without the hemoglobin of red blood cells, 1 L of arterial blood could dissolve and carry only about 3 mL of oxygen (the solubility of oxygen in water). With hemoglobin present, 1 L of blood can dissolve and carry about 250 mL of oxygen.

The transport of oxygen by hemoglobin from the lungs, through the bloodstream, and to the tissues involves a reversible reaction in which the O_2 becomes chemically bonded to hemoglobin:

$$\underset{\text{Hemoglobin}}{Hb(aq)} + O_2(g) \rightleftharpoons \underset{\text{Oxyhemoglobin}}{HbO_2(aq)}$$

In the hemoglobin–oxyhemoglobin reaction, the relative amounts of the two compounds are controlled by the partial pressure of oxygen. High oxygen pressure (in the lungs) causes hemoglobin (Hb) to be converted to oxyhemoglobin (HbO_2). (This is a shift to the right in terms of Le Châtelier's principle; the equilibrium system shifts to decrease the pressure of oxygen.) Decreasing oxygen pressure, on the other hand, accelerates oxygen dissociation from oxyhemoglobin (a shift to the left in terms of Le Châtelier's principle). Decreased oxygen pressure is encountered as arterial blood flows through tissue capillaries and cells.

The partial pressure of oxygen in the lungs is 105 mm Hg. This is sufficient to cause 97% *oxygen saturation;* that is, 97% of the blood's hemoglobin has united with oxygen by the time blood leaves the lung capillaries to return to the heart. The accompanying figure shows the variance in oxygen saturation as a function of oxygen partial pressure. You can see from the figure that oxygen saturation is still high (90%) at an oxygen partial pressure of 60 mm Hg. Thus, a safety zone exists relative to drops in oxygen pressure as a result of doing severe exercise, being at high altitudes, or having heart or lung disease.

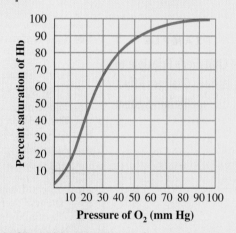

A graph of hemoglobin saturation as a function of oxygen partial pressure.

An individual accustomed to living at sea level will tire more easily at higher altitudes, because the partial pressure of oxygen is lower at the higher altitudes. This translates into fewer molecules of oxyhemoglobin (a shift to the left in terms of Le Châtelier's principle). However, if a person stays at the high altitude for a few days, the body adapts through increased red blood cell (hemoglobin) production. Conversely, individuals living at high altitudes who travel to locations at sea level tire less readily than usual.

Control of oxygen partial pressure is important in the practice of medicine. Increasing the partial pressure of oxygen above that normal for air is often used in the treatment of patients with diseased lungs.

▶ Increasing the pressure associated with an equilibrium system by adding an inert gas (a gas that is not a reactant or a product in the reaction) does not affect the position of the equilibrium.

Pressure changes are usually brought about through volume changes. A pressure increase results from a volume decrease, and a pressure decrease results from a volume increase (Section 7.4). Le Châtelier's principle correctly predicts the direction of the equilibrium position shift resulting from a pressure change only when the pressure change is caused by a change in volume. It does not apply to pressure increases caused by the addition of a nonreactive (inert) gas to the reaction mixture. This addition has no effect on the equilibrium position. The partial pressure (Section 7.8) of each of the gases involved in the reaction remains the same.

According to Le Châtelier's principle, the stress of increased pressure is relieved by decreasing the number of moles of gaseous substances in the system. This is accomplished by the reaction shifting in the direction of fewer moles; that is, it shifts to the side of the equation that contains the fewer moles of gaseous substances. For the reaction

$$2NO_2(g) + 7H_2(g) \rightleftharpoons 2NH_3(g) + 4H_2O(g)$$

an increase in pressure would shift the equilibrium position to the right, because there are 9 moles of gaseous reactants and only 6 moles of gaseous products. On the other hand, the stress of decreased pressure causes an equilibrium system to produce more moles of gaseous substances.

■ Addition of Catalysts

Catalysts cannot change the position of an equilibrium. A catalyst functions by lowering the activation energy for a reaction. It speeds up both the forward and the reverse reactions, so it has no net effect on the position of the equilibrium. However, the lowered activation energy allows equilibrium to be established more quickly than if the catalyst were absent.

EXAMPLE 9.7

Using Le Châtelier's Principle to Predict How Various Changes Affect an Equilibrium System

■ How will the gas-phase equilibrium

$$CH_4(g) + 2H_2S(g) + heat \rightleftharpoons CS_2(g) + 4H_2(g)$$

be affected by each of the following?
a. The removal of $H_2(g)$
b. The addition of $CS_2(g)$
c. An increase in the temperature
d. An increase in the volume of the container (a decrease in pressure)

Solution

a. The equilibrium will *shift to the right,* according to Le Châtelier's principle, in an "attempt" to replenish the H_2 that was removed.
b. The equilibrium will *shift to the left* in an attempt to use up the extra CS_2 that has been placed in the system.
c. Raising the temperature means that heat energy has been added. In an attempt to minimize the effect of this extra heat, the position of the equilibrium will *shift to the right,* the direction that consumes heat; heat is one of the reactants in an endothermic reaction.
d. The system will *shift to the right,* the direction that produces more moles of gaseous substances (an increase of pressure). In this way, the reaction produces 5 moles of gaseous products for every 3 moles of gaseous reactants consumed.

Practice Exercise 9.7

How will the gas-phase equilibrium

$$CO(g) + 3H_2(g) \rightleftharpoons CH_4(g) + H_2O(g) + heat$$

be affected by each of the following?
a. The removal of $CH_4(g)$
b. The addition of $H_2O(g)$
c. A decrease in the temperature
d. A decrease in the volume of the container (an increase in pressure)

CONCEPTS TO REMEMBER

Chemical reaction. A process in which at least one new substance is produced as a result of chemical change.

Combination reaction. A chemical reaction in which a single product is produced from two or more reactants.

Decomposition reaction. A chemical reaction in which a single reactant is converted into two or more simpler substances (elements or compounds).

Single-replacement reaction. A chemical reaction in which an atom or a molecule replaces an atom or a group of atoms from a compound.

Double-replacement reaction. A chemical reaction in which two substances exchange parts with one another and form two different substances.

Combustion reaction. A chemical reaction in which oxygen (usually from air) reacts with a substance with evolution of heat and usually the presence of a flame.

Redox reaction. A chemical reaction in which there is a transfer of electrons from one reactant to another reactant.

Nonredox reaction. A chemical reaction in which there is no transfer of electrons from one reactant to another reactant.

Oxidation number. An oxidation number for an atom is a number that represents the charge that an atom appears to have when the electrons in each bond it is participating in are assigned to the more electronegative of the two atoms involved in the bond. Oxidation numbers are used to identify the electron transfer that occurs in a redox reaction.

Oxidation–reduction terminology. Oxidation is the loss of electrons; reduction is the gain of electrons. An oxidizing agent causes oxidation by accepting electrons from the other reactant. A reducing agent causes reduction by providing electrons for the other reactant to accept.

Collision theory. Collision theory summarizes the conditions required for a chemical reaction to take place. The three basic tenets of collision theory are as follows: (1) Reactant molecules must collide with each other. (2) The colliding reactants must possess a certain minimum of energy. (3) In some cases, colliding reactants must be oriented in a specific way if reaction is to occur.

Exothermic and endothermic chemical reactions. An exothermic chemical reaction releases energy as the reaction occurs. An endothermic chemical reaction requires an input of energy as the reaction occurs.

Reaction rates. Reaction rate is the speed at which reactants are converted to products. Four factors affect the rates of all reactions: (1) the physical nature of the reactants, (2) reactant concentrations, (3) reaction temperature, and (4) the presence of catalysts.

Chemical equilibrium. Chemical equilibrium is the state wherein the rate of the forward reaction is equal to the rate of the reverse reaction. Equilibrium is indicated in chemical equations by writing half-headed arrows pointing in both directions between reactants and products.

Equilibrium constant. The equilibrium constant relates the concentrations of reactants and products at equilibrium. The value of an equilibrium constant is obtained by writing an equilibrium expression and then numerically evaluating it. Equilibrium expressions can be obtained from the balanced chemical equations for reactions.

Equilibrium position. The relative amounts of reactants and products present in a system at equilibrium define the equilibrium position. The equilibrium position is toward the right when a large amount of product is present and is toward the left when a large amount of reactant is present.

Le Châtelier's principle. Le Châtelier's principle states that when a stress (change of conditions) is applied to a system in equilibrium, the system will readjust (change the equilibrium position) in the direction that best reduces the stress imposed on it. Stresses which change an equilibrium position include (1) changes in amount of reactants and/or products, (2) changes in temperature, and (3) changes in pressure.

KEY REACTIONS AND EQUATIONS

1. Combination reaction (Section 9.1)
$$X + Y \longrightarrow XY$$
2. Decomposition reaction (Section 9.1)
$$XY \longrightarrow X + Y$$
3. Single-replacement reaction (Section 9.1)
$$X + YZ \longrightarrow Y + XZ$$
4. Double-replacement reaction (Section 9.1)
$$AX + BY \longrightarrow AY + BX$$
5. Equilibrium expression equation for a general reaction (Section 9.8)
$$wA + xB \rightleftharpoons yC + zD$$
$$K_{eq} = \frac{[C]^y[D]^z}{[A]^w[B]^x}$$

KEY TERMS

Activation energy (9.4)
Catalyst (9.6)
Chemical equilibrium (9.7)
Chemical reaction (9.1)
Collision theory (9.4)
Combination reaction (9.1)
Combustion reaction (9.1)
Decomposition reaction (9.1)

Double-replacement reaction (9.1)
Endothermic reaction (9.5)
Equilibrium constant (9.8)
Equilibrium position (9.8)
Exothermic reaction (9.5)
Le Châtelier's principle (9.9)
Nonoxidation–reduction reaction (9.2)
Oxidation (9.3)

Oxidation number (9.2)
Oxidation–reduction reaction (9.2)
Oxidizing agent (9.3)
Reaction rate (9.6)
Reducing agent (9.3)
Reduction (9.3)
Reversible reaction (9.7)
Single-replacement reaction (9.1)

EXERCISES AND PROBLEMS

The members of each pair of problems in this section test similar material.

■ Types of Chemical Reactions (Section 9.1)

9.1 Classify each of the following reactions as a combination, decomposition, single-replacement, double-replacement, or combustion reaction.
a. $3CuSO_4 + 2Al \longrightarrow Al_2(SO_4)_3 + 3Cu$
b. $K_2CO_3 \longrightarrow K_2O + CO_2$
c. $2AgNO_3 + K_2SO_4 \longrightarrow Ag_2SO_4 + 2KNO_3$
d. $2P + 3H_2 \longrightarrow 2PH_3$

9.2 Classify each of the following reactions as a combination, decomposition, single-replacement, double-replacement, or combustion reaction.
a. $2NaHCO_3 \longrightarrow Na_2CO_3 + CO_2 + H_2O$
b. $2Ag_2CO_3 \longrightarrow 4Ag + 2CO_2 + O_2$
c. $2C_2H_6 + 7O_2 \longrightarrow 4CO_2 + 6H_2O$
d. $Mg + 2HCl \longrightarrow MgCl_2 + H_2$

9.3 Indicate to which of the following types of reactions each of the statements listed applies: *combination, decomposition, single-replacement, double-replacement,* and *combustion.*
a. An element may be a reactant.
b. An element may be a product.
c. A compound may be a reactant.
d. A compound may be a product.

9.4 Indicate to which of the following types of reactions each of the statements listed applies: *combination, decomposition, single-replacement, double-replacement,* and *combustion.*
a. Two reactants are required.
b. Only one reactant is present.
c. Two products are present.
d. Only one product is present.

■ Oxidation Numbers (Section 9.2)

9.5 Determine the oxidation number of
a. Ba in Ba^{2+} b. S in SO_3
c. F in F_2 d. P in PO_4^{3-}

9.6 Determine the oxidation number of
a. Al in Al^{3+} b. N in NO_2
c. O in O_3 d. S in SO_4^{2-}

9.7 Determine the oxidation number of Cr in each of the following chromium-containing species.
a. Cr_2O_3 b. CrO_2 c. CrO_3
d. Na_2CrO_4 e. $BaCrO_4$ f. $BaCr_2O_7$
g. $Na_2Cr_2O_7$ h. CrF_5

9.8 Determine the oxidation number of Cl in each of the following chlorine-containing species.
a. $BeCl_2$ b. $Ba(ClO)_2$ c. ClF_4^+ d. Cl_2O_7
e. NCl_3 f. $AlCl_4^-$ g. ClF h. ClO^-

9.9 What is the oxidation number of each element in each of the following substances?
a. PF_3 b. $NaOH$ c. Na_2SO_4 d. CO_3^{2-}

9.10 What is the oxidation number of each element in each of the following substances?
a. H_2S b. H_2 c. N^{3-} d. MnO_4^-

■ Oxidation–Reduction Reactions (Sections 9.2 and 9.3)

9.11 Classify each of the following reactions as a redox reaction or a nonredox reaction.
a. $2Cu + O_2 \longrightarrow 2CuO$
b. $K_2O + H_2O \longrightarrow 2KOH$

c. $2KClO_3 \longrightarrow 2KCl + 3O_2$
d. $CH_4 + 2O_2 \longrightarrow CO_2 + 2H_2O$

9.12 Classify each of the following reactions as a redox reaction or a nonredox reaction.
a. $2NO + O_2 \longrightarrow 2NO_2$
b. $CO_2 + H_2O \longrightarrow H_2CO_3$
c. $Zn + 2AgNO_3 \longrightarrow Zn(NO_3)_2 + 2Ag$
d. $HNO_3 + NaOH \longrightarrow NaNO_3 + H_2O$

9.13 Identify which substance is oxidized and which substance is reduced in each of the following redox reactions.
a. $N_2 + 3H_2 \longrightarrow 2NH_3$
b. $Cl_2 + 2KI \longrightarrow 2KCl + I_2$
c. $Sb_2O_3 + 3Fe \longrightarrow 2Sb + 3FeO$
d. $3H_2SO_3 + 2HNO_3 \longrightarrow 2NO + H_2O + 3H_2SO_4$

9.14 Identify which substance is oxidized and which substance is reduced in each of the following redox reactions.
a. $2Al + 3Cl_2 \longrightarrow 2AlCl_3$
b. $Zn + CuCl_2 \longrightarrow ZnCl_2 + Cu$
c. $2NiS + 3O_2 \longrightarrow 2NiO + 2SO_2$
d. $3H_2S + 2HNO_3 \longrightarrow 3S + 2NO + 4H_2O$

9.15 Identify which substance is the oxidizing agent and which substance is the reducing agent in each of the redox reactions of Problem 9.13.

9.16 Identify which substance is the oxidizing agent and which substance is the reducing agent in each of the redox reactions of Problem 9.14.

■ Collision Theory (Section 9.4)

9.17 Why are most chemical reactions carried out either in liquid solution or in the gaseous phase?

9.18 Why would a reaction with a high activation energy be expected to be a slow reaction?

9.19 How is activation energy related to the fact that some collisions between reactant molecules do not result in product formation?

9.20 How is collision orientation related to the fact that some collisions between reactant molecules do not result in product formation?

■ Exothermic and Endothermic Reactions (Section 9.5)

9.21 Which of the following reactions are endothermic and which are exothermic?
a. $C_2H_4 + 3O_2 \longrightarrow 2CO_2 + 2H_2O + heat$
b. $N_2 + 2O_2 + heat \longrightarrow 2NO_2$
c. $2H_2O + heat \longrightarrow 2H_2 + O_2$
d. $2KClO_3 \longrightarrow 2KCl + 3O_2 + heat$

9.22 Which of the following reactions are endothermic and which are exothermic?
a. $CaCO_3 + heat \longrightarrow CaO + CO_2$
b. $N_2 + 3H_2 \longrightarrow 2NH_3 + heat$
c. $CO + 3H_2 + heat \longrightarrow CH_4 + H_2O$
d. $2N_2 + 6H_2O + heat \longrightarrow 4NH_3 + 3O_2$

9.23 Sketch an energy diagram graph representing an exothermic reaction, and label the following:
a. Average energy of reactants
b. Average energy of products
c. Activation energy
d. Amount of energy liberated during the reaction

9.24 Sketch an energy diagram graph representing an endothermic reaction, and label the following:
a. Average energy of reactants
b. Average energy of products
c. Activation energy
d. Amount of energy absorbed during the reaction

■ **Factors That Influence Reaction Rates (Section 9.6)**

9.25 Using collision theory, indicate why each of the following factors influences the rate of a reaction.
a. Temperature of reactants
b. Presence of a catalyst

9.26 Using collision theory, indicate why each of the following factors influences the rate of a reaction.
a. Physical nature of reactants
b. Reactant concentrations

9.27 Substances burn more rapidly in pure oxygen than in air. Explain why.

9.28 Milk will sour in a couple of days when left at room temperature, yet it can remain unspoiled for 2 weeks when refrigerated. Explain why.

9.29 Draw an energy diagram graph for an exothermic reaction where no catalyst is present. Then draw an energy diagram graph for the same reaction when a catalyst is present. Indicate the similarities and differences between the two diagrams.

9.30 Draw an energy diagram graph for an endothermic reaction where no catalyst is present. Then draw an energy diagram graph for the same reaction when a catalyst is present. Indicate the similarities and differences between the two diagrams.

9.31 The characteristics of four reactions, each of which involves only two reactants, are given.

Reaction	Activation energy	Temperature	Concentration of reactants
1	low	low	1 mole/L of each
2	high	low	1 mole/L of each
3	low	high	1 mole/L of each
4	low	low	1 mole/L of first reactant and 4 moles/L of second reactant

For each of the following pairs of the preceding reactions, compare the reaction rates when the two reactants are first mixed. Indicate which reaction is faster.
a. 1 and 2   b. 1 and 3   c. 1 and 4   d. 2 and 3

9.32 The characteristics of four reactions, each of which involves only two reactants, are given.

Reaction	Activation energy	Temperature	Concentration of reactants
1	high	low	1 mole/L of each
2	high	high	1 mole/L of each
3	low	low	1 mole/L of first reactant and 4 moles/L of second reactant
4	low	low	4 moles/L of each

For each of the following pairs of the preceding reactions, compare the reaction rates when the two reactants are first mixed. Indicate which reaction is faster.
a. 1 and 2   b. 1 and 3   c. 1 and 4   d. 3 and 4

■ **Chemical Equilibrium (Section 9.7)**

9.33 What condition must be met in order for a system to be in a state of chemical equilibrium?

9.34 What relationship exists between the rates of the forward and reverse reactions for a system in a state of chemical equilibrium?

9.35 Sketch a graph showing how the concentrations of the reactants and products of a typical reversible chemical reaction vary with time.

9.36 Sketch a graph showing how the rates of the forward and reverse reactions for a typical reversible chemical reaction vary with time.

■ **Equilibrium Constants (Section 9.8)**

9.37 Write equilibrium constant expressions for the following reactions.
a. $N_2O_4(g) \rightleftharpoons 2NO_2(g)$
b. $COCl_2(g) \rightleftharpoons CO(g) + Cl_2(g)$
c. $CS_2(g) + 4H_2(g) \rightleftharpoons CH_4(g) + 2H_2S(g)$
d. $2SO_2(g) + O_2(g) \rightleftharpoons 2SO_3(g)$

9.38 Write equilibrium constant expressions for the following reactions.
a. $3O_2(g) \rightleftharpoons 2O_3(g)$
b. $2NOCl(g) \rightleftharpoons 2NO(g) + Cl_2(g)$
c. $4NH_3(g) + 5O_2(g) \rightleftharpoons 4NO(g) + 6H_2O(g)$
d. $CO(g) + H_2O(g) \rightleftharpoons CO_2(g) + H_2(g)$

9.39 Write equilibrium constant expressions for the following reactions.
a. $H_2SO_4(l) \rightleftharpoons SO_3(g) + H_2O(l)$
b. $2Ag(s) + Cl_2(g) \rightleftharpoons 2AgCl(s)$
c. $BaCl_2(aq) + Na_2SO_4(aq) \rightleftharpoons 2NaCl(aq) + BaSO_4(s)$
d. $2Na_2O(s) \rightleftharpoons 4Na(l) + O_2(g)$

9.40 Write equilibrium constant expressions for the following reactions.
a. $2KClO_3(s) \rightleftharpoons 2KCl(s) + 3O_2(g)$
b. $PCl_5(s) \rightleftharpoons PCl_3(l) + Cl_2(g)$
c. $AgNO_3(aq) + NaCl(aq) \rightleftharpoons AgCl(s) + NaNO_3(aq)$
d. $2FeBr_3(s) \rightleftharpoons 2FeBr_2(s) + Br_2(g)$

9.41 Calculate the value of the equilibrium constant for the reaction
$$N_2O_4(g) \rightleftharpoons 2NO_2(g)$$
if the concentrations of the species at equilibrium are $[N_2O_4] = 0.213$ and $[NO_2] = 0.0032$.

9.42 Calculate the value of the equilibrium constant for the reaction
$$N_2(g) + 2O_2(g) \rightleftharpoons 2NO_2(g)$$
if the concentrations of the species at equilibrium are $[N_2] = 0.0013$, $[O_2] = 0.0024$, and $[NO_2] = 0.00065$.

9.43 Use the given K_{eq} value and the terminology in Table 9.2 to describe the relative amounts of reactants and products present in each of the following equilibrium situations.
a. $H_2(g) + Br_2(g) \rightleftharpoons 2HBr(g)$ $K_{eq}(25°C) = 2.0 \times 10^9$
b. $2HCl(g) \rightleftharpoons H_2(g) + Cl_2(g)$ $K_{eq}(25°C) = 3.2 \times 10^{-34}$
c. $SO_2(g) + NO_2(g) \rightleftharpoons NO(g) + SO_3(g)$ $K_{eq}(460°C) = 85.0$
d. $COCl_2(g) \rightleftharpoons CO(g) + Cl_2(g)$ $K_{eq}(395°C) = 0.046$

9.44 Use the given K_{eq} value and the terminology in Table 9.2 to describe the relative amounts of reactants and products present in each of the following equilibrium situations.
a. $2NO(g) \rightleftharpoons N_2(g) + O_2(g)$ $K_{eq}(25°C) = 1 \times 10^{30}$
b. $N_2(g) + 3H_2(g) \rightleftharpoons 2NH_3(g)$ $K_{eq}(25°C) = 1 \times 10^9$
c. $PCl_5(g) \rightleftharpoons PCl_3(g) + Cl_2(g)$ $K_{eq}(127°C) = 1 \times 10^{-2}$
d. $2Na_2O(s) \rightleftharpoons 4Na(l) + O_2(g)$ $K_{eq}(427°C) = 1 \times 10^{-25}$

■ **Le Châtelier's Principle (Section 9.9)**

9.45 For the reaction

$$2Cl_2(g) + 2H_2O(g) \rightleftharpoons 4HCl(g) + O_2(g)$$

determine in what direction the equilibrium will be shifted by each of the following changes.
a. Increase in Cl_2 concentration
b. Increase in O_2 concentration
c. Decrease in H_2O concentration
d. Decrease in HCl concentration

9.46 For the reaction

$$2Cl_2(g) + 2H_2O(g) \rightleftharpoons 4HCl(g) + O_2(g)$$

determine in what direction the equilibrium will be shifted by each of the following changes.
a. Increase in H_2O concentration
b. Increase in HCl concentration
c. Decrease in O_2 concentration
d. Decrease in Cl_2 concentration

9.47 For the reaction

$$C_6H_6(g) + 3H_2(g) \rightleftharpoons C_6H_{12}(g) + heat$$

determine in what direction the equilibrium will be shifted by each of the following changes.
a. Increasing the concentration of C_6H_{12}
b. Decreasing the concentration of C_6H_6
c. Increasing the temperature
d. Decreasing the pressure by increasing the volume of the container

9.48 For the reaction

$$C_6H_6(g) + 3H_2(g) \rightleftharpoons C_6H_{12}(g) + heat$$

determine in what direction the equilibrium will be shifted by each of the following changes.
a. Decreasing the concentration of H_2
b. Increasing the concentration of C_6H_6
c. Decreasing the temperature
d. Increasing the pressure by decreasing the volume of the container

9.49 Consider the following chemical system at equilibrium.

$$CO(g) + H_2O(g) + heat \rightleftharpoons CO_2(g) + H_2(g)$$

For each of the following adjustments of conditions, indicate the effect (shifts left, shifts right, or no effect) on the position of equilibrium.
a. Refrigerating the equilibrium mixture
b. Adding a catalyst to the equilibrium mixture
c. Adding CO to the equilibrium mixture
d. Increasing the size of the reaction container

9.50 Consider the following chemical system at equilibrium.

$$CO(g) + H_2O(g) + heat \rightleftharpoons CO_2(g) + H_2(g)$$

For each of the following adjustments of conditions, indicate the effect (shifts left, shifts right, or no effect) on the position of equilibrium.
a. Heating the equilibrium mixture
b. Increasing the pressure on the equilibrium mixture by adding a nonreactive gas
c. Adding H_2 to the equilibrium mixture
d. Decreasing the size of the reaction container

ADDITIONAL PROBLEMS

9.51 Characterize each of the following reactions using one selection from the choices *redox* and *nonredox* combined with one selection from the choices *combination, decomposition, single-replacement, double-replacement,* and *combustion.*
a. $Zn + Cu(NO_3)_2 \longrightarrow Zn(NO_3)_2 + Cu$
b. $CH_4 + 2O_2 \longrightarrow CO_2 + 2H_2O$
c. $2CuO \longrightarrow 2Cu + O_2$
d. $NaCl + AgNO_3 \longrightarrow AgCl + NaNO_3$

9.52 Classify each of the following reactions as (1) a redox reaction, (2) a nonredox reaction, or (3) "can't classify" because of insufficient information.
a. A combination reaction in which one reactant is an element and the other is a compound
b. A decomposition reaction in which the products are all elements
c. A decomposition reaction in which one of the products is an element
d. A single-replacement reaction in which both of the reactants are compounds

9.53 In each of the following statements, choose the word in parentheses that best completes the statement.
a. The process of reduction is associated with the (loss, gain) of electrons.
b. The oxidizing agent in a redox reaction is the substance that undergoes (oxidation, reduction).

c. Reduction always results in an (increase, decrease) in the oxidation number of an element.
d. A reducing agent in a redox reaction is the substance that contains the element that undergoes an (increase, decrease) in oxidation number.

9.54 Indicate whether each of the following substances undergoes an increase in oxidation number or a decrease in oxidation number in a redox reaction.
a. The oxidizing agent
b. The reducing agent
c. The substance undergoing oxidation
d. The substance undergoing reduction

9.55 Which of the following changes would affect the *value* of a system's equilibrium constant?
a. Removal of a reactant or product from an equilibrium mixture
b. Decrease in the system's total pressure
c. Increase in the system's temperature
d. Addition of a catalyst to the equilibrium mixture

9.56 Write a balanced chemical equation for the totally gaseous equilibrium system that would lead to the following expression for the equilibrium constant.

$$K_{eq} = \frac{[CH_4][H_2S]^2}{[CS_2][H_2]^4}$$

9.57 For which of the following reactions is product formation favored by high temperature?

a. $N_2(g) + 2O_2(g) + heat \rightleftharpoons 2NO_2(g)$

b. $2N_2(g) + 6H_2O(g) + heat \rightleftharpoons 4NH_3(g) + 3O_2(g)$

c. $C_2H_4(g) + 3O_2(g) \rightleftharpoons 2CO_2(g) + 2H_2O(g) + heat$

d. $2KClO_3(s) + heat \rightleftharpoons 2KCl(s) + 3O_2(g)$

9.58 Predict the direction in which each of the following equilibria will shift if the pressure within the system is increased by reducing volume, using the choices *left, right,* and *no effect.*

a. $H_2(g) + C_2N_2(g) \rightleftharpoons 2HCN(g)$

b. $CO(g) + Br_2(g) \rightleftharpoons COBr_2(g)$

c. $CS_2(g) + 4H_2(g) \rightleftharpoons CH_4(g) + 2H_2S(g)$

d. $Ni(s) + 4CO(g) \rightleftharpoons Ni(CO)_4(g)$

ANSWERS TO PRACTICE EXERCISES

9.1 a. combustion   b. combination
c. single-replacement   d. decomposition

9.2 a. N = +4, O = −2   b. K = +1, Cr = +6, O = −2
c. N = −3, H = +1

9.3 nonredox reaction

9.4 a. Al   b. MnO_2 c. MnO_2 d. Al

9.5 a. $K_{eq} = \dfrac{[O_2][HCl]^4}{[H_2O]^2[Cl_2]^2}$; b. $K_{eq} = [NH_3][HCl]$

9.6 0.19

9.7 a. shift to the right   b. shift to the left
c. shift to the right   d. shift to the right

10

Acids, Bases, and Salts

Fish are very sensitive to the acidity of the water present in an aquarium.

Acids, bases, and salts are among the most common and important compounds known. In the form of aqueous solutions, these compounds are key materials in both biochemical systems and the chemical industry. A major ingredient of gastric juice in the stomach is hydrochloric acid. Quantities of lactic acid are produced when the human body is subjected to strenuous exercise. The lye used in making homemade soap contains the base sodium hydroxide. Bases are ingredients in many stomach antacid formulations. The white crystals you sprinkle on your food to make it taste better represent only one of many hundreds of salts that exist.

10.1 Arrhenius Acid–Base Theory

In 1884, the Swedish chemist Svante August Arrhenius (1859–1927) proposed that acids and bases be defined in terms of the chemical species they form when they dissolve in water. An **Arrhenius acid** *is a hydrogen-containing compound that, in water, produces hydrogen ions (H^+ ions).* The acidic species in Arrhenius theory is thus the hydrogen ion. An **Arrhenius base** *is a hydroxide-containing compound that, in water, produces hydroxide ions (OH^- ions).* The basic species in Arrhenius theory is thus the hydroxide ion. For this reason, Arrhenius bases are also called *hydroxide bases*.

Two common examples of Arrhenius acids are HNO_3 (nitric acid) and HCl (hydrochloric acid).

$$HNO_3(l) \xrightarrow{H_2O} H^+(aq) + NO_3^-(aq)$$
$$HCl(g) \xrightarrow{H_2O} H^+(aq) + Cl^-(aq)$$

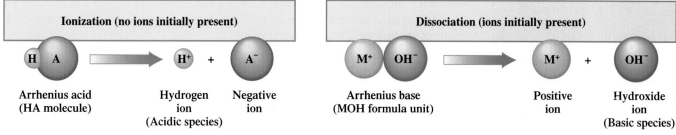

FIGURE 10.1 The difference between the aqueous solution processes of ionization (Arrhenius acids) and dissociation (Arrhenius bases). Ionization is the production of ions from a *molecular compound* that has been dissolved in solution. Dissociation is the production of ions from an *ionic compound* that has been dissolved in solution.

When Arrhenius acids are in the pure state (not in solution), they are covalent compounds; that is, they do not contain H^+ ions. This ion is formed through an interaction between water and the acid when they are mixed. **Ionization** *is the process in which individual positive and negative ions are produced from a molecular compound that is dissolved in solution.*

Two common examples of Arrhenius bases are NaOH (sodium hydroxide) and KOH (potassium hydroxide).

$$NaOH(s) \xrightarrow{H_2O} Na^+(aq) + OH^-(aq)$$

$$KOH(s) \xrightarrow{H_2O} K^+(aq) + OH^-(aq)$$

In direct contrast to acids, Arrhenius bases are ionic compounds in the pure state. When these compounds dissolve in water, the ions separate to yield the OH^- ions. **Dissociation** *is the process in which individual positive and negative ions are released from an ionic compound that is dissolved in solution.* Figure 10.1 contrasts the processes of ionization (acids) and dissociation (bases).

Arrhenius acids have a sour taste, change blue litmus paper to red (see Figure 10.2), and are corrosive to many materials. Arrhenius bases have a bitter taste, change red litmus paper to blue, and are slippery (soapy) to the touch. (The bases themselves are not slippery, but they react with the oils in the skin to form new slippery compounds.)

FIGURE 10.2 Litmus is a vegetable dye obtained from certain lichens found principally in the Netherlands. Paper treated with this dye turns from blue to red in acids (left) and from red to blue in bases (right).

▶ The terms *hydrogen ion* and *proton* are used synonymously in acid–base discussions. Why? The predom-inant hydrogen isotope, ^1_1H, is unique in that no neutrons are present; it consists of a proton and an electron. Thus the ion $^1_1H^+$, a hydrogen atom that has lost its only electron, is simply a proton.

10.2 Brønsted–Lowry Acid–Base Theory

Although it is widely used, Arrhenius acid–base theory has some shortcomings. It is restricted to aqueous solution, and it does not explain why compounds like ammonia (NH_3), which do not contain hydroxide ion, produce a basic water solution.

In 1923, Johannes Nicolaus Brønsted (1879–1947), a Danish chemist, and Thomas Martin Lowry (1874–1936), a British chemist, independently and almost simultaneously proposed broadened definitions for acids and bases—definitions that applied in both aqueous and nonaqueous solutions and that also explained how some nonhydroxide-containing substances, when added to water, produce basic solutions.

A **Brønsted–Lowry acid** *is a substance that can donate a proton (H^+ ion) to some other substance.* A **Brønsted–Lowry base** *is a substance that can accept a proton (H^+ ion) from some other substance.* In short, a Brønsted–Lowry acid is a *proton donor* (or hydrogen ion donor), and a Brønsted–Lowry base is a *proton acceptor* (or hydrogen ion acceptor). The terms *proton* and *hydrogen ion* are used interchangeably in acid–base discussions. Remember that a H^+ ion is a hydrogen atom (proton plus electron) that has lost its electron; hence it is a proton.

Any chemical reaction involving a Brønsted–Lowry acid must also involve a Brønsted–Lowry base. You cannot have one without the other. Proton donation (from an acid) cannot occur unless an acceptor (a base) is present.

Brønsted–Lowry acid–base theory also includes the concept that hydrogen ions in an aqueous solution do not exist in the free state but, rather, react with water to form *hydronium ions*. The attraction between a hydrogen ion and polar water molecules is

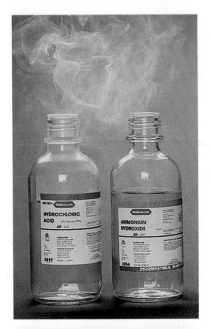

A Brønsted–Lowry base—a proton acceptor—must contain an atom that possesses a pair of nonbonding electrons that can be used in forming a coordinate covalent bond to an incoming proton (from a Brønsted–Lowry acid).

sufficiently strong to bond the hydrogen ion to a water molecule to form a hydronium ion (H_3O^+). The bond between them is a coordinate covalent bond (Section 5.5), because both electrons are furnished by the oxygen atom.

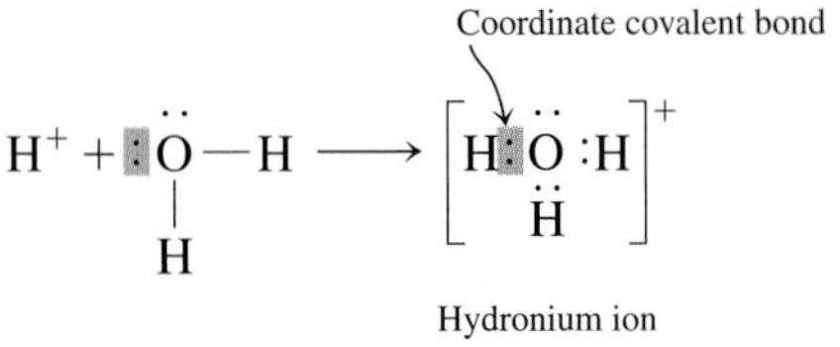

$$H^+ + \ddot{\underset{|}{\ddot{O}}}{-}H \longrightarrow \left[H\!:\!\ddot{\underset{\ddot{H}}{O}}\!:\!H \right]^+$$

Coordinate covalent bond

Hydronium ion

When gaseous hydrogen chloride dissolves in water, it forms hydrochloric acid. This is a simple Brønsted–Lowry acid–base reaction. The chemical equation for this process is

Base: H^+ acceptor   Acid: H^+ donor

$$H_2O(l) + HCl(g) \longrightarrow H_3O^+(aq) + Cl^-(aq)$$

The hydrogen chloride behaves as an acid by donating a proton to a water molecule. Because the water molecule accepts the proton, to become H_3O^+, it is the base.

It is not necessary that a water molecule be one of the reactants in a Brønsted–Lowry acid–base reaction; the reaction does not have to take place in the liquid state. Brønsted–Lowry acid–base theory can be used to describe gas-phase reactions. The white solid haze that often covers glassware in a chemistry laboratory results from the gas-phase reaction between HCl and NH_3:

Base: H^+ acceptor   Acid: H^+ donor

$$NH_3(g) + HCl(g) \longrightarrow NH_4^+(g) + Cl^-(g)$$

This is a Brønsted–Lowry acid–base reaction because the HCl molecules donate protons to the NH_3, forming NH_4^+ and Cl^- ions. These ions instantaneously combine to form the white solid NH_4Cl (see Figure 10.3).

All acids and bases included in Arrhenius theory are also acids and bases according to Brønsted–Lowry theory. However, the converse is not true; some substances that are not considered Arrhenius bases are Brønsted–Lowry bases. Table 10.1 summarizes these two types of acid–base definitions.

■ Conjugate Acid–Base Pairs

For most Brønsted–Lowry acid–base reactions, 100% proton transfer does not occur. Instead, an equilibrium situation (Section 9.7) is reached in which a forward reaction and a reverse reaction occur at the same rate.

The equilibrium mixture for a Brønsted–Lowry acid–base reaction always has two acids and two bases present. Consider the acid–base reaction involving hydrogen fluoride and water:

$$HF(aq) + H_2O(l) \rightleftharpoons H_3O^+(aq) + F^-(aq)$$

For the forward reaction, the HF molecules donate protons to water molecules. Thus the HF is functioning as an acid, and the H_2O is functioning as a base.

$$HF(aq) + H_2O(l) \longrightarrow H_3O^+(aq) + F^-(aq)$$

Acid   Base

TABLE 10.1
Summary of Acid–Base Definitions

Arrhenius acid: hydrogen-containing species that produces H^+ ion in aqueous solution
Arrhenius base: hydroxide-containing species that produces OH^- ion in aqueous solution
Brønsted–Lowry acid: proton (H^+) donor
Brønsted–Lowry base: proton (H^+) acceptor

For the reverse reaction, the one going from right to left, a different picture emerges. Here, H_3O^+ is functioning as an acid (by donating a proton), and F^- behaves as a base (by accepting the proton).

$$H_3O^+(aq) + F^-(aq) \longrightarrow HF(aq) + H_2O(l)$$
$$\text{Acid} \qquad\quad \text{Base}$$

The two acids and two bases involved in a Brønsted–Lowry acid–base equilibrium mixture can be grouped into two conjugate acid–base pairs. A **conjugate acid–base pair** *is two substances in a Brønsted–Lowry acid–base reaction whose chemical formulas differ by one proton (H^+ ion).* The two conjugate acid–base pairs in our example are

> *Conjugate* means "coupled" or "joined together" (as in a pair).

$$\text{Conjugate pair}$$
$$HF(aq) + H_2O(l) \rightleftharpoons H_3O^+(aq) + F^-(aq)$$
$$\text{Acid} \qquad \text{Base} \qquad\quad \text{Acid} \qquad \text{Base}$$
$$\text{Conjugate pair}$$

> Every Brønsted–Lowry acid has a conjugate base, and every Brønsted–Lowry base has a conjugate acid. In general terms, these relationships can be diagrammed as follows:
>
> $$\text{HA} + \text{B} \rightleftharpoons \text{HB}^+ + \text{A}^-$$
> $$\text{Acid} \quad \text{Base} \qquad \text{Conjugate} \quad \text{Conjugate}$$
> $$\text{acid} \qquad\ \text{base}$$

A **conjugate base** *is the species formed when a Brønsted–Lowry acid loses a proton (H^+ ion).* The conjugate base of HF is F^-. A **conjugate acid** *is the species formed when a Brønsted–Lowry base accepts a proton (H^+ ion).* The H_3O^+ ion is the conjugate acid of H_2O. A slash mark separating an acid (on the left) from its conjugate base (on the right) is often used in representing conjugate acid–base pairs: for example, HF/F^- and H_3O^+/H_2O.

EXAMPLE 10.1

Determining the Formula of One Member of a Conjugate Acid–Base Pair When Given the Other Member

■ Write the chemical formula of each of the following.

a. The conjugate base of HSO_4^- b. The conjugate acid of NO_3^-
c. The conjugate base of H_3PO_4 d. The conjugate acid of $HC_2O_4^-$

Solution

a. A conjugate base can always be found by removing one H^+ from a given acid. Removing one H^+ (both the atom and the charge) from HSO_4^- leaves SO_4^{2-}. Thus SO_4^{2-} is the conjugate base of HSO_4^-.
b. A conjugate acid can always be found by adding one H^+ to a given base. Adding one H^+ (both the atom and the charge) to NO_3^- produces HNO_3. Thus HNO_3 is the conjugate acid of NO_3^-.
c. Proceeding as in part **a**, the removal of a H^+ ion from H_3PO_4 produces the $H_2PO_4^-$ ion. Thus $H_2PO_4^-$ is the conjugate base of H_3PO_4.
d. Proceeding as in part **b**, the addition of a H^+ ion to $HC_2O_4^-$ produces the $H_2C_2O_4$ molecule. Thus $H_2C_2O_4$ is the conjugate acid of $HC_2O_4^-$.

Practice Exercise 10.1

Write the chemical formula of each of the following.

a. The conjugate acid of ClO_3^- b. The conjugate base of NH_3
c. The conjugate acid of PO_4^{3-} d. The conjugate base of HS^-

■ **Amphoteric Substances**

Some molecules or ions are able to function as either Brønsted–Lowry acids or bases, depending on the kind of substance with which they react. Such molecules are said to be amphoteric. An **amphoteric substance** *is a substance that can either lose or accept a proton and thus can function as either a Brønsted–Lowry acid or a Brønsted–Lowry base.*

Water is the most common amphoteric substance. Water functions as a base in the first of the following two reactions and as an acid in the second.

> The term *amphoteric* comes from the Greek *amphoteres*, which means "partly one and partly the other." Just as an amphibian is an animal that lives partly on land and partly in the water, an amphoteric substance is sometimes an acid and sometimes a base.

$$HNO_3(aq) + H_2O(l) \rightleftharpoons H_3O^+(aq) + NO_3^-(aq)$$
$$\text{Acid} \qquad\quad \text{Base}$$

$$NH_3(aq) + H_2O(l) \rightleftharpoons NH_4^+(aq) + OH^-(aq)$$
$$\text{Base} \qquad\quad \text{Acid}$$

10.3 Mono-, Di-, and Triprotic Acids

Acids can be classified according to the number of hydrogen ions they can transfer per molecule during an acid–base reaction. A **monoprotic acid** *is an acid that supplies one proton (H^+ ion) per molecule during an acid–base reaction.* Hydrochloric acid (HCl) and nitric acid (HNO_3) are both monoprotic acids.

A **diprotic acid** *is an acid that supplies two protons (H^+ ions) per molecule during an acid–base reaction.* Carbonic acid (H_2CO_3) is a diprotic acid. The transfer of protons for a diprotic acid always occurs in steps. For H_2CO_3, the two steps are

$$H_2CO_3(aq) + H_2O(l) \rightleftharpoons H_3O^+(aq) + HCO_3^-(aq)$$
$$HCO_3^-(aq) + H_2O(l) \rightleftharpoons H_3O^+(aq) + CO_3^{2-}(aq)$$

A few triprotic acids exist. A **triprotic acid** *is an acid that supplies three protons (H^+ ions) per molecule during an acid–base reaction.* Phosphoric acid, H_3PO_4, is the most common triprotic acid. The three proton-transfer steps for this acid are

$$H_3PO_4(aq) + H_2O(l) \rightleftharpoons H_3O^+(aq) + H_2PO_4^-(aq)$$
$$H_2PO_4^-(aq) + H_2O(l) \rightleftharpoons H_3O^+(aq) + HPO_4^{2-}(aq)$$
$$HPO_4^{2-}(aq) + H_2O(l) \rightleftharpoons H_3O^+(aq) + PO_4^{3-}(aq)$$

A **polyprotic acid** *is an acid that supplies two or more protons (H^+ ions) during an acid–base reaction.*

The number of hydrogen atoms present in one molecule of an acid *cannot* always be used to classify the acid as mono-, di-, or triprotic. For example, a molecule of acetic acid contains four hydrogen atoms, and yet it is a monoprotic acid. Only one of the hydrogen atoms in acetic acid is *acidic;* that is, only one of the hydrogen atoms leaves the molecule when it is in solution.

Whether a hydrogen atom is acidic is related to its location in a molecule—that is, to which other atom it is bonded. From a structural viewpoint, the acidic behavior of acetic acid can be represented by the equation

$$\text{H}-\overset{\overset{\displaystyle H}{|}}{\underset{\underset{\displaystyle H}{|}}{\text{C}}}-\overset{\overset{\displaystyle O}{\|}}{\text{C}}-\text{O}-\text{H} + H_2O \rightleftharpoons H_3O^+ + \left[\text{H}-\overset{\overset{\displaystyle H}{|}}{\underset{\underset{\displaystyle H}{|}}{\text{C}}}-\overset{\overset{\displaystyle O}{\|}}{\text{C}}-\text{O}\right]^-$$

Note that one hydrogen atom is bonded to an oxygen atom and the other three hydrogen atoms are bonded to a carbon atom. The hydrogen atom bonded to the oxygen atom is the acidic hydrogen atom; the hydrogen atoms that are bonded to carbon atoms are too tightly held to be removed by reaction with water molecules. Water has very little effect on a carbon–hydrogen bond, because that bond is only slightly polar. On the other hand, the hydrogen bonded to oxygen is involved in a very polar bond because of oxygen's large electronegativity (Section 5.9). Water, which is a polar molecule, readily attacks this bond.

Writing the formula for acetic acid as $HC_2H_3O_2$ instead of $C_2H_4O_2$ emphasizes that there are two different kinds of hydrogen atoms present. One of the hydrogen atoms is acidic and the other three are not. When some hydrogen atoms are acidic and others are not, we write the acidic hydrogens first, thus separating them from the other hydrogen atoms in the formula. Citric acid, the principal acid in citrus fruits (see Figure 10.4), is another example of an acid that contains both acidic and nonacidic hydrogens. Its formula, $H_3C_6H_5O_7$, indicates that three of the eight hydrogen atoms present in a molecule are acidic.

10.4 Strengths of Acids and Bases

Brønsted–Lowry acids vary in their ability to transfer protons and produce hydronium ions in aqueous solution. Acids can be classified as strong or weak on the basis of the extent to which proton transfer occurs in aqueous solution. A **strong acid** *is an acid that transfers 100%, or very nearly 100%, of its protons (H^+ ions) to water in an aqueous solution.* Thus if an acid is strong, nearly all of the acid molecules present give up protons

> ▶ If the double arrows in the equation for a system at equilibrium are of unequal length, the longer arrow indicates the direction in which the equilibrium is displaced.
>
> $\rightleftharpoons$ Equilibrium displaced toward reactants
> $\rightleftharpoons$ Equilibrium displaced toward products

FIGURE 10.4 The sour taste of limes and other citrus fruit is due to the citric acid present in the fruit juice.

TABLE 10.2
Commonly Encountered Strong Acids

HCl	hydrochloric acid
HBr	hydrobromic acid
HI	hydroiodic acid
HNO_3	nitric acid
$HClO_4$	perchloric acid
H_2SO_4	sulfuric acid

▶ Learn the names and formulas of the six commonly encountered strong acids, and then assume that all other acids you encounter are weak unless you are told otherwise.

▶ It is important not to confuse the terms *strong* and *weak* with the terms *concentrated* and *dilute*. *Strong* and *weak* apply to the *extent of proton transfer,* not to the concentration of acid or base. *Concentrated* and *dilute* are relative concentration terms. Stomach acid (gastric juice) is a dilute (not weak) solution of a strong acid (HCl); it is 5% by mass hydrochloric acid.

TABLE 10.3
Commonly Encountered Strong Hydroxide Bases

Group IA hydroxides	Group IIA hydroxides
LiOH	—
NaOH	—
KOH	$Ca(OH)_2$
RbOH	$Sr(OH)_2$
CsOH	$Ba(OH)_2$

▶ HA is a frequently used *general* notation for a monoprotic acid. Similarly, H_2A denotes a diprotic acid.

FIGURE 10.5 A comparison of the number of H_3O^+ ions (the acidic species) present in strong acid and weak acid solutions of the same concentration.

to water. This extensive transfer of protons produces many hydronium ions (the acidic species) within the solution. A **weak acid** *is an acid that transfers only a small percentage of its protons (H$^+$ ions) to water in an aqueous solution.* The extent of proton transfer for weak acids is usually less than 5%.

The vast majority of acids are weak rather than strong. The six most commonly encountered strong acids are listed in Table 10.2.

The extent to which an acid undergoes ionization depends on the molecular structure of the acid; molecular polarity and the strength and polarity of individual bonds are particularly important factors in determining whether an acid is strong or weak.

The difference between a strong acid and a weak acid can also be stated in terms of equilibrium position (Section 9.7). Consider the reaction wherein HA represents the acid and H_3O^+ and A^- are the products from the proton transfer to H_2O. For strong acids, the equilibrium lies far to the right (100% or almost 100%):

$$HA + H_2O \longrightarrow\!\!\!\!\!\rightleftharpoons H_3O^+ + A^-$$

For weak acids, the equilibrium position lies far to the left:

$$HA + H_2O \rightleftharpoons H_3O^+ + A^-$$

Thus, in solutions of strong acids, the predominant species are H_3O^+ and A^-. In solutions of weak acids, the predominant species is HA; very little proton transfer has occurred. The differences between strong and weak acids, in terms of species present in solution, are illustrated in Figure 10.5.

Just as there are strong acids and weak acids, there are also strong bases and weak bases. As with acids, there are only a few strong bases. Strong bases are limited to the hydroxides of Groups IA and IIA listed in Table 10.3. Of the strong bases, only NaOH and KOH are commonly encountered in a chemical laboratory.

Only one of the many weak bases that exist is fairly common—aqueous ammonia. In a solution of ammonia gas (NH_3) in water, small amounts of OH^- ions are produced through the reaction of NH_3 molecules with water.

$$NH_3(g) + H_2O(l) \rightleftharpoons NH_4^+(aq) + OH^-(aq)$$

A solution of aqueous ammonia is sometimes erroneously called ammonium hydroxide. Aqueous ammonia is the preferred designation because most of the NH_3 present has not reacted with water; the equilibrium position lies far to the left. Only a few ammonium ions (NH_4^+) and hydroxide ions (OH^-) are present.

10.5 Ionization Constants for Acids and Bases

The strengths of various acids and bases can be quantified by use of ionization constants, which are forms of equilibrium constants (Section 9.8).

An **acid ionization constant** *is the equilibrium constant for the reaction of a weak acid with water.* For an acid with the general formula HA, the acid ionization constant is

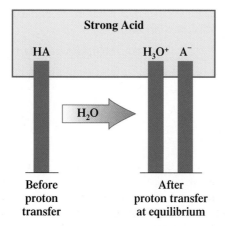

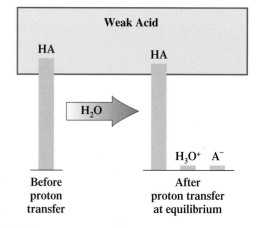

TABLE 10.4
Ionization Constant Values (K_a) and Percent Ionization Values for 1.0 M Solutions, at 24°C, of Selected Weak Acids

Name	Formula	K_a	Percent ionization
phosphoric acid	H_3PO_4	7.5×10^{-3}	8.3
hydrofluoric acid	HF	6.8×10^{-4}	2.6
nitrous acid	HNO_2	4.5×10^{-4}	2.1
acetic acid	$HC_2H_3O_2$	1.8×10^{-5}	0.42
carbonic acid	H_2CO_3	4.3×10^{-7}	0.065
dihydrogen phosphate ion	$H_2PO_4^-$	6.2×10^{-8}	0.025
hydrocyanic acid	HCN	4.9×10^{-10}	0.0022
hydrogen carbonate ion	HCO_3^-	5.6×10^{-11}	0.00075
hydrogen phosphate ion	HPO_4^{2-}	4.2×10^{-13}	0.000065

obtained by writing the equilibrium constant expression for the reaction

$$HA(aq) + H_2O(l) \rightleftharpoons H_3O^+(aq) + A^-(aq)$$

which is

$$K_a = \frac{[H_3O^+][A^-]}{[HA]}$$

The concentration of water is not included in the equilibrium constant expression because water is a pure liquid (Section 9.8). The symbol K_a is used to denote an acid ionization constant.

Table 10.4 gives K_a values, as well as percent ionization values (which can be calculated from K_a values), for selected weak acids. Acid strength increases as the K_a value increases. The actual value of K_a for a given acid must be determined by experimentally measuring the concentrations of HA, H_3O^+, and A^- in the acid solution and then using these values to calculate K_a. Example 10.2 shows how an acid ionization constant value (K_a) can be calculated by using concentration (molarity) and percent ionization data for an acid.

▶ Note the following relationships among acid strength, percent ionization, and K_a magnitude.

- Acid strength increases as percent ionization increases.
- Acid strength increases as the magnitude of K_a increases.
- Percent ionization increases as the magnitude of K_a increases.

EXAMPLE 10.2

Calculating the Acid Ionization Constant for an Acid When Given Its Concentration and Percent Ionization

■ A 0.0100 M solution of an acid, HA, is 15% ionized. Calculate the acid ionization constant for this acid.

Solution

To calculate K_a for the acid, we need the concentrations of H_3O^+, A^-, and HA in the aqueous solution.

The concentration of H_3O^+ will be 15% of the original HA concentration. Thus the concentration of hydronium ion is

$$H_3O^+ = (0.15) \times (0.0100 \text{ M}) = 0.0015 \text{ M}$$

The ionization of a monoprotic acid produces hydronium ions and the conjugate base of the acid (A^- ions) in a 1:1 ratio. Thus the concentration of A^- will be the same as that of hydronium ion—that is, 0.0015 M.

The concentration of HA is equal to the original concentration diminished by that which ionized (15%, or 0.0015 M):

$$HA = 0.0100 \text{ M} - 0.0015 \text{ M} = 0.0085 \text{ M}$$

Substituting these values in the equilibrium expression gives

$$K_a = \frac{[H_3O^+][A^-]}{[HA]} = \frac{[0.0015][0.0015]}{[0.0085]} = 2.6 \times 10^{-4}$$

Practice Exercise 10.2

A 0.100 M solution of an acid, HA, is 6.0% ionized. Calculate the acid ionization constant for this acid.

In Section 10.3, we noted that dissociation of a polyprotic acid occurs in a stepwise manner. In general, each successive step of proton transfer for a polyprotic acid occurs to a lesser extent than the previous step. For the dissociation series

$$H_2CO_3(aq) + H_2O(l) \rightleftharpoons H_3O^+(aq) + HCO_3^-(aq)$$
$$HCO_3^-(aq) + H_2O(l) \rightleftharpoons H_3O^+(aq) + CO_3^{2-}(aq)$$

the second proton is not so easily transferred as the first, because it must be pulled away from a negatively charged particle, HCO_3^-. (Remember that particles with opposite charge attract one another.) Accordingly, HCO_3^- is a weaker acid than H_2CO_3. The K_a values for these two acids (Table 10.4) are 5.6×10^{-11} and 4.3×10^{-7}, respectively.

Base strength follows the same principle as acid strength. Here, however, we deal with a base ionization constant, K_b. A **base ionization constant** *is the equilibrium constant for the reaction of a weak base with water.* The general expression for K_b is

$$K_b = \frac{[BH^+][OH^-]}{[B]}$$

where the reaction is

$$B(aq) + H_2O(l) \rightleftharpoons BH^+(aq) + OH^-(aq)$$

For the reaction involving the weak base NH_3,

$$NH_3(aq) + H_2O(l) \rightleftharpoons NH_4^+(aq) + OH^-(aq)$$

the base ionization constant expression is

$$K_b = \frac{[NH_4^+][OH^-]}{[NH_3]}$$

The K_b value for NH_3, the only common weak base, is 1.8×10^{-5}.

10.6 Salts

To a nonscientist, the term *salt* denotes a white granular substance that is used as a seasoning for food. To the chemist, the term *salt* has a much broader meaning; sodium chloride (table salt) is only one of thousands of salts known to a chemist. A **salt** *is an ionic compound containing a metal or polyatomic ion as the positive ion and a nonmetal or polyatomic ion (except hydroxide) as the negative ion.* (Ionic compounds that contain hydroxide ion are bases rather than salts.)

Much information about salts has been presented in previous chapters, although the term *salt* was not explicitly used in these discussions. Formula writing and nomenclature for binary ionic compounds (salts) were covered in Sections 4.7 and 4.9. Many salts contain polyatomic ions such as nitrate and sulfate; these ions were discussed in Section 4.10. The solubility of ionic compounds (salts) in water was the topic of Section 8.4.

All common soluble salts are *completely* dissociated into ions in solution (Section 8.3). Even if a salt is only slightly soluble, the small amount that does dissolve completely dissociates. Thus the terms *weak* and *strong,* which are used to denote qualitatively the percent dissociation of acids and bases, are not applicable to salts. We do not use the terms *weak salt* and *strong salt.*

Acids, bases, and salts are related in that a salt is one of the products that results from the reaction of an acid with a hydroxide base. This particular type of reaction will be discussed in Section 10.7.

of small numbers that is used to specify molar hydronium ion concentration in an aqueous solution.

The calculation of pH scale values involves the use of logarithms. The **pH** *is the negative logarithm of an aqueous solution's molar hydronium ion concentration.* Expressed mathematically, the definition of pH is

$$pH = -\log[H_3O^+]$$

(The letter *p*, as in pH, means "negative logarithm of.")

■ Integral pH Values

For any hydronium ion concentration expressed in exponential notation in which the coefficient is 1.0, the pH is given directly by the negative of the exponent value of the power of 10:

$$[H_3O^+] = 1.0 \times 10^{-x}$$
$$pH = x$$

Thus, if the hydronium ion concentration is 1.0×10^{-9}, then the pH will be 9.00. This simple relationship between pH and hydronium ion concentration is valid only when the coefficient in the exponential notation expression for the hydronium ion concentration is 1.0.

> The *p* in pH comes from the German word *potenz,* which means "power," as in "power of 10."

> The rule for the number of significant figures in a logarithm is: The number of digits after the decimal place in a logarithm is equal to the number of significant figures in the original number.
>
> $$[H_3O^+] = \underline{6.3} \times 10^{-5}$$
> Two significant figures
>
> $$pH = 4.\underline{20}$$
> Two digits

EXAMPLE 10.4

Calculating the pH of a Solution When Given Its Hydronium Ion or Hydroxide Ion Concentration

■ Calculate the pH for each of the following solutions.

a. $[H_3O^+] = 1.0 \times 10^{-6}$ **b.** $[OH^-] = 1.0 \times 10^{-6}$

Solution

a. Because the coefficient in the exponential expression for the molar hydronium ion concentration is 1.0, the pH can be obtained from the relationships

$$[H_3O^+] = 1.0 \times 10^{-x}$$
$$pH = x$$

The power of 10 is -6 in this case, so the pH will be 6.00.

b. The given quantity involves hydroxide ion rather than hydronium ion. Thus we must calculate the hydronium ion concentration first and then calculate the pH.

$$[H_3O^+] = \frac{1.00 \times 10^{-14}}{1.0 \times 10^{-6}} = 1.0 \times 10^{-8}$$

A solution with a hydronium ion concentration of 1.0×10^{-8} M will have a pH of 8.00.

Practice Exercise 10.4

Calculate the pH for each of the following solutions.

a. $[H_3O^+] = 1.0 \times 10^{-3}$
b. $[OH^-] = 1.0 \times 10^{-8}$

FIGURE 10.10 Most fruits and vegetables are acidic. Tart or sour taste is an indication that such is the case. Nonintegral pH values for selected foods are as shown here.

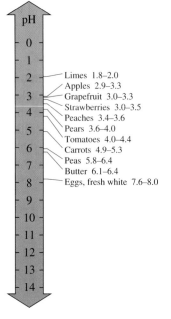

■ Nonintegral pH Values

If the coefficient in the exponential expression for the molar hydronium ion concentration is *not* 1.0, then the pH will have a nonintegral value; that is, it will not be a whole number. For example, consider the following nonintegral pH values.

$$[H_3O^+] = 6.3 \times 10^{-5} \quad pH = 4.20$$
$$[H_3O^+] = 5.3 \times 10^{-5} \quad pH = 4.28$$
$$[H_3O^+] = 2.2 \times 10^{-4} \quad pH = 3.66$$

Figure 10.10 gives nonintegral pH values for selected fruits and vegetables.

The easiest way to obtain nonintegral pH values such as these involves using an electronic calculator that allows for the input of exponential numbers and that has a base-10 logarithm key (LOG).

In using such an electronic calculator, you can obtain logarithm values simply by pressing the LOG key after having entered the number whose log is desired. For pH, you must remember that after obtaining the log value, you must change signs because of the negative sign in the defining equation for pH.

EXAMPLE 10.5

Calculating the pH of a Solution When Given Its Hydronium Ion Concentration

■ Calculate the pH for each of the following solutions.
  a. $[H_3O^+] = 7.23 \times 10^{-8}$
  b. $[H_3O^+] = 5.70 \times 10^{-3}$

Solution

a. Using an electronic calculator, first enter the number 7.23×10^{-8} into the calculator. Then use the LOG key to obtain the logarithm value, -7.1408617. Changing the sign of this number (because of the minus sign in the definition of pH) and adjusting for significant figures yields a pH value of 7.141.
b. Entering the number 5.70×10^{-3} into the calculator and then using the LOG key give a logarithm value of -2.2441251. This value translates into a pH value, after rounding, of 2.244.

Practice Exercise 10.5

Calculate the pH for each of the following solutions.
  a. $[H_3O^+] = 4.44 \times 10^{-11}$
  b. $[H_3O^+] = 8.92 \times 10^{-6}$

■ pH Values and Hydronium Ion Concentration

It is often necessary to calculate the hydronium ion concentration for a solution from its pH value. This type of calculation, which is the reverse of that illustrated in Examples 10.4 and 10.5, is shown in Example 10.6.

EXAMPLE 10.6

Calculating the Molar Hydronium Ion Concentration of a Solution from the Solution's pH

■ The pH of a solution is 6.80. What is the molar hydronium ion concentration for this solution?

Solution

From the defining equation for pH, we have

$$pH = -\log [H_3O^+] = 6.80$$

$$\log [H_3O^+] = -6.80$$

To find $[H_3O^+]$, we need to determine the *antilog* of -6.80.

How an antilog is obtained using a calculator depends on the type of calculator. Many calculators have an antilog function (sometimes labeled INV log) that performs this operation. If this key is present, then

1. Enter the number -6.80. Note that it is the *negative* of the pH that is entered into the calculator.
2. Press the INV log key (or an inverse key and then a log key). The result is the desired hydronium ion concentration.

$$\log [H_3O^+] = -6.80$$

$$\text{antilog} [H_3O^+] = 1.5848931 \times 10^{-7}$$

Rounded off, this value translates into a hydronium ion concentration of 1.6×10^{-7} M.

(continued)

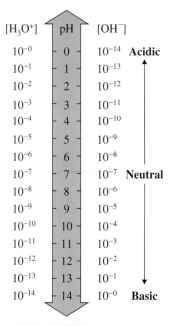

$[H_3O^+]$ pH $[OH^-]$

10^{-0}	0	10^{-14} **Acidic**
10^{-1}	1	10^{-13}
10^{-2}	2	10^{-12}
10^{-3}	3	10^{-11}
10^{-4}	4	10^{-10}
10^{-5}	5	10^{-9}
10^{-6}	6	10^{-8}
10^{-7}	7	10^{-7} **Neutral**
10^{-8}	8	10^{-6}
10^{-9}	9	10^{-5}
10^{-10}	10	10^{-4}
10^{-11}	11	10^{-3}
10^{-12}	12	10^{-2}
10^{-13}	13	10^{-1}
10^{-14}	14	10^{-0} **Basic**

FIGURE 10.11 Relationships among pH values, $[H_3O^+]$, and $[OH^-]$ at 24°C.

Some calculators use a 10^x key to perform the antilog operation. Use of this key is based on the mathematical identity

$$\text{antilog } x = 10^x$$

In our case, this means

$$\text{antilog} -6.80 = 10^{-6.80}$$

If the 10^x key is present, then

1. Enter the number -6.80 (the negative of the pH).
2. Press the function key 10^x. The result is the desired hydronium ion concentration.

$$[H_3O^+] = 10^{-6.80} = 1.6 \times 10^{-7}$$

Practice Exercise 10.6

The pH of a solution is 3.44. What is the molar hydronium ion concentration for this solution?

■ Interpreting pH Values

Identifying an aqueous solution as acidic, basic, or neutral based on pH value is a straightforward process. A **neutral solution** *is an aqueous solution whose pH is 7.0.* An **acidic solution** *is an aqueous solution whose pH is less than 7.0.* A **basic solution** *is an aqueous solution whose pH is greater than 7.0.* The relationships among $[H_3O^+]$, $[OH^-]$, and pH are summarized in Figure 10.11. Note the following trends from the information presented in this figure.

1. The higher the concentration of hydronium ion, the lower the pH value. Another statement of this same trend is that lowering the pH always corresponds to increasing the hydronium ion concentration.
2. A change of 1 unit in pH always corresponds to a tenfold change in hydronium ion concentration. For example,

$$\text{Difference of } 1 \begin{cases} pH = 1.0, \text{ then } [H_3O^+] = 0.1 \text{ M} \\ pH = 2.0, \text{ then } [H_3O^+] = 0.01 \text{ M} \end{cases} \text{ tenfold difference}$$

In a laboratory, solutions of any pH can be created. The range of pH values that are displayed by natural solutions is more limited than that of prepared solutions, but solutions corresponding to most pH values can be found (see Figure 10.12). A pH meter (Figure 10.13) helps chemists determine accurate pH values.

▶ Solutions of low pH are more acidic than solutions of high pH; conversely, solutions of high pH are more basic than solutions of low pH.

FIGURE 10.12 The pH values of selected common liquids. The lower the numerical value of the pH, the more acidic the substance is.

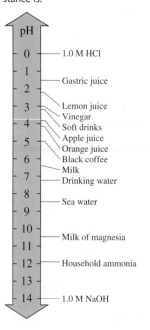

pH

0	1.0 M HCl
1	
2	Gastric juice
3	Lemon juice
4	Vinegar
	Soft drinks
5	Apple juice
	Orange juice
6	Black coffee
	Milk
7	Drinking water
8	
	Sea water
9	
10	Milk of magnesia
11	
12	Household ammonia
13	
14	1.0 M NaOH

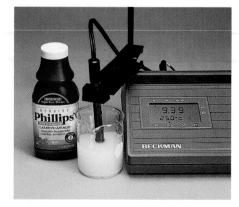

FIGURE 10.13 A pH meter gives an accurate measurement of pH values. The pH of vinegar is 2.32 (left). The pH of milk of magnesia in water is 9.39 (right).

TABLE 10.6
The Normal pH Range of Selected Body Fluids

Type of fluid	pH value
bile	6.8–7.0
blood plasma	7.3–7.5
gastric juices	1.0–3.0
milk	6.6–7.6
saliva	6.5–7.5
spinal fluid	7.3–7.5
urine	4.8–8.4

▶ Like pH, pK_a is a positive number. The lower the pK_a value, the stronger the acid.

The pH values of several human body fluids are given in Table 10.6. Most human body fluids except gastric juices and urine have pH values within one unit of neutrality. Both blood plasma and spinal fluid are always slightly basic.

The Chemistry at a Glance feature on page 250 summarizes what we have said about acids and acidity.

10.10 The pK_a Method for Expressing Acid Strength

In Section 10.5 ionization constants for acids and bases were introduced. These constants give an indication of the strengths of acids and bases. An additional method for expressing the strength of acids is in terms of pK_a units. The definition for pK_a is

$$pK_a = -\log K_a$$

The pK_a for an acid is calculated from K_a in exactly the same way that pH is calculated from hydronium ion concentration.

EXAMPLE 10.7

Calculating the pK_a of an Acid from the Acid's K_a Value

■ Determine the pK_a for acetic acid, $HC_2H_3O_2$, given that K_a for this acid is 1.8×10^{-5}.

Solution

Because the K_a value is 1.8×10^{-5} and $pK_a = -\log K_a$, we have

$$pK_a = -\log(1.8 \times 10^{-5}) = 4.74$$

The logarithm value 4.74 was obtained using an electronic calculator, as explained in Example 10.5.

Practice Exercise 10.7

Determine the pK_a for hydrocyanic acid, HCN, given that K_a for this acid is 4.4×10^{-10}.

10.11 The pH of Aqueous Salt Solutions

The addition of an acid to water produces an acidic solution. The addition of a base to water produces a basic solution. What type of solution is produced when a salt is added to water? Because salts are the products of acid–base neutralizations, a logical supposition would be that salts dissolve in water to produce neutral (pH = 7) solutions. Such is the case for a *few* salts. Aqueous solutions of *most* salts, however, are either acidic or basic rather than neutral. Let us consider why this is so.

When a salt is dissolved in water, it completely ionizes; that is, it completely breaks up into the ions of which it is composed (Section 8.3). For many salts, one or more of the ions so produced are reactive toward water. The ensuing reaction, which is called hydrolysis, causes the solution to have a non-neutral pH. A **hydrolysis reaction** *is the reaction of a salt with water to produce hydronium ion or hydroxide ion or both.*

▶ The term *hydrolysis* comes from the Greek *hydro,* which means "water," and *lysis,* which means "splitting."

■ Types of Salt Hydrolysis

Not all salts hydrolyze. Which ones do and which ones do not? Of those salts that do hydrolyze, which produce acidic solutions and which produce basic solutions? The following guidelines, based on the neutralization "parentage" of a salt—that is, on the acid and base that produce the salt through neutralization—can be used to answer these questions.

1. The salt of a *strong acid* and a *strong base* does not hydrolyze, so the solution is neutral.
2. The salt of a *strong acid* and a *weak base* hydrolyzes to produce an acidic solution.

Acids and Acidic Solutions

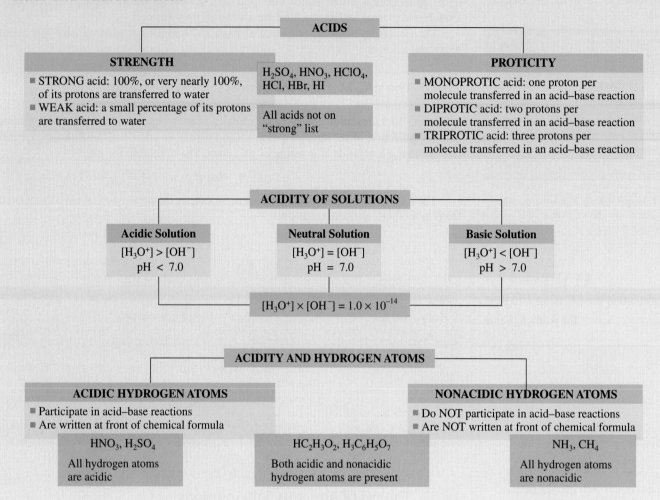

3. The salt of a *weak acid* and a *strong base* hydrolyzes to produce a basic solution.
4. The salt of a *weak acid* and a *weak base* hydrolyzes to produce a slightly acidic, neutral, or slightly basic solution, depending on the relative weaknesses of the acid and base.

These guidelines are summarized in Table 10.7.

The first prerequisite for using these guidelines is the ability to classify a salt into one of the four categories mentioned in the guidelines. This classification is accomplished by writing the neutralization equation (Section 10.7) that produces the salt and then specifying the strength (strong or weak) of the acid and base involved. The "parent" acid and base for the salt are identified by pairing the negative ion of the salt with H^+ (to form the acid) and pairing the positive ion of the salt with OH^- (to form the base). The following two equations illustrate the overall procedure.

TABLE 10.7
Neutralization "Parentage" of Salts and the Nature of the Aqueous Solutions They Form

Type of salt	Nature of aqueous solution	Examples
strong acid–strong base	neutral	$NaCl$, KBr
strong acid–weak base	acidic	NH_4Cl, NH_4NO_3
weak acid–strong base	basic	$NaC_2H_3O_2$, K_2CO_3
weak acid–weak base	depends on the salt	$NH_4C_2H_3O_2$, NH_4NO_2

CHEMICAL CONNECTIONS

Acid Rain: Excess Acidity

Rainfall, even in a pristine environment, has always been and will always be acidic. This acidity results from the presence of carbon dioxide in the atmosphere, which dissolves in water to produce carbonic acid (H_2CO_3), a weak acid.

$$CO_2(g) + H_2O(l) \longrightarrow H_2CO_3(aq)$$

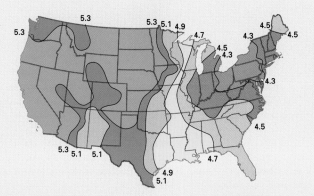

Average annual pH of precipitation in the United States

This reaction produces rainwater with a pH of approximately 5.6–5.7.

Acid rain is a generic term used to describe rainfall (or snowfall) whose pH is lower than the naturally produced value of 5.6. Acid rain has been observed with increasing frequency in many areas of the world. Rainfall with pH values between 4 and 5 is now common, and occasionally rainfall with a pH as low as 2 is encountered. Within the United States, the lowest acid rain pH values are encountered in the northeastern states (see the accompanying map). The maritime provinces of Canada have also been greatly affected.

Acid rain originates from the presence of sulfur oxides (SO_2 and SO_3), and to a lesser extent nitrogen oxides (NO and NO_2), in the atmosphere. After being discharged into the atmosphere, these pollutants can be chemically converted into sulfuric acid (H_2SO_4) and nitric acid (HNO_3) through oxidation processes. Several complicated pathways exist by which these two strong acids are produced. Which pathway is actually taken depends on numerous factors, including the intensity of sunlight and the amount of ammonia present in the atmosphere.

Small amounts of sulfur oxides and nitrogen oxides arise naturally from volcanic activity, lightning, and forest fires, but their major sources are human-related. The major source of sulfur oxide emissions is the combustion of coal associated with power plant operations. (The sulfur content of coal can be a high as 5% by mass.) Automobile exhaust is the major source of nitrogen oxides.

An important factor in determining the impact of acid rain on the environment is the ability of the natural ecosystem to neutralize incoming acidity. Generally speaking, the most sensitive areas overlie crystalline rock, whereas the least sensitive overlie limestone rock. Calcium carbonate and other basic substances associated with limestone rock are good neutralizing agents. Fortunately, low-pH rainfall directly entering lakes and streams does not automatically cause a severe decrease in pH. A large dilution factor accompanies rain falling directly into a large body of water.

The most observable effect of acid rain is the corrosion of building materials. Sulfuric acid (acid rain) readily attacks carbonate-based building materials (limestone, marble); the calcium carbonate is slowly converted into calcium sulfate.

$$CaCO_3(s) + H_2SO_4(aq) \longrightarrow CaSO_4(s) + CO_2(g) + H_2O(l)$$

The $CaSO_4$, which is more soluble than $CaCO_3$, is gradually eroded away. Many stone monuments show distinctly discernible erosion damage.

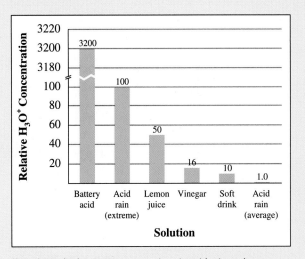

The relative hydronium concentrations in acid rain and some other acidic solutions. Comparisons are based on assigning acid rain of pH 4.5 (a commonly encountered situation) a relative value of 1.0.

$$\boxed{Na}OH + H\boxed{Cl} \longrightarrow H_2O + \boxed{NaCl}$$

Strong base — Strong acid — Strong acid–strong base salt

$$\boxed{K}OH + H\boxed{CN} \longrightarrow H_2O + \boxed{KCN}$$

Strong base — Weak acid — Weak acid–strong base salt

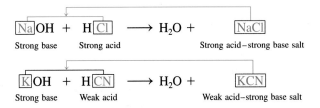

Knowing which acids and bases are strong and which are weak (Section 10.4) is a necessary part of the classification process. Once the salt has been so classified, the guideline that is appropriate for the situation is easily selected.

EXAMPLE 10.8

Predicting Whether a Salt's Aqueous Solution Will Be Acidic, Basic, or Neutral

■ Determine the acid–base "parentage" of each of the following salts, and then use this information to predict whether each salt's aqueous solution is acidic, basic, or neutral

a. Sodium acetate, $NaC_2H_3O_2$
b. Ammonium chloride, NH_4Cl
c. Potassium chloride, KCl
d. Ammonium fluoride, NH_4F

Solution

a. The ions present are Na^+ and $C_2H_3O_2^-$. The "parent" base of Na^+ is NaOH, a strong base. The "parent" acid of $C_2H_3O_2^-$ is $HC_2H_3O_2$, a weak acid. Thus the acid–base neutralization that produces this salt is

$$NaOH + HC_2H_3O_2 \longrightarrow H_2O + NaC_2H_3O_2$$
Strong base   Weak acid   Weak acid–strong base salt

The solution of a weak acid–strong base salt (guideline 3) produces a basic solution.
b. The ions present are NH_4^+ and Cl^-. The "parent" base of NH_4^+ is NH_3, a weak base. The "parent" acid of Cl^- is HCl, a strong acid. This "parentage" will produce a strong acid–weak base salt through neutralization. Such a salt gives an acidic solution upon hydrolysis (guideline 2).
c. The ions present are K^+ and Cl^-. The "parent" base is KOH (a strong base), and the "parent" acid is HCl (a strong acid). The salt produced from neutralization involving this acid–base pair will be a strong acid–strong base salt. Such salts do not hydrolyze. The aqueous solution is neutral (guideline 1).
d. The ions present are NH_4^+ and F^-. Both ions are of weak "parentage"; NH_3 is a weak base, and HF is a weak acid. Thus NH_4F is a weak acid–weak base salt. This is a guideline 4 situation. In this situation, you cannot predict the effect of hydrolysis unless you know the relative strengths of the weak acid and weak base (which is the weaker of the two). HF has a K_a of 6.6×10^{-4} (Table 10.4). NH_3 has a K_b of 1.8×10^{-5} (Section 10.5). Thus, NH_3 is the weaker of the two and will hydrolyze to the greater extent, causing the solution to be acidic.

Practice Exercise 10.8

Predict whether solutions of each of the following salts will be acidic, basic, or neutral.

a. Sodium bromide, NaBr
b. Potassium cyanide, KCN
c. Ammonium iodide, NH_4I
d. Barium chloride, $BaCl_2$

■ **Chemical Equations for Salt Hydrolysis Reactions**

Salt hydrolysis reactions are Brønsted–Lowry acid–base (proton transfer) reactions (Section 10.2). Such reactions are of the following two general types.

1. *Basic hydrolysis:* The reaction of the *negative ion* from a salt with water to produce the ion's conjugate acid and hydroxide ion.

Conjugate acid–base pair

$$CN^- + H_2O \longrightarrow HCN + OH^-$$
Proton   Proton   Weak   Makes solution
acceptor   donor   acid   basic

Conjugate acid–base pair

$$F^- + H_2O \longrightarrow HF + OH^-$$
Proton   Proton   Weak   Makes solution
acceptor   donor   acid   basic

▶ The only *negative ions* that undergo hydrolysis are those of "weak-acid parentage." The driving force for the reaction is the formation of the weak-acid "parent."

CHEMICAL CONNECTIONS

Blood Plasma pH and Hydrolysis

Blood plasma has a slightly basic pH (7.35–7.45), as shown in Table 10.6. The reason for this is related to salt hydrolysis and becomes apparent when the identity of the ions present in blood plasma is specified.

The most abundant positive ion present in blood plasma is Na^+, an ion associated with a strong base (NaOH). Thus it does not hydrolyze. The predominant negative ion present is Cl^-, an ion that comes from a strong acid (HCl). Thus it also does not hydrolyze. Together, these two ions, Na^+ and Cl^-, produce a neutral solution because neither hydrolyzes.

The third most abundant ion in blood plasma is the hydrogen carbonate ion, HCO_3^-, which comes from the weak acid H_2CO_3. Hydrolysis of this ion ties up hydronium ions and leaves hydroxide ions in excess. Thus the plasma fluid has a slightly basic character. Other negative ions present in blood plasma, such as HPO_4^{2-}, also hydrolyze and add to the basic character. However, because of their lower concentrations, their effect on the pH is not so great as that of HCO_3^-.

The pH of numerous other body fluids besides blood plasma is also directly influenced by hydrolysis reactions.

---

2. *Acidic hydrolysis:* The reaction of the *positive ion* from a salt with water to produce the ion's conjugate base and hydronium ion. The most common ion to undergo this type of reaction is the NH_4^+ ion.

▶ The only *positive ions* that undergo hydrolysis are those of "weak-base parentage." The driving force for the reaction is the formation of the weak-base "parent."

$$\overbrace{NH_4^+ + H_2O \longrightarrow NH_3}^{\text{Conjugate acid–base pair}} + H_3O^+$$

Proton donor · Proton acceptor · Weak base · Makes solution acidic

Hydrolysis reactions do not go 100% to completion. They occur only until equilibrium conditions are reached (Section 9.7). At the equilibrium point, solution pH can differ from neutrality by 2 to 4 pH units, as shown in Table 10.8.

10.12 Buffers

A **buffer** *is an aqueous solution containing substances that prevent major changes in solution pH when small amounts of acid or base are added to it.* Buffers are used in a laboratory setting to maintain optimum pH conditions for chemical reactions. Many commercial products contain buffers, which are needed to maintain optimum pH conditions for product behavior. Examples include buffered aspirin (Bufferin) and pH-controlled hair shampoos. Most human body fluids are highly buffered. For example, a buffer system maintains blood's pH at a value close to 7.4, an optimum pH for oxygen transport.

▶ A less common type of buffer involves a weak base and its conjugate acid. We will not consider this type of buffer here.

Buffers contain two active chemical species: (1) a substance to react with and remove added base, and (2) a substance to react with and remove added acid. Typically, a buffer system is composed of a weak acid *and* its conjugate base—that is, a conjugate acid–base pair (Section 10.2). Conjugate acid–base pairs that are commonly employed as buffers include $HC_2H_3O_2/C_2H_3O_2^-$, $H_2PO_4^-/HPO_4^{2-}$, and H_2CO_3/HCO_3^-.

TABLE 10.8
Approximate pH of Selected 0.1 M Aqueous Salt Solutions at 24°C

Name of salt	Formula of salt	pH	Category of salt
ammonium nitrate	NH_4NO_3	5.1	strong acid–weak base
ammonium nitrite	NH_4NO_2	6.3	weak acid–weak base
ammonium acetate	$NH_4C_2H_3O_2$	7.0	weak acid–weak base
sodium chloride	NaCl	7.0	strong acid–strong base
sodium fluoride	NaF	8.1	weak acid–strong base
sodium acetate	$NaC_2H_3O_2$	8.9	weak acid–strong base
ammonium cyanide	NH_4CN	9.3	weak acid–weak base
sodium cyanide	NaCN	11.1	weak acid–strong base

EXAMPLE 10.9

Recognizing Pairs of Chemical Substances That Can Function as a Buffer in Aqueous Solution

■ Predict whether each of the following pairs of substances could function as a buffer system in aqueous solution.

a. HCl and NaCl   **b.** HCN and KCN
c. HCl and HCN   **d.** NaCN and KCN

Solution

Buffer solutions contain either a weak acid and a salt of that weak acid or a weak base and a salt of that weak base.

a. No. We have an acid and the salt of that acid. However, the acid is a strong acid rather than a weak acid.
b. Yes. HCN is a weak acid, and KCN is a salt of that weak acid.
c. No. Both HCl and HCN are acids. No salt is present.
d. No. Both NaCN and KCN are salts. No weak acid is present.

Practice Exercise 10.9

Predict whether each of the following pairs of substances could function as a buffer system in aqueous solution.

a. HCl and NaOH   **b.** $HC_2H_3O_2$ and $KC_2H_3O_2$
c. NaCl and NaCN   **d.** HCN and $HC_2H_3O_2$

As an illustration of buffer action, consider a buffer solution containing approximately equal concentrations of acetic acid (a weak acid) and sodium acetate (a salt of this weak acid). This solution resists pH change by the following mechanisms:

1. When a small amount of a strong acid such as HCl is added to the solution, the newly added H_3O^+ ions react with the acetate ions from the sodium acetate to give acetic acid.

$$H_3O^+ + C_2H_3O_2^- \longrightarrow HC_2H_3O_2 + H_2O$$

Most of the added H_3O^+ ions are tied up in acetic acid molecules, and the pH changes very little.

2. When a small amount of a strong base such as NaOH is added to the solution, the newly added OH^- ions react with the acetic acid (neutralization) to give acetate ions and water.

$$OH^- + HC_2H_3O_2 \longrightarrow C_2H_3O_2^- + H_2O$$

Most of the added OH^- ions are converted to water, and the pH changes only slightly.

The reactions that are responsible for the buffering action in the acetic acid/acetate ion system can be summarized as follows:

$$C_2H_3O_2^- \underset{OH^-}{\overset{H_3O^+}{\rightleftharpoons}} HC_2H_3O_2$$

▶ To resist both increases and decreases in pH effectively, a weak-acid buffer must contain significant amounts of both the weak acid and its conjugate base. If a solution has a large amount of weak acid but very little conjugate base, it will be unable to consume much added acid. Consequently, the pH tends to drop significantly when acid is added. Conversely, a solution that contains a large amount of conjugate base but very little weak acid will provide very little protection against added base. Addition of just a little base will cause a big change in pH.

Note that one member of the buffer pair (acetate ion) removes excess H_3O^+ ion and that the other (acetic acid) removes excess OH^- ion. The buffering action always results in the active species being converted to its partner species.

EXAMPLE 10.10

Writing Equations for Reactions That Occur in a Buffered Solution

■ Write an equation for each of the following buffering actions.

a. The response of $H_2PO_4^-/HPO_4^{2-}$ buffer to the addition of H_3O^+ ions
b. The response of HCN/CN^- buffer to the addition of OH^- ions

Solution

a. The base in a conjugate acid–base pair is the species that responds to the addition of acid. (Recall, from Section 10.2, that the base in a conjugate acid–base pair always has one less hydrogen than the acid.) The base for this reaction is HPO_4^{2-}. The equation for the buffering action is

$$H_3O^+ + HPO_4^{2-} \longrightarrow H_2PO_4^- + H_2O$$

In the buffering response, the base is always converted into its conjugate acid.
b. The acid in a conjugate acid–base pair is the species that responds to the addition of base. The acid for this reaction is HCN. The equation for the buffering action is

$$HCN + OH^- \longrightarrow CN^- + H_2O$$

Water will always be one of the products of buffering action.

Practice Exercise 10.10

Write an equation for each of the following buffering actions.

a. The response of H_2CO_3/HCO_3^- buffer to the addition of H_3O^+ ions
b. The response of $H_2PO_4^-/HPO_4^{2-}$ buffer to the addition of OH^- ions

▶ A common misconception about buffers is that a buffered solution is always a neutral (pH 7.0) solution. This is false. One can buffer a solution at any desired pH. A pH 7.4 buffer will hold the pH of the solution near pH 7.4, whereas a pH 9.3 buffer will tend to hold the pH of a solution near pH 9.3. The pH of a buffer is determined by the degree of weakness of the weak acid used and by the concentrations of the acid and its conjugate base.

A false notion about buffers is that they will hold the pH of a solution *absolutely* constant. The addition of even small amounts of a strong acid or a strong base to any solution, buffered or not, will lead to a change in pH. The important concept is that the shift in pH will be much less when an effective buffer is present (see Table 10.9).

Buffer systems have their limits. If large amounts of H_3O^+ or OH^- are added to a buffer, the buffer capacity can be exceeded; then the buffer system is overwhelmed and the pH changes (Figure 10.14). For example, if large amounts of H_3O^+ were added to the acetate/acetic acid buffer previously discussed, the H_3O^+ ion would react with acetate ion until the acetate was depleted. Then the pH would begin to drop as free H_3O^+ ions accumulated in the solution.

Additional insights into the workings of buffer systems are obtained by considering buffer action within the framework of Le Châtelier's principle and an equilibrium system. Let us again consider an acetic acid/acetate ion buffer system. An equilibrium is established in solution between the acetic acid and the acetate ion.

$$HC_2H_3O_2(aq) + H_2O(l) \rightleftharpoons H_3O^+(aq) + C_2H_3O_2^-(aq)$$

This equilibrium system functions in accordance with *Le Chatelier's principle* (Section 9.9), which states that an equilibrium system, when stressed, will shift its position in such a way as to counteract the stress. Stresses for the buffer will be (1) addition of base (hydroxide ion) and (2) addition of acid (hydronium ion). Further details concerning these two stress situations are as follows.

Addition of base [OH^- ion] *to the buffer.* The addition of base causes the following changes to occur in the solution:

1. The added OH^- ion reacts with H_3O^+ ion, producing water (neutralization).
2. The neutralization reaction produces the stress of *not enough* H_3O^+ ion, because H_3O^+ ion was consumed in the neutralization.

TABLE 10.9

A Comparison of pH Changes in Buffered and Unbuffered Solutions

Unbuffered Solution	
1 liter water	pH = 7.0
1 liter water + 0.01 mole strong base (NaOH)	pH = 12.0
1 liter water + 0.01 mole strong acid (HCl)	pH = 2.0
Buffered Solution	
1 liter buffer^a	pH = 7.2
1 liter buffer^a + 0.01 mole strong base (NaOH)	pH = 7.3
1 liter buffer^a + 0.01 mole strong acid (HCl)	pH = 7.1

^{a}Buffer = equal amounts of 0.1 M HPO_4^{2-} and 0.1 M $H_2PO_4^-$

Buffered Solution	Unbuffered Solution	Buffered Solution	Unbuffered Solution

(a) Before addition of HCl   **(b) After addition of HCl**

FIGURE 10.14 (a) The buffered solution on the left and the unbuffered solution on the right have the same pH (pH 8). They are basic solutions. (b) After the addition of 1 mL of a 0.01 M HCl solution, the pH of the buffered solution has not perceptibly changed, but the unbuffered solution has become acidic, as indicated by the change in the color of the acid–base indicator present.

3. The equilibrium shifts to the right, in accordance with Le Châtelier's principle, to produce more H_3O^+ ion, which maintains the pH close to its original level.

Addition of acid [H_3O^+ ion] *to the buffer.* The addition of acid causes the following changes to occur in the solution:

1. The added H_3O^+ ion increases the overall amount of H_3O^+ ion present.
2. The stress on the system is *too much* H_3O^+ ion.
3. The equilibrium shifts to the left, in accordance with Le Châtelier's principle, consuming most of the excess H_3O^+ ion and resulting in a pH close to the original level.

The Chemistry at a Glance feature on page 257 reviews basic concepts about buffer systems.

10.13 The Henderson–Hasselbalch Equation

Buffers may be prepared from any ratio of concentrations of a weak acid and the salt of its conjugate base. However, a buffer is most effective in counteracting pH change when the acid-to-conjugate-base ratio is 1:1. If a buffer contains considerably more acid than the conjugate base, it is less efficient in handling an acid. Conversely, a buffer with considerably more of the conjugate base than the acid is less efficient in handling added base.

When the concentrations of an acid and its conjugate base are equal in a buffer solution, the solution's hydronium ion concentration is equal to the acid ionization constant of the weak acid—or, stated more concisely, the pH of the solution is equal to the pK_a of the weak acid. The mathematical basis for this equality is as follows:

For the weak acid,

$$K_a = \frac{[H_3O^+][A^-]}{[HA]}$$

If HA and A^- are equal, then they cancel from the equation and we have

$$K_a = [H_3O^+]$$

Taking the negative logarithm of both sides of this equation gives

$$pK_a = pH$$

The relationship between pK_a and pH for buffer solutions where the conjugate acid–base pair concentration ratio is something other than 1:1 is given by the equation

$$pH = pK_a + \log \frac{[A^-]}{[HA]}$$

Buffer Systems

BUFFER SOLUTION
■ A solution that resists major change in pH when small amounts of strong acid or strong base are added ■ A typical buffer system contains a weak acid and its conjugate base
Common biochemical buffer systems are H_2CO_3/HCO_3^- $H_2PO_4^-/HPO_4^{2-}$

Weak Acid
■ The buffer component that reacts with added base ■ Reaction converts it into its conjugate base

Conjugate Base of Weak Acid
■ The buffer component that reacts with added acid ■ Reaction converts it into its conjugate acid

DIAGRAMS OF BUFFER ACTION

$$HA \underset{H^+}{\overset{OH^-}{\rightleftharpoons}} A^-$$

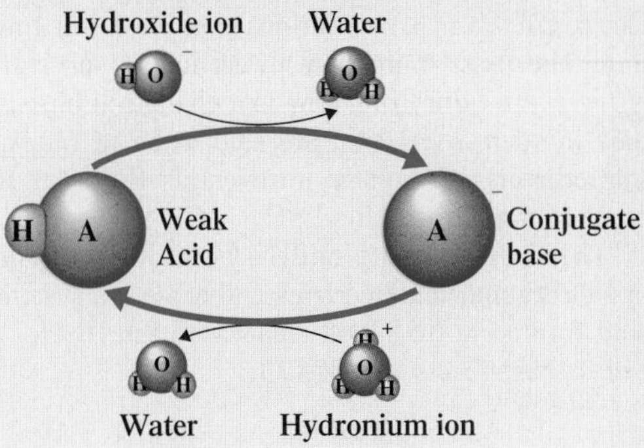

Hydroxide ion   Water

H A Weak Acid   A Conjugate base

Water   Hydronium ion

This equation is called the *Henderson–Hasselbalch equation.* The Henderson–Hasselbalch equation indicates that if there is more A^- than HA in a solution, the pH is higher than pK_a; and if there is more HA than A^-, the pH is lower than pK_a.

EXAMPLE 10.11

Calculating the pH of a Buffer Solution Using the Henderson–Hasselbalch Equation

■ What is the pH of a buffer solution that is 0.5 M in formic acid ($HCHO_2$) and 1.0 M in sodium formate ($NaCHO_2$). The pK_a for formic acid is 3.74.

Solution

The concentrations for the buffering species are

$$HCHO_2 = 0.5\ M \qquad CHO_2^- = 1.0\ M$$

Substituting these values into the Henderson–Hasselbalch equation gives

$$pH = pK_a + \log\frac{[CHO_2^-]}{[HCHO_2]} = 3.74 + \log\frac{1.0}{0.5}$$

$$= 3.74 + \log 2 = 3.74 + 0.30$$

$$= 4.04$$

Practice Exercise 10.11

What is the pH of a buffer solution that is 0.6 M in acetic acid ($HC_2H_3O_2$) and 1.5 M in sodium acetate ($NaC_2H_3O_2$). The pK_a for acetic acid is 4.74.

10.14 Electrolytes

Aqueous solutions in which ions are present are good conductors of electricity, and the greater the number of ions present, the better the solution conducts electricity. Acids, bases, and soluble salts all produce ions in solution; thus they all produce solutions that conduct electricity. All three types of compounds are said to be electrolytes. An **electrolyte** *is a substance whose aqueous solution conducts electricity.* The presence of ions (charged particles) explains the electrical conductivity.

Some substances, such as table sugar (sucrose), glucose, and isopropyl alcohol, do not produce ions in solution. These substances are called nonelectrolytes. A **nonelectrolyte** *is a substance whose aqueous solution does not conduct electricity.*

CHEMICAL CONNECTIONS

Buffering Action in Human Blood

Metabolic processes normally maintain blood pH within the narrow range of 7.35–7.45. Even small departures from blood's normal pH range can cause serious illness, and death can result from pH variations that exceed a few tenths of a unit. The most immediate threat to the survival of a person with severe injury or burns is a change in blood pH; individuals in such a situation are said to be in *shock*. Paramedics immediately administer intravenous fluids in such cases to combat changes in blood pH.

The body's primary buffer system for controlling blood pH is the carbonic acid/hydrogen carbonate ion system. Any excess acid formed in the blood reacts with the HCO_3^- ion, and any excess base reacts with H_2CO_3.

$$H_3O^+ + HCO_3^- \longrightarrow H_2CO_3 + H_2O$$

$$OH^- + H_2CO_3 \longrightarrow HCO_3^- + H_2O$$

The ratio of $[H_2CO_3]$ to $[HCO_3^-]$ in blood is approximately 1 to 10, which means this buffer has a greater ability to interact with acid than with base. Significant amounts of acids (up to 10 moles a day) are produced in the human body as a result of normal metabolic reactions. For example, lactic acid ($HC_3H_5O_2$) is produced in muscle tissue during exercise.

This 1-to-10 ratio of buffering species is also needed to maintain the blood at a pH of 7.4. A 1-to-1 ratio buffer would produce a pH of 6.4. The 1-to-10 ratio is easily maintained.

Carbonic acid concentration is controlled by respiration. Excess H_2CO_3 decomposes to CO_2 and H_2O and is removed from the blood by the lungs.

$$H_2CO_3 \longrightarrow CO_2 + H_2O$$

Hydrogen carbonate ion concentration is controlled by the kidneys. Excess HCO_3^- ion is eliminated from the body through urine.

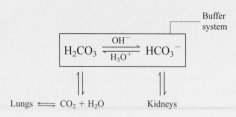

Under certain stress conditions, the blood's buffer systems can be overwhelmed. **Acidosis** *is a body condition in which the pH of blood drops from its normal value of 7.4 to 7.1–7.2.* **Alkalosis** *is a body condition in which the pH of blood increases from its normal value of 7.4 to a value of 7.5.* Both can be life-threatening if not properly taken care of; both can be caused by either metabolic processes or changes in breathing patterns (respiration).

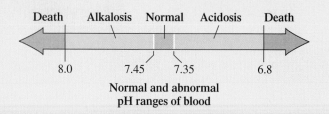

Normal and abnormal pH ranges of blood

Metabolic acidosis is seen in diabetics, who accumulate acidic substances from the metabolism of fats. Excessive loss of bicarbonate ion in cases of severe diarrhea is another cause. A temporary metabolic acidosis condition can result from prolonged intensive exercise. Exercise generates lactic acid (a weak acid) in the muscles. Some of the lactic acid ionizes, and this produces an influx of H_3O^+ ions into the bloodstream.

Metabolic alkalosis is less common than metabolic acidosis. It results from elevated HCO_3^- ion levels. Causes include prolonged vomiting and the side effects of certain drugs that change the concentrations of sodium, potassium, and chloride ions in the blood.

Respiratory acidosis results from higher than normal levels of CO_2 in the blood; inefficient CO_2 removal is usually the origin of this problem. Hypoventilation (a lowered breathing rate), caused by lung diseases such as emphysema and asthma or obstructed air passages, produces respiratory acidosis.

Respiratory alkalosis is caused by hyperventilation (an elevated breathing rate). Causes include hysteria and anxiety (occasioned, for example, by chemistry tests) and the rapid breathing associated with extremely high fevers.

Electrolytes can be divided into two groups—strong electrolytes and weak electrolytes. A **strong electrolyte** *is a substance that completely (or almost completely) ionizes/dissociates into ions in aqueous solution.* Strong electrolytes produce strongly conducting solutions. All strong acids and strong bases and all soluble salts are strong electrolytes. A **weak electrolyte** *is a substance that incompletely ionizes/dissociates into ions in aqueous solution.* Weak electrolytes produce solutions that are intermediate between those containing strong electrolytes and those containing nonelectrolytes in their

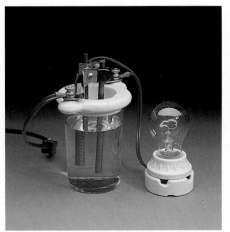
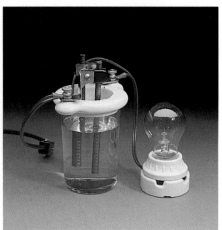

FIGURE 10.15 This simple device can be used to distinguish among strong electrolytes, weak electrolytes, and nonelectrolytes. The light bulb glows strongly for strong electrolytes (left), weakly for weak electrolytes (center), and not at all for non-electrolytes (right).

ability to conduct an electric current. Weak acids, weak bases, and slightly soluble salts constitute the weak electrolytes.

You can determine whether a substance is an electrolyte in solution by testing the ability of the solution to conduct an electric current. A device such as that shown in Figure 10.15 can be used to distinguish among strong electrolytes, weak electrolytes, and nonelectrolytes. If the medium between the electrodes (the solution) is a conductor of electricity, the light bulb glows. A strong glow indicates a strong electrolyte. A faint glow occurs for a weak electrolyte, and there is no glow for a nonelectrolyte.

10.15 Acid–Base Titrations

FIGURE 10.16 A schematic diagram showing the setup used for titration procedures.

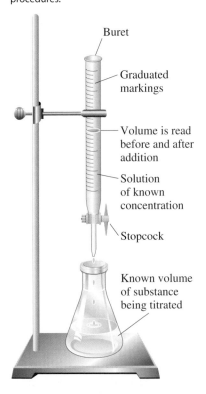

Buret

Graduated markings

Volume is read before and after addition

Solution of known concentration

Stopcock

Known volume of substance being titrated

The analysis of solutions to determine the concentration of acid or base present is performed regularly in many laboratories. Such activity is different from determining a solution's pH value. The pH of a solution gives information about the *concentration* of hydronium ions in solution. Only ionized molecules influence the pH value. The concentration of an acid or a base gives information about the *total number* of acid or base molecules present; both dissociated and undissociated molecules are counted. Thus acid or base concentration is a measure of total acidity or total basicity.

The procedure most frequently used to determine the concentration of an acid or a base solution is an acid–base titration. An **acid–base titration** *is a neutralization reaction in which a measured volume of an acid or a base of known concentration is completely reacted with a measured volume of a base or an acid of unknown concentration.*

Suppose we want to determine the concentration of an acid solution by means of titration. We first measure out a known volume of the acid solution into a beaker or flask. Then we slowly add a solution of base of known concentration to the flask or beaker by means of a buret (Figure 10.16). We continue to add base until all the acid has completely reacted with the added base. The volume of base needed to reach this point is obtained from the buret readings. When we know the original volume of acid, the concentration of the base, and the volume of added base, we can calculate the concentration of the acid, as will be shown in Example 10.12.

In order to complete a titration successfully, we must be able to detect when the reaction between acid and base is complete. Neither the acid nor the base gives any outward sign that the reaction is complete. Thus an indicator is always added to the reaction mixture (Figure 10.17). An **acid–base indicator** *is a compound that exhibits different colors depending on the* pH *of its solution.* Typically, an indicator is one color in basic solutions and another color in acidic solutions. An indicator is selected that

CHEMICAL CONNECTIONS

Electrolytes and Body Fluids

Structurally speaking, there are three types of body fluids: blood plasma, which is the liquid part of the blood; interstitial fluid, which is the fluid in tissues between and around cells; and intracellular fluids, which are the fluids within cells. For every kilogram of its mass, the body contains about 400 mL of intracellular fluid, 160 mL of interstitial fluid, and 40 mL of plasma.

Water is the main component of any type of body fluid. In addition, all body fluids contain electrolytes. It is the electrolytes present in body fluids that govern numerous body processes. The chemical makeup of the three types of body fluids, in terms of electrolytes (ions present), is shown in the accompanying figure.

Chemically, two of the body fluids (plasma and interstitial fluid) are almost identical. Intracellular fluid, on the other hand, shows striking differences. For example, K^+ is the dominant

positive ion in intracellular fluid, and Na^+ dominates in the other two fluids. A similar situation occurs with negative ions. A different ion dominates in intracellular fluid (HPO_4^{2-}) than in the other two fluids (Cl^-).

The electrolytes present in body fluids (1) govern the movement of water between body fluid compartments and (2) maintain acid–base balance within the body fluids. Osmotic pressure, a major factor in controlling water movement, is directly related to electrolyte concentration gradients.

The fact that the presence of ions causes a solution to conduct electricity is of extreme biochemical significance. For example, messages to and from the brain are sent in the form of electrical signals. Ions in intracellular and interstitial fluids are often the carriers of these signals. The presence of electrolytes (ions) is essential to the proper functioning of the human body.

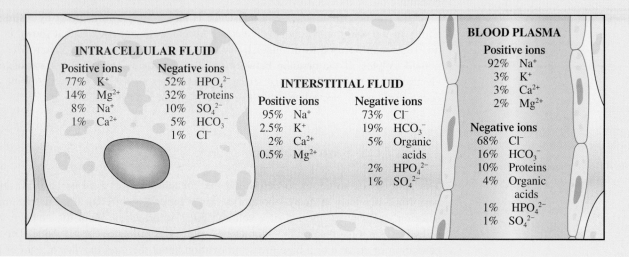

INTRACELLULAR FLUID

Positive ions		Negative ions	
77%	K^+	52%	HPO_4^{2-}
14%	Mg^{2+}	32%	Proteins
8%	Na^+	10%	SO_4^{2-}
1%	Ca^{2+}	5%	HCO_3^-
		1%	Cl^-

INTERSTITIAL FLUID

Positive ions		Negative ions	
95%	Na^+	73%	Cl^-
2.5%	K^+	19%	HCO_3^-
2%	Ca^{2+}	5%	Organic acids
0.5%	Mg^{2+}	2%	HPO_4^{2-}
		1%	SO_4^{2-}

BLOOD PLASMA

Positive ions	
92%	Na^+
3%	K^+
3%	Ca^{2+}
2%	Mg^{2+}

Negative ions	
68%	Cl^-
16%	HCO_3^-
10%	Proteins
4%	Organic acids
1%	HPO_4^{2-}
1%	SO_4^{2-}

FIGURE 10.17 An acid–base titration using an indicator that is yellow in acidic solution and red in basic solution.

changes color at a pH that corresponds as nearly as possible to the pH of the solution when the titration is complete. If the acid and base are both strong, the pH at that point is 7.0. However, because of hydrolysis (Section 10.11), the pH is not 7.0 if a weak acid or weak base is part of the titration system.

Titration of a weak acid by a strong base requires an indicator that changes color above pH 7.0, because the salt formed in the titration will hydrolyze to form a basic solution. Conversely, titration of a weak base by a strong acid requires an indicator that changes color below pH 7.0.

Example 10.12 shows how titration data are used to calculate the molarity of an acid solution of unknown concentration.

EXAMPLE 10.12

Calculating an Unknown Molarity from Acid–Base Titration Data

■ In a sulfuric acid (H_2SO_4)–sodium hydroxide (NaOH) acid–base titration, 17.3 mL of 0.126 M NaOH is needed to neutralize 25.0 mL of H_2SO_4 of unknown concentration. Find the molarity of the H_2SO_4 solution, given that the neutralization reaction that occurs is

$$H_2SO_4(aq) + 2NaOH(aq) \longrightarrow Na_2SO_4(aq) + 2H_2O(l)$$

Solution

First, we calculate the number of moles of H_2SO_4 that reacted with the NaOH. The pathway for this calculation, using dimensional analysis (Section 6.8), is

$$\text{mL of NaOH} \longrightarrow \text{L of NaOH} \longrightarrow \text{moles of NaOH} \longrightarrow \text{moles of } H_2SO_4$$

The sequence of conversion factors that effects this series of unit changes is

$$17.3 \, \text{mL NaOH} \times \left(\frac{10^{-3} \, \text{L NaOH}}{1 \, \text{mL NaOH}} \right) \times \left(\frac{0.126 \, \text{mole NaOH}}{1 \, \text{L NaOH}} \right) \times \left(\frac{1 \, \text{mole } H_2SO_4}{2 \, \text{moles NaOH}} \right)$$

The first conversion factor derives from the definition of a milliliter, the second conversion factor derives from the definition of molarity (Section 8.5), and the third conversion factor uses the coefficients in the balanced chemical equation for the titration reaction (Section 6.7).

The number of moles of H_2SO_4 that react is obtained by combining all the numbers in the dimensional analysis setup in the manner indicated.

$$\left(\frac{17.3 \times 10^{-3} \times 0.126 \times 1}{1 \times 1 \times 2} \right) \, \text{mole } H_2SO_4 = 0.00109 \, \text{mole } H_2SO_4$$

Now that we know how many moles of H_2SO_4 reacted, we calculate the molarity of the H_2SO_4 solution using the definition for molarity.

$$\text{Molarity } H_2SO_4 = \frac{\text{moles } H_2SO_4}{\text{L } H_2SO_4 \text{ solution}} = \frac{0.00109 \, \text{mole}}{0.0250 \, \text{L}}$$

$$= 0.0436 \frac{\text{mole}}{\text{L}}$$

Note that the units in the denominator of the molarity equation must be liters (0.0250) rather than milliliters (25.0).

Practice Exercise 10.12

In a nitric acid (HNO_3)–potassium hydroxide (KOH) acid–base titration, 32.4 mL of 0.352 M KOH is required to neutralize 50.0 mL of HNO_3 of unknown concentration. Find the molarity of the HNO_3 solution, given that the neutralization reaction that occurs is

$$HNO_3(aq) + KOH(aq) \longrightarrow KNO_3(aq) + H_2O(l)$$

CONCEPTS TO REMEMBER

Arrhenius acid–base theory. An Arrhenius acid is a hydrogen-containing compound that, in water, produces hydrogen ions. An Arrhenius base is a hydroxide-containing compound that, in water, produces hydroxide ions.

Brønsted–Lowry acid–base theory. A Brønsted–Lowry acid is any substance that can donate a proton (H^+) to some other substance. A Brønsted–Lowry base is any substance that can accept a proton from some other substance. Proton donation (from an acid) cannot occur unless an acceptor (a base) is present.

Conjugate acids and bases. A conjugate acid–base pair consists of two species that differ by one proton. The conjugate base of an acid is the species that remains when the acid loses a proton. The conjugate acid of a base is the species formed when the base accepts a proton.

Polyprotic acids. Polyprotic acids are acids that can transfer two or more hydrogen ions during an acid–base reaction.

Strengths of acids and bases. Acids can be classified as strong or weak in terms of the extent to which proton transfer occurs in aqueous solution. A strong acid completely transfers its protons to water. A weak acid transfers only a small percentage of its protons to water.

Acid ionization constant. The acid ionization constant quantitatively describes the degree of ionization of an acid. It is the equilibrium constant expression that corresponds to the ionization of the acid.

Salts. Salts are ionic compounds containing a metal or polyatomic ion as the positive ion and a nonmetal or polyatomic ion (except hydroxide ion) as the negative ion. Ionic compounds containing hydroxide ion are bases rather than salts.

Neutralization. Neutralization is the reaction between an acid and a hydroxide base to form a salt and water.

Self-ionization of water. In pure water, a small number of water molecules (1.0×10^{-7} mole/L) donate protons to other water molecules to produce small concentrations (1.0×10^{-7} mole/L) of hydronium and hydroxide ions.

The pH scale. The pH scale is a scale of small numbers that are used to specify molar hydronium ion concentration in an aqueous solution. Mathematically, the pH is the negative logarithm of the hydronium ion concentration. Solutions with a pH lower than 7.0 are acidic, those with a pH higher than 7.0 are basic, and those with a pH equal to 7.0 are neutral.

Hydrolysis of salts. Salt hydrolysis is a reaction in which a salt interacts with water to produce an acidic or a basic solution. Only salts that contain the conjugate base of a weak acid and/or the conjugate acid of a weak base hydrolyze.

Buffer solutions. A buffer solution is a solution that resists pH change when small amounts of acid or base are added to it. The resistance to pH change in most buffers is caused by the presence of a weak acid and a salt of its conjugate base.

Electrolytes. An electrolyte is a substance that forms a solution in water that conducts electricity. Strong acids, strong bases, and soluble salts are strong electrolytes. Weak acids, weak bases, and slightly soluble salts are weak electrolytes.

Acid–base titrations. An acid–base titration is a procedure in which an acid–base neutralization reaction is used to determine an unknown concentration. A measured volume of an acid or a base of known concentration is exactly reacted with a measured volume of a base or an acid of unknown concentration.

KEY REACTIONS AND EQUATIONS

1. Weak-acid equilibrium and acid ionization constant (K_a) expression (Section 10.5)

$$HA(aq) + H_2O(l) \rightleftharpoons H_3O^+(aq) + A^-(aq)$$

$$K_a = \frac{[H_3O^+][A^-]}{[HA]}$$

2. Weak-base equilibrium and base ionization constant (K_b) expression (Section 10.5)

$$B(aq) + H_2O(l) \rightleftharpoons BH^+(aq) + OH^-(aq)$$

$$K_b = \frac{[BH^+][OH^-]}{[B]}$$

3. Ion product constant for water (Section 10.8)

$$[H_3O^+][OH^-] = 1.0 \times 10^{-14}$$

4. Relationship between $[H_3O^+]$ and pH (Section 10.9)

$$pH = -\log [H_3O^+]$$

5. Henderson–Hasselbalch equation (Section 10.13)

$$pH = pK_a + \log \frac{[A^-]}{[HA]}$$

KEY TERMS

Acid–base indicator (10.15)
Acid–base titration (10.15)
Acid ionization constant (10.5)
Acidic solution (10.8 and 10.9)
Amphoteric substance (10.2)
Arrhenius acid (10.1)
Arrhenius base (10.1)
Base ionization constant (10.5)
Basic solution (10.8 and 10.9)
Brønsted–Lowry acid (10.2)
Brønsted–Lowry base (10.2)
Buffer (10.12)

Conjugate acid (10.2)
Conjugate acid–base pair (10.2)
Conjugate base (10.2)
Diprotic acid (10.3)
Dissociation (10.1)
Electrolyte (10.14)
Hydrolysis reaction (10.11)
Ion product constant for water (10.8)
Ionization (10.1)
Monoprotic acid (10.3)
Neutral solution (10.8 and 10.9)

Neutralization reaction (10.7)
Nonelectrolyte (10.14)
pH (10.9)
pH scale (10.9)
Polyprotic acid (10.3)
Salt (10.6)
Strong acid (10.4)
Strong electrolyte (10.14)
Triprotic acid (10.3)
Weak acid (10.4)
Weak electrolyte (10.14)

EXERCISES AND PROBLEMS

The members of each pair of problems in this section test similar material.

■ **Arrhenius Acid–Base Theory (Section 10.1)**

10.1 In Arrhenius acid–base theory, what ion is responsible for the properties of
 a. acidic solutions   b. basic solutions

10.2 What term is used to describe the formation of ions, in aqueous solution, from
 a. a molecular compound   b. an ionic compound

10.3 Classify each of the following as a property of an Arrhenius acid or the property of an Arrhenius base.
 a. Has a sour taste   b. Has a bitter taste

10.4 Classify each of the following as a property of an Arrhenius acid or the property of an Arrhenius base.
 a. Changes the color of blue litmus paper to red
 b. Changes the color of red litmus paper to blue

10.5 Write equations depicting the behavior of the following Arrhenius acids and bases in water.
 a. HI (hydroiodic acid)
 b. HClO (hypochlorous acid)
 c. LiOH (lithium hydroxide)
 d. CsOH (cesium hydroxide)

10.6 Write equations depicting the behavior of the following Arrhenius acids and bases in water.
 a. HBr (hydrobromic acid)
 b. HCN (hydrocyanic acid)
 c. RbOH (rubidium hydroxide)
 d. KOH (potassium hydroxide)

■ **Brønsted–Lowry Acid–Base Theory (Section 10.2)**

10.7 In each of the following reactions, decide whether the underlined species is functioning as a Brønsted–Lowry acid or base.
 a. $\underline{HF} + H_2O \rightarrow H_3O^+ + F^-$
 b. $H_2O + \underline{S^{2-}} \rightarrow HS^- + OH^-$
 c. $H_2O + \underline{H_2CO_3} \rightarrow H_3O^+ + HCO_3^-$
 d. $\underline{HCO_3^-} + H_2O \rightarrow H_3O^+ + CO_3^{2-}$

10.8 In each of the following reactions, decide whether the underlined species is functioning as a Brønsted–Lowry acid or base.
 a. $\underline{HClO_2} + H_2O \rightarrow H_3O^+ + ClO_2^-$
 b. $\underline{OCl^-} + H_2O \rightarrow HOCl + OH^-$
 c. $NH_3 + \underline{HNO_2} \rightarrow NH_4^+ + NO_2^-$
 d. $HCl + \underline{H_2PO_4^-} \rightarrow H_3PO_4 + Cl^-$

10.9 Write equations to illustrate the acid–base reactions that can take place between the following Brønsted–Lowry acids and bases.
 a. Acid: HClO; base: H_2O
 b. Acid: $HClO_4$; base: NH_3
 c. Acid: H_3O^+; base: OH^-
 d. Acid: H_3O^+; base: NH_2^-

10.10 Write equations to illustrate the acid–base reactions that can take place between the following Brønsted–Lowry acids and bases.
 a. Acid: $H_2PO_4^-$; base: NH_3
 b. Acid: H_2O; base: ClO_4^-
 c. Acid: HCl; base: OH^-
 d. Acid: $HC_2H_3O_2$; base: H_2O

10.11 Write the formula of each of the following.
 a. Conjugate base of H_2SO_3
 b. Conjugate acid of CN^-

 c. Conjugate base of $HC_2O_4^-$
 d. Conjugate acid of HPO_4^{2-}

10.12 Write the formula of each of the following.
 a. Conjugate base of NH_4^+
 b. Conjugate acid of OH^-
 c. Conjugate base of H_2S
 d. Conjugate acid of NO_2^-

10.13 For each of the following amphoteric substances, write the two equations needed to describe its behavior in aqueous solution.
 a. HS^- b. HPO_4^{2-} c. NH_3 d. OH^-

10.14 For each of the following amphoteric substances, write the two equations needed to describe its behavior in aqueous solution.
 a. $H_2PO_4^-$ b. HSO_4^- c. $HC_2O_4^-$ d. PH_3

■ **Polyprotic Acids (Section 10.3)**

10.15 Classify the following acids as monoprotic, diprotic, or triprotic.
 a. $HClO_4$ (perchloric acid)   b. $H_2C_2O_4$ (oxalic acid)
 c. $HC_2H_3O_2$ (acetic acid)   d. H_2SO_4 (sulfuric acid)

10.16 Classify the following acids as monoprotic, diprotic, or triprotic.
 a. $HC_4H_7O_2$ (butyric acid)   b. H_3PO_4 (phosphoric acid)
 c. HNO_3 (nitric acid)   d. $H_2C_4H_4O_4$ (succinic acid)

10.17 Write equations showing all steps in the dissociation of citric acid ($H_3C_6H_5O_7$).

10.18 Write equations showing all steps in the dissociation of arsenic acid (H_3AsO_4).

10.19 How many acidic and how many nonacidic hydrogen atoms are present in each of the following molecules?
 a. HNO_3 (nitric acid)   b. $H_2C_2H_4O_2$ (succinic acid)
 c. $HC_4H_7O_2$ (butyric acid)   d. CH_4 (methane)

10.20 How many acidic and how many nonacidic hydrogen atoms are present in each of the following molecules?
 a. H_2CO_3 (carbonic acid)   b. $H_2C_3H_2O_4$ (malonic acid)
 c. NH_3 (ammonia)   d. $HC_3H_5O_2$ (propanoic acid)

10.21 The formula for lactic acid is preferably written as $HC_3H_5O_3$ rather than as $C_3H_6O_3$. Explain why.

10.22 The formula for tartaric acid is preferably written as $H_2C_4H_4O_6$ rather than as $C_4H_6O_6$. Explain why.

10.23 Pyruvic acid, which is produced in metabolic reactions, has the structure

 Would you predict that this acid is a mono-, di-, tri-, or tetraprotic acid? Give your reasoning.

10.24 Oxaloacetic acid, which is produced in metabolic reactions, has the structure

 Would you predict that this acid is a mono-, di-, tri-, or tetraprotic acid? Give your reasoning.

Strengths of Acids and Bases (Section 10.4)

10.25 Classify each of the acids in Problem 10.15 as a strong acid or a weak acid.

10.26 Classify each of the acids in Problem 10.16 as a strong acid or a weak acid.

Ionization Constants for Acids and Bases (Section 10.5)

10.27 Write the acid ionization constant expression for the ionization of each of the following monoprotic acids.
 a. HF (hydrofluoric acid)
 b. $HC_2H_3O_2$ (acetic acid)

10.28 Write the acid ionization constant expression for the ionization of each of the following monoprotic acids.
 a. HCN (hydrocyanic acid)
 b. $HC_6H_7O_6$ (ascorbic acid)

10.29 Write the base ionization constant expression for the ionization of each of the following bases. In each case, the nitrogen atom accepts the proton.
 a. NH_3 (ammonia)
 b. $C_6H_5NH_2$ (aniline)

10.30 Write the base ionization constant expression for the ionization of each of the following bases. In each case, the nitrogen atom accepts the proton.
 a. CH_3NH_2 (methylamine)
 b. $C_2H_5NH_2$ (ethylamine)

10.31 Using the acid ionization constant information given in Table 10.4, indicate which acid is the stronger in each of the following acid pairs.
 a. H_3PO_4 and HNO_2 b. HCN and HF
 c. H_2CO_3 and HCO_3^- d. HNO_2 and HCN

10.32 Using the acid ionization constant information given in Table 10.4, indicate which acid is the stronger in each of the following acid pairs.
 a. H_3PO_4 and $H_2PO_4^-$
 b. H_3PO_4 and H_2CO_3
 c. HPO_4^{2-} and $H_2PO_4^-$
 d. $HC_2H_3O_2$ and HCN

10.33 A 0.00300 M solution of an acid is 12% ionized. Calculate the acid ionization constant K_a.

10.34 A 0.0500 M solution of a base is 7.5% ionized. Calculate the base ionization constant K_b.

Salts (Section 10.6)

10.35 Classify each of the following substances as an acid, a base, or a salt.
 a. HBr   b. NaI   c. NH_4NO_3 d. $Ba(OH)_2$
 e. $AlPO_4$ f. KOH   g. HNO_3 h. $HC_2H_3O_2$

10.36 Classify each of the following substances as an acid, a base, or a salt.
 a. KBr   b. NH_4I c. H_2SO_4 d. $Ba_3(PO_4)_2$
 e. $Ca(OH)_2$ f. HCN   g. NaOH   h. HCl

10.37 Write a balanced equation for the dissociation into ions of each of the following soluble salts in aqueous solution.
 a. $Ba(NO_3)_2$ b. Na_2SO_4 c. $CaBr_2$ d. K_2CO_3

10.38 Write a balanced equation for the dissociation into ions of each of the following soluble salts in aqueous solution.
 a. CaS   b. $BeSO_4$ c. $MgCl_2$ d. $NaC_2H_3O_2$

Acid–Base Neutralization Reactions (Section 10.7)

10.39 Indicate whether each of the following reactions is an acid–base neutralization reaction.

 a. $NaCl + AgNO_3 \rightarrow AgCl + NaNO_3$
 b. $HNO_3 + NaOH \rightarrow NaNO_3 + H_2O$
 c. $HBr + KOH \rightarrow KBr + H_2O$
 d. $H_2SO_4 + Pb(NO_3)_2 \rightarrow PbSO_4 + 2HNO_3$

10.40 Indicate whether each of the following reactions is an acid–base neutralization reaction.
 a. $H_2S + CuSO_4 \rightarrow H_2SO_4 + CuS$
 b. $HCN + LiOH \rightarrow LiCN + H_2O$
 c. $H_2SO_4 + Ba(OH)_2 \rightarrow BaSO_4 + 2H_2O$
 d. $Ni + 2HCl \rightarrow NiCl_2 + H_2$

10.41 Without writing an equation, specify the molecular ratio in which each of the following acid–base pairs will react.
 a. HNO_3 and NaOH   b. H_2SO_4 and NaOH
 c. H_2SO_4 and $Ba(OH)_2$ d. HNO_3 and $Ba(OH)_2$

10.42 Without writing an equation, specify the molecular ratio in which each of the following acid–base pairs will react.
 a. HCl and KOH   b. H_2CO_3 and KOH
 c. HCl and $Ca(OH)_2$ d. H_2CO_3 and $Ca(OH)_2$

10.43 Write a balanced molecular equation to represent each of the following acid–base neutralization reactions.
 a. Acid: HCl; base: NaOH
 b. Acid: HNO_3; base: KOH
 c. Acid: H_2SO_4; base: LiOH
 d. Acid: H_3PO_4; base: $Ba(OH)_2$

10.44 Write a balanced molecular equation to represent each of the following acid–base neutralization reactions.
 a. Acid: HCl; base: LiOH
 b. Acid: HNO_3; base: $Ba(OH)_2$
 c. Acid: H_2SO_4; base: NaOH
 d. Acid: H_3PO_4; base: KOH

10.45 Write a balanced molecular equation for the preparation of each of the following salts, using an acid–base neutralization reaction.
 a. Li_2SO_4 (lithium sulfate)
 b. NaCl (sodium chloride)
 c. KNO_3 (potassium nitrate)
 d. $Ba_3(PO_4)_2$ (barium phosphate)

10.46 Write a balanced molecular equation for the preparation of each of the following salts, using an acid–base neutralization reaction.
 a. $LiNO_3$ (lithium nitrate)
 b. $BaCl_2$ (barium chloride)
 c. K_3PO_4 (potassium phosphate)
 d. Na_2SO_4 (sodium sulfate)

Hydronium Ion and Hydroxide Ion Concentrations (Section 10.8)

10.47 Calculate the molar H_3O^+ ion concentration of a solution if the OH^- ion concentration is
 a. 3.0×10^{-3} M   b. 6.7×10^{-6} M
 c. 9.1×10^{-8} M   d. 1.2×10^{-11} M

10.48 Calculate the molar H_3O^+ ion concentration of a solution if the OH^- ion concentration is
 a. 5.0×10^{-4} M   b. 7.5×10^{-7} M
 c. 2.3×10^{-12} M   d. 1.1×10^{-10} M

10.49 Indicate whether each of the following solutions is acidic, basic, or neutral.
 a. $[H_3O^+] = 1.0 \times 10^{-3}$ b. $[H_3O^+] = 3.0 \times 10^{-11}$
 c. $[OH^-] = 4.0 \times 10^{-6}$ d. $[OH^-] = 2.3 \times 10^{-10}$

10.50 Indicate whether each of the following solutions is acidic, basic, or neutral.

a. $[H_3O^+] = 2.0 \times 10^{-4}$ b. $[H_3O^+] = 2.0 \times 10^{-8}$
c. $[OH^-] = 1.0 \times 10^{-7}$ d. $[OH^-] = 5.0 \times 10^{-9}$

▪ pH Scale (Section 10.9)

10.51 Calculate the pH of the following solutions.
   a. $[H_3O^+] = 1.0 \times 10^{-4}$ b. $[H_3O^+] = 1.0 \times 10^{-11}$
   c. $[OH^-] = 1.0 \times 10^{-3}$ d. $[OH^-] = 1.0 \times 10^{-7}$

10.52 Calculate the pH of the following solutions.
   a. $[H_3O^+] = 1.0 \times 10^{-6}$ b. $[H_3O^+] = 1.0 \times 10^{-2}$
   c. $[OH^-] = 1.0 \times 10^{-9}$ d. $[OH^-] = 1.0 \times 10^{-5}$

10.53 Calculate the pH of the following solutions.
   a. $[H_3O^+] = 2.1 \times 10^{-8}$ b. $[H_3O^+] = 4.0 \times 10^{-8}$
   c. $[OH^-] = 7.2 \times 10^{-11}$ d. $[OH^-] = 7.2 \times 10^{-3}$

10.54 Calculate the pH of the following solutions.
   a. $[H_3O^+] = 3.3 \times 10^{-5}$ b. $[H_3O^+] = 7.6 \times 10^{-5}$
   c. $[OH^-] = 8.2 \times 10^{-10}$ d. $[OH^-] = 8.2 \times 10^{-4}$

10.55 What is the molar hydronium ion concentration in solutions with each of the following pH values?
   a. 2.0   b. 6.0   c. 8.0   d. 10.0

10.56 What is the molar hydronium ion concentration in solutions with each of the following pH values?
   a. 3.0   b. 5.0   c. 9.0   d. 12.0

10.57 What is the molar hydronium ion concentration in solutions with each of the following pH values?
   a. 3.67   b. 5.09   c. 7.35   d. 12.45

10.58 What is the molar hydronium ion concentration in solutions with each of the following pH values?
   a. 2.05   b. 4.88   c. 6.75   d. 11.33

▪ pK_a Values (Section 10.10)

10.59 Calculate the pK_a value for each of the following acids.
   a. Nitrous acid (HNO_2), $K_a = 4.5 \times 10^{-4}$
   b. Carbonic acid (H_2CO_3), $K_a = 4.3 \times 10^{-7}$
   c. Dihydrogen phosphate ion ($H_2PO_4^-$), $K_a = 6.2 \times 10^{-8}$
   d. Sulfurous acid (H_2SO_3), $K_a = 1.5 \times 10^{-2}$

10.60 Calculate the pK_a value for each of the following acids.
   a. Phosphoric acid (H_3PO_4), $K_a = 7.5 \times 10^{-3}$
   b. Hydrofluoric acid (HF), $K_a = 6.8 \times 10^{-4}$
   c. Hydrogen phosphate ion (HPO_4^{2-}), $K_a = 4.2 \times 10^{-13}$
   d. Propanoic acid ($HC_3H_5O_2$), $K_a = 1.3 \times 10^{-5}$

10.61 Acid A has a pK_a value of 4.23, and acid B has a pK_a value of 3.97. Which of the two acids is the stronger?

10.62 Acid A has a pK_a value of 5.71, and acid B has a pK_a value of 5.30. Which of the two acids is the weaker?

▪ Hydrolysis of Salts (Section 10.11)

10.63 Classify each of the following salts as a "strong acid–strong base salt," a "strong acid–weak base salt," a "weak acid–strong base salt," or a "weak acid–weak base salt."
   a. NaCl   b. $KC_2H_3O_2$ c. NH_4Br d. $Ba(NO_3)_2$

10.64 Classify each of the following salts as a "strong acid–strong base salt," a "strong acid–weak base salt," a "weak acid–strong base salt," or a "weak acid–weak base salt."
   a. K_3PO_4 b. $NaNO_3$ c. KCl   d. $Na_2C_2O_4$

10.65 Identify the ion (or ions) present in each of the salts in Problem 10.63 that will undergo hydrolysis in aqueous solution.

10.66 Identify the ion (or ions) present in each of the salts in Problem 10.64 that will undergo hydrolysis in aqueous solution.

10.67 Predict whether solutions of each of the salts in Problem 10.63 will be acidic, basic, or neutral.

10.68 Predict whether solutions of each of the salts in Problem 10.64 will be acidic, basic, or neutral.

▪ Buffers (Section 10.12)

10.69 Predict whether each of the following pairs of substances could function as a buffer system in aqueous solution.
   a. HNO_3 and $NaNO_3$
   b. HF and NaF
   c. KCl and KCN
   d. H_2CO_3 and $NaHCO_3$

10.70 Predict whether each of the following pairs of substances could function as a buffer system in aqueous solution.
   a. HNO_3 and HCl
   b. HNO_2 and KNO_2
   c. $NaC_2H_3O_2$ and $KC_2H_3O_2$
   d. $HC_2H_3O_2$ and $NaNO_3$

10.71 Identify the two "active species" in each of the following buffer systems.
   a. HCN and KCN   b. H_3PO_4 and NaH_2PO_4
   c. H_2CO_3 and $KHCO_3$ d. $NaHCO_3$ and K_2CO_3

10.72 Identify the two "active species" in each of the following buffer systems.
   a. HF and LiF   b. Na_2HPO_4 and KH_2PO_4
   c. K_2CO_3 and $KHCO_3$ d. $NaNO_2$ and HNO_2

10.73 Write an equation for each of the following buffering actions.
   a. The response of a HF/F$^-$ buffer to the addition of H_3O^+ ions
   b. The response of a H_2CO_3/HCO_3^- buffer to the addition of OH^- ions
   c. The response of a HCO_3^-/CO_3^{2-} buffer to the addition of H_3O^+ ions
   d. The response of a $H_3PO_4/H_2PO_4^-$ buffer to the addition of OH^- ions

10.74 Write an equation for each of the following buffering actions.
   a. The response of a HPO_4^{2-}/PO_4^{3-} buffer to the addition of OH^- ions
   b. The response of a HF/F$^-$ buffer to the addition of OH^- ions
   c. The response of a HCN/CN^- buffer to the addition of H_3O^+ ions
   d. The response of a $H_3PO_4/H_2PO_4^-$ buffer to the addition of H_3O^+ ions

▪ The Henderson–Hasselbalch Equation (Section 10.13)

10.75 What is the pH of a buffer that is 0.230 M in a weak acid and 0.500 M in the acid's conjugate base? The pK_a for the acid is 6.72.

10.76 What is the pH of a buffer that is 0.250 M in a weak acid and 0.260 M in the acid's conjugate base? The pK_a for the acid is 5.53.

10.77 What is the pH of a buffer that is 0.150 M in a weak acid and 0.150 M in the acid's conjugate base? The acid's ionization constant in 6.8×10^{-6}.

10.78 What is the pH of a buffer that is 0.175 M in a weak acid and 0.200 M in the acid's conjugate base? The acid's ionization constant in 5.7×10^{-4}.

▪ Electrolytes (Section 10.14)

10.79 Classify each of the following compounds as a strong electrolyte or a weak electrolyte.
   a. H_2CO_3 b. KOH   c. NaCl   d. H_2SO_4

10.80 Classify each of the following compounds as a strong electrolyte or a weak electrolyte.
a. H_3PO_4 b. HNO_3 c. KNO_3 d. $NaOH$

■ **Titration Calculations (Section 10.15)**

10.81 Determine the molarity of a NaOH solution when each of the following amounts of acid neutralizes 25.0 mL of the NaOH solution.
a. 5.00 mL of 0.250 M HNO_3
b. 20.00 mL of 0.500 M H_2SO_4
c. 23.76 mL of 1.00 M HCl
d. 10.00 mL of 0.100 M H_3PO_4

10.82 Determine the molarity of a KOH solution when each of the following amounts of acid neutralizes 25.0 mL of the KOH solution.
a. 5.00 mL of 0.500 M H_2SO_4
b. 20.00 mL of 0.250 M HNO_3
c. 13.07 mL of 0.100 M H_3PO_4
d. 10.00 mL of 1.00 M HCl

ADDITIONAL PROBLEMS

10.83 In which of the following pairs of substances do the two members of the pair constitute a conjugate acid–base pair?
a. HN_3 and N_3^- c. H_2CO_3 and $HClO_3$
b. H_2SO_4 and SO_4^{2-} d. NH_3 and NH_2^-

10.84 For which of the following pairs of acids are both members of the pair of "like strength"—that is, both strong or both weak?
a. HNO_3 and HNO_2 c. H_3PO_4 and $HClO_4$
b. HCl and HBr   d. H_2CO_3 and $H_2C_2O_4$

10.85 Solution A has $[OH^-] = 2.5 \times 10^{-3}$. Solution B has $[H_3O^+] = 3.5 \times 10^{-4}$.
a. Which solution is more acidic?
b. Which solution has the higher pH?

10.86 What would be the pH of a solution that contains 0.1 mole of each of the solutes NaCl, NaOH, and HCl in enough water to give 2.00 L of solution?

10.87 Arrange the following 0.1 M aqueous solutions in order of increasing pH: HCl, HCN, NaOH, and KCl.

10.88 Identify the buffer systems (conjugate acid–base pairs) present in a solution containing equal molar amounts of HCN, KCN, NaCN, and NaCl.

10.89 Both ions in each of the salts NH_4CN and $NH_4C_2H_3O_2$ undergo hydrolysis in aqueous solution. Upon hydrolysis, the first listed salt gives a basic solution and the second a neutral solution. Explain how this is possible.

10.90 How many grams of NaOH are needed to make 875 mL of a solution with a pH of 10.00?

ANSWERS TO PRACTICE EXERCISES

10.1 a. $HClO_3$ b. NH_2^- c. HPO_4^{2-} d. S^{2-}

10.2 3.8×10^{-4}

10.3 1.8×10^{-9} M

10.4 a. 3.00   b. 6.00

10.5 a. 10.353   b. 5.050

10.6 3.6×10^{-4} M

10.7 9.36

10.8 a. neutral   b. basic   c. acidic   d. neutral

10.9 a. no   b. yes   c. no   d. no

10.10 a. $H_3O^+ + HCO_3^- \longrightarrow H_2CO_3 + H_2O$
b. $H_2PO_4^- + OH^- \longrightarrow HPO_4^{2-} + H_2O$

10.11 5.14

10.12 0.228 M HNO_3

11 Nuclear Chemistry

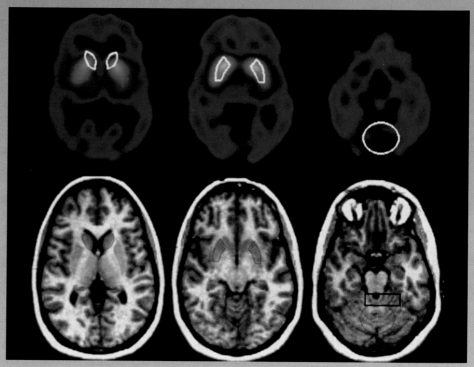

Associated with brain-scan technology is the use of small amounts of radioactive substances.

In this chapter we consider nuclear reactions. It is in the study of such reactions that we encounter the terms *radioactivity, nuclear power plants, nuclear weapons,* and *nuclear medicine.* The electricity produced by a nuclear power plant is generated through the use of heat energy obtained from nuclear reactions. In modern medicine, nuclear reactions are used in the diagnosis and treatment of numerous diseases. Despite some controversy concerning the use of nuclear reactions in weapons and power plants, it is important to remember that it is far more likely that your life will be extended by nuclear medicine than that your life will be taken by nuclear weapons.

11.1 Stable and Unstable Nuclides

A **nuclear reaction** *is a reaction in which changes occur in the nucleus of an atom.* Nuclear reactions are not considered to be ordinary chemical reactions. The governing principles for ordinary chemical reactions deal with the rearrangement of electrons; this rearrangement occurs as the result of electron transfer or electron sharing (Section 4.1). In nuclear reactions, it is nuclei rather than electron arrangements that undergo change.

The term *nuclide* is used extensively in discussions about nuclear reactions. A **nuclide** *is an atom of an element with a specific number of protons and neutrons in its nucleus.* All atoms of a given nuclide must have the same number of protons and the same number of neutrons. The term *nuclide* is used to describe atomic forms of different elements. The species $^{12}_{6}C$ and $^{13}_{6}C$ are *isotopes* of the element carbon. The species $^{12}_{6}C$ and $^{16}_{8}O$ are *nuclides* of different elements.

TABLE 11.5
Some Radionuclides Used in
Radiation Therapy

Nuclide	Half-life	Type of emitter	Use in therapy
cobalt-60	5.3 years	gamma	external source of radiation in treatment of cancer
iodine-131	8 days	beta, gamma	cancer of thyroid
phosphorus-32	14.3 days	beta, gamma	treatment of some types of leukemia and widespread carcinomas
radium-226	1620 years	alpha, gamma	used in implantation cancer therapy
radon-222	3.8 days	alpha, gamma	used in treatment of uterine, cervical, oral, and bladder cancers
yttrium-90	64 hours	beta, gamma	implantation therapy

▶ Abnormal cells are more susceptible to radiation damage than normal cells, because abnormal cells divide more frequently.

This therapy usually causes some radiation sickness because normal cells are also affected, but to a lesser extent. The operating principle here is that abnormal cells are more susceptible to radiation damage than normal cells. Radiation sickness is the price paid for abnormal-cell destruction. Table 11.5 lists some radionuclides that are used in therapy.

11.12 Nuclear Fission and Nuclear Fusion

Our glimpse into the world of nuclear chemistry would not be complete without a brief mention of two additional types of nuclear reactions that are used as sources of energy: nuclear fission and nuclear fusion.

■ Fission Reactions

Nuclear fission *is a nuclear reaction in which a large nucleus (high atomic number) splits into two medium-sized nuclei with an accompanying release of several free neutrons and a large amount of energy.* The most important nucleus that undergoes fission is uranium-235. Bombardment of this nucleus with neutrons causes it to split into two fragments. Characteristics of the uranium-235 fission reaction include the following:

1. There is no unique way in which the uranium-235 nucleus splits. Thus, many different, lighter elements are produced during uranium-235 fission reactions. The following are examples of the ways in which this fission process proceeds.

$$^{235}_{92}\text{U} + {}^{1}_{0}\text{n} \Bigg\langle \begin{array}{l} {}^{135}_{53}\text{I} + {}^{97}_{39}\text{Y} + 4\,{}^{1}_{0}\text{n} \\ {}^{139}_{56}\text{Ba} + {}^{94}_{36}\text{Kr} + 3\,{}^{1}_{0}\text{n} \\ {}^{131}_{50}\text{Sn} + {}^{103}_{42}\text{Mo} + 2\,{}^{1}_{0}\text{n} \\ {}^{139}_{54}\text{Xe} + {}^{95}_{38}\text{Sr} + 2\,{}^{1}_{0}\text{n} \end{array}$$

2. Very large amounts of energy, which are many times greater than that released by ordinary radioactive decay, are emitted during the fission process. It is this large release of energy that makes nuclear fission of uranium-235 the important process that it is. In general, the term *nuclear energy* is used to refer to the energy released during a nuclear fission process. An older term for this energy is *atomic energy*.

3. Neutrons, which are reactants in the fission process, are also produced as products. The number of neutrons produced per fission depends on the way in which the nucleus splits; it ranges from 2 to 4 (as can be seen from the foregoing fission equations). On the average, 2.4 neutrons are produced per fission. The significance of the neutrons that are produced is that they can cause the fission process to continue by colliding with further uranium-235 nuclei. Figure 11.15 shows the chain reaction that can occur once the fission process is started.

FIGURE 11.15 A fission chain reaction is caused by further reaction of the neutrons produced during fission.

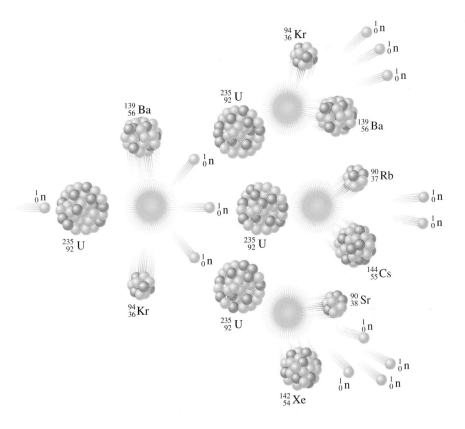

The process of nuclear fission—or "splitting the atom," as it is called in popularized science—can be carried out in both an uncontrolled and a controlled manner. The key to this control lies in what happens to the neutrons produced during fission. Do they react further, causing further fission, or do they escape into the surroundings? If the majority of produced neutrons react further (Figure 11.15), an uncontrolled nuclear reaction (an atomic bomb) results (see Figure 11.16). When only a few neutrons react further (on the average, one per fission), the fission reaction self-propagates in a controlled manner.

The process of *controlled* nuclear fission is the basis for the operation of nuclear power plants that are used to produce electricity (see Figure 11.17). The reaction is controlled with

FIGURE 11.16 Enormous amounts of energy are released in the explosion of a nuclear fission bomb.

FIGURE 11.17 A nuclear power plant. The cooling tower at the Trojan nuclear power plant in Oregon dominates the landscape. The nuclear reactor is housed in the dome-shaped enclosure.

▶ At the high temperature of fusion reactions, electrons completely separate from nuclei. Neutral atoms cannot exist. This high-temperature, gaslike mixture of nuclei and electrons is called a *plasma* and is considered by some scientists to represent a fourth state of matter.

rods that absorb excess neutrons (so that they cannot cause unwanted fissions) and with moderating substances that decrease the speed of the neutrons. The energy produced during the fission process, which appears as heat, is used to operate steam-powered electricity-generating equipment.

■ Fusion Reactions

Another type of nuclear reaction, nuclear fusion, produces even more energy than nuclear fission. **Nuclear fusion** *is a nuclear reaction in which two small nuclei are put together to make a larger one.* This process is essentially the opposite of nuclear fission. In order for fusion to occur, a very high temperature (several hundred million degrees) is required.

Nuclear fusion is the process by which the sun generates its energy (see Figure 11.18). Within the sun, in a three-step reaction, hydrogen-1 nuclei are converted to helium-4 nuclei with the release of extraordinarily large amounts of energy.

The use of nuclear fusion on Earth might seem impossible because of the high temperatures required. It has, however, been accomplished in a hydrogen bomb. In such a weapon, a *fission* bomb is used to achieve the high temperatures needed to start the following process:

$$\ce{^3_1H + ^2_1H \longrightarrow ^4_2He + ^1_0n}$$

The use of nuclear fusion as a *controlled* (peaceful) energy source is a very active area of current scientific research. "Harnessing" this type of nuclear reaction would have numerous advantages:

1. Unlike the by-products of fission reactions, the by-products of fusion reactions are stable (nonradioactive) nuclides. Thus the problem of storing radioactive wastes does not arise.
2. The major fuel under study for controlled fusion is $\ce{^2_1H}$ (called deuterium), a hydrogen isotope that can be readily extracted from ocean water (0.015% of all hydrogen atoms are $\ce{^2_1H}$). Just 0.005 km^3 of ocean water contains enough $\ce{^2_1H}$ to supply the United States with all the energy it needs for 1 year!

However, difficult scientific and engineering problems still remain to be solved before controlled fusion is a reality.

The Chemistry at a Glance feature on page 289 summarizes important concepts about the major types of nuclear reactions that have been considered in this chapter.

FIGURE 11.18 A close-up view of the sun. The process of nuclear fusion maintains the interior of the sun at a temperature of approximately 15 million degrees.

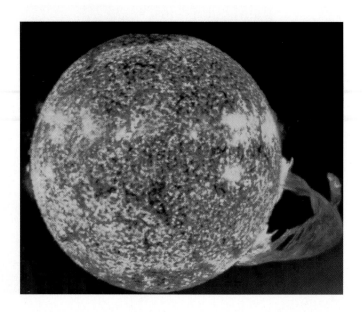

Types of Nuclear Reactions

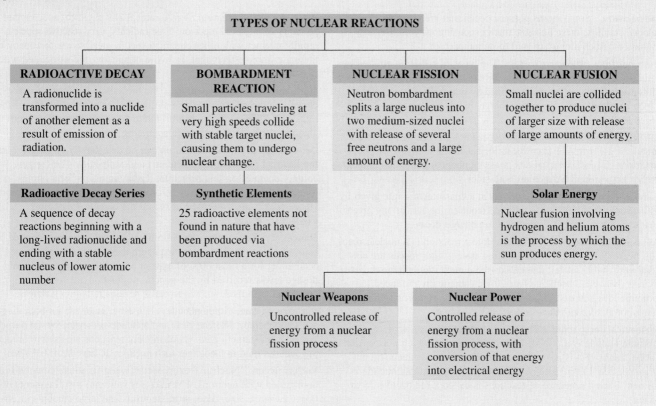

11.13 Nuclear and Chemical Reactions Compared

As the discussions in this chapter have shown, nuclear chemistry is quite different from ordinary chemistry. Many of the laws of chemistry must be modified when we consider nuclear reactions. The major differences between nuclear reactions and ordinary chemical reactions are listed in Table 11.6. This table serves as a summary of many of the concepts presented in this chapter.

TABLE 11.6

Differences Between Nuclear and Chemical Reactions

Chemical reaction	Nuclear reaction
1. Different isotopes of an element have identical chemical properties.	1. Different isotopes of an element have different nuclear properties.
2. The chemical reactivity of an element depends on the element's state of combination (free element, compound, etc.).	2. The nuclear reactivity of an element is independent of the state of chemical combination.
3. Elements retain their identity in chemical reactions.	3. Elements may be changed into other elements during nuclear reactions.
4. Energy changes that accompany chemical reactions are relatively small.	4. Energy changes that accompany nuclear reactions are a number of orders of magnitude larger than those in chemical reactions.
5. Reaction rates are influenced by temperature, pressure, catalysts, and reactant concentrations.	5. Reaction rates are independent of temperature, pressure, catalysts, and reactant concentrations.

CONCEPTS TO REMEMBER

Radioactivity. Some atoms possess nuclei that are unstable. To achieve stability, these unstable nuclei spontaneously emit energy (radiation). Such atoms are said to be radioactive.

Emissions from radioactive nuclei. The types of radiation emitted by naturally occurring radioactive nuclei are alpha, beta, and gamma. These radiations can be characterized by mass and charge values. Alpha particles carry a positive charge, beta particles carry a negative charge, and gamma radiation has no charge.

Balanced nuclear equations. The procedures for balancing nuclear equations are different from those for balancing ordinary chemical equations. In nuclear equations, mass numbers and atomic numbers (rather than atoms) balance on both sides.

Half-life. Every radionuclide decays at a characteristic rate given by its half-life. One half-life is the time required for half of any given quantity of a radioactive substance to undergo decay.

Bombardment reactions. A bombardment reaction is a nuclear reaction in which small particles traveling at very high speeds are collided with stable nuclei; this causes these nuclei to undergo nuclear change (become unstable). Over 2000 synthetically produced radionuclides that do not occur naturally have been produced by using bombardment reactions.

Radioactive decay series. The product of the radioactive decay of an unstable nuclide is a nuclide of another element, which may or may not be stable. If it is not stable, it will decay and produce still another nuclide. Further decay will continue until a stable nuclide is formed. Such a sequence of reactions is called a radioactive decay series.

Chemical Effects of Radiation. Radiation from radioactive decay is ionizing radiation—radiation with enough energy to remove an electron from an atom or molecule. Interaction of ionizing radiation with matter produces ion pairs, with many ion pairs being produced by a single "particle" of radiation. Free-radical formation usually accompanies ion pair formation. A free radical is a chemical species that contains an unpaired electron. Free radicals, very reactive species, rapidly interact with other chemical species nearby, causing many of the undesirable effects that, in living organisms, are associated with radiation exposure.

Biochemical effects of radiation. The biochemical effects of radiation depend on the energy, ionizing ability, and penetrating ability of the radiation. Alpha particles exhibit the greatest ionizing effect, and gamma rays have the greatest penetrating ability.

Detection of radiation. Radiation can be detected by making use of the fact that radiation ionizes atoms and molecules. The Geiger counter operates on this principle. Radiation also affects photographic film in the same way as ordinary light; the film is exposed. Hence film badges are used to record the extent of radiation exposure.

Sources of radiation exposure. Both natural and human-generated sources of low-level radiation exposure exist, with natural sources accounting for almost 80% of exposure (on the average). Current dosage levels received by the general population are very small, compared with those known to cause serious radiation sickness.

Nuclear medicine. Radionuclides are used in medicine for both diagnosis and therapy. The choice of radionuclide is dictated by the purpose as well as the target organ. Bombardment reactions are used to produce the nuclides used in medicine; such nuclides all have short half-lives.

Nuclear fission. Nuclear fission occurs when fissionable nuclides are bombarded with neutrons. The nuclides split into two fragments of about the same size. Also, more neutrons and large amounts of energy are produced. Nuclear fission is the process by which nuclear power plants generate energy.

Nuclear fusion. In nuclear fusion, small nuclei fuse to make heavier nuclei. Nuclear fusion is the process by which the sun generates its energy.

KEY REACTIONS AND EQUATIONS

1. General equation for alpha decay (Section 11.3)
$$_Z^A X \longrightarrow _{Z-2}^{A-4} Y + _2^4 \alpha$$

2. General equation for beta decay (Section 11.3)
$$_Z^A X \longrightarrow _{Z+1}^A Y + _{-1}^0 \beta$$

3. Half-life and amount of undecayed radionuclide (Section 11.4)
$$\begin{pmatrix} \text{Amount of radionuclide} \\ \text{undecayed after } n \text{ half-lives} \end{pmatrix} = \begin{pmatrix} \text{original amount} \\ \text{of radionuclide} \end{pmatrix} \times \left(\frac{1}{2^n} \right)$$

KEY TERMS

Alpha particle (11.2)
Balanced nuclear equation (11.3)
Beta particle (11.2)
Bombardment reaction (11.5)
Daughter nuclide (11.3)
Free radical (11.7)
Gamma ray (11.2)
Half-life (11.4)

Ion pair (11.7)
Ionizing radiation (11.7)
Nonionizing radiation (11.7)
Nuclear equation (11.3)
Nuclear fission (11.12)
Nuclear fusion (11.12)
Nuclear reaction (11.1)
Nuclide (11.1)

Parent nuclide (11.3)
Radioactive decay (11.3)
Radioactive decay series (11.6)
Radioactive nuclide (11.1)
Radioactivity (11.1)
Stable nuclide (11.1)
Transmutation reaction (11.5)
Unstable nuclide (11.1)

EXERCISES AND PROBLEMS

The members of each pair of problems in this section test the same material.

■ **Notation for Nuclides (Section 11.1)**

11.1 Use two different notations to denote each of the following nuclides.
   a. Contains 4 protons, 4 electrons, and 6 neutrons
   b. Contains 11 protons, 11 electrons, and 14 neutrons
   c. Contains 41 protons, 41 electrons, and 55 neutrons
   d. Contains 103 protons, 103 electrons, and 154 neutrons

11.2 Use two different notations to denote each of the following nuclides.
   a. Contains 20 protons, 20 electrons, and 18 neutrons
   b. Contains 37 protons, 37 electrons, and 43 neutrons
   c. Contains 51 protons, 51 electrons, and 74 neutrons
   d. Contains 99 protons, 99 electrons, and 157 neutrons

11.3 Use a notation different from that given to designate each of the following nuclides.
   a. nitrogen-14   b. gold-197
   c. $^{121}_{50}Sn$ d. $^{10}_{5}B$

11.4 Use a notation different from that given to designate each of the following nuclides.
   a. oxygen-17   b. lead-212
   c. $^{92}_{37}Rb$ d. $^{201}_{83}Bi$

■ **The Nature of Radioactivity (Section 11.2)**

11.5 Supply a complete symbol, with superscript and subscript, for each of the following types of radiation.
   a. alpha particle   b. beta particle   c. gamma ray

11.6 Give the charge and mass (in amu) of each of the following types of radiation.
   a. alpha particle   b. beta particle   c. gamma ray

11.7 State the composition of an alpha particle in terms of protons and neutrons.

11.8 What is the relationship between a beta particle and an electron?

■ **Radioactive Decay (Section 11.3)**

11.9 Write balanced nuclear equations for the alpha decay of each of the following nuclides.
   a. $^{200}_{84}Po$ b. curium-240
   c. $^{244}_{96}Cm$ d. uranium-238

11.10 Write balanced nuclear equations for the alpha decay of each of the following nuclides.
   a. $^{229}_{90}Th$ b. bismuth-210
   c. $^{152}_{64}Gd$ d. americium-243

11.11 Write balanced nuclear equations for the beta decay of each of the following nuclides.
   a. $^{10}_{4}Be$ b. carbon-14
   c. $^{21}_{9}F$ d. sodium-25

11.12 Write balanced nuclear equations for the beta decay of each of the following nuclides.
   a. $^{77}_{32}Ge$ b. uranium-235
   c. $^{16}_{7}N$ d. iron-60

11.13 What is the effect on the mass number and atomic number of the parent nuclide when alpha particle decay occurs?

11.14 What is the effect on the mass number and atomic number of the parent nuclide when beta particle decay occurs?

11.15 Supply the missing symbol in each of the following nuclear equations.
   a. $^{34}_{14}Si \rightarrow ^{34}_{15}P + ?$ b. $? \rightarrow ^{28}_{13}Al + ^{0}_{-1}\beta$
   c. $^{252}_{99}Es \rightarrow ^{248}_{97}Bk + ?$ d. $^{204}_{82}Pb \rightarrow ? + ^{4}_{2}\alpha$

11.16 Supply the missing symbol in each of the following nuclear equations.
   a. $? \rightarrow ^{230}_{92}U + ^{4}_{2}\alpha$ b. $^{192}_{78}Pt \rightarrow ? + ^{4}_{2}\alpha$
   c. $^{84}_{35}Br \rightarrow ? + ^{0}_{-1}\beta$ d. $^{10}_{4}Be \rightarrow ^{10}_{5}B + ?$

11.17 Identify the mode of decay for each of the following parent radionuclides, given the identity of the daughter nuclide.
   a. parent = platinum-190; daughter = osmium-186
   b. parent = oxygen-19; daughter = fluorine-19

11.18 Identify the mode of decay for each of the following parent radionuclides, given the identity of the daughter nuclide.
   a. parent = uranium-238; daughter = thorium-234
   b. parent = rhodium-104; daughter = palladium-104

■ **Rate of Radioactive Decay (Section 11.4)**

11.19 Technetium-99 has a half-life of 6.0 hr. What fraction of the technetium-99 atoms in a sample will remain undecayed after the following times?
   a. 12 hr   b. 36 hr
   c. 3 half-lives   d. 6 half-lives

11.20 Copper-66 has a half-life of 5.0 min. What fraction of the copper-66 atoms in a sample will remain undecayed after the following times?
   a. 20 min   b. 30 min
   c. 3 half-lives   d. 8 half-lives

11.21 Determine the half-life of a radionuclide if after 5.4 days the fraction of undecayed nuclides present is
   a. 1/16   b. 1/64   c. 1/256   d. 1/1024

11.22 Determine the half-life of a radionuclide if after 3.2 days the fraction of undecayed nuclides present is
   a. 1/8   b. 1/128   c. 1/32   d. 1/512

11.23 The half-life of sodium-24 is 15.0 hr. How many grams of this nuclide in a 4.00-g sample will remain after 60.0 hr?

11.24 The half-life of strontium-90 is 28 years. How many grams of this nuclide in a 4.00-g sample will remain after 112 years?

■ **Bombardment Reactions (Section 11.5)**

11.25 Approximately how many laboratory-produced radionuclides are known?

11.26 How does the number of laboratory-produced radionuclides compare with the number of naturally occurring nuclides?

11.27 What is the highest-atomic-numbered naturally occurring element?

11.28 What is the highest-atomic-numbered element for which nonradioactive isotopes exist?

11.29 Identify the missing symbol in each of the following equations for bombardment reactions.
   a. $^{24}_{12}Mg + ? \rightarrow ^{27}_{14}Si + ^{1}_{0}n$
   b. $^{27}_{13}Al + ^{2}_{1}H \rightarrow ? + ^{4}_{2}\alpha$
   c. $^{9}_{4}Be + ? \rightarrow ^{12}_{6}C + ^{1}_{0}n$
   d. $^{6}_{3}Li + ? \rightarrow ^{4}_{2}He + ^{3}_{2}He$

11.30 Identify the missing symbol in each of the following equations for bombardment reactions.
a. $? + {}^{4}_{2}\alpha \rightarrow {}^{250}_{99}\text{Es} + {}^{3}_{1}\text{H}$
b. ${}^{14}_{7}\text{N} + {}^{4}_{2}\alpha \rightarrow ? + {}^{1}_{1}\text{H}$
c. ${}^{12}_{6}\text{C} + {}^{2}_{1}\text{H} \rightarrow {}^{13}_{7}\text{N} + ?$
d. ${}^{27}_{13}\text{Al} + ? \rightarrow {}^{30}_{15}\text{P} + {}^{1}_{0}\text{n}$

■ **Radioactive Decay Series (Section 11.6)**

11.31 The uranium-235 decay series terminates with lead-207. Would you expect lead-207 to be a stable or an unstable nuclide? Explain your answer.

11.32 A textbook erroneously indicates that the uranium-235 decay series terminates with radon-222. Explain why such a situation cannot be.

11.33 In the thorium-232 natural decay series, the thorium-232 initially undergoes alpha decay, the resulting daughter emits a beta particle, and the succeeding daughters emit a beta and an alpha particle in that order. Write four nuclear equations, one to represent each of the first four steps in the thorium-232 decay series.

11.34 In the uranium-235 natural decay series, the uranium-235 initially undergoes alpha decay, the resulting daughter emits a beta particle, and the succeeding daughters emit an alpha and a beta particle in that order. Write four nuclear equations, one to represent each of the first four steps in the uranium-235 decay series.

■ **Effects of Radiation (Sections 11.7 and 11.8)**

11.35 What is an ion pair?

11.36 What is a free radical?

11.37 Indicate whether each of the following species is a free radical.
a. H_2O^+ b. H_3O^+ c. OH d. OH^-

11.38 Write an equation that involves water for the formation of the
a. water free radical   b. hydroxyl free radical

11.39 What is the fate of a radiation "particle" that is involved in an ion pair formation reaction?

11.40 What are two possible fates for a free radical produced from a molecule–radiation interaction?

11.41 Contrast the abilities of alpha, beta, and gamma radiations to penetrate a thick sheet of paper.

11.42 Contrast the abilities of alpha, beta, and gamma radiation to penetrate human skin.

11.43 Contrast the velocities with which alpha, beta, and gamma radiations are emitted by nuclei.

11.44 Contrast the ionizing ability of alpha and beta radiations.

■ **Radiation Exposure (Sections 11.8 and 11.10)**

11.45 What would be the expected effect of each of the following short-term, whole-body radiation exposures?
a. 10 rems   b. 150 rems

11.46 What would be the expected effect of each of the following short-term, whole-body radiation exposures?
a. 50 rems   b. 250 rems

11.47 Contrast the radiation exposure that an average American receives from natural sources with that which an average American receives from human-made sources.

11.48 What are the five major sources of low-level radiation exposure, in terms of millirems per year, for the average American?

■ **Detection of Radiation (Section 11.9)**

11.49 Why do technicians who work around radiation usually wear film badges?

11.50 Explain the principle of operation of a Geiger counter.

■ **Nuclear Medicine (Section 11.11)**

11.51 Why are the radionuclides used for diagnostic procedures usually gamma emitters?

11.52 Why do the radionuclides used in diagnostic procedures nearly always have short half-lives?

11.53 Explain how each of the following radionuclides is used in diagnostic medicine.
a. barium-131   b. sodium-24
c. iron-59   d. potassium-42

11.54 Explain how each of the following radionuclides is used in diagnostic medicine.
a. iodine-131   b. phosphorus-32
c. technetium-99   d. chromium-51

11.55 How do the radionuclides used for therapeutic purposes differ from the radionuclides used for diagnostic purposes?

11.56 Contrast the different ways in which cobalt-60 and yttrium-90 are used in radiation therapy.

■ **Nuclear Fission and Nuclear Fusion (Section 11.12)**

11.57 Identify which of the following characteristics apply to the fission process, which to the fusion process, and which to both processes.
a. An extremely high temperature is required to start the process.
b. An example of the process occurs on the sun.
c. Transmutation of elements occurs.
d. Neutrons are needed to start the process.

11.58 Identify which of the following characteristics apply to the fission process, which to the fusion process, and which to both processes.
a. Large amounts of energy are released in the process.
b. Energy released in the process is called *nuclear energy.*
c. The process is now used to generate some electrical power in the United States.
d. A fourth state of matter called *plasma* is encountered in studying this process.

11.59 Identify each of the following nuclear reactions as fission, fusion, or neither.
a. ${}^{3}_{2}\text{He} + {}^{3}_{2}\text{He} \rightarrow {}^{4}_{2}\text{He} + 2\,{}^{1}_{1}\text{H}$
b. ${}^{235}_{92}\text{U} + {}^{1}_{0}\text{n} \rightarrow {}^{144}_{55}\text{Cs} + {}^{90}_{37}\text{Rb} + 2\,{}^{1}_{0}\text{n}$
c. ${}^{209}_{83}\text{Bi} + {}^{4}_{2}\alpha \rightarrow {}^{210}_{85}\text{At} + 3\,{}^{1}_{0}\text{n}$
d. ${}^{238}_{92}\text{U} \rightarrow {}^{234}_{90}\text{Th} + {}^{4}_{2}\alpha$

11.60 Identify each of the following nuclear reactions as fission, fusion, or neither.
a. ${}^{239}_{92}\text{U} \rightarrow {}^{239}_{93}\text{Np} + {}^{0}_{-1}\beta$
b. ${}^{230}_{90}\text{Th} + {}^{1}_{1}\text{p} \rightarrow {}^{223}_{87}\text{Fr} + 2\,{}^{4}_{2}\alpha$
c. ${}^{3}_{1}\text{H} + {}^{2}_{1}\text{H} \rightarrow {}^{4}_{2}\text{He} + {}^{1}_{0}\text{n}$
d. ${}^{235}_{92}\text{U} + {}^{1}_{0}\text{n} \rightarrow {}^{139}_{54}\text{Xe} + {}^{95}_{38}\text{Sr} + 2\,{}^{1}_{0}\text{n}$

ADDITIONAL PROBLEMS

11.61 Write nuclear equations for each of the following radioactive-decay processes.
a. Thallium-206 is formed by beta emission.
b. Palladium-109 undergoes beta emission.
c. Plutonium-241 is formed by alpha emission.
d. Fermium-249 undergoes alpha emission.

11.62 Cobalt-55 has a half-life of 18 hours. How long will it take, in hours, for the following fractions of nuclides in a cobalt-55 sample to decay?
a. 7/8   b. 31/32
c. 63/64   d. 127/128

11.63 Write equations for the following nuclear bombardment processes.
a. Bombardment of a radionuclide with an alpha particle produces curium-242 and one neutron.
b. Bombardment of curium-246 with a small particle produces nobelium-254 and four neutrons.
c. Aluminum-27 is bombarded with an alpha particle and produces a neutron.
d. Bombardment of sodium-23 with hydrogen-2 produces neon-21.

11.64 The second artificially produced element was promethium. Write the equation for the production of ^{143}Pm by the bombardment of ^{142}Nd with neutrons.

11.65 Using Table 11.2 as your source of information, determine for how many of the transuranium elements the most stable isotope has a half-life that is less than 1.0 day.

11.66 How many neutrons are produced as a result of the following fission reaction?

$$\text{Neutron} + \text{U-235} \longrightarrow \text{I-135} + \text{Y-97} + ? \text{ neutrons}$$

11.67 Consider the decay series

$$\text{E} \longrightarrow \text{F} \longrightarrow \text{G} \longrightarrow \text{H}$$

where E, F, and G are radioactive, with half-lives of 10.0 sec, 1.2 min and 12.5 days, respectively, and H is nonradioactive. Starting with 1000 atoms of E, and none of F, G, and H, estimate the numbers of atoms of E, F, G, and H that are present after 50 days.

11.68 Fill in the blanks in the following segment of a radioactive decay series.

$$\underline{\quad?\quad} \xrightarrow{\beta} \underline{\quad?\quad} \xrightarrow{\alpha} {}^{224}\text{Ra} \xrightarrow{\beta} \underline{\quad?\quad}$$

ANSWERS TO PRACTICE EXERCISES

11.1 a. $^{245}_{97}\text{Bk} \rightarrow {}^{4}_{2}\alpha + {}^{241}_{95}\text{Am}$ b. $^{89}_{38}\text{Sr} \rightarrow {}^{0}_{-1}\beta + {}^{89}_{39}\text{Y}$

     c. $^{230}_{90}\text{Th} \rightarrow {}^{4}_{2}\alpha + {}^{226}_{88}\text{Ra}$ d. $^{40}_{19}\text{K} \rightarrow {}^{0}_{-1}\beta + {}^{40}_{20}\text{Ca}$

11.2 0.25 g

11.3 13.4 hr

12 Saturated Hydrocarbons

The historical origins of the terms *organic* and *inorganic* involve the following conceptual pairings:

*org*anic—living *org*anisms

*in*organic—*in*animate materials

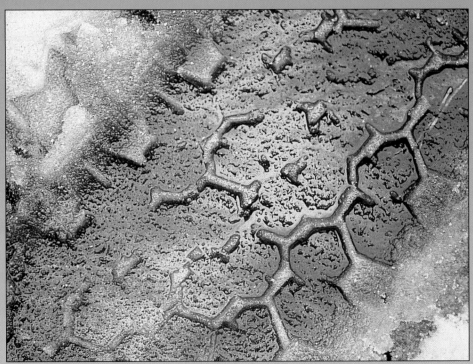

A tire tread in snow with oil (hydrocarbons) causing the iridescent pattern.

This chapter is the first of six that deal with the subject of organic chemistry and organic compounds. Organic compounds are the chemical basis for life itself, as well as an important component of the basis for our current high standard of living. Proteins, carbohydrates, enzymes, and hormones are organic molecules. Organic compounds also include natural gas, petroleum, coal, gasoline, and many synthetic materials such as dyes, plastics, and clothing fibers.

12.1 Organic and Inorganic Compounds

During the latter part of the eighteenth century and the early part of the nineteenth century, chemists began to categorize compounds into two types: organic and inorganic. Compounds obtained from living organisms were called *organic* compounds, and compounds obtained from mineral constituents of the earth were called *inorganic* compounds.

During this early period, chemists believed that a special "vital force," supplied by a living organism, was necessary for the formation of an organic compound. This concept was proved incorrect in 1828 by the German chemist Friedrick Wöhler. Wöhler heated an aqueous solution of two inorganic compounds, ammonium chloride and silver cyanate, and obtained urea (a component of urine).

$$NH_4Cl + AgNCO \longrightarrow \underset{Urea}{(NH_2)_2CO} + AgCl$$

Soon other chemists had successfully synthesized organic compounds from inorganic starting materials. As a result, the vital-force theory was completely abandoned.

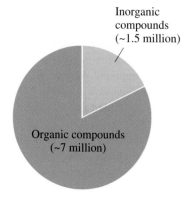

Inorganic
compounds
(~1.5 million)

Organic compounds
(~7 million)

FIGURE 12.1 Sheer numbers is one reason why organic chemistry is a separate field of chemical study. Approximately 7 million organic compounds are known, compared to "just" 1.5 million inorganic compounds.

▶ Some textbooks define organic chemistry as the study of carbon-containing compounds. Almost all carbon-containing compounds qualify as organic compounds. However, the oxides of carbon, carbonates, cyanides, and metallic carbides are classified as inorganic rather than organic compounds. Inorganic carbon compounds involve carbon atoms that are not bonded to hydrogen atoms (CO, CO_2, Na_2CO_3, and so on).

▶ Carbon atoms in organic compounds, in accordance with the octet rule, always form four covalent bonds.

The terms *organic* and *inorganic* continue to be used in classifying compounds, but the definitions of these terms no longer reflect their historical origins. **Organic chemistry** *is the study of hydrocarbons (compounds of carbon and hydrogen) and their derivatives.* Nearly all compounds found in living organisms are still classified as organic compounds, as are many compounds that have been synthesized in the laboratory, and have never been found in a living organism. **Inorganic chemistry** *is the study of all substances other than hydrocarbons and their derivatives.*

In essence, organic chemistry is the study of the compounds of one element (carbon), and inorganic chemistry is the study of the compounds of the other 112 elements. This unequal partitioning occurs because there are approximately 7 million organic compounds and only an estimated 1.5 million inorganic compounds (Figure 12.1). This is an approximately 5:1 ratio between organic and inorganic compounds.

12.2 Bonding Characteristics of the Carbon Atom

Why does the element carbon form 5 times as many compounds as all the other elements combined? The answer is that carbon atoms have the unique ability to bond to each other in a wide variety of ways that involve long chains of carbon atoms or cyclic arrangements (rings) of carbon atoms. Sometimes both chains and rings of carbon atoms are present in the same molecule.

The variety of covalent bonding "behaviors" possible for carbon atoms is related to carbon's electron configuration. Carbon is a member of Group IVA of the periodic table, so carbon atoms possess four valence electrons (Section 4.2). In compound formation, four additional valence electrons are needed to give carbon atoms an octet of valence electrons (the octet rule, Section 4.3). These additional electrons are obtained by electron sharing (covalent bond formation). The sharing of *four* valence electrons requires the formation of *four* covalent bonds.

Carbon can meet this four-bond requirement in three different ways:

1. *By bonding to four other atoms.* This situation requires the presence of four single bonds.

$$-\overset{\mid}{\underset{\mid}{C}}-$$

Four single bonds

2. *By bonding to three other atoms.* This situation requires the presence of two single bonds and one double bond.

$$-\overset{\mid}{C}=$$

Two single bonds and
one double bond

3. *By bonding to two other atoms.* This situation requires the presence of either two double bonds or a triple bond and a single bond.

$$=C= \qquad\qquad -C\equiv$$

Two double bonds　　　One triple bond and
one single bond

12.3 Hydrocarbons and Hydrocarbon Derivatives

The field of organic chemistry encompasses the study of hydrocarbons and hydrocarbon derivatives (Section 12.1). A **hydrocarbon** *is a compound that contains only carbon atoms and hydrogen atoms.* Thousands of hydrocarbons are known. A **hydrocarbon derivative** *is a compound that contains carbon and hydrogen and one or more additional*

▶ The term *saturated* has the general meaning that there is no more room for something. Its use with hydrocarbons comes from early studies in which chemists tried to add hydrogen atoms to various hydrocarbon molecules. Compounds to which no more hydrogen atoms could be added (because they already contained the maximum number) were called saturated, and those to which hydrogen could be added were called unsaturated.

elements. Additional elements commonly found in hydrocarbon derivatives include O, N, S, P, F, Cl, and Br. Millions of hydrocarbon derivatives are known.

Hydrocarbons may be divided into two large classes: saturated and unsaturated. A **saturated hydrocarbon** *is a hydrocarbon in which all carbon–carbon bonds are single bonds.* Saturated hydrocarbons are the simplest type of organic compound. An **unsaturated hydrocarbon** *is a hydrocarbon in which one or more carbon–carbon multiple bonds (double bonds, triple bonds, or both) are present.* In general, saturated and unsaturated hydrocarbons undergo distinctly different chemical reactions.

Saturated hydrocarbons are the subject of this chapter. Unsaturated hydrocarbons are considered in the next chapter. Figure 12.2 summarizes the terminology presented in this section.

12.4 Alkanes: Acyclic Saturated Hydrocarbons

In a saturated hydrocarbon, the carbon atom arrangement may be *acyclic* or *cyclic.* The term *acyclic* means "not cyclic." Examples of acyclic and cyclic carbon atom arrangements are

$$C-C-C-C-C-C$$

Acyclic   Cyclic

In this section we consider acyclic saturated hydrocarbons. Cyclic saturated hydrocarbons are considered in Sections 12.11 and 12.12.

An **alkane** *is a saturated hydrocarbon in which the carbon atom arrangement is acyclic.* Thus an alkane is a hydrocarbon that contains only carbon–carbon single bonds (saturated) and has no rings of carbon atoms (acyclic).

The molecular formulas of all alkanes fit the general formula C_nH_{2n+2}, where n is the number of carbon atoms present. The number of hydrogen atoms present in an alkane is always twice the number of carbon atoms plus two more, as in C_4H_{10}, C_5H_{12}, and C_8H_{18}.

FIGURE 12.2 A summary of classification terms for organic compounds.

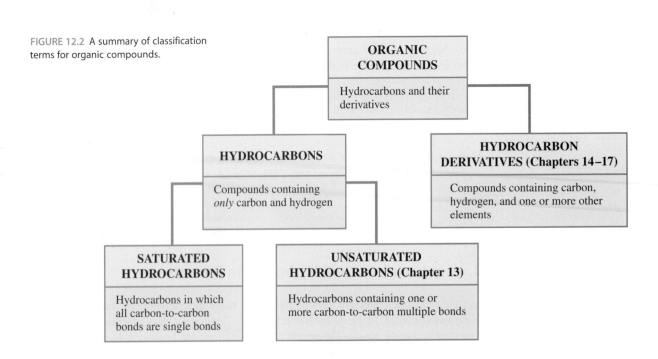

ORGANIC COMPOUNDS

Hydrocarbons and their derivatives

HYDROCARBONS

Compounds containing *only* carbon and hydrogen

HYDROCARBON DERIVATIVES (Chapters 14–17)

Compounds containing carbon, hydrogen, and one or more other elements

SATURATED HYDROCARBONS

Hydrocarbons in which all carbon-to-carbon bonds are single bonds

UNSATURATED HYDROCARBONS (Chapter 13)

Hydrocarbons containing one or more carbon-to-carbon multiple bonds

FIGURE 12.3 Ball-and-stick and space-filling models showing the molecular structures of (a) methane, (b) ethane, and (c) propane, the three simplest alkanes.

(a) Methane   **(b) Ethane**   **(c) Propane**

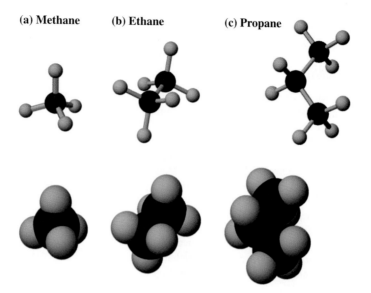

The three simplest alkanes are methane (CH_4), ethane (C_2H_6), and propane (C_3H_8). Ball-and-stick and space-filling models showing the molecular structures of these three alkanes are given in Figure 12.3. Note, from these models, how each carbon atom in each of the models participates in four bonds (Section 12.2). Note also that the geometrical arrangement of atoms about each carbon atom is tetrahedral, an arrangement consistent with the principles of VSEPR theory (Section 5.8). The tetrahedral arrangement of the atoms bonded to alkane carbon atoms is fundamental to understanding the structural aspects of organic chemistry.

12.5 Structural Formulas

The structures of alkanes, as well as other types of organic compounds, are generally represented in two dimensions rather than three (Figure 12.3) because of the difficulty in drawing the latter. These two-dimensional structural representations make no attempt to portray accurately the bond angles or molecular geometry of molecules. Their purpose is to convey information about which atoms in a molecule are bonded to which other atoms.

Two-dimensional structural representations for organic molecules are called structural formulas. A **structural formula** *is a two-dimensional structural representation that shows how the various atoms in a molecule are bonded to each other.* Structural formulas are of two types: expanded structural formulas and condensed structural formulas. An **expanded structural formula** *is a structural formula that shows all atoms in a molecule and all bonds connecting the atoms.* When written out, expanded structural formulas generally occupy a lot of space, and condensed structural formulas represent a shorthand method for conveying the same information. A **condensed structural formula** *is a structural formula that uses groupings of atoms, in which central atoms and the atoms connected to them are written as a group, to convey molecular structural information.* The expanded and condensed structural formulas for methane, ethane, and propane follow.

▶ Structural formulas, whether expanded or condensed, do not show the geometry (shape) of the molecule. That information can be conveyed only by 3-D drawings or models such as those in Figure 12.3.

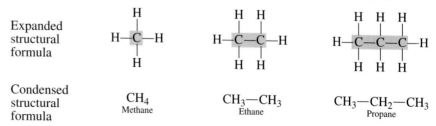

| Expanded structural formula | | | |
| Condensed structural formula | CH_4
 Methane | $CH_3—CH_3$
 Ethane | $CH_3—CH_2—CH_3$
 Propane |

CHEMICAL CONNECTIONS

The Occurrence of Methane

Methane (CH_4), the simplest of all hydrocarbons, is a major component of the atmospheres of Jupiter, Saturn, Uranus, and Neptune but only a minor component of Earth's atmosphere (see the accompanying table). Earth's gravitational field, being weaker than that of the large outer planets, cannot retain enough hydrogen (H_2) in its atmosphere to permit the formation of large amounts of methane; H_2 molecules (the smallest and fastest-moving of all molecules) escape from it into outer space.

The small amount of methane present in Earth's atmosphere comes from terrestrial sources. The decomposition of animal and plant matter in an oxygen-deficient environment—swamps, marshes, bogs, and the sediments of lakes—produces methane. A common name for methane, marsh gas, refers to the production of methane in this manner.

Composition of Earth's Atmosphere (in parts per million by volume)

Major components		Minor components	
nitrogen	780,800	argon	9340
oxygen	209,500	carbon dioxide	314
		neon	18
		helium	5
		methane	2
		krypton	1

Bacteria that live in termites and in the digestive tracts of plant-eating animals have the ability to produce methane from plant materials (cellulose). The methane output of a large cow (belching and flatulence) can reach 20 liters per day.

Methane entering the atmosphere from terrestrial sources presents an environmental problem. Methane is a "greenhouse gas" that contributes to global warming. Methane is 15 to 30 times more efficient than carbon dioxide (the primary greenhouse gas) in trapping heat radiated from Earth. Fortunately, its atmospheric level of 2.0 ppm by volume is much lower than that of carbon dioxide (over 300 ppm).

Methane gas is also found associated with coal and petroleum deposits. Methane associated with coal mines is considered a hazard. If left to accumulate, it can form pockets where air is not present, and asphyxiation of miners can occur. When mixed with air in certain ratios, it can also present an explosion hazard. Methane associated with petroleum deposits is most often recovered, processed, and marketed as *natural gas*. The processed natural gas used in the heating of homes is 85% to 95% methane by volume. Because methane is odorless, an odorant (smelly compound) must be added to the processed natural gas used in home heating. Otherwise, natural gas leaks could not be detected.

Decomposition of plant and animal matter in marshes is a source of methane gas.

The condensed structural formula for propane, CH_3—CH_2—CH_3, is interpreted in the following manner: The first carbon atom is bonded to three hydrogen atoms, and its fourth bond is to the middle carbon atom. The middle carbon atom, besides its bond to the first carbon atom, is also bonded to two hydrogen atoms and to the last carbon atom. The last carbon atom has bonds to three hydrogen atoms in addition to its bond to the middle carbon atom. As is always the case, each carbon atom has four bonds (Section 12.2).

The condensed structural formulas of hydrocarbons in which a long chain of carbon atoms is present are often condensed even more. The formula

$$CH_3-CH_2-CH_2-CH_2-CH_2-CH_2-CH_2-CH_3$$

can be further abbreviated as

$$CH_3-(CH_2)_6-CH_3$$

where parentheses and a subscript are used to denote the number of —CH_2— groups in the chain.

In situations where the focus is solely on the arrangement of carbon atoms in an alkane, *skeletal formulas* that omit the hydrogen atoms are often used.

$$C-C-C-C-C$$
Skeletal formula

means the same as

$$CH_3-CH_2-CH_2-CH_2-CH_3$$
Condensed structural formula

The skeletal formula still represents a unique alkane, because we know that each carbon atom shown must have enough hydrogen atoms attached to it to give the carbon four bonds.

12.6 Structural Isomerism

The molecular formulas CH_4, C_2H_6, and C_3H_8 represent the alkanes methane, ethane, and propane, respectively. Next in the alkane molecular formula sequence is C_4H_{10}, which would be expected to be the molecular formula of the four-carbon alkane. A new phenomenon arises, however, when an alkane has four or more carbon atoms. There is more than one structural formula that is consistent with the molecular formula. Consequently, more than one compound exists with that molecular formula. This situation brings us to the topic of structural isomerism.

> The word *isomer* comes from the Greek *isos*, which means "the same," and *meros*, which means "parts." Isomers have the same parts put together in different ways.

Structural isomers *are compounds with the same molecular formula but different structural formulas—that is, different bonding arrangements between atoms.* Structural isomers, with their differing structural formulas, always have different properties and are different compounds.

There are two four-carbon alkane structural isomers, the compounds *butane* and *isobutane.* Both have the molecular formula C_4H_{10}.

$$CH_3-CH_2-CH_2-CH_3$$
Butane

$$CH_3-CH-CH_3$$
$$|$$
$$CH_3$$
Isobutane

> It is the existence of structural isomers that necessitates the use of structural formulas in organic chemistry. Structural isomers always have the same molecular formula and different structural formulas.

One of these C_4H_{10} isomers, butane, has a chain of four carbon atoms. It is an example of a continuous-chain alkane. A **continuous-chain alkane** *is an alkane in which all carbon atoms are connected in a continuous nonbranching chain.* The other C_4H_{10} isomer, isobutane, has a chain of three carbon atoms with the fourth carbon attached as a branch on the middle carbon of the three-carbon chain. It is an example of a branched-chain alkane. A **branched-chain alkane** *is an alkane in which one or more branches (of carbon atoms) are attached to a continuous chain of carbon atoms.*

There are three structural isomers for alkanes with five carbon atoms (C_5H_{12}):

> Structural isomers of the type we are now considering are also called *skeletal isomers.* Such isomers differ in carbon atom arrangements; that is, the carbon skeletons differ. For alkanes, skeletal isomers are possible whenever more than three carbon atoms are present.

$$CH_3-CH_2-CH_2-CH_2-CH_3$$
Pentane

$$CH_3-CH-CH_2-CH_3$$
$$|$$
$$CH_3$$
Isopentane

$$\begin{array}{c} CH_3 \\ | \\ CH_3-C-CH_3 \\ | \\ CH_3 \end{array}$$
Neopentane

Figure 12.4 shows space-filling models for the three isomeric C_5 alkanes. Note how neopentane, the most branched isomer, has the most compact, most spherical three-dimensional shape.

Table 12.1 contrasts the differing physical properties of the two C_4 alkanes and the three C_5 alkanes.

FIGURE 12.4 Space-filling models for the three isomeric C_5H_{12} alkanes: (a) pentane, (b) isopentane, and (c) neopentane.

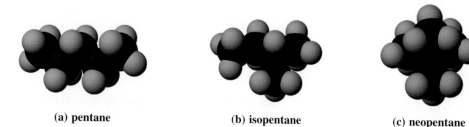

(a) pentane   **(b) isopentane**   **(c) neopentane**

TABLE 12.1
Selected Physical Properties of C_4 and C_5 Alkanes

Alkane	Boiling point	Density (at 20°C)
C_4 alkanes		
butane (C_4H_{10})	−0.5°C	0.579 g/mL
isobutane (C_4H_{10})	−11.6°C	0.549 g/mL
C_5 alkanes		
pentane (C_5H_{12})	36.1°C	0.626 g/mL
isopentane (C_5H_{12})	27.8°C	0.620 g/mL
neopentane (C_5H_{12})	9.5°C	0.614 g/mL

TABLE 12.2
Number of Structural Isomers Possible for Alkanes of Various Carbon-Chain Lengths

Molecular formula	Possible number of structural isomers
CH_4	1
C_2H_6	1
C_3H_8	1
C_4H_{10}	2
C_5H_{12}	3
C_6H_{14}	5
C_7H_{16}	9
C_8H_{18}	18
C_9H_{20}	35
$C_{10}H_{22}$	75
$C_{15}H_{32}$	4,347
$C_{20}H_{42}$	336,319
$C_{25}H_{52}$	36,797,588
$C_{30}H_{62}$	4,111,846,763

The number of possible alkane structural isomers increases dramatically with increasing number of carbon atoms in the alkane, as shown in Table 12.2. Structural isomerism is one of the major reasons for the existence of so many organic compounds.

12.7 Conformations of Alkanes

Rotation about carbon–carbon single bonds is an important property of alkane molecules. Two groups of atoms in an alkane connected by a carbon–carbon single bond can rotate with respect to one another around that bond, much as a wheel rotates around an axle.

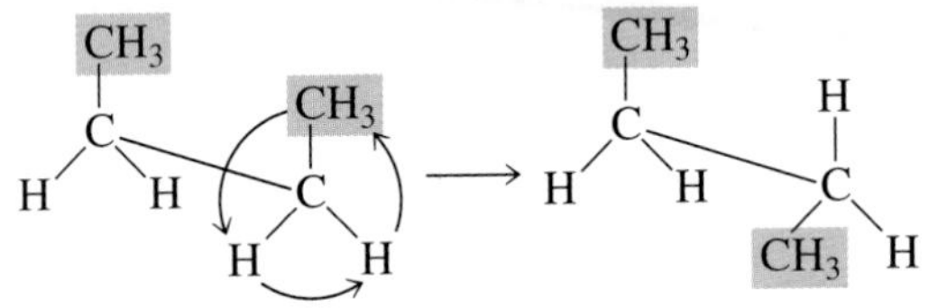

As a result of rotation around single bonds, alkane molecules (except for methane) can exist in infinite numbers of orientations, or conformations. A **conformation** *is the specific three-dimensional arrangement of atoms in an organic molecule at a given instant that results from rotations about carbon–carbon single bonds.*

The following skeletal formulas represent four different conformations for a continuous-chain six-carbon alkane molecule.

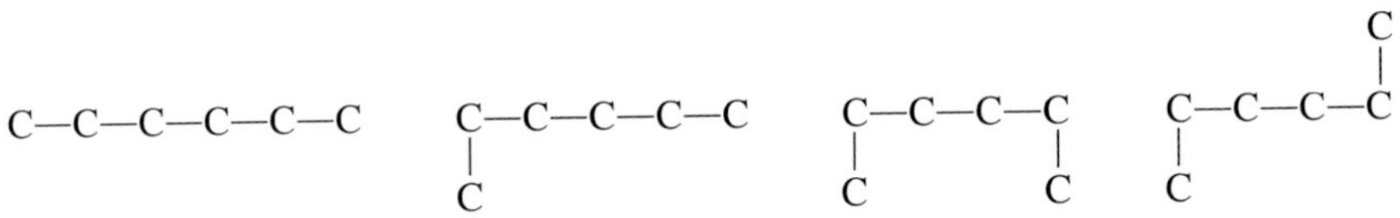

All four skeletal formulas represent the same molecule; that is, they are different conformations of the same molecule. In all four cases, a continuous chain of six carbon atoms is present. In all except the first case, the chain is "bent," but bends do not disrupt the continuity of the chain.

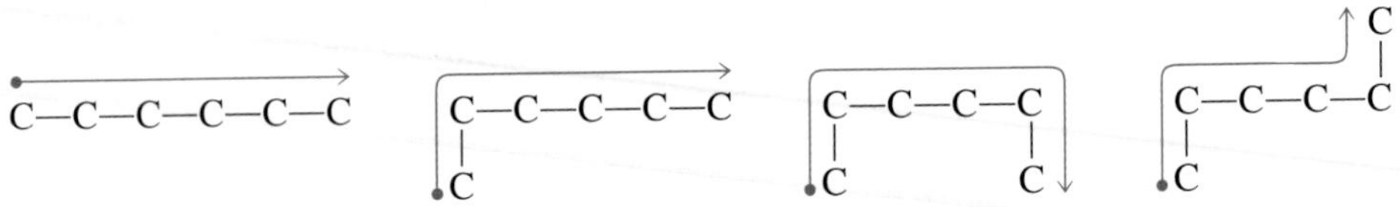

Note that the structures

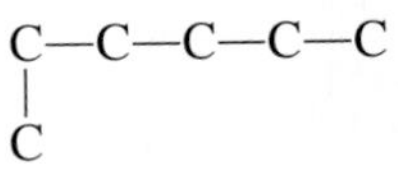

and

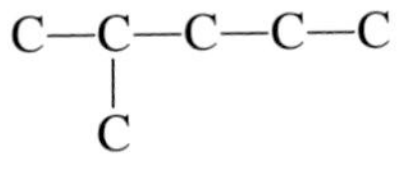

are not two conformations of the same alkane but, rather, represent two different alkanes. The first structure involves a continuous chain of six carbon atoms, and the second

In situations where the focus is solely on the arrangement of carbon atoms in an alkane, *skeletal formulas* that omit the hydrogen atoms are often used.

$$C-C-C-C-C \qquad \text{means the same as} \qquad CH_3-CH_2-CH_2-CH_2-CH_3$$
Skeletal formula   Condensed structural formula

The skeletal formula still represents a unique alkane, because we know that each carbon atom shown must have enough hydrogen atoms attached to it to give the carbon four bonds.

12.6 Structural Isomerism

The molecular formulas CH_4, C_2H_6, and C_3H_8 represent the alkanes methane, ethane, and propane, respectively. Next in the alkane molecular formula sequence is C_4H_{10}, which would be expected to be the molecular formula of the four-carbon alkane. A new phenomenon arises, however, when an alkane has four or more carbon atoms. There is more than one structural formula that is consistent with the molecular formula. Consequently, more than one compound exists with that molecular formula. This situation brings us to the topic of structural isomerism.

Structural isomers *are compounds with the same molecular formula but different structural formulas—that is, different bonding arrangements between atoms.* Structural isomers, with their differing structural formulas, always have different properties and are different compounds.

There are two four-carbon alkane structural isomers, the compounds *butane* and *isobutane.* Both have the molecular formula C_4H_{10}.

$CH_3-CH_2-CH_2-CH_3$ $CH_3-CH-CH_3$

                                    CH_3

Butane   Isobutane

One of these C_4H_{10} isomers, butane, has a chain of four carbon atoms. It is an example of a continuous-chain alkane. A **continuous-chain alkane** *is an alkane in which all carbon atoms are connected in a continuous nonbranching chain.* The other C_4H_{10} isomer, isobutane, has a chain of three carbon atoms with the fourth carbon attached as a branch on the middle carbon of the three-carbon chain. It is an example of a branched-chain alkane. A **branched-chain alkane** *is an alkane in which one or more branches (of carbon atoms) are attached to a continuous chain of carbon atoms.*

There are three structural isomers for alkanes with five carbon atoms (C_5H_{12}):

                                                   CH_3

$CH_3-CH_2-CH_2-CH_2-CH_3$ $CH_3-CH-CH_2-CH_3$ CH_3-C-CH_3

                                      CH_3 CH_3

Pentane   Isopentane   Neopentane

Figure 12.4 shows space-filling models for the three isometric C_5 alkanes. Note how neopentane, the most branched isomer, has the most compact, most spherical three-dimensional shape.

Table 12.1 contrasts the differing physical properties of the two C_4 alkanes and the three C_5 alkanes.

The word *isomer* comes from the Greek *isos*, which means "the same," and *meros*, which means "parts." Isomers have the same parts put together in different ways.

It is the existence of structural isomers that necessitates the use of structural formulas in organic chemistry. Structural isomers always have the same molecular formula and different structural formulas.

Structural isomers of the type we are now considering are also called *skeletal isomers.* Such isomers differ in carbon atom arrangements; that is, the carbon skeletons differ. For alkanes, skeletal isomers are possible whenever more than three carbon atoms are present.

FIGURE 12.4 Space-filling models for the three isomeric C_5H_{12} alkanes: (a) pentane, (b) isopentane, and (c) neopentane.

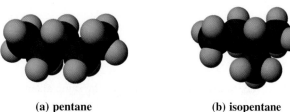

(a) pentane   **(b) isopentane**   **(c) neopentane**

TABLE 12.1

Selected Physical Properties of C_4 and C_5 Alkanes

Alkane	Boiling point	Density (at 20°C)
C_4 alkanes		
butane (C_4H_{10})	−0.5°C	0.579 g/mL
isobutane (C_4H_{10})	−11.6°C	0.549 g/mL
C_5 alkanes		
pentane (C_5H_{12})	36.1°C	0.626 g/mL
isopentane (C_5H_{12})	27.8°C	0.620 g/mL
neopentane (C_5H_{12})	9.5°C	0.614 g/mL

TABLE 12.2

Number of Structural Isomers Possible for Alkanes of Various Carbon-Chain Lengths

Molecular formula	Possible number of structural isomers
CH_4	1
C_2H_6	1
C_3H_8	1
C_4H_{10}	2
C_5H_{12}	3
C_6H_{14}	5
C_7H_{16}	9
C_8H_{18}	18
C_9H_{20}	35
$C_{10}H_{22}$	75
$C_{15}H_{32}$	4,347
$C_{20}H_{42}$	336,319
$C_{25}H_{52}$	36,797,588
$C_{30}H_{62}$	4,111,846,763

The number of possible alkane structural isomers increases dramatically with increasing number of carbon atoms in the alkane, as shown in Table 12.2. Structural isomerism is one of the major reasons for the existence of so many organic compounds.

12.7 Conformations of Alkanes

Rotation about carbon–carbon single bonds is an important property of alkane molecules. Two groups of atoms in an alkane connected by a carbon–carbon single bond can rotate with respect to one another around that bond, much as a wheel rotates around an axle.

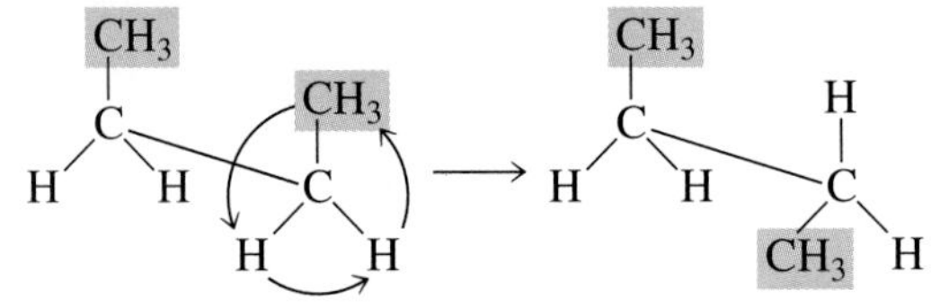

As a result of rotation around single bonds, alkane molecules (except for methane) can exist in infinite numbers of orientations, or conformations. A **conformation** *is the specific three-dimensional arrangement of atoms in an organic molecule at a given instant that results from rotations about carbon–carbon single bonds.*

The following skeletal formulas represent four different conformations for a continuous-chain six-carbon alkane molecule.

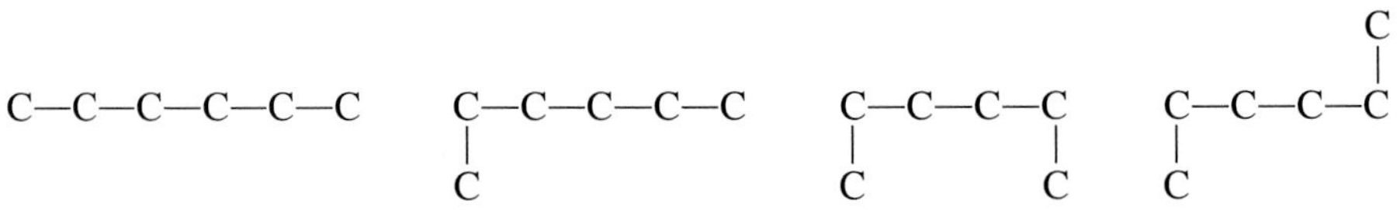

All four skeletal formulas represent the same molecule; that is, they are different conformations of the same molecule. In all four cases, a continuous chain of six carbon atoms is present. In all except the first case, the chain is "bent," but bends do not disrupt the continuity of the chain.

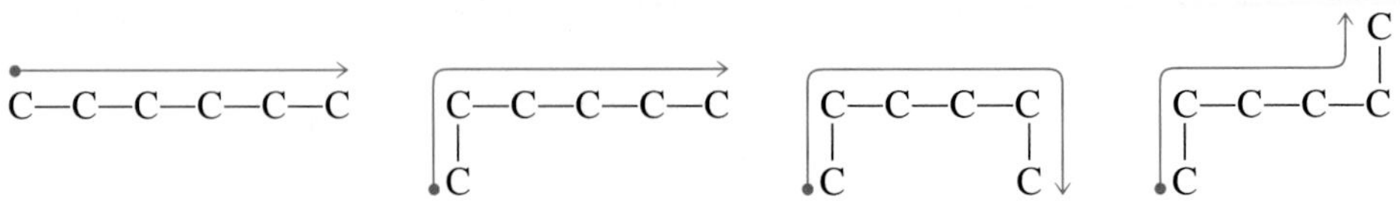

Note that the structures

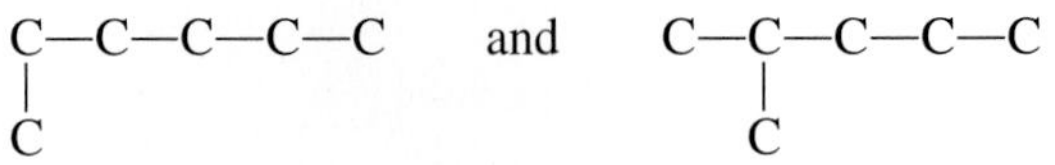

and

are not two conformations of the same alkane but, rather, represent two different alkanes. The first structure involves a continuous chain of six carbon atoms, and the second

You should learn to recognize molecules drawn in several different ways (conformations). Like friends, they can be recognized whether they are sitting, reclining, or standing.

structure involves a continuous chain of five carbon atoms to which a branch is attached. There is no way that you can get a continuous chain of six carbon atoms out of the second structure without "back-tracking," and "back-tracking" is not allowed.

EXAMPLE 12.1

Recognizing Different Conformations of a Molecule and Structural Isomers

■ Determine whether the members of each of the following pairs of structural formulas represent (1) different conformations of the same molecule, (2) different compounds that are structural isomers, or (3) different compounds that are not structural isomers.

a. CH_3—CH_2—CH_2—CH_3 and CH_2—CH_2 | | CH_3 CH_3

b. CH_2—CH_2—CH_3 and CH_2—CH_2—CH_2—CH_3 | | CH_3 CH_3

c. CH_3—CH—CH_3 and CH_3—CH_2—CH_2 | | CH_3 CH_3

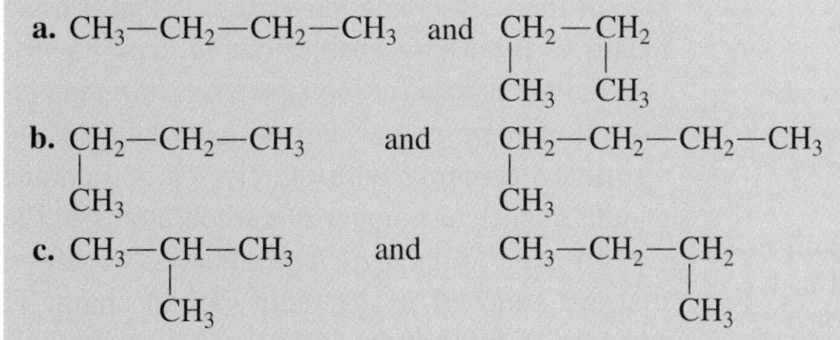

Solution

a. Both molecules have the molecular formula C_4H_{10}. The connectivity of carbon atoms is the same for both molecules: a continuous chain of four carbon atoms. For the second structural formula, we need to go around two corners to get a four-carbon-atom chain, which is fine because of the free rotation associated with single bonds in alkanes.

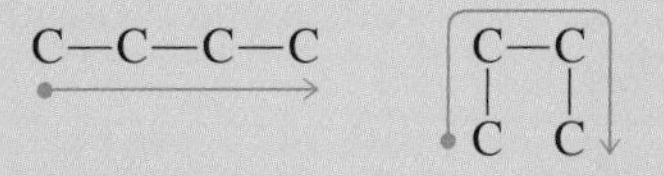

With the same molecular formula and the same connectivity of atoms, these two structural formulas are conformations of the same molecule.

b. The molecular formula of the first compound is C_4H_{10}, and that of the second compound is C_5H_{12}. Thus the two structural formulas represent different compounds that are not structural isomers. Structural isomers must have the same molecular formula.

c. Both molecules have the same molecular formula, C_4H_{10}. The connectivity of atoms is different. In the first case, we have a chain of three carbon atoms with a branch off the chain. In the second case, a continuous chain of four carbon atoms is present.

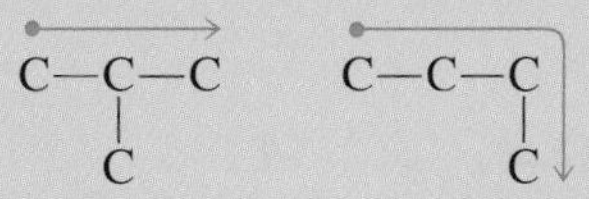

These two structural formulas are those of structural isomers.

Practice Exercise 12.1

Determine whether the members of each of the following pairs of structural formulas represent (1) different conformations of the same molecule, (2) different compounds that are structural isomers, or (3) different compounds that are not structural isomers.

a. CH_3—CH_2—CH_2—CH_2—CH_3 and CH_3—CH_2 | CH_2—CH_2 | CH_3

b. CH_3—CH—CH_2—CH_3 and CH_3—CH—CH_2 | | | CH_3 CH_3 CH_3

c. CH_3—CH—CH_2—CH_3 and CH_2—CH_2—CH_2 | | | CH_3 CH_3 CH_3

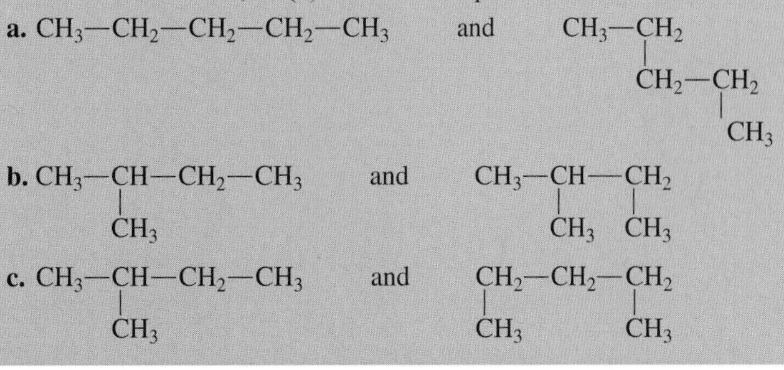

12.8 IUPAC Nomenclature for Alkanes

When relatively few organic compounds were known, chemists arbitrarily named them using what today are called *common names*. These common names gave no information about the structures of the compounds they described. However, as more organic compounds became known, this nonsystematic approach to naming compounds became unwieldy.

▶ IUPAC is pronounced "eye-you-pack."

▶ Continuous-chain alkanes are also frequently called *straight-chain alkanes* and *normal-chain alkanes*.

▶ You need to memorize the prefixes in column 2 of Table 12.3. This is the way to count from 1 to 10 in "organic chemistry language."

Today, formal systematic rules exist for generating names for organic compounds. These rules, which were formulated, and are updated periodically, by the International Union of Pure and Applied Chemistry (IUPAC), are known as *IUPAC rules*. The advantage of the IUPAC naming system is that it assigns each compound a name that not only identifies it but also enables one to draw its structural formula.

IUPAC names for the first ten *continuous-chain* alkanes are given in Table 12.3. Note that all of these names end in *-ane*, the characteristic ending for all alkane names. Note also that beginning with the five-carbon alkane, Greek numerical prefixes are used to denote the actual number of carbon atoms in the continuous chain.

To name *branched-chain* alkanes, we must be able to name the branch or branches that are attached to the main carbon chain. These branches are formally called substituents. A **substituent** *is an atom or group of atoms attached to a chain (or ring) of carbon atoms.* Note that *substituent* is a general term that applies to carbon-chain attachments in all organic molecules, not just alkanes.

For branched-chain alkanes, the substituents are specifically called *alkyl groups*. An **alkyl group** *is the group of atoms that would be obtained by removing a hydrogen atom from an alkane.*

The two most commonly encountered alkyl groups are the two simplest: the one-carbon and two-carbon alkyl groups. Their formulas and names are

$$\text{---CH}_3 \qquad \text{---CH}_2\text{---CH}_3$$

Methyl group   Ethyl group

The extra long bond in these formulas (on the left) denotes the point of attachment to the carbon chain. Note that alkyl groups do not lead a stable, independent existence; that is, they are not molecules. They are always found attached to another entity (usually a carbon chain).

Alkyl groups are named by taking the stem of the name of the alkane that contains the same number of carbon atoms and adding the ending *-yl*. Table 12.4 gives the names for small continuous-chain alkyl groups.

TABLE 12.3

IUPAC Names for the First Ten Continuous-Chain Alkanes[a]

Molecular formula	IUPAC prefix	IUPAC name	Structural formula
CH_4	meth-	methane	CH_4
C_2H_6	eth-	ethane	$CH_3—CH_3$
C_3H_8	prop-	propane	$CH_3—CH_2—CH_3$
C_4H_{10}	but-	butane	$CH_3—CH_2—CH_2—CH_3$
C_5H_{12}	pent-	pentane	$CH_3—CH_2—CH_2—CH_2—CH_3$
C_6H_{14}	hex-	hexane	$CH_3—CH_2—CH_2—CH_2—CH_2—CH_3$
C_7H_{16}	hept-	heptane	$CH_3—CH_2—CH_2—CH_2—CH_2—CH_2—CH_3$
C_8H_{18}	oct-	octane	$CH_3—CH_2—CH_2—CH_2—CH_2—CH_2—CH_2—CH_3$
C_9H_{20}	non-	nonane	$CH_3—CH_2—CH_2—CH_2—CH_2—CH_2—CH_2—CH_2—CH_3$
$C_{10}H_{22}$	dec-	decane	$CH_3—CH_2—CH_2—CH_2—CH_2—CH_2—CH_2—CH_2—CH_2—CH_3$

[a]The IUPAC naming system also includes prefixes for naming continuous-chain alkanes that have more than 10 carbon atoms, but we will not consider them in this text.

TABLE 12.4
Names for the First Six Continuous-Chain Alkyl Groups

Number of carbons	Structural formula	Stem of alkane name	Suffix	Alkyl group name
1	$—CH_3$	meth-	-yl	methyl
2	$—CH_2—CH_3$	eth-	-yl	ethyl
3	$—CH_2—CH_2—CH_3$	prop-	-yl	propyl
4	$—CH_2—CH_2—CH_2—CH_3$	but-	-yl	butyl
5	$—CH_2—CH_2—CH_2—CH_2—CH_3$	pent-	-yl	pentyl
6	$—CH_2—CH_2—CH_2—CH_2—CH_2—CH_3$	hex-	-yl	hexyl

> ▶ The ending -*yl*, as in meth*yl*, eth*yl*, prop*yl*, and but*yl*, appears in the names of all alkyl groups.

We are now ready for the IUPAC rules for naming branched-chain alkanes.

Rule 1: *Identify the longest continuous carbon chain (the parent chain), which may or may not be shown in a straight line, and name the chain.*

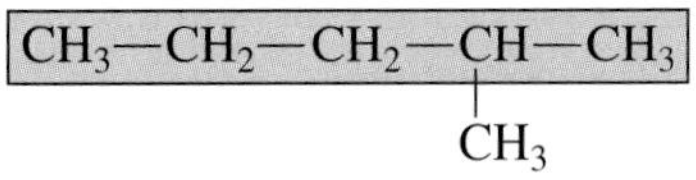

> ▶ An additional guideline for identifying the longest continuous carbon chain: If two different carbon chains in a molecule have the same largest number of carbon atoms, select as the parent chain the one with the larger number of substituents (alkyl groups) attached to the chain.

$$CH_3—CH_2—CH_2—CH—CH_3$$
$$|$$
$$CH_3$$

The parent chain name is *pentane*, because it has five carbon atoms.

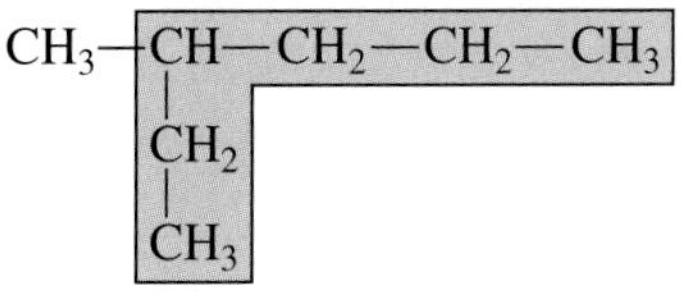

$$CH_3—CH—CH_2—CH_2—CH_3$$
$$|$$
$$CH_2$$
$$|$$
$$CH_3$$

The parent chain name is *hexane*, because it has six carbon atoms.

Rule 2: *Number the carbon atoms in the parent chain from the end of the chain nearest a substituent (alkyl group).*

There always are two ways to number the chain (either from left to right or from right to left). This rule gives the first-encountered alkyl group the lowest possible number.

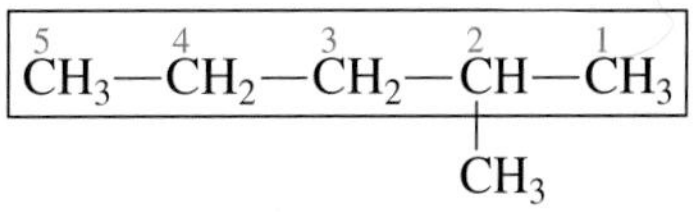

> ▶ Additional guidelines for numbering carbon atom chains:
> 1. If both ends of the chain have a substituent the same distance in, number from the end closest to the second-encountered substituent.
> 2. If there are substituents equidistant from each end of the chain and there is no third substituent to use as the "tie-breaker," begin numbering nearest the substituent that has alphabetical priority—that is, the substituent whose name occurs first in the alphabet.

$$\overset{5}{C}H_3—\overset{4}{C}H_2—\overset{3}{C}H_2—\overset{2}{C}H—\overset{1}{C}H_3$$
$$|$$
$$CH_3$$

Right-to-left numbering system

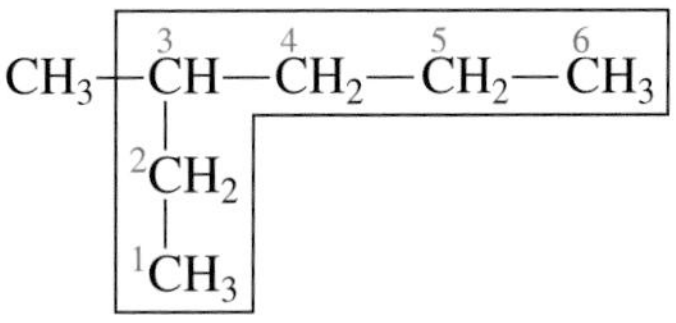

$$CH_3—\overset{3}{C}H—\overset{4}{C}H_2—\overset{5}{C}H_2—\overset{6}{C}H_3$$
$$|$$
$$\overset{2}{C}H_2$$
$$|$$
$$\overset{1}{C}H_3$$

Left-to-right numbering system

Rule 3: *If only one alkyl group is present, name and locate it (by number), and attach the number and name to that of the parent carbon chain.*

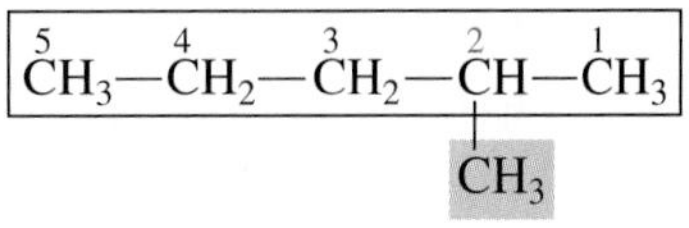

$$\overset{5}{C}H_3—\overset{4}{C}H_2—\overset{3}{C}H_2—\overset{2}{C}H—\overset{1}{C}H_3$$
$$|$$
$$CH_3$$

2-Methylpentane

$$CH_3—\overset{3}{C}H—\overset{4}{C}H_2—\overset{5}{C}H_2—\overset{6}{C}H_3$$
$$|$$
$$\overset{2}{C}H_2$$
$$|$$
$$\overset{1}{C}H_3$$

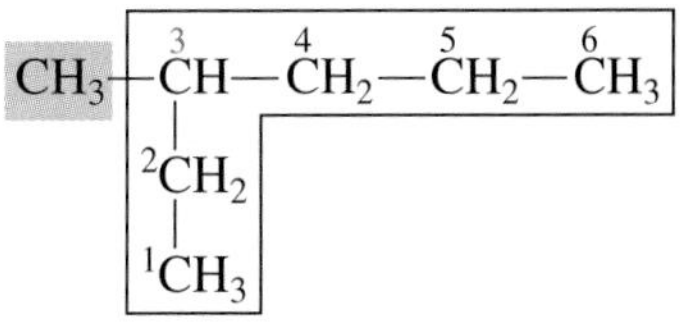

3-Methylhexane

Note that the name is written as one word, with a hyphen between the number (location) and the name of the alkyl group.

Rule 4: *If two or more of the same kind of alkyl group are present in a molecule, indicate the number with a Greek numerical prefix (di-, tri-, tetra-, penta-, and so forth). In addition, a number specifying the location of each identical group must be included. These position numbers, separated by commas, precede the numerical prefix. Numbers are separated from words by hyphens.*

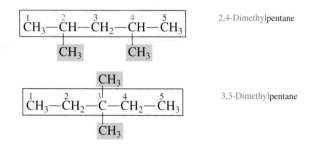

2,4-Dimethylpentane

3,3-Dimethylpentane

There must be as many numbers as there are alkyl groups in the IUPAC name of a branched-chain alkane.

Note that the numerical prefix *di-* must always be accompanied by two numbers, *tri-* by three, and so on, even if the same number is written twice, as in 3,3-dimethylpentane.

Rule 5: *When two kinds of alkyl groups are present on the same carbon chain, number each group separately, and list the names of the alkyl groups in alphabetical order.*

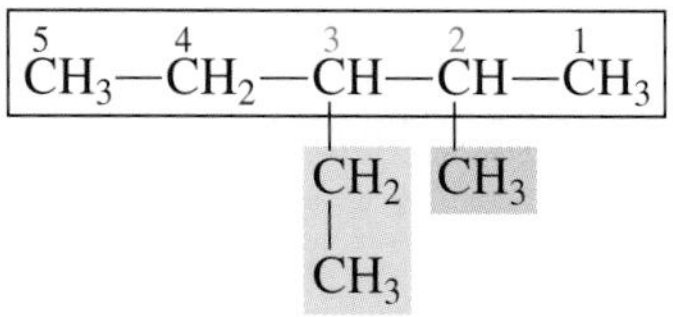

3-Ethyl-2-methylpentane

Note that ethyl is named first in accordance with the alphabetical rule.

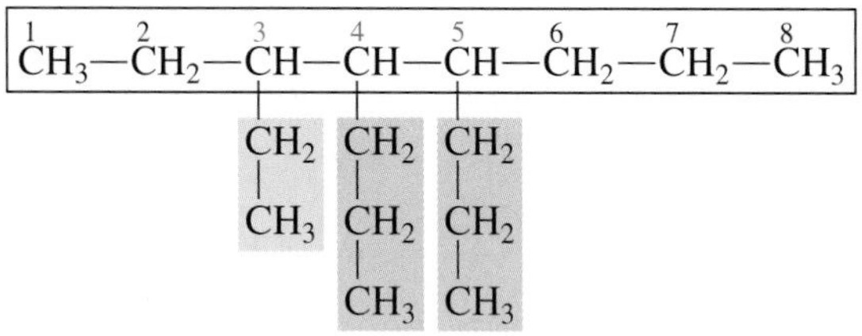

3-Ethyl-4,5-dipropyloctane

Numerical prefixes that designate numbers of alkyl groups, such as *di-*, *tri-*, and *tetra-*, are not considered when determining alphabetical priority for alkyl groups.

Note that the prefix *di-* does not affect the alphabetical order; *ethyl* precedes *propyl*.

Rule 6: *Follow IUPAC punctuation rules, which include the following: (1) Separate numbers from each other by commas. (2) Separate numbers from letters by hyphens. (3) Do not add a hyphen or a space between the last-named substituent and the name of the parent alkane that follows.*

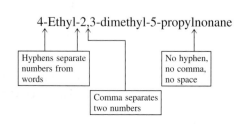

4-Ethyl-2,3-dimethyl-5-propylnonane

Hyphens separate numbers from words

Comma separates two numbers

No hyphen, no comma, no space

EXAMPLE 12.2

Determining IUPAC Names for Branched-Chain Alkanes

■ Give the IUPAC name for each of the following branched-chain alkanes.

a. 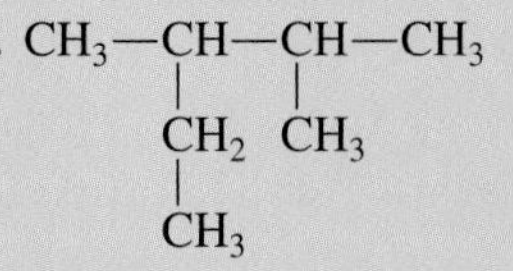 $CH_3-CH-CH-CH_3$
　　　　|　　|
　　　CH_2　CH_3
　　　　|
　　　CH_3

b. 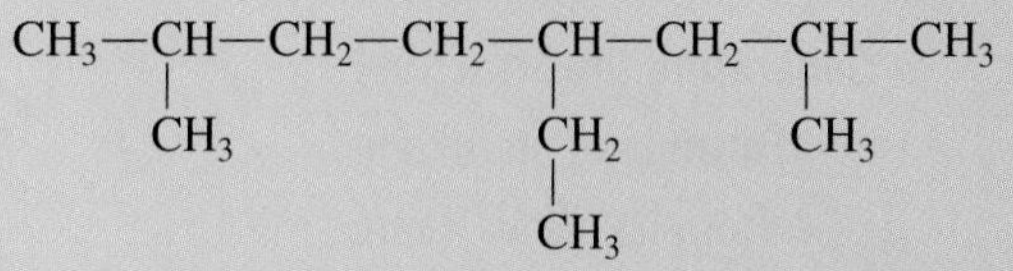 $CH_3-CH-CH_2-CH_2-CH-CH_2-CH-CH_3$
　　　　　|　　　　　　　　　|　　　　|
　　　　CH_3　　　　　　CH_2　CH_3
　　　　　　　　　　　　　　|
　　　　　　　　　　　　CH_3

Solution

a. The longest carbon chain possesses five carbon atoms. Thus the parent-chain name is pentane.

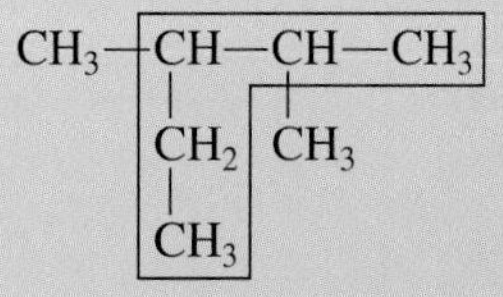

$CH_3-\boxed{CH-CH-CH_3}$
　　　　|　　|
　　　CH_2　CH_3
　　　　|
　　　CH_3

This parent chain is numbered from right to left, because an alkyl substituent is closer to the right end of the chain than to the left end.

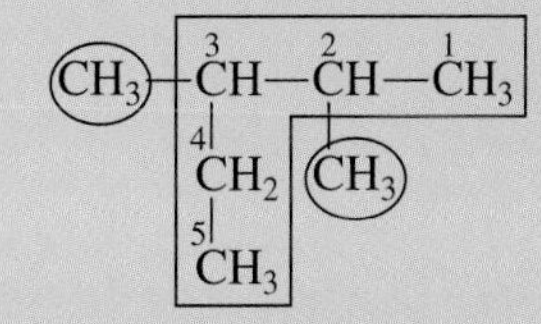

$\boxed{CH_3}-\overset{3}{CH}-\overset{2}{CH}-\overset{1}{CH_3}$
　　　　　　$4|$　　|
　　　　　CH_2　$\boxed{CH_3}$
　　　　　$5|$
　　　　　CH_3

There are two methyl group substituents (circled). One methyl group is located on carbon 2 and the other on carbon 3. The IUPAC name for the compound is 2,3-dimethylpentane.

b. There are eight carbon atoms in the longest carbon chain, so the parent name is octane. There are three alkyl groups present (circled).

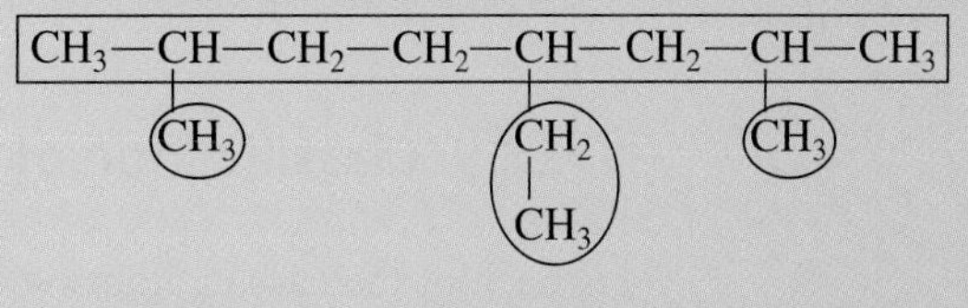

$\boxed{CH_3-CH-CH_2-CH_2-CH-CH_2-CH-CH_3}$
　　　　　　$\boxed{CH_3}$　　　　　$\boxed{CH_2}$　　$\boxed{CH_3}$
　　　　　　　　　　　　　　　|
　　　　　　　　　　　　　CH_3

Selection of the numbering system to be used cannot be made based on the "first-encountered-alkyl-group rule" because an alkyl group is equidistant from each end of the chain. Thus the second-encountered alkyl group is used as the "tie-breaker." It is closer to the right end of the parent chain (carbon 4) than to the left end (carbon 5). Thus we use the right-to-left numbering system.

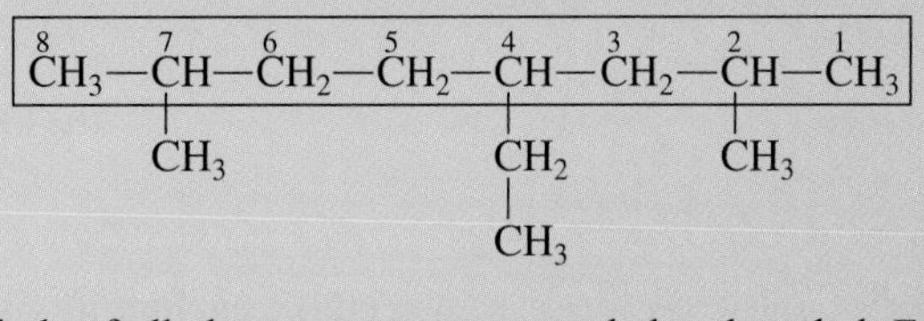

$\boxed{\overset{8}{CH_3}-\overset{7}{CH}-\overset{6}{CH_2}-\overset{5}{CH_2}-\overset{4}{CH}-\overset{3}{CH_2}-\overset{2}{CH}-\overset{1}{CH_3}}$
　　　　　　CH_3　　　　　　CH_2　　CH_3
　　　　　　　　　　　　　　　|
　　　　　　　　　　　　　CH_3

Two different kinds of alkyl groups are present: ethyl and methyl. Ethyl has alphabetical priority over methyl and precedes methyl in the IUPAC name. The IUPAC name is 4-ethyl-2,7-dimethyloctane.

▶ Always compare the total number of carbon atoms in the name with the number of carbon atoms in the structure to make sure they match. The name 4-ethyl-2,7-dimethyloctane indicates the presence of $2 + 2(1) + 8 = 12$ carbon atoms. The structure does have 12 carbon atoms.

Practice Exercise 12.2

Give the IUPAC name for each of the following alkanes.

a. 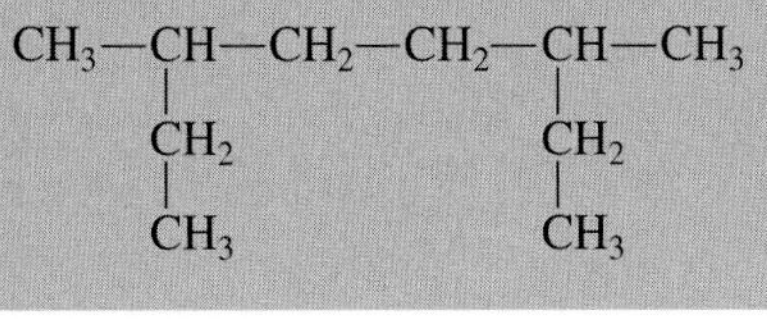 $CH_3-CH-CH_2-CH_2-CH-CH_3$
　　　　　|　　　　　　　　　|
　　　　CH_2　　　　　　CH_2
　　　　　|　　　　　　　　　|
　　　　CH_3　　　　　　CH_3

b. 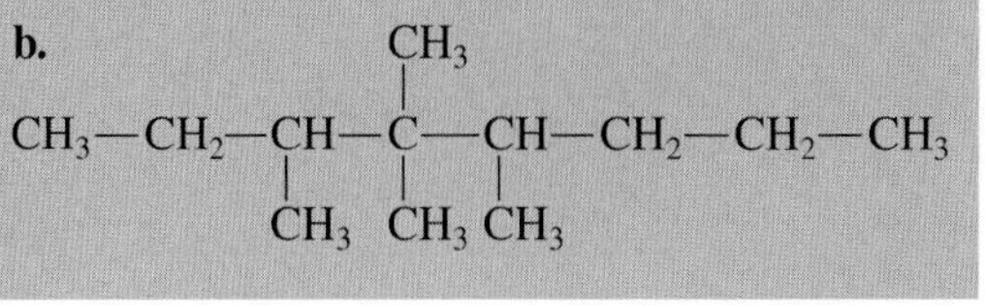
　　　　　　　　　　　　　CH_3
　　　　　　　　　　　　　|
$CH_3-CH_2-CH-C-CH-CH_2-CH_2-CH_3$
　　　　　　　|　|　|
　　　　　CH_3 CH_3 CH_3

After you learn the rules for naming alkanes, it is relatively easy to reverse the procedure and translate the name of an alkane into a structural formula. Example 12.3 shows how this is done.

A few smaller branched alkanes have common names—that is, non-IUPAC names—that still have widespread use. They make use of the prefixes *iso* and *neo,* as in isobutane, isopentane, and neohexane. These prefixes denote particular end-of-chain carbon atom arrangements.

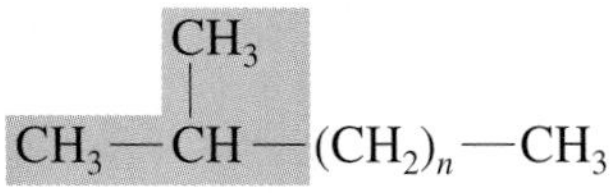

$$CH_3-\overset{\overset{\displaystyle CH_3}{|}}{CH}-(CH_2)_n-CH_3$$

An isoalkane
(e.g., *n* = 1, Isopentane)

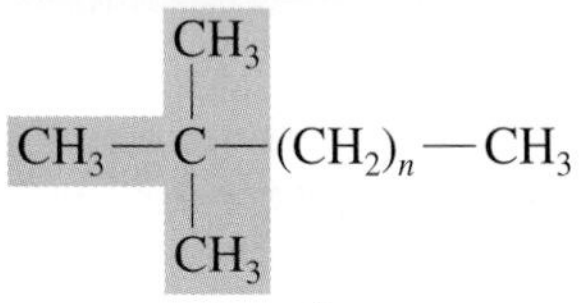

$$CH_3-\overset{\overset{\displaystyle CH_3}{|}}{\underset{\underset{\displaystyle CH_3}{|}}{C}}-(CH_2)_n-CH_3$$

A neoalkane
(e.g., *n* = 1, Neohexane)

■ Draw the condensed structural formula for 3-ethyl-2,3-dimethylpentane.

Solution

Step 1: The name of this compound ends in *pentane,* so the longest continuous chain has five carbon atoms. Draw this chain of five carbon atoms and number it.

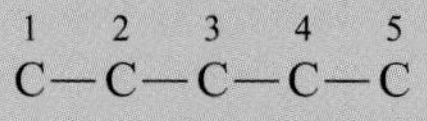

$$\overset{1}{C}-\overset{2}{C}-\overset{3}{C}-\overset{4}{C}-\overset{5}{C}$$

Step 2: Complete the carbon skeleton by attaching alkyl groups as they are specified in the name. An ethyl group goes on carbon 3, and methyl groups are attached to carbons 2 and 3.

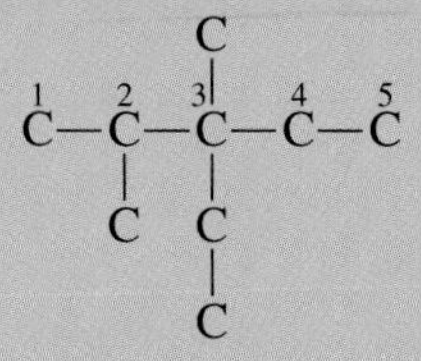

$$\overset{1}{C}-\overset{\overset{}{}}{\underset{\underset{\displaystyle C}{|}}{\overset{2}{C}}}-\overset{\overset{\displaystyle C}{|}}{\underset{\underset{\underset{\displaystyle C}{|}}{\underset{\displaystyle C}{|}}}{\overset{3}{C}}}-\overset{4}{C}-\overset{5}{C}$$

Step 3: Add hydrogen atoms to the carbon skeleton so that each carbon atom has four bonds.

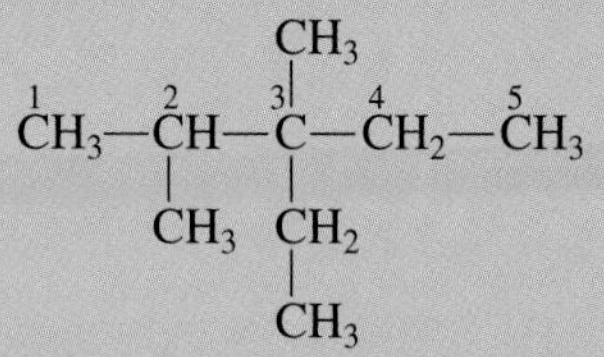

$$\overset{1}{CH_3}-\overset{2}{\underset{\underset{\displaystyle CH_3}{|}}{CH}}-\overset{\overset{\displaystyle CH_3}{|}}{\underset{\underset{\underset{\displaystyle CH_3}{|}}{\underset{\displaystyle CH_2}{|}}}{\overset{3}{C}}}-\overset{4}{CH_2}-\overset{5}{CH_3}$$

Practice Exercise 12.3

Draw the condensed structural formula for 4,5-diethyl-3,4,5-trimethyloctane.

12.9 Classification of Carbon Atoms

Each of the carbon atoms within a hydrocarbon structure can be classified as a primary, secondary, tertiary, or quaternary carbon atom. A **primary carbon atom** *is a carbon atom in an organic molecule that is bonded to only one other carbon atom.* Each of the "end" carbon atoms in the three-carbon propane structure is a primary carbon atom, whereas the middle carbon atom of propane is a secondary carbon atom. A **secondary carbon atom** *is a carbon atom in an organic molecule that is bonded to two other carbon atoms.*

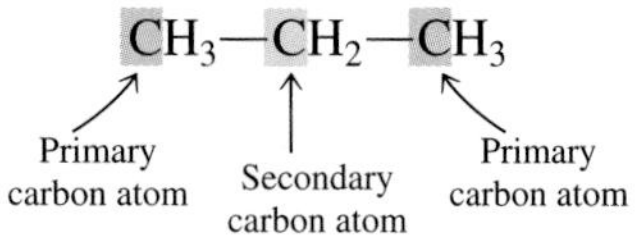

$$CH_3-CH_2-CH_3$$

Primary   Secondary   Primary
carbon atom   carbon atom   carbon atom

A **tertiary carbon atom** *is a carbon atom in an organic molecule that is bonded to three other carbon atoms.* The molecule 2-methylpropane contains a tertiary carbon atom.

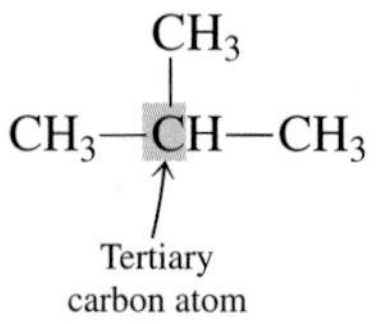

$$CH_3-\overset{\overset{\displaystyle CH_3}{|}}{CH}-CH_3$$

Tertiary
carbon atom

The notations 1°, 2°, 3°, and 4° are often used as designations for the terms *primary, secondary, tertiary,* and *quaternary.* Thus we can write

1° carbon atom
2° carbon atom
3° carbon atom
4° carbon atom

A **quaternary carbon atom** *is a carbon atom in an organic molecule that is bonded to four other carbon atoms.* The molecule 2,2-dimethylpropane contains a quaternary carbon atom.

$$CH_3-\overset{\overset{\displaystyle CH_3}{|}}{\underset{\underset{\displaystyle CH_3}{|}}{C}}-CH_3$$

Quaternary
carbon atom

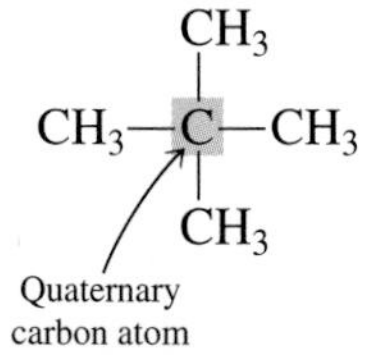

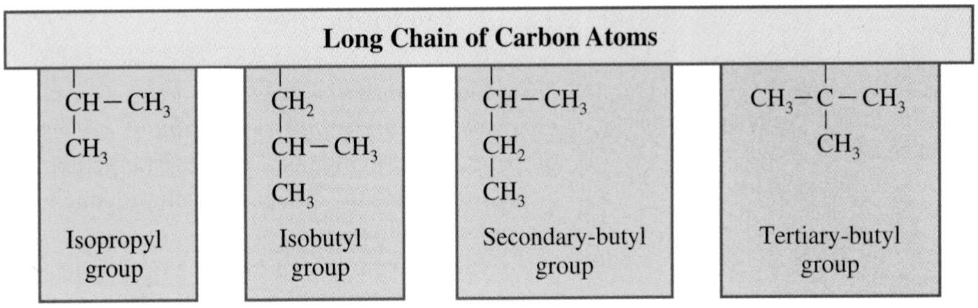

FIGURE 12.5 The four most common branched-chain alkyl groups and their IUPAC names.

Long Chain of Carbon Atoms

$CH-CH_3$	CH_2	$CH-CH_3$	CH_3-C-CH_3
CH_3	$CH-CH_3$	CH_2	CH_3
	CH_3	CH_3	
Isopropyl group	Isobutyl group	Secondary-butyl group	Tertiary-butyl group

12.10 Branched-Chain Alkyl Groups

Just as there are continuous-chain and branched-chain alkanes, there are continuous-chain and branched-chain alkyl groups. Four branched-chain alkyl groups, shown in Figure 12.5, are so common that you should know their names and structures.

For the *sec*-butyl group, the point of attachment of the group to the main carbon chain involves a *secondary* carbon atom. For the *tert*-butyl group, the point of attachment of the group to the main carbon chain involves a *tertiary* carbon atom.

Two examples of alkanes containing branched-chain alkyl groups follow.

> You need to be able to recognize various conformations of branched-chain alkyl groups. For example, these structures all represent an isopropyl group: 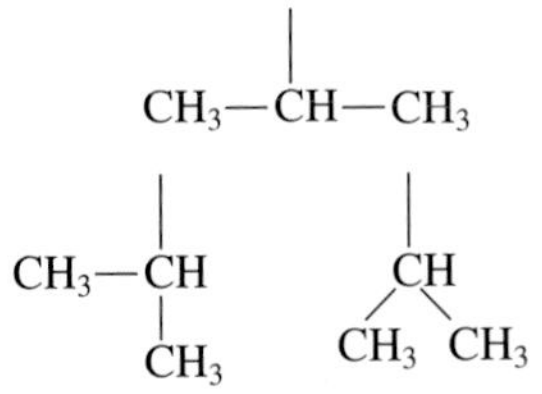
>
> $$CH_3-CH-CH_3$$
>
> $$CH_3-CH \qquad CH$$
> $$\quad\quad CH_3 \qquad CH_3\ CH_3$$
>
> In each case, you have a chain of three carbon atoms with an attachment point (the long bond) involving the middle carbon atom of the chain.

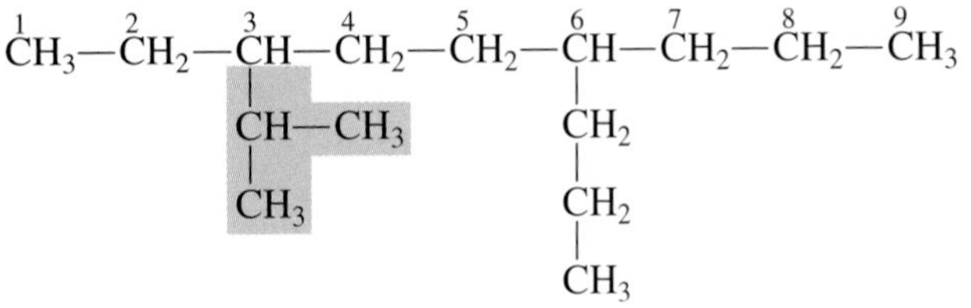

$$\overset{1}{C}H_3-\overset{2}{C}H_2-\overset{3}{C}H-\overset{4}{C}H_2-\overset{5}{C}H_2-\overset{6}{C}H-\overset{7}{C}H_2-\overset{8}{C}H_2-\overset{9}{C}H_3$$
$$CH-CH_3 \qquad\qquad CH_2$$
$$CH_3 \qquad\qquad\qquad CH_2$$
$$CH_3$$

3-Isopropyl-6-propylnonane

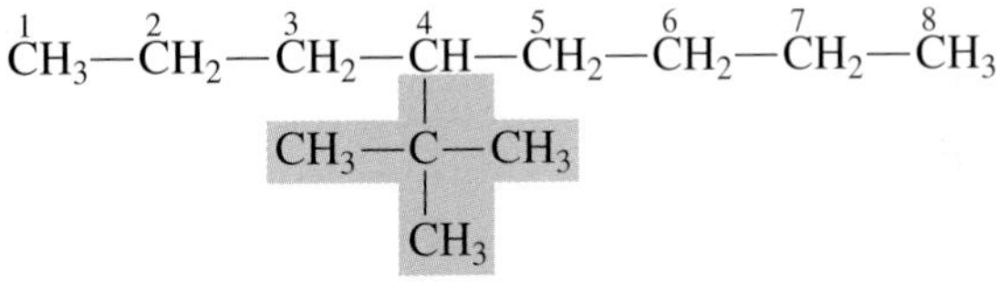

$$\overset{1}{C}H_3-\overset{2}{C}H_2-\overset{3}{C}H_2-\overset{4}{C}H-\overset{5}{C}H_2-\overset{6}{C}H_2-\overset{7}{C}H_2-\overset{8}{C}H_3$$
$$CH_3-C-CH_3$$
$$CH_3$$

4-*tert*-Butyloctane

In IUPAC naming, hyphenated prefixes, such as *sec*- and *tert*-, are not considered when alphabetizing. The prefixes *iso* and *neo* are not hyphenated prefixes and are included when alphabetizing. The following IUPAC name is thus correct:

5-*sec*-Butyl-4-isopropyl-3-methyldecane

■ Complex Branched Alkyl Groups

Complex branched alkyl groups, for which no "simple" name is available, are occasionally encountered. The IUPAC system provision for naming such groups involves selecting the *longest alkyl chain* as the base alkyl group. The base alkyl group is numbered beginning with the carbon atom attached to the main carbon chain. The substituents on the base alkyl group are listed with appropriate numbers, and parentheses are used to set off the name of the complex alkyl group. Two examples of such nomenclature follow.

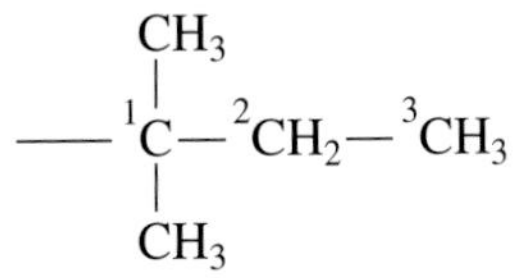

$$CH_3$$
$$\quad-\overset{1}{C}-\overset{2}{C}H_2-\overset{3}{C}H_3$$
$$CH_3$$

(1,1-Dimethylpropyl) group

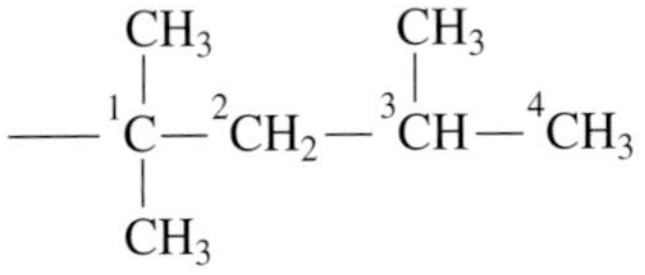

$$CH_3 \qquad CH_3$$
$$\quad-\overset{1}{C}-\overset{2}{C}H_2-\overset{3}{C}H-\overset{4}{C}H_3$$
$$CH_3$$

(1,1,3-Trimethylbutyl) group

12.11 Cycloalkanes

> It takes a minimum of three carbon atoms to form a cyclic arrangement of carbon atoms.

A **cycloalkane** *is a saturated hydrocarbon in which carbon atoms connected to one another in a cyclic (ring) arrangement are present.* The simplest cycloalkane is cyclopropane, which contains a cyclic arrangement of three carbon atoms. Figure 12.6 shows a three-dimensional model of cyclopropane's structure and those of the four-, five-, and six-carbon cycloalkanes.

Cyclopropane's three carbon atoms lie in a flat ring. In all other cycloalkane molecules, some puckering of the ring occurs; that is, the ring systems are nonplanar, as shown in Figure 12.6.

The general formula for cycloalkanes is C_nH_{2n}. Thus a given cycloalkane contains two fewer hydrogen atoms than an alkane with the same number of hydrogen atoms (C_nH_{2n+2}). Butane (C_4H_{10}) and cyclobutane (C_4H_8) are not structural isomers; structural isomers must have the same molecular formula (Section 12.6).

Line-angle drawings are often used to represent cycloalkane structures. A **line-angle drawing** *is an abbreviated structural formula representation in which an angle represents a carbon atom and a line represents a bond.* The line angle drawing for cyclopropane is a triangle, that for cyclobutane a square, that for cyclopentane a pentagon, and that for cyclohexane a hexagon.

Cyclopropane	Cyclobutane	Cyclopentane	Cyclohexane

EXAMPLE 12.4

Generating Molecular Formulas from Line-Angle Drawings

■ Determine the molecular formula for each of the following cycloalkanes and alkanes.

a.

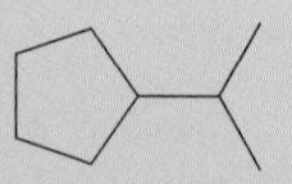

b.

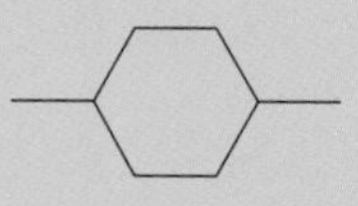

c.

d.

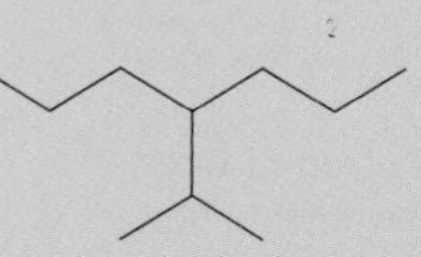

Solution

a. First replace each angle and line terminus with a carbon atom, and then add hydrogens as necessary to give each carbon four bonds. The molecular formula of this compound is C_8H_{16}.

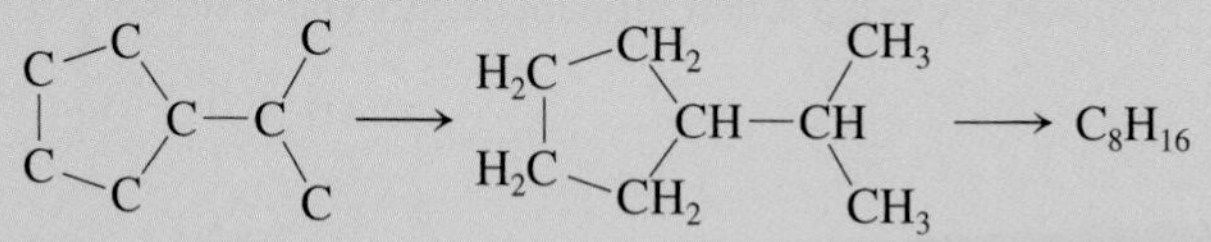

$$C_8H_{16}$$

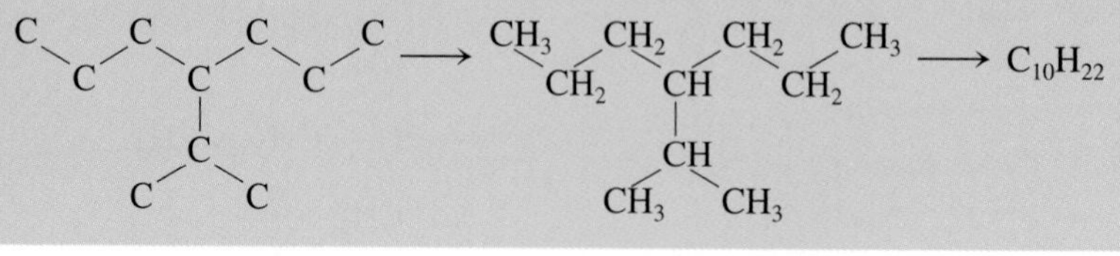

> Line-angle drawings, involving a sawtooth pattern of lines, can be used to denote alkanes as well as cycloalkanes. For alkanes, a carbon atom (with its associated hydrogen atoms) is understood to be present at every point where two lines meet and at the ends of lines. An unbranched chain of eight carbon atoms would be represented as

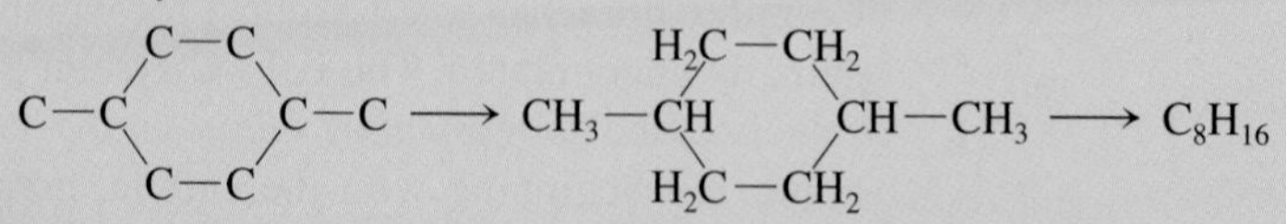

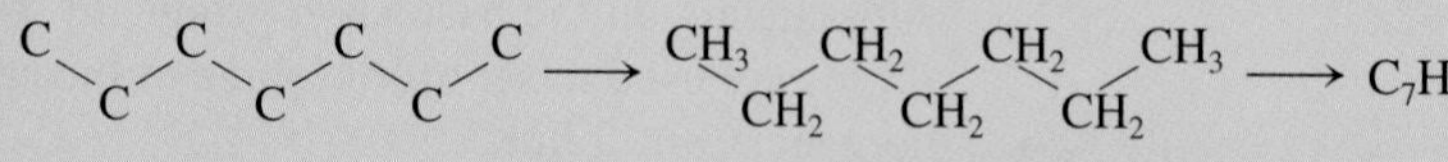

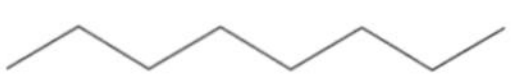

The structural formulas of branched-chain alkanes can also be abbreviated by using this notation. Such notation for 2,3,4-trimethylpentane is

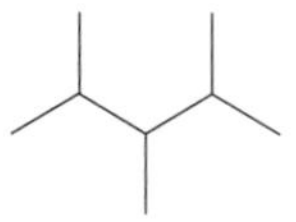

b. Similarly, we have

$$C_8H_{16}$$

c.

$$C_7H_{16}$$

d.

$$C_{10}H_{22}$$

(a) Cyclopropane, C_3H_6

(b) Cyclobutane, C_4H_8

(c) Cyclopentane, C_5H_{10}

(d) Cyclohexane, C_6H_{12}

FIGURE 12.6 Three-dimensional representations of the structures of simple cycloalkanes.

Practice Exercise 12.4
Determine the molecular formula for each of the following cycloalkanes and alkanes.

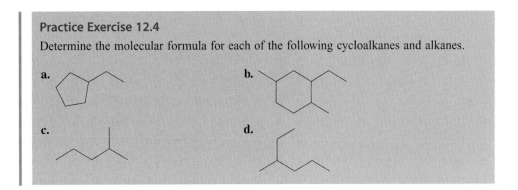

a.

b.

c.

d.

The observed C—C—C bond angles in cyclopropane are 60°, and those in cyclobutane are 90°, values that are considerably smaller than the 109° angle associated with a tetrahedral arrangement of bonds about a carbon atom (Section 5.8). Consequently, cyclopropane and cyclobutane are relatively unstable compounds. Five- and six-membered cycloalkane structures are much more stable, and this structural feature is encountered in many organic molecules.

12.12 IUPAC Nomenclature for Cycloalkanes

IUPAC naming procedures for cycloalkanes are similar to those for alkanes. The ring portion of a cycloalkane molecule serves as the name base, and the prefix *cyclo-* is used to indicate the presence of the ring. Alkyl substituents are named in the same manner as in alkanes. Numbering conventions used in locating substituents include the following:

1. If there is just one ring substituent, it is not necessary to locate it by number.
2. When two ring substituents are present, the carbon atoms in the ring are numbered beginning with the substituent of higher alphabetical priority and proceeding in the direction (clockwise or counterclockwise) that gives the other substituent the lower number.
3. When three or more ring substituents are present, ring numbering begins at the substituent that leads to the lowest set of location numbers. When two or more equivalent numbering sets exist, alphabetical priority among substituents determines the set used.

Example 12.5 illustrates the use of the ring-numbering guidelines.

▶ Cycloalkanes of ring sizes ranging from 3 to over 30 are found in nature, and in principle, there is no limit to ring size. Five-membered rings (cyclopentanes) and six-membered rings (cyclohexanes) are especially abundant in nature.

EXAMPLE 12.5

Determining IUPAC Names for Cycloalkanes

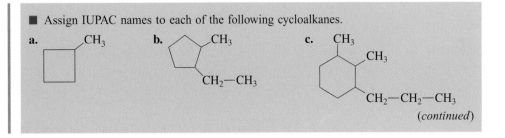

■ Assign IUPAC names to each of the following cycloalkanes.

a.

b.

c.

(continued)

Solution

a. This molecule is a cyclobutane (four-carbon ring) with a methyl substituent. The IUPAC name is simply methylcyclobutane. No number is needed to locate the methyl group, because all four ring positions are equivalent.

b. This molecule is a cyclopentane with ethyl and methyl substituents. The numbers for the carbon atoms that bear the substituents are 1 and 2. On the basis of alphabetical priority, the number 1 is assigned to the carbon atom that bears the ethyl group. The IUPAC name for the compound is 1-ethyl-2-methylcyclopentane.

c. This molecule is a dimethylpropylcyclohexane. Two different 1,2,3 numbering systems exist for locating the substituents. On the basis of alphabetical priority, we use the numbering system that has carbon 1 bearing a methyl group; methyl has alphabetical priority over propyl. Thus the compound name is 1,2-dimethyl-3-propylcyclohexane.

Practice Exercise 12.5

Assign IUPAC names to each of the following cycloalkanes.

a. 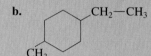   b. 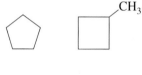   c.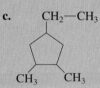

12.13 Isomerism in Cycloalkanes

Structural isomers are possible for cycloalkanes that contain four or more carbon atoms. For example, there are five cycloalkane structural isomers that have the formula C_5H_{10}: one based on a five-membered ring, one based on a four-membered ring, and three based on a three-membered ring. These isomers are

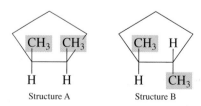

cyclopentane   methylcyclobutane   1,2-dimethyl-cyclopropane   1,1-dimethyl-cyclopropane   ethylcyclopropane

▶ *Cis–trans isomers* have the same molecular formula and the same structural formula. The only difference between them is the orientation of atoms in space. *Structural isomers* have the same molecular formula but different structural formulas.

A second type of isomerism is possible for some *substituted* cycloalkanes. This new type of isomerism is *cis–trans* isomerism. **Cis–trans** isomers *are isomers that have the same molecular and structural formulas but different arrangements of atoms in space because of restricted rotation around bonds.*

In alkanes, there is free rotation about all carbon–carbon bonds (Section 12.7). In cycloalkanes, the ring structure restricts rotation for the carbon atoms in the ring. The consequence of this lack of rotation in a cycloalkane is the creation of "top" and "bottom" positions for the two attachments on each of the ring carbon atoms. This "top–bottom" situation may lead to *cis–trans* isomerism in cycloalkanes with two or more substituents on the ring.

Consider the following two structures for the molecule 1,2-dimethylcyclopentane.

Structure A   Structure B

In structure A, both methyl groups are above the plane of the ring (the "top" side). In structure B, one methyl group is above the plane of the ring (the "top" side) and the other below it (the "bottom" side). Structure A cannot be converted into structure B without breaking bonds. Hence structures A and B are isomers; there are two 1,2-dimethylcyclopentanes. The first isomer is called *cis*-1,2-dimethylcyclopentane and the second *trans*-1,2-dimethylcyclopentane.

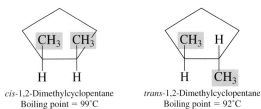

cis-1,2-Dimethylcyclopentane
Boiling point = 99°C

trans-1,2-Dimethylcyclopentane
Boiling point = 92°C

> The Latin *cis* means "on the same side," and the Latin *trans* means "across." Consider the use of the prefix *trans*- in the phrase "transatlantic voyage."

A *cis* **isomer** *is a disubstituted cycloalkane isomer in which substituents on different carbon atoms are on the same side of the carbon ring.* In our current discussion, the restricted rotation barrier is the ring of carbon atoms. A *trans* **isomer** *is a disubstituted cycloalkane isomer in which substituents on different carbon atoms are on opposite sides of the carbon ring.*

Cis–trans isomerism can occur in rings of all sizes. The presence of a substituent on each of two carbon atoms in the ring is the minimum requirement for its occurrence. In biochemistry, we will find that the human body often selectively distinguishes between the *cis* and *trans* isomers of a compound. One isomer will be active in the body and the other inactive.

EXAMPLE 12.6

Identifying and Naming Cycloalkane *Cis–Trans* Isomers

> In cycloalkanes, *cis–trans* isomerism can also be denoted by using wedges and dotted lines. A heavy wedge-shaped bond to a ring structure indicates a bond *above* the plane of the ring, and a broken dotted line indicates a bond *below* the plane of the ring.

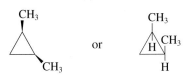

cis-1,2-Dimethylcyclopropane

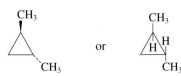

trans-1,2-Dimethylcyclopropane

■ Determine whether *cis–trans* isomerism is possible for each of the following cycloalkanes. If so, then draw structural formulas for the *cis* and *trans* isomers.

a. Methylcyclohexane
b. 1,1-Dimethylcyclohexane
c. 1,3-Dimethylcyclobutane
d. 1-Ethyl-2-methylcyclobutane

Solution

a. *Cis–trans* isomerism is not possible because we do not have two substituents on the ring.
b. *Cis–trans* isomerism is not possible. We have two substituents on the ring, but they are on the same carbon atom. Each of two different carbons must bear substituents.
c. *Cis–trans* isomerism does exist.

cis-1,3-Dimethylcyclobutane   *trans*-1,3-Dimethylcyclobutane

d. *Cis–trans* isomerism does exist.

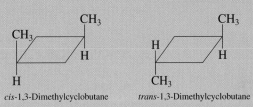

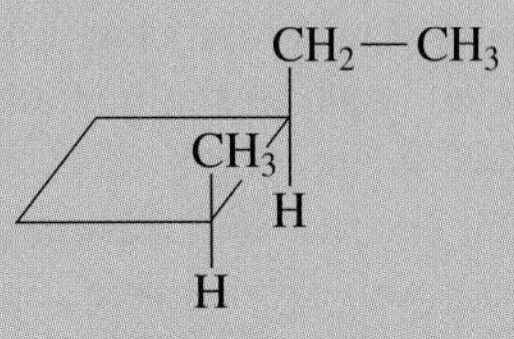

cis-1-Ethyl-2-methylcyclobutane 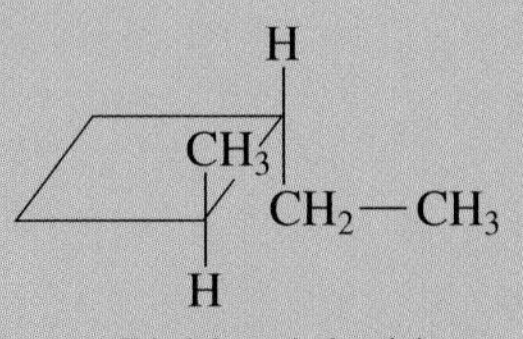   *trans*-1-Ethyl-2-methylcyclobutane

Practice Exercise 12.6

Determine whether *cis–trans* isomerism is possible for each of the following cycloalkanes. If so, then draw structural formulas for the *cis* and *trans* isomers.

a. 1-Ethyl-1-methylcyclopentane
b. Ethylcyclohexane
c. 1,4-Dimethylcyclohexane
d. 1,1-Dimethylcyclooctane

FIGURE 12.7 A rock formation such as this is necessary for the accumulation of petroleum and natural gas.

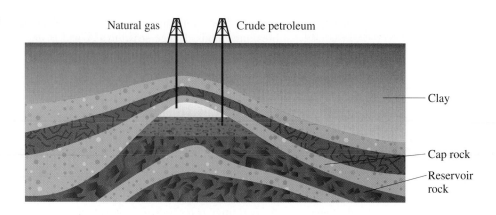

Natural gas   Crude petroleum

Clay

Cap rock

Reservoir rock

12.14 Sources of Alkanes and Cycloalkanes

The word *petroleum* comes from the Latin *petra,* which means "rock," and *oleum,* which means "oil."

Alkanes and cycloalkanes are not "laboratory curiosities" but rather two families of extremely important naturally occurring compounds. Natural gas and petroleum (crude oil) constitute their largest and most important natural source. Deposits of these resources are usually associated with underground dome-shaped rock formations (Figure 12.7). When a hole is drilled into such a rock formation, it is possible to recover some of the trapped hydrocarbons—that is, the natural gas and/or petroleum (Figure 12.8). Note that petroleum and natural gas do not occur in the earth in the form of "liquid pools" but rather are dispersed throughout a porous rock formation.

Unprocessed natural gas contains 50%–90% methane, 1%–10% ethane, and up to 8% higher-molecular-mass alkanes (predominantly propane and butanes). The higher alkanes found in crude natural gas are removed prior to release of the gas into the pipeline distribution systems. Because the removed alkanes can be liquefied by the use of moderate pressure, they are stored as liquids under pressure in steel cylinders and are marketed as bottled gas.

Crude petroleum is a complex mixture of hydrocarbons (both cyclic and acyclic) that can be separated into useful fractions through refining. During refining, the physical separation of the crude into component fractions is accomplished by fractional distillation, a process that takes advantage of boiling-point differences between the components of the crude petroleum. Each fraction contains hydrocarbons within a specific boiling-point range. The fractions obtained from a typical fractionation process are shown in Figure 12.9.

FIGURE 12.8 An oil rig pumping oil from an underground rock formation.

12.15 Physical Properties of Alkanes and Cycloalkanes

In this section, we consider a number of generalizations about the physical properties of alkanes and cycloalkanes.

1. *Alkanes and cycloalkanes are insoluble in water.* Water molecules are polar, and alkane and cycloalkane molecules are nonpolar. Molecules of unlike polarity have limited solubility in one another (Section 8.4). The water insolubility of alkanes makes them good preservatives for metals. They prevent water from reaching the metal surface and causing corrosion. They also have biological functions as protective coatings (see Figure 12.10).

2. *Alkanes and cycloalkanes have densities lower than that of water.* Alkane and cycloalkane densities fall in the range 0.6 g/mL to 0.8 g/mL, compared with water's density of 1.0 g/mL. When alkanes and cycloalkanes are mixed with water, two layers form (because of insolubility), with the hydrocarbon layer on top (because of its lower density). This density difference between alkanes/cycloalkanes and water explains why oil spills in aqueous environments spread so quickly. The *floating* oil follows the movement of the water.

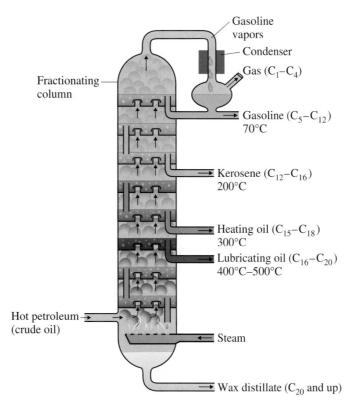

FIGURE 12.9 The complex hydrocarbon mixture present in petroleum is separated into simpler mixtures by means of a fractionating column.

FIGURE 12.10 The insolubility of alkanes in water is used to advantage by many plants, which produce unbranched long-chain alkanes that serve as protective coatings on leaves and fruits. Such protective coatings minimize water loss for plants. Apples can be "polished" because of the long-chain alkane coating on their skin, which involves the unbranched alkanes $C_{27}H_{56}$ and $C_{29}H_{60}$. The leaf wax of cabbage and broccoli is mainly unbranched $C_{29}H_{60}$.

3. *The boiling points of continuous-chain alkanes and cycloalkanes increase with an increase in carbon chain length or ring size.* For continuous-chain alkanes, the boiling point increases roughly 30°C for every carbon atom added to the chain. This trend, shown in Figure 12.11, is the result of increasing London force strength (Section 7.13). London forces become stronger as molecular surface area increases. Short, continuous-chain alkanes (1 to 4 carbon atoms) are gases at room temperature. Continuous-chain alkanes containing 5 to 17 carbon atoms are liquids, and alkanes that have carbon chains longer than this are solids at room temperature.

Branching on a carbon chain lowers the boiling point of an alkane. A comparison of the boiling points of unbranched alkanes and their 2-methyl-branched isomers is given in Figure 12.11. Branched alkanes are more compact, with smaller surface areas than their straight-chain isomers.

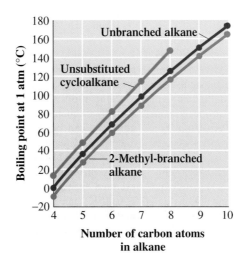

FIGURE 12.11 Trends in normal boiling points for continuous-chain alkanes, 2-methyl branched alkanes, and unsubstituted cycloalkanes as a function of the number of carbon atoms present. For a series of alkanes or cycloalkanes, melting point increases as carbon chain length increases.

Unbranched Alkanes			
C_1	C_3	C_5	C_7
C_2	C_4	C_6	C_8

Unsubstituted Cycloalkanes			
✕	C_3	C_5	C_7
✕	C_4	C_6	C_8

☐ Gas ☐ Liquid

FIGURE 12.12 A physical-state summary for unbranched alkanes and unsubstituted cycloalkanes at room temperature and pressure.

Cycloalkanes have higher boiling points than their noncyclic counterparts with the same number of carbon atoms (Figure 12.11). These differences are due in large part to cyclic systems having more rigid and more symmetrical structures. Cyclopropane and cyclobutane are gases at room temperature, and cyclopentane through cyclooctane are liquids at room temperature. Figure 12.12 is a physical-state summary for unbranched alkanes or unsubstituted cycloalkanes with 8 or fewer carbon atoms.

The alkanes and cycloalkanes whose boiling points are compared in Figure 12.11 constitute *homologous series* of organic compounds. In a homologous series, the members differ structurally only in the number of —CH_2— groups present. Members exhibit gradually changing physical properties and usually have very similar chemical properties.

The existence of homologous series of organic compounds gives organization to organic chemistry in the same way that the periodic table gives organization to the chemistry of the elements. Knowing something about a few members of a homologous series usually enables us to deduce the properties of other members in the series.

CHEMICAL CONNECTIONS

The Physiological Effects of Alkanes

The simplest alkanes (methane, ethane, propane, and butane) are gases at room temperature and pressure. Methane and ethane are difficult to liquefy, so they are usually handled as compressed gases. Propane and butane are easily liquefied at room temperature under a moderate pressure. They are stored in low-pressure cylinders in a liquefied form. These four gases are colorless, odorless, and nontoxic, and they have limited physiological effects. The danger in inhaling them lies in potential suffocation due to lack of oxygen. The major immediate danger associated with a natural gas leak is the potential formation of an explosive air–alkane mixture rather than the formation of a toxic air–alkane mixture.

The C_5 to C_8 alkanes, of which there are many isomeric forms, are free-flowing, nonpolar, volatile liquids. They are the primary constituents of gasoline. These compounds are not particularly toxic, but gasoline should not be swallowed because (1) some of the additives present are harmful and (2) liquid alkanes can damage lung tissue because of physical rather than chemical effects. Physical effects include the dissolving of lipid molecules of cell membranes (see Chapter 19), causing pneumonia-like symptoms. Liquid alkanes can also affect the skin for related reasons. These alkanes dissolve natural body oils, causing the skin to dry out. (This "drying out" effect is easily noticed when paint thinner, a mixture of hydrocarbons, is used to remove paint from the hands.)

In direct contrast to liquid alkanes, solid alkanes are used to protect the skin. Pharmaceutical-grade *petrolatum* and *mineral oil* (also called *liquid petrolatum*), obtained as products from petroleum distillation, have such a function. Petrolatum is a mixture of C_{25} to C_{30} alkanes, and mineral oil involves alkanes in the C_{18} to C_{24} range.

Petrolatum (Vaseline is a well-known brand name) is a semi-solid hydrocarbon mixture that is useful both as a skin softener and as a skin protector. Many moisturizing hand lotions and some medicated salves contain petrolatum. Neither water nor water solutions (for example, urine) will penetrate protective

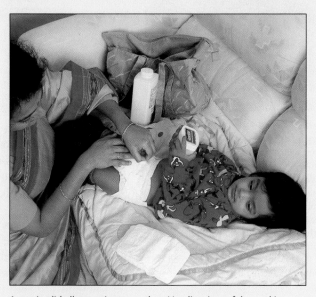

A semi-solid alkane mixture, such as Vaseline, is useful as a skin protector because neither water nor water solutions will penetrate a coating of it. Here, Vaseline is applied to a baby's bottom as a protection against diaper rash.

petrolatum coatings. This explains why petrolatum products protect a baby's bottom from diaper rash.

Mineral oil is often used to replace natural skin oils washed away by frequent bathing and swimming. Too much mineral oil, however, can be detrimental; it will dissolve nonpolar skin materials. Mineral oil has some use as a laxative; it effectively softens and lubricates hard stools. When taken by mouth, it passes through the gastrointestinal tract unchanged and is excreted chemically intact. Loss of fat-soluble vitamins (A, D, E, and K) can occur if mineral oil is consumed while these vitamins are in the digestive tract. Using a mineral oil enema instead avoids this drawback.

12.16 Chemical Properties of Alkanes and Cycloalkanes

The term *paraffins* is an older name for the alkane family of compounds. This name comes from the Latin *parum affinis,* which means "little activity." That is a good summary of the general chemical properties of alkanes.

Alkanes are the least reactive type of organic compound. They can be heated for long periods of time in strong acids and bases with no appreciable reaction. Strong oxidizing agents and reducing agents have little effect on alkanes.

Alkanes are not absolutely unreactive. Two important reactions that they undergo are combustion, which is reaction with oxygen, and halogenation, which is reaction with halogens.

■ Combustion

A **combustion reaction** *is a reaction between a substance and oxygen (usually from air) that proceeds with the evolution of heat and light (usually as a flame).* Alkanes readily undergo combustion when ignited. When sufficient oxygen is present to support total combustion, carbon dioxide and water are the products.

$$CH_4 + 2O_2 \longrightarrow CO_2 + 2H_2O + energy$$

$$2C_6H_{14} + 19O_2 \longrightarrow 12CO_2 + 14H_2O + energy$$

The exothermic nature (Section 9.5) of alkane combustion reactions explains the extensive use of alkanes as fuels. Natural gas, used in home heating, is predominantly methane. Propane is used in home heating in rural areas and in gas barbecue units (see Figure 12.13). Butane fuels portable camping stoves. Gasoline is a complex mixture of many alkanes and other types of hydrocarbons.

Incomplete combustion can occur if insufficient oxygen is present during the combustion process. When this is the case, some carbon monoxide (CO) and/or elemental carbon are reaction products along with carbon dioxide (CO_2). In a chemical laboratory setting, incomplete combustion is often observed. The appearance of deposits of carbon black (soot) on the bottom of glassware is physical evidence that incomplete combustion is occurring. The problem is that the air-to-fuel ratio for the Bunsen burner is not correct. It is too rich; it contains too much fuel and not enough oxygen (air).

FIGURE 12.13 Propane fuel tank on a home barbecue unit.

■ Halogenation

The halogens are the elements in Group VIIA of the periodic table: fluorine (F_2), chlorine (Cl_2), bromine (Br_2), and iodine (I_2) (Section 3.4). A **halogenation reaction** *is a reaction between a substance and a halogen in which one or more halogen atoms are incorporated into molecules of the substance.*

Halogenation of an alkane produces a hydrocarbon derivative in which one or more halogen atoms have been substituted for hydrogen atoms. An example of an alkane halogenation reaction is

$$\underset{\begin{matrix}|\\H\end{matrix}}{\overset{\begin{matrix}H & H\\|&|\end{matrix}}{H-C-C-H}} + Br_2 \xrightarrow[\text{light}]{\text{Heat or}} \underset{\begin{matrix}|\\H\end{matrix}}{\overset{\begin{matrix}H & H\\|&|\end{matrix}}{H-C-C-}}Br + HBr$$

Alkane halogenation is an example of a substitution reaction, a type of reaction that occurs often in organic chemistry. A **substitution reaction** *is a reaction in which part of a small reacting molecule replaces an atom or a group of atoms on a hydrocarbon or hydrocarbon derivative.* A diagrammatic representation of a substitution reaction is shown in Figure 12.14.

A *general* equation for the substitution of a single halogen atom for one of the hydrogen atoms of an alkane is

$$\underset{\text{Alkane}}{R-H} + \underset{\text{Halogen}}{X_2} \xrightarrow[\text{light}]{\text{Heat or}} \underset{\begin{matrix}\text{Halogenated}\\\text{alkane}\end{matrix}}{R-X} + \underset{\begin{matrix}\text{Hydrogen}\\\text{halide}\end{matrix}}{H-X}$$

FIGURE 12.14 In an alkane substitution reaction, an incoming atom or group of atoms (represented by the orange sphere) replaces a hydrogen atom in the alkane molecule.

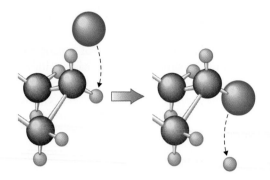

▶ Occasionally, it is useful to represent alkyl groups in a nonspecific way. The symbol R is used for this purpose. Just as *city* is a generic term for Chicago, New York, or San Francisco, the symbol R is a generic designation for any alkyl group. The symbol R comes from the German word *radikal*, which means, in a chemical context, "molecular fragment."

Note the following features of this general equation:

1. The notation R—H is a general formula for an alkane. R— in this case represents an alkyl group. Addition of a hydrogen atom to an alkyl group produces the parent hydrocarbon of the alkyl group.
2. The notation R—X on the product side is the general formula for a halogenated alkane. X is the general symbol for a halogen atom.
3. Reaction conditions are noted by placing these conditions on the equation arrow that separates reactants from products. Halogenation of an alkane requires the presence of heat or light.

(The symbol **R** is used frequently in organic chemistry and will be encountered in numerous generalized formulas in subsequent chapters; it always represents a generalized organic group in a structural formula. An R group can be an alkyl group—methyl, ethyl, propyl, etc.—or any number of other organic groups. Consider the symbol R to represent the **R**est of an organic molecule, which is not specifically specified because it is not the focal point of the discussion occurring at that time.)

In halogenation of an alkane, the alkane is said to undergo *fluorination, chlorination, bromination,* or *iodination,* depending on the identity of the halogen reactant. Chlorination and bromination are the two widely used alkane halogenation reactions. Fluorination reactions generally proceed too quickly to be useful, and iodination reactions go too slowly.

Halogenation usually results in the formation of a mixture of products rather than a single product. More than one product results because more than one hydrogen atom on an alkane can be replaced with halogen atoms. To illustrate this concept, let us consider the chlorination of methane, the simplest alkane.

Methane and chlorine, when heated to a high temperature or in the presence of light, react as follows:

$$CH_4 + Cl_2 \xrightarrow[\text{light}]{\text{Heat or}} CH_3Cl + HCl$$

The reaction does not stop at this stage, however, because the chlorinated methane product can react with additional chlorine to produce polychlorinated products.

$$CH_3Cl + Cl_2 \xrightarrow[\text{light}]{\text{Heat or}} CH_2Cl_2 + HCl$$

$$CH_2Cl_2 + Cl_2 \xrightarrow[\text{light}]{\text{Heat or}} CHCl_3 + HCl$$

$$CHCl_3 + Cl_2 \xrightarrow[\text{light}]{\text{Heat or}} CCl_4 + HCl$$

By controlling the reaction conditions and the ratio of chlorine to methane, it is possible to *favor* formation of one or another of the possible chlorinated methane products.

Properties of Alkanes and Cycloalkanes

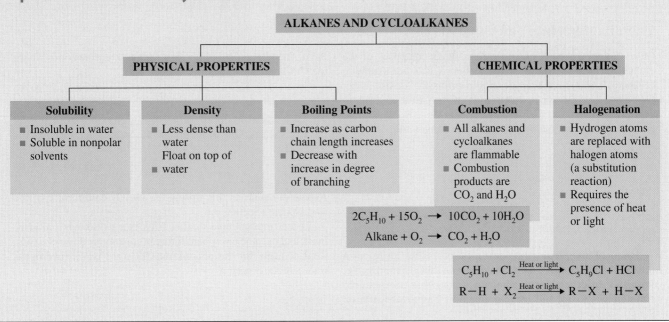

ALKANES AND CYCLOALKANES

PHYSICAL PROPERTIES

CHEMICAL PROPERTIES

Solubility
- Insoluble in water
- Soluble in nonpolar solvents

Density
- Less dense than water
  Float on top of
- water

Boiling Points
- Increase as carbon chain length increases
- Decrease with increase in degree of branching

Combustion
- All alkanes and cycloalkanes are flammable
- Combustion products are CO_2 and H_2O

Halogenation
- Hydrogen atoms are replaced with halogen atoms (a substitution reaction)
- Requires the presence of heat or light

$$2C_5H_{10} + 15O_2 \rightarrow 10CO_2 + 10H_2O$$
$$\text{Alkane} + O_2 \rightarrow CO_2 + H_2O$$

$$C_5H_{10} + Cl_2 \xrightarrow{\text{Heat or light}} C_5H_9Cl + HCl$$
$$R-H + X_2 \xrightarrow{\text{Heat or light}} R-X + H-X$$

> The contrast between IUPAC and common names for halogenated hydrocarbons is as follows:
>
> IUPAC (one word)
> boxed: haloalkane
> chloromethane
> Common (two words)
> boxed: alkyl halide
> methyl chloride

> An alternative designation for a halogenated alkane is *alkyl halide*.

> Halogenated alkane boiling points are generally higher than those of the corresponding alkane. An important factor contributing to this effect is the polarity of carbon–halogen bonds, which results in increased dipole–dipole interactions.

The chemical properties of cycloalkanes are similar to those of alkanes. Cycloalkanes readily undergo combustion as well as chlorination and bromination. With unsubstituted cycloalkanes, monohalogenation produces a single product, because all hydrogen atoms present in the cycloalkane are equivalent to one another.

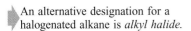

Figure 12.15 shows space-filling models of the four ethyl halides. Note the large differences in size associated with the various halogen atoms. Fluorine is the smallest and iodine the largest.

The Chemistry at a Glance feature on this page summarizes the properties of alkanes and cycloalkanes.

12.17 Nomenclature and Properties of Halogenated Alkanes

A **halogenated alkane** *is an alkane derivative in which one or more halogen atoms are present.* Similarly, a **halogenated cycloalkane** *is a cycloalkane derivative in which one or more halogen atoms are present.* Produced by halogenation reactions (Section 12.16), these two types of compounds represent the first class of hydrocarbon derivatives (Section 12.3) that we formally consider in this text.

FIGURE 12.15 Space-filling models of the four ethyl halides.

Ethyl fluoride

Ethyl chloride

Ethyl bromide

Ethyl iodide

CHEMICAL CONNECTIONS

Chlorofluorocarbons and the Ozone Layer

Chlorofluorocarbons (CFCs) are compounds composed of the elements chlorine, fluorine, and carbon. CFCs are synthetic compounds that have been developed primarily for use as refrigerants. The two most widely used of the CFCs are trichlorofluoromethane and dichlorodifluoromethane. Both of these compounds are marketed under the trade name Freon.

Cl
|
Cl—C—F
|
Cl
Trichlorofluoromethane
(Freon-11)

F
|
Cl—C—F
|
Cl
Dichlorodifluoromethane
(Freon-12)

Freon-11 and Freon-12 possess ideal properties for use as a refrigerant gas. Both are inert, nontoxic, and easily compressible. Prior to their development, ammonia was used in refrigeration. Ammonia is toxic, and leaking ammonia-based refrigeration units have been fatal.

We now know that CFCs contribute to a serious environmental problem: destruction of the stratospheric (high-altitude) ozone that we commonly call the ozone layer. Once released into the atmosphere, CFCs persist for long periods without reaction. Consequently, they slowly drift upward in the atmosphere, finally reaching the stratosphere.

It is in the stratosphere, the location of the "ozone layer," that environmental problems occur. At these high altitudes, the CFCs are exposed to ultraviolet light (from the sun), which activates

them. The ultraviolet light breaks carbon–chlorine bonds within the CFCs, releasing chlorine atoms.

$$CCl_2F_2 + \text{ultraviolet light} \longrightarrow CClF_2 + Cl$$

The Cl atoms so produced (called atomic chlorine) are extremely reactive species. One of the molecules with which they react is ozone (O_3).

$$Cl + O_3 \longrightarrow ClO + O_2$$

A reaction such as this upsets the O_3–O_2 equilibrium in the stratosphere (Section 9.6).

Recent international treaties (this is a worldwide problem) limit, and in some cases ban, future production and use of CFCs. Replacements for the phased-out CFCs are HFCs (hydrogen-fluorocarbons) such as

F   F
| |
F—C—C—H
| |
F   H
1,1,1,2-Tetrafluoroethane

Haloalkanes with some carbon–hydrogen bonds are more reactive than CFCs and are generally destroyed at lower altitudes before they reach the stratosphere. Unfortunately, however, their refrigeration properties are not as good as those of the CFCs.

■ Nomenclature of Halogenated Alkanes

The IUPAC rules for naming halogenated alkanes are similar to those for naming branched alkanes, with the following modifications:

1. Halogen atoms, treated as substituents on a carbon chain, are called *fluoro-, chloro-, bromo-,* and *iodo-.*
2. When a carbon chain bears both a halogen and an alkyl substituent, the two substituents are considered of equal rank in determining the numbering system for the chain. The chain is numbered from the end closer to a substituent, whether it be a *halo-* or an alkyl group.
3. Alphabetical priority determines the order in which all substituents present are listed.

▶ Some halogenated alkanes have densities that are greater than that of water, a situation not common for organic compounds. Chloroalkanes containing two or more chlorine atoms, bromoalkanes, and iodoalkanes are all more dense than water.

The following names are derived using these rule adjustments.

CH₃—CH—CH—CH₃
        | |
        Cl   CH₃
2-Chloro-3-methylbutane

CH₃—CH—CH₂—CH₂
        | |
        Br   Cl
3-Bromo-1-chlorobutane

CH₂—CH₃
|
F
1-Ethyl-2-fluorocyclohexane

A common (non-IUPAC) system also exists for naming simple halogenated alkanes. These common names have two parts. The first part is the name of the hydrocarbon portion of the molecule (the alkyl group). The second part (as a separate word) identifies the halogen portion, which is named as if it were an ion (chloride, bromide, and so on), even though no ions are present (all bonds are covalent bonds). The following examples contrast the IUPAC names and the common names (in parentheses) of selected halogenated alkanes.

$$CH_3—CH_2—Cl \qquad CH_3—CH_2—CH_2—Br \qquad CH_3—\underset{\underset{Cl}{|}}{CH}—CH_3$$

Chloroethane   1-Bromopropane   2-Chloropropane
(ethyl chloride)   (propyl bromide)   (isopropyl chloride)

TABLE 12.5
Additional Common Names for Selected Polyhalogenated Methanes

CH_2Cl_2	methylene chloride
$CHCl_3$	chloroform
CCl_4	carbon tetrachloride
CCl_3F	Freon-11
CCl_2F_2	Freon-12

A number of polyhalogenated methanes have acquired additional common names that are usually not clearly related to their structures. Table 12.5 gives five important examples of this additional nomenclature. The compounds Freon-11 and Freon-12, the last two entries in Table 12.5, are examples of chlorofluorocarbons (CFCs). CFCs are synthetic compounds that have been heavily used as refrigerants and as air conditioning chemicals. We now know that CFCs are factors in the destruction of stratospheric (high-altitude) ozone.

CONCEPTS TO REMEMBER

Carbon atom bonding characteristics. Carbon atoms in organic compounds must have four bonds.

Types of hydrocarbons. Hydrocarbons, binary compounds of carbon and hydrogen, are of two types: saturated and unsaturated. In saturated hydrocarbons, all carbon–carbon bonds are single bonds. Unsaturated hydrocarbons have one or more carbon–carbon multiple bonds—double bonds, triple bonds, or both.

Alkanes. Alkanes are saturated hydrocarbons in which the carbon atom arrangement is that of an unbranched or branched chain. The formulas of all alkanes can be represented by the general formula C_nH_{2n+2}, where n is the number of carbon atoms present.

Structural formulas. Structural formulas are two-dimensional representations of the arrangement of the atoms in molecules. These formulas give complete information about the arrangement of the atoms in a molecule but not the spatial orientation of the atoms. Two types of structural formulas are commonly encountered: expanded and condensed.

Structural isomerism. Structural isomers are two or more compounds that have the same molecular formula but different structural formulas—that is, different arrangements of atoms within the molecule.

Conformations. Conformations are differing orientations of the same molecule made possible by free rotation about single bonds in the molecule.

Alkane nomenclature. The IUPAC name for an alkane is based on the longest continuous chain of carbon atoms in the molecule. A group of carbon atoms attached to the chain is an alkyl group. Both the position and the identity of the alkyl group are prefixed to the name of the longest carbon chain.

Cycloalkanes. Cycloalkanes are saturated hydrocarbons in which at least one cyclic arrangement of carbon atoms is present. The formulas of all cycloalkanes can be represented by the general formula C_nH_{2n}, where n is the number of carbon atoms present.

Cycloalkane nomenclature. The IUPAC name for a cycloalkane is obtained by placing the prefix *cyclo-* before the alkane name that corresponds to the number of carbon atoms in the ring. Alkyl groups attached to the ring are located by using a ring-numbering system.

Cis–trans isomerism. For certain disubstituted cycloalkanes, *cis–trans* isomers exist. *Cis–trans* isomers are compounds that have the same molecular and structural formulas but different arrangements of atoms in space because of restricted rotation about bonds.

Natural sources of saturated hydrocarbons. Natural gas and petroleum are the largest and most important natural sources of both alkanes and cycloalkanes.

Physical properties of saturated hydrocarbons. Saturated hydrocarbons are not soluble in water and have lower densities than water. Melting and boiling points increase with increasing carbon chain length or ring size.

Chemical properties of saturated hydrocarbons. Two important reactions that saturated hydrocarbons undergo are combustion and halogenation. In combustion, saturated hydrocarbons burn in air to produce CO_2 and H_2O. Halogenation is a substitution reaction in which one or more hydrogen atoms of the hydrocarbon are replaced by halogen atoms.

Halogenated alkanes. Halogenated alkanes are hydrocarbon derivatives in which one or more halogen atoms have replaced hydrogen atoms of the alkane.

Halogenated alkane nomenclature. Halogenated alkanes are named by using the rules that apply to branched-chain alkanes, with halogen substituents being treated the same as alkyl groups.

KEY REACTIONS AND EQUATIONS

1. Combustion (rapid reaction with O_2) of alkanes (Section 12.16)

$$\text{Alkane} + O_2 \longrightarrow CO_2 + H_2O$$

2. Halogenation of alkanes (Section 12.16)

$$R—H + X_2 \xrightarrow[\text{light}]{\text{Heat or}} R—X + H—X$$

KEY TERMS

Alkane (12.4)
Alkyl group (12.8)
Branched-chain alkane (12.6)
Cis isomer (12.13)
Cis–trans isomers (12.13)
Combustion reaction (12.16)
Condensed structural formula (12.5)
Conformation (12.7)
Continuous-chain alkane (12.6)
Cycloalkane (12.11)

Expanded structural formula (12.5)
Halogenated alkane (12.17)
Halogenated cycloalkane (12.17)
Halogenation reaction (12.16)
Hydrocarbon (12.3)
Hydrocarbon derivative (12.3)
Inorganic chemistry (12.1)
Line-angle drawing (12.1)
Organic chemistry (12.1)
Primary carbon atom (12.9)

Quaternary carbon atom (12.9)
Saturated hydrocarbon (12.3)
Secondary carbon atom (12.9)
Structural formula (12.5)
Structural isomers (12.6)
Substituent (12.8)
Substitution reaction (12.16)
Tertiary carbon atom (12.9)
Trans isomer (12.13)
Unsaturated hydrocarbon (12.3)

EXERCISES AND PROBLEMS

The members of each pair of problems in this section test similar material.

■ **Organic and Inorganic Compounds (Section 12.1)**

12.1 Indicate whether each of the following statements is true or false.
   a. The number of organic compounds exceeds the number of inorganic compounds by a factor of about 2.
   b. Chemists now believe that a special "vital force" is needed to form an organic compound.
   c. Historically, the *org-* of the term *organic* was conceptually paired with the *org-* in the term *living organism*.
   d. Most but not all compounds found in living organisms are organic compounds.

12.2 Indicate whether each of the following statements in true or false.
   a. Over 5 million organic compounds have been characterized.
   b. The number of known organic compounds and the number of known inorganic compounds are approximately the same.
   c. In essence, organic chemistry is the study of the compounds of one element.
   d. Numerous organic compounds are known that do not occur in living organisms.

■ **Bonding Characteristics of the Carbon Atom (Section 12.2)**

12.3 Indicate whether each of the following situations meet or do not meet the "bonding requirement" for carbon atoms.
   a. Two single bonds and a double bond
   b. A single bond and two double bonds
   c. Three single bonds and a triple bond
   d. A double bond and a triple bond

12.4 Indicate whether each of the following situations meet or do not meet the "bonding requirement" for carbon atoms.
   a. Four single bonds
   b. Three single bonds and a double bond
   c. Two double bonds and two single bonds
   d. Two double bonds

■ **Hydrocarbons and Hydrocarbon Derivatives (Section 12.3)**

12.5 What is the difference between a *hydrocarbon* and a *hydrocarbon derivative*?

12.6 Contrast hydrocarbons and hydrocarbon derivatives in terms of number of compounds that are known.

12.7 What is the difference between a *saturated hydrocarbon* and an *unsaturated hydrocarbon*?

12.8 What structural feature is present in an unsaturated hydrocarbon that is not present in a saturated hydrocarbon?

12.9 Classify each of the following hydrocarbons as saturated or unsaturated.

12.10 Classify each of the following hydrocarbons as saturated or unsaturated.

■ **Formulas for Alkanes (Section 12.4)**

12.11 Using the general formula for an alkane, derive the following for specific alkanes.
   a. Number of hydrogen atoms present when 8 carbon atoms are present
   b. Number of carbon atoms present when 10 hydrogen atoms are present
   c. Number of carbon atoms present when 41 total atoms are present
   d. Total number of covalent bonds present in the molecule when 7 carbon atoms are present

12.12 Using the general formula for an alkane, derive the following for specific alkanes.
   a. Number of carbon atoms present when 14 hydrogen atoms are present
   b. Number of hydrogen atoms present when 6 carbon atoms are present
   c. Number of hydrogen atoms present when 32 total atoms are present

d. Total number of covalent bonds present in the molecule when 16 hydrogen atoms are present

Structural Formulas (Section 12.6)

12.13 Convert each of the following expanded structural formulas into a condensed structural formula.

a.
$$H-\overset{\overset{\displaystyle H}{|}}{\underset{\underset{\displaystyle H}{|}}{C}}-\overset{\overset{\displaystyle H}{|}}{\underset{\underset{\displaystyle H}{|}}{C}}-\overset{\overset{\displaystyle H}{|}}{\underset{\underset{\displaystyle H}{|}}{C}}-\overset{\overset{\displaystyle H}{|}}{\underset{\underset{\displaystyle H}{|}}{C}}-H$$

b.
$$H-\overset{H}{\underset{H}{C}}-\overset{H}{\underset{H}{C}}-\overset{H}{\underset{\underset{H-\overset{H}{\underset{H}{C}}-H}{|}}{C}}-\overset{H}{\underset{H}{C}}-\overset{H}{\underset{H}{C}}-H$$

c. (expanded structural formula)

d.
$$H-\overset{H}{\underset{H}{C}}-\overset{H}{\underset{H}{C}}-\overset{H}{\underset{\underset{H-\overset{H}{\underset{H-\overset{H}{\underset{H}{C}}-H}{C}}-H}{|}}{C}}-\overset{H}{\underset{H}{C}}-\overset{H}{\underset{H}{C}}-H$$

12.14 Convert each of the following expanded structural formulas into a condensed structural formula.

a.
$$H-\overset{H}{\underset{H}{C}}-\overset{H}{\underset{H}{C}}-\overset{H}{\underset{H}{C}}-\overset{H}{\underset{H}{C}}-\overset{H}{\underset{H}{C}}-\overset{H}{\underset{H}{C}}-H$$

b. (expanded structural formula)

c. (expanded structural formula)

d. (expanded structural formula)

12.15 The following structural formulas for alkanes are incomplete in that the hydrogen atoms attached to each carbon are not shown. Complete each of these formulas by writing in the correct number of hydrogen atoms attached to each carbon atom. That is, rewrite each of these formulas as a condensed structural formula such as $CH_3-CH_2-CH_3$.

a. C—C—C—C
      |
      C

b. C—C—C—C—C—C
      | | |
      C C C

c. C—C—C—C—C—C

d.
$$C-\overset{\overset{\displaystyle C}{|}}{\underset{\underset{\displaystyle C}{|}}{C}}-C-C$$

12.16 The following structural formulas for alkanes are incomplete in that the hydrogen atoms attached to each carbon are not shown. Complete each of these formulas by writing in the correct number of hydrogen atoms attached to each carbon atom. That is, rewrite each of these formulas as a condensed structural formula such as $CH_3-CH_2-CH_3$.

a. C—C—C—C—C
     | |
     C   C

b.
$$C-\overset{\overset{\displaystyle C}{|}}{\underset{\underset{\displaystyle C}{|}}{C}}-C$$

c. C—C—C—C—C

d. C—C—C—C—C
       | |
       C C
           |
           C

12.17 Draw the indicated type of formula for the following alkanes.
  a. The expanded structural formula for a continuous-chain alkane with the formula C_5H_{12}
  b. The expanded structural formula for $CH_3-(CH_2)_6-CH_3$
  c. The condensed structural formula, using parentheses for the $-CH_2-$ groups, for the continuous-chain alkane $C_{10}H_{22}$
  d. The molecular formula for the alkane $CH_3-(CH_2)_4-CH_3$

12.18 Draw the indicated type of formula for the following alkanes.
  a. The expanded structural formula for a continuous-chain alkane with the molecular formula C_6H_{14}
  b. The condensed structural formula, using parentheses for the $-CH_2-$ groups, for the straight-chain alkane $C_{12}H_{26}$
  c. The molecular formula for the alkane $CH_3-(CH_2)_6-CH_3$
  d. The expanded structural formula for $CH_3-(CH_2)_3-CH_3$

Structural Isomers and Molecular Conformations (Sections 12.6 and 12.7)

12.19 For each of the following pairs of structures, determine whether they are
  1. Different conformations of the same molecule
  2. Different compounds that are structural isomers

3. Different compounds that are not structural isomers

a. 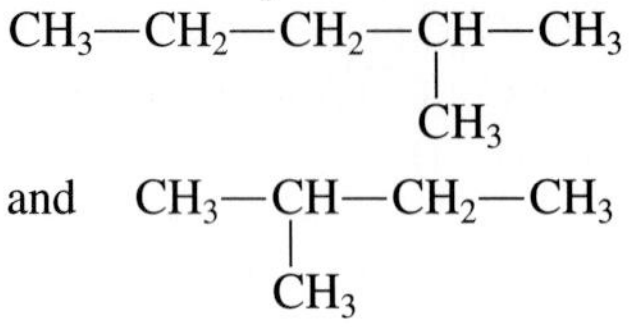 CH₃—CH₂—CH₂—CH—CH₃
                     |
                     CH₃

and   CH₃—CH—CH₂—CH₃
                |
                CH₃

b. 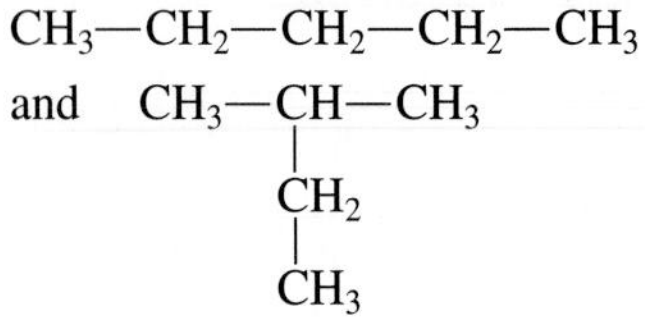 CH₃—CH₂—CH₂—CH₂—CH₃

and   CH₃—CH—CH₃
                |
                CH₂
                |
                CH₃

c. 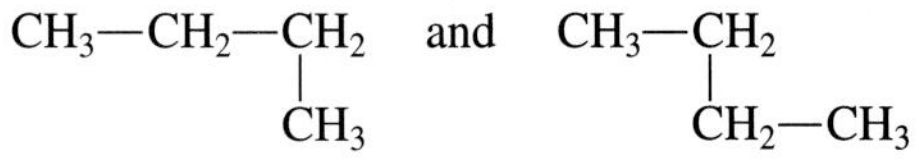 CH₃—CH₂—CH₂ and CH₃—CH₂
               | |
               CH₃   CH₂—CH₃

d. 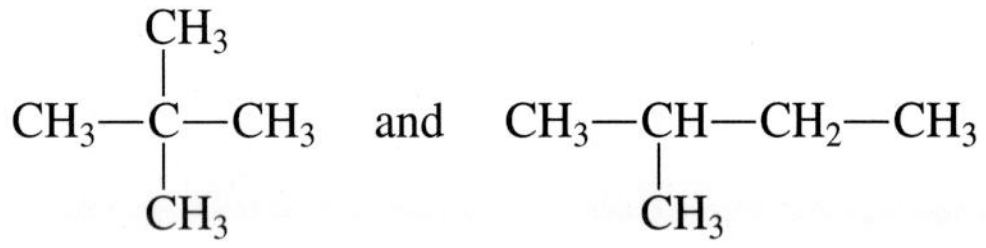   CH₃
          |
CH₃—C—CH₃ and CH₃—CH—CH₂—CH₃
      | |
      CH₃   CH₃

12.20 For each of the following pairs of structures, determine whether they are

1. Different conformations of the same molecule
2. Different compounds that are structural isomers
3. Different compounds that are not structural isomers

a. 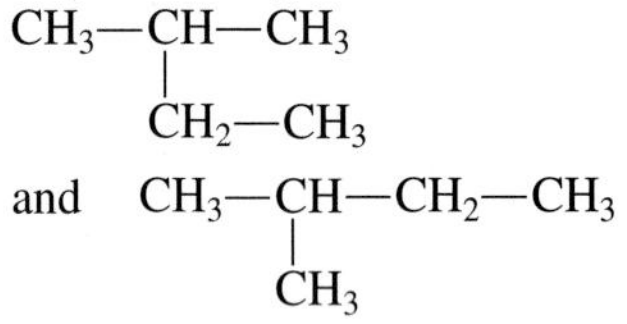 CH₃—CH—CH₃
         |
         CH₂—CH₃

and   CH₃—CH—CH₂—CH₃
                |
                CH₃

b. 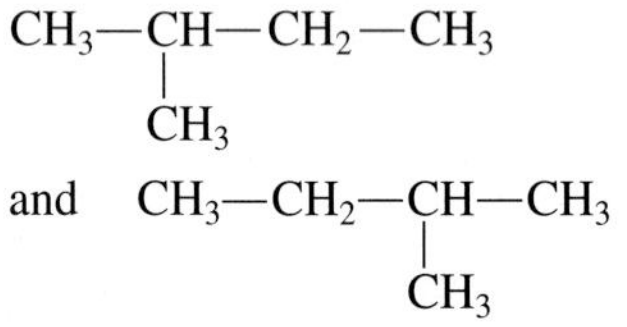 CH₃—CH—CH₂—CH₃
         |
         CH₃

and   CH₃—CH₂—CH—CH₃
                   |
                  CH₃

c. 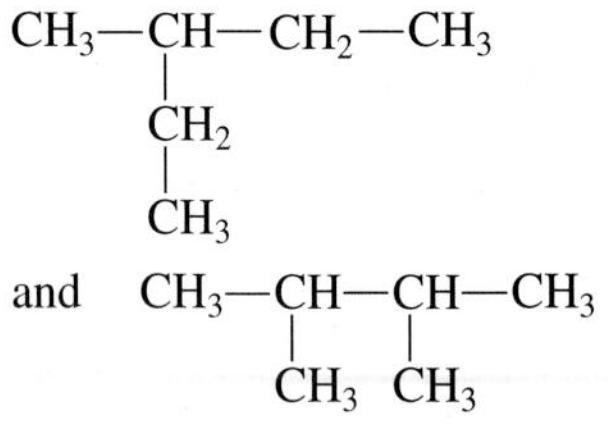 CH₃—CH—CH₂—CH₃
         |
         CH₂
         |
         CH₃

and   CH₃—CH—CH—CH₃
                | |
                CH₃ CH₃

d. 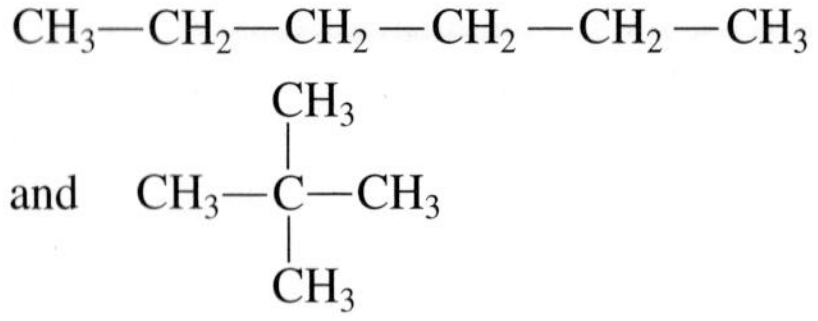 CH₃—CH₂—CH₂—CH₂—CH₂—CH₃
                     CH₃
                     |
and   CH₃—C—CH₃
                 |
                 CH₃

■ IUPAC Nomenclature for Alkanes (Section 12.8)

12.21 The first step in naming an alkane is to identify the longest continuous chain of carbon atoms. For each of the following skeletal carbon arrangements, how many carbon atoms are present in the longest continuous chain?

a.

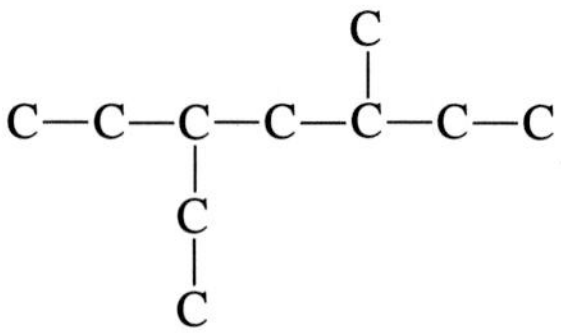

b.

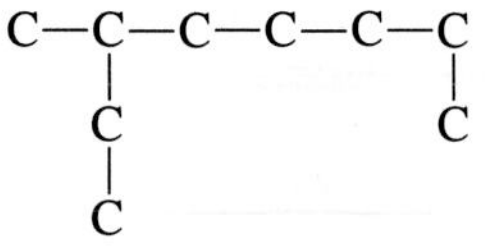

c.

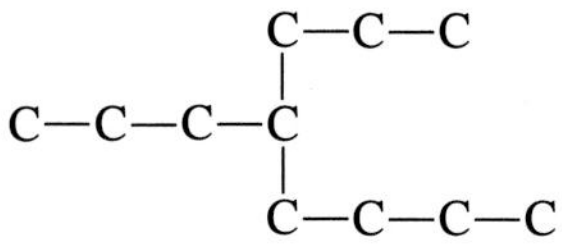

d.

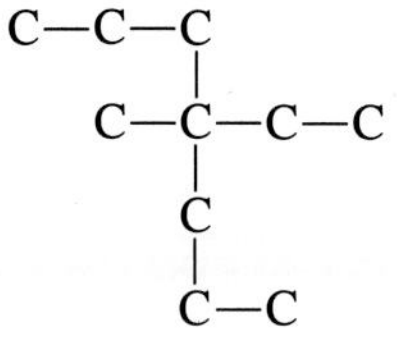

12.22 The first step in naming an alkane is to identify the longest continuous chain of carbon atoms. For each of the following skeletal carbon arrangements, how many carbon atoms are present in the longest continuous chain?

a.

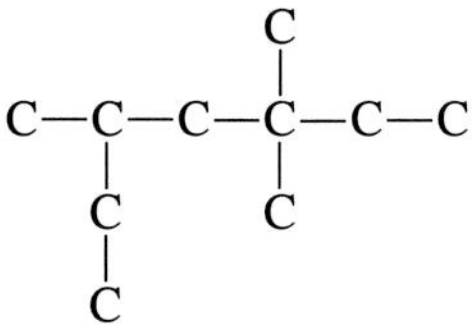

b.

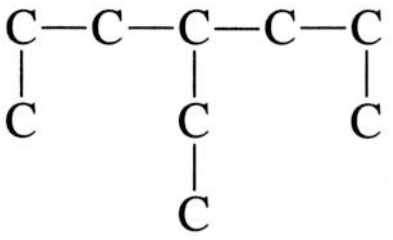

c.

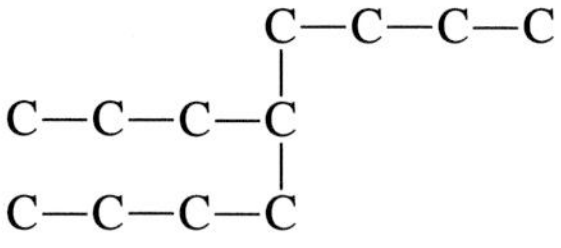

d.

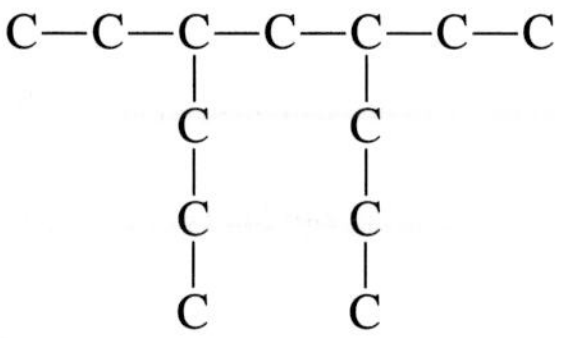

12.23 Give the IUPAC name for each of the following alkanes.

a. 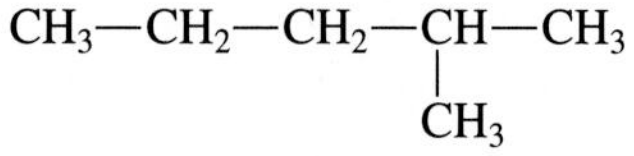 CH₃—CH₂—CH₂—CH—CH₃
                         |
                        CH₃

b. 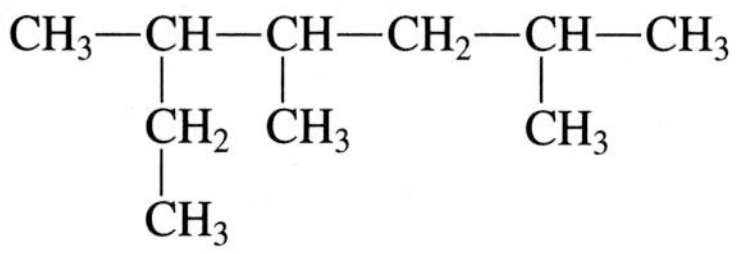 CH₃—CH—CH—CH₂—CH—CH₃
          | | |
          CH₂ CH₃   CH₃
          |
          CH₃

c.

$$CH_3-CH-\overset{\overset{\displaystyle CH_3}{|}}{\underset{\underset{\displaystyle CH_2}{|}}{\underset{|}{C}}}-CH_2-CH_3$$
$$CH_3 \quad CH_2$$
$$CH_3$$

d. $CH_3-CH-CH-CH-CH_2-CH_3$
$\quad\quad\;\; CH_3 \;\; CH_2 \;\; CH_3$
$\quad\quad\quad\quad\quad CH_3$

e. $CH_3-(CH_2)_8-CH_3$

f. $CH_2-CH_2-CH-CH_2-CH_2$
$\quad CH_3 \quad\quad CH_2 \quad CH_3$
$\quad\quad\quad\quad\quad CH_2$
$\quad\quad\quad\quad\quad CH_3$

12.24 Give the IUPAC name for each of the following alkanes.

a. $CH_3-CH-CH-CH_2-CH_3$
$\quad\quad\;\; CH_3 \;\; CH_3$

b.
$$CH_3-\overset{\overset{\displaystyle CH_3}{|}}{\underset{\underset{\displaystyle CH_3}{|}}{C}}-CH_2-\overset{\overset{\displaystyle CH_3}{|}}{\underset{\underset{\displaystyle CH_3}{|}}{C}}-CH_3$$

c. $CH_3-CH-CH_2-CH-CH_3$
$\quad\quad\;\; CH_3 \quad\quad CH_2$
$\quad\quad\quad\quad\quad\;\; CH_3$

d. $CH_3-CH_2-CH_2-CH-CH_3$
$\quad\quad\quad\quad\quad\;\; CH_3-CH_2-CH_3$

e. $CH_3-(CH_2)_7-CH_3$

f.
$\quad\quad\quad\quad\quad\quad\quad\quad CH_3$
$CH_3-(CH_2)_3-CH-CH-(CH_2)_3-CH_3$
$\quad\quad\quad\quad\quad\quad CH_2$
$\quad\quad\quad\quad\quad\quad CH_2$
$\quad\quad\quad\quad\quad\quad CH_3$

12.25 Two different carbon chains of eight atoms can be located in the following alkane.

$\quad\quad\quad\quad\quad\quad\quad CH_2-CH_2-CH_3$
$CH_3-CH_2-CH_2-CH_2-CH-CH-CH_2-CH_3$
$\quad\quad\quad\quad\quad\quad\quad\quad\quad CH_3$

Which of these chains should be used in determining the IUPAC name for the alkane? Explain your answer.

12.26 Two different carbon chains of seven atoms can be located in the following alkane.

$CH_3-CH-CH_2-CH-CH_2-CH-CH_3$
$\quad\quad CH_3 \quad\quad CH_2 \quad\quad CH_3$
$\quad\quad\quad\quad\quad\; CH_2-CH_3$

Which of these chains should be used in determining the IUPAC name for the alkane? Explain your answer.

12.27 Draw a condensed structural formula for each of the following alkanes.
a. 2-Methylbutane
b. 3,4-Dimethylhexane
c. 3-Ethyl-3-methylpentane
d. 2,3,4,5-Tetramethylheptane
e. 3,5-Diethyloctane
f. 4-Propylnonane

12.28 Draw a condensed structural formula for each of the following alkanes.
a. 3-Methylhexane
b. 2,4-Dimethylhexane
c. 5-Propyldecane
d. 2,3,4-Trimethyloctane
e. 3-Ethyl-3-methylheptane
f. 3,3,4,4-Tetramethylheptane

12.29 For each of the alkanes in Problem 12.27, determine (a) the number of alkyl groups present and (b) the number of substituents present.

12.30 For each of the alkanes in Problem 12.28 determine (a) the number of alkyl groups present and (b) the number of substituents present.

12.31 Explain why the name given for each of the following alkanes is not the correct IUPAC name. Then give the correct IUPAC name for the compound.
a. 4-Methylpentane
b. 2-Ethyl-2-methylpropane
c. 2,3,3-Trimethylbutane
d. 1,2,2-Trimethylpentane
e. 3-Methyl-4-ethylhexane
f. 2-Methyl-4-methylhexane

12.32 Explain why the name given for each of the following alkanes is not the correct IUPAC name. Then give the correct IUPAC name for the compound.
a. 1,6-Dimethylhexane
b. 2-Ethylpentane
c. 3,3,4-Trimethylpentane
d. 2-Ethyl-4-methylhexane
e. 3-Ethyl-4-ethylhexane
f. 3,4,5,5,6-Pentamethylhexane

■ **Classification of Carbon Atoms (Section 12.9)**

12.33 For each of the alkane structures in Problem 12.23, give the number of (a) primary, (b) secondary, (c) tertiary, and (d) quaternary carbon atoms present.

12.34 For each of the alkane structures in Problem 12.24, give the number of (a) primary, (b) secondary, (c) tertiary, and (d) quaternary carbon atoms present.

■ **Branched-Chain Alkyl Groups (Section 12.10)**

12.35 Give the name of the branched alkyl group attached to each of the following carbon chains, where the carbon chain is denoted by a horizontal line.

a. _____
$\quad\;\; CH-CH_3$
$\quad\;\; CH_3$

b. _____
$\quad\;\; CH_2$
$CH_3-CH-CH_3$

c. _____
$CH_3-CH-CH_3$

d. _____
$\quad\;\; CH-CH_3$
$\quad\;\; CH_2$
$\quad\;\; CH_3$

12.36 Give the name of the branched alkyl group attached to each of the following carbon chains, where the carbon chain is denoted by a horizontal line.

a.
$$CH_2$$
$$|$$
$$CH—CH_3$$
$$|$$
$$CH_3$$

b.
$$CH_3—C—CH_3$$
$$|$$
$$CH_3$$

c.
$$CH_3—CH_2—CH—CH_3$$

d.
$$CH$$
$$CH_3 \quad CH_3$$

12.37 Draw condensed structural formulas for the following branched alkanes.
a. 5-(*sec*-Butyl)decane
b. 4,4-Diisopropyloctane
c. 5-Isobutyl-2,3-dimethylnonane
d. 4-(*tert*-Butyl)-3-methylheptane

12.38 Draw condensed structural formulas for the following branched alkanes.
a. 5-Isobutylnonane
b. 4,4-Di(*sec*-butyl)decane
c. 4-(*tert*-Butyl)-3,3-diethylheptane
d. 4-Isopropyl-2,3-dimethyloctane

■ **Cycloalkanes (Sections 12.11 through 12.13)**

12.39 Using the general formula for a cycloalkane, derive the following for specific cycloalkanes.
a. Number of hydrogen atoms present when 8 carbon atoms are present
b. Number of carbon atoms present when 12 hydrogen atoms are present
c. Number of carbon atoms present when a total of 15 atoms are present
d. Number of covalent bonds present when 5 carbon atoms are present

12.40 Using the general formula for a cycloalkane, derive the following for specific cycloalkanes.
a. Number of hydrogen atoms present when 4 carbon atoms are present
b. Number of carbon atoms present when 6 hydrogen atoms are present
c. Number of hydrogen atoms present when a total of 18 atoms are present
d. Number of covalent bonds present when 8 hydrogen atoms are present

12.41 What is the molecular formula for each of the following cycloalkane molecules?

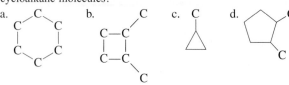

12.42 What is the molecular formula for each of the following cycloalkane molecules?

12.43 Assign an IUPAC name to each of the cycloalkanes in Problem 12.41.

12.44 Assign an IUPAC name to each of the cycloalkanes in Problem 12.42.

12.45 What is wrong with each of the following attempts to name a cycloalkane using IUPAC rules?
a. Dimethylcyclohexane
b. 3,4-Dimethylcyclohexane
c. 1-Ethylcyclobutane
d. 2-Ethyl-1-methylcyclopentane

12.46 What is wrong with each of the following attempts to name a cycloalkane using IUPAC rules?
a. Dimethylcyclopropane
b. 1-Methylcyclohexane
c. 2,5-Dimethylcyclobutane
d. 1-Propyl-2-ethylcyclohexane

12.47 Write structural formulas, with line-angle drawings denoting ring structure, for the following cycloalkanes.
a. Propylcyclobutane
b. Isopropylcyclobutane
c. *cis*-1,2-Diethylcyclohexane
d. *trans*-1-Ethyl-3-propylcyclopentane

12.48 Write structural formulas, with line-angle drawings denoting ring structure, for the following cycloalkanes.
a. Butylcyclopentane
b. Isobutylcyclopentane
c. *cis*-1,3-Diethylcyclopentane
d. *trans*-1-Ethyl-4-methylcyclohexane

12.49 Determine whether *cis–trans* isomerism is possible for each of the following cycloalkanes. If it is, then draw structural formulas for the *cis* and *trans* isomers.
a. Isopropylcyclobutane
b. 1,2-Diethylcyclopropane
c. 1-Ethyl-1-propylcyclopentane
d. 1,3-Dimethylcyclohexane

12.50 Determine whether *cis–trans* isomerism is possible for each of the following cycloalkanes. If it is, then draw structural formulas for the *cis* and *trans* isomers.
a. *sec*-Butylcyclohexane
b. 1-Ethyl-3-methylcyclobutane
c. 1,1-Dimethylcyclohexane
d. 1,3-Dipropylcyclopentane

■ **Sources of Alkanes and Cycloalkanes (Section 12.14)**

12.51 What physical property of hydrocarbons is the basis for the fractional distillation process for separating hydrocarbons?

12.52 Describe the process by which crude petroleum is separated into simpler mixtures (fractions).

■ **Physical Properties of Alkanes and Cycloalkanes (Section 12.15)**

12.53 Which member in each of the following pairs of compounds has the higher boiling point?
a. Hexane and octane
b. Cyclobutane and cyclopentane
c. Pentane and 1-methylbutane
d. Pentane and cyclopentane

12.54 Which member in each of the following pairs of compounds has the higher boiling point?
a. Methane and ethane
b. Cyclohexane and hexane

c. Butane and methylpropane
d. Pentane and 2,2-dimethylpropane

12.55 For which of the following pairs of compounds do both members of the pair have the same physical state (solid, liquid, or gas) at room temperature and pressure?
a. Ethane and hexane
b. Cyclopropane and butane
c. Octane and 3-methyloctane
d. Pentane and decane

12.56 For which of the following pairs of compounds do both members of the pair have the same physical state (solid, liquid, or gas) at room temperature and pressure?
a. Methane and butane
b. Cyclobutane and cyclopentane
c. Hexane and 2,3-dimethylbutane
d. Pentane and octane

■ **Chemical Properties of Alkanes and Cycloalkanes (Section 12.16)**

12.57 Write the formulas of the products from the complete combustion of each of the following alkanes or cycloalkanes.
a. C_3H_8　　　　　　b. Butane
c. Cyclobutane　　　　d. CH_3—$(CH_2)_{15}$—CH_3

12.58 Write the formulas of the products from the complete combustion of each of the following alkanes or cycloalkanes.
a. C_4H_{10}　　　　　b. 2-Methylpentane
c. Cyclopentane　　　d. CH_3—$(CH_2)_7$—CH_3

12.59 Write molecular formulas for all the possible halogenated hydrocarbon products from the bromination of methane.

12.60 Write molecular formulas for all the possible halogenated hydrocarbon products from the fluorination of methane.

12.61 Write structural formulas for all the possible halogenated hydrocarbon products from the monochlorination of the following alkanes or cycloalkanes.
a. Ethane　　　　　　b. Butane
c. 2-Methylpropane　　d. Cyclopentane

12.62 Write structural formulas for all the possible halogenated hydrocarbon products from the monobromination of the following alkanes or cycloalkanes.
a. Propane　　　　　　b. Pentane
c. 2-Methylbutane　　　d. Cyclohexane

■ **Nomenclature of Halogenated Alkanes (Section 12.17)**

12.63 Give both IUPAC and common names to each of the following halogenated hydrocarbons.
a. CH_3—I　　　　　b. CH_3—CH_2—CH_2—Cl

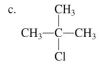
c. 　　　　　　　　　d.

12.64 Give both IUPAC and common names to each of the following halogenated hydrocarbons.
a. CH_3—CH_2—CH_2—CH_2—Br　　b. CH_3—CH—Cl
　　　　　　　　　　　　　　　　　　　　　　　　 |
　　　　　　　　　　　　　　　　　　　　　　　　CH_3

c. 　　　　　　　　　　　　　　d.

12.65 Draw structural formulas for the following halogenated hydrocarbons.
a. Trichloromethane
b. 1,2-Dichloro-1,1,2,2-tetrafluoroethane
c. Isopropyl bromide
d. *trans*-1-Bromo-3-chlorocyclopentane

12.66 Draw structural formulas for the following halogenated hydrocarbons.
a. Trifluorochloromethane
b. Pentafluoroethane
c. Isobutyl chloride
d. *cis*-1,2-Dichlorocyclohexane

■ **ADDITIONAL PROBLEMS**

12.67 Answer the following questions about the unbranched alkane with seven carbon atoms.
a. How many hydrogen atoms are present?
b. How many carbon–carbon bonds are present?
c. How many carbon atoms have two hydrogen atoms bonded to them?
d. How many total covalent bonds are present?
e. Is the alkane a solid, a liquid, or a gas at room temperature?
f. Is the alkane less dense or more dense than water?
g. Is the alkane soluble or insoluble in water?
h. Is the alkane flammable or nonflammable in air?

12.68 Indicate whether the members of each of the following pairs of compounds are structural isomers.
a. Hexane and 2-methylhexane
b. Hexane and 2,2-dimethylbutane
c. Hexane and methylcyclopentane
d. Hexane and cyclohexane

12.69 Draw structural formulas for the following compounds.
a. *trans*-1,4-Difluorocyclohexane
b. *cis*-1-Chloro-2-methylcyclobutane

c. *tert*-Butyl bromide
d. Isobutyl iodide

12.70 What is the IUPAC name of the compound obtained by attaching a *tert*-butyl group to carbon 4 and a *sec*-butyl group to carbon 5 of the alkane 2,2-dimethylheptane?

12.71 Give the molecular formula for each of the following compounds.
a. an 18-carbon alkane
b. a 7-carbon cycloalkane
c. a 7-carbon difluorinated alkane
d. a 6-carbon dibrominated cycloalkane

12.72 Draw the structural formula of each of the following compounds.
a. Neooctane　　　　　b. Isobutane
c. Methylene chloride　　d. Freon-12

12.73 Classify each of the following molecular formulas as representing an alkane, a cycloalkane, a halogenated alkane, or a halogenated cycloalkane.
a. C_6H_{14}　　　　　b. $C_6H_{10}Cl_2$
c. $C_4H_8Cl_2$　　　　d. C_4H_8

12.74 Write skeletal formulas (without hydrogen atoms) and assign IUPAC names to all saturated hydrocarbon structural isomers (ignore *cis–trans* isomers) with the following molecular formulas.
  a. C_6H_{12} (twelve isomers are possible)
  b. C_6H_{14} (five isomers are possible)
  c. $C_3H_6Br_2$ (four isomers are possible)

12.75 Assign an IUPAC name to each of the following hydrocarbons, whose structural formulas are given using line-angle drawings.

a.

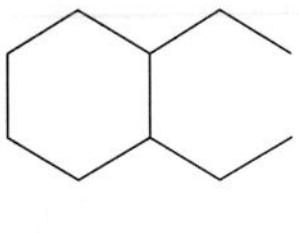

b.

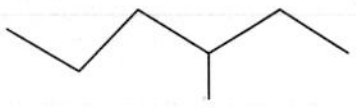

c.

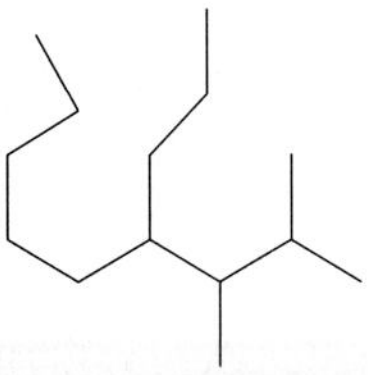

d.

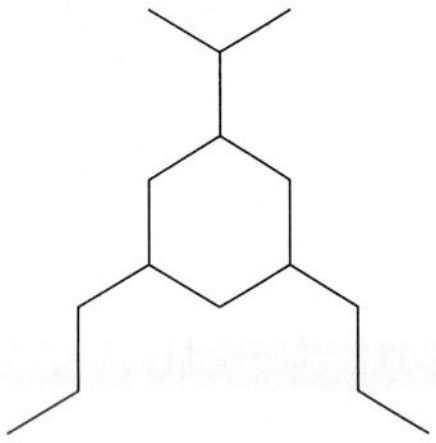

12.76 Assign an IUPAC name to the following alkane which has a five-carbon branched alkyl group.

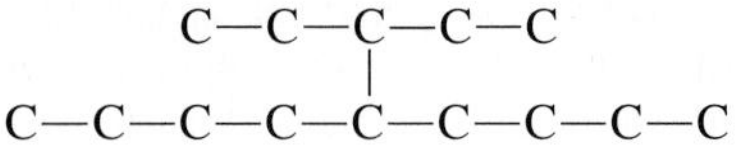

$$C-C-C-C-C$$
$$C-C-C-C-C-C-C-C-C$$

ANSWERS TO PRACTICE EXERCISES

12.1 a. different conformations   b. different conformations
  c. structural isomers

12.2 a. 3,6-dimethyloctane   b. 3,4,4,5-tetramethyloctane

12.3

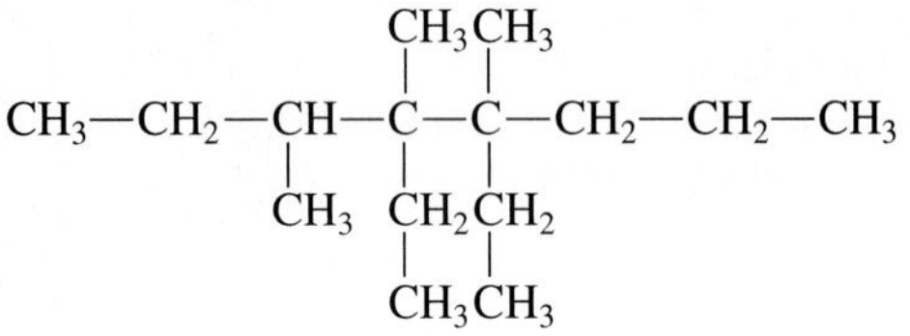

$$CH_3-CH_2-CH-\underset{\underset{CH_3}{|}}{\overset{\overset{CH_3}{|}}{C}}-\underset{\underset{\underset{\underset{CH_3CH_3}{|}}{CH_3CH_2}}{|}}{\overset{\overset{CH_3}{|}}{C}}-CH_2-CH_2-CH_3$$

12.4 a. C_7H_{14} b. $C_{10}H_{20}$ c. C_6H_{14} d. C_7H_{16}

12.5 a. methylcyclopropane   b. 1-ethyl-4-methylcyclohexane
  c. 4-ethyl-1,2-dimethylcyclopentane

12.6 a. not possible   b. not possible
  c.

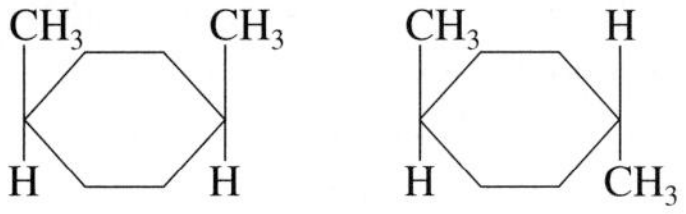

| CH₃ | CH₃ | CH₃ | H | d. not possible |

Cis isomer   *Trans* isomer

13

Unsaturated Hydrocarbons

When acetylene, an unsaturated hydrocarbon, is burned with oxygen in an oxyacetylene welding torch, a temperature high enough to cut metals is produced.

Two general types of hydrocarbons exist, *saturated* and *unsaturated*. Saturated hydrocarbons, discussed in the previous chapter, include the alkanes and cycloalkanes. All bonds in saturated hydrocarbons are single bonds. Unsaturated hydrocarbons, the topic of this chapter, contain one or more carbon–carbon multiple bonds. There are three classes of unsaturated hydrocarbons: the *alkenes*, the *alkynes*, and the *aromatic hydrocarbons*, all of which we will consider.

13.1 Unsaturated Hydrocarbons

An **unsaturated hydrocarbon** *is a hydrocarbon in which one or more carbon–carbon multiple bonds (double bonds, triple bonds, or both) are present.* Unsaturated hydrocarbons have *physical* properties similar to those of saturated hydrocarbons. However, their *chemical* properties are much different. Unsaturated hydrocarbons are chemically more reactive than their saturated counterparts. The increased reactivity of unsaturated hydrocarbons is related to the presence of the carbon–carbon multiple bond(s) in such compounds. These multiple bonds serve as locations where chemical reactions can occur.

Whenever a specific portion of a molecule governs its chemical properties, that portion of the molecule is called a functional group. A **functional group** *is the part of an organic molecule where most of its chemical reactions occur.* Carbon–carbon multiple bonds are the functional group for an unsaturated hydrocarbon.

The study of various functional groups and their respective reactions provides the organizational structure for organic chemistry. Each of the organic chemistry chapters

> The field of organic chemistry is organized in terms of functional groups.

▶ Alkanes and cycloalkanes (Chapter 12) lack functional groups; as a result, they are relatively unreactive.

that follow introduces new functional groups that characterize families of hydrocarbon derivatives.

Unsaturated hydrocarbons are subdivided into three groups on the basis of the type of multiple bond(s) present: (1) *alkenes,* which contain one or more carbon–carbon double bonds, (2) *alkynes,* which contain one or more carbon–carbon triple bonds, and (3) *aromatic hydrocarbons,* which exhibit a special type of "delocalized" bonding that involves a six-membered carbon ring (to be discussed in Section 13.10).

We begin our consideration of unsaturated hydrocarbons with a discussion of alkenes. Information about alkynes and aromatic hydrocarbons then follows.

13.2 Characteristics of Alkenes and Cycloalkenes

An **alkene** *is an acyclic unsaturated hydrocarbon that contains one or more carbon–carbon double bonds.* The alkene functional group is, thus, a C=C group. Note the close similarity between the family names *alkene* and *alkane* (Section 12.4); they differ only in their endings: *-ene* versus *-ane.* The *-ene* ending means a double bond is present.

▶ The general formula for an alkene with one double bond, C_nH_{2n}, is the same as that for a cycloalkane (Section 12.11). Thus such alkenes and cycloalkanes with the same number of carbon atoms are isomeric with one another.

The simplest type of alkene contains only one carbon–carbon double bond. Such compounds have the general formula C_nH_{2n}. Thus alkenes with one double bond have two fewer hydrogen atoms than do alkanes (C_nH_{2n+2}).

The two simplest alkenes are ethene (C_2H_4) and propene (C_3H_6).

$$CH_2{=}CH_2 \qquad CH_2{=}CH{-}CH_3$$
$$\text{Ethene} \qquad\qquad \text{Propene}$$

Comparing the geometrical shape of ethene with that of methane (the simplest alkane) reveals a major difference. The arrangement of bonds about the carbon atom in methane is tetrahedral (Section 12.4), whereas the carbon atoms in ethene have a trigonal planar arrangement of bonds; that is, they form a flat, triangle-shaped arrangement (see Figure 13.1). The two carbon atoms participating in a double bond and the four other atoms attached to these two carbon atoms always lie in a plane with a trigonal planar arrangement of atoms about each carbon atom of the double bond. Such an arrangement of atoms is consistent with the principles of VSEPR theory (Section 5.7).

▶ An older but still widely used name for alkenes is *olefins,* pronounced "oh-la-fins." The term *olefin* means "oil-forming." Many alkenes react with Cl_2 to form "oily" compounds.

A **cycloalkene** *is a cyclic unsaturated hydrocarbon that contains one or more carbon–carbon double bonds within the ring system.* Cycloalkenes in which there is only one double bond have the general formula C_nH_{2n-2}. This general formula reflects the loss of four hydrogen atoms from that of an alkane (C_nH_{2n+2}). Note that two hydrogen atoms are lost because of the double bond and two because of the ring structure.

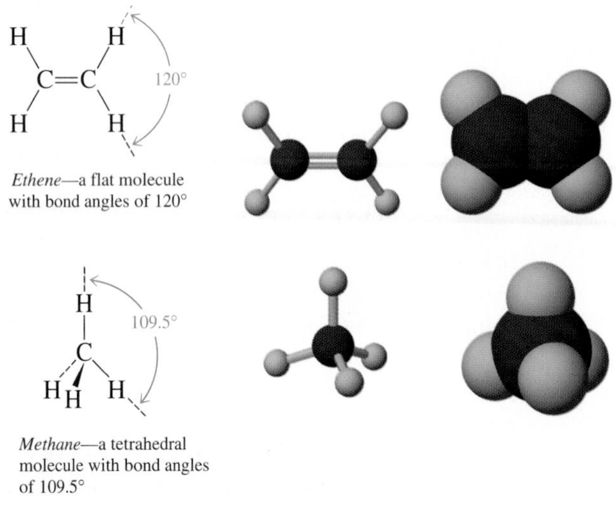

FIGURE 13.1 Three-dimensional representations of the structures of ethene and methane. In ethene, the atoms are in a flat (planar) rather than a tetrahedral arrangement. Bond angles are 120°.

Ethene—a flat molecule with bond angles of 120°

Methane—a tetrahedral molecule with bond angles of 109.5°

The simplest cycloalkene is the compound cyclopropene (C_3H_4), a three-membered carbon ring system containing one double bond.

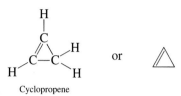

Cyclopropene

Alkenes with more than one carbon–carbon double bond are relatively common. When two double bonds are present, the compounds are often called *dienes*. Cycloalkenes that contain more than one double bond are possible but are not common.

13.3 Names for Alkenes and Cycloalkenes

The rules previously presented for naming alkanes and cycloalkanes (Sections 12.8 and 12.12) can be used, with some modification, to name alkenes and cycloalkenes.

1. Replace the alkane suffix *-ane* with the suffix *-ene,* which is used to indicate the presence of a carbon–carbon double bond.
2. Select as the parent carbon chain the longest continuous chain of carbon atoms that *contains both carbon atoms of the double bond.* For example, select

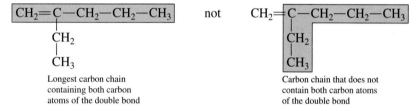

Longest carbon chain containing both carbon atoms of the double bond

Carbon chain that does not contain both carbon atoms of the double bond

> Carbon–carbon double bonds take precedence over alkyl groups and halogen atoms in determining the direction in which the parent carbon chain is numbered.

3. Number the parent carbon chain beginning at the end nearest the double bond.

$$\overset{1}{C}H_3-\overset{2}{C}H=\overset{3}{C}H-\overset{4}{C}H_2-\overset{5}{C}H_3 \quad \text{not} \quad \overset{5}{C}H_3-\overset{4}{C}H=\overset{3}{C}H-\overset{2}{C}H_2-\overset{1}{C}H_3$$

If the double bond is equidistant from both ends of the parent chain, begin numbering from the end closer to a substituent.

$$\overset{4}{C}H_3-\overset{3}{C}H=\overset{2}{C}H-\overset{1}{C}H_2 \quad \text{not} \quad \overset{1}{C}H_3-\overset{2}{C}H=\overset{3}{C}H-\overset{4}{C}H_2$$
$$\qquad\qquad\qquad | \qquad\qquad\qquad\qquad\qquad\qquad |$$
$$\qquad\qquad\qquad Cl \qquad\qquad\qquad\qquad\qquad\qquad Cl$$

> A number is not needed to specify double bond position in ethene and propene, because there is only one way of positioning the double bond in these molecules.

4. Give the position of the double bond in the chain as a *single* number, which is the lower-numbered carbon atom participating in the double bond. This number is placed immediately before the name of the parent carbon chain.

$$\overset{1}{C}H_3-\overset{2}{C}H=\overset{3}{C}H-\overset{4}{C}H_3 \qquad \overset{1}{C}H_2=\overset{2}{C}H-\overset{3}{C}H-\overset{4}{C}H_3$$
$$\qquad\qquad\qquad\qquad\qquad\qquad\qquad\qquad\qquad | $$
$$\qquad\qquad\qquad\qquad\qquad\qquad\qquad\qquad\quad CH_3$$

2-Butene   3-Methyl-1-butene

> Line-angle drawings for the simpler noncyclic 1-alkenes are as follows:
>
> Propene
>
> 1-Butene
>
> 1-Pentene
>
> 1-Hexene

5. Use the suffixes *-diene, -triene, -tetrene,* and so on when more than one double bond is present in the molecule. A separate number must be used to locate each double bond.

$$\overset{1}{C}H_2=\overset{2}{C}H-\overset{3}{C}H=\overset{4}{C}H_2 \qquad \overset{1}{C}H_2=\overset{2}{C}H-\overset{3}{C}H-\overset{4}{C}H=\overset{5}{C}H_2$$
$$\qquad\qquad\qquad\qquad\qquad\qquad\qquad\qquad\qquad | $$
$$\qquad\qquad\qquad\qquad\qquad\qquad\qquad\qquad\quad CH_3$$

1,3-Butadiene   3-Methyl-1,4-pentadiene

6. Do not use a number to locate the double bond in unsubstituted cycloalkenes with only one double bond, because that bond is assumed to be between carbons 1 and 2.

7. In substituted cycloalkenes with only one double bond, the double-bonded carbon atoms are numbered 1 and 2 in the direction (clockwise or counterclockwise) that gives the first-encountered substituent the lower number. Again, no number is used in the name to locate the double bond.

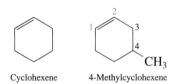

Cyclohexene   4-Methylcyclohexene

8. In cycloalkenes with more than one double bond within the ring, assign one double bond the numbers 1 and 2 and the other double bonds the lowest numbers possible.

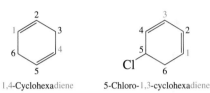

1,4-Cyclohexadiene   5-Chloro-1,3-cyclohexadiene

EXAMPLE 13.1

Assigning IUPAC Names to Alkenes and Cycloalkenes

■ Assign IUPAC names to the following alkenes and cycloalkenes.

a. $CH_3—CH=CH—CH_2—CH_2—CH_3$ **b.** $CH_3—CH_2—C=CH_2$
$\qquad\qquad\qquad\qquad\qquad\qquad\qquad\qquad\qquad\qquad\quad |$
$\qquad\qquad\qquad\qquad\qquad\qquad\qquad\qquad\qquad\qquad\ CH_2$
$\qquad\qquad\qquad\qquad\qquad\qquad\qquad\qquad\qquad\qquad\quad |$
$\qquad\qquad\qquad\qquad\qquad\qquad\qquad\qquad\qquad\qquad\ CH_3$

c.

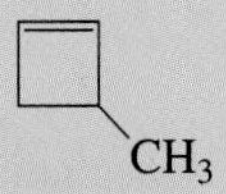

d.

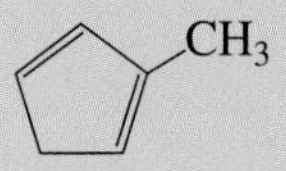

Solution

a. The carbon chain in this hexene is numbered from the end closest to the double bond.

$$\overset{1}{C}H_3—\overset{2}{C}H=\overset{3}{C}H—\overset{4}{C}H_2—\overset{5}{C}H_2—\overset{6}{C}H_3$$

The complete IUPAC name is 2-hexene.

b. The longest carbon chain containing *both* carbons of the double bond has four carbon atoms. Thus we have a butene.

$$\boxed{CH_3—CH_2—C=CH_2}$$
$$|$$
$$CH_2$$
$$|$$
$$CH_3$$

The chain is numbered from the end closest to the double bond. The complete IUPAC name is 2-ethyl-1-butene.

c. This compound is a methylcyclobutene. The numbers 1 and 2 are assigned to the carbon atoms of the double bond, and the ring is numbered clockwise, which results in a carbon 3 location for the methyl group. (Counterclockwise numbering would have placed the methyl group on carbon 4.) The complete IUPAC name of the cycloalkene is 3-methylcyclobutene. The double bond is understood to involve carbons 1 and 2.

d. A ring system containing five carbon atoms, two double bonds, and a methyl substituent on the ring is called a methylcyclopentadiene. Two different numbering systems produce the same locations (carbons 1 and 3) for the double bonds.

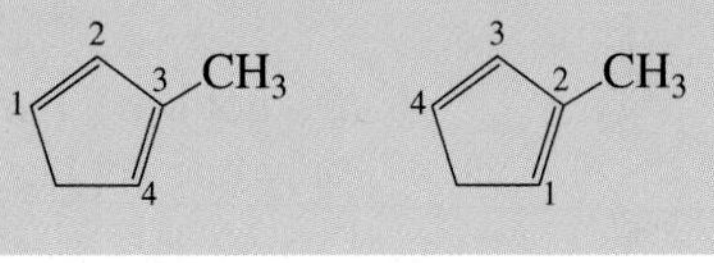

The counterclockwise numbering system assigns the lower number to the methyl group. The complete IUPAC name of the compound is 2-methyl-1,3-cyclopentadiene.

Practice Exercise 13.1

Assign IUPAC names to the following alkenes and cycloalkenes.

a. $CH_3—CH{=}CH—CH_2—CH—CH_3$
                                 CH_3

b.

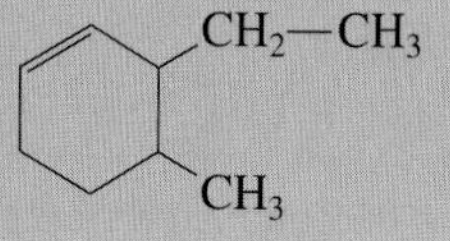

c. $CH_2{=}CH—CH{=}CH_2$

d.

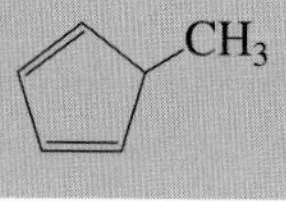

■ Common Names (Non-IUPAC Names)

The simpler members of most families of organic compounds, including alkenes, have common names in addition to IUPAC names. In many cases these common (non-IUPAC) names are used almost exclusively for the compounds. It would be nice if such common names did not exist, but they do. We have no choice but to memorize these names; fortunately, there are not many of them.

The two simplest alkenes, ethene and propene, have common names that you should be familiar with. They are ethylene and propylene, respectively.

$$CH_2{=}CH_2 \qquad CH_2{=}CH—CH_3$$

Ethylene   Propylene

■ Alkenes as Substituents

Just as there are *alkanes* and *alkyl groups* (Section 12.8), there are *alkenes* and *alkenyl groups*. An **alkenyl group** *is a noncyclic hydrocarbon substituent in which a carbon–carbon double bond is present*. The three most frequently encountered alkenyl groups are the one-, two-, and three-carbon entities, which may be named using IUPAC nomenclature (methylidene, ethenyl, and 2-propenyl) or with common names (methylene, vinyl, and allyl).

$$CH_2{=} \qquad\qquad CH_2{=}CH{\longrightarrow} \qquad\qquad CH_2{=}CH—CH_2{\longrightarrow}$$

Methylene group   Vinyl group   Allyl group
(IUPAC name: methylidene group)   (IUPAC name: ethenyl group)   (IUPAC name: 2-propenyl group)

The use of these alkenyl group names in actual compound nomenclature is illustrated in the following examples.

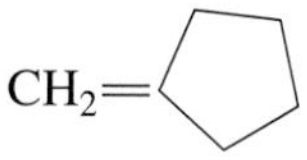

$$CH_2{=}\!\!\bigtriangleup \qquad\qquad CH_2{=}CH—Cl \qquad CH_2{=}CH—CH_2—Br$$

Methylene cyclopentane   Vinyl chloride   Allyl bromide
(IUPAC name: methylidenecyclopentane)   (IUPAC name: chloroethene)   (IUPAC name: 3-bromopropene)

13.4 Isomerism in Alkenes

Structural isomerism is possible for alkenes, just as it was for alkanes (Section 12.6). Figure 13.2 compares structural isomer possibilities for four- and five-carbon alkane and alkene systems. The potential for structural isomers is greater for alkenes than for alkanes, because there is more than one location where the double bond can be placed in systems containing four or more carbon atoms.

Cis–trans isomerism (Section 12.13) is possible for some alkenes. Such isomerism results from the structural rigidity associated with carbon–carbon double bonds: Unlike

---

Despite the universal acceptance and precision of the IUPAC nomenclature system, some alkenes (those of low molecular mass) are known almost exclusively by common names.

Structural isomers that differ only in the location of the functional group are often called positional isomers. *The compounds 1-butene and 2-butene are positional isomers.*

$$CH_2{=}CH—CH_2—CH_3$$
1-Butene

$$CH_3—CH{=}CH—CH_3$$
2-Butene

CHEMICAL CONNECTIONS

Ethene: A Plant Hormone and High-Volume Industrial Chemical

Ethene (ethylene), the simplest unsaturated hydrocarbon (C_2H_4), is a colorless, flammable gas with a slightly sweet odor. It occurs naturally in *small* amounts in plants, where it functions as a plant hormone. A few parts per million ethene (less than 10 parts per million) stimulates the fruit-ripening process.

The commercial fruit industry uses ethene's ripening property to advantage. Bananas, tomatoes, and some citrus fruits are picked green to prevent spoiling and bruising during transportation to markets. At its destination, the fruit is exposed to small amounts of ethene gas, which stimulates the ripening process.

Despite having no large natural source, ethene is an exceedingly important industrial chemical. Indeed, industrial production of ethene exceeds that of every other organic compound. Petrochemicals (substances found in natural gas and petroleum) are the starting materials for ethene production.

In one process, ethane (from natural gas) is *dehydrogenated* at a high temperature to produce ethene.

$$CH_3-CH_3 \xrightarrow{750°C} CH_2{=}CH_2 + H_2$$
$$\text{Ethane} \qquad\qquad \text{Ethene}$$

In another process, called *thermal cracking,* hydrocarbons from petroleum are heated to a high temperature in the absence of air (to prevent combustion), which causes the cleavage of carbon–carbon bonds. Ethene is one of the smaller molecules produced by this process.

Industrially produced ethene serves as a starting material for the production of many plastics and fibers. Almost one-half of ethene production is used in the production of the well-known plastic polyethylene (Section 13.8). Polyvinyl chloride (PVC) and polystyrene (styrofoam) are two other important ethene-based materials. About one-sixth of ethene production is converted to ethylene glycol, the principal component of most brands of antifreeze for automobile radiators (Section 14.4).

Ethene is the hormone that causes tomatoes to ripen.

FIGURE 13.2 A comparison of structural isomerism possibilities for four- and five-carbon alkane and alkene systems

Four-Carbon Alkanes (two isomers)	Four-Carbon Alkenes (three isomers)	Five-Carbon Alkanes (three isomers)	Five-Carbon Alkenes (five isomers)
$CH_3-CH_2-CH_2-CH_3$ Butane	$CH_2{=}CH-CH_2-CH_3$ 1-Butene	$CH_3-CH_2-CH_2-CH_2-CH_3$ Pentane	$CH_2{=}CH-CH_2-CH_2-CH_3$ 1-Pentene
$CH_3-\underset{\underset{\displaystyle CH_3}{\mid}}{CH}-CH_3$ 2-Methylpropane	$CH_3-CH{=}CH-CH_3$ 2-Butene	$CH_3-\underset{\underset{\displaystyle CH_3}{\mid}}{CH}-CH_2-CH_3$ 2-Methylbutane	$CH_3-CH{=}CH-CH_2-CH_3$ 2-Pentene
	$CH_2{=}\underset{\underset{\displaystyle CH_3}{\mid}}{C}-CH_3$ 2-Methylpropene	$CH_3-\underset{\underset{\displaystyle CH_3}{\overset{\overset{\displaystyle CH_3}{\mid}}{C}}}{C}-CH_3$ 2,2-Dimethylpropane	$CH_2{=}\underset{\underset{\displaystyle CH_3}{\mid}}{C}-CH_2-CH_3$ 2-Methyl-1-butene
			$CH_2{=}CH-\underset{\underset{\displaystyle CH_3}{\mid}}{CH}-CH_3$ 3-Methyl-1-butene
			$CH_3-\underset{\underset{\displaystyle CH_3}{\mid}}{C}{=}CH-CH_3$ 2-Methyl-2-butene

FIGURE 13.3 *Cis–trans* isomers: Different representations of the *cis* and *trans* isomers of 2-butene.

cis-2-Butene
boiling point = 4°C
density = 0.62 g/ml

trans-2-Butene
boiling point = 1°C
density = 0.60 g/ml

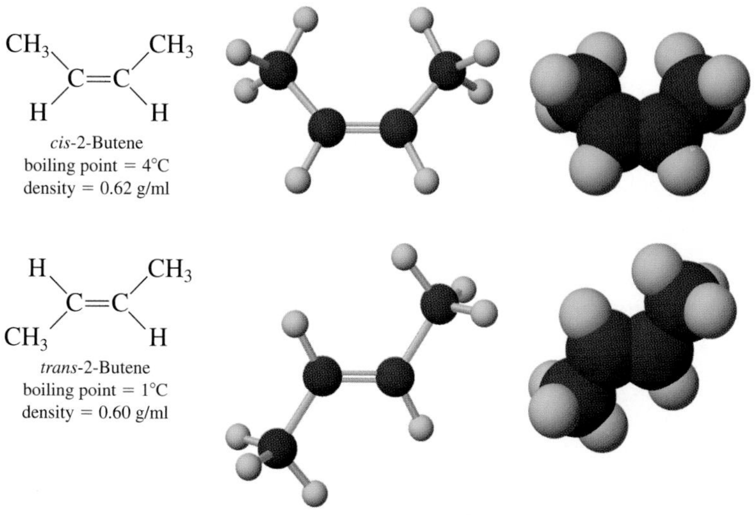

the situation in alkanes, where free rotation about carbon–carbon single bonds is possible (Section 12.7), no rotation about carbon–carbon double bonds (or carbon–carbon triple bonds) can occur.

To determine whether an alkene has *cis* and *trans* isomers, draw the alkene structure in a manner that emphasizes the four attachments to the double-bonded carbon atoms.

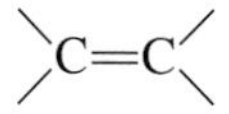

▶ The double bond of alkenes, like the ring of cycloalkanes, imposes rotational restrictions.

If *each* of the two carbons of the double bond has two *different* groups attached to it, *cis* and *trans* isomers exist.

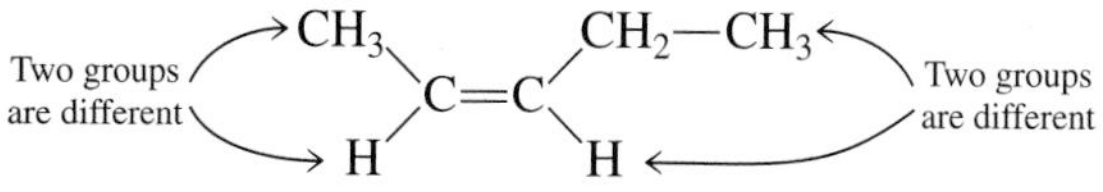

▶ Recall, from Section 12.13, that *cis* means "on the same side" and *trans* means "across."

The simplest alkene for which *cis* and *trans* isomers exist is 2-butene.

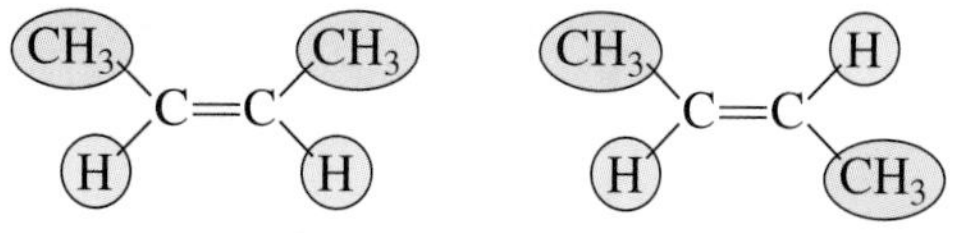

Structure A
(*cis*-2-butene)

Structure B
(*trans*-2-butene)

Structure A is the *cis* isomer and structure B is the *trans* isomer. In alkene chemistry, a **cis isomer** *is an alkene isomer in which two similar (or identical) double-bond substituents are on the same side of the double bond.* In structure A, both methyl groups are on the same side of the double bond. A **trans isomer** *is an alkene isomer in which two similar (or identical) double-bond substituents are on opposite sides of the double bond.* In structure B, the methyl groups are on opposite sides of the double bond. The only way to convert structure A to structure B is to break the double bond. At room temperature, such bond breaking does not occur. Hence these two structures represent two different compounds (*cis–trans* isomers) that differ in boiling point, density, and so on. Figure 13.3 shows three-dimensional representations of the *cis* and *trans* isomers of 2-butene.

Cis–trans isomerism is not possible when one of the double-bonded carbons bears two identical groups. Thus neither 1-butene nor 2-methylpropene is capable of existing in *cis* and *trans* forms.

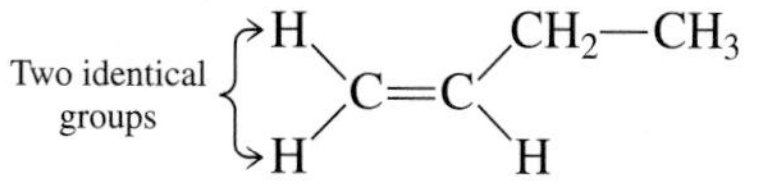

1-Butene

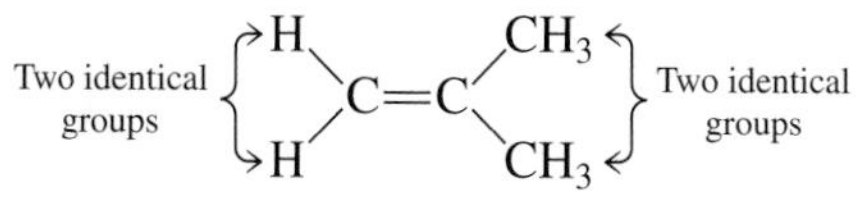

2-Methylpropene

When alkenes contain more than one double bond, *cis–trans* considerations are more complicated. Orientation about each double bond must be considered independently of that at other sites. For example, for the molecule 2,4-heptadiene (two double bonds) there are four different *cis–trans* isomers (*trans–trans, trans–cis, cis–trans,* and *cis–cis*). The structures of two of these isomers are

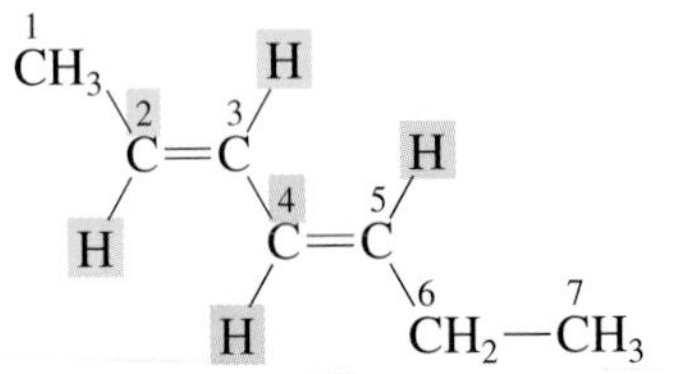

trans-trans-2,4-Heptadiene 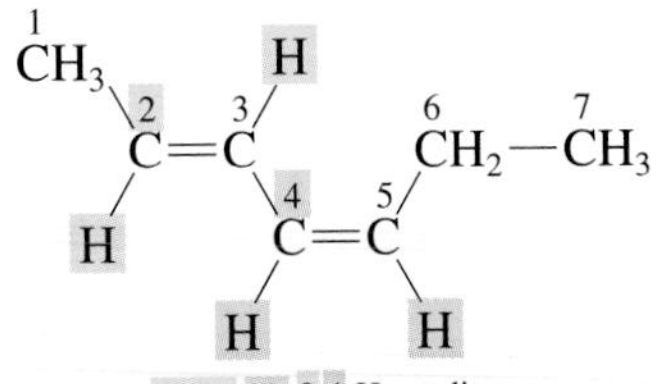   *trans-cis*-2,4-Heptadiene

EXAMPLE 13.2

Determining Whether *Cis–Trans* Isomerism Is Possible in Substituted Alkenes

■ Determine whether each of the following substituted alkenes can exist in *cis–trans* isomeric forms.

a. 1-Bromo-1-chloroethene   **b.** 2-Chloro-2-butene

Solution

a. The condensed structural formula for this compound is

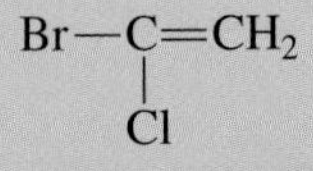

$$Br-C=CH_2$$
$$|$$
$$Cl$$

Redrawing this formula to emphasize the four attachments to the double-bonded carbon atoms gives

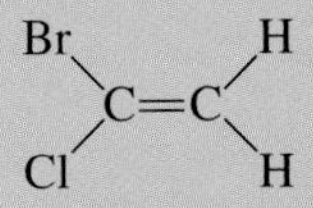

The carbon atom on the right has two identical attachments. Hence *cis–trans* isomerism is not possible.

b. The condensed structural formula for this compound is

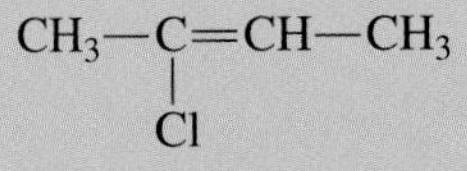

$$CH_3-C=CH-CH_3$$
$$|$$
$$Cl$$

Redrawing this formula to emphasize the four attachments to the double-bonded carbon atoms gives

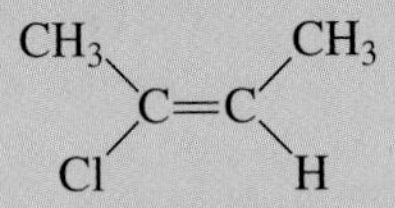

Because both carbon atoms of the double bond bear two different attachments, *cis–trans* isomers are possible.

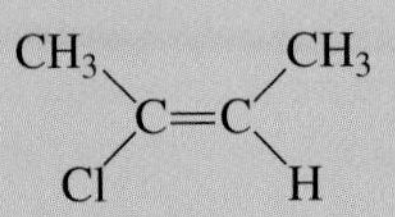

cis-2-Chloro-2-butene 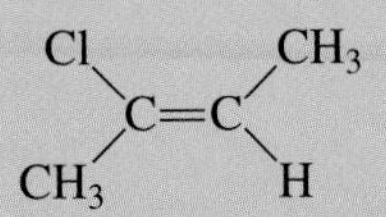   *trans*-2-Chloro-2-butene

Practice Exercise 13.2

Determine whether each of the following substituted alkenes can exist in *cis–trans* isomeric forms.

a. 1-Chloropropene   **b.** 2-Chloropropene

CHEMICAL CONNECTIONS

Cis–Trans Isomerism and Vision

Cis–trans isomerism plays an important role in many biochemical processes, including the reception of light by the retina of the eye. Within the retina, microscopic structures called rods and cones contain a compound called *retinal,* which absorbs light. Retinal contains a carbon chain with five carbon–carbon double bonds, four in a *trans* configuration and one in a *cis* configuration. This arrangement of double bonds gives retinal a shape that fits the protein *opsin,* to which it is attached, as shown in the accompanying diagram.

When light strikes retinal, the *cis* double bond is converted to a *trans* double bond. The resulting *trans*-retinal no longer fits the protein opsin and is subsequently released. Accompanying this release is an electrical impulse, which is sent to the brain. Receipt of such impulses by the brain is what enables us to see.

In order to trigger nerve impulses again, *trans*-retinal must be converted back to *cis*-retinal. This occurs in the membranes of the rods and cones, where enzymes change *trans*-retinal back into *cis*-retinal.

13.5 Naturally Occurring Alkenes

Alkenes are abundant in nature. Many important biological molecules are characterized by the presence of carbon–carbon double bonds within their structure. Two important types of naturally occurring substances to which alkenes contribute are pheromones and terpenes.

■ Pheromones

A **pheromone** *is a compound used by insects (and some animals) to transmit a message to other members of the same species.* Pheromones are often alkenes or alkene derivatives. The biological activity of alkene-type pheromones is usually highly dependent on whether the double bonds present are in a *cis* or a *trans* arrangement (Section 13.4).

The sex attractant of the female silkworm is a 16-carbon alkene derivative containing an —OH group. Two double bonds are present, *trans* at carbon 10 and *cis* at carbon 12.

$$CH_3-(CH_2)_2 \overset{13}{\underset{}{}} \overset{12}{\underset{}{}} \overset{11}{\underset{}{}} \overset{10}{\underset{}{}} (CH_2)_8-CH_2-OH$$

This compound is 10 billion times more effective in eliciting a response from the male silkworm than the 10-*cis*–12-*trans* isomer and 10 trillion times more effective than the isomer wherein both bonds are in a *trans* configuration.

Insect sex pheromones are useful in insect control. A small amount of synthetically produced sex pheromone is used to lure male insects of a single species into a trap (see Figure 13.4). The trapped males are either killed or sterilized. Releasing sterilized males

FIGURE 13.4 The application of sex pheromones in insect control involves using a small amount of synthetically produced pheromone to lure a particular insect into a trap. This is accomplished without harming other "beneficial" insects.

CHEMICAL CONNECTIONS

Carotenoids: A Source of Color

Carotenoids are the most widely distributed of the substances that give color to our world; they occur in flowers, fruits, plants, insects, and animals. These compounds are terpenes (Section 13.5) in which eight isoprene units are present. Structural formulas for two members of the carotenoid family, β-carotene and lycopene, follow.

β-Carotene

Lycopene

Present in both of these carotenoid structures is a *conjugated* system of 11 double bonds. (Conjugated double bonds are double bonds separated from each other by one single bond.) Color is frequently caused by the presence of compounds that contain extended conjugated-double-bond systems. When *visible* light strikes these compounds, certain wavelengths of the visible light are absorbed by the electrons in the conjugated-bond system. The unabsorbed wavelengths of visible light are reflected and are perceived as color.

The molecule β-carotene is responsible for the yellow-orange color in carrots, apricots, and yams. The yellowish tint of animal fat comes from β-carotene present in animal diets. The yellow-orange color of autumn leaves comes from β-carotene. Leaves contain chlorophyll (green pigment) and β-carotene (yellow pigment) in a ratio of approximately 3 to 1. The yellow-orange β-carotene color is masked by the chlorophyll until autumn, when the chlorophyll molecules decompose, as a result of lower temperatures and less sunlight, and are not replaced.

The molecule lycopene is the red pigment in tomatoes, paprika, and watermelon. Lycopene's structure differs from that of β-carotene in that the two rings in β-carotene have been opened. The ripening of a green tomato involves the gradual decomposition of chlorophyll with an associated unmasking of the red color of the lycopene present. A green pepper becomes red after ripening for the same reason.

The following table gives the carotenoid levels present in selected fruits and vegetables (in terms of parts per million on a dry-weight basis.)

carrots	54	spinach	26–76
tomatoes	51	peas	3–7
apricots	35	lemons	2–3
peaches	27	apples	1–5

Carotenoids are synthesized only by plants. They reach animal tissues via feed, however, and can be modified and deposited therein. The chicken egg yolk, which is colored by carotenoids, is an important example of such transfer.

Carotenoids, molecules that contain eight isoprene units, are responsible for the yellow-orange color of autumn leaves.

The following word associations are important to remember:

alkane—substitution reaction

alkene—addition reaction

An analogy can be drawn to a basketball team. When a *substitution* is made, one player leaves the game as another enters. The number of players on the court remains at five per team. If *addition* were allowed during a basketball game, two players could enter the game and no one would leave; there would be seven players per team on the court rather than five.

Addition reactions can be classified as symmetrical or unsymmetrical. A **symmetrical addition reaction** *is an addition reaction in which identical atoms (or groups of atoms) are added to each carbon of a carbon–carbon multiple bond.* An **unsymmetrical addition reaction** *is an addition reaction in which different atoms (or groups of atoms) are added to the carbon atoms of a carbon–carbon multiple bond.*

■ Symmetrical Addition Reactions

The two most common examples of symmetrical addition reactions are hydrogenation and halogenation.

A **hydrogenation reaction** *is an addition reaction in which H_2 is incorporated into molecules of an organic compound.* In alkene hydrogenation a hydrogen atom is added to each carbon atom of a double bond. It is accomplished by heating the alkene and H_2 in the presence of a catalyst (usually Ni or Pt).

▶ Hydrogenation of an alkene requires a catalyst. No reaction occurs if the catalyst is not present.

$$CH_2{=}CH{-}CH_3 + H_2 \xrightarrow[\substack{150°C \\ 12-15\ atm \\ pressure}]{Ni\ or\ Pt} \overset{\overset{H}{|}\ \overset{H}{|}}{CH_2{-}CH{-}CH_3}$$

Propene   Propane

The identity of the catalyst used in hydrogenation is specified by writing it above the arrow in the chemical equation for the hydrogenation. In general terms, hydrogenation of an alkene can be written as

$$\underset{\text{Alkene}}{\diagup\text{C}{=}\text{C}\diagdown} + H_2 \xrightarrow[\substack{Heat, \\ pressure}]{Ni\ or\ Pt} \underset{\text{Alkane}}{\overset{\overset{H}{|}\ \overset{H}{|}}{-\text{C}{-}\text{C}-}}$$

▶ The Chemical Connection feature "Trans Fatty Acids and Blood Cholesterol Levels" in Chapter 19 addresses health issues relative to consumption of partially hydrogenated products.

The hydrogenation of vegetable oils is a very important commercial process today. Vegetable oils from sources such as soybeans and cottonseeds are composed of long-chain organic molecules that contain several double bonds. When these oils are hydrogenated, they are converted to low-melting solids that are used in margarines and shortenings.

A **halogenation reaction** *is an addition reaction in which a halogen (F_2, Cl_2, Br_2 or I_2) is incorporated into molecules of an organic compound.* In alkene halogenation a halogen atom is added to each carbon atom of a double bond. Chlorination (Cl_2) and bromination (Br_2) are the two halogenation processes most commonly encountered. No catalyst is needed.

$$CH_3{-}CH{=}CH{-}CH_3 + Cl_2 \longrightarrow \overset{\overset{Cl}{|}\ \overset{Cl}{|}}{CH_3{-}CH{-}CH{-}CH_3}$$

2-Butene   2,3-Dichlorobutane

In general terms, halogenation of an alkene can be written as

$$\underset{\text{Alkene}}{\diagup\text{C}{=}\text{C}\diagdown} + \underset{\text{Halogen}}{X_2} \longrightarrow \underset{\text{Dihalogenated alkane}}{\overset{\overset{X}{|}\ \overset{X}{|}}{-\text{C}{-}\text{C}-}} \quad (X = Cl, Br)$$

Bromination is often used to test for the presence of carbon–carbon double bonds in organic substances. Bromine in water or carbon tetrachloride is reddish brown. The dibromo compound(s) formed from the symmetrical addition of bromine to an organic compound is(are) colorless. Thus the decolorization of a Br_2 solution indicates the presence of carbon–carbon double bonds (see Figure 13.9).

FIGURE 13.9 A bromine in water solution is reddish brown (left). When a small amount of such a solution is added to an unsaturated hydrocarbon, the added solution is decolorized as the bromine adds to the hydrocarbon to form colorless dibromo compounds (right).

■ Unsymmetrical Addition Reactions

Two important types of unsymmetrical addition reactions are hydrohalogenation and hydration.

A **hydrohalogenation reaction** *is an addition reaction in which a hydrogen halide (HCl, HBr, or HI) is incorporated into molecules of an organic compound.* In alkene hydrohalogenation one carbon atom of a double bond receives a halogen atom and the other carbon atom receives a hydrogen atom. Hydrohalogenation reactions require no catalyst. For *symmetrical* alkenes, such as ethene, only one product results from hydrohalogenation.

$$\underset{\text{Ethene}}{CH_2{=}CH_2} + H{-}Cl \longrightarrow \underset{\text{Chloroethane}}{CH_2{-}CH_2} \overset{\boxed{H} \quad \boxed{Cl}}{\phantom{CH_2{-}CH_2}}$$

A **hydration reaction** *is an addition reaction in which H_2O is incorporated into molecules of an organic compound.* In alkene hydration one carbon atom of a double bond receives a hydrogen atom and the other carbon atom receives an $-OH$ group. Alkene hydration requires a small amount of H_2SO_4 (sulfuric acid) as a catalyst. For *symmetrical* alkenes, only one product results from hydration.

$$\underset{\text{Ethene}}{CH_2{=}CH_2} + H{-}OH \xrightarrow{H_2SO_4} \underset{\text{An alcohol}}{CH_2{-}CH_2} \overset{\boxed{H} \quad \boxed{OH}}{\phantom{CH_2{-}CH_2}}$$

In this equation, the water (H_2O) is written as $H-OH$ to emphasize how this molecule adds to the double bond. Note also that the product of this hydration reaction contains an $-OH$ group. Hydrocarbon derivatives of this type are called *alcohols.* Such compounds are the subject of Chapter 14.

When the alkene involved in a hydrohalogenation or hydration reaction is itself *unsymmetrical,* more than one product is possible. (An unsymmetrical alkene is one in which the two carbon atoms of the double bond are not equivalently substituted.) For example, the addition of HCl to propene (an unsymmetrical alkene) could produce either 1-chloropropane or 2-chloropropane, depending on whether the H from the HCl attaches itself to carbon 2 or carbon 1.

$$\underset{\text{Propene}}{CH_2{=}CH{-}CH_3} + HCl \longrightarrow \underset{\text{1-Chloropropane}}{CH_2{-}CH{-}CH_3} \overset{\boxed{Cl} \quad H}{\phantom{CH_2{-}CH{-}CH_3}}$$

or

$$\underset{\text{Propene}}{CH_2{=}CH{-}CH_3} + HCl \longrightarrow \underset{\text{2-Chloropropane}}{CH_2{-}CH{-}CH_3} \overset{H \quad \boxed{Cl}}{\phantom{CH_2{-}CH{-}CH_3}}$$

When two isomeric products are possible, one product often predominates. The dominant product can be predicted by using Markovnikov's rule, named after the Russian chemist Vladimir V. Markovnikov (see Figure 13.10). **Markovnikov's rule** states that *when an unsymmetrical molecule of the form* HQ *adds to an unsymmetrical alkene, the hydrogen atom from the* HQ *becomes attached to the unsaturated carbon atom that already has the most hydrogen atoms.* Thus the major product in our example involving propene is 2-chloropropane.

> ◢ The addition of water to carbon–carbon double bonds occurs in many biochemical reactions that take place in the human body—for example, in the citric acid cycle (Section 23.6) and in the oxidation of fatty acids (Section 25.4).

FIGURE 13.10 Vladimir Vasilevich Markovnikov (1837–1904). A professor of chemistry at several Russian universities, Markovnikov (pronounced Mar-cove-na-coff) synthesized rings containing four carbon atoms and seven carbon atoms, thereby disproving the notion of the day that carbon could form only five- and six-membered rings.

> ◢ Two catchy summaries of Markovnikov's rule are "Hydrogen goes where hydrogen is" and "The rich get richer" (in terms of hydrogen).

E X A M P L E   1 3 . 3

Predicting Products in Alkene Addition Reactions Using Markovnikov's Rule

■ Using Markovnikov's rule, predict the predominant product in each of the following addition reactions.

a. $CH_3{-}CH_2{-}CH_2{-}CH{=}CH_2 + HBr \longrightarrow$

b. [cyclopentene ring]$-CH_3 + HCl \longrightarrow$

c. $CH_3{-}CH{=}CH{-}CH_2{-}CH_3 + HBr \longrightarrow$

Solution

a. The hydrogen atom will add to carbon 1, because carbon 1 already contains more hydrogen atoms than carbon 2. The predominant product of the addition will be 2-bromopentane.

$$CH_3-CH_2-CH_2-\overset{②}{C}H=\overset{①}{C}H_2 + HBr \longrightarrow CH_3-CH_2-CH_2-\overset{\overset{\displaystyle Br}{|}}{C}H-\overset{\overset{\displaystyle H}{|}}{C}H_2$$

b. Carbon 1 of the double bond does not have any H atoms directly attached to it. Carbon 2 of the double bond has one H atom (H atoms are not shown in the structure but are implied) attached to it. The H atom from the HCl will add to carbon 2, giving 1-chloro-1-methylcyclopentane as the product.

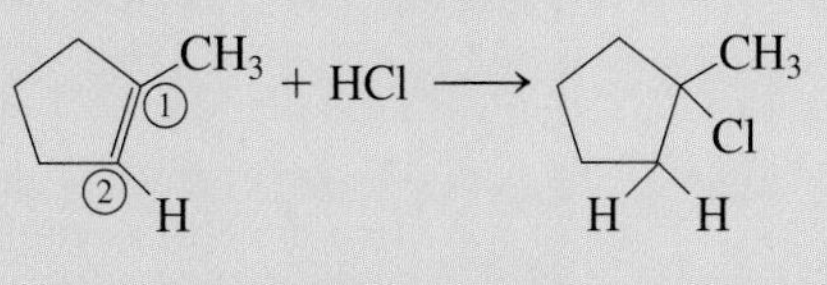

c. Each carbon atom of the double bond in this molecule has one hydrogen atom. Thus Markovnikov's rule does not favor either carbon atom. The result is two isomeric products that are formed in almost equal quantities.

$$CH_3-\overset{\overset{\displaystyle }{|}}{\underset{\underset{\displaystyle Br}{|}}{C}}H-CH_2-CH_2-CH_3 \quad \text{and} \quad CH_3-CH_2-\overset{\overset{\displaystyle }{|}}{\underset{\underset{\displaystyle Br}{|}}{C}}H-CH_2-CH_3$$

2-Bromopentane   3-Bromopentane

Practice Exercise 13.3

Using Markovnikov's rule, predict the predominant product in each of the following addition reactions.

a. $CH_2=CH-CH_2-CH_3 + HCl \rightarrow$

b.

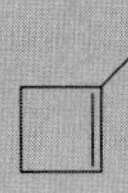

In compounds that contain more than one carbon–carbon double bond, such as dienes and trienes, addition can occur at each of the double bonds. In the complete hydrogenation of a diene and in that of a triene, the amounts of hydrogen needed are twice as much and three times as much, respectively, as that needed for the hydrogenation of an alkene with one double bond.

$$CH_2=CH-CH_2-CH_2-CH_2-CH_3 + H_2 \xrightarrow{Ni} CH_3-(CH_2)_4-CH_3$$

$$CH_2=CH-CH=CH-CH_2-CH_3 + 2H_2 \xrightarrow{Ni} CH_3-(CH_2)_4-CH_3$$

$$CH_2=CH-CH=CH-CH=CH_2 + 3H_2 \xrightarrow{Ni} CH_3-(CH_2)_4-CH_3$$

EXAMPLE 13.4

Predicting Reactants and Products in Alkene Addition Reactions

■ Supply the structural formula of the missing substance in each of the following addition reactions.

a. $CH_3-CH_2-CH=CH_2 + H_2O \xrightarrow{H_2SO_4}$?

b.
$$? + Br_2 \rightarrow$$

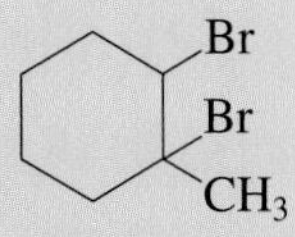

c.

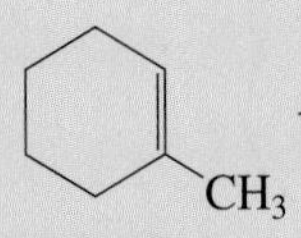

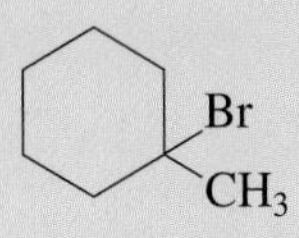

d. $CH_3-CH=CH-CH=CH_2 + 2H_2 \xrightarrow{Ni}$?

(continued)

Solution

a. This is a hydration reaction. Using Markovnikov's rule, we determine that the H will become attached to carbon 1, which has more hydrogen atoms than carbon 2, and that the —OH group will be attached to carbon 2.

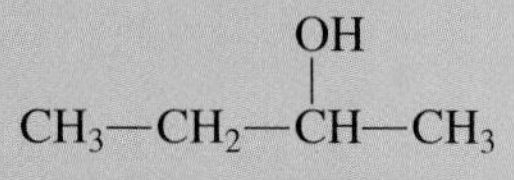

$$\underset{\text{OH}}{CH_3-CH_2-\overset{\mid}{CH}-CH_3}$$

b. The reactant alkene will have to have a double bond between the two carbon atoms that bromine atoms are attached to in the product.

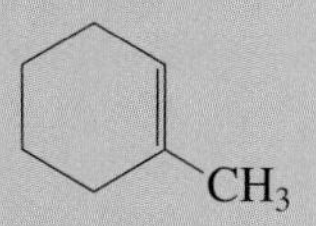

c. The small reactant molecule that adds to the double bond is HBr. The added Br atom from the HBr is explicitly shown in the product's structural formula, but the added H atom is not shown.

d. Hydrogen will add at each of the double bonds. The product hydrocarbon is pentane.

$$CH_3-CH_2-CH_2-CH_2-CH_3$$

Practice Exercise 13.4

Supply the structural formula of the missing substance in each of the following addition reactions.

a. $CH_3-CH_2-CH=CH_2 + HBr \rightarrow ?$

b. 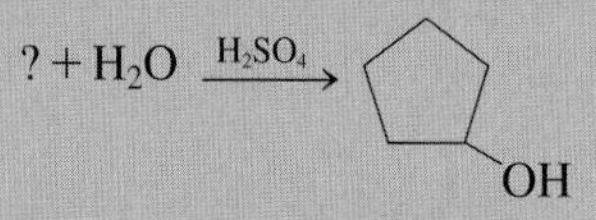 $? + H_2O \xrightarrow{H_2SO_4}$

c.

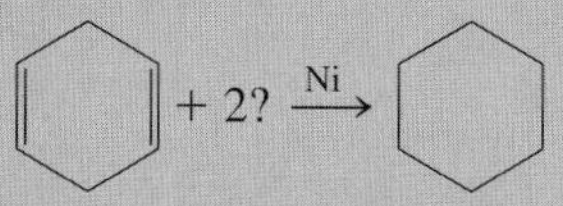

d. 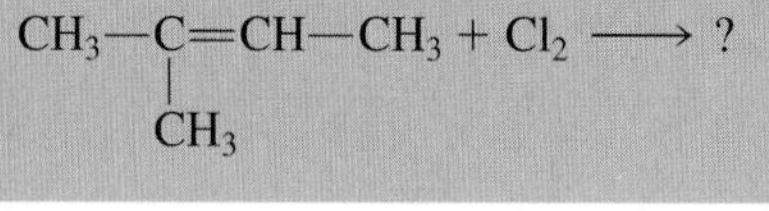 $CH_3-\underset{\underset{CH_3}{\mid}}{C}=CH-CH_3 + Cl_2 \longrightarrow ?$

13.8 Polymerization of Alkenes: Addition Polymers

▶ The word *polymer* comes from the Greek *poly*, which means "many," and *meros*, which means "parts."

A **polymer** *is a large molecule formed by the repetitive bonding together of many smaller molecules.* The smaller repeating units of a polymer are called *monomers*. A **monomer** *is the small molecule that is the structural repeating unit in a polymer.* The process by which a polymer is made is called *polymerization*. A **polymerization reaction** *is a reaction in which the repetitious combining of many small molecules (monomers) produces a very large molecule (the polymer).* With appropriate catalysts, simple alkenes and simple substituted alkenes readily undergo polymerization.

The type of polymer that alkenes and substituted alkenes form is an *addition polymer*. An **addition polymer** *is a polymer in which the monomers simply "add together" with no other products formed besides the polymer.* Addition polymerization is similar to the addition reactions described in Section 13.7 except that there is no reactant other than the alkene or substituted alkene.

▶ We will consider polymer types other than addition polymers in Sections 15.10 and 16.18.

The simplest alkene addition polymer has ethylene (ethene) as the monomer. With appropriate catalysts, ethylene readily adds to itself to produce polyethylene.

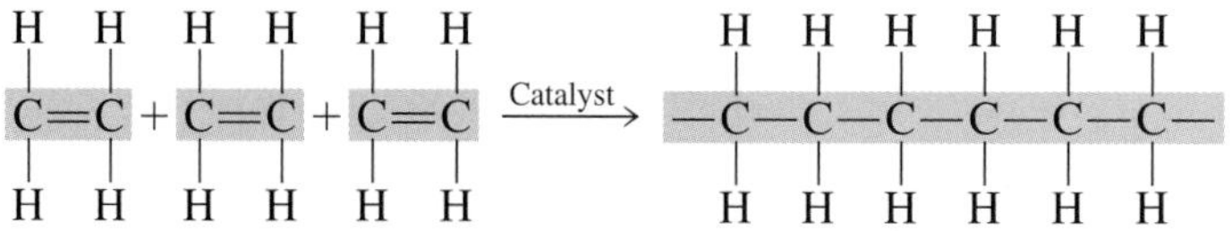

Polyethylene

An *exact* formula for a polymer such as polyethylene cannot be written, because the length of the carbon chain varies from polymer molecule to polymer molecule. In recognition of this "inexactness" of formula, the notation used for denoting polymer formulas is independent of carbon chain length. We write the formula of the simplest repeating unit (the monomer with the double bond changed to a single bond) in parentheses and then add the subscript n after the parentheses, with n being understood to represent a very large number. Using this notation, we have, for the formula of polyethylene,

$$\left(\begin{array}{cc} \overset{\textstyle H}{\underset{\textstyle |}{|}} & \overset{\textstyle H}{\underset{\textstyle |}{|}} \\ -C-C- \\ \underset{\textstyle H}{|} & \underset{\textstyle H}{|} \end{array}\right)_n$$

This notation clearly identifies the basic repeating unit found in the polymer.

■ Substituted-ethene Addition Polymers

Many substituted alkenes undergo polymerization similar to that of ethene when they are treated with the proper catalyst. For a monosubstituted-ethene monomer, the general polymerization equation is

$$H_2C=\overset{\textstyle Z}{\underset{\textstyle |}{C}}H \xrightarrow{\text{Polymerization}} +CH_2-\overset{\textstyle Z}{\underset{\textstyle |}{C}}H)_n$$

Variation in the substituent group Z can change polymer properties dramatically, as is shown by the entries in Table 13.1, a listing of monomers for the five ethene-based polymers *polyethylene, polypropylene, poly(vinyl chloride) (PVC), Teflon,* and *polystyrene,* along with several uses of each. Figure 13.11 depicts the preparation of polystyrene. Figure 13.12 contrasts the structures of polyethylene, polypropylene, and polyvinylchloride as depicted in space-filling models.

The properties of an ethene-based polymer depend not only on monomer identity but also on the average size (length) of polymer molecules and on the extent of polymer branching. For example, polyethylene has both high-density (HDPE) and low-density (LDPE) forms. HDPE, which has nonbranched polymer molecules, is a rigid material used in threaded bottle caps, toys, bottles, and milk jugs. LDPE, which has highly branched polymer molecules, is a flexible material used in plastic bags, plastic film, and squeeze bottles (see Figure 13.13). Objects made of HDPE hold their shape in boiling water, whereas objects made of LDPE become severely deformed at this temperature.

■ Butadiene-based Addition Polymers

When dienes such as 1,3-butadiene are used as the monomers in addition polymerization reactions, the resulting polymers contain double bonds and are thus still unsaturated.

$$CH_2=CH-CH=CH_2 \xrightarrow{\text{Polymerization}} +CH_2-CH=CH-CH_2)_n$$

1,3-Butadiene   Polybutadiene

FIGURE 13
vested in N

FIGURE 13.11 Preparation of polystyrene. When styrene, $C_6H_5-CH=CH_2$, is heated with a catalyst (benzoyl peroxide), it yields a viscous liquid. After some time, this liquid sets to a hard plastic (sample shown at left).

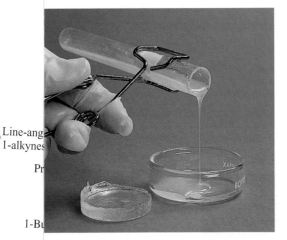

Line-ang
1-alkynes

Pr

1-Bu

1-Penty

1-Hexyne

FIGURE 13.15
of ethyne (acet
The molecule is
angles are 180°

FIGURE 13.12 Line-angle drawings and space-filling models of segments of the ethene-based polymers (a) polyethylene, (b) polypropylene, and (c) poly (vinyl chloride).

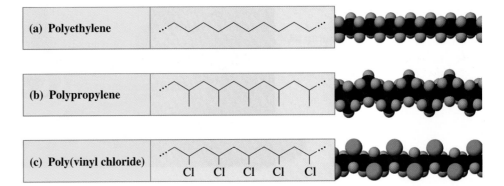

(a) **Polyethylene**

(b) **Polypropylene**

(c) **Poly(vinyl chloride)**   Cl   Cl   Cl   Cl   Cl

TABLE 13.
Some Com
from Ethen

CHEMISTRY AT A GLANCE

Chemical Reactions of Alkenes

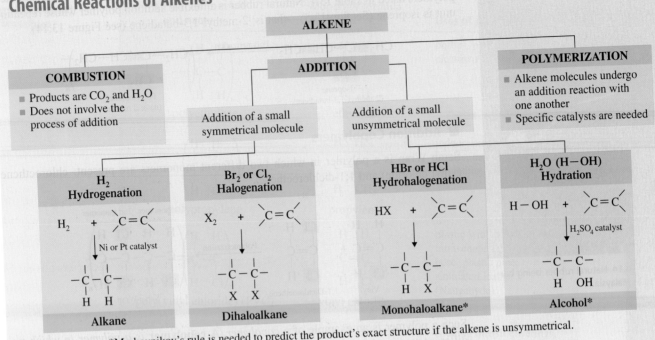

*Markovnikov's rule is needed to predict the product's exact structure if the alkene is unsymmetrical.

Cycloalkynes, molecules that contain a triple bond as part of a ring structure, are known, but they are not common. Because of the 180° angle associated with a triple bond, a ring system containing a triple bond has to be quite large. The smallest cycloalkyne that has been isolated is cyclooctyne.

Early cars had carbide headlights that produced acetylene by the action of slowly dripping water on calcium carbide. This same type of lamp, which was also used by miners, is still often used by spelunkers (cave explorers).

FIGURE 13.13 E
of polyethylene. F
are strong and rig
for wire) contain
polyethylene). Pol
are very flexible (
ing materials) con
polyethylene).

FIGURE 13.16 A physical-state summary for unbranched 1-alkynes at room temperature and pressure.

Unbranched 1-Alkynes		
✕	C_3	C_5	C_7
C_2	C_4	C_6	C_8

☐ Gas ☐ Liquid

■ IUPAC Nomenclature for Alkynes

The rules for naming alkynes are identical to those used to name alkenes (Section 13.3), except the ending *-yne* is used instead of *-ene.* Consider the following structures and their IUPAC names.

$$\overset{4}{CH_3}-\overset{3}{CH}-\overset{2}{C}\equiv\overset{1}{CH}$$
$$\underset{CH_3}{|}$$
3-Methyl-1-butyne

$$\overset{1}{CH_3}-\overset{2}{CH_2}-\overset{3}{C}\equiv\overset{4}{C}-\overset{5}{CH_2}-\overset{6}{\underset{|}{C}}-\overset{7}{CH_3}$$
6,6-Dimethyl-3-heptyne

$$\overset{1}{CH}\equiv\overset{2}{C}-\overset{3}{CH_2}-\overset{4}{CH_2}-\overset{5}{CH_2}-\overset{6}{C}\equiv\overset{7}{CH}$$
1,6-Heptadiyne

Common names for simple alkynes are based on the name *acetylene,* as shown in the following examples.

$$CH\equiv CH \qquad CH_3-C\equiv CH \qquad CH_3-C\equiv C-CH_3$$
Acetylene   Methylacetylene   Dimethylacetylene

■ Physical and Chemical Properties of Alkynes

The physical properties of alkynes are similar to those of alkenes and alkanes. In general, alkynes are insoluble in water but soluble in organic solvents, have densities less than that of water, and have boiling points that increase with molecular mass. Low-molecular-mass alkynes are gases at room temperature. Figure 13.16 is a physical-state summary for unbranched 1-alkynes with eight or fewer carbon atoms.

The triple-bond functional group of alkynes behaves chemically quite similarly to the double-bond functional group of alkenes. Thus there are many parallels between alkene chemistry and alkyne chemistry. The same substances that add to double bonds (H_2, HCl,

Cl_2, and so on) also add to triple bonds. However, two molecules of a specific reactant can add to a triple bond, as contrasted to the addition of one molecule of reactant to a double bond. In triple-bond addition, the first molecule converts the triple bond into a double bond, and the second molecule then converts the double bond into a single bond. For example, propyne reacts with H_2 to form propene first and then to form propane.

$$CH\equiv C-CH_3 \xrightarrow[Ni]{H_2} CH_2\equiv CH-CH_3 \xrightarrow[Ni]{H_2} CH_3=CH_2-CH_3$$

An alkyne	An alkene	An alkane
(propyne)	(propene)	(propane)

Students often ask whether it is possible to have hydrocarbons in which both double and triple bonds are present. The answer is yes. Immediately, another question is asked. How do we name such compounds? Such compounds are called *alkenynes*. An example is

$$CH\equiv C-CH=CH_2$$

1-Buten-3-yne

A double bond has priority over a triple bond in numbering the chain when numbering systems are equivalent. Otherwise, the chain is numbered from the end closest to a multiple bond.

13.10 Aromatic Hydrocarbons

Aromatic hydrocarbons are the third class of unsaturated hydrocarbons; the alkenes and alkynes (previously considered) are the other two classes. An **aromatic hydrocarbon** *is an unsaturated cyclic hydrocarbon that does not readily undergo addition reactions*. This reaction behavior, which is very different from that of alkenes and alkynes, explains the separate classification for aromatic hydrocarbons.

The fact that even though they are unsaturated compounds, aromatic hydrocarbons do not readily undergo addition reactions suggests that the bonding present in this type of compound must differ significantly from that in alkenes and alkynes. Such is indeed the case.

Let's look at the bonding present in *benzene,* the simplest aromatic hydrocarbon, to explore this new type of bonding situation and to also characterize the aromatic hydrocarbon functional group. Benzene, a flat, symmetrical molecule with a molecular formula of C_6H_6 (see Figure 13.17), has a structural formula that is often formalized as that of a cyclohexatriene—in other words, as a structural formula that involves a six-membered carbon ring in which three double bonds are present.

This structure is one of two equivalent structures that can be drawn for benzene that differ only in the locations of the double bonds (1,3,5 positions versus 2,4,6 positions):

FIGURE 13.17 Space-filling and ball-and-stick models for the structure of benzene.

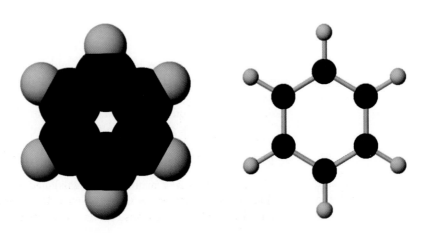

Neither of these conventional structures, however, is totally correct. Experimental evidence indicates that all of the carbon–carbon bonds in benzene are equivalent (identical), and these structures imply three bonds of one type (double bonds) and three bonds of a different type (single bonds).

The equivalent nature of the carbon–carbon bonds in benzene is addressed by considering the correct bonding structure for benzene to be an *average* of the two "triene" structures. Related to this "average"-structure situation is the concept that electrons associated with the ring double bonds are not held between specific carbon atoms; instead, they are free to move "around" the carbon ring. Thus the true structure for benzene, an intermediate between that represented by the two "triene" structures, is a situation in which all carbon–carbon bonds are equivalent; they are neither single nor double bonds but something in between. Placing a double-headed arrow between the conventional structures that are averaged to obtain the true structure is one way to denote the average structure.

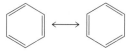

An alternative notation for denoting the bonding in benzene—a notation that involves a single structure—is

In this "circle-in-the-ring" structure for benzene, the circle denotes the electrons associated with the double bond that move "around" the ring. Each carbon atom in the ring can be considered to participate in three conventional (localized) bonds (two C—C bonds and one C—H bond) and in one *delocalized bond* (the circle) that involves all six carbon atoms. A **delocalized bond** *is a covalent bond in which electrons are shared among more than two atoms.* This delocalized bond is what causes benzene and its derivatives to be resistant to addition reactions, a property normally associated with unsaturation in a molecule.

The structure represented by the notation

is called an *aromatic ring system,* and it is the functional group present in aromatic compounds. An **aromatic ring system** *is a highly unsaturated carbon ring system in which both localized and delocalized bonds are present.*

13.11 Names for Aromatic Hydrocarbons

Replacement of one or more of the hydrogen atoms on benzene with other groups produces benzene derivatives. Compounds with alkyl groups or halogen atoms attached to the benzene ring are commonly encountered. We consider first the naming of benzene derivatives with one substituent, then the naming of those with two substituents, and finally the naming of those with three or more substituents.

■ Benzene Derivatives with One Substituent

The IUPAC system of naming monosubstituted benzene derivatives uses the name of the substituent as a prefix to the name benzene. Examples of this type of nomenclature include

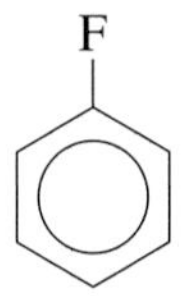

Fluorobenzene 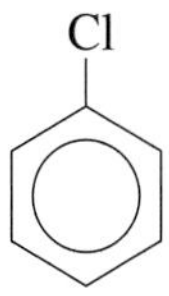   Chlorobenzene 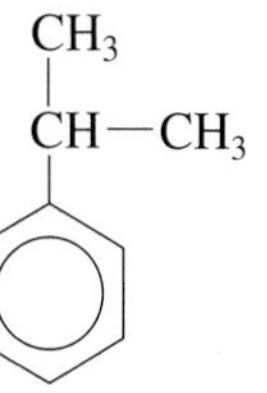   Isopropylbenzene 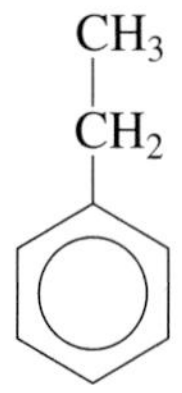   Ethylbenzene

A few monosubstituted benzenes have names wherein the substituent and the benzene ring taken together constitute a new parent name. Two important examples of such nomenclature with hydrocarbon substituents are

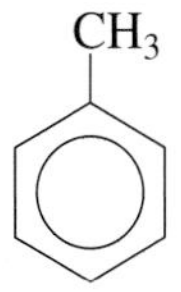

Toluene
(not methylbenzene)

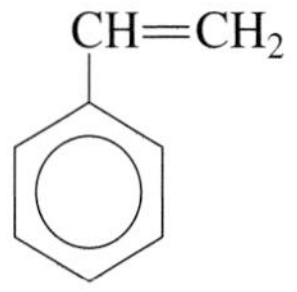

Styrene
(not vinylbenzene)

Both of these compounds are industrially important chemicals.

Monosubstituted benzene structures are often drawn with the substituent at the "12 o'clock" position, as in the previous structures. However, because all the hydrogen atoms in benzene are equivalent, it does not matter at which carbon of the ring the substituted group is located. Each of the following formulas represents toluene.

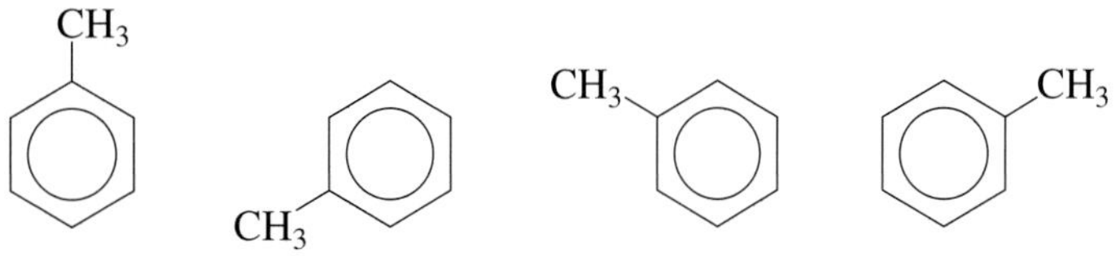

For monosubstituted benzene rings that have a group attached that is not easily named as a substituent, the benzene ring is often treated as a group attached to this substituent. In this reversed approach, the benzene ring attachment is called a *phenyl* group, and the compound is named according to the rules for naming alkanes, alkenes, and alkynes.

The word *phenyl* comes from "phene," a European term used during the 1800s for benzene. The word is pronounced *fen*-nil.

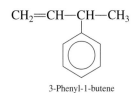

$CH_2\!=\!CH\!-\!CH\!-\!CH_3$

3-Phenyl-1-butene

■ Benzene Derivatives with Two Substituents

When two substituents, either the same or different, are attached to a benzene ring, three isomeric structures are possible.

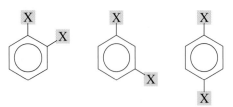

To distinguish among these three isomers, we must specify the positions of the substituents relative to one another. This can be done in either of two ways: by using numbers or by using nonnumerical prefixes.

▶ *Cis–trans* isomerism is not possible for disubstituted benzenes. All 12 atoms of benzene are in the same plane—that is, benzene is a flat molecule. When a substituent group replaces an H atom, the atom that bonds the group to the ring is also in the plane of the ring.

When numbers are used, the three isomeric dichlorobenzenes have the first-listed set of names:

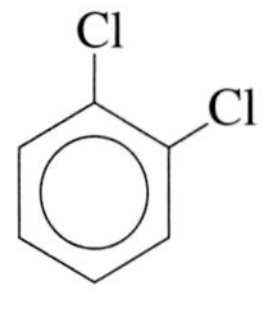

1,2-Dichlorobenzene
(*ortho*-dichlorobenzene)

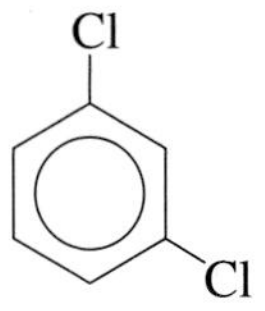

1,3-Dichlorobenzene
(*meta*-dichlorobenzene)

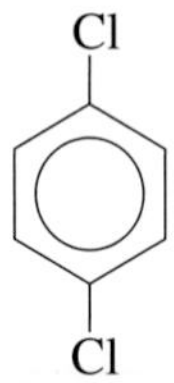

1,4-Dichlorobenzene
(*para*-dichlorobenzene)

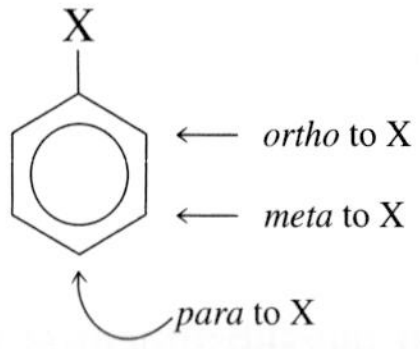

The prefix system uses the prefixes *ortho-*, *meta-*, and *para-* (abbreviated *o-*, *m-*, and *p-*).

Ortho- means 1,2 disubstitution; the substituents are on adjacent carbon atoms.
Meta- means 1,3 disubstitution; the substituents are one carbon removed from each other.
Para- means 1,4 disubstitution; the substituents are two carbons removed from each other (on opposite sides of the ring).

▶ Learn the meaning of the prefixes *ortho-*, *meta-*, and *para-*. These prefixes are extensively used in naming disubstituted benzenes.

⟵ *ortho* to X

⟵ *meta* to X

para to X

When prefixes are used, the three isomeric dichlorobenzenes have the second-listed set of names above.

When one of the two substituents in a disubstituted benzene imparts a special name to the compound (as, for example, toluene), the compound is named as a derivative of that parent molecule. The special substituent is assumed to be at ring position 1.

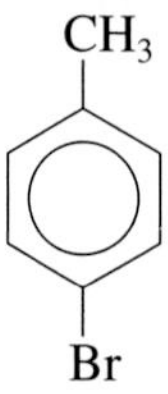

▶ The use of *ortho-*, *meta-*, and *para-* in place of position numbers is reserved for disubstituted benzenes. The system is never used with cyclohexanes or other ring systems.

4-Bromotoluene
(not 1-bromo-4-methylbenzene)

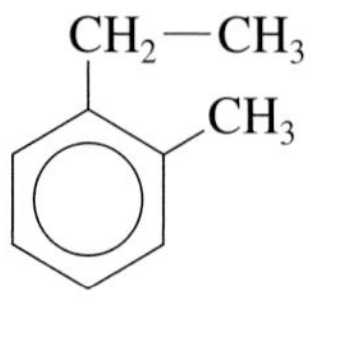

2-Ethyltoluene
(not 1-ethyl-2-methylbenzene)

When neither substituent group imparts a special name, the substituents are cited in alphabetical order before the ending *-benzene*. The carbon of the benzene ring bearing the substituent with alphabetical priority becomes carbon 1.

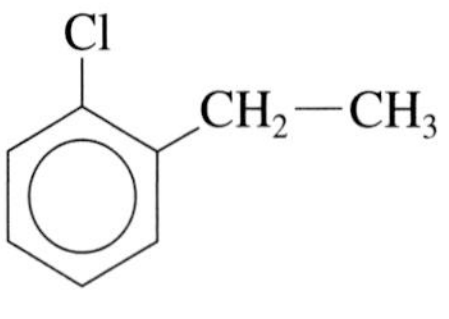

1-Chloro-2-ethylbenzene
(not 2-chloro-1-ethylbenzene)

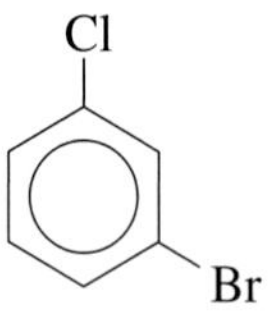

1-Bromo-3-chlorobenzene
(not 3-bromo-1-chlorobenzene)

▶ When parent names such as *toluene* and *xylene* are used, additional substituents present cannot be the same as those included in the parent name. If such is the case, name the compound as a substituted benzene. The compound

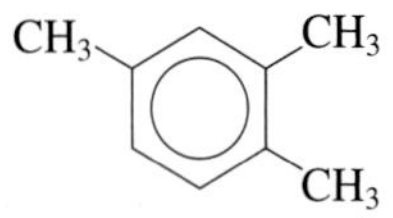

is named as a trimethylbenzene and not as a methylxylene or a dimethyltoluene.

A benzene ring bearing two methyl groups is a situation that generates a new special base name. Such compounds (there are three isomers) are not named as dimethylbenzenes or as methyl toluenes. They are called xylenes.

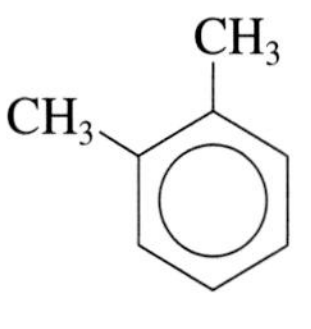

o-Xylene

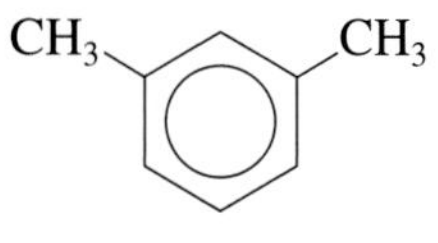

m-Xylene

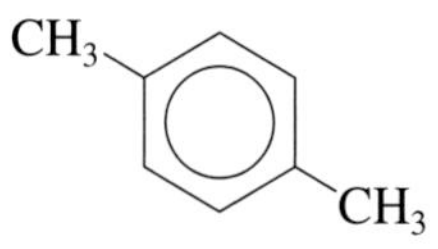

p-Xylene

The xylenes are good solvents for grease and oil and are used for cleaning microscope slides and optical lenses and for removing wax from skis.

■ Benzene Derivatives with Three or More Substituents

When more than two groups are present on the benzene ring, their positions are indicated with *numbers*. The ring is numbered in such a way as to obtain the lowest possible numbers for the carbon atoms that have substituents. If there is a choice of numbering systems (two systems give the same lowest set), then the group that comes first alphabetically is given the lower number.

1,2,4-Tribromobenzene   1-Bromo-3,5-dichlorobenzene

EXAMPLE 13.5

Assigning IUPAC Names to Benzene Derivatives

■ Assign IUPAC names to the following benzene derivatives.

a. Cl, CH₂—CH₃

b. Br, Cl, CH₂—CH₃

c. CH₃—CH—CH—CH₃ with Br and phenyl

d. CH₃, Cl

Solution

a. No substituents that will change the parent name from benzene are present on the ring. Alphabetical priority dictates that the chloro group is on carbon 1 and the ethyl group on carbon 3. The compound is named 1-chloro-3-ethylbenzene (or *m*-chloroethylbenzene).

b. Again, no substituents that will change the parent name from benzene are present on the ring. Alphabetical priority among substituents dictates that the bromo group is on carbon 1, the chloro group on carbon 3, and the ethyl group on carbon 5. The compound is named 1-bromo-3-chloro-5-ethylbenzene.

c. This compound is named with the benzene ring treated as a substituent—that is, as a phenyl group. The compound is named 2-bromo-3-phenylbutane.

d. The methyl group present on the benzene ring changes the parent name from benzene to toluene. Carbon 1 bears the methyl group. Numbering clockwise, we obtain the name 2-chlorotoluene. (See Figure 13.18)

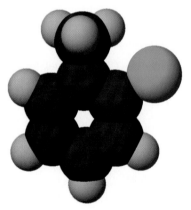

FIGURE 13.18 Space-filling model for the compound 2-chlorotoluene.

Practice Exercise 13.5

Assign IUPAC names to the following benzene derivatives.

a. Br, CH₂—CH₂—CH₃

b. CH₂—CH₂—CH₃, Cl

c. CH₃—CH₂—CH₂—CH—CH₂—CH₃ with phenyl

d. Br, Cl, Cl

13.12 Aromatic Hydrocarbons: Physical Properties and Sources

In general, aromatic hydrocarbons resemble other hydrocarbons in physical properties. They are insoluble in water, are good solvents for other nonpolar materials, and are less dense than water.

Benzene, monosubstituted benzenes, and many disubstituted benzenes are liquids at room temperature. Benzene itself is a colorless, flammable liquid that burns with a sooty flame because of incomplete combustion.

At one time, coal tar was the main source of aromatic hydrocarbons. Petroleum is now the primary source of such compounds. At high temperatures, with special catalysts, saturated hydrocarbons obtained from petroleum can be converted to aromatic hydrocarbons. The production of toluene from heptane is representative of such a conversion.

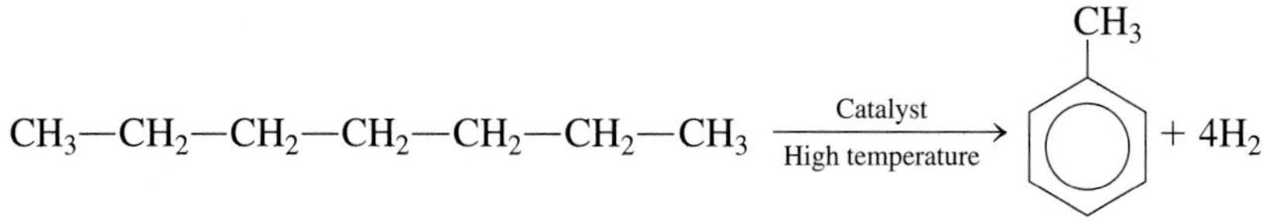

$$CH_3-CH_2-CH_2-CH_2-CH_2-CH_2-CH_3 \xrightarrow[\text{High temperature}]{\text{Catalyst}} \underset{}{\bigcirc}\!\!-CH_3 + 4H_2$$

> Two common situations in which a person can be exposed to low-level benzene vapors are
> 1. Inhaling of gasoline vapors while refueling an automobile. Gasoline contains about 2% (v/v) benzene.
> 2. Being around a cigarette smoker. Benzene is a combustion product present in cigarette smoke. For smokers themselves, inhaled cigarette smoke is a serious benzene-exposure source.

Benzene was once widely used as an organic solvent. Such use has been discontinued because benzene's short- and long-term toxic effects are now recognized. Benzene inhalation can cause nausea and respiratory problems.

13.13 Chemical Reactions of Aromatic Hydrocarbons

We have noted that aromatic hydrocarbons do not readily undergo the addition reactions characteristic of other unsaturated hydrocarbons. An addition reaction would require breaking up the delocalized bonding (Section 13.10) present in the ring system.

If benzene is so unresponsive to addition reactions, what reactions does it undergo? Benzene undergoes *substitution* reactions. As you recall from Section 12.16, substitution reactions are characterized by different atoms or groups of atoms replacing hydrogen atoms in a hydrocarbon molecule. Two important types of substitution reactions for benzene and other aromatic hydrocarbons are alkylation and halogenation.

1. *Alkylation:* An alkyl group (R—) from an alkyl chloride (R—Cl) substitutes for a hydrogen atom on the benzene ring. A catalyst, $AlCl_3$, is needed for alkylation.

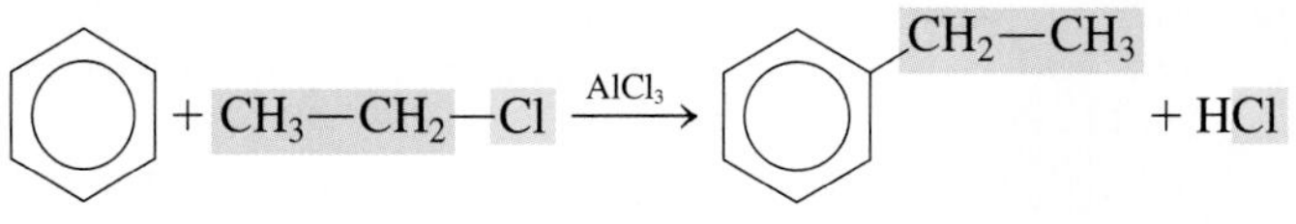

In general terms, the alkylation of benzene can be written as

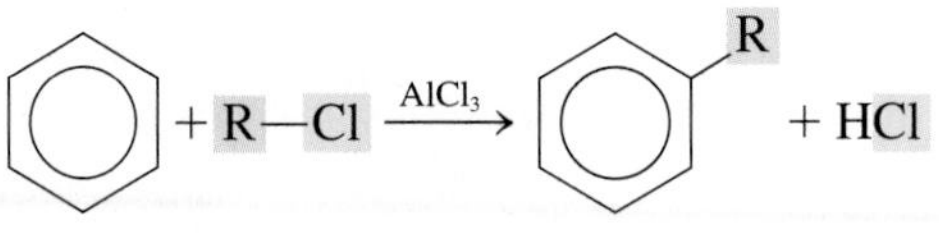

> Alkylation, the reaction that attaches an alkyl group to an aromatic ring, is also known as a *Friedel–Crafts reaction,* named after Charles Friedel and James Mason Crafts, the French and American chemists responsible for its discovery in 1877.

Alkylation is the most important industrial reaction of benzene.

2. *Halogenation* (bromination or chlorination): A hydrogen atom on a benzene ring can be replaced by bromine or chlorine if benzene is treated with Br_2 or Cl_2 in the presence of a catalyst. The catalyst is usually $FeBr_3$ for bromination and $FeCl_3$ for chlorination.

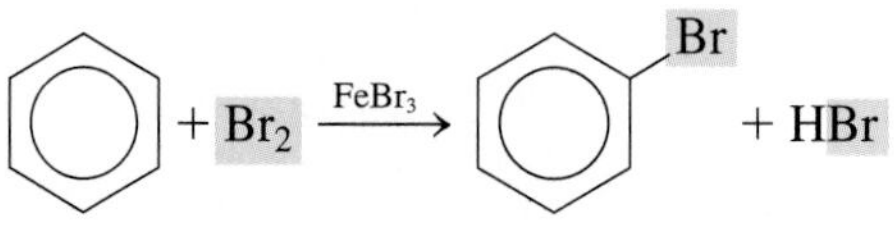

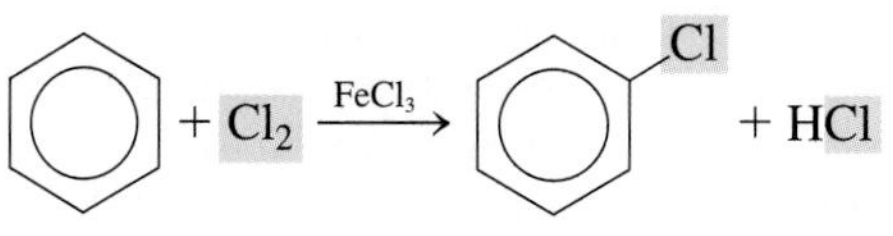

CHEMICAL CONNECTIONS

Fused-Ring Aromatic Hydrocarbons and Cancer

A number of fused-ring aromatic hydrocarbons are known to be carcinogens—that is, to cause cancer. Three of the most potent carcinogens are

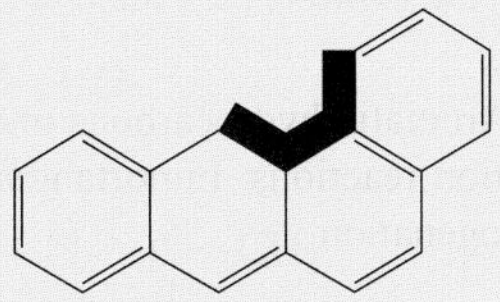

1,2-Benzanthracene

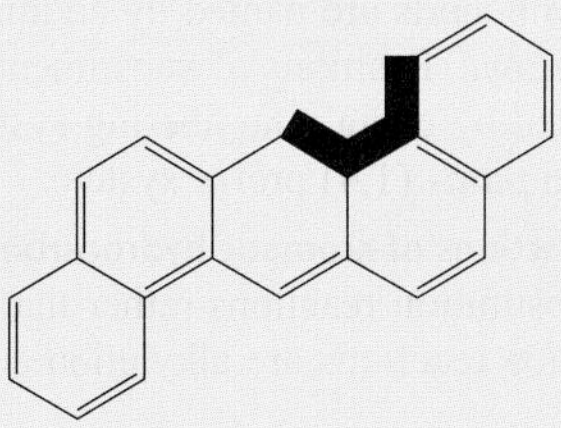

1,2,5,6-Dibenzanthracene

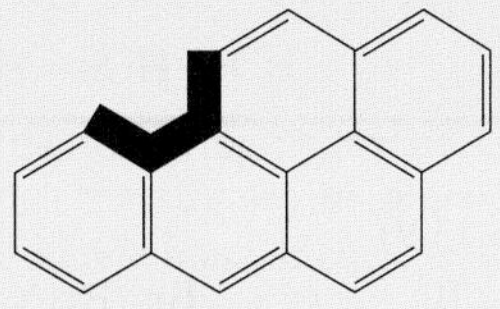

3,4-Benzpyrene

Very small amounts of these substances, when applied to the skin of mice, cause cancer.

Carcinogenic fused-ring aromatic hydrocarbons share some structural features. They all contain four or more fused rings, and they all have the same "angle" in the series of rings (the dark area in the structures shown). Fused-ring aromatic hydrocarbons are often formed when hydrocarbon materials are heated to high temperatures. These resultant compounds are present in low concentrations in tobacco smoke, in automobile exhaust, and sometimes in burned (charred) food. The charred portions of a well-done steak cooked over charcoal are a likely source.

Angular, fused-ring hydrocarbon systems are believed to be partially responsible for the high incidence of lung and lip cancer among cigarette smokers, because tobacco smoke contains 3,4-benzpyrene. The more a person smokes, the greater his or her risk of developing cancer.

We now know that the high incidence of lung cancer in British chimney sweeps (documented over 200 years ago) was caused by fused-ring hydrocarbon compounds present in the chimney soot that the sweeps inhaled regularly.

Aromatic halogenation differs from alkane halogenation (Section 12.16) in that light is not required to initiate aromatic halogenation.

13.14 Fused-Ring Aromatic Compounds

Benzene and its substituted derivatives are not the only type of aromatic hydrocarbon that exists. Another large class of aromatic hydrocarbons is the fused-ring aromatic hydrocarbons. A **fused-ring aromatic hydrocarbon** *is an aromatic hydrocarbon whose structure contains two or more rings fused together.* Two carbon rings that share a pair of carbon atoms are said to be *fused.*

The three simplest fused-ring aromatic compounds are naphthalene, anthracene, and phenanthrene. All three are solids at room temperature.

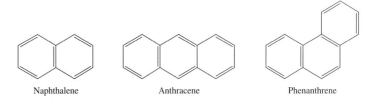

Naphthalene   Anthracene   Phenanthrene

CONCEPTS TO REMEMBER

Unsaturated hydrocarbons. An unsaturated hydrocarbon is a hydrocarbon that contains one or more carbon–carbon multiple bonds. Three main classes of unsaturated hydrocarbons exist: alkenes, alkynes, and aromatic hydrocarbons.

Alkenes and cycloalkenes. An alkene is an acyclic unsaturated hydrocarbon in which one or more carbon–carbon double bonds are present. A cycloalkene is a cyclic unsaturated hydrocarbon that contains one or more carbon–carbon double bonds within the ring system.

Alkene nomenclature. Alkenes and cycloalkenes are given IUPAC names using rules similar to those for alkanes and cycloalkanes, except that the ending *-ene* is used. Also, the double bond takes precedence both in selecting and in numbering the main chain or ring.

Cis–trans isomers. Because rotation about a carbon–carbon double bond is restricted, some alkenes exist in two isomeric (*cis–trans*) forms. *Cis–trans* isomerism is possible when each carbon of the double bond is attached to two different groups.

Physical pr
physical pr
than water,

Addition re
Cl_2, Br_2, H
ble bond. V
metrical, th
carbon ato
of H atom

Addition p
monomers
Many fami
are additio

Alkynes an
hydrocarb
They are n

1. Haloger

2. Hydrog

3. Hydroh

4. Hydrat

5. Hydrog

Addition
Addition
Alkene (1
Alkenyl g
Alkyne (
Aromatic
Aromatic
***Cis* isome**
Copolym

13.14 The following names are *incorrect* by IUPAC rules. Determine the correct IUPAC name for each compound.
a. 2-Methyl-4-pentene
b. 3-Methyl-2,4-pentadiene
c. 3-Methyl-3-cyclopentene
d. 1,2-Dimethyl-3-cyclohexene

■ Isomerism in Alkenes (Section 13.4)

13.15 Draw skeletal formulas showing only carbon atoms, and give the IUPAC names for the 13 possible alkene structural isomers with the formula C_6H_{12}. (Three of the structural isomers are hexenes, six are methylpentenes, three are dimethylbutenes, and one is an ethylbutene.)

13.16 Draw skeletal formulas showing only carbon atoms, and give the IUPAC names for the 16 possible alkadiene structural isomers with the formula C_6H_{10}. (Six of the structural isomers are hexadienes, eight are methylpentadienes, one is a dimethylbutadiene, and one is an ethylbutadiene.)

13.17 For each molecule, tell whether *cis–trans* isomers exist. If they do, draw the two isomers and label them as *cis* and *trans*.
a. $CH_2{=}CH{-}CH_3$ b. $CH_2{=}CH{-}CH_2{-}Cl$ (Cl below)

c. $CH_3{-}C{=}CH{-}CH_3$ (with CH_3 below C)   d. 3-Hexene

e. 4-Methyl-2-pentene   f. 1,2-Dimethylcyclopentane

13.18 For each molecule, tell whether *cis–trans* isomers exist. If they do, draw the two isomers and label them as *cis* and *trans*.
a. $CH_3{-}CH_2{-}CH{=}CH_2$ b. $CH_3{-}CH_2{-}C{=}CH_2$ (with Cl below C)

c. $CH_3{-}CH_2{-}CH{=}CH$ (with Cl below)   d. 2-Pentene

e. 1,2-Dichloroethene   f. 1,3-Dichlorocyclobutane

13.19 Assign an IUPAC name to each of the following molecules. Include the prefix *cis-* or *trans-* when appropriate.

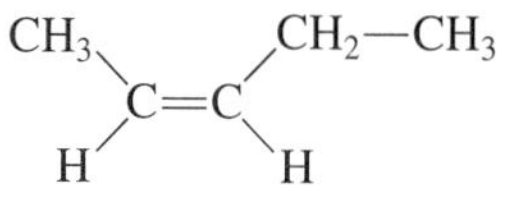

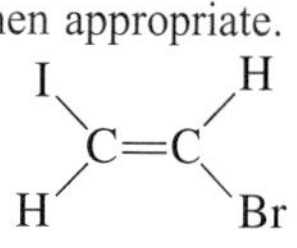

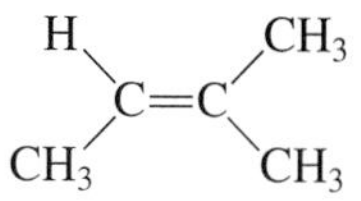

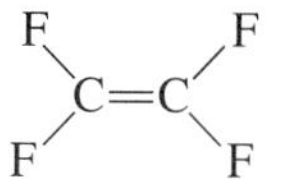

13.20 Assign an IUPAC name to each of the following molecules. Include the prefix *cis-* or *trans-* when appropriate.

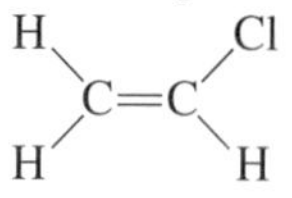

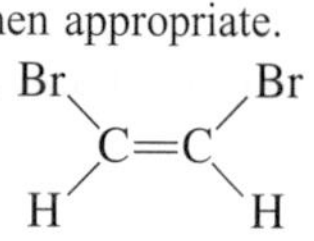

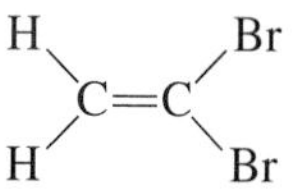

13.21 Draw a structural formula for each of the following compounds.
a. *trans*-3-Methyl-3-hexene   b. *cis*-2-Pentene
c. *trans*-5-Methyl-2-heptene   d. *trans*-1,3-Pentadiene

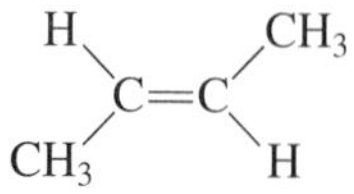

13.22 Draw a structural formula for each of the following compounds.
a. *trans*-2-Hexene   b. *cis*-4-Methyl-2-pentene
c. *cis*-1-Chloro-1-pentene   d. *cis*-1,3-Pentadiene

■ Naturally Occurring Alkenes (Section 13.5)

13.23 What is a pheromone?

13.24 What is a terpene?

13.25 Why is the number of carbon atoms in a terpene always a multiple of the number 5?

13.26 What is the structural relationship between β-carotene and vitamin A?

■ Physical Properties of Alkenes (Section 13.6)

13.27 Indicate whether each of the following alkenes would be expected to be a solid, a liquid, or a gas at room temperature and pressure?
a. Propene   b. 1-Pentene   c. 1-Octene   d. Cyclopentene

13.28 Indicate whether each of the following statements is true or false.
a. 1-butene has a density greater than that of water.
b. 1-butene has a higher boiling point than 1-hexene.
c. 1-butene is flammable but 1-hexene is not.
d. Both 1-pentene and cyclopentene are gases at room temperature and pressure.

■ Alkene Addition Reactions (Section 13.7)

13.29 Which of the following reactions are addition reactions?
a. $C_4H_8 + Cl_2 \rightarrow C_4H_8Cl_2$
b. $C_6H_6 + Cl_2 \rightarrow C_6H_5Cl + HCl$
c. $C_3H_6 + HCl \rightarrow C_3H_7Cl$
d. $C_7H_{16} \rightarrow C_7H_8 + 4H_2$

13.30 Which of the following reactions are addition reactions?
a. $C_3H_6 + Cl_2 \rightarrow C_3H_6Cl_2$
b. $C_8H_{10} \rightarrow C_8H_8 + H_2$
c. $C_6H_6 + C_2H_5Cl \rightarrow C_8H_{10} + HCl$
d. $C_4H_8 + HCl \rightarrow C_4H_9Cl$

13.31 Write a chemical equation, showing reactants, products, and catalysts needed (if any), for the reaction of ethene with each of the following substances.
a. Cl_2 b. HCl   c. H_2 d. HBr

13.32 Write a chemical equation, showing reactants, products, and catalysts needed (if any), for the reaction of ethene with each of the following substances.
a. H_2O b. Br_2 c. HI   d. I_2

13.33 Write a chemical equation, showing reactants, products, and catalysts needed (if any), for the reaction of propene with each of the reactants in Problem 13.31. Use Markovnikov's rule as needed.

13.34 Write a chemical equation, showing reactants, products, and catalysts needed (if any), for the reaction of propene with each of the reactants in Problem 13.32. Use Markovnikov's rule as needed.

13.35 Supply the structural formula of the product in each of the following alkene addition reactions.
a. $CH_3{-}CH{=}CH{-}CH_3 + Cl_2 \rightarrow$?
b. $CH_3{-}C{=}CH_2 + HBr \rightarrow$? (with CH_3 below C)
c. $CH_3{-}CH_2{-}CH{=}CH_2 + HCl \rightarrow$?
d. 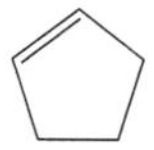 (cyclopentane) $+ H_2 \xrightarrow[\text{catalyst}]{\text{Ni}}$?
e. 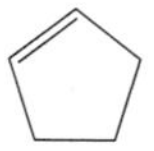 (cyclopentane) $+ H_2 \xrightarrow[\text{catalyst}]{\text{no}}$?
f. 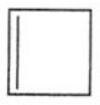 (cyclobutane) $+ H_2O \xrightarrow{H_2SO_4}$?

13.36 Supply the structural formula of the product in each of the following alkene addition reactions.

a. $CH_3-CH_2-CH=CH_2 + Cl_2 \rightarrow$?

b. 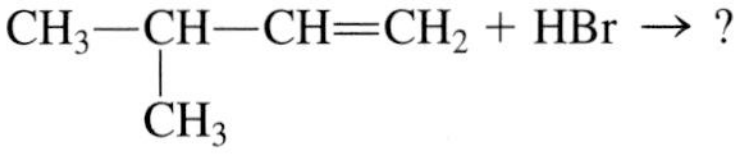 $CH_3-CH-CH=CH_2 + HBr \rightarrow$?
$\quad\quad\quad |$
$\quad\quad CH_3$

c. 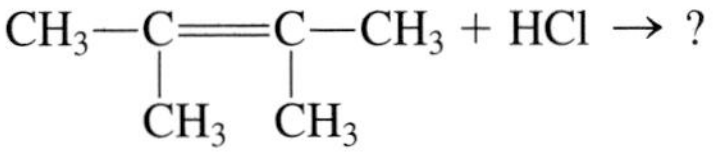 $CH_3-C=C-CH_3 + HCl \rightarrow$?
$\quad\quad\quad | \quad\; |$
$\quad\quad CH_3 \; CH_3$

d. 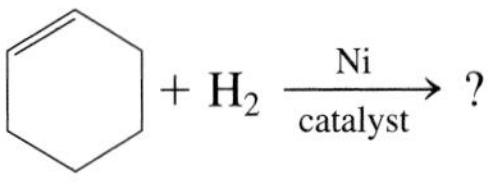
 $+ H_2 \xrightarrow[\text{catalyst}]{\text{Ni}}$?

e. 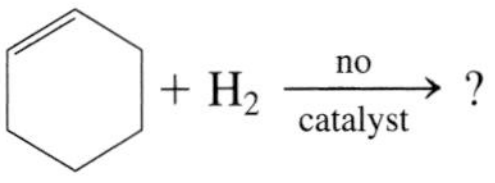
$+ H_2 \xrightarrow[\text{catalyst}]{\text{no}}$?

f. $CH_3-CH=CH_2 + H_2O \xrightarrow{H_2SO_4}$?

13.37 What reactant would you use to prepare each of the following compounds from cyclohexene?

a. 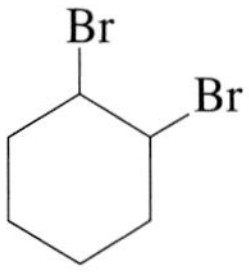 Br, Br

b.

c. 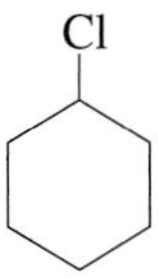 Cl

d. 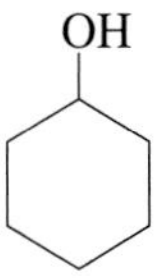 OH

13.38 What reactant would you use to prepare each of the following compounds from cyclopentene?

a.

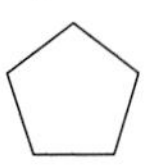

b. 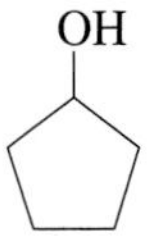 OH

c. 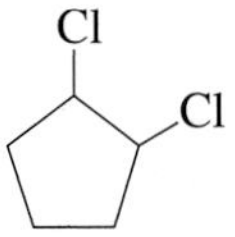 Cl, Cl

d. 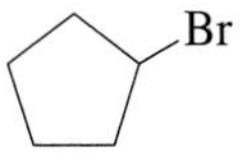 Br

13.39 How many molecules of H_2 gas will react with 1 molecule of each of the following unsaturated hydrocarbons?

a. $CH_3-CH=CH-CH=CH-CH_3$

b. 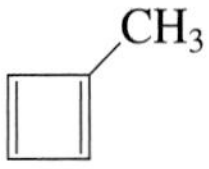
$\quad CH_3$

c. 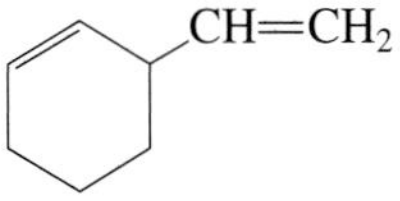 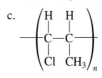 CH=CH_2

d. 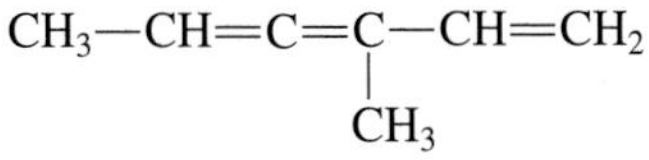 $CH_3-CH=C=C-CH=CH_2$
$\quad\quad\quad\quad\quad\quad |$
$\quad\quad\quad\quad\quad CH_3$

13.40 How many molecules of H_2 gas will react with 1 molecule of each of the following unsaturated hydrocarbons?

a. $CH_3-CH=CH-CH_3$

b. 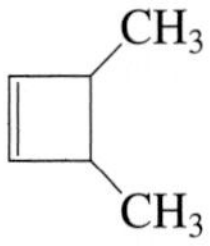
$\quad\quad CH_3$

$\quad\quad CH_3$

c. 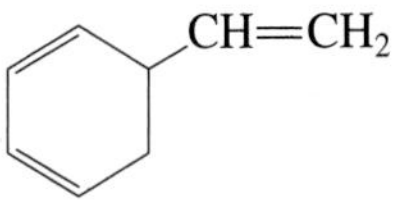 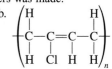 CH=CH_2

d. 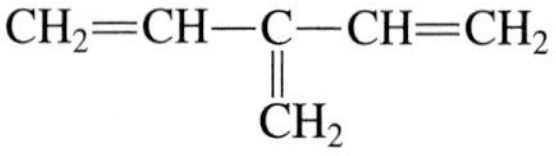 $CH_2=CH-C-CH=CH_2$
$\quad\quad\quad\quad\;\; ||$
$\quad\quad\quad\quad CH_2$

■ **Polymerization of Alkenes (Section 13.8)**

13.41 Draw the structural formula of the monomer(s) from which each of the following polymers was made.

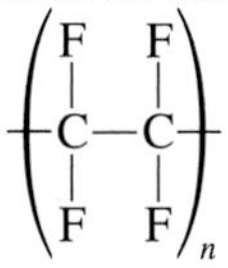

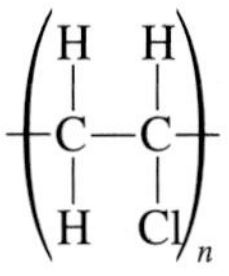

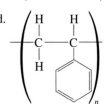

13.42 Draw the structural formula of the monomer(s) from which each of the following polymers was made.

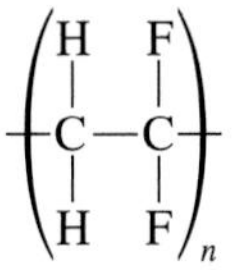

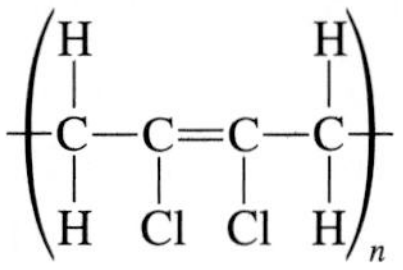

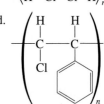

13.43 Draw the "start" (the first three repeating units) of the structural formula of the addition polymers made from the following monomers.

a. Ethylene
b. Vinyl chloride
c. 1,2-Dichloroethene
d. 1-Chloroethene

13.44 Draw the "start" (the first three repeating units) of the structural formula of the addition polymers made from the following monomers.

a. Propylene
b. 1,1,2,2-Tetrafluoroethene
c. 2-Methyl-1-propene
d. 1,2-Dichloroethylene

■ **Alkynes (Section 13.9)**

13.45 Assign an IUPAC name to each of the following unsaturated hydrocarbons.

a. $CH_3-CH_2-CH_2-CH_2-C\equiv CH$

b. 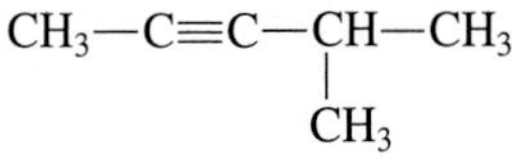 $CH_3-C\equiv C-CH-CH_3$
$\quad\quad\quad\quad\quad\quad |$
$\quad\quad\quad\quad\quad CH_3$

c.
$\quad\quad\quad CH_3$
$\quad\quad\quad |$
$CH_3-C-C\equiv C-CH_2-CH_2-CH_3$
$\quad\quad\quad |$
$\quad\quad\quad CH_3$

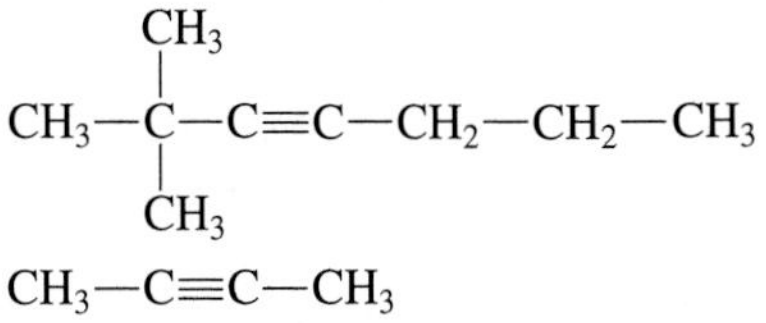

d. $CH_3-C\equiv C-CH_3$

e. CH≡C—CH—C≡C—CH₃
         |
        CH₃

f. CH₃—C≡C—CH—CH₃
           |
        CH₂—CH₃

13.46 Assign an IUPAC name to each of the following unsaturated hydrocarbons.

a. CH₃—CH—C≡CH
      |
     CH₃

b. CH₃—C≡C—CH₃

c. CH₃—CH—C≡C—CH—CH₃
      | |
     CH₃   CH₃

d. CH₃—CH—CH₂—C
        | ‖
      CH₂   CH
      |
     CH₃

e. CH≡C—C≡CH

f. CH₃—CH₂—CH—C≡CH
           |
        CH₂—CH₂—CH₃

13.47 Supply the structural formula of the product in each of the following alkyne addition reactions.

a. CH≡CH + 2H₂ $\xrightarrow{Ni}$?

b. CH₃—C≡CH + 2Br₂ ⟶ ?

c. CH₃—C≡CH + 2HBr ⟶ ?

d. CH≡CH + 1HCl ⟶ ?

e. 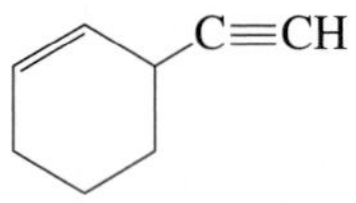 + 3H₂ $\xrightarrow{Ni}$?

f. CH₃—CH₂—C≡CH + 1HBr ⟶ ?

13.48 Supply the structural formula of the product in each of the following alkyne addition reactions.

a. CH₃—C≡C—CH₃ + 2Br₂ ⟶ ?

b. CH₃—C≡C—CH₃ + 2HBr ⟶ ?

c. CH≡C—CH₂—CH₃ + 1H₂ $\xrightarrow{Ni}$?

d. CH≡C—CH₃ + 1HCl ⟶ ?

e. 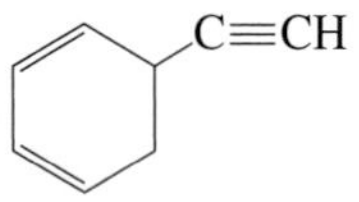 + 4H₂ $\xrightarrow{Ni}$?

f. CH₃—CH₂—C≡CH + 1HBr ⟶ ?

■ **Nomenclature for Aromatic Compounds (Section 13.11)**

13.49 Assign an IUPAC name to each of the following disubstituted benzenes. Use numbers rather than prefixes to locate the substituents on the benzene ring.

a. 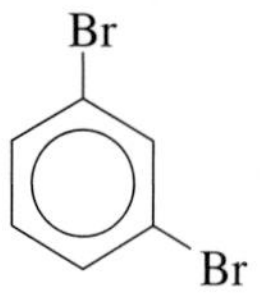
b.

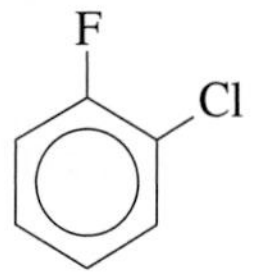

c. 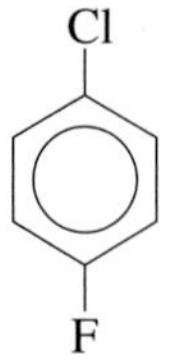
d.

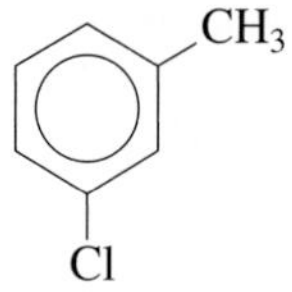

e. 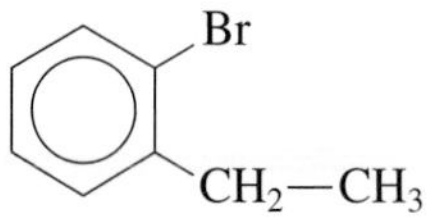
f.

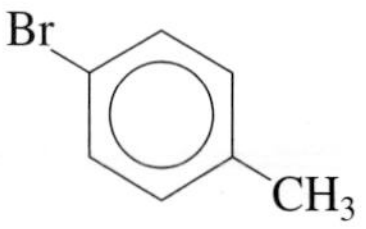

13.50 Assign an IUPAC name to each of the following disubstituted benzenes. Use numbers rather than prefixes to locate the substituents on the benzene ring.

a. 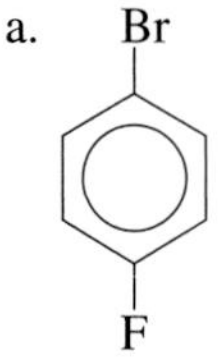
b.

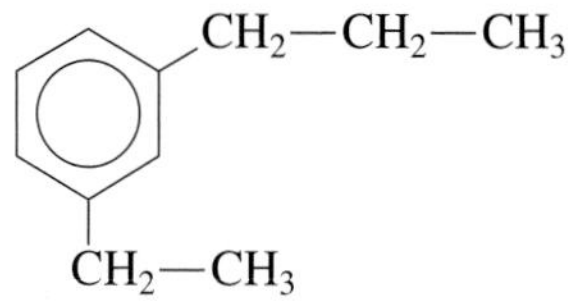

c. 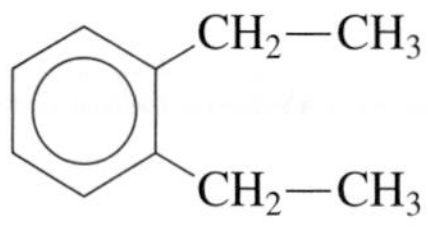
d.

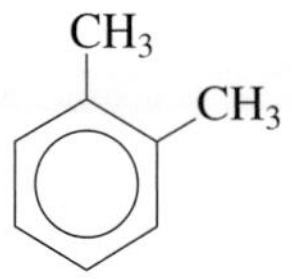

e. 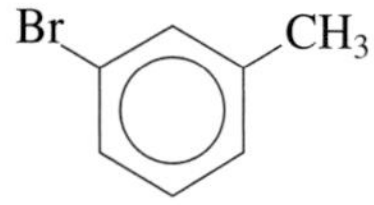
f.

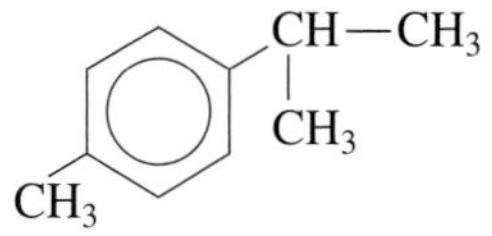

13.51 To each of the compounds in Problem 13.49, assign an IUPAC name in which the substituents on the benzene ring are located using the *ortho-*, *meta-*, *para-* prefix system.

13.52 To each of the compounds in Problem 13.50, assign an IUPAC name in which the substituents on the benzene ring are located using the *ortho-*, *meta-*, *para-* prefix system.

13.53 Assign an IUPAC name to each of the following substituted benzenes.

a. 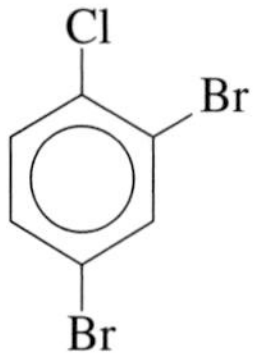
b.

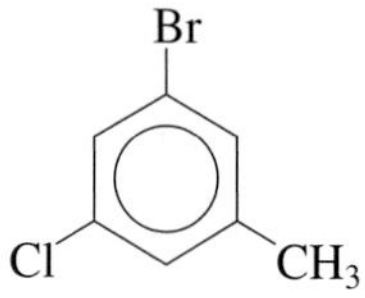

c. 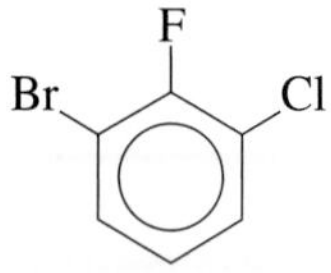
d.

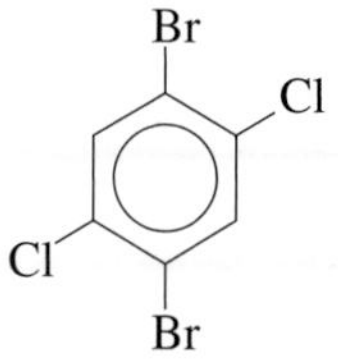

13.54 Assign an IUPAC name to each of the following substituted benzenes.

a. 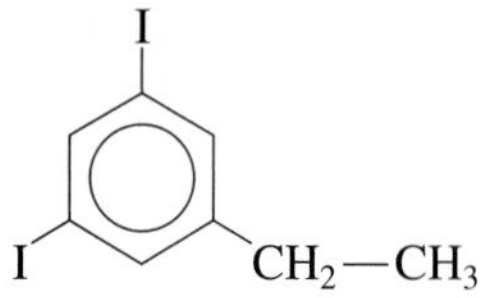
b.

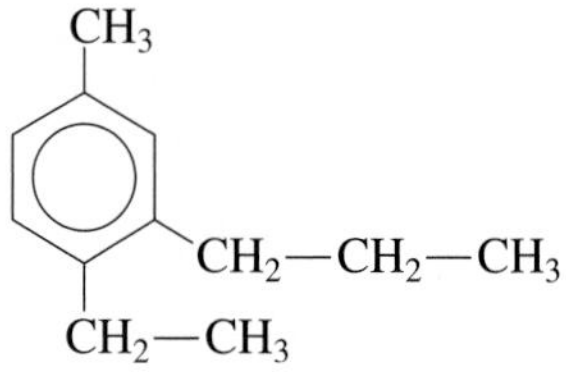

c.

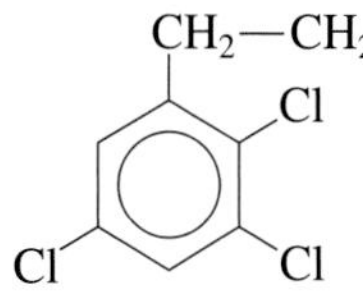

CH$_2$—CH$_2$—CH$_3$ d.

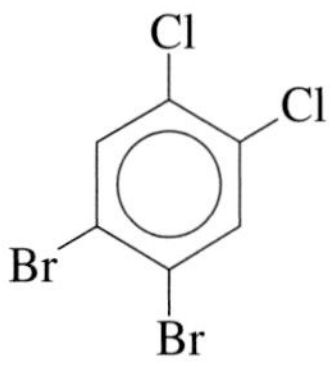

13.55 Assign an IUPAC name to each of the following compounds, in which the benzene ring is treated as a substituent.
a. 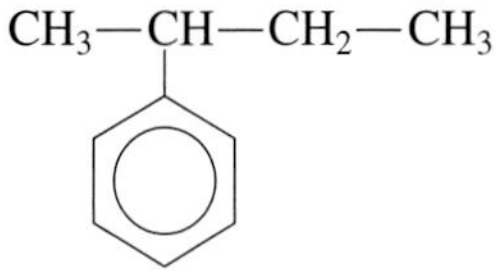 CH$_3$—CH—CH$_2$—CH$_3$

b. 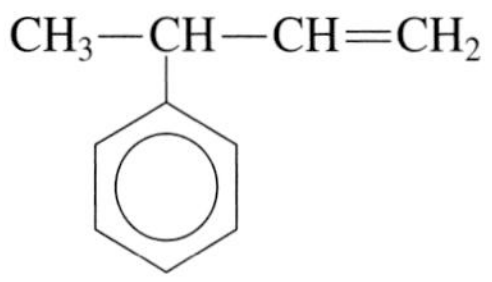 CH$_3$—CH—CH=CH$_2$

c. 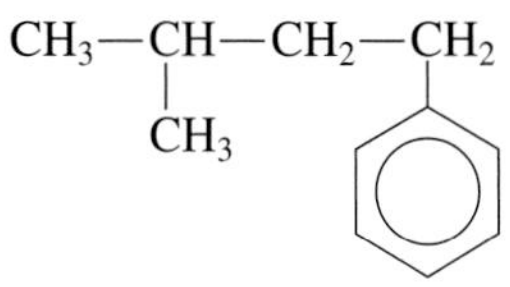 CH$_3$—CH—CH$_2$—CH$_2$
   |
   CH$_3$

d. 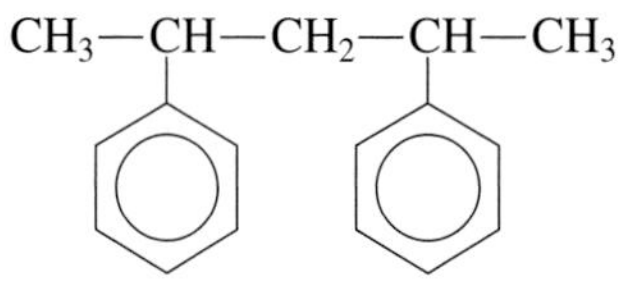 CH$_3$—CH—CH$_2$—CH—CH$_3$

13.56 Assign an IUPAC name to each of the following compounds, in which the benzene ring is treated as a substituent.
a. 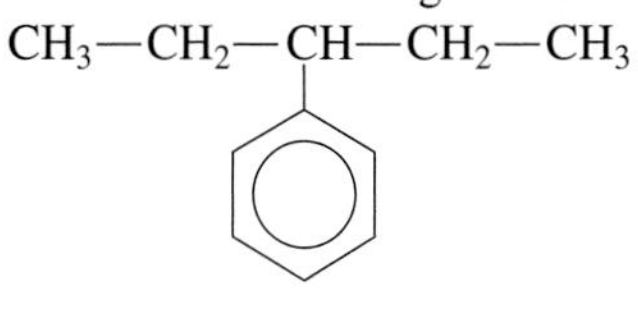 CH$_3$—CH$_2$—CH—CH$_2$—CH$_3$

b. 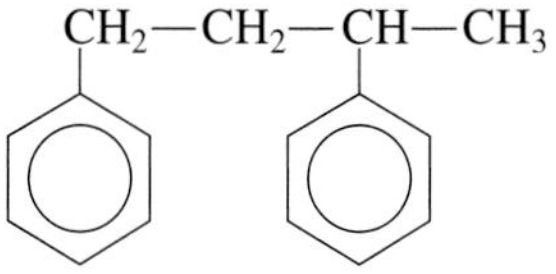 CH$_2$—CH$_2$—CH—CH$_3$

c. 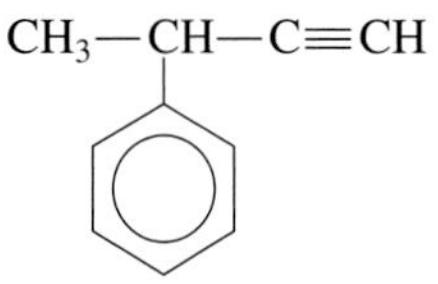 CH$_3$—CH—C≡CH

d. 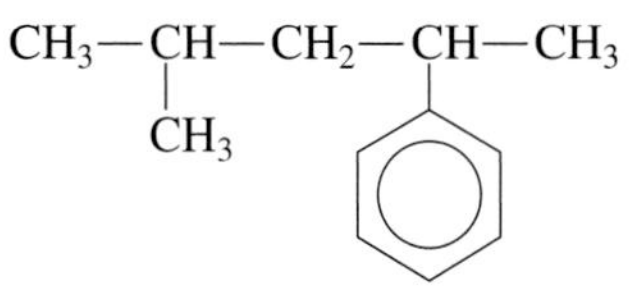 CH$_3$—CH—CH$_2$—CH—CH$_3$
   |
   CH$_3$

13.57 Write a structural formula for each of the following compounds.

a. 1,3-Diethylbenzene
b. *o*-Xylene
c. *p*-Ethyltoluene
d. Phenylbenzene
e. 1,2-Diphenylethane
f. 3-Methyl-3-phenylpentane

13.58 Write a structural formula for each of the following compounds.
a. *o*-Ethylpropylbenzene   b. *m*-Xylene
c. 2-Bromotoluene   d. 2-Phenylpropane
e. Isopropylbenzene   f. Triphenylmethane

■ **Chemical Reactions of Aromatic Hydrocarbons (Section 13.13)**

13.59 For each of the following classes of compounds, indicate whether addition or substitution is the most characteristic reaction for the class.
a. Alkanes   b. Dienes
c. Alkylbenzenes   d. Cycloalkenes

13.60 For each of the following classes of compounds, indicate whether addition or substitution is the most characteristic reaction for the class.
a. Alkynes   b. Cycloalkanes
c. Aromatic hydrocarbons   d. Saturated hydrocarbons

13.61 Complete the following reaction equations by supplying the formula of the missing reactant or product.
a.

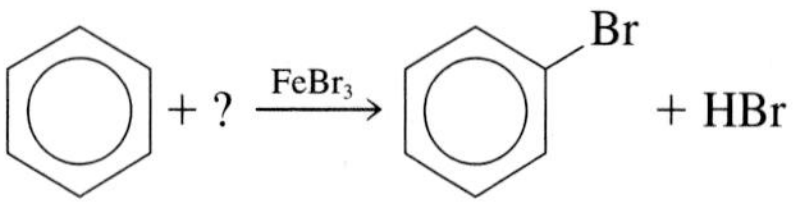

b.

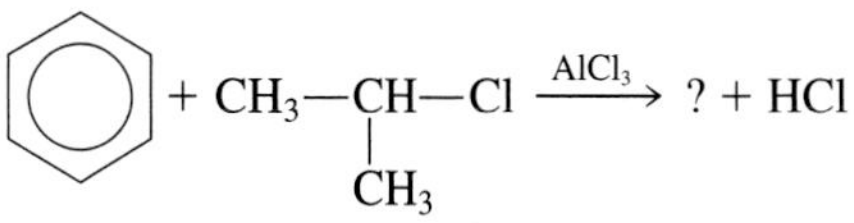

c.

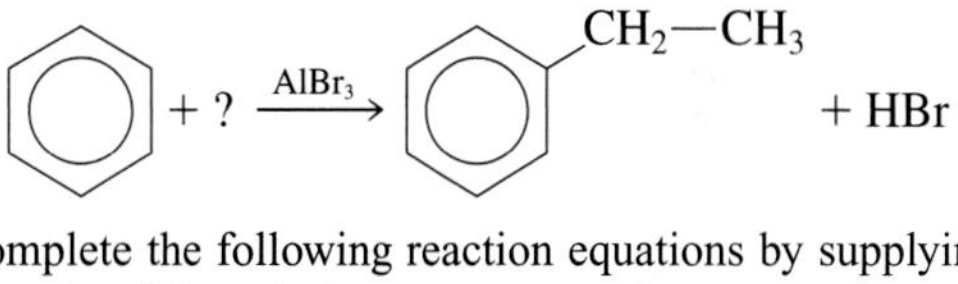

13.62 Complete the following reaction equations by supplying the formula of the missing reactant, product, or catalyst.
a.

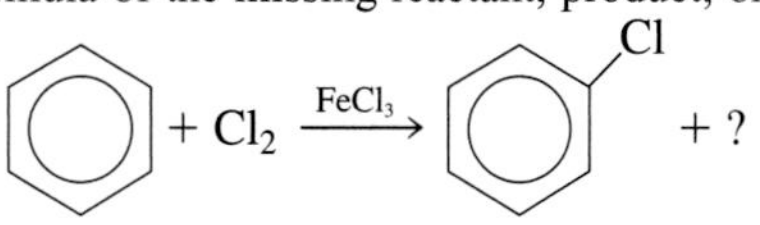

b.

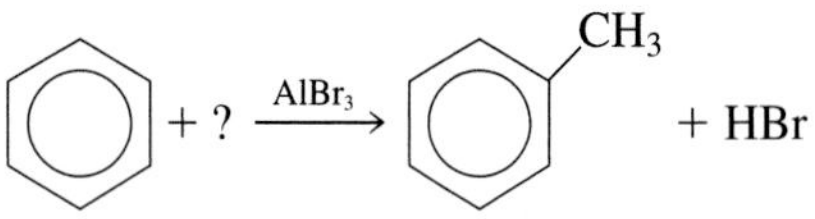

c.

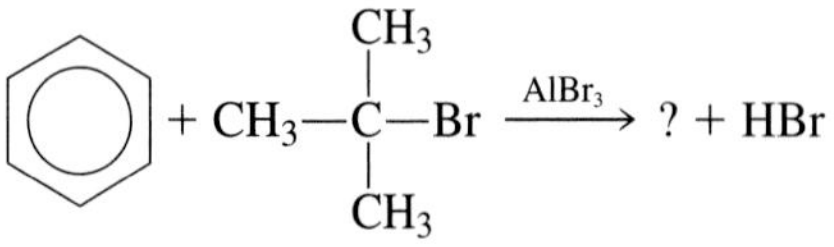

ADDITIONAL PROBLEMS

13.63 What is the molecular formula for the simplest compound of each of the following types?
a. Alkene with one multiple bond
b. Cycloalkene with one multiple bond
c. Alkyne with one multiple bond
d. Alkane

13.64 Indicate whether the hydrocarbon listed first in each of the following pairs of hydrocarbons contains (1) more hydrogen atoms, (2) the same number of hydrogen atoms, or (3) fewer hydrogen atoms than the hydrocarbon listed second.
a. Propane and propene
b. Propene and propyne
c. Propene and cyclopropene
d. Propyne and cyclopropene

13.65 Indicate whether each of the following pairs of hydrocarbons are *structural* isomers?
a. Propene and cyclopropene
b. 1-Pentene and 2-pentene
c. *cis*-2-Butene and *trans*-2-butene
d. Cyclobutene and 2-butyne

13.66 Contrast the compounds cyclohexane, cyclohexene, and benzene in terms of each of the following:
a. Number of carbon atoms present
b. Number of hydrogen atoms present
c. Whether they undergo substitution or addition reactions
d. Whether they are a solid, a liquid, or a gas at room temperature and pressure

13.67 Draw a condensed structural formula for each of the following unsaturated hydrocarbons.
a. 5-Methyl-2-hexyne
b. 1-Chloro-2-butene
c. 5,6-Dimethyl-2-heptyne

d. 3-Isopropyl-1-hexene
e. 1,6-Heptadiene
f. 3-Methyl-1,4-pentadiyne

13.68 How many molecules of H_2 will react with 1 molecule of each of the compounds in Problem 13.67 when the appropriate catalyst is present?

13.69 Draw a condensed structural formula for each of the following compounds.
a. Vinylbenzene
b. Allyl chloride
c. Propylacetylene
d. Dipropylacetylene
e. *o*-Xylene
f. *m*-Phenyltoluene

13.70 The compound 2-methyl-1-propene is a well-known substance. The compound 2,2-dimethyl-1-propene does not exist. Explain why this is so.

13.71 The compound 1,2-dichlorocyclohexane exists in *cis–trans* forms. However, *cis–trans* isomerism is not possible for the compound 1,2-dichlorobenzene. Explain why this is so.

13.72 Hydrocarbons with the formula C_5H_{10} can be either alkenes or cycloalkanes. Draw the nine possible structural isomers that fit this formula; five are alkenes and four are cycloalkanes. Then indicate which of these nine isomers exist in *cis–trans* forms.

13.73 There are eight isomeric substituted benzenes that have the formula C_9H_{12}. What are the IUPAC names for these eight structural isomers?

13.74 How many different compounds are there that fit each of the following descriptions?
a. Bromochlorobenzenes
b. Trichlorobenzenes
c. Dibromodichlorobenzenes
d. Monobromoanthracenes

ANSWERS TO PRACTICE EXERCISES

13.1 a. 5-methyl-3-hexene
b. 3-ethyl-4-methylcyclohexene
c. 1,3-butadiene
d. 5-methyl-1,3-cyclopentadiene

13.2 a. yes   b. no

13.3 a. 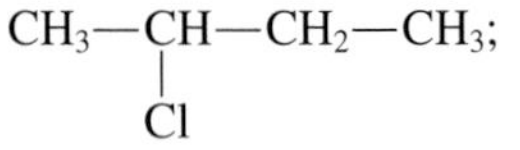 CH_3—CH—CH_2—CH_3;
            |
            Cl

b. 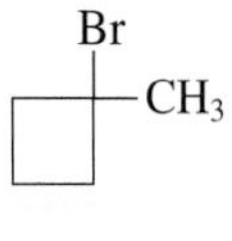 Br $\rule{0.4cm}{0pt}$—CH_3

13.4 a. 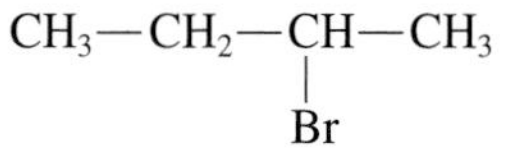 CH_3—CH_2—CH—CH_3 b. 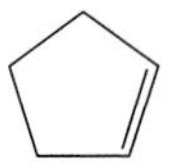
                      |
                      Br

c. H_2 d. 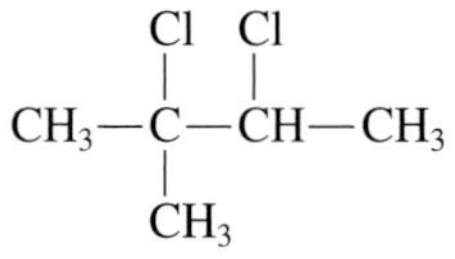
            Cl Cl
            | |
CH_3—C—CH—CH_3
            |
            CH_3

13.5 a. 1-bromo-3-propylbenzene (or *m*-bromopropylbenzene)
b. 1-chloro-4-propylbenzene (or *p*-chloropropylbenzene)
c. 3-phenylhexane
d. 4-bromo-1,2-dichlorobenzene

14

Alcohols, Phenols, and Ethers

The physiological effects of poison ivy are caused by certain phenol compounds present in the leaves.

This chapter is the first of three that consider hydrocarbon derivatives with *oxygen-containing functional groups*. Many biochemically important molecules contain carbon atoms bonded to oxygen atoms.

In this chapter we consider hydrocarbon derivatives whose functional groups contain one oxygen atom participating in two single bonds (alcohols, phenols, and ethers). Chapter 15 focuses on derivatives whose functional groups have one oxygen atom participating in a double bond (aldehydes and ketones), and in Chapter 16 we examine functional groups that contain two oxygen atoms, one participating in single bonds and the other in a double bond (carboxylic acids and esters).

14.1 Bonding Characteristics of Oxygen Atoms in Organic Compounds

An understanding of the bonding characteristics of the oxygen atom is a prerequisite to our study of compounds with oxygen-containing functional groups. Normal bonding behavior for oxygen atoms in such functional groups is the formation of two covalent bonds. Oxygen is a member of Group VIA of the periodic table and thus possesses six valence electrons. To complete its octet by electron sharing, an oxygen atom can form either two single bonds or a double bond.

Two single bonds   One double bond

361

FIGURE 14.1 Space-filling models for the three simplest unbranched-chain alcohols: methyl alcohol, ethyl alcohol, and propyl alcohol.

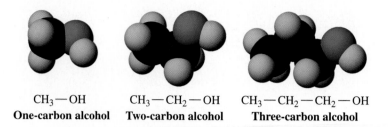

$CH_3—OH$
One-carbon alcohol $CH_3—CH_2—OH$
Two-carbon alcohol $CH_3—CH_2—CH_2—OH$
Three-carbon alcohol

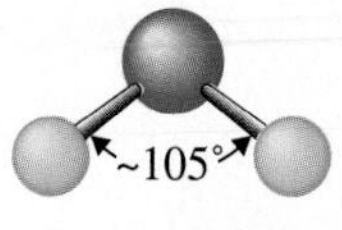

Water (HOH) ~105°

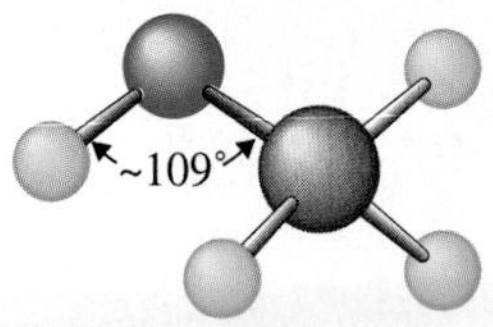

Methyl alcohol (CH₃OH) ~109°

FIGURE 14.2 The similar shapes of water and methanol. Methyl alcohol may be viewed structurally as an alkyl derivative of water.

▶ The hydroxyl group (—OH) should not be confused with the *hydroxide ion* (OH⁻) that we have encountered previously. Alcohols are not *hydroxides.* Hydroxides are ionic compounds that contain the OH⁻ polyatomic ion (Section 4.10). Alcohols are not ionic compounds. In an alcohol, the —OH group, which is not an ion, is *covalently* bonded to a saturated carbon atom.

Thus, in organic chemistry, carbon forms four bonds, hydrogen forms one bond, and oxygen forms two bonds.

$$—\overset{|}{\underset{|}{C}}—\qquad H—\qquad :\overset{..}{\underset{..}{O}}—$$

4 valence electrons, 4 covalent bonds, no nonbonding electron pairs

1 valence electron, 1 covalent bond, no nonbonding electron pairs

6 valence electrons, 2 covalent bonds, 2 nonbonding electron pairs

14.2 Structural Characteristics of Alcohols

We begin our discussion of hydrocarbon derivatives containing a single oxygen atom by considering *alcohols,* substances with the generalized formula

$$R—OH$$

An **alcohol** *is an organic compound in which an —OH group is bonded to a saturated carbon atom.*

Saturated carbon atom $—\overset{|}{\underset{|}{C}}—OH$ Alcohol functional group

The —OH group, the functional group that is characteristic of an alcohol, is called a *hydroxyl group.* A **hydroxyl group** *is the —OH functional group.*

Examples of structural formulas for alcohols include

$$CH_3—OH \qquad CH_3—CH_2—OH \qquad CH_3—CH_2—CH_2—OH$$

Space-filling models for these three alcohols, the simplest alcohols possible that have unbranched carbon chains, are given in Figure 14.1.

Alcohols may be viewed structurally as being alkyl derivatives of water in which a hydrogen atom has been replaced by an alkyl group.

$$H—\overset{..}{\underset{..}{O}}—H \qquad R—\overset{..}{\underset{..}{O}}—H$$
Water   An alcohol

Figure 14.2 shows the similarity in oxygen bond angles for water and $CH_3—OH$, the simplest alcohol.

Alcohols may also be viewed structurally as hydroxyl derivatives of alkanes in which a hydrogen atom has been replaced by a hydroxyl group

$$R—H \qquad R—OH$$
An alkane   An alcohol

14.3 Nomenclature for Alcohols

Common names exist for alcohols with simple (generally C_1 through C_4) alkyl groups. The word *alcohol,* as a separate word, is placed after the name of the alkyl or cycloalkyl group present.

Line-angle drawings for selected simple alcohols:

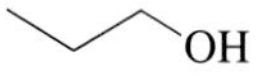

OH

Propyl alcohol
(1-propanol)

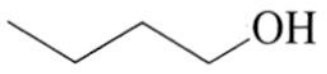

OH

Butyl alcohol
(1-butanol)

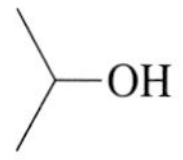

OH

Isopropyl alcohol
(2-propanol)

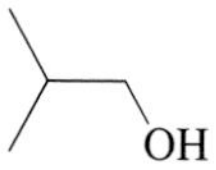

OH

Isobutyl alcohol
(2-methyl-1-propanol)

CH_3—OH
Methyl alcohol

CH_3—CH_2—OH
Ethyl alcohol

CH_3—CH_2—CH_2—OH
Propyl alcohol

CH_3—CH—OH
|
CH_3
Isopropyl alcohol

OH

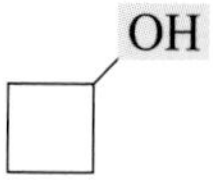

Cyclobutyl alcohol

IUPAC rules for naming alcohols that contain a single hydroxyl group follow.

Rule 1: *Name the longest carbon chain to which the hydroxyl group is attached.* The chain name is obtained by dropping the final -*e* from the alkane name and adding the suffix -*ol*.

Rule 2: *Number the chain starting at the end nearest the hydroxyl group, and use the appropriate number to indicate the position of the —OH group.* (In numbering of the longest carbon chain, the hydroxyl group has priority over double and triple bonds, as well as over alkyl, cycloalkyl, and halogen substituents.)

Rule 3: *Name and locate any other substituents present.*

Rule 4: *In alcohols where the —OH group is attached to a carbon atom in a ring, the hydroxyl group is assumed to be on carbon 1.*

EXAMPLE 14.1

Determining IUPAC Names for Alcohols

■ Name the following alcohols, utilizing IUPAC nomenclature rules.

a.
CH_3
|
CH_3—CH_2—C—CH_2—CH_2—CH_3
|
OH

b. CH_3—CH_2—CH—CH_2—CH_3
|
CH_2—OH

c.
CH_3

CH_3——OH

In the naming of alcohols with *unsaturated* carbon chains, two endings are needed: one for the double or triple bond and one for the hydroxyl group. The -*ol* suffix always comes last in the name; that is, unsaturated alcohols are named as *alkenols* or *alkynols*.

$\overset{3}{C}H_2$=$\overset{2}{C}H$—$\overset{1}{C}H_2$—OH
2-Propen-1-ol
(common name: allyl alcohol)

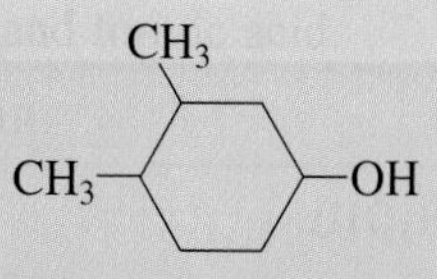

Solution

a. The longest carbon chain that contains the alcohol functional group has six carbons. When we change the -*e* to -*ol*, hexane becomes *hexanol*. Numbering the chain from the end nearest the —OH group identifies carbon number 3 as the location of both the —OH group and a methyl group. The complete name is 3-methyl-3-hexanol.

CH_3
|
$\overset{1}{C}H_3$—$\overset{2}{C}H_2$—$\overset{3}{C}$—$\overset{4}{C}H_2$—$\overset{5}{C}H_2$—$\overset{6}{C}H_3$
|
OH

b. The longest carbon chain containing the —OH group has four carbon atoms. It is numbered from the end closest to the —OH group as follows:

CH_3—CH_2—$\overset{2}{C}H$—$\overset{3}{C}H_2$—$\overset{4}{C}H_3$
|
$\overset{1}{C}H_2$—OH

The base name is 1-butanol. The complete name is 2-ethyl-1-butanol.

(continued)

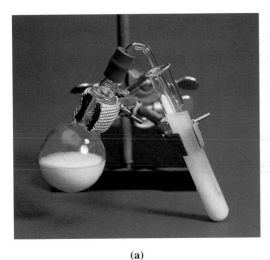 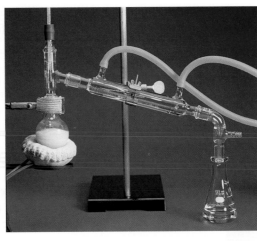 

(a) (b) (c)

FIGURE 14.4 An experimental setup for preparing ethyl alcohol by fermentation. (a) A small amount of yeast has been added to the aqueous sugar solution in the flask. Yeast enzymes catalyze the decomposition of sugar to ethanol and carbon dioxide, CO_2. The CO_2 is bubbling through lime water, $Ca(OH)_2$, producing calcium carbonate, $CaCO_3$. (b) More concentrated ethanol is produced from the solution in the flask by collecting the fraction that boils at about 78°C. (c) Concentrated ethanol (50% v/v) burns when it is ignited.

Long-term excessive use of ethyl alcohol may cause undesirable effects such as cirrhosis of the liver, loss of memory, and strong physiological addiction. Links have also been established between certain birth defects and the ingestion of ethyl alcohol by women during pregnancy (fetal alcohol syndrome).

Ethyl alcohol can be produced by yeast fermentation of sugars found in plant extracts (see Figure 14.4). The synthesis of ethyl alcohol in this manner, from grains such as corn, rice, and barley, is the reason why ethyl alcohol is often called *grain alcohol.*

Fermentation is the process by which ethyl alcohol for alcoholic beverages is produced. The maximum concentration of ethyl alcohol obtainable by fermentation is about 18% (v/v), because yeast enzymes cannot function in stronger alcohol solutions. Alcoholic beverages with a higher concentration of alcohol than this are prepared by either distillation or fortification with alcohol obtained by the distillation of another fermentation product. Table 14.2 lists the alcohol content of common alcoholic beverages and of selected common household products and over-the-counter drug products.

Denatured alcohol is ethyl alcohol that has been rendered unfit to drink by the addition of small amounts of toxic substances (denaturing agents). Almost all of the ethyl alcohol used for industrial purposes is denatured alcohol.

Most ethyl alcohol used in industry is prepared from ethene via a hydration reaction (Section 13.7).

$$CH_2{=}CH_2 + H_2O \xrightarrow{\text{Catalyst}} CH_3{-}CH_2{-}OH$$

The reaction produces a product that is 95% alcohol and 5% water. In applications where water does interfere with use, the mixture is treated with a dehydrating agent to produce 100% ethyl alcohol. Such alcohol, with all traces of water removed, is called *absolute alcohol.*

▶ The alcohol content of strong alcoholic beverages is often stated in terms of proof. *Proof* is twice the percentage of alcohol. This system dates back to the seventeenth century and is based on the fact that a 50% (v/v) alcohol–water mixture will burn. Its flammability was *proof* that a liquor had not been watered down.

▶ The "medicinal" odor associated with doctors' offices is usually that of isopropyl alcohol.

■ Isopropyl Alcohol (2-Propanol)

Isopropyl alcohol is one of two three-carbon monohydroxy alcohols; the other is propyl alcohol. A 70% isopropyl alcohol–30% water solution is marketed as *rubbing alcohol.* Isopropyl alcohol's rapid evaporation rate creates a dramatic cooling effect when it is applied to the skin, hence its use for alcohol rubs to combat high body temperature.

TABLE 14.2
Ethyl Alcohol Content (volume percent) of Common Alcoholic Beverages, Household Products, and Over-the-Counter Drugs

Product type	Product	Volume % ethyl alcohol
Alcoholic Beverages	Beer	3.2–9
	Wine (unfortified)	12
	Brandy	40–45
	Whiskey	45–55
	Rum	45
Flavorings	Vanilla extract	35
	Almond extract	50
Cough and Cold Remedies	Pertussin Plus	25
	Nyquil	25
	Dristan	12
	Vicks 44	10
	Robitussin, DM	1.4
Mouthwashes	Listerine	25
	Scope	18
	Colgate 100	17
	Cepacol	14
	Lavoris	5

Isopropyl alcohol has a bitter taste. Its toxicity is twice that of ethyl alcohol but it causes few fatalities because it often induces vomiting and thus doesn't stay down long enough to be fatal. In the body it is oxidized to acetone.

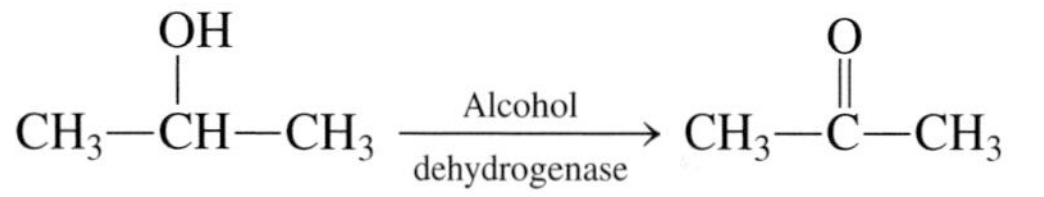

$$CH_3-\underset{\underset{OH}{|}}{CH}-CH_3 \xrightarrow[\text{dehydrogenase}]{\text{Alcohol}} CH_3-\underset{\overset{O}{\|}}{C}-CH_3$$

Isopropyl alcohol   Acetone

Large amounts (about 150 mL) of ingested isopropyl alcohol can be fatal; death occurs from paralysis of the central nervous system.

■ Ethylene Glycol (1,2-Ethanediol) and Propylene Glycol (1,2-Propanediol)

Ethylene glycol and propylene glycol are the two simplest alcohols possessing two —OH groups. Besides being diols, they are also classified as glycols. A **glycol** *is a diol in which the two —OH groups are on adjacent carbon atoms.*

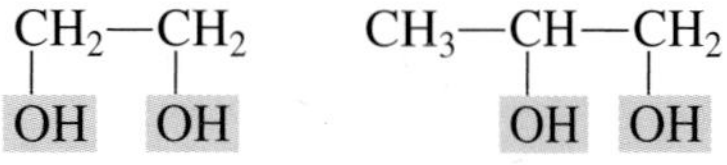

$$\underset{\underset{OH}{|}\;\;\underset{OH}{|}}{CH_2-CH_2} \qquad \underset{\;\;\;\;\underset{OH}{|}\;\;\underset{OH}{|}}{CH_3-CH-CH_2}$$

Ethylene glycol   Propylene glycol

> The ethylene glycol and propylene glycol used in antifreeze formulations are colorless and odorless; the color and odor of antifreezes come from additives for rust protection and the like.

> Ethylene glycol and propylene glycol are synthesized from ethylene and propylene, respectively, hence their common names.

Both of these glycols are colorless, odorless, high-boiling liquids that are completely miscible with water. Their major uses are as the main ingredient in automobile "year-round" antifreeze and airplane "de-icers" (Figure 14.5) and as a starting material for the manufacture of polyester fibers (Section 17.17).

Ethylene glycol is extremely toxic when ingested. In the body, liver enzymes oxidize it to oxalic acid.

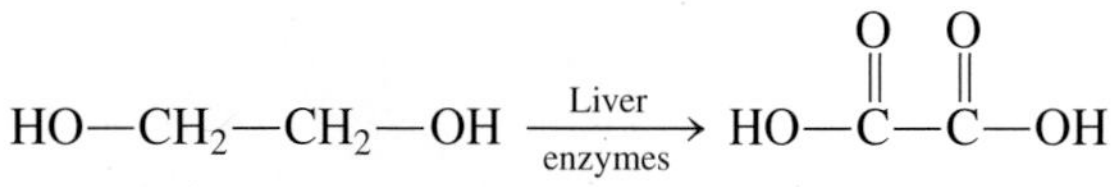

$$HO-CH_2-CH_2-OH \xrightarrow[\text{enzymes}]{\text{Liver}} HO-\underset{\overset{O}{\|}}{C}-\underset{\overset{O}{\|}}{C}-OH$$

Ethylene glycol   Oxalic acid

Oxalic acid, as a calcium salt, crystallizes in the kidneys, which leads to renal problems.

FIGURE 14.5 Space-filling molecular model for the dihydroxy alcohol *ethylene glycol,* which is the major ingredient in airplane "de-icers."

$$CH_2-CH_2$$
$$\;\;|\;\;\;\;\;|$$
$$OH\;\;\;OH$$

Propylene glycol, on the other hand, is essentially nontoxic and has been used as a solvent for drugs. Like ethylene glycol, it is oxidized by liver enzymes; however, pyruvic acid, its oxidation product, is a compound normally found in the human body, being an intermediate in carbohydrate metabolism (Chapter 24).

$$CH_3-CH-CH_2 \xrightarrow[\text{enzymes}]{\text{Liver}} CH_3-\overset{O}{\overset{||}{C}}-\overset{O}{\overset{||}{C}}-OH$$
$$\qquad\;|\quad\;\;|$$
$$\quad\;\;OH\;\;OH$$

Propylene glycol   Pyruvic acid

■ Glycerol (1,2,3-Propanetriol)

Glycerol is a clear, thick liquid that has the consistency of honey. Its molecular structure involves three —OH groups on three different carbon atoms.

$$CH_2-CH-CH_2$$
$$\;\;|\qquad|\qquad|$$
$$OH\;\;\;OH\;\;\;OH$$

Glycerol is normally present in the human body because it is a product of fat metabolism. It is present, in combined form, in all animal fats and vegetable oils (Section 19.4). In some Arctic species, glycerol functions as a "biological antifreeze" (see Figure 14.6).

Because glycerol has a great affinity for water vapor (moisture), it is often added to pharmaceutical preparations such as skin lotions and soap. Florists sometimes use glycerol on cut flowers to help retain water and maintain freshness. Its lubricative properties also make it useful in shaving creams and in applications such as glycerin suppositories for rectal administration of medicines. It is used in candies and icings as a retardant for preventing sugar crystallization.

FIGURE 14.6 Space-filling molecular model for the trihydroxy alcohol *glycerol,* which is often called biological antifreeze. For survival in Arctic and northern winters, many fish and insects, including the common housefly, produce large amounts of glycerol that dissolve in their blood, thereby lowering the freezing point of the blood.

$$CH_2-CH-CH_2$$
$$\;\;|\qquad|\qquad|$$
$$OH\;\;\;OH\;\;\;OH$$

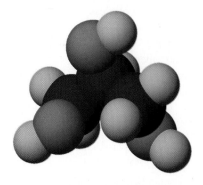

CHEMICAL CONNECTIONS

Menthol: A Useful Naturally Occurring Terpene Alcohol

Menthol is a naturally occurring terpene (Section 13.5) alcohol with a pleasant, minty odor. Its IUPAC name is 2-isopropyl-5-methylcyclohexanol.

In the pure state, menthol is a white crystalline solid with a melting point of 41°C to 43°C. It can be obtained from peppermint oil and can also be prepared synthetically.

Topical application of menthol to the skin causes a refreshing, cooling sensation followed by a slight burning-and-prickling sensation. Its mode of action is that of a *differential* anesthetic. It stimulates the receptor cells in the skin that normally respond to cold to give a sensation of coolness that is unrelated to body temperature. (This cooling sensation is particularly noticeable in the respiratory tract when low concentrations of menthol are inhaled.) At the same time as cooling is perceived, menthol can depress the nerves for pain reception.

Numerous products contain menthol.

■ Throat sprays and lozenges containing menthol temporarily soothe inflamed mucous surfaces of the nose and throat. Lozenges contain 2–20 milligrams of menthol per wafer.
■ Cough drops and cigarettes of the "mentholated" type use menthol for its counterirritant effect.
■ Pre-electric shave preparations and aftershave lotions often contain menthol. A concentration of only 0.1% (m/v) gives ample cooling to allay the irritation of a "close" shave.
■ Many dermatologic preparations contain menthol as an antipruritic (anti-itching agent).
■ Chest-rub preparations containing menthol include Ben Gay [7% (m/v)] and Mentholatum [6% (m/v)].
■ Artificial mint flavors have menthol as an ingredient. Several toothpastes and mouthwashes use menthol as a flavoring agent.

14.5 Physical Properties of Alcohols

Alcohol molecules have both polar and nonpolar character. The hydroxyl groups present are polar, and the alkyl (R) group present is nonpolar.

Nonpolar portion ↘   ↙ Polar portion

$$(CH_3—CH_2—CH_2)—(OH)$$

FIGURE 14.7 Space-filling molecular models showing the nonpolar (green) and polar (pink) parts of methanol and 1-octanol. (a) The polar hydroxyl functional group dominates the physical properties of methanol. The molecule is completely soluble in water (polar) but only partially so in hexane (nonpolar). (b) Conversely, the nonpolar portion of 1-octanol dominates its physical properties; it is infinitely soluble in hexane and has limited solubility in water.

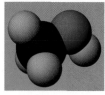

$CH_3—OH$
Nonpolar   Polar

(a) Methanol

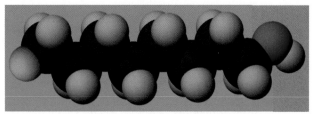

$CH_3CH_2CH_2CH_2CH_2CH_2CH_2CH_2—OH$
Nonpolar   Polar

(b) 1- Octanol

FIGURE 14.8 (a) Boiling points and (b) solubilities in water of selected 1-alcohols.

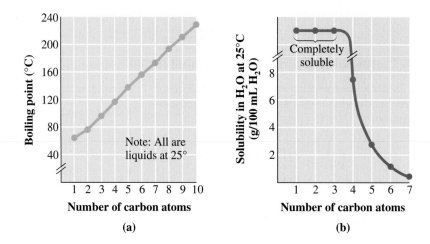

The physical properties of an alcohol depend on whether the polar or the nonpolar portion of its structure "dominates." Factors that determine this include the *length* of the nonpolar carbon chain present and the *number* of polar hydroxyl groups present (see Figure 14.7).

Boiling Points and Water Solubilities

Figure 14.8a shows that the boiling point for unbranched-chain alcohols with an —OH group on an end carbon, 1-alcohols, increases as the length of the carbon chain increases. This trend results from increasing London forces (Sections 7.13 and 12.15) with increasing carbon chain length. Alcohols with more than one hydroxyl group present have significantly higher boiling points (bp) than their monohydroxy counterparts.

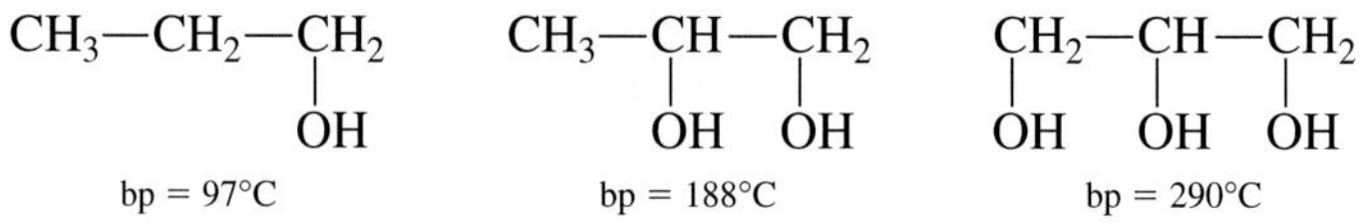

$$CH_3—CH_2—CH_2 \atop \qquad\qquad\ \ |\ \atop \qquad\qquad OH$$
bp = 97°C

$$CH_3—CH—CH_2 \atop \qquad\ \ |\quad\ \ |\ \atop \qquad\ OH\quad OH$$
bp = 188°C

$$CH_2—CH—CH_2 \atop \ \ |\quad\ \ |\quad\ \ |\ \atop OH\quad OH\quad OH$$
bp = 290°C

This boiling-point trend is related to increased hydrogen bonding between alcohol molecules (to be discussed shortly). Figure 14.9 is a physical-state summary for unbranched 1-alcohols and unsubstituted cycloalcohols with eight or fewer carbon atoms.

Small monohydroxy alcohols are soluble in water in all proportions. As carbon chain length increases beyond three carbons, solubility in water rapidly decreases (Figure 14.8b) because of the increasingly nonpolar character of the alcohol. Alcohols with two —OH groups present are more soluble in water than their counterparts with only one —OH group. Increased hydrogen bonding is responsible for this. Diols containing as many as seven carbon atoms show appreciable solubility in water.

Alcohols and Hydrogen Bonding

A comparison of the properties of alcohols with their alkane counterparts (Table 14.3) shows that

1. Alcohols have *higher* boiling points than alkanes of similar molecular mass.
2. Alcohols have much *higher* solubility in water than alkanes of similar molecular mass.

The differences in physical properties between alcohols and alkanes are related to hydrogen bonding. Because of their hydroxyl group(s), alcohols can participate in hydrogen bonding, whereas alkanes cannot. Hydrogen bonding between alcohol molecules (see Figure 14.10) is similar to that which occurs between water molecules (Section 7.13).

FIGURE 14.9 A physical-state summary for unbranched 1-alcohols and unsubstituted cycloalcohols at room temperature and pressure.

Unbranched 1-Alcohols			
C_1	C_3	C_5	C_7
C_2	C_4	C_6	C_8

Unsubstituted Cycloalcohols			
✕	C_3	C_5	C_7
✕	C_4	C_6	C_8

▢ Liquid

TABLE 14.3
A Comparison of Selected Physical Properties of Alcohols with Alkane Counterparts of Similar Molecular Mass

Type of compound	Compound	Structure	Molecular mass	Boiling point (°C)	Solubility in water
alkane	ethane	CH₃—CH₃	30	−89	slight solubility
alcohol	methanol	CH₃—OH	32	65	unlimited solubility
alkane	propane	CH₃—CH₂—CH₃	44	−42	slight solubility
alcohol	ethanol	CH₃—CH₂—OH	46	78	unlimited solubility
alkane	butane	CH₃—CH₂—CH₂—CH₃	58	−1	slight solubility
alcohol	1-propanol	CH₃—CH₂—CH₂—OH	60	97	unlimited solubility
alcohol	2-propanol	CH₃—CH—OH 	 CH₃	60	83	unlimited solubility

Extra energy is needed to overcome alcohol–alcohol hydrogen bonds before alcohol molecules can enter the vapor phase. Hence alcohol boiling points are higher than those for the corresponding alkanes (where no hydrogen bonds are present).

Alcohol molecules can also hydrogen-bond to water molecules (see Figure 14.11). The formation of such hydrogen bonds explains the solubility of small alcohol molecules in water. As the alcohol chain length increases, alcohols become more alkane-like (non-polar), and solubility decreases.

14.6 Preparation of Alcohols

A general method for preparing alcohols—the hydration of alkenes—was discussed in the previous chapter (Section 13.7). Alkenes react with water (an unsymmetrical addition agent) in the presence of sulfuric acid (the catalyst) to form an alcohol. Markovnikov's rule is used to determine the predominant alcohol product.

Another method of synthesizing alcohols involves the addition of H₂ to a carbon–oxygen double bond (a carbonyl group, C=O). (The carbonyl group is a functional group that will be discussed in detail in Chapter 15.) A carbonyl group behaves very much

Alcohols are intermediate products in the metabolism of both carbohydrates (Chapter 24) and fats (Chapter 25). In these metabolic processes, both addition of water to a carbon–carbon double bond and addition of hydrogen to a carbon–oxygen double bond lead to the introduction of the alcohol functional group into a biomolecule.

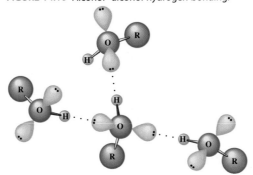
FIGURE 14.10 Alcohol–alcohol hydrogen bonding.

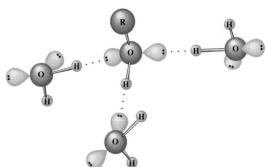
FIGURE 14.11 Alcohol–water hydrogen bonding.

like a carbon–carbon double bond when it reacts with H_2 under the proper conditions. As a result of H_2 addition, the oxygen of the carbonyl group is converted to an —OH group.

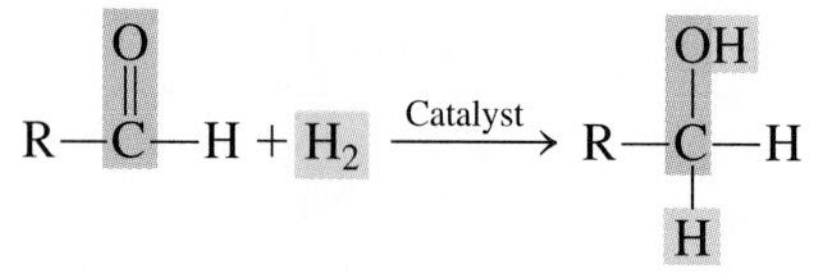

$$R-\overset{\overset{\displaystyle O}{\|}}{C}-H + H_2 \xrightarrow{\text{Catalyst}} R-\overset{\overset{\displaystyle OH}{|}}{\underset{\underset{\displaystyle H}{|}}{C}}-H$$

Aldehyde (Section 15.2)   Alcohol

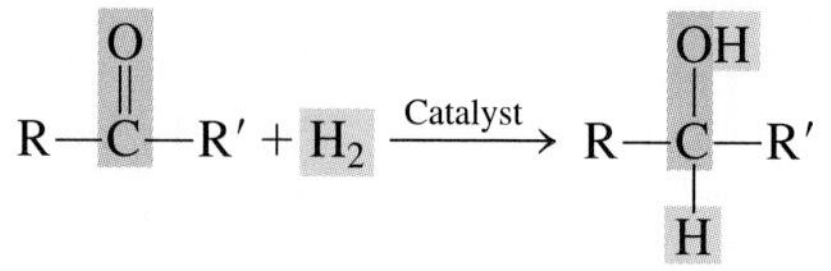

$$R-\overset{\overset{\displaystyle O}{\|}}{C}-R' + H_2 \xrightarrow{\text{Catalyst}} R-\overset{\overset{\displaystyle OH}{|}}{\underset{\underset{\displaystyle H}{|}}{C}}-R'$$

Ketone (Section 15.2)   Alcohol

14.7 Reactions of Alcohols

Of the many reactions that alcohols undergo, we consider four in this section: (1) combustion, (2) dehydration, (3) oxidation, and (4) halogenation.

■ Combustion

As we have seen in the previous two chapters, hydrocarbons of all types undergo combustion in air to produce carbon dioxide and water. Alcohols are also flammable; as with hydrocarbons, the combustion products are carbon dioxide and water. Methyl alcohol is the fuel of choice for racing cars (Section 14.4). Oxygenated gasoline, which is used in winter in many areas of the United States because it burns "cleaner," contains ethyl alcohol as one of the "oxygenates."

■ Intramolecular Alcohol Dehydration

A **dehydration reaction** *is a reaction in which the components of water (H and OH) are removed from a single reactant or from two reactants (H from one and OH from the other).* In *intramolecular* dehydration, both water components are removed from the same molecule.

Reaction conditions for the intramolecular dehydration of an alcohol are a temperature of 180°C and the presence of sulfuric acid (H_2SO_4) as a catalyst. The dehydration product is an alkene.

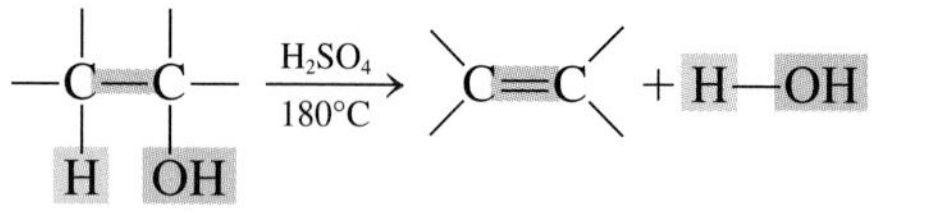

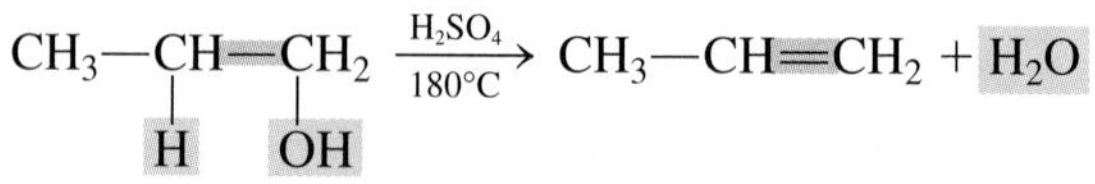

$$-\overset{|}{\underset{\underset{\displaystyle H}{|}}{C}}-\overset{|}{\underset{\underset{\displaystyle OH}{|}}{C}}- \xrightarrow[180°C]{H_2SO_4} \overset{}{C}=C + H-OH$$

$$CH_3-\overset{|}{\underset{\underset{\displaystyle H}{|}}{CH}}-\overset{|}{\underset{\underset{\displaystyle OH}{|}}{CH_2}} \xrightarrow[180°C]{H_2SO_4} CH_3-CH=CH_2 + H_2O$$

Intramolecular alcohol dehydration is an example of an *elimination reaction* (see Figure 14.12), as contrasted to a substitution reaction (Section 12.16) and an addition reaction (Section 13.7). An **elimination reaction** *is a reaction in which two groups or two atoms on neighboring carbon atoms are removed, or eliminated, from a molecule, leaving a multiple bond between the carbon atoms.*

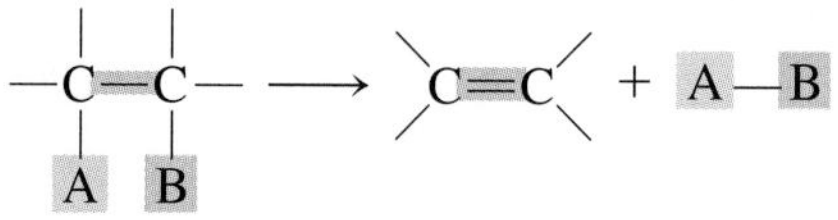

$$-\overset{|}{\underset{\underset{\displaystyle A}{|}}{C}}-\overset{|}{\underset{\underset{\displaystyle B}{|}}{C}}- \longrightarrow \overset{}{C}=C + A-B$$

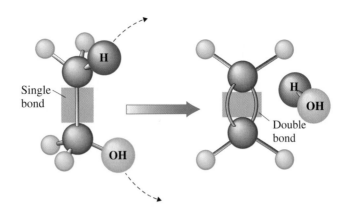

FIGURE 14.12 In an intramolecular alcohol dehydration the components of water (H and OH) are removed from neighboring carbon atoms with the resultant introduction of a double bond into the molecule.

What occurs in an elimination reaction is the reverse of what occurs in an addition reaction.

Dehydration of an alcohol can result in the production of more than one alkene product. This happens when there is more than one neighboring carbon atom from which hydrogen loss can occur. Dehydration of 2-butanol produces two alkenes.

▶ Dehydration of alcohols to form carbon–carbon double bonds occurs in several metabolic pathways in living systems, such as the citric acid cycle (Section 23.6) and the fatty acid spiral (Section 25.4). In these biochemical dehydrations, enzymes serve as catalysts instead of acids, and the reaction temperature is 37°C instead of the elevated temperatures required in the laboratory.

$$CH_2\!-\!CH\!-\!CH\!-\!CH_3 \xrightarrow[180°C]{H_2SO_4}$$

$$\underset{H}{|}\quad\underset{OH}{|}\quad\underset{H}{|}$$

2-Butanol

Removal produces 1-butene   Removal produces 2-butene

$$\overset{①}{CH_2}\!\!=\!\!\overset{②}{CH}\!-\!\overset{③}{CH}\!-\!\overset{④}{CH_3} + \overset{①}{CH_2}\!-\!\overset{②}{CH}\!\!=\!\!\overset{③}{CH}\!-\!\overset{④}{CH_3} + H_2O$$

$$\underset{H}{|}\qquad\qquad\underset{H}{|}$$

1-Butene   2-Butene

The dominant product can be predicted using Zaitsev's rule, named after the Russian chemist Alexander Zaitsev. **Zaitsev's rule** states that *the major product in an intramolecular alcohol dehydration reaction is the alkene that has the greatest number of alkyl groups attached to the carbon atoms of the double bond.* In the preceding reaction, 2-butene (with two alkyl groups) is favored over 1-butene (with one alkyl group).

▶ Alexander Zaitsev (1841–1910), a nineteenth-century Russian chemist, studied at the University of Paris and then returned to his native Russia to become a professor of chemistry at the University of Kazan. His surname is pronounced "zait-zeff."

▶ An alternative way of expressing Zaitsev's rule is "Hydrogen atom loss, during intramolecular alcohol dehydration to form an alkene, will occur preferentially from the carbon atom (adjacent to the hydroxyl-bearing carbon) that already has the fewest hydrogen atoms."

Two alkyl groups on double-bonded carbons $\boxed{CH_3}\!-\!CH\!\!=\!\!CH\!-\!\boxed{CH_3}$

2-Butene

$CH_2\!\!=\!\!CH\!-\!\boxed{CH_2\!-\!CH_3}$ One alkyl group on double-bonded carbons

1-Butene

Alkene formation via intramolecular alcohol *dehydration* is the "reverse reaction" to the reaction for preparing an alcohol through *hydration* of an alkene (Section 14.6). This relationship can be diagrammed as follows:

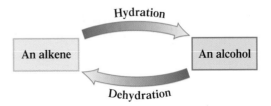

Hydration

An alkene   An alcohol

Dehydration

This "reverse reaction" situation illustrates the fact that many organic reactions can go both forward or backwards, depending on reaction conditions. Noting relationships such as this helps in keeping track of the numerous reactions that hydrocarbon derivatives undergo.

■ Intermolecular Alcohol Dehydration

At a lower temperature (140°C) than that required for alkene formation (180°C), an *inter*molecular rather than an *intra*molecular alcohol dehydration process can occur to produce an ether—a compound with the general structure R—O—R (Section 14.3). In ether formation, two alcohol molecules interact, an H atom being lost from one and an —OH group from the other. The resulting "leftover" portions of the two alcohol molecules join to form the ether. This reaction, which gives useful yields only for primary alcohol reactants (2° and 3° alcohols yield predominantly alkenes), can be written as

$$-\overset{|}{\underset{|}{C}}-O-H + H-O-\overset{|}{\underset{|}{C}}- \xrightarrow[140°C]{H_2SO_4} -\overset{|}{\underset{|}{C}}-O-\overset{|}{\underset{|}{C}}- + H-O-H$$

$$CH_3-CH_2-O-H + H-O-CH_2-CH_3 \xrightarrow[140°C]{H_2SO_4}$$
Ethanol   Ethanol

$$CH_3-CH_2-O-CH_2-CH_3 + H_2O$$

The preceding reaction is an example of *condensation*. A **condensation reaction** *is a reaction in which two molecules combine to form a larger one while liberating a small molecule, usually water.* In this case, two alcohol molecules combine to give an ether and water.

EXAMPLE 14.2

Predicting the Reactant in an Alcohol Dehydration Reaction When Given the Product

▶ The following is a summary of products obtained from alcohol dehydration reactions using H_2SO_4 as a catalyst.

Primary alcohol $\underset{\xrightarrow{140°C}}{\overset{180°C}{\rightleftarrows}}$ alkene / ether

Secondary alcohol $\underset{\xrightarrow{140°C}}{\overset{180°C}{\rightleftarrows}}$ alkene / alkene

Tertiary alcohol $\underset{\xrightarrow{140°C}}{\overset{180°C}{\rightleftarrows}}$ alkene / alkene

---

■ Identify the alcohol reactant needed to produce each of the following compounds as the *major product* of an alcohol dehydration reaction.

a. Alcohol $\xrightarrow[180°C]{H_2SO_4}$ $CH_3-CH=CH-CH_3$

b. Alcohol $\xrightarrow[180°C]{H_2SO_4}$ $CH_2=CH-\underset{\underset{CH_3}{|}}{CH}-CH_3$

c. Alcohol $\xrightarrow[140°C]{H_2SO_4}$ $CH_3-\underset{\underset{CH_3}{|}}{CH}-CH_2-O-CH_2-\underset{\underset{CH_3}{|}}{CH}-CH_3$

Solution

a. Both carbon atoms of the double bond are equivalent to each other. Add an H atom to one carbon atom of the double bond and an OH group to the other carbon atom of the double bond. It does not matter which goes where; you get the same molecule either way.

$$CH_3-\underset{\underset{OH}{|}}{CH}-\underset{\underset{H}{|}}{CH}-CH_3 \quad \text{or} \quad CH_3-\underset{\underset{H}{|}}{CH}-\underset{\underset{OH}{|}}{CH}-CH_3$$

b. There are two possible parent alcohols: one with an —OH group on carbon 1 and the other with an —OH group on carbon 2.

$$\underset{\underset{OH}{|}}{CH_2}-CH_2-\underset{\underset{CH_3}{|}}{CH}-CH_3 \quad \text{or} \quad CH_3-\underset{\underset{OH}{|}}{CH}-\underset{\underset{CH_3}{|}}{CH}-CH_3$$

Using the reverse of Zaitsev's rule, we find that the hydrogen atom will go back on the double-bonded carbon that bears the most alkyl groups.

Zero alkyl groups   One alkyl group

$$\underset{\underset{\text{OH atom} \quad \text{H atom}}{}}{CH_2=\overset{}{CH}-\underset{\underset{CH_3}{|}}{CH}-CH_3} \longrightarrow \underset{\underset{OH}{|}}{CH_2}-CH_2-\underset{\underset{CH_3}{|}}{CH}-CH_3$$

c. This is an ether. The primary alcohol from which the ether was formed will have the same alkyl group present as is in the ether. Thus the alcohol is

$$CH_3-CH-CH_2-OH$$
$$|$$
$$CH_3$$

Practice Exercise 14.2

Identify the starting alcohol from which each of the following products was obtained by an alcohol dehydration reaction.

a. Alcohol $\xrightarrow[140°C]{H_2SO_4}$ $CH_2=CH-CH_2-CH_3$

b. Alcohol $\xrightarrow[140°C]{H_2SO_4}$ $CH_3-\underset{\underset{CH_3}{|}}{C}=\underset{\underset{CH_3}{|}}{C}-CH_3$

c. Alcohol $\xrightarrow[140°C]{H_2SO_4}$ $CH_3-CH_2-CH_2-O-CH_2-CH_2-CH_3$

■ Oxidation

Before discussing alcohol oxidation reactions, we consider (1) a new method for recognizing when oxidation and reduction have occurred in a chemical reaction and (2) an alcohol classification system to "sort out" reaction products when an alcohol is oxidized.

The processes of oxidation and reduction were considered in Section 9.2 in the context of inorganic, rather than organic, reactions. Oxidation numbers were used to characterize oxidation–reduction processes. This same technique could be used in characterizing oxidation–reduction processes involving organic compounds, but it is not. Formal use of the oxidation number rules with organic compounds is usually cumbersome because of the many carbon and hydrogen atoms present; often, fractional oxidation numbers for carbon result.

A better approach for organic redox reactions is to use the following set of operational rules instead of oxidation numbers.

1. A carbon atom in an organic compound is considered *oxidized* if it *loses hydrogen atoms* or *gains oxygen atoms* in a redox reaction.
2. A carbon atom in an organic compound is considered *reduced* if it *gains hydrogen atoms* or *loses oxygen atoms* in a redox reaction.

Note that these operational definitions for oxidation and reduction are "opposites." This is just as it should be; oxidation and reduction are "opposite" processes.

Some alcohols readily undergo oxidation with mild oxidizing agents; others are resistant to oxidation with these same oxidizing agents. An alcohol classification system helps in determining whether oxidation will occur and what the oxidation product will be.

Alcohols are classified as primary (1°), secondary (2°), or tertiary (3°), depending on the number of carbon atoms bonded to the carbon atom that bears the hydroxyl group. A **primary alcohol** *is an alcohol in which the hydroxyl-bearing carbon atom is bonded to only one other carbon atom.* A **secondary alcohol** *is an alcohol in which the hydroxyl-bearing carbon atom is bonded to two other carbon atoms.* A **tertiary alcohol** *is an alcohol in which the hydroxyl-bearing carbon atom is bonded to three other carbon atoms.* Chemical reactions of alcohols often depend on alcohol class (1°, 2°, or 3°).

▶ Pronounce 1° as "primary," 2° as "secondary," and 3° as "tertiary."

▶ Methyl alcohol, CH_3—OH, an alcohol in which the hydroxyl-bearing carbon atom is attached to three hydrogen atoms, does not fit any of the alcohol classification definitions. It is usually grouped with the primary alcohols because its reactions are similar to those of these alcohols.

$$CH_3-\underset{\underset{H}{|}}{\overset{\overset{H}{|}}{C}}-OH \qquad CH_3-\underset{\underset{H}{|}}{\overset{\overset{CH_3}{|}}{C}}-OH \qquad CH_3-\underset{\underset{CH_3}{|}}{\overset{\overset{CH_3}{|}}{C}}-OH$$

$$\text{1° Alcohol} \qquad\qquad \text{2° Alcohol} \qquad\qquad \text{3° Alcohol}$$

Primary and secondary alcohols, but not tertiary alcohols, readily undergo oxidation in the presence of mild oxidizing agents to produce compounds that contain a carbon–oxygen double bond (aldehydes, ketones, and carboxylic acids). A number of different oxidizing agents can be used for the oxidation, including potassium permanganate ($KMnO_4$), potassium dichromate ($K_2Cr_2O_7$), and chromic acid (H_2CrO_4).

The net effect of the action of a mild oxidizing agent on a primary or secondary alcohol is the removal of two hydrogen atoms from the alcohol. One hydrogen comes from the —OH group, the other from the carbon atom to which the —OH group is attached. This H removal generates a carbon–oxygen double bond.

An alcohol   Compound containing a carbon–oxygen double bond

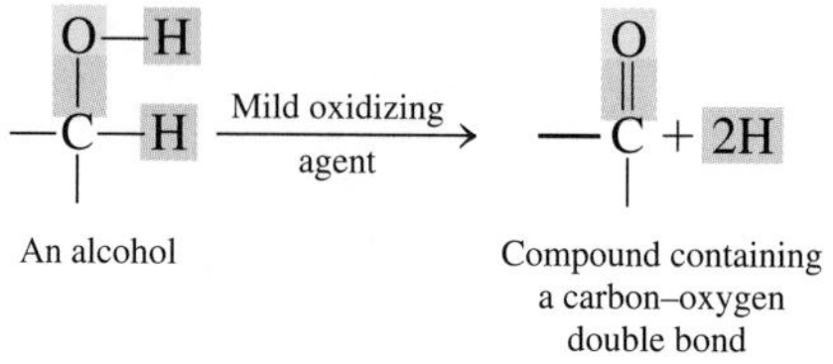

The two "removed" hydrogen atoms combine with oxygen supplied by the oxidizing agent to give H_2O.

Primary and secondary alcohols, the two types of oxidizable alcohols, yield different products upon oxidation. A 1° alcohol produces an *aldehyde* that is often then further oxidized to a *carboxylic acid,* and a 2° alcohol produces a *ketone.*

$$\text{Primary alcohol} \xrightarrow[\text{ox. agent}]{\text{Mild}} \text{aldehyde} \xrightarrow[\text{ox. agent}]{\text{Mild}} \text{carboxylic acid}$$

$$\text{Secondary alcohol} \xrightarrow[\text{ox. agent}]{\text{Mild}} \text{ketone}$$

$$\text{Tertiary alcohol} \xrightarrow[\text{ox. agent}]{\text{Mild}} \text{no reaction}$$

The general reaction for the oxidation of a primary alcohol is

1° Alcohol   Aldehyde   Carboxylic acid

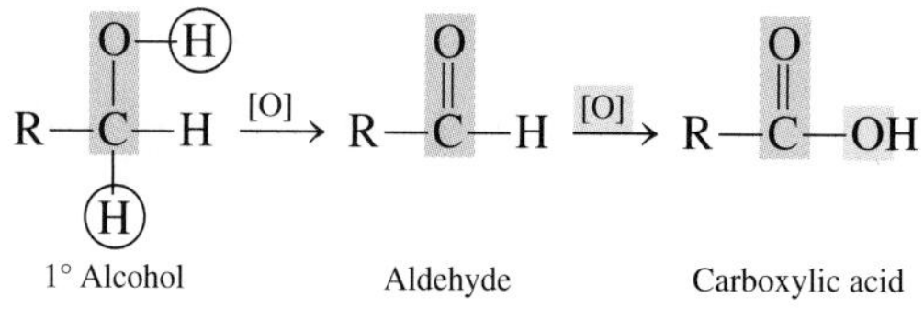

In this equation, the symbol [O] represents the mild oxidizing agent. The immediate product of the oxidation of a primary alcohol is an aldehyde. Because aldehydes themselves are readily oxidized by the same oxidizing agents that oxidize alcohols, aldehydes are further converted to carboxylic acids. A specific example of a primary alcohol oxidation reaction is

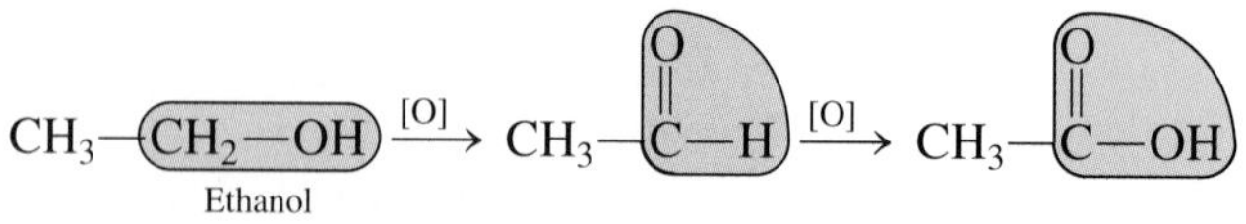

Ethanol

This specific oxidation reaction—that of ethanol—is the basis for the "breathalyzer test" used by law enforcement officers to determine whether an automobile driver is "drunk" (see Figure 14.13).

The general reaction for the oxidation of a secondary alcohol is

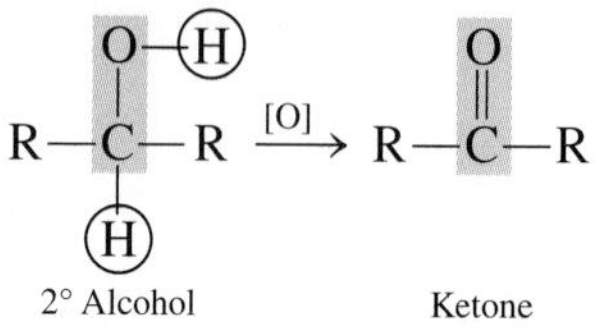

2° Alcohol   Ketone

◀ FIGURE 14.13 The oxidation of ethanol is the basis for the "breathalyzer test" that law enforcement officers use to determine whether an individual suspected of driving under the influence (DUI) has a blood alcohol level exceeding legal limits.

The DUI suspect is required to breathe into an apparatus containing a solution of potassium dichromate ($K_2Cr_2O_7$). The unmetabolized alcohol in the person's breath is oxidized by the dichromate ion ($Cr_2O_7{}^{2-}$), and the extent of the reaction gives a measure of the amount of alcohol present.

The dichromate ion is a yellow-orange color in solution. As oxidation of the alcohol proceeds, the dichromate ions are converted to Cr^{3+} ions, which have a green color in solution. The intensity of the green color that develops is measured and is proportional to the amount of ethanol in the suspect's breath, which in turn has been shown to be proportional to the person's blood alcohol level.

As with primary alcohols, oxidation involves the removal of two hydrogen atoms. Unlike aldehydes, ketones are resistant to further oxidation. A specific example of the oxidation of a secondary alcohol is

$$CH_3-\overset{\text{OH}}{\underset{|}{C}}H-CH_3 \xrightarrow{[O]} CH_3-\overset{\text{O}}{\overset{\|}{C}}-CH_3$$

Tertiary alcohols do not undergo oxidation with mild oxidizing agents. This is because they do not have hydrogen on the —OH-bearing carbon atom.

$$R-\overset{\text{OH}}{\underset{\underset{R}{|}}{\overset{|}{C}}}-R \xrightarrow{[O]} \text{no reaction}$$

3° Alcohol

■ Halogenation

Alcohols undergo halogenation reactions in which a halogen atom is substituted for the hydroxyl group, producing an alkyl halide. Alkyl halide production in this manner is superior to alkyl halide production through halogenation of an alkane (Section 12.17) because mixtures of products are *not* obtained. A single product is produced in which the halogen atom is found only where the —OH group was originally located.

Several different halogen-containing reactants, including phosphorus trihalides (PX_3; X is Cl or Br), are useful in producing alkyl halides from alcohols.

$$3R-OH + PX_3 \xrightarrow{\Delta} 3R-X + H_3PO_3$$

Note that heating of the reactants (Δ) is required.

The Chemistry at a Glance feature on page 378 summarizes the reaction chemistry of alcohols.

14.8 Polymeric Alcohols

It is possible to synthesize polymeric alcohols with structures similar to those of substituted polyethylenes (Section 13.8). Two of the simplest such compounds are poly(vinyl alcohol) (PVA) and poly(ethylene glycol) (PEG).

$$\left(\begin{array}{cc} H & H \\ | & | \\ -C-C- \\ | & | \\ H & OH \end{array}\right)_n \qquad \left(\begin{array}{cc} H & H \\ | & | \\ -C-C- \\ | & | \\ OH & OH \end{array}\right)_n$$

PVA   PEG

Poly(vinyl alcohol) is a tough, whitish polymer that can be formed into strong films, tubes, and fibers that are highly resistant to hydrocarbon solvents. Unlike most organic polymers, PVA is water-soluble. PVA has oxygen-barrier properties under dry conditions that are superior to those of any other polymer. PVA can be rendered insoluble in water, if needed, by use of chemical agents that cross-link individual polymer strands.

Aqueous solutions of PEG are very viscous (thick) because of the great solubility of PEG in water. PEG is used as an additive in many shampoos. It contributes little to the cleansing action of the shampoo, but it gives the shampoo texture or richness.

Summary of Reactions Involving Alcohols

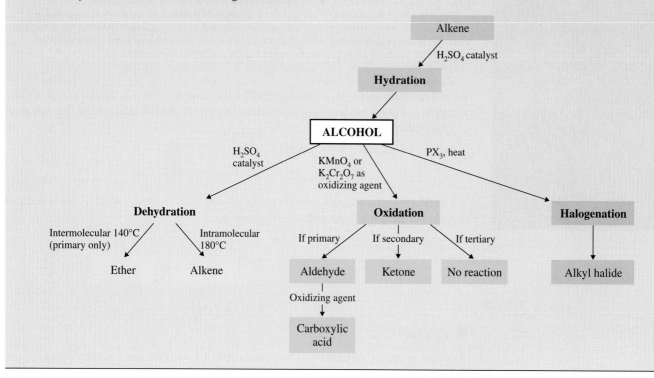

14.9 Structural Characteristics of Phenols

▶ The generic term *aryl group* (Ar) is the aromatic counterpart of the nonaromatic general term *alkyl group* (R).

A **phenol** *is an organic compound in which an —OH group is attached to a carbon atom that is part of an aromatic carbon ring system.*

The general formula for phenols is Ar—OH, where Ar represents an *aryl group*. An **aryl group** *is an aromatic carbon ring system from which one hydrogen atom has been removed.*

A hydroxyl group is thus the functional group for both phenols and alcohols. The reaction chemistry for phenols is sufficiently different from that for nonaromatic alcohols (Section 14.7) to justify discussing these compounds separately. Remember that phenols contain a "benzene ring" and that the chemistry of benzene is much different from that of other unsaturated hydrocarbons (Section 13.13).

The following are examples of compounds classified as phenols.

14.10 Nomenclature for Phenols

Besides being the name for a family of compounds, *phenol* is also the IUPAC-approved name for the simplest member of the phenol family of compounds.

Phenol

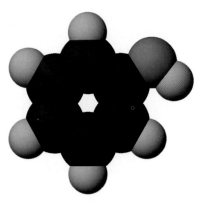

FIGURE 14.14 A space-filling model for *phenol*, a compound that has an —OH group bonded directly to a benzene (aromatic) ring.

A space-filling model for the compound *phenol* is shown in Figure 14.14. The name *phenol* is derived from a combination of the terms *phenyl* and alcoh*ol*.

The IUPAC rules for naming phenols are simply extensions of the rules used to name benzene derivatives with hydrocarbon or halogen substituents (Section 13.11). The parent name is phenol. Ring numbering always begins with the hydroxyl group and proceeds in the direction that gives the lower number to the next carbon atom bearing a substituent. The numerical position of the hydroxyl group is not specified in the name because it is 1 by definition.

3-Chlorophenol
(or *meta*-chlorophenol)

4-Ethyl-2-methylphenol

2,5-Dibromophenol

Methyl and hydroxy derivatives of phenol have IUPAC-accepted common names. Methylphenols are called cresols. The name *cresol* applies to all three isomeric methylphenols.

ortho-Cresol

meta-Cresol

para-Cresol

> Several neurotransmitters in the human body (Section 17.9), including norepinephrine, epinephrine (adrenaline), and dopamine, are catechol derivatives.

For hydroxyphenols, each of the three isomers has a different common name.

Catechol

Resorcinol

Hydroquinone

14.11 Physical and Chemical Properties of Phenols

> An *antiseptic* is a substance that kills microorganisms on living tissue. A *disinfectant* is a substance that kills microorganisms on inanimate objects.

Phenols are generally low-melting solids or oily liquids at room temperature. Most of them are only slightly soluble in water. Many phenols have antiseptic and disinfectant properties. The simplest phenol, phenol itself, is a colorless solid with a medicinal odor. Its melting point is 41°C, and it is more soluble in water than are most other phenols.

We have previously noted that the chemical properties of phenols are significantly different from those of alcohols (Section 14.9). The similarities and differences between these two reaction chemistries are as follows:

1. Both alcohols and phenols are flammable.
2. Dehydration is a reaction of alcohols but not of phenols; phenols cannot be dehydrated.

3. Both 1° and 2° alcohols are oxidized by mild oxidizing agents. Tertiary (3°) alcohols and phenols do not react with the oxidizing agents that cause 1° and 2° alcohol oxidation. Phenols can be oxidized by stronger oxidizing agents.
4. Both alcohols and phenols undergo halogenation in which the hydroxyl group is replaced by a halogen atom in a substitution reaction.

Acidity of Phenols

One of the most important properties of phenols is their acidity. Unlike alcohols, phenols are weak acids in solution. As acids, phenols have K_a values (Section 10.5) of about 10^{-10}. Such K_a values are lower than those of most weak inorganic acids (10^{-5} to 10^{-10}). The acid ionization reaction for phenol itself is

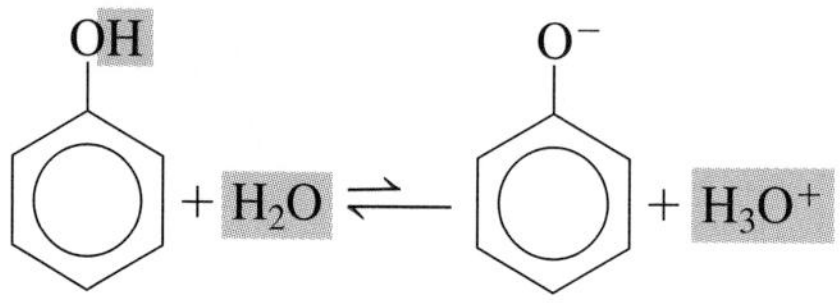

Phenol   Phenoxide ion

Note that the negative ion produced from the ionization is called the phenoxide ion. When phenol itself is reacted with sodium hydroxide (a base), the salt sodium phenoxide in produced.

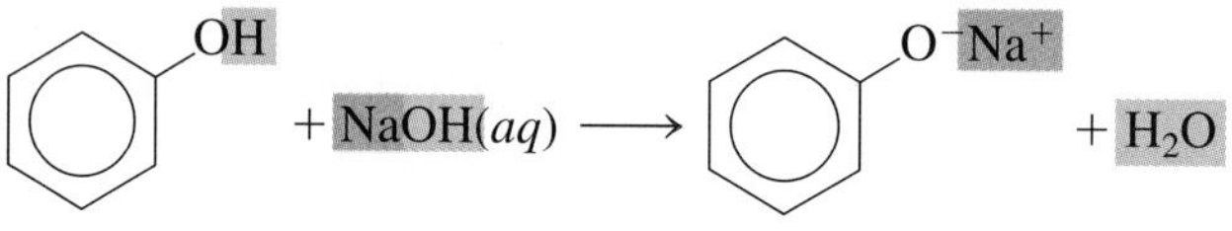

Phenol   Sodium phenoxide

14.12 Occurrence of and Uses for Phenols

Dilute (2%) solutions of phenol have long been used as antiseptics. Concentrated phenol solutions, however, can cause severe skin burns. Today, phenol has been largely replaced by more effective phenol derivatives such as 4-hexylresorcinol. The compound 4-hexylresorcinol is an ingredient in many mouthwashes and throat lozenges.

The "parent" name for a benzene ring bearing two hydroxyl groups "meta" to each other is resorcinol (Section 14.10).

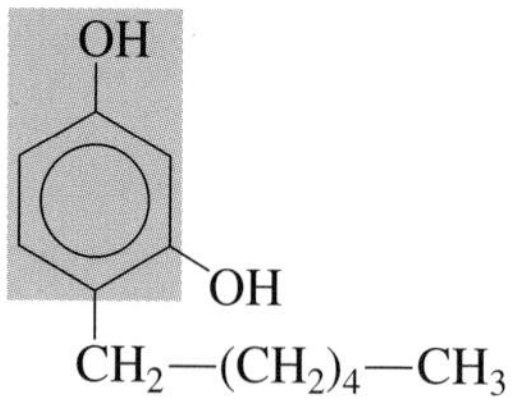

4-Hexylresorcinol

The phenol derivatives *o*-phenylphenol and 2-benzyl-4-chlorophenol are the active ingredients in Lysol, a disinfectant for walls, floors, and furniture in homes and hospitals.

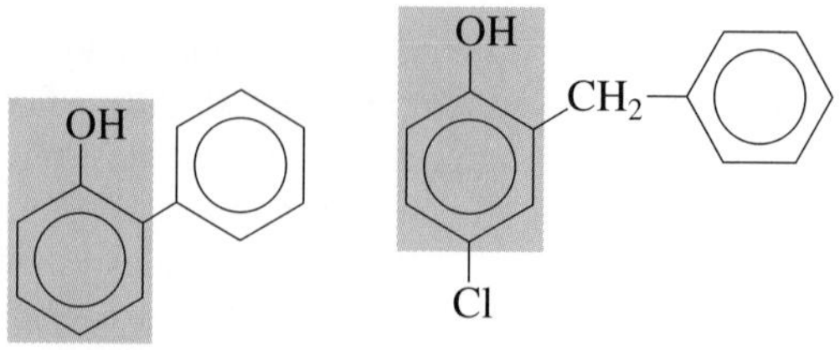

o-Phenylphenol   2-Benzyl-4-chlorophenol

A number of phenols possess antioxidant activity. An **antioxidant** *is a substance that protects other substances from being oxidized by being oxidized itself in preference to the other substances.* An antioxidant has a greater affinity for a particular oxidizing agent than do the substances the antioxidant is "protecting"; the antioxidant, therefore, reacts with the oxidizing agent first. Many foods sensitive to air are protected from oxidation through the use of phenolic additives. Two commercial phenolic antioxidant food additives are BHA (butylated hydroxy anisole) and BHT (butylated hydroxy toluene) (see Figure 14.15).

FIGURE 14.15 Many commercially baked goods contain the antioxidants BHA and BHT to help prevent spoilage.

▶ Within the human body, natural dietary antioxidants also offer protection against undesirable oxidizing agents. They include vitamin C (section 21.13), beta-carotene (Section 21,14), vitamin E (Section 21.14), and flavonoids (Sec. 23.11).

BHA (2 isomers)

BHT

A naturally occurring phenolic antioxidant that is important in the functioning of the human body is vitamin E (Section 21.14).

Vitamin E

A number of phenols found in plants are used as flavoring agents and/or antibacterials. Included among these phenols are

Thymol

Eugenol

Isoeugenol

Vanillin

Thymol, obtained from the herb thyme, possesses both flavorant and antibacterial properties. It is used as an ingredient in several mouthwash formulations.

Eugenol is responsible for the flavor of cloves. Dentists traditionally used clove oil as an antiseptic because of eugenol's presence; they use it to a limited extent even today.

FIGURE 14.16 Nutmeg tree fruit. A phenolic compound, isoeugenol, is responsible for the odor associated with nutmeg.

Water (HOH)

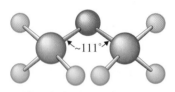

Dimethyl ether (CH₃OCH₃)

FIGURE 14.17 The similar shapes of water and dimethyl ether. Diethyl ether may be viewed structurally as a dialkyl derivative of water.

Structural isomers that differ in the type of functional group present, such as ethyl alcohol and dimethyl ether, are often called *functional group isomers.*

Isoeugenol, which differs in structure from eugenol only in the location of the double bond in the hydrocarbon side chain, is responsible for the odor associated with nutmeg (see Figure 14.16).

Vanillin, which gives vanilla its flavor, is extracted from the dried seed pods of the vanilla orchid. Natural supplies of vanillin are inadequate to meet demand for this flavoring agent. Synthetic vanillin is produced by oxidation of eugenol. Vanillin is an unusual substance in that even though its odor can be perceived at extremely low concentrations, the strength of its odor does not increase greatly as its concentration is increased.

Certain phenols exert profound physiological effects. For example, the irritating constituents of poison ivy and poison oak are derivatives of catechol (Section 14.10). These skin irritants have 15-carbon alkyl side chains with varying degrees of unsaturation (zero to three double bonds).

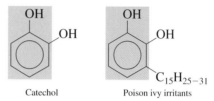

Catechol   Poison ivy irritants

14.13 Structural Characteristics of Ethers

An **ether** *is an organic compound in which an oxygen atom is bonded to two carbon atoms by single bonds.* In an ether, the carbon atoms that are attached to the oxygen atom can be part of alkyl, cycloalkyl, or aryl groups. Examples of ethers include

$$CH_3-O-CH_3 \qquad CH_3-CH_2-O-\bigcirc \qquad CH_3-O-\bigcirc$$

The two groups attached to the oxygen atom of an ether can be the same (first structure), but they need not be so (second and third structures).

All ethers contain a C—O—C unit, which is the ether functional group.

Ether functional group
$$---\overbrace{C-O-C}---$$

Generalized formulas for ethers, which depend on the types of groups attached to the oxygen atom (alkyl or aryl), include R—O—R, R—O—R′ (where R′ is an alkyl group different from R), R—O—Ar, and Ar—O—Ar.

Structurally, an ether can be visualized as a derivative of water in which both hydrogen atoms have been replaced by hydrocarbon groups (see Figure 14.17). Note that unlike alcohols and phenols, ethers do not possess a hydroxyl (—OH) group.

$$H-\overset{..}{\underset{..}{O}}-H \qquad\qquad R-\overset{..}{\underset{..}{O}}-R$$

Water   An ether

Alcohols and ethers with the same number of carbon atoms and the same degree of saturation are structural isomers. For example, a three-carbon alcohol and a three-carbon ether both have the molecular formula C_3H_8O and are thus structural isomers (see Figure 14.18.)

14.14 Nomenclature for Ethers

Common names for ethers are formed by naming the two hydrocarbon groups attached to the oxygen atom and adding the word *ether.* The hydrocarbon groups are listed in alphabetical order. When both hydrocarbon groups are the same, the prefix *di-* is placed before the name of the hydrocarbon group. In this system, ether names consist of two or three separate words.

FIGURE 14.18 Alcohols and ethers with the same number of carbon atoms and the same degree of saturation are structural isomers, as is illustrated here for propyl alcohol and ethyl methyl ether.

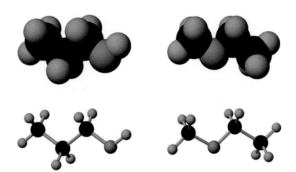

Propyl alcohol (C_3H_8O)   **Ethyl methyl ether (C_3H_8O)**

▶ Line-angle drawings for selected simple ethers:

Methyl ethyl ether
(methoxyethane)

Diethyl ether
(ethoxyethane)

Dipropyl ether
(1-propoxypropane)

$CH_3—O—CH_2—CH_3$ $CH_3—CH_2—O—\bigcirc$ $CH_3—O—CH_3$

Ethyl methyl ether   Ethyl phenyl ether   Dimethyl ether

In the IUPAC nomenclature system, ethers are named as substituted hydrocarbons. The smaller hydrocarbon attachment and the oxygen atom are called an *alkoxy group,* and this group is considered a substituent on the larger hydrocarbon group. An **alkoxy group** *is an —OR group, an alkyl (or aryl) group attached to an oxygen atom.* Simple alkoxy groups include the following:

$CH_3—O—$ $CH_3—CH_2—O—$ $CH_3—CH_2—CH_2—O—$
Methoxy group   Ethoxy group   Propoxy group

The general symbol for an alkoxy group is —O—R (or —OR).

The steps in naming an ether using the IUPAC system are

1. Select the longest carbon chain and use its name as the base name.
2. Change the *-yl* ending of the other hydrocarbon group to *-oxy* to obtain the alkoxy group name; *methyl* becomes *methoxy,* *ethyl* becomes *ethoxy,* etc.
3. Place the alkoxy name, with a locator number, in front of the base chain name.

Here are some examples of IUPAC ether nomenclature, with the alkoxy groups present highlighted in each structure:

▶ It is possible to have compounds that contain both ether and alcohol functional groups, such as

$$CH_3—CH—CH_2—CH_2—O—CH_3$$
$$\quad\quad |$$
$$\quad\quad OH$$

4-Methoxy-2-butanol

(The alcohol functional group has higher priority in IUPAC nomenclature, so the compound is named as an alcohol rather than as an ether.)

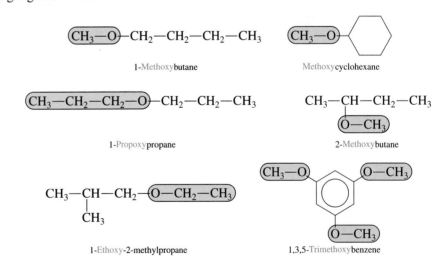

$\boxed{CH_3—O}—CH_2—CH_2—CH_2—CH_3$
1-Methoxybutane

$\boxed{CH_3—O}—\bigcirc$
Methoxycyclohexane

$\boxed{CH_3—CH_2—CH_2—O}—CH_2—CH_2—CH_3$
1-Propoxypropane

$CH_3—CH—CH_2—CH_3$
$\quad\quad |$
$\quad\boxed{O—CH_3}$
2-Methoxybutane

$CH_3—CH—CH_2—\boxed{O—CH_2—CH_3}$
$\quad\quad |$
$\quad\quad CH_3$
1-Ethoxy-2-methylpropane

1,3,5-Trimethoxybenzene

▶ The compound responsible for the characteristic odor of anise and fennel is *anethole,* an allyl derivative of anisole.

$O—CH_3$

$CH_2—CH=CH_2$

The simplest aromatic ether involves a methoxy group attached to a benzene ring. This ether goes by the common name *anisole.*

$\bigcirc—O—CH_3$

Anisole

CHEMICAL CONNECTIONS

Ethers as General Anesthetics

For many people, the word *ether* evokes thoughts of hospital operating rooms and anesthesia. This response derives from the *former* large-scale use of diethyl ether as a general anesthetic. In 1846, the Boston dentist William Morton was the first to demonstrate publicly the use of diethyl ether as a surgical anesthetic.

In many ways, diethyl ether is an ideal general anesthetic. It is relatively easy to administer, it is readily made in pure form, and it causes excellent muscle relaxation. There is less danger of an overdose with diethyl ether than with almost any other anesthetic, because there is a large gap between the effective level for anesthesia and the lethal dose.

Despite these ideal properties, diethyl ether is rarely used today because of two drawbacks: (1) It causes nausea and

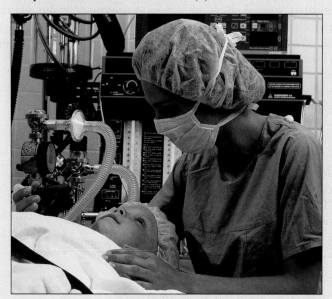

irritation of the respiratory passage, and (2) it is a highly flammable substance, forming explosive mixtures with air, which can be set off by a spark.

By the 1930s, nonether anesthetics had been developed that solved the problems of nausea and irritation. They also, however, were extremely flammable compounds. The simple hydrocarbon cyclopropane was the most widely used of these newer compounds.

It was not until the late 1950s and early 1960s that nonflammable general anesthetics became available. Anesthetic nonflammability was achieved by incorporating halogen atoms into anesthetic molecules. Three of the most used of these "halogenated" anesthetics are enflurane, isoflurane, and halothane.

Enflurane and isoflurane, which are structural isomers, are hexahalogenated ethers.

Enflurane

Isoflurane

With these compounds, induction of anesthesia can be achieved in less than 10 minutes with an inhaled concentration of 3% in oxygen.

Halothane, which is potent at relatively low doses, and whose effects wear off quickly, is a hexahalogenated ethane derivative rather than an ether.

Halothane

Derivatives of anisole are named as substituted anisoles, in a manner similar to that for substituted phenols (Section 14.10). Anisole derivatives were encountered in Section 14.12 when considering antioxidant food additives: BHAs are both a phenol and an anisole.

In the United States, the hydrocarbon ethene, $CH_2{=}CH_2$, is the organic chemical produced in the largest quantities. An ether, methyl *tert*-butyl ether (MTBE), is the number-two organic chemical in terms of production.

The contrast between IUPAC and common names for ethers is as follows:

IUPAC (one word)

alkoxyalkane

2-methoxybutane

Common (three or two words)

alkyl alkyl ether

ethyl methyl ether

or

dialkyl ether

dipropyl ether

Methyl *tert*-butyl ether (MTBE)

Since 1979, small amounts of MTBE have been used in gasoline to raise octane levels, but in recent years it has been added in much larger amounts as a clean-burning oxygenate to produce EPA-mandated reformulated gasoline used to improve air quality in polluted areas. The amount of MTBE in gasoline is likely to be reduced in the future in response to a growing problem: contamination of water supplies by small amounts of MTBE from leaking gasoline tanks and from spills. MTBE in the water supplies is not a health-and-safety issue at this time, but its presence does affect taste and odor in contaminated supplies.

Compounds with ether functional groups occur in a variety of plants. The phenolic flavoring agents eugenol, isoeugenol, and vanillin (Section 14.12) are also ethers; each has a methoxy substituent on the ring.

14.15 Physical and Chemical Properties of Ethers

The boiling points of ethers are similar to those of alkanes of comparable molecular mass and are much lower than those of alcohols of comparable molecular mass.

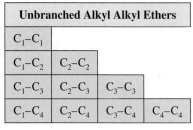

Unbranched Alkyl Alkyl Ethers			
C_1–C_1			
C_1–C_2	C_2–C_2		
C_1–C_3	C_2–C_3	C_3–C_3	
C_1–C_4	C_2–C_4	C_3–C_4	C_4–C_4

☐ Gas ▨ Liquid

FIGURE 14.19 A physical-state summary for unbranched alkyl alkyl ethers at room temperature and pressure.

Alkane CH_3—CH_2—CH_2—CH_2—CH_3 Mol. mass = 72 amu
bp = 36°C

Ether CH_3—CH_2—O—CH_2—CH_3 Mol. mass = 74 amu
bp = 35°C

Alcohol CH_3—CH_2—CH_2—CH_2—OH   Mol. mass = 74 amu
bp = 117°C

The much higher boiling point of the alcohol results from hydrogen bonding between alcohol molecules. Ether molecules, like alkanes, cannot hydrogen-bond to one another. Ether oxygen atoms have no hydrogen atom attached directly to them. Figure 14.19 is a physical-state summary for unbranched alkyl alkyl ethers where the alkyl groups range in size from C_1 to C_4.

Ethers, in general, are more soluble in water than are alkanes of similar molecular mass, because ether molecules are able to form hydrogen bonds with water (Figure 14.20).

Ethers have water solubilities similar to those of alcohols of the same molecular mass. For example, diethyl ether and butyl alcohol have the same solubility in water. Because ethers can also hydrogen-bond to alcohols, alcohols and ethers tend to be mutually soluble. Nonpolar substances tend to be more soluble in ethers than in alcohols because ethers have no hydrogen-bonding network that has to be broken up for solubility to occur.

Two chemical properties of ethers are especially important.

1. *Ethers are flammable.* Special care must be exercised in laboratories where ethers are used. Diethyl ether, whose boiling point of 35°C is only a few degrees above room temperature, is a particular flash-fire hazard.
2. *Ethers react slowly with oxygen from the air to form unstable hydroperoxides and peroxides.*

R—O—O—H   R—O—O—R
Hydroperoxide   Peroxide

> The term *ether* comes from the Latin *aether,* which means "to ignite." This name is given to these compounds because of their high vapor pressure at room temperature, which makes them very flammable.

FIGURE 14.20 Ether–water hydrogen bonding.

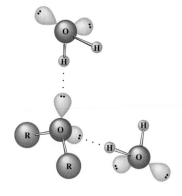

Such compounds, when concentrated, represent an explosion hazard and must be removed before *stored* ethers are used.

Ethers are unreactive toward acids, bases, and oxidizing agents. Like alkanes, they do undergo halogenation reactions.

The general chemical unreactivity of ethers, coupled with the fact that most organic compounds are ether-soluble, makes ethers excellent solvents in which to carry out organic reactions. Their relatively low boiling points simplify their separation from the reaction products.

A chemical reaction for the preparation of ethers has been previously considered. In Section 14.7 we noted that the intermolecular dehydration of a primary alcohol will produce an ether. Although additional methods exist for ether preparation, we will not consider them in this text.

14.16 Cyclic Ethers

Cyclic ethers contain ether functional groups as part of a ring system. Some examples of such cyclic ethers, along with their common names, follow.

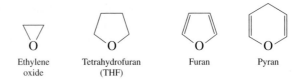

Ethylene   Tetrahydrofuran   Furan   Pyran
oxide   (THF)

Ethylene oxide is used extensively in industrial chemical processes. THF is a particularly useful solvent in that it dissolves many organic compounds and yet is miscible with water. In carbohydrate chemistry (Chapter 18), we will encounter many cyclic structures that are polyhydroxy derivatives of the five-membered (furan) and six-membered (pyran) cyclic ether systems. These carbohydrate derivatives are called *furanoses* and *pyranoses*, respectively (Section 18.10).

Vitamin E (Section 14.11) and THC (the active ingredient in marijuana; page 387) have structures in which a cyclic ether component is present.

Cyclic ethers are our first encounter with heterocyclic organic compounds. A **heterocyclic organic compound** *is a cyclic organic compound in which one or more of the carbon atoms in the ring have been replaced with atoms of other elements.* The hetero atom is usually oxygen or nitrogen.

We have just seen that *cyclic ethers*—compounds in which the ether functional group is part of a ring system—exist. In contrast, cyclic alcohols—compounds in which the alcohol functional group is part of a ring system—do not exist. To incorporate an alcohol functional group into a ring system would require an oxygen atom with three bonds, and oxygen atoms form only two bonds.

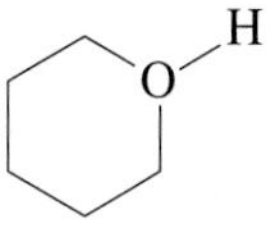

The oxygen atom in this structure has three bonds, which is not possible.

Compounds such as

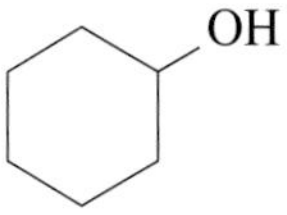

OH   OH

and

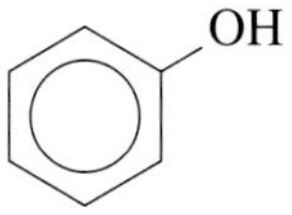

which do exist, are not cyclic alcohols in the sense in which we are using the term, because the alcohol functional group is attached to a ring system rather than part of it.

14.17 Sulfur Analogs of Alcohols and Ethers

Many organic compounds containing oxygen have sulfur analogs, in which a sulfur atom has replaced an oxygen atom. Sulfur is in the same group of the periodic table as oxygen, so the two elements have similar electron configurations (Section 3.8).

Thiols, the sulfur analogs of alcohols, contain —SH functional groups instead of —OH functional groups. The thiol functional group is called a *sulfhydryl group.* A **sulfhydryl group** *is the —SH functional group.* A **thiol** *is an organic compound in which a sulfhydryl group is present.* An older term used for thiols is *mercaptans.* Contrasting the

The root *thio-* indicates that a sulfur atom has replaced an oxygen atom in a compound. It originates from the Greek *theion,* meaning "brimstone," which is an older name for the element sulfur.

CHEMICAL CONNECTIONS

Marijuana: The Most Commonly Used Illicit Drug

Prepared from the leaves, flowers, seeds, and small stems of a hemp plant called *Cannabis sativa,* marijuana, which is also called pot or grass, is the most commonly used illicit drug in the United States. The most active ingredient of the many in marijuana is the molecule *tetrahydrocannabinol,* called THC for short. Three different functional groups are present in a THC molecule; it is a phenol, a cyclic ether, and a cycloalkene.

Tetrahydrocannabinol

The THC content of marijuana varies considerably. Most marijuana sold in the North American illegal drug market has a THC content of 1% to 2%.

Marijuana has a pharmacology unlike that of any other drug. A marijuana "high" is a combination of sedation, tranquiliza-

tion, and mild hallucination. THC readily penetrates the brain. The portions of the brain that involve memory and motor control contain the receptor sites where THC molecules interact. Even moderate doses of marijuana cause short-term memory loss. Marijuana unquestionably impairs driving ability, even after ordinary social use. THC readily crosses the placental barrier and reaches the fetus. Heavy marijuana users experience inflammation of the bronchi, sore throat, and inflamed sinuses. Increased heart rate, to as high as a dangerous 160 beats per minute, can occur with marijuana use.

The onset of action of THC is usually within minutes after smoking begins, and peak concentration in plasma occurs in 10 to 30 minutes. Unless more is smoked, the effects seldom last longer than 2 to 3 hours. Because THC is only slightly soluble in water, it tends to be deposited in fatty tissues. Unlike alcohol, THC persists in the bloodstream for several days, and the products of its breakdown remain in the blood for as long as 8 days.

New research indicates that physical dependence on THC can develop. Drug withdrawal symptoms are seen in some individuals who have been exposed repeatedly to high doses.

general structures for alcohols and thiols, we have

$$R-\underset{\text{An alcohol}}{\underbrace{OH}} \quad \text{and} \quad R-\underset{\text{A thiol}}{\underbrace{SH}}$$

Hydroxyl group   Sulfhydryl group

Thiols are named in the same way as alcohols in the IUPAC system, except that the *-ol* becomes *-thiol*. The prefix *thio-* indicates the substitution of a sulfur atom for an oxygen atom in a compound.

$$\underset{\text{2-Butanol}}{CH_3-\underset{\underset{OH}{|}}{CH}-CH_2-CH_3} \qquad \underset{\text{2-Butanethiol}}{CH_3-\underset{\underset{SH}{|}}{CH}-CH_2-CH_3}$$

As in the case of diols and triols, the *-e* at the end of the alkane name is also retained for thiols.

Common names for thiols are based on use of the term *mercaptan,* the older name for thiols. The name of the alkyl group present (as a separate word) precedes the word *mercaptan.*

$$\underset{\text{Ethyl mercaptan}}{CH_3-CH_2-SH} \qquad \underset{\text{Isopropyl mercaptan}}{CH_3-\underset{\underset{CH_3}{|}}{CH}-SH}$$

▶ Even though thiols have a higher molecular mass than alcohols with the same number of carbon atoms, they have much lower boiling points, because they do not exhibit hydrogen bonding as alcohols do.

Two important properties of thiols are low boiling points (because of lack of hydrogen bonding) and a strong, disagreeable odor. The familiar odor of natural gas results from the addition of thiols to the gas. The exceptionally low threshold of detection for thiols enables consumers to smell a gas leak long before the gas reaches dangerous levels. Thiols are responsible for the odor of skunks (Figure 14.21). Many of the strong odors associated with foods are caused by small thiol molecules (Table 14.4).

FIGURE 14.21 Thiols are responsible for the strong odor of "essence of skunk." Their odor is an effective defense mechanism.

TABLE 14.4

Selected Thiols That Are Present in Foods

Flavor or odor of	Structures of compounds involved
oysters	CH_3—SH
cheddar cheese	CH_3—SH

freshly chopped onions	CH_3—CH_2—CH_2
	               │
	               SH

garlic	CH_2=CH—CH_2
	              │
	              SH

▸ Disulfide common names are similar to ether common names, with *disulfide* used in place of *ether*.

$$CH_3—S—S—CH_2—CH_3$$
ethyl methyl disulfide

▸ Bacteria in the mouth interact with saliva and leftover food to produce such compounds as hydrogen sulfide, methanethiol (a thioalcohol), and dimethyl sulfide (a thioether). These compounds, which have odors detectable in air at concentrations of parts per billion, are responsible for "morning breath."

FIGURE 14.22 A comparison involving space-filling models for dimethyl ether and dimethyl sulfide (dimethyl thioether). A sulfur atom is much larger than an oxygen atom. This results in a carbon–sulfur bond being weaker than a carbon–oxygen bond.

Thiols are easily oxidized but yield different products than their alcohol analogs. Thiols form *disulfides*. Each of two thiol groups loses a hydrogen atom, thus linking the two sulfur atoms together via a disulfide group, —S—S—.

$$R—SH + HS—R \xrightarrow{\text{Oxidation}} R—S—S—R + 2H$$
A disulfide

Reversal of this reaction, a reduction process, is also readily accomplished. Breaking of the disulfide bond regenerates two thiol molecules.

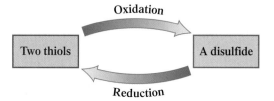

These two "opposite reactions" are of biological importance in the area of protein chemistry. Disulfide bonds formed from the interaction of two —SH groups contribute in a major way to protein structure (Chapter 20).

Sulfur analogs of ethers exist. They are known as thioethers (or sulfides). A **thioether** *is an organic compound in which a sulfur atom is bonded to two carbon atoms by single bonds.* The generalized formula for a thioether is R—S—R. Like thiols, thioethers (or sulfides) have strong characteristic odors.

Thioethers are named in the same way as ethers, with *sulfide* used in place of *ether* in common names and *alkylthio* used in place of *alkoxy* in IUPAC names.

$$CH_3—S—CH_3$$
Dimethyl sulfide
(methylthiomethane)

Methyl phenyl sulfide
(methylthiobenzene)

4-(ethylthio)-2-Methyl-2-pentene

In general, thiols are more reactive than their alcohol counterparts, and thioethers are more reactive than their corresponding ethers. The larger size of a sulfur atom compared to an oxygen atom (see Figure 14.22) results in a carbon–sulfur covalent bond that is weaker than a carbon–oxygen covalent bond. An added factor is that sulfur's electronegativity (2.5) is significantly lower than that of oxygen (3.5).

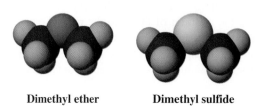

Dimethyl ether   **Dimethyl sulfide**

CONCEPTS TO REMEMBER

Alcohols. Alcohols are organic compounds that contain an —OH group attached to a saturated carbon atom. The general formula for an alcohol is R—OH, where R is an alkyl group.

Nomenclature of alcohols. The IUPAC names of simple alcohols end in *-ol,* and their carbon chains are numbered to give precedence to the location of the —OH group. Alcohol common names contain the word *alcohol* preceded by the name of the alkyl group.

Physical properties of alcohols. Alcohol molecules hydrogen-bond to each other and to water molecules. They thus have higher-than-normal boiling points, and the low-molecular-mass alcohols are soluble in water.

Classifications of alcohols. Alcohols are classified on the basis of the number of carbon atoms bonded to the carbon attached to the —OH group. In primary alcohols, the —OH group is bonded to a carbon atom bonded to only one other C atom. In secondary alcohols, the —OH-containing C atom is attached to two other C atoms. In tertiary alcohols, it is attached to three other C atoms.

Alcohol dehydration. Alcohols can be dehydrated in the presence of sulfuric acid to form alkenes or ethers. At 180°C, an alkene is produced; at 140°C, primary alcohols produce an ether.

Alcohol oxidation. Oxidation of primary alcohols first produces an aldehyde, which is then further oxidized to a carboxylic acid. Secondary alcohols are oxidized to ketones, and tertiary alcohols are resistant to oxidation.

Phenols. Phenols have the general formula Ar—OH, where Ar represents an aryl group derived from an aromatic compound. Phenols are named as derivatives of the parent compound phenol, using the conventions for aromatic hydrocarbon nomenclature.

Properties of Phenols. Phenols are generally low-melting solids; most are only slightly soluble in water. The chemical reactions of phenols are significantly different from those of alcohols, even though both types of compounds possess hydroxyl groups. Phenols are more resistant to oxidation and do not undergo dehydration. Phenols have acidic properties, whereas alcohols do not.

Ethers. The general formula for an ether is R—O—R′, where R and R′ are alkyl, cycloalkyl, or aryl groups. In the IUPAC system, ethers are named as alkoxy derivatives of alkanes. Common names are obtained by giving the R group names in alphabetical order and adding the word *ether.*

Properties of ethers. Ethers have lower boiling points than alcohols because ether molecules do not hydrogen-bond to each other. Ethers are slightly soluble in water because water forms hydrogen bonds with ethers.

Thioethers (sulfides). Thioethers (sulfides) are the sulfur analogs of ethers. They have the general formula R—S—R.

Thiols and disulfides. Thiols are the sulfur analogs of alcohols. They have the general formula R—SH. The —SH group is called the sulfhydryl group. Oxidation of thiols forms disulfides, which have the general formula R—S—S—R. The most distinctive physical property of thiols is their foul odor.

KEY REACTIONS AND EQUATIONS

1. Intramolecular dehydration of alcohols to give alkenes (Section 14.7)

2. Intermolecular dehydration of primary alcohols to give ethers (Section 14.7)

$$R{-}O{-}H + H{-}O{-}R \xrightarrow[140°C]{H_2SO_4} R{-}O{-}R + H_2O$$

3. Oxidation of a primary alcohol to give an aldehyde and then a carboxylic acid (Section 14.7)

4. Oxidation of a secondary alcohol to give a ketone (Section 14.7)

5. Attempted oxidation of a tertiary alcohol, which gives no reaction (Section 14.7)

6. Production of an alkyl halide from an alcohol by substitution using PX₃ (X is Cl or Br) (Section 14.7)

$$R{-}OH \xrightarrow[\Delta]{PX_3} R{-}X$$

7. Oxidation of a thiol to give a disulfide (Section 14.17)

$$R{-}SH + HS{-}R \xrightarrow{[O]} R{-}S{-}S{-}R + 2H$$

8. Reduction of a disulfide to give a thiol (Section 14.17)

$$R{-}S{-}S{-}R + 2H \longrightarrow R{-}SH + HS{-}R$$

KEY TERMS

Alcohol (14.2)
Alkoxy group (14.14)
Antioxidant (14.12)
Aryl group (14.9)
Condensation reaction (14.7)
Dehydration reaction (14.7)
Elimination reaction (14.7)

Ether (14.13)
Glycol (14.4)
Heterocyclic organic compound (14.16)
Hydroxyl group (14.2)
Phenol (14.9)
Primary alcohol (14.7)
Secondary alcohol (14.7)

Sulfhydryl group (14.17)
Tertiary alcohol (14.7)
Thioether (14.17)
Thiol (14.17)
Zaitsev's rule (14.7)

EXERCISES AND PROBLEMS

The members of each pair of problems in this section test similar material.

▪ Bonding Characteristics of Oxygen (Section 14.1)

14.1 In organic compounds, how many covalent bonds does each of the following types of atoms form?
a. Oxygen   b. Hydrogen
c. Carbon   d. A halogen

14.2 In organic compounds, which of the following bonding behaviors does an oxygen atom exhibit?
a. One single bond   b. Two single bonds
c. One double bond   d. Two double bonds

▪ Structural Characteristics of Alcohols (Section 14.2)

14.3 What is the generalized formula for an alcohol?

14.4 What is the name of the functional group that characterizes an alcohol?

14.5 Contrast, in general terms, the structures of an alcohol and water.

14.6 Contrast, in general terms, the structures of an alcohol and an alkane.

▪ Nomenclature for Alcohols (Section 14.3)

14.7 Assign an IUPAC name to each of the following alcohols.

a.
$$CH_3-CH_2-CH_2-\overset{\overset{\displaystyle OH}{|}}{CH}-CH_3$$

b. CH_3-CH_2-OH

c.
$$CH_3-\overset{\overset{\displaystyle CH_3}{|}}{CH}-\overset{\overset{\displaystyle OH}{|}}{CH}-CH_3$$

d.
$$CH_3-CH_2-CH_2-\overset{\overset{\displaystyle |}{\underset{\underset{\displaystyle OH}{|}}{CH_2}}}{CH}-CH_2-CH_3$$

e.
$$CH_3-CH_2-\overset{\overset{\displaystyle |}{\underset{\underset{\displaystyle CH_3}{|}}{CH}}}{CH}-OH$$

f.
$$CH_3-\overset{\overset{\displaystyle CH_3}{|}}{\underset{\underset{\displaystyle CH_3}{|}}{C}}-CH_2-CH_2-OH$$

14.8 Assign an IUPAC name to each of the following alcohols.

a.
$$CH_3-CH_2-\overset{\overset{\displaystyle OH}{|}}{CH}-CH_2-CH_3$$

b.
$$CH_3-CH_2-\overset{\overset{\displaystyle OH}{|}}{CH}-\overset{\overset{\displaystyle CH_3}{|}}{CH}-CH_3$$

c.
$$CH_3-CH_2-\overset{\overset{\displaystyle |}{\underset{\underset{\displaystyle CH_2-CH_2-CH_2-OH}{|}}{CH}}}{CH}-CH_2-CH_2-CH_3$$

d.
$$CH_3-\overset{\overset{\displaystyle |}{\underset{\underset{\displaystyle CH_3}{|}}{CH}}}{CH}-\overset{\overset{\displaystyle |}{\underset{\underset{\displaystyle CH_3}{|}}{CH}}}{CH}-CH_2-OH$$

e.
$$CH_3-\overset{\overset{\displaystyle OH}{|}}{\underset{\underset{\displaystyle CH_3}{|}}{C}}-CH_2-CH_3$$

f.
$$CH_3-CH_2-\overset{\overset{\displaystyle OH}{|}}{CH}-\overset{\overset{\displaystyle CH_3}{|}}{\underset{\underset{\displaystyle CH_2-CH_3}{|}}{C}}-\overset{\overset{\displaystyle |}{\underset{\underset{\displaystyle }{}}{CH}}}{CH}-CH_3$$

14.9 Write a structural formula for each of the following alcohols.
a. 3-Pentanol   b. 3-Ethyl-3-hexanol
c. 2-Methyl-1-propanol   d. 4-Methyl-2-pentanol
e. 2-Phenyl-2-propanol   f. 2-Methylcyclobutanol

14.10 Write a structural formula for each of the following alcohols.
a. 2-Heptanol   b. 2,2-Dimethyl-1-hexanol
c. 2-Methyl-2-heptanol   d. 3-Ethyl-2-pentanol
e. 3-Phenyl-1-butanol   f. 3,5-Dimethylcyclohexanol

14.11 Write a structural formula for, and assign an IUPAC name to, each of the following alcohols.
a. Pentyl alcohol   b. Propyl alcohol
c. Isobutyl alcohol   d. *sec*-Butyl alcohol

14.12 Write a structural formula for, and assign an IUPAC name to, each of the following alcohols.
a. Butyl alcohol   b. Hexyl alcohol
c. Isopropyl alcohol   d. *tert*-Butyl alcohol

14.13 Assign an IUPAC name to each of the following polyhydroxy alcohols.

a.
$$\overset{\overset{\displaystyle }{\underset{\underset{\displaystyle OH}{|}}{CH_2}}}{CH_2}-\overset{\overset{\displaystyle }{\underset{\underset{\displaystyle OH}{|}}{CH}}}{CH}-CH_3$$

b. 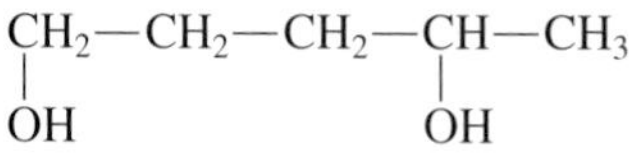

c. 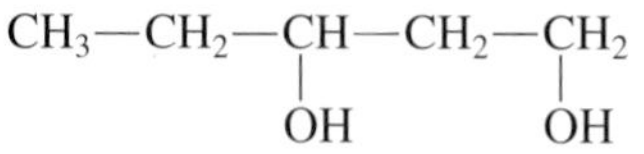

d. 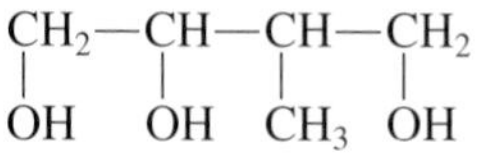

14.14 Assign an IUPAC name to each of the following polyhydroxy alcohols.

a. 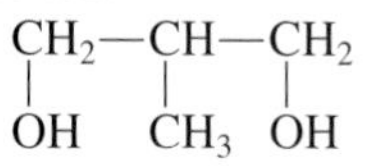

b. 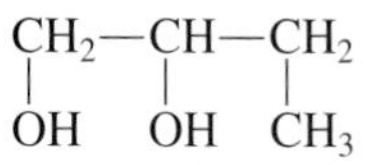

c. 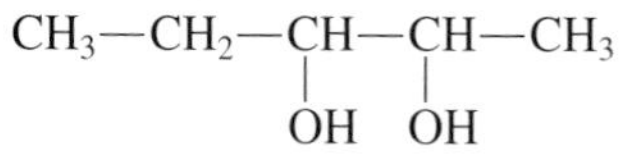

d. 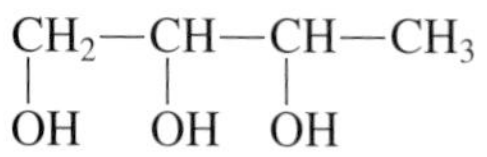

14.15 Utilizing IUPAC rules, name each of the following compounds. Don't forget to use *cis-* and *trans-* prefixes (Section 12.13) where needed.

a. 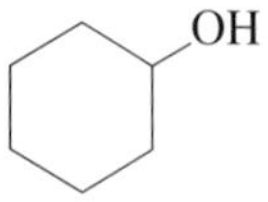

b. 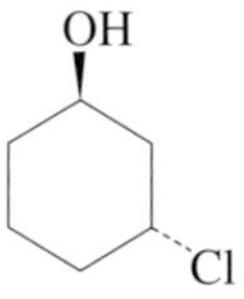

c. 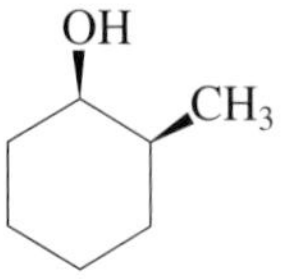

d. 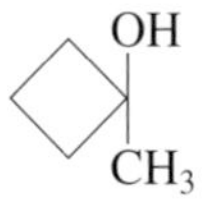

14.16 Utilizing IUPAC rules, name each of the following compounds. Don't forget to use *cis-* and *trans-* prefixes (Section 12.13) where needed.

a. 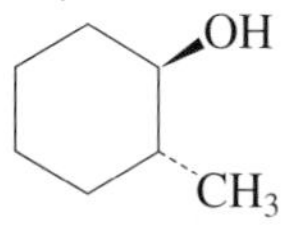

b. 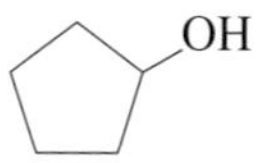

c. 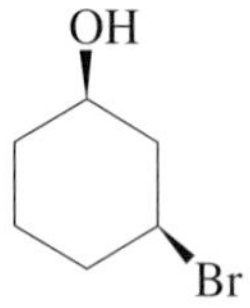

d. 

14.17 Write a structural formula for each of the following *unsaturated* alcohols.

a. 4-Penten-2-ol b. 1-Pentyn-3-ol
c. 3-Methyl-3-buten-2-ol d. *cis*-2-Buten-1-ol

14.18 Write a structural formula for each of the following *unsaturated* alcohols.

a. 1-Penten-3-ol b. 3-Butyn-1-ol
c. 2-Methyl-3-buten-1-ol d. *trans*-3-Penten-1-ol

14.19 Each of the following alcohols is named incorrectly. However, the names give correct structural formulas. Draw structural formulas for the compounds, and then write the correct IUPAC name for each alcohol.

a. 2-Ethyl-1-propanol b. 2,4-Butanediol
c. 2-Methyl-3-butanol d. 1,4-Cyclopentanediol

14.20 Each of the following alcohols is named incorrectly. However, the names give correct structural formulas. Draw structural formulas for the compounds, and then write the correct IUPAC name for each alcohol.

a. 3-Ethyl-2-butanol b. 3,4-Pentanediol
c. 3-Methyl-3-butanol d. 1,1-Dimethyl-1-butanol

Important Common Alcohols (Section 14.4)

14.21 What does each of the following terms mean?

a. Absolute alcohol b. Grain alcohol
c. Rubbing alcohol d. Drinking alcohol

14.22 What does each of the following terms mean?

a. Wood alcohol b. Denaturated alcohol
c. 70-Proof alcohol d. "Alcohol"

14.23 Give the name of the alcohol that fits each of the following descriptions.

a. Thick liquid that has the consistency of honey
b. Often produced via a fermentation process
c. Used as a race car fuel
d. Industrially produced from CO and H_2

14.24 Give the name of the alcohol that fits each of the following descriptions.

a. Sometimes used as a skin coolant for the human body
b. Antifreeze ingredient
c. Active ingredient in alcoholic beverages
d. Moistening agent in many cosmetics

Physical Properties of Alcohols (Section 14.5)

14.25 Explain why the boiling points of alcohols are much higher than those of alkanes with similar molecular masses.

14.26 Explain why the water solubilities of alcohols are much higher than those of alkanes with similar molecular masses.

14.27 Which member of each of the following pairs of compounds would you expect to have the higher boiling point?

a. 1-Butanol and 1-heptanol
b. Butane and 1-propanol
c. Ethanol and 1,2-ethanediol

14.28 Which member of each of the following pairs of compounds would you expect to have the higher boiling point?

a. 1-Octanol and 1-pentanol
b. Pentane and 1-butanol
c. 1,3-Propanediol and 1-propanol

14.29 Which member of each of the following pairs of compounds would you expect to be more soluble in water?

a. Butane and 1-butanol
b. 1-Octanol and 1-pentanol
c. 1,2-Butanediol and 1-butanol

14.30 Which member of each of the following pairs of compounds would you expect to be more soluble in water?

a. 1-Pentanol and 1-butanol
b. 1-Propanol and 1-hexanol
c. 1,2,3-Propanetriol and 1-hexanol

14.31 Determine the maximum number of hydrogen bonds that can form between an ethanol molecule and

a. other ethanol molecules b. water molecules
c. methanol molecules d. 1-propanol molecules

14.32 Determine the maximum number of hydrogen bonds that can form between a methanol molecule and

a. other methanol molecules b. water molecules
c. 1-propanol molecules d. 2-propanol molecules

■ **Preparation of Alcohols (Section 14.6)**

14.33 Write the structure of the expected predominant organic product formed in each of the following reactions.

a. $CH_2{=}CH_2 + H_2O \xrightarrow{H_2SO_4}$

b.
$$CH_3{-}CH_2{-}\overset{\displaystyle O}{\overset{\|}{C}}{-}H + H_2 \xrightarrow{Catalyst}$$

c. $CH_3{-}CH_2{-}\underset{\underset{\displaystyle CH_3}{|}}{C}{=}CH_2 + H_2O \xrightarrow{H_2SO_4}$

d.
$$CH_3{-}CH_2{-}\overset{\displaystyle O}{\overset{\|}{C}}{-}CH_2{-}CH_3 + H_2 \xrightarrow{Catalyst}$$

14.34 Write the structure of the expected predominant organic product formed in each of the following reactions.

a. $CH_3{-}CH{=}CH{-}CH_3 + H_2O \xrightarrow{H_2SO_4}$

b.
$$CH_3{-}CH_2{-}\overset{\displaystyle O}{\overset{\|}{C}}{-}CH_3 + H_2 \xrightarrow{Catalyst}$$

c.
$$CH_3{-}\overset{\displaystyle O}{\overset{\|}{C}}{-}H + H_2 \xrightarrow{Catalyst}$$

d. $CH_3{-}\underset{\underset{\displaystyle CH_3}{|}}{CH}{-}CH{=}CH{-}CH_3 + H_2O \xrightarrow{H_2SO_4}$

■ **Reactions of Alcohols (Section 14.7)**

14.35 Draw the structure of the organic product expected to be predominant when each of the following alcohols is dehydrated using sulfuric acid at the temperature indicated.

a. $CH_3{-}\underset{\underset{\displaystyle OH}{|}}{CH}{-}CH_3 \xrightarrow[180°C]{H_2SO_4}$

b. $CH_3{-}\underset{\underset{\displaystyle CH_3}{|}}{CH}{-}CH_2{-}OH \xrightarrow[180°C]{H_2SO_4}$

c. $CH_3{-}\underset{\underset{\displaystyle CH_3}{|}}{CH}{-}OH \xrightarrow[140°C]{H_2SO_4}$

d. $CH_3{-}CH_2{-}CH_2{-}OH \xrightarrow[180°C]{H_2SO_4}$

14.36 Draw the structure of the organic product expected to be predominant when each of the following alcohols is dehydrated using sulfuric acid at the temperature indicated.

a. $CH_3{-}\underset{\underset{\displaystyle CH_3}{|}}{CH}{-}CH_2{-}OH \xrightarrow[140°C]{H_2SO_4}$

b. $CH_3{-}\underset{\underset{\displaystyle CH_3}{|}}{CH}{-}CH_2{-}OH \xrightarrow[180°C]{H_2SO_4}$

c. $CH_3{-}\underset{\underset{\displaystyle OH}{|}}{CH}{-}CH_2{-}CH_3 \xrightarrow[180°C]{H_2SO_4}$

d. $CH_3{-}\underset{\underset{\displaystyle OH}{|}}{CH}{-}CH_2{-}CH_3 \xrightarrow[140°C]{H_2SO_4}$

14.37 Identify the alcohol reactant from which each of the following products was obtained by an alcohol dehydration reaction.

a. Alcohol $\xrightarrow[180°C]{H_2SO_4} CH_3{-}CH{=}\underset{\underset{\displaystyle CH_3}{|}}{C}{-}CH_3$

b. Alcohol $\xrightarrow[180°C]{H_2SO_4} CH_3{-}CH{=}CH_2$

c. Alcohol $\xrightarrow[140°C]{H_2SO_4} CH_3{-}CH_2{-}O{-}CH_2{-}CH_3$

d. Alcohol $\xrightarrow[140°C]{H_2SO_4} CH_3{-}\underset{\underset{\displaystyle CH_3}{|}}{CH}{-}O{-}\underset{\underset{\displaystyle CH_3}{|}}{CH}{-}CH_3$

14.38 Identify the alcohol reactant from which each of the following products was obtained by an alcohol dehydration reaction.

a. Alcohol $\xrightarrow[180°C]{H_2SO_4} CH_2{=}\underset{\underset{\displaystyle CH_3}{|}}{C}{-}CH_2{-}CH_3$

b. Alcohol $\xrightarrow[180°C]{H_2SO_4} CH_3{-}CH_2{-}CH{=}CH_2$

c. Alcohol $\xrightarrow[140°C]{H_2SO_4} CH_3{-}O{-}CH_3$

d. Alcohol $\xrightarrow[140°C]{H_2SO_4} CH_3{-}CH_2{-}O{-}CH_2{-}CH_3$

14.39 Draw the structure of the alcohol that could be used to prepare each of the following compounds in an oxidation reaction.

a.
$$CH_3{-}CH_2{-}\overset{\displaystyle O}{\overset{\|}{C}}{-}CH_3$$

b.
$$CH_3{-}CH_2{-}\overset{\displaystyle O}{\overset{\|}{C}}{-}OH$$

c.
$$CH_3{-}CH_2{-}\overset{\displaystyle O}{\overset{\|}{C}}{-}H$$

d.

14.40 Draw the structure of the alcohol that could be used to prepare each of the following compounds in an oxidation reaction.

a.
$$CH_3{-}\underset{\underset{\displaystyle CH_3}{|}}{CH}{-}\overset{\displaystyle O}{\overset{\|}{C}}{-}H$$

b.
$$CH_3{-}\underset{\underset{\displaystyle CH_3}{|}}{CH}{-}\overset{\displaystyle O}{\overset{\|}{C}}{-}CH_3$$

c.
$$CH_3{-}\underset{\underset{\displaystyle CH_3}{|}}{CH}{-}CH_2{-}\overset{\displaystyle O}{\overset{\|}{C}}{-}OH$$

d.

14.41 Draw the structure of the expected predominant organic product formed in each of the following reactions.

a. $CH_3-CH_2-CH_2-OH \xrightarrow[\Delta]{PCl_3}$

b.
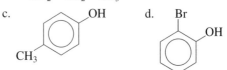
(cyclopentane with CH_3) $\xrightarrow[180°C]{H_2SO_4}$
OH

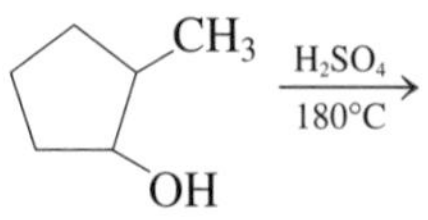

c. $CH_3-\underset{\underset{OH}{|}}{CH}-CH_2-CH_3 \xrightarrow{K_2Cr_2O_7}$

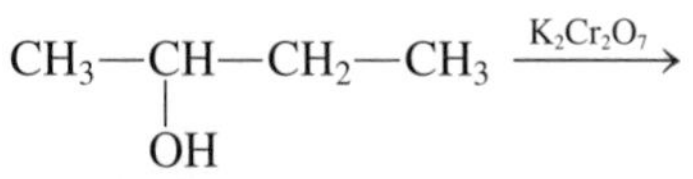

d. 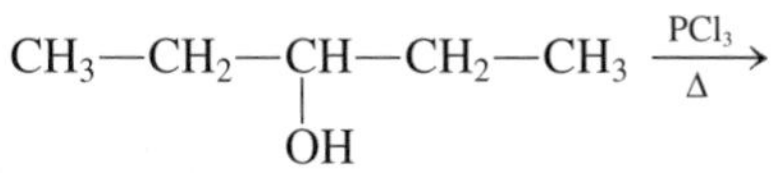 $CH_3-CH_2-\underset{\underset{OH}{|}}{CH}-CH_2-CH_3 \xrightarrow[\Delta]{PCl_3}$

e. $CH_3-CH_2-OH \xrightarrow[140°C]{H_2SO_4}$

f. 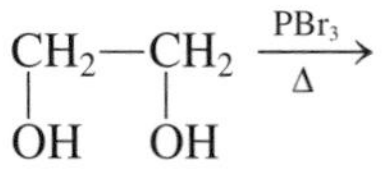 $\underset{\underset{OH}{|}}{CH_2}-\underset{\underset{OH}{|}}{CH_2} \xrightarrow[\Delta]{PBr_3}$

14.42 Draw the structure of the expected predominant organic product formed in each of the following reactions.

a. $CH_3-CH_2-CH_2-OH \xrightarrow[\Delta]{PBr_3}$

b. 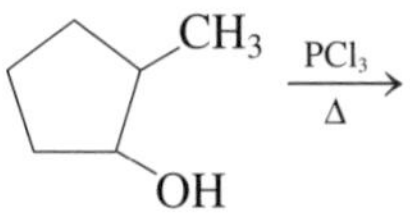 (cyclopentane with CH_3) $\xrightarrow[\Delta]{PCl_3}$
OH

c. 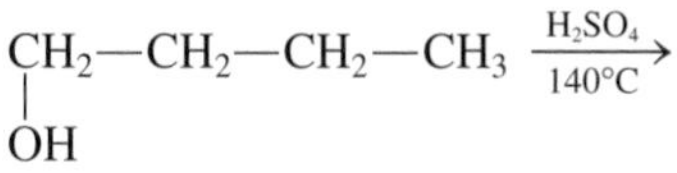 $\underset{\underset{OH}{|}}{CH_2}-CH_2-CH_2-CH_3 \xrightarrow[140°C]{H_2SO_4}$

d. 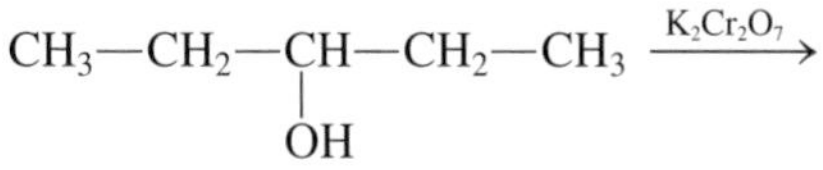 $CH_3-CH_2-\underset{\underset{OH}{|}}{CH}-CH_2-CH_3 \xrightarrow{K_2Cr_2O_7}$

e. 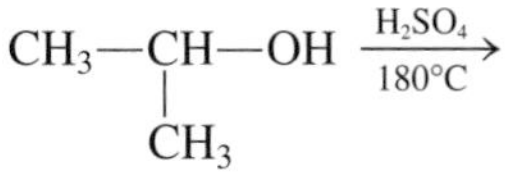 $CH_3-\underset{\underset{CH_3}{|}}{CH}-OH \xrightarrow[180°C]{H_2SO_4}$

f. 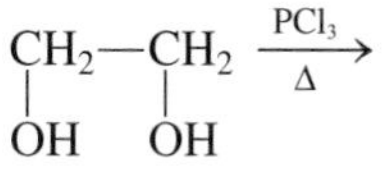 $\underset{\underset{OH}{|}}{CH_2}-\underset{\underset{OH}{|}}{CH_2} \xrightarrow[\Delta]{PCl_3}$

■ **Polymeric Alcohols (Section 14.8)**

14.43 Draw a structural representation for the polymeric alcohol PEG [poly(ethylene glycol)].

14.44 Draw a structural representation for the polymeric alcohol PVA [poly(vinyl alcohol)].

■ **Structural Characteristics of and Nomenclature for Phenols (Sections 14.9 and 14.10)**

14.45 Explain why the first of the following two compounds is a phenol and the second is not.

CH_3-⟨benzene ring⟩$-OH$ CH_3-⟨benzene ring⟩$-CH_2-OH$

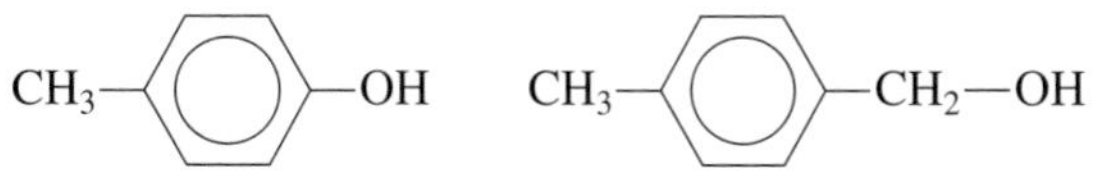

14.46 Explain why the first of the following two compounds is a phenol and the second is not.

CH_3-⟨benzene ring⟩$-OH$ CH_3-⟨cyclohexane ring⟩$-OH$

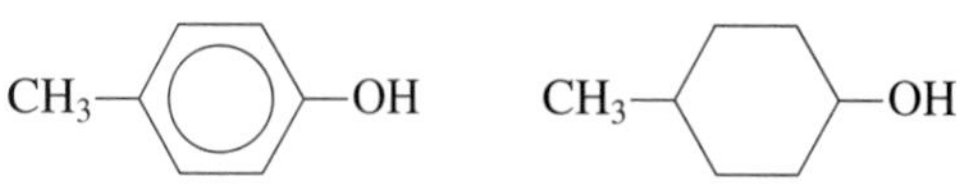

14.47 Name the following phenols.

a. 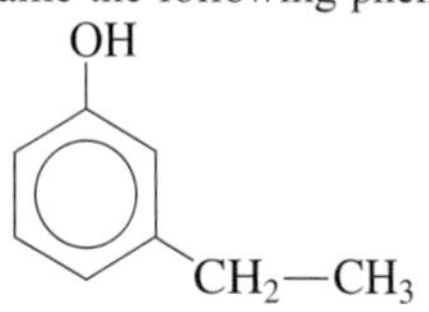 OH (benzene ring with CH_2-CH_3)

b. 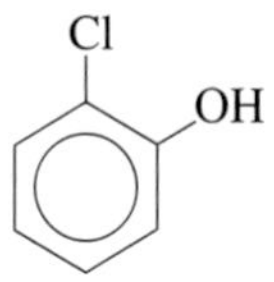 Cl (benzene ring with OH)

c. 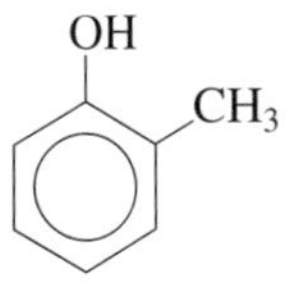 OH (benzene ring with CH_3)

d. 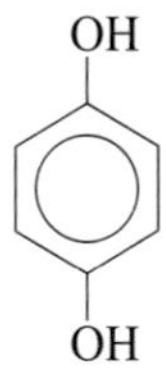 OH (benzene ring with OH)

e. 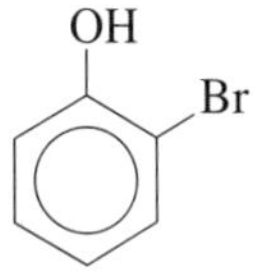 OH (benzene ring with Br)

f. 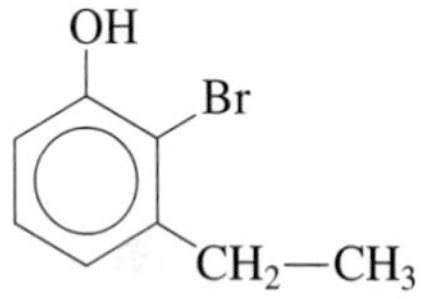 OH (benzene ring with Br and CH_2-CH_3)

14.48 Name the following phenols.

a. 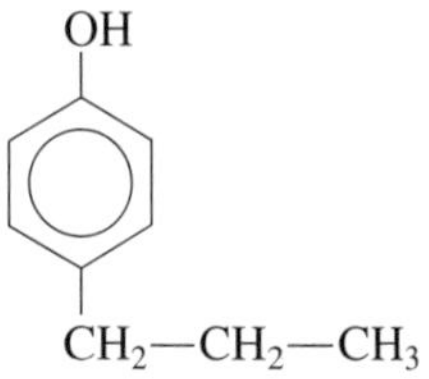 OH (benzene ring with $CH_2-CH_2-CH_3$)

b. 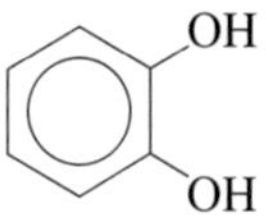 OH (benzene ring with OH)

c. OH (benzene ring with CH_3)

d. Br (benzene ring with OH)

e. 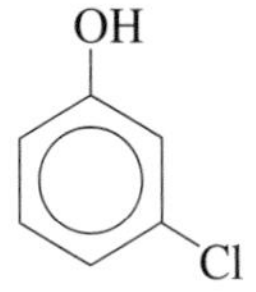 OH (benzene ring with Cl)

f. 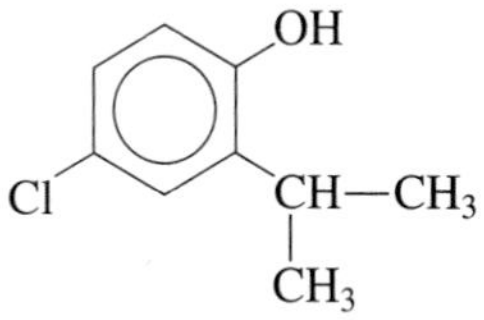 OH (benzene ring with Cl and $CH-CH_3$, CH_3)

14.49 Draw a structural formula for each of the following phenols.
a. 4-Chlorophenol
b. 2-Ethylphenol
c. 2,4-Dibromophenol
d. *m*-Cresol
e. Resorcinol
f. 2,6-Diethyl-4-methylphenol

14.50 Draw a structural formula for each of the following phenols.
a. 3-Bromophenol
b. *m*-Ethylphenol
c. Hydroquinone
d. *o*-Cresol
e. Catechol
f. 2,6-Dichlorophenol

■ **Properties and Uses of Phenols (Sections 14.11 and 14.12)**

14.51 Phenolic compounds are frequently used as antiseptics and disinfectants. What is the difference between an antiseptic and a disinfectant?

14.52 Phenolic compounds are frequently used as antioxidants. What is an antioxidant?

14.53 Phenols are weak acids. Write an equation for the acid dissociation of the compound phenol.

14.54 How does the acidity of phenols compare with that of inorganic weak acids?

■ **Structural Characteristics of and Nomenclature for Ethers (Sections 14.13 and 14.14)**

14.55 Indicate whether each of the following structural notations denotes an ether.
a. R—O—R
b. R—O—H
c. Ar—O—R
d. Ar—O—Ar

14.56 What is the difference in meaning associated with each of the following pairs of notations?
a. R—O—R and R—O—H
b. Ar—O—R and Ar—O—Ar
c. Ar—O—H and Ar—O—R
d. R—O—R and Ar—O—Ar

14.57 Assign an IUPAC name to each of the following ethers.
a. CH₃—O—CH₂—CH₂—CH₃
b. CH₃—CH₂—CH₂—O—CH₂—CH₃
c. CH₃—CH—CH₃
　　　　|
　　　　O—CH₃
d.

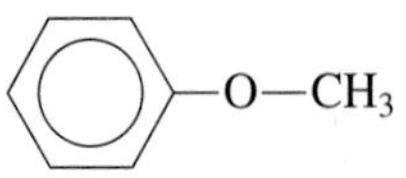

e.

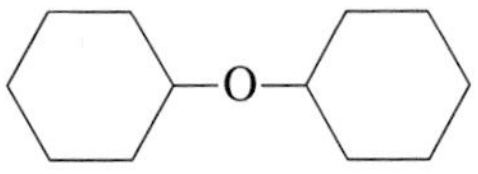

f. 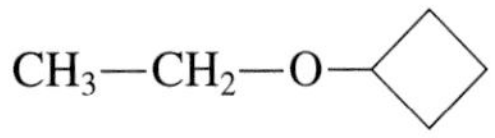 CH₃—CH₂—O—◇

14.58 Assign an IUPAC name to each of the following ethers.
a. CH₃—CH₂—O—CH₂—CH₃
b. CH₃—CH—O—CH₃
　　　　|
　　　　CH₃
c. CH₃—CH₂—CH—CH₃
　　　　　　|
　　　　　　O—CH₂—CH₃
d.

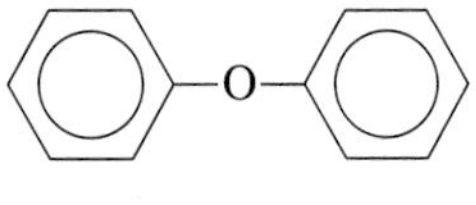

e. 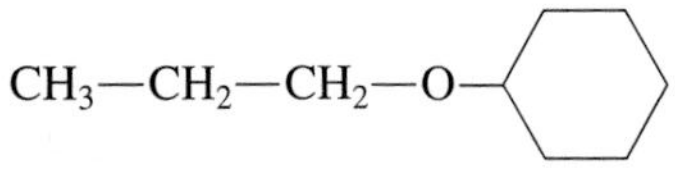 CH₃—CH₂—CH₂—O—◯

f.

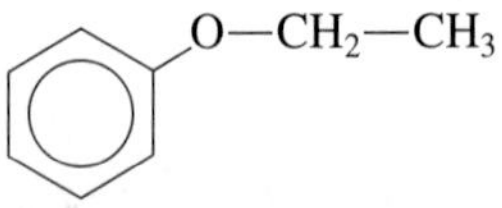

14.59 Assign a common name to each of the ethers in Problem 14.57.

14.60 Assign a common name to each of the ethers in Problem 14.58.

14.61 Draw the structure of each of the following ethers.
a. Isopropyl propyl ether
b. Ethyl phenyl ether
c. 3-Methylanisole
d. 2-Ethoxypentane
e. Ethoxycyclobutane
f. 1-Methoxy-2,2-dimethylpropane

14.62 Draw the structure of each of the following ethers.
a. Butyl methyl ether
b. Anisole
c. Phenyl propyl ether
d. 3-Propoxyheptane
e. 1,3-Dimethoxybenzene
f. 3-Methoxy-2-methylhexane

■ **Properties of Ethers (Section 14.15)**

14.63 Dimethyl ether and ethanol have the same molecular mass. Dimethyl ether is a gas at room temperature, and ethanol is a liquid at room temperature. Explain these observations.

14.64 Compare the solubility in water of ethers and alcohols that have similar molecular masses.

14.65 What are the two chemical hazards associated with ether use?

14.66 How do the chemical reactivities of ethers compare with those of
a. alkanes
b. alcohols

14.67 Explain why ether molecules cannot hydrogen-bond to each other.

14.68 How many hydrogen bonds can form between a single ether molecule and water molecules?

■ **Cyclic Ethers (Section 14.16)**

14.69 Classify each of the following molecular structures as that of a cyclic ether, a noncyclic ether, or a nonether.
a. 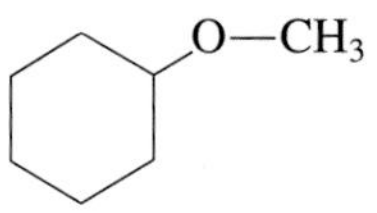 O—CH₃
b. 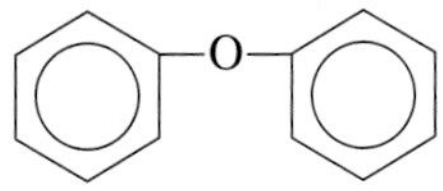 ◯—O—◯
c. 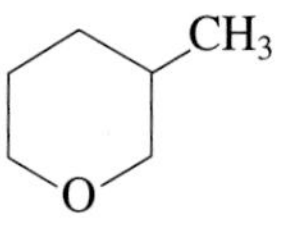 CH₃
d. 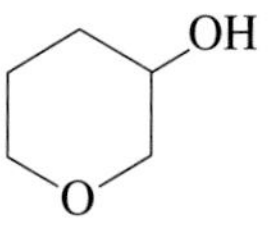 OH
e. 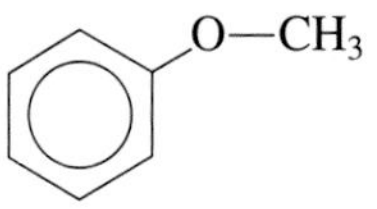 O—CH₃
f. 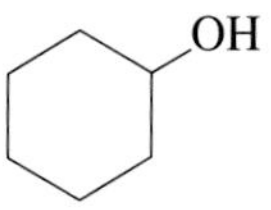 OH

14.70 Classify each of the following molecular structures as that of a cyclic ether, a noncyclic ether, or a nonether.
a. 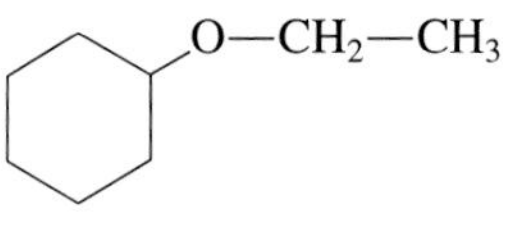 O—CH₂—CH₃
b. ◯—O—◯
c. 
d. 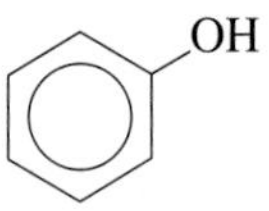 OH
e. 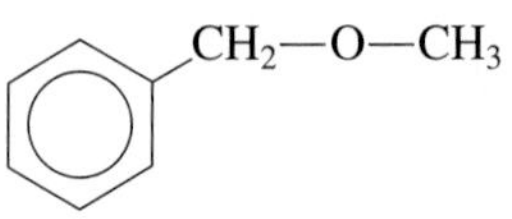 CH₂—O—CH₃
f.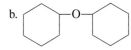

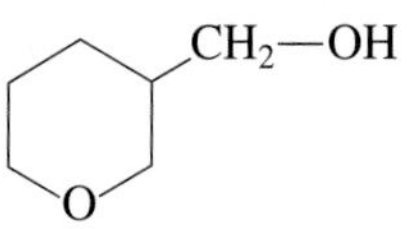

■ **Sulfur Analogs of Alcohols and Ethers (Section 14.17)**

14.71 Contrast the general structural formulas for a thioalcohol and an alcohol.

14.72 Contrast the general structural formulas for a thioether and an ether.

14.73 Draw a structural formula for each of the following thiols.
a. Methanethiol
b. 2-Propanethiol
c. 1-Butanethiol

d. 3-Methyl-1-pentanethiol
e. Cyclopentanethiol
f. 1,2-Ethanedithiol

14.74 Draw a structural formula for each of the following thiols.
  a. 1-Propanethiol
  b. Ethanethiol
  c. 1,3-Pentanedithiol
  d. 3-Methyl-3-pentanethiol
  e. 2-Methylcyclopentanethiol
  f. 2,2-Dimethyl-1-hexanethiol

14.75 Assign a common name to each of the following thiols.
  a. CH_3—SH
  b. CH_3—CH_2—CH_2—SH
  c. CH_3—CH_2—CH—SH
                         |
                        CH_3
  d. CH_3—CH—CH_2—SH
              |
             CH_3

14.76 Assign a common name to each of the following thiols.
  a. CH_3—CH_2—SH
  b. CH_3—CH—SH
            |
           CH_3

  c. CH_3—CH_2—CH_2—CH_2—SH

              CH_3
              |
  d. CH_3—C—SH
              |
              CH_3

14.77 Contrast the products that result from the oxidation of an alcohol and the oxidation of a thiol.

14.78 Write the formulas for the sulfur-containing organic products of the following reactions.

  a. $2CH_3$—CH_2—SH $\xrightarrow[\text{agent}]{\text{Oxidizing}}$

  b. CH_3—CH_2—S—S—CH_2—CH_3 $\xrightarrow[\text{agent}]{\text{Reducing}}$

14.79 Assign both an IUPAC name and a common name to each of the following thioethers.

a. CH_3—CH_2—S—CH_3
b. CH_3—CH—S—CH_3
          |
         CH_3

c. 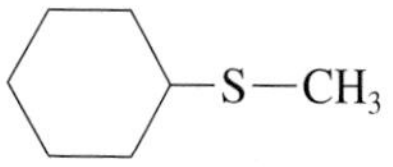
  ⬡—S—CH_3

d. 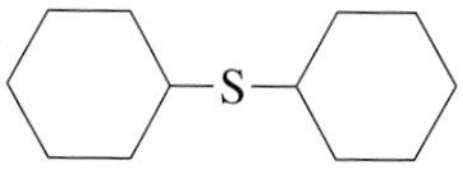
  ⬡—S—⬡

e. CH_2=CH—CH_2—S—CH_3
f. CH_3—CH_2—CH—CH_3
                  |
                  S
                  |
                 CH_3

14.80 Assign both an IUPAC name and a common name to each of the following thioethers.
  a. CH_3—CH_2—S—CH_2—CH_3
  b. CH_3—CH—S—CH_2—CH_3
            |
           CH_3

  c. 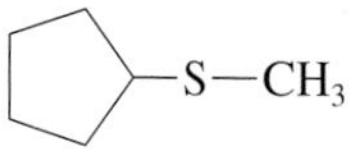
    ⬠—S—CH_3

  d. 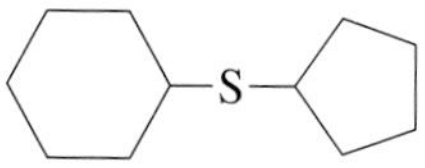
    ⬡—S—⬠

  e. CH_2=CH—CH_2—S—CH_2—CH_3
  f. CH_3—CH—CH_3
            |
            S
            |
           CH_3

ADDITIONAL PROBLEMS

14.81 Classify each of the alcohols in Problem 14.8 as a primary, secondary, or tertiary alcohol.

14.82 Classify each of the alcohol functional groups in the compounds in Problem 14.14 as a primary-, secondary-, or tertiary-alcohol functional group.

14.83 Assign an IUPAC name to each of the following compounds.
  a. 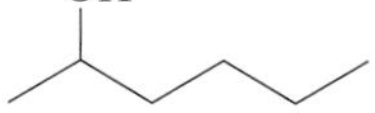 OH
  b. 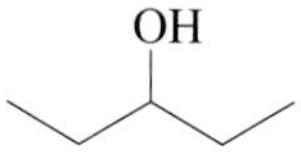 OH
  c. 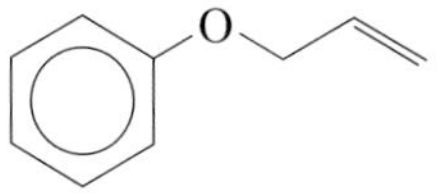 ⬡—O—allyl
  d. 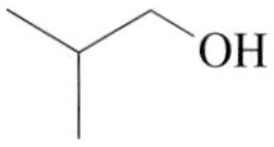 OH
  e. 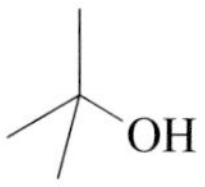 OH (tert-butyl)
  f. 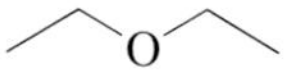 ethyl ether

14.84 Draw structural formulas for the eight isomeric alcohols and six isomeric ethers that have the molecular formula $C_5H_{10}O$.

14.85 Three isomeric pentanols with unbranched carbon chains exist. Which of these, upon dehydration at 180°C, yields only 1-pentene as a product?

14.86 A mixture of methanol, 1-propanol, and H_2SO_4 (catalyst) is heated to 140°C. After reaction, the solution contains three different ethers. Draw a structural formula for each of the ethers.

14.87 Which of the terms *ether, alcohol, diol, thiol, thioether, thioalcohol, disulfide, sulfide,* and *peroxide* characterize(s) each of the following compounds? Note that more than one term may apply to a given compound.
  a. CH_3—S—S—CH_3
  b. CH_3—CH_2—SH
  c. CH_3—CH—CH_2—CH_3
            |
            OH
  d. CH_3—CH_2—O—O—CH_3
  e. HO—CH_2—CH_2—SH
  f. CH_3—O—CH_2—S—CH_3

14.88 Assign IUPAC names to the following compounds.
    a. $HS-CH_2-CH_2-SH$
    b. $CH_3-O-CH_2-CH_2-CH_2-OH$
    c. $HO-CH_2-CH_2-CH_3$

    d. $CH_3-CH_2-O-CH_2-CH_2-O-CH_2-CH_3$
    e. $CH_3-S-CH_2-CH_3$
    f. $CH_3-O-CH_2-CH_2-S-CH_2-CH_3$

ANSWERS TO PRACTICE EXERCISES

14.1 a. 2,5-dimethyl-3-hexanol   b. 3-methyl-1-pentanol
    c. 1,2-dimethylcyclopentanol

14.2 a.
$$CH_2-CH_2-CH_2-CH_3$$
$$|$$
$$OH$$

b.
$$\begin{array}{c} \quad\quad\quad OH \\ \quad\quad\quad | \\ CH_3-C-CH-CH_3 \\ \quad | \quad | \\ \quad CH_3\ CH_3 \end{array}$$

c. $CH_3-CH_2-CH_2-OH$

15 Aldehydes and Ketones

Benzaldehyde is the main flavor component in almonds. Aldehydes and ketones are responsible for the odor and taste of numerous nuts and spices.

In this chapter, we continue our discussion of hydrocarbon derivatives that contain the element oxygen. The functional groups we considered in the previous chapter (alcohols, phenols, and ethers) have the common feature of carbon–oxygen *single* bonds. Carbon–oxygen *double* bonds are also possible in hydrocarbon derivatives. We will now consider the simplest types of compounds that contain this structural feature: aldehydes and ketones.

15.1 The Carbonyl Group

Both aldehydes and ketones contain a carbonyl functional group. A **carbonyl group** *is a carbon atom double bonded to an oxygen atom.* The structural representation for a carbonyl group is

$$\underbrace{\diagdown C\!=\!\ddot{O}\,:}$$

Carbonyl group

Carbon–oxygen and carbon–carbon double bonds differ in a major way. A carbon–oxygen double bond is *polar*, and a carbon–carbon double bond is *nonpolar*. The electronegativity (Section 5.9) of oxygen (3.5) is much greater than that of carbon (2.5). Hence the carbon–oxygen double bond is polarized, the oxygen atom acquiring a partial negative charge (δ^-) and the carbon atom acquiring a partial positive charge (δ^+).

$$\diagdown \overset{\delta^+}{C}\!=\!\overset{\delta^-}{O} \qquad \text{or} \qquad \diagdown \overset{+}{C}\!=\!\overset{\longrightarrow}{O}$$

Polar nature of carbon–oxygen double bond

15.2 Structure of Aldehydes and Ketones

▶ The word *carbonyl* is pronounced "carbon-EEL."

▶ The difference in electronegativity between oxygen and carbon causes a carbon–oxygen double bond to be polar.

▶ The word *aldehyde* is pronounced "AL-da-hide."

An **aldehyde** *is an organic compound in which a carbonyl carbon atom is bonded to at least one hydrogen atom.* The remaining group attached to the carbonyl carbon atom can be hydrogen, an alkyl group (R), a cycloalkyl group, or an aryl group (Ar).

The aldehyde functional group, the structural feature common to all the preceding compounds, is

Linear notations for an aldehyde functional group and for an aldehyde itself are —CHO and RCHO, respectively. Note that the ordering of the symbols H and O in these notations is HO, not OH (which denotes a hydroxyl group).

Cyclic aldehyde structures are not possible. For a carbonyl carbon atom to be part of a ring system, it would have to form two bonds to ring atoms. This is not possible, because three of the four bonds from an aldehyde carbonyl carbon atom *must* go to oxygen and hydrogen. This means that an aldehyde group must always be at the end of a carbon chain.

▶ Remember that for general *condensed* functional group structures such as RCHO, carbon always has four bonds and hydrogen always has only one. In RCHO, you know one of carbon's bonds goes to the R group and one to H; therefore, two bonds must go to O.

▶ The word *ketone* is pronounced "KEY-tone."

A **ketone** *is an organic compound in which a carbonyl carbon atom is bonded to two carbon groups.* The groups containing these bonded carbon atoms may be alkyl, cycloalkyl, or aryl.

The ketone functional group, the structural feature common to all the preceding compounds, is

The general condensed formula for a ketone is RCOR, in which the oxygen atom is understood to be double-bonded to the carbonyl carbon at the left of it in the formula.

Unlike aldehydes, ketones can form cyclic structures, such as

▶ In an aldehyde, the carbonyl group is always located at the end of a hydrocarbon chain.

$$CH_3-CH_2-CH_2-CH_2-\boxed{C-H}$$
with O double-bonded to C

In a ketone, the carbonyl group is always at a nonterminal (interior) position on the hydrocarbon chain.

$$CH_3-CH_2-CH_2-\boxed{C}-CH_2-CH_3$$
with O double-bonded to C

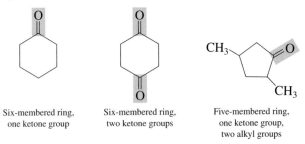

Six-membered ring, one ketone group

Six-membered ring, two ketone groups

Five-membered ring, one ketone group, two alkyl groups

Cyclic ketones are *not* heterocyclic ring systems as were cyclic ethers (Section 14.16).

▶ This is the second time we have encountered functional group isomerism. Alcohols and ethers (Section 14.14) also can exhibit such isomerism.

Aldehydes and ketones with the same number of carbon atoms and the same degree of saturation are isomeric (functional group isomerism). For example, there are both an aldehyde and a ketone that have the molecular formula C_3H_6O. (See Figure 15.1.)

$$CH_3-CH_2-\overset{\overset{\displaystyle O}{\|}}{C}-H \qquad \text{and} \qquad CH_3-\overset{\overset{\displaystyle O}{\|}}{C}-CH_3$$

An aldehyde   A ketone

FIGURE 15.1 Aldehydes and ketones with the same number of carbon atoms and the same degree of saturation are structural isomers, as is illustrated here for the three-carbon aldehyde (propanal) and the three-carbon ketone (propanone). Both have the molecular formula C_3H_6O.

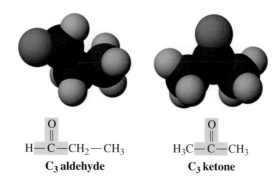

$$\underset{\textbf{C}_3\textbf{ aldehyde}}{H-\overset{\overset{\displaystyle O}{\|}}{C}-CH_2-CH_3}$$

$$\underset{\textbf{C}_3\textbf{ ketone}}{H_3C-\overset{\overset{\displaystyle O}{\|}}{C}-CH_3}$$

15.3 Nomenclature for Aldehydes

The IUPAC rules for naming aldehydes are as follows:

1. Select as the parent carbon chain the longest chain that *includes* the carbon atom of the carbonyl group.
2. Name the parent chain by changing the *-e* ending of the corresponding alkane name to *-al*.
3. Number the parent chain by assigning the number 1 to the carbonyl carbon atom of the aldehyde group.
4. Determine the identity and location of any substituents, and append this information to the front of the parent chain name.

▶ The carbonyl carbon atom in an aldehyde cannot have any number but 1, so we do not have to include this number in the aldehyde's IUPAC name.

EXAMPLE 15.1

Determining IUPAC Names for Aldehydes

▶ Line-angle drawings for the simpler unbranched-chain aldehydes:

$$\underset{\text{Methanal}}{\overset{\overset{\displaystyle O}{\|}}{H\quad\quad H}}$$

$$\underset{\text{Ethanal}}{\overset{\overset{\displaystyle O}{\|}}{\quad\quad H}}$$

$$\underset{\text{Propanal}}{\overset{\displaystyle O}{\diagup\diagdown_{H}}}$$

$$\underset{\text{Butanal}}{\overset{\displaystyle O}{\diagup\diagdown_{H}}}$$

▶ Be careful about the endings *-al* and *-ol*. They are easily confused. The suffix *-al* (pronounced like the man's name Al) denotes an aldehyde; the suffix *-ol* (pronounced like the *ol* in old) denotes an alcohol.

■ Assign IUPAC names to the following aldehydes.

a.
$$CH_3-CH_2-CH_2-CH_2-\overset{\overset{\displaystyle O}{\|}}{C}-H$$

b.
$$CH_3-\underset{\underset{\displaystyle CH_3}{|}}{CH}-CH_2-\overset{\overset{\displaystyle O}{\|}}{C}-H$$

c.
$$CH_3-CH_2-CH_2-\underset{\underset{\underset{\displaystyle CH_3}{|}}{\underset{\displaystyle CH_2}{|}}}{CH}-\overset{\overset{\displaystyle O}{\|}}{C}-H$$

d.
$$CH_3-CH_2-\underset{\underset{\displaystyle OH}{|}}{CH}-CH_2-\overset{\overset{\displaystyle O}{\|}}{C}-H$$

Solution

a. The parent chain name comes from pentane. Remove the *-e* ending and add the aldehyde suffix *-al*. The name becomes *pentanal*. The location of the carbonyl carbon atom need not be specified, because this carbon atom is always number 1. The complete name is simply *pentanal*.

b. The parent chain name is *butanal*. To locate the methyl group, we number the carbon chain beginning with the carbonyl carbon atom. The complete name of the aldehyde is *3-methylbutanal*.

$$\overset{4}{C}H_3-\overset{3}{\underset{\underset{\displaystyle CH_3}{|}}{C}H}-\overset{2}{C}H_2-\overset{\overset{\overset{\displaystyle O}{\|}}{}}{\overset{1}{C}}-H$$

(continued)

▶ When a compound contains more than one type of functional group, the suffix for only one of them can be used as the ending of the name. The IUPAC rules establish priorities that specify which suffix is used. For the functional groups we have discussed up to this point in the text, the IUPAC priority system is

Increasing priority ↑
aldehyde
ketone
alcohol
alkene
alkyne
alkoxy ⎤
alkyl ⎬ Equal-priority substituents (listed in alphabetical order)
halogen ⎦

aldehyde–ether

$$CH_3-O-CH_2-\overset{\overset{\displaystyle O}{\|}}{C}-H$$

2-methoxyethanal

aldehyde–alkene

$$CH_2=CH-\overset{\overset{\displaystyle O}{\|}}{C}-H$$

2-propenal

▶ The common names for simple aldehydes illustrate a second method for counting from one to four: *form-, acet-, propion-,* and *butyr-.* We will use this method again in the next chapter when we consider the common names for carboxylic acids and esters. (The first method for counting from one to four, with which you are now thoroughly familiar, is *meth-, eth-, prop-,* and *but-,* as in methane, ethane, propane, and butane.)

▶ The contrast between IUPAC names and common names for aldehydes is as follows:

IUPAC (one word)
⎡alkanal⎤
pentanal
Common (one word)
⎡(prefix) aldehyde*⎤
butyraldehyde

*The common-name prefixes are related to natural sources for carboxylic acids with the same number of carbon atoms (see Section 16.3).

▶ Aromatic aldehydes are not cyclic aldehydes (which do not exist). The carbonyl carbon atom in an aromatic aldehyde is not part of the ring system.

c. The longest chain containing the carbonyl carbon atom is five carbons long, giving a parent chain name of *pentanal.* An ethyl group is present on carbon 2. Thus the complete name is *2-ethylpentanal.*

$$\overset{5}{C}H_3-\overset{4}{C}H_2-\overset{3}{C}H_2-\overset{2}{C}H-\overset{1}{\overset{\overset{\displaystyle O}{\|}}{C}}-H$$
$$\underset{\underset{\displaystyle CH_3}{|}}{CH_2}$$

d. This is a hydroxyaldehyde, with the hydroxyl group located on carbon 3.

$$\overset{5}{C}H_3-\overset{4}{C}H_2-\overset{3}{C}H-\overset{2}{C}H_2-\overset{1}{\overset{\overset{\displaystyle O}{\|}}{C}}-H$$
$$\underset{\displaystyle OH}{|}$$

The complete name of the compound is *3-hydroxypentanal.* An aldehyde functional group has priority over an alcohol functional group in IUPAC nomenclature. An alcohol group named as a substituent is a *hydroxy* group.

Practice Exercise 15.1

Assign IUPAC names to the following aldehydes.

a.
$$CH_3-\underset{\underset{\displaystyle CH_3}{|}}{CH}-\overset{\overset{\displaystyle O}{\|}}{C}-H$$

b.
$$CH_3-CH_2-\underset{\underset{\displaystyle CH_2-CH_2-CH_3}{|}}{CH}-\overset{\overset{\displaystyle O}{\|}}{C}-I$$

c.
$$CH_3-\underset{\underset{\displaystyle Cl}{|}}{CH}-\underset{\underset{\displaystyle Cl}{|}}{CH}-\overset{\overset{\displaystyle O}{\|}}{C}-H$$

d.
$$CH_3-(CH_2)_4-\overset{\overset{\displaystyle O}{\|}}{C}-H$$

Unbranched aldehydes with four or fewer carbon atoms have common names:

$$H-\overset{\overset{\displaystyle O}{\|}}{C}-H$$
Formaldehyde

$$CH_3-\overset{\overset{\displaystyle O}{\|}}{C}-H$$
Acetaldehyde

$$CH_3-CH_2-\overset{\overset{\displaystyle O}{\|}}{C}-H$$
Propionaldehyde

$$CH_3-CH_2-CH_2-\overset{\overset{\displaystyle O}{\|}}{C}-H$$
Butyraldehyde

Unlike the common names for alcohols and ethers, the common names for aldehydes are *one* word rather than two or three.

In assigning common names to branched aldehyde systems, we name and locate substituents (branches) in the same way as in the IUPAC system.

$$\overset{4}{C}H_3-\overset{3}{\underset{\underset{\displaystyle CH_3}{|}}{C}H}-\overset{2}{C}H_2-\overset{1}{\overset{\overset{\displaystyle O}{\|}}{C}}-H$$
3-Methylbutyraldehyde

$$Cl-\overset{3}{C}H_2-\overset{2}{C}H_2-\overset{1}{\overset{\overset{\displaystyle O}{\|}}{C}}-H$$
3-Chloropropionaldehyde

In the IUPAC system, aromatic aldehydes—compounds in which an aldehyde group is attached to a benzene ring—are named as derivatives of benzaldehyde, the parent compound (see Figure 15.2 on page 402).

Propanone is the simplest possible ketone. One- and two-carbon ketones cannot exist. A minimum of three carbon atoms is required for a ketone: one C atom for the carbonyl group and one C atom for each of the groups attached to the carbonyl carbon atom. No locator number is needed in the name propanone, because there is only one possible location for the double bond.

Line-angle drawings for the simpler unbranched-chain 2-ketones:

2-Propanone

2-Butanone

2-Pentanone

2-Hexanone

Benzaldehyde

3-Chloro-5-methylbenzaldehyde

4-Hydroxybenzaldehyde

The last of these compounds is named as a benzaldehyde rather than as a phenol because the aldehyde group has priority over the hydroxyl group in the IUPAC naming system.

15.4 Nomenclature for Ketones

Assigning IUPAC names to ketones is similar to naming aldehydes except that the ending *-one* is used instead of *-al*. The rules for IUPAC ketone nomenclature follow.

1. Select as the parent carbon chain the longest carbon chain that *includes* the carbon atom of the carbonyl group.
2. Name the parent chain by changing the *-e* ending of the corresponding alkane name to *-one*.
3. Number the carbon chain such that the carbonyl carbon atom receives the lowest possible number. The position of the carbonyl carbon atom is noted by placing a number immediately before the name of the parent chain.
4. Determine the identity and location of any substituents, and append this information to the front of the parent chain name.
5. Cyclic ketones are named by assigning the number 1 to the carbon atom of the carbonyl group. The ring is then numbered to give the lowest number(s) to the atom(s) bearing substituents.

EXAMPLE 15.2

Determining IUPAC Names for Ketones

In IUPAC nomenclature, the ketone functional group has precedence over all groups we have discussed so far except the aldehyde group. When both aldehyde and ketone groups are present in the same molecule, the ketone group is named as a substituent (the *oxo-* group).

$$CH_3-\overset{O}{\overset{\|}{C}}-CH_2-CH_2-\overset{O}{\overset{\|}{C}}-H$$

4-Oxopentanal

■ Assign IUPAC names to the following ketones.

a.
$$CH_3-\overset{O}{\overset{\|}{C}}-CH_2-CH_2-CH_3$$

b.
$$CH_3-\overset{}{\underset{\underset{CH_3}{|}}{CH}}-\overset{}{\underset{\underset{CH_3}{|}}{CH}}-\overset{O}{\overset{\|}{C}}-CH_3$$

c.

d.

Solution

a. The parent chain name is *pentanone*. We number the chain from the end closest to the carbonyl carbon atom. Locating the carbonyl carbon at carbon 2 completes the name: *2-pentanone*.

b. The longest carbon chain of which the carbonyl carbon is a member is five carbons long. The parent chain name is again *pentanone*. There are two methyl groups attached, and the numbering system is from right to left.

$$\overset{5}{C}H_3-\overset{4}{\underset{\underset{CH_3}{|}}{C}H}-\overset{3}{\underset{\underset{CH_3}{|}}{C}H}-\overset{2}{\overset{O}{\overset{\|}{C}}}-\overset{1}{C}H_3$$

The complete name for the compound is *3,4-dimethyl-2-pentanone*.

(continued)

c. The base name is *cyclohexanone*. The methyl group is bonded to carbon 2 because we begin numbering at the carbonyl carbon; the name is *2-methylcyclohexanone*.

d. This ketone has a base name of *cyclopentanone*. Numbering clockwise from the carbonyl carbon atom locates the bromo group on carbon 3. The complete name is *3-bromocyclopentanone*.

Practice Exercise 15.2

Assign IUPAC names to the following ketones.

a.

$$CH_3—CH_2—\overset{\overset{\displaystyle O}{\|}}{C}—CH_2—CH_3$$

b.

$$CH_3—\underset{\underset{\displaystyle CH_3}{|}}{CH}—\overset{\overset{\displaystyle O}{\|}}{C}—\underset{\underset{\displaystyle CH_3}{|}}{CH}—CH_3$$

c.

d.

Common names for ketones are constructed by giving, in alphabetical order, the names of the alkyl or aryl groups attached to the carbonyl functional group and then adding the word *ketone*. Unlike aldehyde common names, which are one word, those for ketones are two or three words.

$$CH_3—\overset{\overset{\displaystyle O}{\|}}{C}—CH_2—CH_3 \qquad CH_3—CH_2—\overset{\overset{\displaystyle O}{\|}}{C}—\bigcirc \qquad CH_3—\underset{\underset{\displaystyle CH_3}{|}}{CH}—\overset{\overset{\displaystyle O}{\|}}{C}—CH_3$$

Ethyl methyl ketone   Cyclohexyl ethyl ketone   Isopropyl methyl ketone

Three ketones have additional common names besides those obtained with the preceding procedures. These three ketones are

$$CH_3—\overset{\overset{\displaystyle O}{\|}}{C}—CH_3 \qquad \bigcirc—\overset{\overset{\displaystyle O}{\|}}{C}—CH_3 \qquad \bigcirc—\overset{\overset{\displaystyle O}{\|}}{C}—\bigcirc$$

Acetone   Acetophenone   Benzophenone
(dimethyl ketone)   (methyl phenyl ketone)   (diphenyl ketone)

Acetophenone is the simplest aromatic ketone.

15.5 Selected Common Aldehydes and Ketones

Formaldehyde, the simplest aldehyde, with only one carbon atom, is manufactured on a large scale by the oxidation of methanol.

$$CH_3—OH \xrightarrow[600°C-700°C]{Ag\ catalyst} H—\overset{\overset{\displaystyle O}{\|}}{C}—H + H_2$$

Its major use is in the manufacture of polymers (Section 15.10). At room temperature and pressure, formaldehyde is an irritating gas. Bubbling this gas through water produces *formalin*, an aqueous solution containing 37% formaldehyde by mass or 40% by volume.

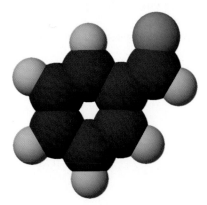

FIGURE 15.2 Space-filling model for benzaldehyde, the simplest aromatic aldehyde.

▶ The procedure for coining common names for ketones is the same as that used for ether common names (Section 14.15).

▶ The contrast between IUPAC names and common names for ketones is as follows:

IUPAC (one word)

    alkanone
    2-butanone

Common (three or two words)

    alkyl alkyl ketone
    ethyl methyl ketone
    or
    dialkyl ketone
    dipropyl ketone

CHEMICAL CONNECTIONS

Lachrymatory Aldehydes and Ketones

A lachrymator, pronounced "lack-ra-mater," is a compound that causes the production of tears. A number of aldehydes and ketones have lachrymatic properties.

Two lachrymal ketones are 2-chloroacetophenone and bromoacetone.

2-Chloroacetophenone
(2-chloro-1-phenylethanone)

Bromoacetone
(1-bromopropanone)

2-Chloroacetophenone is a component of the tear gas used by police and the military. It is also the active ingredient in MACE canisters now marketed for use by individuals to protect themselves from attackers. The compound bromoacetone has been used as a chemical war gas.

Smoke contains compounds that cause the eyes to tear. A predominant lachrymator in wood smoke is formaldehyde, the one carbon aldehyde. The smoke associated with an outdoor barbecue contains the unsaturated aldehyde *acrolein*.

Acrolein
(propenal)

Acrolein is formed as fats that are present in meat break down when heated. (Besides being a lachrymator, acrolein is responsible for the "pleasant" odor associated with the process of barbecuing meat.)

The lachrymatory compound associated with onions is a derivative of thiopropionaldehyde.

Thiopropionaldehyde
(propanethial)

Thiopropionaldehyde-S-oxide
(propanethial-S-oxide)
[lachrymator in chopped onions]

Onions do not cause tear production until they are chopped or sliced. The onion cells damaged by these actions release the enzyme *allinase*, which converts an odorless compound naturally present in onions to the lachrymatic compound.

Scientists are not sure why thiopropionaldehyde-S-oxide causes tear production, but it is known that it is an unstable molecule that is readily broken up by water into propanal, hydrogen sulfide, and sulfuric acid.

$$CH_3-CH_2-\overset{\overset{\displaystyle S}{\|}}{C}-H \xrightarrow{H_2O}$$

$$CH_3-CH_2-\overset{\overset{\displaystyle O}{\|}}{C}-H + H_2S + H_2SO_4$$

Sulfuric acid may be responsible for the eye irritation.

Many people can peel onions under water, or peel cold ones from the refrigerator, without crying. Water washes away the soluble lachrymator and also breaks it down chemically. If the onion is cold, the enzymatic reaction making the lachrymator is slower, so less is formed. Also the vapor pressure of the lachrymator is greatly reduced at the lower temperature, so its concentration in air is reduced.

▶ The oxidation of methanol to formaldehyde has been previously mentioned (Section 14.7). Ingested methanol is oxidized in the human body to formaldehyde, and it is the formaldehyde that causes blindness.

(This represents the solubility limit of formaldehyde gas in water.) Very little free formaldehyde gas is actually present in formalin; most of it reacts with water, producing methylene glycol.

$$H-\overset{\overset{\displaystyle O}{\|}}{C}-H + H-O-H \longrightarrow HO-CH_2-OH$$

Formalin is used for preserving biological specimens (see Figure 15.3); anyone who has experience in a biology laboratory is familiar with the pungent odor of formalin. Formalin

FIGURE 15.3 Formalin is used to preserve biological specimens. This coelacanth, a "prehistoric" fish, is preserved in formaldehyde.

is also the most widely used preservative chemical in embalming fluids used by morticians. Its mode of action involves reaction with protein molecules in a manner that links the protein molecules together; the result is a "hardening" of the protein.

Acetone, a colorless, volatile liquid with a pleasant, mildly "sweet" odor, is the simplest ketone and is also the ketone used in largest volume in industry. Acetone is an excellent solvent because it is miscible with both water and nonpolar solvents. Acetone is the main ingredient in gasoline treatments that are designed to solubilize water in the gas tank and allow it to pass through the engine in miscible form. Acetone can also be used to remove water from glassware in the laboratory. And it is a major component of some nail polish removers.

Small amounts of acetone are produced in the human body in reactions related to obtaining energy from fats. Normally, such acetone is degraded to CO_2 and H_2O. Diabetic people produce larger amounts of acetone, not all of which can be degraded. The presence of acetone in urine is a sign of diabetes. In severe diabetes, the odor of acetone can be detected on the person's breath.

■ Naturally Occurring Aldehydes and Ketones

Aldehydes and ketones occur widely in nature. Naturally occurring compounds of these types, with higher molecular masses, usually have pleasant odors and flavors and are often used for these properties in consumer products (perfumes, air fresheners, and the like). Table 15.1 gives the structures and uses for selected naturally occurring aldehydes and ketones. The unmistakable odor of melted butter is largely due to the four-carbon diketone butanedione (see Figure 15.4).

Many important steroid hormones (Chapter 19) are ketones, including testosterone, the hormone that controls the development of male sex characteristics; progesterone, the hormone secreted at the time of ovulation in females; and cortisone, a hormone from the adrenal glands that is used medicinally to relieve inflammation.

Testosterone

FIGURE 15.4 The delightful odor of melted butter is largely due to butanedione.

Progesterone

Cortisone

TABLE 15.1

Selected Aldehydes and Ketones Whose Uses Are Based on Their Odor or Flavor

Aldehydes

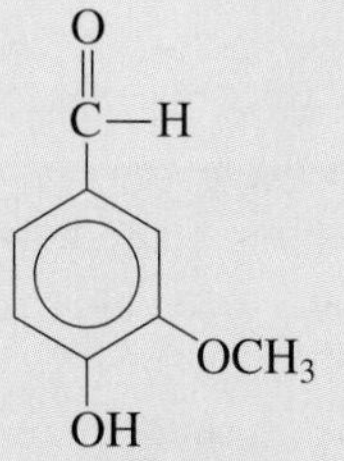

Vanillin
(vanilla flavoring)

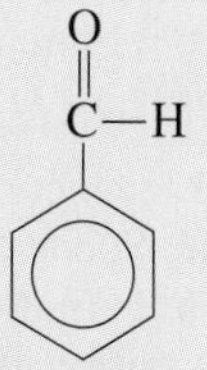

Benzaldehyde
(almond flavoring)

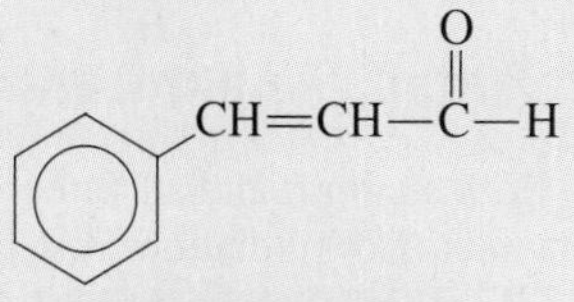

Cinnamaldehyde
(cinnamon flavoring)

Ketones

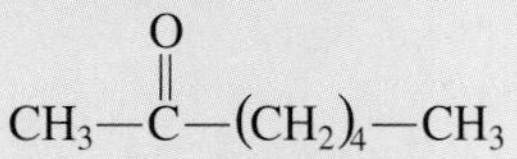

$CH_3-\overset{\overset{\displaystyle O}{\|}}{C}-(CH_2)_4-CH_3$

2-Heptanone
(clove flavoring)

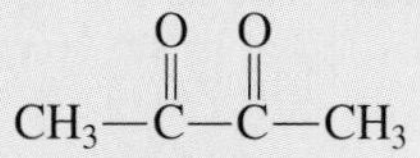

$CH_3-\overset{\overset{\displaystyle O}{\|}}{C}-\overset{\overset{\displaystyle O}{\|}}{C}-CH_3$

Butanedione
(butter flavoring)

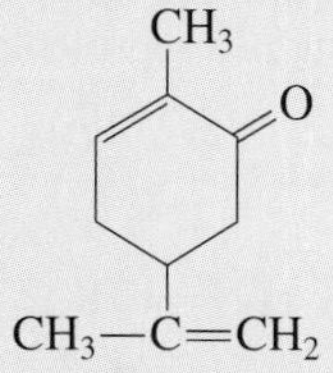

Carvone
(spearmint flavoring)

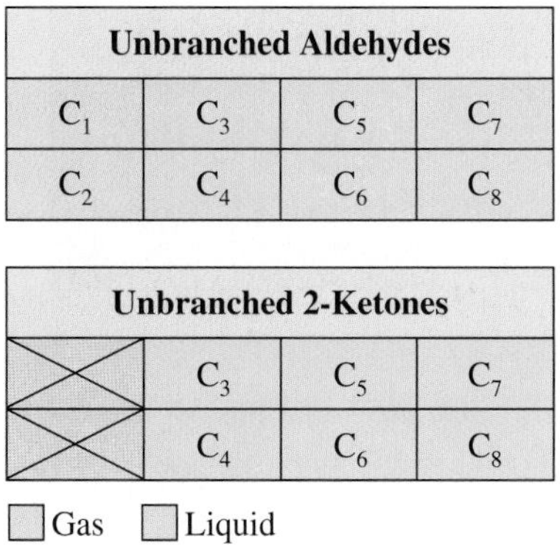

Unbranched Aldehydes			
C_1	C_3	C_5	C_7
C_2	C_4	C_6	C_8

Unbranched 2-Ketones			
✕	C_3	C_5	C_7
✕	C_4	C_6	C_8

☐ Gas   ☐ Liquid

FIGURE 15.5 A physical-state summary for unbranched aldehydes and unbranched 2-ketones at room temperature and pressure.

▶ The ordering of boiling points for carbonyl compounds (aldehydes and ketones), alcohols, and alkanes of similar molecular mass is

Alcohols > carbonyl compounds > alkanes

15.6 Physical Properties of Aldehydes and Ketones

The C_1 and C_2 aldehydes are gases at room temperature. The C_3 through C_{11} straight-chain saturated aldehydes are liquids, and the higher aldehydes are solids (Figure 15.5). The presence of alkyl groups tends to lower both boiling points and melting points, as does the presence of unsaturation in the carbon chain. Lower-molecular-mass ketones are colorless liquids at room temperature (Figure 15.5).

The boiling points of aldehydes and ketones are intermediate between those of alcohols and alkanes of similar molecular mass. Aldehydes and ketones have higher boiling points than alkanes because of dipole–dipole attractions between molecules. Carbonyl group polarity (Section 15.1) makes these dipole–dipole interactions possible. Aldehydes and ketones have lower boiling points than the corresponding alcohols, because no hydrogen bonding occurs as it does with alcohols. Dipole–dipole attractions are weaker forces than hydrogen bonds (Section 7.13). Table 15.2 provides boiling-point information for selected aldehydes, alcohols, and alkanes.

TABLE 15.2

Boiling Points of Some Alkanes, Aldehydes, and Alcohols of Similar Molecular Mass

Type of compound	Compound	Structure	Molecular mass	Boiling point (8C)
alkane	ethane	CH_3-CH_3	30	−89
aldehyde	methanal	$H-CHO$	30	−21
alcohol	methanol	CH_3-OH	32	65
alkane	propane	$CH_3-CH_2-CH_3$	44	−42
aldehyde	ethanal	CH_3-CHO	44	20
alcohol	ethanol	CH_3-CH_2-OH	46	78
alkane	butane	$CH_3-CH_2-CH_2-CH_3$	58	−1
aldehyde	propanal	CH_3-CH_2-CHO	58	49
alcohol	1-propanol	$CH_3-CH_2-CH_2-OH$	60	97

CHEMICAL CONNECTIONS

Suntan, Sunburn, and Ketones

Both sunburn and suntan are caused by the interaction of ultraviolet (UV) radiation from the sun with skin. Sudden high-level UV radiation exposure can cause the skin to burn. Steady low-level exposure to UV radiation can have a different effect—a suntan.

Human skin has a built-in defense system to protect itself against ultraviolet radiation. The brown-colored skin pigment called melanin acts as a protective barrier by absorbing and scattering UV light. Structurally, melanin is a polymeric substance involving many interconnected *cyclic ketone* units. The following is a representation of a portion of its structure.

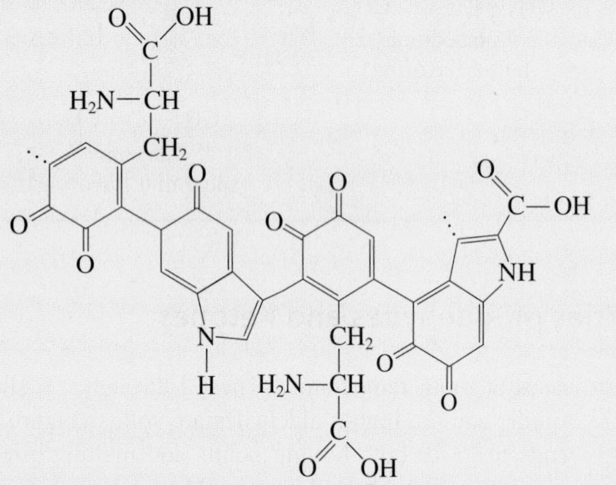

A dark-skinned person has more melanin molecules in the upper layers of the skin (and more protection against sunburn) than a light-skinned person.

When melanin-producing cells deep in the skin are exposed to UV radiation, melanin production increases. The presence of this extra melanin in the skin gives the skin an appearance that we call a "tan." The larger the melanin molecules so produced, the deeper the tan. People who tan readily have skin that can produce a large amount of melanin.

When a person experiences a sunburn, the skin peels. When peeling occurs, any tan that has been built up (excess melanin) sloughs off with the dead skin. Thus the tanning process must begin anew.

Sunscreen products contain substances that absorb UV radiation in a manner similar to that of melanin. Sunscreen use enables sunbathers to control UV radiation dosage and thus avoid sunburn (too much UV radiation) and yet develop a suntan (a small amount of UV radiation). Light-skinned people must use stronger sunscreen protection, because they have less natural protection (melanin). The UV-absorbing agents in many sunscreen formulations are benzophenone derivatives such as

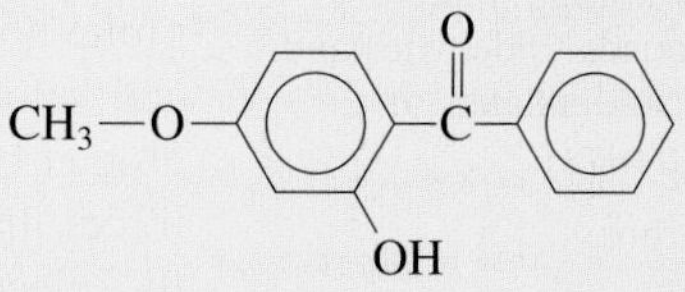

Oxybenzone

Studies show that sunbathing, especially when sunburn results, ages the skin prematurely and increases the risk of skin cancer.

Water molecules can hydrogen-bond with aldehyde and ketone molecules (Figure 15.6). This hydrogen bonding causes low-molecular-mass aldehydes and ketones to be water-soluble. As the hydrocarbon portions get larger, the water solubility of aldehydes and ketones decreases. Table 15.3 gives data on solubility in water for selected aldehydes and ketones.

Although low-molecular-mass aldehydes have pungent, penetrating, unpleasant odors, higher-molecular-mass aldehydes (above C_8) are more fragrant, especially benzaldehyde derivatives. Ketones generally have pleasant odors, and several are used in perfumes and air fresheners.

FIGURE 15.6 Low-molecular-mass aldehydes and ketones are soluble in water because of hydrogen bonding.

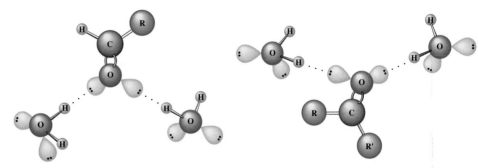

(a) Aldehyde–Water Hydrogen Bonding   **(b) Ketone–Water Hydrogen Bonding**

15.7 Preparation of Aldehydes and Ketones

Aldehydes and ketones can be produced by the oxidation of primary and secondary alcohols, respectively, using mild oxidizing agents such as $KMnO_4$ or $K_2Cr_2O_7$ (Section 14.7).

$$R-\underset{\underset{H}{|}}{\overset{\overset{OH}{|}}{C}}-H \xrightarrow{\text{Oxidation}} R-\overset{\overset{O}{\|}}{C}-H \qquad R-\underset{\underset{H}{|}}{\overset{\overset{OH}{|}}{C}}-R' \xrightarrow{\text{Oxidation}} R-\overset{\overset{O}{\|}}{C}-R'$$

Primary alcohol   Aldehyde   Secondary alcohol   Ketone

When this type of reaction is used for aldehyde preparation, reaction conditions must be sufficiently mild to avoid further oxidation of the aldehyde to a carboxylic acid (Section 14.7). Ketones do not undergo the further oxidation that aldehydes do.

In the oxidation of an alcohol to an aldehyde or a ketone, the alcohol molecule loses H atoms. Recall that the *loss of H atoms* by an organic molecule is one of the operational definitions for the process of *oxidation* (Section 14.7).

▶ The term *aldehyde* stems from *al*cohol *dehyd*rogenation, indicating that aldehydes are related to alcohols by the loss of hydrogen.

EXAMPLE 15.3

Predicting Products in Alcohol Oxidation Reactions

■ Draw the structure of the aldehyde or ketone formed from the oxidation of each of the following alcohols. Assume that reaction conditions are sufficiently mild that any aldehydes produced are not oxidized further.

a. $CH_3-CH_2-CH_2-OH$

b. $CH_3-\underset{\underset{OH}{|}}{CH}-CH_3$

c. ⬡—OH

d. $CH_3-\underset{\underset{CH_3}{|}}{\overset{\overset{CH_3}{|}}{C}}-OH$

(continued)

TABLE 15.3
Water Solubility (g/100 g H_2O) for Various Aldehydes and Ketones

Number of carbon atoms	Aldehyde	Water solubility of aldehyde	Ketone	Water solubility of ketone
1	methanal	very soluble		
2	ethanal	infinite		
3	propanal	16	propanone	infinite
4	butanal	7	2-butanone	26
5	pentanal	4	2-pentanone	5
6	hexanal	1	2-hexanone	1.6
7	heptanal	0.1	2-heptanone	0.4
8	octanal	insoluble	2-octanone	insoluble

Solution

a. This is a primary alcohol that will give the aldehyde *propanal* as the oxidation product.

$$CH_3-CH_2-\overset{\displaystyle O}{\overset{\|}{C}}-H$$

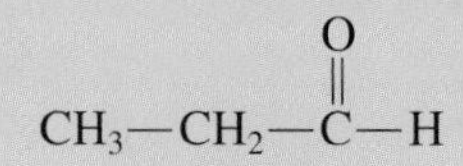

Propanal

b. This is a secondary alcohol. Upon oxidation, secondary alcohols are converted to ketones.

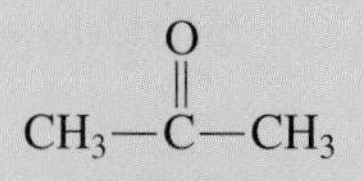

$$CH_3-\overset{\displaystyle O}{\overset{\|}{C}}-CH_3$$

Propanone

c. This cyclic alcohol is a secondary alcohol; hence a ketone is the oxidation product.

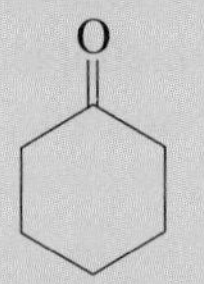

Cyclohexanone

d. This is a tertiary alcohol. Tertiary alcohols do not undergo oxidation (Section 14.7).

Practice Exercise 15.3

Draw the structure of the aldehyde or ketone formed from the oxidation of each of the following alcohols. Assume that reaction conditions are sufficiently mild that any aldehydes produced are not oxidized further to carboxylic acids.

a. 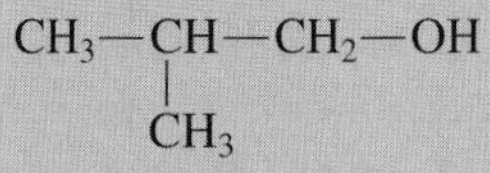 $CH_3-\overset{\displaystyle |}{\underset{\displaystyle CH_3}{CH}}-CH_2-OH$ **b.** 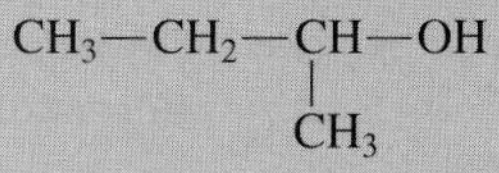 $CH_3-CH_2-\overset{\displaystyle |}{\underset{\displaystyle CH_3}{CH}}-OH$

c. 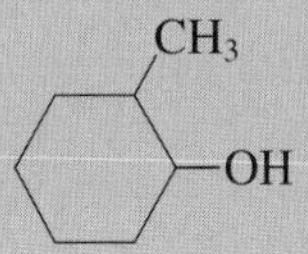 CH₃ / —OH

d. 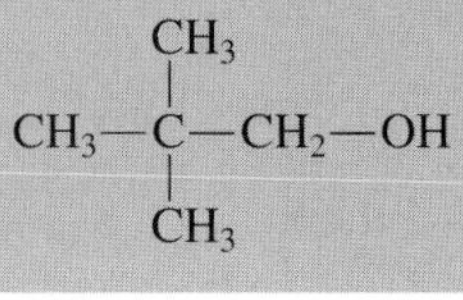 $CH_3-\overset{\displaystyle CH_3}{\underset{\displaystyle CH_3}{\overset{\displaystyle |}{\underset{\displaystyle |}{C}}}}-CH_2-OH$

15.8 Oxidation and Reduction of Aldehydes and Ketones

■ Oxidation of Aldehydes and Ketones

Aldehydes readily undergo oxidation to carboxylic acids (Section 15.7), and ketones are resistant to oxidation.

$$R-\overset{\displaystyle O}{\overset{\|}{C}}-H \xrightarrow{[O]} R-\overset{\displaystyle O}{\overset{\|}{C}}-OH \qquad\qquad R-\overset{\displaystyle O}{\overset{\|}{C}}-R \xrightarrow{[O]} \text{no reaction}$$

Aldehyde   Carboxylic acid   Ketone

In aldehyde oxidation, the aldehyde gains an oxygen atom (supplied by the oxidizing agent). *Gain of oxygen* is one of the operational definitions for the process of *oxidation* (Section 14.7).

Among the mild oxidizing agents that convert aldehydes into carboxylic acids is oxygen in air. Thus aldehydes must be protected from air. When an aldehyde is prepared from oxidation of a primary alcohol (Section 15.7), it is usually removed from the reaction mixture immediately to prevent it from being further oxidized to a carboxylic acid.

Because both aldehydes and ketones contain carbonyl groups, we might expect similar oxidation reactions for the two types of compounds. Oxidation of an aldehyde

CHEMICAL CONNECTIONS

Diabetes, Aldehyde Oxidation, and Glucose Testing

Diabetes mellitus is a disease that involves the hormone insulin, a substance necessary to control blood-sugar (glucose) levels. There are two forms of diabetes. In one form, the pancreas does not produce insulin at all. Patients with this condition require injections of insulin to control glucose levels. In the second form, the body cannot make proper use of insulin. Patients with this form of diabetes can often control glucose levels through their diet but may require medication. If the blood-sugar level of a diabetic becomes too high, serious kidney damage can result.

Normal urine does not contain glucose. When the kidneys become overloaded with glucose (the blood-glucose level is too high), glucose is excreted in the urine. Benedict's test (Section 15.8) can be used to detect glucose in urine, because glucose has an aldehyde group present in its structure.

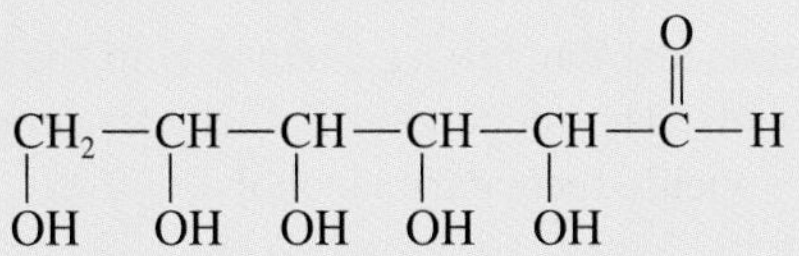

$$CH_2-CH-CH-CH-CH-\overset{\displaystyle O}{\overset{\displaystyle \|}{C}}-H$$
$$\ \ |\quad\ \ |\quad\ \ |\quad\ \ |\quad\ \ |$$
$$OH\quad OH\quad OH\quad OH\quad OH$$

Glucose

A urine glucose test is carried out using either plastic test strips coated with Benedict's solution or Clinitest tablets (a convenient solid form of Benedict's reagent). A few drops of urine are added to the plastic strip or tablet, and the degree of coloration is used to estimate the blood-glucose level. The solution turns greenish at a low glucose level, then yellow-orange, and finally a dark orange-red.

Tests are also available for directly measuring glucose concentration in blood. These tests involve placing a drop of blood (from a finger prick) on a plastic strip containing a dye and an enzyme that will oxidize glucose's aldehyde group. A two-step reaction sequence occurs. First, the enzyme causes glucose oxidation to a carboxylic acid, with hydrogen peroxide (H_2O_2) being another product of the reaction.

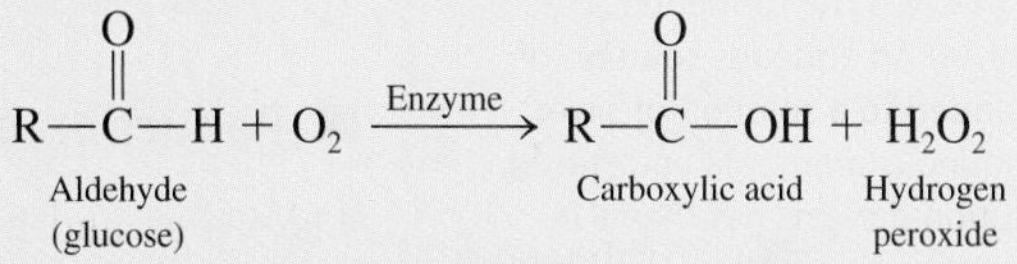

$$R-\overset{\displaystyle O}{\overset{\displaystyle \|}{C}}-H + O_2 \xrightarrow{\text{Enzyme}} R-\overset{\displaystyle O}{\overset{\displaystyle \|}{C}}-OH + H_2O_2$$

Aldehyde (glucose) — Carboxylic acid — Hydrogen peroxide

Then the H_2O_2 reacts with the dye to produce a colored product. The amount of color produced, measured by comparison with a color chart or by an electronic monitor, is proportional to the blood-glucose concentration.

---

involves breaking a carbon–hydrogen bond, and oxidation of a ketone involves breaking a carbon–carbon bond. The former is much easier to accomplish than the latter. For ketones to be oxidized, strenuous reaction conditions must be employed.

Several tests, based on the ease with which aldehydes are oxidized, have been developed for distinguishing between aldehydes and ketones, for detecting the presence of aldehyde groups in sugars (carbohydrates), and for measuring the amounts of sugars present in a solution. The most widely used of these tests are the Tollens test and Benedict's test.

The Tollens test, also called the silver mirror test, involves a solution that contains silver nitrate ($AgNO_3$) and ammonia (NH_3) in water. When Tollens solution is added to an aldehyde, Ag^+ ion (the oxidizing agent) is reduced to silver metal, which deposits on the inside of the test tube, forming a silver mirror. The appearance of this silver mirror (see Figure 15.7) is a positive test for the presence of the aldehyde group.

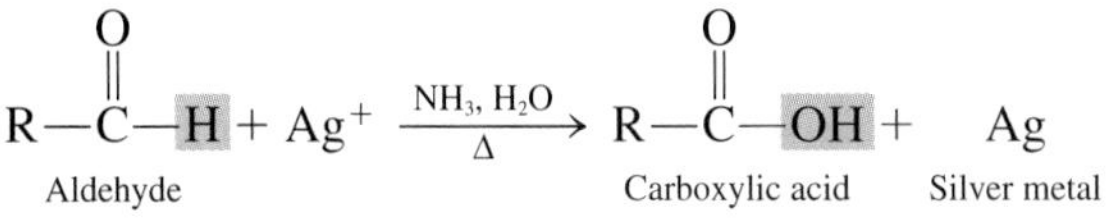

$$R-\overset{\displaystyle O}{\overset{\displaystyle \|}{C}}-H + Ag^+ \xrightarrow[\Delta]{NH_3,\ H_2O} R-\overset{\displaystyle O}{\overset{\displaystyle \|}{C}}-OH + Ag$$

Aldehyde — Carboxylic acid — Silver metal

The Ag^+ ion will not oxidize ketones.

Benedict's test is similar to the Tollens test in that a metal ion is the oxidizing agent. With this test, Cu^{2+} ion is reduced to Cu^+ ion, which precipitates from solution as Cu_2O (a brick red solid; Figure 15.8).

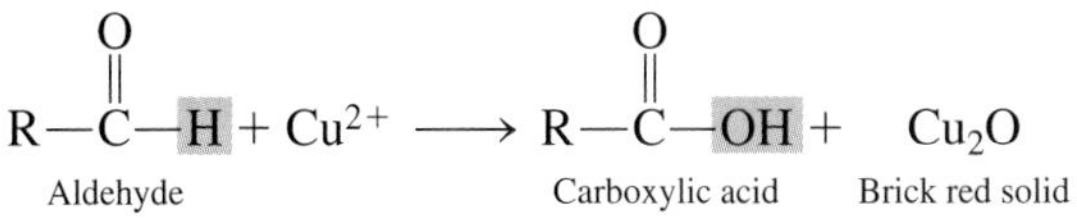

$$R-\overset{\displaystyle O}{\overset{\displaystyle \|}{C}}-H + Cu^{2+} \longrightarrow R-\overset{\displaystyle O}{\overset{\displaystyle \|}{C}}-OH + Cu_2O$$

Aldehyde — Carboxylic acid — Brick red solid

Benedict's solution is made by dissolving copper sulfate, sodium citrate, and sodium carbonate in water.

$$\text{Alcohol + aldehyde} \rightleftharpoons \text{hemiacetal}$$

$$\text{Alcohol + ketone} \rightleftharpoons \text{hemiacetal}$$

This situation makes isolation of the hemiacetal difficult; in practice, it usually cannot be done.

An important exception to this difficulty with isolation is the case where the —OH and $\text{C}{=}\text{O}$ functional groups that react to form the hemiacetal come from the *same* molecule. This produces a *cyclic* hemiacetal rather than a noncyclic one, and cyclic acetals are more stable than the noncyclic ones and can be isolated.

Illustrative of intramolecular hemiacetal formation is the reaction

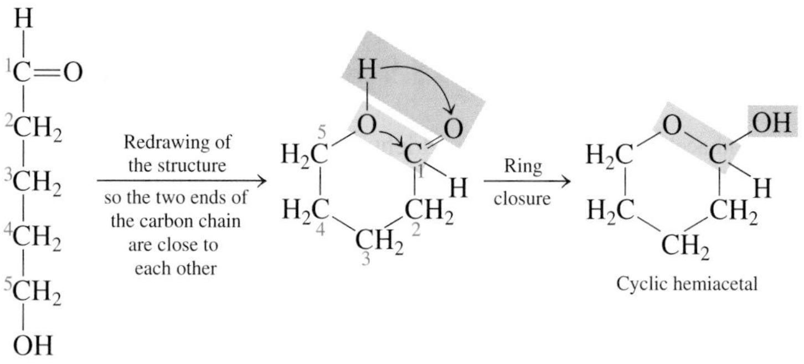

Cyclic hemiacetals are very important compounds in carbohydrate chemistry, the topic of Chapter 18.

EXAMPLE 15.4

Recognizing Hemiacetal Structures

■ Indicate whether each of the following compounds is a hemiacetal.

a. 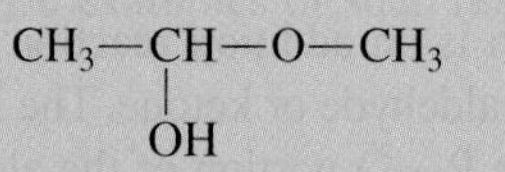 CH₃—CH—O—CH₃
          |
          OH

b. 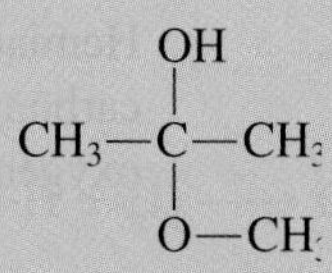   OH
             |
    CH₃—C—CH₃
             |
            O—CH₃

c.   CH₃
                      |
    CH₃—CH—CH—O—CH₃
          |
          OH

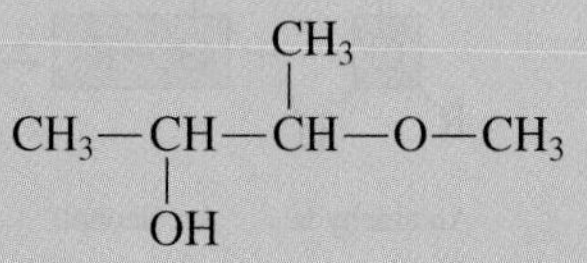

d.

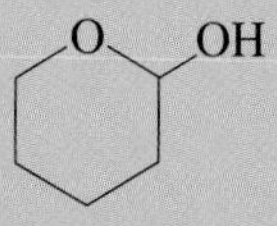

Solution

In each part, we will be looking for the following structural feature: the presence of an —OH group and an —OR group attached to the same carbon atom.

a. We have an —OH group and an —OR group attached to the same carbon atom. The compound is a *hemiacetal*.
b. We have an —OH group and an —OR group attached to the same carbon atom. The compound is a *hemiacetal*.
c. The —OH and —OR groups present in this molecule are attached to *different* carbon atoms. Therefore, the molecule is *not a hemiacetal*.
d. We have a ring carbon atom bonded to two oxygen atoms: one oxygen atom in an —OH substituent and the other oxygen atom bonded to the rest of the ring (the same as an R group). This is a *hemiacetal*.

Practice Exercise 15.4

Indicate whether each of the following compounds is a hemiacetal.

a. 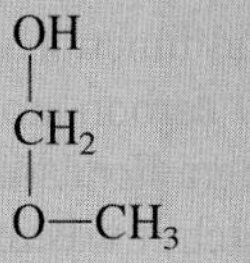 OH
   |
   CH₂
   |
   O—CH₃

b. 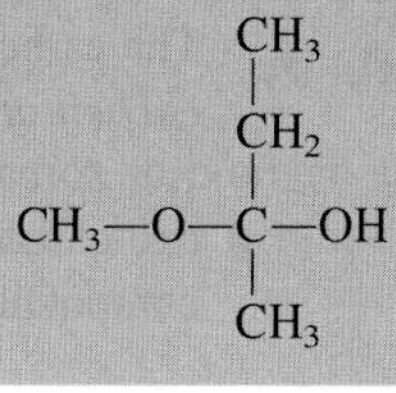   CH₃
             |
             CH₂
             |
    CH₃—O—C—OH
             |
             CH₃

c.

$$CH_3—O—CH \overset{\overset{\displaystyle CH_3}{|}}{\underset{\underset{\displaystyle CH_3}{|}}{\underset{\displaystyle HO—CH}{|}}}$$

d.

> This is our second encounter with condensation reactions. The first encounter involved intermolecular alcohol dehydration (Section 14.7).

■ Acetal Formation

If a small amount of acid catalyst is added to a hemiacetal reaction mixture, then the hemiacetal reacts with a second alcohol molecule, in a condensation reaction, to form an acetal.

$$R_1—\overset{\overset{\displaystyle OH}{|}}{\underset{\underset{\displaystyle H}{|}}{C}}—OR_2 + R_3—OH \underset{}{\overset{H^+}{\rightleftharpoons}} R_1—\overset{\overset{\displaystyle OR_3}{|}}{\underset{\underset{\displaystyle H}{|}}{C}}—OR_2 + H—OH$$

A hemiacetal   An alcohol   An acetal

> Acetals have two alkoxy groups (—OR) attached to the same carbon atom.

An **acetal** *is an organic compound in which a carbon atom is bonded to two alkoxy groups (—OR).* The functional group for an acetal is thus

$$—\overset{\overset{\displaystyle OR}{|}}{\underset{\underset{\displaystyle OR}{|}}{C}}—OR$$

A specific example of acetal formation from a hemiacetal is

$$CH_3—\overset{\overset{\displaystyle OH}{|}}{\underset{\underset{\displaystyle O—CH_3}{|}}{CH}} + CH_3—CH_2—OH \overset{H^+}{\rightleftharpoons} CH_3—\overset{\overset{\displaystyle O—CH_2—CH_3}{|}}{\underset{\underset{\displaystyle O—CH_3}{|}}{CH}} + H—OH$$

Note that acetal formation does not involve addition to a carbon–oxygen double bond as hemiacetal formation does; no double bond is present in either of the reactants involved in acetal formation. Acetal formation involves a substitution reaction; the —OR′ group of the alcohol replaces the —OH group on the hemiacetal.

Figure 15.9 show molecular models for acetaldehyde (the two-carbon aldehyde) and the hemiacetal and acetal formed when this aldehyde reacts with ethyl alcohol.

■ Acetal Hydrolysis

> In Section 24.1, we will find that the enzyme-catalyzed hydrolysis of acetals is an important step in the digestion of carbohydrates.

Acetals, unlike hemiacetals, are easily isolated from reaction mixtures. They are stable in basic solution but undergo *hydrolysis* in acidic solution. A **hydrolysis reaction** *is the reaction of a compound with H_2O, in which the compound splits into two or more fragments as the elements of water (H— and —OH) are added to the compound.* The

FIGURE 15.9 Molecular models for acetaldehyde and its hemiacetal and acetal formed by reaction with ethyl alcohol.

Acetaldehyde   Acetaldehyde hemiacetal   Acetaldehyde acetal
with ethyl alcohol   with ethyl alcohol

products of acetal hydrolysis are the aldehyde or ketone and alcohols that originally reacted to form the acetal.

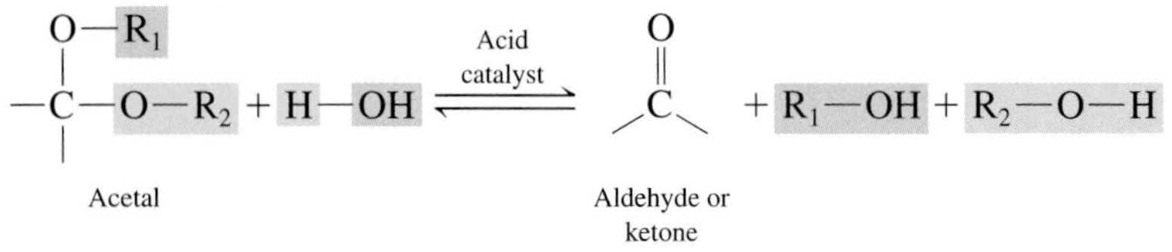

$$\underset{\text{Acetal}}{-\overset{\overset{\displaystyle O-R_1}{|}}{\underset{|}{C}}-O-R_2} + H-OH \overset{\text{Acid catalyst}}{\rightleftharpoons} \underset{\substack{\text{Aldehyde or}\\\text{ketone}}}{\overset{\overset{\displaystyle O}{\|}}{C}} + R_1-OH + R_2-O-H$$

For example,

$$\underset{\underset{\displaystyle CH_3}{|}}{\overset{\overset{\displaystyle O-CH_2-CH_3}{|}}{CH_3-C-O-CH_3}} + H-OH \overset{\text{Acid catalyst}}{\rightleftharpoons} CH_3-\overset{\overset{\displaystyle O}{\|}}{C}-CH_3 + CH_3-OH + CH_3-CH_2-OH$$

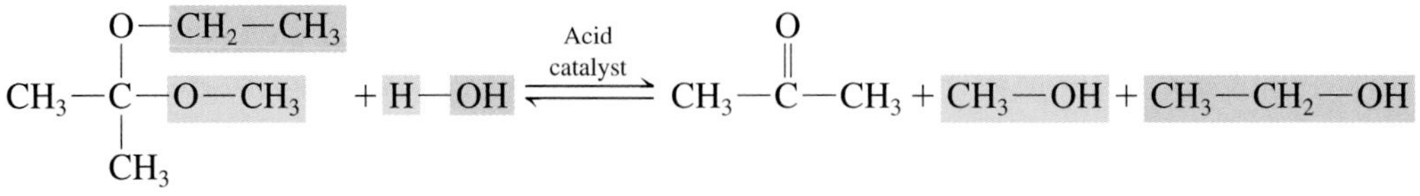

The carbonyl hydrolysis product is an aldehyde if the acetal carbon atom has a hydrogen atom attached directly to it, and it is a ketone if no hydrogen attachment is present. In the preceding example, the carbonyl product is a ketone because the two additional acetal carbon atom attachments are methyl groups.

EXAMPLE 15.5

Predicting Products in Acetal Hydrolysis Reactions

■ Draw the structures of the aldehyde (or ketone) and the two alcohols produced when the following acetals undergo hydrolysis in acidic solution.

a.
$$CH_3-CH_2-\overset{\overset{\displaystyle O-CH_3}{|}}{\underset{\underset{\displaystyle O-CH_2-CH_3}{|}}{CH}}$$

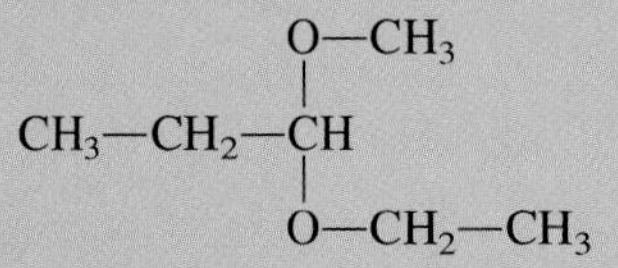

b.
$$CH_3-\overset{\overset{\displaystyle CH_3}{|}}{\underset{\underset{\displaystyle CH_3-C-CH_3}{\underset{|}{O}}}{C}}-O-\overset{\overset{\displaystyle }{}}{\underset{\underset{\displaystyle CH_3}{|}}{CH}}-CH_3$$
$$\overset{\overset{\displaystyle CH_3}{|}}{\underset{\underset{\displaystyle CH_3}{|}}{CH_3-C-CH_3}}$$

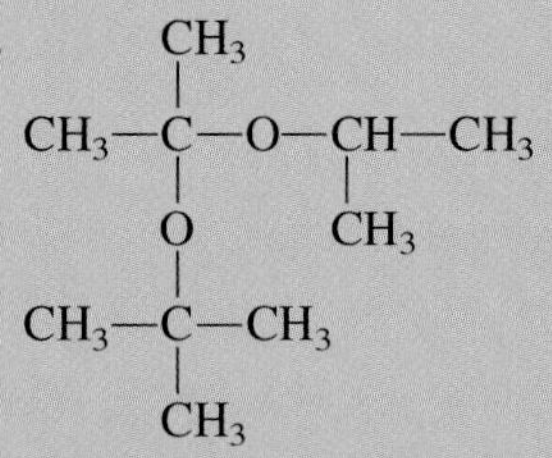

Solution

a. Each of the alkoxy (—OR) groups present will be converted into an alcohol during the hydrolysis. Because the acetal carbon atom has a H attachment, the remainder of the molecule becomes an aldehyde, with the carbon atom to which the alkoxy groups were attached becoming the carbonyl carbon atom.

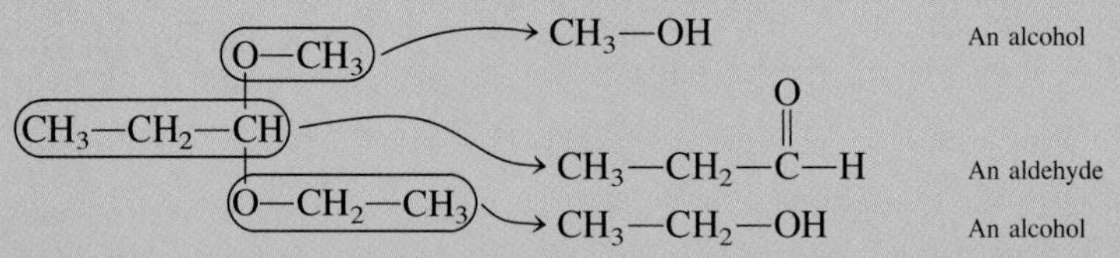

b. Again, each of the alkoxy groups present will be converted into an alcohol during the hydrolysis. Because the acetal carbon atoms lacks a H attachment, the remainder of the molecule becomes a ketone.

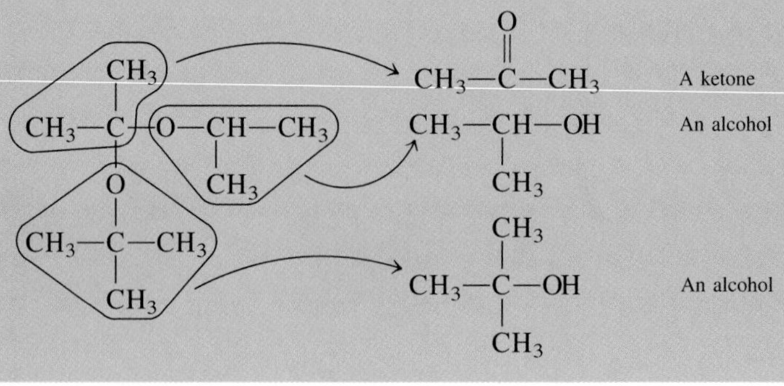

Reactions Involving Aldehydes and Ketones

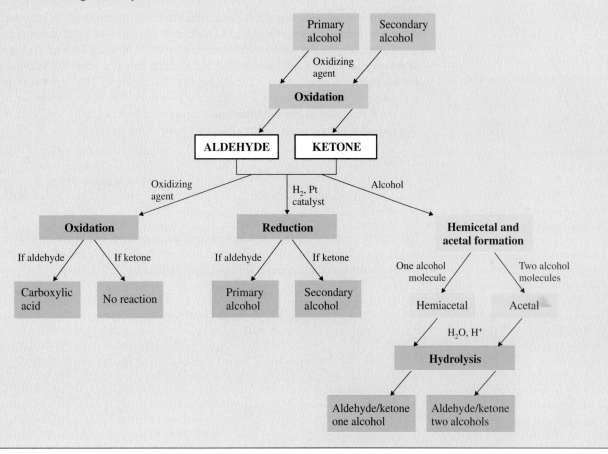

Practice Exercise 15.5

Draw the structures of the aldehyde (or ketone) and the two alcohols produced when the following acetal undergoes hydrolysis in acidic solution.

$$CH_3-CH_2-CH_2-O-\overset{\overset{\displaystyle CH_3}{|}}{\underset{\underset{\displaystyle CH_3}{|}}{C}}-O-CH_2-CH_3$$

■ **Nomenclature for Hemiacetals and Acetals**

A "descriptive" type of common nomenclature that includes the terms *hemiacetal* and *acetal* as well as the name of the carbonyl compound (aldehyde or ketone) produced in the hydrolysis of the hemiacetal or acetal is commonly used in describing such compounds. Two examples of such nomenclature are

$$C-C-\overset{\overset{\displaystyle OH}{|}}{\underset{\underset{\displaystyle H}{|}}{C}}-O-C \qquad C-\overset{\overset{\displaystyle O-C-C}{|}}{\underset{\underset{\displaystyle C}{|}}{C}}-O-C-C$$

Methyl hemiacetal of propanal   Diethyl acetal of propanone

The Chemistry at a Glance feature on this page summarizes reactions that involve aldehydes and ketones.

15.10 Formaldehyde-Based Polymers

Many types of organic compounds can serve as reactants (monomers) for polymerization reactions, including ethylenes (Section 13.8), alcohols (Section 14.8), and carbonyl compounds.

Formaldehyde, the simplest aldehyde, is a prolific "polymer former." As representative of its polymer reactions, let us consider the reaction between formaldehyde and phenol to form a phenol–formaldehyde network polymer (see Figure 15.10). A **network polymer** *is a polymer in which monomers are connected in a three-dimensional cross-linked network.*

As a first step in the polymerization process, a formaldehyde molecule reacts with two phenol molecules with the production of one water molecule.

Each of the phenol groups in this new species can form two more —CH_2— bridges, producing a large polymeric species, a section of which has the following structure. Each —CH_2— bridge comes from a formaldehyde molecule.

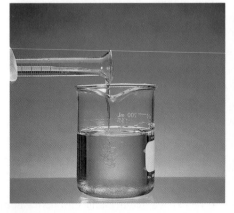

FIGURE 15.10 When a mixture of phenol and formaldehyde dissolved in acetic acid is treated with concentrated hydrochloric acid, a cross-linked phenol–formaldehyde network polymer is formed.

FIGURE 15.11 Bakelite jewelry in use during the 1930–1950 time period.

The first synthetic plastic, Bakelite, produced in 1907, was a phenol–formaldehyde polymer (Figure 15.11). One of its first uses was in the manufacture of billiard balls. Modern phenol–formaldehyde polymers, called phenolics, are adhesives used in the production of plywood and particle board.

15.11 Sulfur-Containing Carbonyl Groups

The introduction of sulfur into a carbonyl group produces two different classes of compounds, depending on whether the sulfur atom replaces the carbonyl oxygen atom or the carbonyl carbon atom.

Replacement of the carbonyl oxygen atom with sulfur produces *thiocarbonyl compounds*—thioaldehydes (thials) and thioketones (thiones)—the simplest of which are

$$
\begin{array}{cc}
\overset{\displaystyle S}{\underset{\displaystyle \parallel}{}} & \overset{\displaystyle S}{\underset{\displaystyle \parallel}{}} \\
H-C-H & CH_3-C-CH_3 \\
\text{Thioformaldehyde} & \text{Thioacetone} \\
\text{(methanethial)} & \text{(propanethione)}
\end{array}
$$

Thiocarbonyl compounds, such as these, are unstable and readily decompose.

Replacement of the carbonyl carbon atom with sulfur produces *sulfoxides,* compounds that are much more stable than thiocarbonyl compounds. The oxidation of a thioether (sulfide) [Section 14.17] constitutes the most common route to a sulfoxide.

$$
\underset{\text{Thioether}}{R-S-R} \xrightarrow{\text{[O]}} \underset{\text{Sulfoxide}}{R-\overset{\displaystyle O}{\overset{\displaystyle \parallel}{S}}-R}
$$

A highly interesting sulfoxide is DMSO (dimethyl sulfoxide), a sulfur analog of acetone, the simplest ketone.

$$
\underset{\text{DMSO}}{CH_3-\overset{\displaystyle O}{\overset{\displaystyle \parallel}{S}}-CH_3} \qquad \underset{\text{Acetone}}{CH_3-\overset{\displaystyle O}{\overset{\displaystyle \parallel}{C}}-CH_3}
$$

DMSO is an odorless liquid with unusual properties. Because of the presence of the polar sulfur–oxide bond, DMSO is miscible with water and also quite soluble in less polar organic solvents. When rubbed on the skin, DMSO has remarkable penetrating power and is quickly absorbed into the body, where it relieves pain and inflammation. For many years it has been heralded as a "miracle drug" for arthritis, sprains, burns, herpes, infections, and high blood pressure. However, the FDA has steadfastly refused to approve it for general medical use. For example, the FDA says that DMSO's powerful penetrating action could cause an insecticide on a gardener's skin to be carried accidentally into his or her bloodstream. Another complication is that DMSO is reduced in the body to dimethyl sulfide, a compound with a strong garlic-like odor that soon appears on the breath.

$$CH_3—\overset{\overset{\displaystyle O}{\|}}{S}—CH_3 \xrightarrow{\text{Reduction}} CH_3—S—CH_3$$

The FDA has approved DMSO for use in certain bladder conditions and as a veterinary drug for topical use in nonbreeding dogs and horses. For example, DMSO is used as an anti-inflammatory rub for race horses.

CONCEPTS TO REMEMBER

The carbonyl group. A carbonyl group consists of a carbon atom bonded to an oxygen atom through a double bond. Aldehydes and ketones are compounds that contain a carbonyl functional group. The carbonyl carbon in an aldehyde has at least one hydrogen attached to it, and the carbonyl carbon in a ketone has no hydrogens attached to it.

Nomenclature of aldehydes and ketones. The IUPAC names of aldehydes and ketones are based on the longest carbon chain that contains the carbonyl group. The chain numbering is done from the end that results in the lowest number for the carbonyl group. The names of aldehydes end in *-al,* those of ketones in *-one.*

Physical properties of aldehydes and ketones. The boiling points of aldehydes and ketones are intermediate between those of alcohols and alkanes. The polarity of the carbonyl groups enables aldehyde and ketone molecules to interact with each other through dipole–dipole interactions. They cannot, however, hydrogen bond to each other. Lower-molecular-mass aldehydes and ketones are soluble in water.

Preparation of aldehydes and ketones. Oxidation of primary and secondary alcohols, using mild oxidizing agents, produces aldehydes and ketones, respectively.

Oxidation and reduction of aldehydes and ketones. Aldehydes are easily oxidized to carboxylic acids; ketones do not readily undergo oxidation. Reduction of aldehydes and ketones produces primary and secondary alcohols, respectively.

Hemiacetals and acetals. A characteristic reaction of aldehydes and ketones is the addition of an alcohol across the carbonyl double bond to produce hemiacetals. The reaction of a second alcohol molecule with a hemiacetal produces an acetal.

KEY REACTIONS AND EQUATIONS

1. Oxidation of an aldehyde to give a carboxylic acid (Section 15.8)

$$R—\overset{\overset{\displaystyle O}{\|}}{C}—H \xrightarrow{[O]} R—\overset{\overset{\displaystyle O}{\|}}{C}—OH$$

2. Attempted oxidation of a ketone (Section 15.8)

$$R—\overset{\overset{\displaystyle O}{\|}}{C}—R' \xrightarrow{[O]} \text{no reaction}$$

3. Reduction of an aldehyde to give a primary alcohol (Section 15.8)

$$R—\overset{\overset{\displaystyle O}{\|}}{C}—H + H_2 \xrightarrow{\text{Catalyst}} R—\overset{\overset{\displaystyle OH}{|}}{\underset{\underset{\displaystyle H}{|}}{C}}—H$$

4. Reduction of a ketone to give a secondary alcohol (Section 15.8)

$$R—\overset{\overset{\displaystyle O}{\|}}{C}—R' + H_2 \xrightarrow{\text{Catalyst}} R—\overset{\overset{\displaystyle OH}{|}}{\underset{\underset{\displaystyle H}{|}}{C}}—R'$$

5. Addition of an alcohol to an aldehyde to form a hemiacetal and then an acetal (Section 15.9)

$$R_1—\overset{\overset{\displaystyle O}{\|}}{C}—H + R_2—O—H \rightleftharpoons \overset{H^+}{} R_1—\overset{\overset{\displaystyle OH}{|}}{\underset{\underset{\displaystyle H}{|}}{C}}—OR_2$$

Aldehyde   Hemiacetal

$$R_1—\overset{\overset{\displaystyle OH}{|}}{\underset{\underset{\displaystyle H}{|}}{C}}—OR_2 + R_3—OH \overset{H^+}{\rightleftharpoons} R_1—\overset{\overset{\displaystyle OR_3}{|}}{\underset{\underset{\displaystyle H}{|}}{C}}—OR_2 + H_2O$$

Hemiacetal   Acetal

6. Hydrolysis of an acetal to yield an aldehyde and two alcohols (Section 15.9)

$$R_1—\overset{\overset{\displaystyle OR_3}{|}}{\underset{\underset{\displaystyle H}{|}}{C}}—OR_2 + H_2O \overset{H^+}{\rightleftharpoons} R_1—\overset{\overset{\displaystyle O}{\|}}{C}—H + R_2—OH + R_3—OH$$

KEY TERMS

Acetal (15.9)
Aldehyde (15.2)
Carbonyl group (15.1)

Hemiacetal (15.9)
Hydrolysis reaction (15.9)

Ketone (15.2)
Network polymer (15.10)

EXERCISES AND PROBLEMS

The members of each pair of problems in this section test similar material.

■ **The Carbonyl Group (Section 15.1)**

15.1 Indicate which of the following compounds contain a carbonyl group.

a. 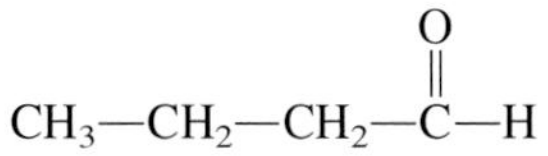
$$CH_3-CH_2-CH_2-\overset{\overset{\displaystyle O}{\|}}{C}-H$$

b. CH_3-O-CH_3

c. 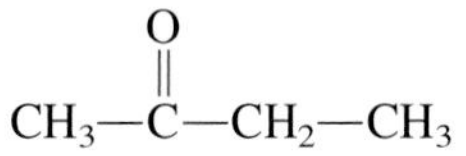
$$CH_3-\overset{\overset{\displaystyle O}{\|}}{C}-CH_2-CH_3$$

d. 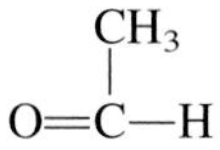
$$\overset{\overset{\displaystyle CH_3}{|}}{O=C-H}$$

e. $CH_3-O-CH_2-O-CH_3$

f. 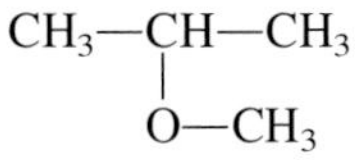
$$CH_3-\overset{\overset{\displaystyle }{|}}{\underset{\underset{\displaystyle O-CH_3}{|}}{CH}}-CH_3$$

15.2 Indicate which of the following compounds contain a carbonyl group.

a. 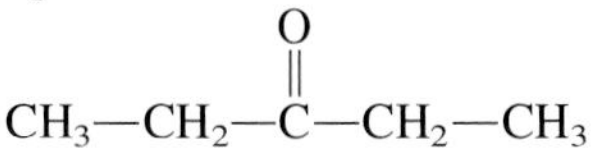
$$CH_3-CH_2-\overset{\overset{\displaystyle O}{\|}}{C}-CH_2-CH_3$$

b. $CH_3-CH_2-O-CH_2-CH_3$

c. $CH_3-CH_2-CH_2-O-H$

d. 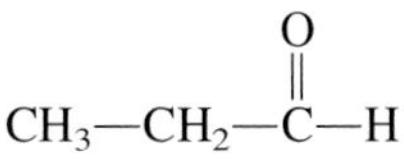
$$CH_3-CH_2-\overset{\overset{\displaystyle O}{\|}}{C}-H$$

e. $CH_3-O-CH_2-CH_2-O-H$

f. 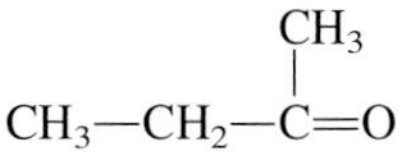
$$CH_3-CH_2-\overset{\overset{\displaystyle CH_3}{|}}{C}=O$$

15.3 What are the similarities and differences between the bonding in a carbon–oxygen double bond and that in a carbon–carbon double bond?

15.4 Use δ^+ and δ^- notation to show the polarity in a carbon–oxygen double bond.

■ **Structure of Aldehydes and Ketones (Section 15.2)**

15.5 Classify each of the following structures as an aldehyde, a ketone, or neither.

a. 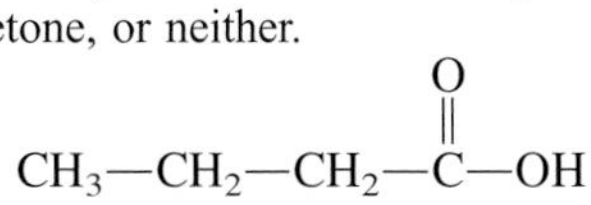
$$CH_3-CH_2-CH_2-\overset{\overset{\displaystyle O}{\|}}{C}-OH$$

b.
$$CH_3-CH_2-CH_2-CH_2-\overset{\overset{\displaystyle O}{\|}}{C}-H$$

c.
$$CH_3-\overset{\overset{\displaystyle O}{\|}}{C}-CH_3$$

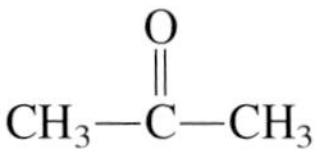

d. $CH_3-O-CH_2-CH_3$

e. 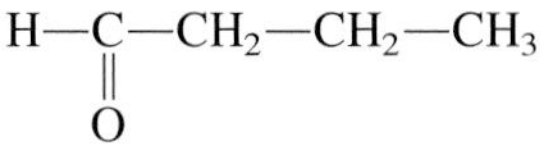
$$H-\overset{\overset{\displaystyle }{}}{\underset{\underset{\displaystyle O}{\|}}{C}}-CH_2-CH_2-CH_3$$

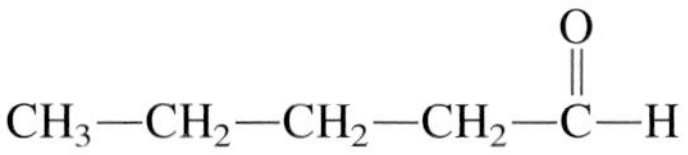

f. CH_3-CHO

15.6 Classify each of the following structures as an aldehyde, a ketone, or neither.

a. 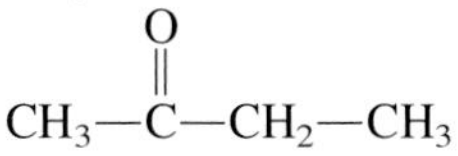
$$CH_3-\overset{\overset{\displaystyle O}{\|}}{C}-CH_2-CH_3$$

b.
$$CH_3-CH_2-CH_2-\overset{\overset{\displaystyle O}{\|}}{C}-O-CH_3$$

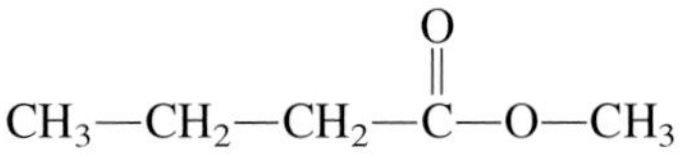

c.
$$CH_3-CH_2-\overset{\overset{\displaystyle }{}}{\underset{\underset{\displaystyle CH_3}{|}}{C}}=O$$

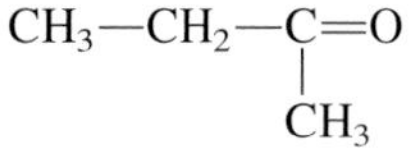

d. 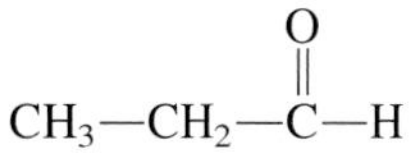
$$CH_3-CH_2-\overset{\overset{\displaystyle O}{\|}}{C}-H$$

e. 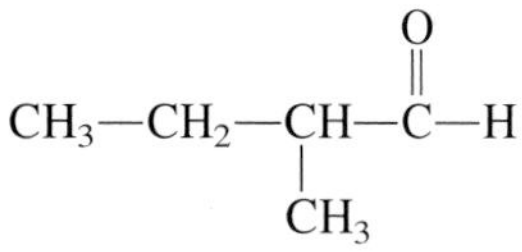
$$CH_3-CH_2-\overset{\overset{\displaystyle }{}}{\underset{\underset{\displaystyle CH_3}{|}}{CH}}-\overset{\overset{\displaystyle O}{\|}}{C}-H$$

f. 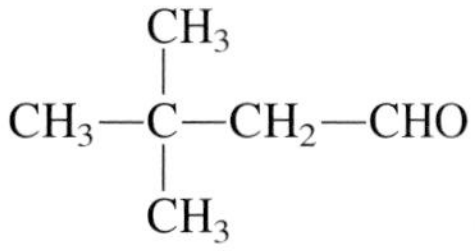
$$CH_3-\overset{\overset{\displaystyle CH_3}{|}}{\underset{\underset{\displaystyle CH_3}{|}}{C}}-CH_2-CHO$$

15.7 Draw the structures of the two simplest aldehydes and the two simplest ketones.

15.8 One- and two-carbon ketones do not exist. Explain why.

15.9 Classify each of the following structures as an aldehyde, a ketone, or neither.

a.

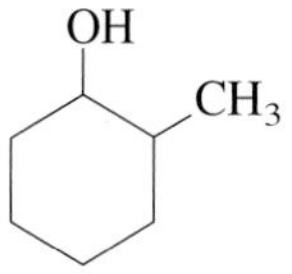

b.

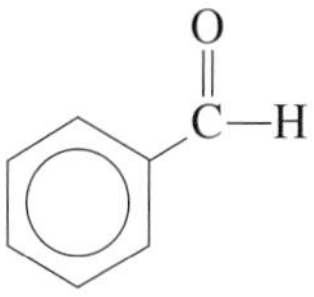

c.

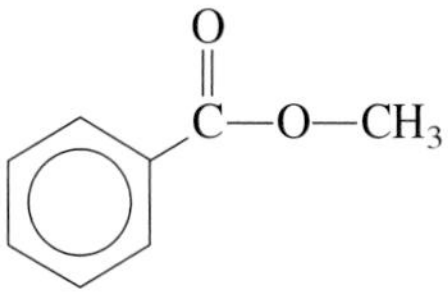

d.

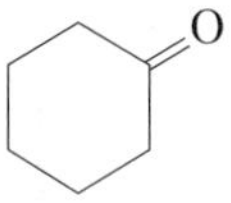

e.

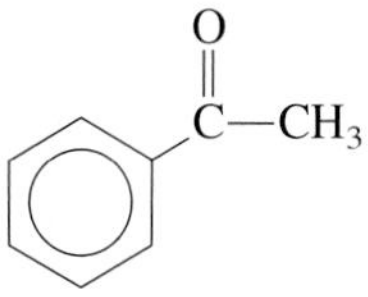

f.

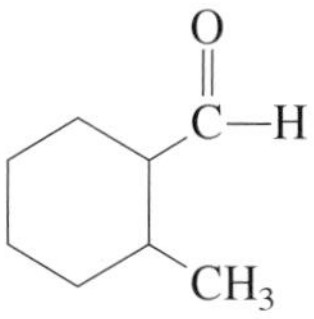

15.10 Classify each of the following structures as an aldehyde, a ketone, or neither.

a.

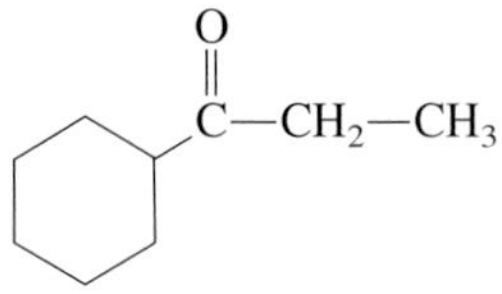

b.

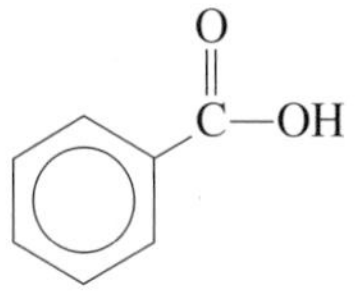

c.

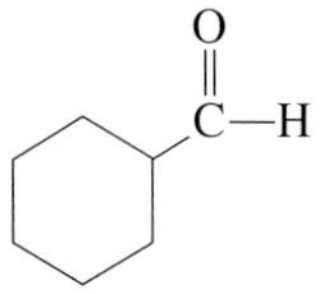

d.

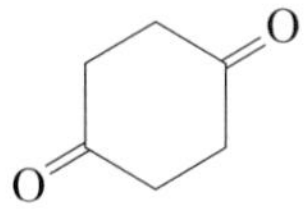

e.

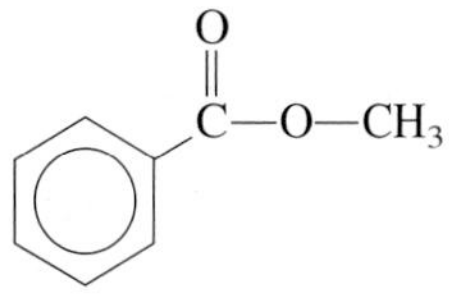

f.

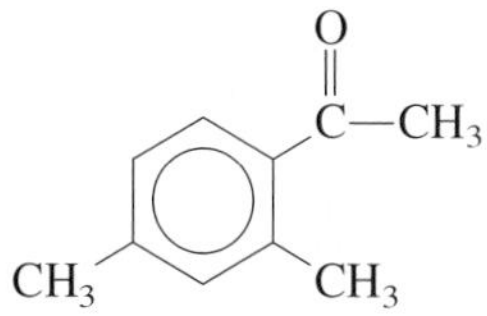

■ Nomenclature for Aldehydes (Section 15.3)

15.11 Assign an IUPAC name to each of the following aldehydes.

a.

$$CH_3-CH_2-CH_2-\overset{\displaystyle O}{\overset{\|}{C}}-H$$

b.

$$CH_3-CH_2-\underset{\underset{\displaystyle CH_3}{|}}{CH}-\overset{\displaystyle O}{\overset{\|}{C}}-H$$

c.

$$CH_3-\underset{\underset{\displaystyle CH_2-CH_2-CH_3}{|}}{CH}-CH_2-CH_2-\overset{\displaystyle O}{\overset{\|}{C}}-H$$

d.

e. CH_3-CH_2-CHO

f.

$$CH_3-\underset{\underset{\displaystyle CH_3}{|}}{\overset{\overset{\displaystyle CH_3}{|}}{C}}-CH_2-\overset{\displaystyle O}{\overset{\|}{C}}-H$$

15.12 Assign an IUPAC name to each of the following aldehydes.

a.

$$CH_3-\underset{\underset{\displaystyle CH_3}{|}}{CH}-CH_2-CH_2-\overset{\displaystyle O}{\overset{\|}{C}}-H$$

b.

$$CH_3-CH_2-\underset{\underset{\displaystyle CH_2}{|}}{\underset{\underset{\displaystyle CH_3}{|}}{CH}}-\overset{\displaystyle O}{\overset{\|}{C}}-H$$

c.

$$\underset{\underset{\displaystyle CH_3}{|}}{CH_2}-CH_2-\underset{\underset{\displaystyle CH_2-CH_2-CH_3}{|}}{CH}-CH_2-\overset{\displaystyle O}{\overset{\|}{C}}-H$$

d.

e.

$$CH_3-CH_2-\underset{\underset{\displaystyle CH_3}{|}}{\overset{\overset{\displaystyle CH_3}{|}}{C}}-CH_2-\overset{\displaystyle O}{\overset{\|}{C}}-H$$

f. $CH_3-CH_2-CH_2-CHO$

15.13 Draw a structural formula for each of the following aldehydes.
a. 3-Methylpentanal   b. 2-Ethylhexanal
c. 3,4-Dimethylheptanal   d. 2,2-Dichloropropanal
e. 2,4,5-Trimethylheptanal   f. 4-Hydroxy-2-methyloctanal

15.14 Draw a structural formula for each of the following aldehydes.
a. 2-Methylpentanal   b. 4-Ethylhexanal
c. 3,3-Dimethylhexanal   d. 2,3-Dibromopropanal
e. 2-Bromo-4-methylhexanal   f. 2,4-Dichloroheptanal

15.15 Draw a structural formula for each of the following aldehydes.
a. Formaldehyde   b. Propionaldehyde
c. Chloroacetaldehyde   d. 2-Chlorobutyraldehyde
e. *o*-Methylbenzaldehyde   f. 2,4-Dimethylbenzaldehyde

15.16 Draw a structural formula for each of the following aldehydes.
a. Acetaldehyde   b. Butyraldehyde
c. 2-Chloropropionaldehyde   d. 2-Methylbutyraldehyde
e. Benzaldehyde   f. *p*-Bromobenzaldehyde

15.17 Assign a common name to each of the following aldehydes.

a.

$$CH_3-CH_2-\overset{\displaystyle O}{\overset{\|}{C}}-H$$

b.

$$CH_3-\underset{\underset{\displaystyle CH_3}{|}}{CH}-\overset{\displaystyle O}{\overset{\|}{C}}-H$$

c.

$$\underset{\underset{\displaystyle CH_3}{|}}{CH_2}-CH_2-\overset{\displaystyle O}{\overset{\|}{C}}-H$$

d.

$$\underset{\underset{\displaystyle Cl}{|}}{Cl}-CH-\overset{\displaystyle O}{\overset{\|}{C}}-H$$

e.

f.

15.18 Assign a common name to each of the following aldehydes.

a.

$$CH_3-CH_2-CH_2-\overset{\displaystyle O}{\overset{\|}{C}}-H$$

b.

$$CH_3-CH_2-\underset{\underset{\displaystyle CH_3}{|}}{CH}-\overset{\displaystyle O}{\overset{\|}{C}}-H$$

c.

$$CH_3-\underset{\underset{\displaystyle CH_3}{|}}{CH}-CH_2-\overset{\displaystyle O}{\overset{\|}{C}}-H$$

d.

$$\underset{\underset{\displaystyle Cl}{|}}{Cl}-\overset{\overset{\displaystyle Cl}{|}}{C}-\overset{\displaystyle O}{\overset{\|}{C}}-H$$

e.

f.

■ Nomenclature for Ketones (Section 15.4)

15.19 Using IUPAC nomenclature, name each of the following ketones.

a.

$$CH_3-CH_2-\overset{\displaystyle O}{\overset{\|}{C}}-CH_3$$

b.

$$CH_3-\underset{\underset{\displaystyle CH_3}{|}}{CH}-\underset{\underset{\displaystyle CH_3}{|}}{CH}-\overset{\displaystyle O}{\overset{\|}{C}}-\underset{\underset{\displaystyle CH_3}{|}}{CH}-CH_3$$

c.

$$CH_3-\underset{\underset{\displaystyle CH_3}{|}}{CH}-CH_2-CH_2-\overset{\displaystyle O}{\overset{\|}{C}}-CH_2-CH_3$$

d.

$$CH_3-CH_2-\underset{\underset{\displaystyle CH_2}{|}}{CH_2}\underset{\underset{\displaystyle CH_2}{|}}{\quad}CH_2-\overset{\displaystyle O}{\overset{\|}{C}}-CH_3$$

e.

$$Cl{-}CH_2{-}CH_2{-}\overset{\displaystyle O}{\overset{\|}{C}}{-}CH_2{-}CH_2{-}Cl$$

f.

$$CH_3{-}CH_2{-}\overset{\displaystyle O}{\overset{\|}{C}}{-}\underset{\underset{\displaystyle Cl}{|}}{CH}{-}Cl$$

15.20 Using IUPAC nomenclature, name each of the following ketones.

a.

$$CH_3{-}\overset{\displaystyle O}{\overset{\|}{C}}{-}CH_2{-}CH_2{-}CH_2{-}CH_3$$

b. $$CH_3{-}\underset{\underset{\displaystyle CH_3}{|}}{CH}{-}\overset{\displaystyle O}{\overset{\|}{C}}{-}CH_3$$

c.

$$CH_3{-}CH_2{-}\overset{\displaystyle O}{\overset{\|}{C}}{-}CH_2{-}\underset{\underset{\underset{\displaystyle CH_3}{|}}{CH_2}}{\underset{|}{CH}}{-}CH_3$$

d.

$$CH_3{-}\underset{\underset{\displaystyle Cl}{|}}{CH}{-}\overset{\displaystyle O}{\overset{\|}{C}}{-}\underset{\underset{\displaystyle Br}{|}}{CH}{-}CH_3$$

e.

$$CH_3{-}\underset{\underset{\displaystyle Cl}{|}}{CH}{-}\overset{\displaystyle O}{\overset{\|}{C}}{-}CH_2{-}\underset{\underset{\displaystyle Cl}{|}}{CH_2}$$

f.

$$\underset{\underset{\displaystyle CH_3}{|}}{CH_2}{-}CH_2{-}\underset{\underset{\displaystyle CH_3}{|}}{CH}{-}\overset{\displaystyle O}{\overset{\|}{C}}{-}\underset{\underset{\displaystyle CH_3}{|}}{CH_2}$$

15.21 Using IUPAC nomenclature, name each of the following ketones.

a.

b.

c.

d.

15.22 Using IUPAC nomenclature, name each of the following ketones.

a.

b.

c.

d.

15.23 Draw a structural formula for each of the following ketones.
a. 3-Methyl-2-pentanone   b. 3-Hexanone

c. Cyclobutanone   d. 2,4-Dimethyl-3-pentanone
e. Chloropropanone   f. 1,3-Dichloropropanone

15.24 Draw a structural formula for each of the following ketones.
a. 2-Methyl-3-pentanone   b. 2-Pentanone
c. 2,2-Dimethyl-4-octanone   d. Bromopropanone
e. 1,1-Dibromopropanone   f. Cyclopentanone

15.25 Draw a structural formula for each of the following ketones.
a. Diethyl ketone   b. Acetone
c. Isopropyl propyl ketone   d. Chloromethyl methyl ketone
e. Acetophenone   f. Methyl phenyl ketone

15.26 Draw a structural formula for each of the following ketones.
a. Dimethyl ketone   b. Phenyl propyl ketone
c. Methyl *tert*-butyl ketone   d. Dichloromethyl ethyl ketone
e. Benzophenone   f. Diphenyl ketone

■ **Physical Properties of Aldehydes and Ketones (Section 15.6)**

15.27 Aldehydes and ketones have higher boiling points than alkanes of similar molecular mass. Explain why.

15.28 Aldehydes and ketones have lower boiling points than alcohols of similar molecular mass. Explain why.

15.29 How many hydrogen bonds can form between an acetone molecule and water molecules?

15.30 How many hydrogen bonds can form between an acetaldehyde molecule and water molecules?

15.31 Would you expect ethanal or octanal to be more soluble in water? Explain your answer.

15.32 Would you expect ethanal or octanal to have the more fragrant odor? Explain your answer.

■ **Preparation of Aldehydes and Ketones (Section 15.7)**

15.33 Draw the structure of the aldehyde or ketone formed from oxidation of each of the following alcohols. Assume that reaction conditions are sufficiently mild that any aldehydes produced are not oxidized further to carboxylic acids.
a. $CH_3{-}CH_2{-}CH_2{-}CH_2{-}CH_2{-}OH$

b. $$CH_3{-}CH_2{-}\underset{\underset{\displaystyle CH_3}{|}}{CH}{-}OH$$

c.

$$CH_3{-}\underset{\underset{\displaystyle CH_3}{|}}{\overset{\overset{\displaystyle CH_3}{|}}{C}}{-}CH_2{-}CH_2{-}OH$$

d. $$CH_3{-}CH_2{-}\underset{\underset{\displaystyle OH}{|}}{CH}{-}CH_2{-}CH_3$$

e.

f.

15.34 Draw the structure of the aldehyde or ketone formed from oxidation of each of the following alcohols. Assume that reaction conditions are sufficiently mild that any aldehydes formed are not oxidized further to carboxylic acids.
a. $$CH_3{-}CH_2{-}\underset{\underset{\displaystyle CH_3}{|}}{CH}{-}CH_2{-}OH$$

TABLE 16.1

Common Names for the First Six Monocarboxylic Acids

Length of carbon chain	Structural formula	Common name[a]	IUPAC name
C_1 monoacid	H—COOH	formic acid	methanoic acid
C_2 monoacid	CH_3—COOH	acetic acid	ethanoic acid
C_3 monoacid	CH_3—CH_2—COOH	propionic acid	propanoic acid
C_4 monoacid	CH_3—$(CH_2)_2$—COOH	butyric acid	butanoic acid
C_5 monoacid	CH_3—$(CH_2)_3$—COOH	valeric acid	pentanoic acid
C_6 monoacid	CH_3—$(CH_2)_4$—COOH	caproic acid	hexanoic acid

[a]The mnemonic "*F*rogs *a*re *p*olite, *b*eing *v*ery *c*ourteous" is helpful in remembering, in order, the first letters of the common names of these six simple saturated monocarboxylic acids.

▶ There is a connection between acetic acid and sourdough bread. The yeast used in leavening the dough for this bread is a type that cannot metabolize the sugar maltose as most yeasts do. Consequently, bacteria that thrive on maltose become abundant in the dough. These bacteria produce acetic acid and lactic acid from the maltose, and the dough becomes *sour* (acidic); hence the name *sourdough* bread.

▶ The alpha-carbon atom in a carboxylic acid is the carbon atom to which the carboxyl group is attached. It is never the carboxyl carbon atom itself.

▶ Greek letters are *never* used for specifying substituent location in the IUPAC system.

FIGURE 16.3 "Drug-sniffing" dogs used by narcotics agents can find hidden heroin by detecting the odor of acetic acid (vinegar odor). Acetic acid is a by-product of the final step in illicit heroin production, and trace amounts remain in the heroin.

Acetic acid is the most widely used of all carboxylic acids. Its primary use is as an *acidulant*—a substance that gives the proper acidic conditions for a chemical reaction. In the pure state, acetic acid is a colorless liquid with a sharp odor (see Figure 16.3). Vinegar is a 4%–8% (v/v) acetic acid solution; its characteristic odor comes from the acetic acid present. Pure acetic acid is often called *glacial* acetic acid because it freezes on a moderately cold day (f.p. = 17°C), producing icy-looking crystals.

When using common names for carboxylic acids, we designate the positions (locations) of substituents by using letters of the Greek alphabet rather than numbers. The first four letters of the Greek alphabet are alpha (α), beta (β), gamma (γ), and delta (δ). The alpha-carbon atom is carbon 2, the beta-carbon atom is carbon 3, and so on.

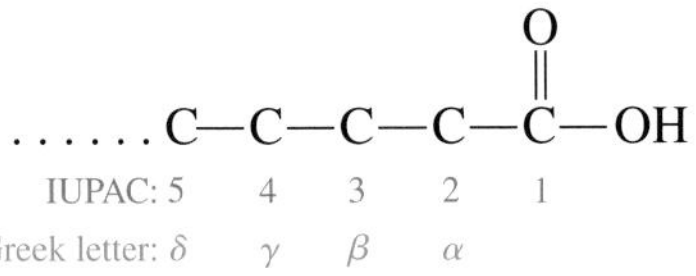

$$\ldots\ldots\overset{}{C}-\overset{}{C}-\overset{}{C}-\overset{}{C}-\overset{\displaystyle O}{\overset{\|}{C}}-OH$$

IUPAC: 5   4   3   2   1

Greek letter: δ γ β α

With the Greek-letter system, the compound

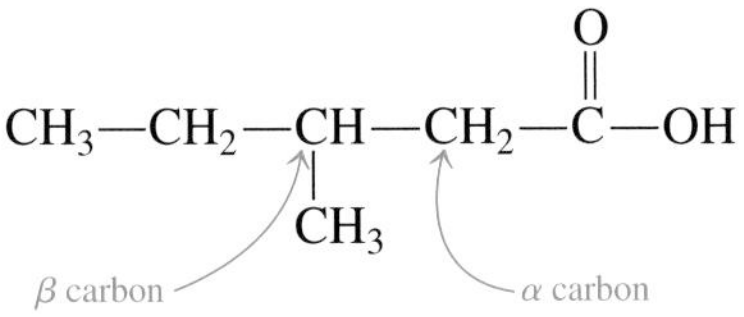

$$CH_3-CH_2-\underset{\underset{\displaystyle CH_3}{|}}{CH}-CH_2-\overset{\displaystyle O}{\overset{\|}{C}}-OH$$

β carbon α carbon

would be called β-methylvaleric acid.

■ Dicarboxylic Acids

Common names for the first six dicarboxylic acids are given in Table 16.2. Oxalic acid, the simplest dicarboxylic acid, is found in plants of the genus *Oxalis,* which includes rhubarb and spinach, and in cabbage (see Figure 16.4). This acid and its salts are poisonous in *high* concentrations. The amount of oxalic acid present in spinach, cabbage, and

TABLE 16.2

Common Names for the First Six Dicarboxylic Acids

Length of carbon chain	Structural formula	Common name[a]	IUPAC name
C_2 diacid	HOOC—COOH	oxalic acid	ethanedioic acid
C_3 diacid	HOOC—CH_2—COOH	malonic acid	propanedioic acid
C_4 diacid	HOOC—$(CH_2)_2$—COOH	succinic acid	butanedioic acid
C_5 diacid	HOOC—$(CH_2)_3$—COOH	glutaric acid	pentanedioic acid
C_6 diacid	HOOC—$(CH_2)_4$—COOH	adipic acid	hexanedioic acid
C_7 diacid	HOOC—$(CH_2)_5$—COOH	pimelic acid	heptanedioic acid

[a]The mnemonic "*O*h *m*y, *s*uch *g*ood *a*pple *p*ie" is helpful in remembering, in order, the first letters of the common names of these six simple dicarboxylic acids.

rhubarb is not harmful. Oxalic acid is used to remove rust, bleach straw and leather, and remove ink stains. Succinic and glutaric acid and their derivatives play important roles in biological reactions that occur in the human body (Section 23.6).

EXAMPLE 16.2

Generating the Structural Formulas of Carboxylic Acids from Their Names

The contrast between IUPAC names and common names for mono- and dicarboxylic acids is as follows:

Monocarboxylic Acids

    IUPAC (two words)

        | alkanoic acid |

    Common (two words)

        | (prefix)ic acid* |

Dicarboxylic Acids

    IUPAC (two words)

        | alkanedioic acid |

    Common (two words)

        | (prefix)ic acid* |

*The common-name prefixes are related to natural sources for the acids.

FIGURE 16.4 The C_2 dicarboxylic acid, oxalic acid, contributes to the tart taste of rhubarb stalks.

$$HO-\overset{\overset{\displaystyle O}{\|}}{C}-\overset{\overset{\displaystyle O}{\|}}{C}-OH$$

■ Draw a structural formula for each of the following carboxylic acids.

a. Caproic acid   **b.** Glutaric acid
c. α-Phenylsuccinic acid   **d.** β-Chlorobutyric acid

Solution

a. Caproic acid is the six-carbon unsubstituted monocarboxylic acid. Its structural formula is

$$CH_3-CH_2-CH_2-CH_2-CH_2-\overset{\overset{\displaystyle O}{\|}}{C}-OH$$

b. Glutaric acid is the five-carbon unsubstituted dicarboxylic acid, with a carboxyl group at each end of the carbon chain.

$$HO-\overset{\overset{\displaystyle O}{\|}}{C}-CH_2-CH_2-CH_2-\overset{\overset{\displaystyle O}{\|}}{C}-OH$$

c. Succinic acid is the four-carbon unsubstituted dicarboxylic acid. A phenyl group (Section 13.11) is present on the alpha-carbon atom.

$$HO-\overset{\overset{\displaystyle O}{\|}}{C}-\overset{\alpha}{C}H-\overset{\beta}{C}H_2-\overset{\overset{\displaystyle O}{\|}}{C}-OH$$

d. Butyric acid is the four-carbon unsubstituted monocarboxylic acid. A chloro group is attached to the beta-carbon atom (carbon 3).

$$\overset{\gamma}{C}H_3-\overset{\beta}{C}H-\overset{\alpha}{C}H_2-\overset{\overset{\displaystyle O}{\|}}{C}-OH$$
$$\quad\quad |$$
$$\quad\quad Cl$$

Practice Exercise 16.2

Draw a structural formula for each of the following carboxylic acids.

a. Adipic acid   **b.** β-Chlorovaleric acid
c. Malonic acid   **d.** Phenylacetic acid

16.4 Polyfunctional Carboxylic Acids

Polyfunctional carboxylic acids are carboxylic acids that contain one or more additional functional groups besides the carboxyl group. Such acids occur naturally in many fruits, are important in the normal functioning of the human body (metabolism), and find use in over-the-counter skin-care products and in prescription drugs. Three commonly encountered types of polyfunctional carboxylic acids are *unsaturated* acids, *hydroxy* acids, and *keto* acids.

$$C-C=C-COOH \qquad C-C-\overset{\overset{\displaystyle OH}{|}}{C}-COOH \qquad C-\overset{\overset{\displaystyle O}{\|}}{C}-C-COOH$$

An unsaturated acid   A hydroxy acid   A keto acid

More information about these types of polyfunctional acids follows.

CHEMICAL CONNECTIONS

Nonprescription Pain Relievers Derived from Propanoic Acid

Consumers are faced with a shelf-full of choices when looking for an over-the-counter medicine to treat aches, pains, and fever. The vast majority of brands available, however, represent only four chemical formulations. Besides the long-available aspirin and acetaminophen, consumers can now purchase products that contain ibuprofen and naproxen.

These two newer entrants into the over-the-counter pain-reliever market are derivatives of propanoic acid, the three-carbon monocarboxylic acid.

Ibuprofen, marketed under the brand names Advil, Motrin-IB, and Nuprin, was cleared by the FDA in 1984 for nonprescription sales. Numerous studies have shown that nonprescription-strength ibuprofen relieves minor pain and fever as well as aspirin or acetaminophen. Like aspirin, ibuprofen reduces inflammation. (Prescription-strength ibuprofen has extensive use as an anti-inflammatory agent for the treatment of rheumatoid arthritis.) There is evidence that ibuprofen is more effective than either aspirin or acetaminophen in reducing dental pain and menstrual pain. Both aspirin and ibuprofen can cause stomach bleeding in some people, although ibuprofen seems to cause fewer problems. Ibuprofen is more expensive than either aspirin or acetaminophen.

Naproxen, marketed under the brand names Aleve and Anaprox, was cleared by the FDA in 1994 for nonprescription use. The effects of naproxen last longer in the body (8–12 hr per dose) than the effects of ibuprofen (4–6 hr per dose) and of

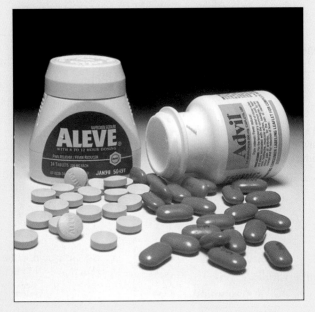

aspirin and acetaminophen (4 hr per dose). Naproxen is more likely to cause slight intestinal bleeding and stomach upset than is ibuprofen. It is also not recommended for use by children under 12.

$$CH_3\!-\!CH_2\!-\!COOH$$

Propanoic acid

Ibuprofen

Naproxen

■ Unsaturated Acids

The simplest *unsaturated monocarboxylic* acid is propenoic acid (acrylic acid), a substance used in the manufacture of several polymeric materials. Two forms exist for the simplest *unsaturated dicarboxylic* acid, butenedioic acid. The two isomers have separate common names, fumaric acid (*trans*) and maleic acid (*cis*), a naming procedure seldom encountered.

> An unsaturated monocarboxylic acid with the structure
>
> $$CH_3\!-\!CH_2\!-\!CH_2\!-\!C\!=\!CH\!-\!COOH$$
> $$\qquad\qquad\qquad\quad |$$
> $$\qquad\qquad\qquad\ CH_3$$
>
> 3-Methyl-2-hexenoic acid
>
> has been found to be largely responsible for "body odor." It is produced by skin bacteria, particularly those found in armpits.

$$CH_2\!=\!CH\!-\!COOH$$

Acrylic acid

Maleic acid
(*cis* isomer)

Fumaric acid
(*trans* isomer)

Some antihistamines (Section 17.9) are salts of maleic acid. The addition of small amounts of maleic acid to fats and oils prevents them from becoming rancid. Fumaric acid is a *metabolic acid*. Metabolic acids are intermediate compounds in the metabolic reactions (Section 23.1) that occur in the human body. More information about metabolic acids is presented in the next section.

■ Hydroxy Acids

Four of the simpler *hydroxy* acids are

C—COOH   C—C—COOH   HOOC—C—C—COOH   HOOC—C—C—COOH
| | | | |
OH   OH   OH   OH OH
Glycolic acid   Lactic acid   Malic acid   Tartaric acid

Malic and tartaric acids are derivatives of succinic acid, the four-carbon unsubstituted diacid (Section 16.3).

Hydroxy acids occur naturally in many foods. Glycolic acid is present in the juice from sugar cane and sugar beets. Lactic acid is present in sour milk, sauerkraut, and dill pickles. Both malic acid and tartaric acid occur naturally in fruits. The sharp taste of apples (fruit of trees of the genus *Malus*) is due to malic acid. Tartaric acid is particularly abundant in grapes (Figure 16.5). It is also a component of tartar sauce and an acidic ingredient in many baking powders. Lactic and malic acids are also *metabolic acids* (Section 16.5).

Citric acid, perhaps the best known of all carboxylic acids, is a hydroxy acid with a structural feature we have not previously encountered. It is a hydroxy *tri*carboxylic acid. Besides there being acid groups at both ends of a carbon chain, a third acid group is present as a substituent on the chain. An acid group as a substituent is called a *carboxy* group. Thus citric acid is a hydroxycarboxy diacid.

> The IUPAC name for citric acid is 2-hydroxy-1,2,3-propanetrioic acid.

OH
|
HOOC—C—C—C—COOH
|
COOH
Citric acid

Citric acid gives citrus fruits their "sharp" taste; lemon juice contains 4%–8% citric acid, and orange juice is about 1% citric acid. Citric acid is used widely in beverages and in foods. In jams, jellies, and preserves, it produces tartness and pH adjustment to optimize conditions for gelation. In fresh salads, citric acid prevents enzymatic browning reactions, and in frozen fruits, it prevents deterioration of color and flavor. Addition of citric acid to seafood retards microbial growth by lowering pH. Citric acid is also a metabolic acid (Section 16.5).

■ Keto Acids

Keto acids, as the designation implies, contain a carboxyl group within a carbon chain. Pyruvic acid, with three carbon atoms, is the simplest keto acid that can exist.

O
‖
C—C—COOH
Pyruvic acid

In the pure state, pyruvic acid is a liquid with an odor resembling that of vinegar (acetic acid; Section 16.3). Pyruvic acid is a metabolic acid (Section 16.5).

FIGURE 16.5 Tartaric acid, the dihydroxy derivative of succinic acid, is particularly abundant in ripe grapes.

CHEMICAL CONNECTIONS

Carboxylic Acids and Skin Care

A number of carboxylic acids are used as "skin-care acids." Heavily advertised at present are cosmetic products that contain *alpha-hydroxy* acids, carboxylic acids in which a hydroxyl group is attached to the acid's alpha carbon atom (the carbon atom attached to the carboxyl group carbon atom). Such cosmetic products address problems such as dryness, flaking, and itchiness of the skin and are highly promoted for removing wrinkles.

Alpha-hydroxy acids "work" by loosening the cells of the outer layer of skin (the epidermis) and by accelerating the flaking off of dead skin. The result is healthier-looking skin.

The alpha-hydroxy acids most commonly found in cosmetic products are glycolic acid and lactic acid, the two simplest alpha-hydroxy acids.

Tretinoin has an *all-trans* double-bond configuration. Accutane has a *cis* double bond in the position closest to the carboxyl group, with the rest of the double bonds in *trans* configurations.

Both acids are naturally occurring substances. Glycolic acid occurs in sugar cane and sugar beets, and lactic acid occurs in sour milk.

The use of alpha-hydroxy acids in cosmetics is considered safe at acid concentrations of less than 10%; higher concentrations can cause skin irritation, burning, and stinging. (Lactic acid becomes a prescription drug at concentrations of 12% or more.) One drawback of the cosmetic use of alpha-hydroxy products is that such use can increase the skin's sensitivity to the ultraviolet light component of sunlight; it is this component that causes sunburn. Individuals who are using "alpha-hydroxys" should apply a sunscreen whenever they go outside for an extended period of time.

Glycolic acid, at higher concentrations than that found in cosmetics, is used by dermatologists for the "spot" removal of *keratoses* (precancerous lesions and/or patches of darker, thickened skin).

Polyunsaturated carboxylic acids are used extensively in the treatment of severe acne. The prescription drugs Tretinoin and Accutane are such compounds.

Two skin-care products containing alpha-hydroxy acids.

16.5 "Metabolic" Acids

Numerous polyfunctional acids, including some mentioned in the previous section, are intermediates in the metabolic reactions that occur in the human body as food is processed. There are eight such acids that will appear repeatedly in the biochemistry chapters of this text.

Interestingly, these eight key metabolic intermediates are derived from only three of the simple carboxylic acids. These three simple acids and the metabolic acids related to them are

propionic acid (3-carbon monoacid): lactic, glyceric, and pyruvic acids
succinic acid (4-carbon diacid): fumaric, oxaloacetic, and malic acids
glutaric acid (5-carbon diacid): α-ketoglutaric and citric acids

Metabolic acids derived from the diacids succinic and glutaric are encountered in the citric acid cycle (Section 23.6), a series of reactions in which C_2 units obtained from all

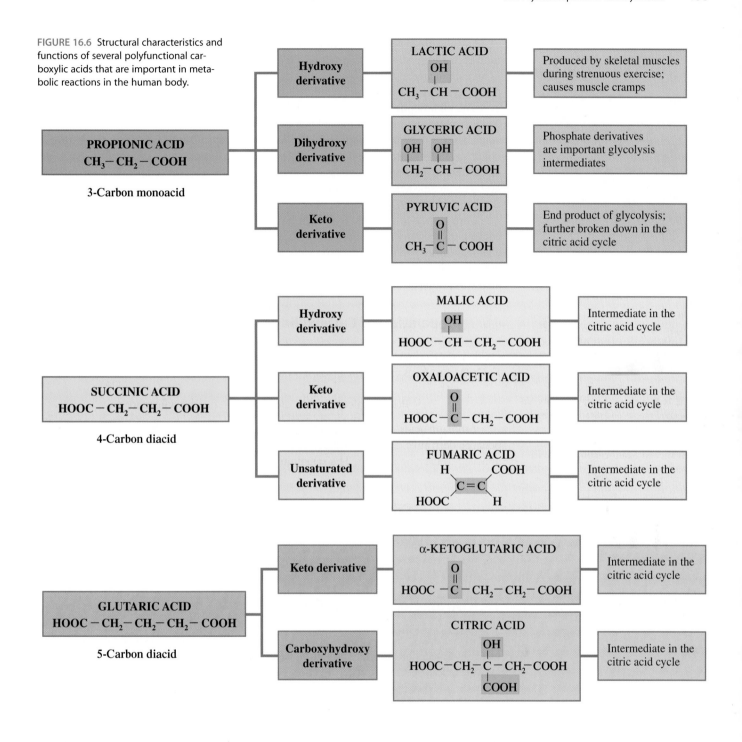

FIGURE 16.6 Structural characteristics and functions of several polyfunctional carboxylic acids that are important in metabolic reactions in the human body.

types of foods are further processed for the purpose of obtaining energy. Glyceric and pyruvic acid (propionic acid derivatives) are encountered in glycolysis (Section 24.2), a series of reactions in which glucose is processed. Lactic acid (a propionic derivative) is a by-product of strenuous exercise (Section 24.3). Figure 16.6 gives further details about the eight "metabolic" acids.

16.6 Physical Properties of Carboxylic Acids

Carboxylic acids are the most *polar* organic compounds we have discussed so far. Both the carbonyl part ($C=O$) and the hydroxyl part (—OH) of the carboxyl functional group are polar. The result is very high melting and boiling points for carboxylic acids, the highest of any type of organic compound yet considered (Figure 16.7).

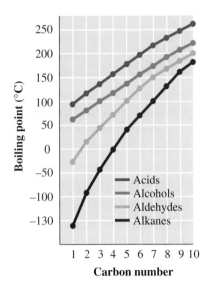

FIGURE 16.7 The boiling points of mono-carboxylic acids compared to those of other types of compounds. All compounds in the comparison have unbranched carbon chains.

FIGURE 16.8 A physical-state summary for unbranched mono- and dicarboxylic acids at room temperature and pressure.

Unbranched Monocarboxylic Acids			
C_1	C_3	C_5	C_7
C_2	C_4	C_6	C_8

Unbranched Dicarboxylic Acids			
✕	C_3	C_5	C_7
C_2	C_4	C_6	C_8

☐ Liquid ☐ Solid

Unsubstituted saturated monocarboxylic acids containing up to nine carbon atoms are liquids that have strong, sharp odors (Figure 16.8). Acids with 10 or more carbon atoms in an unbranched chain are waxy solids that are odorless (because of low volatility). Aromatic carboxylic acids, as well as dicarboxylic acids, are also odorless solids.

The high boiling points of carboxylic acids indicate the presence of strong intermolecular attractive forces. A unique hydrogen-bonding arrangement, shown in Figure 16.9, contributes to these attractive forces. A given carboxylic acid molecule forms two hydrogen bonds to another carboxylic acid molecule, producing a "complex" known as a *dimer*. Because dimers have twice the mass of a single molecule, a higher temperature is needed to boil a carboxylic acid than would be needed for similarly sized aldehyde and alcohol molecules where dimerization does not occur.

Carboxylic acids readily hydrogen-bond to water molecules. Such hydrogen bonding contributes to water solubility for short-chain carboxylic acids. The unsubstituted C_1 to C_4 monocarboxylic acids are completely miscible with water. Solubility then rapidly decreases with carbon number, as shown in Figure 16.10. Short-chain dicarboxylic acids are also water-soluble. In general, aromatic acids are not water-soluble.

16.7 Preparation of Carboxylic Acids

Oxidation of primary alcohols or aldehydes, using an oxidizing agent such as CrO_3 or $K_2Cr_2O_7$, produces carboxylic acids, a process that we examined in Sections 14.7 and 15.8.

$$\text{Primary alcohol} \xrightarrow{\text{[O]}} \text{aldehyde} \xrightarrow{\text{[O]}} \text{carboxylic acid}$$

Aromatic acids can be prepared by oxidizing a carbon side chain (alkyl group) on a benzene derivative. In this process, all the carbon atoms of the alkyl group except the one attached to the ring are lost. The remaining carbon becomes part of a carboxyl group.

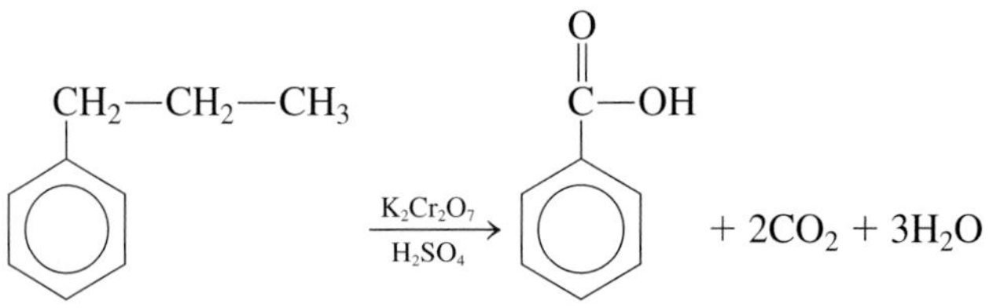

$$\text{C}_6\text{H}_5-\text{CH}_2-\text{CH}_2-\text{CH}_3 \xrightarrow[\text{H}_2\text{SO}_4]{\text{K}_2\text{Cr}_2\text{O}_7} \text{C}_6\text{H}_5-\overset{\displaystyle O}{\overset{\|}{\text{C}}}-\text{OH} + 2\text{CO}_2 + 3\text{H}_2\text{O}$$

16.8 Acidity of Carboxylic Acids

Carboxylic acids, as the name implies, are *acidic*. When a carboxylic acid is placed in water, hydrogen ion transfer (proton transfer; Section 10.2) occurs to produce hydronium ion (the acidic species in water; Section 10.2) and carboxylate ion.

$$\text{R}-\text{COOH} + \text{H}_2\text{O} \longrightarrow \underset{\substack{\text{Hydronium} \\ \text{ion}}}{\text{H}_3\text{O}^+} + \underset{\substack{\text{Carboxylate} \\ \text{ion}}}{\text{R}-\text{COO}^-}$$

FIGURE 16.9 A given carboxylic acid molecule can form two hydrogen bonds to another carboxylic acid molecule, producing a "dimer"—a complex with a mass twice that of a single molecule.

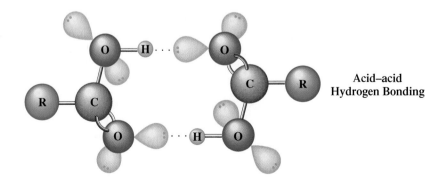

Acid–acid Hydrogen Bonding

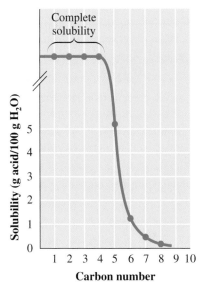

FIGURE 16.10 The solubility in water of saturated unbranched-chain carboxylic acids.

At normal human body pH values (pH = 7.35 to 7.45), most carboxylic acids exist as carboxylate ions. Acetic acid is in the form of acetate ion, pyruvic acid is in the form of pyruvate ion, lactic acid is in the form of lactate ion, and so on.

Carboxylic acid salt formation involves an acid–base neutralization reaction (Section 10.7).

A **carboxylate ion** *is the negative ion produced when a carboxylic acid loses one or more acidic hydrogen atoms.*

Carboxylate ions formed from monocarboxylic acids always carry a -1 charge; only one acidic hydrogen atom is present in such molecules. Dicarboxylic acids, which possess two acidic hydrogen atoms (one in each carboxyl group), can produce carboxylate ions bearing a -2 charge.

Carboxylate ions are named by dropping the *-ic acid* ending from the name of the parent acid and replacing it with *-ate.*

$$CH_3-\overset{O}{\overset{\|}{C}}-OH + H_2O \longrightarrow H_3O^+ + CH_3-\overset{O}{\overset{\|}{C}}-O^-$$

Acetic acid (ethanoic acid)   Acetate ion (ethanoate ion)

$$HO-\overset{O}{\overset{\|}{C}}-\overset{O}{\overset{\|}{C}}-OH + 2H_2O \longrightarrow 2H_3O^+ + {}^-O-\overset{O}{\overset{\|}{C}}-\overset{O}{\overset{\|}{C}}-O^-$$

Oxalic acid (ethanedioic acid)   Oxalate ion (ethanedioate ion)

Carboxylic acids are weak acids (Section 10.4). The extent of proton transfer is usually less than 5%; that is, an equilibrium situation exists in which the equilibrium lies far to the left.

$$R-COOH + H_2O \rightleftharpoons H_3O^+ + R-COO^-$$

More than 95% of molecules in this form   Less than 5% of molecules in this form

Table 16.3 gives K_a values (Section 10.5) and percent ionization in 0.100 M solution for selected monocarboxylic acids.

16.9 Carboxylic Acid Salts

In a manner similar to that of inorganic acids (Section 10.6), carboxylic acids react completely with strong bases to produce water and a carboxylic acid salt.

$$CH_3-\overset{O}{\overset{\|}{C}}-OH + NaOH \longrightarrow CH_3-\overset{O}{\overset{\|}{C}}-O^- Na^+ + H_2O$$

Carboxylic acid   Strong base   Carboxylic acid salt   Water

A **carboxylic acid salt** *is an ionic compound in which the negative ion is a carboxylate ion.*

Carboxylic acid salts are named similarly to other ionic compounds (Section 4.9): *The positive ion is named first, followed by a separate word giving the name of the negative ion.* The salt formed in the preceding reaction contains sodium ions and acetate ions (from acetic acid); hence the salt's name is sodium acetate.

TABLE 16.3
Acid Strength for Selected Monocarboxylic Acids

Acid	K_a	Percent ionization (0.100 M solution)
Formic	1.8×10^{-4}	4.2%
Acetic	1.8×10^{-5}	1.3%
Propionic	1.3×10^{-5}	1.2%
Butyric	1.5×10^{-5}	1.2%
Valeric	1.5×10^{-5}	1.2%
Caproic	1.4×10^{-5}	1.2%

EXAMPLE 16.3

Writing Equations for the Formation of Carboxylic Acid Salts

■ Using an acid–base neutralization reaction, write a chemical equation for the formation of each of the following carboxylic acid salts.

a. Sodium propionate   **b.** Potassium oxalate

Solution

a. This salt contains sodium ion (Na^+) and propionate ion, the three-carbon monocarboxylate ion.

$$CH_3-CH_2-\overset{\displaystyle O}{\overset{\displaystyle \|}{C}}-O^-Na^+$$

From a neutralization standpoint, the sodium ion's source is the base sodium hydroxide, NaOH, and the negative ion's source is the acid propanoic acid. The acid–base neutralization equation is

$$CH_3-CH_2-\overset{\displaystyle O}{\overset{\displaystyle \|}{C}}-OH + NaOH \longrightarrow CH_3-CH_2-\overset{\displaystyle O}{\overset{\displaystyle \|}{C}}-O^-Na^+ + H_2O$$

Propionic acid   Sodium   Sodium propionate   Water
                                hydroxide

b. This salt contains potassium ions (K^+) whose source would be the base potassium hydroxide, KOH. The salt also contains oxalate ions, whose source would be the acid oxalic acid.

$$HO-\overset{\displaystyle O}{\overset{\displaystyle \|}{C}}-\overset{\displaystyle O}{\overset{\displaystyle \|}{C}}-OH + 2KOH \longrightarrow K^+{}^-O-\overset{\displaystyle O}{\overset{\displaystyle \|}{C}}-\overset{\displaystyle O}{\overset{\displaystyle \|}{C}}-O^-K^+ + 2H_2O$$

Oxalic acid   Potassium   Potassium oxalate   Water
                              hydroxide

Note that two molecules of base are needed to react completely with one molecule of acid because the acid is a dicarboxylic acid.

Practice Exercise 16.3

Using an acid–base neutralization reaction, write a chemical equation for the formation of each of the following carboxylic acid salts.

a. Sodium formate   **b.** Potassium malonate

Converting a carboxylic acid salt back to a carboxylic acid is very simple. React the salt with a solution of a strong acid such as hydrochloric acid (HCl) or sulfuric acid (H_2SO_4).

$$CH_3-\overset{\displaystyle O}{\overset{\displaystyle \|}{C}}-O^-Na^+ + HCl \longrightarrow CH_3-\overset{\displaystyle O}{\overset{\displaystyle \|}{C}}-OH + NaCl$$

Sodium acetate   Hydrochloric   Acetic acid   Sodium
                            acid   chloride

The interconversion reactions between carboxylic acid salts and their "parent" carboxylic acids are so easy to carry out that organic chemists consider these two types of compounds interchangeable.

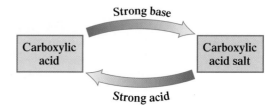

FIGURE 16.11 Propionates, salts of propionic acid, extend the shelf life of bread by preventing the formation of mold.

▶ The solubility of benzoic acid in water at 25°C is 3.4 g/L. The solubility of sodium benzoate, the sodium salt of benzoic acid in water at 25°C is 550 g/L.

■ Uses for Carboxylic Acid Salts

The solubility of carboxylic acid salts in water is much greater than that of the carboxylic acids from which they are derived. Drugs and medicines that contain acid groups are usually marketed as the sodium or potassium salt of the acid. This greatly enhances the solubility of the medication, increasing the ease of its absorption by the body.

Many *antimicrobials,* compounds used as food preservatives, are carboxylic acid salts. Particularly important are the salts of benzoic, sorbic, and propionic acids.

$$COOH$$

Benzoic acid

$$CH_3-CH=CH-CH=CH-COOH$$

Sorbic acid
(2,4-hexadienoic acid)

$$CH_3-CH_2-COOH$$

Propionic acid

The benzoate salts of sodium and potassium are effective against yeast and mold in beverages, jams and jellies, pie fillings, ketchup, and syrups. Concentrations of up to 0.1% (m/m) benzoate are found in such products.

Sodium benzoate

Potassium benzoate

Sodium and potassium sorbates inhibit mold and yeast growth in dairy products, dried fruits, sauerkraut, and some meat and fish products. Sorbate preservative concentrations range from 0.02% to 0.2% (m/m).

$$CH_3-CH=CH-CH=CH-\overset{O}{\overset{\|}{C}}-O^-\ Na^+$$

Sodium sorbate

$$CH_3-CH=CH-CH=CH-\overset{O}{\overset{\|}{C}}-O^-\ K^+$$

Potassium sorbate

Calcium and sodium propionates are used in baked products and also in cheese foods and spreads (see Figure 16.11). Benzoates and sorbates cannot be used in yeast-leavened baked goods, because they affect the activity of the yeast.

$$\left(CH_3-CH_2-\overset{O}{\overset{\|}{C}}-O^-\right)_2 Ca^{2+}$$

Calcium propionate

$$CH_3-CH_2-\overset{O}{\overset{\|}{C}}-O^-\ Na^+$$

Sodium propionate

Carboxylate salts do not directly kill microorganisms present in food. Rather, they prevent further growth and proliferation of these organisms by increasing the pH of the foods in which they are used.

16.10 Structure of Esters

An **ester** *is a carboxylic acid derivative in which the —OH portion of the carboxyl group has been replaced with an —OR group.*

$$R-\overset{O}{\overset{\|}{C}}-O-H$$

Carboxylic acid

$$R-\overset{O}{\overset{\|}{C}}-O-R$$

Ester

The ester functional group is thus

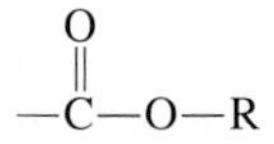

$$-\overset{\displaystyle O}{\overset{\displaystyle \|}{C}}-O-R$$

In linear form, the ester functional group can be represented as —COOR or —CO$_2$R.

The simplest ester, which has two carbon atoms, has a hydrogen atom attached to the ester functional group.

$$H-\overset{\displaystyle O}{\overset{\displaystyle \|}{C}}-O-CH_3$$

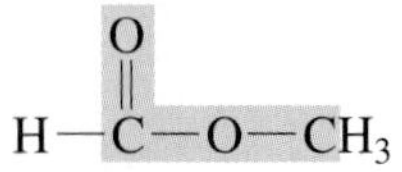

Note that the two carbon atoms present are not bonded to each other.

There are two three-carbon esters.

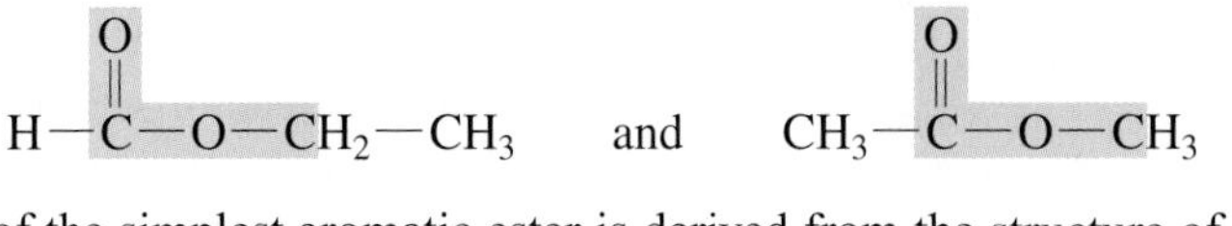

$$H-\overset{\displaystyle O}{\overset{\displaystyle \|}{C}}-O-CH_2-CH_3 \quad \text{and} \quad CH_3-\overset{\displaystyle O}{\overset{\displaystyle \|}{C}}-O-CH_3$$

The structure of the simplest aromatic ester is derived from the structure of benzoic acid, the simplest aromatic carboxylic acid.

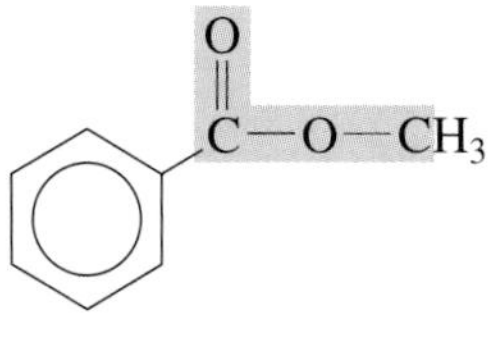

$$\text{(benzene ring)}-\overset{\displaystyle O}{\overset{\displaystyle \|}{C}}-O-CH_3$$

Note that the difference between a carboxylic acid and an ester is a "H versus R" relationship.

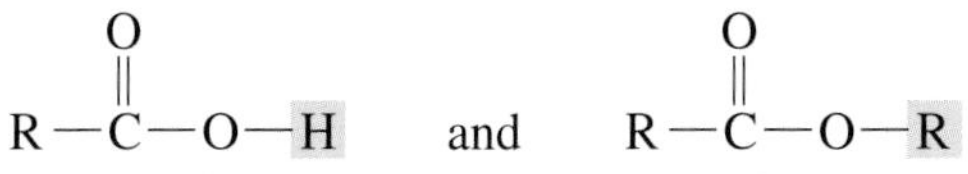

$$R-\overset{\displaystyle O}{\overset{\displaystyle \|}{C}}-O-H \quad \text{and} \quad R-\overset{\displaystyle O}{\overset{\displaystyle \|}{C}}-O-R$$
$$\text{Acid} \qquad\qquad\qquad \text{Ester}$$

We have encountered this "H versus R" relationship several times before in our study of hydrocarbon derivatives. The Chemistry at a Glance feature on page 441 summarizes the "H versus R" relationships we have encountered so far.

16.11 Preparation of Esters

Esters are produced through *esterification*. An **esterification reaction** *is the reaction of a carboxylic acid with an alcohol (or phenol) to produce an ester.* A strong acid catalyst (generally H$_2$SO$_4$) is needed for esterification.

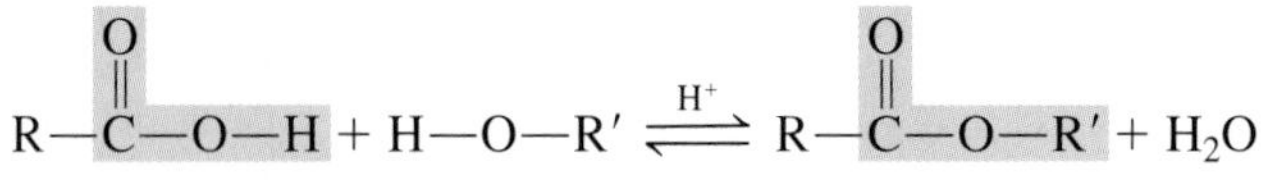

$$R-\overset{\displaystyle O}{\overset{\displaystyle \|}{C}}-O-H + H-O-R' \overset{H^+}{\rightleftharpoons} R-\overset{\displaystyle O}{\overset{\displaystyle \|}{C}}-O-R' + H_2O$$
$$\text{Carboxylic acid} \qquad \text{Alcohol} \qquad\qquad \text{Ester} \qquad \text{Water}$$

In the esterification process, a —OH group is lost from the carboxylic acid, a —H atom is lost from the alcohol, and water is formed as a by-product. The net effect of this reaction is substitution of the —OR group of the alcohol for the —OH group of the acid.

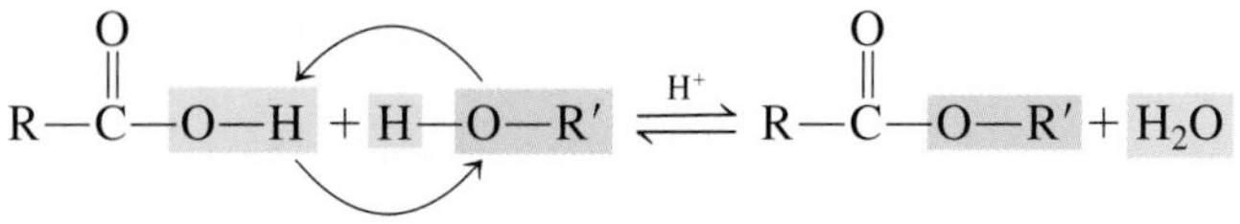

> Esterification is a *condensation reaction*. This is the third time we have encountered this type of reaction. The first encounter involved intermolecular alcohol dehydration (Section 14.7) and the second encounter involved the preparation of acetals (Section 15.9).

$$R-\overset{\displaystyle O}{\overset{\displaystyle \|}{C}}-O-H + H-O-R' \overset{H^+}{\rightleftharpoons} R-\overset{\displaystyle O}{\overset{\displaystyle \|}{C}}-O-R' + H_2O$$

A specific example of esterification is the reaction of acetic acid with methyl alcohol.

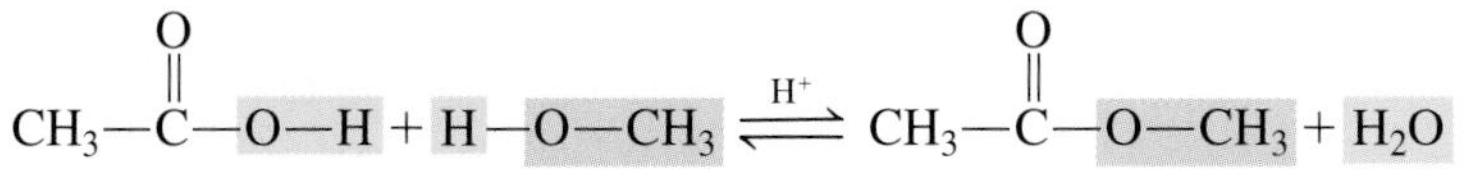

> Studies show that in ester formation, the hydroxyl group of the acid (not of the alcohol) becomes part of the water molecule.

$$CH_3-\overset{\displaystyle O}{\overset{\displaystyle \|}{C}}-O-H + H-O-CH_3 \overset{H^+}{\rightleftharpoons} CH_3-\overset{\displaystyle O}{\overset{\displaystyle \|}{C}}-O-CH_3 + H_2O$$

Summary of the "H versus R" Relationship for Pairs of Hydrocarbon Derivatives

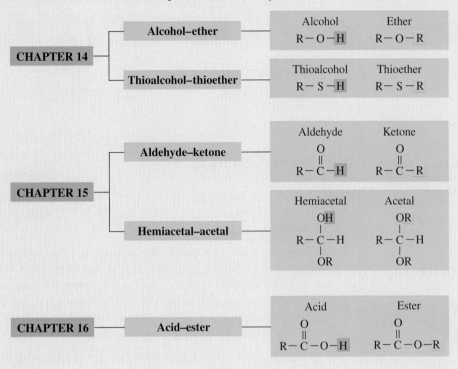

Esterification reactions are equilibrium processes, with the position of equilibrium (Section 9.8) usually favoring products only slightly. That is, at equilibrium, substantial amounts of both reactants and products are present. The amount of ester formed can be increased by using an excess of alcohol or by constantly removing one of the products. According to Le Châtelier's principle (Section 9.9), either of these techniques will shift the position of equilibrium to the right (the product side of the equation). This equilibrium problem explains the use of the "double-arrow notation" in all the esterification equations in this section.

It is often useful to think of the structure of an ester in terms of its "parent" alcohol and acid molecules; the ester has an acid part and an alcohol part.

$$\underset{\substack{\text{Acid} \\ \text{part}}}{R-\overset{\overset{\textstyle O}{\|}}{C}} \underset{\substack{\text{Alcohol} \\ \text{part}}}{O-R'}$$

> Esters and carboxylic acids with the same number of carbon atoms and the same degree of saturation are structural isomers (functional group isomers). This is the third time we have encountered functional group isomerism. The other examples are alcohol–ether and aldehyde–ketone isomers.

In this context, it is easy to identify the acid and alcohol from which a given ester can be produced; just add a —OH group to the acid part of the ester and a —H atom to the alcohol part to generate the parent molecules.

$$CH_3-CH_2-\overset{\overset{\textstyle O}{\|}}{C}\!\!\mid\!\!O-CH_2-CH_2-CH_3$$

+OH   +H

$$\underset{\text{"Parent" acid}}{CH_3-CH_2-\overset{\overset{\textstyle O}{\|}}{C}-\boxed{OH}} \qquad \underset{\text{"Parent" alcohol}}{\boxed{H}-O-CH_2-CH_2-CH_3}$$

Hydroxy acids—compounds which contain both a hydroxyl and a carboxyl group (Section 16.4)—have the capacity to undergo intermolecular esterification to form *lactones* (cyclic esters). Such internal esterification easily takes place in situations where a five- or six-membered ring can be formed.

> **Cyclic esters formed from hydroxyacids are called *lactones*.**

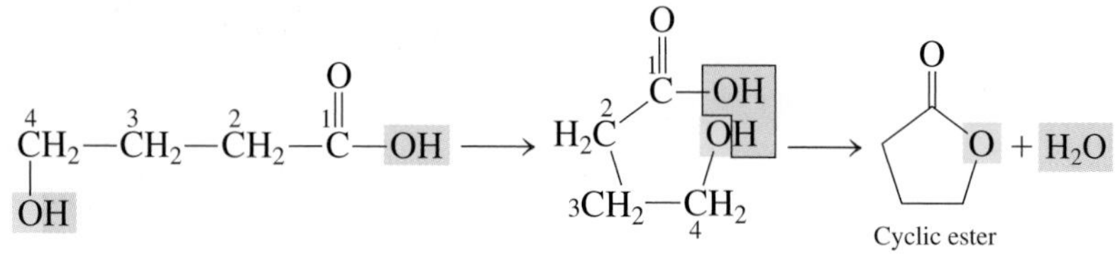

16.12 Nomenclature for Esters

> **Salts and esters of carboxylic acids are named in the same way. The name of the positive ion (in the case of a salt) or the name of the organic group attached to the single-bonded oxygen of the carbonyl group (in the case of an ester) precedes the name of the acid. The *-ic acid* part of the name of the acid is converted to *-ate*.**

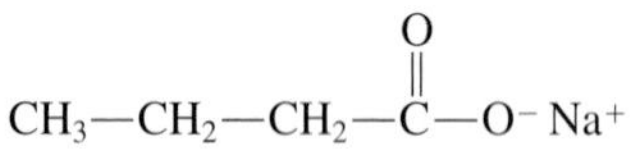

Visualizing esters as having an "alcohol part" and an "acid part" (Section 16.11) is the key to naming them in both the common and the IUPAC systems of nomenclature. The rules are as follows:

1. The name for the alcohol part of the ester appears first and is followed by a *separate word* giving the name for the acid part of the ester.
2. The name for the alcohol part of the ester is simply the name of the R group (alkyl, cycloalkyl, or aryl) present in the —OR portion of the ester.
3. The name for the acid part of the ester is obtained by dropping the *-ic acid* ending for the acid's name and adding the suffix *-ate*.

$$CH_3-CH_2-CH_2-\overset{\overset{\textstyle O}{\|}}{C}-O^-\ Na^+$$

IUPAC: Sodium butanoate
Common: Sodium butyrate

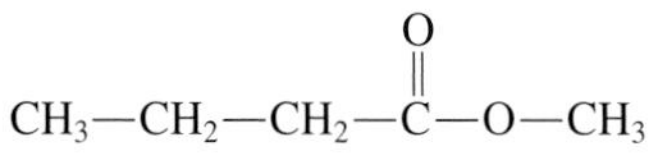

$$CH_3-CH_2-CH_2-\overset{\overset{\textstyle O}{\|}}{C}-O-CH_3$$

IUPAC: Methyl butanoate
Common: Methyl butyrate

Consider the ester derived from ethanoic acid (acetic acid) and methanol (methyl alcohol). Its name will be *methyl ethanoate* (IUPAC) or *methyl acetate* (common); see Figure 16.12.

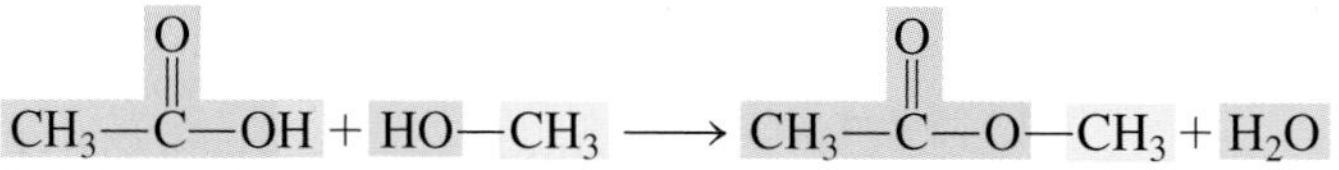

IUPAC: Ethanoic acid   Methanol   Methyl ethanoate
Common: Acetic acid   Methyl alcohol   Methyl acetate

Dicarboxylic acids can form diesters, with each of the carboxyl groups undergoing esterification. An example of such a molecule and how it is named is

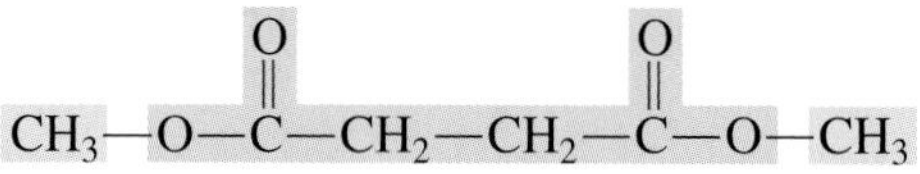

IUPAC: Dimethyl butanedioate
Common: Dimethyl succinate

Further examples of ester nomenclature, for compounds in which substituents are present, are

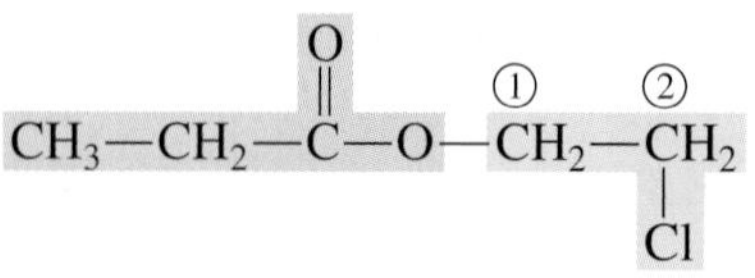

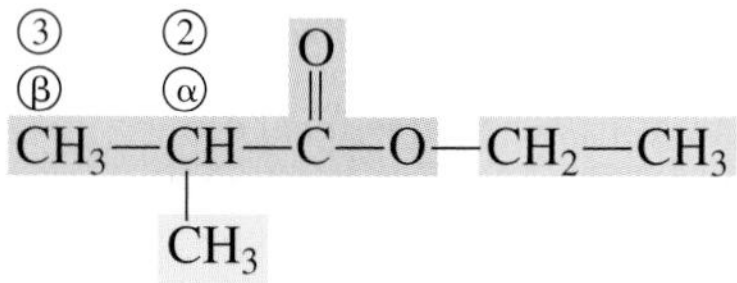

IUPAC: 2-Chloroethyl propanoate   Ethyl 2-methylpropanoate
Common: 2-Chloroethyl propionate   Ethyl α-methylpropionate

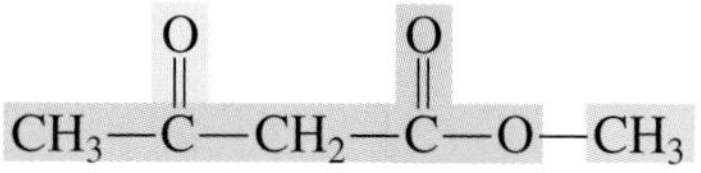

IUPAC: Methyl 3-oxobutanoate
Common: Methyl 3-oxobutyrate

FIGURE 16.12 Space-filling models for the methyl and ethyl esters of acetic acid.

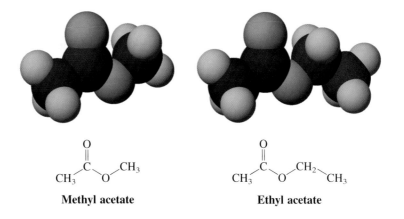

Methyl acetate   **Ethyl acetate**

EXAMPLE 16.4

Determining IUPAC and Common Names for Esters

Line-angle drawings for the simpler unbranched-chain methyl esters:

Methyl methanoate

Methyl ethanoate

Methyl propanoate

Methyl butanoate

■ Assign both IUPAC and common names to the following esters.

a.
$$CH_3—CH_2—\overset{\overset{\displaystyle O}{\|}}{C}—O—CH_2—CH_3$$

b.
$$CH_3—\underset{\underset{\displaystyle CH_3}{|}}{CH}—CH_2—\overset{\overset{\displaystyle O}{\|}}{C}—O—CH_3$$

c.
$$\overset{\overset{\displaystyle O}{\|}}{C}—O—CH_2—CH_2—CH_3$$

Solution

a. The name *ethyl* characterizes the alcohol part of the molecule. The name of the acid is propanoic acid (IUPAC) or propionic acid (common). Deleting the *-ic acid* ending and adding *-ate* gives the name *ethyl propanoate* (IUPAC) or *ethyl propionate* (common).

b. The name of the alcohol part of the molecule is methyl (from methanol or methyl alcohol). The name of the five-carbon acid is 3-methylbutanoic acid or β-methylbutyric acid. Hence the ester name is *methyl 3-methylbutanoate* (IUPAC) or *methyl β-methylbutyrate* (common).

c. The name *propyl* characterizes the alcohol part of the molecule. The acid part of the molecule is derived from benzoic acid (both IUPAC and common name). Hence the ester name in both systems is *propyl benzoate.*

Practice Exercise 16.4

Assign both IUPAC and common names to the following esters.

a.
$$CH_3—\overset{\overset{\displaystyle O}{\|}}{C}—O—CH_2—CH_3$$

b.
$$CH_3—CH_2—CH_2—CH_2—\overset{\overset{\displaystyle O}{\|}}{C}—O—CH_3$$

c.
$$H—\overset{\overset{\displaystyle O}{\|}}{C}—O—CH_2—CH_2—CH_3$$

The contrast between IUPAC names and common names for unbranched esters of carboxylic acids is as follows:

IUPAC (two words)

alkyl alkanoate

methyl propanoate

Common (two words)

alkyl (prefix)ate*

methyl acetate

*The common-name prefixes are related to natural sources for the "parent" carboxylic acids.

Lactones (cyclic esters) are named by replacing the *-oic* ending of the parent hydroxycarboxylic acid name with *-olide* and identifying the hydroxyl-bearing carbon by number.

CHEMICAL CONNECTIONS

Aspirin

Aspirin, an ester of salicylic acid (Section 16.13), is a drug that has the ability to decrease pain (analgesic properties), to lower body temperature (antipyretic properties), and to reduce inflammation (anti-inflammatory properties). It is most frequently taken in tablet form, and the tablet usually contains 325 mg of aspirin held together with an inert starch binder.

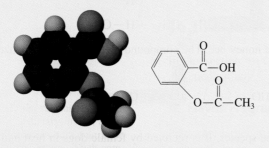

After ingestion, aspirin undergoes hydrolysis to produce salicylic acid and acetic acid. Salicylic acid is the active ingredient of aspirin—the substance that has analgesic, antipyretic, and anti-inflammatory effects.

Salicylic acid is capable of irritating the lining of the stomach, inducing a small amount of bleeding. Breaking (or chewing) an aspirin tablet, rather than taking it whole, reduces the chance of bleeding by eliminating drug concentration on one part of the stomach lining. Buffered aspirin products contain alkaline chemicals (such as aluminum glycinate or aluminum hydroxide) to neutralize the acidity of the aspirin when it contacts the stomach lining.

Aspirin—that is, salicylic acid—inhibits the synthesis of a class of hormones called prostaglandins (Section 19.13), molecules that cause pain, fever, and inflammation when present in the bloodstream in higher-than-normal levels. Salicylic acid's mode of action is irreversible inhibition (Section 21.7) of *cyclooxygenase,* an enzyme necessary for the production of prostaglandins.

Recent studies show that aspirin also increases the time it takes blood to coagulate (clot). For blood to coagulate, platelets must first be able to aggregate, and prostaglandins (which aspirin inhibits) appear to be necessary for platelet aggregation to occur. One study suggests that healthy men can cut their risk of heart attacks nearly in half by taking one baby aspirin per day (81 mg compared to the 325 mg in a regular tablet). Aspirin acts by making the blood less likely to clot. Heart attacks usually occur when clots form in the coronary arteries, cutting off blood supply to the heart.

Aspirin manufacturers indicate that "low dose" (81 mg) aspirin tablets now represent 23% of the total market for aspirin tablets. In 2001, about 26 million people in America regularly took aspirin for "heart health"—up from 7 million in 1997.

Aspirin   + H₂O $\xrightarrow{H^+}$ Salicylic acid   + CH₃—C—OH   Acetic acid

Oil of wintergreen, also called methyl salicylate, is used in skin rubs and liniments to help decrease the pain of sore muscles. It is absorbed through the skin, where it is hydrolyzed to produce salicylic acid. Salicylic acid, as with aspirin, is the actual pain reliever.

16.14 Physical Properties of Esters

Ester molecules cannot form hydrogen bonds to each other because they do not have a hydrogen atom bonded to an oxygen atom. Consequently, the boiling points of esters are much lower than those of alcohols and carboxylic acids of comparable molecular mass. Esters are more like ethers in their physical properties. Table 16.5 gives boiling-point data for compounds of similar molecular mass that contain different functional groups.

Water molecules can hydrogen-bond to esters through the oxygen atoms present in the ester functional group (Figure 16.14).

TABLE 16.5
Boiling Points of Compounds of Similar Molecular Mass That Contain Different Functional Groups

Name	Functional-group class	Molecular mass	Boiling point (°C)
diethyl ether	ether	74	34
ethyl formate	ester	74	54
methyl acetate	ester	74	57
butanal	aldehyde	72	76
1-butanol	alcohol	74	118
propionic acid	acid	74	141

Because of such hydrogen bonding, low-molecular-mass esters are soluble in water. Solubility rapidly decreases with increasing carbon chain length; borderline solubility situations are reached when three to five carbon atoms are in a chain.

Low- and intermediate-molecular-mass esters are usually colorless liquids at room temperature (see Figure 16.15). Most have pleasant odors (Section 16.13).

16.15 Chemical Reactions of Esters

The most important reaction of esters involves breaking the carbon–oxygen single bond that holds the "alcohol part" and the "acid part" of the ester together. This reaction process is called either ester hydrolysis or ester saponification, depending on reaction conditions.

■ Ester Hydrolysis

▶ This is our second encounter with hydrolysis reactions. The first encounter involved the hydrolysis of acetals (Section 15.9).

▶ The breaking of a bond within a molecule and the attachment of the components of water to the fragments are characteristics of all hydrolysis reactions.

In ester hydrolysis, an ester reacts with water, producing the carboxylic acid and alcohol from which the ester was formed.

$$R-\overset{\overset{\displaystyle O}{\|}}{C}-O-R' + H-OH \xrightarrow{H^+} R-\overset{\overset{\displaystyle O}{\|}}{C}-OH + R'-O-H$$

$$CH_3-\overset{\overset{\displaystyle O}{\|}}{C}-O-CH_3 + H-OH \xrightarrow{H^+} CH_3-\overset{\overset{\displaystyle O}{\|}}{C}-OH + CH_3-O-H$$

Methyl acetate   Water   Acetic acid   Methyl alcohol

Ester hydrolysis requires the presence of a strong-acid catalyst or enzymes. Ester hydrolysis is the reverse of esterification (Section 16.11), the formation of an ester from a carboxylic acid and an alcohol.

Esterification

| Carboxylic acid + Alcohol | | Ester |

Ester hydrolysis

FIGURE 16.14 Ester–water hydrogen bonding.

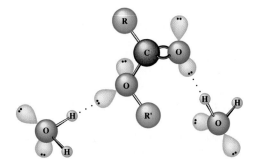

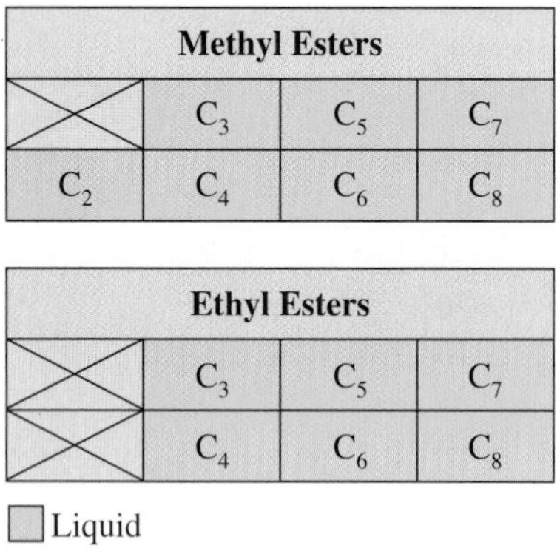

Methyl Esters			
✕	C_3	C_5	C_7
C_2	C_4	C_6	C_8

Ethyl Esters			
✕	C_3	C_5	C_7
✕	C_4	C_6	C_8

☐ Liquid

FIGURE 16.15 A physical-state summary for methyl and ethyl esters of unbranched-chain carboxylic acids at room temperature and pressure.

▶ In both ester hydrolysis and ester saponification, an alcohol is produced. Under acidic conditions (ester hydrolysis), the other product is a carboxylic acid. Under basic conditions (ester saponification), the other product is a carboxylic acid salt.

■ Ester Saponification

A **saponification reaction** *is the hydrolysis of an organic compound, under basic conditions, in which a carboxylic acid salt is one of the products.* Esters, amides (Section 17.16), and fats and oils (Section 19.6) all undergo saponification reactions.

In ester saponification either NaOH and KOH is used as the base and the saponification products are an alcohol and a carboxylic acid salt. (Any carboxylic acid product formed is converted to its salt because of the basic reaction conditions.)

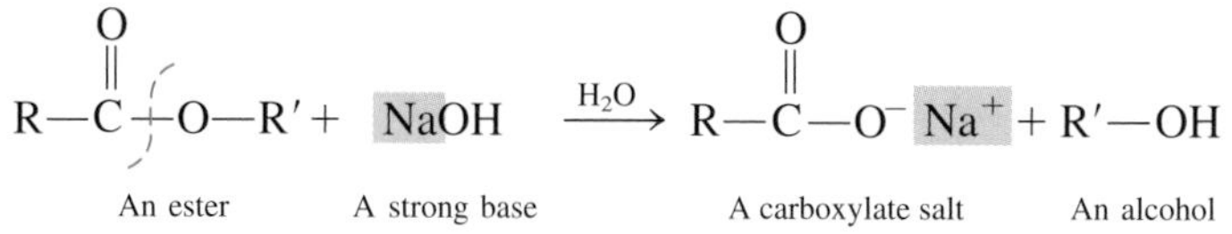

$$
\underset{\text{An ester}}{R-\overset{\overset{\displaystyle O}{\|}}{C}-O-R'} + \underset{\text{A strong base}}{NaOH} \xrightarrow{H_2O} \underset{\text{A carboxylate salt}}{R-\overset{\overset{\displaystyle O}{\|}}{C}-O^-\;Na^+} + \underset{\text{An alcohol}}{R'-OH}
$$

A specific example of ester saponification is

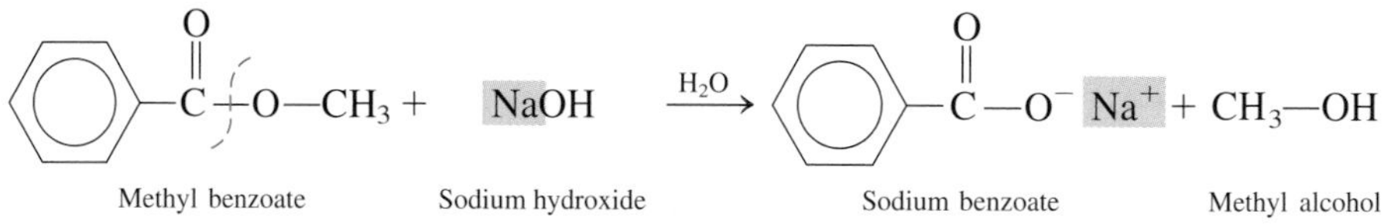

$$
\underset{\text{Methyl benzoate}}{\text{Ph}-\overset{\overset{\displaystyle O}{\|}}{C}-O-CH_3} + \underset{\text{Sodium hydroxide}}{NaOH} \xrightarrow{H_2O} \underset{\text{Sodium benzoate}}{\text{Ph}-\overset{\overset{\displaystyle O}{\|}}{C}-O^-\;Na^+} + \underset{\text{Methyl alcohol}}{CH_3-OH}
$$

EXAMPLE 16.5

Structural Equations for Reactions That Involve Esters

■ Write structural equations for each of the following reactions.

a. Hydrolysis, with an acidic catalyst, of ethyl acetate
b. Saponification, with NaOH, of methyl formate
c. Esterification of propionic acid using isopropyl alcohol

Solution

a. Hydrolysis, under acidic conditions, cleaves an ester to produce its "parent" carboxylic acid and alcohol.

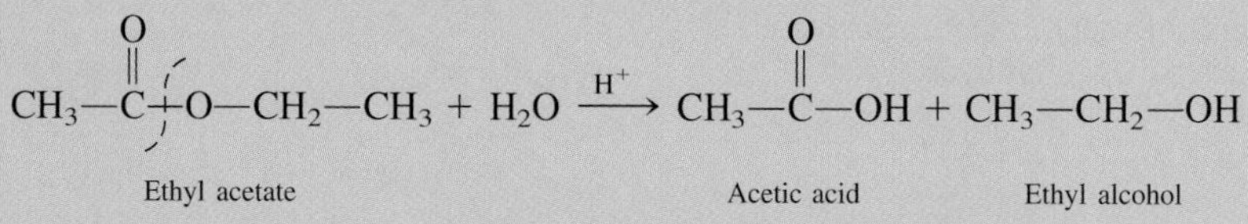

$$
\underset{\text{Ethyl acetate}}{CH_3-\overset{\overset{\displaystyle O}{\|}}{C}-O-CH_2-CH_3} + H_2O \xrightarrow{H^+} \underset{\text{Acetic acid}}{CH_3-\overset{\overset{\displaystyle O}{\|}}{C}-OH} + \underset{\text{Ethyl alcohol}}{CH_3-CH_2-OH}
$$

b. Saponification cleaves an ester to produce its "parent" alcohol and the *salt* of its "parent" carboxylic acid.

$$
\underset{\text{Methyl formate}}{H-\overset{\overset{\displaystyle O}{\|}}{C}-O-CH_3} + \underset{\text{Sodium hydroxide}}{NaOH} \xrightarrow{H_2O} \underset{\text{Sodium formate}}{H-\overset{\overset{\displaystyle O}{\|}}{C}-O^-\;Na^+} + \underset{\text{Methyl alcohol}}{CH_3-OH}
$$

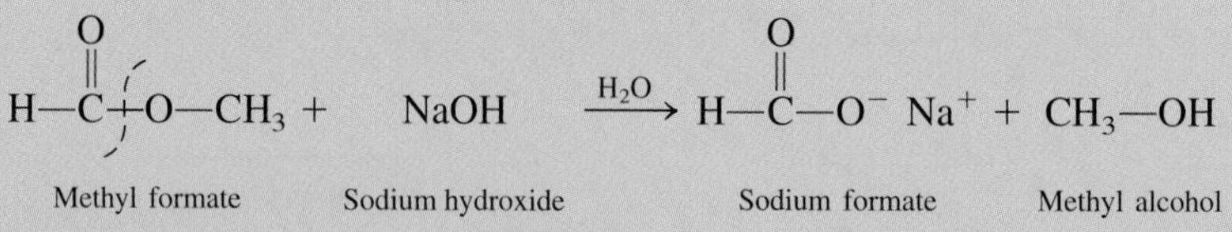

c. Esterification is the reaction in which a carboxylic acid and an alcohol react to produce an ester.

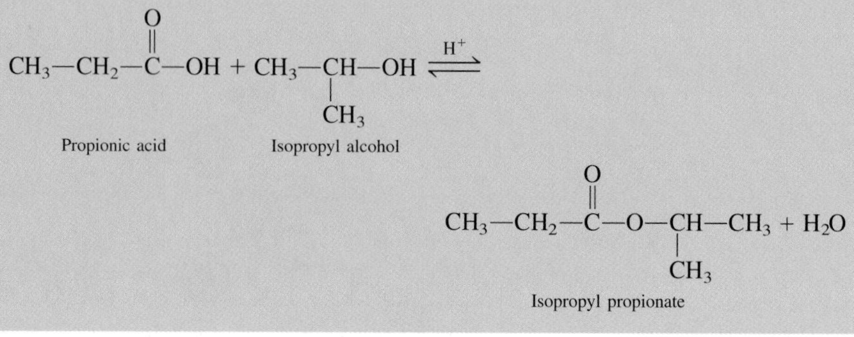

$$
\underset{\text{Propionic acid}}{CH_3-CH_2-\overset{\overset{\displaystyle O}{\|}}{C}-OH} + \underset{\text{Isopropyl alcohol}}{CH_3-\underset{\underset{\displaystyle CH_3}{|}}{CH}-OH} \rightleftharpoons
$$

$$
\underset{\text{Isopropyl propionate}}{CH_3-CH_2-\overset{\overset{\displaystyle O}{\|}}{C}-O-\underset{\underset{\displaystyle CH_3}{|}}{CH}-CH_3} + H_2O
$$

Practice Exercise 16.5

Write structural equations for each of the following reactions.

a. Hydrolysis, with an acidic catalyst, of propyl propanoate
b. Saponification, with KOH, of ethyl propanoate
c. Esterification of acetic acid with propyl alcohol

The Chemistry at a Glance feature on page 450 summarizes reactions that involve carboxylic acids and esters.

16.16 Sulfur Analogs of Esters

Just as alcohols react with carboxylic acids to produce esters, thiols (Section 14.13) react with carboxylic acids to produce thioesters. A **thioester** *is a sulfur-containing analog of an ester in which an —SR group has replaced the —OR group.*

$$CH_3-\overset{\overset{\text{O}}{\|}}{C}-OH + CH_3-CH_2-S-H \longrightarrow CH_3-\overset{\overset{\text{O}}{\|}}{C}-S-CH_2-CH_3 + H_2O$$

A carboxylic acid   A thiol   A thioester

The thioester methyl thiobutanoate is used as an artificial flavoring agent. It generates the taste we call strawberry.

$$CH_3-CH_2-CH_2-\overset{\overset{\text{O}}{\|}}{C}-S-CH_3$$

Methyl thiobutanoate

The most important naturally occurring thioester is acetyl coenzyme A, whose abbreviated structure is

$$CH_3-\overset{\overset{\text{O}}{\|}}{C}-S-CoA$$

Acetyl coenzyme A

Coenzyme A, the parent molecule for acetyl coenzyme A, is a large complex *thiol* whose structure, for simplicity, is usually abbreviated as CoA—S—H. The formation of acetyl coenzyme A (acetyl CoA) from coenzyme A can be envisioned as a thioesterification reaction between acetic acid and coenzyme A.

$$CH_3-\overset{\overset{\text{O}}{\|}}{C}-OH + CoA-S-H \longrightarrow CH_3-\overset{\overset{\text{O}}{\|}}{C}-S-CoA + H_2O$$

Acetic acid   Coenzyme A   Acetyl coenzyme A
                (a thiol)   (acetyl CoA)

Acetyl coenzyme A plays a central role in the metabolic cycles through which the body obtains energy to "run itself" (Section 23.6).

The complete structure of acetyl CoA is given in Section 23.3.

16.17 Polyesters

A **condensation polymer** *is a polymer formed by reacting difunctional monomers to give a polymer and some small molecule (such as water) as a by-product of the process. Polyesters* are an important type of condensation polymer. A **polyester** *is a*

16.18 Esters of Inorganic Acids

Inorganic acids such as sulfuric, phosphoric, and nitric acids react with alcohols to form esters in a manner similar to that for carboxylic acids.

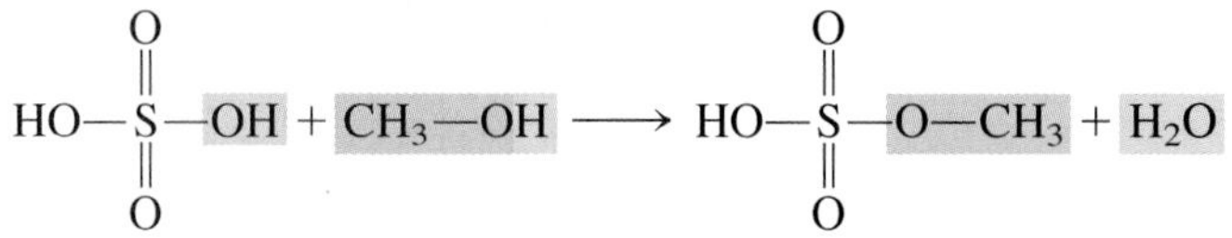

Sulfuric acid
(H_2SO_4)

Methyl ester of
sulfuric acid

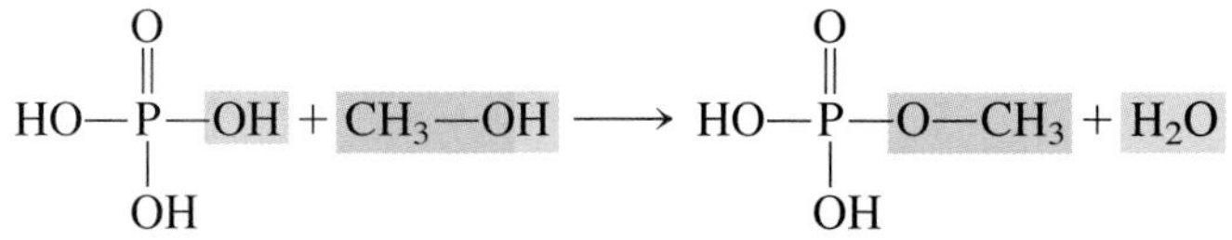

Phosphoric acid
(H_3PO_4)

Methyl ester of
phosphoric acid

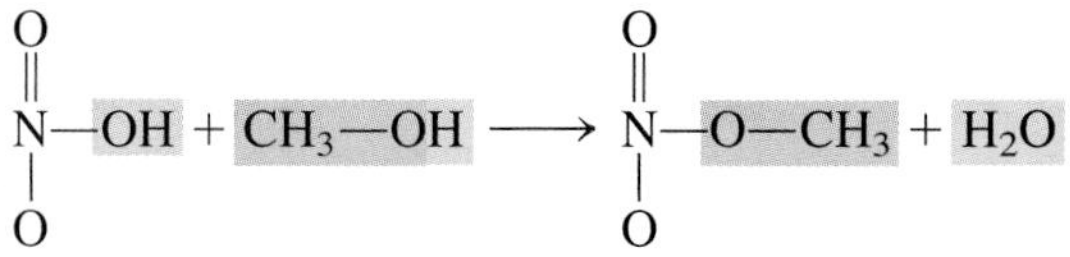

Nitric acid
(HNO_3)

Methyl ester of
nitric acid

▶ Esters of inorganic acids undergo hydrolysis reactions in a manner similar to that for esters of carboxylic acids (Section 16.15).

The most important inorganic esters, from a biochemical standpoint, are those of phosphoric acid—that is, phosphate esters. A **phosphate ester** *is an organic compound formed by reaction of an alcohol with phosphoric acid.* Because phosphoric acid has three hydroxyl groups, it can form mono-, di-, and triesters by reaction with one, two, and three molecules of alcohol, respectively.

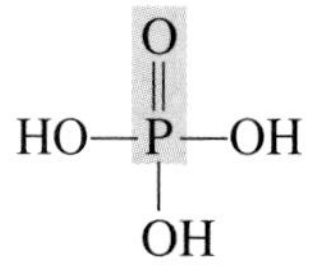

Phosphoric acid

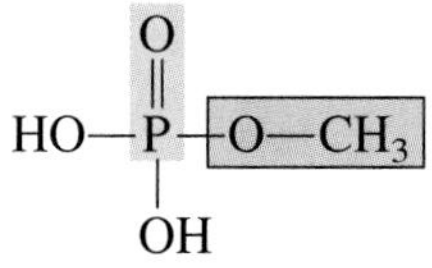

Monoester (one —OR group)

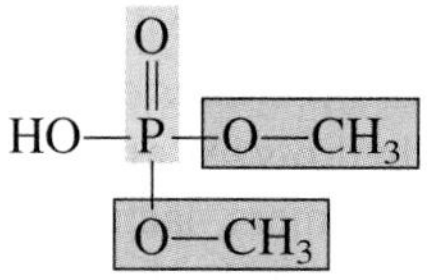

Diester (two —OR groups)

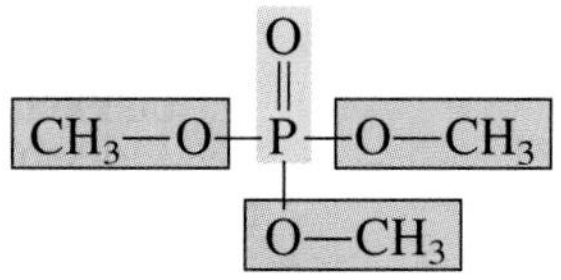

Triester (three —OR groups)

Phosphoric acid molecules have the ability to react with each other to produce diphosphoric acid and triphosphoric acid.

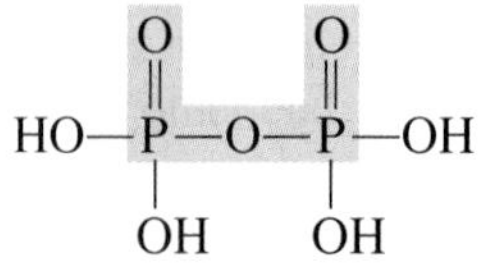

Diphosphoric acid

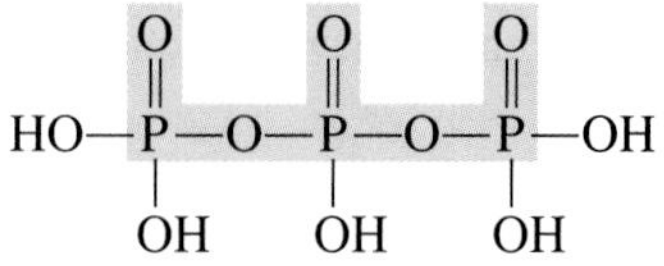

Triphosphoric acid

These acids also undergo esterification reactions with alcohols, producing species such as those shown below.

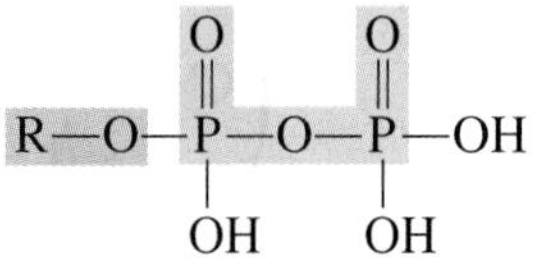

Diphosphate monoester

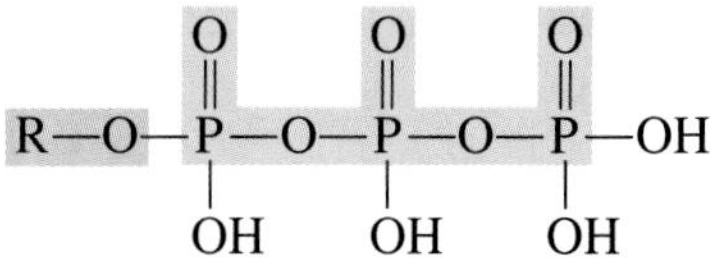

Triphosphate monoester

We will encounter esters of these types in Chapter 23 when we consider the biochemical production of energy in the human body. Adenosine diphosphate (ADP) and adenosine triphosphate (ATP) are important examples of such compounds.

CHEMICAL CONNECTIONS

Nitroglycerin: An Inorganic Triester

The reaction of one molecule of glycerol (a trihydroxy-alcohol) with three molecules of nitric acid produces the trinitrate ester called nitroglycerin.

$$
\begin{array}{l}
CH_2{-}OH \\
| \\
CH{-}OH \\
| \\
CH_2{-}OH
\end{array}
+ 3HO{-}NO_2 \longrightarrow
\begin{array}{l}
CH_2{-}O{-}NO_2 \\
| \\
CH{-}O{-}NO_2 \\
| \\
CH_2{-}O{-}NO_2
\end{array}
+ 3H_2O
$$

Besides being a component of dynamite explosives, nitroglycerin has medicinal value. It is used in treating patients with angina pectoris—sharp chest pains caused by an insufficient supply of oxygen reaching heart muscle. Its effect on the human body is that of a vasodilator, a substance that increases blood flow by relaxing constricted muscles around blood vessels.

Nitroglycerin medication is available in several forms: (1) as a liquid diluted with alcohol to render it nonexplosive, (2) as a liquid adsorbed to a tablet for convenience of sublingual (under the tongue) administration, (3) in ointments for topical use, and (4) as "skin patches" that release the drug continuously through the skin over a 24-hr period. Nitroglycerin is rapidly absorbed through the skin, enters the bloodstream, and finds its way to heart muscle within seconds.

In the pure state, nitroglycerin is a shock-sensitive liquid that can decompose to produce large volumes of gases (N_2, CO_2, H_2O, and O_2). When used in dynamite, it is adsorbed on clay-like materials, giving products that will not explode without a formal ignition system.

Another compound used for the same medicinal purposes as nitroglycerin is isopentyl nitrite. It is a monoester involving nitrous acid (HNO_2) and isopentyl alcohol (3-methyl-1-butanol).

$$
\begin{array}{l}
CH_3{-}CH{-}CH_2{-}CH_2{-}OH + HO{-}NO \longrightarrow \\
\qquad\quad | \\
\qquad\;\; CH_3
\end{array}
$$

$$
\begin{array}{l}
CH_3{-}CH{-}CH_2{-}CH_2{-}O{-}NO + H_2O \\
\qquad\quad | \\
\qquad\;\; CH_3
\end{array}
$$

CONCEPTS TO REMEMBER

The carboxyl group.   The functional group present in carboxylic acids is the carboxyl group. A carboxyl group is composed of a hydroxyl group bonded to a carbonyl carbon atom. It thus contains two oxygen atoms directly bonded to the same carbon atom.

Nomenclature of carboxylic acids.   The IUPAC name for a monocarboxylic acid is formed by replacing the final *-e* of the hydrocarbon parent name with *-oic acid*. As with previous IUPAC nomenclature, the longest carbon chain containing the functional group is identified, and it is numbered starting with the carboxyl carbon atom. Common-name usage is more prevalent for carboxylic acids than for any other type of organic compound.

Types of carboxylic acids.   Carboxylic acids are classified by the number of carboxyl groups present (monocarboxylic, dicarboxylic, etc.), by the degree of saturation (saturated, unsaturated, aromatic), and by additional functional groups present (hydroxy, keto, etc.).

Physical properties of carboxylic acids.   Low-molecular-mass carboxylic acids are liquids at room temperature and have sharp or unpleasant odors. Long-chain acids are waxlike solids. The carboxyl group is polar and forms hydrogen bonds to other carboxyl groups or other molecules. Thus carboxylic acids have relatively high boiling points, and those with lower molecular masses are soluble in water.

Preparation of carboxylic acids.   Carboxylic acids are synthesized through oxidation of primary alcohols or aldehydes using strong oxidizing agents. Aromatic carboxylic acids can be prepared by oxidizing a carbon side chain on a benzene derivative using a strong oxidizing agent.

Acidity of carboxylic acids.   Soluble carboxylic acids behave as weak acids, donating protons to water molecules. The portion of the acid molecule left after proton loss is called a carboxylate ion.

Carboxylic acid salts.   Carboxylic acids are neutralized by bases to produce carboxylic acid salts. Such salts are usually more soluble in water than are the acids from which they were derived. Carboxylic acid salts are named by changing the *-ic* ending of the acid to *-ate*.

Esters.   Esters are formed by the reaction of an acid with an alcohol. In such reactions, the —OR group from the alcohol replaces the —OH group in the carboxylic acid. Esters are polar compounds, but they cannot form hydrogen bonds to each other. Therefore, their boiling points are lower than those of alcohols and acids of similar molecular mass.

Nomenclature of esters.   An ester is named as an alkyl (from the name of the alcohol reactant) carboxylate (from the name of the acid reactant).

Reactions of esters.   Esters can be converted back to carboxylic acids and alcohols under either acidic or basic conditions. Under acidic conditions, the process is called hydrolysis, and the products are the acid and alcohol. Under basic conditions, the process is called saponification, and the products are the acid salt and alcohol.

Thioesters.   Thioesters are sulfur-containing analogs of esters in which a —SR group has replaced the —OR group.

Polyesters.   Polyesters are polymers in which the monomers (diacids and dialcohols) are joined through ester linkages.

Esters of inorganic acids.   Alcohols can react with inorganic acids, such as nitric, sulfuric, and phosphoric acids, to form esters. Phosphate esters are an important class of biochemical compounds.

16.51 For each of the following esters, draw the structural formula of the "parent" acid and the "parent" alcohol.

a.
$$CH_3-CH_2-\overset{\overset{\displaystyle O}{\|}}{C}-O-CH_2-CH_3$$

b.
$$CH_3-CH_2-CH_2-\overset{\overset{\displaystyle O}{\|}}{C}-O-CH_3$$

c.
$$CH_3-O-\overset{\overset{\displaystyle O}{\|}}{C}-CH_2-CH_2-CH_3$$

d.
$$CH_3-\overset{\overset{\displaystyle O}{\|}}{C}-O-\bigcirc$$

e.
$$\bigcirc-\overset{\overset{\displaystyle O}{\|}}{C}-O-CH_3$$

f.
$$CH_3-\overset{\overset{\displaystyle Cl}{|}}{C}H-\overset{\overset{\displaystyle O}{\|}}{C}-O-CH_2-CH_3$$

16.52 For each of the following esters, draw the structural formula of the "parent" acid and the "parent" alcohol.

a.
$$CH_3-\overset{\overset{\displaystyle O}{\|}}{C}-O-CH_2-CH_3$$

b.
$$CH_3-CH_2-\overset{\overset{\displaystyle O}{\|}}{C}-O-CH_3$$

c.
$$CH_3-O-\overset{\overset{\displaystyle O}{\|}}{C}-CH_2-CH_3$$

d.
$$\bigcirc-\overset{\overset{\displaystyle O}{\|}}{C}-O-CH_3$$

e.
$$CH_3-\overset{\overset{\displaystyle CH_3}{|}}{C}H-CH_2-\overset{\overset{\displaystyle O}{\|}}{C}-O-\bigcirc$$

f.
$$CH_3-\overset{}{C}H-CH_2-\overset{\overset{\displaystyle O}{\|}}{C}-O-CH_3$$
$$\underset{\displaystyle \bigcirc}{|}$$

■ **Nomenclature for Esters (Section 16.12)**

16.53 Assign an IUPAC name to each of the following esters.

a.
$$CH_3-CH_2-\overset{\overset{\displaystyle O}{\|}}{C}-O-CH_3$$

b.
$$H-\overset{\overset{\displaystyle O}{\|}}{C}-O-CH_3$$

c.
$$CH_3-\overset{\overset{\displaystyle O}{\|}}{C}-O-CH_3$$

d.
$$CH_3-CH_2-CH_2-O-\overset{\overset{\displaystyle O}{\|}}{C}-CH_3$$

e.
$$CH_3-CH_2-\overset{\overset{\displaystyle O}{\|}}{C}-O-\overset{\overset{\displaystyle CH_3}{|}}{C}H-CH_3$$

f.
$$\bigcirc-\overset{\overset{\displaystyle O}{\|}}{C}-O-CH_2-CH_3$$

16.54 Assign an IUPAC name to each of the following esters.

a.
$$CH_3-\overset{\overset{\displaystyle O}{\|}}{C}-O-CH_2-CH_2-CH_2-CH_3$$

b.
$$CH_3-CH_2-CH_2-\overset{\overset{\displaystyle O}{\|}}{C}-O-CH_3$$

c.
$$CH_3-CH_2-\overset{\overset{\displaystyle O}{\|}}{C}-O-CH_2-CH_2-CH_3$$

d.
$$CH_3-CH_2-O-\overset{\overset{\displaystyle O}{\|}}{C}-H$$

e.
$$CH_3-\overset{\overset{\displaystyle CH_3}{|}}{C}H-\overset{\overset{\displaystyle CH_3}{|}}{C}H-\overset{\overset{\displaystyle O}{\|}}{C}-O-CH_2-CH_3$$

f.
$$CH_3-CH_2-\overset{\overset{\displaystyle O}{\|}}{C}-O-\bigcirc$$

16.55 Assign a common name to each of the esters in Problem 16.53.

16.56 Assign a common name to each of the esters in Problem 16.54.

16.57 Draw a structural formula for each of the following esters.
a. Methyl formate   b. Propyl acetate
c. Octyl decanoate   d. Ethyl phenylacetate
e. Isopropyl acetate
f. 2-Bromopropyl ethanoate

16.58 Draw a structural formula for each of the following esters.
a. Ethyl butyrate   b. Butyl ethanoate
c. 2-Methylpropyl formate
d. Ethyl 2-methylpropanoate
e. Methyl valerate   f. Phenyl benzoate

16.59 Assign IUPAC names to the esters that are produced from the reaction of the following carboxylic acids and alcohols.
a. Acetic acid and ethanol
b. Ethanoic acid and methanol
c. Butyric acid and ethyl alcohol
d. Lactic acid and propyl alcohol
e. 1-Pentanol and pentanoic acid
f. 2-Butanol and caproic acid

16.60 Assign IUPAC names to the esters that are produced from the reaction of the following carboxylic acids and alcohols.
a. Ethanoic acid and propyl alcohol
b. Acetic acid and 1-pentanol
c. Acetic acid and 2-pentanol
d. Methyl alcohol and butyric acid
e. Ethanol and benzoic acid
f. Pyruvic acid and methyl alcohol

■ **Physical Properties of Esters (Section 16.14)**

16.61 Explain why ester molecules cannot form hydrogen bonds to each other.

16.62 How many hydrogen bonds can form between a methyl acetate molecule and two water molecules?

16.63 Explain why esters have lower boiling points than carboxylic acids of comparable molecular mass.

16.64 Explain why esters are less soluble in water than carboxylic acids of comparable molecular mass.

■ **Chemical Reactions of Esters (Section 16.15)**

16.65 Write the structural formulas of the reaction products when each of the following esters is hydrolyzed under acidic conditions.

a.
$$CH_3—CH_2—\overset{\overset{\displaystyle O}{\|}}{C}—O—CH_2—CH_3$$

b.
$$CH_3—\overset{\overset{\displaystyle O}{\|}}{C}—O—CH_2—CH_3$$

c.
$$CH_3—\overset{\overset{\displaystyle CH_3}{|}}{CH}—\overset{\overset{\displaystyle O}{\|}}{C}—O—\bigcirc$$

d. Methyl butanoate
e. Ethyl formate
f. Isopropyl benzoate

16.66 Write the structural formulas of the reaction products when each of the following esters is hydrolyzed under acidic conditions.

a.
$$H—\overset{\overset{\displaystyle O}{\|}}{C}—O—CH_2—CH_2—CH_3$$

b.
$$CH_3—\overset{\overset{\displaystyle CH_3}{|}}{CH}—\overset{\overset{\displaystyle O}{\|}}{C}—O—\overset{\overset{\displaystyle CH_3}{|}}{CH}—CH_3$$

c.
$$\bigcirc\overset{\overset{\displaystyle O}{\|}}{C}—O—\bigcirc$$

d. Ethyl valerate
e. Butyl butyrate
f. Pentyl benzoate

16.67 Write the structural formulas of the reaction products when each of the esters in Problem 16.65 is saponified using sodium hydroxide.

16.68 Write the structural formulas of the reaction products when each of the esters in Problem 16.66 is saponified using sodium hydroxide.

16.69 Draw structures of the reaction products in the following chemical reactions.

a.
$$CH_3—\overset{\overset{\displaystyle CH_3}{|}}{CH}—\overset{\overset{\displaystyle O}{\|}}{C}—O—CH_2—CH_3 + H_2O \xrightarrow{H^+}$$

b.
$$CH_3—\overset{\overset{\displaystyle CH_3}{|}}{CH}—\overset{\overset{\displaystyle O}{\|}}{C}—O—CH_2—CH_3 + NaOH \xrightarrow{H_2O}$$

c.
$$H—\overset{\overset{\displaystyle O}{\|}}{C}—O—CH_2—CH_2—CH_2—CH_3 + H_2O \xrightarrow{H^+}$$

d.
$$CH_3—\overset{\overset{\displaystyle O}{\|}}{C}—O—CH_2—\overset{\overset{\displaystyle CH_3}{|}}{CH}—\overset{\overset{\displaystyle CH_3}{|}}{CH}—CH_3 + NaOH \xrightarrow{H_2O}$$

16.70 Draw structures of the reaction products in the following chemical reactions.

a.
$$CH_3—\overset{\overset{\displaystyle CH_3}{|}}{CH}—CH_2—\overset{\overset{\displaystyle O}{\|}}{C}—O—CH_3 + H_2O \xrightarrow{H^+}$$

b.
$$CH_3—\overset{\overset{\displaystyle CH_3}{|}}{CH}—CH_2—\overset{\overset{\displaystyle O}{\|}}{C}—O—CH_3 + NaOH \xrightarrow{H_2O}$$

c.
$$CH_3—CH_2—\overset{\overset{\displaystyle O}{\|}}{C}—O—(CH_2)_5—CH_3 + H_2O \xrightarrow{H^+}$$

d.
$$CH_3—(CH_2)_5—\overset{\overset{\displaystyle O}{\|}}{C}—O—CH_2—CH_3 + NaOH \xrightarrow{H_2O}$$

■ **Sulfur Analogs of Esters (Section 16.16)**

16.71 Draw the structures of the thioesters formed as a result of each of the following reactions between carboxylic acids and thiols.

a.
$$CH_3—\overset{\overset{\displaystyle O}{\|}}{C}—OH + CH_3—CH_2—SH \rightarrow$$

b.
$$CH_3—(CH_2)_8—\overset{\overset{\displaystyle O}{\|}}{C}—OH + CH_3—SH \rightarrow$$

c.
$$\bigcirc COOH + CH_3—\overset{\overset{\displaystyle CH_3}{|}}{CH}—SH \rightarrow$$

d.
$$H—\overset{\overset{\displaystyle O}{\|}}{C}—OH + CH_3—CH_2—CH_2—SH \rightarrow$$

16.72 Draw the structures of the thioesters formed as a result of each of the following reactions between carboxylic acids and thiols.

a.
$$CH_3—CH_2—\overset{\overset{\displaystyle O}{\|}}{C}—OH + CH_3—CH_2—SH \rightarrow$$

b. $CH_3—CH_2—CH_2—COOH + CH_3—SH \rightarrow$

c.
$$CH_3—\overset{\overset{\displaystyle O}{\|}}{C}—OH + CH_3—CH_2—\overset{\overset{\displaystyle}{\underset{\overset{|}{CH_3}}{CH}}}—SH \rightarrow$$

d.
$$\bigcirc—COOH + \bigcirc—SH \rightarrow$$

■ **Polyesters (Section 16.17)**

16.73 Write the structure (two repeating units) of the polyester polymer formed from oxalic acid and 1,3-propanediol.

16.74 Write the structure (two repeating units) of the polyester polymer formed from malonic acid and ethylene glycol.

16.75 Draw the structural formulas of the monomers needed to form the following polyester.

$$\left(O—(CH_2)_3—O—\overset{\overset{\displaystyle O}{\|}}{C}—(CH_2)_2—\overset{\overset{\displaystyle O}{\|}}{C}—O\right)_n$$

16.76 Draw the structural formulas of the monomers needed to form the following polyester.

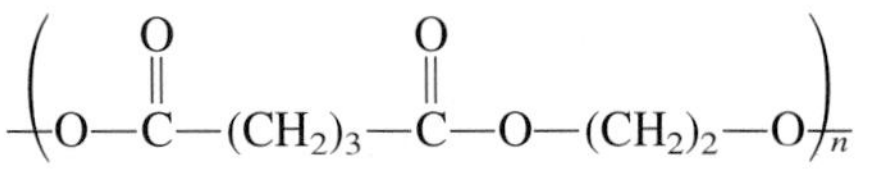

$$\left(\!-O\!-\!\overset{\displaystyle O}{\overset{\|}{C}}\!-\!(CH_2)_3\!-\!\overset{\displaystyle O}{\overset{\|}{C}}\!-\!O\!-\!(CH_2)_2\!-\!O\!-\!\right)_{\!n}$$

■ **Esters of Inorganic Acids (Section 16.18)**

16.77 Draw the structures of the esters formed by reacting the following substances.
   a. 1 molecule methanol and 1 molecule phosphoric acid
   b. 2 molecules methanol and 1 molecule phosphoric acid
   c. 1 molecule methanol and 1 molecule nitric acid
   d. 1 molecule ethylene glycol and 2 molecules nitric acid

16.78 Draw the structures of the esters formed by reacting the following substances.
   a. 1 molecule ethanol and 1 molecule phosphoric acid
   b. 2 molecules methanol and 1 molecule sulfuric acid
   c. 1 molecule ethylene glycol and 1 molecule nitric acid
   d. 1 molecule glycerol and 3 molecules nitric acid

16.79 Phosphoric acid can form triesters but sulfuric acid cannot. Explain why.

16.80 Sulfuric acid can form diesters but nitric acid cannot. Explain why.

ADDITIONAL PROBLEMS

16.81 With the help of Figure 16.6 and IUPAC naming rules, specify the number of carbon atoms present and the number of carboxyl groups present in each of the following carboxylic acids.
   a. Oxalic acid   b. Heptanoic acid
   c. *cis*-3-Heptenoic acid   d. Citric acid
   e. Pyruvic acid   f. Dichloroethanoic acid

16.82 Malonic, maleic, and malic acids are dicarboxylic acids with similar-sounding names. How do the structures of these acids differ from each other?

16.83 The general molecular formula for an alkane is C_nH_{2n+2}. What is the general molecular formula for an unsaturated unsubstituted monocarboxylic acid containing one carbon–carbon double bond?

16.84 Draw structural formulas and give IUPAC names for all possible saturated unsubstituted monocarboxylic acids that contain six carbon atoms. There are eight isomers.

16.85 Monocarboxylic acids and esters with the same number of carbon atoms and the same degree of unsaturation are isomeric. Give IUPAC names for all possible esters that are isomeric with butanoic acid. There are four isomers.

16.86 Assign IUPAC names to the following compounds.
   a. 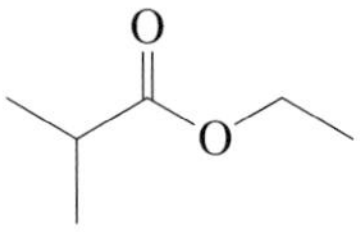
   b. 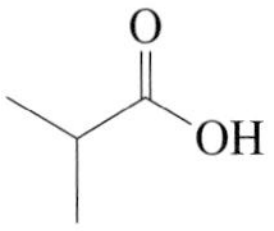
   c. 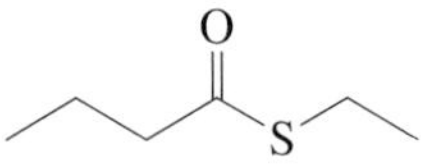
   d.

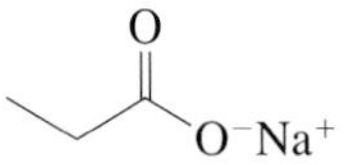

16.87 A sample of ethyl alcohol is divided into two portions. Portion A is added to an aqueous solution of a strong oxidizing agent and allowed to react. The organic product of this reaction is mixed with portion B of the ethyl alcohol. A trace of acid is added and the solution is heated. What is the structure of the final product of this reaction scheme?

16.88 For each of the following reactions, draw the structure(s) of the organic product(s).
   a. 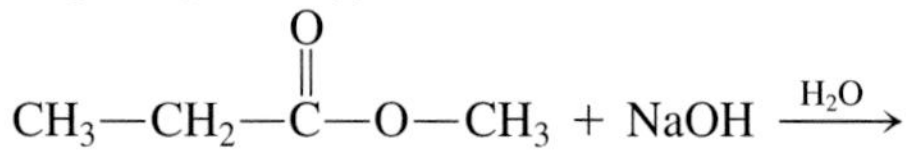
   $$CH_3\!-\!CH_2\!-\!\overset{\displaystyle O}{\overset{\|}{C}}\!-\!O\!-\!CH_3 + NaOH \xrightarrow{H_2O}$$
   b.
   $$CH_3\!-\!CH_2\!-\!\overset{\displaystyle O}{\overset{\|}{C}}\!-\!OH + CH_3\!-\!SH \longrightarrow$$
   c.
   $$CH_3\!-\!\overset{\displaystyle O}{\overset{\|}{C}}\!-\!OH + NaOH \longrightarrow$$
   d. 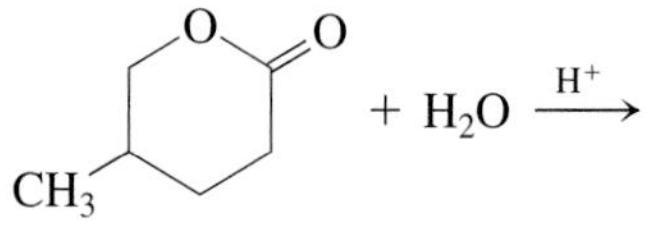
   $$+ H_2O \xrightarrow{H^+}$$

ANSWERS TO PRACTICE EXERCISES

16.1   a. propanoic acid   b. 2,2-dimethylpropanoic acid
       c. 2-ethylpentanoic acid

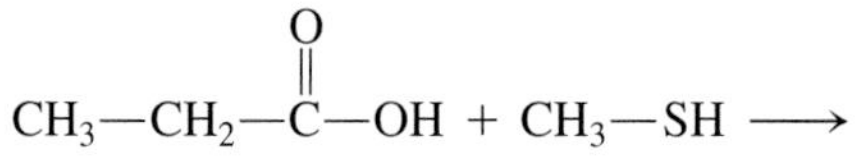

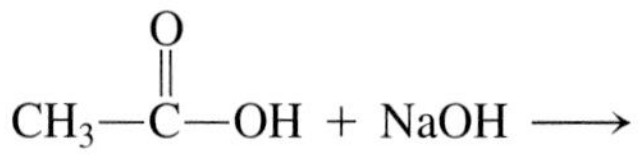

16.2   a. 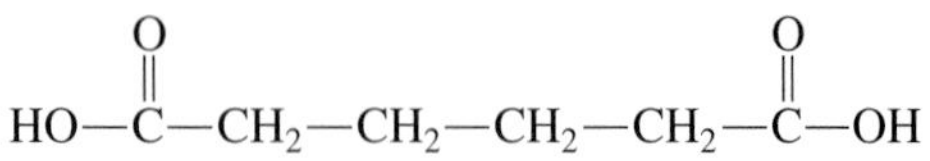
       $$HO\!-\!\overset{\displaystyle O}{\overset{\|}{C}}\!-\!CH_2\!-\!CH_2\!-\!CH_2\!-\!CH_2\!-\!\overset{\displaystyle O}{\overset{\|}{C}}\!-\!OH$$
       b.
       $$CH_3\!-\!CH_2\!-\!\underset{\underset{\displaystyle Cl}{|}}{CH}\!-\!CH_2\!-\!\overset{\displaystyle O}{\overset{\|}{C}}\!-\!OH$$
       c.
       $$HO\!-\!\overset{\displaystyle O}{\overset{\|}{C}}\!-\!CH_2\!-\!\overset{\displaystyle O}{\overset{\|}{C}}\!-\!OH$$
       d. 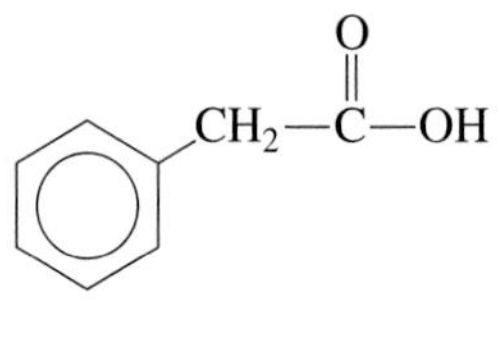
       $$\overset{\displaystyle O}{\overset{\|}{CH_2\!-\!C}}\!-\!OH$$

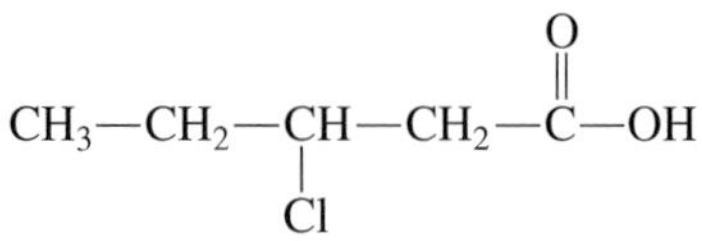

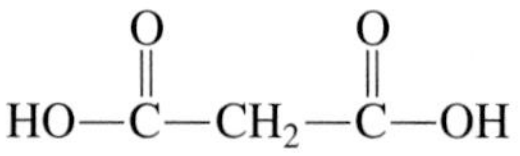

16.3 a.

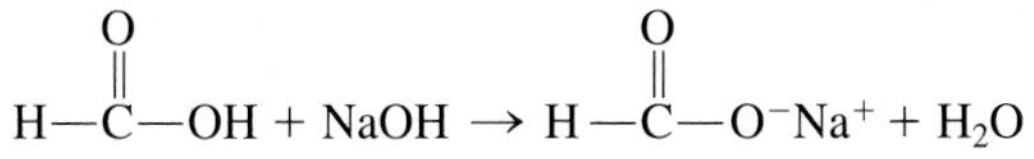

$$\text{H}-\overset{\overset{\displaystyle O}{\|}}{\text{C}}-\text{OH} + \text{NaOH} \rightarrow \text{H}-\overset{\overset{\displaystyle O}{\|}}{\text{C}}-\text{O}^-\text{Na}^+ + \text{H}_2\text{O}$$

b.

$$\text{HO}-\overset{\overset{\displaystyle O}{\|}}{\text{C}}-\text{CH}_2-\overset{\overset{\displaystyle O}{\|}}{\text{C}}-\text{OH} + 2\text{KOH} \rightarrow$$

$$\text{K}^+{}^-\text{O}-\overset{\overset{\displaystyle O}{\|}}{\text{C}}-\text{CH}_2-\overset{\overset{\displaystyle O}{\|}}{\text{C}}-\text{O}^-\text{K}^+ + 2\text{H}_2\text{O}$$

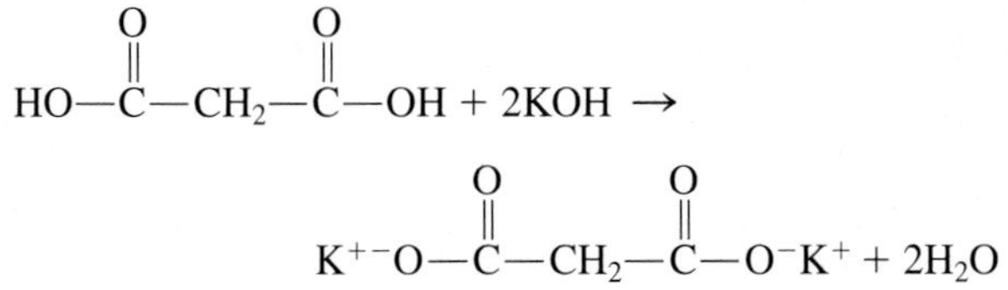

16.4 a. ethyl ethanoate (IUPAC), ethyl acetate (common)
b. methyl pentanoate (IUPAC); methyl valerate (common)
c. propyl methanoate (IUPAC); propyl formate (common)

16.5 a.

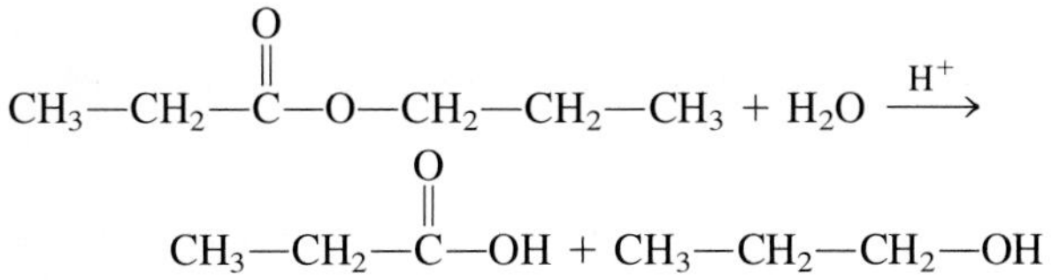

$$\text{CH}_3-\text{CH}_2-\overset{\overset{\displaystyle O}{\|}}{\text{C}}-\text{O}-\text{CH}_2-\text{CH}_2-\text{CH}_3 + \text{H}_2\text{O} \xrightarrow{\text{H}^+}$$

$$\text{CH}_3-\text{CH}_2-\overset{\overset{\displaystyle O}{\|}}{\text{C}}-\text{OH} + \text{CH}_3-\text{CH}_2-\text{CH}_2-\text{OH}$$

b.

$$\text{CH}_3-\text{CH}_2-\overset{\overset{\displaystyle O}{\|}}{\text{C}}-\text{O}-\text{CH}_2-\text{CH}_3 + \text{KOH} \xrightarrow{\text{H}_2\text{O}}$$

$$\text{CH}_3-\text{CH}_2-\overset{\overset{\displaystyle O}{\|}}{\text{C}}-\text{O}^-\text{K}^+ + \text{CH}_3-\text{CH}_2-\text{OH}$$

c.

$$\text{CH}_3-\overset{\overset{\displaystyle O}{\|}}{\text{C}}-\text{OH} + \text{CH}_3-\text{CH}_2-\text{CH}_2-\text{OH} \underset{}{\overset{\text{H}^+}{\rightleftarrows}}$$

$$\text{CH}_3-\overset{\overset{\displaystyle O}{\|}}{\text{C}}-\text{O}-\text{CH}_2-\text{CH}_2-\text{CH}_3 + \text{H}_2\text{O}$$

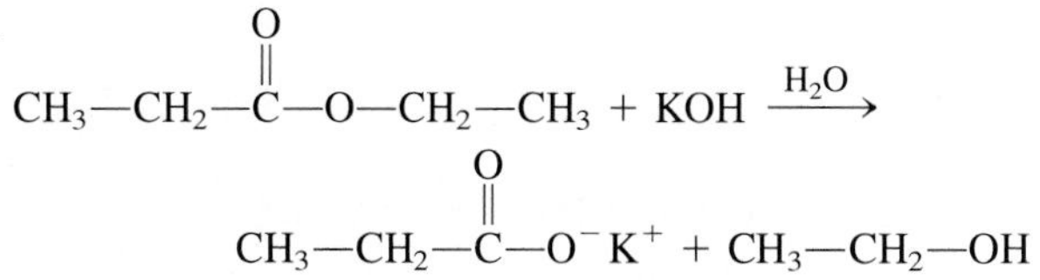

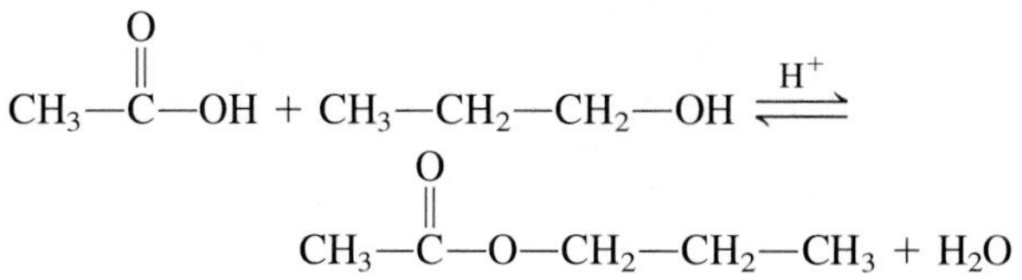

17 Amines and Amides

Parachutist with a parachute made of the polyamide nylon.

The four most abundant elements in living organisms are carbon, hydrogen, oxygen, and nitrogen. In previous chapters, we have discussed compounds containing the first three of these elements. Alkanes, alkenes, alkynes, and aromatic hydrocarbons are all carbon–hydrogen compounds. The carbon–hydrogen–oxygen compounds we have discussed include alcohols, phenols, ethers, aldehydes, ketones, carboxylic acids, and esters. We now extend our discussion to organic compounds that contain the element nitrogen.

Two types of organic nitrogen-containing compounds are the focus of this chapter: amines and amides. Amines are carbon–hydrogen–nitrogen compounds, and amides contain oxygen in addition to these elements. Amines and amides occur widely in nature in living organisms. Many of these naturally occurring compounds are very active physiologically. In addition, numerous drugs used for the treatment of mental illness, hay fever, heart problems, and other physical disorders are amines or amides.

17.1 Bonding Characteristics of Nitrogen Atoms in Organic Compounds

An understanding of the bonding characteristics of the nitrogen atom is a prerequisite to our study of amines and amides. Nitrogen is a member of Group VA of the periodic table; it has five valence electrons (Section 4.2) and will form three covalent bonds to complete its octet of electrons (Section 4.3). Thus, in organic chemistry, carbon forms four bonds (Section 12.2), nitrogen forms three bonds, and oxygen forms two bonds (Section 14.1).

$$-\overset{|}{\underset{|}{C}}- \qquad -\overset{\cdot\cdot}{\underset{|}{N}}- \qquad :\overset{\cdot\cdot}{\underset{}{O}}-$$

4 valence electrons	5 valence electrons	6 valence electrons
4 covalent bonds	3 covalent bonds	2 covalent bonds
no nonbonding	1 nonbonding	2 nonbonding
electron pairs	electron pair	electron pairs

17.2 Structure and Classification of Amines

> Amines bear the same relationship to ammonia that alcohols and ethers bear to water (Sections 14.2 and 14.13).

An **amine** *is an organic derivative of ammonia (NH_3) in which one or more alkyl, cyclo-alkyl, or aryl groups are attached to the nitrogen atom.* Amines are classified as primary (1°), secondary (2°), or tertiary (3°) on the basis of how many hydrocarbon groups are bonded to the ammonia nitrogen atom (see Figure 17.1). A **primary amine** *is an amine in which the nitrogen atom is bonded to one hydrocarbon group and two hydrogen atoms.* The generalized formula for a primary amine is RNH_2. A **secondary amine** *is an amine in which the nitrogen atom is bonded to two hydrocarbon groups and one hydrogen atom.* The generalized formula for a secondary amine is R_2NH. A **tertiary amine** *is an amine in which the nitrogen atom is bonded to three hydrocarbon groups and no hydrogen atoms.* The generalized formula for a tertiary amine is R_3N.

The basis for the amine primary-secondary-tertiary classification system differs from that for alcohols (Section 14.7).

1. For alcohols we look at how many R groups are on a *carbon* atom, the hydroxyl-bearing carbon atom.
2. For amines we look at how many R groups are on the *nitrogen* atom.

Tert-butyl alcohol is a *tertiary* alcohol, whereas *tert*-butylamine is a *primary* amine.

Tertiary carbon atom →
$$CH_3-\overset{\overset{\displaystyle CH_3}{|}}{\underset{\underset{\displaystyle CH_3}{|}}{C}}-OH$$
tert-Butyl alcohol (a tertiary alcohol)

Tertiary carbon atom →   Primary nitrogen atom →
$$CH_3-\overset{\overset{\displaystyle CH_3}{|}}{\underset{\underset{\displaystyle CH_3}{|}}{C}}-NH_2$$
tert-Butylamine (a primary amine)

FIGURE 17.1 Classification of amines is related to the number of R groups attached to the nitrogen atom.

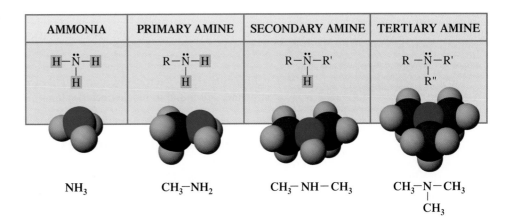

AMMONIA	PRIMARY AMINE	SECONDARY AMINE	TERTIARY AMINE
$H-\overset{\cdot\cdot}{\underset{\underset{H}{\|}}{N}}-H$	$R-\overset{\cdot\cdot}{\underset{\underset{H}{\|}}{N}}-H$	$R-\overset{\cdot\cdot}{\underset{\underset{H}{\|}}{N}}-R'$	$R-\overset{\cdot\cdot}{\underset{\underset{R''}{\|}}{N}}-R'$
NH_3	CH_3-NH_2	$CH_3-NH-CH_3$	$CH_3-\underset{\underset{\displaystyle CH_3}{\|}}{N}-CH_3$

The functional group present in a primary amine, the —NH$_2$ group, is called an *amino* group. An **amino group** *is the —NH$_2$ functional group.* Secondary and tertiary amines possess substituted amino groups.

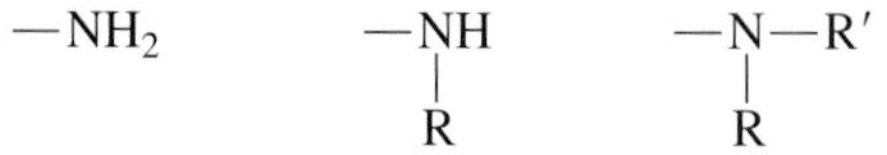

$$-NH_2 \qquad -NH \qquad -N-R'$$
$$\quad\quad\quad\quad | \qquad\quad\quad |$$
$$\quad\quad\quad\quad R \qquad\quad\quad R$$

Amino group   Monosubstituted   Disubstituted
amino group   amino group

EXAMPLE 17.1

Classifying Amines as Primary, Secondary, or Tertiary

■ Classify each of the following amines as a primary, secondary, or tertiary amine.

a. 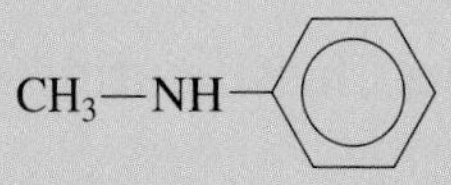
CH$_3$—NH—⬡

b. 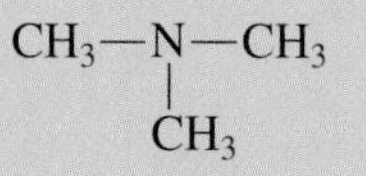 CH$_3$—N—CH$_3$
$\qquad\qquad\quad$ |
$\qquad\qquad\quad$ CH$_3$

c. 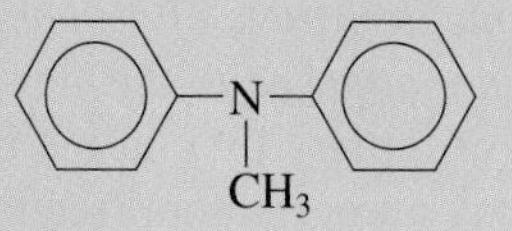
⬡—N—⬡
$\quad$ |
$\quad$ CH$_3$

d. 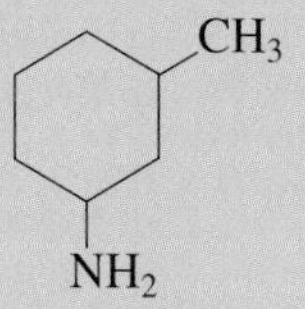
⬡ with CH$_3$ and NH$_2$

Solution

The number of carbon atoms directly bonded to the nitrogen atom determines the amine classification.

a. This is a secondary amine, because the nitrogen is bonded to both a methyl group and a phenyl group.
b. Here we have a tertiary amine, because the nitrogen atom is bonded to three methyl groups.
c. This is also a tertiary amine; the nitrogen atom is bonded to two phenyl groups and a methyl group.
d. This is a primary amine. The nitrogen atom is bonded to only one carbon atom.

Practice Exercise 17.1

Classify each of the following amines as a primary, secondary, or tertiary amine.
a. CH$_3$—CH$_2$—CH$_2$—NH$_2$ **b.** CH$_3$—NH—CH$_2$—CH$_3$
c. 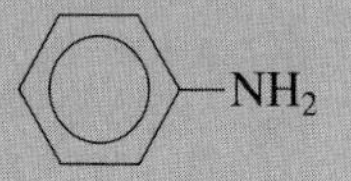
⬡—NH$_2$
d. 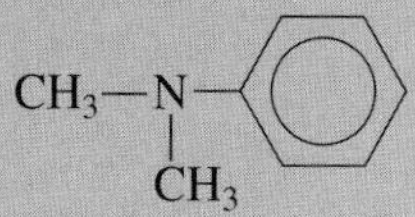
CH$_3$—N—⬡
$\qquad\qquad$ |
$\qquad\qquad$ CH$_3$

▶ Line-angle drawings for selected primary, secondary, and tertiary amines.

1° ⌇⌇⌇ NH$_2$

1° ⌇⌇ with NH$_2$

1° ⌇⌇ with NH$_2$

2° ⌇ NH ⌇

3° ⌇ N ⌇

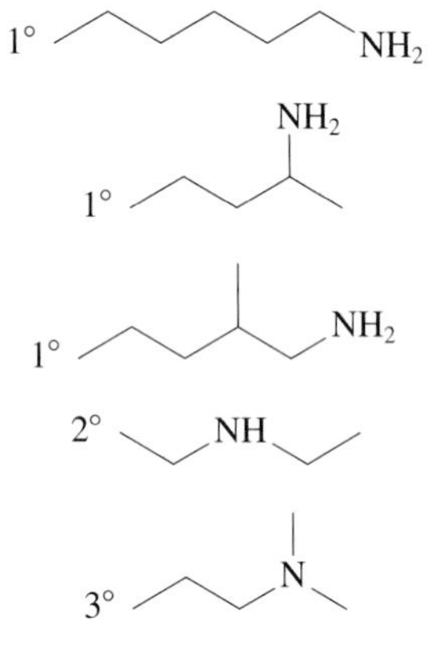

Cyclic amines exist. Such compounds are always either secondary or tertiary amines.

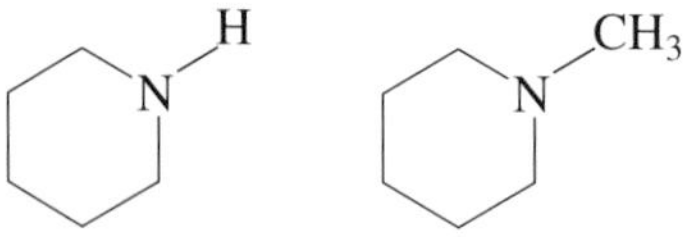

⬡ N—H

⬡ N—CH$_3$

2° Cyclic amine   3° Cyclic amine

Cyclic amines are heterocyclic compounds (Section 14.16). Numerous cyclic amine compounds are found in biochemical systems (Section 17.8).

17.3 Nomenclature for Amines

Both common and IUPAC names are extensively used for amines. In the common system of nomenclature, amines are named by listing the alkyl group or groups attached to the nitrogen atom in alphabetical order and adding the suffix -*amine;* all of this appears as

The common names of amines, like those of aldehydes, are written as a single word, which is different from the common names of alcohols (two words), ethers (two or three words), ketones (two or three words), acids (two words), and esters (two words).

one word. Prefixes such as *di-* and *tri-* are added when identical groups are bonded to the nitrogen atom.

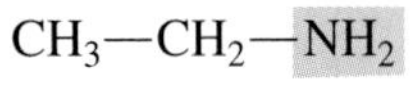

$$CH_3—CH_2—\boxed{NH_2} \qquad CH_3—\boxed{NH}—CH_3$$

Ethylamine    Dimethylamine 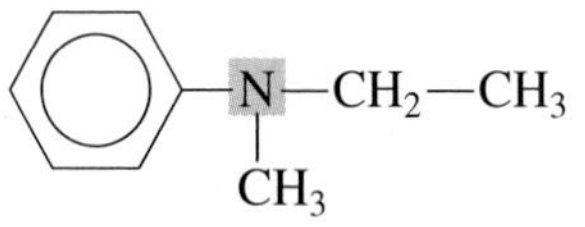   Ethylmethylphenylamine

The IUPAC rules for naming amines are similar to those for alcohols (Section 14.3). Alcohols are named as *alkanols* and amines are named as *alkanamines*. IUPAC rules for naming *primary* amines are as follows:

1. Select as the parent carbon chain the longest chain to which the nitrogen atom is attached.
2. Name the parent chain by changing the *-e* ending of the corresponding alkane name to *-amine*.
3. Number the parent chain from the end nearest the nitrogen atom.
4. The position of attachment of the nitrogen atom is indicated by a number in front of the parent chain name.
5. The identity and location of any substituents are appended to the front of the parent chain name.

IUPAC nomenclature for primary amines is similar to that for alcohols, except that the suffix is *-amine* rather than *-ol*. An —NH$_2$ group, like an —OH group, has priority in numbering the parent carbon chain.

$$CH_3—\underset{\underset{NH_2}{|}}{CH}—CH_2—CH_3 \qquad CH_3—\underset{\underset{CH_3}{|}}{CH}—CH_2—CH_2—\boxed{NH_2}$$

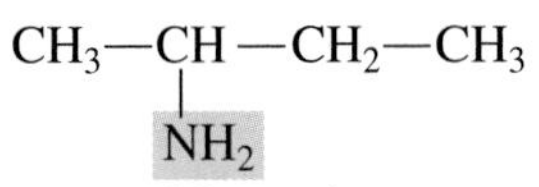

2-Butanamine 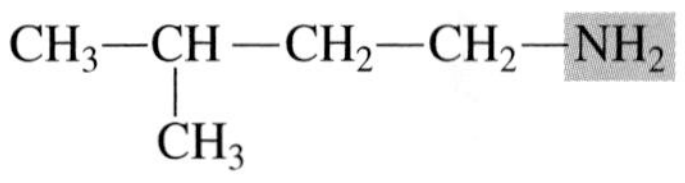   3-Methyl-1-butanamine

In diamines, the final *-e* of the carbon chain name is retained for ease of pronunciation. Thus the base name for a four-carbon chain bearing two amino groups is butan*e*diamine.

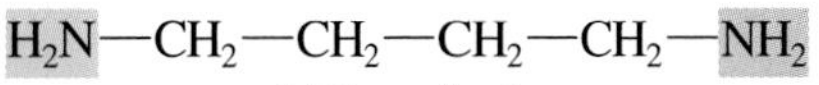

$$\boxed{H_2N}—CH_2—CH_2—CH_2—CH_2—\boxed{NH_2}$$

1,4-Butanediamine

In secondary and tertiary amines, the base name involves the longest carbon chain attached to the nitrogen atom. The names of the other groups attached to the nitrogen are appended to the front of the base name, and *N-* or *N,N-* prefixes are used to indicate that these groups are attached to the nitrogen atom rather than to the base carbon chain.

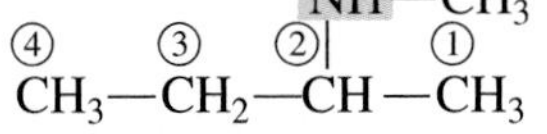

$$\underset{④}{CH_3}—\underset{③}{CH_2}—\underset{②}{\overset{\overset{NH—CH_3}{|}}{CH}}—\underset{①}{CH_3}$$

N-methyl-2-butanamine

$$CH_3—\underset{\underset{\underset{N-methyl-2-butanamine}{}}{\overset{\overset{CH_3}{|}}{N}}}{}—\underset{①}{CH_2}—\underset{②}{CH_2}—\underset{③}{CH_3}$$

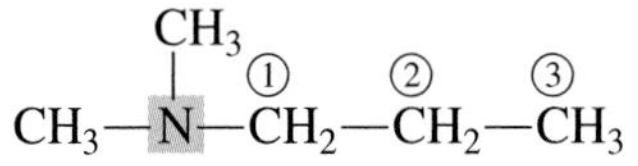

N,N-dimethyl-1-propanamine

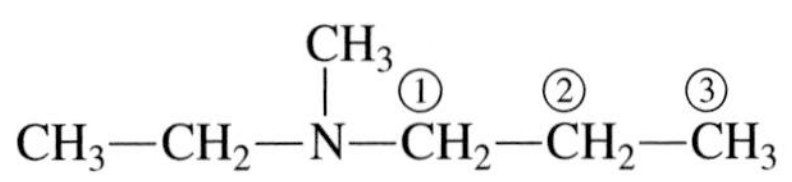

$$CH_3—CH_2—\underset{\overset{|}{CH_3}}{N}—\underset{①}{CH_2}—\underset{②}{CH_2}—\underset{③}{CH_3}$$

N-ethyl-*N*-methyl-1-propanamine

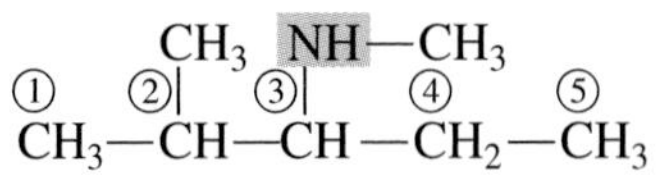

$$\underset{①}{CH_3}—\underset{②}{\overset{\overset{CH_3}{|}}{CH}}—\underset{③}{\overset{\overset{NH—CH_3}{|}}{CH}}—\underset{④}{CH_2}—\underset{⑤}{CH_3}$$

2,*N*-dimethyl-3-pentanamine

In IUPAC nomenclature, the amino group has a priority just below that of an alcohol. The priority list for functional groups is

carboxylic acid ↑
aldehyde
ketone   Increasing
alcohol   priority
amine

In amines where additional functional groups are present, the amine group is treated as a substituent. As a substituent, an —NH$_2$ group is called an *amino* group.

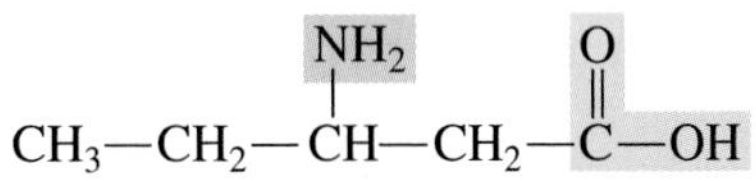

$$CH_3—CH_2—\underset{\overset{|}{NH_2}}{CH}—CH_2—\overset{\overset{O}{||}}{C}—OH$$

3-Aminopentanoic acid

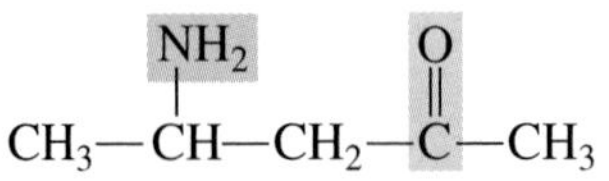

$$CH_3—\underset{\overset{|}{NH_2}}{CH}—CH_2—\overset{\overset{O}{||}}{C}—CH_3$$

4-Amino-2-pentanone

$$CH_3—NH—\underset{③}{CH_2}—\underset{②}{CH_2}—\underset{①}{CH_2}—OH$$

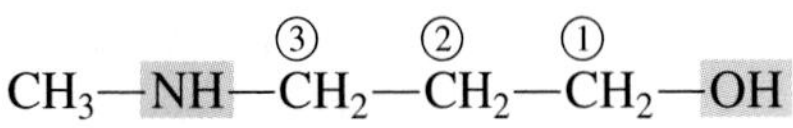

3-(*N*-methylamino)-1-propanol

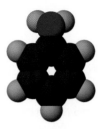

FIGURE 17.2 Space-filling model of aniline, the simplest aromatic amine. Aromatic amines, including aniline, are generally toxic; they are readily absorbed through the skin.

▶ A benzene ring with both an amino group and a methyl group as substituents is called *toluidine*. This name is a combination of the names *toluene* and *aniline*.

The simplest aromatic amine, a benzene ring bearing an amino group, is called *aniline* (Figure 17.2). Other simple aromatic amines are named as derivatives of aniline.

Aniline   *m*-Chloroaniline   2,3-Dichloroaniline

In secondary and tertiary aromatic amines, the additional group or groups attached to the nitrogen atom are located using a capital *N-*.

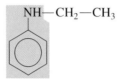

N-ethylaniline   *N,N*-dimethylaniline   3,*N*-dimethylaniline

EXAMPLE 17.2

Determining IUPAC Names for Amines

▶ The contrast between IUPAC names and common names for primary, secondary, and tertiary amines is as follows:

Primary Amines

IUPAC (one word)

alkanamine

Common (one word)

alkylamine

Secondary Amines

IUPAC (one word)

N-alkylalkanamine

Common (one word)

alkylalkylamine

Tertiary Amines

IUPAC (one word)

N-alkyl-*N*-alkylalkanamine

Common (one word)

alkylalkylalkylamine

■ Assign IUPAC names to each of the following amines.

a. $CH_3-CH_2-NH-(CH_2)_4-CH_3$

b. Br—⬡—NH_2

c. $H_2N-CH_2-CH_2-NH_2$

d. $CH_3-N-CH_2-CH_3$ with CH_3 below N

Solution

a. The longest carbon chain has five carbons. The name of the compound is *N-ethyl-1-pentanamine*.

b. This compound is named as a derivative of aniline: *4-bromoaniline* (or *p*-bromoaniline). The carbon in the ring to which the —NH_2 is attached is carbon 1.

c. Two —NH_2 groups are present in this molecule. The name is *1,2-ethanediamine*.

d. This is a tertiary amine in which the longest carbon chain has two carbons (ethane). The base name is thus *ethanamine*. We also have two methyl groups attached to the nitrogen atom. The name of the compound is *N,N-dimethylethanamine*.

Practice Exercise 17.2

Assign IUPAC names to each of the following amines.

a. $CH_3-CH_2-CH-CH_2-CH_2-CH_3$ with NH_2 below the CH

b. $CH_3-CH_2-CH_2-NH-CH_2-CH_2-CH_3$

c. CH_3-N-CH_3 with CH_3 below N

d. ⬡—$NH-CH_3$

17.4 Physical Properties of Amines

The methylamines (mono-, di-, and tri-) and ethylamine are gases at room temperature and have ammonia-like odors. Most other amines are liquids (see Figure 17.3), and many have odors resembling that of raw fish. A few amines, particularly diamines, have strong, disagreeable odors. The foul odor arising from dead fish and decaying flesh is due to amines released by the bacterial decomposition of protein. Two of these "odoriferous" compounds are the diamines putrescine and cadaverine.

Unbranched Primary Amines			
C_1	C_3	C_5	C_7
C_2	C_4	C_6	C_8

☐ Gas ☐ Liquid

FIGURE 17.3 A physical-state summary for unbranched primary amides at room temperature and room pressure.

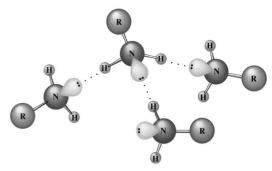

FIGURE 17.4 Amine–amine hydrogen bonding.

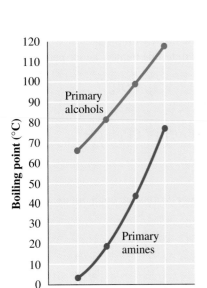

FIGURE 17.5 A comparison of boiling points of unbranched primary amines and unbranched primary alcohols.

$$H_2N\!\!-\!\!(CH_2)_4\!\!-\!\!NH_2 \qquad H_2N\!\!-\!\!(CH_2)_5\!\!-\!\!NH_2$$

Putrescine
(1,4-butanediamine)

Cadaverine
(1,5-pentanediamine)

The simpler amines are irritating to the skin, eyes, and mucous membranes and are toxic by ingestion. Aromatic amines are generally toxic (see Figure 17.3). Many are readily absorbed through the skin and affect both the blood and the nervous system.

The boiling points of amines are intermediate between those of alkanes and alcohols of similar molecular mass. They are higher than alkane boiling points, because hydrogen bonding is possible between amine molecules but not between alkane molecules. Intermolecular hydrogen bonding of amines involves the hydrogen atoms and nitrogen atoms of the amino groups (Figure 17.4).

The boiling points of amines are lower than those of corresponding alcohols (Figure 17.5), because N···H hydrogen bonds are weaker than O···H hydrogen bonds. [The difference in hydrogen-bond strength results from electronegativity differences; nitrogen is less electronegative than oxygen (Section 5.9).]

Amines with fewer than six carbon atoms are infinitely soluble in water. This solubility results from hydrogen bonding between the amines and water. Even tertiary amines are water-soluble, because the amine nitrogen atom has a nonbonding electron pair that can form a hydrogen bond with a hydrogen atom of water (Figure 17.6).

17.5 Basicity of Amines

Amines, like ammonia, are weak bases. In Section 10.4 we learned that ammonia's weak-base behavior results from its accepting a proton (H^+) from water to produce ammonium ion (NH_4^+) and hydroxide ion (OH^-).

$$\ddot{N}H_3 \; + \; HOH \; \rightleftharpoons \; NH_4^+ \; + \; OH^-$$

Ammonia Ammonium ion Hydroxide ion

Amines behave in a similar manner.

$$CH_3\!\!-\!\!\ddot{N}H_2 \; + \; HOH \; \rightleftharpoons \; CH_3\!\!-\!\!\overset{+}{N}H_3 \; + \; OH^-$$

Methylamine Methylammonium ion Hydroxide ion

The result of the interaction of an amine with water is a basic solution containing substituted ammonium ions and hydroxide ions. A **substituted ammonium ion** *is an ammonium ion in which one or more alkyl, cycloalkyl, or aryl groups have been substituted for hydrogen atoms.*

▶ Tertiary amines have lower boiling points than primary and secondary amines, because intermolecular hydrogen bonding is not possible in tertiary amines. Such amines have no hydrogen atoms directly bonded to the nitrogen atom.

▶ Amines, like ammonia, have a pair of unshared electrons on the nitrogen atom present. These unshared electrons can accept a hydrogen ion from water. Thus both amines and ammonia produce basic aqueous solutions.

FIGURE 17.6 Amine–water hydrogen bonding.

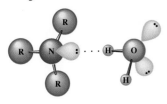

Three important generalizations apply to substituted ammonium ions.

1. *Substituted ammonium ions are charged species rather than neutral molecules.*
2. *The nitrogen atom in an ammonium ion or a substituted ammonium ion participates in four bonds.* In a neutral compound, nitrogen atoms form only three bonds. Four bonds about a nitrogen atom are possible, however, when the species is a positive ion.
3. *Substituted ammonium ions have common names derived from the names of the "parent" amines.* Replacement of the word *amine* in the name of the "parent" amine with the words *ammonium ion* generates the name of the substituted ammonium ion. The following two examples illustrate this nomenclature pattern.

$$CH_3-CH_2-NH_2 \xrightarrow{H_2O} CH_3-CH_2-\overset{+}{N}H_3 + OH^-$$
Ethylamine   Ethylammonium ion

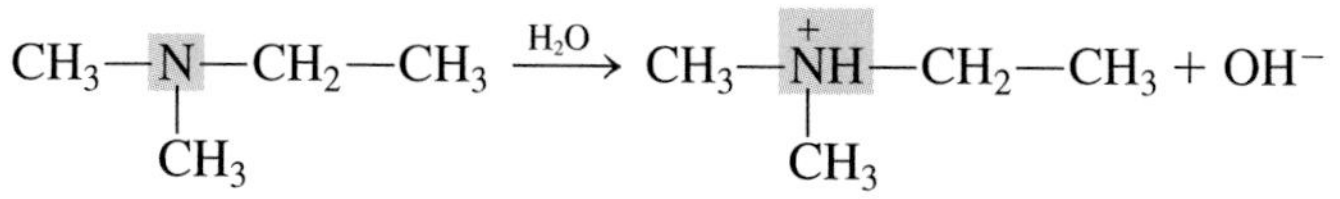

$$CH_3-\underset{\underset{CH_3}{|}}{N}-CH_2-CH_3 \xrightarrow{H_2O} CH_3-\underset{\underset{CH_3}{|}}{\overset{+}{N}H}-CH_2-CH_3 + OH^-$$
Ethyldimethylamine   Ethyldimethylammonium ion

▶ Substituted ammonium ions always contain one more hydrogen atom than their "parent" amine. They also always carry a +1 charge, whereas the "parent" amine is a neutral molecule.

Aromatic amines also exhibit basic behavior in water. With such compounds, the positive ion formed is called a substituted *anilinium ion*.

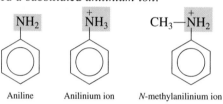

Aniline   Anilinium ion   *N*-methylanilinium ion

EXAMPLE 17.3

Determining Names for Substituted Ammonium and Substituted Anilinium Ions

■ Name the following substituted ammonium or substituted anilinium ions.

a. $CH_3-CH_2-\overset{+}{N}H_2-CH_2-CH_3$

b. 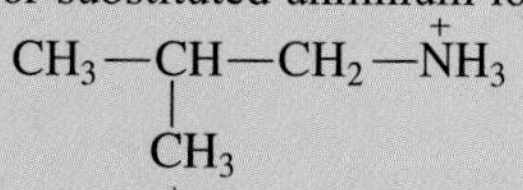 $CH_3-\underset{\underset{CH_3}{|}}{CH}-CH_2-\overset{+}{N}H_3$

c. 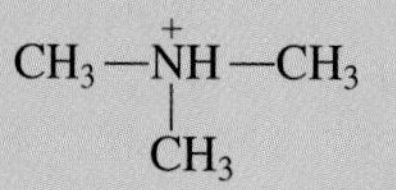 $CH_3-\underset{\underset{CH_3}{|}}{\overset{+}{N}H}-CH_3$

d. 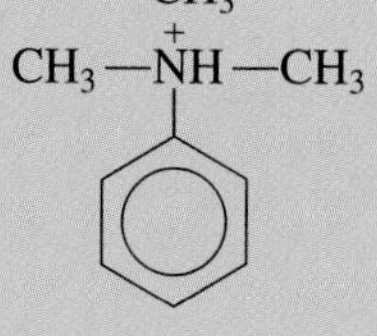 $CH_3-\overset{+}{N}H-CH_3$ (attached to benzene ring)

Solution

a. The parent amine is diethylamine. Replacing the word *amine* in the parent name with *ammonium ion* generates the name of the ion, *diethylammonium ion*.
b. The parent amine is isobutylamine. The name of the ion is *isobutylammonium ion*.
c. The parent amine is trimethylamine. The name of the ion is *trimethylammonium ion*.
d. The parent name is *N,N*-dimethylaniline. Replacing the word *aniline* in the parent name with *anilinium* ion generates the name of the ion, *N,N*-dimethylanilinium ion.

Practice Exercise 17.3

Name the following substituted ammonium or substituted anilinium ions.

a. $CH_3-CH_2-\overset{+}{N}H_2-CH_3$

b. 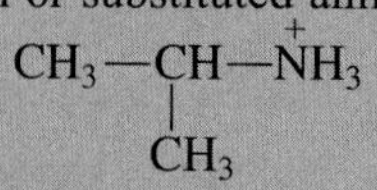 $CH_3-\underset{\underset{CH_3}{|}}{CH}-\overset{+}{N}H_3$

c. 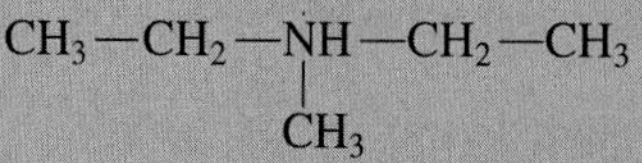 $CH_3-CH_2-\underset{\underset{CH_3}{|}}{\overset{+}{N}H}-CH_2-CH_3$

d. 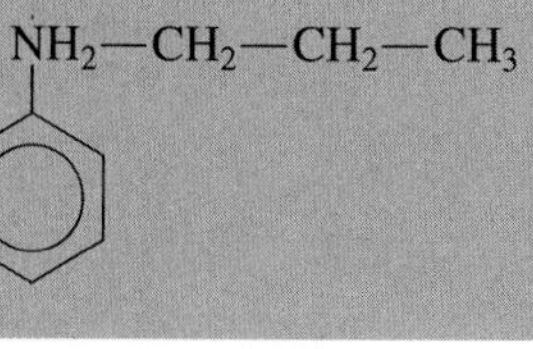 $NH_2-CH_2-CH_2-CH_3$ (attached to benzene ring)

17.6 Amine Salts

▶ The cocaine molecule is both an amine and an ester; one amine and two ester functional groups are present. As an illegal street drug, cocaine is consumed as a water-soluble amine salt and in a water-insoluble, free-base form (nonsalt form—freed of the base required to make the salt). Cocaine hydrochloride, the amine salt, is a white powder that is snorted or injected intravenously. Free-base cocaine is heated and its vapors are inhaled. Cocaine users and dealers call the water-insoluble form of the drug "crack." "Snow" and "coke" are street names for the water-soluble form of the drug.

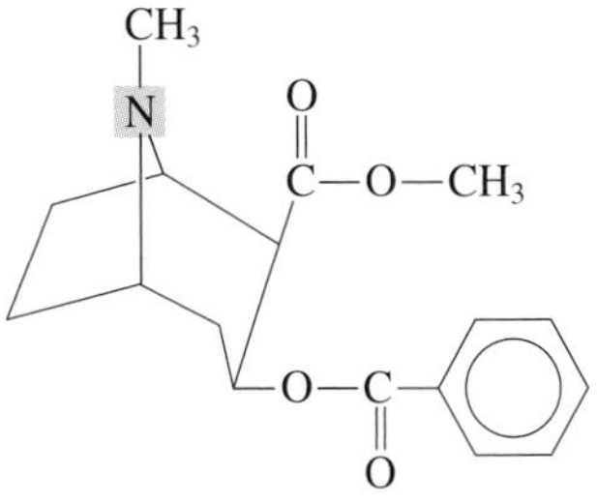

Cocaine
(an amine)

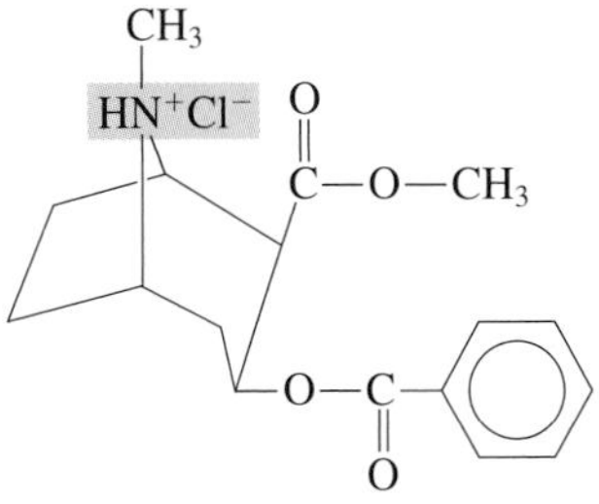

Cocaine hydrochloride
(an amine salt)

The reaction of an acid with a base (neutralization) produces a salt (Section 10.7). Because amines are bases (Section 17.5), their reaction with an acid produces a salt, an amine salt.

$$CH_3-\ddot{N}H_2 + (H)-Cl \longrightarrow CH_3-\overset{+}{N}H_3\,Cl^-$$
Amine   Acid   Amine salt

Aromatic amines react with acids in a similar manner.

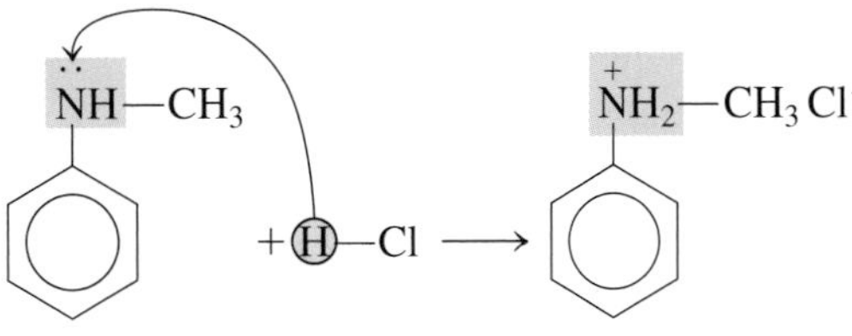

NH—CH_3 + (H)—Cl ⟶ $\overset{+}{N}H_2$—CH_3 Cl⁻
Amine   Acid   Amine salt

An **amine salt** *is an ionic compound in which the positive ion is a mono-, di-, or trisubstituted ammonium ion (RNH_3^+, $R_2NH_2^+$, or R_3NH^+) and the negative ion comes from an acid.* Amine salts can be obtained in crystalline form (odorless, white crystals) by evaporating the water from the acidic solutions in which amine salts are prepared.

Amine salts are named using standard nomenclature procedures for ionic compounds (Section 4.9). The name of the positive ion, the substituted ammonium or anilinium ion, is given first and is followed by a separate word for the name of the negative ion.

$$CH_3-CH_2-\overset{+}{N}H_3\,Cl^- \qquad CH_3-\overset{+}{N}H_2-CH_3\,Br^-$$
Ethylammonium chloride   Dimethylammonium bromide

An older naming system for amine salts, still used in the pharmaceutical industry, treats amine salts as amine–acid complexes rather than as ionic compounds. In this system, the amine salt made from dimethylamine and hydrochloric acid is named and represented as

$$CH_3-\underset{\underset{CH_3}{|}}{N}H \cdot HCl \qquad \text{rather than as} \qquad CH_3-\underset{\underset{CH_3}{|}}{\overset{+}{N}}H_2\,Cl^-$$
Dimethylamine hydrochloride   Dimethylammonium chloride

Many medication labels refer to hydrochlorides or hydrogen sulfates (from sulfuric acid), indicating that the medications are in a water-soluble ionic (salt) form.

Many higher-molecular-mass amines are water-insoluble; however, virtually all amine salts are water-soluble. Thus amine salt formation, like carboxylic acid salt formation (Section 16.9), provides a means for converting water-insoluble compounds into water-soluble compounds. Many drugs that contain amine functional groups are administered to patients in the form of amine salts because of their increased solubility in water in this form.

Many people unknowingly use acids to form amine salts when they put vinegar or lemon juice on fish. Such action converts amines in fish (often smelly compounds) to salts, which are odorless.

The process of forming amine salts with acids is an easily reversed process. Treating an amine salt with a strong base such as NaOH regenerates the "parent" amine.

$$CH_3-\overset{+}{N}H_3\,Cl^- + NaOH \longrightarrow CH_3-NH_2 + NaCl + H_2O$$
Amine salt   Base   Amine

The "opposite nature" of the processes of amine salt formation from an amine and the regeneration of the amine from its amine salt can be diagrammed as follows:

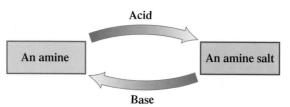

■ Write the structures of the products that form when each of the following reactions involving amines or amine salts takes place.

a. $CH_3-NH-CH_3 + HCl \longrightarrow$

b.

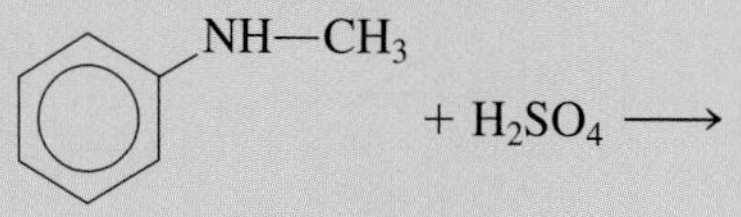

$$\text{(ring)}-NH-CH_3 + H_2SO_4 \longrightarrow$$

c. $CH_3-\overset{+}{N}H_2-CH_3\ Cl^- + NaOH \longrightarrow$

Solution

a. The reactants are an amine and a strong acid. Their interaction produces an amine salt.

$$CH_3-\overset{\cdot\cdot}{N}H-CH_3 + \textcircled{H}Cl \longrightarrow CH_3-\overset{+}{N}H_2-CH_3\ Cl^-$$

b. Again, we have the reaction of an amine with a strong acid. A hydrogen ion is transferred from the acid to the amine.

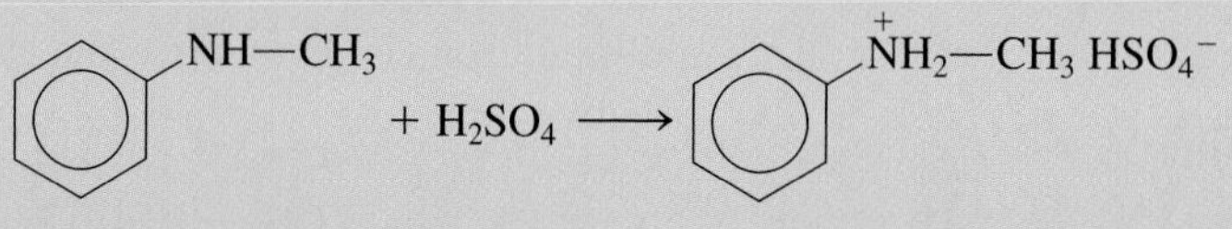

$$\text{(ring)}-NH-CH_3 + H_2SO_4 \longrightarrow \text{(ring)}-\overset{+}{N}H_2-CH_3\ HSO_4^-$$

c. The reactants are an amine salt and a strong base. Their interaction regenerates the "parent" amine.

$$CH_3-\overset{+}{N}H_2-CH_3\ Cl^- + NaOH \longrightarrow CH_3-NH-CH_3 + NaCl + H_2O$$

Practice Exercise 17.4

Write the structures of the products formed in the following reactions.

a. $CH_3-CH_2-\underset{\underset{CH_3}{|}}{N}-CH_3 + HCl \longrightarrow$

b. $CH_3-CH_2-NH_2 + H_2SO_4 \longrightarrow$

c. $CH_3-CH_2-\underset{\underset{CH_3}{|}}{\overset{+}{N}H}-CH_3\ Br^- + NaOH \longrightarrow$

17.7 Preparation of Amines and Quaternary Ammonium Salts

Several methods exist for preparing amines. We consider only one: alkylation in the presence of base. Generalized equations for the alkylation process are

$$\text{Ammonia + alkyl halide} \xrightarrow{\text{Base}} 1° \text{ amine}$$

$$1° \text{ Amine + alkyl halide} \xrightarrow{\text{Base}} 2° \text{ amine}$$

$$2° \text{ Amine + alkyl halide} \xrightarrow{\text{Base}} 3° \text{ amine}$$

$$3° \text{ Amine + alkyl halide} \xrightarrow{\text{Base}} \text{quaternary ammonium salt}$$

Alkylation under basic conditions is actually a two-step process. In the first step, using 1° amine preparation as an example, an amine salt is produced.

$$NH_3 + R{-}X \longrightarrow R{-}\overset{+}{N}H_3 \ X^-$$

The second step, which involves the base (NaOH) present, converts the amine salt to free amine.

$$R{-}\overset{+}{N}H_3 \ X^- + NaOH \longrightarrow RNH_2 + NaX + H_2O$$

A specific example of the production of a primary amine from ammonia is the reaction of ethyl bromide with ammonia to produce ethylamine. The chemical equation (with both steps combined) is

$$NH_3 + CH_3{-}CH_2{-}Br + NaOH \longrightarrow CH_3{-}CH_2{-}NH_2 + NaBr + H_2O$$

If the newly formed primary amine produced in an ammonia alkylation reaction is not quickly removed from the reaction mixture, then the nitrogen atom of the amine may react with further alkyl halide molecules, giving, in succession, secondary and tertiary amines.

$$NH_3 \xrightarrow[OH^-]{RX} RNH_2 \xrightarrow[OH^-]{RX} R_2NH \xrightarrow[OH^-]{RX} R_3N$$

Primary   Secondary   Tertiary
amine   amine   amine

Examples of the production of a 2° amine and a 3° amine via alkylation are

$$CH_3{-}CH_2{-}NH_2 + CH_3{-}Br + NaOH \longrightarrow CH_3{-}CH_2{-}NH{-}CH_3 + NaBr + H_2O$$

Primary amine   Alkyl halide   Base   Secondary amine

$$CH_3{-}CH_2{-}NH{-}CH_3 + CH_3{-}Br + NaOH \longrightarrow CH_3{-}CH_2{-}\underset{\underset{CH_3}{|}}{N}{-}CH_3 + NaBr + H_2O$$

Secondary amine   Alkyl halide   Base   Tertiary amine

Tertiary amines react with alkyl halides in the presence of a strong base to produce a quaternary ammonium salt. A **quaternary ammonium salt** *is an ammonium salt in which all four groups attached to the nitrogen atom of the ammonium ion are hydrocarbon groups.*

$$R{-}\underset{\underset{R}{|}}{N}{-}R + R{-}X \xrightarrow{OH^-} R{-}\overset{\overset{R}{|}}{\underset{\underset{R}{|}}{N^+}}{-}R \ X^-$$

Quaternary ammonium salts differ from amine salts in that addition of strong base does not convert quaternary ammonium salts back to their "parent" amines; there is no hydrogen atom on the nitrogen with which the OH⁻ can react. Quaternary ammonium salts are colorless, odorless, crystalline solids that have high melting points and are usually water-soluble.

Quaternary ammonium salts are named in the same way as amine salts (Section 17.6), taking into account that four organic groups are attached to the nitrogen atom rather than a lesser number of groups. Figure 17.7 contrasts the structures of an ammonium ion and a tetramethyl ammonium ion.

FIGURE 17.7 Space-filling models showing that the ammonium ion (NH_4^+) has a tetrahedral structure, as does the quaternary ammonium ion, in which four methyl groups are present [$(CH_3)_4N^+$].

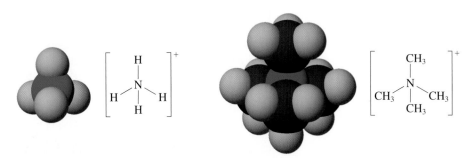

Compounds that contain quaternary ammonium ions are important in biochemical systems. Choline and acetylcholine are two important quaternary ammonium *ions* present in the human body. Choline has important roles in both fat transport and growth regulation. Acetylcholine is involved in the transmission of nerve impulses.

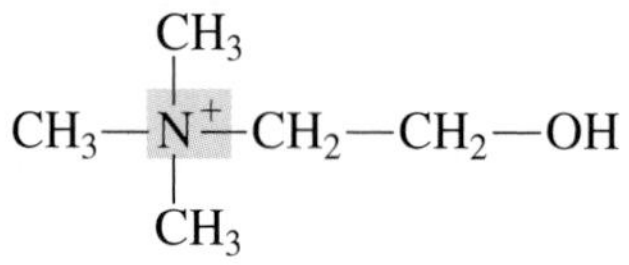

Choline 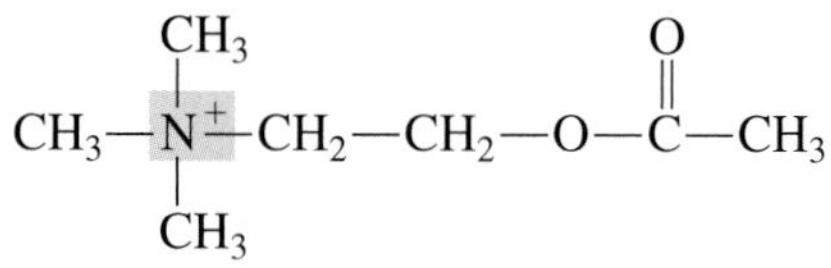   Acetylcholine

17.8 Heterocyclic Amines

▶ Heterocyclic amines are the first heterocyclic compounds we have encountered that have nitrogen heteroatoms. In previous chapters, we have discussed heterocyclic compounds with oxygen as the heteroatom: cyclic ethers (Section 14.16); the cyclic forms of hemiacetals and acetals (Section 15.9); and cyclic esters (Section 16.10).

A **heterocyclic amine** *is an organic compound in which nitrogen atoms of amine groups are part of either an aromatic or a nonaromatic ring system.* Heterocyclic amines are the most common type of heterocyclic organic compound (Section 14.16). Figure 17.4 gives structures for a number of "key" unsubstituted heterocyclic amines. These compounds are the "parent" compounds for numerous derivatives that are important in medicinal, agricultural, food, and industrial chemistry, as well as in the functioning of the human body.

Study of the heterocyclic amine structures in Figure 17.8 shows that (1) ring systems may be saturated, unsaturated, or aromatic, (2) more than one nitrogen atom may be present in a given ring, and (3) fused ring systems often occur.

▶ Heterocyclic amines often have strong odors, some agreeable and others disagreeable. The "pleasant" aroma of many heat-treated foods is caused by heterocyclic amines formed during the heat treatment. The compounds responsible for the pervasive odors of popped popcorn and hot roasted peanuts are heterocyclic amines.

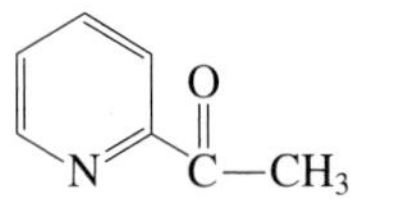

The two most widely used central nervous system stimulants in our society, caffeine and nicotine, are heterocyclic amine derivatives. Caffeine's structure is based on a purine ring system. Nicotine's structure contains one pyridine ring and one pyrrolidine ring.

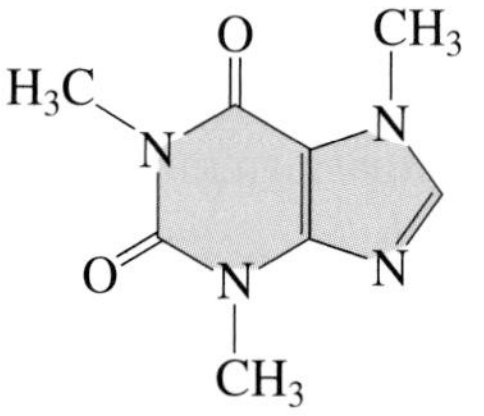

Caffeine 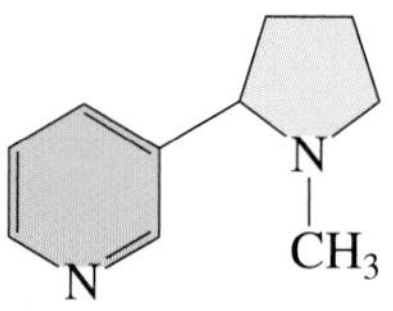   Nicotine

Methyl-2-pyridyl ketone
(odor of popcorn)

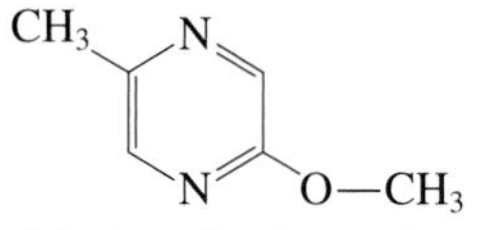

2-Methoxy-5-methylpyrazine
(odor of peanuts)

A large cyclic structure built on four pyrrole rings (Figure 17.8), called a porphyrin, is important in the chemistry of living organisms. Porphyrins form metal ion complexes in which the metal ion is located in the middle of the large ring structure. *Heme,* an iron–porphyrin complex present in the red blood pigment hemoglobin, is responsible for oxygen transport in the human body.

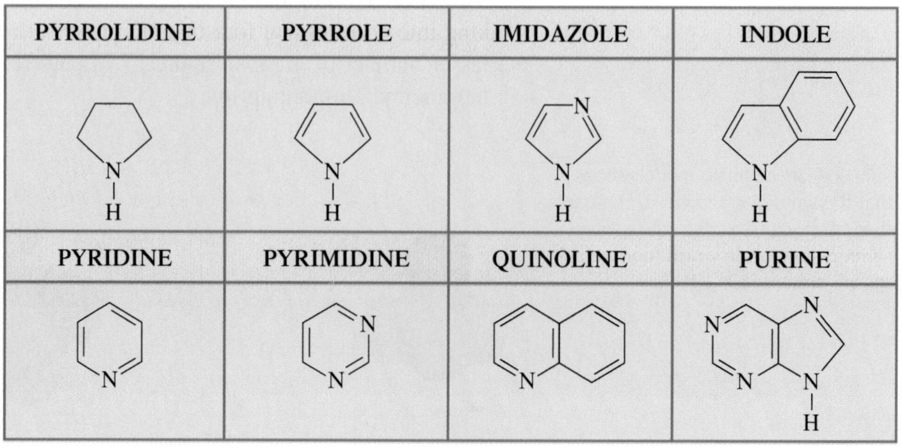

FIGURE 17.8 Structural formulas for selected heterocyclic amines that serve as "parent" molecules for more complex amine derivatives.

PYRROLIDINE	PYRROLE	IMIDAZOLE	INDOLE
PYRIDINE	PYRIMIDINE	QUINOLINE	PURINE

CHEMICAL CONNECTIONS

Caffeine: The Most Widely Used Central Nervous System Stimulant

Caffeine, found in coffee beans and tea leaves, is the most widely used nonprescription central nervous system stimulant today. Structurally, it is a derivative of the heterocyclic amine purine (see Figure 17.8).

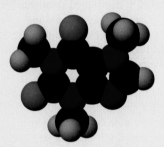

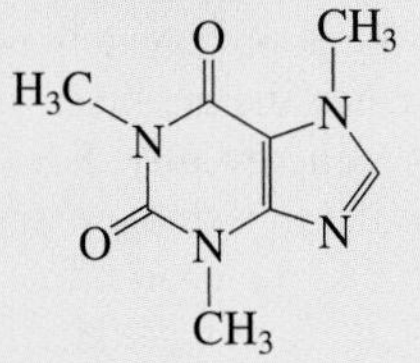

Caffeine

In addition to stimulating the central nervous system, caffeine can produce a variety of other effects. It increases heartbeat and basal metabolic rate, promotes secretion of stomach acid, and steps up production of urine. The overall effect an individual experiences is usually interpreted as a "lift."

Caffeine is mildly addicting. People who ordinarily consume substantial amounts of caffeine-containing beverages or drugs experience withdrawal symptoms if caffeine is eliminated. Such symptoms include headache and depression for a period of several days. As a result of caffeine dependence, many people need a cup of coffee before they feel good each morning.

Current scientific thought holds that caffeine's mode of action in the body is exerted through a chemical substance called cyclic adenosine monophosphate (cyclic AMP). Caffeine inhibits an enzyme that ordinarily breaks down cyclic AMP to its inactive end product. The resulting increase in cyclic AMP leads to increased glucose production within cells and thus makes available more energy to allow higher rates of cellular activity.

Coffee is the major source of caffeine for most Americans. However, substantial amounts of caffeine may be consumed in soft drinks, tea, and numerous nonprescription medications, including combination pain relievers (Anacin, Midol, Empirin), cold remedies (Dristan, Triaminicin), and antisleep agents (No Doz, Vivarin).

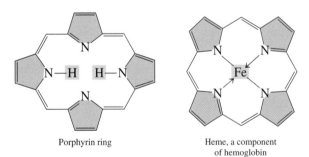

Porphyrin ring

Heme, a component of hemoglobin

17.9 Selected Biochemically Important Amines

■ Neurotransmitters

A **neurotransmitter** *is a chemical substance that is released at the end of a nerve, travels across the synaptic gap between the nerve and another nerve, and then bonds to a receptor site on the other nerve, triggering a nerve impulse.* Figure 17.9 shows schematically how neurotransmitters function.

The most important neurotransmitters in the human body are acetylcholine (Section 17.7) and the amines norepinephrine, dopamine, and serotonin.

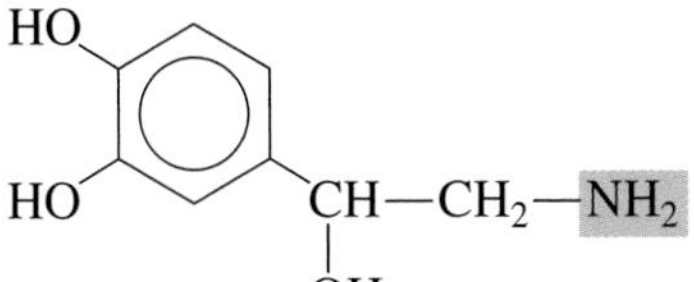

Norepinephrine

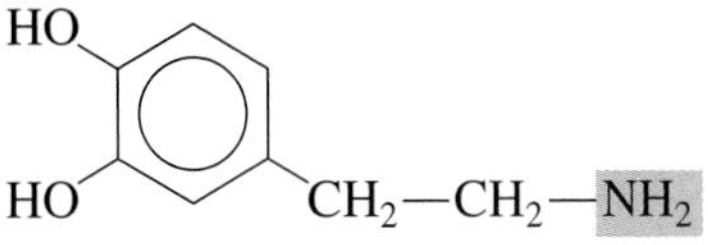

Dopamine

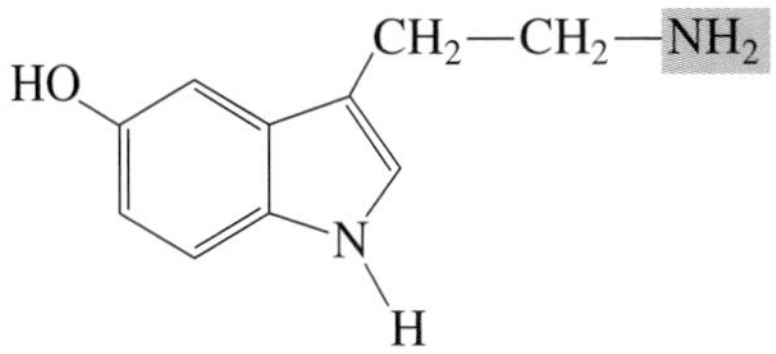

Serotonin

CHEMICAL CONNECTIONS

Nicotine Addiction: A Widespread Example of Drug Dependence

Next to caffeine, nicotine is the most widely used central nervous system stimulant in our society. It is found in smoking tobacco and chewing tobacco. Structurally, nicotine contains both a pyrrolidine and a pyridine ring system (see Figure 17.4); these rings are connected through a carbon–carbon bond rather than fused.

Nicotine

Nicotine is readily and completely absorbed from the stomach after oral administration and from the lungs upon inhalation. Its mild effect on the central nervous system is rather transient. After an initial response, depression follows. Most cigarettes contain between 0.5 mg and 2.0 mg of nicotine, of which approximately 20% (between 0.1 mg and 0.4 mg) will actually be inhaled and absorbed into the bloodstream.

Nicotine is quickly distributed throughout the body, rapidly penetrating the brain, all body organs, and, in general, all body fluids. There is now strong evidence that cigarette smoking affects a developing fetus.

Nicotine induces both physiological and psychological dependence. Withdrawal from nicotine is accompanied by headache, stomach pain, irritability, and insomnia. Full withdrawal can take six months or longer.

In large doses nicotine is a potent poison. Nicotine poisoning causes vomiting, diarrhea, nausea, and abdominal pain. Death occurs from respiratory paralysis. The lethal dose of the drug is considered to be approximately 60 mg.

Nicotine can influence the action of prescription drugs that the smoker is taking concurrently. For example, Darvon can lose its pain-killing effect in a smoker, and tranquilizers such as Valium may work with diminished effect. The carcinogenic effects associated with cigarette smoking are caused not by nicotine but by other substances present in tobacco, including fused-ring aromatic hydrocarbons (see the Chemical Connections feature on page 353) and radon-222 decay products (see the Chemical Connections feature on page 277).

▶ The names of the amine neurotransmitters are pronounced nor-ep-in-NEFF-rin, SER-oh-tone-in, and DOPE-a-mean.

Norepinephrine, a compound secreted by the adrenal glands into the blood, helps maintain muscle tone in the blood vessels.

Dopamine is found in the brain. A deficiency of this neurotransmitter results in Parkinson's disease, a degenerative neurological disease. Administration of dopamine to a patient does not relieve the symptoms of this disease, because dopamine in the blood cannot cross the blood–brain barrier. The drug L-dopa, which can pass through the blood–

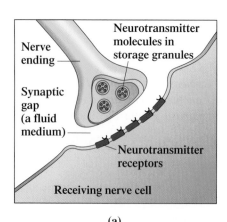

(a)

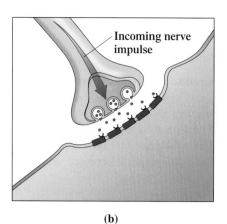

(b)

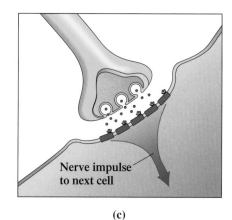

(c)

FIGURE 17.9 Neurotransmitters are chemical messengers between nerve cells. Neurotransmitters released from one nerve cell stimulate (activate) an adjacent nerve cell. (a) Before the conduction of a nerve impulse. (b) An incoming nerve impulse triggers the release of neurotransmitter molecules. (c) Neurotransmitters bind to receptor sites, activating the receptor nerve cell.

Prozac, the most widely prescribed drug for mental depression, inhibits the reuptake of serotonin, thus maintaining serotonin levels. Chemically, Prozac is a methyl propyl secondary amine derivative with three fluorine atoms present in the structure. The element fluorine is very seldom encountered in biochemical molecules.

$$F_3C-\text{(ring)}-CH-(CH_2)_2-NH-CH_3$$

Ephedrin, pronounced "eh-FEH-drin," is a substance extracted from the Asian plant ephedra that is used in many products sold as weight-loss aids and energy boosters at nutrition stores, in supermarkets and on the Internet. Its chemical structure resembles that of epinephrine (adrenaline).

$$\text{(ring)}-CH-CH-NH$$
$$\qquad\quad OH\quad CH_3\ CH_3$$

Ephedrin

Within the body, it behaves as a stimulant to the heart and the central nervous system. Users say it is a good "energy booster." Concerns exist about the safety of its use, particularly with people who have hypertension and other cardiovascular problems.

Currently, its use is controversial in the realm of sports. It is banned in the Olympics and in college sports. It is banned in professional football, but not in professional basketball or baseball.

Meanwhile, it is bought by millions of Americans who take billions of doses; it is readily available in capsule, drink, and chewing gum form.

Antihistamines are drugs that counteract, to some extent, the effects of histamine release in the body. Antihistamines share a common structural feature with histamine—an ethanamine chain.

$$-CH_2-CH_2-N\big\langle$$

This structure allows antihistamines to occupy receptor sites in nerves normally occupied by histamine, thus blocking histamine from occupying the nerve sites.

brain barrier, does give relief from Parkinson's symptoms. Inside brain cells, enzymes catalyze the conversion of L-dopa to dopamine.

$$HO-\text{(ring)}-CH_2-CH-NH_2 \quad\xrightarrow{\text{Enzymes}}\quad HO-\text{(ring)}-CH_2-CH_2-NH_2 \ +\ CO_2$$
$$\qquad\qquad\qquad\quad COOH$$

L-Dopa · · · Dopamine

Serotonin, also a brain chemical, is involved in sleep, sensory perception, and the regulation of body temperature. Serotonin deficiency has been implicated in mental illness. Treatment of mental depression can involve the use of drugs that help maintain serotonin at normal levels by preventing its breakdown within the brain.

■ Epinephrine

Epinephrine, also known as adrenaline, has some neurotransmitter functions but is more important as a central nervous system stimulant. Produced by the adrenal glands, epinephrine differs in structure from norepinephrine in that a methyl group substituent is present on the amine nitrogen atom.

$$HO-\text{(ring)}-CH-CH_2-NH$$
$$\qquad\qquad\quad OH\qquad\quad CH_3$$

Epinephrine

Pain, excitement, and fear trigger the release of large amounts of epinephrine into the bloodstream. The effect is increased blood glucose levels, which in turn increase blood pressure, rate and force of heart contraction, and muscular strength. These changes cause the body to function at a "higher" level. Epinephrine is often called the "fight or flight" hormone.

■ Histamine

The heterocyclic amine *histamine* is responsible for the unpleasant effects felt by individuals susceptible to hay fever and various pollen allergies.

$$H_2N-CH_2-CH_2-\text{(imidazole ring, N, N-H)}$$

Histamine

Histamine is naturally present in the human body in a "stored" form; it is part of more complex molecules. A number of situations can trigger the release of histamine. Activators include (1) contact with pollen, dust, and other allergens, (2) substances released from damaged cells, and (3) contact with chemicals to which an individual has become sensitized.

The presence of "free" histamine causes the symptoms associated with hay fever, such as watery eyes and stuffy nose, and many of the symptoms associated with the common cold. A group of substances called *antihistamines* can be taken as medication to counteract the effects of the histamine.

CHEMICAL CONNECTIONS

Amphetamines: Central Nervous System Stimulants

Amphetamines are a set of powerful *synthetic* amines that function as central nervous system stimulants. Benzedrin, also called simply amphetamine, is the parent compound for the amphetamine family. Important amphetamine derivatives include methamphetamine (prescribed as an antidepressant), methoxyamphetamine (prescribed as a bronchodilator), and isoproterenol (prescribed for emphysema and asthma).

Amphetamine
(benzedrine)

Methamphetamine
(methedrine)

Methoxyamphetamine

Isoproterenol

Structurally, these compounds are all related to adrenaline and mimic its stimulant effects.

Generally, amphetamines increase both heart rate and respiratory rate. They also reduce fatigue and diminish hunger by raising the glucose level in blood. At one time, they were widely used as appetite suppressants in the treatment of obesity, but because of many adverse effects, their use in weight control has diminished.

Hyperactive children (so overactive that they cannot sit still or concentrate) can benefit from amphetamines. Paradoxically, these CNS system stimulants calm hyperactive patients. (Recent research suggests that the mode of action involves increasing serotonin levels, with resulting inhibition of aggressive and impulsive behavior.) Ritalin (methylphenidate), an amphetamine-like drug used for treating hyperactive children, is one of the most widely prescribed stimulant drugs in the United States. Its

structure is similar enough to that of amphetamine that, pharmacologically, its effects are virtually identical to those of amphetamine.

Amphetamine

Methylphenidate
(Ritalin)

A great problem with amphetamines is the diversion of large quantities of these relatively inexpensive drugs into the illegal drug market. In this illegal drug market, *speed, splash,* and *crank* are names for powdered forms of methamphetamine; *ice* is a pure, crystallized form of methamphetamine that is smoked. *STP* is methoxyamphetamine. Abuse of these drug can produce severe physiological reactions. Once the drugs wear off, the user tends to "crash" into a state of physical and mental exhaustion. Withdrawal produces fatigue and profound and prolonged sleep.

Methylenedioxymethamphetamine (also known as MDMA and as Ecstasy) is an illegal hallucinogenic amphetamine derivative whose use is rapidly increasing; it now ranks behind only alcohol and marijuana in use by teens. It is typically sold as a white powder, and it can be inhaled, injected, or swallowed. Animal studies indicate that Ecstasy use can damage the brain's serotonin cells; serotonin is involved in appetite, sleep, mood regulation, memory, and sexual function. One of the most dangerous aspects of the use of this particular drug is that it can damage brain cells without any warning that the damage is taking place.

Methylenedioxymethamphetamine
(MDMA, Ecstasy)

17.10 Alkaloids

▶ The name *alkaloid,* which means "like a base," reflects the fact that alkaloids react with acids. Such is expected behavior for substances with amine functional groups, because amines are weak bases.

People in various parts of the world have known for centuries that physiological effects can be obtained by eating or chewing the leaves, roots, or bark of certain plants. Over 5000 different compounds that are physiologically active have been isolated from such plants. Nearly all of these compounds, which are collectively called alkaloids, contain amine functional groups. An **alkaloid** *is a nitrogen-containing organic compound extracted from plant material.*

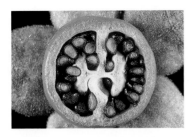

FIGURE 17.10 Fruit of the belladonna plant; the alkaloid atropine is obtained from this plant.

Three well-known compounds that we have considered previously are alkaloids. They are nicotine (tobacco plant), caffeine (coffee beans and tea leaves), and cocaine (coca plant).

A number of alkaloids are currently used in medicine. Quinine, which occurs in cinchona bark, is used to treat malaria. Atropine, which is isolated from the belladonna plant, is used to dilate the pupil of the eye in patients undergoing eye examinations (Figure 17.10). Atropine is also used as a preoperative drug to relax muscles and reduce the secretion of saliva in surgical patients.

Quinine

Atropine

An extremely important family of alkaloids is the narcotic painkillers, a class of drugs derived from the resin (opium) of the oriental poppy plant (Figure 17.11). The most important drugs obtained from opium are morphine and codeine. Synthetic modification of morphine produces the illegal drug heroin.

These three compounds have similar chemical structures.

Morphine

Codeine

Heroin

FIGURE 17.11 Oriental poppy plants, the source of several narcotic painkillers, including morphine.

Morphine is one of the most effective painkillers known; its painkilling properties are about a hundred times greater than those of aspirin. Morphine acts by blocking the process in the brain that interprets pain signals coming from the peripheral nervous system. The major drawback to the use of morphine is that it is addictive.

Codeine is a methylmorphine. Almost all codeine used in modern medicine is produced by methylating the more abundant morphine. Codeine is less potent than morphine, having a painkilling effect about one-sixth that of morphine.

Heroin is a synthetic compound, the diacetyl ester of morphine; it is produced from morphine. This chemical modification increases painkilling potency; heroin has more than three times the painkilling effect of morphine. However, heroin is so addictive that it has no accepted medical use in the United States.

17.11 Structure and Classification of Amides

An **amide** *is a carboxylic acid derivative in which the carboxyl —OH group has been replaced with an amino or a substituted amino group.* The amide functional group is thus

$$-\overset{\overset{\displaystyle O}{\|}}{C}-NH_2 \quad \text{or} \quad -\overset{\overset{\displaystyle O}{\|}}{C}-NH-R \quad \text{or} \quad -\overset{\overset{\displaystyle O}{\|}}{C}-\underset{\underset{\displaystyle R}{|}}{N}-R$$

depending on the degree of substitution.

Amides, like amines, can be classified as primary (1°), secondary (2°), or tertiary (3°), depending on how many carbon atoms are attached to the nitrogen atom.

> Primary, secondary, and tertiary amides are also called unsubstituted, monosubstituted, and disubstituted amides, respectively.

$$R-\overset{\overset{\displaystyle O}{\|}}{C}-NH_2 \qquad R-\overset{\overset{\displaystyle O}{\|}}{C}-NH-R' \qquad R-\overset{\overset{\displaystyle O}{\|}}{C}-\underset{\underset{\displaystyle R''}{|}}{N}-R'$$

Primary amide   Secondary amide   Tertiary amide

A **primary amide** *is an amide in which two hydrogen atoms are bonded to the amide nitrogen atom.* Such amides are also called *unsubstituted* amides. A **secondary amide** *is an amide in which an alkyl (or aryl) group and a hydrogen atom are bonded to the amide nitrogen atom. Monosubstituted* amide is another name for this type of amide. A **tertiary amide** *is an amide in which two alkyl (or aryl) groups and no hydrogen atoms are bonded to the amide nitrogen atom.* Such amides are *disubstituted* amides.

Note that the difference between a 1° amide and a 2° amide is "H versus R" and that the difference between a 2° amide and a 3° amide is again "H versus R." These "H versus R" relationships are the same relationships that exist between 1° and 2° amines and 2° and 3° amines (Section 17.2), as is summarized in Figure 17.12.

The simplest amide has a hydrogen atom attached to an unsubstituted amide functional group.

$$H-\overset{\overset{\displaystyle O}{\|}}{C}-NH_2$$

Next in complexity are amides in which a methyl group is present. There are two of them, one with the methyl group attached to the carbon atom and the other with the methyl group attached to the nitrogen atom.

FIGURE 17.12 Primary, secondary, and tertiary amines and amides and the "H versus R" relationship.

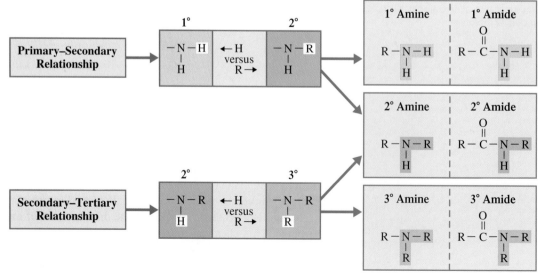

Line-angle drawings for selected primary, secondary, and tertiary amides:

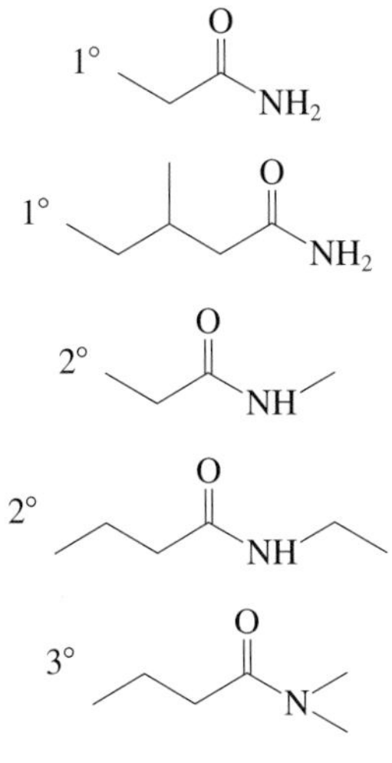

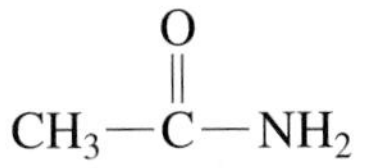

1°

1°

2°

2°

3°

$$CH_3-\overset{\overset{\displaystyle O}{\|}}{C}-NH_2 \quad \text{and} \quad H-\overset{\overset{\displaystyle O}{\|}}{C}-NH-CH_3$$

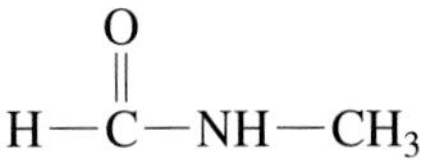

The first of these structures is a 1° amide, and the second structure is a 2° amide. The structure of the simplest aromatic amide involves a benzene ring to which an unsubstituted amide functional group is attached.

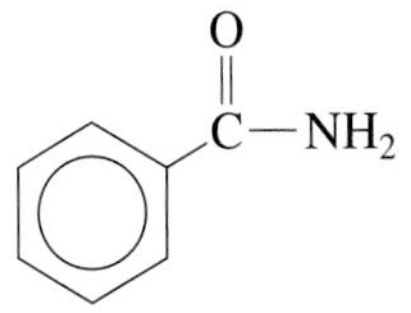

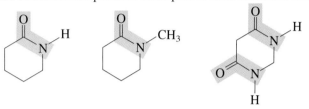

Cyclic amide structures are possible. Examples of such structures include

Cyclic amides are called *lactams,* a term that parallels the use of the term *lactones* for cyclic esters (Section 16.11).

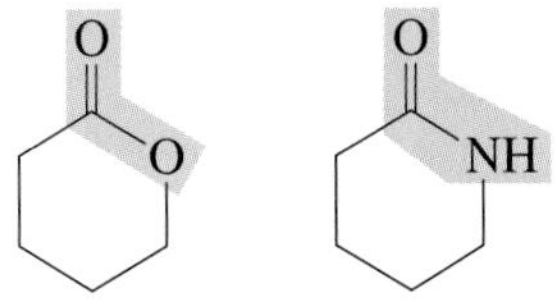

A lactone
(a cyclic ester)

A lactam
(a cyclic amide)

The members of the penicillin family of antibiotics (Section 21.10) have structures that contain a four-membered lactam ring.

17.12 Nomenclature for Amides

For nomenclature purposes (both IUPAC and common), amides are considered to be derivatives of carboxylic acids. Hence their names are based on the name of the parent carboxylic acid. (A similar procedure was used for naming esters; Section 16.12). The rules are as follows:

1. The ending of the name of the carboxylic acid is changed from *-ic acid* (common) or *-oic acid* (IUPAC) to *-amide.* For example, *benzoic acid* becomes *benzamide.*
2. The names of groups attached to the nitrogen (2° and 3° amides) are appended to the front of the base name, using an *N-* prefix as a locator.

Selected primary amide IUPAC names (with the common name in parentheses) are

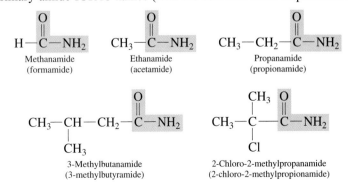

Acrylamide (2-propenamide), the simplest unsaturated amide, has the structure

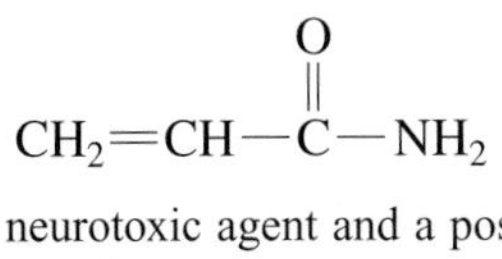

$$CH_2=CH-\overset{\overset{\displaystyle O}{\|}}{C}-NH_2$$

It is a neurotoxic agent and a possible human carcinogen.

Surprisingly, in 2002, low concentrations of acrylamide were found in potato chips, french fries, and other starchy foods prepared at high temperatures (greater than 120°C). Its possible source is the reaction between the amino acid asparagine (present in food proteins; Section 20.2) and carbohydrate sugars (Section 18.8) present in food.

Human risk studies are now underway concerning acrylamide presence in fried and some baked foods. No traces of acrylamide have been found in uncooked or boiled foods.

Nomenclature for secondary and tertiary amides, amides with substituted amino groups, involves use of the prefix *N-*, a practice we previously encountered with amine nomenclature (Section 17.3).

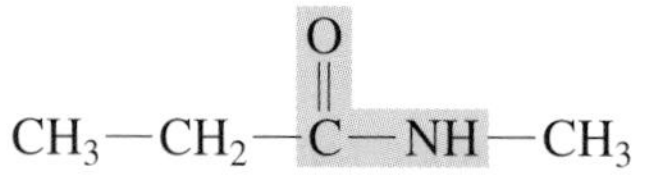

$$CH_3-CH_2-\overset{\overset{\displaystyle O}{\|}}{C}-NH-CH_3 \qquad CH_3-\overset{\overset{\displaystyle O}{\|}}{C}-\overset{\overset{\displaystyle CH_3}{|}}{N}-CH_3$$

N-Methylpropanamide
(*N*-methylpropionamide)

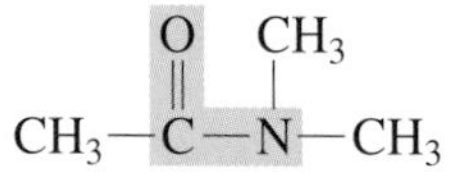

N,N-Dimethylethanamide
(*N,N*-dimethylacetamide)

The simplest aromatic amide, a benzene ring bearing an unsubstituted amide group, is called *benzamide.* Other aromatic amines are named as benzamide derivatives.

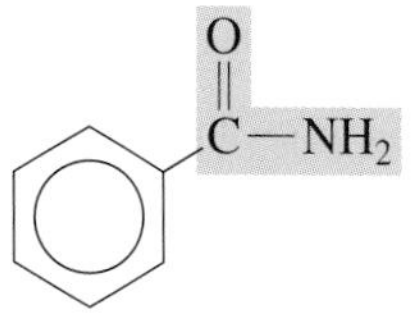

Benzamide 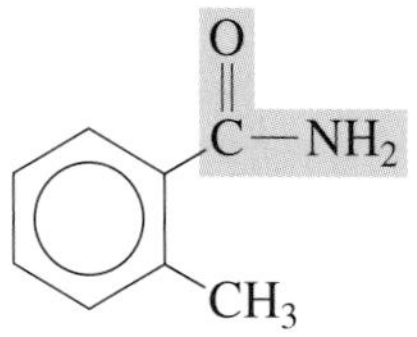   2-Methylbenzamide 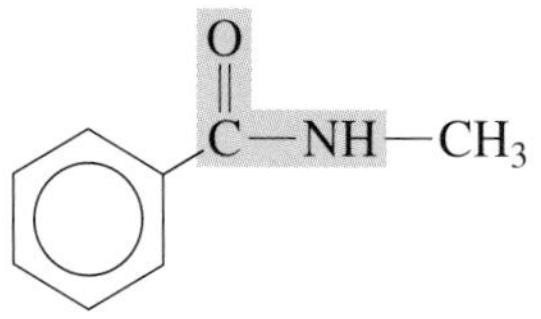   *N*-Methylbenzamide

---

EXAMPLE 17.5

Determining IUPAC and Common Names for Amides

▶ The contrast between IUPAC names and common names for unbranched unsubstituted amides is as follows:

IUPAC (one word)

alkanamide

ethanamide

Common (one word)

(prefix)amide*

acetamide

*The common-name prefixes are related to natural sources for the acids.

■ Assign both common and IUPAC names to each of the following amides.

a.
$$CH_3-CH_2-CH_2-\overset{\overset{\displaystyle O}{\|}}{C}-NH_2$$

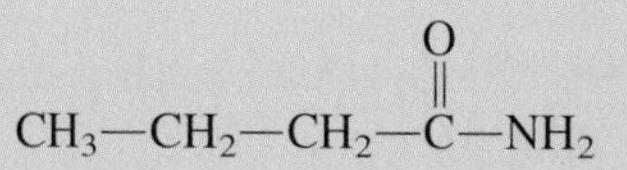

b.
$$CH_3-\overset{\overset{\displaystyle Br}{|}}{CH}-\overset{\overset{\displaystyle O}{\|}}{C}-NH-CH_3$$

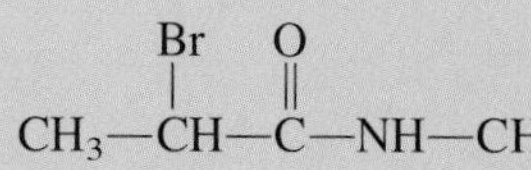

c.

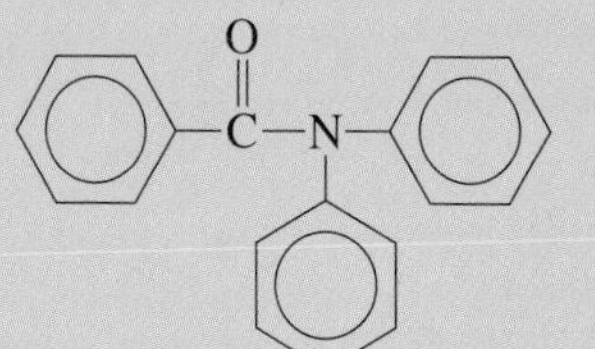

Solution

a. The parent acid for this amide is butyric acid (common) or butanoic acid (IUPAC). The common name for this amide is *butyramide,* and the IUPAC name is *butanamide.*
b. The common and IUPAC names of the acid are very similar; they are propionic acid and propanoic acid, respectively. The common name is *2-bromo-N-methylpropionamide,* and the IUPAC name is *2-bromo-N-methylpropanamide.* The prefix *N-* must be used with the methyl group to indicate that it is attached to the nitrogen atom.
c. In both the common and IUPAC systems of nomenclature, the name of the parent acid is the same: benzoic acid. The name of the amide is *N,N-diphenylbenzamide.*

Practice Exercise 17.5

Assign both common and IUPAC names to each of the following amides.

a.
$$CH_3-CH_2-\overset{\overset{\displaystyle O}{\|}}{\underset{\underset{\displaystyle Br}{|}}{CH}}-\overset{\overset{\displaystyle O}{\|}}{C}-NH_2$$

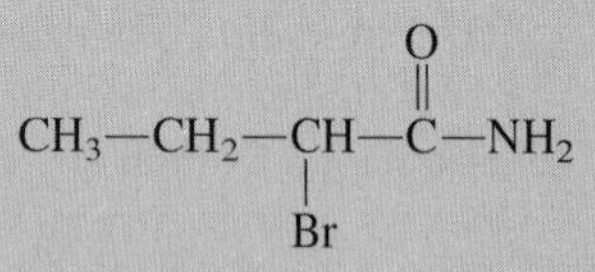

b.
$$CH_3-\overset{\overset{\displaystyle O}{\|}}{C}-NH-CH_3$$

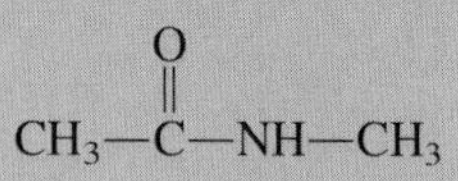

c.

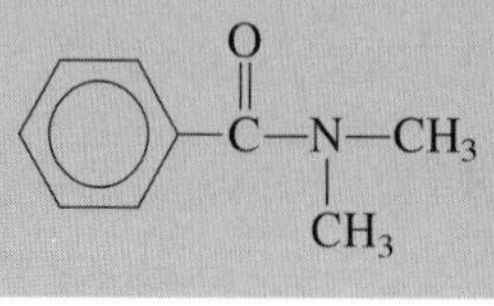

17.13 Selected Amides and Their Uses

The simplest naturally occurring amide is urea, a water-soluble white solid produced in the human body from carbon dioxide and ammonia through a complex series of metabolic reactions (Section 26.4).

$$CO_2 + 2NH_3 \longrightarrow (H_2N)_2CO + H_2O$$

Methanamide
(a primary amide)

N-Methyl methamide
(a secondary amide)

N,N-Dimethyl methamide
(a tertiary amide)

FIGURE 17.13 Space-filling models for the simplest primary, secondary, and tertiary amides.

Lidocaine (xylocaine), a substance commonly administered by injection as a dental anesthetic, is a synthetic molecule that contains both amide and amine functional groups.

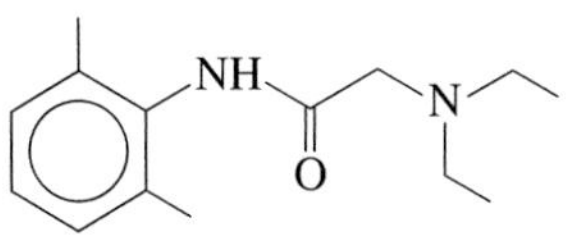

Another well known local anesthetic is procaine (novocaine). Its structure contains two amine groups and an ester group but no amide group.

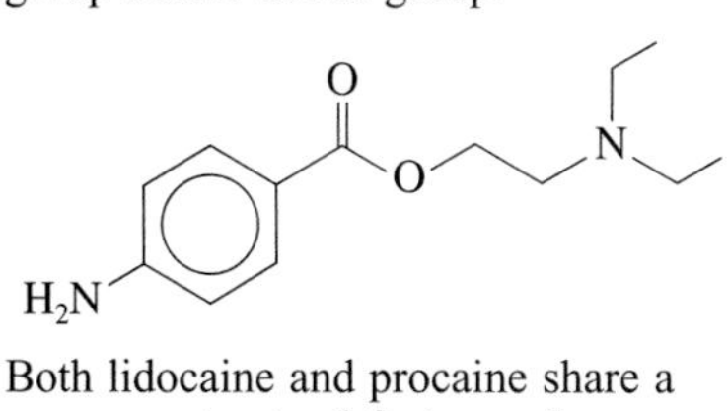

Both lidocaine and procaine share a common structural feature—the presence of a diethyl amino group (on the right side of each structure).

Urea is a one-carbon diamide. Its molecular structure is

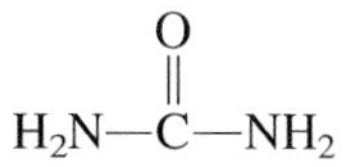

$$H_2N-\overset{\overset{\textstyle O}{\|}}{C}-NH_2$$

Urea formation is the human body's primary method for eliminating "waste" nitrogen. The kidneys remove urea from the blood and provide for its excretion in urine. With malfunctioning kidneys, urea concentrations in the body can build to toxic levels—a condition called *uremia.*

Melatonin is a hormone that is synthesized by the pineal gland and that regulates the sleep–wake cycle in humans. Melatonin levels within the body increase in evening hours and then decrease as morning approaches. High melatonin levels are associated with longer and more sound sleeping. The concentration of this hormone in the blood decreases with age; a six-year-old has a blood melatonin concentration over five times that of an 80-year-old. This is one reason why young children have less trouble sleeping than senior citizens. As a prescription drug, melatonin is used to treat insomnia and jet lag.

Structurally, melatonin is a polyfunctional amide; amine and ether groups are also present.

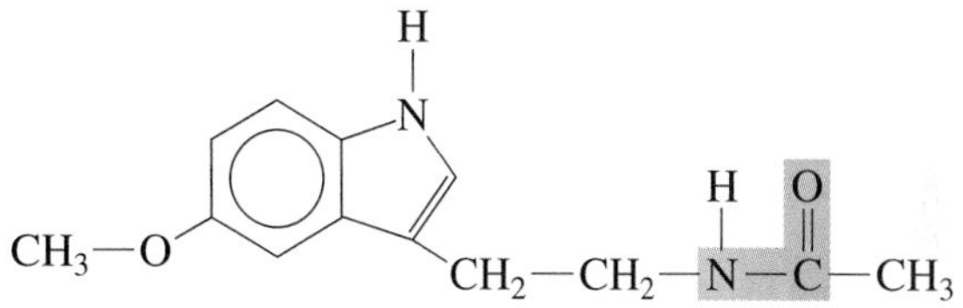

A number of synthetic amides exhibit physiological activity and are used as drugs in the human body. Foremost among them, in terms of use, is acetaminophen, which in 1992 replaced aspirin as the top-selling over-the-counter pain reliever. Acetaminophen is a derivative of acetamide (see the Chemical Connections feature on page 483).

Barbiturates, which are cyclic amide compounds, are a heavily used group of prescription drugs that cause relaxation (tranquilizers), sleep (sedatives), and death (overdoses). All barbiturates are derivatives of barbituric acid, a cyclic amide that was first synthesized from urea and malonic acid.

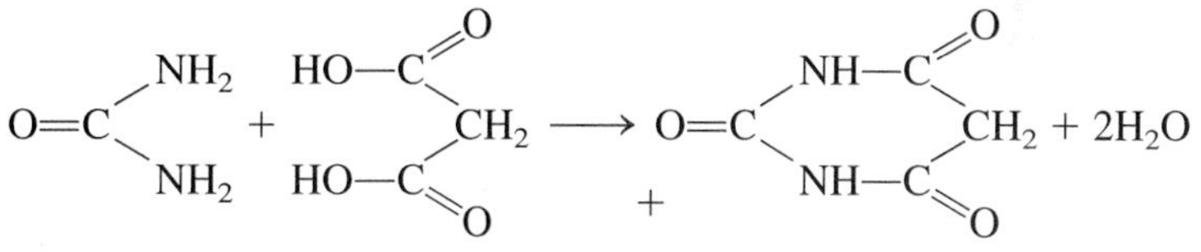

Urea   Malonic acid   Barbituric acid

(The researcher who first synthesized this compound named it after his girlfriend Barbara.)

17.14 Properties of Amides

Amides do not exhibit basic properties in solution as amines do (Section 17.5). Although the nitrogen atom present in amides has a nonbonding pair of electrons, as in amines, these electrons are not available for bonding to a H^+ ion. The reason for this is related to the polarity of the carbonyl portion (—C=O) of the amide functional group.

Methanamide and its *N*-methyl and *N,N*-dimethyl derivatives (the simplest 1°, 2°, and 3° amides, respectively), are all liquids at room temperature; molecular models for these three simple amides are given in Figure 17.13. All unbranched primary amides, except methanamide, are solids at room temperature (Figure 17.14), as are most other amides. In many cases, the amide melting point is even higher than that of the corresponding carboxylic acid. The high melting points result from the numerous intermolecular hydrogen-bonding possibilities that exist between amide H atoms and carbonyl O atoms. Figure

Unbranched Primary Amides			
C$_1$	C$_3$	C$_5$	C$_7$
C$_2$	C$_4$	C$_6$	C$_8$

☐ Liquid ☐ Solid

FIGURE 17.14 A physical-state summary for unbranched primary amines at room temperature and pressure.

17.15 shows selected hydrogen-bonding interactions that are possible among several primary amide molecules.

Fewer hydrogen-bonding possibilities exist for 2° amides, because the nitrogen atom now has only one hydrogen atom; hence lower melting points are the rule for such amides. Still lower melting points are observed for 3° amides, because no hydrogen bonding is possible. The disubstituted *N,N*-dimethylacetamide has a melting point of −20°C, which is about 100°C lower than that of the unsubstituted acetamide.

Amides of low molecular mass, up to five or six carbon atoms, are soluble in water. Again, numerous hydrogen-bonding possibilities exist between water and the amide. Even disubstituted amides can participate in such hydrogen bonding.

$$
\begin{array}{c}
\downarrow \\
\ddot{O}\!:\!\leftarrow \\
\parallel \quad \swarrow \\
R\!-\!C\!-\!\overset{\ddots}{N}\!-\!H \leftarrow \\
\mid \\
H \\
\uparrow
\end{array}
\qquad
\begin{array}{l}
\text{Arrows denote sites} \\
\text{where hydrogen bonding} \\
\text{to water can occur.}
\end{array}
$$

17.15 Preparation of Amides

The reaction of a carboxylic acid with ammonia or a 1° or 2° amine produces an amide, provided that the reaction is carried out at an elevated temperature (greater than 100°C).

$$\text{Ammonia + carboxylic acid} \xrightarrow[\text{temp.}]{\text{High}} \text{1° amide}$$

$$\text{1° Amine + carboxylic acid} \xrightarrow[\text{temp.}]{\text{High}} \text{2° amide}$$

$$\text{2° Amine + carboxylic acid} \xrightarrow[\text{temp.}]{\text{High}} \text{3° amide}$$

If the preceding reactions are run at room temperature (25°), no amide formation occurs; instead an acid–base reaction occurs in which a carboxylic acid salt is produced. This acid–base reaction when a 1° amine is the reactant is

$$
\underset{\text{Acid}}{
\begin{array}{c}
O \\
\parallel \\
R\!-\!C\!-\!OH
\end{array}}
+
\underset{\substack{\text{Primary} \\ \text{amine}}}{
\begin{array}{c}
H \\
\mid \\
H\!-\!N\!-\!R
\end{array}}
\xrightarrow[\text{temp.}]{\text{Room}}
\underset{\text{Carboxylate salt}}{
\begin{array}{cc}
H & O \\
\mid & \parallel \\
H\!-\!\overset{+}{N}\!-\!R & R\!-\!C\!-\!O^{-} \\
\mid \\
H
\end{array}}
$$

FIGURE 17.15 Amide–amide hydrogen bonding.

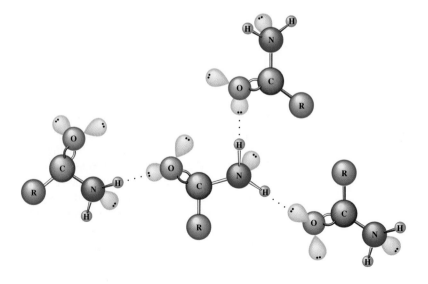

CHEMICAL CONNECTIONS

Acetaminophen: A Substituted Amide

Often called the aspirin substitute, acetaminophen is the most widely used of all nonprescription pain relievers, accounting for over half of this market. Acetaminophen is a derivative of acetamide in which a hydroxyphenyl group has replaced one of the amide hydrogens.

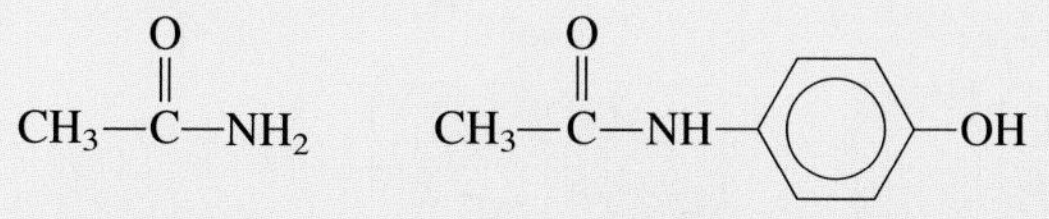

Acetamide Acetaminophen

Acetaminophen is the active ingredient in Tylenol, Datril, Tempra, and Anacin-3. Excedrin, which contains both acetaminophen and aspirin, is a combination pain reliever.

Acetaminophen is often used as an aspirin substitute because it has no irritating effect on the intestinal tract and yet has comparable analgesic and antipyretic effects. Unlike aspirin, however, it is not effective against inflammation and is of limited use for the aches and pains of arthritis. Also, acetaminophen does not inhibit platelet aggregation and therefore is not useful for preventing vascular clotting.

Acetaminophen is available in a liquid form that is used extensively for small children and other patients who have difficulty taking solid tablets. The extensive use of acetaminophen for children has a drawback; it is the drug most often involved in childhood poisonings. An overdose of acetaminophen is toxic to the liver.

Acetaminophen's mode of action in the body is similar to that of aspirin—inhibition of prostaglandin synthesis.

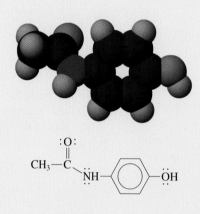

General structural equations for 1°, 2°, and 3° amide production from carboxylic acids are

$$R—\overset{O}{\overset{\|}{C}}—OH + H—\overset{H}{\overset{|}{N}}—H \xrightarrow{100°C} R—\overset{O}{\overset{\|}{C}}—\overset{H}{\overset{|}{N}}—H + H_2O$$

Carboxylic acid; Ammonia; Primary amide

$$R—\overset{O}{\overset{\|}{C}}—OH + H—\overset{H}{\overset{|}{N}}—R \xrightarrow{100°C} R—\overset{O}{\overset{\|}{C}}—\overset{H}{\overset{|}{N}}—R + H_2O$$

Carboxylic acid; Primary amine; Secondary amide

$$R—\overset{O}{\overset{\|}{C}}—OH + H—\overset{R}{\overset{|}{N}}—R \xrightarrow{100°C} R—\overset{O}{\overset{\|}{C}}—\overset{R}{\overset{|}{N}}—R + H_2O$$

Carboxylic acid; Secondary amine; Tertiary amide

These reactions are called *amidification* reactions. An **amidification reaction** *is the reaction of a carboxylic acid with an amine (or ammonia) to produce an amide.* In amidification, a —OH group is lost from the carboxylic acid, a —H atom is lost from the ammonia or amine, and water is formed as a by-product. Amidification reactions are thus *condensation* reactions.

> This is the fourth time we have encountered *condensation* reactions. Esterification (Section 16.10), acetal formation (Section 15.9), and intermolecular alcohol dehydration (Section 14.7) were the other three condensation situations.

Two specific amidification reactions, in which a 2° amide and a 3° amide are produced, respectively, are

$$CH_3—CH_2—CH_2—\overset{O}{\overset{\|}{C}}—OH + H—\overset{H}{\overset{|}{N}}—CH_3 \xrightarrow{\Delta} CH_3—CH_2—CH_2—\overset{O}{\overset{\|}{C}}—\overset{H}{\overset{|}{N}}—CH_3 + H_2O$$

Butanoic acid; Methylamine (1° amine); *N*-methylbutanamide (a 2° amine)

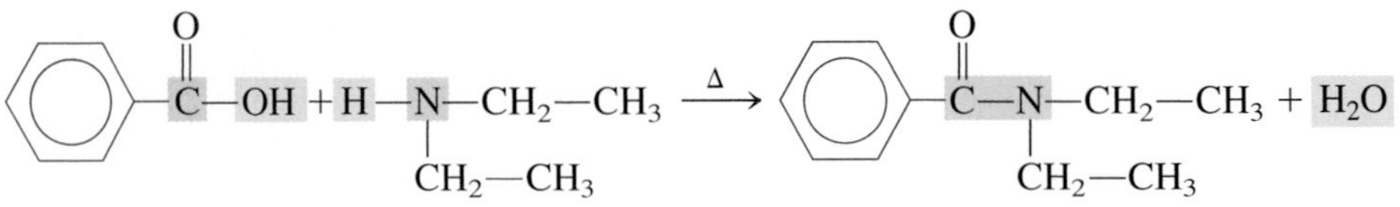

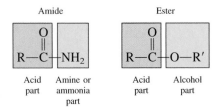

Benzoic acid   Diethylamine   *N,N*-diethylbenzamide
               (2° amine)   (a 3° amine)

Just as it is useful to think of the structure of an ester (Section 16.10) in terms of an "acid part" and an "alcohol part," it is useful to think of an amide in terms of an "acid part" and an "amine (or ammonia) part."

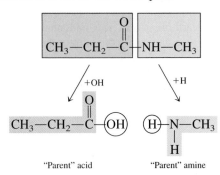

In this context, it is easy to identify the parent acid and amine from which a given amide can be produced; to generate the parent molecules, just add an —OH group to the acid part of the amide and an H atom to the amine part.

$$CH_3-CH_2-\overset{\overset{\displaystyle O}{\|}}{C}+NH-CH_3$$

+OH   +H

$$CH_3-CH_2-\overset{\overset{\displaystyle O}{\|}}{C}-(OH)\quad (H)-\overset{\overset{\displaystyle}{}}{\underset{\underset{\displaystyle H}{|}}{N}}-CH_3$$

"Parent" acid   "Parent" amine

EXAMPLE 17.6

Predicting Reactants Needed to Prepare Specific Amides

■ What carboxylic acid and amine (or ammonia) are needed to prepare each of the following amides?

a.
$$CH_3-\overset{\overset{\displaystyle O}{\|}}{C}-NH-CH_2-CH_3$$

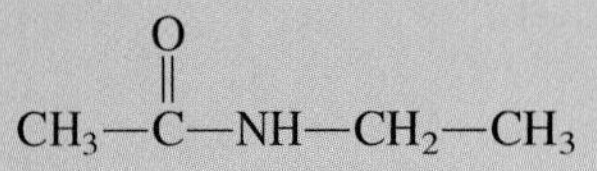

b.
$$CH_3-CH_2-\overset{\overset{\displaystyle O}{\|}}{C}-NH_2$$

c.
$$CH_3-CH_2-\overset{\overset{\displaystyle O}{\|}}{C}-\underset{\underset{\displaystyle CH_3}{|}}{N}-CH_3$$

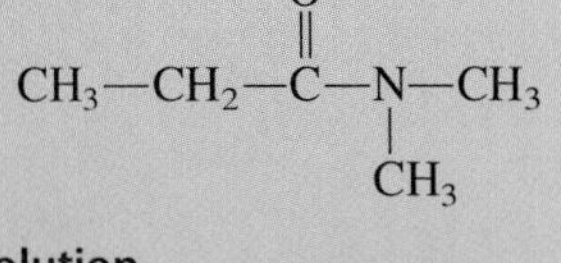

Solution

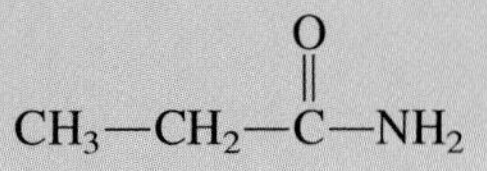

a. Viewing the molecule as having an acid part and an amine part, we obtain

$$CH_3-\overset{\overset{\displaystyle O}{\|}}{C}+NH-CH_2-CH_3$$

Acid part   Amine part

Adding an —OH group to the acid part and a H atom to the amine part, we obtain the "parent" molecules, which are

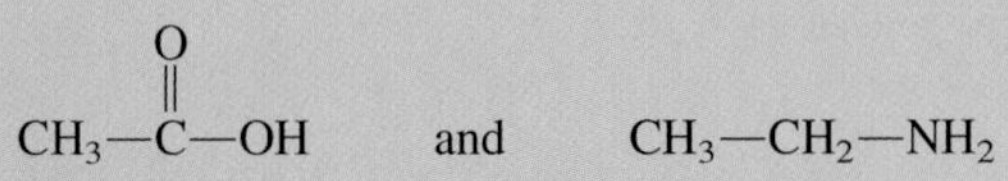

$$CH_3-\overset{\overset{\displaystyle O}{\|}}{C}-OH \quad \text{and} \quad CH_3-CH_2-NH_2$$

b. Proceeding as in part **a**, we find that the "parent" acid and amine molecules are, respectively,

$$CH_3-CH_2-\overset{\overset{\displaystyle O}{\|}}{C}-OH \quad \text{and} \quad NH_3$$

c. Proceeding again as in part **a,** we find that the "parent" acid and amine molecules are, respectively,

$$CH_3-CH_2-\overset{\overset{\displaystyle O}{\|}}{C}-OH \quad \text{and} \quad CH_3-NH-CH_3$$

Practice Exercise 17.6

What carboxylic acid and amine (or ammonia) are needed to prepare each of the following amides?

a.
$$CH_3-\overset{\overset{\displaystyle O}{\|}}{C}-\underset{\underset{\displaystyle CH_3}{|}}{N}-CH_2-CH_3$$

b.
$$CH_3-CH_2-\overset{\overset{\displaystyle O}{\|}}{C}-NH-CH_3$$

c. 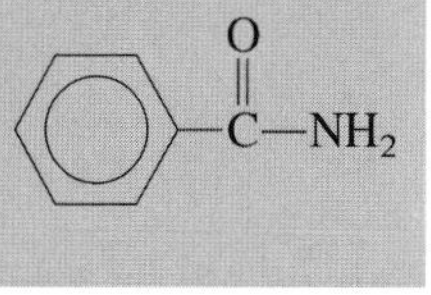
$$\langle\bigcirc\rangle-\overset{\overset{\displaystyle O}{\|}}{C}-NH_2$$

17.16 Hydrolysis of Amides

As was the case with esters (Section 16.15), the most important reaction of amides is hydrolysis. In amide hydrolysis, the bond between the carbonyl carbon atom and the nitrogen is broken, and free acid and free amine are produced. Amide hydrolysis is catalyzed by acids, bases, or certain enzymes; sustained heating is also often required.

$$\underset{\text{Amide}}{R-\overset{\overset{\displaystyle O}{\|}}{C}-NH-R'} + H_2O \overset{\Delta}{\longrightarrow} \underset{\substack{\text{Carboxylic} \\ \text{acid}}}{R-\overset{\overset{\displaystyle O}{\|}}{C}-OH} + \underset{\text{Amine}}{R'-NH_2}$$

Acidic or basic hydrolysis conditions have an effect on the products. *Acidic* conditions convert the product amine to an amine salt (Section 17.6). *Basic* conditions convert the product acid to a carboxylic acid salt (Section 16.9).

▶ Amide hydrolysis under basic conditions is also called amide saponification, just as ester hydrolysis under basic conditions is called ester saponification (Section 16.16).

$$\underset{\text{Acidic hydrolysis of an amide}}{R-\overset{\overset{\displaystyle O}{\|}}{C}-NH-R'} + H_2O + \boxed{HCl} \overset{\Delta}{\longrightarrow} \underset{\text{Carboxylic acid}}{R-\overset{\overset{\displaystyle O}{\|}}{C}-OH} + \underset{\text{Amine salt}}{R'-\overset{+}{N}H_3\,Cl^-}$$

$$\underset{\text{Basic hydrolysis of an amide}}{R-\overset{\overset{\displaystyle O}{\|}}{C}-NH-R'} + \boxed{NaOH} \overset{\Delta}{\longrightarrow} \underset{\text{Carboxylic acid salt}}{R-\overset{\overset{\displaystyle O}{\|}}{C}-O^-\,Na^+} + \underset{\text{Amine}}{R'-NH_2}$$

EXAMPLE 17.7

Predicting the Products of Amide Hydrolysis Reactions

■ Draw structural formulas for the organic products of each of the following amide hydrolysis reactions. Be sure to take into account whether the hydrolysis occurs under neutral, acidic, or basic conditions.

a.
$$CH_3-CH_2-\overset{\overset{\displaystyle O}{\|}}{C}-NH-CH_3 + H_2O \overset{\Delta}{\longrightarrow}$$

b.
$$CH_3-CH_2-\overset{\overset{\displaystyle O}{\|}}{C}-NH-CH_2-CH_3 + H_2O \overset{\Delta}{\underset{HCl}{\longrightarrow}}$$

c.
$$CH_3-\overset{\overset{\displaystyle O}{\|}}{C}-\underset{\underset{\displaystyle CH_3}{|}}{N}-CH_3 + H_2O \overset{\Delta}{\underset{NaOH}{\longrightarrow}}$$

(continued)

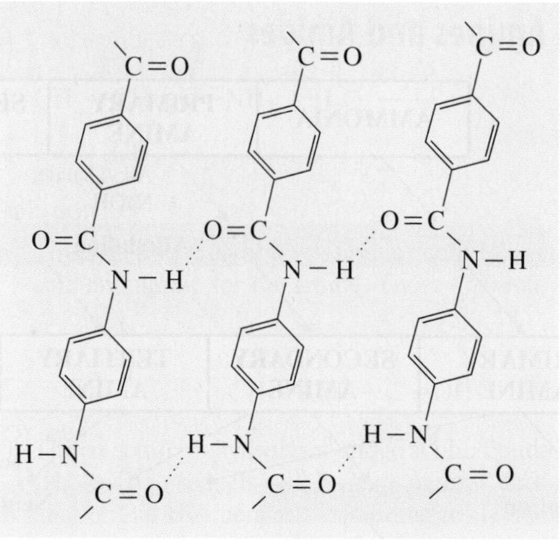

FIGURE 17.17 A regular hydrogen-bonding pattern among Kevlar polymer strands contributes to the great strength of this polymer.

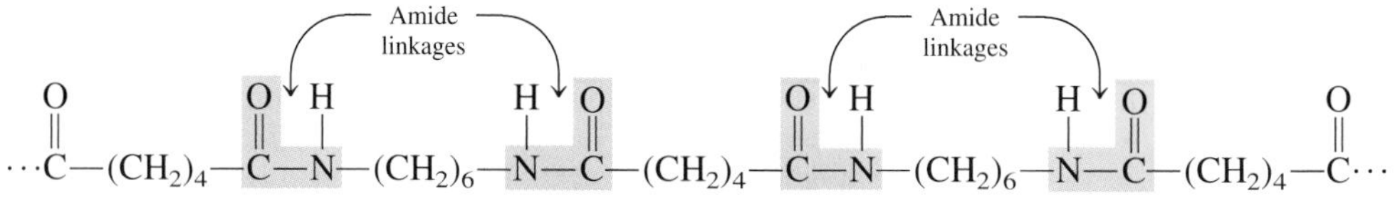

A portion of the polyamide nylon 66

Additional stiffness and toughness are imparted to polyamides if aromatic rings are present in the polymer "backbone." The polyamide Kevlar is now used in place of steel in bullet-resistant vests. The polymeric repeating unit in Kevlar is

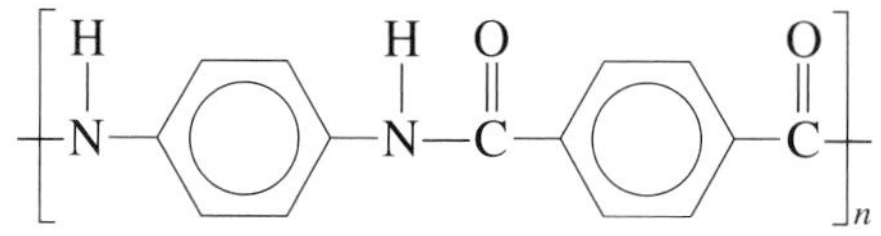

Kevlar

A uniform system of hydrogen bonds that holds polymer chains together accounts for the "amazing" strength of Kevlar (see Figure 17.17).

Nomex is a polymer whose structure is a variation of that of Kevlar. With Nomex, the monomers are *meta* isomers rather than *para* isomers. Nomex is used in flame-resistant clothing for fire fighters and race car drivers (Figure 17.18).

Silk and wool are examples of *naturally occurring* polyamide polymers. Silk and wool are proteins, and proteins are polyamide polymers. Because much of the human body is protein material, much of the human body is polyamide polymer. The monomers for proteins are amino acids, difunctional molecules containing both amino and carboxyl groups. Here are some representative structures for amino acids, of which there are many (Section 20.2).

$$H_2N-CH_2-COOH \qquad H_2N-CH-COOH \qquad H_2N-CH-COOH$$

FIGURE 17.18 Fire fighters with flame-resistant clothing containing Nomex.

Polyurethanes are polymers related to polyesters and polyamides. The backbone of a polyurethane polymer contains aspects of both ester and amide functional groups. The following is a portion of the structure of a typical polyurethane polymer.

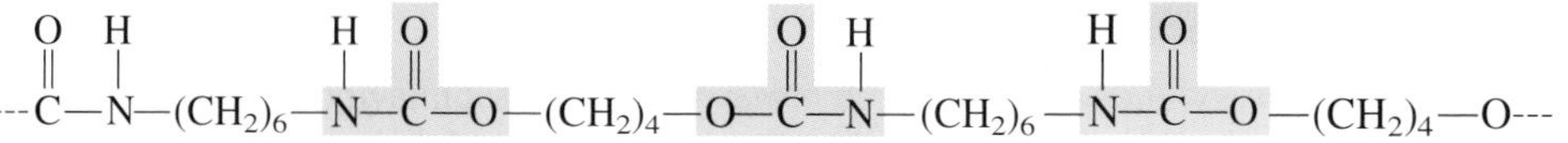

Foam rubber in furniture upholstery, packaging materials, life preservers, elastic fibers, and many other products contain polyurethane polymers (Figure 17.19).

18 Carbohydrates

Carbohydrates in the form of cotton and linen may be woven into clothing materials.

Beginning with this chapter on carbohydrates, we will focus almost exclusively on biochemistry, the chemistry of living systems. Like organic chemistry, biochemistry is a vast subject, and we can discuss only a few of its facets. Our approach to biochemistry will be similar to our approach to organic chemistry. We will devote individual chapters to each of the major classes of biochemical compounds, which are carbohydrates, lipids, proteins, and nucleic acids. Then we will examine the major types of chemical reactions in living organisms. In this first "biochapter," carbohydrates are considered.

The same functional groups found in organic compounds are also present in biochemical compounds. Usually, however, there is greater structural complexity associated with biochemical compounds as a result of polyfunctionality; several different functional groups are present. Often biochemical compounds interact with each other, within cells, to form larger structures. But the same chemical principles and chemical reactions associated with the various organic functional groups that we have studied apply to these larger biochemical structures as well.

18.1 Biochemistry—An Overview

Biochemistry *is the study of the chemical substances found in living organisms and the chemical interactions of these substances with each other.* Biochemistry is a field in which new discoveries are made almost daily about how cells manufacture the molecules needed for life and how the chemical reactions by which life is maintained occur. The knowledge explosion that has occurred in the field of biochemistry during the last decades of the twentieth century is truly phenomenal.

FIGURE 18.1 Mass composition data for the human body in terms of major types of biochemical substances.

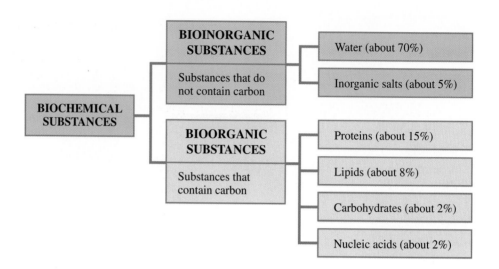

▶ As isolated compounds, bioinorganic and bioorganic substances have no life in and of themselves. Yet when these substances are gathered together in a cell, their chemical interactions are able to sustain life.

▶ It is estimated that more than half of all organic carbon atoms are found in the carbohydrate materials of plants.

▶ Human uses for carbohydrates of the plant kingdom extend beyond food. Carbohydrates in the form of cotton and linen are used as clothing. Carbohydrates in the form of wood are used for shelter and heating and in making paper.

FIGURE 18.2 Most of the matter in plants, except water, is carbohydrate material. Photosynthesis, the process by which carbohydrates are made, requires sunlight.

A **biochemical substance** *is a chemical substance found within a living organism.* Biochemical substances are divided into two groups: bioinorganic substances and bioorganic substances. *Bioinorganic substances* include water and inorganic salts. *Bioorganic substances* include carbohydrates, lipids, proteins, and nucleic acids. Figure 18.1 gives an approximate mass composition for the human body in terms of types of biochemical substances present.

Although we tend to think of the human body as made up of organic substances, bioorganic molecules make up only about one-fourth of body mass. The bioinorganic substance water constitutes over two-thirds of the mass of the human body, and another 4%–5% of body mass comes from inorganic salts (Section 10.6).

18.2 Occurrence and Functions of Carbohydrates

Carbohydrates are the most abundant class of bioorganic molecules on planet Earth. Although their abundance in the human body is relatively low (Section 18.1), carbohydrates constitute about 75% by mass of dry plant materials (see Figure 18.2).

Green (chlorophyll-containing) plants produce carbohydrates via *photosynthesis.* In this process, carbon dioxide from the air and water from the soil are the reactants, and sunlight absorbed by chlorophyll is the energy source.

$$CO_2 + H_2O + \text{solar energy} \xrightarrow[\text{Plant enzymes}]{\text{Chlorophyll}} \text{carbohydrates} + O_2$$

Plants have two main uses for the carbohydrates they produce. In the form of *cellulose,* carbohydrates serve as structural elements, and in the form of *starch,* they provide energy reserves for the plants.

Dietary intake of plant materials is the major carbohydrate source for humans and animals. The average human diet should ideally be about two-thirds carbohydrate by mass. Carbohydrates have the following functions in humans:

1. Carbohydrate oxidation provides energy.
2. Carbohydrate storage, in the form of glycogen, provides a short-term energy reserve.
3. Carbohydrates supply carbon atoms for the synthesis of other biochemical substances (proteins, lipids, and nucleic acids).
4. Carbohydrates form part of the structural framework of DNA and RNA molecules.
5. Carbohydrates linked to lipids (Chapter 19) are structural components of cell membranes.
6. Carbohydrates linked to proteins (Chapter 20) function in a variety of cell–cell and cell–molecule recognition processes.

18.3 Classification of Carbohydrates

Most simple carbohydrates have empirical formulas that fit the general formula $C_nH_{2n}O_n$. An early observation by scientists that this general formula can also be written as $C_n(H_2O)_n$ is the basis for the term *carbohydrate*—that is, "hydrate of carbon." It is now known that this hydrate viewpoint is not correct, but the term *carbohydrate* still persists. Today the term is used to refer to an entire family of compounds, only some of which have the formula $C_nH_{2n}O_n$.

A **carbohydrate** *is a polyhydroxy aldehyde, a polyhydroxy ketone, or a compound that yields polyhydroxy aldehydes or polyhydroxy ketones upon hydrolysis.* The carbohydrate glucose is a polyhydroxy aldehyde, and the carbohydrate fructose is a polyhydroxy ketone.

Glucose
(a polyhydroxy aldehyde)

Fructose
(a polyhydroxy ketone)

A striking structural feature of carbohydrates is the large number of functional groups present. In glucose and fructose there is a functional group attached to each carbon atom.

Carbohydrates are classified on the basis of molecular size as monosaccharides, oligosaccharides, and polysaccharides.

A **monosaccharide** *is a carbohydrate that contains a single polyhydroxy aldehyde or polyhydroxy ketone unit.* Monosaccharides cannot be broken down into simpler units by hydrolysis reactions. Both glucose and fructose are monosaccharides. Naturally occurring monosaccharides have from three to seven carbon atoms; five- and six-carbon species are especially common. Pure monosaccharides are water-soluble, white, crystalline solids.

An **oligosaccharide** *is a carbohydrate that contains two to ten monosaccharide units covalently bonded to each other.* Disaccharides are the most common type of oligosaccharide. A **disaccharide** *is a carbohydrate that contains two monosaccharide units covalently bonded to each other.* Like monosaccharides, disaccharides are crystalline, water-soluble substances. Sucrose (table sugar) and lactose (milk sugar) are disaccharides.

Within the human body, oligosaccharides are often found associated with proteins and lipids in complexes that have both structural and regulatory functions. Free oligosaccharides, other than disaccharides, are seldom encountered in biochemical systems.

Complete hydrolysis of an oligosaccharide produces monosaccharides. Upon hydrolysis, a disaccharide produces two monosaccharides, a trisaccharide three monosaccharides, a hexasaccharide six monosaccharides, and so on.

A **polysaccharide** *is a polymeric carbohydrate that contains many monosaccharide units covalently bonded to each other.* Polysaccharides often contain several thousand monosaccharide units. Both cellulose and starch are polysaccharides. We encounter these two substances everywhere. The paper on which this book is printed is mainly cellulose, as are the cotton in our clothes and the wood in our houses. Starch is a component of many types of foods, including bread, pasta, potatoes, rice, corn, beans, and peas.

18.4 Chirality: Handedness in Molecules

Monosaccharides are the simplest type of carbohydrate. Before considering specific structures for and specific reactions of monosaccharides, we will consider an important general structural property called *handedness,* which most monosaccharides exhibit. Most monosaccharides exist in two forms: a "left-handed" form and a "right-handed" form.

The term *monosaccharide* is pronounced "mon-oh-SACK-uh-ride."

The *oligo* in the term *oligosaccharides* comes from the Greek *oligos,* which means "small" or "few." The term *oligosaccharide* is pronounced "OL-ee-go-SACK-uh-ride."

Types of carbohydrates are related to each other through hydrolysis.

Polysaccharides
↓ Hydrolysis
Oligosaccharides
↓ Hydrolysis
Monosaccharides

The property of handedness is not restricted to carbohydrates. It is a general phenomenon found in all classes of organic compounds.

FIGURE 18.3 The mirror image of the right hand is the left hand. Conversely, the mirror image of the left hand is the right hand.

Left   Right

Mirror image of left hand
is in the back of the mirror

These two forms are related to each other in the same way your left and right hands are related to each other. That relationship is that of *mirror images.* Figure 18.3 shows this mirror-image relationship for human hands.

■ Mirror Images

The concept of *mirror images* is the key to understanding molecular handedness. All objects, including all molecules, have mirror images. A **mirror image** *is the reflection of an object in a mirror.* Objects can be divided into two classes on the basis of their mirror images: objects with *superimposable* mirror images and objects with *nonsuperimposable* mirror images. **Superimposable mirror images** *are images that coincide at all points when the images are laid upon each other.* A dinner plate with no design features has superimposable mirror images. **Nonsuperimposable mirror images** *are images where not all points coincide when the images are laid upon each other.* Human hands are non-superimposable mirror images, as Figure 18.4 shows; note in this figure that the two thumbs point in opposite directions and that the fingers do not align correctly. Like human hands, all objects with nonsuperimposable mirror images exist in "left-handed" and "right-handed" forms.

The term *chiral* (rhymes with *spiral*) comes from the Greek word *cheir,* which means "hand." Chiral objects are said to possess "handedness."

■ Chirality

Of particular concern to us is the "handedness concept" as it applies to molecules. Not all molecules possess handedness. What, then, is the molecular structural feature that generates "handedness?" Any organic molecule that contains a carbon atom with four *different* groups attached to it in a tetrahedral orientation possesses handedness. Such a carbon atom is called a *chiral center.* A **chiral center** *is an atom in a molecule that has four different groups tetrahedrally bonded to it.*

A molecule that contains a chiral center is said to be *chiral.* A **chiral molecule** *is a molecule whose mirror images are not superimposable.* Chiral molecules have handedness. An **achiral molecule** *is a molecule whose mirror images are superimposable.* Achiral molecules do not possess handedness.

A trisubstituted methane molecule, such as bromochloroiodomethane, is the simplest example of a chiral organic molecule.

FIGURE 18.4 A person's left and right hands are not superimposable upon each other.

$$Br-\underset{\underset{I}{|}}{\overset{\overset{H}{|}}{C}}-Cl$$

bromochloroiodomethane

Note the four different groups attached to the carbon atom present: —H, —Br, —Cl, and —I. Figure 18.5a shows the nonsuperimposability of the two mirror-image forms of this molecule.

The simplest example of a chiral monosaccharide molecule is the three-carbon monosaccharide called glyceraldehyde.

FIGURE 18.5 Examples of simple molecules that are chiral. (a) The mirror-image forms of the molecule bromochloroiodomethane are nonsuperimposable. (b) The mirror-image forms of the molecule glyceraldehyde are nonsuperimposable.

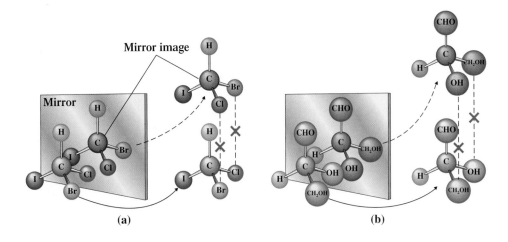

(a)   (b)

$$CHO$$
$$H-C-OH$$
$$CH_2OH$$

glyceraldehyde

> There are a few chiral molecules known that do not have a chiral center. Such exceptions are not important for the applications that we will make of the chirality concept in this text.

The four different groups attached to the carbon atom at the chiral center in this molecule are —H, —OH, —CHO, and —CH₂OH. The nonsuperimposability of the two mirror-image forms of glyceraldehyde is shown in Figure 18.5b.

Chiral centers within molecules are often denoted by a small asterisk. Note the chiral centers in the following molecules.

$$CH_3-CH_2-\overset{H}{\underset{OH}{*C}}-CH_3 \qquad H-\overset{Cl}{\underset{I}{*C}}-CH_3 \qquad CH_3-CH_2-CH_2-\overset{CH_3}{\underset{H}{*C}}-CH_2-CH_3$$

2-Butanol   1-Chloro-1-iodoethane   3-Methylhexane

EXAMPLE 18.1

Identifying Chiral Centers in Molecules

■ Indicate whether the circled carbon atom in each of the following molecules is a chiral center.

a. $CH_3-\textcircled{C}H-CH_2-CH_3$
    $\qquad\qquad\underset{Cl}{|}$

b. $CH_3-CH_2-\overset{O}{\overset{\|}{\textcircled{C}}}-CH_3$

c. $CH_3-CH_2-\textcircled{C}H-OH$
    $\qquad\qquad\qquad\underset{|}{CH_2}$
    $\qquad\qquad\qquad\underset{|}{CH_3}$

d.
$$\begin{array}{c} Br \\ | \\ \textcircled{C}H \\ H_2C \qquad CH_2 \\ | \qquad\qquad | \\ HC \qquad\quad CH_2 \\ Br \qquad CH_2 \end{array}$$

Solution

a. This is a chiral center. The four different groups attached to the carbon atom are —CH₃, —Cl, —CH₂—CH₃, and —H.

b. No chiral center is present. The carbon atom is attached to only *three* groups.

c. No chiral center is present. Two of the groups attached to the carbon atom are identical.

d. The chirality rules for ring carbon atoms are the same as those for acyclic carbon atoms. A chiral center is present. Two of the groups are —H and —Br. The third group, obtained by proceeding clockwise around the ring, is —CH₂—CH₂—CH₂. The fourth group, obtained by proceeding counterclockwise around the ring, is —CH₂—CHBr—CH₂.

(continued)

Practice Exercise 18.1

Indicate whether the circled carbon atom in each of the following molecules is a chiral center.

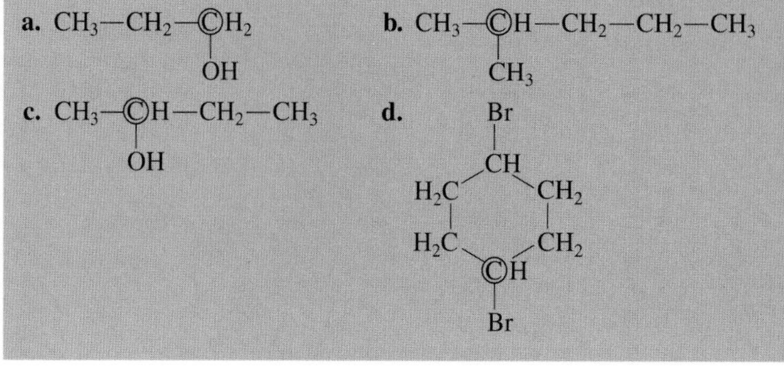

a. CH_3—CH_2—$\textcircled{C}H_2$
　　　　　　　　　|
　　　　　　　　OH

b. CH_3—$\textcircled{C}H$—CH_2—CH_2—CH_3
　　　　　|
　　　CH_3

c. CH_3—$\textcircled{C}H$—CH_2—CH_3
　　　　　|
　　　　OH

d.
　　　　　Br
　　　　　|
　　　　CH
　H_2C　　CH_2
　H_2C　　CH_2
　　　$\textcircled{C}H$
　　　　|
　　　　Br

> Remember the meaning of the structural notations —CHO and —CH_2OH.
>
> 　　　　　　O
> 　　　　　‖
> —CHO　means　—C—H
>
> 　　　　　　　H
> 　　　　　　　|
> —CH_2OH　means　—C—OH
> 　　　　　　　|
> 　　　　　　　H

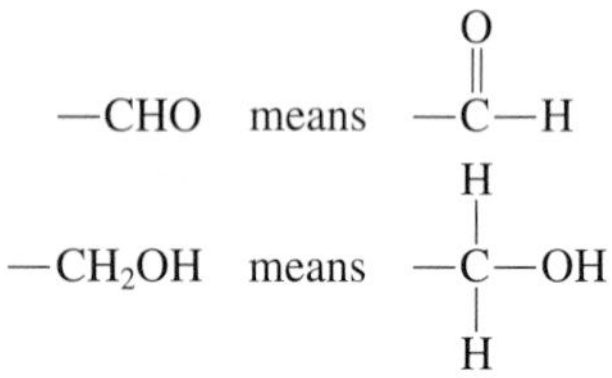

Organic molecules, especially monosaccharides, may contain more than one chiral center. For example, the following monosaccharide has two chiral centers.

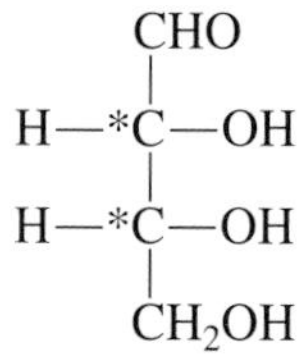

$$
\begin{array}{c}
CHO \\
H-\overset{*}{C}-OH \\
H-\overset{*}{C}-OH \\
CH_2OH
\end{array}
$$

What is the importance of the handedness that we have been discussing? In human body chemistry, right-handed and left-handed forms of a molecule often elicit different responses within the body. Sometimes both forms are biologically active, each form giving a different response; sometimes both elicit the same response, but one form's response is many times greater than that of the other; and sometimes only one of the two forms is biochemically active. For example, studies show that the body's response to the right-handed form of the hormone epinephrine (Section 17.9) is 20 times greater than its response to the left-handed form.

Naturally occurring monosaccharides are almost always "right-handed." Plants, our dietary source for carbohydrates, produce only right-handed monosaccharides. Interestingly, when we consider protein chemistry (Chapter 20), we will find that amino acids, the building blocks for proteins, are always left-handed molecules.

18.5 Stereoisomerism: Enantiomers and Diastereomers

The left- and right-handed forms of a chiral molecule are isomers. They are not *structural isomers,* the type of isomerism that we encountered repeatedly in the organic chemistry chapters of the text, but rather are *stereoisomers.* **Stereoisomers** *are isomers whose atoms are connected in the same way but which differ in the orientation of these atoms in space.* By contrast, atoms are connected to each other in different ways in structural isomers (Section 12.6).

There are two major structural features that generate *stereoisomerism:* (1) the presence of a chiral center in a molecule and (2) the presence of "structural rigidity" in a molecule. Structural rigidity is caused by restricted rotation about chemical bonds. It is the basis for *cis–trans* isomerism, a phenomenon found in some substituted cycloalkanes (Section 12.13) and some alkenes (Section 13.4). Thus handedness is our second encounter with stereoisomerism. (When we discussed *cis–trans* isomerism, we did not mention that it is a form of stereoisomerism.)

> The term *enantiomer* comes from the Greek *enantios,* which means "opposite." It is pronounced "en-AN-tee-o-mer."

Stereoisomers can be subdivided into two types: *enantiomers* and *diastereomers* (Figure 18.5). **Enantiomers** *are stereoisomers whose molecules are nonsuperimposable*

FIGURE 18.6 (a) Enantiomers are stereoisomers whose molecules are non-superimposable mirror images of each other, as in left-handed and right-handed forms of a molecule. (b) Diastereomers are stereoisomers whose molecules are not mirror images of each other.

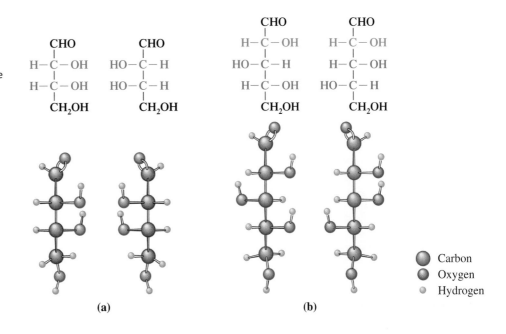

(a)   (b)

Carbon
Oxygen
Hydrogen

mirror images of each other. Left- and right-handed forms of a molecule with a single chiral center are enantiomers.

> Some textbooks use the term *diastereoisomers* instead of *diastereomers*. The pronunciation for *diastereomer* is "dye-a-STEER-ee-o-mer."

Diastereomers *are stereoisomers whose molecules are not mirror images of each other. Cis–trans* isomers (of both the alkene and the cycloalkane types) are diastereomers. Molecules that contain more than one chiral center can also exist in diastereomeric as well as enantiomeric forms, as is shown in Figure 18.6.

Figure 18.7 shows the "thinking pattern" involved in using the terms *stereoisomers, enantiomers,* and *diastereomers.*

FIGURE 18.7 A summary of the "thought process" used in classifying molecules as enantiomers or diastereomers.

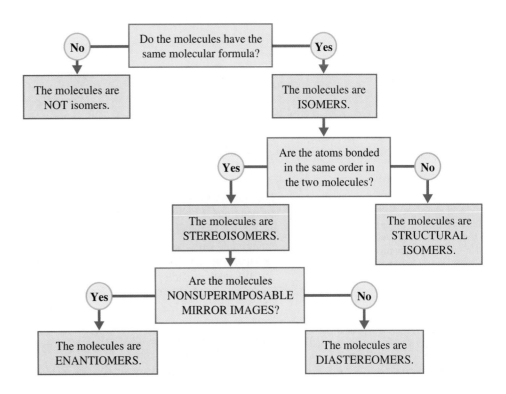

18.6 Designating Handedness Using Fischer Projections

Fischer projections carry the name of their originator, the German chemist Hermann Emil Fischer (see Figure 18.8).

FIGURE 18.8 The German chemist Hermann Emil Fischer (1852–1919), the developer of the two-dimensional system for specifying chirality, was one of the early greats in organic chemistry. He made many fundamental discoveries about carbohydrates, proteins, and other natural products. In 1902 he was awarded the second Nobel Prize in chemistry.

Drawing *three*-dimensional representations of chiral molecules, as in Figure 18.6, can be both time-consuming and awkward. Fischer projections represent a method for giving molecular chirality specifications in *two* dimensions. A **Fischer projection** *is a two-dimensional structural notation for showing the spatial arrangement of groups about chiral centers in molecules.*

In a Fischer projection, a chiral center is represented as the intersection of vertical and horizontal lines. The atom at the chiral center, which is almost always carbon, is not explicitly shown.

The tetrahedral arrangement of the four groups attached to the atom at the chiral center is governed by the following conventions: (1) Vertical lines from the chiral center represent bonds to groups directed into the printed page. (2) Horizontal lines from the chiral center represent bonds to groups directed out of the printed page.

Fischer projection

In Fischer projections for monosaccharides (the simplest type of carbohydrate; Section 18.2), the monosaccharide *carbon chain is* positioned vertically with the carbonyl group (aldehyde or ketone) at or near the top. The smallest monosaccharide that has a chiral center is the compound glyceraldehyde (2,3-dihydroxypropanal; Section 18.4).

$$CH_2-*CH-\overset{\overset{\displaystyle O}{\|}}{C}-H$$
$$\quad | \qquad |$$
$$OH \quad OH$$

The Fischer projections for the two enantiomers of glyceraldehyde are

L-Glyceraldehyde   D-Glyceraldehyde

The D and L designations for the handedness of the two members of an enantiomeric pair come from the Latin words *dextro,* which means "right," and *levo,* which means "left."

The handedness (right and left) of these two enantiomers is specified by using the designations D and L. The enantiomer with the chiral center —OH group on the right in the Fischer projection is by definition the right-handed isomer (D-glyceraldehyde), and the enantiomer with the chiral center —OH group on the left in the Fischer projection is by definition the left-handed isomer (L-glyceraldehyde).

We now consider Fischer projections for the compound 2,3,4-trihydroxybutanal, a monosaccharide with four carbons and *two* chiral centers.

$$CH_2-*CH-*CH-\overset{\overset{\displaystyle O}{\|}}{C}-H$$
$$\quad | \qquad | \qquad |$$
$$OH \quad OH \quad OH$$

There are four stereoisomers for this compound—two pairs of enantiomers.

$$
\begin{array}{cccc}
\text{CHO} & \text{CHO} & \text{CHO} & \text{CHO} \\
\text{H} \!-\!\! \text{OH} & \text{HO} \!-\!\! \text{H} & \text{HO} \!-\!\! \text{H} & \text{H} \!-\!\! \text{OH} \\
\text{H} \!-\!\! \text{OH} & \text{HO} \!-\!\! \text{H} & \text{H} \!-\!\! \text{OH} & \text{HO} \!-\!\! \text{H} \\
\text{CH}_2\text{OH} & \text{CH}_2\text{OH} & \text{CH}_2\text{OH} & \text{CH}_2\text{OH}
\end{array}
$$

First enantiomeric pair   Second enantiomeric pair

> Any given molecular structure can have only one mirror image. Hence enantiomers always come in pairs; there can never be more than two.

In the first enantiomeric pair, both chiral-center —OH groups are on the same side of the Fischer projection, and in the second enantiomeric pair, the chiral-center —OH groups are on opposite sides of the Fischer projection. These are the only —OH group arrangements possible.

The D,L system used to designate the handedness of glyceraldehyde enantiomers is extended to monosaccharides with more than one chiral center in the following manner. The carbon chain is numbered, starting at the carbonyl group end of the molecule, and the highest-numbered chiral center is used to determine D or L configuration.

$$
\begin{array}{cccc}
\text{CHO} & \text{CHO} & \text{CHO} & \text{CHO} \\
\text{H}\overset{2}{-}\text{OH} & \text{HO}\overset{2}{-}\text{H} & \text{HO}\overset{2}{-}\text{H} & \text{H}\overset{2}{-}\text{OH} \\
\text{H}\overset{3}{-}\text{OH} & \text{HO}\overset{3}{-}\text{H} & \text{H}\overset{3}{-}\text{OH} & \text{HO}\overset{3}{-}\text{H} \\
\overset{4}{\text{CH}_2\text{OH}} & \overset{4}{\text{CH}_2\text{OH}} & \overset{4}{\text{CH}_2\text{OH}} & \overset{4}{\text{CH}_2\text{OH}} \\
\textbf{A} & \textbf{B} & \textbf{C} & \textbf{D} \\
\text{D isomer} & \text{L isomer} & \text{D isomer} & \text{L isomer}
\end{array}
$$

The D,L nomenclature gives the configuration (handedness) only at the highest-numbered chiral center. The configuration at other chiral centers in a molecule is accounted for by assigning a different common name to each pair of D,L enantiomers. In our present example, compounds A and B (the first enantiomeric pair) are D-erythrose and L-erythrose; compounds C and D (the second enantiomeric pair) are D-threose and L-threose.

What is the relationship between compounds A and C in our present example? They are diastereomers (Section 18.5), stereoisomers that are not mirror images of each other. Other diastereomeric pairs in our example are A and D, B and C, and B and D. The members of each of these four pairs are epimers. **Epimers** *are diastereomers whose molecules differ only in the configuration at one chiral center.*

> Diastereomers that have two chiral centers must have the same handedness (both left or both right) at one chiral center and opposite handedness (one left and one right) at the other chiral center.

EXAMPLE 18.2

Drawing Fischer Projections for Monosaccharides

■ Draw a Fischer projection for the enantiomer of each of the following monosaccharides.

a.
$$
\begin{array}{c}
\text{CHO} \\
\text{H} \!-\!\! \text{OH} \\
\text{H} \!-\!\! \text{OH} \\
\text{H} \!-\!\! \text{OH} \\
\text{CH}_2\text{OH}
\end{array}
$$

b.
$$
\begin{array}{c}
\text{CH}_2\text{OH} \\
\text{C}\!=\!\text{O} \\
\text{HO} \!-\!\! \text{H} \\
\text{H} \!-\!\! \text{OH} \\
\text{CH}_2\text{OH}
\end{array}
$$

Solution

Given the Fischer projection of one member of an enantiomeric pair, we draw the other enantiomer's Fischer projection by reversing the substituents that are in horizontal positions *at each chiral center.*

(continued)

a. Three chiral centers are present in this polyhydroxy aldehyde. Reversing the positions of the —H and —OH groups at each chiral center produces the Fischer projection of the other enantiomer.

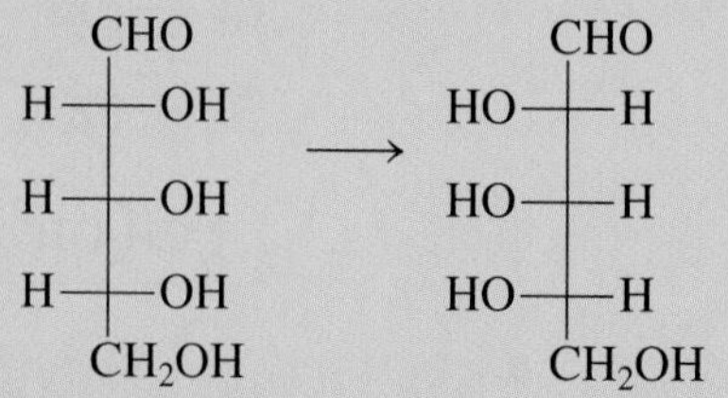

$$
\begin{array}{ccc}
\text{CHO} & & \text{CHO} \\
\text{H}-\!\!\!-\text{OH} & & \text{HO}-\!\!\!-\text{H} \\
\text{H}-\!\!\!-\text{OH} & \longrightarrow & \text{HO}-\!\!\!-\text{H} \\
\text{H}-\!\!\!-\text{OH} & & \text{HO}-\!\!\!-\text{H} \\
\text{CH}_2\text{OH} & & \text{CH}_2\text{OH}
\end{array}
$$

The given enantiomer   The other enantiomer

b. This monosaccharide is a polyhydroxy ketone with two chiral centers. Reversing the positions of the —H and —OH groups at both chiral centers generates the Fischer projection of the other enantiomer.

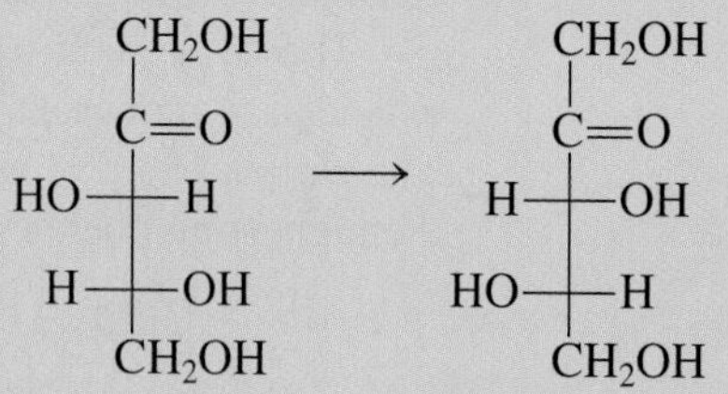

$$
\begin{array}{ccc}
\text{CH}_2\text{OH} & & \text{CH}_2\text{OH} \\
\text{C}=\text{O} & & \text{C}=\text{O} \\
\text{HO}-\!\!\!-\text{H} & \longrightarrow & \text{H}-\!\!\!-\text{OH} \\
\text{H}-\!\!\!-\text{OH} & & \text{HO}-\!\!\!-\text{H} \\
\text{CH}_2\text{OH} & & \text{CH}_2\text{OH}
\end{array}
$$

The given enantiomer   The other enantiomer

Practice Exercise 18.2

Draw a Fischer projection for the enantiomer of each of the following monosaccharides.

a.
$$
\begin{array}{c}
\text{CHO} \\
\text{H}-\!\!\!-\text{OH} \\
\text{HO}-\!\!\!-\text{H} \\
\text{HO}-\!\!\!-\text{H} \\
\text{CH}_2\text{OH}
\end{array}
$$

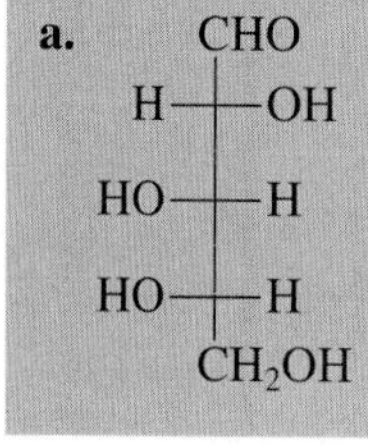

b.
$$
\begin{array}{c}
\text{CH}_2\text{OH} \\
\text{C}=\text{O} \\
\text{H}-\!\!\!-\text{OH} \\
\text{H}-\!\!\!-\text{OH} \\
\text{CH}_2\text{OH}
\end{array}
$$

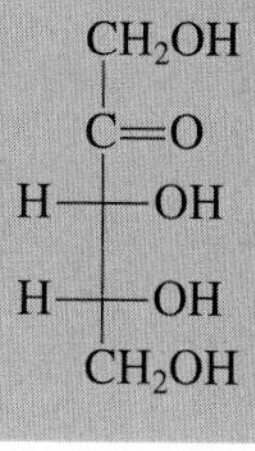

EXAMPLE 18.3

Classifying Monosaccharides as D or L Enantiomers

■ Classify each of the following monosaccharides as a D enantiomer or an L enantiomer.

a.
$$
\begin{array}{c}
{}^1\text{CHO} \\
\text{HO}-\!\!\!-^2\!\!\!-\text{H} \\
\text{H}-\!\!\!-^3\!\!\!-\text{OH} \\
\text{H}-\!\!\!-^4\!\!\!-\text{OH} \\
{}^5\text{CH}_2\text{OH}
\end{array}
$$

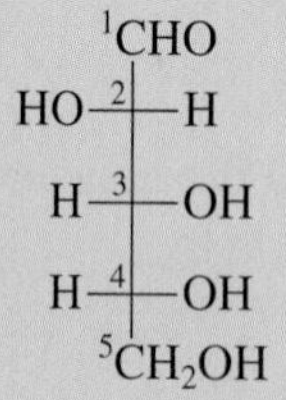

b.
$$
\begin{array}{c}
{}^1\text{CH}_2\text{OH} \\
{}^2\text{C}=\text{O} \\
\text{H}-\!\!\!-^3\!\!\!-\text{OH} \\
\text{HO}-\!\!\!-^4\!\!\!-\text{H} \\
\text{HO}-\!\!\!-^5\!\!\!-\text{H} \\
{}^6\text{CH}_2\text{OH}
\end{array}
$$

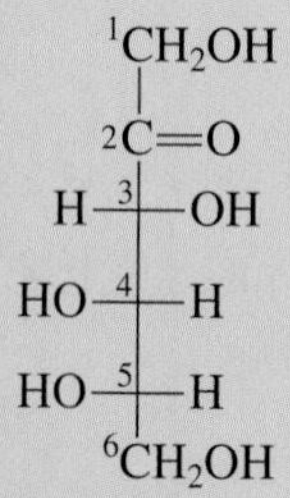

Solution

D or L configuration for a monosaccharide is determined by the highest-numbered chiral center, the one farthest from the carbonyl carbon atom.

a. The highest-numbered chiral center, which involves carbon 4, has the —OH group on the right. Thus, this monosaccharide is a D enantiomer.

b. The highest-numbered chiral center, which involves carbon 5, has the —OH group on the left. Thus, this monosaccharide is an L enantiomer.

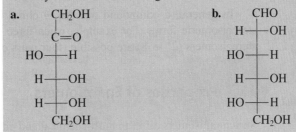
EXAMPLE 18.4

Recognizing Enantiomers and Diastereomers

■ Characterize each of the following pairs of structures as enantiomers, diastereomers, or neither enantiomers nor diastereomers.

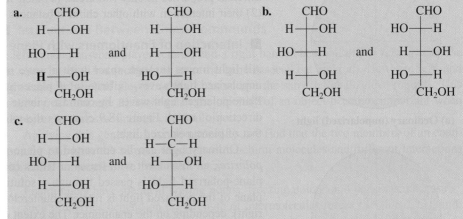

Solution

a. These two structures represent *diastereomers*—the arrangement of —H and —OH substituents is identical for at least one chiral center, whereas the arrangement of —H and —OH substituents at remaining chiral centers is that of mirror images. The —H and —OH substituent arrangement is the same at the first chiral center and is that of mirror images at the second and third chiral centers.

b. These two structures represent *enantiomers*—a mirror-image substituent relationship exists between the two isomers at *every* chiral center.

c. These two structures are *neither enantiomers nor diastereomers*. The connectivity of atoms differs in the two structures at carbon 2. Stereoisomers (enantiomers and diastereomers) must have the same connectivity throughout both structures. (The two structures are not even structural isomers because the first structure contains one more oxygen atom than the second.)

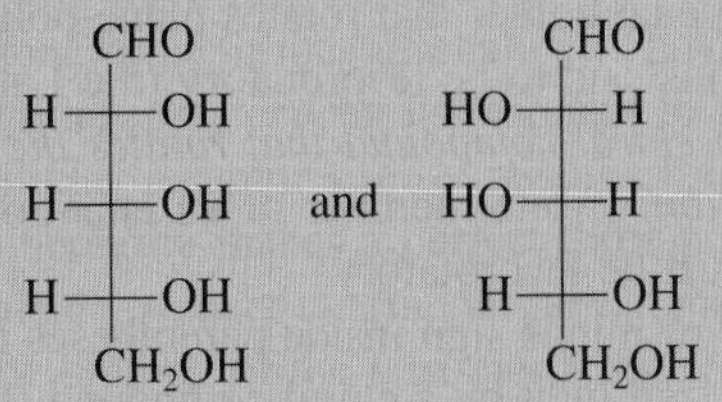

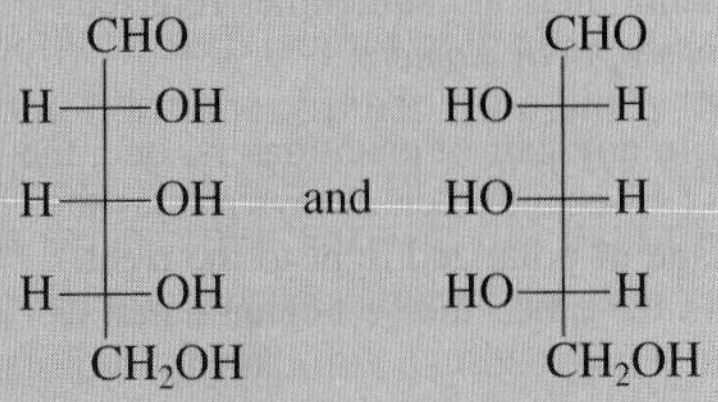

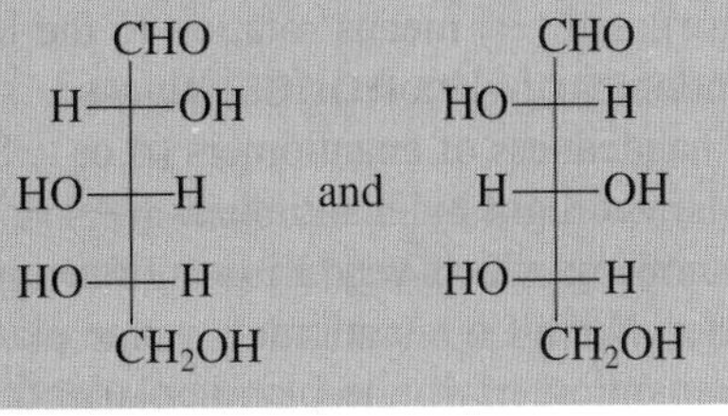

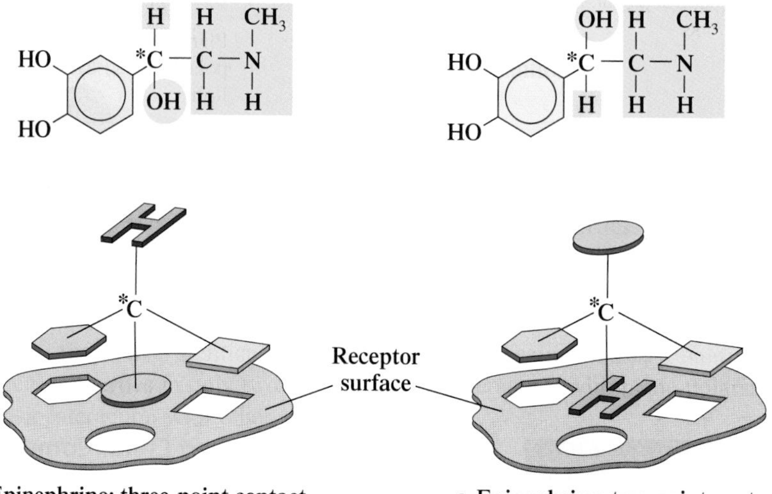
HO
HO
H H CH3
*C C N
OH H H
OH H CH3
*C C N
H H H
*C
Receptor
surface

D-Epinephrine: three-point contact, positive response

L-Epinephrine: two-point contact, much smaller response

FIGURE 18.12 D-Epinephrine binds to the receptor at three points, whereas the biochemically weaker L-epinephrine binds at only two sites.

found in nature. A three-carbon monosaccharide is called a *triose,* and those that contain four, five, and six carbon atoms are called *tetroses*, *pentoses*, and *hexoses*, respectively.

Monosaccharides are classified as *aldoses* or *ketoses* on the basis of type of carbonyl group (Section 15.1) present. An **aldose** *is a monosaccharide that contains an aldehyde functional group.* Aldoses are polyhydroxy aldehydes. A **ketose** *is a monosaccharide that contains a ketone functional group.* Ketoses are polyhydroxy ketones.

Monosaccharides are often classified by both their number of carbon atoms and their functional group. A six-carbon monosaccharide with an aldehyde functional group is an *aldohexose;* a five-carbon monosaccharide with a ketone functional group is a *ketopentose.*

The Latin word for "sugar" is *saccharum,* from whence comes the term *saccharide.*

Monosaccharides are also often called sugars. Hexoses are six-carbon sugars, pentoses five-carbon sugars, and so on. The word *sugar* is associated with "sweetness," and most (but not all) monosaccharides have a sweet taste. The designation *sugar* is also applied to disaccharides, many of which also have a sweet taste. Thus **sugar** *is a general designation for either a monosaccharide or a disaccharide.*

EXAMPLE 18.5

Classifying Monosaccharides on the Basis of Structural Characteristics

■ Classify each of the following monosaccharides according to both the number of carbon atoms and the type of carbonyl group present.

a.

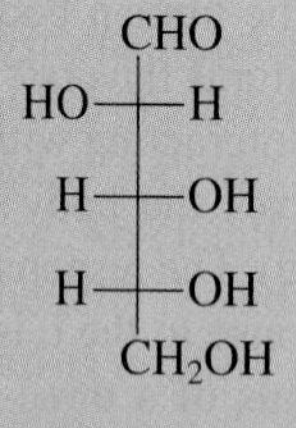
CHO
HO H
H OH
H OH
CH2OH

b.

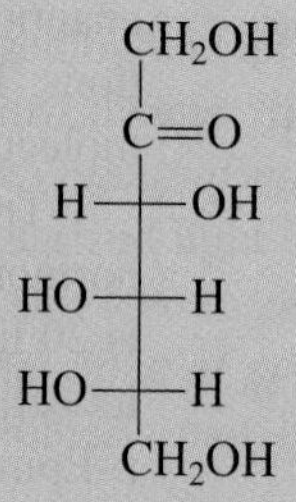
CH2OH
C=O
H OH
HO H
HO H
CH2OH

c.

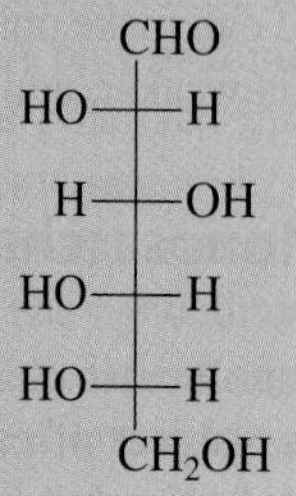
CHO
HO H
H OH
HO H
HO H
CH2OH

d.

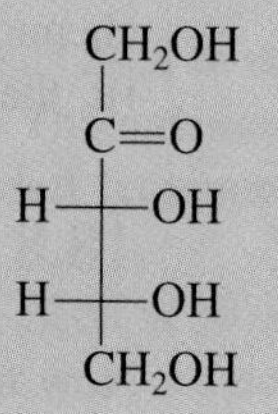
CH2OH
C=O
H OH
H OH
CH2OH

(continued)

Solution

a. An aldehyde functional group is present as well as five carbon atoms. This monosaccharide is thus an *aldopentose*.

b. This monosaccharide contains a ketone group and six carbon atoms, so it is a *ketohexose*.

c. Six carbon atoms and an aldehyde group in a monosaccharide are characteristic of an *aldohexose*.

d. This monosaccharide is a *ketopentose*.

Practice Exercise 18.5

Classify each of the following monosaccharides according to both the number of carbon atoms and the type of carbonyl group present.

a.
```
    CH2OH
     |
    C=O
HO——H
 H——OH
 H——OH
    CH2OH
```

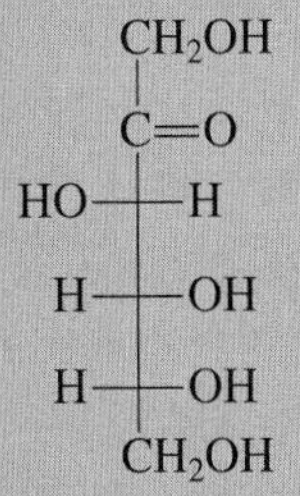

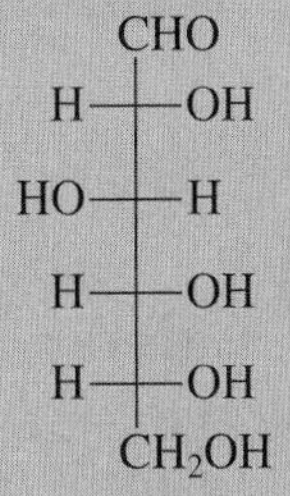

b.
```
    CHO
 H——OH
HO——H
 H——OH
 H——OH
    CH2OH
```

c.
```
    CHO
 H——OH
 H——OH
    CH2OH
```

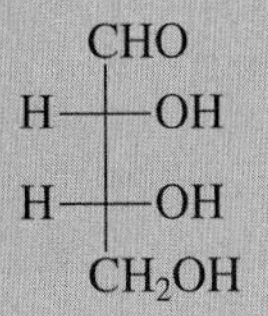

d.
```
    CH2OH
     |
    C=O
 H——OH
HO——H
    CH2OH
```

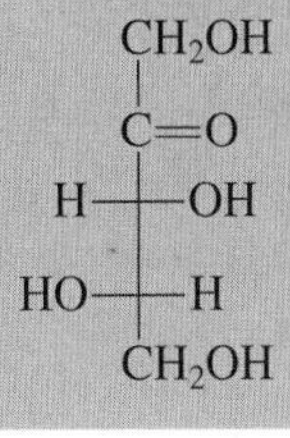

▶ Nearly all naturally occurring monosaccharides are D isomers. These D monosaccharides are important energy sources for the human body. L Monosaccharides, which can be produced in the laboratory, cannot be used by the body as energy sources. Body enzymes are specific for D isomers.

In terms of carbon atoms, trioses are the smallest monosaccharides that can exist. There are two such compounds, one an aldose (glyceraldehyde) and the other a ketose (dihydroxyacetone).

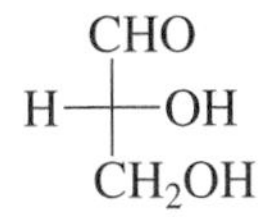

```
    CHO              CH2OH
 H——OH                |
    CH2OH            C=O
                      |
                     CH2OH
 D-Glyceraldehyde   Dihydroxyacetone
```

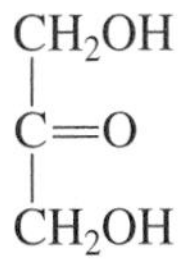

These two triose structures serve as the reference points for consideration of the structures of aldoses and ketoses that contain more carbon atoms.

The structures of all D aldoses containing three, four, five, and six carbon atoms are given in Figure 18.13. Figure 18.13 starts with the triose glyceraldehyde at the top and proceeds downward through the tetroses, pentoses, and hexoses. The number of possible aldoses doubles each time an additional carbon atom is added, because the new carbon atom is a chiral center. Glyceraldehyde has one chiral center, the tetroses two chiral centers, the pentoses three chiral centers, and the hexoses four chiral centers.

In aldose structures such as those shown in Figure 18.13, the chiral center farthest from the aldehyde group determines the D or L designation for the aldose. The configurations about the other chiral centers present are accounted for by assigning a different common name to each set of D and L enantiomers. (Only the D isomer is shown in Figure 18.13; the L isomer is the mirror image of the structure shown.)

A major difference between glyceraldehyde and dihydroxyacetone is that the latter does not possess a chiral carbon atom. Thus D and L forms are not possible for dihydroxyacetone. This reduces by half (compared with aldoses) the number of stereoisomers possible for ketotetroses, ketopentoses, and ketohexoses. An aldohexose has four chiral

▶ All monosaccharides have names that end in *-ose* except the trioses glyceraldehyde and dihydroxyacetone.

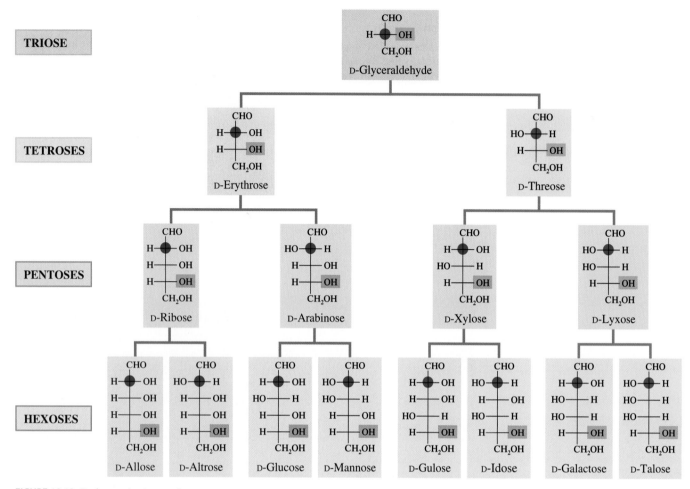

FIGURE 18.13 Fischer projections and common names for D aldoses containing three, four, five, and six carbon atoms. The new chiral-center carbon atom added in going from triose to tetrose to pentose to hexose is marked in color. This new chiral center can have the hydroxyl group at the right or left in the Fischer projection, which doubles the number of stereoisomers. The hydroxyl group that specifies the D configuration is highlighted in red.

▶ You should memorize the structures of the six monosaccharides considered in this section.

carbon atoms, but a ketohexose has only three. Figure 18.14 gives the projection formulas and common names for the D forms of ketoses containing three, four, five, and six carbon atoms.

18.9 Biochemically Important Monosaccharides

Of the many monosaccharides, six that are particularly important in the functioning of the human body are the trioses D-glyceraldehyde and dihydroxyacetone and the D forms of glucose, galactose, fructose, and ribose. Glucose and galactose are aldohexoses, fructose is a ketohexose, and ribose is an aldopentose. All six of these monosaccharides are water-soluble, white, crystalline solids.

■ D-Glyceraldehyde and Dihydroxyacetone

The simplest of the monosaccharides, these two trioses are important intermediates in the process of glycolysis (Section 24.2), a series of reactions whereby glucose is converted into two molecules of pyruvate. D-Glyceraldehyde is a chiral molecule but dihydroxyacetone is not.

FIGURE 18.14 Fischer projections and common names for D ketoses containing three, four, five, and six carbon atoms. The new chiral-center carbon atom added in going from triose to tetrose to pentose to hexose is marked in color. This new chiral center can have the hydroxyl group at the right or left in the Fischer projection, which doubles the number of stereoisomers. The hydroxyl group that specifies the D configuration is highlighted in red.

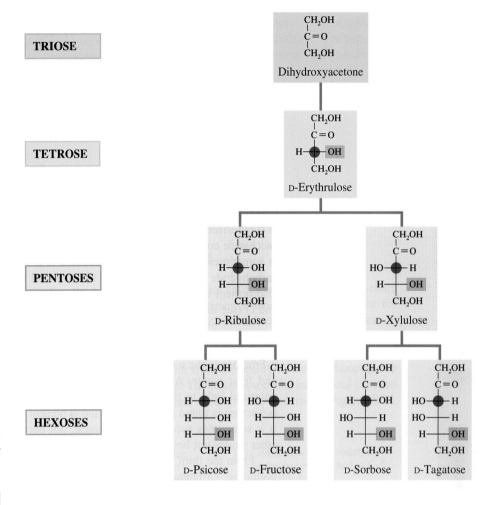

FIGURE 18.15 A 5% (m/v) glucose solution is often used in hospitals as an intravenous source of nourishment for patients who cannot take food by mouth. The body can use it as an energy source without digesting it.

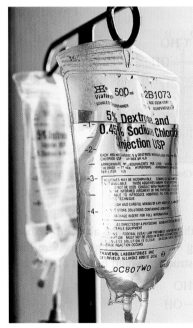

D-Glucose tastes sweet, is nutritious, and is an important component of the human diet. L-Glucose, on the other hand, is tasteless, and the body cannot use it.

■ D-Glucose

Of all monosaccharides, D-glucose is the most abundant in nature and the most important from a human nutritional standpoint. Its Fischer projection is

$$
\begin{array}{c}
\text{CHO} \\
\text{H}-\!\!\!-\text{OH} \\
\text{HO}-\!\!\!-\text{H} \\
\text{H}-\!\!\!-\text{OH} \\
\text{H}-\!\!\!-\text{OH} \\
\text{CH}_2\text{OH}
\end{array}
$$

D-Glucose

Ripe fruits, particularly ripe grapes (20%–30% glucose by mass), are a good source of glucose, which is often referred to as *grape sugar*. Two other names for D-glucose are *dextrose* and *blood sugar*. The name *dextrose* draws attention to the fact that the optically active D-glucose, in aqueous solution, rotates plane-polarized light to the right. The term *blood sugar* draws attention to the fact that blood contains dissolved glucose. The concentration of glucose in human blood is fairly constant; it is in the range of 70–100 mg per 100 mL of blood. Cells use this glucose as a primary energy source (see Figure 18.15).

■ D-Galactose

A comparison of the Fischer projections for D-galactose and D-glucose shows that these two compounds differ only in the configuration of the —OH group and —H group on carbon 4.

In situations where α or β configuration does not matter, the —OH group on carbon 1 is placed in a horizontal position, and a wavy line is used as the bond that connects it to the ring.

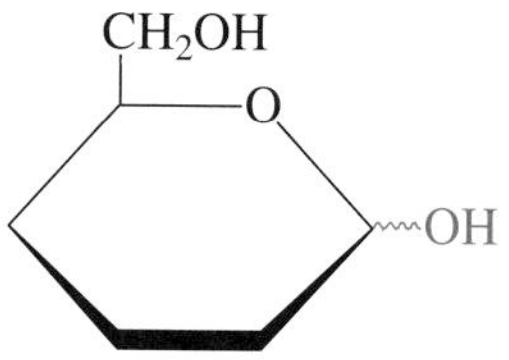

The specific identity of a monosaccharide is determined by the positioning of the other —OH groups in the Haworth projection. Any —OH group at a chiral center that is to the right in a Fischer projection formula points down in the Haworth projection. Any group to the left in a Fischer projection points up in the Haworth projection. The following is a matchup between Haworth projection and Fischer projection.

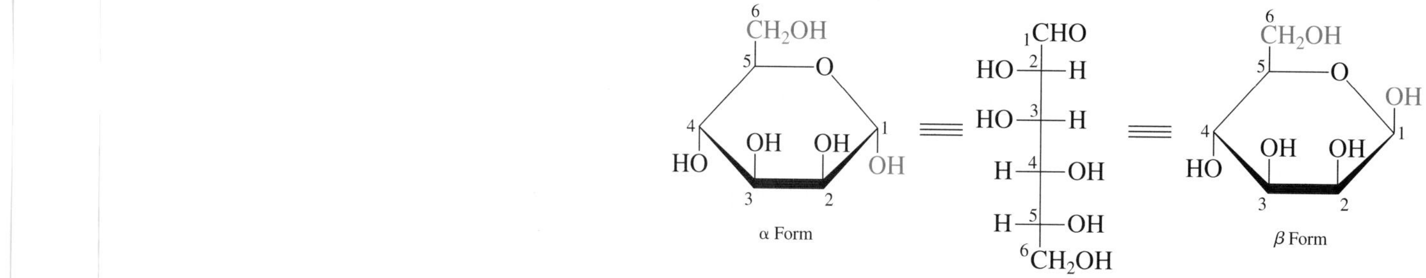

Comparison of this Fischer projection with those given in Figure 18.13 reveals that the monosaccharide is D-mannose.

18.12 Reactions of Monosaccharides

Five important reactions of monosaccharides are oxidation to acidic sugars, reduction to sugar alcohols, glycoside formation, phosphate ester formation, and amino sugar formation. In considering these reactions, we will use glucose as the monosaccharide reactant. Remember, however, that other aldoses, as well as ketoses, undergo similar reactions.

■ Oxidation to Produce Acidic Sugars

The redox chemistry of monosaccharides is closely linked to that of the alcohol and aldehyde functional groups. This latter redox chemistry, which we considered in Chapters 14 and 15, is summarized in the following diagram.

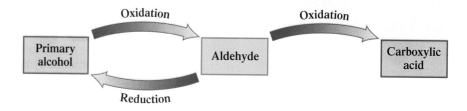

Monosaccharide oxidation can yield three different types of *acidic sugars*. The oxidizing agent used determines the product.

   Weak oxidizing agents, such as Tollens and Benedict's solutions (Section 15.8), oxidize the aldehyde end of an aldose to give an *aldonic acid*. Oxidation of the aldehyde end of glucose produces gluconic acid, and oxidation of the aldehyde end of galactose produces galactonic acid. The structures involved in the glucose reaction are

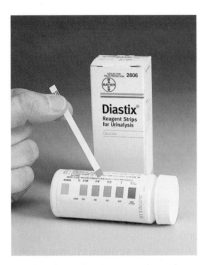

FIGURE 18.19 The glucose content of urine can be determined by dipping a plastic strip treated with oxidizing agents into the urine sample and comparing the color change of the strip to a color chart that indicates glucose concentration.

$$
\begin{array}{ccc}
\boxed{\text{CHO}} & & \boxed{\text{COOH}} \\
\text{H}{-}\text{OH} & & \text{H}{-}\text{OH} \\
\text{HO}{-}\text{H} & \xrightarrow[\text{oxidizing agent}]{\text{Weak}} & \text{HO}{-}\text{H} \\
\text{H}{-}\text{OH} & & \text{H}{-}\text{OH} \\
\text{H}{-}\text{OH} & & \text{H}{-}\text{OH} \\
\text{CH}_2\text{OH} & & \text{CH}_2\text{OH} \\
\text{D-Glucose} & & \text{D-Gluconic acid}
\end{array}
$$

Because aldoses act as reducing agents in such reactions, they are called *reducing sugars*. With Tollens solution, glucose reduces Ag^+ ion to Ag, and with Benedict's solution, glucose reduces Cu^{2+} ion to Cu^+ ion (see Section 15.8). A **reducing sugar** *is a carbohydrate that gives a positive test with Tollens and Benedict's solutions.*

Under the basic conditions associated with Tollens and Benedict's solutions, ketoses are also reducing sugars. In this situation the ketose undergoes a structural rearrangement that produces an aldose, and the aldose then reacts. Thus all monosaccharides, both aldoses and ketoses, are reducing sugars.

Tollens and Benedict's solutions can be used to test for glucose in urine, a symptom of diabetes. For example, using Benedict's solution, we observe that if no glucose is present in the urine (a normal condition), the Benedict's solution remains blue. The presence of glucose is indicated by the formation of a red precipitate. Testing for the presence of glucose in urine is such a common laboratory procedure that much effort has been put into the development of easy-to-use test methods (Figure 18.19).

Strong oxidizing agents can oxidize both ends of a monosaccharide at the same time (the carbonyl group and the terminal primary alcohol group) to produce a dicarboxylic acid. Such polyhydroxy dicarboxylic acids are known as *aldaric acids*. For glucose, this oxidation produces glucaric acid.

$$
\begin{array}{ccc}
\boxed{\text{CHO}} & & \boxed{\text{COOH}} \\
\text{H}{-}\text{OH} & & \text{H}{-}\text{OH} \\
\text{HO}{-}\text{H} & \xrightarrow[\text{oxidizing agent}]{\text{Strong}} & \text{HO}{-}\text{H} \\
\text{H}{-}\text{OH} & & \text{H}{-}\text{OH} \\
\text{H}{-}\text{OH} & & \text{H}{-}\text{OH} \\
\boxed{\text{CH}_2\text{OH}} & & \boxed{\text{COOH}} \\
\text{D-Glucose} & & \text{D-Glucaric acid}
\end{array}
$$

Although it is difficult to do in the laboratory, in biochemical systems enzymes can oxidize the primary alcohol end of an aldose such as glucose, without oxidation of the aldehyde group, to produce an *alduronic acid*. For glucose, such an oxidation produces D-glucuronic acid.

$$
\begin{array}{ccc}
\text{CHO} & & \text{CHO} \\
\text{H}{-}\text{OH} & & \text{H}{-}\text{OH} \\
\text{HO}{-}\text{H} & \xrightarrow{\text{Enzymes}} & \text{HO}{-}\text{H} \\
\text{H}{-}\text{OH} & & \text{H}{-}\text{OH} \\
\text{H}{-}\text{OH} & & \text{H}{-}\text{OH} \\
\boxed{\text{CH}_2\text{OH}} & & \boxed{\text{COOH}} \\
\text{D-Glucose} & & \text{D-Glucuronic acid}
\end{array}
$$

■ Reduction to Produce Sugar Alcohols

The carbonyl group present in a monosaccharide (either an aldose or a ketose) can be reduced to a hydroxyl group, using hydrogen as the reducing agent. For aldoses and ketoses, the product of the reduction is the corresponding polyhydroxy alcohol, which is sometimes called a *sugar alcohol.* For example, the reduction of D-glucose gives D-glucitol.

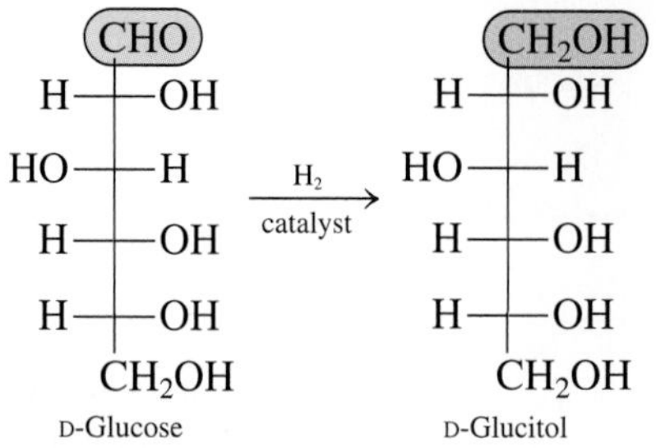

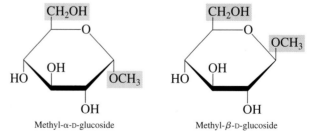

CHO
H——OH
HO——H
H——OH
H——OH
CH$_2$OH
D-Glucose

$\xrightarrow[\text{catalyst}]{H_2}$

CH$_2$OH
H——OH
HO——H
H——OH
H——OH
CH$_2$OH
D-Glucitol

D-Glucitol is also known by the common name D-sorbitol. Hexahydroxy alcohols such as D-sorbitol have properties similar to those of the trihydroxy alcohol *glycerol* (Section 14.4). These alcohols are used as moisturizing agents in foods and cosmetics because of their affinity for water. D-Sorbitol is also used as a sweetening agent in chewing gum; bacteria that cause tooth decay cannot use polyalcohols as food sources, as they can glucose and many other monosaccharides.

> ▶ D-Sorbitol accumulation in the eye is a major factor in the formation of cataracts due to diabetes.

■ Glycoside Formation

> ▶ Remember, from Section 15.9, that acetals have two —OR groups attached to the same carbon atom.

In Section 15.9 we learned that hemiacetals can react with alcohols in acid solution to produce acetals. Because the cyclic forms of monosaccharides are hemiacetals, they react with alcohols to form acetals, as is illustrated here for the reaction of β-D-glucose with methyl alcohol.

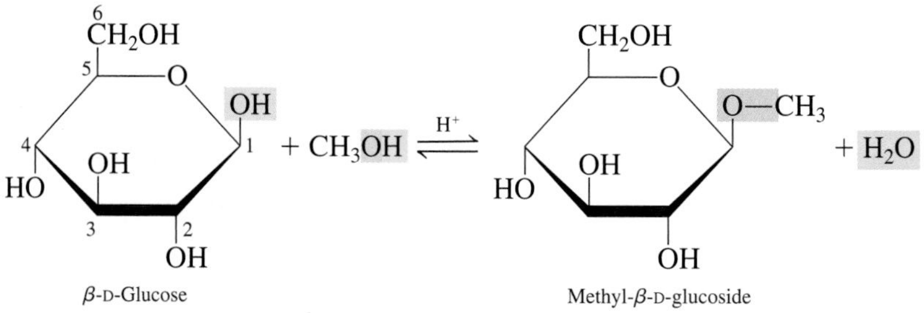

β-D-Glucose + CH$_3$OH $\xrightleftharpoons{\ \text{H}^+\ }$ Methyl-β-D-glucoside + H$_2$O

The general name for monosaccharide acetals is *glycoside.* A **glycoside** *is an acetal formed from a cyclic monosaccharide by replacement of the hemiacetal carbon —OH group with an —OR group.* More specifically, a glycoside produced from glucose is called a glucoside, that from galactose is called a galactoside, and so on. Glycosides, like the hemiacetals from which they are formed, can exist in both α and β forms. Glycosides are named by listing the alkyl or aryl group attached to the oxygen, followed by the name of the monosaccharide involved, with the suffix *-ide* appended to it.

Methyl-α-D-glucoside   Methyl-β-D-glucoside

■ Phosphate Ester Formation

The hydroxyl groups of a monosaccharide can react with inorganic oxyacids to form inorganic esters (Section 16.18). Phosphate esters, formed from phosphoric acid and various monosaccharides, are commonly encountered in biochemical systems. For example,

CHEMICAL CONNECTIONS

Blood Types and Monosaccharides

Human blood is classified into four types: A, B, AB, and O. If a blood transfusion is necessary and the patient's own blood is not available, the donor's blood must be matched to that of the patient. Blood of one type cannot be given to a recipient with blood of another type unless the two types are compatible. A transfusion of the wrong blood type can cause the blood cells to form clumps, a potentially fatal reaction. The following table shows compatibility relationships. People with type O blood are universal donors, and those with type AB blood are universal recipients.

Human Blood Group Compatibilities

	Recipient Blood Type			
Donor blood type	A	B	AB	O
A	+	−	+	−
B	−	+	+	−
AB	−	−	+	−
O	+	+	+	+

+ = compatible; − = incompatible

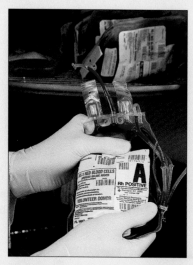

A unit of blood obtained from a blood bank.

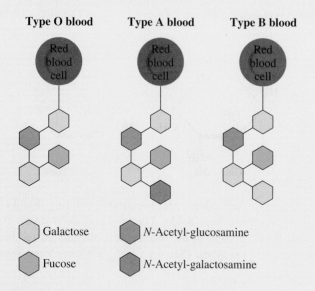

Type O blood **Type A blood** **Type B blood**

Galactose

Fucose

N-Acetyl-glucosamine

N-Acetyl-galactosamine

In the United States, sampling studies show that 41% of the population has type A blood, 10% type B, 4% type AB, and 45% type O.

The biochemical basis for the various blood types involves monosaccharides. The plasma membranes of red blood cells carry biochemical markers made up of monosaccharides. Four monosaccharides are involved in the "marking system." One is the simple monosaccharide D-galactose and the other three are monosaccharide derivatives. Two of these are *N*-acetyl amino derivatives (Section 18.12), those of D-glucose and D-galactose. The third is L-fucose (6-deoxy-L-galactose), an L-galactose derivative in which the oxygen atom at carbon 6 has been removed (converting the

—CH_2OH group to a —CH_3 group). The L configuration of this derivative is unusual in that L-monosaccharides are seldom found in the human body. The arrangement of these monosaccharides in the biochemical marker determines blood type.

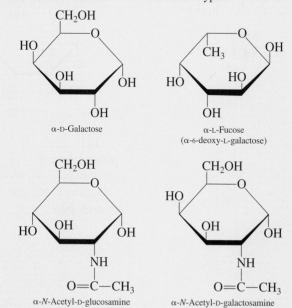

α-D-Galactose

α-L-Fucose
(α-6-deoxy-L-galactose)

α-*N*-Acetyl-D-glucosamine

α-*N*-Acetyl-D-galactosamine

Note that all of the biochemical markers have a common structural portion that involves four monosaccharide units. Type A markers differ from type O markers in that an *N*-acetyl galactosamine unit is also present. In type B markers, a second galactose unit is present. Type AB blood contains both type A and type B markers.

specific enzymes in the human body catalyze the esterification of the carbonyl group (carbon 1) and the primary alcohol group (carbon 6) in glucose to produce the compounds glucose 1-phosphate and glucose 6-phosphate, respectively.

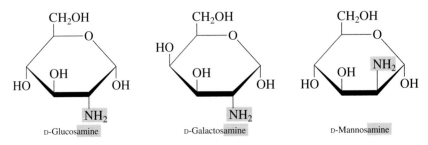

α-D-Glucose 1-phosphate   α-D-Glucose 6-phosphate

These phosphate esters of glucose are stable in aqueous solution and play important roles in the metabolism of carbohydrates.

■ Amino Sugar Formation

> An *acetyl group* has the structure
>
> $$CH_3-\overset{\overset{\displaystyle O}{\|}}{C}-$$
>
> It can be considered to be derived from acetic acid by removal of the —OH portion of that structure.
>
> $$CH_3-\overset{\overset{\displaystyle O}{\|}}{C}-\boxed{OH} \rightarrow$$
>
> Acetic acid

If one of the hydroxyl groups of a monosaccharide is replaced with an amino group, an amino sugar is produced. In naturally occurring amino sugars, of which there are three common ones, the amino group replaces the carbon 2 hydroxyl group. The three common natural amino sugars are

CH₂OH   CH₂OH   CH₂OH

D-Glucosamine   D-Galactosamine   D-Mannosamine

Amino sugars and their *N*-acetyl derivatives are important building blocks of polysaccharides found in cartilage (Section 18.15). The *N*-acetyl derivatives of D-glucosamine and D-galactosamine are present in the biochemical markers on red blood cells, which distinguish the various blood types. (See the Chemical Connections feature on page 523.)

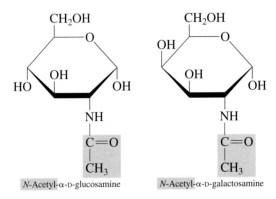

N-Acetyl-α-D-glucosamine   *N*-Acetyl-α-D-galactosamine

The Chemistry at a Glance feature on page 525 summarizes the "sugar terminology" associated with the common types of monosaccharides and monosaccharides derivatives that we have considered so far in this chapter.

"Sugar Terminology" Associated with Monosaccharides and Their Derivatives

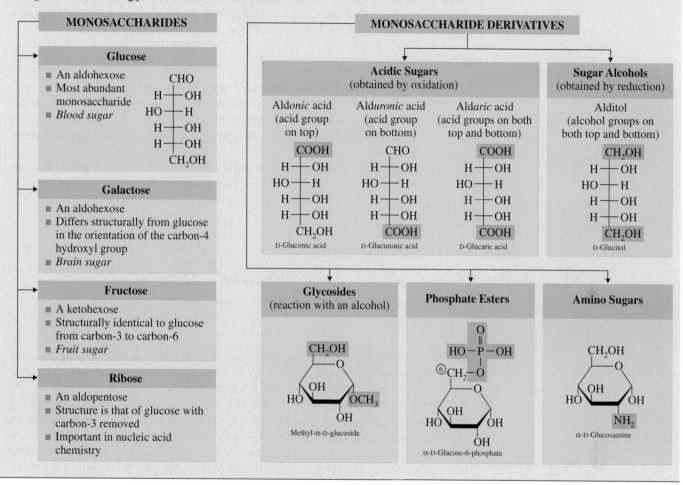

18.13 Disaccharides

A monosaccharide that has cyclic forms (hemiacetal forms) can react with an alcohol to form a glycoside (acetal), as we noted in Section 18.12. This same type of reaction can be used to produce a *disaccharide,* a carbohydrate in which two monosaccharides are bonded together (Section 18.3). In disaccharide formation, one of the monosaccharide reactants functions as a hemiacetal, and the other functions as an alcohol.

Monosaccharide + monosaccharide ⟶ disaccharide + H₂O

$$\text{(Functioning as a hemiacetal)} \qquad \text{(Functioning as an alcohol)} \qquad \text{(Glycoside)}$$

The bond that links the two monosaccharides of a disaccharide (glycoside) together is called a glycosidic linkage. A **glycosidic linkage** *is the bond in a disaccharide resulting from the reaction between the hemiacetal carbon atom —OH group of one monosaccharide and a —OH group on the other monosaccharide.* It is always a carbon–oxygen–carbon bond in a disaccharide.

We now examine the structures and properties of four important disaccharides: maltose, cellobiose, lactose, and sucrose. As we consider details of the structures of these compounds, we will find that the configuration (α or β) at carbon 1 of the reacting monosaccharides that functions as a hemiacetal is of prime importance.

■ Maltose

Maltose, often called *malt sugar,* is produced whenever the polysaccharide starch (Section 18.14) breaks down, as happens in plants when seeds germinate and in human beings during starch digestion. It is a common ingredient in baby foods and is found in malted milk. Malt (germinated barley that has been baked and ground) contains maltose; hence the name *malt sugar*.

Structurally, maltose is made up of two D-glucose units, one of which must be α-D-glucose. The formation of maltose from two glucose molecules is as follows:

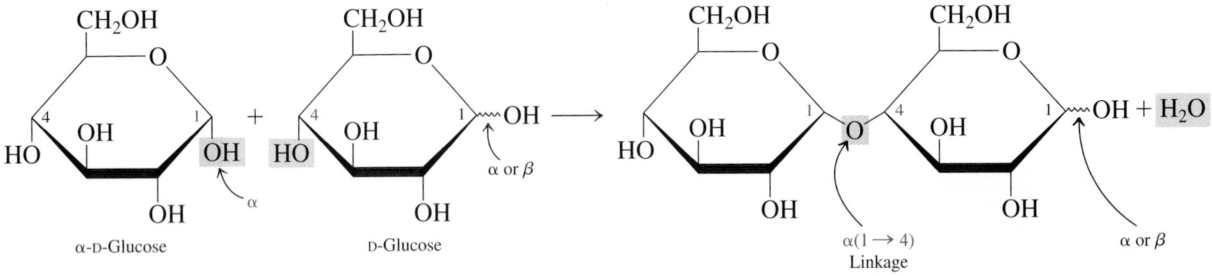

The glycosidic linkage between the two glucose units is called an $\alpha(1 \rightarrow 4)$ linkage. The two —OH groups that form the linkage are attached, respectively, to carbon 1 of the first glucose unit (in an α configuration) and to carbon 4 of the second.

Maltose is a reducing sugar (Section 18.12), because the glucose unit on the right has a hemiacetal carbon atom (C-1). Thus this glucose unit can open and close; it is in equilibrium with its open-chain aldehyde form (Section 18.10). This means there are actually three forms of the maltose molecule: α-maltose, β-maltose, and the open-chain form. Structures for these three maltose forms are shown in Figure 18.20. In the solid state, the β form is dominant.

The most important chemical reaction of maltose is that of hydrolysis. Hydrolysis of D-maltose whether in a laboratory flask or in a living organism, produces two molecules of D-glucose. Acidic conditions or the enzyme *maltase* is needed for the hydrolysis to occur.

$$\text{D-Maltose} + H_2O \xrightarrow{\text{H}^+ \text{ or maltase}} 2 \text{ D-Glucose}$$

■ Cellobiose

Cellobiose is produced as an intermediate in the hydrolysis of the polysaccharide cellulose (Section 18.14). Like maltose, cellobiose contains two D-glucose monosaccharide units. It differs from maltose in that one of the D-glucose units—the one functioning as a hemiacetal—must have a β configuration instead of the α configuration for maltose. This change in configuration results in a $\beta(1 \rightarrow 4)$ glycosidic linkage.

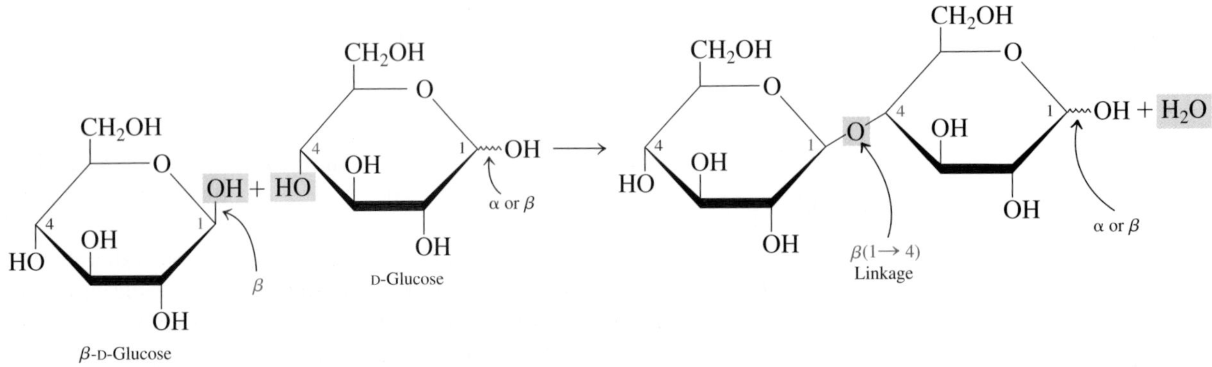

FIGURE 18.20 The three forms of maltose present in aqueous solution.

α-Maltose

β-Maltose

Open-chain aldehyde form

Like maltose, cellobiose is a reducing sugar, has three isomeric forms in aqueous solution, and upon hydrolysis produces two D-glucose molecules.

$$\text{D-Cellobiose} + H_2O \xrightarrow{\text{H}^+ \text{ or cellobiase}} 2 \text{ D-glucose}$$

Despite these similarities, maltose and cellobiose have different biochemical behaviors. These differences are related to the stereochemistry of their glycosidic linkages. Maltase, the enzyme that breaks the glucose–glucose $\alpha(1 \rightarrow 4)$ linkage present in maltose, is found both in the human body and in yeast. Consequently, maltose is digested easily by humans and is readily fermented by yeast. Both the human body and yeast lack the enzyme cellobiase needed to break the glucose–glucose $\beta(1 \rightarrow 4)$ linkage of cellobiose. Thus cellobiose cannot be digested by humans or fermented by yeast.

■ Lactose

In both maltose and cellobiose, the monosaccharide units present are identical—two glucose units in each case. However, the two monosaccharide units in a disaccharide need not be identical. *Lactose* is made up of a β-D-galactose unit and a D-glucose unit joined by a $\beta(1 \rightarrow 4)$ glycosidic linkage.

$\beta(1 \rightarrow 4)$ Glycosidic linkage

β-D-Galactose   D-Glucose

Lactose

> It is important to distinguish between the structural notation used for an $\alpha(1 \rightarrow 4)$ glycosidic linkage and that used for a $\beta(1 \rightarrow 4)$ glycosidic linkage.

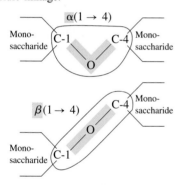

$\alpha(1 \rightarrow 4)$

Mono-saccharide — C-1 ⟍ C-4 — Mono-saccharide
O

$\beta(1 \rightarrow 4)$ — C-4 — Mono-saccharide
O
Mono-saccharide — C-1

> The α form of lactose is sweeter to the taste and more soluble in water than the β form. The β form can be found in ice cream that has been stored for a long time; it crystallizes and gives the ice cream a gritty texture.

The glucose hemiacetal center is unaffected when galactose bonds to glucose in the formation of lactose, so lactose is a reducing sugar (the glucose ring can open to give an aldehyde).

Lactose is the major sugar found in milk. This accounts for its common name, *milk sugar*. Enzymes in mammalian mammary glands take glucose from the bloodstream and synthesize lactose in a four-step process. Epimerization (Section 18.9) of glucose yields galactose, and then the $\beta(1 \rightarrow 4)$ linkage forms between a galactose and a glucose unit.

CHEMICAL CONNECTIONS

Lactose Intolerance and Galactosemia

Lactose is the principal carbohydrate in milk. Human mother's milk obtained by nursing infants contains 7%–8% lactose, almost double the 4%–5% lactose found in cow's milk.

For many people, the digestion and absorption of lactose are a problem. This problem, called *lactose intolerance,* is a condition in which people lack the enzyme *lactase,* which is needed to hydrolyze lactose to galactose and glucose.

$$\text{Lactose} + \text{H}_2\text{O} \xrightarrow{\text{Lactase}} \text{Glucose} + \text{galactose}$$

Deficiency of lactase can be caused by a genetic defect, by physiological decline with age, or by injuries to the mucosa lining the intestines. When lactose molecules remain in the intestine undigested, they attract water to themselves, causing fullness, discomfort, cramping, nausea, and diarrhea. Bacterial fermentation of the lactose further along the intestinal tract produces acid (lactic acid) and gas, adding to the discomfort.

The level of the enzyme lactase in humans varies with age. Most children have sufficient lactase during the early years of their life when milk is a much-needed source of calcium in their diet. In adulthood, the enzyme level decreases, and lactose intolerance develops. This explains the change in milk-drinking habits of many adults. Some researchers estimate that as many as one of three adult Americans exhibits a degree of lactose intolerance.

The level of the enzyme lactase in humans varies widely among ethnic groups, indicating that the trait is genetically determined (inherited). The occurrence of lactose intolerance is lowest among Scandinavians and other northern Europeans and highest among native North Americans, Southeast Asians, Africans, and Greeks. The estimated prevalence of lactose intolerance is as follows:

80% Asian Americans	60% Inuits
80% Native Americans	50% Hispanics
75% African Americans	20% Caucasians
70% Mediterranean peoples	10% Northern Europeans

After lactose has been degraded into glucose and galactose, the galactose has to be converted into glucose before it can be used by cells. In humans, the genetic condition called *galactosemia* is caused by the absence of one or more of the enzymes needed for this conversion. In people with this condition, galactose and its toxic metabolic derivative galactitol (dulcitol) accumulate in the blood.

If not treated, galactosemia can cause mental retardation in infants and even death. Treatment involves exclusion of milk and milk products from the diet.

◀ The enzyme needed to break the $\beta(1 \rightarrow 4)$ linkage in lactose is different from the one needed to break the $\beta(1 \rightarrow 4)$ linkage in cellobiose. Because the two disaccharides have slightly different structures, different enzymes are required—*lactase* for lactose and *cellobiase* for cellobiose.

Lactose is an important ingredient in commercially produced infant formulas that are designed to simulate mother's milk. Souring of milk is caused by the conversion of lactose to lactic acid by bacteria in the milk. Pasteurization of milk is a quick-heating process that kills most of the bacteria and retards the souring process.

Lactose can be hydrolyzed by acid or by the enzyme *lactase,* forming an equimolar mixture of galactose and glucose.

$$\text{D-Lactose} + \text{H}_2\text{O} \xrightarrow{\text{H}^+ \text{ or lactase}} \text{D-galactose} + \text{D-glucose}$$

In the human body, the galactose so produced is then converted to glucose by other enzymes. The genetic condition *lactose intolerance,* an inability of the human digestive system to hydrolyze lactose, is considered in the Chemical Connections feature on this page.

FIGURE 18.21 Space-filling model of the disaccharide sucrose. Average per capita consumption of sucrose in the United States is approximately 100 pounds per year. Two-thirds of this is sucrose that is added to food for extra sweetening.

▶ The glycosidic linkage in sucrose is very different from that in maltose, cellobiose, and lactose. The linkages in the latter three compounds can be characterized as "head-to-tail" linkages—that is, the front end (carbon 1) of one monosaccharide is linked to the back end (carbon 4) of the other monosaccharide. Sucrose has a "head-to-head" glycosidic linkage; the front ends of the two monosaccharides (carbon 1 for glucose and carbon 2 for fructose) are linked.

▶ The term *invert sugar* comes from the observation that the direction of rotation of plane-polarized light (Section 18.7) changes from positive (clockwise) to negative (counterclockwise) when sucrose is hydrolyzed to invert sugar. The rotation is $+66°$ for sucrose. The *net* rotation for the invert sugar mixture of fructose ($-92°$) and glucose ($+52°$) is $-40°$.

FIGURE 18.22 Honeybees and many other insects possess an enzyme called *invertase* that hydrolyzes sucrose to invert sugar. Thus honey is predominantly a mixture of D-glucose and D-fructose with some unhydrolyzed sucrose. Honey also contains flavoring agents obtained from the particular flowers whose nectars are collected. Whether a person eats monosaccharides individually, as in honey, or linked together, as in sucrose, they end up the same way in the human body: as glucose and fructose.

■ Sucrose

Sucrose, common *table sugar*, is the most abundant of all disaccharides and occurs throughout the plant kingdom. It is produced commercially from the juice of sugar cane and sugar beets. Sugar cane contains up to 20% by mass sucrose, and sugar beets contain up to 17% by mass sucrose. Figure 18.21 shows a molecular model for sucrose.

The two monosaccharide units present in a D-sucrose molecule are α-D-glucose and β-D-fructose. The glycosidic linkage is not a $(1 \rightarrow 4)$ linkage, as was the case for maltose, cellobiose, and lactose. It is instead an $\alpha,\beta(1 \rightarrow 2)$ glycosidic linkage. The —OH group on carbon 2 of D-fructose (the hemiacetal carbon) reacts with the —OH group on carbon 1 of D-glucose (the hemiacetal carbon).

Sucrose, unlike maltose, cellobiose, and lactose, is a *nonreducing sugar*. No hemiacetal is present in the molecule, because the glycosidic linkage involves the reducing ends of both monosaccharides. Sucrose, in the solid state and in solution, exists in only one form—there are no α and β isomers, and an open-chain form is not possible.

Sucrase, the enzyme needed to break the $\alpha,\beta(1 \rightarrow 2)$ linkage in sucrose, is present in the human body. Hence sucrose is an easily digested substance. Sucrose hydrolysis (digestion) produces an equimolar mixture of glucose and fructose called *invert sugar* (see Figure 18.22).

$$\text{D-Sucrose} + H_2O \xrightarrow{\;H^+ \text{ or sucrase}\;} \underbrace{\text{D-glucose} + \text{D-fructose}}_{\text{Invert sugar}}$$

CHEMICAL CONNECTIONS

Artificial Sweeteners

Because of the high caloric value of sucrose, it is often difficult to satisfy a demanding "sweet tooth" with sucrose without adding pounds to the body frame or inches to the waistline. Artificial sweeteners, which provide virtually no calories, are now used extensively as a solution to the "sucrose problem."

Three artificial sweeteners that have been widely used are saccharin, sodium cyclamate, and aspartame.

Saccharin

Sodium cyclamate

Aspartame

All three of these artificial sweeteners have received much publicity because of concern about their safety.

Saccharin is the oldest of the artificial sweeteners, having been in use for over 100 years. Questions about its safety arose in 1977 from a study that suggested that large doses of saccharin caused bladder tumors in rats. As a result, the FDA proposed banning saccharin, but public support for its use caused Congress to impose a moratorium on the ban. In 1991, on the basis of many further studies, the FDA withdrew its proposal to ban saccharin.

Sodium cyclamate, approved by the FDA in 1949, dominated the artificial sweetener market for 20 years. In 1969, principally on the basis of one study suggesting that it caused cancer in laboratory animals, the FDA banned its use. Further studies have shown that neither sodium cyclamate nor its metabolites cause cancer in animals. Reapproval of sodium cyclamate has been suggested, but there has been little action by the FDA. Interestingly, Canada has approved sodium cyclamate use but banned the use of saccharin.

Aspartame (Nutra-Sweet), approved by the FDA in 1981, is used in both the United States and Canada and accounts for three-fourths of current artificial sweetener use. It tastes like sucrose but is 180 times sweeter. It provides 4 kcal/g, as does sucrose, but because so little is used, its calorie contribution is negligible. Aspartame has quickly found its way into almost every diet food on the market today.

The safety of aspartame lies with its hydrolysis products: the amino acids aspartic acid and phenylalanine. These amino acids are identical to those obtained from digestion of proteins. The only danger aspartame poses is that it contains phenylalanine, an amino acid that can lead to mental retardation among young children suffering from PKU (phenylketonuria). Labels on all products containing aspartame warn phenylketonurics of this potential danger.

Sucralose, a derivative of sucrose, is a new low-calorie entry into the artificial sweetener market. It is synthesized from sucrose by substitution of three chlorine atoms for hydroxyl groups.

An advantage of sucralose over aspartame is that it is heat-stable and can therefore be used in cooked food. Aspartame loses its sweetness when heated. Sucralose is 600 times sweeter than sucrose and has a similar taste. It is calorie-free because it cannot be hydrolyzed as it passes through the digestive tract.

Name	Type	Sweetness
lactose	disaccharide	16
glucose	monosaccharide	74
sucrose	disaccharide	100
fructose	monosaccharide	173
sodium cyclamate	noncarbohydrate	3,000
aspartame	noncarbohydrate	15,000
saccharin	noncarbohydrate	35,000
sucralose	disaccharide derivative	60,000

When sucrose is cooked with acid-containing foods such as fruits or berries, partial hydrolysis takes place, forming some invert sugar. Jams and jellies prepared in this manner are actually sweeter than the pure sucrose added to the original mixture, because one-to-one mixtures of glucose and fructose taste sweeter than sucrose. The Chemical Connections feature on this page discusses alternatives to sucrose use.

18.14 Polysaccharides

A **polysaccharide** *is a polymer that contains many monosaccharide units bonded to each other by glycosidic linkages.* The number of monosaccharide units varies with the polysaccharide from a few hundred to hundreds of thousands. In some polysaccharides, the monosaccharides are bonded together in a linear (unbranched) chain. In others, there is extensive branching of the chain (Figure 18.23).

Unlike monosaccharides and most disaccharides, polysaccharides are not sweet and do not test positive in Tollens and Benedict's solutions. They have limited water solubility because of their size. However, the —OH groups present can individually become hydrated by water molecules. The result is usually a thick colloidal suspension of the polysaccharide in water. Polysaccharides, such as flour and cornstarch, are often used as thickening agents in sauces, desserts, and gravy.

Although there are many naturally occurring polysaccharides, in this section we will focus on only four of them: cellulose, starch, glycogen, and chitin. All play vital roles in living systems—cellulose and starch in plants, glycogen in humans and other animals, and chitin in arthropods.

> In nutrition discussions, monosaccharides and disaccharides are called *simple carbohydrates,* and polysaccharides are called *complex carbohydrates.*

■ Cellulose

Cellulose is the most abundant polysaccharide. It is the structural component of the cell walls of plants. Approximately half of all the carbon atoms in the plant kingdom are contained in cellulose molecules. Structurally, cellulose is a linear (unbranched) D-glucose polymer in which the glucose units are linked by $\beta(1 \rightarrow 4)$ glycosidic bonds.

Typically, cellulose chains contain about 5000 glucose units, which gives macromolecules with molecular masses of about 900,000 amu. Cotton is almost pure cellulose (95%), and wood is about 50% cellulose.

Even though it is a glucose polymer, cellulose is not a source of nutrition for human beings. Humans lack the enzymes capable of catalyzing the hydrolysis of $\beta(1 \rightarrow 4)$ linkages in cellulose. Even grazing animals lack the enzymes necessary for cellulose digestion. However, the intestinal tracts of animals such as horses, cows, and sheep contain

FIGURE 18.23 Some polysaccharides have a linear chain structure (a); others have a branched-chain structure (b).

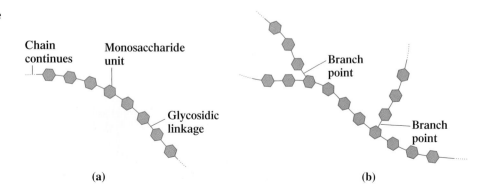

FIGURE 18.24 A sandwich such as this is high in dietary fiber; that is, it is a cellulose-rich "meal."

bacteria that produce *cellulase,* an enzyme that can hydrolyze cellulose $\beta(1 \rightarrow 4)$ linkages and produce free glucose from cellulose. Thus grasses and other plant materials are a source of nutrition for grazing animals. The intestinal tracts of termites contain the same microorganisms, which enable termites to use wood as their source of food. Microorganisms in the soil can also metabolize cellulose, which makes possible the biodegradation of dead plants.

Despite its nondigestibility, cellulose is still an important component of a balanced diet. It serves as dietary fiber. Dietary fiber provides the digestive tract with "bulk" that helps move food through the intestinal tract and facilitates the excretion of solid wastes. Cellulose readily absorbs water, leading to softer stools and frequent bowel action. Links have been found between the length of time stools spend in the colon and possible colon cancer.

High-fiber food may also play a role in weight control. Obesity is not seen in parts of the world where people eat large amounts of fiber-rich foods (see Figure 18.24). Many of the weight-loss products on the market are composed of bulk-inducing fibers such as methylcellulose.

Some dietary fibers bind lipids such as cholesterol (Section 19.9) and carry them out of the body with the feces. This lowers blood lipid concentrations and, possibly, the risk of heart and artery disease.

About 20–35 grams of dietary fiber daily is a desirable intake. This is two to three times higher than the average intake of Americans.

■ Starch

Starch, like cellulose, is a polysaccharide containing only glucose units. It is the energy-storage polysaccharide in plants. If excess glucose enters a plant cell, it is converted to starch and stored for later use. When the cell cannot get enough glucose from outside the cell, it hydrolyzes starch to release glucose.

Iodine is often used to test for the presence of starch in solution. Starch-containing solutions turn a dark blue-black when iodine is added (see Figure 18.25). As starch is broken down through acid or enzymatic hydrolysis to glucose monomers, the blue-black color disappears.

Two different polyglucose polysaccharides can be isolated from most starches: amylose and amylopectin. *Amylose,* a straight-chain glucose polymer, usually accounts for 15%–20% of the starch; *amylopectin,* a branched glucose polymer, accounts for the remaining 80%–85% of the starch.

In amylose's non-branched structure, the glucose units are connected by $\alpha(1 \rightarrow 4)$ glycosidic linkages.

▶ Amylose and cellulose are both linear chains of D-glucose molecules. They are stereoisomers that differ in the configuration at carbon 1 of each D-glucose unit. In amylose, α-D-glucose is present; in cellulose, β-D-glucose.

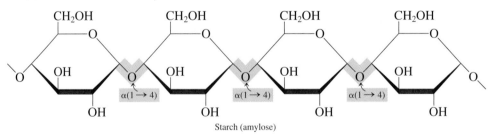

Starch (amylose)

FIGURE 18.25 Use of iodine to test for starch. Starch-containing solutions turn dark blue-black when iodine is added.

The number of glucose units present in an amylose chain depends on the source of the starch; 300–500 monomer units are usually present.

Amylopectin, the other polysaccharide in starch has a high degree of branching in its polyglucose structure. A branch occurs about once every 25–30 glucose units. The branch points involve $\alpha(1 \rightarrow 6)$ linkages (Figure 18.26). Because of the branching, amylopectin has a larger average molecular mass than the linear amylose. The average molecular mass of amylose is 50,000 amu or more; it is 300,000 or more for amylopectin.

All of the glycosidic linkages in starch (both amylose and amylopectin) are of the α type. In amylose, they are all $(1 \rightarrow 4)$; in amylopectin, both $(1 \rightarrow 4)$ and $(1 \rightarrow 6)$ linkages are present. Because both types of α linkages can be broken through hydrolysis within the human digestive tract (with the help of the enzyme *amylase*), starch has nutritional value for humans. The starches present in potatoes and cereal grains (wheat, rice, corn, etc.) account for approximately two-thirds of the world's food consumption.

■ Glycogen

Glycogen, like cellulose and starch, is a polysaccharide containing only glucose units. It is the glucose storage polysaccharide in humans and animals. Its function is thus similar to that of starch in plants, and it is sometimes referred to as *animal starch*. Liver cells and muscle cells are the storage sites for glycogen in humans.

Glycogen has a structure similar to that of amylopectin; all glycosidic linkages are of the α type, and both $(1 \rightarrow 4)$ and $(1 \rightarrow 6)$ linkages are present. Glycogen and amylopectin differ in the number of glucose units between branches and in the total number of glucose units present in a molecule. Glycogen is about three times more highly branched than amylopectin, and it is much larger, with a molar mass of up to 3,000,000 amu.

When excess glucose is present in the blood (normally from eating too much starch), the liver and muscle tissue convert the excess glucose to glycogen, which is then stored in these tissues. Whenever the glucose blood level drops (from exercise, fasting, or normal activities), some stored glycogen is hydrolyzed back to glucose. These two opposing processes are called *glycogenesis* and *glycogenolysis,* the formation and decomposition of glycogen, respectively.

$$\text{Glucose} \underset{\text{Glycogenolysis}}{\overset{\text{Glycogenesis}}{\rightleftharpoons}} \text{glycogen}$$

Glycogen is an ideal storage form for glucose. The large size of these macromolecules prevents them from diffusing out of cells. Also, conversion of glucose to glycogen reduces osmotic pressure (Section 9.9). Cells would burst because of increased osmotic

▶ The glucose polymers amylose, amylopectin, and glycogen compare as follows in molecular size and degree of branching.

Amylose:	Up to 1000 glucose units; no branching
Amylopectin:	Up to 100,000 glucose units; branch points every 24–30 glucose units
Glycogen:	Up to 1,000,000 glucose units; branch points every 8–12 glucose units

▶ The amount of stored glycogen in the human body is relatively small. Muscle tissue is approximately 1% glycogen, liver tissue 2%–3%. However, this amount is sufficient to take care of normal-activity glucose demands for about 15 hours. During strenuous exercise, glycogen supplies can be exhausted rapidly. At this point, the body begins to oxidize fat as a source of energy.

Many marathon runners eat large quantities of starch foods the day before a race. This practice, called *carbohydrate loading,* maximizes body glycogen reserves.

FIGURE 18.26 Two perspectives on the structure of the polysaccharide amylopectin. (a) Molecular structure of amylopectin. (b) An overview of the branching that occurs in the amylopectin structure. Each circle is a glucose unit.

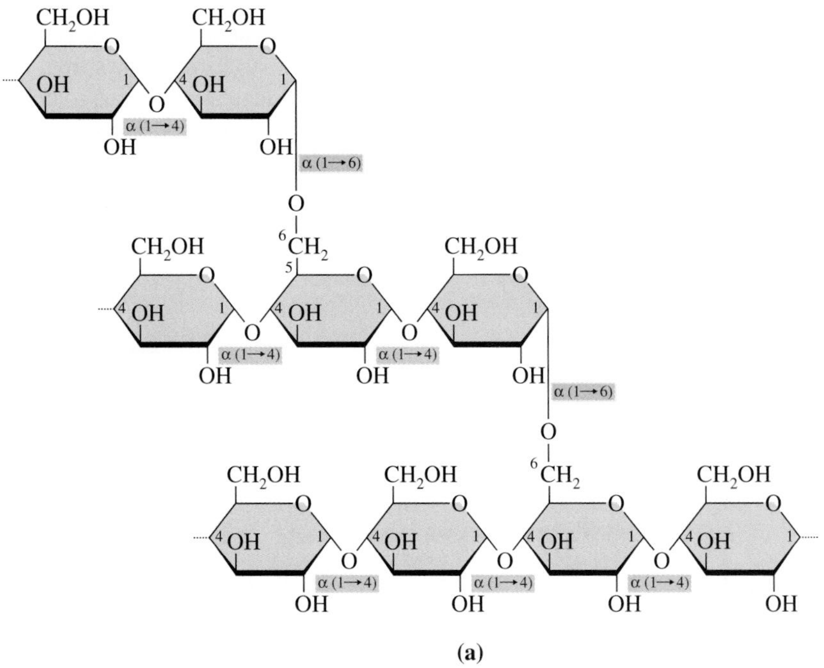

An $\alpha\,(1 \rightarrow 6)$ linkage is present in the amylopectin structure at each branch point.

(a)

(b)

FIGURE 18.27 The small, dense particles within this electron micrograph of a liver cell are glycogen granules.

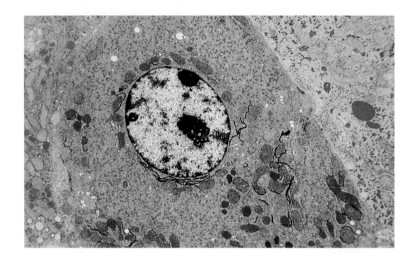

pressure if all of the glucose in glycogen were present in cells in free form. High concentrations of glycogen in a cell sometimes precipitate or crystallize into *glycogen granules*. These granules are discernible in photographs of cells under electron microscope magnification (Figure 18.27).

■ Chitin

The word *chitin* is pronounced "kye-ten"; it rhymes with *Titan.*

Chitin is a polysaccharide that is similar to cellulose in both function and structure. Its function is to give rigidity to the exoskeletons of crabs, lobsters, shrimp, insects, and other arthropods (see Figure 18.28). It also occurs in the cell walls of fungi.

Structurally, chitin is a linear polymer (no branching) with all $\beta(1 \rightarrow 4)$ glycosidic linkages, as is cellulose. Chitin differs from cellulose in that the monosaccharide present is an *N*-acetyl amino derivative of D-glucose (Section 18.12). Figure 18.29 contrasts the structures of chitin and cellulose.

The Chemistry at a Glance feature on page 535 summarizes the types of glycosidic linkages present in commonly encountered glucose-containing di- and polysaccharides.

FIGURE 18.28 Chitin, a linear $\beta(1 \rightarrow 4)$ polysaccharide, produces the rigidity in the exoskeletons of crabs and other arthropods.

18.15 Mucopolysaccharides

A **mucopolysaccharide** *is a polysaccharide that occurs in connective tissue associated with joints in animals and humans.* Its function is primarily that of lubrication, which is necessary if movement is to occur (see Figure 18.30). The name *muco*polysaccharide comes from the highly viscous, gelatinous (mucus-like) consistency of these substances in aqueous solution.

Unlike all of the polysaccharides we have discussed up to this point, mucopolysaccharides are *hetero*polysaccharides rather than *homo*polysaccharides. A **homopoly-**

FIGURE 18.29 The structures of cellulose (a) and chitin (b). In both substances, all glycosidic linkages are of the $\beta(1 \rightarrow 4)$ type.

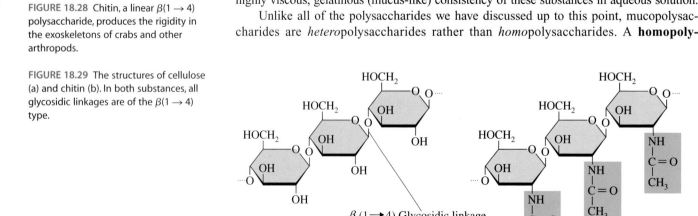

β (1 → 4) Glycosidic linkage

(a)   (b)

Types of Glycosidic Linkages for Common Glucose-Containing Di- and Polysaccharides

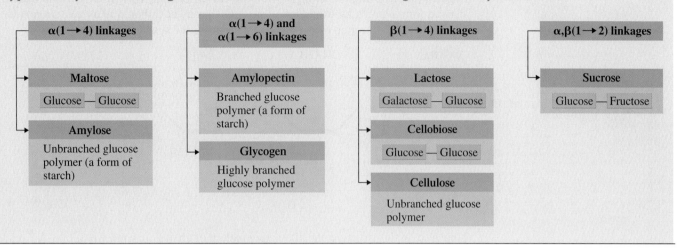

α(1 → 4) linkages	α(1 → 4) and α(1 → 6) linkages	β(1 → 4) linkages	α,β(1 → 2) linkages
Maltose Glucose — Glucose	**Amylopectin** Branched glucose polymer (a form of starch)	**Lactose** Galactose — Glucose	**Sucrose** Glucose — Fructose
Amylose Unbranched glucose polymer (a form of starch)	**Glycogen** Highly branched glucose polymer	**Cellobiose** Glucose — Glucose	
		Cellulose Unbranched glucose polymer	

saccharide *is a polysaccharide in which only one type of monosaccharide monomer is present.* Cellulose, starch, glycogen, and chitin are all homopolysaccharides. A **heteropolysaccharide** *is a polysaccharide in which more than one (usually two) type of monosaccharide monomer is present.*

One of the most common mucopolysaccharides is hyaluronic acid, a heteropolysaccharide in which the following two glucose derivatives alternate in the structure.

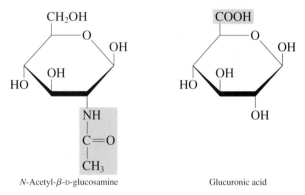

N-Acetyl-β-D-glucosamine   Glucuronic acid

FIGURE 18.30 Mucopolysaccharides associated with the connective tissue of joints give hurdlers such as these the flexibility needed to accomplish their task.

Both of these monosaccharide derivatives have been encountered previously. *N*-acetyl-*β*-D-glucosamine is the repeating unit in chitin (see Section 18.14). Glucuronic acid is derived from glucose by oxidation of the —OH group at carbon 6 to an acid group (see Section 18.12). A section of the polymeric structure for hyaluronic acid is

Note also, in this structure, the alternating pattern of glycosidic bond types, $\beta(1 \rightarrow 3)$ and $\beta(1 \rightarrow 4)$.

18.16 Glycolipids and Glycoproteins

Prior to 1960, the biochemistry of carbohydrates was thought to be rather simple. These compounds served (1) as energy sources for plants, humans, and animals and (2) as structural materials for plants and arthropods.

Research since that time has shown that carbohydrates (oligosaccharides) attached through glycosidic linkages to lipids (Chapter 19) and proteins (Chapter 20), which are called *glycolipids* and *glycoproteins*, respectively, have a wide variety of cellular functions.

It is now known that carbohydrate units on cell surfaces (glycolipids and glycoproteins) govern how individual cells interact with other cells, with invading bacteria, and with viruses. In the human reproductive process, fertilization begins with the binding of a sperm to a specific oligosaccharide on the surface of an egg. The oligosaccharide markers on cell surfaces that are the basis for blood types were considered in the Chemical Connections feature on page 523.

The prefix *glyco-*, used in the terms *glycolipid* and *glycoprotein*, is derived from the Greek word *glykys*, which means "sweet." Most monosaccharides and disaccharides have a sweet taste.

18.17 Dietary Considerations and Carbohydrates

Foods high in carbohydrate content constitute over 50% of the diet of most people of the world—rice in Asia, corn in South America, cassava (a starchy root vegetable) in parts of Africa, the potato and wheat in North America, and so on. Current nutritional recommendations support such a situation; a balanced diet should ideally be about 60% carbohydrate.

Nutritionists usually subdivide dietary carbohydrates into the categories *simple* and *complex*. A **simple carbohydrate** *is a dietary monosaccharide or disaccharide.* Simple carbohydrates are usually sweet to the taste and are commonly referred to as sugars (Section 18.8). A **complex carbohydrate** *is a dietary polysaccharide.* The main complex carbohydrates are starch and cellulose, substances not generally sweet to the taste.

Simple carbohydrates provide 20% of the energy in the U.S. diet. Half of this energy content comes from *natural sugars* (milk, fresh fruits) and the other half from *refined sugars* added to foods. Despite claims to the contrary, refined sugars are chemically and structurally no different from the sugars naturally present in foods. The only difference is that the refined sugar is in a pure form, whereas natural sugars are part of mixtures of substances obtained from a plant source.

Refined sugars are often said to provide *empty Calories* because they provide energy but few other nutrients. Natural sugars, on the other hand, are accompanied by nutrients. A tablespoon of sucrose (table sugar) provides 50 Calories of energy just as a small

CHEMICAL CONNECTIONS

"Good and Bad Carbs": The Glycemic Index

The glycemic index (GI) is a dietary carbohydrate rating system that indicates how fast a particular carbohydrate is broken down into glucose (through hydrolysis) and the level of blood glucose that results. Its focal point is thus blood glucose levels.

Slow generation of glucose, a modest rise in blood glucose, and a smooth return to normal blood glucose levels are desir-able. A rapid increase (surge) in blood glucose levels, with a re-sulting overcorrection (from excess insulin production) that drops glucose levels below normal, is undesirable. Low-GI foods promote the first of these two effects, and high-Gl foods promote the latter.

Selected examples of GI ratings for foods follow.

Low-glycemic foods (less than 55)		Intermediate-glycemic foods (55 to 70)		High-glycemic foods (over 70)	
Low-fat yogurt, artificially sweetened	14	Brown rice	55	Corn chips	72
Grapefruit	25	Popcorn, sweet corn	55	Watermelon**	72
Kidney beans	27	Long-grain white rice	56	Cheerios**	74
Low-fat yogurt, sugar sweetened	33	Mini shredded wheats	58	French fries	76
Apple, pear	38	Cheese pizza	60	Rice cakes	82
Spaghetti	41	Coca-Cola*	63	Pretzels	83
Orange	44	Raisins	64	Cornflakes	84
Low-fat ice cream*	50	Table sugar*	65	Baked potato**	85
Potato chips*	54	White bread	70	Dried dates**	103

*High in a lot of "empty Calories," so eat sparingly. Otherwise, they'll crowd out more nutritious foods.

**Don't avoid these healthful foods. Instead, combine them with low-glycemic choices.

Considerations relative to use of GI values such as these include the following:

1. At least one low-GI food should be part of each meal.
2. Fruits, vegetables, and legumes tend to have low-GI ratings.
3. Whole-grain foods, substances high in fiber, tend to have slower digestion rates.
4. High-GI rated foods should still be consumed, but as part of meals that also contain low-GI foods.

orange does. The small orange, however, also supplies vitamin C, potassium, calcium, and fiber; table sugar provides no other nutrients.

The major dietary source for complex carbohydrates in the U.S. diet is grains, a source of both starch and fiber as well as of protein, vitamins, and minerals. The pulp of a potato provides starch, and the skin provides fiber. Vegetables such as broccoli and green beans are low in starch but high in fiber.

A developing concern about dietary intake of carbohydrates involves *how fast* a given dietary carbohydrate is broken down to generate glucose within the human body. The term *glycemic effect* refers to how quickly carbohydrates are digested (broken down into glucose), how high blood glucose levels rise, and how quickly blood glucose levels return to normal. A measurement system called the *glycemic index (GI)* has been developed for rating foods in terms of their glycemic effect. The Chemical Connections feature on this page discusses this topic further.

CONCEPTS TO REMEMBER

Biochemistry. Biochemistry is the study of the chemical substances found in living systems and the chemical interactions of these sub-stances with each other.

Carbohydrates. Carbohydrates are polyhydroxy aldehydes, polyhy-droxy ketones, or compounds that yield such substances upon hy-drolysis. Plants contain large quantities of carbohydrates produced via photosynthesis.

Carbohydrate classification. Carbohydrates are classified into three groups: monosaccharides, oligosaccharides, and polysaccharides.

Chirality and achirality. A chiral object is not identical to its mirror image. An achiral object is identical to its mirror image.

Chiral center. A chiral center is an atom in a molecule that has four different groups tetrahedrally bonded to it. Molecules that contain a single chiral center exist in a left-handed and a right-handed form.

Stereoisomerism. The atoms of stereoisomers are connected in the same way but are arranged differently in space. The major causes of stereoisomerism in molecules are structural rigidity and the presence of a chiral center.

Enantiomers and diastereomers. Two types of stereoisomers exist: enantiomers and diastereomers. Enantiomers have structures that are nonsuperimposable mirror images of each other. Enantiomers have identical achiral properties but different chiral properties. Diastereomers have structures that are not mirror images of each other.

Fischer projections. Fischer projections are two-dimensional structural formulas used to depict the three-dimensional shapes of molecules with chiral centers.

Chirality of monosaccharides. Monosaccharides are classified as D or L stereoisomers on the basis of the configuration of the chiral center farthest from the carbonyl group.

Optical activity. Chiral compounds are optically active—that is, they rotate the plane of polarized light. Enantiomers rotate the plane of polarized light in opposite directions. The prefix (+) indicates that the compound rotates the plane of polarized light in a clockwise direction, whereas compounds that rotate the plane of polarized light in a counterclockwise direction have the prefix (−).

Classification of monosaccharides. Monosaccharides are classified as aldoses or ketoses on the basis of the type of carbonyl group present. They are further classified as trioses, tetroses, pentoses, etc. on the basis of the number of carbon atoms present.

Important monosaccharides. Important monosaccharides include glucose, galactose, fructose, and ribose. Glucose and galactose are aldohexoses, fructose is a ketohexose, and ribose is an aldopentose.

Cyclic monosaccharides. Cyclic monosaccharides form through an intramolecular reaction between the carbonyl group and an alcohol group of an open-chain monosaccharide. These cyclic forms predominate in solution.

Reactions of monosaccharides. Five important reactions of monosaccharides are (1) oxidation to an acidic sugar, (2) reduction to a sugar alcohol, (3) glycoside formation, (4) phosphate ester formation, and (5) amino sugar formation.

Disaccharides. Disaccharides are glycosides formed from the linkage of two monosaccharides. The most important disaccharides are maltose, cellobiose, lactose, and sucrose. Each of these has at least one glucose unit in its structure.

Polysaccharides. Polysaccharides are polymers of monosaccharides. Cellulose and starch are polymers of glucose, but the linkages between glucose units are different. The α linkages in starch are easily broken by enzymatic hydrolysis in humans, thus starch is a source of dietary energy. The β linkages in cellulose cannot be broken by enzymatic hydrolysis in humans, so cellulose provides no usable energy in our diets. Glycogen is a highly branched glucose polymer utilized for energy storage within the body.

KEY REACTIONS AND EQUATIONS

1. Monosaccharide oxidation (Section 18.12)

   Aldose or ketose + weak oxidizing agent $\longrightarrow$ acidic sugar

2. Monosaccharide reduction (Section 18.12)

   Aldose or ketose + H$_2$ $\xrightarrow{\text{Catalyst}}$ sugar alcohol

3. Glycoside (acetal) formation (Section 18.12)

   Cyclic monosaccharide + alcohol $\longrightarrow$ glycoside (acetal) + H$_2$O

4. Monosaccharide ester formation (Section 18.12)

   Monosaccharide + oxyacid $\longrightarrow$ ester + H$_2$O

5. Hydrolysis of disaccharide (Section 18.13)

   Disaccharide + H$_2$O $\xrightarrow{\text{Catalyst}}$ two monosaccharides

6. Hydrolysis of maltose (Section 18.13)

   D-Maltose + H$_2$O $\xrightarrow{\text{H}^+ \text{ or maltase}}$ 2 D-glucose

7. Hydrolysis of cellobiose (Section 18.13)

   D-Cellobiose + H$_2$O $\xrightarrow{\text{H}^+ \text{ or cellobiase}}$ 2 D-glucose

8. Hydrolysis of lactose (Section 18.13)

   D-Lactose + H$_2$O $\xrightarrow{\text{H}^+ \text{ or lactose}}$ D-galactose + D-glucose

9. Hydrolysis of sucrose (Section 18.13)

   D-Sucrose + H$_2$O $\xrightarrow{\text{H}^+ \text{ or sucrase}}$ D-fructose + D-glucose

10. Complete hydrolysis of polysaccharide (Section 18.14)

    Polysaccharide + H$_2$O $\xrightarrow{\text{H}^+ \text{ or enzymes}}$ many monosaccharides

11. Complete hydrolysis of starch (Section 18.14)

    Starch + H$_2$O $\xrightarrow{\text{H}^+ \text{ or enzymes}}$ many D-glucose

12. Complete hydrolysis of glycogen (Section 18.14)

    Glycogen + H$_2$O $\xrightarrow{\text{H}^+ \text{ or enzymes}}$ many D-glucose

---

KEY TERMS

Achiral molecule (18.4)
Aldose (18.8)
Biochemical substance (18.1)
Biochemistry (18.1)
Carbohydrate (18.3)
Chiral center (18.4)
Chiral molecule (18.4)
Complex carbohydrate (18.17)

Dextrorotatory compound (18.7)
Diastereomers (18.5)
Disaccharide (18.3)
Enantiomers (18.5)
Epimers (18.6)
Fischer projection (18.6)
Glycoside (18.12)
Glycosidic linkage (18.13)

Haworth projection (18.11)
Heteropolysaccharide (18.15)
Homopolysaccharide (18.15)
Ketose (18.8)
Levorotatory compound (18.7)
Mirror image (18.4)
Monosaccharide (18.3)
Mucopolysaccharide (18.15)

Nonsuperimposable mirror image (18.4)   Polysaccharide (18.3, 18.14)   Stereoisomers (18.5)
Oligosaccharide (18.3)   Reducing sugar (18.12)   Sugar (18.8)
Optically active compound (18.7)   Simple carbohydrate (18.17)   Superimposable mirror image (18.4)

EXERCISES AND PROBLEMS

The members of each pair of problems in this section test similar material.

■ **Biochemical Substances (Section 18.1)**

18.1 Define the term *biochemistry*.

18.2 What are the two general groups of biochemical substances?

18.3 What are the four major types of bioorganic substances?

18.4 For each of the following pairs of bioorganic substances, indicate which member of the pair is more abundant in the human body.
   a. Proteins and nucleic acids
   b. Proteins and carbohydrates
   c. Lipids and carbohydrates
   d. Lipids and nucleic acids

■ **Occurrence of Carbohydrates (Section 18.2)**

18.5 Write a general chemical equation for photosynthesis.

18.6 What role does chlorophyll play in photosynthesis?

18.7 What are the two major functions of carbohydrates in the plant kingdom?

18.8 What are the six major functions of carbohydrates in the human body?

■ **Structural Characteristics of Carbohydrates (Section 18.3)**

18.9 Define the term *carbohydrate*.

18.10 What functional group is present in all carbohydrates?

18.11 Explain the difference between
   a. a monosaccharide and an oligosaccharide
   b. a disaccharide and a tetrasaccharide

18.12 Explain the difference between
   a. an oligosaccharide and a polysaccharide
   b. a trisaccharide and an oligosaccharide

■ **Chirality (Section 18.4)**

18.13 Explain what the term *superimposable* means.

18.14 Explain what the term *nonsuperimposable* means.

18.15 In each of the following lists of objects, identify those objects that are chiral.
   a. Nail, hammer, screwdriver, drill bit
   b. Your hand, your foot, your ear, your nose
   c. The words TOT, TOOT, POP, PEEP

18.16 In each of the following lists of objects, identify those objects that are chiral.
   a. Baseball cap, glove, shoe, scarf
   b. Pliers, scissors, spoon, fork
   c. The words MOM, DAD, AHA, WAX

18.17 Indicate whether the circled carbon atom in each of the following molecules is a chiral center.
   a. CH_3—ⒸH_2—OH   b. CH_3—ⒸH—OH
                                              |
                                             CH_3

   c. CH_3—ⒸH—OH   d. CH_3—CH_2—ⒸH—OH
              | |
             Cl CH_3

18.18 Indicate whether the circled carbon atom in each of the following molecules is a chiral center.
   a. CH_3—ⒸH_2—NH_2 b. CH_3—ⒸH—CH_3
                                                    |
                                                  NH_2

   c. CH_3—ⒸH—NH_2 d. CH_3—ⒸH—NH_2
              | |
             CH_3 Cl

18.19 Use asterisks to show the chiral center(s) in the following structures.
   a.   H H
            | |
     Cl—C—C—Br
            | |
           H Cl

   b.   H Cl H
            | | |
     Br—C—C—C—Cl
            | | |
           H Br Br

   c.   O
                                  ‖
     CH_2—CH—CH—CH—C—H
      | | | |
     OH   OH   OH   OH

   d. CH_2—CH—CH—CH—CH—CH_2
       | | | | | |
      OH   OH   OH   OH   OH   OH

18.20 Use asterisks to show the chiral center(s) in the following structures.
   a.   H Cl
            | |
     Cl—C—C—Cl
            | |
           Br Br

   b.   H H H
            | | |
     Cl—C—C—C—Cl
            | | |
           Br OH Br

   c.   O
                                 ‖
     CH_3—CH—CH—CH—C—H
              | | |
             OH   OH   OH

   d. CH_2—CH—CH—CH—CH_2
       | | | | |
      OH   OH   OH   OH   OH

18.21 How many chiral centers are present in each of the following molecular structures?

   a.   b.

   c.   d.

18.22 How many chiral centers are present in each of the following molecular structures?

   a.   b.

pairs of monosaccharides. More than one term may apply in a given situation.
a. D-Glucose and D-ribose
b. D-Fructose and dihydroxyacetone
c. D-Glyceraldehyde and dihydroxyacetone
d. D-Galactose and D-ribose

18.47 Draw the Fischer projection for each of the following monosaccharides.
a. D-Glucose
b. D-Glyceraldehyde
c. D-Fructose
d. L-Galactose

18.48 Draw the Fischer projection for each of the following monosaccharides.
a. D-Galactose
b. D-Ribose
c. Dihydroxyacetone
d. L-Glucose

18.49 To which of the common monosaccharides does each of the following terms apply?
a. Levulose
b. Grape sugar
c. Brain sugar

18.50 To which of the common monosaccharides does each of the following terms apply?
a. Dextrose
b. Fruit sugar
c. Blood sugar

■ **Cyclic Forms of Monosaccharides (Section 18.10)**

18.51 The intermolecular reaction that produces the cyclic forms of monosaccharides involves functional groups on which two carbon atoms in the case of each of the following?
a. D-Glucose
b. D-Galactose
c. D-Fructose
d. D-Ribose

18.52 How many carbon atoms and how many oxygen atoms are present in the ring portion of the cyclic forms of each of the following monosaccharides?
a. D-Glucose
b. D-Galactose
c. D-Fructose
d. D-Ribose

18.53 What is the structural difference between the alpha and beta forms of D-glucose?

18.54 What is the structural difference between the alpha forms of D-glucose and D-galactose?

18.55 Fructose contains six carbon atoms, and ribose has only five carbon atoms. Why do both of these monosaccharides have cyclic forms that involve a five-membered ring?

18.56 Fructose and glucose both contain six carbon atoms. Why do the cyclic forms of fructose have a five-membered ring instead of the six-membered ring found in the cyclic forms of glucose?

18.57 The structure of glucose is sometimes written in an open-chain form and sometimes as a cyclic hemiacetal structure. Explain why either form is acceptable.

18.58 When pure α-D-glucose is dissolved in water, β-D-glucose and α-D-glucose are both soon present. Explain how this is possible.

■ **Haworth Projection Formulas (Section 18.11)**

18.59 Identify each of the following structures as an α-D-monosaccharide or a β-D-monosaccharide.

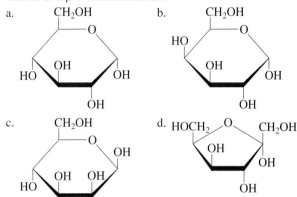

18.60 Identify each of the following structures as an α-D-monosaccharide or a β-D-monosaccharide.

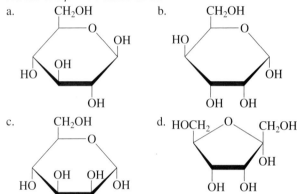

18.61 Identify whether each of the structures in Problem 18.59 is that of a hemiacetal.

18.62 Identify whether each of the structures in Problem 18.60 is that of a hemiacetal.

18.63 Draw the open-chain form for each of the monosaccharides in Problem 18.59.

18.64 Draw the open-chain form for each of the monosaccharides in Problem 18.60.

18.65 Using the information in Figures 18.13 and 18.14, assign a name to each of the monosaccharides in Problem 18.59.

18.66 Using the information in Figures 18.13 and 18.14, assign a name to each of the monosaccharides in Problem 18.60.

18.67 Draw the Haworth projection for each of the following monosaccharides.
a. α-D-Galactose   b. β-D-Galactose
c. α-L-Galactose   d. β-L-Galactose

18.68 Draw the Haworth projection for each of the following monosaccharides.
a. α-D-Mannose   b. β-D-Mannose
c. α-L-Mannose   d. β-L-Mannose

■ **Reactions of Monosaccharides (Section 18.12)**

18.69 Which of the following monosaccharides is a *reducing sugar*?
a. D-Glucose   b. D-Galactose
c. D-Fructose   d. D-Ribose

18.70 Which of the following monosaccharides will give a positive test with Benedict's solution?
a. D-Glucose   b. D-Galactose
c. D-Fructose   d. D-Ribose

18.71 In terms of oxidation and reduction, explain what occurs to both D-glucose and Tollens solution when they react with each other.

18.72 Describe the chemical reaction that is used to detect glucose in urine that involves Benedict's solution.

18.73 Draw structures for the following compounds.
a. Galactonic acid   b. Galactaric acid
c. Galacturonic acid   d. Galactitol

18.74 Draw structures for the following compounds.
a. Mannonic acid   b. Mannaric acid
c. Mannuronic acid   d. Mannitol

18.75 Indicate whether each of the following structures is that of a glycoside.

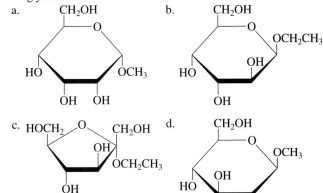

18.76 Indicate whether each of the following structures is that of a glycoside.

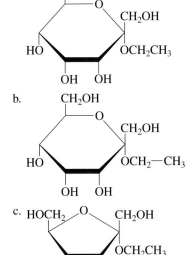

18.77 For each structure in Problem 18.75, identify the configuration at the acetal carbon atom as α or β.

18.78 For each structure in Problem 18.76, identify the configuration at the acetal carbon atom as α or β.

18.79 Identify the alcohol needed to produce each of the compounds in Problem 18.75 by reaction of the alcohol with the appropriate monosaccharide.

18.80 Identify the alcohol needed to produce each of the compounds in Problem 18.76 by reaction of the alcohol with the appropriate monosaccharide.

18.81 What is the difference in meaning between the terms *glycoside* and *glucoside?*

18.82 What is the difference in meaning between the terms *glycoside* and *galactoside?*

18.83 Draw structures for the following compounds.
a. Ethyl-β-D-glucoside
b. Methyl-α-D-galactoside

18.84 Draw structures for the following compounds.
a. Ethyl-α-D-galactoside
b. Methyl-β-D-glucoside

18.85 Draw structures for the following compounds.
a. α-D-Galactose 6-phosphate
b. *N*-Acetyl-α-D-galactosamine

18.86 Draw structures for the following compounds.
a. α-D-Mannose 6-phosphate
b. *N*-Acetyl-α-D-mannosamine

■ **Disaccharides (Section 18.13)**

18.87 What monosaccharides are produced from the hydrolysis of the following disaccharides?
a. Sucrose   b. Maltose
c. Lactose   d. Cellobiose

18.88 What type of glycosidic linkage [α(1 → 4), etc.] is present in each of the following disaccharides?
a. Sucrose   b. Maltose
c. Lactose   d. Cellobiose

18.89 Explain why lactose is a reducing sugar.

18.90 Explain why sucrose is not a reducing sugar.

18.91 Indicate whether each of the following disaccharides gives a positive or a negative Benedict's test.
a. Sucrose   b. Maltose
c. Lactose   d. Cellobiose

18.92 Indicate whether each of the following disaccharides gives a positive or a negative Tollens test.
a. Maltose   b. Lactose
c. Cellobiose   d. Sucrose

18.93 What type of glycosidic linkage [α(1 → 4), etc.] is present in each of the following disaccharides?

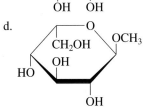

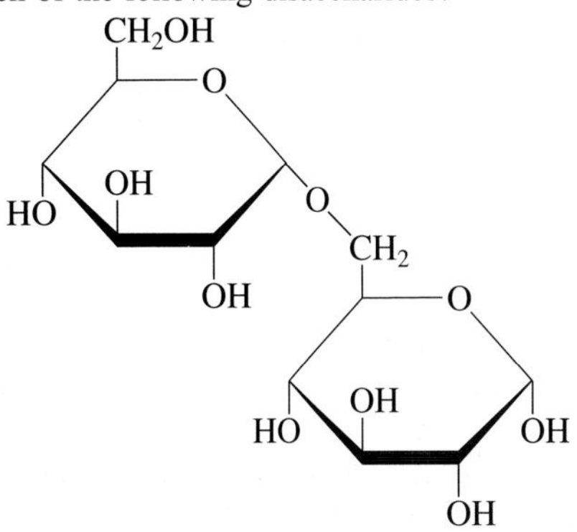

b.

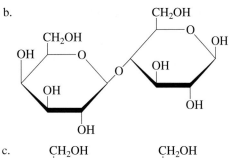

c.

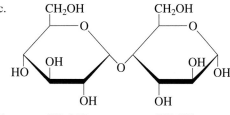

d.

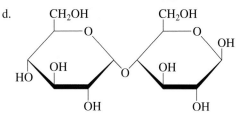

d.

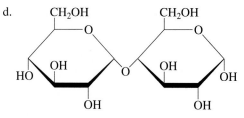

18.95 For each of the structures in Problem 18.93, specify whether the disaccharide is in an α configuration or a β configuration, or neither.

18.96 For each of the structures in Problem 18.94, specify whether the disaccharide is in an α configuration or a β configuration, or neither.

18.97 Identify each of the structures in Problem 18.93 as a reducing sugar or a nonreducing sugar.

18.98 Identify each of the structures in Problem 18.94 as a reducing sugar or a nonreducing sugar.

18.99 Using the information in Figures 18.13 and 18.14, assign a name to each monosaccharide present in each of the structures in Problem 18.93.

18.100 Using the information in Figures 18.13 and 18.14, assign a name to each monosaccharide present in each of the structures in Problem 18.94.

18.94 What type of glycosidic linkage [α(1 → 4), etc.] is present in each of the following disaccharides?

a.

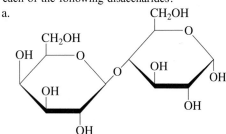

b.

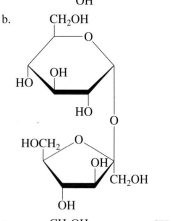

c.

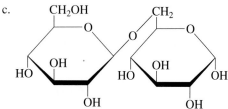

■ **Polysaccharides (Section 18.14)**

18.101 Describe the structural differences and similarities between the following pairs of polysaccharides.
a. Glycogen and amylopectin
b. Amylose and cellulose

18.102 Describe the structural differences and similarities between the following pairs of polysaccharides.
a. Amylose and glycogen
b. Amylose and amylopectin

18.103 Match each of the following structural characteristics to the polysaccharides amylopectin, amylose, glycogen, cellulose, and chitin. A specific characteristic may apply to more than one polysaccharide.
a. Contains both α(1 → 4) and α(1 → 6) glycosidic linkages
b. Composed of glucose monosaccharide units
c. Composed of unbranched molecular chains
d. Contains only β(1 → 4) glycosidic linkages

18.104 Match each of the following structural characteristics to the polysaccharides amylopectin, amylose, glycogen, cellulose, and chitin. A specific characteristic may apply to more than one polysaccharide.
a. Contains acetal linkages between monosaccharide units
b. Contains only α(1 → 4) glycosidic linkages
c. Monosaccharide units are derivatives of glucose
d. Composed of highly branched molecular chains

18.105 Why, when both contain D-glucose, can humans digest starch but not cellulose?

18.106 What is the difference between plant starch and animal starch?

■ **Mucopolysaccharides (Section 18.15)**

18.107 In what location in the human body are mucopolysaccharides present?

18.108 Describe the mucopolysaccharide hyaluronic acid in terms of
a. monosaccharide derivatives present
b. types of glycosidic linkages present

18.109 What do the structures of chitin and hyaluronic acid have in common?

18.110 Contrast the glycosidic linkages present in chitin and hyaluronic acid.

■ **Dietary Considerations and Carbohydrates (Section 18.17)**

18.111 In a dietary context, what is the difference between a *simple* carbohydrate and a *complex* carbohydrate?

18.112 In a dietary context, what is the difference between a *natural* sugar and a *refined* sugar?

18.113 In a dietary context, what are *empty* Calories?

18.114 In a dietary context, what is the *glycemic effect?*

ADDITIONAL PROBLEMS

18.115 Indicate whether each of the following compounds is chiral or achiral.
   a. 1-Chloro-2-methylpentane   b. 2-Chloro-2-methylpentane
   c. 2-Chloro-3-methylpentane   d. 3-Chloro-2-methylpentane

18.116 Which of the following compounds is (are) optically active?

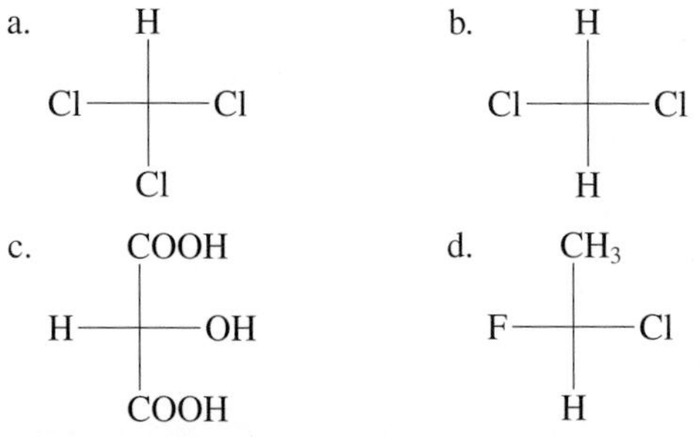

18.117 In which of the following pairs of monosaccharides do both members of the pair contain the same number of carbon atoms?
   a. Glyceraldehyde and glucose
   b. Dihydroxyketone and ribose
   c. Ribose and deoxyribose
   d. Glyceraldehyde and dihydroxyketone

18.118 Draw Fischer projections for the four stereoisomers of the molecule

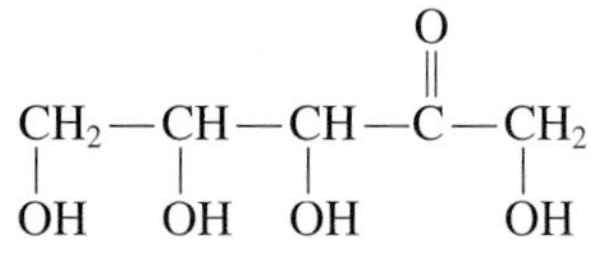

$$CH_2-CH-CH-\overset{\overset{\displaystyle O}{\|}}{C}-CH_2$$
$$\quad | \qquad | \qquad | \qquad\qquad |$$
$$OH\quad OH\quad OH\qquad\quad OH$$

18.119 What is the alkane of lowest molecular mass that is a chiral compound?

18.120 What monosaccharide(s) is (are) obtained from the hydrolysis of each of the following?
   a. Sucrose   b. Glycogen
   c. Starch   d. Amylose

18.121 Classify each of the following carbohydrates as a homopolysaccharide or a heteropolysaccharide.
   a. Chitin   b. Amylopectin
   c. Hyaluronic acid   d. Glycogen

18.122 List the reactant(s) necessary to effect the following chemical changes.

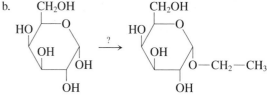

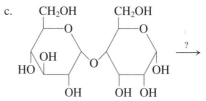

ANSWERS TO PRACTICE EXERCISES

18.1 a. not a chiral center   b. not a chiral center   c. chiral center
   d. not a chiral center

18.2 a.

CHO
HO——H
H——OH
H——OH
CH₂OH

   b.

CH₂OH
C=O
HO——H
HO——H
CH₂OH

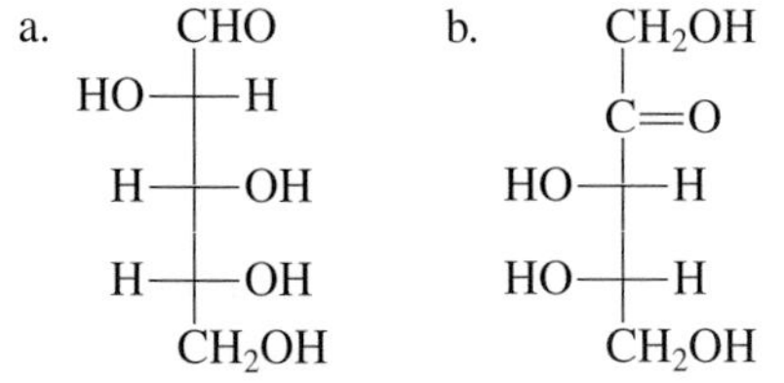

18.3 a. D enantiomer   b. L entantiomer

18.4 a. diastereomers   b. enantiomers   c. diastereomers

18.5 a. ketohexose   b. aldohexose   c. aldotetrose   d. ketopentose

19

Lipids

Fats and oils are the most widely occurring types of lipids. Thick layers of fat help insulate polar bears against the effects of low temperatures.

There are four major classes of bioorganic substances: carbohydrates, lipids, proteins, and nucleic acids (Section 18.1). In the previous chapter we considered the first of these classes, carbohydrates. We now turn our attention to the second of the bioorganic classes, the compounds we call lipids.

Lipids known as fats provide a major way of storing chemical energy and carbon atoms in the body. Fats also surround and insulate vital body organs, providing protection from mechanical shock and preventing excessive loss of heat energy. Phospholipids, glycolipids, and cholesterol (a lipid) are the basic components of cell membranes. Several cholesterol derivatives function as chemical messengers (hormones) within the body.

19.1 Structure and Classification of Lipids

Unlike carbohydrates and most other classes of compounds, lipids do not have a common structural feature that serves as the basis for defining such compounds. Instead, their characterization is based on solubility characteristics. A **lipid** *is an organic compound found in living organisms that is insoluble (or only sparingly soluble) in water but soluble in nonpolar organic solvents.* When a biochemical material (human, animal, or plant tissue) is homogenized in a blender and mixed with a nonpolar organic solvent, the substances that dissolve in the solvent are the lipids.

Figure 19.1 shows the structural diversity that is associated with lipid molecules. Some are esters, some are amides, and some are alcohols; some are acyclic, some are cyclic, and some are polycyclic. The common thread that ties all of the compounds of Figure 19.1 together is solubility rather than structure. All are insoluble in water.

FIGURE 19.1 The structural formulas of these types of lipids illustrate the great structural diversity among lipids. The defining parameter for lipids is solubility rather than structure.

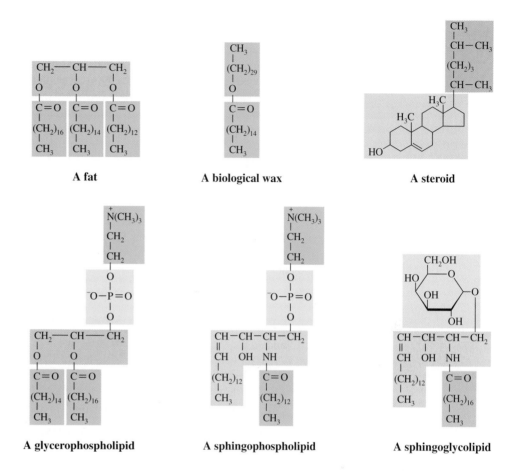

A fat

A biological wax

A steroid

A glycerophospholipid

A sphingophospholipid

A sphingoglycolipid

For purposes of study, we will divide lipids into five categories on the basis of lipid function:

1. **Energy-storage lipids** (triacylglycerols)
2. **Membrane lipids** (phospholipids, sphingoglycolipids, and cholesterol)
3. **Emulsification lipids** (bile acids)
4. **Messenger lipids** (steroid hormones and eicosanoids)
5. **Protective-coating lipids** (biological waxes)

Our entry point into a discussion of these five general types of lipids is a consideration of molecules called *fatty acids.* Fatty acids are structural components of all the lipids that we consider in this chapter except cholesterol, bile acids, and steroid hormones. Familiarity with the structural characteristics and physical properties of fatty acids makes it easier to understand the behavior of the many fatty-acid-containing lipids found in the human body.

19.2 Fatty Acids: Lipid Building Blocks

A **fatty acid** *is a naturally occurring monocarboxylic acid.* Because of the pathway by which they are biosynthesized (Section 25.7), fatty acids nearly always contain an even number of carbon atoms and have a carbon chain that is unbranched. In terms of carbon chain length, fatty acids are characterized as *long-chain fatty acids* (C_{12} to C_{26}), *medium-chain fatty acids* (C_6 to C_{10}), or *short-chain fatty acids* (C_4 and C_6). Fatty acids are rarely found free in nature but rather occur as part of the structure of more complex lipid molecules.

■ Saturated and Unsaturated Fatty Acids

The carbon chain of a fatty acid may or may not contain carbon–carbon double bonds. On the basis of this consideration, fatty acids are classified as saturated fatty acids (SFAs), monounsaturated fatty acids (MUFAs), or polyunsaturated fatty acids (PUFAs).

A **saturated fatty acid** *is a fatty acid with a carbon chain in which all carbon–carbon bonds are single bonds.* The structural formula for the 16-carbon SFA is

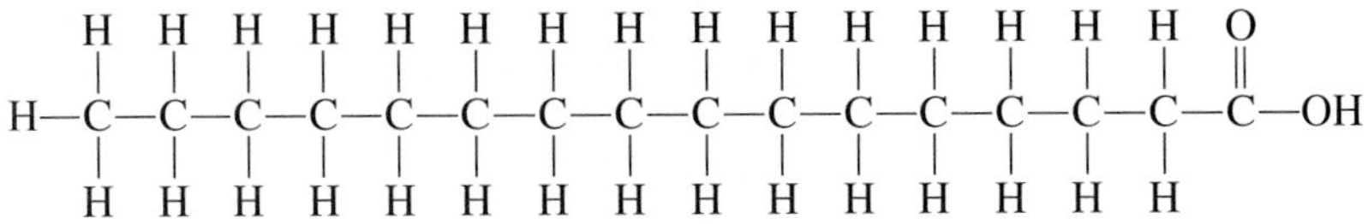

IUPAC name: hexadecanoic acid
Common name: palmitic acid

The structural formula for a fatty acid is usually written in a more condensed form than the preceding structural formula. Two alternative structural notations for palmitic acid are

$$CH_3—(CH_2)_{14}—\overset{\overset{\displaystyle O}{\|}}{C}—OH$$

and

(We first encountered "line-angle notation" in Section 12.11.)

A **monounsaturated fatty acid** *is a fatty acid with a carbon chain in which one carbon–carbon double bond is present.* In biochemically important MUFAs, the configuration about the double bond is nearly always *cis* (Section 14.5). Different ways of depicting the structure of a MUFA follow.

> More than 500 different fatty acids have been isolated from the lipids of microorganisms, plants, animals, and humans. These fatty acids differ from one another in the length of their carbon chains, their degree of unsaturation (number of double bonds), and the positions of the double bonds in the chains.

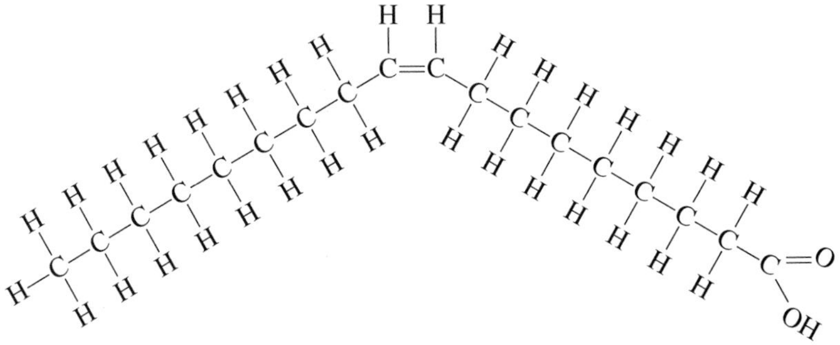

$$CH_3—(CH_2)_7—CH{=}CH—(CH_2)_7—\overset{\overset{\displaystyle O}{\|}}{C}—OH$$

IUPAC name: *cis*-9-octadecenoic acid
Common name: oleic acid

The first of these structures correctly emphasizes that the presence of a *cis* double bond in the carbon chain puts a rigid 30° bend in the chain. Such a bend affects the physical properties of a fatty acid, as we will see in Section 19.3.

A **polyunsaturated fatty acid** *is a fatty acid with a carbon chain in which two or more carbon–carbon double bonds are present.* Up to six double bonds are found in biochemically important PUFAs.

Fatty acids are nearly always referred to using their common names. IUPAC names for fatty acids, although easily constructed, are usually quite long. These two types of names for an 18-carbon PUFA containing *cis* double bonds in the 9 and 12 positions are as follows:

IUPAC name: *cis,cis*-9,12-octadecadienoic acid
Common name: linoleic acid

■ Unsaturated Fatty Acids and Double-Bond Position

A numerically based shorthand system exists for specifying key structural parameters for fatty acids. In this system, two numbers separated by a colon are used to specify the number of carbon atoms and the number of carbon–carbon double bonds present. The notation 18:0 denotes a C_{18} fatty acid with no double bonds, whereas the notation 18:2 signifies a C_{18} fatty acid in which two double bonds are present.

To specify double-bond positioning within the carbon chain of an unsaturated fatty acid, the preceding notation is expanded by adding the Greek capital letter delta (Δ) followed by one or more superscript numbers. The notation $18:3(\Delta^{9,12,15})$ denotes a C_{18} PUFA with three double bonds at locations between carbons 9 and 10, 12 and 13, and 15 and 16.

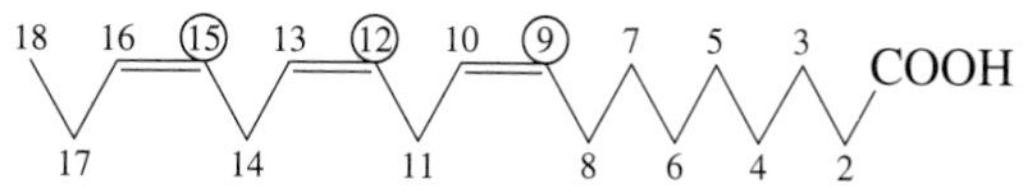

MUFAs are usually Δ^9 acids, and the first two additional double bonds in PUFAs are generally at the Δ^{12} and Δ^{15} locations. [A notable exception to this generalization is the biochemically important arachidonic acid, a PUFA with the structural parameters $20:4(\Delta^{5,8,11,14})$]. Denoting double-bond locations using this "delta notation" always assumes a numbering system in which the carboxyl carbon atom is C-1.

Several different "families" of unsaturated fatty acids exist. These family relationships become apparent when double-bond position is specified relative to the methyl (noncarboxyl) end of the fatty acid carbon chain. Double-bond positioning determined in this manner is denoted by using the Greek lower-case letter omega (ω). An **omega-3 fatty acid** *is an unsaturated fatty acid with its endmost double bond three carbon atoms away from its methyl end.* An example of an omega-3 fatty acid is

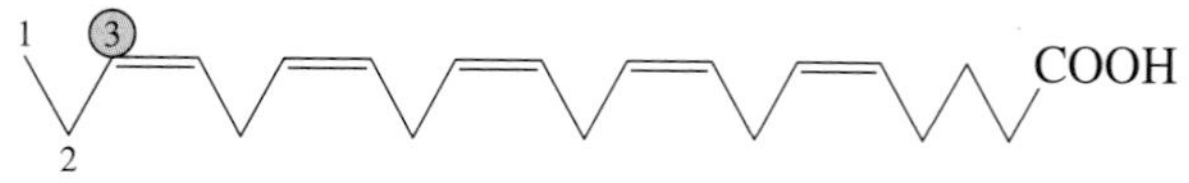

An **omega-6 fatty acid** *is an unsaturated fatty acid with its endmost double bond six carbon atoms away from its methyl end.*

The following three acids all belong to the omega-6 fatty acid family.

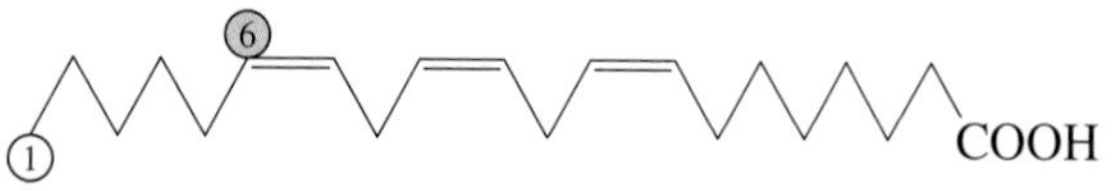

The structural feature common to these omega-6 fatty acids is highlighted with color in the preceding structural formulas. All the members of an omega family of fatty acids have structures in which the same "methyl end" is present.

Table 19.1 gives the names and structures of the fatty acids most commonly encountered as building blocks in biochemically important lipid structures, as well as the "delta" and "omega" notations for the acids.

TABLE 19.1
Selected Fatty Acids of Biological Importance

Structure notation		Common name	Structure
Saturated Fatty Acids			
12:0		lauric acid	 COOH
14:0		myristic acid	 COOH
16:0		palmitic acid	 COOH
18:0		stearic acid	 COOH
20:0		arachidic acid	 COOH
Monounsaturated Fatty Acids			
16:1 Δ^9	ω-7	palmitoleic acid	 COOH
18:1 Δ^9	ω-9	oleic acid	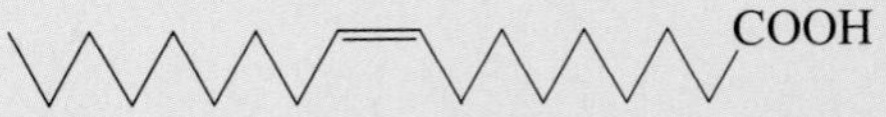 COOH
Polyunsaturated Fatty Acids			
18:2 $\Delta^{9,12}$	ω-6	linoleic acid	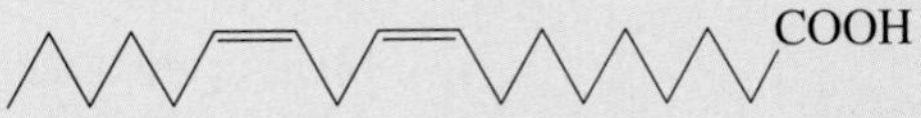 COOH
18:3 $\Delta^{9,12,15}$	ω-3	linolenic acid	 COOH
20:4 $\Delta^{5,8,11,14}$	ω-6	arachidonic acid	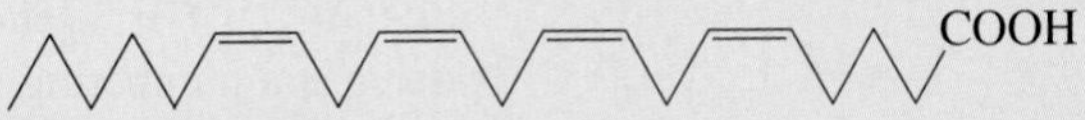 COOH
20:5 $\Delta^{5,8,11,14,17}$	ω-3	EPA (eicosapentaenoic acid)	 COOH
22:6 $\Delta^{4,7,10,13,16,19}$	ω-3	DHA (docosahexaenoic acid)	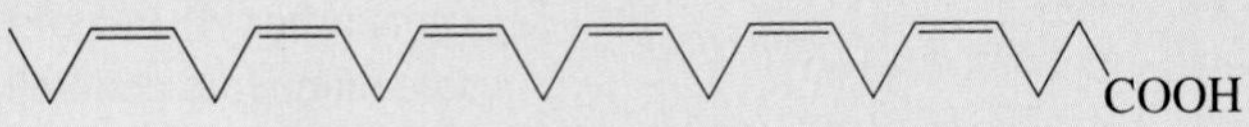 COOH

EXAMPLE 19.1

Classifying Fatty Acids on the Basis of Structural Characteristics

■ Classify the fatty acid with the following structural formula in the ways indicated.

COOH

a. What is the type designation (SFA, MUFA, or SUFA) for this fatty acid?
b. On the basis of carbon chain length and degree of unsaturation, what is the numerical shorthand designation for this fatty acid?
c. To which "omega" family of fatty acids does this fatty acid belong?
d. What is the "delta" designation for the carbon chain double-bond locations for this fatty acid?

Solution

a. Two carbon–carbon double bonds are present in this molecule, which makes it a *polyunsaturated fatty acid (PUFA)*.
b. Eighteen carbon atoms and two carbon–carbon double bonds are present. The shorthand numerical designation for this fatty acid is thus *18:2*.
c. Counting from the methyl end of the carbon chain, the first double bond encountered involves carbons 6 and 7. This fatty acid belongs to the *omega-6* family of fatty acids.
d. Counting from the carboxyl end of the carbon chain, with C-1 being the carboxyl group, the double-bond locations are 9 and 12. This is a $\Delta^{9,12}$ fatty acid.

Practice Exercise 19.1

Classify the fatty acid with the following structural formula in the ways indicated.

a. What is the type designation (SFA, MUFA, or SUFA) for this fatty acid?
b. On the basis of carbon chain length and degree of unsaturation, what is the numerical shorthand designation for this fatty acid?
c. To which "omega" family of fatty acids does this fatty acid belong?
d. What is the "delta" designation for the carbon chain double-bond location for this fatty acid?

19.3 Physical Properties of Fatty Acids

The physical properties of fatty acids, and of lipids that contain them, are largely determined by the length and degree of unsaturation of the fatty acid carbon chain.

Water solubility for fatty acids is a direct function of carbon chain length; solubility decreases as carbon chain length increases. Short-chain fatty acids have a slight solubility in water. Long-chain fatty acids are essentially insoluble in water. The slight solubility of short-chain fatty acids is related to the polarity of the carboxyl group present. In longer-chain fatty acids, the nonpolar nature of the hydrocarbon chain completely dominates solubility considerations.

Melting points for fatty acids are strongly influenced by both carbon chain length and degree of unsaturation (number of double bonds present). Figure 19.2 shows melting-point variation as a function of both of these variables. As carbon chain length increases, melting point increases. This trend is related to the greater surface area associated with a longer carbon chain and to the increased opportunities that this greater surface area affords for intermolecular attractions between fatty acid molecules.

A trend of particular significance is that saturated fatty acids have higher melting points than unsaturated fatty acids with the same number of carbon atoms. The greater the degree of unsaturation, the greater the reduction in melting points. Figure 19.2 shows this effect for the 18-carbon acids with zero, one, two, and three double bonds. Long-chain saturated fatty acids tend to be solids at room temperature, whereas long-chain unsaturated fatty acids tend to be liquids at room temperature.

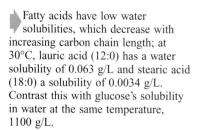

Fatty acids have low water solubilities, which decrease with increasing carbon chain length; at 30°C, lauric acid (12:0) has a water solubility of 0.063 g/L and stearic acid (18:0) a solubility of 0.0034 g/L. Contrast this with glucose's solubility in water at the same temperature, 1100 g/L.

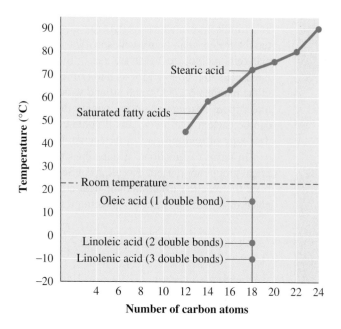

FIGURE 19.2 The melting point of a fatty acid depends on the length of the carbon chain and on the number of double bonds present in the carbon chain.

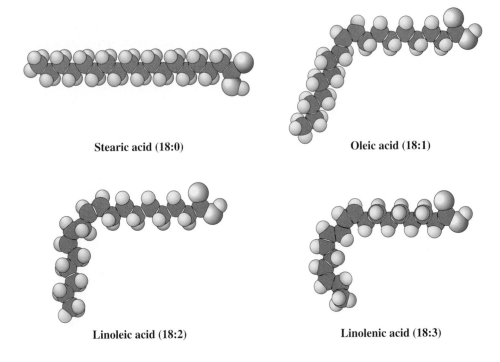

Stearic acid (18:0)

Oleic acid (18:1)

Linoleic acid (18:2)

Linolenic acid (18:3)

The decreasing melting point associated with increasing degree of unsaturation in fatty acids is explained by decreased molecular attractions between carbon chains. The double bonds in unsaturated fatty acids, which generally have the *cis* configuration, produce "bends" in the carbon chains of these molecules (see Figure 19.3). These "bends" prevent unsaturated fatty acids from packing together as tightly as saturated fatty acids. The greater the number of double bonds, the less efficient the packing. As a result, unsaturated fatty acids always have fewer intermolecular attractions, and therefore lower melting points, than their saturated counterparts.

19.4 Energy-Storage Lipids: Triacylglycerols

With the notable exception of nerve cells, human cells store small amounts of energy-providing materials for use when energy demand is high. The most widespread energy-storage material within cells is the carbohydrate glycogen (Section 18.14); it is present in small amounts in most cells.

Lipids known as triacylglycerols also function within the body as energy-storage materials. Rather than being widespread, triacylglycerols, are concentrated primarily in special cells (adipocytes) that are nearly filled with the material. Adipose tissue containing these cells is found in various parts of the body: under the skin, in the abdominal cavity, in the mammary glands, and around various organs (see Figure 19.4). Triacylglycerols are much more efficient at storing energy than is glycogen, because large quantities of them can be packed into a very small volume. These energy-storage lipids are the most abundant type of lipid present in the human body.

In terms of functional groups present, triacylglycerols are triesters; three ester functional groups are present. Recall, from Section 16.13, that an ester is a compound produced from the reaction of an alcohol with a carboxylic acid. The alcohol involved in triacylglycerol formation is always glycerol, a three-carbon alcohol with three hydroxyl groups.

FIGURE 19.4 An electron micrograph of adipocytes, the body's triacylglycerol-storing cells. Note the bulging spherical shape.

$$CH_2—OH$$
$$|$$
$$CH—OH$$
$$|$$
$$CH_2—OH$$
Glycerol

Fatty acids are the carboxylic acids involved in triacylglycerol formation. In the esterification producing a triacylglycerol, a single molecule of glycerol reacts with three fatty acid molecules; each of the three hydroxyl groups present is esterified. Figure 19.5 shows the triple esterification reaction that occurs between glycerol and three molecules of stearic acid (18:0); note the production of three molecules of water as a by-product of the reaction.

Two general ways to represent the structure of a triacylglycerol are

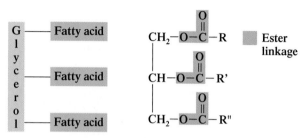

> Triacylglycerols do not actually contain glycerol and three fatty acids, as the block diagram for a triacylglycerol implies. They actually contain a glycerol *residue* and three fatty acid *residues*. In the formation of the triacylglycerol, three molecules of water have been removed from the structural components of the triacylglycerol, leaving residues of the reacting molecules.

The first representation, a block diagram, shows the four subunits present in the structure: glycerol and three fatty acids. The second representation, a general structural formula, shows the three ester linkages present in a triacylglycerol. Each of the fatty acids is attached to glycerol through an ester linkage.

Formally defined, a **triacylglycerol** *is a lipid formed by esterification of three fatty acids to a glycerol molecule.* Within the name *triacylglycerol* is the term *acyl.* An **acyl group** *is the portion of a carboxylic acid that remains after the —OH group is removed from the carbonyl carbon atom.* The structural representation for an acyl group is

$$R - \overset{\overset{\textstyle O}{\textstyle \|}}{C} -$$

An acyl group

Thus, as the name implies, triacylglycerol molecules contain three fatty acid residues (three acyl groups) attached to a glycerol residue. An older name that is still frequently used for a triacylglycerol is *triglyceride.*

The triacylglycerol produced from glycerol and three molecules of stearic acid (as in Figure 19.5) is an example of a simple triacylglycerol. A **simple triacylglycerol** *is a triester formed from the esterification of glycerol with three identical fatty acid molecules.* If the reacting fatty acid molecules are not all identical, then the result is a mixed triacylglycerol. A **mixed triacylglycerol** *is a triester formed from the esterification of glycerol with more than one kind of fatty acid molecule.* Figure 19.6 shows the structure of a mixed triacylglycerol in which one fatty acid is saturated, another monounsaturated, and the third polyunsaturated. Naturally occurring *simple* triacylglycerols are rare. Most biochemically important triacylglycerols are *mixed* triacylglycerols.

FIGURE 19.5 Structure of the simple triacylglycerol produced from the triple esterification reaction between glycerol and three molecules of stearic acid (18:0 acid). Three molecules of water are a by-product of this reaction.

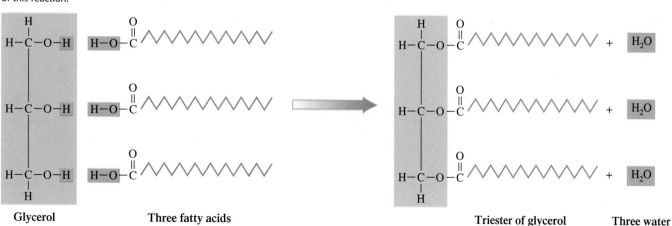

Glycerol　　　　Three fatty acids　　　　　　　　Triester of glycerol　　Three water molecules

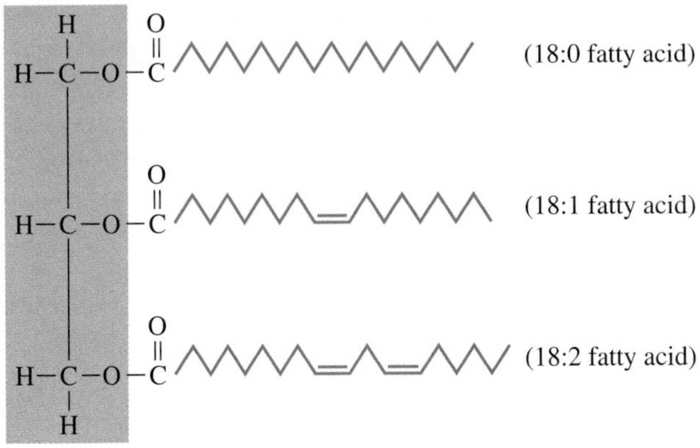

FIGURE 19.6 Structure of a mixed triacylglycerol in which three different fatty acid residues are present.

(18:0 fatty acid)

(18:1 fatty acid)

(18:2 fatty acid)

■ Fats and Oils

Fats are naturally occurring complex mixtures of triacylglycerol molecules in which many different kinds of triacylglycerol molecules are present. *Oils* are also naturally occurring complex mixtures of triacylglycerol molecules in which there are many different kinds of triacylglycerol molecules present. Given that both are triacylglycerol mixtures, what distinguishes a fat from an oil? The answer is physical state at room temperature. A **fat** *is a triacylglycerol mixture that is a solid or a semi-solid at room temperature (25°C)*. Generally, fats are obtained from animal sources. An **oil** *is a triacylglycerol mixture that is a liquid at room temperature (25°C)*. Generally, oils are obtained from plant sources. Because they are mixtures, no fat or oil can be represented by a single specific chemical formula. Many different fatty acids are represented in the triacylglycerol molecules present in the mixture. The actual composition of a fat or oil varies even for the species from which it is obtained. Composition depends on both dietary and climatic factors. For example, fat obtained from corn-fed hogs has a different overall composition than fat obtained from peanut-fed hogs. Flax seed grown in warm climates gives oil with a different composition from that obtained from flax seed grown in colder climates.

Additional generalizations and comparisons between fats and oils follow.

▶ *Petroleum oils* (Section 12.14) are structurally different from *lipid oils.* The former are mixtures of alkanes and cycloalkanes. The latter are mixtures of triesters of glycerol.

1. Fats are composed largely of triacylglycerols in which saturated fatty acids predominate, although some unsaturated fatty acids are present. Such triacylglycerols can pack closely together because of the "linearity" of their fatty acid chains (Figure 19.7a), causing the higher melting points associated with fats. Oils contain triacylglycerols with larger amounts of mono- and polyunsaturated fatty acids than those in fats. Such triacylglycerols cannot pack as tightly together because of "bends" in their fatty acid chains (Figure 19.7b). The result is lower melting points.

2. Fats are generally obtained from animals; hence the term *animal fat.* Although fats are solids at room temperature, the warmer body temperature of the living animal keeps the fat somewhat liquid (semi-solid) and thus allows for movement. Oils

FIGURE 19.7 Representative triacylglycerols from (a) a fat and (b) an oil.

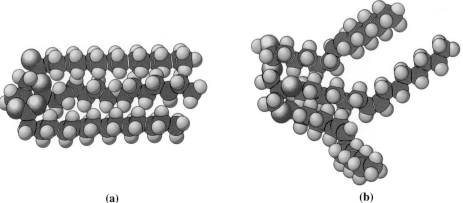

(a)   (b)

FIGURE 19.8 Percentages of saturated, monounsaturated, and polyunsaturated fatty acids in the triacylglycerols of various dietary fats and oils.

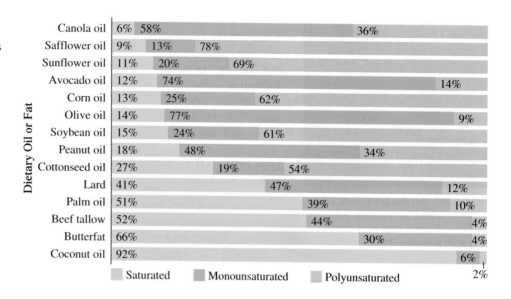

Dietary Oil or Fat

	Saturated	Monounsaturated	Polyunsaturated
Canola oil	6%	58%	36%
Safflower oil	9%	13%	78%
Sunflower oil	11%	20%	69%
Avocado oil	12%	74%	14%
Corn oil	13%	25%	62%
Olive oil	14%	77%	9%
Soybean oil	15%	24%	61%
Peanut oil	18%	48%	34%
Cottonseed oil	27%	19%	54%
Lard	41%	47%	12%
Palm oil	51%	39%	10%
Beef tallow	52%	44%	4%
Butterfat	66%	30%	4%
Coconut oil	92%		6% · 2%

▶ All oils, even polyunsaturated oils, contain some *saturated* fatty acids. All fats, even highly saturated fats, contain some *unsaturated* fatty acids.

typically come from plants, although there are also fish oils. A fish would have some serious problems if its triacylglycerols "solidified" when it encountered cold water.

3. Pure fats and pure oils are colorless, odorless, and tasteless. The tastes, odors, and colors associated with dietary plant oils are caused by small amounts of other naturally occurring substances present in the plant that have been carried along during processing. The presence of these "other" compounds is usually considered desirable.

Figure 19.8 gives the percentages of saturated, monounsaturated, and polyunsaturated fatty acids found in common dietary oils and fats. In general, a higher degree of fatty acid unsaturation is associated with oils than with fats. A notable exception to this generalization is coconut oil, which is highly saturated. This oil is a liquid not because it contains many double bonds within the fatty acids but because it is rich in *shorter-chain* fatty acids, particularly lauric acid (12:0).

19.5 Dietary Considerations and Triacylglycerols

In the past two decades, considerable research has been carried out concerning the role of dietary factors as a cause of disease (obesity, diabetes, cancer, hypertension, and atherosclerosis). Numerous studies have shown that *in general,* nations whose citizens have high dietary intakes of triacylglycerols (fats and oils) tend to have higher incidences of heart disease and certain types of cancers. This is the reason for concern that the typical American diet contains too much fat and the call for Americans to reduce their total dietary fat intake.

Contrary to the general trend, however, there are several areas of the world where high dietary fat intake does not translate into high risks for cardiovascular disease, obesity, and certain types of cancers. These exceptions, which include some Mediterranean countries and the Inuit people of Greenland, suggest that relationships between dietary triacylglycerol intake and risk factors for disease involve more than simply the *total amount* of triacylglycerols consumed.

■ "Good Fats" Versus "Bad Fats"

In dietary discussions, the term *fat* is used as a substitute for the term *triacylglycerol.* Thus a dietary fat can be either a "fat" or an "oil." Ongoing studies indicate that both the *type of dietary fat* consumed and the *amount of dietary fat* consumed are important factors in determining human body responses to dietary fat. Current dietary fat recommendations are that people limit their total fat intake to 30% of total calories—with up to 15% coming from monounsaturated fat, up to 10% from polyunsaturated fat, and less than 10% from saturated fats.

Artificial Fat Substitutes

Artificial sweeteners (sugar substitutes) have been an accepted part of the diet of most people for many years. New since the 1990s are artificial fats—substances that create the sensations of "richness" of taste and "creaminess" of texture in food without the negative effects associated with dietary fats (heart disease and obesity).

Food scientists have been trying to develop fat substitutes since the 1960s. Now available for consumer use are two types of fat substitutes: *calorie-reduced* fat substitutes and *calorie-free* substitutes. They differ in their chemical structures and therefore in how the body handles them.

Simplesse, the best-known calorie-reduced fat substitute, received FDA marketing approval in 1990. It is made from the protein of fresh egg whites and milk by a procedure called microparticulation. This procedure produces tiny, round protein particles so fine that the tongue perceives them as a fluid rather than as the solid they are. Their fineness creates a sensation of smoothness, richness, and creaminess on the tongue.

In the body, Simplesse is digested and absorbed, contributing to energy intake. But 1 g of Simplesse provides 1.3 cal, compared with the 9 cal provided by 1 g of fat. Simplesse is used only to replace fats in *formulated* foods such as salad dressings, cheeses, sour creams, and other dairy products. Simplesse is unsuitable for frying or baking, because it turns rubbery or rigid (gels) when heated. Consequently, it is not available for home use.

Olestra, a sucrose polyester and the best-known calorie-free fat substitute, received FDA marketing approval in 1996. It is produced by heating soybean oil with sucrose in the presence of methyl alcohol. The resulting product has a structure somewhat similar to that of a triacylglycerol, sucrose takes the place of the glycerol molecule, and six to eight fatty acids are attached to it rather than the three in a triacylglycerol. Unlike triacylglycerols, however, olestra cannot be hydrolyzed by the body's digestive enzymes and therefore passes through the digestive tract undigested.

Olestra looks, feels, and tastes like dietary fat and can substitute for fats and oils in foods such as shortenings, oils, margarines, snacks, ice creams, and other desserts. It has the same cooking properties as fats and oils.

In the digestive tract, Olestra interferes with the absorption of both dietary and body-produced cholesterol; thus it may lower total cholesterol levels. A problem with its use is that it also reduces the absorption of the fat-soluble vitamins A, D, E, and K. To avoid such depletion, Olestra is fortified with these vitamins. Another problem with Olestra use is that in some individuals it can cause gastrointestinal irritation and/or diarrhea. All products containing Olestra must carry the following label: "Olestra may cause abdominal cramping and loose stools. Olestra inhibits the absorption of some vitamins and other nutrients. Vitamins A, D, E, and K have been added."

■ Hydrogenation

Hydrogenation is a reaction we encountered in Section 13.7. It involves hydrogen addition across carbon–carbon multiple bonds, which increases the degree of saturation as some double bonds are converted to single bonds. With this change, there is a corresponding increase in the melting point of the substance.

Hydrogenation involving just one carbon–carbon bond within a fatty acid residue of a triacylglycerol can be diagrammed as follows:

$\cdots CH_2 - CH_2 - CH = CH - CH_2 - CH_2 \cdots + H_2 \longrightarrow \cdots CH_2 - CH_2 - CH_2 - CH_2 - CH_2 - CH_2 \cdots$

Portion of an unsaturated fatty acid residue in a triacylglycerol containing one double bond

The double bond has been converted to a single bond; the degree of saturation has increased

The structural equation for the complete hydrogenation of a triacylglycerol in which all three fatty acid residues are oleic acid (18:1) is

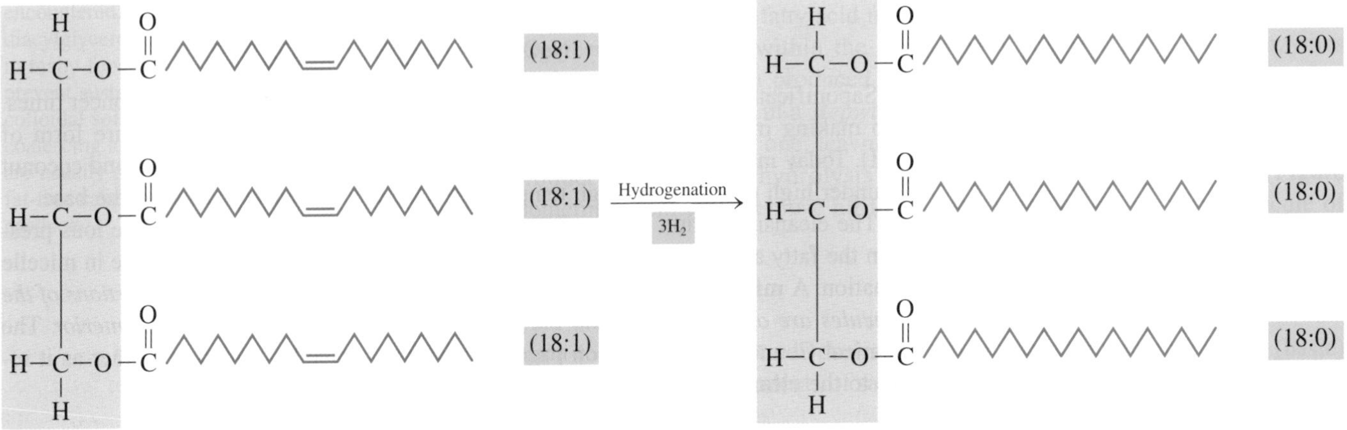

The Cleansing Action of Soap

The cleansing action of soap is directly related to the structure of the carboxylate ions present in soap as fatty acid salts. Their structure is such that they exhibit a "dual polarity." The hydrocarbon portion of the carboxylate ion is nonpolar, and the carboxyl portion is polar. This dual polarity for the fatty acid salt *sodium stearate,* which is representative of all fatty acid salts present in soap, is as follows:

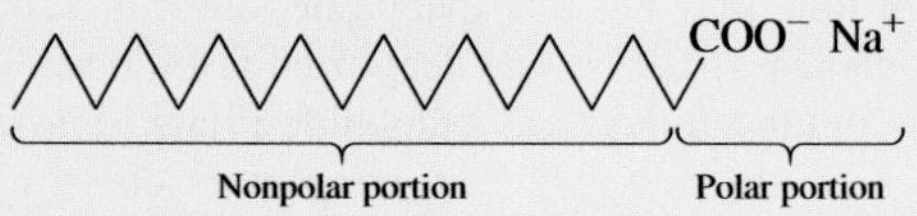

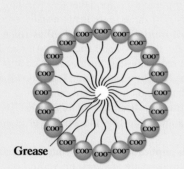

Soap solubilizes oily and greasy materials in the following manner: The nonpolar portion of the carboxylate ion dissolves in the nonpolar oil or grease, but the polar carboxyl portion maintains its solubility in the polar water.

The penetration of the oil or grease by the nonpolar end of the carboxylate ion is followed by the formation of micelles (see the accompanying diagram). The carboxyl groups (the micelle exterior) and water molecules are attracted to each other, causing the solubilizing of the micelle.

The micelles do not combine into larger drops, because their surfaces are all negatively charged, and like charges repel each other. The water-soluble micelles are subsequently rinsed away, leaving a material devoid of oil and grease.

For many cleansing purposes, synthetic detergents have largely replaced soaps. The basis of the cleansing action of synthetic detergents is very similar to that of soaps, because their structures are very similar. The structure of the sodium salt of a benzene sulfonic acid is typical of the types of molecules used in detergents.

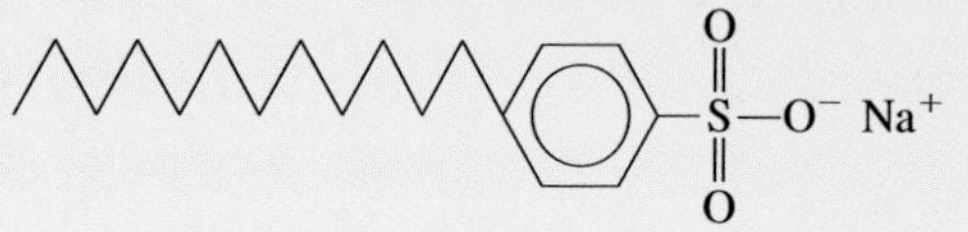

Many food products are produced via partial hydrogenation. In partial hydrogenation some, but not all, of the double bonds present are converted into single bonds. In this manner, liquids (usually plant oils) are converted into semi-solid materials.

Peanut butter is produced from peanut oil through partial hydrogenation. Solid cooking shortenings and stick margarine are produced from liquid plant oils through partial hydrogenation. Soft-spread margarines are also partial-hydrogenation products. Here, the extent of hydrogenation is carefully controlled to make the margarine soft at refrigerator temperatures (4°C). Concern has arisen about food products obtained from hydrogenation processes, because the hydrogenation process itself converts some *cis* double bonds within fatty acid residues into *trans* double bonds. The Chemical Connections feature on page 562 explores this issue further.

■ Oxidation

The carbon–carbon double bonds present in the fatty acid residues of a triacylglycerol are subject to oxidation with molecular oxygen (from air) as the oxidizing agent. Such oxidation breaks the carbon–carbon bonds, producing both aldehyde and carboxylic acid products.

$$-CH{=}CH- \xrightarrow{\text{Oxidation}} \underset{\substack{\text{Short-chain}\\ \text{aldehydes}}}{-\overset{\overset{\displaystyle O}{\|}}{C}-H + H-\overset{\overset{\displaystyle O}{\|}}{C}-} \xrightarrow{\text{Oxidation}} \underset{\substack{\text{Short-chain}\\ \text{carboxylic acids}}}{-\overset{\overset{\displaystyle O}{\|}}{C}-OH + HO-\overset{\overset{\displaystyle O}{\|}}{C}-}$$

Unsaturated fatty acids

▶ Antioxidants are compounds that are easily oxidized. When added to foods, they are more easily oxidized than the food. Thus they prevent the food from being oxidized (see Section 14.12).

The short-chain aldehydes and carboxylic acids so produced often have objectionable odors, and fats and oils containing them are said to have become *rancid*. To avoid this unwanted oxidation process, commercially prepared foods containing fats and oils nearly always contain *antioxidants*—substances that are more easily oxidized than the food. Two naturally occurring antioxidants are vitamin C (Section 21.13) and vitamin E (Section

Trans Fatty Acids and Blood Cholesterol Levels

All current dietary recommendations stress reducing saturated fat intake. In accordance with such recommendations, many people have switched from butter to margarine and now use partially hydrogenated vegetable oils rather than animal fat for cooking. However, recent studies suggest that partially hydrogenated products also play a role in raising blood cholesterol levels. Why would this be so?

During hydrogenation (Section 19.6) some of the *unreacted* double bonds that are naturally present in the *cis* configuration are isomerized to the *trans* configuration. Such *cis–trans* conversions can dramatically alter the overall shape of the fatty acid. In the following diagram, note how conversion of a *cis,cis*-18:2 fatty acid to a *trans,trans*-18:2 fatty acid affects molecular shape. The *trans,trans*-18:2 fatty acid has a shape very much like that of an 18:0 saturated fatty acid (the structure on the right).

Studies show that fatty acids with *trans* double bonds (an unnatural situation) affect blood cholesterol levels in a manner similar to saturated fatty acids.

Trans fatty acids (*trans* fat) make up approximately 5% of the fat intake in the typical diet in the United States, and the amount of *trans* fat a person consumes depends on the amount of fat eaten and on the types of foods selected. The best ex-ample of a *trans*-fat food may be stick margarine, but it is also found in crackers, cookies, pastries, and deep-fried fast foods. Spreadable margarine in tubs, though, contains little if any *trans* fat.

Until recently, the only way consumers could determine whether a food included *trans* fat was to look for "hydrogenated" on the list of ingredients. A food that lists partially hydrogenated oils among its first three ingredients usually contains substantial amounts of *trans* fatty acids as well as some saturated fat. In the year 2000, the U.S. Food and Drug Administration (FDA) adopted rules requiring that the *trans*-fat content of a food be included in the Nutrition Facts panel found on all food products. When *trans* fatty acids are present, an asterisk (or other symbol) placed after the heading "Saturated Fat" refers the consumer to a footnote that lists the amount of grams of *trans* fat in the product.

The health implications of *trans* fatty acids remain an area of active research; many answers are yet to be found. At present, eating saturated fatty acids still appears to pose greater health risks than eating *trans* fatty acids. However, consuming excessive amounts of either type of fatty acid is unwise.

18:2 *(cis, cis)*   **18:2** *(trans, trans)*   **18:0**

21.14). Two synthetic oxidation inhibitors are BHA and BHT (Section 14.12). In the presence of air, antioxidants, rather than food, are oxidized.

Perspiration, generated by strenuous exercise or by "hot and muggy" climatic conditions, contains numerous triacylglycerols (oils). Rapid oxidation of these oils, promoted by microorganisms on the skin, generates the body odor that accompanies most "sweaty" people (see Figure 19.11).

FIGURE 19.11 The oils (triacylglycerols) present in skin perspiration rapidly undergo oxidation. The oxidation products, short-chain aldehydes and short-chain carboxylic acids, often have strong odors.

■ Using words rather than structural formulas, characterize the products formed when the following triacylglycerol undergoes the reactions listed.

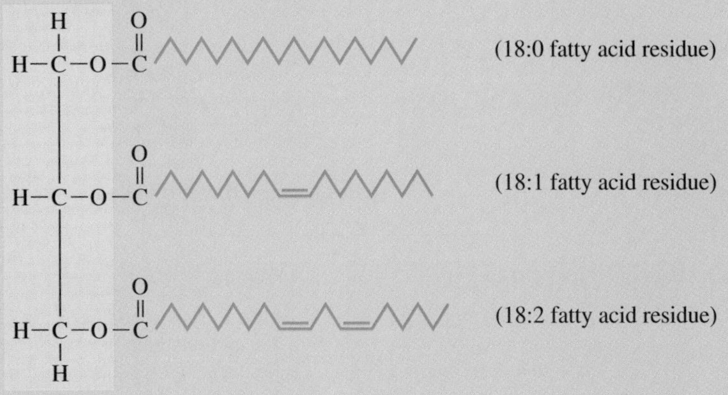

(18:0 fatty acid residue)

(18:1 fatty acid residue)

(18:2 fatty acid residue)

a. Complete hydrolysis
b. Complete saponification using NaOH
c. Complete hydrogenation

Solution

a. When a triacylglycerol undergoes complete hydrolysis, there are four products: glycerol and three fatty acids. For the given triacylglycerol the products are *glycerol, an 18:0 fatty acid, an 18:1 fatty acid,* and *an 18:2 fatty acid.*
b. When a triacylglycerol undergoes complete saponification, there are four products: glycerol and three fatty acid salts. For the given triacylglycerol, with NaOH as the base involved in the saponification, the products are *glycerol, the sodium salt of the 18:0 fatty acid, the sodium salt of the 18:1 fatty acid,* and *the sodium salt of the 18:2 fatty acid.*
c. Complete hydrogenation will change the given triacylglycerol into a *triacylglycerol in which all three fatty acid residues are 18:0 fatty acid residues.* That is, all of the fatty acid residues are completely saturated (there are no carbon–carbon double bonds).

Practice Exercise 19.2

Using words rather than structural formulas, characterize the products formed when the following triacylglycerol undergoes the reactions listed.

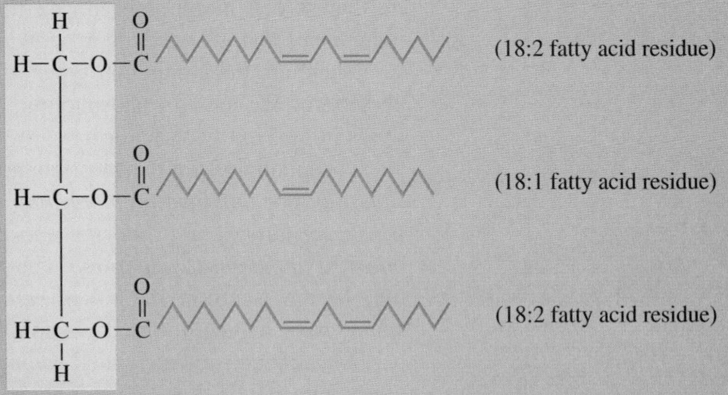

(18:2 fatty acid residue)

(18:1 fatty acid residue)

(18:2 fatty acid residue)

a. Complete hydrolysis
b. Complete saponification using NaOH
c. Complete hydrogenation

The Chemistry at a Glance on page 564 contains a summary of the terminology used in characterizing the properties of the fatty acid residues that are part of the structure of triacylglycerols (fats and oils).

Classification Schemes for Fatty Acid Residues Present in Triacylglycerols

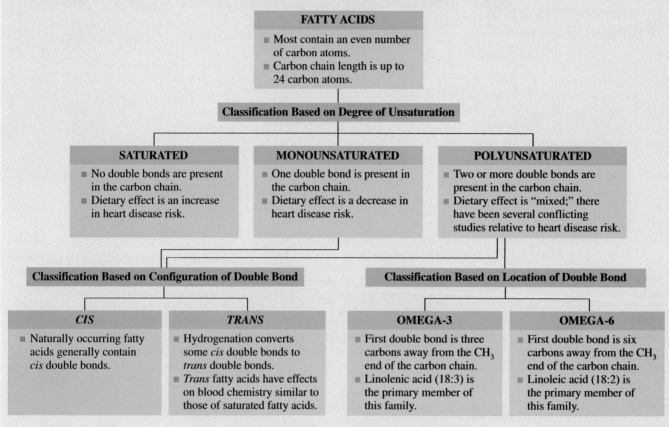

19.7 Membrane Lipids: Phospholipids

All cells are surrounded by a membrane that confines their contents. Up to 80% of the mass of a cell membrane can be lipid materials; the rest is primarily protein. It is membranes that give cells their individuality by separating them from their environment.

There are three common types of membrane lipids: phospholipids, glycolipids, and cholesterol. We consider phospholipids in this section and the other two types of membrane lipids in the next two sections.

Phospholipids are the most abundant type of membrane lipid. A **phospholipid** *is a lipid that contains one or more fatty acids, a phosphate group, a platform molecule to which the fatty acid(s) and the phosphate group are attached, and an alcohol that is attached to the phosphate group.* The platform molecule on which a phospholipid is built may be the 3-carbon alcohol *glycerol* or a more complex C_{18} aminodialcohol called *sphingosine*. Glycerol-based phospholipids are called *glycerophospholipids,* and those based on sphingosine are called *sphingophospholipids*. The general block diagrams for a glycerophospholipid and a sphingophospholipid are as follows:

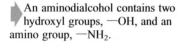

An aminodialcohol contains two hydroxyl groups, —OH, and an amino group, —NH$_2$.

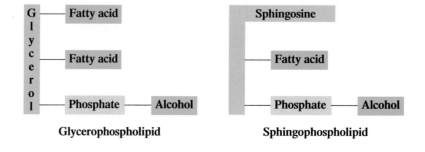

■ Glycerophospholipids

A **glycerophospholipid** *is a lipid that contains two fatty acids and a phosphate group esterified to a glycerol molecule and an alcohol esterified to the phosphate group.* All attachments (bonds) between groups in a glycerophospholipid are ester linkages, a situation similar to that in triacylglycerols (Section 19.4). However, glycerophospholipids have four ester linkages as contrasted to three ester linkages in triacylglycerols.

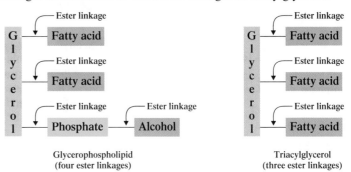

Glycerophospholipid
(four ester linkages)

Triacylglycerol
(three ester linkages)

Because of the ester linkages present, glycerophospholipids undergo hydrolysis and saponification reactions in a manner similar to that for triacylglycerols (Section 19.6). There will be five reaction products, however, instead of the four for triacylglycerols.

The alcohol attached to the phosphate group in a glycophospholipid is usually one of three amino alcohols: choline, ethanolamine, or serine. The structures of these three amino alcohols, given in terms of the charged forms (Sections 17.8 and 18.6) that they adopt in neutral solution, are

$$HO-CH_2-CH_2-\overset{+}{N}(CH_3)_3 \qquad HO-CH_2-CH_2-\overset{+}{N}H_3 \qquad HO-CH_2-CH-\overset{+}{N}H_3$$
$$\qquad\qquad\qquad\qquad\qquad\qquad\qquad\qquad\qquad\qquad\qquad\qquad\qquad\qquad\qquad\qquad | $$
$$\qquad\qquad\qquad\qquad\qquad\qquad\qquad\qquad\qquad\qquad\qquad\qquad\qquad\qquad\qquad\qquad COO^-$$

Choline
(a quaternary ammonium ion)

Ethanolamine
(positive-ion form)

Serine
(two ionic groups present)

Glycerophospholipids containing these three amino alcohols are respectively known as phosphatidylcholines, phosphatidylethanolamines, and phosphatidylserines. The fatty acid, glycerol, and phosphate portions of a glycerophospholipid structure constitute a *phosphatidyl* group.

Although the general structural features of glycerophospholipids are similar in many respects to those of triacylglycerols, these two types of lipids have quite different biochemical functions. Triacylglycerols serve as storage molecules for metabolic fuel. Glycerophospholipids function almost exclusively as components of cell membranes (Section 19.10) and are not stored. A major structural difference between the two types of lipids, that of polarity, is related to their differing biochemical functions. Triacylglycerols are a nonpolar class of lipids, whereas glycerophospholipids are polar. In general, membrane lipids have polarity associated with their structures.

Further consideration of general phosphoacylglycerol structure reveals an additional structural characteristic of most membrane lipids. Let us consider a phosphatidylcholine containing stearic and oleic acids to illustrate this additional feature. The chemical structure of this molecule is shown in Figure 19.12a.

A molecular model for this compound, which gives the orientation of groups in space, is illustrated in Figure 19.12b. There are two important things to notice about this model: (1) There is a "head" part, the choline and phosphate, and (2) there are two "tails," the two fatty acid carbon chains. The head part is polar. The two tails, the carbon chains, are nonpolar.

All glycerophospholipids have structures similar to that shown in Figure 19.12. All have a "head" and two "tails." A simplified representation for this structure uses a circle to represent the polar head and two wavy lines to represent the nonpolar tails.

▶ Glycerophospholipids have a *hydrophobic* ("water-hating") portion, the nonpolar fatty acid groups, and a *hydrophilic* ("water-loving") portion, the polar head group.

Polar head group,

Nonpolar chains

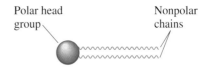

EXERCISES AND PROBLEMS

The members of each pair of problems in this section test similar material.

■ Structure and Classification of Lipids (Section 19.1)

19.1 What characteristic do all lipids have in common?

19.2 What structural feature, if any, do all lipid molecules have in common? Explain your answer.

19.3 Would you expect lipids to be soluble or insoluble in each of the following solvents?
a. H_2O (polar)
b. CH_3—CH_2—O—CH_2—CH_3 (nonpolar)
c. CH_3—OH (polar)
d. CH_3—CH_2—CH_2—CH_2—CH_3 (nonpolar)

19.4 Would you expect lipids to be soluble or insoluble in each of the following solvents?
a. CH_3—$(CH_2)_7$—CH_3 (nonpolar)
b. CH_3—Cl (polar)
c. CCl_4 (nonpolar)
d. CH_3—CH_2—OH (polar)

19.5 In terms of biochemical function, what are the five major categories of lipids?

19.6 What is the biochemical function of each of the following types of lipids?
a. Triacylglycerols   b. Bile acids
c. Sphingoglycolipids   d. Eicosanoids

■ Fatty Acids (Sections 19.2 and 19.3)

19.7 Classify each of the following fatty acids as long-chain, medium-chain, or short-chain.
a. Myristic (14:0)   b. Caproic (6:0)
c. Arachidic (20:0)   d. Capric (10:0)

19.8 Classify each of the following fatty acids as long-chain, medium-chain, or short-chain.
a. Lauric (12:0)   b. Oleic (18:1)
c. Butyric (4:0)   d. Stearic (18:0)

19.9 Classify each of the following fatty acids as saturated, monounsaturated, or polyunsaturated.
a. Stearic (18:0)   b. Linolenic (18:3)
c. Docosahexaenoic (22:6)   d. Oleic (18:1)

19.10 Classify each of the following fatty acids as saturated, monounsaturated, or polyunsaturated.
a. Palmitic (16:0)   b. Linoleic (18:2)
c. Arachidonic (20:4)   d. Palmitoleic (16:1)

19.11 Structurally, what is the difference between a SFA and a MUFA?

19.12 Structurally, what is the difference between a MUFA and a PUFA?

19.13 With the help of Table 19.1, classify each of the acids in Problem 19.9 as an omega-3 acid, an omega-6 acid, or neither an omega-3 nor an omega-6 acid.

19.14 With the help of Table 19.1, classify each of the acids in Problem 19.10 as an omega-3 acid, an omega-6 acid, or neither an omega-3 nor an omega-6 acid.

19.15 Draw the condensed structural formula for the fatty acid whose numerical shorthand designation is 18:2 ($\Delta^{9,12}$)

19.16 Draw the condensed structural formula for the fatty acid whose numerical shorthand designation is 20:4 ($\Delta^{5,8,11,14}$)

19.17 Why does the introduction of double bonds into a fatty acid molecule lower its melting point?

19.18 What effect does a *cis* double bond have on the shape of a fatty acid molecule?

19.19 In each of the following pairs of fatty acids, select the fatty acid that has the lower melting point.
a. 18:0 acid and 18:1 acid   b. 18:2 acid and 18:3 acid
c. 14:0 acid and 16:0 acid   d. 18:1 acid and 20:0 acid

19.20 In each of the following pairs of fatty acids, select the fatty acid that has the higher melting point.
a. 14:0 acid and 18:0 acid   b. 20:4 acid and 20:5 acid
c. 18:3 acid and 20:3 acid   d. 16:0 acid and 16:1 acid

19.21 Using the structural information given in Table 19.1, assign an IUPAC name to each of the following fatty acids.
a. Myristic acid   b. Palmitoleic acid

19.22 Using the structural information given in Table 19.1, assign an IUPAC name to each of the following fatty acids.
a. Stearic acid   b. Linolenic acid

■ Triacylglycerols (Section 19.4)

19.23 What are the four structural subunits that contribute to the structure of a triacylglycerol?

19.24 Draw the general block diagram for a triacylglycerol.

19.25 How many different kinds of functional groups are present in a triacylglycerol in which all three fatty acid residues come from saturated fatty acids?

19.26 How many different kinds of functional groups are present in a triacylglycerol in which all three fatty acid residues come from unsaturated fatty acids?

19.27 Draw the condensed structural formula of a triacylglycerol formed from glycerol and three molecules of palmitic acid.

19.28 Draw the condensed structural formula of a triacylglycerol formed from glycerol and three molecules of stearic acid.

19.29 Draw block diagram structures for the four different triacylglycerols that can be produced from glycerol, stearic acid, and linolenic acid.

19.30 Draw block diagram structures for the three different triacylglycerols that can be produced from glycerol, palmitic acid, stearic acid, and linolenic acid.

19.31 Identify the fatty acids present in each of the following triacylglycerols.

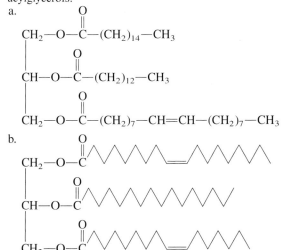

19.32 Identify the fatty acids present in each of the following triacylglycerols.

a.

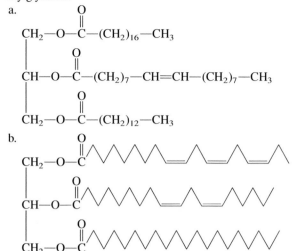

b.

19.33 For each of the acyl groups present in the triacylglycerol of Problem 19.31a, indicate how many carbon atoms are present and how many oxygen atoms are present.

19.34 For each of the acyl groups present in the triacylglycerol of Problem 19.32a, indicate how many carbon atoms are present and how many oxygen atoms are present.

19.35 What is the difference in meaning, if any, between the members of each of the following pairs of terms?
a. Triacylglycerol and triglyceride
b. Triacylglycerol and fat
c. Triacylglycerol and mixed triacylglycerol
d. Fat and oil

19.36 What is the difference in meaning, if any, between the members of each of the following pairs of terms?
a. Triacylglycerol and oil
b. Triacylglycerol and simple triacylglycerol
c. Simple triacylglycerol and mixed triacylglycerol
d. Triglyceride and fat

■ **Dietary Considerations and Triacylglycerols (Section 19.5)**

19.37 In a dietary context, indicate whether each of the following pairings of concepts is correct?
a. "Saturated fat" and "good fat"
b. "Polyunsaturated fat" and "bad fat"

19.38 In a dietary context, indicate whether each of the following pairings of concepts is correct?
a. "Monounsaturated fat" and "good fat"
b. "Saturated fat" and "good and bad fat"

19.39 In a dietary context, which of the following pairings of concepts is correct?
a. "Cold-water fish" and "high in omega-3 fatty acids"
b. "Fatty fish" and "low in omega-3 fatty acids"

19.40 In a dietary context, which of the following pairings of concepts is correct?
a. "Warm-water fish" and "low in omega-3 fatty acids"
b. "Fish and chips" and "high in omega-3 fatty acids"

19.41 In a dietary context, classify each of the following fatty acids as an essential fatty acid or as a nonessential fatty acid.
a. Lauric acid (12:0)   b. Linoleic acid (18:2)
c. Myristic acid (14:0)   d. Palmitoleic (16:1)

19.42 In a dietary context, classify each of the following fatty acids as an essential fatty acid or as a nonessential fatty acid.

a. Stearic acid (18:0)   b. Linolenic acid (18:3)
c. Oleic acid (18:1)   d. Arachidic acid (20:0)

■ **Chemical Reactions of Triacylglycerols (Section 19.5)**

19.43 Name, in general terms, the products of the complete
a. hydrolysis of a fat   b. saponification of an oil

19.44 Name, in general terms, the products of the complete
a. saponification of a fat   b. hydrolysis of an oil

19.45 Draw condensed structural formulas for all products you would obtain from the complete hydrolysis of the following triacylglycerol.

$$CH_2-O-\overset{\overset{\displaystyle O}{\|}}{C}-(CH_2)_{14}-CH_3$$
$$CH-O-\overset{\overset{\displaystyle O}{\|}}{C}-(CH_2)_{12}-CH_3$$
$$CH_2-O-\overset{\overset{\displaystyle O}{\|}}{C}-(CH_2)_7-CH=CH-(CH_2)_7-CH_3$$

19.46 Draw condensed structural formulas for all products you would obtain from the complete hydrolysis of the following triacylglycerol.

$$CH_2-O-\overset{\overset{\displaystyle O}{\|}}{C}-(CH_2)_{16}-CH_3$$
$$CH-O-\overset{\overset{\displaystyle O}{\|}}{C}-(CH_2)_{12}-CH_3$$
$$CH_2-O-\overset{\overset{\displaystyle O}{\|}}{C}-(CH_2)_6-(CH_2-CH=CH)_3-CH_2-CH_3$$

19.47 With the help of Table 19.1, determine the names of each of the products obtained in Problem 19.45.

19.48 With the help of Table 19.1, determine the names of each of the products obtained in Problem 19.46.

19.49 Draw condensed structural formulas for all products you would obtain from the saponification with NaOH of the triacylglycerol in Problem 19.45.

19.50 Draw condensed structural formulas for all products you would obtain from the saponification with KOH of the triacylglycerol in Problem 19.46.

19.51 With the help of Table 19.1, name each of the products obtained in Problem 19.49.

19.52 With the help of Table 19.1, name each of the products obtained in Problem 19.50.

19.53 Why can only unsaturated triacylglycerols undergo hydrogenation?

19.54 A food package label lists an oil as "partially hydrogenated." What does this mean?

19.55 How many molecules of H_2 will react with one molecule of the following triacylglycerol?

$$CH_2-O-\overset{\overset{\displaystyle O}{\|}}{C}-(CH_2)_6-(CH_2-CH=CH)_2-(CH_2)_4-CH_3$$
$$CH-O-\overset{\overset{\displaystyle O}{\|}}{C}-(CH_2)_7-CH=CH-(CH_2)_7-CH_3$$
$$CH_2-O-\overset{\overset{\displaystyle O}{\|}}{C}-(CH_2)_6-(CH_2-CH=CH)_3-CH_2-CH_3$$

19.56 How many molecules of H_2 will react with one molecule of the following triacylglycerol?

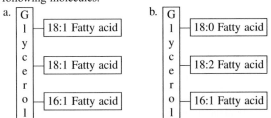

19.57 Draw block diagram structures for all possible products of the partial hydrogenation, with two molecules of H_2, of the following molecules.

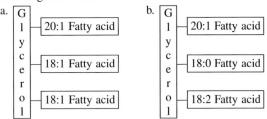

19.58 Draw block diagram structures for all possible products of the partial hydrogenation, with two molecules of H_2, of the following molecules.

a.
G l y c e r o l	20:1 Fatty acid
	18:1 Fatty acid
	18:1 Fatty acid

b.
G l y c e r o l	20:1 Fatty acid
	18:0 Fatty acid
	18:2 Fatty acid

19.59 Why do animal fats and vegetable oils become rancid when exposed to moist, warm air?

19.60 Why are the compounds BHA and BHT often added to foods that contain fats and oils?

■ **Phospholipids (Section 19.7)**

19.61 What are the two common types of platform molecules for a phospholipid?

19.62 How many fatty acid residues are present in a phospholipid?

19.63 Draw the general block diagram for a glycerophospholipid.

19.64 Draw the general block diagram for a sphingophospholipid.

19.65 Draw the structures of the three amino alcohols commonly esterified to the phosphate group in a glycerophospholipid.

19.66 What structural subunits are present in a phosphatidyl group?

19.67 Sphingophospholipids have a "head and two tails" structure. Give the chemical identity of the head and of each of the two tails.

19.68 Glycerophospholipids have a "head and two tails" structure. Give the chemical identity of the head and of each of the two tails.

19.69 Which portion of the structure of a phospholipid has hydrophobic characteristics?

19.70 Which portion of the structure of a phospholipid has hydrophilic characteristics?

19.71 Indicate how many ester linkages are present in the structure of a
a. glycerophospholipid   b. sphingophospholipid

19.72 Indicate how many amide linkages are present in the structure of a
a. glycerophospholipid   b. sphingophospholipid

19.73 Structurally, what is the difference between a lecithin and a phosphatidylserine?

19.74 Structurally, what is the difference between a lecithin and a sphingomyelin?

■ **Sphingoglycolipids (Section 19.8)**

19.75 Draw the general block diagram for a sphingoglycolipid.

19.76 How many of each of the following types of linkages are present in a sphingoglycolipid?
a. Ester linkages   b. Amide linkages
c. Glycosidic linkages

19.77 How does the general structure of a sphingoglycolipid differ from that of a sphingophospholipid?

19.78 Structurally, what is the difference between a cerebroside and a ganglioside?

■ **Cholesterol (Section 19.9)**

19.79 Draw and number the fused hydrocarbon ring system characteristic of all steroids.

19.80 What positions in the steroid nucleus are particularly likely to bear substituents?

19.81 Describe the structure of cholesterol in terms of substituents attached to the steroid nucleus.

19.82 Structurally, what is considered the "head" of a cholesterol molecule?

19.83 In a dietary context, what is the difference between "good cholesterol" and "bad cholesterol"?

19.84 In a dietary context, how do HDL and LDL differ in function?

■ **Cell Membranes (Section 19.10)**

19.85 What are the three major types of lipids present in cell membranes?

19.86 What is the structural characteristic common to all the non-steroid lipids present in cell membranes?

19.87 What is a lipid bilayer?

19.88 What is the basic structure of a cell membrane?

19.89 What is the function of unsaturation in the hydrocarbon tails of membrane lipids?

19.90 What function does cholesterol serve when it is present in cell membranes?

19.91 What is the difference between *passive transport* and *facilitated transport*?

19.92 What is the difference between *facilitated transport* and *active transport*?

19.93 Match each of the following statements related to membrane transport processes to the appropriate term: *passive transport, facilitated transport, active transport.* More than one term may apply in a given situation.
a. Movement across the membrane is against the concentration gradient.
b. Proteins serve as "gates."
c. Expenditure of cellular energy is required.
d. Movement across the membrane is from a high to a low concentration.

19.94 Match each of the following statements related to membrane transport processes to the appropriate term: *passive transport,* *facilitated transport, active transport.* More than one term may apply in a given situation.

  a. Movement across the membrane is with the concentration gradient.
  b. Proteins serve as "pumps."
  c. Expenditure of cellular energy is not required.
  d. Movement across the membrane is from a low to a high concentration.

■ **Bile Acids (Section 19.11)**

19.95 Describe the structural differences between a bile acid and cholesterol.

19.96 Describe the structural differences between cholic acid and a deoxycholic acid.

19.97 Describe the structural differences between glycocholic acid and taurocholic acid.

19.98 Describe the structural differences between glycocholic acid and glyco-7-deoxycholic acid.

19.99 What is the medium through which bile acids are supplied to the small intestine?

19.100 What is the chemical composition of bile?

19.101 At what location in the body are bile acids stored until needed?

19.102 What is the chemical composition of the majority of gallstones?

■ **Steroid Hormones (Section 19.12)**

19.103 What are the two major classes of steroid hormones?

19.104 Describe the general function of each of the following types of steroid hormones.

  a. Estrogens   b. Androgens
  c. Progestins   d. Mineralocorticoids

19.105 How do the sex hormones estradiol and testosterone differ in structure?

19.106 What functional groups are present in each of the following steroid hormones?

  a. Estradiol   b. Testosterone
  c. Progesterone   d. Cortisone

■ **Eicosanoids (Section 19.13)**

19.107 What is the major structural difference between a prostaglandin and its parent fatty acid?

19.108 What is the major structural difference between a leukotriene and its parent fatty acid?

19.109 List six physiological processes that are regulated by eicosanoids.

19.110 What is the biochemical basis for the effectiveness of aspirin in decreasing inflammation?

■ **Biological Waxes (Section 19.14)**

19.111 Draw the general block diagram for a biological wax.

19.112 Draw the condensed structural formula of a wax formed from palmitic acid (Table 19.1) and cetyl alcohol, CH_3—$(CH_2)_{14}$—CH_2—OH.

19.113 What is the difference between a biological wax and a mineral wax?

19.114 Biological waxes have a "head and two tails" structure. Give the chemical identity of the head and of the two tails?

ADDITIONAL PROBLEMS

19.115 Classify each of the following types of lipids as (1) glycerol-based, (2) sphingosine-based, or (3) neither glycerol-based nor sphingosine-based.

  a. Bile acids   b. Fats
  c. Thromboxanes   d. Gangliosides
  e. Waxes   f. Leukotrienes

19.116 Indicate whether each of the lipid types in Problem 19.115 has a "head and two tails" structure.

19.117 Identify the type of lipid that fits each of the following "structural component" characterizations.

  a. Sphingosine + fatty acid + phosphoric acid + choline
  b. Glycerol + three fatty acids
  c. Fused-ring system with three 6-membered rings and one 5-membered ring
  d. 20-carbon fatty acid + three conjugated double bonds
  e. 20-carbon fatty acid + cyclopentane ring
  f. Sphingosine + fatty acid + monosaccharide

19.118 Classify each of the following types of lipids as (1) an energy-storage lipid, (2) a membrane lipid, (3) an emulsification lipid, (4) a messenger lipid, or (5) a protective-coating lipid.

  a. Fats   b. Cholic acid
  c. Cholesterol   d. Estrogens
  e. Sphingomyelins   f. Prostaglandins

19.119 Which of the terms *glycerolipid, sphingolipid,* and *phospholipid* apply to each of the following lipids. More than one term may apply in a given situation.

  a. Triacylglycerol   b. Sphingoglycolipid
  c. Glycerophospholipid   d. Sphingophospholipid

19.120 Specify the numbers of ester linkages, amide linkages, and glycosidic linkages present in each of the following types of lipids.

  a. Oils   b. Lecithins
  c. Sphinogomyelins   d. Biological Waxes
  e. Cerebrosides   f. Phosphatidylcholines

19.121 Indicate whether each of the following types of lipids contain a "steroid nucleus" as part of its structure.

  a. Prostaglandins   b. Cortisone
  c. Cholesterol   d. Bile acids
  e. Estrogens   f. Leukotrienes

ANSWERS TO PRACTICE EXERCISES

19.1 a. MUFA (monounsaturated fatty acid)
  b. 12:1 fatty acid
  c. omega-3 fatty acid (ω-3)
  d. delta-9 fatty acid (Δ^9)

19.2 a. four products: glycerol and three fatty acids
  b. four products: glycerol and three fatty acid salts
  c. one product: a triacylglycerol in which all fatty acid residues are 18:0 residues

20 Proteins

The word protein comes from the Greek *proteios*, which means "of first importance." This reflects the key role that proteins play in life processes.

The protein made by spiders to produce a web is a form of silk that can be exceptionally strong.

In this chapter we consider the third of the bioorganic classes of molecules (Section 18.1), the compounds called proteins. An extraordinary number of different proteins, each with a different function, exist in the human body. A typical human cell contains about 9000 different kinds of proteins, and the human body contains about 100,000 different proteins. Proteins are needed for the synthesis of enzymes, certain hormones, and some blood components; for the maintenance and repair of existing tissues; for the synthesis of new tissue; and sometimes for energy.

20.1 Characteristics of Proteins

Next to water, proteins are the most abundant substances in nearly all cells—they account for about 15% of a cell's overall mass (Section 18.1) and for almost half of a cell's dry mass. All proteins contain the elements carbon, hydrogen, oxygen, and nitrogen; most also contain sulfur. The presence of nitrogen in proteins sets them apart from carbohydrates and lipids, which generally do not contain nitrogen. The average nitrogen content of proteins is 15.4% by mass. Other elements, such as phosphorus and iron, are essential constituents of certain specialized proteins. Casein, the main protein of milk, contains phosphorus, an element very important in the diet of infants and children. Hemoglobin, the oxygen-transporting protein of blood, contains iron.

A **protein** *is a biochemical polymer in which the monomer units are amino acids.* Thus the starting point for a discussion of proteins is an understanding of the structures and chemical properties of amino acids.

20.2 Amino Acids: The Building Blocks for Proteins

An **amino acid** *is an organic compound that contains both an amino* ($-NH_2$) *group and a carboxyl* ($-COOH$) *group.* The amino acids found in proteins are always α-amino acids. An **α-amino acid** *is an amino acid in which the amino group and the carboxyl group are attached to the α-carbon atom.* The carbon atom to which these two groups are attached is called the α-*carbon atom.* The general structural formula for an α-amino acid is

The R group present in an α-amino acid is called the amino acid *side chain.* The nature of this side chain distinguishes α-amino acids from each other. Side chains vary in size, shape, charge, acidity, functional groups present, hydrogen-bonding ability, and chemical reactivity.

Over 700 different naturally occurring amino acids are known, but only 20 of them, called standard amino acids, are normally present in proteins. A **standard amino acid** *is one of the 20 α-amino acids normally found in proteins.* The structures of the 20 standard amino acids are given in Table 20.1. Within Table 20.1, amino acids are grouped according to side-chain polarity. In this system there are four categories: (1) nonpolar amino acids, (2) polar neutral amino acids, (3) polar acidic amino acids, and (4) polar basic amino acids. This classification system gives insights into how various types of amino acid side chains help determine the properties of proteins (Section 20.11).

A **nonpolar amino acid** *is an amino acid that contains one amino group, one carboxyl group, and a nonpolar side chain.* When incorporated into a protein, such amino acids are *hydrophobic* ("water-fearing"); that is, they are not attracted to water molecules. They are generally found in the interior of proteins, where there is limited contact with water. There are eight nonpolar amino acids.

The three types of polar amino acids have varying degrees of affinity for water. Within a protein, such amino acids are said to be *hydrophilic* ("water-loving"). Hydrophilic amino acids are often found on the surfaces of proteins.

A **polar neutral amino acid** *is an amino acid that contains one amino group, one carboxyl group, and a side chain that is polar but neutral.* In solution at physiological pH, the side chain of a polar neutral amino acid is neither acidic nor basic. There are seven polar neutral amino acids.

A **polar acidic amino acid** *is an amino acid that contains one amino group and two carboxyl groups, the second carboxyl group being part of the side chain.* In solution at physiological pH, the side chain of a polar acidic amino acid bears a negative charge; the side-chain carboxyl group has lost its acidic hydrogen atom. There are two polar acidic amino acids: aspartic acid and glutamic acid.

A **polar basic amino acid** *is an amino acid that contains two amino groups and one carboxyl group, the second amino group being part of the side chain.* In solution at physiological pH, the side chain of a polar basic amino acid bears a positive charge; the nitrogen atom of the amino group has accepted a proton (basic behavior; Section 17.5). There are three polar basic amino acids: lysine, arginine, and histidine.

The names of the standard amino acids are often abbreviated using three-letter codes. Except in four cases, these abbreviations are the first three letters of the amino acid's name. In addition, a new one-letter code for amino acid names is currently gaining popularity (particularly in computer applications). Both sets of abbreviations are used extensively in describing peptides and proteins, which contain tens and hundreds of amino acid units. Both types of abbreviations are given in Table 20.1.

▶ In an α-amino acid, the carboxyl group and the amino group are attached to the same carbon atom.

▶ The nature of the side chain (R group) distinguishes α-amino acids from each other, both physically and chemically.

▶ The nonpolar amino acid *proline* has a structural feature not found in any other standard amino acid. Its side chain, a propyl group, is bonded to both the α-carbon atom and the amino nitrogen atom, giving a cyclic side chain.

Proline

▶ A variety of functional groups are present in the side chains of the 20 standard amino acids: six have alkyl groups (Section 12.8), three have aromatic groups (Section 13.10), two have sulfur-containing groups (Section 14.17), two have hydroxyl (alcohol) groups (Section 14.2), three have amino groups (Section 17.2), two have carboxyl groups (Section 16.1), and two have amide groups (Section 17.11).

TABLE 20.1

The 20 Standard Amino Acids, Grouped According to Side-Chain Polarity. Below each amino acid's structure are its name (with pronunciation), its three-letter abbreviation, and its one-letter abbreviation.

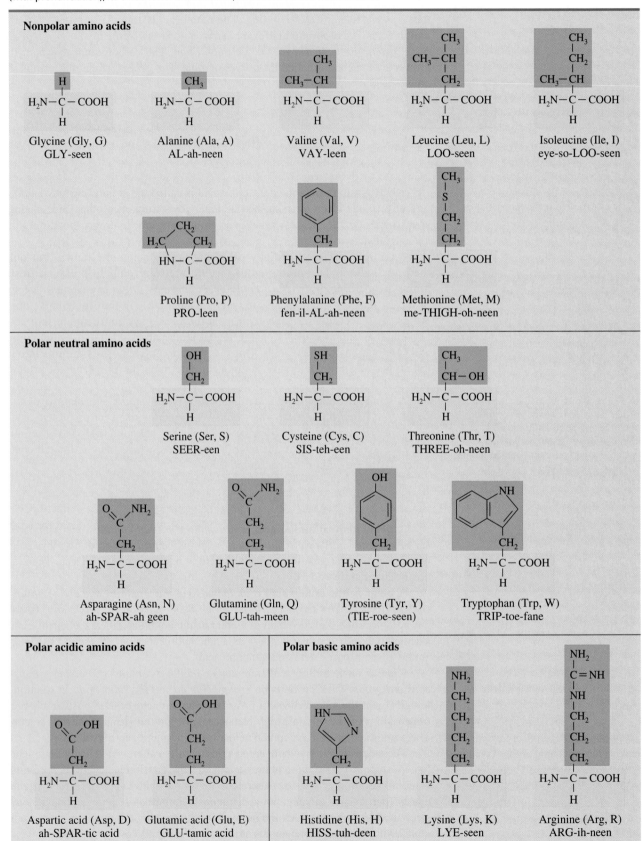

Nonpolar amino acids

Glycine (Gly, G)
GLY-seen

Alanine (Ala, A)
AL-ah-neen

Valine (Val, V)
VAY-leen

Leucine (Leu, L)
LOO-seen

Isoleucine (Ile, I)
eye-so-LOO-seen

Proline (Pro, P)
PRO-leen

Phenylalanine (Phe, F)
fen-il-AL-ah-neen

Methionine (Met, M)
me-THIGH-oh-neen

Polar neutral amino acids

Serine (Ser, S)
SEER-een

Cysteine (Cys, C)
SIS-teh-een

Threonine (Thr, T)
THREE-oh-neen

Asparagine (Asn, N)
ah-SPAR-ah geen

Glutamine (Gln, Q)
GLU-tah-meen

Tyrosine (Tyr, Y)
(TIE-roe-seen)

Tryptophan (Trp, W)
TRIP-toe-fane

Polar acidic amino acids

Aspartic acid (Asp, D)
ah-SPAR-tic acid

Glutamic acid (Glu, E)
GLU-tamic acid

Polar basic amino acids

Histidine (His, H)
HISS-tuh-deen

Lysine (Lys, K)
LYE-seen

Arginine (Arg, R)
ARG-ih-neen

CHEMICAL CONNECTIONS

The Essential Amino Acids

All of the amino acids in Table 20.1 are necessary constituents of human protein. Adequate amounts of 11 of the 20 amino acids can be synthesized from carbohydrates and lipids in the body if a source of nitrogen is also available. Because the human body is incapable of producing 9 of these 20 acids fast enough or in sufficient quantities to sustain normal growth, these 9 amino acids, called essential amino acids, must be obtained from food. An **essential amino acid** *is an amino acid needed in the human body that must be obtained from dietary sources because it cannot be synthesized within the body from other substances in adequate amounts.* The following table lists the *essential* amino acids for humans.

The Essential Amino Acids for Humans

arginine[a]	methionine
histidine	phenylalanine
isoleucine	threonine
leucine	tryptophan
lysine	valine

[a]Arginine is required for growth in children but is not required by adults.

The human body can synthesize small amounts of some of the essential amino acids, but not enough to meet its needs, especially in the case of growing children.

A **complete dietary protein** *is a protein that contains all the essential amino acids in the same relative amounts in which the human body needs them.* A complete dietary protein may or may not contain all the nonessential amino acids. Most animal proteins, including casein from milk and proteins found in meat, fish, and eggs are complete proteins, although gelatin is an exception (it lacks tryptophan). Proteins from plants (vegetables, grains, and legumes) have quite diverse amino acid patterns and some tend to be limiting in one or more essential amino acids. Some plant proteins (for example, corn protein) are far from complete. Others (for example, soy protein) are complete. Thus vegetarians must eat a variety of plant foods to obtain all of the essential amino acids in appropriate quantities.

The following table lists the essential amino acid deficiencies associated with selected vegetables and grains.

Amino Acids Missing in Selected Vegetables and Grains

Food source	Amino acid deficiency
soy	none
wheat, rice, oats	lysine
corn	lysine, tryptophan
beans	methionine, tryptophan
peas	methionine
almonds, walnuts	lysine, tryptophan

20.3 Chirality and Amino Acids

> Glycine, the simplest of the standard amino acids, is achiral. All of the other standard amino acids are chiral.

Four different groups are attached to the α-carbon atom in all of the standard amino acids except glycine, where the R group is a hydrogen atom.

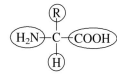

This means that the structures of 19 of the 20 standard amino acids possess a chiral center (Section 18.4) at this location, so enantiomeric forms (left- and right-handed forms; Section 18.5) exist for each of these amino acids.

> Because only L amino acids are constituents of proteins, the enantiomer designation of L or D will be omitted in subsequent amino acid and protein discussions. It is understood that it is the L isomer that is always present.

With few exceptions (in some bacteria), the amino acids found in nature and in proteins are L isomers. Thus, as is the case with monosaccharides (Section 18.8), nature favors one mirror-image form over the other. Interestingly, for amino acids the L isomer is the preferred form, whereas for monosaccharides the D isomer is preferred.

The rules for drawing Fischer projections (Section 18.6) for amino acid structures follow.

1. The —COOH group is put at the top of the projection, the R group at the bottom. This positions the carbon chain vertically.
2. The —NH₂ group is in a horizontal position. Positioning it on the left denotes the L isomer, and positioning it on the right denotes the D isomer.

FIGURE 20.1 Designation of handedness in standard amino acid structures involves aligning the carbon chain vertically and looking at the position of the horizontally aligned —NH$_2$ group. The L form has the —NH$_2$ group on the left, and the D form has the —NH$_2$ group on the right.

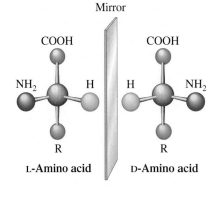

L-Amino acid   D-Amino acid

Figure 20.1 shows molecular models that illustrate the use of these rules. Fischer projections for both enantiomers of the amino acids alanine and serine follow.

▶ Two of the 19 chiral standard amino acids, isoleucine and threonine, possess two chiral centers (see Table 20.1). With two chiral centers present, four stereoisomers are possible for these amino acids. However, only one of the L isomers is found in proteins.

L-Alanine   D-Alanine   L-Serine   D-Serine

20.4 Acid–Base Properties of Amino Acids

In pure form, amino acids are white crystalline solids with relatively high decomposition points. (Most amino acids decompose before they melt.) Also most amino acids are *not* very soluble in water because of strong intermolecular forces within their crystal structures. Such properties are those often exhibited by compounds in which charged species are present. Studies of amino acids confirm that they are charged species both in the solid state and in solution. Why is this so?

Both an acidic group (—COOH) and a basic group (—NH$_2$) are present on the same carbon in an α-amino acid.

▶ In drawing amino acid structures, where handedness designation is not required, the placement of the four groups about the α-carbon atom is arbitrary. From this point on in the text, we will draw amino acid structures such that the —COOH group is on the left, the —NH$_2$ group is on the right, the R group points down, and the H atom points up. Drawing amino acids in this "arrangement" makes it easier to draw structures where amino acids are linked together to form longer amino acid chains.

Basic group → H$_2$N—C—COOH ← Acidic group

In Section 16.8, we learned that in neutral solution, carboxyl groups have a tendency to lose protons (H$^+$), producing a negatively charged species:

$$—COOH \longrightarrow —COO^- + H^+$$

In Section 17.5, we learned that in neutral solution, amino groups have a tendency to accept protons (H$^+$), producing a positively charged species:

$$—NH_2 + H^+ \longrightarrow —\overset{+}{N}H_3$$

Consistent with the behavior of these groups, in neutral solution, the —COOH group of an amino acid donates a proton to the —NH$_2$ of the same amino acid. We can characterize this behavior as an *internal* acid–base reaction. The net result is that in neutral solution, amino acid molecules have the structure

H$_3\overset{+}{N}$—C—COO$^-$

Strong intermolecular forces between the positive and negative centers of zwitterions are the cause of the high melting points of amino acids.

Such a molecule is known as a zwitterion, from the German term meaning "double ion." A **zwitterion** *is a molecule that has a positive charge on one atom and a negative charge on another atom, but which has no net charge.* Note that the net charge on a zwitterion is zero even though parts of the molecule carry charges. In solution and also in the solid state, α-amino acids are zwitterions.

Zwitterion structure changes when the pH of a solution containing an amino acid is changed from neutral either to acidic (low pH) by adding an acid such as HCl or to basic (high pH) by adding a base such as NaOH. In an acidic solution, the zwitterion accepts a proton (H^+) to form a positively charged ion.

From this point on in the text, the structures of amino acids will be drawn in their zwitterion form unless information given about the pH of the solution indicates otherwise.

$$\underset{\substack{\text{Zwitterion (no net charge)}}}{\overset{\displaystyle H}{\underset{\displaystyle R}{H_3\overset{+}{N}-\overset{|}{\underset{|}{C}}-COO^-}}} + H_3O^+ \longrightarrow \underset{\substack{\text{Positively charged ion}}}{\overset{\displaystyle H}{\underset{\displaystyle R}{H_3\overset{+}{N}-\overset{|}{\underset{|}{C}}-COOH}}} + H_2O$$

In basic solution, the $-\overset{+}{N}H_3$ of the zwitterion loses a proton, and a negatively charged species is formed.

$$\underset{\substack{\text{Zwitterion}\\\text{(no net charge)}}}{\overset{\displaystyle H}{\underset{\displaystyle R}{H_3\overset{+}{N}-\overset{|}{\underset{|}{C}}-COO^-}}} + OH^- \longrightarrow \underset{\substack{\text{Negatively charged ion}}}{\overset{\displaystyle H}{\underset{\displaystyle R}{H_2N-\overset{|}{\underset{|}{C}}-COO^-}}} + H_2O$$

Thus, in solution, three different amino acid forms can exist (zwitterion, negative ion, and positive ion). The three species are actually in equilibrium with each other, and the equilibrium shifts with pH change. The overall equilibrium process can be represented as follows:

$$\underset{\substack{\text{Acidic solution}\\\text{(low pH)}}}{\overset{\displaystyle H}{\underset{\displaystyle R}{H_3\overset{+}{N}-\overset{|}{\underset{|}{C}}-COOH}}} \underset{H_3O^+}{\overset{OH^-}{\rightleftharpoons}} \underset{\substack{\text{Neutral solution}\\\text{(pH = 7.0)}}}{\overset{\displaystyle H}{\underset{\displaystyle R}{H_3\overset{+}{N}-\overset{|}{\underset{|}{C}}-COO^-}}} \underset{H_3O^+}{\overset{OH^-}{\rightleftharpoons}} \underset{\substack{\text{Basic solution}\\\text{(high pH)}}}{\overset{\displaystyle H}{\underset{\displaystyle R}{H_2N-\overset{|}{\underset{|}{C}}-COO^-}}}$$

In acidic solution, the positively charged species on the left predominates; nearly neutral solutions have the middle species (the zwitterion) as the dominant species; in basic solution, the negatively charged species on the right predominates.

EXAMPLE 20.1

Determining Amino Acid Form in Solutions of Various pH

■ Draw the structural form of the amino acid alanine that predominates in solution at each of the following pH values.

a. pH = 1.0
b. pH = 7.0
c. pH = 11.0

Solution

At low pH, both amino and carboxyl groups are protonated. At high pH, both groups have lost their protons. At neutral pH, the zwitterion is present.

a.
$$\overset{\displaystyle H}{\underset{\displaystyle CH_3}{H_3\overset{+}{N}-\overset{|}{\underset{|}{C}}-COOH}}$$
pH = 1.0
(net charge of +1)

b.
$$\overset{\displaystyle H}{\underset{\displaystyle CH_3}{H_3\overset{+}{N}-\overset{|}{\underset{|}{C}}-COO^-}}$$
pH = 7.0
(no net charge)

c.
$$\overset{\displaystyle H}{\underset{\displaystyle CH_3}{H_2N-\overset{|}{\underset{|}{C}}-COO^-}}$$
pH = 11.0
(net charge of −1)

(continued)

Guidelines for amino acid form as a function of solution pH follow.

Low pH: All acid groups are protonated (—COOH). All amino groups are protonated ($-\overset{+}{N}H_3$).
High pH: All acid groups are deprotonated (—COO^-). All amino groups are deprotonated (—NH_2).
Neutral pH: All acid groups are deprotonated (—COO^-). All amino groups are protonated ($-\overset{+}{N}H_3$).

The term *protonated* denotes gain of a H^+ ion, and the term *deprotonated* denotes loss of a H^+ ion.

Practice Exercise 20.1

Draw the structural form of the amino acid valine that predominates in solution at each of the following pH values.

a. pH = 7.0
b. pH = 12.0
c. pH = 2.0

The previous discussion assumed that the side chain (R group) of an amino acid remains unchanged in solution as the pH is varied. This is the case for neutral amino acids but not for acidic or basic ones. For these latter compounds, the side chain can also acquire a charge, because it contains an amino or a carboxyl group that can, respectively, gain or lose a proton.

Because of the extra site that can be protonated or deprotonated, acidic and basic amino acids have four charged forms in solution. These four forms for aspartic acid, one of the acidic amino acids, are

Low-pH form (+1 charge) | Moderately-low-pH form (no net charge) (zwitterion) | Neutral-pH form (−1 net charge) | High-pH form (−2 net charge)

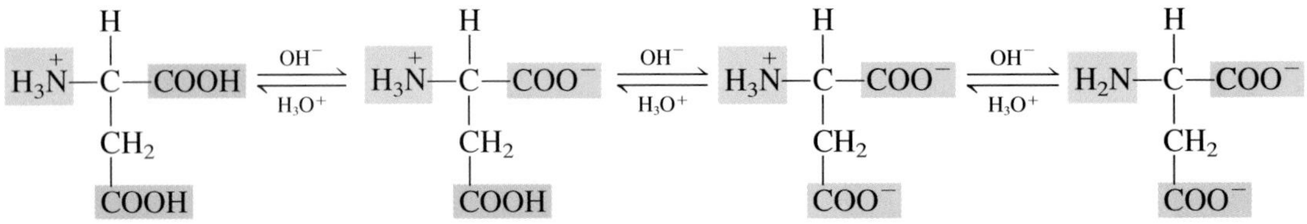

The existence of two low-pH forms for aspartic acid results from the two carboxyl groups being deprotonated at different pH values. For basic amino acids, two high-pH forms exist because deprotonation of the amino groups does not occur simultaneously. The side-chain amino group deprotonates before the α-amino group.

Isoelectric Points and Electrophoresis

The amounts of the various forms of an amino acid—zwitterion, negative ion(s), and positive ion(s)—that are present in an aqueous solution of the amino acid vary with pH. For each amino acid, there is a pH value, called the *isoelectric point,* at which nearly all molecules are present in zwitterion form. The **isoelectric point** *is the pH at which the concentration of the zwitterion is at a maximum in an amino acid solution.* Every amino acid has a different isoelectric point. Fifteen of the 20 amino acids, those with nonpolar or polar neutral side chains (Table 20.1), have isoelectric points in the range of 4.8–6.3. The three basic amino acids have higher isoelectric points (His = 7.59, Lys = 9.74, Arg = 10.76), and the two acidic amino acids have lower ones (Asp = 2.77, Glu = 3.22).

The isoelectric point of an amino acid is measured by observing its behavior in an electric field. In an electric field, a charged molecule is attracted to (migrates toward) the electrode of opposite charge. At a high pH, an amino acid has a net negative charge and migrates toward the positive electrode. At a low pH, the opposite is true; with a net positive charge, the amino acid migrates toward the negative electrode. At the isoelectric point, migration does not occur because the zwitterions present have no net charge.

Mixtures of amino acids in solution can be separated by using their different migration patterns at various pH values. This type of analytical separation is called electrophoresis. **Electrophoresis** *is the process of separating charged molecules on the basis of their migration toward charged electrodes associated with an electric field.*

FIGURE 20.2 Separation, at a pH of 5.5, of the three amino acids Lys, Phe, and Glu using electrophoresis.

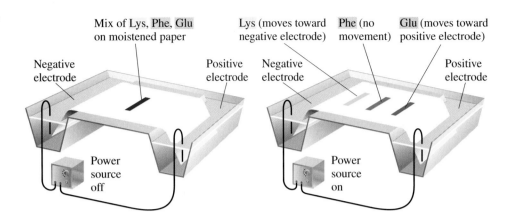

Figure 20.2 schematically shows the separation of the amino acids Lys, Phe, and Glu by electrophoresis. At a pH of 5.5, these amino acids exist in the following forms:

$$
\begin{array}{ccc}
\text{Lys} & \text{Phe} & \text{Glu} \\
(+1 \text{ charge}) & (\text{no net charge}) & (-1 \text{ charge})
\end{array}
$$

When a current is applied, Phe does not move (it has no net charge); Lys, because of its positive charge, migrates toward the negative electrode; and Glu, with a negative charge, moves toward the positive electrode.

Proteins, which are amino acid polymers (Section 20.8), also have isoelectric points and also can be separated via electrophoresis techniques.

EXAMPLE 20.2

Migration Patterns of Amino Acids at Various pH Values in an Electric Field

■ Predict the direction of migration (if any) toward the positively or negatively charged electrode for the following amino acids in solutions of the specified pH. Write "isoelectric" if no migration occurs.

a. Lysine at pH = 7.0
b. Glutamic acid at pH = 7.0
c. Serine at pH = 1.0

Solution

a. Lysine at pH = 7.0

$$
\begin{array}{c}
\text{H} \\
| \\
H_3\overset{+}{N}\text{—C—COO}^- \\
| \\
CH_2 \\
| \\
CH_2 \\
| \\
CH_2 \\
| \\
CH_2 \\
| \\
^+NH_3
\end{array}
$$

Net positive charge (2 "+" and 1 "−"); migrates toward negatively charged electrode

(continued)

b. Glutamic acid at pH = 7.0

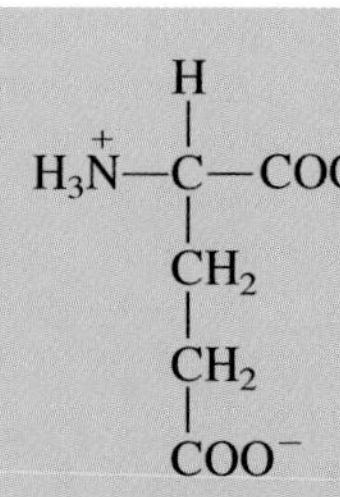

$$\overset{H}{\underset{\underset{\underset{COO^-}{|}}{\underset{\underset{CH_2}{|}}{\overset{+}{H_3N}-C-COO^-}}}{}}$$

Net negative charge (1 "+" and 2 "−"); migrates toward positively charged electrode

c. Serine at pH = 1.0

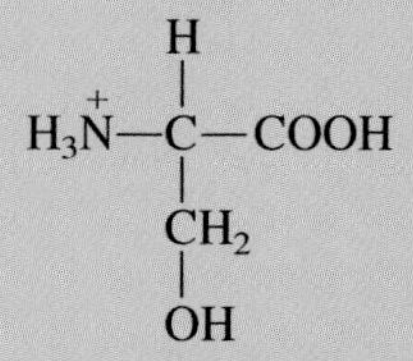

$$\overset{H}{\underset{\underset{\underset{OH}{|}}{\underset{CH_2}{|}}}{\overset{+}{H_3N}-C-COOH}}$$

One positive charge; migrates toward negatively charged electrode

Practice Exercise 20.2

Predict the direction of migration (if any) toward the positively or negatively charged electrode for the following amino acids in solutions of the specified pH. Write "isoelectric" if no migration occurs.

a. Lysine at pH = 12.0   **b.** Glutamic acid at pH = 2.0   **c.** Serine at pH = 7.0

20.5 Cysteine: A Chemically Unique Amino Acid

Cysteine is the only standard amino acid (Table 20.1) that has a side chain that contains a sulfhydryl group (—SH group; Section 14.17). The presence of this sulfhydryl group imparts to cysteine a chemical property that is unique among the standard amino acids. Cysteine, in the presence of mild oxidizing agents, readily *dimerizes,* that is, reacts with another cysteine molecule to form a dicysteine molecule. (A *dimer* is a molecule that is made up of two like subunits.) In dicysteine, the two cysteine residues are linked via a covalent disulfide bond.

$$\overset{+}{H_3N}-CH-COO^- + \overset{+}{H_3N}-CH-COO^- \longrightarrow$$

CH₂   CH₂

SH   SH

Cysteine   Cysteine   Dicysteine

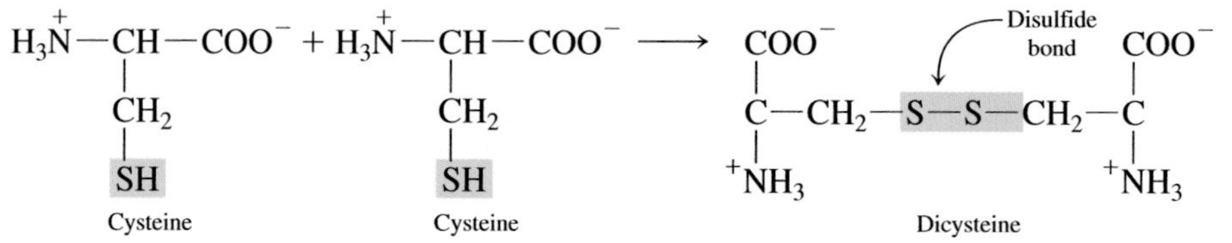

The covalent disulfide bond of dicysteine is readily broken, using reducing agents, to regenerate two cysteine molecules. This oxidation–reduction behavior involving sulfhydryl groups and disulfide bonds was previously encountered in Section 14.17 when the reactions of thioalcohols were considered.

$$-SH + HS- \underset{\text{Reduction}}{\overset{\text{Oxidation}}{\rightleftarrows}} -S-S- + 2H$$

As we shall see in Section 20.10, the formation of disulfide bonds between cysteine residues present in protein molecules has important consequences relative to protein structure and protein shape.

20.6 Peptide Formation

In Section 17.14, we learned that a carboxylic acid and an amine can react to produce an amide. The general equation for this reaction is

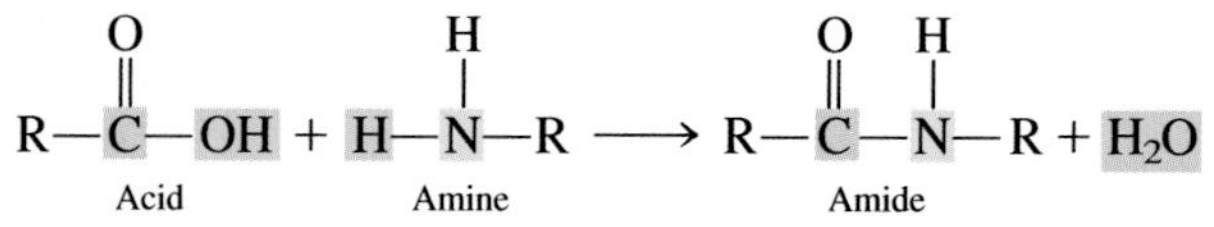

$$\underset{\text{Acid}}{R-\overset{\overset{O}{\|}}{C}-OH} + \underset{\text{Amine}}{H-\overset{\overset{H}{|}}{N}-R} \longrightarrow \underset{\text{Amide}}{R-\overset{\overset{O}{\|}}{C}-\overset{\overset{H}{|}}{N}-R} + H_2O$$

Two amino acids can combine in a similar way—the carboxyl group of one amino acid interacts with the amino group of the other amino acid. The products are a molecule of water and a molecule containing the two amino acids linked by an amide bond.

$$H_3\overset{+}{N}-\underset{R_1}{\overset{\overset{\displaystyle H}{|}}{C}}-COO^- + H_3\overset{+}{N}-\underset{R_2}{\overset{\overset{\displaystyle H}{|}}{C}}-COO^- \longrightarrow H_3\overset{+}{N}-\underset{R_1}{\overset{\overset{\displaystyle H}{|}}{C}}-\overset{\overset{\displaystyle O}{||}}{C}-\overset{\overset{\displaystyle H}{|}}{N}-\underset{R_2}{\overset{\overset{\displaystyle H}{|}}{C}}-COO^- + H_2O$$

Amide bond

Removal of the elements of water from the reacting carboxyl and amino groups and the ensuing formation of the amide bond are better visualized when expanded structural formulas for the reacting groups are used.

$$-\overset{\overset{\displaystyle O}{||}}{C}-O + H-\underset{\underset{\displaystyle H}{|}}{\overset{\overset{\displaystyle H}{|}}{N}}- \longrightarrow -\overset{\overset{\displaystyle O}{||}}{C}-\overset{\overset{\displaystyle H}{|}}{N}- + H_2O$$

Carboxyl group $(-COO^-)$ Amino group $(H_3\overset{+}{N}-)$ Amide bond

> Peptide bond formation is an example of a condensation reaction.

In amino acid chemistry, amide bonds that link amino acids together are given the specific name of peptide bond. A **peptide bond** *is a covalent bond between the carboxyl group of one amino acid and the amino group of another amino acid.*

Under proper conditions, many amino acids can bond together to give chains of amino acids containing numerous peptide bonds. For example, four peptide bonds are present in a chain of five amino acids.

| Amino acid | Amino acid | Amino acid | Amino acid | Amino acid |

Peptide bond   Peptide bond   Peptide bond   Peptide bond

Chains of amino acids are known as peptides. A **peptide** *is a sequence of amino acids in which the amino acids are joined together through amide (peptide) bonds.* A compound containing two amino acids joined by a peptide bond is specifically called a *dipeptide;* three amino acids in a chain constitute a *tripeptide;* and so on. The name *oligopeptide* is loosely used to refer to peptides with 10 to 20 amino acid residues, and the name *polypeptide* is used to refer to larger peptides. A **polypeptide** *is a long chain of amino acids, each joined to the next by a peptide bond.*

> A peptide chain has *directionality* because its two ends are different. There are an N-terminal end and a C-terminal end. By convention, the direction of the peptide chain is always
>
> N-terminal end $\longrightarrow$ C-terminal end
>
> The N-terminal end is always on the left, and the C-terminal end is always on the right.

In all peptides, the amino acid at one end of the amino acid sequence has a free $H_3\overset{+}{N}$ group, and the amino acid at the other end of the sequence has a free COO^- group. The end with the free $H_3\overset{+}{N}$ group is called the *N-terminal end,* and the end with the free COO^- group is called the *C-terminal end.* By convention, the sequence of amino acids in a peptide is written with the N-terminal end amino acid on the left. The individual amino acids within a peptide chain are called *amino acid residues.*

The structural formula for a polypeptide may be written out in full, or the sequence of amino acids present may be indicated by using the standard three-letter amino acid abbreviations. The abbreviated formula for the tripeptide.

N-terminal end $\longrightarrow$ $H_3\overset{+}{N}-\underset{\underset{\displaystyle H}{|}}{\overset{\overset{\displaystyle H}{|}}{C}}-\overset{\overset{\displaystyle O}{||}}{C}-\overset{\overset{\displaystyle H}{|}}{N}-\underset{\underset{\displaystyle CH_3}{|}}{\overset{\overset{\displaystyle H}{|}}{C}}-\overset{\overset{\displaystyle O}{||}}{C}-\overset{\overset{\displaystyle H}{|}}{N}-\underset{\underset{\displaystyle CH_2-OH}{|}}{\overset{\overset{\displaystyle H}{|}}{C}}-COO^-$ $\longleftarrow$ C-terminal end

Glycine   Alanine   Serine

CHEMICAL CONNECTIONS

Substitutes for Human Insulin

In humans, an insufficient production of insulin results in the disease *diabetes mellitus*. Treatment of this disease involves giving the patient extra insulin via subcutaneous injection. For many years, because of the limited availability of human insulin, most insulin used by diabetics was obtained from the pancreases of slaughter-house animals. Such animal insulin, primarily from cows and pigs, was used by most diabetics without serious side effects because it is structurally very similar to human insulin. Immunological reactions gradually do increase over time, however, because the animal insulin is foreign to the human body.

A comparison of the primary structure of human insulin with pig and cow insulins shows differences at only 4 of the 51 amino acid positions: positions 8, 9, and 10 on chain A and position 30 on chain B (see Figure 20.9 and the following table).

Species	Chain A			Chain B
	#8	**#9**	**#10**	**#30**
human	Thr	Ser	Ile	Thr
pig (porcine)	Thr	Ser	Ile	Ala
cow (bovine)	Ala	Ser	Val	Ala

The dependence of diabetics on animal insulin has declined because of the availability of human insulin produced by genetically engineered bacteria (Section 22.14). These bacteria carry a gene that directs the synthesis of human insulin. Such bacteria-produced insulin is fully functional. All diabetics now have the choice of using human insulin or using animal insulin. Many still continue to use the animal insulin because it is cheaper.

▶ When a protein consists of more than one polypeptide chain, as in insulin, each chain is called a *subunit* of the protein.

▶ The primary structure of a protein is the *sequence* of amino acids in a protein chain—that is, the order in which the amino acids are connected to each other.

cows, pigs, sheep, and horses are very similar both to each other and to human insulin. Until recently, this similarity was particularly important for diabetics who required supplemental injections of insulin. (See the Chemical Connections feature on this page.)

An analogy is often drawn between the primary structure of proteins and words. Words, which convey information, are formed when the 26 letters of the English alphabet are properly sequenced. Proteins are formed from proper sequences of the 20 standard amino acids. The proper sequence of letters in a word is necessary for it to make sense, just as the proper sequence of amino acids is necessary to make biochemically active protein. Furthermore, the letters that form a word are written from left to right, as are amino acids in protein formulas. As any dictionary of the English language will document, a tremendous variety of words can be formed by different letter sequences. Imagine the number of amino acid sequences possible for a large protein. There are 1.55×10^{66} sequences possible for the 51 amino acids found in insulin! From these possibilities, the body reliably produces only *one*, illustrating the remarkable precision of life processes. From the simplest bacterium to the human brain cell, only those amino acid sequences needed by the cell are produced. The fascinating process of protein biosynthesis and the way in which genes in DNA direct this process will be discussed in Chapter 22.

20.10 Secondary Structure of Proteins

Secondary protein structure *is the arrangement in space adopted by the backbone portion of a protein.* The two most common types of secondary structure are the *alpha helix* (α helix) and the *beta pleated sheet* (β pleated sheet). The type of interaction responsible for both of these types of secondary structure is hydrogen bonding (Section 7.13) between a carbonyl oxygen atom of a peptide linkage and the hydrogen atom of an amino group of another peptide linkage farther along the backbone. This hydrogen-bonding interaction is diagrammed as follows:

$$\diagup N\!-\!H\cdots\cdots O\!=\!C\diagdown$$

■ The Alpha Helix

The *alpha helix* (α helix) structure resembles a coiled helical spring, with the coil configuration maintained by hydrogen bonds between $>$N—H and $>$C$=$O groups of every fourth amino acid, as is shown diagrammatically in Figure 20.5.

FIGURE 20.5 Four representations of the α helix protein secondary structure. (a) Arrangement of protein backbone with no detail shown. (b) Backbone arrangement with hydrogen-bonding interactions shown. (c) Backbone atomic detail shown, as well as hydrogen-bonding interactions. (d) Top view of an α helix showing that amino acid side chains (R groups) point away from the long axis of the helix.

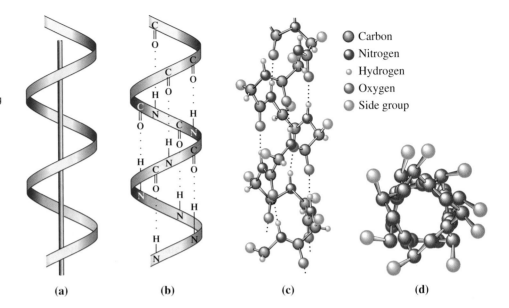

Carbon
Nitrogen
Hydrogen
Oxygen
Side group

(a)   (b)   (c)   (d)

Proteins have varying amounts of α-helical secondary structure, ranging from a few percent to nearly 100%. In an α helix, all of the amino acid side chains (R groups) lie outside the helix; there is not enough room for them in the interior. Figure 20.5d illustrates this situation.

■ The Beta Pleated Sheet

> The hydrogen bonding present in an α helix is *intra*molecular. In a β pleated sheet, the hydrogen bonding can be *inter*molecular (between two different chains) or *intra*molecular (a single chain folding back on itself).

The *beta pleated sheet* secondary structure involves amino acid chains that have almost completely extended backbones. Hydrogen bonds form between oxygen and hydrogen peptide linkage atoms that are either in different parts of a single chain that folds back on itself (intrachain bonds) or between atoms in different polypeptide chains in those proteins that contain more than one chain (interchain bonds). (Many proteins exist in which several polypeptide chains are present.) Figure 20.6a shows a representation of the β pleated sheet structure that occurs when portions of two different polypeptide chains are aligned parallel to each other (interchain bonds). The term *pleated sheet* arises from the repeated zigzag pattern in the structure (Figure 20.6b). Note how in a pleated sheet structure the amino acid side chains are positioned above and below the plane of the sheet.

FIGURE 20.6 Two representations of the β pleated sheet protein structure. (a) A representation emphasizing the hydrogen bonds between protein chains. (b) A representation emphasizing the pleats and the location of the R groups.

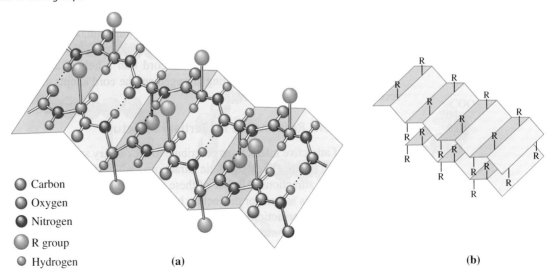

Carbon
Oxygen
Nitrogen
R group
Hydrogen

(a)   (b)

CHEMICAL CONNECTIONS

Protein Structure and the Color of Meat

The meat that humans eat is composed primarily of muscle tissue. The major proteins present in such muscle tissue are *myosin* and *actin,* which lie in alternating layers and which slide past each other during muscle contraction. Contraction is temporarily maintained through interactions between these two types of proteins.

Structurally, myosin consists of a rodlike coil of two alpha helices (fibrous protein) with two globular protein heads. It is the "head portions" of myosin that interact with the actin.

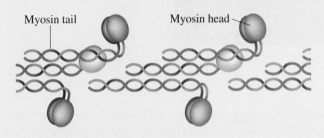

Myosin tail   Myosin head

Structurally, actin has the appearance of two filaments spiraling about one another (see diagram below). Each circle in this structural diagram represents a monomeric unit of actin (called globular actin). The monomeric actin units associate to form a long polymer (called fibrous actin). Each identical monomeric actin unit is a globular protein containing many amino acid residues.

The chemical process associated with muscle contraction (interaction between myosin and actin) requires molecular oxygen. The oxygen storage protein *myoglobin* (Section 20.14) is the oxygen source. The amount of myoglobin present in a muscle

is determined by how the muscle is used. Heavily used muscles require larger amounts of myoglobin than infrequently used muscles require.

The amount of myoglobin present in muscle tissue is a major determiner of the color of the muscle tissue. Myoglobin molecules have a red color when oxygenated and a purple color when deoxygenated. Thus, heavily worked muscles have a darker color than infrequently used muscles.

The different colors of meat reflect the concentration of myoglobin in the muscle tissue. In turkeys and chickens, which walk around a lot but rarely fly, the leg meat is dark, the breast meat is white. On the other hand, game birds that do fly a lot have dark breast meat. In general, game animals (which use all of their muscles regularly) tend to have darker meat than domesticated animals.

All land animals and birds need to support their own weight. Fish, on the other hand, are supported by water as they swim, which reduces the need for myoglobin oxygen support. Hence fish tend to have lighter flesh. Fish that spend most of their time lying at the bottom of a body of water have the lightest (whitest) flesh of all. Salmon flesh contains additional pigments that give it its characteristic "orange-pink" color.

Meat, when cooked, turns brown as the result of changes in myoglobin structure caused by the heat; the iron atom in the heme unit of myoglobin (Section 20.14) becomes oxidized. When meat is heavily salted with preservatives (NaCl, $NaNO_2$, or the like), as in the preparation of ham, the myoglobin picks up nitrite ions, and its color changes to pink.

20.15 Protein Hydrolysis

When a protein or polypeptide in a solution of strong acid or strong base is heated, the peptide bonds of the amino acid chain are hydrolyzed and free amino acids are produced. The hydrolysis reaction is the reverse of the formation reaction for a peptide bond. Amine and carboxylic acid functional groups are regenerated.

Let us consider the hydrolysis of the tripeptide Ala–Gly–Cys under acidic conditions. Complete hydrolysis produces one unit each of the amino acids alanine, glycine, and cysteine. The equation for the hydrolysis is

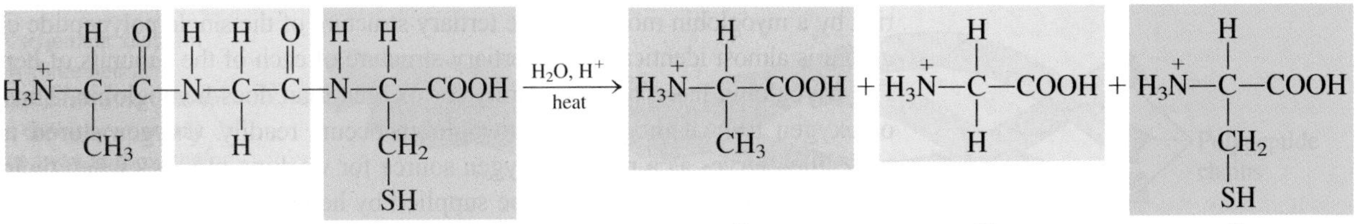

FIGURE 20.17 Protein denaturation involves loss of the protein's three-dimensional structure. Complete loss of such structure produces a random-coil protein strand.

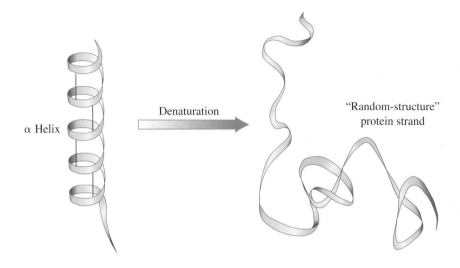

α Helix → Denaturation → "Random-structure" protein strand

Note that the product amino acids in this reaction are written in positive-ion form because of the acidic reaction conditions.

Protein hydrolysis produces free amino acids. This process is the reverse of protein synthesis, where free amino acids are combined.

Protein digestion (Section 26.1) is simply enzyme-catalyzed hydrolysis of ingested protein. The free amino acids produced from this process are absorbed through the intestinal wall into the bloodstream and transported to the liver. Here they become the raw materials for the synthesis of new protein tissue. Also, the hydrolysis of cellular proteins to amino acids is an ongoing process, as the body resynthesizes needed molecules and tissue.

20.16 Protein Denaturation

A consequence of protein denaturation, the partial or complete loss of a protein's three-dimensional structure, is loss of biological activity for the protein.

Protein denaturation *is the partial or complete disorganization of a protein's characteristic three-dimensional shape as a result of disruption of its secondary, tertiary, and quaternary structural interactions.* Because the biochemical function of a protein depends on its three-dimensional shape, the result of denaturation is loss of biochemical activity. Protein denaturation does not affect the primary structure of a protein.

Although some proteins lose all of their three-dimensional structural characteristics upon denaturation (Figure 20.17), most proteins maintain some three-dimensional structure. For a few small proteins, it is possible to find conditions under which the effects of denaturation can be reversed; this restoration process, in which the protein is "refolded," is called *renaturation.* For most proteins, however, denaturation is irreversible.

Loss of water solubility is a frequent physical consequence of protein denaturation. The precipitation out of biochemical solution of denatured protein is called *coagulation.*

A most dramatic example of protein denaturation occurs when egg white (a concentrated solution of the protein albumin) is poured onto a hot surface. The clear albumin solution immediately changes into a white solid with a jelly-like consistency (see Figure 20.18). A similar process occurs when hamburger juices encounter a hot surface. A brown, jelly-like solid forms.

When protein-containing foods are cooked, protein denaturation occurs. Such "cooked" protein is more easily digested because it is easier for digestive enzymes to "work on" denatured (unraveled) protein. Cooking foods also kills microorganisms through protein denaturation. For example, ham and bacon can harbor parasites that cause trichinosis. Cooking the ham or bacon denatures parasite protein.

FIGURE 20.18 Heat denatures the protein in egg white, producing a white jellylike solid. The primary structure of the protein remains intact, but all higher levels of protein structure are disrupted.

In surgery, heat is often used to seal small blood vessels. This process is called *cauterization.* Small wounds can also be sealed by cauterization. Heat-induced denaturation is used in sterilizing surgical instruments and in canning foods; bacteria are destroyed when the heat denatures their protein.

The body temperature of a patient with fever may rise to 102°F, 103°F, or even 104°F without serious consequences. A temperature above 106°F (41°C) is extremely dangerous,

FIGURE 20.19 Storage room for cheese; during storage cheese "matures" as bacteria and enzymes ferment the cheese, giving it a stronger flavor.

for at this level, the enzymes of the body begin to be inactivated. Enzymes, which function as catalysts for almost all body reactions, are protein. Inactivation of enzymes, through denaturation, can have lethal effects on body chemistry.

The effect of ultraviolet radiation from the sun, a nonionizing radiation (Section 11.7), is similar to that of heat. Denatured skin proteins cause most of the problems associated with sunburn.

A curdy precipitate of casein, the principal protein in milk, is formed in the stomach when the hydrochloric acid of gastric juice denatures the casein. The curdling of milk that takes place when milk sours or cheese is made (see Figure 20.19) results from the presence of lactic acid, a by-product of bacterial growth. Yogurt is prepared by growing lactic-acid-producing bacteria in skim milk. The coagulated denatured protein gives yogurt its semi-solid consistency.

Serious eye damage can result from eye tissue contact with acids or bases, when irreversibly denatured and coagulated protein causes a clouded cornea. This reaction is part of the basis for the rule that students wear protective eyewear in the chemistry laboratory.

Alcohols are an important type of denaturing agent. Denaturation of bacterial protein takes place when isopropyl or ethyl alcohol is used as a disinfectant—hence the common practice of swabbing the skin with alcohol before giving an injection. Interestingly, pure isopropyl or ethyl alcohol is less effective than the commonly used 70% alcohol solution. Pure alcohol quickly denatures and coagulates the bacterial surface, thereby forming an effective barrier to further penetration by the alcohol. The 70% solution denatures more slowly and allows complete penetration to be achieved before coagulation of the surface proteins takes place.

The process of giving a person a "hair permanent" involves protein denaturation through the use of reducing agents and oxidizing agents (see the Chemical Connections feature on page 615).

Table 20.5 is a listing of selected physical and chemical agents that cause protein denaturation. The effectiveness of a given denaturing agent depends on the type of protein upon which it is acting.

TABLE 20.5
Selected Physical and Chemical Denaturing Agents

Denaturing agent	Mode of action
heat	disrupts hydrogen bonds by making molecules vibrate too violently; produces coagulation, as in the frying of an egg
microwave radiation	causes violent vibrations of molecules that disrupt hydrogen bonds
ultraviolet radiation	operates very similarly to the action of heat (e.g., sunburning)
violent whipping or shaking	causes molecules in globular shapes to extend to longer lengths, which then entangle (e.g., beating egg white into meringue)
detergent	affects hydrogen bonds and salt linkages
organic solvents (e.g., ethanol, 2-propanol, acetone)	interfere with hydrogen bonds, because these solvents also can form hydrogen bonds; quickly denature proteins in bacteria, killing them (e.g., the disinfectant action of 70% ethanol)
strong acids and bases	disrupt hydrogen bonds and salt bridges; prolonged action leads to actual hydrolysis of peptide bonds
salts of heavy metals (e.g., salts of Hg^{2+}, Ag^+, Pb^{2+})	metal ions combine with —SH groups and form poisonous salts
reducing agents	oxidize disulfide linkages to produce —SH groups

Denaturation and Human Hair

The process used in waving hair—that is, in a hair permanent—involves reversible denaturation. Hair is protein in which many disulfide (—S—S—) linkages occur as part of its tertiary structure; 16%–18% of hair is the amino acid cysteine. It is these disulfide linkages that give hair protein its overall shape. When a permanent is administered, hair is first treated with a reducing agent (ammonium thioglycolate) that breaks the disulfide linkages in the hair, producing two sulfhydryl (—SH) groups:

$$\text{Disulfide bridges} \xrightarrow{\text{Reducing agents}} \text{sulfhydryl groups}$$

The "reduced" hair, whose tertiary structure has been disrupted, is then wound on curlers to give it a new configuration.

Finally, the reduced and rearranged hair is treated with an oxidizing agent (potassium bromate) to form disulfide linkages at new locations within the hair:

$$\text{Sulfhydryl groups} \xrightarrow{\text{Oxidizing agents}} \text{re-formed disulfide bridges}$$

The new shape and curl of the hair are maintained by the newly formed disulfide bonds and the resulting new tertiary structure accompanying their formation. Of course, as new hair grows in, the "permanent" process has to be repeated.

20.17 Glycoproteins

A **glycoprotein** *is a conjugated protein that contains carbohydrates or carbohydrate derivatives in addition to amino acids.* The carbohydrate content of glycoproteins is variable (from a few percent up to 85%), but it is fixed for any specific glycoprotein.

Glycoproteins include a number of very important substances; two of these, collagen and immunoglobulins, are described in this section. Many of the proteins in cell membranes (lipid bilayers; see Section 19.10) are actually glycoproteins. The blood group markers of the ABO system (see the Chemical Connections feature on page 523 in Chapter 18) are also glycoproteins in which the carbohydrate content can reach 85%.

■ Collagen

▶ Nonstandard amino acids consist of amino acid residues that have been chemically modified after their incorporation into a protein (as is the case with 4-hydroxyproline and 5-hydroxylysine) and amino acids that occur in living organisms but are not found in proteins.

The fibrous protein collagen (Section 20.14) qualifies as a *glycoprotein* because carbohydrate units are present in its structure. A structural feature of collagen not considered in Section 20.14 is the presence of the *nonstandard* amino acids 4-hydroxyproline (5%) and 5-hydroxylysine (1%)—derivatives of the standard amino acids proline and lysine (Table 20.1).

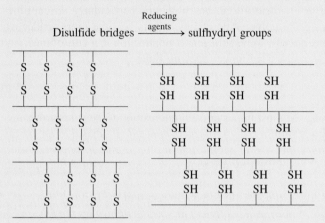

▶ The primary biochemical function of vitamin C involves the hydroxylation of proline and lysine during collagen formation. These hydroxylation processes require the enzymes *proline hydroxylase* and *lysine hydroxylase*. These enzymes can function only in the presence of vitamin C.

The presence of carbohydrate units (mostly glucose, galactose, and their disaccharides) attached by glycosidic linkages (Section 18.13) to collagen at its 5-hydroxylysine residues causes collagen to be classified as a *glycoprotein*. The function of the carbohydrate groups in collagen is related to cross-linking; they direct the assembly of collagen triple helices into more complex aggregations called *collagen fibrils*.

When collagen is boiled in water, under basic conditions, it is converted to the water-soluble protein gelatin. This process involves both denaturation (Section 20.16) and hydrolysis (Section 20.15). Heat acts as a denaturant, causing rupture of the hydrogen bonds supporting collagen's triple-helix structure. Regions in the amino acid chains where proline and hydroxyproline concentrations are high are particularly susceptible to hydrolysis, which breaks up the polypeptide chains. Meats become more tender when cooked because of the conversion of some collagen to gelatin. Tougher cuts of meat (more cross-linking), such as stew meat, need longer cooking times.

■ Immunoglobulins

Immunoglobulins are among the most important and interesting of the soluble proteins in the human body. An **immunoglobulin** *is a glycoprotein molecule produced by an organism as a protective response to the invasion of microorganisms or foreign molecules.* Different classes of immunoglobulins, identified by differing carbohydrate content and molecular mass, exist.

Immunoglobulins serve as *antibodies* to combat invasion of the body by *antigens*. An **antigen** *is a foreign substance, such as a bacterium or virus, that invades the human body.* An **antibody** *is a biochemical molecule that counteracts a specific antigen.* The immune system of the human body has the capability to produce immunoglobulins that respond to several thousand different antigens.

All types of immunoglobulin molecules have much the same basic structure, which includes the following features:

1. Four polypeptide chains are present: two identical heavy (H) chains and two identical light (L) chains.
2. The H chains, which usually contain 400–500 amino acid residues, are approximately twice as long as the L chains.
3. Both the H and L chains have constant and variable regions. The constant regions have the same amino acid sequence from immunoglobulin to immunoglobulin, and the variable regions have a different amino acid sequence in each immunoglobulin.
4. The carbohydrate content of various immunoglobulins varies from 1% to 12% by mass.

FIGURE 20.20 This schematic diagram shows the structure of an immunoglobulin. Two heavy (H) polypeptide chains and two light (L) polypeptide chains are cross-linked by disulfide bridges. The purple areas are the constant amino acid regions, and the areas shown in red are the variable amino acid regions of each chain. Carbohydrate molecules attached to the heavy chains aid in determining the destinations of immunoglobulins in the tissues.

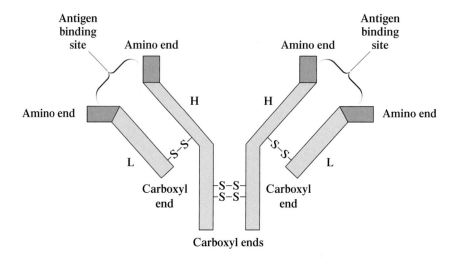

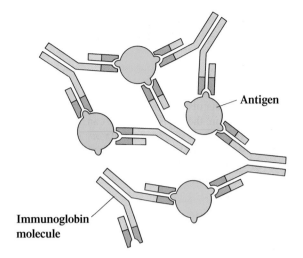

FIGURE 20.21 In this immunoglobulin–antigen complex, note that more than one immunoglobulin molecule can attach itself to a given antigen. Also, any given immunoglobulin has only two sites where antigen can bind.

Antigen

Immunoglobin molecule

5. The secondary and tertiary structures are similar for all immunoglobulins. They involve a Y-shaped conformation (Figure 20.20) with disulfide linkages between H and L chains stabilizing the structure.

The interaction of an immunoglobulin molecule with an antigen occurs at the "tips" (upper-most part) of the Y structure. These tips are the variable-composition region of the immunoglobulin structure. It is here that the antigen binds specifically, and it is here that the amino acid sequence differs from one immunoglobulin to another.

Each immunoglobulin has two identical active sites and can thus bind to two molecules of the antigen it is "designed for." The action of many such immunoglobulins of a given type in concert with each other creates an *antigen–antibody complex* that precipitates from solution (Figure 20.21). Eventually, an invading antigen can be eliminated from the

CHEMICAL CONNECTIONS

Cyclosporine: An Antirejection Drug

The survival rate for patients undergoing human organ transplant operations such as heart, liver, or kidney replacement, has risen dramatically since the late 1980s. This increased success coincides with the introduction of a new drug for controlling transplant rejection by a patient's own immune system. This new immunosuppressive agent (antirejection drug) is cyclosporine a substance obtained from a particular type of soil fungus.

The primary structure of cyclosporine is that of a cyclic peptide containing 11 amino acid units. Ten of these are amino acids with simple side chains (four or fewer carbon atoms). The eleventh amino acid, which is the key to cyclosporine's pharmacological activity, had not been previously reported. This novel amino acid has a 7-carbon branched, unsaturated, hydroxylated side chain with the structure.

$$-CH-CH-CH_2-CH=CH-CH_3$$
$$\quad | \quad \ | $$
$$\quad OH \quad CH_3$$

The diagram shows the amino acid sequence within the cyclosporine ring. Seven of the amino acid units, denoted by asterisks in the preceding structure, have their nitrogen atom

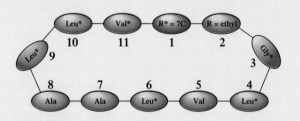

methylated; that is, a methyl group has replaced the hydrogen atom. This unique structural feature makes cyclosporine water-insoluble but fat-soluble.

The fat solubility of cyclosporine allows it to cross cell membranes readily and to be widely distributed in the body. It is administered either intravenously or orally. Because of its low water solubility, the drug is supplied in olive oil for oral administration. Cyclosporine has a narrow therapeutic index. When the blood concentration is too low, inadequate immunosuppression occurs. On the other hand, a high cyclosporine concentration can lead to kidney problems.

CHEMICAL CONNECTIONS

Lipoproteins and Heart Attack Risk

The lipoproteins present in blood serum are classified according to their density, which is related to the fractions of protein and lipid present. The more protein in the lipoprotein, the higher its density. On a density basis, there are three general categories of blood serum lipoproteins: (1) very-low-density lipoprotein (VLDL), (2) low-density lipoprotein (LDL), and (3) high-density lipoprotein (HDL). Characteristics of these three types of lipoproteins are given in the following table.

Type of lipoprotein	Density range (g/mL)	Approximate percent-by-mass protein
VLDL	1.006–1.019	5
LDL	1.019–1.063	25
HDL	1.063–1.21	50

The various types of lipoproteins have different functions. VLDLs are the principal carriers of triacylglycerols in the blood. (As cells remove triacylglycerols as needed from VLDLs, the VLDLs become LDLs.) Both LDLs and HDLs are involved in cholesterol transport. LDLs carry approximately 80% of this substance, and HDLs the remainder.

Of significance, LDLs and HDLs carry cholesterol for different purposes. LDLs carry cholesterol to cells for their use, whereas HDLs carry excess cholesterol away from cells to the liver for processing and excretion from the body.

Studies show that LDL levels correlate *directly* with heart disease, whereas HDL levels correlate *inversely* with heart disease risk. Thus HDL is sometimes referred to as "good" cholesterol (*H*DL = *H*ealthy) and LDL as "bad" cholesterol (*L*DL = *L*ess healthy).

The goal of dietary measures to slow the advance of atherosclerosis is to reduce LDL cholesterol levels. Reduction in the dietary intake of saturated fat appears to be a key action (see the Chemical Connections feature on page 562 in Chapter 19).

High HDL levels are desirable, because they give the body an efficient means of removing excess cholesterol. Low HDL levels can result in excess cholesterol depositing within the circulatory system.

In general, women have higher HDL levels than men—an average of 55 mg per 100 mL of blood serum versus 45 mg per 100 mL. This may explain in part why proportionately fewer women have heart attacks than men. Nonsmokers have uniformly higher HDL levels than smokers. Exercise on a regular basis tends to increase HDL levels. This discovery has increased the popularity of walking and running exercise. It appears that genetics also plays a role in establishing HDL as well as other lipoprotein concentrations in the blood.

A person's *total* blood cholesterol level does not necessarily correlate with that individual's real risk for heart and blood vessel disease. A better measure is the *cholesterol ratio,* which is defined as

$$\text{Cholesterol ratio} = \frac{\text{total cholesterol}}{\text{HDL cholesterol}}$$

For example, if a person's total cholesterol is 200 and his or her HDL is 45, then the cholesterol ratio would be 4.4. According to the accompanying guidelines for interpreting cholesterol ratio values, this indicates an average risk for heart disease.

What Your Cholesterol Ratio Means

Ratio	Heart disease risk
6.0	high
5.0	above average
4.5	average
4.0	below average
3.0	low

body through such precipitation. The bonding of an antigen to the variable region of an immunoglobulin occurs through dipole–dipole interactions and hydrogen bonds rather than covalent bonds.

The importance of immunoglobulins is amply and tragically demonstrated by the effects of AIDS (acquired immunodeficiency syndrome). The AIDS virus upsets the body's normal production of immunoglobulins and leaves the body susceptible to what would otherwise not be debilitating and deadly infections.

Individuals who receive organ transplants must be given drugs to suppress the production of immunoglobulins against foreign proteins in the new organ, thus preventing rejection of the organ. The major reason for the increasing importance of organ transplants is the successful development of drugs that can properly manipulate the body's immune system (see the Chemical Connections feature on cyclosporine on page 617).

Many reasons exist for a mother to breast-feed a newborn infant. One of the most important is immunoglobulins. During the first two or three days of lactation, the breasts produce *colostrum,* a premilk substance containing immunoglobulins from the mother's blood. Colostrum helps protect the newborn infant from those infections to which the mother has developed immunity. These diseases are the ones in her environment—precisely those the infant needs protection from. Breast milk, once it is produced, is a source of immunoglobulins for the infant for a short time. (After the first week of nursing, immunoglobulin concentrations in the milk decrease rapidly.) Infant formula used as a substitute for breast milk is almost always nutritionally equivalent, but it does not contain immunoglobulins.

20.18 Lipoproteins

A **lipoprotein** *is a conjugated protein that contains lipids in addition to amino acids.* The major function of such proteins is to help suspend lipids and transport them through the bloodstream. Lipids, in general, are insoluble in blood (an aqueous medium) because of their nonpolar nature (Section 19.1).

The presence or absence of various types of lipoproteins in the blood appears to have implications for the health of the heart and blood vessels. Lipoprotein levels in the blood are now used as an indicator of heart attack risk (see the Chemical Connections feature on page 618).

CONCEPTS TO REMEMBER

Protein. A protein is a polymer in which the monomer units are amino acids.

α-Amino acid. An α-amino acid is an amino acid in which the amino group and the carboxyl group are attached to the same carbon atom.

Standard amino acid. A standard amino acid is one of the 20 α-amino acids that are normally present in protein.

Amino acid classifications. Amino acids are classified as nonpolar, polar neutral, polar basic, or polar acidic, depending on the nature of the side chain (R group) present.

Chirality of amino acids. Amino acids found in proteins are always left-handed (L isomer).

Zwitterion. A zwitterion is a molecule that has a positive charge on one atom and a negative charge on another atom. In neutral solution and in the solid state, amino acids exist as zwitterions. For amino acids in solution, the isoelectric point is the pH at which the concentration of zwitterions is a maximum.

Disulfide bond formation. The amino acid cysteine readily dimerizes; the —SH groups of two cysteine molecules interact to form a covalent disulfide bond.

Peptide bond. A peptide bond is an amide bond involving the carboxyl group of one amino acid and the amino group of another amino acid. In a protein, the amino acids are linked to each other through peptide bonds.

Biochemically important peptides. Numerous small peptides are biologically active. Their functions include hormonal action, neurotransmission functions, and antioxidant activity.

Primary protein structure. The primary structure of a protein is the sequence of amino acids present in the peptide chain or chains of the protein.

Secondary protein structure. The secondary structure of a protein is the arrangement in space of the backbone portion of the protein. The two major types of protein secondary structure are the α helix and the β pleated sheet.

Tertiary protein structure. The tertiary structure of a protein is the overall three-dimensional shape that results from the attractive forces among amino acid side chains (R groups).

Quaternary protein structure. The quaternary structure of a protein involves the associations among various polypeptide chains present in the protein.

Simple and conjugated proteins. A simple protein contains only amino acids. A conjugated protein contains one or more additional chemical components, called prosthetic groups, in addition to amino acids.

Fibrous and globular proteins. Fibrous proteins are generally insoluble in water and have a long, thin, fibrous shape. α Keratin and collagen are important fibrous proteins. Globular proteins are generally soluble in water and have a roughly spherical or globular overall shape. Hemoglobin and myoglobin are important globular proteins.

Protein hydrolysis. Protein hydrolysis is a chemical reaction in which peptide bonds within a protein are broken through reaction with water. Complete hydrolysis produces free amino acids.

Protein denaturation. Protein denaturation is the partial or complete disorganization of a protein's characteristic three-dimensional shape as a result of disruption of its secondary, tertiary, and quaternary structural interactions.

Glycoproteins. Glycoproteins are conjugated proteins that contain carbohydrates or carbohydrate derivatives in addition to amino acids. Collagen and immunoglobulins are important glycoproteins.

Lipoproteins. Lipoproteins are conjugated proteins that are composed of both lipids and amino acids. Lipoproteins are classified on the basis of their density.

KEY REACTIONS AND EQUATIONS

1. Formation of a zwitterion at pH 7 (Section 20.4)

$$H_2N-CH-COOH \longrightarrow \overset{+}{H_3N}-CH-COO^-$$
$$\quad\quad | \quad\quad\quad\quad\quad\quad | $$
$$\quad\quad R \quad\quad\quad\quad\quad\quad R$$

2. Conversion of a zwitterion to a positive ion in acidic solution (Section 20.4)

$$\overset{+}{H_3N}-CH-COO^- + H_3O^+ \longrightarrow \overset{+}{H_3N}-CH-COOH + H_2O$$
$$\quad\quad | \quad\quad\quad\quad\quad\quad\quad\quad\quad\quad | $$
$$\quad\quad R \quad\quad\quad\quad\quad\quad\quad\quad\quad\quad R$$

3. Conversion of a zwitterion to a negative ion in basic solution (Section 20.4)

$$\overset{+}{H_3N}-CH-COO^- + OH^- \longrightarrow H_2N-CH-COO^- + H_2O$$
$$\quad\quad | \quad\quad\quad\quad\quad\quad\quad\quad\quad\quad | $$
$$\quad\quad R \quad\quad\quad\quad\quad\quad\quad\quad\quad\quad R$$

4. Formation of a peptide bond (Section 20.6)

$$\overset{+}{H_3N}-CH-COO^- + \overset{+}{H_3N}-CH-COO^- \longrightarrow$$
$$\quad\quad | \quad\quad\quad\quad\quad\quad | $$
$$\quad\quad R \quad\quad\quad\quad\quad\quad R'$$

$$\overset{+}{H_3N}-CH-\overset{\overset{O}{\|}}{C}-\overset{\overset{H}{|}}{N}-CH-COO^- + H_2O$$
$$\quad\quad | \quad\quad\quad\quad\quad\quad\quad | $$
$$\quad\quad R \quad\quad\quad\quad\quad\quad\quad R'$$

5. Hydrolysis of a protein in acidic solution (Section 20.15)

$$\text{Protein} + H_2O \xrightarrow{H^+} \text{smaller peptides} \xrightarrow{H^+} \text{amino acids}$$

6. Denaturation of a protein (Section 20.16)

$$\begin{matrix}\text{Protein with} \\ 1°, 2°, \text{and} \\ 3° \text{ structure}\end{matrix} \xrightarrow[\text{agent}]{\text{Denaturing}} \begin{matrix}\text{Protein with } 1° \\ \text{structure only}\end{matrix}$$

KEY TERMS

α-amino acid (20.2)
Amino acid (20.2)
Antibody (20.17)
Antigen (20.17)
Complete dietary protein (20.2)
Conjugated protein (20.13)
Electrophoresis (20.4)
Essential amino acid (20.2)
Fibrous protein (20.14)
Globular protein (20.14)
Glycoprotein (20.17)

Immunoglobulin (20.17)
Isoelectric point (20.4)
Lipoprotein (20.18)
Nonpolar amino acid (20.2)
Peptide (20.6)
Peptide bond (20.6)
Polar acidic amino acid (20.2)
Polar basic amino acid (20.2)
Polar neutral amino acid (20.2)
Polypeptide (20.6)
Primary protein structure (20.9)

Prosthetic group (20.13)
Protein (20.1 and 20.8)
Protein denaturation (20.16)
Quaternary protein structure (20.12)
Secondary protein structure (20.10)
Simple protein (20.13)
Standard amino acid (20.2)
Tertiary protein structure (20.11)
Zwitterion (20.4)

EXERCISES AND PROBLEMS

The members of each pair of problems in this section test similar material.

■ **Amino Acid Structural Characteristics (Section 20.2)**

20.1 Which of the following structures represent *α*-amino acids?

a.
$$\overset{\overset{H}{|}}{H_2N-C-COOH}$$
$$\quad\quad |$$
$$\quad\quad CH_3$$

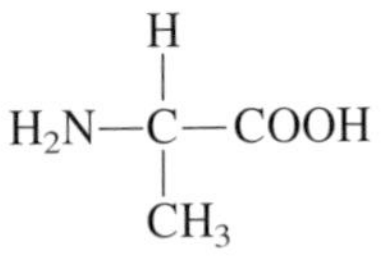

b.
$$\overset{\overset{H\ \ COOH}{|\ \ \ |}}{H_2N-C-CH_2}$$
$$\quad\quad |$$
$$\quad\quad CH_3$$

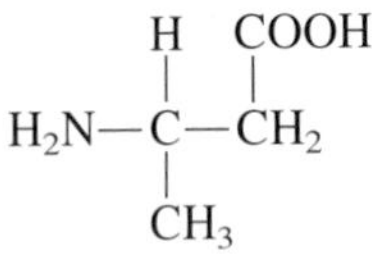

c.
$$\overset{\overset{H}{|}}{H_2N-CH_2-C-COOH}$$
$$\quad\quad\quad\quad |$$
$$\quad\quad\quad\quad H$$

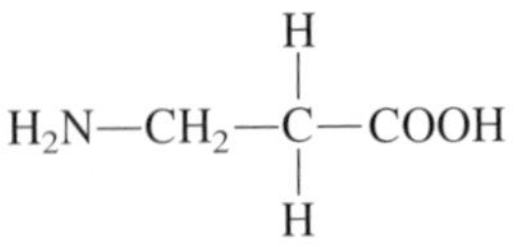

d.
$$\overset{\overset{H}{|}}{H_2N-C-COOH}$$
$$\quad\quad |$$
$$\quad\quad CH_2$$
$$\quad\quad |$$
$$\quad\quad CH_3$$

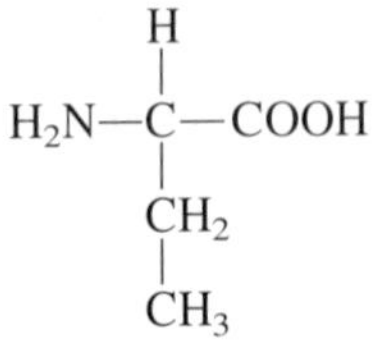

20.2 What is the significance of the prefix *α* in the designation *α*-amino acid?

20.3 What is the major structural difference among the various standard amino acids?

20.4 On the basis of polarity, what are the four types of side chains found in the standard amino acids?

20.5 With the help of Table 20.1, determine which of the standard amino acids have a side chain with the following characteristics.
a. Contains an aromatic group
b. Contains the element sulfur
c. Contains a carboxyl group
d. Contains a hydroxyl group

20.6 With the help of Table 20.1, determine which of the standard amino acids have a side chain with the following characteristics.
a. Contains only carbon and hydrogen
b. Contains an amino group
c. Contains an amide group
d. Contains more than four carbon atoms

20.7 What is the distinguishing characteristic of a polar basic amino acid?

20.8 What is the distinguishing characteristic of a polar acidic amino acid?

20.9 In what way is the structure of the amino acid proline different from that of the other 19 standard amino acids?

20.10 Which two of the standard amino acids are structural isomers?

■ **Amino Acid Nomenclature (Section 20.2)**

20.11 What amino acids do these abbreviations stand for?
a. Ala   b. Leu   c. Met   d. Trp

20.12 What amino acids do these abbreviations stand for?
a. Asp   b. Cys   c. Phe   d. Val

20.13 Which four standard amino acids have three-letter abbreviations that are not the first three letters of their common names?

20.14 What are the three-letter abbreviations for the three polar basic amino acids?

20.15 Classify each of the following amino acids as nonpolar, polar neutral, polar acidic, or polar basic.
a. Asn   b. Glu   c. Pro   d. Ser

20.16 Classify each of the following amino acids as nonpolar, polar neutral, polar acidic, or polar basic.
a. Gly   b. Thr   c. Tyr   d. His

■ **Chirality and Amino Acids (Section 20.3)**

20.17 To which family of mirror-image isomers do nearly all naturally occurring amino acids belong?

20.18 In what way is the structure of glycine different from that of the other 19 common amino acids?

20.19 Draw Fischer projection formulas for the following amino acids.
a. L-Serine   b. D-Serine   c. D-Alanine   d. L-Leucine

20.20 Draw Fischer projection formulas for the following amino acids.
a. L-Cysteine   b. D-Cysteine   c. L-Alanine   d. L-Valine

■ **Acid–Base Properties of Amino Acids (Section 20.4)**

20.21 Amino acids are solids at room temperature with relatively high decomposition points. Explain why.

20.22 Amino acids exist as zwitterions in the solid state. Explain why.

20.23 Draw the zwitterion structure for each of the following amino acids.
a. Leucine   b. Isoleucine   c. Cysteine   d. Glycine

20.24 Draw the zwitterion structure for each of the following amino acids.
a. Serine   b. Methionine   c. Threonine   d. Phenylalanine

20.25 Draw the structure of serine at each of the following pH values.
a. 7.0   b. 1.0   c. 12.0   d. 3.0

20.26 Draw the structure of glycine at each of the following pH values.
a. 7.0   b. 13.0   c. 2.0   d. 11.0

20.27 Explain what is meant by the term *isoelectric point*.

20.28 Most amino acids have isoelectric points between 5.0 and 6.0, but the isoelectric point of lysine is 9.7. Explain why lysine has such a high value for its isoelectric point.

20.29 Glutamic acid exists in two low-pH forms instead of the usual one. Explain why.

20.30 Arginine exists in two high-pH forms instead of the usual one. Explain why.

20.31 Predict the direction of movement of each of the following amino acids in a solution at the pH value specified under the influence of an electric field. Indicate the direction as toward the positive electrode or toward the negative electrode. Write "isoelectric" if no net movement occurs.
a. Alanine at pH = 12.0   b. Valine at pH = 7.0
c. Aspartic acid at pH = 1.0   d. Arginine at pH = 13.0

20.32 Predict the direction of movement of each of the following amino acids in a solution at the pH value specified under the influence of an electric field. Indicate the direction as toward the positive electrode or toward the negative electrode. Write "isoelectric" if no net movement occurs.
a. Alanine at pH = 2.0   b. Valine at pH = 12.0
c. Aspartic acid at pH = 13.0   d. Arginine at pH = 1.0

20.33 A direct current was passed through a solution containing valine, histidine, and aspartic acid at a pH of 6.0. One amino acid migrated to the positive electrode, one migrated to the negative electrode, and one did not migrate to either electrode. Which amino acids went where?

20.34 A direct current was passed through a solution containing alanine, arginine, and glutamic acid at a pH of 6.0. One amino acid migrated to the positive electrode, one migrated to the negative electrode, and one did not migrate to either electrode. Which amino acids went where?

■ **Cysteine and Disulfide Bonds (Section 20.5)**

20.35 When two cysteine molecules dimerize, what happens to the R groups present?

20.36 What chemical reaction involving the cysteine molecule produces a disulfide bond?

■ **Peptide Formation (Section 20.6)**

20.37 What two functional groups are involved in the formation of a peptide bond?

20.38 What is meant by the N-terminal end and the C-terminal end of a peptide?

20.39 Write out the full structure of the tripeptide Val–Phe–Cys.

20.40 Write out the full structure of the tripeptide Glu–Ala–Leu.

20.41 Explain why the notations Ser–Cys and Cys–Ser represent two different molecules rather than the same molecule.

20.42 Explain why the notations Ala–Gly–Val–Ala and Ala–Val–Gly–Ala represent two different molecules rather than the same molecule.

20.43 There are a total of six different amino acid sequences for a tripeptide containing one molecule each of serine, valine, and glycine. Using three-letter abbreviations for the amino acids, draw the six possible sequences of amino acids.

20.44 There are a total of six different amino acid sequences for a tetrapeptide containing two molecules each of serine and valine. Using three-letter abbreviations for the amino acids, draw the six possible sequences of amino acids.

20.45 Identify the amino acids contained in each of the following tripeptides.

a.

b.

20.46 Identify the amino acids contained in each of the following tripeptides.

a.

$$H_3\overset{+}{N}-CH_2-\underset{\underset{}{}}{\overset{O}{\overset{\|}{C}}}-\underset{}{\overset{H}{\overset{|}{N}}}-\underset{\underset{\underset{CH_3}{|}}{\overset{|}{CH-CH_3}}}{CH}-\underset{}{\overset{O}{\overset{\|}{C}}}-\underset{}{\overset{H}{\overset{|}{N}}}-CH_2-COO^-$$

b.

$$H_3\overset{+}{N}-\underset{\underset{OH}{\overset{|}{CH_2}}}{CH}-\underset{}{\overset{O}{\overset{\|}{C}}}-\underset{}{\overset{H}{\overset{|}{N}}}-\underset{\underset{\underset{CH_3}{|}}{\overset{|}{CH-CH_3}}}{\underset{\overset{|}{CH_2}}{CH}}-\underset{}{\overset{O}{\overset{\|}{C}}}-\underset{}{\overset{H}{\overset{|}{N}}}-\underset{\underset{\underset{COO^-}{|}}{\overset{|}{CH_2}}}{\overset{\overset{|}{CH_2}}{CH}}-COO^-$$

20.47 How many peptide bonds are present in each of the molecules in Problem 20.45?

20.48 How many peptide bonds are present in each of the molecules in Problem 20.46?

20.49 What are the two repeating units present in the "backbone" of a peptide?

20.50 For a peptide, describe
a. the regularly repeating part of its structure.
b. the variable part of its structure.

■ **Biochemically Important Small Peptides (Section 20.7)**

20.51 Contrast the structures of the protein hormones oxytocin and vasopressin in terms of
a. what they have in common.
b. how they differ.

20.52 Contrast the protein hormones oxytocin and vasopressin in terms of their biochemical functions.

20.53 Contrast the binding-site locations in the brain for enkephalins and the prescription painkillers morphine and codeine.

20.54 Contrast the structures of the peptide neurotransmitters Met-enkephalin and Leu-enkephalin in terms of
a. what they have in common.
b. how they differ.

20.55 What is the unusual structural feature present in the molecule glutathione?

20.56 What is the major biological function of glutathione?

■ **Levels of Protein Structure (Sections 20.8–20.12)**

20.57 What is primary protein structure?

20.58 Two proteins with the same amino acid composition do not have to have the same primary structure. Explain why.

20.59 What are the two common types of secondary protein structure?

20.60 Hydrogen bonding between which functional groups stabilizes protein secondary structure arrangements?

20.61 The β pleated sheet secondary structure can be formed through either intramolecular hydrogen bonding or intermolecular hydrogen bonding. Explain why.

20.62 The α helix secondary structure always involves intramolecular hydrogen bonding and never involves intermolecular hydrogen bonding. Explain why.

20.63 Can more than one type of secondary structure be present in the same protein molecule? Explain your answer.

20.64 What is meant by the statement that a section of a protein has a "random structure" arrangement?

20.65 What is the difference between the types of hydrogen bonding that occur in secondary and tertiary protein structures?

20.66 State the four types of attractive interactions that give rise to tertiary protein structure.

20.67 Specify the nature of each of the following tertiary-structure interactions, using the choices hydrophobic, electrostatic, hydrogen bonding, and disulfide bond.
a. Phenylalanine and leucine
b. Arginine and glutamic acid
c. Two cysteines
d. Serine and tyrosine

20.68 Specify the nature of each of the following tertiary-structure interactions, using the choices hydrophobic, electrostatic, hydrogen bonding, and disulfide bond.
a. Lysine and aspartic acid
b. Threonine and tyrosine
c. Alanine and valine
d. Leucine and isoleucine

20.69 What is an oligomeric protein?

20.70 What is quaternary protein structure?

■ **Simple and Conjugated Proteins (Section 20.13)**

20.71 What is the major difference between a simple protein and a conjugated protein?

20.72 What is the prosthetic group in each of the following types of conjugated proteins?
a. Lipoproteins   b. Glycoproteins
c. Phosphoproteins   d. Metalloproteins

■ **Fibrous and Globular Proteins (Section 20.14)**

20.73 Contrast fibrous and globular proteins in terms of
a. solubility characteristics in water.
b. general biological function.

20.74 Contrast fibrous and globular proteins in terms of
a. general secondary structure.
b. relative abundance within the human body.

20.75 Classify each of the following proteins as a globular protein or a fibrous protein.
a. α Keratin   b. Collagen
c. Hemoglobin   d. Myoglobin

20.76 What is the major biochemical function for each of the following proteins?
a. α Keratin   b. Collagen
c. Hemoglobin   d. Myoglobin

20.77 Contrast the structures of the proteins α keratin and collagen.

20.78 Contrast the structures of the proteins myoglobin and hemoglobin.

■ **Protein Hydrolysis (Section 20.15)**

20.79 Will hydrolysis of the dipeptides Ala–Val and Val–Ala yield the same products? Explain your answer.

20.80 A shampoo bottle lists "partially hydrolyzed protein" as one of its ingredients. What is the difference between partially hydrolyzed protein and completely hydrolyzed protein?

20.81 Drugs that are proteins, such as insulin, must always be injected rather than taken by mouth. Explain why.

20.82 Which structural levels of a protein are affected by hydrolysis?

20.83 Identify the primary structure of a hexapeptide containing six different amino acids if the following smaller peptides are among the partial-hydrolysis products:
Ala–Gly, His–Val–Arg, Ala–Gly–Met, and Gly–Met–His.

20.84 Identify the primary structure of a hexapeptide containing five different amino acids if the following smaller peptides are among the partial-hydrolysis products: Gly–Cys, Ala–Ser, Ala–Gly, and Cys–Val–Ala.

20.85 How many different di- and tripeptides could be present in a solution of partially hydrolyzed Ala–Gly–Ser–Tyr?

20.86 How many different di- and tripeptides could be present in a solution of partially hydrolyzed Ala–Gly–Ala–Gly?

■ **Protein Denaturation (Section 20.16)**

20.87 Which structural levels of a protein are affected by denaturation?

20.88 Suppose a sample of protein is completely hydrolyzed and another sample of the same protein is denatured. Compare the final products of these processes.

20.89 In what way is the protein in a cooked egg the same as that in a raw egg?

20.90 Why is 70% ethanol rather than pure ethanol preferred for use as an antiseptic agent?

■ **Glycoproteins (Section 20.17)**

20.91 What two nonstandard amino acids are present in collagen?

20.92 Where are the carbohydrate units located in collagen?

20.93 What is the function of the carbohydrate groups present in collagen?

20.94 What is the role of vitamin C in the biosynthesis of collagen?

20.95 What is the difference between an antigen and an antibody?

20.96 What is an immunoglobulin?

20.97 Describe the structural features of a typical immunoglobulin molecule.

20.98 Describe the process by which blood immunoglobulins help protect the body from invading bacteria and viruses.

■ **Lipoproteins (Section 20.18)**

20.99 What is the major biochemical function of lipoproteins?

20.100 What is the basis for the classification of blood serum lipoproteins into groups?

ADDITIONAL PROBLEMS

20.101 State whether each of the following statements applies to primary, secondary, tertiary, or quaternary protein structure.
a. A disulfide bond forms between two cysteine residues in different protein chains.
b. A salt bridge forms between amino acids with acidic and basic side chains.
c. Hydrogen bonding between carbonyl oxygen atoms and nitrogen atoms of amino groups causes a peptide to coil into a helix.
d. Peptide linkages hold amino acids together in a polypeptide chain.

20.102 What is the common name for each of the following IUPAC-named standard amino acids?
a. 2-Aminopropanoic acid
b. 2-Amino-4-methylbutanoic acid
c. 2-Amino-3-hydroxybutanoic acid
d. 2-Aminobutanedioic acid

20.103 What is the net charge at a pH of 1.0 for each of the following peptides?
a. Val–Ala–Leu   b. Tyr–Trp–Thr
c. Asp–Asp–Glu–Gly   d. His–Arg–Ser–Ser

20.104 What is the net charge at a pH of 13.0 for each of the peptides in Problem 20.103?

20.105 The amino acid isoleucine possesses two chiral centers. Draw Fischer projection formulas for the four stereoisomers that are possible for this amino acid.

20.106 Indicate how many structurally isomeric tetrapeptides are possible for a tetrapeptide in which
a. four different amino acids are present
b. three different amino acids are present
c. two different amino acids are present

20.107 Draw the structures of the three hydrolysis products obtained when the tripeptide in part a of Problem 20.45 undergoes hydrolysis under
a. low-pH (acidic) conditions
b. high-pH (basic) conditions

20.108 Classify each of the following proteins as a simple protein, a conjugated protein, a glycoprotein, a lipoprotein, a fibrous protein, or a globular protein. More than one classification may apply to a given protein.
a. α Keratin   b. Hemoglobin
c. Myoglobin   d. Collagen

ANSWERS TO PRACTICE EXERCISES

20.1 a., b., c.

20.2 a. toward positively charged electrode
b. toward negatively charged electron
c. isoelectric

20.3

21 | Enzymes and Vitamins

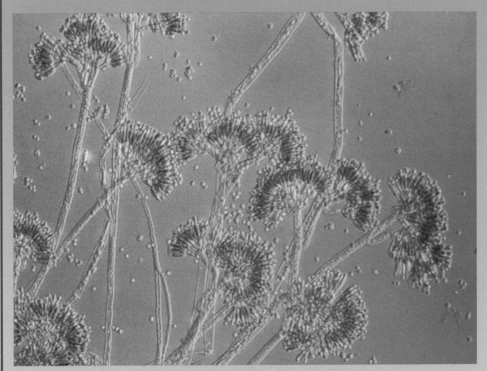

Penicillium mold which produces the penicillin antibiotic.

In this chapter we consider two topics: enzymes and vitamins. Enzymes govern all chemical reactions in living orgnisms. They are specialized proteins that, with fascinating precision and selectivity, catalyze biochemical reactions that store and release energy, make pigments in our hair and eyes, digest the food we eat, synthesize cellular building materials, and protect us by repairing cellular damage or clotting our blood. Enzymes are sensitive to their environment, responding quickly to changes in the cell. The deficiency or excess of particular enzymes can cause certain diseases or signal problems such as heart attacks and other organ damage. Our knowledge of protein structure (Chapter 20) can help us appreciate and better understand how enzymes function in living cells.

Vitamins, which are necessary components of a healthful diet, play important roles in cellular metabolism. In most cases, they function as enzyme cofactors or carriers of functional groups during biosynthesis.

21.1 General Characteristics of Enzymes

An **enzyme** *is an organic compound that acts as a catalyst for a biochemical reaction.* Each cell in the human body contains thousands of different enzymes, because almost every reaction in a cell requires its own specific enzyme. Enzymes cause cellular reactions to occur millions of times faster than corresponding uncatalyzed reactions. As catalysts (Section 9.6), enzymes are not consumed during the reaction but merely help the reaction occur more rapidly.

The word *enzyme* comes from the Greek words *en,* which means "in," and *zyme,* which means "yeast." Long before their chemical nature was understood, yeast enzymes were used in the production of bread and alcoholic beverages. The action of yeast on sugars

FIGURE 21.1 Bread dough rises as a result of the action of yeast enzymes.

produces the carbon dioxide gas that causes bread to rise (see Figure 21.1) Fermentation of sugars in fruit juices with the same yeast enzymes produces alcoholic beverages.

Most enzymes are globular proteins (Section 20.14). Some are simple proteins, consisting entirely of amino acid chains. Others are conjugated proteins, containing additional chemical components (Section 21.3). Until the 1980s, it was thought that *all* enzymes were proteins. A few enzymes are now known that are made of ribonucleic acids (RNA; Section 22.7) and that catalyze cellular reactions involving nucleic acids. In this chapter, we will consider only enzymes that are proteins.

Enzymes undergo all the reactions of proteins, including *denaturation* (Section 20.16). Slight alterations in pH, temperature, or other protein denaturants affect enzyme activity dramatically. Good cooks realize that overheating yeast kills the action of the yeast. A person suffering from a high fever (greater than 106°F) runs the risk of denaturing certain enzymes. The biochemist must exercise extreme caution in handling enzymes to avoid the loss of their activity. Even vigorous shaking of an enzyme solution can destroy enzyme activity.

Enzymes differ from nonbiochemical (laboratory) catalysts in that their activity is usually regulated by other substances present in the cell in which they are found. Most laboratory catalysts need to be removed from a reaction mixture to stop their catalytic action; this is not so with enzymes. In some cases, if a certain chemical is needed in the cell, the enzyme responsible for its production is activated by other cellular components. When a sufficient quantity has been produced, the enzyme is then deactivated. In other situations, the cell may produce more or less enzyme as required. Because different enzymes are required for nearly all cellular reactions, certain necessary reactions can be accelerated or decelerated without affecting the rest of the cellular chemistry.

21.2 Nomenclature and Classification of Enzymes

Enzymes are most commonly named by using a system that attempts to provide information about the *function* (rather than the structure) of the enzyme. Type of reaction catalyzed and *substrate* identity are focal points for the nomenclature. A **substrate** *is the reactant in an enzyme-catalyzed reaction.* The substrate is the substance upon which the enzyme "acts."

Three important aspects of the enzyme-naming process are the following:

1. The suffix -*ase* identifies a substance as an enzyme. Thus ure*ase,* sucr*ase,* and lip*ase* are all enzyme designations. The suffix -*in* is still found in the names of some of the first enzymes studied, many of which are digestive enzymes. Such names include *trypsin, chymotrypsin,* and *pepsin.*
2. The type of reaction catalyzed by an enzyme is often noted with a prefix. An *oxidase* enzyme catalyzes an oxidation reaction, and a *hydrolase* enzyme catalyzes a hydrolysis reaction.
3. The identity of the substrate is often noted in addition to the type of reaction. Enzyme names of this type include *glucose oxidase, pyruvate carboxylase,* and *succinate dehydrogenase.* Infrequently, the substrate but not the reaction type is given, as in the names *urease* and *lactase.* In such names, the reaction involved is hydrolysis; *urease* catalyzes the hydrolysis of urea, *lactase* the hydrolysis of lactose.

> Enzymes, the most efficient catalysts known, increase the rates of biochemical reactions by factors of up to 10^{20} over uncatalyzed reactions. Nonenzymatic catalysts, on the other hand, typically enhance the rate of a reaction by factors of 10^2 to 10^4.

EXAMPLE 21.1

Predicting Enzyme Function from an Enzyme's Name

■ Predict the function of the following enzymes.

a. Cellulase **b.** Sucrase
c. L-Amino acid oxidase **d.** Aspartate aminotransferase

Solution

a. Cellulase catalyzes the hydrolysis of cellulose.
b. Sucrase catalyzes the hydrolysis of the disaccharide sucrose.
c. L-Amino acid oxidase catalyzes the oxidation of L-amino acids.
d. Aspartate aminotransferase catalyzes the transfer of an amino group from aspartate to a different molecule.

(continued)

Practice Exercise 21.1

Predict the function of the following enzymes.

a. Maltase　　　　　　　　　　　**b.** Lactate dehydrogenase

c. Fructose oxidase　　　　　　　**d.** Maleate isomerase

Enzymes are grouped into six major classes on the basis of the types of reactions they catalyze.

1. An **oxidoreductase** *is an enzyme that catalyzes oxidation–reduction reactions.*
2. A **transferase** *is an enzyme that catalyzes the transfer of a functional group from one molecule to another.*
3. A **hydrolase** *is an enzyme that catalyzes hydrolysis reactions in which the addition of a water molecule to a bond, causes the bond to break.*
4. A **lyase** *is an enzyme that catalyzes the addition of a group to a double bond or the removal of a group to form a double bond in a manner that does not involve hydrolysis or oxidation.*
5. An **isomerase** *is an enzyme that catalyzes the rearrangement of functional groups within a molecule, converting the molecule into another molecule isomeric with it.*
6. A **ligase** *is an enzyme the catalyzes the bonding together of two molecules into one, with the participation of ATP.*

Within each of these six main classes of enzymes there are subclasses. Table 21.1 gives further information about enzyme subclass terminology.

TABLE 21.1
Main Classes and Subclasses of Enzymes

Main classes	Selected subclasses	Type of reaction catalyzed
oxidoreductases	oxidases	oxidation of a substrate
	reductases	reduction of a substrate
	dehydrogenases	introduction of double bond (oxidation) by formal removal of two H atoms from substrate, the H being accepted by a coenzyme
transferases	transaminases	transfer of an amino group between substrates
	kinases	transfer of a phosphate group between substrates
hydrolases	lipases	hydrolysis of ester linkages in lipids
	proteases	hydrolysis of amide linkages in proteins
	nucleases	hydrolysis of sugar–phosphate ester bonds in nucleic acids
	carbohydrases	hydrolysis of glycosidic bonds in carbohydrates
	phosphatases	hydrolysis of phosphate–ester bonds
lyases	dehydratases	removal of H_2O from substrate
	decarboxylases	removal of CO_2 from substrate
	deaminases	removal of NH_3 from substrate
	hydratases	addition of H_2O to a substrate
isomerases	racemases	conversion of D to L isomer, or vice versa
	mutases	conversion of one structural isomer into another
ligases	synthetases	formation of new bond between two substrates, with participation of ATP
	carboxylases	formation of new bond between substrate and CO_2, with participation of ATP

21.3 Enzyme Structure

Enzymes can be divided into two general structural classes: simple enzymes and conjugated enzymes. A **simple enzyme** *is an enzyme composed only of protein (amino acid chains).* A **conjugated enzyme** *is an enzyme that has a nonprotein part in addition to a protein part.* By itself, neither the protein part nor the nonprotein portion of a conjugated enzyme has catalytic properties. An **apoenzyme** *is the protein part of a conjugated enzyme.* A **cofactor** *is the nonprotein part of a conjugated enzyme.* It is the combination of apoenzyme with cofactor that produces a biochemically active enzyme. A **holoenzyme** *is the biochemically active conjugated enzyme produced from an apoenzyme and a cofactor.*

$$\text{apoenzyme} + \text{cofactor} = \text{holoenzyme}$$

Why do many apoenzymes need cofactors? Cofactors provide additional chemically reactive functional groups besides those present in the amino acid side chains of apoenzymes.

A cofactor is generally either a small organic molecule or an inorganic ion (usually a metal ion). A **coenzyme** *is a small organic molecule that serves as a cofactor in a conjugated enzyme.* Many vitamins (Section 21.12) have coenzyme functions in the human body.

Typical inorganic ion cofactors include Zn^{2+}, Mg^{2+}, Mn^{2+}, and Fe^{2+}. The nonmetallic Cl^{-} ion occasionally acts as a cofactor. Dietary minerals are an important source of inorganic ion cofactors.

21.4 Models of Enzyme Action

Explanations of *how* enzymes function as catalysts in biochemical systems are based on the concepts of an enzyme active site and enzyme–substrate complex formation.

■ Enzyme Active Site

Studies show that only a small portion of an enzyme molecule, called the active site, participates in the interaction with a substrate or substrates during a reaction. The **active site** *is the relatively small part of an enzyme's structure that is actually involved in catalysis.*

The active site in an enzyme is a three-dimensional entity formed by groups that come from different parts of the protein chain(s); these groups are brought together by the folding and bending (secondary and tertiary structure; Sections 20.10 and 20.11) of the protein. The active site is usually a "crevicelike" location in the enzyme (see Figure 21.2).

▶ It is possible for an enzyme to have more than one active site.

FIGURE 21.2 The active site of an enzyme is usually a crevicelike region formed as a result of the protein's secondary and tertiary structural characteristics.

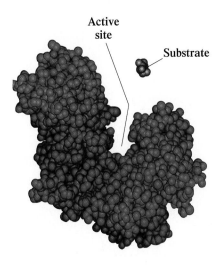

Active site

Substrate

■ Enzyme–Substrate Complex

Catalysts offer an alternative pathway with lower activation energy through which a reaction can occur (Section 9.6). In enzyme-controlled reactions, this alternative pathway involves the formation of an enzyme–substrate complex as an intermediate species in the reaction. An **enzyme–substrate complex** *is the intermediate reaction species that is formed when a substrate binds to the active site of an enzyme.* Within the enzyme–substrate complex, the substrate encounters more favorable reaction conditions than if it were free. The result is faster formation of product.

■ Lock-and-Key Model

To account for the highly specific way an enzyme recognizes a substrate and binds it to the active site, researchers have proposed several models. The simplest of these models is the lock-and-key model.

In the lock-and-key model, the active site in the enzyme has a fixed, rigid geometrical conformation. Only substrates with a complementary geometry can be accommodated at such a site, much as a lock accepts only certain keys. Figure 21.3 illustrates the lock-and-key concept of substrate–enzyme interaction.

■ Induced-Fit Model

The lock-and-key model explains the action of numerous enzymes. It is, however, too restrictive for the action of many other enzymes. Experimental evidence indicates that many enzymes have flexibility in their shapes. They are not rigid and static; there is constant change in their shape. The induced-fit model is used for this type of situation.

The induced-fit model allows for small changes in the shape or geometry of the active site of an enzyme to accommodate a substrate. A good analogy is the changes that occur in the shape of a glove when a hand is inserted into it. The induced fit is a result of the enzyme's flexibility; it adapts to accept the incoming substrate. This model, illustrated in Figure 21.4, is a more thorough explanation for the active-site properties of an enzyme because it includes the specificity of the lock-and-key model coupled with the flexibility of the enzyme protein.

The forces that draw the substrate into the active site are many of the same forces that maintain tertiary structure in the folding of polypeptide chains. Electrostatic interactions, hydrogen bonds, and hydrophobic interactions all help attract and bind substrate molecules. For example, a protonated (positively charged) amino group in a substrate could be attracted and held at the active site by a negatively charged aspartate or glutamate residue. Alternatively, cofactors such as positively charged metal ions often help bind substrate molecules. Figure 21.5 is a schematic representation of the amino acid R group interactions that bind a substrate to an enzyme active site.

> The lock-and-key model is more than just a "shape fit." In addition, there are weak binding forces (R group interactions) between parts.

FIGURE 21.3 The lock-and-key model for enzyme activity. Only a substrate whose shape and chemical nature are complementary to those of the active site can interact with the enzyme.

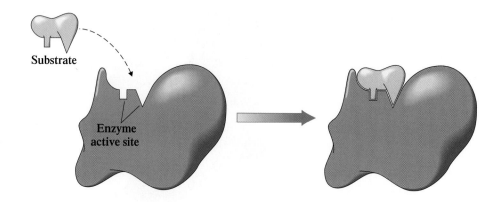

FIGURE 21.4 The induced-fit model for enzyme activity. The enzyme active site, although not exactly complementary in shape to that of the substrate, is flexible enough that it can adapt to the shape of the substrate.

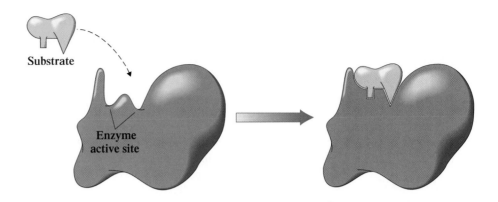

21.5 Enzyme Specificity

Enzymes exhibit different levels of selectivity, or specificity, for substrates. The degree of enzyme specificity is determined by the active site. Some active sites accommodate only one particular compound, whereas others can accommodate a "family" of closely related compounds. Types of enzyme specificity include

1. *Absolute Specificity.* Such specificity means an enzyme will catalyze a particular reaction for *only one* substrate. This most restrictive of all specificities is not common. Urease is an enzyme with absolute specificity.
2. *Stereochemical Specificity.* Such specificity means an enzyme can distinguish between stereoisomers. Chirality is inherent in an active site, because amino acids are chiral compounds. L-Amino-acid oxidase will catalyze reactions of L-amino acids but not of D-amino acids.
3. *Group Specificity.* Such specificity involves structurally similar compounds that have the same functional groups. Carboxypeptidase is group-specific; it cleaves amino acids, one at a time, from the carboxyl end of the peptide chain.
4. *Linkage Specificity.* Such specificity involves a particular type of bond, irrespective of the structural features in the vicinity of the bond. Phosphatases hydrolyze phosphate–ester bonds in all types of phosphate esters. Linkage specificity is the most general of the specificities considered.

FIGURE 21.5 A schematic diagram representing amino acid R group interactions that bind a substrate to an enzyme active site. The R group interactions that maintain the three-dimensional structure of the enzyme (secondary and tertiary structure) are also shown.

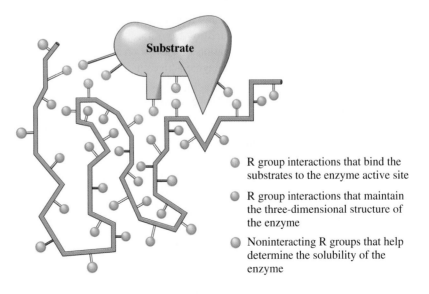

- R group interactions that bind the substrates to the enzyme active site
- R group interactions that maintain the three-dimensional structure of the enzyme
- Noninteracting R groups that help determine the solubility of the enzyme

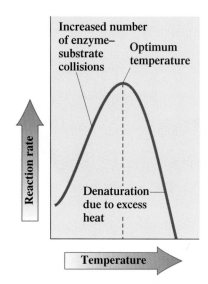

FIGURE 21.6 A graph showing the effect of temperature on the rate of an enzymatic reaction.

21.6 Factors That Affect Enzyme Activity

Enzyme activity *is a measure of the rate at which an enzyme converts substrate to products in a biochemical reaction.* Four factors affect enzyme activity: temperature, pH, substrate concentration, and enzyme concentration.

■ Temperature

Temperature is a measure of the kinetic energy (energy of motion) of molecules. Higher temperatures mean molecules are moving faster and colliding more frequently. This concept applies to collisions between substrate molecules and enzymes. As the temperature of an enzymatically catalyzed reaction increases, so does the rate (velocity) of the reaction.

However, when the temperature increases beyond a certain point, the increased energy begins to cause disruptions in the tertiary structure of the enzyme; denaturation is occurring. Change in tertiary structure at the active site impedes catalytic action, and the enzyme activity quickly decreases as the temperature climbs past this point (Figure 21.6). The temperature that produces maximum activity for an enzyme is known as the optimum temperature for that enzyme. **Optimum temperature** *is the temperature at which an enzyme exhibits maximum activity.* For human enzymes, the optimum temperature is often 37°C, normal body temperature.

■ pH

The pH of an enzyme's environment can affect its activity. This is not surprising, because the *charge* on acidic and basic amino acids (Section 20.2) located at the active site depends on pH. Small changes in pH (less than one unit) can result in enzyme denaturation (Section 20.16) and subsequent loss of catalytic activity.

Most enzymes exhibit maximum activity over a very narrow pH range. Only within this narrow pH range do the enzyme's amino acids exist in properly charged forms (Section 20.4). **Optimum pH** *is the pH at which an enzyme exhibits maximum activity.* Figure 21.7 shows the effect of pH on an enzyme's activity. Biochemical buffers help maintain the optimum pH for an enzyme.

Each enzyme has a characteristic optimum pH, which usually falls within the physiological pH range of 7.0–7.5. Notable exceptions to this generalization are the digestive enzymes pepsin and trypsin. Pepsin, which is active in the stomach, functions best at a pH of 2.0. On the other hand, trypsin, which operates in the small intestine, functions best at a pH of 8.0. The amino acid sequences present in pepsin and trypsin are those needed such that the R groups present can maintain protein tertiary structure (Section 20.11) at low (2.0) and high (8.0) pH values, respectively.

A variation from normal pH can also affect substrates, causing either protonation or deprotonation of groups on the substrate. The interaction between the altered substrate and the enzyme active site may be less efficient than normal—or even impossible.

■ Substrate Concentration

When the concentration of an enzyme is kept constant and the concentration of substrate is increased, the enzyme activity pattern shown in Figure 21.8 is obtained. This activity pattern is called a *saturation curve*. Enzyme activity increases up to a certain substrate concentration and thereafter remains constant.

What limits enzymatic activity to a certain maximum value? As substrate concentration increases, the point is eventually reached where enzyme capabilities are used to their maximum extent. The rate remains constant from this point on (Figure 21.8). Each substrate must occupy an enzyme active site for a finite amount of time, and the products must leave the site before the cycle can be repeated. When each enzyme molecule is working at full capacity, the incoming substrate molecules must "wait their turn" for an empty active site. At this point, the enzyme is said to be under saturating conditions.

FIGURE 21.7 A graph showing the effect of pH on the rate of an enzymatic reaction.

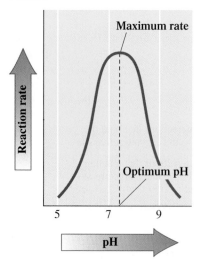

Enzymatic Browning: Discoloration of Fruits and Vegetables

Everyone is familiar with the way fruits such as apples, pears, peaches, apricots, and bananas and vegetables such as potatoes quickly turn brown when their tissue is exposed to oxygen. Such oxygen exposure occurs when the food is sliced or bitten into or when it has sustained bruises, cuts or other injury to the peel. This "browning reaction" is related to the work of an enzyme called phenolase (or polyphenoloxidase), a conjugated enzyme in which copper is present.

Phenolase is classified as an oxidoreductase. The substrates for phenolase are phenolic compounds present in the tissues of the fruits and vegetables. Phenolase hydroxylates monophenols to *o*-diphenols and oxidizes *o*-diphenols to *o*-quinones. The *o*-quinones then enter into a number of other reactions, which produce the "undesirable" brown discolorations. Quinone formation is enzyme- and oxygen-dependent.

Once the quinones have formed, the subsequent reactions occur spontaneously and no longer depend on the presence of phenolase or oxygen.

Enzymatic browning can be prevented or slowed in several ways. Immersing the "injured" food (for example, apple slices) in cold water slows the browning process. The lower temperature decreases enzyme activity, and the water limits the enzyme's access to oxygen. Refrigeration slows enzyme activity even more, and boiling temperatures destroy (denature) the enzyme. A long-used method for preventing browning involves lemon juice. Phenolase works very slowly in the acidic environment created by the lemon juice's presence. In addition, the vitamin C (ascorbic acid) present in lemon juice functions as an antioxidant. It is more easily oxidized than the phenolic-derived compounds, and its oxidation products are colorless.

At left, a freshly cut apple. Brownish oxidation products form in a few minutes (at right).

The rate at which an enzyme accepts and releases substrate molecules at substrate saturation is given by its turnover number. An enzyme's **turnover number** *is the number of substrate molecules transformed per minute by one molecule of enzyme under optimum conditions of temperature, pH, and saturation.* Table 21.2 gives turnover numbers for selected enzymes. Some enzymes have a much faster mode of operation than others.

■ Enzyme Concentration

Because enzymes are not consumed in the reactions they catalyze, the cell usually keeps the number of enzymes low compared with the number of substrate molecules. This is efficient; the cell avoids paying the energy costs of synthesizing and maintaining a large work force of enzyme molecules. Thus, in general, the concentration of substrate in a reaction is much higher than that of the enzyme.

If the amount of substrate present is kept constant and the enzyme concentration is increased, the reaction rate increases because more substrate molecules can be accommodated

■ Vitamin C

Vitamin C, which has the simplest structure of the 13 vitamins, exists in two active forms in the human body: an oxidized form and a reduced form.

$$CH_2-CH \quad \underset{Reduction}{\overset{Oxidation}{\rightleftharpoons}} \quad CH_2-CH \quad + 2H$$

Ascorbic acid
(reduced)

Dehydroascorbic acid
(oxidized)

> Vitamin C, the best known of all vitamins, was the first vitamin to be discovered (1928), the first to be structurally characterized (1933), and the first to be synthesized in the laboratory (1933). Laboratory production of vitamin C, which exceeds 80 million pounds per year, is greater than the combined production of all the other vitamins. In addition to its use as a vitamin supplement, synthetic vitamin C is used as a food additive (preservative), a flour additive, and an animal feed additive.

> Why is vitamin C called *ascorbic acid* when there is no carboxyl group (acid group) present in its structure? Vitamin C is a cyclic ester in which a carbon 1 carboxyl group has reacted with a carbon 4 hydroxyl group, forming the ring structure.

> Other naturally occurring dietary antioxidants include glutathione (Section 20.7), vitamin E (Section 21.14), beta-carotene (Section 21.14), and flavonoids (Section 23.11).

Humans, monkeys, apes, and guinea pigs are among the relatively few species that require dietary sources of vitamin C. Other species synthesize vitamin C from carbohydrates. Vitamin C's biosynthesis involves L-gulonic acid, an acid derivative of the monosaccharide L-gulose (see Figure 18.13). L-Gulonic acid is reduced by the enzyme *lactonase* to give a cyclic ester (lactone, Section 16.13); ring closure involves carbons 1 and 4. An *oxidase* then introduces a double bond into the ring, producing L-ascorbic acid.

L-Gulonic acid $\xrightarrow{\text{Lactonase}}$ γ-L-Gulonolactone $\xrightarrow{\text{Oxidase}}$ L-Ascorbic acid

The four —OH groups present in vitamin C's reduced form are suggestive of its biosynthetic monosaccharide (*polyhydroxy* aldehyde) origins. Its chemical name, L-ascorbic acid, correctly indicates that vitamin C is a weak acid. Although no carboxyl group is present, the carbon 3 hydroxyl group hydrogen atom exhibits acidic behavior as a result of its attachment to an unsaturated carbon atom.

The most completely characterized role of vitamin C is its function as a cosubstrate in the formation of the structural protein collagen (Section 20.17), which makes up much of the skin, ligaments, and tendons and also serves as the matrix on which bone and teeth are formed. Specifically, biosynthesis of the amino acids hydroxyproline and hydroxylysine (important in binding collagen fibers together) from proline and lysine requires the presence of both vitamin C and iron. Iron serves as a cofactor in the reaction, and vitamin C maintains iron in the oxidation state that allows it to function. In this role, vitamin C is functioning as a *specific* antioxidant.

Vitamin C also functions as a *general* antioxidant (Section 14.12) for water-soluble substances in the blood and other body fluids. Its antioxidant properties are also beneficial for several other vitamins. The active form of vitamin E is regenerated by vitamin C, and it also helps keep the active form of folate (a B vitamin) in its reduced state. Because of its antioxidant properties, vitamin C is often added to foods as a preservative.

Vitamin C is also involved in the metabolism of several amino acids that end up being converted to the hormones norepinephrine and thyroxine. The adrenal glands contain a higher concentration of vitamin C than any other organ in the body.

An intake of 100 mg/day of vitamin C saturates all body tissues with the compound. After the tissues are saturated, all additional vitamin C is excreted. The RDA for vitamin C varies from country to country. It is 30 mg/day in Great Britain, 60 mg/day in the United States and Canada, and 75 mg/day in Germany. A variety of fruits and vegetables have a relatively high vitamin C content (see Figure 21.18).

FIGURE 21.18 Rows of cabbage plants. Although many people think citrus fruits (50 mg per 100 g) are the best source of vitamin C, peppers (128 mg per 100 g), cauliflower (70 mg per 100 g), strawberries (60 mg per 100 g), and spinach or cabbage (60 mg per 100 g) are all richer in vitamin C.

■ Vitamin B

Pronunciation guidelines for the standard names of the B vitamins:

THIGH-ah-min
RYE-boh-flay-vin
NIGH-a-sin
FOLL-ate
PAN-toe-THEN-ick acid
BY-oh-tin

There are eight B vitamins. Our discussion of them involves four topics: nomenclature, function, structural characteristics, and dietary sources.

Much confusion exists about the B vitamins' names. Many have "number" names as well as "word" names (often several). The *preferred* names for the B vitamins (alternative names in parenthesis) are

1. **Thiamin** (vitamin B_1)
2. **Riboflavin** (vitamin B_2)
3. **Niacin** (nicotinic acid, nicotinamide, vitamin B_3)
4. **Vitamin B_6** (pyridoxine, pyridoxal, pyridoxamine)
5. **Folate** (folic acid)
6. **Vitamin B_{12}** (cobalamin)
7. **Pantothenic acid** (vitamin B_5)
8. **Biotin**

B vitamin structure is very diverse. The only common thread among structures is that all structures, except that of pantothenic acid, involve heterocyclic nitrogen ring systems. The element sulfur is present in two structures (thiamin and biotin), and vitamin B_{12} contains a metal atom (cobalt). (Biotin does not contain a tin atom, as the name might imply.) Table 21.5 gives structural forms for the eight B vitamins. Note that for two B vitamins (niacin and vitamin B_6), more than one form of the vitamin exists.

The major function of B vitamins within the human body is as components of coenzymes. Unlike vitamin C, all of the B vitamins must be chemically modified before they become functional within the coenzymes. For example, thiamine is converted to thiamine pyrophosphate (TPP), which then serves as the coenzyme in several reactions involving carbohydrate metabolism.

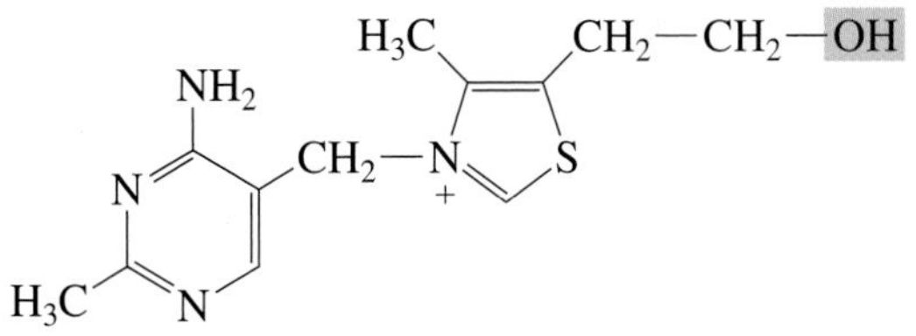

Thiamine (vitamin B_1)

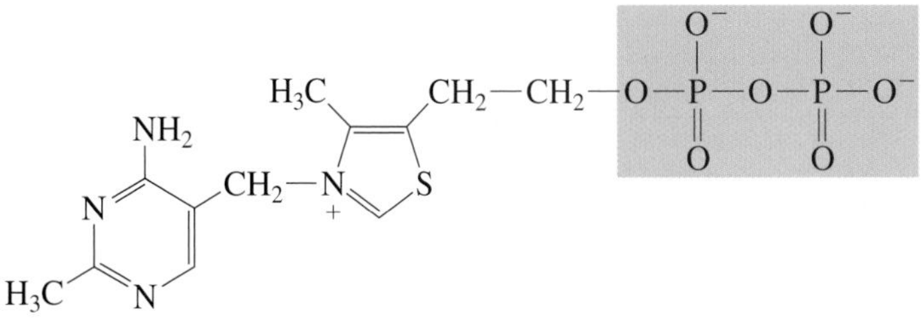

Thiamine pyrophosphate (TPP)

Another example of chemical modification for a B vitamin is the conversion of folate to tetrahydrofolate (THF).

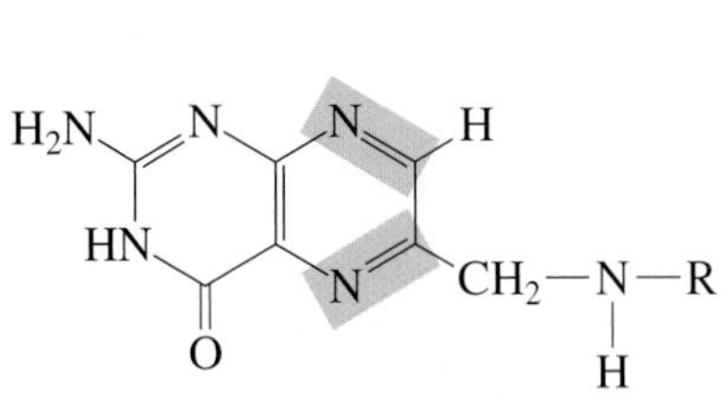

Folate

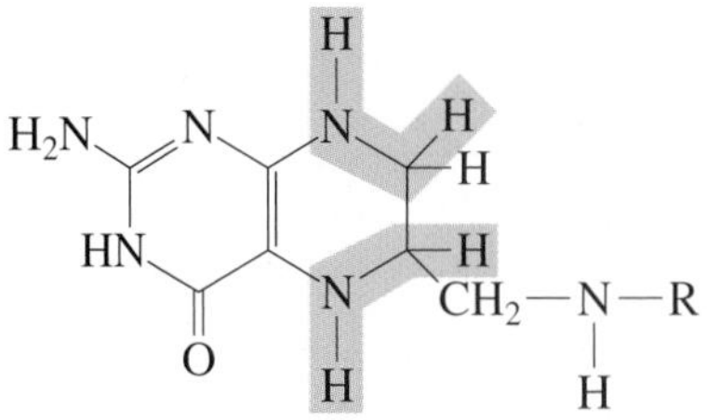

Tetrahydrofolate (THF)

Table 21.6 lists selected important coenzymes that involve B vitamins and indicates how these enzymes function. In general, coenzymes serve as temporary carriers of atoms or functional groups in redox and group transfer reactions.

An ample supply of the B vitamins can be obtained from normal dietary intake as long as a variety of foods are consumed. A certain food may be a better source of a particular B vitamin than others; however, there are multiple sources for each of the B vitamins, as Table 21.7 shows. Note, from Table 21.7, that fruits, in general, are very poor sources of B vitamins and that only certain vegetables are good B vitamin sources. Vitamin B_{12} is

TABLE 21.5
Structures of the Eight B Vitamins

Thiamin	Riboflavin

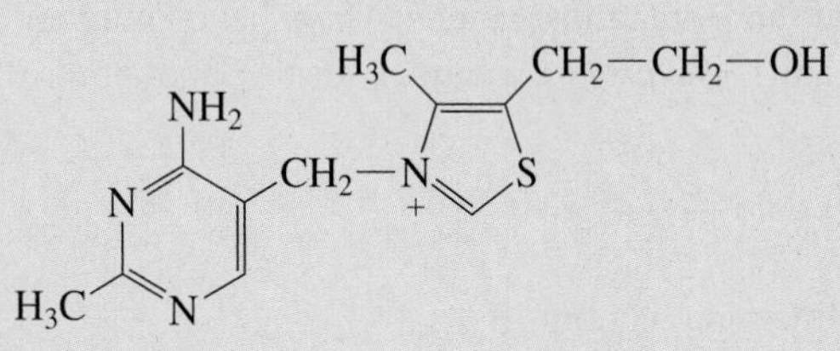

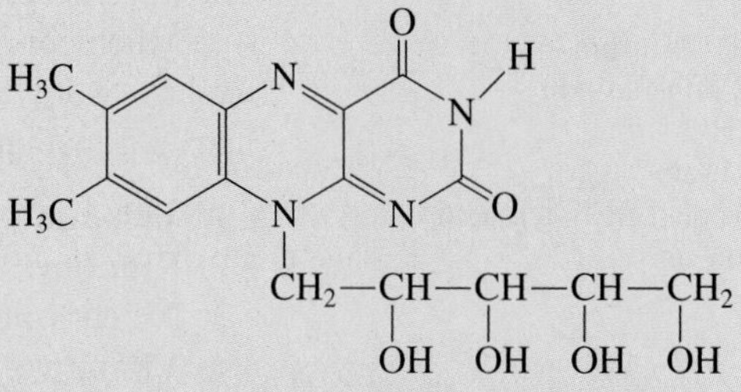

Niacin (two forms)	Pantothenic Acid
Nicotinic acid Nicotinamide	

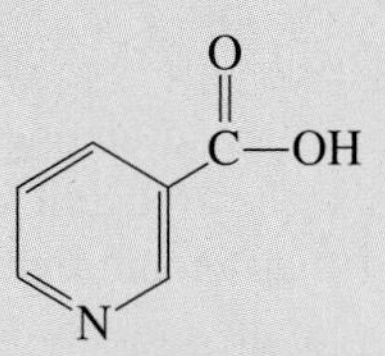

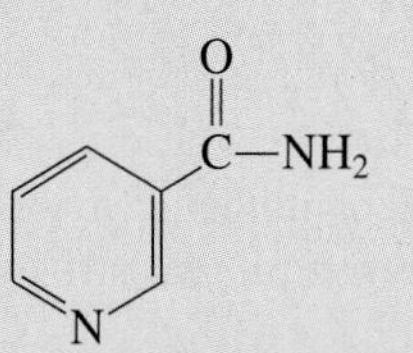

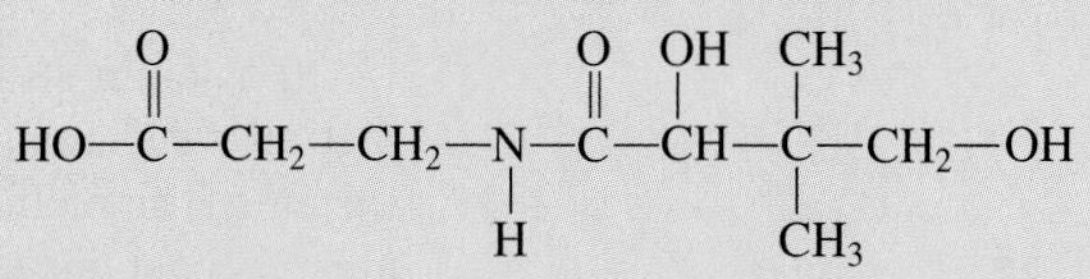

Vitamin B$_6$ (three forms)

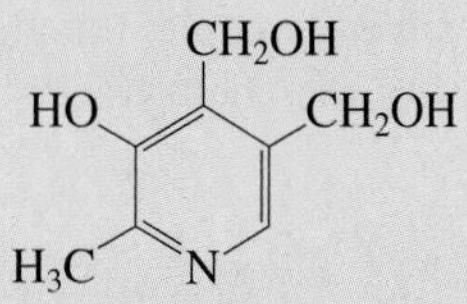

Pyridoxine 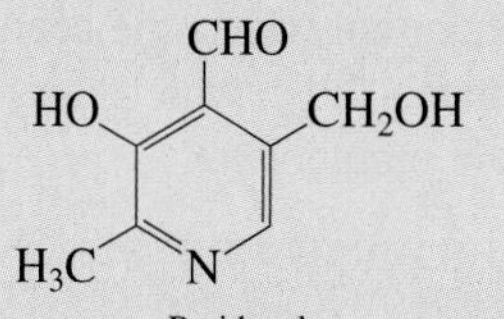   Pyridoxal 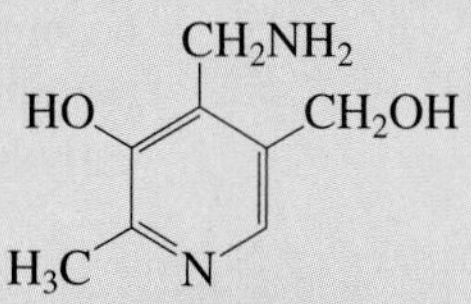   Pyridoxamine

Folate

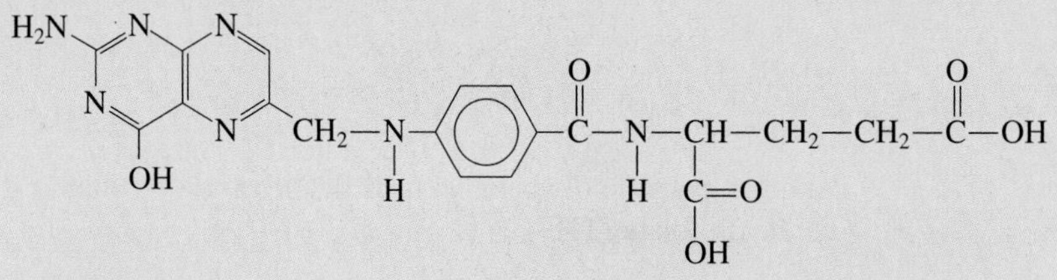

Vitamin B$_{12}$	Biotin

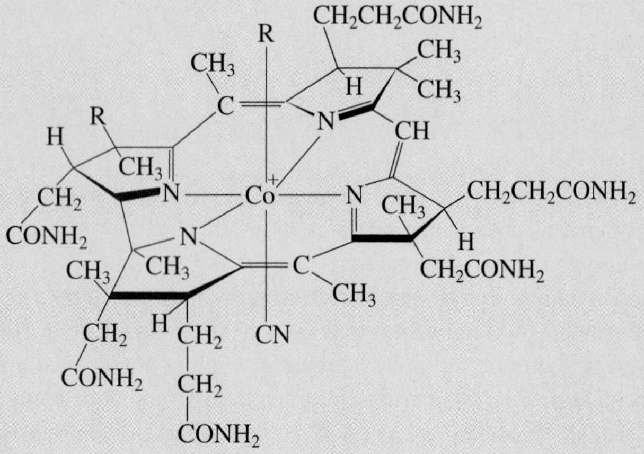

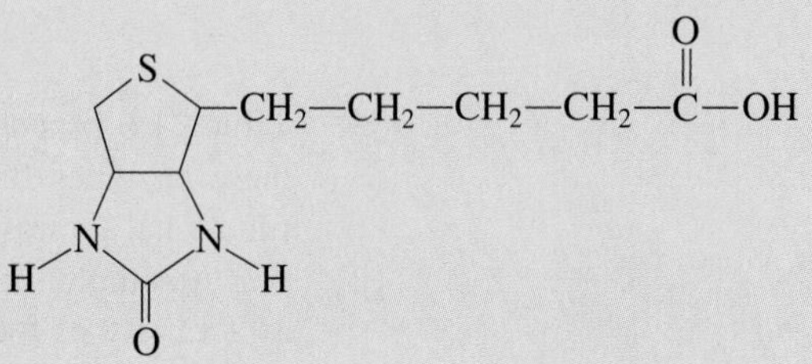

TABLE 21.6
Selected Important Coenzymes in Which B Vitamins Are Present

B Vitamin	Coenzymes	Groups Transferred
thiamine	thiamine pyrophosphate (TPP)	aldehydes
riboflavin	flavin mononucleotide (FMN) flavin adenine dinucleotide (FAD)	hydrogen atoms
niacin	nicotinamide adenine dinucleotide (NAD^+) nicotinamide adenine dinucleotide phosphate ($NADP^+$)	hydride ion (H^-)
vitamin B_6	pyridoxal-5-phosphate (PLP)	amino groups
folate	tetrahydrofolate (THF)	one-carbon groups other than CO_2
vitamin B_{12}	5'-deoxyadenosylcobalamine	alkyl groups, hydrogen atoms
pantothenic acid	coenzyme A (CoA) acyl carrier protein (ACP)	acyl groups
biotin	biocytin	carbon dioxide

TABLE 21.7
A Summary of Dietary Sources of B Vitamins

Vitamin	Vegetable group	Fruit group	Bread, cereal, rice, and pasta group	Milk, yogurt, and cheese group	Meat, poultry, fish, dry beans, eggs, and nuts group			
					Meats	Dry Beans	Eggs	Nuts and seeds
thiamin		watermelon	whole and enriched grains		pork, organ meats	legumes		sunflower seeds
riboflavin	mushrooms, asparagus, broccoli, leafy greens		whole and enriched grains	milk, cheeses	liver, red meat, poultry, fish	legumes	eggs	
niacin	mushrooms, asparagus, potato		whole and enriched grains, wheat bran		tuna, chicken, beef, turkey	legumes, peanuts		sunflower seeds
vitamin B_6	broccoli, spinach, potato, squash	bananas, watermelon	whole wheat, brown rice		chicken, fish, pork, organ meats	soybeans		sunflower seeds
folate	mushrooms, leafy greens, broccoli, asparagus, corn	oranges	fortified grains		organ meats (muscle meats are poor sources)	legumes		sunflower seeds, nuts
vitamin B_{12}				milk products	beef, poultry, fish, shellfish		egg yolk	
pantothenic acid	mushrooms, broccoli, avocados		whole grains		meat	legumes	egg yolk	
biotin			fortified cereals	yogurt	liver (muscle meats are poor sources)	soybeans	egg yolk	nuts

unique among the vitamins in being found almost exclusively in food derived from animals. Legislation that dates back to the 1940s requires that all grain products that cross state lines be enriched in thiamin, riboflavin, and niacin. Folate was added to the legislated enrichment list in 1996, when research showed that folate was essential in the prevention of certain birth defects.

21.14 Fat-Soluble Vitamins

There are four fat-soluble vitamins, denoted by the letters A, D, E, and K. Many of the functions of the fat-soluble vitamins involve processes that occur in cell membranes. The structures of the fat-soluble vitamins are more hydrocarbon-like, with fewer functional groups than the water-soluble vitamins. Their structures as a whole are nonpolar, which enhances their solubility in cell membranes.

■ Vitamin A

Normal dietary intake provides a person with both *preformed* and *precursor forms* (provitamin forms) of vitamin A. Preformed vitamin A forms are called *retinoids*. The retinoids include retin*al,* retin*ol,* and retin*oic acid*.

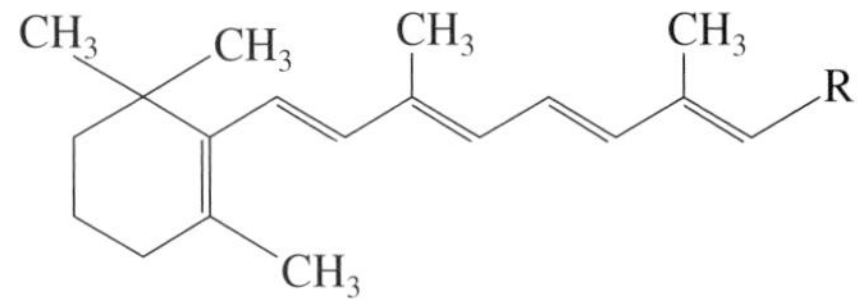

> Beta-carotene is a deep yellow (almost orange) compound. If a plant food is white or colorless, it possesses little or no vitamin A activity. Potatoes, pasta, and rice are foods in this category.

R = CH$_2$—OH (Retinol)
R = CHO (Retinal)
R = COOH (Retinoic acid)

Retinoids

Foods derived from animals, including egg yolks and dairy products, provide compounds (retinyl esters) that are easily hydrolyzed to retinoids in the intestine.

Foods derived from plants provide carotenoids (see the Chemical Connections feature "Carotenoids: A Source of Color" on page 338 in Chapter 13), which serve as precursor forms of vitamin A. The major carotenoid with vitamin A activity is beta-carotene (β-carotene), which can be cleaved to yield two molecules of vitamin A.

> The retinoids are terpenes (Section 13.5) in which four isoprene units are present. Beta-carotene has an eight-unit terpene structure.

Cleavage at this point can yield two molecules of vitamin A

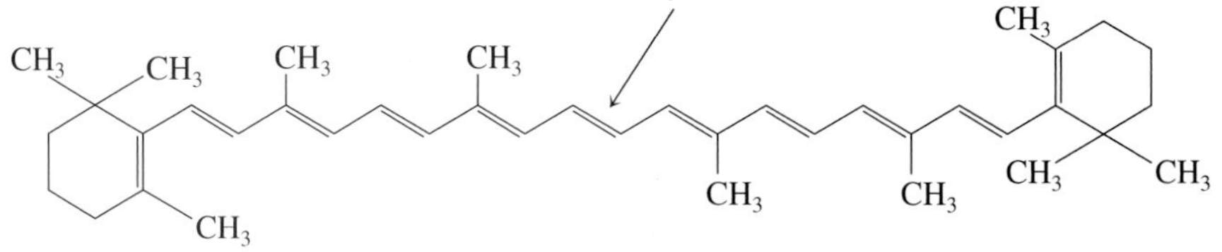

Beta-carotene, a precursor for vitamin A

Beta-carotene is a yellow to red-orange pigment plentiful in carrots, squash, cantaloupe, apricots, and other yellow vegetables and fruits, as well as in leafy green vegetables (where the yellow pigment is masked by green chlorophyll).

Vitamin A has four major functions in the body.

> Beta-carotene cleavage does not always occur in the "middle" of the molecule, so only one molecule of vitamin A is produced. Furthermore, not all beta-carotene is converted to vitamin A, and its absorption is not so efficient as that of vitamin A itself. It is estimated that 6 mg of beta-carotene is needed to produce 1 mg of retinol. Unconverted beta-carotene serves as an antioxidant (see Section 13.5), a role independent of its conversion to vitamin A.

1. *Vision.* In the eye, vitamin A combines with the protein opsin to form the visual pigment rhodopsin (see the Chemical Connections feature "*Cis–Trans* Isomerism and Vision" on page 335 in Chapter 13). Rhodopsin participates in the conversion of light energy into nerve impulses that are sent to the brain. Although vitamin A's involvement in the process of vision is its best-known function ("Eat your carrots and you'll see better"), only 0.1% of the body's vitamin A is found in the eyes.

2. *Regulating Cell Differentiation.* Cell differentiation is the process whereby immature cells change in structure and function to become specialized cells. For

example, some immature bone marrow cells differentiate into white blood cells and others into red blood cells. In the cellular differentiation process, vitamin A binds to protein receptors; these vitamin A–protein receptor complexes then bind to regulatory regions of DNA molecules.

3. *Maintenance of the Health of Epithelial Tissues.* Epithelial tissue covers outer body surfaces as well as lining internal cavities and tubes. It includes skin and the linings of the mouth, stomach, lungs, vagina, and bladder. Lack of vitamin A causes such surfaces to become drier and harder than normal. Vitamin A's role here is related to cellular differentiation involving mucus-secreting cells.

4. *Reproduction and Growth.* In men, vitamin A participates in sperm development. In women, normal fetal development during pregnancy requires vitamin A. Again vitamin A's role is related to cellular differentiation processes.

■ Vitamin D

The two most important members of the vitamin D family of molecules are vitamin D_3 (cholecalciferol) and vitamin D_2 (ergocalciferol). Vitamin D_3 is produced in the skin of humans and animals by the action of sunlight (ultraviolet light) on its precursor molecule, the cholesterol derivative 7-dehydrocholesterol (a normal metabolite of cholesterol found in the skin). Absorption of light energy induces breakage of the 9, 10 carbon bond; a spontaneous isomerization (shifting of double bonds) then occurs (see Figure 21.19).

Vitamin D_2 (ergocalciferol) differs from vitamin D_3 only in the side-chain structure. It is produced from the plant sterol ergosterol through the action of light.

Vitamin D_3 (cholecalciferol) is sometimes called the "sunshine vitamin" because of its synthesis in the skin by sunlight irradiation.

FIGURE 21.19 The quantity of vitamin D synthesized by exposure of the skin to sunlight (ultraviolet radiation) varies with latitude, the length of exposure time, and skin pigmentation. (Darker-skinned people synthesize less vitamin D because the pigmentation filters out ultraviolet light.)

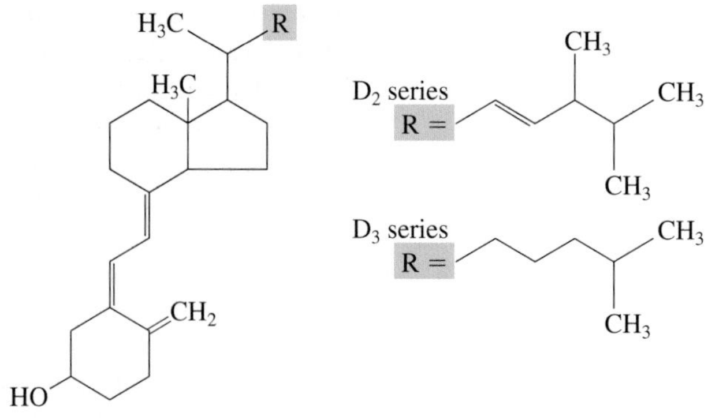

Vitamin D

Both the cholecalciferol and the ergocalciferol forms of vitamin D must undergo two further hydroxylation steps before the vitamin D becomes fully functional. The first step, which occurs in the liver, adds a —OH group to carbon 25. The second step, which occurs in the kidneys, adds a —OH group to carbon 1.

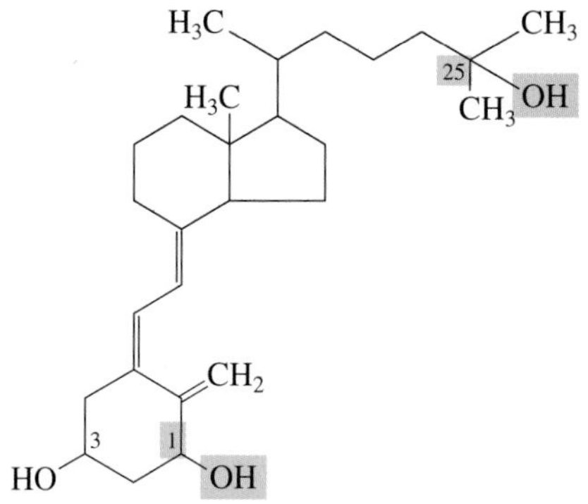

1,25-Dihydroxyvitamin D_3

Only a few foods, including liver, fatty fish (such as salmon), and egg yolks, are good natural sources of vitamin D. Such vitamin D is vitamin D_3. Foods fortified with vitamin D include milk and margarine. The rest of the body's vitamin D supplies are made within the body (skin) with the help of sunlight.

The principal function of vitamin D is to maintain normal blood levels of calcium ion and phosphate ion so that bones can absorb these ions. Vitamin D stimulates absorption of these ions from the gastrointestinal tract and aids in their retention by the kidneys. Vitamin D triggers the deposition of calcium salts into the organic matrix of bones by activating the biosynthesis of calcium-binding proteins.

▶ Milk is enriched in vitamin D by exposure to ultraviolet light. Cholesterol in milk is converted to cholecalciferol (vitamin D) by ultraviolet light.

■ Vitamin E

There are four forms of vitamin E: alpha-, beta-, delta-, and gamma-tocopherol. These forms differ from each other structurally in what substituents (—CH₃ or —H) are present at two positions on an aromatic ring.

▶ The word *tocopherol* is pronounced "tuh-KOFF-er-ol."

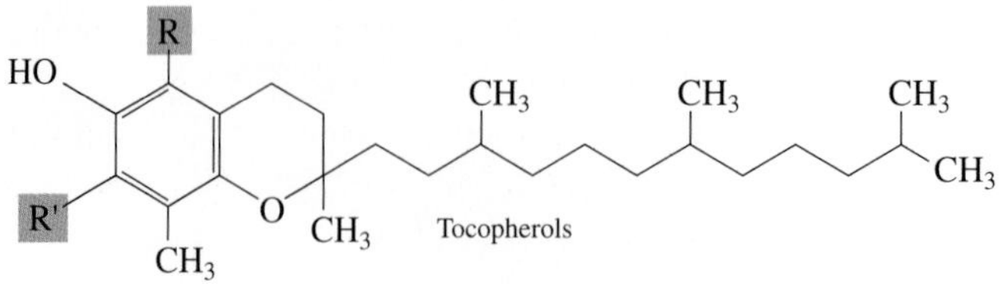

Tocopherols

	R'	R"
α	CH₃	CH₃
β	CH₃	H
γ	H	CH₃
δ	H	H

The tocopherol form with the greatest biological activity is alpha-tocopherol, the vitamin E form in which methyl groups are present at both the R and R′ positions on the aromatic ring.

Vitamin E is unique among the vitamins in that antioxidant activity is its *principal* biochemical role.

Plant oils (margarine, salad dressings, and shortenings), green and leafy vegetables, and whole-grain products are sources of vitamin E.

The primary function of vitamin E in the body is as an antioxidant—a compound that protects other compounds from oxidation by being oxidized itself. Vitamin E is particularly important in preventing the oxidation of polyunsaturated fatty acids (Section 19.2) in membrane lipids. It also protects vitamin A from oxidation. After vitamin E is "spent" as an antioxidant, its antioxidant function can be restored by vitamin C.

A most important location in the human body where vitamin E exerts its antioxidant effect is the lungs, where exposure of cells to oxygen (and air pollutants) is greatest. Both red and white blood cells that pass through the lungs, as well as the cells of the lung tissue itself, benefit from vitamin E's protective effect.

Infants, particularly premature infants, do not have a lot of vitamin E, which is passed from the mother to the infant only in the last weeks of pregnancy. Often, premature infants require oxygen supplementation for the purpose of controlling respiratory distress. In such situations, vitamin E is administered to the infant along with oxygen, to give antioxidant protection.

Vitamin E has also been found to be involved in the conversion of arachidonic acid (20:4) to prostaglandins (Section 19.13.)

Vitamin K

Like the other fat-soluble vitamins, vitamin K occurs in several forms. *Menaquinones* (vitamin K_1) are found in fish oils and meats and are synthesized by bacteria, including those in the human intestinal tract. *Phylloquinones* (vitamin K_2) are found in plants. Typically, about half of the body's vitamin K comes from the diet and half from synthesis by intestinal bacteria. Menaquinones are the form found in vitamin K supplements.

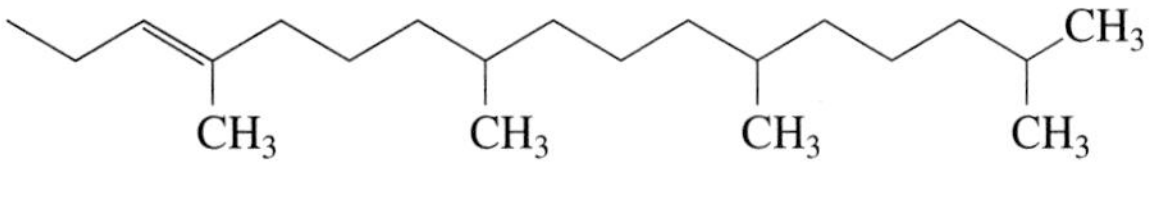

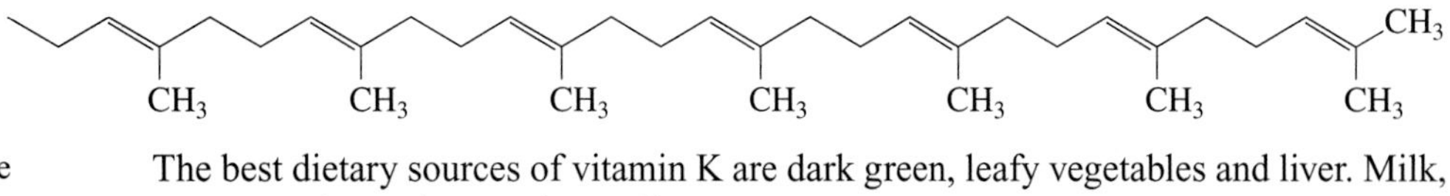

Vitamin K_1 and vitamin K_2 differ structurally in the length and degree of unsaturation of a side chain.

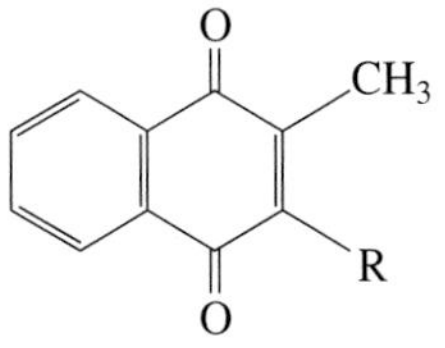

Vitamin K

K_1: R =

K_2: R =

All of the fat-soluble vitamins share a common structural feature; they all have terpenelike structures. That is, they are all made up of five-carbon isoprene units (Section 13.5). No common structural pattern exists for the water-soluble vitamins. On the other hand, the water-soluble vitamins have *functional* uniformity, whereas the fat-soluble vitamins have diverse functions.

The best dietary sources of vitamin K are dark green, leafy vegetables and liver. Milk, meat, eggs, and cereals contain smaller amounts.

Vitamin K is essential to the blood-clotting process. Over a dozen different proteins and the mineral calcium are involved in the formation of a blood clot. Vitamin K is essential for the formation of prothrombin and at least five other proteins involved in the regulation of blood clotting. Vitamin K is sometimes given to presurgical patients to ensure adequate prothrombin levels and prevent hemorrhaging.

Vitamin K is also required for the biosynthesis of several other proteins found in the plasma, bone, and kidney.

CONCEPTS TO REMEMBER

Enzyme. Enzymes are highly specialized protein molecules that act as biochemical catalysts. Enzymes have common names that provide information about their function rather than their structure. The suffix *-ase* is characteristic of most enzyme names.

Enzyme structure. Simple enzymes are composed only of protein (amino acids). Conjugated enzymes have a nonprotein portion (cofactor) in addition to a protein portion (apoenzyme). Cofactors may be small organic molecules (coenzymes) or inorganic ions.

Enzyme classification. There are six classes of enzymes based on function: oxidoreductases, transferases, hydrolases, lyases, isomerases, and ligases.

Enzyme active site. An enzyme active site is the relatively small part of the enzyme that is actually involved in catalysis. It is where substrate binds to the enzyme.

Lock-and-key model of enzyme activity. The active site in an enzyme has a fixed, rigid geometrical conformation. Only substrates with a complementary geometry can be accommodated at the active site.

Induced-fit model of enzyme activity. The active site in an enzyme can undergo small changes in geometry in order to accommodate a series of related substrates.

Enzyme activity. Enzyme activity is a measure of the rate at which an enzyme converts substrate to products. Four factors that affect enzyme activity are temperature, pH, substrate concentration, and enzyme concentration.

Enzyme inhibition. An enzyme inhibitor slows or stops the normal catalytic function of an enzyme by binding to it. Three modes of inhibition are reversible competitive inhibition, reversible noncompetitive inhibition, and irreversible inhibition.

Allosteric enzyme. An allosteric enzyme is an enzyme with two or more protein chains and two kinds of binding sites (for substrate and regulator).

Zymogen. A zymogen is an inactive precursor of a proteolytic enzyme; the zymogen is activated by a chemical reaction that removes or adds to part of its structure.

Vitamin. A vitamin is an organic compound necessary in small amounts for the normal growth of humans and some animals. Vitamins must be obtained from dietary sources, because they cannot be synthesized in the body.

Water-soluble vitamins. Vitamin C and the eight B vitamins are the water-soluble vitamins. Vitamin C is essential for the proper formation of bones and teeth and is also an important antioxidant. All eight B vitamins function as coenzymes.

Fat-soluble vitamins. The four fat-soluble vitamins are vitamins A, D, E, and K. The best-known function of vitamin A is its role in vision. Vitamin D is essential for the proper use of calcium and phosphorus to form bones and teeth. The primary function of vitamin E is as an antioxidant. Vitamin K is essential in the regulation of blood clotting.

KEY REACTIONS AND EQUATIONS

1. Conversion of an apoenzyme to an active enzyme (Section 21.3)

   Apoenzyme + cofactor ⟶ holoenzyme (active enzyme)

2. Mechanism of enzyme action (Section 21.4)

   Enzyme + substrate ⟶ enzyme–substrate complex
   Enzyme–substrate complex ⟶ enzyme + product

3. Enzyme inhibition (Section 21.7)

   Enzyme + inhibitor ⟶ inactive enzyme

4. Conversion of a zymogen to an active enzyme (Section 21.9)

   Zymogen $\xrightarrow{\text{Enzyme}}$ active enzyme + peptide fragment

KEY TERMS

Active site (21.4)
Allosteric enzyme (21.8)
Antibiotic (21.10)
Apoenzyme (21.3)
Coenzyme (21.3)
Cofactor (21.3)
Competitive enzyme inhibitor (21.7)
Conjugated enzyme (21.3)
Enzyme (21.1)
Enzyme activity (21.6)
Enzyme inhibitor (21.7)

Enzyme–substrate complex (21.4)
Feedback control (21.8)
Holoenzyme (21.3)
Hydrolase (21.2)
Irreversible enzyme inhibitor (21.7)
Isoenzymes (21.11)
Isomerase (21.2)
Ligase (21.2)
Lyase (21.2)
Noncompetitive enzyme inhibitor (21.7)
Optimum pH (21.6)

Optimum temperature (21.6)
Oxidoreductase (21.2)
Proteolytic enzyme (21.9)
Simple enzyme (21.3)
Substrate (21.2)
Transferase (21.2)
Turnover number (21.6)
Vitamin (21.12)
Zymogen (21.9)

EXERCISES AND PROBLEMS

The members of each pair of problems in this section test similar material.

■ **Importance of Enzymes (Section 21.1)**

21.1 What is the general role of enzymes in the human body?

21.2 Why does the body need so many different enzymes?

21.3 List two ways in which enzymes differ from inorganic laboratory catalysts.

21.4 Occasionally, we refer to the "delicate" nature of enzymes. Explain why this adjective is appropriate.

■ **Enzyme Nomenclature (Section 21.2)**

21.5 Which of the following substances are enzymes?
a. Sucrase   b. Galactose
c. Trypsin   d. Xylulose reductase

21.6 Which of the following substances are enzymes?
a. Sucrose   b. Pepsin
c. Glutamine synthetase   d. Cellulase

21.7 Predict the function of each of the following enzymes.
a. Pyruvate carboxylase
b. Alcohol dehydrogenase
c. L-Amino acid reductase
d. Maltase

21.8 Predict the function of each of the following enzymes.
a. Cytochrome oxidase
b. *Cis–trans* isomerase
c. Succinate dehydrogenase
d. Lactase

21.9 Suggest a name for an enzyme that catalyzes each of the following reactions.
a. Hydrolysis of sucrose
b. Decarboxylation of pyruvate
c. Isomerization of glucose
d. Removal of hydrogen from lactate

21.10 Suggest a name for an enzyme that catalyzes each of the following reactions.
a. Hydrolysis of lactose
b. Oxidation of nitrite
c. Decarboxylation of citrate
d. Reduction of oxalate

21.11 Give the name of the substrate on which each of the following enzymes acts.
a. Pyruvate carboxylase
b. Galactase
c. Alcohol dehydrogenase
d. L-Amino acid reductase

21.12 Give the name of the substrate on which each of the following enzymes acts.
a. Cytochrome oxidase   b. Lactase
c. Succinate dehydrogenase   d. Tyrosine kinase

21.13 To which of the six major classes of enzymes does each of the following belong?
a. Mutase   b. Dehydratase
c. Carboxylase   d. Kinase

21.14 To which of the six major classes of enzymes does each of the following belong?
a. Protease   b. Racemase
c. Dehydrogenase   d. Synthase

21.15 To which of the six major classes of enzymes does the enzyme that catalyzes each of the following reactions belong?
a. A *cis* double bond is converted to a *trans* double bond.
b. An alcohol is dehydrated to form a compound with a double bond.
c. An amino group is transferred from one substrate to another.
d. An ester linkage is hydrolyzed.

21.16 To which of the six major classes of enzymes does the enzyme that catalyzes each of the following reactions belong?
a. An L isomer is converted to a D isomer.
b. A phosphate group is transferred from one substrate to another.
c. An amide linkage is hydrolyzed.
d. Hydrolysis of a carbohydrate to monosaccharides occurs.

21.17 Identify the enzyme needed in each of the following reactions as an isomerase, a decarboxylase, a dehydrogenase, a lipase, or a phosphatase.

a.
$$CH_3-\overset{\overset{\displaystyle O}{\|}}{C}-COOH \rightarrow CH_3-\overset{\overset{\displaystyle O}{\|}}{C}-H + CO_2$$

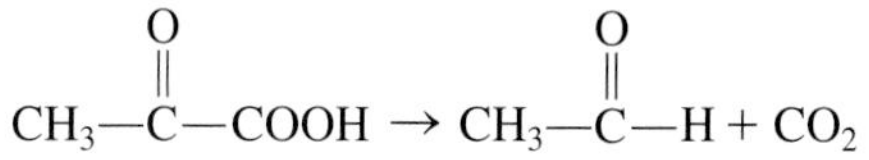

b.
$$\begin{array}{c} CH_2-O-\overset{\overset{\displaystyle O}{\|}}{C}-R \\ | \\ CH-O-\overset{\overset{\displaystyle O}{\|}}{C}-R + 3H_2O \rightarrow \\ | \\ CH_2-O-\overset{\overset{\displaystyle O}{\|}}{C}-R \end{array} \quad \begin{array}{c} CH_2-OH \\ | \\ CH-OH \; + \; 3R-COOH \\ | \\ CH_2-OH \end{array}$$

c.
$$\overset{+}{H_3}N-\underset{\underset{\underset{OPO_3^{2-}}{|}}{\overset{|}{CH_2}}}{CH}-COO^- + H_2O \rightarrow$$

$$\overset{+}{H_3}N-\underset{\underset{\underset{OH}{|}}{\overset{|}{CH_2}}}{CH}-COO^- + HPO_4^{2-}$$

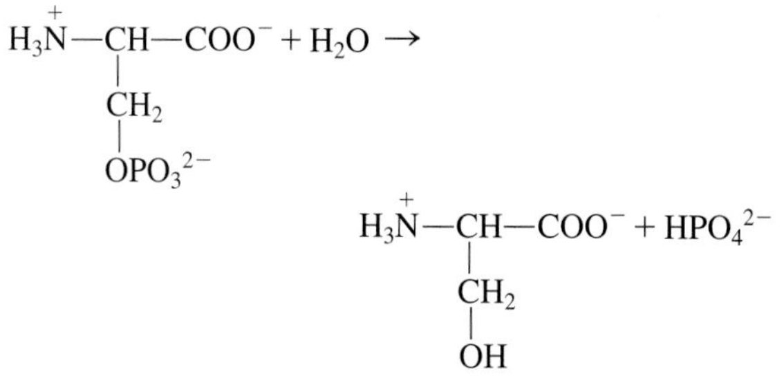

d.
$$CH_3-\underset{\underset{OH}{|}}{CH}-COOH + NAD^+ \rightarrow$$
$$CH_3-\overset{\overset{\displaystyle O}{\|}}{C}-COOH + NADH + H^+$$

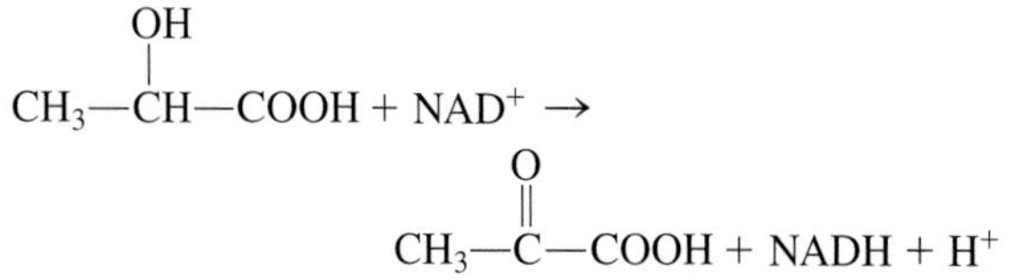

21.18 Identify the enzyme needed in each of the following reactions as an isomerase, a decarboxylase, a dehydrogenase, a protease, or a phosphatase.

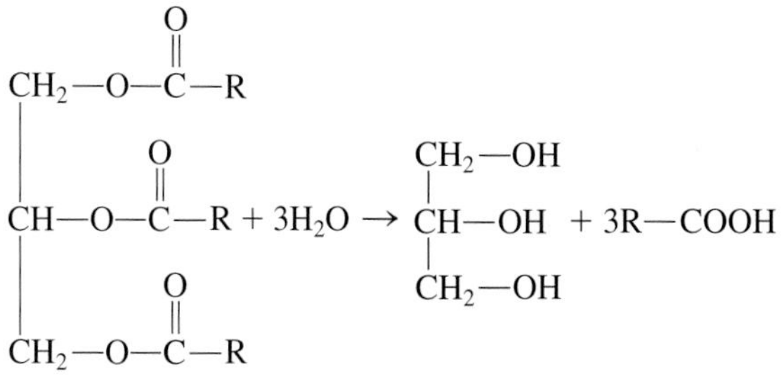

a.
$$CH_3-\overset{\overset{\displaystyle O}{\|}}{C}-\overset{\overset{\displaystyle O}{\|}}{C}-OH \rightarrow CH_3-\overset{\overset{\displaystyle O}{\|}}{C}-H + CO_2$$

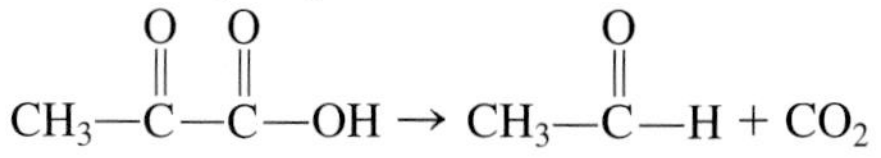

b.

$$
\begin{array}{ccc}
\text{CHO} & & \text{CH}_2\text{OH} \\
| & & | \\
\text{HC—OH} & & \text{C}=\text{O} \\
| & & | \\
\text{HO—CH} & \rightarrow & \text{HO—CH} \qquad + 2\text{H} \\
| & & | \\
\text{HC—OH} & & \text{HC—OH} \\
| & & | \\
\text{HO—CH} & & \text{HO—CH} \\
| & & | \\
\text{CH}_2\text{OPO}_3^{2-} & & \text{CH}_2\text{OPO}_3^{2-}
\end{array}
$$

c.

$$
\text{HO}-\overset{\overset{\displaystyle O}{\|}}{\text{C}}-\text{CH}_2-\text{CH}_2-\overset{\overset{\displaystyle O}{\|}}{\text{C}}-\text{OH} \rightarrow
$$

$$
\text{HO}-\overset{\overset{\displaystyle O}{\|}}{\text{C}}-\text{CH}=\text{CH}-\overset{\overset{\displaystyle O}{\|}}{\text{C}}-\text{OH} + 2\text{H}
$$

d.

$$
\overset{+}{\text{H}_3}\text{N}-\underset{\underset{\displaystyle \text{CH}_3}{|}}{\text{CH}}-\overset{\overset{\displaystyle O}{\|}}{\text{C}}-\text{NH}-\underset{\underset{\displaystyle \text{CH}_3}{|}}{\text{CH}}-\text{COO}^- + \text{H}_2\text{O} \rightarrow
$$

$$
2\ \overset{+}{\text{H}_3}\text{N}-\underset{\underset{\displaystyle \text{CH}_3}{|}}{\text{CH}}-\text{COO}^-
$$

■ Enzyme Structure (Section 21.3)

21.19 Indicate whether each of the following phrases describes a simple or a conjugated enzyme.
  a. An enzyme that has both a protein and a nonprotein portion
  b. An enzyme that requires Mg^{2+} ion for activity
  c. An enzyme in which only amino acids are present
  d. An enzyme in which a cofactor is present

21.20 Indicate whether each of the following phrases describes a simple or a conjugated enzyme.
  a. An enzyme that contains a carbohydrate portion
  b. An enzyme that contains only protein
  c. A holoenzyme
  d. An enzyme that has a vitamin as part of its structure

21.21 What is the difference between a cofactor and a coenzyme?

21.22 All coenzymes are cofactors, but not all cofactors are coenzymes. Explain this statement.

21.23 Why are cofactors present in most enzymes?

21.24 What is the difference between an apoenzyme and a holoenzyme?

■ Models of Enzyme Action (Section 21.4)

21.25 What is an enzyme active site?

21.26 What is an enzyme–substrate complex?

21.27 How does the lock-and-key model of enzyme action explain the highly specific way some enzymes select a substrate?

21.28 How does the induced-fit model of enzyme action explain the broad specificities of some enzymes?

21.29 What types of forces hold a substrate at an enzyme active site?

21.30 The forces that hold a substrate at an enzyme active site are not covalent bonds. Explain why not.

■ Enzyme Specificity (Section 21.5)

21.31 Define the following terms dealing with enzyme specificity.
  a. Absolute specificity
  b. Linkage specificity

21.32 Define the following terms dealing with enzyme specificity.
  a. Group specificity
  b. Stereochemical specificity

21.33 Which type(s) of enzyme specificity are best accounted for by the lock-and-key model of enzyme action?

21.34 Which type(s) of enzyme specificity are best accounted for by the induced-fit model of enzyme action?

21.35 Which enzyme in each of the following pairs would be more limited in its catalytic scope?
  a. An enzyme that exhibits absolute specificity or an enzyme that exhibits group specificity
  b. An enzyme that exhibits stereochemical specificity or an enzyme that exhibits linkage specificity

21.36 Which enzyme in each of the following pairs would be more limited in its catalytic scope?
  a. An enzyme that exhibits linkage specificity or an enzyme that exhibits absolute specificity
  b. An enzyme that exhibits group specificity or an enzyme that exhibits stereochemical specificity

■ Factors That Affect Enzyme Activity (Section 21.6)

21.37 Temperature affects enzymatic reaction rates in two ways. An increase in temperature can accelerate the rate of a reaction or it can stop the reaction. Explain each of these effects.

21.38 Define the optimum temperature for an enzyme.

21.39 Explain why all enzymes do not possess the same optimum pH.

21.40 Why does an enzyme lose activity when the pH is drastically changed from the optimum pH?

21.41 Draw a graph that shows the effect of increasing substrate concentration on the rate of an enzyme-catalyzed reaction (at constant temperature, pH, and enzyme concentration).

21.42 Draw a graph that shows the effect of increasing enzyme concentration on the rate of an enzyme-catalyzed reaction (at constant temperature, pH, and substrate concentration).

21.43 In an enzyme-catalyzed reaction, all of the enzyme active sites are saturated by substrate molecules at a certain substrate concentration. What happens to the rate of the reaction when the substrate concentration is doubled?

21.44 What is an enzyme turnover number?

■ Enzyme Inhibition (Section 21.7)

21.45 In competitive inhibition, can both the inhibitor and the substrate bind to an enzyme at the same time? Explain your answer.

21.46 Compare the sites where competitive and noncompetitive inhibitors bind to enzymes.

21.47 Indicate whether each of the following statements describes a reversible competitive inhibitor, a reversible noncompetitive inhibitor, or an irreversible inhibitor. More than one answer may apply.
  a. Both inhibitor and substrate bind at the active site on a random basis.
  b. The inhibitor effect cannot be reversed by the addition of more substrate.
  c. Inhibitor structure does not have to resemble substrate structure.
  d. The inhibitor can bind to the enzyme at the same time as substrate.

21.48 Indicate whether each of the following statements describes a reversible competitive inhibitor, a reversible noncompetitive

inhibitor, or an irreversible inhibitor. More than one answer may apply.
a. It bonds covalently to the enzyme active site.
b. The inhibitor effect can be reversed by the addition of more substrate.
c. Inhibitor structure must be somewhat similar to that of substrate.
d. The inhibitor cannot bind to the enzyme at the same time as substrate.

■ **Regulation of Enzyme Activity (Sections 21.8 and 21.9)**

21.49 What is an allosteric enzyme?

21.50 What is a regulator molecule?

21.51 What is feedback control?

21.52 What is the difference between positive and negative feedback to an allosteric enzyme?

21.53 What is the general relationship between zymogens and proteolytic enzymes?

21.54 What, if any, is the difference in meaning between the terms *zymogen* and *proenzyme*?

21.55 Why are proteolytic enzymes always produced in an inactive form?

21.56 What is the mechanism by which most zymogens are activated?

■ **Antibiotics That Inhibit Enzyme Activity (Section 21.10)**

21.57 By what mechanism do sulfa drugs kill bacteria?

21.58 By what mechanism do penicillins kill bacteria?

21.59 Why is penicillin toxic to bacteria but not to higher organisms?

21.60 What amino acid in transpeptidase forms a covalent bond to penicillin?

21.61 What situation has made Cipro a prominent antibiotic?

21.62 Describe the structure of Cipro in terms of common and "unusual" functional groups present.

■ **Vitamins (Sections 21.12–21.14)**

21.63 What is a vitamin?

21.64 List a way in which vitamins differ from carbohydrates, fats, and proteins (the major classes of food).

21.65 Indicate whether each of the following is a fat-soluble or a water-soluble vitamin.
a. Vitamin K   b. Vitamin B_{12}
c. Vitamin C   d. Thiamin

21.66 Indicate whether each of the following is a fat-soluble or a water-soluble vitamin.
a. Vitamin A   b. Vitamin B_6
c. Vitamin E   d. Riboflavin

21.67 Indicate whether each of the vitamins in Problem 21.65 would be likely or unlikely to be toxic when consumed in excess.

21.68 Indicate whether each of the vitamins in Problem 21.66 would be likely or unlikely to be toxic when consumed in excess.

21.69 Describe the two most completely characterized roles of vitamin C in the body.

21.70 Structurally, how do the oxidized and reduced forms of vitamin C differ?

21.71 What is the dominant function within the human body of the B vitamins as a group?

21.72 With the help of Table 21.6, identify the B vitamin or vitamins to which each of the following characterizations applies.
a. Is part of the coenzymes NAD and NADP
b. Is part of coenzyme A
c. Is part of THF
d. Is part of TPP

21.73 With the help of Table 21.6, indicate whether each of the following B vitamins exists in more than one structural form.
a. Folate   b. Niacin
c. Vitamin B_6 d. Biotin

21.74 With the help of Table 21.6, identify the B vitamin or vitamins to which each of the following characterizations applies.
a. Contain a heterocyclic nitrogen ring system
b. Has the most complex structure
c. Contains a metal atom as part of its structure
d. Contains sulfur as part of its structure

21.75 Describe the structural differences among the three retinoic forms of vitamin A.

21.76 What is the relationship between the plant pigment beta-carotene and vitamin A?

21.77 What is *cell differentiation* and how does vitamin A participate in this process?

21.78 List four major functions for vitamin A in the human body.

21.79 How do vitamin D_2 and vitamin D_3 differ in structure?

21.80 In terms of source, how do vitamin D_2 and vitamin D_3 differ?

21.81 What is the principal function of vitamin D in the human body?

21.82 Why is vitamin D often called the sunshine vitamin?

21.83 Which form of tocopherol (vitamin E) exhibits the greatest biochemical activity?

21.84 How do the various forms of tocopherol differ in structure?

21.85 What is the principal function of vitamin E in the human body?

21.86 Why is vitamin E often given to premature infants that are on oxygen therapy?

21.87 How do vitamin K_1 and vitamin K_2 differ in structure?

21.88 In terms of source, how do vitamin K_1 and vitamin K_2 differ?

21.89 How are *menaquinones, phylloquinones,* and vitamin K related?

21.90 What is the principal function of vitamin K in the human body?

ADDITIONAL PROBLEMS

21.91 Explain the difference, if any, between the following types of enzymes.
a. Apoenzyme and proenzyme
b. Simple enzyme and allosteric enzyme
c. Coenzyme and isoenzyme
d. Conjugated enzyme and holoenzyme

21.92 Identify the functional groups present in a molecule of each of the following vitamins.
a. Vitamin C   b. Vitamin A (retinol)
c. Vitamin D   d. Vitamin K

21.93 Indicate whether each of the following vitamins functions as a coenzyme.

a Vitamin C   b. Vitamin A
c. Vitamin D   d. Niacin
e. Riboflavin   f. Biotin

21.94 Which vitamin has each of the following functions?
   a. Water-soluble antioxidant
   b. Fat-soluble antioxidant
   c. Involved in the process of calcium deposition in bone
   d. Involved in the blood-clotting process
   e. Involved in cell differentiation
   f. Involved in vision
   g. Involved in collagen formation
   h. Involved in prostaglandin formation

21.95 What general kinds of reactions do the following types of enzymes catalyze?
   a. Oxidoreductases   b. Lyases
   c. Isomerases   d. Ligases
   e. Hydrolases   f. Transferases

21.96 Explain what is meant by the equation

$$E + S \rightleftharpoons ES \rightarrow E + P$$

given that ES stands for enzyme–substrate complex.

21.97 Alcohol dehydrogenase catalyzes the conversion of ethanol to acetaldehyde. This enzyme, in its active state, consists of a protein molecule and a zinc ion. On the basis of this information, identify the following for this chemical system.
   a. Substrate   b. Cofactor
   c. Apoenzyme   d. Holoenzyme

21.98 Each of the following is an abbreviation for an enzyme used in the diagnosis and/or treatment of heart attacks. What does each abbreviation stand for?
   a. TPA   b. LDH
   c. CPK   d. AST

ANSWER TO PRACTICE EXERCISE

21.1 a. hydrolysis of maltose
   b. removal of hydrogen from lactate ion
   c. oxidation of fructose
   d. rearrangement (isomerization) of maleate ion

22

Nucleic Acids

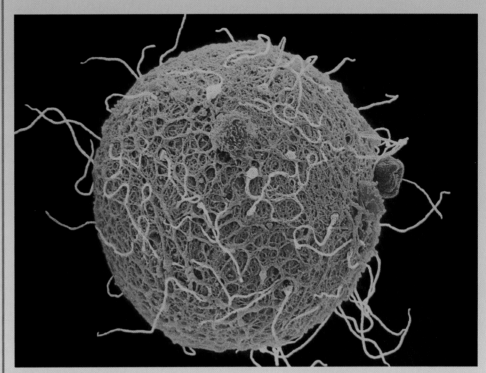

Human egg and sperm.

A most remarkable property of living cells is their ability to produce exact replicas of themselves. Furthermore, cells contain all the instructions needed for making the complete organism of which they are a part. The molecules within a cell that are responsible for these amazing capabilities are nucleic acids.

The Swiss physiologist Friedrich Miescher (1844–1895) discovered nucleic acids in 1869 while studying the nuclei of white blood cells. The fact that they were initially found in cell nuclei and are acidic accounts for the name *nucleic acid*. Although we now know that nucleic acids are found throughout a cell, not just in the nucleus, the name is still used for such materials.

22.1 Types of Nucleic Acids

Two types of nucleic acids are found within cells of higher organisms: *deoxyribonucleic acid* (DNA) and *ribonucleic acid* (RNA). Nearly all the DNA is found within the cell nucleus. Its primary function is the storage and transfer of genetic information. This information is used (indirectly) to control many functions of a living cell. In addition, DNA is passed from existing cells to new cells during cell division. RNA occurs in all parts of a cell. It functions primarily in synthesis of proteins, the molecules that carry out essential cellular functions. The structural distinctions between DNA and RNA molecules are considered in Section 22.2.

All nucleic acid molecules are polymers. A **nucleic acid** *is a polymer in which the monomer units are nucleotides.* Thus the starting point for a discussion of nucleic acids is an understanding of the structures and chemical properties of nucleotides.

▶ It was not until 1944, 75 years after the discovery of nucleic acids, that scientists obtained the first evidence that these molecules are responsible for the storage and transfer of genetic information.

655

22.2 Nucleotides: Building Blocks of Nucleic Acids

Proteins are polypeptides, many carbohydrates are polysaccharides, and nucleic acids are polynucleotides.

A **nucleotide** *is a three-subunit molecule in which a pentose sugar is bonded to both a phosphate group and a nitrogen-containing heterocyclic base.* With a three-subunit structure, nucleotides are more complex monomers than the monosaccharides of polysaccharides (Section 18.8) and the amino acids of proteins (Section 20.2). A block structural diagram for a nucleotide is

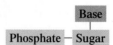

■ Pentose Sugars

The sugar unit of a nucleotide is either the pentose *ribose* or the pentose *2-deoxyribose.*

The systems for numbering the atoms in the pentose and nitrogen-containing base subunits of a nucleotide are important and will be used extensively in later sections of this chapter. The convention is that

1. Pentose ring atoms are designated with *primed* numbers.
2. Nitrogen-containing base ring atoms are designated with *unprimed* numbers.

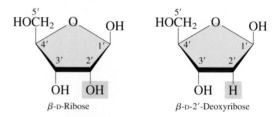

β-D-Ribose β-D-2'-Deoxyribose

Structurally, the only difference between these two sugars occurs at carbon 2′. The —OH group present on this carbon in ribose becomes a —H atom in 2-deoxyribose. (The prefix *deoxy-* means "without oxygen.")

RNA and DNA differ in the identity of the sugar unit in their nucleotides. In RNA the sugar unit is *r*ibose—hence the *R* in RNA. In DNA the sugar unit is 2-*d*eoxyribose—hence the *D* in DNA.

■ Nitrogen-Containing Heterocyclic Bases

A pyrimidine derivative that we have encountered previously is the B vitamin thiamin (see Section 21.13).

Caffeine, the most widely used nonprescription central nervous system stimulant, is the 3,7-dimethyl-2,6-dioxo derivative of purine (Section 17.8).

Five nitrogen-containing heterocyclic bases are nucleotide components. Three of them are derivatives of pyrimidine (Section 17.8), a monocyclic base with a six-membered ring, and two are derivatives of purine (Section 17.8), a bicyclic base with fused five- and six-membered rings.

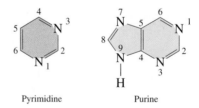

Pyrimidine   Purine

Both of these heterocyclic compounds are bases because they contain amine functional groups (secondary or tertiary), and amine functional groups exhibit basic behavior (proton acceptors; Section 17.5).

The three pyrimidine derivatives found in nucleotides are thymine (T), cytosine (C), and uracil (U).

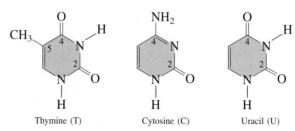

Thymine (T)   Cytosine (C)   Uracil (U)

Thymine is the 5-methyl-2,4-dioxo derivative, cytosine the 4-amino-2-oxo derivative, and uracil the 2,4-dioxo derivative of pyrimidine.

The two purine derivatives found in nucleotides are adenine (A) and guanine (G).

Adenine (A)   Guanine (G)

FIGURE 22.1 Space-filling model of the molecule adenine, a nucleotide present in both DNA and RNA.

Adenine is the 6-amino derivative of purine, and guanine is the 2-amino-6-oxo purine derivative. A space-filling model for adenine is shown in Figure 22.1.

Adenine, guanine, and cytosine are found in both DNA and RNA. Uracil is found only in RNA, and thymine usually occurs only in DNA. Figure 22.2 summarizes the occurrences of nitrogen-containing heterocyclic bases in nucleic acids.

■ Phosphate

Phosphate, the third component of a nucleotide, is derived from phosphoric acid (H_3PO_4). Under cellular pH conditions, the phosphoric acid loses two of its hydrogen atoms to give a -2-charged hydrogen phosphate ion.

$$O{=}\overset{\overset{\textstyle OH}{|}}{\underset{\underset{\textstyle OH}{|}}{P}}{-}OH \;\rightleftharpoons\; O{=}\overset{\overset{\textstyle O^-}{|}}{\underset{\underset{\textstyle O^-}{|}}{P}}{-}OH + 2H^+$$

Phosphoric acid   Hydrogen phosphate ion

■ Nucleotide Formation

The formation of a nucleotide from sugar, base, and phosphate can be visualized as occurring in the following manner:

Base

Phosphate   Sugar   Nucleotide

Important characteristics of this combining of three molecules into one molecule (the nucleotide) are that

1. Condensation, with formation of a water molecule, occurs at two locations: between sugar and base and between sugar and phosphate.
2. The base is always attached at the C-1′ position of the sugar. For purine bases, attachment is through N-9; for pyrimidine bases, N-1 is involved. The C-1′ carbon atom of the ribose unit is always in a β configuration (Section 18.10), and the bond connecting the sugar and base is a β-N-glycosidic linkage (Section 18.13).

FIGURE 22.2 Two purine bases and three pyrimidine bases are found in the nucleotides present in nucleic acids.

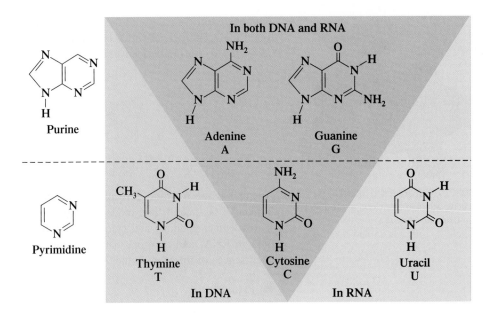

To remember which two of the five nucleotide bases are the purine derivatives (fused rings), use the phrase "pure silver" and the chemical symbol for silver, which is Ag.

pure Ag

purine A and G

3. The phosphate group is attached to the sugar at the C-5′ position through a phosphate–ester linkage.

There are four possible RNA nucleotides, differing in the base present (A, C, G, or U) and four possible DNA nucleotides, differing in the base present (A, C, G, or T).

■ Nucleotide Nomenclature

The common names and abbreviations for the eight nucleotides of DNA and RNA molecules are given in Table 22.1. It is important to be familiar with them, because they are frequently encountered in biochemistry.

We can make several generalizations about the nomenclature given in Table 22.1.

1. All of the names end in 5′-monophosphate, which signifies the presence of a phosphate group attached to the 5′ carbon atom of ribose or deoxyribose. (In Chapter 23 we will encounter nucleotides that contain two or three phosphate groups—diphosphates and triphosphates.)
2. Preceding the monophosphate ending is the name of the base present in a modified form. The suffix -osine is used with purine bases, the suffix -idine with pyrimidine bases.
3. The prefix deoxy- at the start of the name signifies that the sugar present is deoxyribose. When no prefix is present, the sugar is ribose.

TABLE 22.1
The Names of the Eight Nucleotides Found in DNA and RNA

Base	Sugar	Nucleotide name	Nucleotide abbreviation
DNA Nucleotides			
adenine	deoxyribose	deoxyadenosine 5′-monophosphate	dAMP
guanine	deoxyribose	deoxyguanosine 5′-monophosphate	dGMP
cytosine	deoxyribose	deoxycytidine 5′-monophosphate	dCMP
thymine	deoxyribose	deoxythymidine 5′-monophosphate	dTMP
RNA Nucleotides			
adenine	ribose	adenosine 5′-monophosphate	AMP
guanine	ribose	guanosine 5′-monophosphate	GMP
cytosine	ribose	cytidine 5′-monophosphate	CMP
uracil	ribose	uridine 5′-monophosphate	UMP

Use of Synthetic Nucleic Acid Bases in Medicine

Many hundreds of modified nucleic acid bases have been prepared in laboratories, and their effects on nucleic acid synthesis investigated. Several of them are now in clinical use as drugs for controlling, at the cellular level, cancers and other related disorders.

The theory behind the use of these modified bases involves their masquerading as legitimate nucleic acid building blocks. The enzymes associated with the DNA replication process (Section 22.5) incorporate the modified bases into growing nucleic acid chains. The presence of these "pseudonucleotides" in the chain stops further growth of the chain, thus interfering with nucleic acid synthesis.

Examples of drugs now in use include 5-fluorouracil, which is employed against a variety of cancers, especially those of the breast and digestive tract, and 6-mercaptopurine, which is used in the treatment of leukemia.

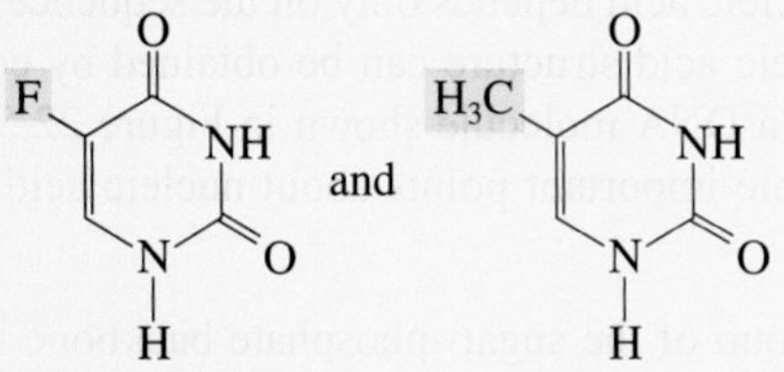

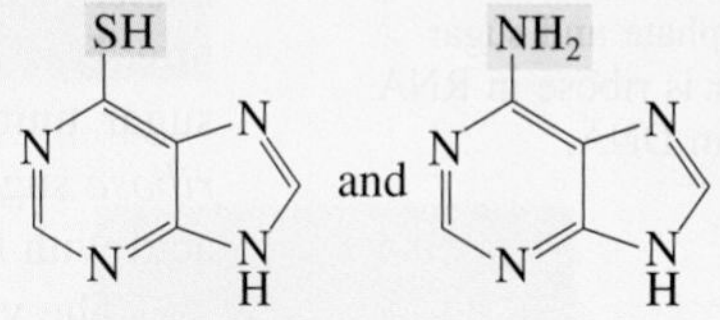

6-Mercaptopurine (a modified adenine)   and   Adenine

The rapidly dividing cells that are characteristic of cancer require large quantities of DNA. Anticancer drugs based on modified nucleic acid bases block DNA synthesis and therefore block the increase in the number of cancer cells. Cancer cells are generally affected to a greater extent than normal cells because of this rapid growth. Eventually, the normal cells are affected to such a degree that use of the drugs must be discontinued. 5-Fluorouracil inhibits the formation of thymine-containing nucleotides required for DNA synthesis. 6-Mercaptopurine, which substitutes for adenine, inhibits the synthesis of nucleotides that incorporate adenine and guanine.

5-Fluorouracil (a modified thymine)   and   Thymine

4. The abbreviations in Table 22.1 for the nucleotides come from the one-letter symbols for the bases (A, C, G, T, and U), the use of MP for monophosphate, and a lower-case *d* at the start of the abbreviation whenever deoxyribose is the sugar.

The use of synthetic nucleic acid bases in medicine is considered in the Chemical Connections feature on this page.

22.3 Primary Nucleic Acid Structure

Nucleotides are related to nucleic acids in the same way that amino acids are related to proteins.

Nucleic acids are polymers in which the repeating units, the monomers, are nucleotides (Section 22.2). The nucleotide units within a nucleic acid molecule are linked to each other through sugar–phosphate bonds. The resulting molecular structure (Figure 22.3) involves a chain of alternating sugar and phosphate groups with a base group protruding from the chain at regular intervals.

We can now define, in terms of structure, the two major types of nucleic acids: ribonucleic acids and deoxyribonucleic acids (Section 22.1). A **ribonucleic acid (RNA)** *is a nucleotide polymer in which each of the monomers contains ribose, a phosphate group, and one of the heterocyclic bases adenine, cytosine, guanine, or uracil.* Two changes to this definition generate the deoxyribonucleic acid definition; deoxyribose replaces ribose

FIGURE 22.3 The general structure of a nucleic acid in terms of nucleotide subunits.

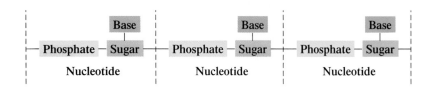

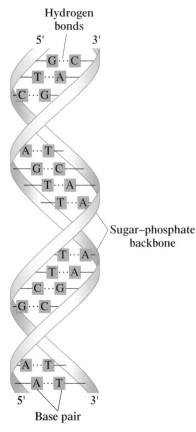

Hydrogen bonds

Sugar–phosphate backbone

Base pair

FIGURE 22.7 A schematic drawing of the DNA double helix that emphasizes the hydrogen bonding between bases on the two chains.

▶ The α helix secondary structure of proteins involves *one* polypeptide chain; the double-helix secondary structure of DNA involves *two* polynucleotide chains. In the α helix of proteins, the R groups are on the *outside* of the helix; in the double helix of DNA, the bases are on the

▶ The *antiparallel* nature of the two polynucleotide chains in the DNA double helix means that there are a 5′ end and a 3′ end at both ends of the double helix.

▶ A mnemonic device for recalling base-pairing combinations in DNA and RNA involves listing the base abbreviations in alphabetical order. Then the first and last bases pair, and so do the middle two bases.

DNA: A C G T

RNA: A C G U

Another way to remember these base-pairing combinations is to note that AT spells a word and that C and G look very much alike.

2. The differences among various nucleic acids and among various proteins are related to the order in which groups are attached to the backbones (nitrogen bases in nucleic acids and amino acid R groups in proteins).
3. Both nucleic acid polymer chains and protein polymer chains have directionality; for nucleic acids there are a 5′ end and a 3′ end, and for proteins there are an N-terminal end and a C-terminal end.

22.4 The DNA Double Helix

Like proteins, nucleic acids have secondary, or three-dimensional, structure as well as primary structure. The secondary structures of DNAs and RNAs differ, and we will discuss them separately.

The amounts of the bases A, T, G, and C present in DNA molecules were the key to determination of the general three-dimensional structure of DNA molecules. Base composition data for DNA molecules from many different organisms revealed a definite pattern of base occurrence. The amounts of A and T were always equal, and the amounts of C and G were always equal, as were the amounts of total purines and total pyrimidines.

The relative amounts of these base pairs in DNA vary depending on the life form from which the DNA is obtained. (Each animal or plant has a unique base composition.) However, the relationships

$$\%A = \%T \qquad \text{and} \qquad \%C = \%G$$

always hold true. For example, human DNA contains 30% adenine, 30% thymine, 20% guanine, and 20% cytosine.

In 1953, an explanation for the base composition patterns associated with DNA molecules was proposed by the American microbiologist James Watson and the English biophysicist Francis Crick. Their model, which has now been validated in numerous ways, involves a double-helix structure that accounts for the equality of bases present, as well as for all other known DNA structural data.

The DNA double helix involves two polynucleotide strands coiled around each other in a manner somewhat like a spiral staircase. The sugar–phosphate backbones of the two polynucleotide strands can be thought of as being the outside banisters of the spiral staircase (see Figure 22.7). The bases (side chains) of each backbone extend inward toward the bases of the other strand. The two strands are connected by *hydrogen bonds* (Section 7.13) between their bases. Additionally, the two strands of the double helix are *antiparallel*—that is, they run in opposite directions. One strand runs in the 5′-to-3′ direction, and the other is oriented in the 3′-to-5′ direction.

■ Base Pairing

A physical restriction, the size of the interior of the DNA double helix, limits the base pairs that can hydrogen-bond to one another. Only pairs involving one small base (a pyrimidine) and one large base (a purine) correctly "fit" within the helix interior. There is not enough room for two large purine bases to fit opposite each other (they overlap), and two small pyrimidine bases are too far apart to hydrogen-bond to one another effectively. Of the four possible purine–pyrimidine combinations (A–T, A–C, G–T, and G–C), hydrogen-bonding possibilities are *most favorable* for the A–T and G–C pairings, and these two combinations are the *only two* that normally occur in DNA. Figure 22.8 shows the specific hydrogen-bonding interactions for the four possible purine–pyrimidine base-pairing combinations.

The pairing of A with T and that of G with C are said to be *complementary*. A and T are complementary bases, as are G and C. **Complementary bases** *are pairs of bases in a nucleic acid structure that can hydrogen-bond to each other*. The fact that complementary base pairing occurs in DNA molecules explains, very simply, why the amounts of the bases A and T present are always equal, as are the amounts of G and C.

FIGURE 22.8 Hydrogen-bonding possibilities are more favorable when A–T and G–C base pairing occurs than when A–C and G–T base pairing occurs. (a) Two and three hydrogen bonds can form, respectively, between A–T and G–C base pairs. These combinations are present in DNA molecules. (b) Only one hydrogen bond can form between G–T and A–C base pairs. These combinations are not present in DNA molecules.

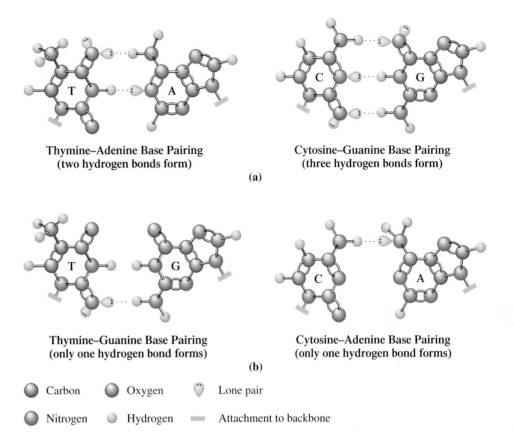

Thymine–Adenine Base Pairing
(two hydrogen bonds form)

Cytosine–Guanine Base Pairing
(three hydrogen bonds form)

(a)

Thymine–Guanine Base Pairing
(only one hydrogen bond forms)

Cytosine–Adenine Base Pairing
(only one hydrogen bond forms)

(b)

Carbon   Oxygen   Lone pair

Nitrogen   Hydrogen   Attachment to backbone

▶ Hydrogen bonding is responsible for the secondary structure (double helix) of DNA. Hydrogen bonding is also responsible for secondary structure in proteins (Section 20.10).

▶ The two strands of DNA in a double helix are complementary. This means that if you know the order of bases in one strand, you can predict the order of bases in the other strand.

Hydrogen bonding holds the two strands of the DNA double helix together. Although hydrogen bonds are relatively weak forces, each DNA molecule contains so many base pairs that the hydrogen-bonding attractions are sufficient in magnitude, collectively, to prevent the two entwined DNA strands from separating spontaneously under normal physiological conditions.

The two strands of DNA in a double helix are *not identical*—they are complementary. **Complementary DNA strands** *are strands of DNA in a double helix with base pairing such that each base is located opposite its complementary base.* Wherever G occurs in one strand, there is a C in the other strand; wherever T occurs in one strand, there is an A in the other strand. An important ramification of this complementary relationship is that knowing the base sequence of one strand of DNA enables us to predict the base sequence of the complementary strand.

In specifying the base sequence of a segment of a strand of DNA (or RNA), we list the bases in sequential order (using their one-letter abbreviations) in the direction from the 5′ end to the 3′ end of the segment.

$$5' \text{ A–A–G–C–T–A–G–C–T–T–A–C–T } 3'$$

EXAMPLE 22.1

Predicting Base Sequence in a Complementary DNA Strand

■ Predict the sequence of bases in the DNA strand that is complementary to the single DNA strand shown.

$$5' \text{ C–G–A–A–T–C–C–T–A } 3'$$

Solution

Because only A forms a complementary base pair with T, and only G with C, the complementary strand is as follows:

Given: 5′ C–G–A–A–T–C–C–T–T–A 3′

Complementary strand: 3′ G–C–T–T–A–G–G–A–T 5′

(continued)

Note the reversal of the numbering of the ends of the complementary strand compared to the given strand. This is due to the antiparallel nature of the two strands in a DNA double helix.

Practice Exercise 22.1

Predict the sequence of bases in the DNA strand complementary to the single DNA strand shown.

5′ A–A–T–G–C–A–G–C–T 3′

22.5 Replication of DNA Molecules

DNA molecules are the carriers of the genetic information within a cell; that is, they are the molecules of heredity. Each time a cell divides, an exact copy of the DNA of the parent cell is needed for the new daughter cell. The process by which new DNA molecules are generated is DNA replication. **DNA replication** *is the biochemical process by which DNA molecules produce exact duplicates of themselves.* The key concept in understanding DNA replication is the base pairing associated with the DNA double helix.

■ DNA Replication Overview

To understand DNA replication, we must regard the two strands of the DNA double helix as a pair of *templates,* or patterns. During replication, the strands separate. Each can then act as a template for the synthesis of a new, complementary strand. The result is two daughter DNA molecules with base sequences identical to those of the parent double helix. Let us consider details of this replication.

Under the influence of the enzyme *DNA helicase,* the DNA double helix unwinds, and the hydrogen bonds between complementary bases are broken. This unwinding process, as shown in Figure 22.9, is somewhat like opening a zipper.

The bases of the separated strands are no longer connected by hydrogen bonds. They can pair with *free* individual nucleotides present in the cell's nucleus. As shown in Figure 22.9, the base pairing always involves C pairing with G and A pairing with T. After the free nucleotides have formed hydrogen bonds with the old strand (the template), the enzyme *DNA polymerase* verifies that the base pairing is correct and catalyzes the formation of new phosphodiester linkages between nucleotides (represented by colored ribbons in Figure 22.9).

Each of the two daughter molecules of double-stranded DNA formed in the DNA replication process contains one strand from the original parent molecule and one newly formed strand.

FIGURE 22.9 In DNA replication, the two strands of the DNA double helix unwind, the separated strands serving as templates for the formation of new DNA strands. Free nucleotides pair with the complementary bases on the separated strands of DNA. This process ultimately results in the complete replication of the DNA molecule.

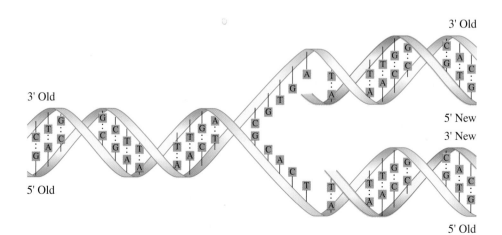

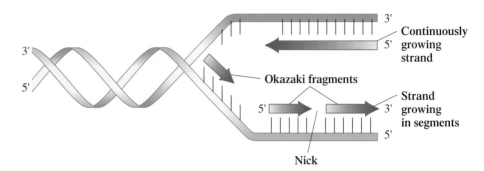

FIGURE 22.10 Because the enzyme *DNA polymerase* can act only in the 5′-to-3′ direction, one strand (top) grows continuously in the direction of the unwinding, and the other strand grows in segments in the opposite direction. The segments in this latter chain are then connected by a different enzyme, *DNA ligase*.

■ The Replication Process in Finer Detail

Though simple in principle, the DNA replication process has many intricacies.

1. The enzyme *DNA polymerase* can operate on a forming DNA daughter strand only in the 5′-to-3′ direction. Because the two strands of parent DNA run in opposite directions (one is 5′ to 3′ and the other 3′ to 5′; Section 22.4), only one strand can grow continuously in the 5′-to-3′ direction. The other strand must be formed in short segments, called *Okazaki fragments* (after their discoverer, Reiji Okazaki), as the DNA unwinds (see Figure 22.10). The breaks or gaps in this daughter strand are called *nicks*. To complete the formation of this strand, the Okazaki fragments are connected by action of the enzyme *DNA ligase.*

2. The process of DNA unwinding does not have to begin at an end of the DNA molecule. It may occur at any location within the molecule. Indeed, studies show that unwinding usually occurs at several interior locations simultaneously and that DNA replication is bidirectional for these locations; that is, it proceeds in both directions from the unwinding sites. As shown in Figure 22.11, the result of this multiple-site replication process is formation of "bubbles" of newly synthesized DNA. The bubbles grow larger and eventually coalesce, giving rise to two complete daughter DNAs. Multiple-site replication enables large DNA molecules to be replicated rapidly.

FIGURE 22.11 DNA replication usually occurs at multiple sites within a molecule, and the replication is bidirectional from these sites.

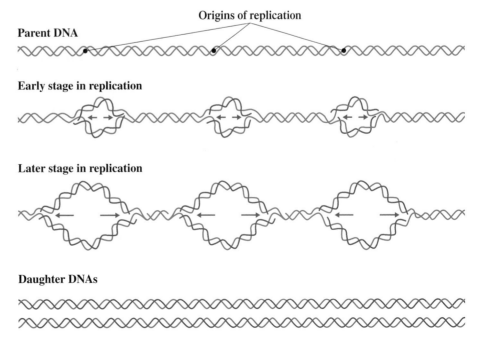

FIGURE 22.12 Identical twins share identical physical characteristics because they received identical DNA from their parents.

Chromosomes are *nucleoproteins.* They are a combination of nucleic acid (DNA) and various proteins.

■ Chromosomes

Once the DNA within a cell has been replicated, it interacts with specific proteins in the cell called *histones* to form structural units that provide the most stable arrangement for the long DNA molecules. These histone–DNA complexes are called *chromosomes.* A **chromosome** *is an individual DNA molecule bound to a group of proteins.* Typically, a chromosome is about 15% by mass DNA and 85% by mass protein.

Cells from different kinds of organisms have different numbers of chromosomes. A normal human has 46 chromosomes per cell, a mosquito 6, a frog 26, a dog 78, and a turkey 82.

Chromosomes occur in matched (*homologous*) pairs. The 46 chromosomes of a human cell constitute 23 homologous pairs. One member of each homologous pair is derived from a chromosome inherited from the father, and the other is a copy of one of the chromosomes inherited from the mother. Homologous chromosomes have similar, but not identical, DNA base sequences; both code for the same traits but for different forms of the trait (for example, blue eyes versus brown eyes). Offspring are like their parents, but they are different as well; part of their DNA came from one parent and part from the other parent. Occasionally, identical twins are born (see Figure 22.12). Such twins have received identical DNA from their parents.

The Chemistry at a Glance feature on page 667 summarizes the steps in DNA replication.

22.6 Overview of Protein Synthesis

We saw in the previous section how the replication of DNA makes it possible for a new cell to contain the same genetic information as its parent cell. We will now consider how the genetic information contained in a cell is expressed in cell operation. This brings us to the topic of protein synthesis. The synthesis of proteins (skin, hair, enzymes, hormones, and so on) is under the direction of DNA molecules. It is this role of DNA that establishes the similarities between parent and offspring that we regard as hereditary characteristics.

We can divide the overall process of protein synthesis into two steps. The first step is called transcription and the second translation. The following diagram summarizes the relationship between transcription and translation.

$$\boxed{\text{DNA}} \xrightarrow{\text{Transcription}} \boxed{\text{RNA}} \xrightarrow{\text{Translation}} \boxed{\text{protein}}$$

Before discussing the details of transcription and translation, we need to learn more about RNA molecules. We will be particularly concerned with differences between RNA and DNA and among various types of RNA molecules.

DNA Replication

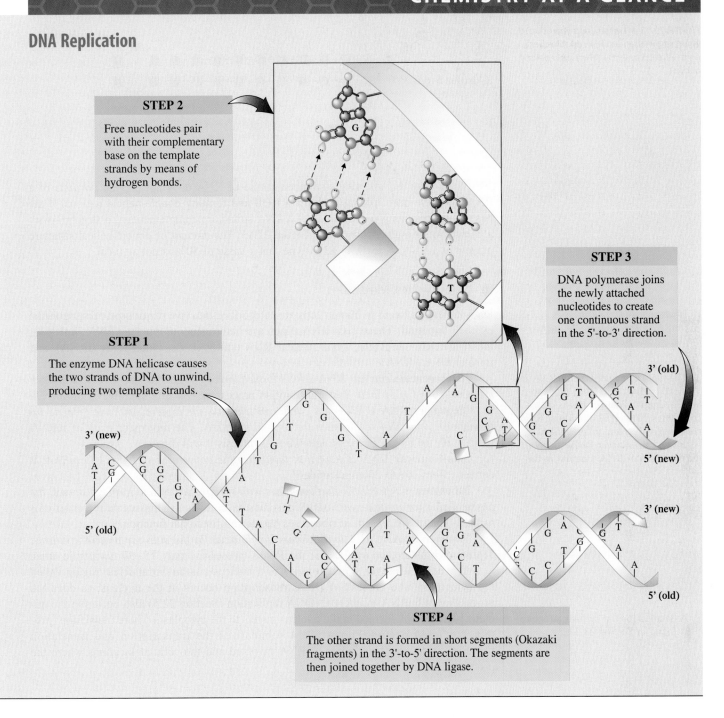

STEP 2

Free nucleotides pair with their complementary base on the template strands by means of hydrogen bonds.

STEP 1

The enzyme DNA helicase causes the two strands of DNA to unwind, producing two template strands.

STEP 3

DNA polymerase joins the newly attached nucleotides to create one continuous strand in the 5'-to-3' direction.

STEP 4

The other strand is formed in short segments (Okazaki fragments) in the 3'-to-5' direction. The segments are then joined together by DNA ligase.

22.7 Ribonucleic Acids

Four major differences exist between RNA molecules and DNA molecules.

1. The sugar unit in the backbone of RNA is ribose; it is deoxyribose in DNA.
2. The base thymine found in DNA is replaced by uracil in RNA (Figure 22.1). Uracil, instead of thymine, pairs with (forms hydrogen bonds with) adenine in RNA.
3. RNA is a single-stranded molecule; DNA is double-stranded (double helix). Thus RNA, unlike DNA, does not contain equal amounts of specific bases.
4. RNA molecules are much smaller than DNA molecules, ranging from 75 nucleotides to a few thousand nucleotides.

▶ The bases thymine (T) and uracil (U) have similar structures. Thymine is a methyluracil (Section 22.2). The hydrogen-bonding patterns (Figure 22.7) for the A–U base pair (RNA) and the A–T base pair (DNA) are identical.

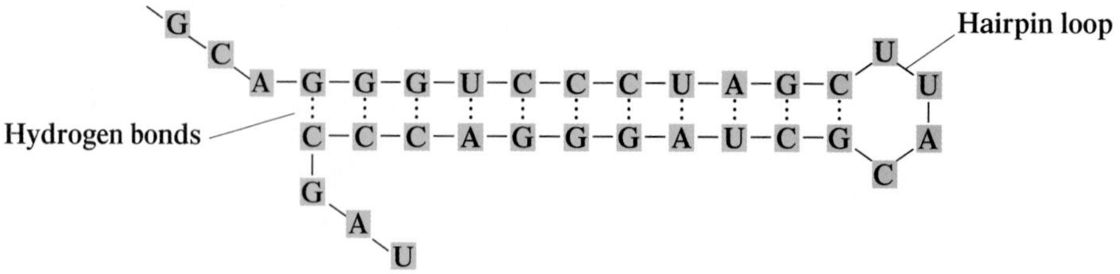

FIGURE 22.13 A hairpin loop is produced when single-stranded RNA doubles back on itself and complementary base pairing occurs.

```
        G                                                    U
         C                                              Hairpin loop
          A—G—G—G—U—C—C—C—U—A—G—C        U
                                                        U
Hydrogen bonds —— C—C—C—A—G—G—G—A—U—C—G        A
          G                                            C
           A
            U
```

We should note that the single-stranded nature of RNA does not prevent *portions* of a RNA molecule from folding back upon itself and forming double-helical regions. If the base sequences along two portions of a RNA strand are complementary, a structure with a hairpin loop results, as shown in Figure 22.13. The amount of double-helical structure present in a RNA varies with RNA type, but a value of 50% is not atypical.

■ Types of RNA Molecules

▶ Heterogeneous nuclear RNA (hnRNA) also goes by the name *primary transcript RNA* (ptRNA).

RNA molecules found in human cells are categorized into five major types, distinguished by their function. These five RNA types are heterogeneous nuclear RNA (hnRNA), messenger RNA (mRNA), small nuclear RNA (snRNA), ribosomal RNA (rRNA), and transfer RNA (tRNA).

Heterogeneous nuclear RNA *is RNA formed directly by DNA transcription from which messenger RNA is formed.* Post-transcription processing converts the hnRNA to mRNA.

Messenger RNA *is RNA that carries instructions for protein synthesis (genetic information) from DNA to the sites for protein synthesis.* The molecular mass of mRNA varies with the length of the protein whose synthesis it will direct.

▶ The most abundant type of RNA in a cell is ribosomal RNA (75% to 80% by mass). Transfer RNA constitutes 10%–15% of cellular RNA; messenger RNA and its precursor, heterogeneous nuclear RNA, make up the 5%–10% of RNA material in the cell.

Small nuclear RNA *is RNA that facilitates the conversion of hnRNA to mRNA.* It contains from 100 to 200 nucleotides.

Ribosomal RNA *is RNA that combines with specific proteins to form ribosomes, the physical sites for protein synthesis.* Ribosomes have molecular masses on the order of 3 million. The rRNA present in ribosomes has no informational function.

Transfer RNA *is RNA that delivers amino acids to the sites for protein synthesis.* Transfer RNAs are the smallest of the RNAs, possessing only 75–90 nucleotide units.

At a nondetail level, a cell consists of a nucleus and an extranuclear region called the cytoplasm. The process of DNA transcription occurs in the nucleus, as does the processing of hnRNA to mRNA. [DNA replication (Section 22.5) also occurs in the nucleus.] The mRNA formed in the nucleus travels to the cytoplasm where translation (protein synthesis) occurs. Figure 22.14 summarizes the transcription and translation processes in terms of the types of RNA involved and the cellular locations where the processes occur.

▶ A detailed look at cellular structure is found in Section 23.2.

FIGURE 22.14 An overview of types of RNA in terms of cellular locations where they are encountered and processes in which they are involved.

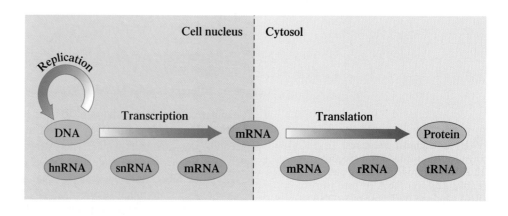

22.8 Transcription: RNA Synthesis

Transcription *is the process by which DNA directs the synthesis of mRNA molecules that carry the coded information needed for protein synthesis.* Messenger RNA production via transcription is actually a "two-step" process in which a hnRNA molecule is initially produced and then is "edited" to yield the desired mRNA molecule. The mRNA molecule so produced then functions as the carrier of the information needed to direct protein synthesis.

Within a strand of a DNA molecule are instructions for the synthesis of numerous hnRNA/mRNA molecules. During transcription, a DNA molecule unwinds, under enzyme influence, at the particular location where the appropriate base sequence is found for the hnRNA/mRNA of concern, and the "exposed" base sequence is transcribed. A short segment of a DNA strand so transcribed, which contains instructions for the formation of a particular hnRNA/mRNA, is called a *gene*. A **gene** *is a segment of a DNA strand that contains the base sequence for the production of a specific hnRNA/mRNA molecule.*

In humans, most genes are composed of 1000–3500 nucleotide units. Hundreds of genes can exist along a DNA strand. Information concerning the total number of genes and the total number of nucleotide base pairs present in human DNA was "updated" in 2001. In that year, the initial results from the *Human Genome Project,* a decade-long internationally based research project to determine the location and base sequence of each of the genes in the human genome, were announced. A **genome** *is all of the genetic material in the chromosomes of an organism.*

The DNA found in a human cell contains an estimated 2.9 billion nucleotide base pairs and approximately 30,000–40,000 genes. This estimate of 30,000–40,000 genes in the human genome is significantly lower than the 100,000 genes that were previously thought to be present. Initial Human Genome Project results also indicate that the base pairs present in these 30,000–40,000 genes constitute only a very small percentage (2%) of the total number of base pairs present in the chromosomes of the human genome.

■ Steps in the Transcription Process

The mechanics of transcription are in many ways similar to those of DNA replication. Four steps are involved.

1. A *portion* of the DNA double helix unwinds, exposing some bases (a gene). The unwinding process is governed by the enzyme *RNA polymerase* rather than by *DNA helicase* (replication enzyme).
2. Free *ribo*nucleotides align along *one* of the exposed strands of DNA bases, the *template* strand, forming new base pairs. In this process, U rather than T aligns with A in the base-pairing process. Because ribonucleotides rather than deoxyribonucleotides are involved in the base pairing, ribose, rather than deoxyribose, becomes incorporated into the new nucleic acid backbone.
3. RNA polymerase links the aligned ribonucleotides.
4. Transcription ends when the RNA polymerase enzyme encounters a sequence of bases that is "read" as a stop signal. The newly formed RNA molecule and the RNA polymerase enzyme are released, and the DNA then rewinds to re-form the original double helix.

Figure 22.15 shows the overall process of transcription of DNA to form RNA.

▶ In DNA–RNA base pairing, the complementary base pairs are

DNA		RNA
A	——	U
G	——	C
C	——	G
T	——	A

RNA molecules contain the base U instead of the base T.

EXAMPLE 22.2

Base Pairing Associated with the Transcription Process

■ From the base sequence 5′ A–T–G–C–C–A 3′ in a DNA template strand, determine the base sequence in the RNA synthesized from the DNA template strand.

Solution

A RNA molecule cannot contain the base T. The base U is present instead. Therefore, U–A base pairing will occur instead of T–A base pairing. The other base-pairing

(continued)

FIGURE 22.15 The transcription of DNA to form RNA involves an unwinding of a portion of the DNA double helix. Only one strand of the DNA is copied during transcription.

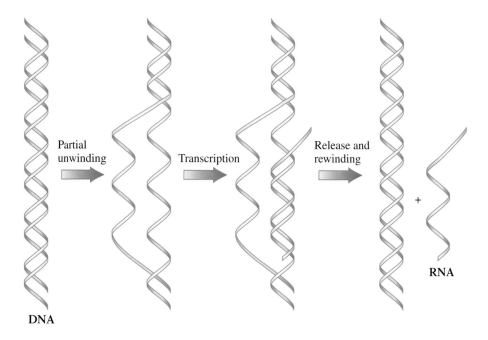

Partial unwinding

Transcription

Release and rewinding

+

RNA

DNA

combination, G–C, remains the same. The RNA product of the transcription process will therefore be

DNA template: 5' A–T–G–C–C–A 3'

RNA molecule: 3' U–A–C–G–G–U 5'

Note that the direction of the RNA strand is antiparallel to that of the DNA template. This will always be the case during transcription.

It is standard procedure, when writing and reading base sequences for nucleic acids (both DNAs and RNAs), always to specify base sequence in the 5' ⟶ 3' direction unless otherwise directed. Thus

3' U–A–C–G–G–U 5'   becomes   5' U–G–G–C–A–U 3'

Practice Exercise 22.2

From the base sequence 5' T–A–A–C–C–T 3' in a DNA template strand, determine the base sequence in the RNA synthesized from the DNA template strand.

■ Post-Transcription Processing: Formation of mRNA

The RNA produced from a gene through transcription is hnRNA, the precursor for mRNA. The conversion of hnRNA to mRNA involves *post-transcription processing* of the hnRNA. In this processing, certain portions of the hnRNA are deleted and the retained parts are then spliced together. This process leads us to the concepts of *exons* and *introns.*

It is now known that not all bases in a gene convey genetic information. Instead, a gene is *segmented;* it has portions called *exons* that contain genetic information and portions called *introns* that do not convey genetic information.

An **exon** *is a gene segment that conveys (codes for) genetic information. Ex*ons are DNA segments that help *ex*press a genetic message. An **intron** *is a gene segment that does not convey (code for) genetic information. In*trons are DNA segments that *in*terrupt a genetic message. A gene consists of alternating exon and intron segments (Figure 22.16).

Both the exons and the introns of a gene are transcribed during production of *heterogeneous nuclear RNA* (hnRNA). The hnRNA is then "edited," under the direction

FIGURE 22.16 Heterogeneous nuclear RNA contains both exons and introns. Messenger RNA is heterogeneous nuclear RNA from which the introns have been excised.

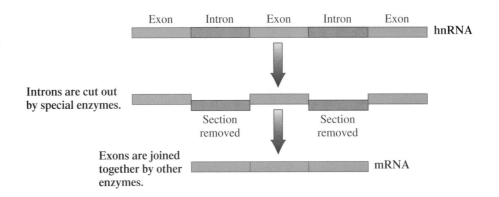

of enzymes, to remove the introns. The remaining exons are joined together to form a shortened RNA strand that carries the genetic information of the transcribed gene. This "edited" RNA is the messenger RNA (mRNA) that serves as a blueprint for protein assembly. Much is yet to be learned about introns and why they are present in genes; investigating their function is an active area of biochemical research.

Splicing *is the process of joining the exons of a hnRNA molecule together to form a mRNA molecule.* The splicing process involves snRNA molecules, the most recent of the RNA types to be discovered. This type of RNA is never found "free" in a cell. A snRNA molecule is always found complexed with proteins in particles called *small nuclear ribonucleoprotein particles,* which are usually called *snRNPs* (pronounced "snurps"). A **small nuclear ribonucleoprotein particle** *is a complex formed from a snRNA molecule and several proteins.* "Snurps" always further collect together into larger complexes called *spliceosomes.* A **spliceosome** *is a large assembly of snRNA molecules and proteins involved in the conversion of hnRNA molecules to mRNA molecules.*

■ Alternative Splicing

As we have previously noted, the initial results of the Human Genome Project reduced the estimated number of genes present in the human genome from 100,000 to 30,000–40,000. Rather than simplifying our understanding of how genes work, this reduced estimate in number of genes actually complicates it considerably.

Prior to the announcement of the Human Genome Project's results, biochemistry had largely embraced the "one-protein-one-gene" concept. It was generally assumed that each type of protein had "its own" gene that carried the instructions for its synthesis. This no longer seems plausible, because the estimated number of different proteins present in the human body now significantly exceeds the estimated number of genes.

Intensive research is now centered on the concept of *alternative splicing.* **Alternative splicing** *is a process by which several different proteins that are variations of a basic structural motif can be produced from a single gene.* In alternative splicing, a hnRNA molecule with multiple exons present is spliced in several different ways. Figure 22.17 shows the four alternative splicing patterns that can occur when a hnRNA contains four exons, two of which are *alternative exons.*

FIGURE 22.17 An hnRNA molecule containing four exons, two of which (B and C) are alternative exons, can be spliced in four different ways, producing four different proteins. Proteins can be produced with neither, either, or both of the alternative exons present.

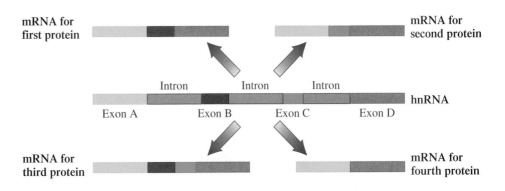

22.9 The Genetic Code

The nucleotide (base) sequence of a mRNA molecule is the informational part of such a molecule. This base sequence in a given mRNA determines the amino acid sequence for the protein synthesized under that mRNA's direction.

How can the base sequence of a mRNA molecule (which involves only *4* different bases—A, C, G, and U) encode enough information to direct proper sequencing of *20* amino acids in proteins? If each base encoded for a particular standard amino acid, then only 4 amino acids would be specified out of the 20 needed for protein synthesis, a clearly inadequate number. If two-base sequences were used to code amino acids, then there would be $4^2 = 16$ possible combinations, so 16 amino acids could be represented uniquely. This is still an inadequate number. If three-base sequences were used to code for amino acids, there would be $4^3 = 64$ possible combinations, which is more than enough combinations for uniquely specifying each of the 20 standard amino acids found in proteins.

Research has verified that sequences of three nucleotides in mRNA molecules specify the amino acids that go into synthesis of a protein. Such three-nucleotide sequences are called codons. A **codon** *is a three-nucleotide sequence in a mRNA molecule that codes for a specific amino acid.*

Which amino acid is specified by which codon? (We have 64 codons to choose from.) Researchers deciphered codon–amino acid relationships by adding different *synthetic* mRNA molecules (whose base sequences were known) to cell extracts and then determining the structure of any newly formed protein. After many such experiments, researchers finally matched all 64 possible codons with their functions in protein synthesis. It was found that 61 of the 64 codons formed by various combinations of the bases A, C, G, and U were related to specific amino acids; the other 3 combinations were termination codons ("stop" signals) for protein synthesis. Collectively, these relationships between three-nucleotide sequences in mRNA and amino acid identities are known as the genetic code. The **genetic code** *is the assignment of the 64 mRNA codons to specific amino acids (or stop signals).* The determination of this code during the early 1960s is one of the most remarkable of twentieth-century scientific achievements. The 1968 Noble Prize in chemistry was awarded to Marshall Nirenberg and Har Gobind Khovana for their work in illuminating how mRNA encodes for proteins.

The complete genetic code is given in Table 22.2. Examination of this table indicates that the genetic code has several remarkable features.

> There is a rough correlation between the number of codons for a particular amino acid and that amino acid's frequency of occurrence in proteins. For example, the two amino acids that have a single codon, Met and Trp, are two of the least common amino acids in proteins.

1. *The genetic code is highly degenerate; that is, many amino acids are designated by more than one codon.* Three amino acids (Arg, Leu, and Ser) are represented by six codons. Two or more codons exist for all other amino acids except Met and Trp, which have only a single codon. Codons that specify the same amino acid are called *synonyms.*

2. *There is a pattern to the arrangement of synonyms in the genetic code table.* All synonyms for an amino acid fall within a single box in Table 22.2, unless there are more than four synonyms, where two boxes are needed. The significance of the "single box" pattern is that with synonyms, the first two bases of the codon are the same—they differ only in the third base. For example, the four synonyms for the amino acid Pro are CCU, CCC, CCA, and CCG.

3. *The genetic code is almost universal.* Although Table 22.2 does not show this feature, studies of many organisms indicate that with minor exceptions, the code is the same in all of them. The same codon specifies the same amino acid whether the cell is a bacterial cell, a corn plant cell, or a human cell.

4. *An initiation codon exists.* The existence of "stop" codons (UAG, UAA, and UGA) suggests the existence of "start" codons. There is one initiation codon. Besides coding for the amino acid methionine, the codon AUG functions as an initiator of protein synthesis when it occurs as the first codon in an amino acid sequence.

TABLE 22.2
The Universal Genetic Code. The code is composed of 64 three-nucleotide sequences (codons), which can be read from the table. The left-hand column indicates the nucleotide base found in the first (5′) position of the codon. The nucleotides in the second (middle) position of the codon are in the middle columns. The right-hand column indicates the nucleotide found in the third (3′) position. Thus the codon ACG encodes for the amino acid Thr, and the codon GGG encodes for the amino acid Gly.

First position (5′ end)	Second Position				Third position (3′ end)
	U	C	A	G	
U	Phe	Ser	Tyr	Cys	U
	Phe	Ser	Tyr	Cys	C
	Leu	Ser	Stop	Stop	A
	Leu	Ser	Stop	Trp	G
C	Leu	Pro	His	Arg	U
	Leu	Pro	His	Arg	C
	Leu	Pro	Gln	Arg	A
	Leu	Pro	Gln	Arg	G
A	Ile	Thr	Asn	Ser	U
	Ile	Thr	Asn	Ser	C
	Ile	Thr	Lys	Arg	A
	Met	Thr	Lys	Arg	G
G	Val	Ala	Asp	Gly	U
	Val	Ala	Asp	Gly	C
	Val	Ala	Glu	Gly	A
	Val	Ala	Glu	Gly	G

EXAMPLE 22.3

Using the Genetic Code and mRNA Codons to Predict Amino Acid Sequences

■ Using the genetic code in Table 22.2, determine the sequence of amino acids encoded by the mRNA codon sequence

$$\text{5′ GCC–AUG–GUA–AAA–UGC–GAC–CCA 3′}$$

Solution

Matching the codons with the amino acids, using Table 22.2, yields

mRNA: 5′ GCC–AUG–GUA–AAA–UGC–GAC–CCA 3′

Peptide: Ala   Met   Val   Lys   Cys   Asp   Pro

Practice Exercise 22.3

Using the genetic code in Table 22.2, determine the sequence of amino acids encoded by the mRNA codon sequence

$$\text{5′ CAU–CCU–CAC–ACU–GUU–UGU–UGG 3′}$$

EXAMPLE 22.4

Relating Protein Amino Acid Sequence to Base Sequence on a DNA Template Strand

■ Sections A, C, and E of the following base sequence section of a DNA template strand are exons, and sections B and D are introns.

DNA   5′ ATT – CGT – TGT – TTT – CCC – AGT – GCC 3′
              A   B   C   D   E

a. What is the structure of the hnRNA transcribed from this template?
b. What is the structure of the mRNA obtained by splicing the hnRNA?
c. What polypeptide amino acid sequence will be synthesized using the mRNA?

Solution

a. The base sequence in the hnRNA will be complementary to that of the template DNA, except that U is used in the RNA instead of T. The hnRNA will have a directionality antiparallel to that of the DNA sequence.

hnRNA   3′ UAA – GCA – ACA – AAA – GGG – UCA – CGG 5′

(continued)

b. In the splicing process, introns are removed and the exons combined to give the mRNA.

mRNA   3′ UAA – ACA – AAA – CGG 5′

c. The codons in a mRNA must be read in the 5′-to-3′ direction to use the genetic code correctly to determine the sequence of amino acids in the peptide. Rewriting the mRNA structure in the 5′-to-3′ direction gives

mRNA   5′ GGC – AAA – ACA – AUU 3′

The Universal Genetic Code (Table 22.2), reveals that this mRNA sequence codes for the amino acid sequence

Gly–Lys–Thr–Ile

Practice Exercise 22.4

Sections A, C, and E of the following base sequence section of a DNA template strand are exons, and sections B and D are introns.

DNA   5′ CGC – CGT – AGT – TGG – CCC – GGA – GGA 3′
           A   B   C   D   E

a. What is the structure of the hnRNA transcribed from this template?
b. What is the structure of the mRNA obtained by splicing the hnRNA?
c. What polypeptide amino acid sequence will be synthesized using the mRNA?

22.10 Anticodons and tRNA Molecules

The amino acids used in protein synthesis do not directly interact with the codons of a mRNA molecule. Instead, tRNA molecules function as intermediaries that deliver amino acids to the mRNA. At least one type of tRNA molecule exists for each of the 20 amino acids found in proteins.

All tRNA molecules have the same general shape, and this shape is crucial to how they function. Figure 22.18a shows the general *two-dimensional* "cloverleaf" shape of a tRNA molecule, a shape produced by the molecule's folding and twisting into regions of

FIGURE 22.18 A tRNA molecule. The amino acid attachment site is at the open end of the cloverleaf (the 3′ end), and the anticodon is located in the hairpin loop opposite the open end.

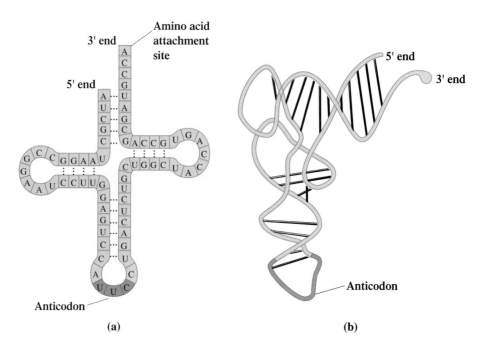

(a)   (b)

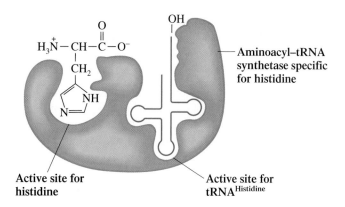

FIGURE 22.19 An aminoacyl–tRNA synthetase has an active site for tRNA and a binding site for the particular amino acid that is to be attached to that tRNA.

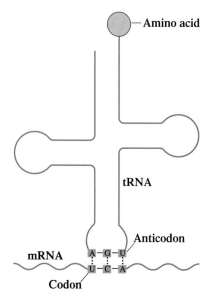

FIGURE 22.20 The interaction between anticodon (tRNA) and codon (mRNA), which involves complementary base pairing, governs the proper placement of amino acids in a protein.

parallel strands and regions of hairpin loops. (The actual three-dimensional shape of a tRNA molecule involves considerable additional twisting of the "cloverleaf" shape—Figure 22.18b.)

Two features of the tRNA structure are of particular importance.

1. The 3′ end of the open part of the cloverleaf structure is where an amino acid becomes *covalently* bonded to the tRNA molecule through an ester bond. Each of the different tRNA molecules is specifically recognized by an *aminoacyl synthetase* enzyme. These enzymes also recognize the one kind of amino acid that "belongs" with the particular tRNA and facilitates its bonding to the tRNA (see Figure 22.19).

2. The loop *opposite* the open end of the cloverleaf is the site for a sequence of three bases called an anticodon. An **anticodon** *is a three-nucleotide sequence on a tRNA molecule that is complementary to a codon on a mRNA molecule.*

The interaction between the anticodon of the tRNA and the codon of the mRNA leads to the proper placement of an amino acid into a growing peptide chain during protein synthesis. This interaction, which involves complementary base pairing, is shown in Figure 22.20.

22.11 Translation: Protein Synthesis

Translation *is the process by which the codes within mRNA molecules are deciphered and a particular protein molecule is synthesized.*

The substances needed for the translation phase of protein synthesis are mRNA molecules, tRNA molecules, amino acids, ribosomes, and a number of different enzymes. A **ribosome** *is a rRNA–protein complex that serves as the site for the translation phase of protein synthesis.* Ribosomes have structures involving two subunits—a large subunit and a small subunit (see Figure 22.21). Each subunit is approximately 65% rRNA and 35% protein. The rRNA helps maintain the structure of the ribosome and also provides sites where mRNA can attach itself to the ribosome.

FIGURE 22.21 Ribosomes, which contain both rRNA and protein, have structures that contain two subunits. One subunit is much larger than the other.

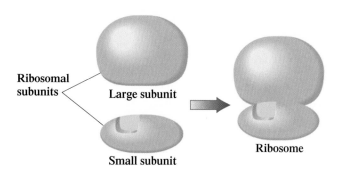

There are five general steps to the translation process: (1) activation of tRNA, (2) initiation, (3) elongation, (4) termination, and (5) post-translational processing.

■ Activation of tRNA

There are two steps involved in tRNA activation. First, an amino acid interacts with an activator molecule (ATP; Section 23.3) to form a highly energetic complex. This complex then reacts with the appropriate tRNA molecule to produce an *activated tRNA molecule,* a tRNA molecule that has an amino acid covalently bonded to it at its 3′ end through an ester linkage.

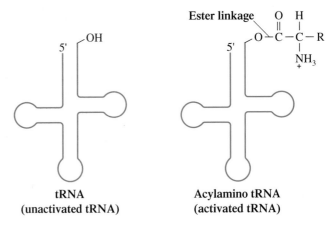

tRNA
(unactivated tRNA)

Acylamino tRNA
(activated tRNA)

■ Initiation

The initiation of protein synthesis begins when mRNA attaches itself to the surface of a small ribosomal subunit such that its first codon, which is always the initiating codon AUG, occupies a site called the P site (peptidyl site.) (See Figure 22.22a.) An activated tRNA molecule with anticodon complementary to the codon AUG attaches itself, through complementary base pairing, to the AUG codon (Figure 22.22b). The resulting complex then interacts with a large ribosomal subunit to complete the formation of an initiation complex (Figure 22.22c).

When functioning as the initiating codon, AUG codes for a derivative of methionine, *N*-formyl methionine, rather than for methionine itself. This derivative, which is designated as f-Met, has the structure

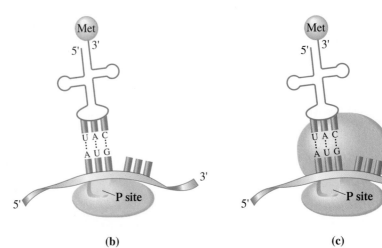

FIGURE 22.22 Initiation of protein synthesis begins with the formation of an initiation complex.

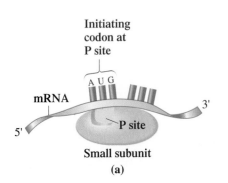

(a)

(b)

(c)

■ Elongation

Next to the P site in an mRNA–ribosome complex is a second binding site called the A site (aminoacyl site). (See Figure 22.23a.) At this second site the next mRNA codon is exposed, and a tRNA with the appropriate anticodon binds to it (Figure 22.23b). With amino acids in place at both the P and the A sites, the enzyme *peptidyl transferase* effects the linking of the P site amino acid to the A site amino acid to form a dipeptide. Such peptide bond formation leaves the tRNA at the P site empty and the tRNA at the A site bearing the dipeptide (Figure 22.23c).

The empty tRNA at the P site now leaves that site and is free to pick up another molecule of its specific amino acid. Simultaneously with the release of tRNA from the P site, the ribosome shifts along the mRNA. This shift puts the newly formed dipeptide at the P site, and the third codon of mRNA is now available, at site A, to accept a tRNA molecule whose anticodon complements this codon (see Figure 22.23d). The movement of a ribosome along a mRNA molecule is called *translocation*. **Translocation** *is the part of translation in which a ribosome moves down a mRNA molecule three base positions (one codon) so that a new codon can occupy the ribosomal A site.*

Now a repetitious process begins. The third codon, now at the A site, accepts an incoming tRNA with its accompanying amino acid; and then the entire dipeptide at the P site is transferred and bonded to the A site amino acid to give a tripeptide (see Figure 22.23e). The empty tRNA at the P site is released, the ribosome shifts along the mRNA, and the process continues.

▶ In elongation, the polypeptide chain grows one amino acid at a time.

FIGURE 22.23 The process of translation that occurs during protein synthesis. The anticodons of tRNA molecules are paired with the codons of an mRNA molecule to bring the appropriate amino acids into sequence for protein formation.

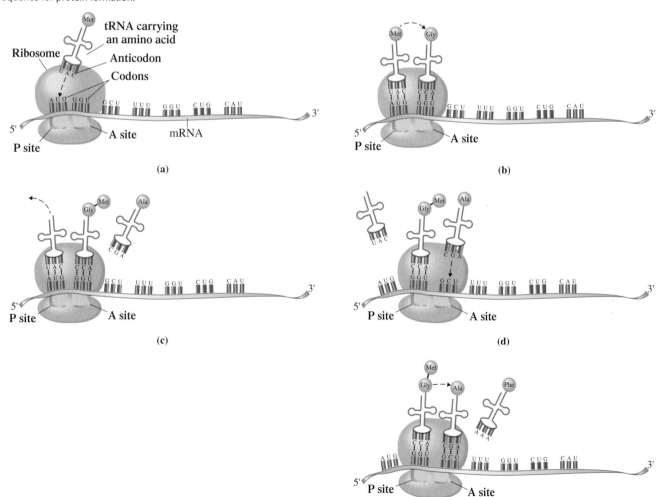

CHEMICAL CONNECTIONS

Antibiotics That Inhibit Bacterial Protein Synthesis

Some antibiotics work because they inhibit protein synthesis in bacteria but not in humans. They inhibit one specific enzyme or another in the bacterial ribosomes. These antibiotics are useful in treating disease and in studying protein synthesis mechanisms in bacteria. The accompanying table lists a few of the most commonly encountered antibiotics and their modes of action relative to protein synthesis.

Antibiotic	Biological action	Antibiotic	Biological action
chloramphenicol	inhibits an important enzyme (peptidyl transferase) in the large ribosomal subunit	streptomycin	inhibits initiation of protein synthesis and also causes the mRNA codons to be read incorrectly
erythromycin	binds to the large subunit and stops the ribosome from moving along the mRNA from one codon to the next	tetracycline	binds to the small ribosomal subunit and inhibits the binding of incoming tRNA molecules
puromycin	induces premature polypeptide chain termination		

■ Termination

The polypeptide continues to grow by way of translocation until all necessary amino acids are in place and bonded to each other. Appearance in the mRNA codon sequence of one of the three stop codons (UAA, UAG, or UGA) terminates the process. No tRNA has an anticodon that can base-pair with these stop codons. The polypeptide is then cleaved from the tRNA through hydrolysis.

■ Post-Translation Processing

Some modification of proteins usually occurs after translation. For example, most proteins do not have f-Met (the initiation codon) as their first amino acid. Cleavage of N-terminal f-Met is part of post-translation processing. Formation of S—S bonds between cysteine units is another example of post-translation processing.

■ Efficiency of mRNA Utilization

Many ribosomes can move simultaneously along a single mRNA molecule (Figure 22.24). In this highly efficient arrangement, many identical protein chains can be synthesized almost at the same time from a single strand of mRNA. This multiple use of mRNA molecules reduces the amount of resources and energy that the cell expends to synthesize needed protein. Such complexes of several ribosomes and mRNA are called polyribosomes or polysomes. A **polysome** *is a complex of mRNA and several ribosomes.*

The Chemistry at a Glance feature on page 679 summarizes the steps in protein synthesis.

FIGURE 22.24 Several ribosomes can simultaneously proceed along a single strand of mRNA one after another. Such a complex of mRNA and ribosomes is called a polysome.

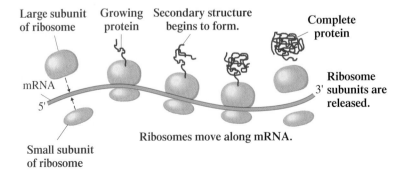

Protein Synthesis

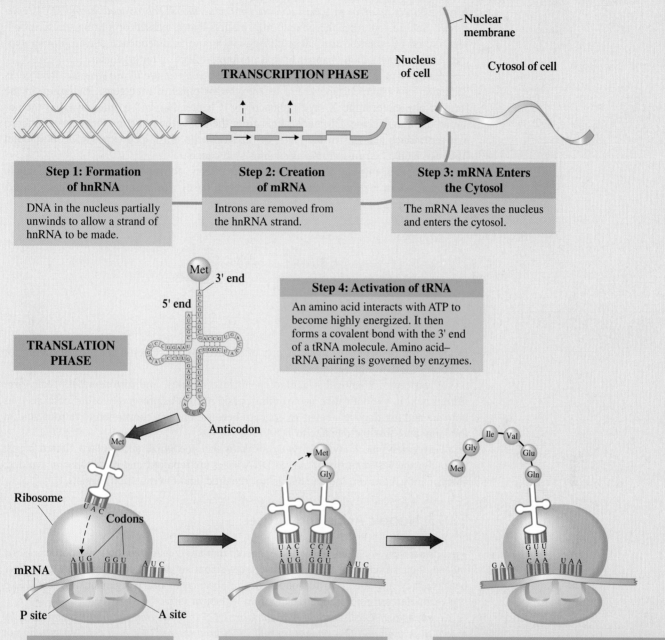

TRANSCRIPTION PHASE

Nuclear membrane

Nucleus of cell

Cytosol of cell

Step 1: Formation of hnRNA

DNA in the nucleus partially unwinds to allow a strand of hnRNA to be made.

Step 2: Creation of mRNA

Introns are removed from the hnRNA strand.

Step 3: mRNA Enters the Cytosol

The mRNA leaves the nucleus and enters the cytosol.

Met
3' end
5' end

TRANSLATION PHASE

Step 4: Activation of tRNA

An amino acid interacts with ATP to become highly energized. It then forms a covalent bond with the 3' end of a tRNA molecule. Amino acid–tRNA pairing is governed by enzymes.

Anticodon

Met

Ribosome

Codons

mRNA

P site   A site

Met
Gly

Ile   Val
Gly   Glu
Met   Gln

Step 5: Initiation

The mRNA attaches to a ribosome so that the first codon (AUG) is at the P site. A tRNA carrying *N*-formyl methionine, an amino acid derivative, attaches to the first codon.

Step 6: Elongation

Another tRNA with the second amino acid binds at the A site. The methionine derivative transfers from the P site to the A site. The ribosome shifts to the next codon, making its A site available for the tRNA carrying the third amino acid.

Steps 7 and 8: Termination and Post-Translation Processing

The polypeptide chain continues to lengthen until a stop codon appears on the mRNA. The new protein is cleaved from the last tRNA.

During post-translation processing, cleavage of f-Met (the initiation codon) usually occurs. S—S bonds between Cys units also can form.

22.12 Mutations

A **mutation** *is an error in base sequence in a gene that is reproduced during DNA replication.* Such errors alter the genetic information that is passed on during transcription. The altered information can cause changes in amino acid sequence during protein synthesis. Sometimes, such changes have a profound effect on an organism.

A **mutagen** *is a substance or agent that causes a change in the structure of a gene.* Radiation and chemical agents are two important types of mutagens. Radiation, in the form of ultraviolet light, X rays, radioactivity (Chapter 11), and cosmic rays, has the potential to be mutagenic. Ultraviolet light from the sun is the radiation that causes sunburn and can induce changes in the DNA of the skin cells. Sustained exposure to ultraviolet light can lead to serious problems such as skin cancer.

Chemical agents can also have mutagenic effects. Nitrous acid (HNO_2) is a mutagen that causes deamination of heterocyclic nitrogen bases. For example, HNO_2 can convert cytosine to uracil.

NH$_2$ structure $\xrightarrow{HNO_2}$ structure

Cytosine   Uracil

Deamination of a cytosine that was part of an mRNA codon would change the codon; for example, CGG would become UGG.

A variety of chemicals—including nitrites, nitrates, and nitrosamines—can form nitrous acid in the body. The use of nitrates and nitrites as preservatives in foods such as bologna and hot dogs is a cause of concern because of their conversion to nitrous acid in the body and possible damage to DNA.

Fortunately, the body has *repair enzymes* that recognize and replace altered bases. Normally, the vast majority of altered DNA bases are repaired, and mutations are avoided. Occasionally, however, the damage is not repaired, and the mutation persists.

22.13 Nucleic Acids and Viruses

Viruses are very small disease-causing agents that are considered the lowest order of life. Indeed, their structure is so simple that some scientists do not consider them truly alive because they are unable to reproduce in the absence of other organisms. Figure 22.25 shows an electron microscope image of an influenza virus.

A **virus** *is a small particle that contains DNA or RNA (but not both) surrounded by a coat of protein and that cannot reproduce without the aid of a host cell.* Viruses do not possess the nucleotides, enzymes, amino acids, and other molecules necessary to replicate their nucleic acid or to synthesize proteins. To reproduce, viruses must invade the cells of another organism and cause these host cells to carry out the reproduction of the virus. Such an invasion disrupts the normal operation of cells, causing diseases within the host organism. The only function of a virus is reproduction; viruses do not generate energy.

There is no known form of life that is not subject to attack by viruses. Viruses attack bacteria, plants, animals, and humans. Many human diseases are of viral origin. Among them are the common cold, mumps, measles, smallpox, rabies, influenza, infectious mononucleosis, hepatitis, and AIDS.

Viruses most often attach themselves to the outside of specific cells in a host organism. An enzyme within the protein overcoat of the virus catalyzes the breakdown of the cell membrane, opening a hole in the membrane. The virus then injects its DNA or RNA into the cell. Once inside, this nucleic acid material is mistaken by the host cell for its own, whereupon that cell begins to translate and/or transcribe the viral nucleic acid. When all the virus components have been synthesized by the host cell, they assemble

FIGURE 22.25 An electron microscope image of an influenza virus.

automatically to form many new virus particles. Within 20 to 30 minutes after a single molecule of viral nucleic acid enters the host cell, hundreds of new virus particles have formed. So many are formed that they eventually burst the host cell and are free to infect other cells.

If a virus contains DNA, the host cell replicates the viral DNA in a manner similar to the way it replicates its own DNA. The newly produced viral DNA then proceeds to make the proteins needed for the production of protein coats for additional viruses.

A RNA-containing virus is called a *retrovirus.* Once inside a host, such viruses first make viral DNA. This *reverse* synthesis is governed by the enzyme *reverse transcriptase.* The template is the viral RNA rather than DNA. The viral DNA so produced then produces additional viral DNA and the proteins necessary for the protein coats.

The AIDS (acquired immunodeficiency syndrome) virus is an example of a retrovirus. This virus has an affinity for a specific type of white blood cell called a *helper T cell,* which is an important part of the body's immune system. When helper T cells are unable to perform their normal functions as a result of such viral infection, the body becomes more susceptible to infection and disease.

A **vaccine** *is a preparation containing an inactive or weakened form of a virus or bacterium.* The antibodies produced by the body against these specially modified viruses or bacteria effectively act against the naturally occurring active forms as well. Thanks to vaccination programs, many diseases, such as polio and mumps (caused by RNA-containing viruses) and smallpox and yellow fever (caused by DNA-containing viruses), are now seldom encountered.

> Viral infections are more difficult to treat than bacterial infections because viruses, unlike bacteria, replicate inside cells. It is difficult to design drugs that prevent the replication of the virus that do not also affect the normal activities of the host cells.

22.14 Recombinant DNA and Genetic Engineering

Increased knowledge about how DNA molecules function under various chemical conditions has opened the door to the field of technology called *genetic engineering* or *biotechnology.* Techniques now exist whereby a "foreign" gene can be added to an organism, and the organism will produce the protein associated with the added gene.

As an example of benefits that can come from genetic engineering, consider the case of human insulin. For many years, because of the very limited availability of human insulin, the insulin used by diabetics was obtained from the pancreases of slaughterhouse animals. Such insulin is structurally very similar to human insulin (see the Chemical Connections feature "Substitutes for Human Insulin" in Chapter 20) and can be substituted for it. Today, diabetics can also choose to use "real" human insulin produced by genetically altered bacteria. Such "genetically engineered" bacteria are grown in large numbers, and the insulin they produce is harvested in a manner similar to the way some antibiotics are obtained from cultured microorganisms. Human growth hormone is another substance that is now produced by genetically altered bacteria.

Genetic engineering procedures involve a type of DNA called *recombinant DNA.* **Recombinant DNA** *is DNA that contains genetic material from two different organisms that has been combined into one DNA molecule.* Let us examine the theory and procedures used in obtaining recombinant DNA through genetic engineering.

The bacterium *E. coli,* which is found in the intestinal tract of humans and animals, is the organism most often used in recombinant DNA experiments. Yeast cells are also used, with increasing frequency, in this research.

In addition to their chromosomal DNA, *E. coli* (and other bacteria) contain DNA in the form of small, circular, double-stranded molecules called *plasmids.* These plasmids, which carry only a few genes, replicate independently of the chromosome. Also, they are transferred relatively easily from one cell to another. Plasmids from *E. coli* are used in recombinant DNA work.

The procedure used to obtain *E. coli* cells that contain recombinant DNA involves the following steps (see Figure 22.26).

Step 1: **Cell membrane dissolution.** *E. coli* cells of a specific strain are placed in a solution that dissolves cell membranes, thus releasing the contents of the cells.

FIGURE 22.26 Recombinant DNA is made by inserting a gene obtained from DNA of one organism into the DNA from another kind of organism.

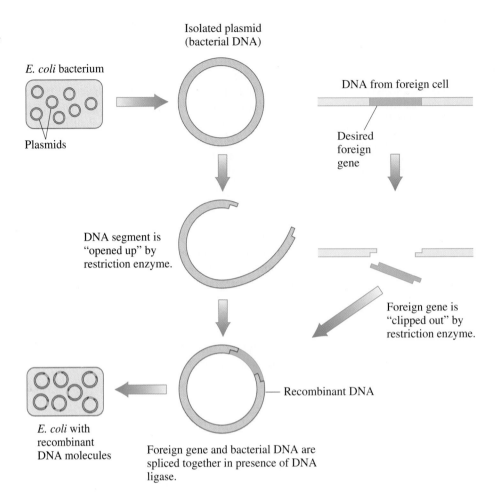

Isolated plasmid (bacterial DNA)

E. coli bacterium

DNA from foreign cell

Plasmids

Desired foreign gene

DNA segment is "opened up" by restriction enzyme.

Foreign gene is "clipped out" by restriction enzyme.

Recombinant DNA

E. coli with recombinant DNA molecules

Foreign gene and bacterial DNA are spliced together in presence of DNA ligase.

Step 2:   **Isolation of plasmid fraction.** The released cell components are separated into fractions, one fraction being the plasmids. The isolated plasmid fraction is the material used in further steps.

Step 3:   **Cleavage of plasmid DNA.** A special enzyme, called a *restriction enzyme,* is used to cleave the double-stranded DNA of a circular plasmid. The result is a linear (noncircular) DNA molecule.

Step 4:   **Gene removal from another organism.** The same restriction enzyme is then used to remove a desired gene from a chromosome of another organism.

Step 5:   **Gene–plasmid splicing.** The gene (from Step 4) and the opened plasmid (from Step 3) are mixed in the presence of the enzyme *DNA ligase,* which splices the two together. This splicing, which attaches one end of the gene to one end of the opened plasmid and attaches the other end of the gene to the other end of the plasmid, results in an altered circular plasmid (the recombinant DNA).

Step 6:   **Uptake of recombinant DNA.** The altered plasmids (recombinant DNA) are placed in a live *E. coli* culture, where they are taken up by the *E. coli* bacteria. The *E. coli* culture into which the plasmids are placed need not be identical to that from which the plasmids were originally obtained.

We noted that in Step 3, the conversion of a circular plasmid into a linear DNA molecule requires a restriction enzyme. A **restriction enzyme** *is an enzyme that recognizes specific base sequences in DNA and cleaves the DNA in a predictable manner at these sequences.* The discovery of restriction enzymes made genetic engineering possible.

Restriction enzymes occur naturally in numerous types of bacterial cells. Their function is to protect the bacteria from invasion by foreign DNA by catalyzing the cleavage of the invading DNA. The term *restriction* relates to such enzymes placing a "restriction" on the type of DNA allowed into the bacterial cells.

FIGURE 22.27 Cleavage pattern resulting from the use of a restriction enzyme that cleaves DNA between G and A bases in the 5′-to-3′ direction in the sequence G–A–A–T–T–C. The double-helix structure is not cut straight across.

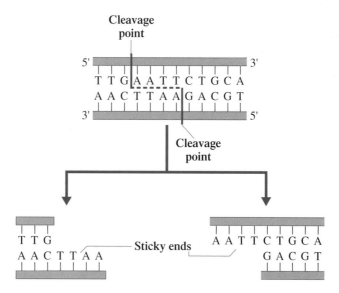

To understand how a restriction enzyme works, let us consider one that cleaves DNA between G and A bases in the 5′-to-3′ direction in the sequence G–A–A–T–T–C. This enzyme will cleave the double-helix structure of a DNA molecule in the manner shown in Figure 22.27.

Note that the double helix is not cut straight across; the individual strands are cut at different points, giving a staircase cut. (Both cuts must be between G and A in the 5′-to-3′ direction.) This staircase cut leaves unpaired bases on each cut strand. These ends with unpaired bases are called "sticky ends" because they are ready to "stick to" (pair up with) a complementary section of DNA if they can find one.

If the same restriction enzyme used to cut a plasmid is also used to cut a gene from another DNA molecule, the sticky ends of the gene will be complementary to those of the plasmid. This enables the plasmid and gene to combine readily, forming a new, modified plasmid molecule. This modified plasmid molecule is called recombinant DNA. In addition to the newly spliced gene, the recombinant DNA plasmid contains all of the genes and characteristics of the original plasmid. Figure 22.28 shows diagrammatically the match between sticky ends that occurs when plasmid and gene combine.

FIGURE 22.28 The "sticky ends" of the cut plasmid and the cut gene are complementary and combine to form recombinant DNA.

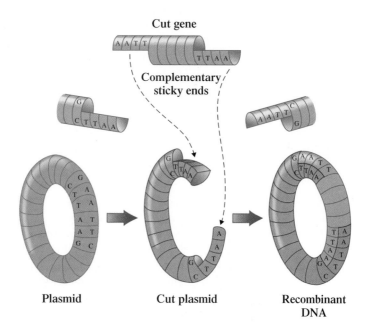

Step 6 involves inserting the recombinant DNA (modified plasmids) back into *E. coli* cells. The process is called transformation. **Transformation** *is the process of incorporating recombinant DNA into a host cell.*

The transformed cells then reproduce, resulting in large numbers of identical cells called clones. **Clones** *are cells with identical DNA that have descended from a single cell.* Within a few hours, a single genetically altered bacterial cell can give rise to thousands of clones. Each clone has the capacity to synthesize the protein directed by the foreign gene it carries.

Researchers are not limited to selection of naturally occurring genes for transforming bacteria. Chemists have developed nonenzymatic methods of linking nucleotides together such that they can construct artificial genes of any sequence they desire. In fact, benchtop instruments are now available that can be programmed by a microprocessor to synthesize any DNA base sequence *automatically.* The operator merely enters a sequence of desired bases, starts the instrument, and returns later to obtain the product. Such flexibility in manufacturing DNA has opened many doors, accelerated the pace of recombinant DNA research, and redefined the term *designer genes!*

22.15 The Polymerase Chain Reaction

▶ PCR temperature conditions are higher than those in the human body. This is possible because the DNA polymerase used was isolated from an organism that lives in the "hot pots" of Yellowstone National Park at temperatures of 70°C–75°C.

The **polymerase chain reaction (PCR)** *is a method for rapidly producing multiple copies of a DNA nucleotide sequence.* Billions of copies of a specific DNA sequence (gene) can be produced in a few hours via this reaction. The PCR is easy to carry out, requiring only a few chemicals, a container, and a source of heat. (In actuality, the PCR process is now completely automated.)

By means of the PCR process, DNA that is available only in very small quantities can be amplified to quantities large enough to analyze. The PCR process, devised in 1983, has become a valuable tool for diagnosing diseases and detecting pathogens in the body. It is now used in the prenatal diagnosis of a number of genetic disorders, including muscular dystrophy and cystic fibrosis, and in the identification of bacterial pathogens. It is also the definitive way to detect the AIDS virus.

The PCR process has also proved useful in certain types of forensic investigations. A DNA sample may be obtained from a single drop of blood or semen or a single strand of hair at a crime scene and amplified by the PCR process. A forensic chemist can then compare the amplified samples with DNA samples taken from suspects. Work with DNA in the forensic area is often referred to as *DNA fingerprinting.*

▶ After n cycles of the PCR process, the amount of DNA will have increased 2^n times.

2^{10} is approximately 1000.

2^{20} is approximately 1,000,000.

Twenty-five cycles of the PCR can be carried out in an hour in a process that is fully automated.

DNA polymerase, an enzyme present in all living organisms, is a key substance in the PCR process. It can attach additional nucleotides to a short starter nucleotide chain, called a *primer,* when the primer is bound to a complementary strand of DNA that functions as a template. The original DNA is heated to separate its strands, and then primers, DNA polymerase, and deoxyribonucleotides are added so that the polymerase can replicate the original strand. The process is repeated until, in a short time, millions of copies of the original DNA have been made.

Figure 22.29 shows diagrammatically, in very simplified terms, the basic steps in the PCR process.

22.16 DNA Sequencing

DNA sequencing *is a method by which the base sequence in a DNA molecule (or a portion of it) is determined.* Discovered in 1977, this is the process that made the Human Genome Project (Section 22.8) possible. Today, thanks to computer technology, sequencing a nucleic acid is a fairly routine, fully automated process.

The key concept in DNA sequencing is the *selective interruption* of polynucleotide synthesis. This interruption of synthesis, which is caused to occur at every possible nucleotide site, depends on the presence of 2′,3′-*dideoxy*ribonucleotide triphosphates

FIGURE 22.29 The basic steps, in simplified terms, of the polymerase chain reaction process. Each cycle of the polymerase chain reaction doubles the number of copies of the target DNA sequence.

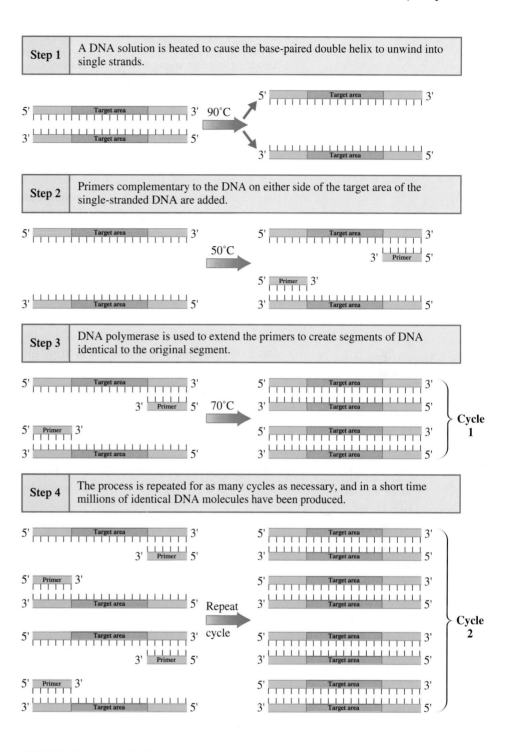

| Step 1 | A DNA solution is heated to cause the base-paired double helix to unwind into single strands. |

| Step 2 | Primers complementary to the DNA on either side of the target area of the single-stranded DNA are added. |

| Step 3 | DNA polymerase is used to extend the primers to create segments of DNA identical to the original segment. |

| Step 4 | The process is repeated for as many cycles as necessary, and in a short time millions of identical DNA molecules have been produced. |

(ddNTPs) in the synthesis mixture. Such compounds are synthetic analogs of the standard deoxyribonucleotide triphosphates in which both the $2'$ and the $3'$ hydroxy groups of deoxyribose have been replaced by hydrogen substituents.

The basic steps involved in the DNA sequencing process are as follows:

Step 1: Cleavage using restriction enzymes. Restriction enzymes (Section 22.14) are used to cleave a DNA molecule, which is too large to be sequenced as a whole, into smaller fragments (100–200 base pairs). These smaller fragments are the DNA actually sequenced. By later identifying the points of overlap among the fragments sequenced, it is possible to determine the base sequence of the entire original DNA molecule.

Step 2: Separation into individual components. The mixture of small DNA fragments generated by the restriction enzymes is separated into individual components. Each component type is then sequenced independently. Separation of the fragment mixture is accomplished via gel electrophoresis techniques (Section 20.4).

Step 3: Separation into single strands. Using chemical methods, a given DNA fragment is separated into its two strands, and one strand is then used as a template to create complementary strands of varying length via the "interruption of synthesis" process.

Step 4: Addition of primer to the single strands. The single-stranded DNA to be sequenced (the template) is mixed with short polynucleotides that serve as a primer for the complementary strand (see Figure 22.30a).

Step 5: Separation of the reaction mixture into four parts. The mixture of template DNA and primer is divided into four portions, and four parallel synthesis reactions are carried out. Each reaction mixture contains all four deoxyribonucleotide triphosphates: dATP, dCTP, dGTP, and dTTP (Section 22.2). Each test tube also contains a unique ingredient—one of the ddNTPs that has been

FIGURE 22.30 A schematic diagram of selected steps in the DNA sequencing procedure for the 10-base DNA segment 5′ AGCAGCTGGT 3′.

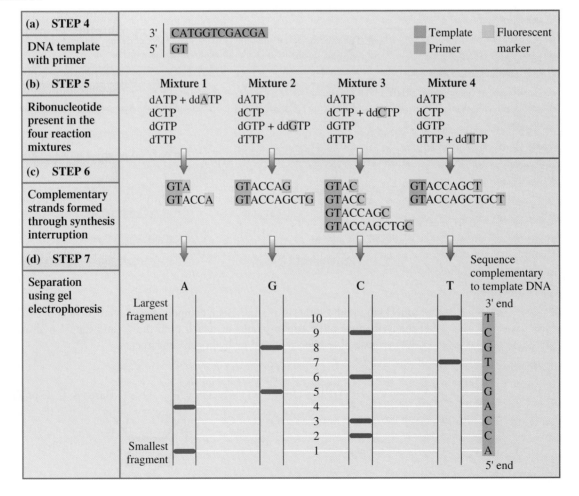

labeled with a fluorescent material that can be detected using instrumentation (see Figure 22.30b).

Step 6: **Polynucleotide synthesis with interruption.** As DNA complementary strand synthesis proceeds, nucleotides from the solution are added to the growing polynucleotide chain. Elongation of the growing chain takes place without complication until a ddNTP is incorporated into the chain. Synthesis stops at this point because a ddNTP lacks a hydroxyl group at carbon 3 and hence cannot participate in a 3′-to-5′ phosphodiester linkage, a necessary requirement for chain elongation (Section 22.3). Thus the portion of the reaction mixture that contains ddATP will be a mixture of all possible lengths of DNA complementary strands that terminate in ddA. Similarly, all of the complementary strands in the portion that contains ddGTP will terminate in ddG, and so on (see Figure 22.30c).

Step 7: **Identification of the reaction mixture components.** The newly synthesized complementary DNA strands of the four portions of the reaction mixture are then subjected to gel electrophoresis. Smaller DNA fragments move more rapidly through the gel than do larger ones, which is the basis for the separation. Fluorescence from the four differently marked ddNTPs present in the complementary strands is the basis for identification; the labeling pattern observed indicates the sequence of bases. Figure 22.30d shows the gel separation of the complementary nucleotide strands for the 10-base DNA segment shown in Figure 22.30a.

CONCEPTS TO REMEMBER

Nucleic acids. Nucleic acids are polymeric molecules in which the repeating units are nucleotides. Cells contain two kinds of nucleic acids—deoxyribonucleic acids (DNA) and ribonucleic acids (RNA). The major biological functions of DNA and RNA are, respectively, transfer of genetic information and synthesis of proteins.

Nucleotides. Nucleotides, the monomers of nucleic acid polymers, are molecules composed of a pentose sugar bonded to both a phosphate group and a nitrogen-containing heterocyclic base. The pentose sugar must be either ribose or deoxyribose. Five nitrogen-containing bases are found in nucleotides: adenine (A), guanine (G), cytosine (C), thymine (T), and uracil (U).

Primary nucleic acid structure. The "backbone" of a nucleic acid molecule is a constant alternating sequence of sugar and phosphate groups. Each sugar unit has a nitrogen-containing base attached to it.

Secondary DNA structure. A DNA molecule exists as two polynucleotide chains coiled around each other in a double-helix arrangement. The double helix is held together by hydrogen bonding between complementary pairs of bases. Only two base-pairing combinations occur: A with T, and C with G.

DNA replication. DNA replication occurs when the two strands of a parent DNA double helix separate and act as templates for the synthesis of new chains using the principle of complementary base pairing.

Chromosome. A chromosome is a cell structure that consists of an individual DNA molecule bound to a group of proteins.

RNA molecules. Five important types of RNA molecules, distinguished by their function, are ribosomal RNA (rRNA), messenger RNA (mRNA), heterogeneous nuclear RNA (hnRNA), transfer RNA (tRNA), and small nuclear RNA (snRNA).

Transcription. Transcription is the process in which the genetic information encoded in the base sequence of DNA is copied into RNA molecules.

Complementary bases. Complementary bases are specific pairs of bases in nucleic acid structures that hydrogen-bond to each other.

Gene. A gene is a portion of a DNA molecule that contains the base sequences needed for the production of a specific hnRNA/mRNA molecule. Genes are segmented, with portions called exons that contain genetic information and portions called introns that do not convey genetic information.

Codon. A codon is a three-nucleotide sequence in mRNA that codes for a specific amino acid needed during the process of protein synthesis.

Genetic code. The genetic code consists of all the mRNA codons that specify either a particular amino acid or the termination of protein synthesis.

Anticodon. An anticodon is a three-nucleotide sequence in tRNA that binds to a complementary sequence (a codon) in mRNA.

Translation. Translation is the stage of protein synthesis in which the codons in mRNA are translated into amino acid sequences of new proteins. Translation involves interactions between the codons of mRNA and the anticodons of tRNA.

Mutations. Mutations are changes in the base sequence in DNA molecules.

Recombinant DNA. Recombinant DNA molecules are synthesized by splicing a segment of DNA, usually a gene, from one organism into the DNA of another organism.

Polymerase chain reaction. The polymerase chain reaction is a method for rapidly producing many copies of a DNA sequence.

DNA sequencing. DNA sequencing is a multistep process for determining the sequence of bases in a DNA segment.

KEY REACTIONS AND EQUATIONS

1. Formation of a nucleotide (Section 22.2)

Pentose sugar (ribose or deoxyribose) + phosphate group
+ nitrogen-containing heterocyclic base ⟶

Phosphate — Sugar
|
Base   + 2H$_2$O

2. Formation of a nucleic acid (Section 22.3)

Many deoxyribose-containing nucleotides ⟶ DNA
Many ribose-containing nucleotides ⟶ RNA

3. Protein synthesis (Section 22.6)

DNA $\xrightarrow{\text{Transcription}}$ RNA $\xrightarrow{\text{Translation}}$ Protein

---

KEY TERMS

Alternative splicing (22.8)
Anticodon (22.10)
Chromosome (22.5)
Clones (22.14)
Codon (22.9)
Complementary bases (22.4)
Complementary DNA strands (22.4)
Deoxyribonucleic acid (22.3)
DNA replication (22.5)
DNA sequencing (22.16)
Exon (22.8)
Gene (22.8)
Genetic code (22.9)
Genome (22.8)

Heterogeneous nuclear RNA (22.7)
Intron (22.8)
Messenger RNA (22.7)
Mutagen (22.12)
Mutation (22.12)
Nucleic acid (22.1)
Nucleotide (22.2)
Polymerase chain reaction (22.15)
Polysome (22.11)
Primary nucleic acid structure (22.3)
Recombinant DNA (22.14)
Restriction enzyme (22.14)
Ribonucleic acid (RNA) (22.3)
Ribosomal RNA (22.7)

Ribosome (22.11)
Small nuclear ribonucleoprotein particle (22.8)
Small nuclear RNA (22.7)
Spliceosome (22.8)
Splicing (22.8)
Transcription (22.8)
Transfer RNA (22.7)
Transformation (22.14)
Translation (22.11)
Translocation (22.11)
Vaccine (22.13)
Virus (22.13)

EXERCISES AND PROBLEMS

The members of each pair of problems in this section test similar material.

■ **Nucleotides (Section 22.2)**

22.1 What is the structural difference between the pentose sugars ribose and 2-deoxyribose?

22.2 What are the names of the pentose sugars present, respectively, in DNA and RNA molecules?

22.3 Characterize each of the following nitrogen-containing bases as a purine derivative or a pyrimidine derivative.
a. Thymine
b. Cytosine
c. Adenine
d. Guanine

22.4 Characterize each of the following nitrogen-containing bases as a component of (1) both DNA and RNA, (2) DNA but not RNA, or (3) RNA but not DNA.
a. Adenine
b. Thymine
c. Uracil
d. Cytosine

22.5 How many different choices are there for each of the following subunits in the specified type of nucleotide?
a. Pentose sugar subunit in DNA nucleotides
b. Nitrogen-containing base subunit in RNA nucleotides
c. Phosphate subunit in DNA nucleotides

22.6 How many different choices are there for each of the following subunits in the specified type of nucleotide?
a. Pentose sugar subunit in RNA nucleotides
b. Nitrogen-containing base subunit in DNA nucleotides
c. Phosphate subunit in RNA nucleotides

22.7 Which nitrogen-containing base is present in each of the following nucleotides?
a. AMP
b. dGMP
c. dTMP
d. UMP

22.8 Which nitrogen-containing base is present in each of the following nucleotides?
a. GMP
b. dAMP
c. CMP
d. dCMP

22.9 Which pentose sugar is present in each of the nucleotides in Problem 22.7?

22.10 Which pentose sugar is present in each of the nucleotides in Problem 22.8?

22.11 Characterize as true or false each of the following statements about the given nucleotide.

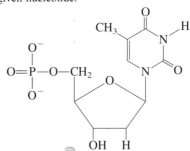

a. The nitrogen-containing base is a purine derivative.
b. The phosphate group is attached to the sugar unit at carbon 3′.
c. The sugar unit is ribose.
d. The nucleotide could be a component of both DNA and RNA.

22.12 Characterize as true or false each of the following statements about the given nucleotide.

a. The sugar unit is 2-deoxyribose.
b. The sugar unit is attached to the nitrogen-containing base at nitrogen 3.
c. The nitrogen-containing base is a pyrimidine derivative.
d. The nucleotide could be a component of both DNA and RNA.

22.13 Draw the structures of the three products produced when the nucleotide in Problem 22.11 undergoes hydrolysis.

22.14 Draw the structures of the three products produced when the nucleotide in Problem 22.12 undergoes hydrolysis.

■ **Primary Nucleic Acid Structure (Section 22.3)**

22.15 What are the two repeating subunits present in the *backbone* portion of a nucleic acid?

22.16 To which type of subunit in a nucleic acid *backbone* are the nitrogen-containing bases attached?

22.17 What distinguishes various DNA molecules from each other?

22.18 What distinguishes various RNA molecules from each other?

22.19 What is the difference between a nucleic acid's 3′ end and its 5′ end?

22.20 In the lengthening of a polynucleotide chain, which type of nucleotide subunit would bond to the 3′ end of the polynucleotide chain?

22.21 What are the nucleotide subunits that participate in a nucleic acid 3′,5′-phosphodiester linkage?

22.22 How many 3′,5′-phosphodiester linkages are present in a tetranucleotide segment of a nucleic acid?

22.23 Draw the structure of the dinucleotide product obtained by combining the nucleotides of Problems 22.11 and 22.12 such that the Problem 22.11 nucleotide is the 5′ end of the dinucleotide.

22.24 Draw the structure of the dinucleotide product obtained by combining the nucleotides of Problems 22.11 and 22.12 such that the Problem 22.11 nucleotide is the 3′ end of the dinucleotide.

■ **The DNA Double Helix (Section 22.4)**

22.25 Describe the DNA double helix in terms of
a. general shape
b. what is on the outside of the helix and what is within the interior of the helix

22.26 Describe the DNA double helix in terms of
a. the directionality of the two polynucleotide chains present
b. a comparison of the total number of nitrogen-containing bases present in each of the two polynucleotide chains

22.27 The base content of a particular DNA molecule is 36% thymine. What is the percentage of each of the following bases in the molecule?
a. Adenine   b. Guanine   c. Cytosine

22.28 The base content of a particular DNA molecule is 24% guanine. What is the percentage of each of the following bases in the molecule?
a. Adenine   b. Cytosine   c. Thymine

22.29 In terms of hydrogen bonding, a G–C base pair is more stable than an A–T base pair. Explain why this is so.

22.30 What structural consideration prevents the following bases from forming complementary base pairs?
a. A and G   b. T and C

22.31 What is the relationship between the total number of purine bases (A and G) and the total number of pyrimidine bases (C and T) present in a DNA double helix?

22.32 The base composition for one of the strands of a DNA double helix is 19% A, 34% C, 28% G, and 19% T. What is the percent base composition for the other strand of the DNA double helix?

22.33 Identify the 3′ and 5′ ends of the DNA base sequence TAGCC.

22.34 The two-base DNA sequences TA and AT represent different dinucleotides. Explain why this is so.

22.35 Using the concept of complementary base pairing, write the complementary DNA strands, with their 5′ and 3′ ends labeled, for each of the following DNA base sequences.
a. 5′ ACGTAT 3′   b. 5′ TTACCG 3′
c. 3′ GCATAA 5′   d. AACTGG

22.36 Using the concept of complementary base pairing, write the complementary DNA strands, with their 5′ and 3′ ends labeled, for each of the following DNA base sequences.
a. 5′ CCGGTA 3′   b. 5′ CACAGA 3′
c. 3′ TTTAGA 5′   d. CATTAC

22.37 How many total hydrogen bonds would exist between the DNA strand 5′ AGTCCTCA 3′ and its complementary strand?

22.38 How many total hydrogen bonds would exist between the DNA strand 5′ CCTAGGAT 3′ and its complementary strand?

■ **Replication of DNA Molecules (Section 22.5)**

22.39 What is the function of the enzyme *DNA helicase* in the DNA replication process?

22.40 What are two functions of the enzyme *DNA polymerase* in the DNA replication process?

22.41 In the replication of a DNA molecule, two daughter molecules, Q and R, are formed. The following base sequence is part of the newly formed strand in daughter molecule Q.

5′ ACTTAG 3′

Indicate the corresponding base sequence in
a. the newly formed strand in daughter molecule R
b. the "parent" strand in daughter molecule Q
c. the "parent" strand in daughter molecule R

22.42 In the replication of a DNA molecule, two daughter molecules, S and T, are formed. The following base sequence is part of the "parent" strand in daughter molecule S.

5′ TTCAGAG 3′

Indicate the corresponding base sequence in
a. the newly formed strand in daughter molecule T
b. the newly formed strand in daughter molecule S
c. the "parent" strand in daughter molecule T

22.43 During DNA replication, one of the newly formed strands grows continuously, whereas the other grows in segments that are later connected together. Explain why this is so.

22.44 DNA replication is most often a bidirectional process. Explain why this is so.

22.45 What is a chromosome?

22.46 Chromosomes are nucleoproteins. Explain.

■ **RNA Molecules (Section 22.7)**

22.47 What are the four major differences between RNA molecules and DNA molecules?

22.48 What are the names and abbreviations for the five major types of RNA molecules?

22.49 State whether each of the following phrases applies to hnRNA, mRNA, tRNA, rRNA, or snRNA.
a. Material from which messenger RNA is made
b. Delivers amino acids to protein synthesis sites
c. Smallest of the RNAs in terms of nucleotide units present
d. Also goes by the designation ptRNA

22.50 State whether each of the following phrases applies to hnRNA, mRNA, tRNA, rRNA, or snRNA.
a. Associated with a series of proteins in a complex structure
b. Contains genetic information needed for protein synthesis
c. Most abundant type of RNA in a cell
d. Involved in the editing of hnRNA molecules

22.51 For each of the following types of RNA, indicate whether the predominant cellular location for the RNA is the nuclear region, the extranuclear region, or both the nuclear and the extranuclear regions.
a. hnRNA   b. tRNA   c. rRNA   d. mRNA

22.52 Indicate whether each of the following processes occurs in the nuclear or the extranuclear region of a cell?
a. DNA transcription
b. Processing of hnRNA to mRNA
c. mRNA translation (protein synthesis)
d. DNA replication

■ **Transcription: RNA Synthesis (Section 22.8)**

22.53 What serves as a template in the process of *transcription*?

22.54 What is the initial product of the *transcription* process?

22.55 What are two functions of the enzyme *RNA polymerase* in the transcription process?

22.56 What is a *gene*?

22.57 What are the complementary base pairs in DNA–RNA interactions?

22.58 In DNA–DNA interactions there are two complementary base pairs, and in DNA–RNA interactions there are three complementary base pairs. Explain.

22.59 Write the base sequence of the hnRNA formed by transcription of the following DNA base sequence.

5′ ATGCTTA 3′

22.60 Write the base sequence of the hnRNA formed by transcription of the following DNA base sequence.

5′ TAGTGAT 3′

22.61 From what DNA base sequence was the following hnRNA sequence transcribed?

5′ UUCGCAG 3′

22.62 From what DNA base sequence was the following hnRNA sequence transcribed?

5′ GCUUAUC 3′

22.63 What is the relationship between an exon and a gene?

22.64 What is the relationship between an intron and a gene?

22.65 What mRNA base sequence would be obtained from the following portion of a gene?

| exon | intron | exon |

5′ TCAG–TAGC–TTCA 3′

22.66 What mRNA base sequence would be obtained from the following portion of a gene?

| intron | exon | intron |

5′ TTAC–AACG–GCAT 3′

22.67 In the process of splicing, which type of RNA
a. undergoes the splicing?
b. is present in the spliceosomes?

22.68 What is the difference between snRNA and snRNPs?

22.69 What is *alternative splicing*?

22.70 How many different mRNAs can be produced from an hnRNA that contains three exons, one of which is an "alternative" exon?

■ **The Genetic Code (Section 22.9)**

22.71 What is a codon?

22.72 On what type of RNA molecule are codons found?

22.73 Using the information in Table 22.2, determine what amino acid is coded for by each of the following codons.
a. CUU   b. AAU   c. AGU   d. GGG

22.74 Using the information in Table 22.2, determine what amino acid is coded for by each of the following codons.
a. GUA   b. CCC   c. CAC   d. CCA

22.75 Using the information in Table 22.2, determine the synonyms, if any, of each of the codons in Problem 22.73.

22.76 Using the information in Table 22.2, determine the synonyms, if any, of each of the codons in Problem 22.74.

22.77 Explain why the base sequence ATC could not be a codon.

22.78 Explain why the base sequence AGAC could not be a codon.

22.79 Predict the sequence of amino acids coded by the mRNA sequence

5′ AUG–AAA–GAA–GAC–CUA 3′

22.80 Predict the sequence of amino acids coded by the mRNA sequence

5′ GGA–GGC–ACA–UGG–GAA 3′

■ **Anticodons and tRNA Molecules (Section 22.10)**

22.81 Describe the general structure of a tRNA molecule.

22.82 Where is the anticodon site on a tRNA molecule?

22.83 By what type of bond is an amino acid attached to a tRNA molecule?

22.84 What principle governs the codon–anticodon interaction that leads to proper placement of amino acids in proteins?

22.85 What is the anticodon that would interact with each of the following codons?
a. AGA   b. CGU   c. UUU   d. CAA

22.86 What is the anticodon that would interact with each of the following codons?
a. CCU   b. GUA   c. AUC   d. GCA

22.87 Identify the amino acid associated with each of the codons in Problem 22.85.
a. UGG   b. GAC   c. GGA   d. AGA

22.88 Identify the amino acid associated with each of the codons in Problem 22.86.
a. UGU   b. ACG   c. AGU   d. CAC

■ Translation: Protein Synthesis (Section 22.11)

22.89 What are the five major steps in translation (protein synthesis)?

22.90 What is a ribosome, and what role do ribosomes play in protein synthesis?

22.91 In the growth step of protein synthesis, at which site in the ribosome does new peptide bond formation actually take place?

22.92 What two changes occur at a ribosome during protein synthesis immediately after peptide bond formation?

22.93 Write a possible mRNA base sequence that would lead to the production of this pentapeptide. (There is more than one correct answer.)

Gly–Ala–Cys–Val–Tyr

22.94 Write a possible mRNA base sequence that would lead to the production of this pentapeptide. (There is more than one correct answer.)

Lys–Met–Thr–His–Phe

■ Mutations (Section 22.12)

22.95 For the codon sequence

5′ GGC–UAU–AGU–AGC–CCC 3′

write the amino acid sequence produced in each of the following ways.
a. Translation proceeds in a normal manner.
b. A mutation changes CCC to CCU.
c. A mutation changes CCC to ACC.

22.96 For the codon sequence

5′ GGA–AUA–UGG–UUC–CUA 3′

write the amino acid sequence produced in each of the following ways.
a. Translation proceeds in a normal manner.
b. A mutation changes GGA to GGG.
c. A mutation changes GGA to CGA.

■ Viruses and Vaccines (Section 22.13)

22.97 Describe the general structure of a virus.

22.98 What is the only function of a virus?

22.99 What is the most common method by which viruses invade cells?

22.100 Why must a virus infect another organism in order to reproduce?

■ Recombinant DNA and Genetic Engineering (Section 22.14)

22.101 How does recombinant DNA differ from normal DNA?

22.102 Give two reasons why bacterial cells are used for recombinant DNA procedures.

22.103 What role do plasmids play in recombinant DNA procedures?

22.104 Describe what occurs when a particular restriction enzyme operates on a segment of double-stranded DNA.

22.105 Describe what happens during transformation.

22.106 How are plasmids obtained from *E. coli* bacteria?

22.107 A particular restriction enzyme will cleave DNA between A and A in the sequence AAGCTT in the 5′-to-3′ direction. Draw a diagram showing the structural details of the "sticky ends" that result from cleavage of the following DNA segment.

22.108 A particular restriction enzyme will cleave DNA between A and A in the sequence AAGCTT in the 5′-to-3′ direction. Draw a diagram showing the structural details of the "sticky ends" that result from cleavage of the following DNA segment.

5′ ┌─────────────────┐ 3′
   G G A A G C T T A
   C C T T C G A A T
3′ └─────────────────┘ 5′

■ Polymerase Chain Reaction (Section 22.15)

22.109 What is the function of the polymerase chain reaction?

22.110 What is the function of the enzyme *DNA polymerase* in the PCR process?

22.111 What is a *primer* and what is its function in the PCR process?

22.112 What are the four types of substances needed to carry out the PCR process?

■ DNA Sequencing (Section 22.16)

22.113 How do the notations dATP and ddATP differ in meaning?

22.114 What role do dideoxynucleotides play in the DNA sequencing process?

22.115 Assume that the "dark lines" in the first and second columns of Figure 22.30d are interchanged, while the other labels stay the same.
a. Given this change, what would be the sequence of bases in the DNA fragment under study?
b. Given this change, what would be the sequence of bases in the original DNA fragment that was to be sequenced?

22.116 Assume that the "dark lines" in the second and fourth columns of Figure 22.30d are interchanged, while the other labels stay the same.
a. Given this change, what would be the sequence of bases in the DNA fragment under study?
b. Given this change, what would be the sequence of bases in the original DNA fragment that was to be sequenced?

ADDITIONAL PROBLEMS

22.117 With the help of the structures given in Section 22.2, describe the structural differences between the following pairs of nucleotide bases.
a. Thymine and uracil   b. Adenine and guanine

22.118 The following is a sequence of bases for an exon portion of a strand of a gene.

5′ CATACAGCCTGGAAGCTA 3′

a. What is the sequence of bases on the strand of DNA complementary to this segment?
b. What is the sequence of bases on the mRNA molecule synthesized from this strand?
c. What codons are present on the mRNA molecule from part b?
d. What anticodons will be found on the tRNA molecules that interact with the codons from part c?
e. What is the sequence of amino acids in the peptide formed using these protein synthesis instructions?

22.119 Which of these RNA types, (1) mRNA, (2) hnRNA, (3) rRNA, or (4) tRNA, is most closely associated with each of the following terms?
a. Codon   b. Anticodon
c. Intron   d. Amino acid carrier

22.120 Which of these processes, (1) translation phase of protein synthesis, (2) transcription phase of protein synthesis, (3) replication of DNA, or (4) formation of recombinant DNA, is associated with each of the following events?
a. Complete unwinding of a DNA molecule occurs.
b. Partial unwinding of a DNA molecule occurs.
c. An mRNA–ribosome complex is formed.
d. Okazaki fragments are formed.

22.121 Which of these base-pairing situations, (1) between two DNA segments, (2) between two RNA segments, (3) between a DNA segment and an RNA segment, or (4) between a codon and an anticodon, fits each of the following base-pairing sequences? More than one response may apply to a given base-pairing situation.

a. A G T   b. A C T
    U C A   T G A

c. A G U   d. C C G
    U C A   G G C

22.122 Which of these characterizations, (1) found in DNA but not RNA, (b) found in RNA but not DNA, (3) found in both DNA and RNA, or (4) not found in DNA or RNA, fits each of the following mono-, di- or trinucleotides?
a. 5′ dAMP–dAMP 3′
b. 5′ AMP–AMP–CMP 3′
c. 5′ dAMP–CMP 3′
d. 5′ GGA 3′

22.123 Suppose that 28% of the nucleotides of a DNA molecule are deoxythymidine 5′-monophosphate, and during replication the relative amounts of available nucleotide bases are 22% A, 28% T, 22% C, and 28% G. What base would be depleted first in the replication process?

22.124 On the basis of the initial results of the Human Genome Project concerning the DNA present in a human cell
a. How many base-pairs are present in the DNA?
b. How many genes are present in the DNA?
c. What percentage of the base-pairs are accounted for by the genes?

ANSWERS TO PRACTICE EXERCISES

22.1 3′ T–T–A–C–G–T–C–G–A 5′
22.2 3′ A–U–U–G–G–A 5′ which becomes 5′ A–G–G–U–U–A 3′
22.3 His–Pro–His–Thr–Val–Cys–Trp

22.4 a. 3′ GCG–GCA–UCA–ACC–GGG–CCU–CCU 5′
b. 3′ GCG–ACC–CCU–CCU 5′ which becomes
5′ UCC–UCC–CCA–GCG 3′
c. Ser–Ser–His–Gly

23

Biochemical Energy Production

The energy consumed by these scarlet ibises in flight is generated by numerous sequences of biochemical reactions that occur within their bodies.

This chapter is the first of four dealing with the chemical reactions that occur in a living organism. In this first chapter, we consider those molecules that are repeatedly encountered in biological reactions, as well as those reactions that are common to the processing of carbohydrates, lipids, and proteins. The three following chapters consider the reactions associated uniquely with carbohydrate, lipid, and protein processing, respectively.

23.1 Metabolism

Metabolism *is the sum total of all the biochemical reactions that take place in a living organism.* Human metabolism is quite remarkable. An average human adult whose weight remains the same for 40 years processes about 6 *tons* of solid food and 10,000 gallons of water, during which time the composition of the body is essentially constant. Just as we must put gasoline in a car to make it go or plug in a kitchen appliance to make it run, we also need a source of energy to think, breathe, exercise, or work. As we have seen in previous chapters, even the simplest living cell is continually carrying on energy-demanding processes such as protein synthesis, DNA replication, RNA transcription, and membrane transport.

Metabolic reactions fall into one of two subtypes: catabolism and anabolism. **Catabolism** *is all metabolic reactions in which large biochemical molecules are broken down to smaller ones.* Catabolic reactions usually release energy. The reactions involved in the oxidation of glucose are catabolic. **Anabolism** *is all metabolic reactions in which*

Catabolism is pronounced ca-TAB-o-lism, and *anabolism* is pronounced an-ABB-o-lism. *Catabolic* is pronounced CAT-a-bol-ic, and *anabolic* is pronounced AN-a-bol-ic.

FIGURE 23.1 The processes of catabolism and anabolism are opposite in nature. The first usually produces energy, and the second usually consumes energy.

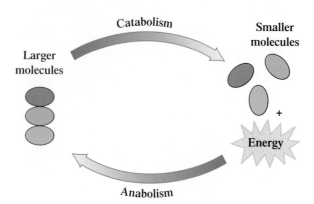

small biochemical molecules are joined together to form larger ones. Anabolic reactions usually require energy in order to proceed. The synthesis of proteins from amino acids is an anabolic process. Figure 23.1 contrasts catabolic and anabolic processes.

The metabolic reactions that occur in a cell are usually organized into sequences called *metabolic pathways.* A **metabolic pathway** *is a series of consecutive biochemical reactions used to convert a starting material into an end product.* Such pathways may be *linear,* in which a series of reactions generates a final product, or *cyclic,* in which a series of reactions regenerates the first reactant.

Linear metabolic pathway: A $\longrightarrow$ B $\longrightarrow$ C $\longrightarrow$ D

Cyclic metabolic pathway:

The major metabolic pathways for all life forms are similar. This enables scientists to study metabolic reactions in simpler life forms and use the results to help understand the corresponding metabolic reactions in more complex organisms, including humans.

23.2 Metabolism and Cell Structure

Knowledge of the major structural features of a cell is a prerequisite to understanding *where* metabolic reactions take place.

Cells are of two types: prokaryotic and eukaryotic. *Prokaryotic cells* have no nucleus and are found only in bacteria. The DNA that governs the reproduction of prokaryotic cells is usually a single circular molecule found near the center of the cell in a region called the *nucleoid.* A **eukaryotic cell** *is a cell in which the DNA is found in a membrane-enclosed nucleus.* Cells of this type, which are found in all higher organisms, are about 1000 times larger than bacterial cells. Our focus in the remainder of this section will be on eukaryotic cells, the type present in humans. Figure 23.2 shows the general internal structure of a eukaryotic cell. Note the key components shown: the outer membrane, nucleus, cytosol, ribosomes, lysosomes, and mitochondria.

The **cytoplasm** *is the water-based material of a eukaryotic cell that lies between the nucleus and the outer membrane of the cell.* Within the cytoplasm are several kinds of small structures called *organelles.* An **organelle** *is a minute structure within the cytoplasm of a cell that carries out a specific cellular function.* The organelles are surrounded by the *cytosol.* The **cytosol** *is the water-based fluid part of the cytoplasm of a cell.*

Three important types of organelles are ribosomes, lysosomes, and mitochondria. We have considered ribosomes before; they are the sites where protein synthesis occurs (Section 22.11) A **lysosome** *is an organelle that contains hydrolytic enzymes needed for cellular rebuilding, repair, and degradation.* Some lysosome enzymes hydrolyze proteins to amino acids; others hydrolyze polysaccharides to monosaccharides. Bacteria and viruses

The term *eukaryotic,* pronounced you-KAHR-ee-ah-tic, is from the Greek *eu,* meaning "true," and *karyon,* meaning "nucleus." The term *prokaryotic,* which contains the Greek *pro,* meaning "before," literally means "before the nucleus."

Eukaryotic and prokaryotic cells differ in that the former contain a well-defined nucleus, set off from the rest of the cell by a membrane.

FIGURE 23.2 A schematic representation of a eukaryotic cell with selected internal components identified.

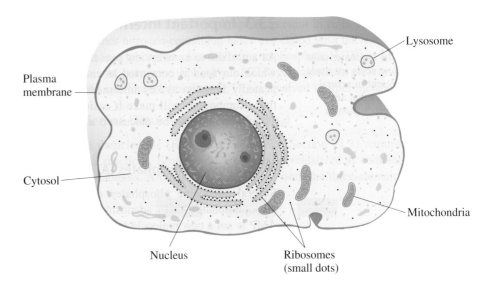

Lysosome

Plasma membrane

Cytosol

Mitochondria

Nucleus

Ribosomes (small dots)

▶ A protective mechanism exists to prevent lysosome enzymes from destroying the cell in which they are found if they should be accidently released (via membrane rupture or leakage). The optimum pH (Section 21.6) for lysosome enzyme activity is 4.8. The cytoplasmic pH of 7.0–7.3 renders them inactive.

▶ *Mitochondria*, pronounced my-toe-KON-dree-ah, is plural. The singular form of the term is *mitochondrion*. The threadlike shape of mitochondria is responsible for this organelle's name; *mitos* is Greek for "thread," and *chondrion* is Greek for "granule."

"trapped" by the body's immune system (Section 20.17) are degraded and destroyed by enzymes from lysosomes.

A **mitochondrion** *is an organelle that is responsible for the generation of most of the energy for a cell.* Much of the discussion of this chapter deals with the energy-producing chemical reactions that occur within mitochondria. Further details of mitochondrion structure will help us understand more about how these reactions occur.

Mitochondria are sausage-shaped organelles containing both an *outer membrane* and a *multifolded inner membrane* (see Figure 23.3). The outer membrane, which is about 50% lipid and 50% protein, is freely permeable to small molecules. The inner membrane, which is about 20% lipid and 80% protein, is highly impermeable to most substances. The nonpermeable nature of the inner membrane divides a mitochondrion into two separate compartments—an interior region called the *matrix* and the region between the inner and outer membranes, called the *intermembrane space*. The folds of the inner membrane that protrude into the matrix are called *cristae*.

The invention of high-resolution electron microscopes allowed researchers to see the interior structure of the mitochondrion more clearly and led to the discovery, in 1962, of small spherical knobs attached to the cristae called *ATP synthase complexes*. As their name implies, these relatively small knobs, which are located on the matrix side of the inner membrane, are responsible for ATP synthesis, and their association with the inner membrane is critically important for this task. More will be said about ATP in the next section.

FIGURE 23.3 (a) A schematic representation of a mitochondrion, showing key features of its internal structure. (b) An electron micrograph of a single mitochondrial crista, showing the ATP synthase knobs extending into the matrix.

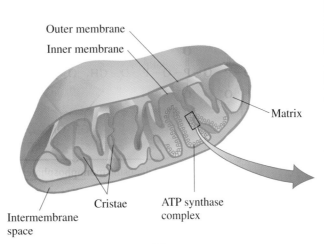

Outer membrane

Inner membrane

Matrix

Intermembrane space

Cristae

ATP synthase complex

(a)

(b)

▶ The first stage of biochemical energy production, digestion, is not considered part of metabolism because it is extracellular. Metabolic processes are intracellular.

Stage 1: The first stage, *digestion,* begins in the mouth (saliva contains starch-digesting enzymes), continues in the stomach (gastric juices), and is completed in the small intestine (the majority of digestive enzymes and bile salts). The end products of digestion—glucose and other monosaccharides from carbohydrates, amino acids from proteins, and fatty acids and glycerol from fats and oils—are small enough to pass across intestinal membranes and into the blood, where they are transported to the body's cells.

Stage 2: The second stage, *acetyl group formation,* involves numerous reactions, some of which occur in the cytosol of cells and some in cellular mitochondria. The small molecules from digestion are further oxidized during this stage. Primary products include two-carbon acetyl units (which become attached to coenzyme A to give acetyl CoA) and the reduced coenzyme NADH.

Stage 3: The third stage, the *citric acid cycle,* occurs inside mitochondria. Here acetyl groups are oxidized to produce CO_2 and energy. Some of the energy released by these reactions is lost as heat, and some is carried by the reduced coenzymes NADH and $FADH_2$ to the fourth stage. The CO_2 that we exhale as part of the breathing process comes primarily from this stage.

Stage 4: The fourth stage, the *electron transport chain and oxidative phosphorylation,* also occurs inside mitochondria. NADH and $FADH_2$ supply the "fuel" (hydrogen ions and electrons) needed for the production of ATP molecules, the primary energy carriers in metabolic pathways. Molecular O_2, inhaled via breathing, is converted to H_2O in this stage.

The reactions in stages 3 and 4 are the same for all types of foods (carbohydrates, fats, proteins). These reactions constitute the common metabolic pathway. The **common metabolic pathway** *is the sum total of the biochemical reactions of the citric acid cycle, the electron transport chain, and oxidative phosphorylation.* The remainder of this chapter deals with the common metabolic pathway. The reactions of stages 1 and 2 of biochemical energy production differ for different types of foodstuffs. They are discussed in Chapters 24–26, which cover the metabolism of carbohydrates, fats (lipids), and proteins, respectively.

The Chemistry at a Glance feature on page 703 summarizes the four general stages in the process of production of biochemical energy from ingested food. This diagram is a *very simplified* version of the "energy generation" process that occurs in the human body, as will become clear from the discussions presented in later sections of this chapter, which give further details of the process.

23.6 The Citric Acid Cycle

The **citric acid cycle** *is the series of biochemical reactions in which the acetyl portion of acetyl CoA is oxidized to carbon dioxide and the reduced coenzymes $FADH_2$ and NADH are produced.* This cycle, stage 3 of biochemical energy production, gets its name from the first intermediate product in the cycle, citric acid. It is also known as the *Krebs cycle,* after its discoverer Hans Adolf Krebs (see Figure 23.7), and as the *tricarboxylic acid cycle,* in reference to the three carboxylate groups present in citric acid. Figure 23.8 shows the compounds produced in all eight steps of the citric acid cycle.

We shall now consider the individual steps of the cycle in detail. As we go through these steps, we will observe two important types of reactions: (1) oxidation, which produces NADH and $FADH_2$, and (2) decarboxylation, wherein a carbon chain is shortened by the removal of a carbon atom as CO_2.

■ Reactions of the Citric Acid Cycle

Step 1: *Formation of Citrate.* Acetyl CoA, the two-carbon degradation product of carbohydrates, fats, and proteins (Section 23.5), enters the cycle by combining

FIGURE 23.7 Hans Adolf Krebs (1900–1981), a German-born British biochemist, received the 1953 Nobel Prize in medicine for establishing the relationships among the different compounds in the cycle that carries his name, the Krebs cycle.

Simplified Summary of the Four Stages of Biochemical Energy Production

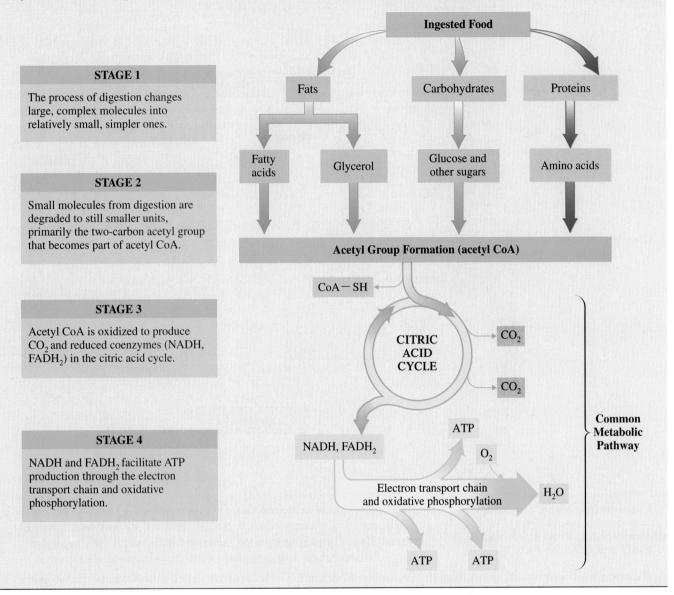

STAGE 1

The process of digestion changes large, complex molecules into relatively small, simpler ones.

STAGE 2

Small molecules from digestion are degraded to still smaller units, primarily the two-carbon acetyl group that becomes part of acetyl CoA.

STAGE 3

Acetyl CoA is oxidized to produce CO_2 and reduced coenzymes (NADH, $FADH_2$) in the citric acid cycle.

STAGE 4

NADH and $FADH_2$ facilitate ATP production through the electron transport chain and oxidative phosphorylation.

▶ The formation of citryl CoA is a condensation reaction (Section 14.7) because a new carbon–carbon bond is formed.

with the four-carbon keto dicarboxylate species oxaloacetate. This results in the transfer of the acetyl group from coenzyme A to oxaloacetate, producing the C_6 citrate species and free coenzyme A.

FIGURE 23.8 The citric acid cycle. Details of the numbered steps are given in the text.

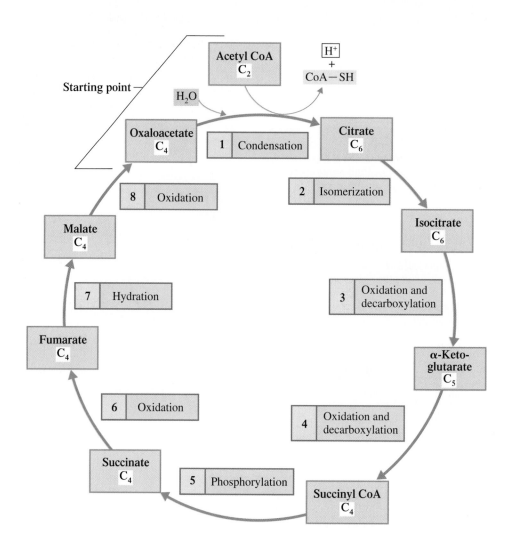

A *synthase* is an enzyme that makes a new covalent bond during a reaction without the direct involvement of an ATP molecule.

The formation of citryl CoA involves addition of acetyl CoA to the carbon–oxygen double bond. A hydrogen atom of the acetyl —CH₃ group adds to the oxygen atom of the double bond, and the remainder of the acetyl CoA adds to the carbon atom of the double bond.

There are two parts to the reaction: (1) the condensation of acetyl CoA and oxaloacetate to form citryl CoA, a process catalyzed by the enzyme *citrate synthase,* and (2) hydrolysis of the thioester bond in citryl CoA to produce CoA—SH and citrate, also catalyzed by the enzyme *citrate synthase.*

Step 2: *Formation of Isocitrate.* Citrate is converted to its less symmetrical isomer isocitrate in an isomerization process that involves a dehydration followed by a hydration, both catalyzed by the enzyme *aconitase.* The net result of these reactions is that the —OH group from citrate is moved to a different carbon atom.

Citrate is a tertiary alcohol and isocitrate a secondary alcohol. Tertiary alcohols are not readily oxidized; secondary alcohols are easier to oxidize (Section 14.7). The next step in the cycle involves oxidation.

All acids found in the citric acid cycle exist as negative ions (carboxylate ions; Section 16.8) at cellular pH.

Step 3: *Oxidation of Isocitrate and Formation of CO_2.* This step involves oxidation–reduction (the first of four redox reactions in the citric acid cycle) and decarboxylation. The reactants are a NAD^+ molecule and isocitrate. The reaction, catalyzed by *isocitrate dehydrogenase,* is complex: (1) Isocitrate is oxidized to a ketone (oxalosuccinate) by NAD^+, releasing two hydrogens. (2) One hydrogen and two electrons are transferred to NAD^+ to form NADH; the remaining hydrogen ion (H^+) is released. (3) The oxalosuccinate remains bound to the enzyme and undergoes decarboxylation (loses CO_2), which produces the C_5 α-ketoglutarate (a keto dicarboxylate species).

Isocitrate   Oxalosuccinate α-Ketoglutarate

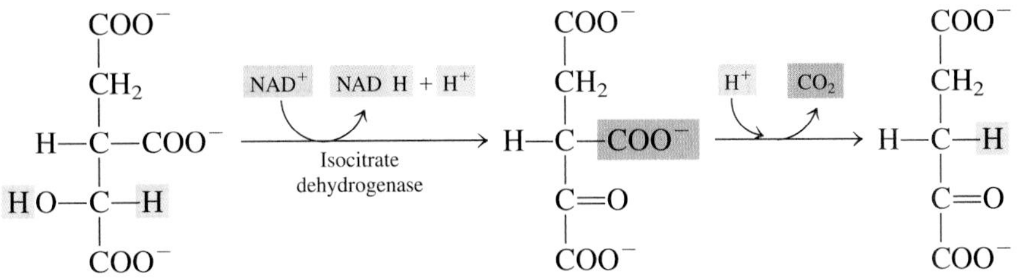

This step yields the first molecules of CO_2 and NADH in the cycle.

The CO_2 molecules produced in Steps 3 and 4 of the citric acid cycle are the CO_2 molecules we exhale in the process of respiration.

Step 4: *Oxidation of α-Ketoglutarate and Formation of CO_2.* This second redox reaction of the cycle involves one molecule each of NAD^+, CoA—SH, and α-ketoglutarate. The catalyst is an aggregate of three enzymes called the *α-ketoglutarate dehydrogenase* complex. As in Step 3, both oxidation and decarboxylation occur. There are three products: CO_2, NADH, and the C_4 species succinyl CoA.

α-Ketoglutarate   Succinyl CoA

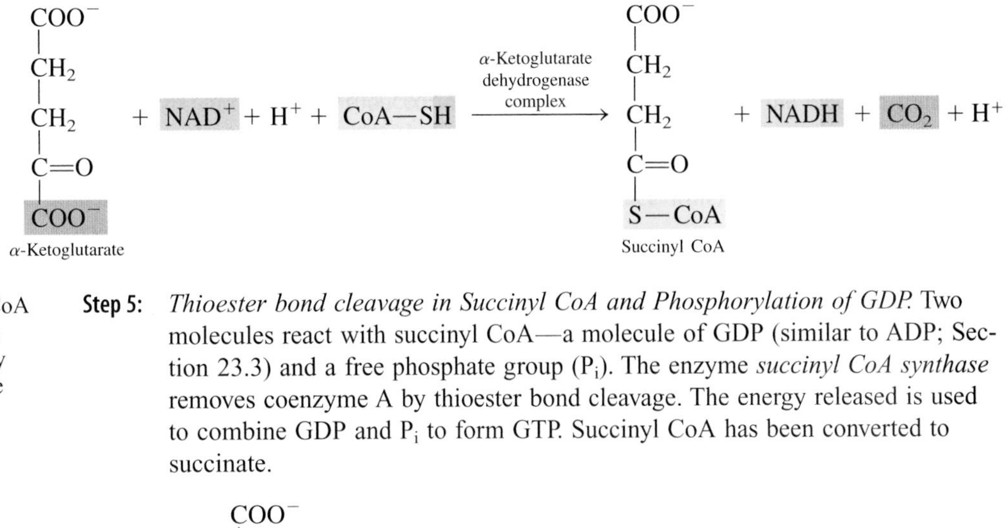

The thioester bond in succinyl CoA is a strained bond. Its hydrolysis releases energy, which is trapped by GTP formation. The function of the GTP produced is similar to that of ATP: to store energy in the form of a high-energy phosphate bond (Section 23.4).

Step 5: *Thioester bond cleavage in Succinyl CoA and Phosphorylation of GDP.* Two molecules react with succinyl CoA—a molecule of GDP (similar to ADP; Section 23.3) and a free phosphate group (P_i). The enzyme *succinyl CoA synthase* removes coenzyme A by thioester bond cleavage. The energy released is used to combine GDP and P_i to form GTP. Succinyl CoA has been converted to succinate.

Succinyl CoA   Succinate

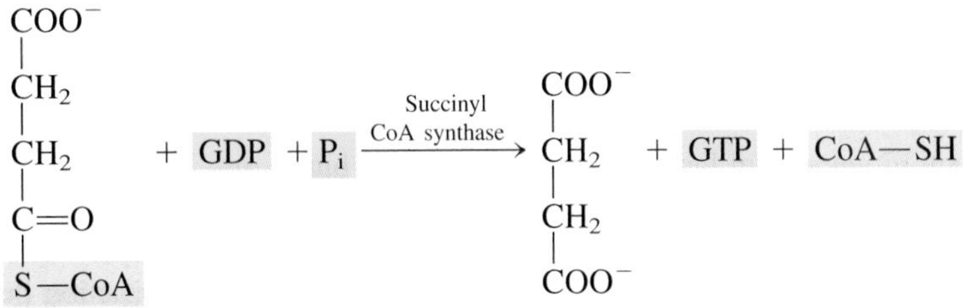

Steps 6 through 8 of the citric acid cycle involve a sequence of functional group changes that we have encountered several times in the organic sections of the text. The reaction sequence is

$$\text{Alkane} \xrightarrow[\text{(dehydrogenation)}]{\text{① Oxidation}} \text{alkene} \xrightarrow{\text{② Hydration}} \text{secondary alcohol} \xrightarrow[\text{(dehydrogenation)}]{\text{③ Oxidation}} \text{ketone}$$

Step 6: *Oxidation of Succinate.* This is the third redox reaction of the cycle. The enzyme involved is *succinate dehydrogenase,* and the oxidizing agent is FAD

Cyanide Poisoning

Inhalation of hydrogen cyanide gas (HCN) or ingestion of solid potassium cyanide (KCN) rapidly inhibits the electron transport chain in all tissues, making cyanide one of the most potent and rapidly acting poisons known. The attack point for the cyanide ion (CN^-) is cytochrome oxidase, the last fixed-site enzyme in the electron transport chain. Cyanide inactivates this enzyme by bonding itself to the Fe^{3+} in the enzyme's heme portions. As a result, Fe^{3+} is unable to transfer electrons to oxygen, blocking the cell's use of oxygen. Death results from tissue asphyxiation,

particularly of the central nervous system. Cyanide also binds to the heme group in hemoglobin, blocking oxygen transport in the bloodstream.

One treatment for cyanide poisoning is to administer various nitrites (NO_2^-), which oxidize the iron atoms of hemoglobin to Fe^{3+}. This form of hemoglobin helps draw CN^- back into the bloodstream, where it can be converted to thiocyanate (SCN^-) by thiosulfate ($S_2O_3^{2-}$), which is administered along with the nitrite (see the accompanying figure).

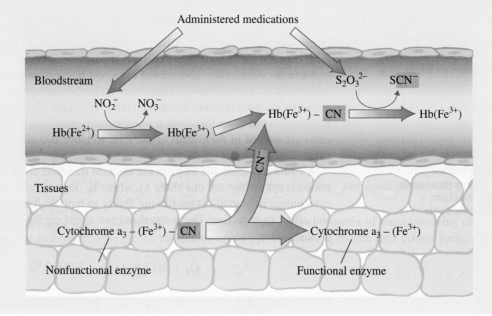

oxidative phosphorylation has been—and still is—one of the most challenging problems in biochemistry.

One concept central to the oxidative phosphorylation process is that of *coupled reactions*. **Coupled reactions** *are pairs of biochemical reactions that occur concurrently in which energy released by one reaction is used in the other reaction.* Oxidative phosphorylation and the oxidation reactions of the electron transport chain are coupled systems.

The interdependence (coupling) of ATP synthesis with the reactions of the ETC is related to the movement of protons (H^+ ions) across the inner mitochondrial membrane. The three fixed enzyme complexes involved in the ETC chain have a second function besides electron transfer down the chain. They also serve as "proton pumps," transferring protons from the matrix side of the inner mitochondrial membrane to the intermembrane space (Figure 23.12).

Some of the H^+ ions crossing the inner mitochondrial membrane come from the reduced electron carriers, and some come from the matrix; the details of how the H^+ ions cross the inner mitochondrial membrane are not fully understood.

For every two electrons passed through the ETC, four protons cross the inner mitochondrial membrane at the first fixed enzyme site, two at the second fixed enzyme site, and four more at the third fixed enzyme site. This proton flow causes a buildup of H^+ ions (protons) in the intermembrane space; this high concentration of protons becomes the basis for ATP synthesis (Figure 23.13).

▶ *Oxidative phosphorylation* is not the only process by which ATP is produced in cells. A second process, *substrate phosphorylation* (Section 24.2), can also be an ATP source. However, the amount of ATP produced by this second process is much less than that produced by oxidative phosphorylation.

FIGURE 23.12 A second function for the fixed enzyme sites involved in the electron transport chain is that of acting as proton pumps. For every two electrons passed through the ETC, 10 H$^+$ ions are transferred from the mitochondrial matrix to the intermembrane space.

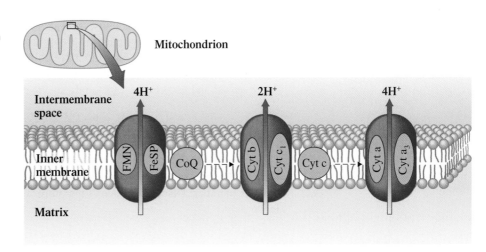

The difference in H$^+$ ion concentration between the two sides of the inner mitochondrial membrane causes a pH difference of about 1.4 units. A pH difference of 1.4 units means that the intermembrane space, the more acidic region, has 25 times more protons than the matrix.

Some of the energy released at each of the three fixed enzyme complex sites is consumed in the movement of H$^+$ ions across the inner membrane from the matrix into the intermembrane space. Movement of ions from a region of lower concentration (the matrix) to one of higher concentration (the intermembrane space) requires the expenditure of energy because it opposes the natural tendency, as exhibited in the process of osmosis (Section 8.8), to equalize concentrations.

The "proton flow" explanation for ATP–ETC coupling is formally called chemiosmotic coupling. **Chemiosmotic coupling** *is an explanation for the coupling of ATP synthesis with electron transport chain reactions that requires a proton gradient across the inner mitochondrial membrane.* The main concepts in this explanation for coupling follow.

1. The result of the pumping of protons from the mitochondrial matrix across the inner mitochondrial membrane is a higher concentration of protons in the intermembrane space than in the matrix. This concentration difference constitutes an *electrochemical (proton) gradient.* A chemical gradient exists whenever a substance has a higher concentration in one region than in another. Because the proton has an electrical charge (is an ion), an electrical gradient also exists. Potential energy (Section 7.2) is always associated with an electrochemical gradient.
2. A spontaneous flow of protons from the region of high concentration to the region of low concentration occurs because of the electrochemical gradient. This proton flow is not through the membrane itself (it is not permeable to H$^+$ ions) but rather through enzyme complexes called *ATP synthases* located on the inner mitochondrial membrane (Section 23.2). This proton flow through the ATP synthases "powers" the synthesis of ATP. ATP synthases are thus the *coupling factors* that link the processes of oxidative phosphorylation and the electron transport chain.

FIGURE 23.13 Formation of ATP accompanies the flow of protons from the intermembrane space back into the mitochondrial matrix. The proton flow results from an electrochemical gradient across the inner mitochondrial membrane.

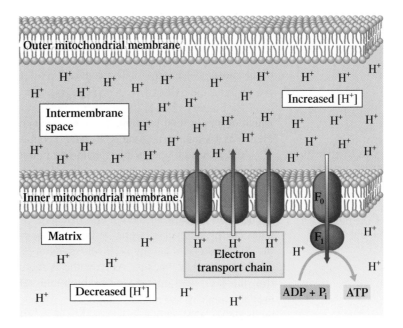

3. ATP synthase has two subunits, the F_0 and F_1 subunits (Figure 23.13). The F_0 part of the synthase is the channel for proton flow, whereas the formation of ATP takes place in the F_1 subunit. As protons return to the mitochondrial matrix through the F_0 subunit, the potential energy associated with the electrochemical gradient is released and used in the F_1 subunit for the synthesis of ATP.

$$\boxed{ADP} + P_i \xrightarrow{\text{ATP synthase}} \boxed{ATP} + H_2O$$

The Chemistry at a Glance feature on page 715 brings together into one diagram the three processes that constitute the common metabolic pathway: the citric acid cycle, the electron transport chain, and oxidative phosphorylation. These three processes operate together. Discussing them separately, as we have done, is a matter of convenience only.

23.9 ATP Production for the Common Metabolic Pathway

For each mole of NADH oxidized in the ETC, 2.5 moles of ATP are formed. $FADH_2$, which does not enter the ETC at its start, produces only 1.5 moles of ATP per mole of $FADH_2$ oxidized. $FADH_2$'s entrance point into the chain, coenzyme Q, is beyond the site of the first fixed enzyme complex site—that is, beyond the first "proton-pumping" site. Hence fewer ATP molecules are produced than for NADH.

The energy yield, in terms of ATP production, can now be totaled for the common metabolic pathway (Section 23.5). Every acetyl CoA entering the CAC produces three NADH, one $FADH_2$, and one GTP (which is equivalent in energy to ATP; Section 23.6). Thus 10 molecules of ATP are produced for each acetyl CoA catabolized.

$$3 \boxed{NADH} \longrightarrow 7.5 \boxed{ATP}$$
$$1 \boxed{FADH_2} \longrightarrow 1.5 \boxed{ATP}$$
$$1 \boxed{GTP} \longrightarrow \underline{\quad 1 \boxed{ATP}}$$
$$10 \boxed{ATP}$$

23.10 The Importance of ATP

The cycling of ATP and ADP in metabolic processes is the principal medium for energy exchange in biochemical processes. The conversion

$$\boxed{ATP} \longrightarrow \boxed{ADP} + P_i$$

powers life processes (the biosynthesis of essential compounds, muscle contraction, nutrient transport, and so on). The conversion

$$P_i + \boxed{ADP} \longrightarrow \boxed{ATP}$$

which occurs in food catabolism cycles, regenerates the ATP expended in cell operation. Figure 23.14 summarizes the ATP–ADP cycling process.

ATP is a high-energy phosphate compound (Section 23.4). Its hydrolysis to ADP produces an *intermediate* amount of free energy (-7.5 kcal/mole; Table 23.1) compared with hydrolysis energies for other organophosphate compounds (Table 23.1). Of major importance, the energy derived from ATP hydrolysis is a *biochemically useful* amount of energy. It is larger than the amount of energy needed by compounds to which ATP donates energy, and yet it is smaller than that available in compounds used to form ATP. If the ATP hydrolysis energy were *unusually high,* the body would not be able to convert ADP back to ATP, because ATP synthesis requires an energy input equal to or greater than the hydrolysis energy, and such an unusually high amount of energy would not be available.

▶ Biochemistry textbooks published before the mid-1990s make the following statements:

1 NADH produces 3 ATP in the ETC.

1 $FADH_2$ produces 2 ATP in the ETC.

As more has been learned about the electron transport chain and oxidative phosphorylation, these numbers have had to be reduced. The overall conversion process is more complex than was originally thought, and not as much ATP is produced.

▶ ATP molecules in cells have a high turnover rate. Normally, a given ATP molecule in a cell does not last more than a minute before it is converted to ADP. The concentration of ATP in a cell varies from 0.5 to 2.5 milligram per milliliter of cell fluid.

Summary of the Common Metabolic Pathway

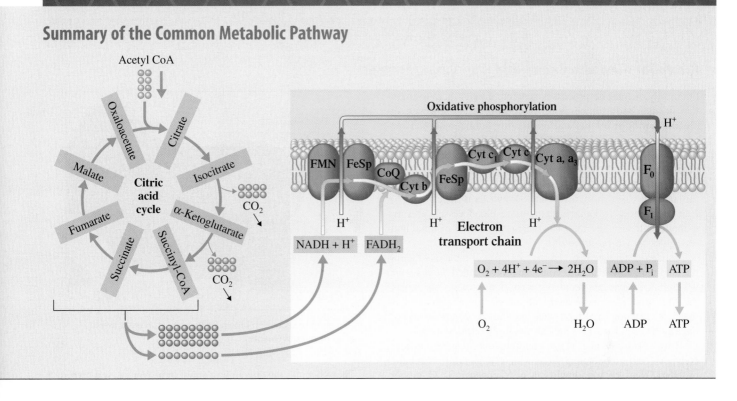

23.11 Non-ETC Oxygen-Consuming Reactions

The electron transport chain/oxidative phosphorylation phase of metabolism consumes more than 90% of the oxygen taken into the human body via respiration. What happens to the remainder of the inspired O_2?

As a normal part of metabolic chemistry, significant amounts of this remaining O_2 is converted into several highly reactive oxygen species (ROS). Among these ROS are hydrogen peroxide (H_2O_2), superoxide ion (O_2^-), and hydroxyl radical (OH). The latter two of these substances are free radicals, substances that contain an unpaired electron (Section 11.7). Reactive oxygen species have beneficial functions within the body, but they can also cause problems if they are not eliminated when they are no longer needed.

FIGURE 23.14 The interconversion of ATP and ADP is the principal medium for energy exchange in biochemical processes.

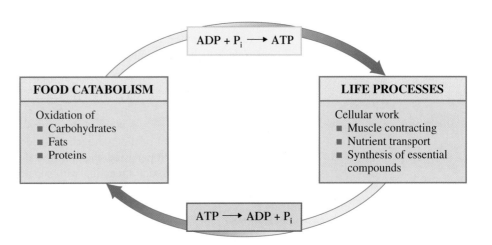

CHEMICAL CONNECTIONS

Brown Fat, Newborn Babies, and Hibernating Animals

Ordinarily, metabolic processes generate enough heat to maintain normal body temperature. In certain cases, however, including newborn infants and hibernating animals, normal metabolism is not sufficient to meet the body's heat requirements. In these cases, a supplemental method of heat generation, which involves *brown fat tissue,* occurs.

Brown fat tissue, as the name implies, is darker in color than ordinary fat tissue, which is white. Brown fat is specialized for heat production. It contains many more blood vessels and mitochondria than white fat. (The increased number of mitochondria gives brown fat its color.)

Another difference between the two types of fat is that the mitochondria in brown fat cells contain a protein called *thermogenin,* which functions as an *uncoupling agent.* This protein "uncouples" the ATP production associated with the electron transport chain. The ETC reactions still take place, but the energy that would ordinarily be used for ATP synthesis is simply released as heat.

Brown fat tissue is of major importance for newborn infants. Newborns are immediately faced with a temperature regulation problem. They leave an environment of constant 37°C temperature and enter a much colder environment (25°C). A supply of *active* brown fat, present at birth, helps the baby adapt to the cooler environment.

Very limited amounts of brown fat are present in most adults. However, stores of brown fat increase in adults who are regularly exposed to cold environs. Thus the production of brown fat is one of the body's mechanisms for adaptation to cold.

Hibernating bears rely on brown fat tissue to help meet their bodies' heat requirements.

Thermogenin, the uncoupling agent in brown fat, is a protein bound to the inner mitochondrial membrane. When activated, it functions as a proton channel through the inner membrane. The proton gradient produced by the electron transport chain is dissipated through this "new" proton channel, and less ATP synthesis occurs because the normal proton channel, ATP synthase, has been bypassed. The energy of the proton gradient, no longer useful for ATP synthesis, is released as heat.

White blood cells have a significant concentration of superoxide free radicals. Here, these free radicals aid in the destruction of invading bacteria and viruses. Their formation reaction is

$$2O_2 + NADPH \longrightarrow 2O_2^- + NADP^+ + H^+$$

(NADP is a phosphorylated version of the coenzyme NADH; see Section 24.8.)

Superoxide ion that is not needed is eliminated from cells in a two-step process governed by the enzymes *superoxide dismutase* and *catalase,* two of the most rapidly working enzymes known (see Table 21.2). In the first step, superoxide ion is converted to hydrogen peroxide, which is then, in the second step, converted to H_2O.

$$2O_2^- + 2H^+ \xrightarrow{\text{superoxide dismutase}} H_2O_2 + O_2$$
$$2H_2O_2 \xrightarrow{\text{catalase}} 2H_2O + O_2$$

Immediate destruction of the hydrogen peroxide produced in the first of these two steps is critical, because if it persisted, then unwanted production of hydroxyl radical would occur via hydrogen peroxide's reaction with superoxide ion.

$$H_2O_2 + O_2^- + H^+ \longrightarrow H_2O + O_2 + OH$$

Hydroxyl radicals quickly react with other substances by taking an electron from them. Such action usually causes bond breaking. Lipids in cell membranes are particularly vulnerable to such attack by hydroxyl radicals.

CHEMICAL CONNECTIONS

Flavonoids: An Important Class of Dietary Antioxidants

Numerous studies indicate that diets high in fruits and vegetables are associated with a "healthy lifestyle." One reason for this is that fruits and vegetables contain compounds called *phytochemicals*. Phytochemicals are compounds found in plants that have biological activity in the human body even though they have no nutritional value. The functions that phytochemicals perform in the human body include antioxidant activity, cancer inhibition, cholesterol regulation, and anti-inflammatory activity.

Each fruit and vegetable is a unique package of phytochemicals, so consuming a wide variety of fruits and vegetables provides the body with the broadest spectrum of benefits. In such a situation, many phytochemicals are consumed in *small* amounts. This approach is much safer than taking supplemental doses of particular phytochemicals; in larger doses, some phytochemicals are toxic.

A major group of phytochemicals are the *flavonoids,* of which over 4000 individual compounds are known. All flavonoids are antioxidants (Section 14.12), but some are stronger antioxidants than others, depending on their molecular structure. About 50 flavonoids are present in foods and in beverages derived from plants (tea leaves, grapes, oranges, and so on).

The core *flavonoid* structure is

Both aromatic and cyclic ether ring systems are present. Of particular importance as antioxidants in foods are flavonoids known as *flavones* and *flavonols,* flavonoids whose core structures are enhanced by the presence of ketone and/or hydroxyl groups and a double bond in the oxygen-containing ring system.

Flavones   Flavonols

The formation of flavone and flavonol compounds depends normally on the action of light, so in general the highest concentration of these compounds occurs in leaves or in the skins of fruits, whereas only traces are found in parts of plants that grow below the ground. The common onion is, however, a well-known exception to this generalization.

The most widespread flavonoid in food is the flavonol *quercetin.*

It is predominant in fruits, vegetables, and the leaves of various vegetables. In fruits, apples contain the highest amounts of quercetin, the majority of it being found in the outer tissues (skin, peel). A small peeled apple contains about 5.7 mg of the antioxidant vitamin C. But the same amount of apple *with the skin* contains flavonoids and other phytochemicals that have the effect of 1500 mg of vitamin C. Onions are also major dietary sources of quercetin.

In addition to their antioxidant benefits, flavonoids may also help fight bacterial infections. Recent studies indicate that flavonoids can stop the growth of some strains of drug-resistant bacteria.

It is estimated that 5% of the ROS escape destruction through normal channels (superoxide dimutase and catalase). Operating within a cell is a backup system—a network of antioxidants—to deal with this problem. Participating in this antioxidant network are glutathione (Section 20.7), vitamin C (Section 21.13), and beta-carotene and vitamin E (Section 21.14), as well as other compounds obtained from plants through dietary intake. Particularly important in this latter category are compounds called *flavonoids* (see the Chemical Connections feature on this page). The vitamin antioxidants as well as the other antioxidants present prevent oxidative damage by reacting with the harmful oxidizing agents before they can react with other biologically important substances.

Reactive oxygen species can also be formed in the body as the result of external influences such as polluted air, cigarette smoke, and radiation exposure (including solar radiation). Vitamin C is particularly active against such free-radical damage.

CONCEPTS TO REMEMBER

Metabolism. Metabolism is the sum total of all the biochemical reactions that take place in a living organism. Metabolism consists of catabolism and anabolism. Catabolic biochemical reactions involve the breakdown of large molecules into smaller fragments. Anabolic biochemical reactions synthesize large molecules from smaller ones.

Mitochondria. Mitochondria are membrane-enclosed subcellular structures that are the site of energy production in the form of ATP molecules. Enzymes for both the citric acid cycle and the electron transport chain are housed in the mitochondria.

Important coenzymes. Three very important coenzymes involved in catabolism are NAD^+, FAD, and CoA. NAD^+ and FAD are oxidizing agents that participate in the oxidation reactions of the citric acid cycle. They transport hydrogen atoms and electrons from the citric acid cycle to the electron transport chain. CoA interacts with acetyl groups produced from food degradation to form acetyl CoA. Acetyl CoA is the "fuel" for the citric acid cycle.

High-energy compounds. A high-energy compound liberates a larger-than-normal amount of free energy upon hydrolysis, because structural features in the molecule contribute to repulsive strain in one or more bonds. Most high-energy biochemical molecules contain phosphate groups.

Common catabolic pathway. The common catabolic pathway includes the reactions of the citric acid cycle and those of the electron transport chain and oxidative phosphorylation. The degradation products from all types of foods (carbohydrates, fats,

and proteins) participate in the reactions of the common metabolic pathway.

Citric acid cycle. The citric acid cycle is a cyclic series of eight reactions that oxidize the acetyl portion of acetyl CoA, resulting in the production of two molecules of CO_2. The complete oxidation of one acetyl group produces three molecules of NADH, one of $FADH_2$, and one of GTP besides the CO_2.

Electron transport chain. The electron transport chain is a series of reactions that passes electrons from NADH and $FADH_2$ to molecular oxygen. Each electron carrier that participates in the chain has an increasing affinity for electrons. Upon accepting the electrons and hydrogen ions, the O_2 is reduced to H_2O.

Oxidative phosphorylation. Oxidative phosphorylation is the biochemical process by which ATP is synthesized from ADP as the result of a proton gradient across the inner mitochondrial membrane. Oxidative phosphorylation is coupled to the reactions of the electron transport chain.

Chemiosmotic coupling. Chemiosmotic coupling explains how the energy needed for ATP synthesis is obtained. Synthesis takes place because of a flow of protons across the inner mitochondrial membrane.

Importance of ATP. ATP is the link between energy production and energy use in cells. The conversion of ATP to ADP powers life processes, and the conversion of ADP back to ATP regenerates the energy expended in cell operation.

KEY REACTIONS AND EQUATIONS

1. Oxidation by NAD^+ (Section 23.3)
$$NAD^+ + 2H^+ + 2e^- \longrightarrow NADH + H^+$$

2. Oxidation by FAD (Section 23.3)
$$FAD + 2H^+ + 2e^- \longrightarrow FADH_2$$

3. The citric acid cycle (Section 23.6)
$$\text{Acetyl CoA} + 3NAD^+ + FAD + GDP + P_i + 2H_2O \longrightarrow$$
$$2CO_2 + CoA + 3NADH + 2H^+ + FADH_2 + GTP$$

4. The electron transport chain (Section 23.7)
$$NADH + H^+ \longrightarrow NAD^+ + 2H^+ + 2e^-$$
$$FADH_2 \longrightarrow FAD + 2H^+ + 2e^-$$
$$O_2 + 4H^+ + 4e^- \longrightarrow 2H_2O$$

5. Oxidative phosphorylation (Section 23.8)
$$ADP + P_i \xrightarrow[\text{from ETC}]{\text{Energy}} ATP$$

KEY TERMS

Acetyl group (23.3)
Anabolism (23.1)
Catabolism (23.1)
Chemiosmotic coupling (23.8)
Citric acid cycle (23.6)
Common metabolic pathway (23.5)
Coupled reactions (23.8)

Cytochrome (23.7)
Cytoplasm (23.2)
Cytosol (23.2)
Electron transport chain (23.7)
Eukaryotic cell (23.2)
High-energy compound (23.4)
Lysosome (23.2)

Metabolic pathway (23.1)
Metabolism (23.1)
Mitochondrion (23.2)
Organelle (23.2)
Oxidative phosphorylation (23.8)

EXERCISES AND PROBLEMS

The members of each pair of problems in this section test similar material.

■ Metabolism (Section 23.1)

23.1 Classify anabolism and catabolism as synthetic or degradative processes.

23.2 Classify anabolism and catabolism as energy-producing or energy-consuming processes.

23.3 What is a metabolic pathway?

23.4 What is the difference between a linear and a cyclic metabolic pathway?

23.5 What general characteristics are associated with a catabolic pathway?

23.6 What general characteristics are associated with an anabolic pathway?

■ **Cell Structure (Section 23.2)**

23.7 List several differences between prokaryotic cells and eukaryotic cells.

23.8 What kinds of organisms have prokaryotic cells and what kinds have eukaryotic cells?

23.9 What is an organelle?

23.10 What is the general function of each of the following types of organelles?
a. Ribosome   b. Lysosome   c. Mitochondrion

23.11 In a mitochondrion, what separates the matrix from the intermembrane space?

23.12 In what major way do the inner and outer mitochondrial membranes differ?

23.13 What is the intermembrane space of a mitochondrion?

23.14 Where are ATP synthase complexes located in a mitochondrion?

■ **Intermediate Compounds in Metabolic Pathways (Section 23.3)**

23.15 What does each letter in ATP stand for?

23.16 What does each letter in ADP stand for?

23.17 Draw a block diagram structure for ATP.

23.18 Draw a block diagram structure for ADP.

23.19 What is the structural difference between ATP and AMP?

23.20 What is the structural difference between ADP and AMP?

23.21 What is the structural difference between ATP and GTP?

23.22 What is the structural difference between ATP and CTP?

23.23 In terms of hydrolysis, what is the relationship between ATP and ADP?

23.24 In terms of hydrolysis, what is the relationship between ADP and AMP?

23.25 What does each letter in FAD stand for?

23.26 What does each letter in NAD^+ stand for?

23.27 Draw a block diagram structure for FAD based on the presence of an ADP core (three-block diagram).

23.28 Draw a block diagram structure for FAD based on the presence of two nucleotides (six-block diagram).

23.29 Draw a block diagram structure for NAD^+ based on the presence of two nucleotides (six-block diagram).

23.30 Draw a block diagram structure for NAD^+ based on the presence of an ADP core (three-block diagram).

23.31 Which part of an NAD^+ molecule is the active participant in redox reactions?

23.32 Which part of an FAD molecule is the active participant in redox reactions?

23.33 Give the letter designation for
a. the reduced form of FAD   b. the oxidized form of NADH

23.34 Give the letter designation for
a. the oxidized form of $FADH_2$
b. the reduced form of NAD^+

23.35 Name the vitamin B molecule that is part of the structure of
a. NAD^+ b. FAD

23.36 In which of the following molecules is the vitamin B portion the "active" portion in redox processes?
a. NAD^+ b. FAD

23.37 Draw the three-block diagram structure for coenzyme A.

23.38 Which part of a coenzyme A molecule is the active participant in a redox reaction?

■ **High-Energy Phosphate Compounds (Section 23.4)**

23.39 What is a high-energy compound?

23.40 What factors contribute to a strained bond in high-energy phosphate compounds.

23.41 What does the designation P_i denote?

23.42 What does the designation PP_i denote?

23.43 With the help of Table 23.1, determine which compound in each of the following pairs of phosphate-containing compounds releases more free energy upon hydrolysis?
a. ATP and phosphoenolpyruvate
b. Creatine phosphate and ADP
c. Glucose 1-phosphate and 1,3-diphosphoglycerate
d. AMP and glycerol 3-phosphate

23.44 With the help of Table 23.1, determine which compound in each of the following pairs of phosphate-containing compounds releases more free energy upon hydrolysis?
a. ATP and creatine phosphate
b. Glucose 1-phosphate and glucose 6-phosphate
c. ADP and AMP
d. Phosphoenolpyruvate and PP_i

■ **Biochemical Energy Production (Section 23.5)**

23.45 Describe the four general stages of the process by which biochemical energy is obtained from food.

23.46 Of the four general stages of biochemical energy production from food, which are part of the common metabolic pathway?

■ **The Citric Acid Cycle (Section 23.6)**

23.47 What are two other names for the citric acid cycle?

23.48 What is the basis for the name *citric acid cycle*?

23.49 What is the "fuel" for the citric acid cycle?

23.50 What are the products of the citric acid cycle?

23.51 Consider the reactions that occur during *one turn* of the citric acid cycle in answering each of the following questions.
a. How many CO_2 molecules are formed?
b. How many molecules of $FADH_2$ are formed?
c. How many times is a secondary alcohol oxidized?
d. How many times does water add to a carbon–carbon double bond?

23.52 Consider the reactions that occur during *one turn* of the citric acid cycle in answering each of the following questions.
a. How many molecules of NADH are formed?
b. How many GTP molecules are formed?
c. How many decarboxylation reactions occur?
d. How many oxidation–reduction reactions occur?

23.53 There are eight steps in the citric acid cycle. List those steps that involve
a. oxidation
b. isomerization
c. hydration

23.54 There are eight steps in the citric acid cycle. List those steps that involve
a. oxidation and decarboxylation
b. phosphorylation
c. condensation

23.55 There are four C_4 dicarboxylic acid species in the citric acid cycle. What are their names and structures?

23.56 There are two keto carboxylic acid species in the citric acid cycle. What are their names and structures?

23.57 What type of reaction occurs in the citric acid cycle whereby a C_6 compound is converted to a C_5 compound?

23.58 What type of reaction occurs in the citric acid cycle whereby a C_5 compound is converted to a C_4 compound?

23.59 Identify the oxidized coenzyme (NAD^+ or FAD) that participates in each of the following citric acid cycle reactions.
a. Isocitrate $\longrightarrow$ α-ketoglutarate
b. Succinate $\longrightarrow$ fumarate

23.60 Identify the oxidized coenzyme (NAD^+ or FAD) that participates in each of the following citric acid cycle reactions.
a. Malate $\longrightarrow$ oxaloacetate
b. α-Ketoglutarate $\longrightarrow$ succinyl CoA

23.61 List the two citric acid cycle intermediates involved in the reaction governed by each of the following enzymes. List the reactant first.
a. Isocitrate dehydrogenase
b. Fumarase
c. Malate dehydrogenase
d. Aconitase

23.62 List the two citric acid cycle intermediates involved in the reaction governed by each of the following enzymes. List the reactant first.
a. α-Ketoglutarate dehydrogenase
b. Succinate dehydrogenase
c. Citrate synthase
d. Succinyl CoA synthase

■ **The Electron Transport Chain (Section 23.7)**

23.63 By what other name is the electron transport chain known?

23.64 Give a one-sentence summary of what occurs during the reactions known as the electron transport chain.

23.65 What is the final electron acceptor of the electron transport chain?

23.66 Which substances generated in the citric acid cycle participate in the electron transport chain?

23.67 What do the following abbreviations stand for?
a. FMN   b. Cyt   c. FeSP   d. NADH

23.68 What do the following abbreviations stand for?
a. CoQ   b. $FMNH_2$ c. $FADH_2$ d. $CoQH_2$

23.69 Classify each of the following electron transport chain electron carriers as mobile or fixed-site. For fixed-site electron carriers, further specify the location as first, second, or third fixed site.
a. Cyt c   b. Cyt b   c. FMN   d. CoQ

23.70 Classify each of the following electron transport chain electron carriers as mobile or fixed-site. For fixed-site electron carriers, further specify the location as first, second, or third fixed site.
a. Cyt c_1 b. FeSP   c. $CoQH_2$ d. Cyt a_3

23.71 Put the following substances in the correct order of their participation in the electron transport chain: cyt c_1, cyt a_3, FeSP, and NADH.

23.72 Put the following substances in the correct order of their participation in the electron transport chain: CoQ, $FADH_2$, FMN, and cyt b.

23.73 Indicate whether each of the following changes represents oxidation or reduction.
a. $CoQH_2 \longrightarrow CoQ$
b. $NAD^+ \longrightarrow NADH$
c. Cyt c (Fe^{2+}) $\longrightarrow$ cyt c (Fe^{3+})
d. Cyt b (Fe^{3+}) $\longrightarrow$ cyt b (Fe^{2+})

23.74 Indicate whether each of the following changes represents oxidation or reduction.
a. $FADH_2 \longrightarrow FAD$
b. FMN $\longrightarrow$ $FMNH_2$
c. Fe(III)SP $\longrightarrow$ Fe(II)SP
d. Cyt c_1 (Fe^{3+}) $\longrightarrow$ cyt c_1 (Fe^{2+})

23.75 Fill in the missing substances in the following electron transport chain reaction sequences.

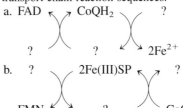

23.76 Fill in the missing substances in the following electron transport chain reaction sequences.

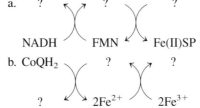

23.77 Fill in the blanks in the following electron transport chain reactions.
a. $FMNH_2 + 2Fe(III)SP \longrightarrow$ _____ + _____
b. $FADH_2 +$ _____ $\longrightarrow$ FAD + _____

23.78 Fill in the blanks in the following electron transport chain reactions.
a. Cyt b (Fe^{2+}) + cyt c_1 (Fe^{3+}) $\longrightarrow$ _____ + _____
b. NADH + _____ $\longrightarrow$ $NAD^+ +$ _____

■ **Oxidative Phosphorylation (Section 23.8)**

23.79 What is oxidative phosphorylation?

23.80 What are coupled reactions?

23.81 The coupling of ATP synthesis with the reactions of the ETC is related to the movement of what chemical species across the inner mitochrondial membrane?

23.82 At what enzyme location(s) in the electron transport chain does proton pumping occur?

23.83 At what mitochondrial location does H^+ ion buildup occur as the result of proton pumping?

23.84 How many protons cross the inner mitochondrial membrane for every two electrons that are passed through the electron transport chain?

23.85 What is the name of the enzyme that catalyzes ATP production during oxidative phosphorylation?

23.86 What is the location of the enzyme that uses stored energy in a proton gradient to drive the reaction that produces ATP.

23.87 How is the proton gradient associated with chemiosmotic coupling dissipated during ATP synthesis?

23.88 What are the "starting materials" from which ATP is synthesized as the proton gradient associated with chemiosmotic coupling is dissipated?

■ **ATP Production (Section 23.9)**

23.89 How many ATP molecules are formed for each NADH molecule that enters the electron transport chain?

23.90 How many ATP molecules are formed for each FADH$_2$ molecule that enters the electron transport chain?

23.91 NADH and FADH$_2$ molecules do not yield the same number of ATP molecules. Explain why.

23.92 What is the energy yield, in terms of ATP molecules, from one turn of the citric acid cycle, assuming that the products of the cycle enter the electron transport chain?

■ **Non-ETC Oxygen-Consuming Reactions (Section 23.11)**

23.93 What does the designation ROS stand for?

23.94 Give the chemical formula for each of the following.
a. Superoxide ion   b. Hydroxyl radical

23.95 Give the chemical equation for the reaction by which
a. superoxide ion is generated within cells
b. superoxide ion is converted to hydrogen peroxide within cells

23.96 Give the chemical equation for the reaction by which
a. hydrogen peroxide is converted to desirable products within cells
b. hydrogen peroxide is converted to an undesirable product within cells

---

ADDITIONAL PROBLEMS

23.97 Classify each of the following substances as (1) a reactant in the citric acid cycle, (2) a reactant in the electron transport chain, or (3) a reactant in both the CAC and the ETC.
a. NAD$^+$ b. NADH   c. O$_2$
d. H$_2$O   e. Fumarate f. Cytochrome a

23.98 Classify each of the following substances as (1) a product in the citric acid cycle, (2) a product in the electron transport chain, or (3) a product in both the CAC and the ETC.
a. FADH$_2$ b. FAD   c. CO$_2$
d. H$_2$O   e. Malate   f. Flavin mononucleotide

23.99 Which of these substances, (1) ATP, (2) CoA, (3) FAD, and (4) NAD$^+$, contain the following subunits of structure? More than one choice may apply in a given situation.
a. Contains two ribose subunits
b. Contains two phosphate subunits
c. Contains one adenine subunit
d. Contains one ribitol subunit

23.100 Characterize, in terms of number of carbon atoms present, each of the following citric acid cycle changes as (a) a C$_6$ to C$_6$ change, (b) a C$_6$ to C$_5$ change, (c) a C$_5$ to C$_4$ change, or (d) a C$_4$ to C$_4$ change.
a. Citrate to isocitrate   b. Succinate to fumarate
c. Malate to oxaloacetate
d. Isocitrate to α-ketoglutarate

23.101 At which of these locations, (1) first fixed enzyme site, (2) second fixed enzyme site, (3) third fixed enzyme site, or (4) mobile enzyme site, does each of the following ETC reactions occur?
a. FADH$_2$ + CoQ   b. NADH + FMN
c. cyt a$_3$ + O$_2$ d. FeSP + CoQ

23.102 In what way are the processes of the citric acid cycle and the electron transport chain interrelated?

23.103 Where within a cell does each of the following take place?
a. Citric acid cycle
b. Electron transport chain and oxidative phosphorylation

23.104 One of the oxidation steps that occurs when lipids are metabolized is
Would you expect this reaction to require FAD or NAD$^+$ as the oxidizing agent?

$$R-CH_2-CH_2-\overset{\displaystyle O}{\overset{\displaystyle \|}{C}}-S-CoA \longrightarrow R-CH=CH-\overset{\displaystyle O}{\overset{\displaystyle \|}{C}}-S-CoA$$

23.105 In oxidative phosphorylation, what is oxidized and what is phosphorylated?

23.106 Which atom in cytochromes undergoes oxidation and reduction in the electron transport chain?

24 Carbohydrate Metabolism

Carbohydrates are the major energy source for animals as well as human beings.

In this chapter we explore the relationship between carbohydrate metabolism and energy production in cells. The molecule glucose is the focal point of carbohydrate metabolism. Commonly called blood sugar, glucose is supplied to the body via the circulatory system and, after being absorbed by the cell, can be either oxidized to yield energy or stored as glycogen for future use. When sufficient oxygen is present, glucose is totally oxidized to CO_2 and H_2O. However, in the absence of oxygen, glucose is only partially oxidized to lactic acid. Besides supplying energy needs, glucose and other six-carbon sugars can be converted into a variety of different sugars (C_3, C_4, C_5, and C_7) needed for biosynthesis. Some of the oxidative steps in carbohydrate metabolism also produce NADH and NADPH, sources of reductive power in cells.

24.1 Digestion and Absorption of Carbohydrates

Digestion *is the biochemical process by which food molecules, through hydrolysis, are broken down into simpler chemical units that can be used by cells for their metabolic needs.* Digestion is the first stage in the processing of food products.

The digestion of carbohydrates begins in the mouth, where the enzyme *salivary α-amylase* catalyzes the hydrolysis of α-glycosidic linkages (Section 18.13) in starch from plants and glycogen from meats to produce smaller polysaccharides and the disaccharide maltose.

Only a small amount of carbohydrate digestion occurs in the mouth, because food is swallowed so quickly. Although the food mass remains longer in the stomach, very little further carbohydrate digestion occurs there either, because *salivary α-amylase* is inactivated

▶ *Salivary α-amylase* is a constituent of saliva, the fluid secreted by the salivary glands. Saliva is 99% water plus small amounts of several inorganic ions and organic molecules. Saliva secretion can be triggered by the taste, smell, sight, and even thought of food. Average saliva output is about 1.5 L per day.

FIGURE 24.1 A section of the small intestine, showing its folds and the villi that cover the inner surface of the folds. Villi greatly increase the inner intestinal surface area.

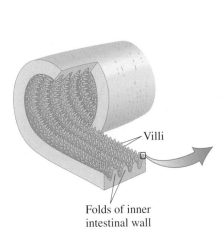 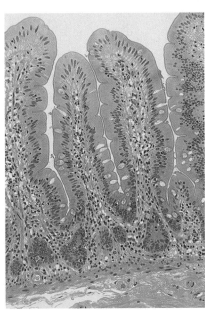

Villi

Folds of inner
intestinal wall

by the acidic environment of the stomach, and the stomach's own secretions do not contain any carbohydrate-digesting enzymes.

The primary site for carbohydrate digestion is within the lumen of the small intestine, where α-amylase, this time secreted by the pancreas, again begins to function. The *pancreatic α-amylase* breaks down polysaccharide chains into shorter and shorter segments until the disaccharide maltose (two glucose units; Section 18.13) and glucose itself are the dominant species.

The final step in carbohydrate digestion occurs on the outer membranes of intestinal mucosal cells, where the enzymes that convert disaccharides to monosaccharides are located. The important disaccharidase enzymes are *maltase, sucrase,* and *lactase.* These enzymes convert, respectively, maltose to two glucose units, sucrose to one glucose and one fructose unit, and lactose to one glucose and one galactose unit (Section 18.13). (The disaccharides sucrose and lactose present in food are not digested until they reach this point.)

The three major breakdown products from carbohydrate digestion are thus glucose, galactose, and fructose. These monosaccharides are absorbed into the bloodstream through the intestinal wall. The folds of the intestinal wall are lined with fingerlike projections called *villi,* which are rich in blood capillaries (Figure 24.1). Absorption is by *active transport* (Section 19.10), which, unlike passive transport, is an energy-requiring process. In this case, ATP is needed. Protein carriers mediate the passage of the monosaccharides through cell membranes. Figure 24.2 summarizes the different phases in the digestive process for carbohydrates.

After their absorption into the bloodstream, monosaccharides are transported to the liver, where fructose and galactose are rapidly converted into compounds that are metabolized by the same pathway as glucose. Thus the central focus of carbohydrate metabolism is the pathway by which glucose is further processed, a pathway called *glycolysis* (Section 24.2)—a series of ten reactions, each of which involves a different enzyme.

24.2 Glycolysis

Glycolysis *is the metabolic pathway by which glucose (a C_6 molecule) is converted into two molecules of pyruvate (a C_3 molecule).* This metabolic pathway functions in almost all cells.

The conversion of glucose to pyruvate is an oxidation process in which no molecular oxygen is utilized. The oxidizing agent is the coenzyme NAD^+. Metabolic pathways

▶ The term *glycolysis,* pronounced "gligh-KOLL-ih-sis," comes from the Greek *glyco,* meaning "sweet," and *lysis,* meaning "breakdown."

▶ *Pyruvate,* pronounced "PIE-roo-vate," is the carboxylate ion (Section 16.9) produced when pyruvic acid (a three-carbon keto acid) loses its acidic hydrogen atom.

$$
\begin{array}{ccc}
\text{CH}_3 & & \text{CH}_3 \\
| & & | \\
\text{C}=\text{O} & \longrightarrow & \text{C}=\text{O} \quad + \text{H}^+ \\
| & & | \\
\text{COOH} & & \text{COO}^- \\
\text{Pyruvic acid} & & \text{Pyruvate ion}
\end{array}
$$

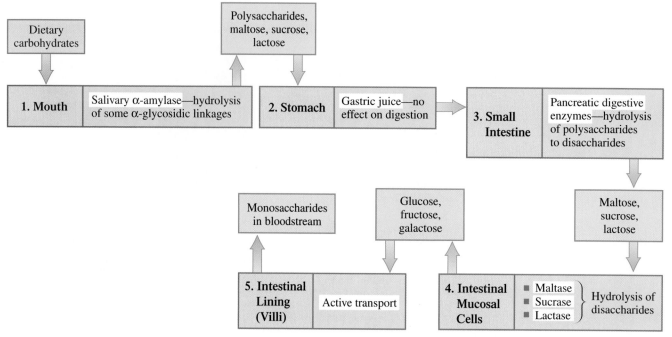

FIGURE 24.2 Summary of carbohydrate digestion in the human body.

> *Anaerobic* is pronounced "AN-air-ROE-bic." *Aerobic* is pronounced "air-ROE-bic."

> Glycolysis is also called the *Embden–Meyerhof pathway* after the German chemists Gustav Embden and Otto Meyerhof, who discovered many of the details of the pathway.

> A *kinase* is an enzyme that catalyzes the transfer of a phosphoryl group (PO_3^{2-}) from ATP (or some other high-energy phosphate compound) to a substrate.

in which molecular oxygen is not a participant are called *anaerobic* pathways. Pathways that require molecular oxygen are called *aerobic* pathways. Glycolysis is an anaerobic pathway.

Glycolysis is a ten-step process (compared to the eight steps of the citric acid cycle; Section 23.6) in which every step is enzyme-catalyzed. Figure 24.3 gives an overview of glycolysis. There are two stages in the overall process, a *six-carbon stage* (Steps 1–3) and a *three-carbon stage* (Steps 4–10). All of the enzymes needed for glycolysis are present in the cell cytosol (Section 23.2), which is where glycolysis takes place. Details of the individual steps within the glycolysis pathway are now considered.

■ Six-Carbon Stage of Glycolysis (Steps 1–3)

The intermediates of the six-carbon stage of glycolysis are all either *glucose* or *fructose* derivatives in which phosphate groups are present.

Step 1: *Formation of Glucose 6-phosphate.* Glycolysis begins with the phosphorylation of glucose to yield glucose 6-phosphate, a glucose molecule with a phosphate group attached to the hydroxyl oxygen on carbon 6 (the carbon atom outside the ring). The phosphate group is from an ATP molecule. *Hexokinase,* an enzyme that requires Mg^{2+} ion for its activity, catalyzes the reaction.

The symbol ⓟ is a shorthand notation for a PO_3^{2-} unit.

Glucose → Glucose 6-phosphate

This reaction requires energy, which is provided by the breakdown of an ATP molecule. This energy expenditure will be recouped later in the cycle.

FIGURE 24.3 An overview of glycolysis.

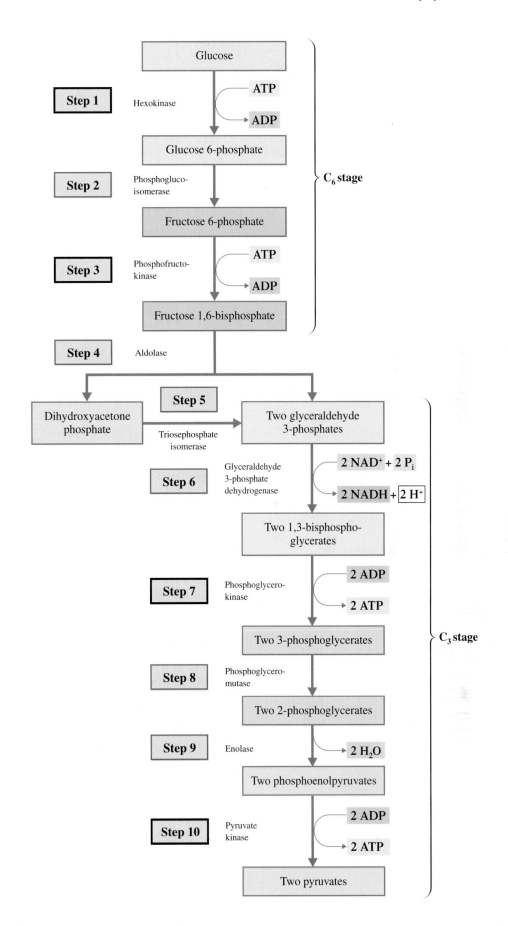

Phosphorylation of glucose provides a way of "trapping" glucose within a cell. Glucose can cross cell membranes, but glucose 6-phosphate cannot.

Step 2: *Formation of Fructose 6-phosphate.* Glucose 6-phosphate is isomerized to fructose 6-phosphate by *phosphoglucoisomerase.*

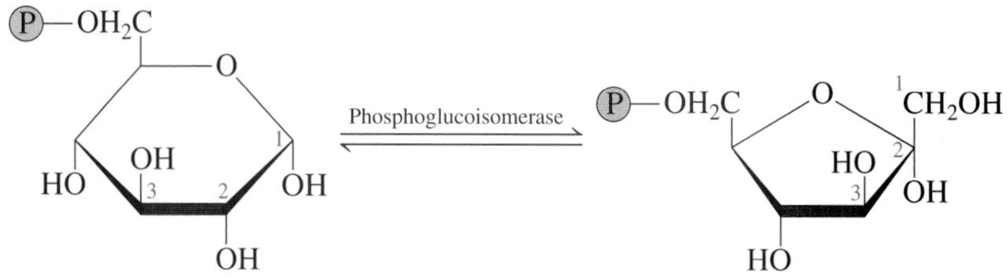

Glucose 6-phosphate — Phosphoglucoisomerase → Fructose 6-phosphate

The net result of this change is that carbon 1 of glucose is no longer part of the ring structure. [Glucose, an aldose, forms a six-membered ring, and fructose, a ketose, forms a five-membered ring (Section 18.10); both sugars, however, contain six carbon atoms.]

Step 3: *Formation of Fructose 1,6-bisphosphate.* This step, like Step 1, is a phosphorylation reaction and therefore requires the expenditure of energy. ATP is the source of the phosphate and the energy. The enzyme involved, *phosphofructokinase,* is another enzyme that requires Mg^{2+} ion for its activity. The fructose molecule now contains two phosphate groups.

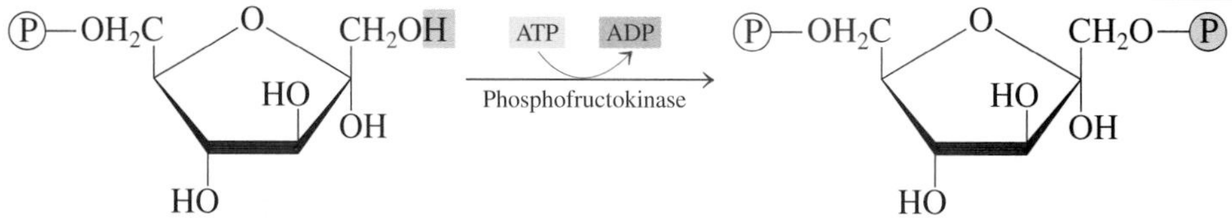

> Step 3 of glycolysis commits the original glucose molecule to the glycolysis pathway. Glucose 6-phosphate (Step 1) and fructose 6-phosphate (Step 2) can enter other metabolic pathways, but fructose 1,6-bisphosphate can enter only glycolysis.

Fructose 6-phosphate — ATP ADP Phosphofructokinase → Fructose 1,6-bisphosphate

■ Three-Carbon Stage of Glycolysis (Steps 4–10)

All intermediates in the three-carbon stage of glycolysis are phosphorylated derivatives of *dihydroxyacetone, glyceraldehyde, glycerate,* or *pyruvate,* which in turn are derivatives of either glycerol or acetone. Figure 24.4 shows the structural relationships among these molecules.

Step 4: *Formation of Triose Phosphates.* In this step, the reacting C_6 species is split into two C_3 (triose) species. Because fructose 1,6-bisphosphate, the molecule being split, is unsymmetrical, the two trioses produced are not identical. One product is dihydroxyacetone phosphate, and the other is glyceraldehyde 3-phosphate. *Aldolase* is the enzyme that catalyzes this reaction. A better understanding of the structural relationships between reactant and products is obtained if the fructose 1,6-bisphosphate is written in its open-chain form (Section 18.10) rather than in its cyclic form.

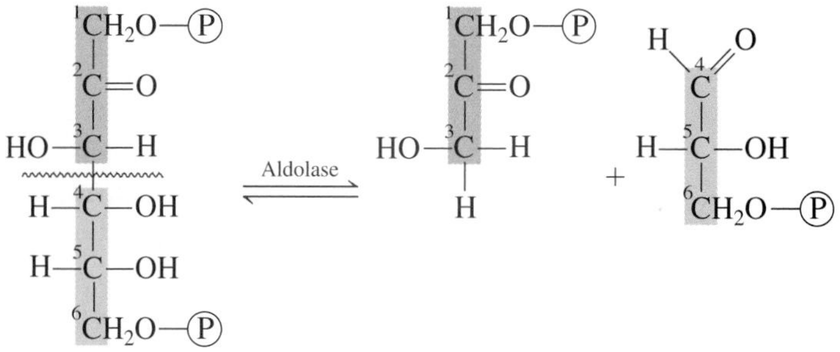

Fructose 1,6-bisphosphate (open-chain form) — Aldolase → Dihydroxyacetone phosphate + Glyceraldehyde 3-phosphate

FIGURE 24.4 Structural relationships among glycerol and acetone and the C_3 intermediates in the process of glycolysis.

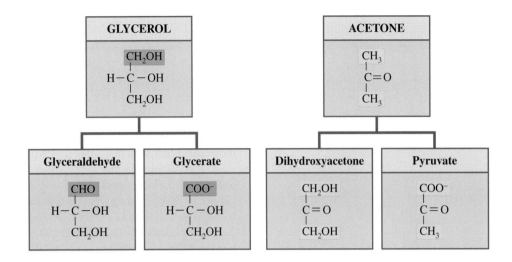

Step 5: *Isomerization of Triose Phosphates.* Only one of the two trioses produced in Step 4, glyceraldehyde 3-phosphate, is a glycolysis intermediate. Dihydroxyacetone phosphate, the other triose, can, however, be readily converted into glyceraldehyde 3-phosphate. Dihydroxyacetone phosphate (a ketose) and glyceraldehyde 3-phosphate (an aldose) are isomers, and the isomerization process from ketose to aldose is catalyzed by the enzyme *triosephosphate isomerase.*

Step 6: *Formation of 1,3-Bisphosphoglycerate.* In a reaction catalyzed by *glyceraldehyde 3-phosphate dehydrogenase,* a phosphate group is added to glyceraldehyde 3-phosphate to produce 1,3-bisphosphoglycerate. The hydrogen of the aldehyde group becomes part of NADH.

▶ Keep in mind that from Step 6 onward, two molecules of each of the C_3 compounds take part in every reaction for each original C_6 glucose molecule.

The newly added phosphate group in 1,3-bisphosphoglycerate is a high-energy phosphate group (Section 23.4). A high-energy phosphate group is produced when a phosphate group is attached to a carbon atom that is also participating in a carbon–carbon or carbon–oxygen double bond.

Note that a molecule of the reduced coenzyme NADH is a product of this reaction and also that the source of the added phosphate is inorganic phosphate (P_i).

Step 7: *Formation of 3-Phosphoglycerate.* In this step, the diphosphate species just formed is converted back to a monophosphate species. This is an ATP-producing step in which the C-1 phosphate group of 1,3-bisphosphoglycerate (the high-energy phosphate) is transferred to an ADP molecule to form the ATP. The enzyme involved is *phosphoglycerokinase.*

1,3-Bisphosphoglycerate   3-Phosphoglycerate

Remember that two ATP molecules are produced for each original glucose molecule, because both C_3 molecules produced from the glucose react.

ATP production in this step involves substrate-level phosphorylation. **Substrate-level phosphorylation** *is the biochemical process by which a high-energy phosphate group from an intermediate compound (substrate) is directly transferred to ADP to produce ATP.* Substrate-level phosphorylation differs from oxidative phosphorylation (Section 23.8) in that the latter process involves the transfer of free phosphate ions in solution (P_i) to ADP molecules to form ATP.

> A *mutase* is an enzyme that effects the shift of a phosphoryl group (PO_3^{2-}) from one oxygen atom to another within a molecule.

Step 8: *Formation of 2-Phosphoglycerate.* In this isomerization step, the phosphate group of 3-phosphoglycerate is moved from carbon 3 to carbon 2. The enzyme *phosphoglyceromutase* catalyzes the exchange of the phosphate group between the two carbons.

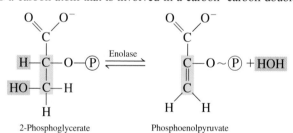

3-Phosphoglycerate   2-Phosphoglycerate

> An *enol* (from *ene* + *ol*), as in phospho*enol*pyruvate, is a compound in which a —OH group is attached to a carbon atom involved in a carbon–carbon double bond. Note that in phosphoenolpyruvate, the —OH group has been phosphorylated.

Step 9: *Formation of Phosphoenolpyruvate.* This is an alcohol dehydration reaction that proceeds with the enzyme *enolase,* another Mg^{2+}-requiring enzyme. The result is another compound containing a high-energy phosphate group; the phosphate group is attached to a carbon atom that is involved in a carbon–carbon double bond.

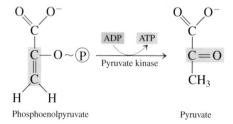

2-Phosphoglycerate   Phosphoenolpyruvate

Step 10: *Formation of Pyruvate.* In this step, substrate-level phosphorylation again occurs. Phosphoenolpyruvate transfers its high-energy phosphate group to an ADP molecule to produce ATP and pyruvate.

Phosphoenolpyruvate   Pyruvate

The enzyme involved, *pyruvate kinase,* requires both Mg^{2+} and K^+ ions for its activity. Again, because two C_3 molecules are reacting, two ATP molecules are produced.

ATP molecules are involved in Steps 1, 3, 7, and 10 of glycolysis. Considering these steps collectively shows that there is a net gain of two ATP molecules for every glucose

TABLE 24.1
ATP Production and Consumption
During Glycolysis

Step	Reaction	ATP change per glucose
1	Glucose → glucose 6-phosphate	−1
3	Fructose 6-phosphate → fructose 1,6-bisphosphate	−1
7	2(1,3-Bisphosphoglycerate → 3-phosphoglycerate)	+2
10	2(Phosphoenolpyruvate → pyruvate)	+2
		Net +2

molecule converted into two pyruvates (Table 24.1). Though useful, this is a small amount of ATP compared to that generated in oxidative phosphorylation (Section 23.8).

The net overall equation for the process of glycolysis is

$$\text{Glucose} + 2\text{NAD}^+ + 2\text{ADP} + 2\text{P}_i \longrightarrow$$
$$2 \text{ pyruvate} + 2\text{NADH} + 2\text{ATP} + 2\text{H}^+ + 2\text{H}_2\text{O}$$

■ Entry of Galactose and Fructose into Glycolysis

The breakdown products from carbohydrate digestion are glucose, fructose, and galactose (Section 24.1). Both fructose and galactose are converted, in the liver, to intermediates that enter into the glycolysis pathway.

The entry of fructose into the glycolytic pathway involves phosphorylation by ATP to produce fructose 1-phosphate, which is then split into two trioses—glyceraldehyde and dihydroxyacetone phosphate. Dihydroxyacetone phosphate enters glycolysis directly; glyceraldehyde must be phosphorylated by ATP to glyceraldehyde 3-phosphate before it enters the pathway (see Figure 24.5).

The entry of galactose into the glycolytic pathway begins with its conversion to glucose 1-phosphate (a four-step sequence), which is then converted to glucose 6-phosphate, a glycolysis intermediate (see Figure 24.5).

■ Regulation of Glycolysis

Glycolysis, like all metabolic pathways, must have control mechanisms associated with it. In glycolysis, the control points are Steps 1, 3, and 10 (see Figure 24.3).

Step 1, the conversion of glucose to glucose 6-phosphate, involves the enzyme *hexokinase*. This particular enzyme is inhibited by glucose 6-phosphate, the substance produced by its action (feedback inhibition; Section 21.9).

At Step 3, where fructose 6-phosphate is converted to fructose 1,6-bisphosphate by the enzyme *phosphofructokinase,* high concentrations of ATP and citrate inhibit enzyme activity. A high ATP concentration, which is characteristic of a state of low energy consumption, thus stops glycolysis at the fructose 6-phosphate stage. This stoppage also causes increases in glucose 6-phosphate stores, because glucose 6-phosphate is in equilibrium with fructose 6-phosphate.

The third control point involves the last step of glycolysis, the conversion of phosphoenolpyruvate to pyruvate. *Pyruvate kinase,* the enzyme needed at this point, is inhibited by high ATP concentrations. Both pyruvate kinase (Step 10) and phosphofructokinase (Step 3) are allosteric enzymes (Section 21.9).

24.3 Fates of Pyruvate

The production of pyruvate from glucose (glycolysis) occurs in a similar manner in most cells. In contrast, the fate of the pyruvate so produced varies with cellular conditions and the nature of the organism. Three common fates for pyruvate are of prime importance: conversion into acetyl CoA, into lactate, and into ethanol (see Figure 24.6).

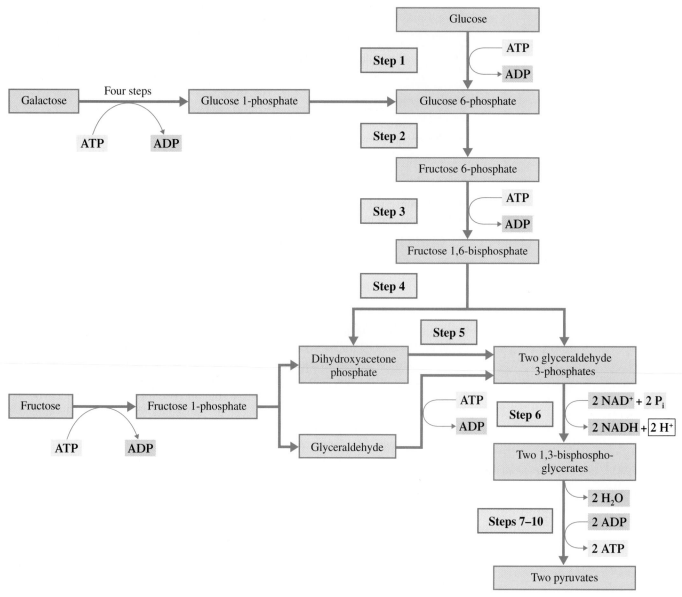

FIGURE 24.5 Entry points for fructose and galactose into the glycolysis pathway.

A key concept in considering these fates of pyruvate is the need for a continuous supply of NAD^+ for glycolysis. As glucose is oxidized to pyruvate in glycolysis, NAD^+ is reduced to NADH.

$$\text{Glucose} + \boxed{2NAD^+} \xrightarrow{ 2ADP + 2P_i 2ATP} 2\ \text{pyruvate} + \boxed{2NADH} + 2H^+$$

It is significant that each pathway of pyruvate metabolism includes provisions for regeneration of NAD^+ from NADH so that glycolysis can continue.

■ Oxidation to Acetyl CoA

Under *aerobic* (oxygen-rich) conditions, pyruvate is oxidized to acetyl CoA. Pyruvate formed in the cytosol through glycolysis crosses the two mitochondrial membranes and enters the mitochondrial matrix, where the oxidation takes place. The overall reaction, in simplified terms, is

$$\underset{\text{Pyruvate}}{CH_3-\overset{\overset{\displaystyle O}{\|}}{C}-COO^-} + \boxed{CoA-SH} + \boxed{NAD^+} \xrightarrow[\text{complex}]{\text{Pyruvate dehydrogenase}} \underset{\text{Acetyl CoA}}{CH_3-\overset{\overset{\displaystyle O}{\|}}{C}-\boxed{S-CoA}} + \boxed{NADH} + \boxed{CO_2}$$

FIGURE 24.6 The three common fates of pyruvate generated by glycolysis.

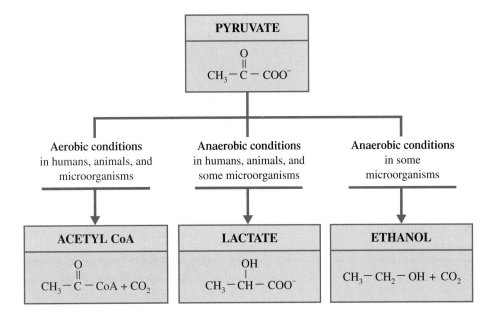

> Not all acetyl CoA produced from pyruvate enters the citric acid cycle. Particularly when high levels of acetyl CoA are produced (from excess ingestion of dietary carbohydrates), some acetyl CoA is used as the starting material for the production of the fatty acids needed for fat (triacylglycerol) formation (Section 25.7).

This reaction, which involves both oxidation and decarboxylation (CO_2 is produced), is far more complex than the simple stoichiometry of the equation suggests. The enzyme complex involved contains three different enzymes, each with numerous subunits. The overall reaction process involves four separate steps and requires NAD^+, CoA—SH, FAD, and two other coenzymes (lipoic acid and thiamin pyrophosphate, the latter derived from the B vitamin thiamin).

Most acetyl CoA molecules produced from pyruvate enter the citric acid cycle. Citric acid cycle operations change more NAD^+ to its reduced form, NADH. The NADH from glycolysis, from the conversion of pyruvate to acetyl CoA, and from the citric acid cycle enters the electron transport chain directly or indirectly (Section 23.7). In the ETC, electrons from NADH are transferred to O_2, and the NADH is changed back to NAD^+. The NAD^+ needed for glycolysis, pyruvate–acetyl CoA conversion, and the citric acid cycle can be regenerated.

The net overall reaction for processing one glucose molecule to two molecules of acetyl CoA is

$$\text{Glucose} + \boxed{\text{2ADP}} + 2P_i + \boxed{4NAD^+} + 2CoA-SH \longrightarrow$$
$$2\text{acetyl CoA} + 2CO_2 + \boxed{\text{2ATP}} + \boxed{\text{4NADH}} + 4H^+ + 2H_2O$$

■ **Fermentation Processes**

When the body becomes oxygen deficient (anaerobic conditions), such as during strenuous exercise, the electron transport chain process slows down because its last step is dependent on oxygen. The result of this "slowing down" is a buildup in NADH concentration (it is not being consumed so fast) and a decreased amount of available NAD^+ (it is not being produced so fast). Decreased NAD^+ concentration then negatively affects the rate of glycolysis. An alternative method for conversion of NADH to NAD^+—a method that does not require oxygen—is needed if glycolysis is to continue, it being the only available source of *new* ATP under these conditions.

Fermentation processes solve this problem. **Fermentation** *is a biochemical process by which NADH is oxidized to* NAD^+ *without the need for oxygen.* We consider here two fermentation processes: lactate fermentation and ethanol fermentation.

■ **Lactate Fermentation**

Lactate fermentation *is the enzymatic anaerobic reduction of pyruvate to lactate.* The sole purpose of this process is the conversion of NADH to NAD^+. The lactate so formed is converted back to pyruvate when aerobic conditions are again established in a cell (Section 24.6).

Working muscles often produce lactate. If strenuous exercise is continued too long, the buildup of lactate in the muscles reaches a point beyond which fermentation cannot continue. This slows glycolysis and new ATP production, and the muscle action can no longer continue (fatigue and exhaustion; see the accompanying Chemical Connections feature on page 733) until oxygen supplies are re-established.

The equation for lactate formation from pyruvate is

$$CH_3-\underset{\text{Pyruvate}}{\overset{\overset{\displaystyle O}{\|}}{C}}-COO^- + \boxed{NAD\ H} + \boxed{H^+} \xrightarrow[\text{dehydrogenase}]{\text{Lactate}} CH_3-\underset{\text{Lactate}}{\overset{\overset{\displaystyle O\ \boxed{H}}{|}}{C\ \boxed{H}}}-COO^- + \boxed{NAD^+}$$

▶ Red blood cells have no mitochondria and therefore always form lactate as the end product of glycolysis.

When the reaction for conversion of pyruvate to lactate is added to the net glycolysis reaction (Section 24.2), an overall reaction for the conversion of glucose to lactate is obtained.

$$Glucose + \boxed{2ADP} + 2P_i \longrightarrow Lactate + \boxed{2ATP} + 2H_2O$$

Note that NADH and NAD^+ do not appear in this equation, even though the process cannot proceed without them. The NADH generated during glycolysis (Step 6) is consumed in the conversion of pyruvate to lactate. Thus there is no net oxidation–reduction in the conversion of glucose to lactate.

■ Ethanol Fermentation

▶ With bread and other related products obtained using yeast, the ethanol produced by fermentation evaporates during baking.

Under anaerobic conditions, several simple organisms, including yeast, possess the ability to regenerate NAD^+ through ethanol, rather than lactate, production. Such a process is called alcohol fermentation. **Ethanol fermentation** *is the enzymatic anaerobic conversion of pyruvate to ethanol and carbon dioxide.* Ethanol fermentation involving yeast causes bread and related products to rise as a result of CO_2 bubbles being released during baking. Beer, wine, and other alcoholic drinks are produced by ethanol fermentation of the sugars in grain and fruit products.

The first step in conversion of pyruvate to ethanol is a decarboxylation reaction to produce acetaldehyde.

$$CH_3-\underset{\text{Pyruvate}}{\overset{\overset{\displaystyle O}{\|}}{C}}-\boxed{COO^-} + H^+ \xrightarrow[\text{decarboxylase}]{\text{Pyruvate}} CH_3-\underset{\text{Acetaldehyde}}{\overset{\overset{\displaystyle O}{\|}}{C}}-H + \boxed{CO_2}$$

The second step involves acetaldehyde reduction to produce ethanol.

$$CH_3-\underset{\text{Acetaldehyde}}{\overset{\overset{\displaystyle O}{\|}}{C}}-H + \boxed{NAD\ H} + \boxed{H^+} \xrightarrow[\text{dehydrogenase}]{\text{Alcohol}} CH_3-\underset{\text{Ethanol}}{\overset{\overset{\displaystyle O\ \boxed{H}}{|}}{\underset{\underset{\displaystyle H}{|}}{C}}}-H + \boxed{NAD^+}$$

The overall equation for the conversion of pyruvate to ethanol (the sum of the two steps) is

$$CH_3-\underset{\text{Pyruvate}}{\overset{\overset{\displaystyle O}{\|}}{C}}-COO^- + 2H^+ + \boxed{NADH} \xrightarrow{\text{Two steps}} CH_3-CH_2-\underset{\text{Ethanol}}{OH} + \boxed{NAD^+} + CO_2$$

An overall reaction for the production of ethanol from glucose is obtained by combining the reaction for the conversion of pyruvate with the net reaction for glycolysis (Section 24.2).

$$Glucose + \boxed{2ADP} + 2P_i \longrightarrow 2Ethanol + 2CO_2 + \boxed{2ATP} + 2H_2O$$

Again note that NADH and NAD^+ do not appear in the final equation; they are both generated and consumed.

CHEMICAL CONNECTIONS

Lactate Accumulation

During strenuous exercise, conditions in muscle cells can change from aerobic to anaerobic as the oxygen supply becomes inadequate to meet demand. Such conditions cause pyruvate to be converted to lactate rather than acetyl CoA. (Lactate production can also be high at the start of strenuous exercise before the delivery of oxygen is stepped up via an increased respiration rate.)

The resulting lactate begins to accumulate in the cytosol of cells where it is produced. Some lactate diffuses out of the cells into the blood, where it contributes to a slight decrease in blood pH. This lower pH triggers fast breathing, which helps supply more oxygen to the cells.

Lactate accumulation is the cause of muscle pain and cramping during prolonged, strenuous exercise. As a result of such cramping, muscles may be stiff and sore the next day. Regular, hard exercise increases the efficiency with which oxygen is delivered to the body. Thus athletes can function longer than nonathletes under aerobic conditions without lactate production.

Lactate accumulation can also occur in heart muscle if it experiences decreased oxygen supply (from artery blockage). The heart muscle experiences cramps and stops beating (cardiac arrest). Massage of heart muscle often reduces such cramps, just as it does for skeletal muscle, and it is sometimes possible to start the heart beating again by using such a technique.

Premature infants born with underdeveloped lungs are often given increased amounts of oxygen to minimize lactate accumulation. They are also given bicarbonate (HCO_3^-) solution to counteract the acidity change in blood that accompanies lactate buildup.

Strenuous muscular activity can result in lactate accumulation.

Figure 24.7 summarizes the relationship between the fates of pyruvate and the regeneration of NAD^+ from NADH.

24.4 ATP Production for the Complete Oxidation of Glucose

We now assemble energy production figures for glycolysis, oxidation of pyruvate to acetyl CoA, the citric acid cycle, and the electron transport chain. The result, with one added piece of information, gives the ATP yield for the *complete* oxidation of one molecule of glucose.

The new piece of information involves the NADH produced during Step 6 of glycolysis. This NADH, produced in the cytosol, cannot *directly* participate in the electron transport chain, because mitochondria are impermeable to NADH (and NAD^+). A transport

FIGURE 24.7 All three of the common fates of pyruvate from glycolysis provide for the regeneration of NAD^+ from NADH.

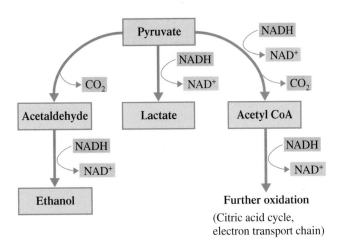

system shuttles the electrons from NADH, but not NADH itself, across the membrane. This shuttle involves dihydroxyacetone phosphate (a glycolysis intermediate) and glycerol 3-phosphate.

The first step in the shuttle is the cytosolic reduction of dihydroxyacetone phosphate by NADH to produce glycerol 3-phosphate and NAD^+ (see Figure 24.8). Glycerol 3-phosphate then crosses the outer mitochondrial membrane, where it is reoxidized to dihydroxyacetone phosphate. The oxidizing agent is FAD rather than NAD^+. The regenerated dihydroxyacetone phosphate diffuses out of the mitochondrion and returns to the cytosol for participation in another "turn" of the shuttle. The $FADH_2$ coproduced in the mitochondrial reaction can participate in the electron transport chain reactions. The net reaction of this shuttle process is

$$\underset{\text{(cytosolic)}}{NADH} + H^+ + \underset{\text{(mitochondrial)}}{FAD} \longrightarrow \underset{\text{(cytosolic)}}{NAD^+} + \underset{\text{(mitochondrial)}}{FADH_2}$$

The consequence of this reaction is that only 1.5 rather than 2.5 molecules of ATP are formed for each cytosolic NADH, because $FADH_2$ yields one less ATP than does NADH in the electron transport chain.

Table 24.2 shows ATP production for the complete oxidation of a molecule of glucose. The final number is 30 ATP, 26 of which come from the oxidative phosphorylation associated with the electron transport chain. This total of 30 ATP for complete oxidation contrasts markedly with a total of 2 ATP for oxidation of glucose to lactate and 2 ATP for oxidation of glucose to ethanol. Neither of these latter processes involves the citric acid cycle or the electron transport chain. Thus the aerobic oxidation of glucose is 15 times more efficient in the production of ATP than the anaerobic lactate and ethanol processes.

The production of 30 ATP molecules per glucose (Table 24.2) is for those cells where the dihydroxyacetone phosphate–glycerol 3-phosphate shuttle operates (skeletal muscle and nerve cells). In certain other cells, particularly heart and liver cells, a more complex shuttle system called the malate–aspartate shuttle functions. In this shuttle, 2.5 ATP molecules result from 1 cytosolic NADH, which changes the total ATP production to 32 molecules per glucose.

The net overall reaction for the *complete* metabolism (oxidation) of a glucose molecule is the simple equation

$$\text{Glucose} + 6O_2 + 30ADP + 30P_i \longrightarrow 6CO_2 + 6H_2O + 30ATP$$

FIGURE 24.8 The dihydroxyacetone phosphate–glycerol 3-phosphate shuttle.

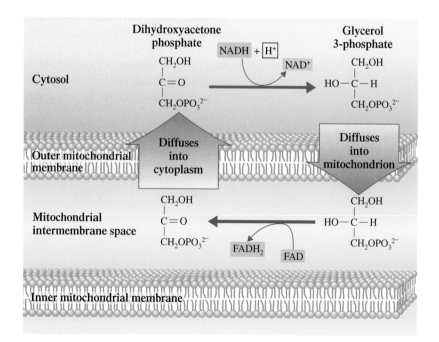

TABLE 24.2
Production of ATP from the Complete Oxidation of One Glucose Molecule in a Skeletal Muscle Cell

Reaction	Comments	Yield of ATP
Glycolysis		
glucose → glucose 6-phosphate	consumes 1 ATP	−1
glucose 6-phosphate → fructose 1,6-bisphosphate	consumes 1 ATP	−1
2(glyceraldehyde 3-phosphate → 1,3-bisphosphoglycerate)	each produces 1 cytosolic NADH	—
2(1,3-bisphosphoglycerate → 3-phosphoglycerate)	each produces 1 ATP	+2
2(phosphoenolpyruvate → pyruvate)	each produces 1 ATP	+2
Oxidation of Pyruvate		
2(pyruvate → acetyl CoA + CO_2)	each produces 1 NADH	—
Citric Acid Cycle		
2(isocitrate → α-ketoglutarate + CO_2)	each produces 1 NADH	—
2(α-ketoglutarate → succinyl CoA + CO_2)	each produces 1 NADH	—
2(succinyl CoA → succinate)	each produces 1 GTP	+2
2(succinate → fumarate)	each produces 1 $FADH_2$	—
2(malate → oxaloacetate)	each produces 1 NADH	—
Electron Transport Chain and Oxidative Phosphorylation		
2 cytosolic NADH formed in glycolysis	each produces 1.5 ATP	+3
2 NADH formed in the oxidation of pyruvate	each produces 2.5 ATP	+5
2 $FADH_2$ formed in the citric acid cycle	each produces 1.5 ATP	+3
6 NADH formed in the citric acid cycle	each produces 2.5 ATP	+15
	Net production of ATP	+30

Note that substances such as NADH, NAD^+, and $FADH_2$ are not part of this equation. Why? They cancel out—that is, they are consumed in one step (reactant) and regenerated in another step (product). Note also what the net equation does not acknowledge: the many dozens of reactions that are needed to generate the 30 molecules of ATP. The Chemistry at a Glance feature on page 737 summarizes the glucose metabolism process in terms of stages or phases and also gives the total number of reactions involved in each stage. The processing of 1 glucose molecule requires 157 chemical reactions; it is astounding how complex, and yet how simple, human body chemistry can be.

24.5 Glycogen Synthesis and Degradation

Glycogen, a branched polymeric form of glucose (Section 18.14), is the storage form of carbohydrates in humans and animals. It is found primarily in muscle and liver tissue. In muscles it is the source of glucose needed for glycolysis. In the liver, it is the source of glucose needed to maintain normal glucose levels in the blood.

■ Glycogenesis

Glycogenesis *is the metabolic pathway by which glycogen is synthesized from glucose.* Glycogenesis involves three reactions (steps).

Step 1: *Formation of Glucose 1-phosphate.* The starting material for this step is not glucose itself but rather glucose 6-phosphate (available from the first step of glycolysis). The enzyme *phosphoglucomutase* effects the change from a 6-phosphate to a 1-phosphate.

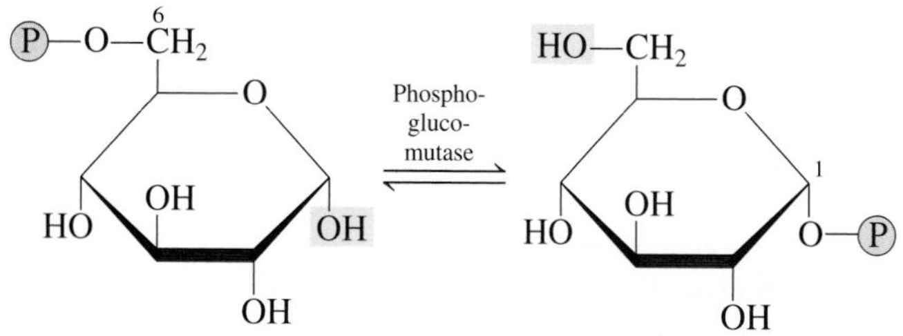

Glucose 6-phosphate   Glucose 1-phosphate

Step 2: *Formation of UDP-glucose.* Glucose 1-phosphate from Step 1 must be activated before it can be added to a growing glycogen chain. The activator is the high-energy compound UTP (uridine triphosphate). The UTP is hydrolyzed to UMP and PP_i, and then the PP_i is further hydrolyzed to $2P_i$. The UMP that is formed bonds to the glucose 1-phosphate to form UDP-glucose.

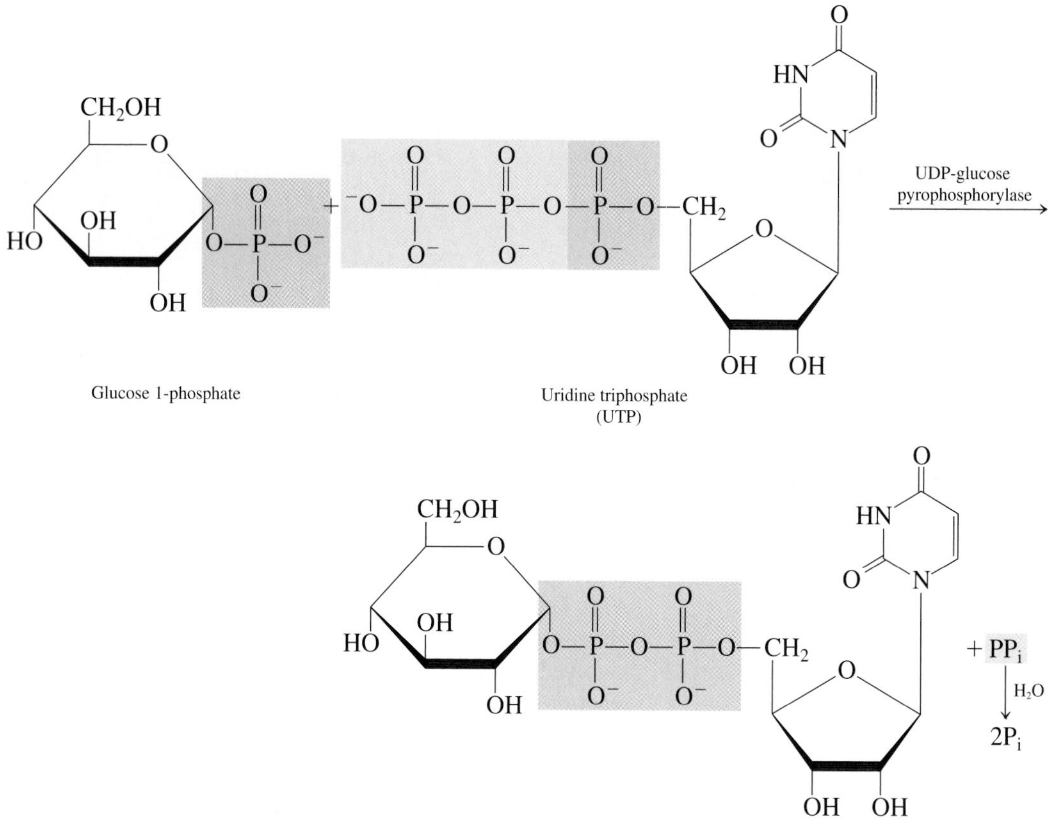

Glucose 1-phosphate   Uridine triphosphate
                                              (UTP)

Uridine diphosphate glucose (UDP-glucose)

Step 3: *Glucose Transfer to a Glycogen Chain.* The glucose unit of UDP-glucose is then attached to the end of a glycogen chain.

$$UDP\text{-glucose} + (glucose)_n \longrightarrow (glucose)_{n+1} + UDP$$

Glucogen chain   Glycogen with an additional glucose unit

In a subsequent reaction, the UDP produced in Step 3 is converted back to UTP, which can then react with another glucose 1-phosphate (Step 2). The conversion reaction requires ATP.

$$UDP + ATP \longrightarrow UTP + ADP$$

Adding a single glucose unit to a growing glycogen chain requires the investment of two ATP molecules: one in the formation of glucose 6-phosphate and one in the regeneration of UTP.

Reactions Involved in the Complete Oxidation of One Glucose Molecule

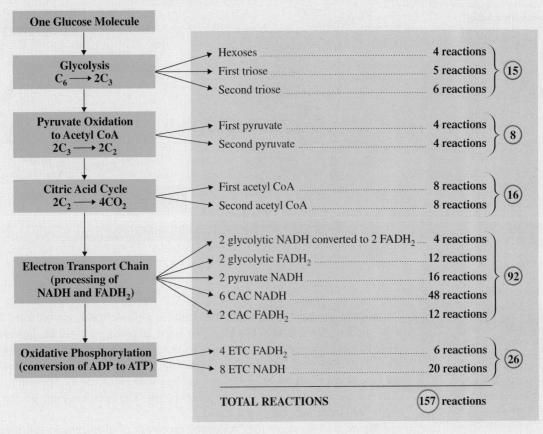

One Glucose Molecule		
Glycolysis $C_6 \longrightarrow 2C_3$	Hexoses .... 4 reactions	
	First triose .... 5 reactions	15
	Second triose .... 6 reactions	
Pyruvate Oxidation to Acetyl CoA $2C_3 \longrightarrow 2C_2$	First pyruvate .... 4 reactions	8
	Second pyruvate .... 4 reactions	
Citric Acid Cycle $2C_2 \longrightarrow 4CO_2$	First acetyl CoA .... 8 reactions	16
	Second acetyl CoA .... 8 reactions	
Electron Transport Chain (processing of NADH and FADH$_2$)	2 glycolytic NADH converted to 2 FADH$_2$ .... 4 reactions	
	2 glycolytic FADH$_2$ .... 12 reactions	
	2 pyruvate NADH .... 16 reactions	92
	6 CAC NADH .... 48 reactions	
	2 CAC FADH$_2$ .... 12 reactions	
Oxidative Phosphorylation (conversion of ADP to ATP)	4 ETC FADH$_2$ .... 6 reactions	26
	8 ETC NADH .... 20 reactions	
TOTAL REACTIONS	157 reactions	

■ Glycogenolysis

Glycogenolysis *is the metabolic pathway by which free glucose units are obtained from glycogen.* This process is not simply the reverse of glycogen synthesis (glycogenesis), because it does not require UTP or UDP molecules. Glycogenolysis, like glycogenesis, is a three-step process.

▶ A *phosphorylase* is an enzyme that catalyzes the cleavage of a bond by P_i (in contrast to hydrolysis, which refers to bond cleavage by water), such as removal of a glucose unit from glucogen to give glucose 1-phosphate.

Step 1: *Phosphorylation of a Glucose Residue.* The enzyme *glycogen phosphorylase* effects the removal of an end glucose unit from a glycogen molecule as glucose 1-phosphate.

$$(\text{Glucose})_n + P_i \longrightarrow (\text{glucose})_{n-1} + \text{glucose 1-phosphate}$$
Glycogen   Glycogen with one fewer glucose unit

Step 2: *Glucose 1-phosphate Isomerization.* The enzyme *phosphoglucomutase* catalyzes the isomerization process whereby the phosphate group of glucose 1-phosphate is moved to the carbon 6 position.

$$\text{Glucose 1-phosphate} \rightleftharpoons \text{glucose 6-phosphate}$$

This process is the reverse of the first step of glycogenesis.

Step 3: *Hydrolysis of Glucose 6-phosphate to Glucose.* The reaction for this step is

$$\text{Glucose 6-phosphate} + H_2O \longrightarrow \text{glucose} + P_i$$

▶ A *phosphatase* is an enzyme that effects the removal of a phosphate group (P_i) from a molecule, such as converting glucose 6-phosphate to glucose, with H_2O as the attacking species.

The enzyme needed for this reaction, *glucose 6-phosphatase,* is found only in the liver, kidneys, and intestine. Thus *complete* glycogenolysis occurs only in these tissues.

FIGURE 24.9 The processes of glycogenesis and glycogenolysis contrasted. The intermediate glucose–UDP is part of glycogenesis but not of glycogenolysis.

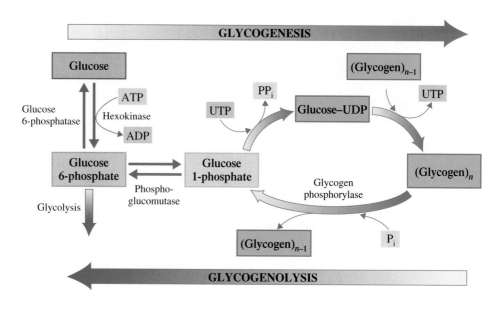

▶ The fact that glycogen synthesis (glycogenesis) and glycogen degradation (glycogenolysis) are not totally reverse processes has significance. In fact, it is almost always the case in biochemistry that "opposite" biosynthetic and degradative pathways differ in some steps. This allows for separate control of the pathways.

Muscle and brain cells, which lack glucose 6-phosphatase, cannot form free glucose from glucose 6-phosphate. The first two steps of glycogenolysis do, however, occur in these tissues. The glucose 6-phosphate so produced can contribute to energy production through glycolysis and the common metabolic pathway, because glucose 6-phosphate is the first intermediate in the glycolytic pathway (see Figure 24.3). Thus brain and muscle cells can use glycogen for energy production only. The liver, however, has the capacity to supply additional glucose to the blood.

Figure 24.9 contrasts the "opposite" processes of glycogenesis and glycogenolysis. Both processes involve the intermediate glucose 6-phosphate. Glucose–UDP is unique to glycogenesis.

24.6 Gluconeogenesis

Gluconeogenesis *is the metabolic pathway by which glucose is synthesized from noncarbohydrate materials.* Glycogen stores in muscle and liver tissue are depleted within 12–18 hours from fasting or in even less time from heavy work or strenuous exercise. Without gluconeogenesis, the brain, which is dependent on glucose as a fuel, would have problems functioning if food intake were restricted for even one day.

The noncarbohydrate starting materials for gluconeogenesis are lactate (from hardworking muscles and from red blood cells), glycerol (from triacylglycerol hydrolysis), and certain amino acids (from dietary protein hydrolysis or from muscle protein during starvation). About 90% of gluconeogenesis takes place in the liver. Hence gluconeogenesis helps to maintain normal blood-glucose levels in times of inadequate dietary carbohydrate intake (such as between meals).

The processes of gluconeogenesis (pyruvate to glucose) and glycolysis (glucose to pyruvate) are not exact opposites. The most obvious difference between these two processes is that 11 compounds are involved in gluconeogenesis and only 10 in glycolysis. Why the difference? The last step of glycolysis is the conversion of the high-energy compound phosphoenolpyruvate to pyruvate. The reverse of this process, which is the beginning of gluconeogenesis, cannot be accomplished in a single step because of the large energy difference between the two compounds and the slow rate of the reaction. Instead, a two-step process by way of oxaloacetate is required to effect the change, and this adds an extra compound to the gluconeogenesis pathway (see Figure

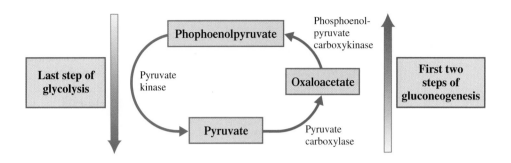

FIGURE 24.10 The "opposite" processes of gluconeogenesis (pyruvate to glucose) and glycolysis (glucose to pyruvate) are not exact opposites. The reversal of the last step of glycolysis requires two steps in gluconeogenesis. Therefore, gluconeogenesis has 11 steps, whereas glycolysis has only 10 steps.

24.10). Both an ATP molecule and a GTP molecule are needed to drive this two-step process.

The oxaloacetate intermediate in this two-step process provides a connection to the citric acid cycle. In the first step of this cycle, oxaloacetate combines with acetyl CoA. If energy rather than glucose is needed, then oxaloacetate can go directly into the citric acid cycle.

As is shown in Figure 24.11, there are two other locations where gluconeogenesis and glycolysis differ. In Steps 9 and 11 of gluconeogenesis (Steps 1 and 3 of glycolysis), the reactant–product combinations match between pathways. However, different enzymes are required for the forward and reverse processes. The new enzymes for gluconeogenesis are *fructose 1,6-bisphosphatase* and *glucose 6-phosphatase*.

The overall net reaction for gluconeogenesis is

$$2 \text{ Pyruvate} + 4\text{ATP} + 2\text{GTP} + 2\text{NADH} + 2H_2O \longrightarrow$$
$$\text{glucose} + 4\text{ADP} + 2\text{GDP} + 6P_i + 2\text{NAD}^+$$

Thus to reconvert pyruvate to glucose requires the expenditure of 4 ATP and 2 GTP. Whenever gluconeogenesis occurs, it is at the expense of other ATP-producing metabolic processes.

Glycolysis has a net production of 2 ATP (Section 24.2). Gluconeogenesis has a net expenditure of 4 ATP and 2 GTP, which is equivalent to the expenditure of 6 ATP.

■ The Cori Cycle

Gluconeogenesis using lactate as a source of pyruvate is particularly important because of lactate formation during strenuous exercise. The lactate so produced (Section 24.3) diffuses from muscle cells into the blood, where it is transported to the liver. Here the enzyme lactate dehydrogenase (the same enzyme that catalyzes lactate formation in muscle) converts lactate back to pyruvate.

FIGURE 24.11 The pathway for gluconeogenesis is similar, but not identical, to the pathway for glycolysis.

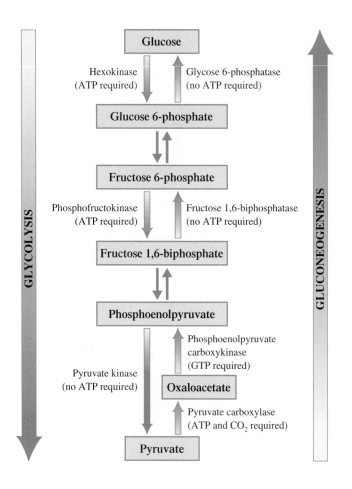

The newly formed pyruvate is then converted via gluconeogenesis to glucose, which enters the bloodstream and goes to the muscles. This cyclic process, which is called the Cori cycle, is diagrammed in Figure 24.12. The **Cori cycle** *is a cyclic biochemical process in which glucose is converted to lactate in muscle tissue, the lactate is reconverted to glucose in the liver, and the glucose is returned to the muscle tissue.*

24.7 Terminology for Glucose Metabolic Pathways

In the preceding three sections we considered the processes of glycolysis, glycogenesis, glycogenolysis, and gluconeogenesis. Because of their like-sounding names, keeping the terminology for these four processes "straight" is often a problem. Figure 24.13 shows the relationships among these processes. Note that the glycogen degradation pathways

▶ The Cori cycle is named in honor of Gerty and Carl Cori, the husband-and-wife team who discovered it. They were awarded a Nobel Prize in 1947, the third husband-and-wife team to be so recognized. Marie and Pierre Curie were the first, Irene and Frederic Joliot-Curie the second.

FIGURE 24.12 The Cori cycle. Lactate, formed from glucose under anaerobic conditions in muscle cells, is transferred to the liver, where it is reconverted to glucose, which is then transferred back to the muscle cells.

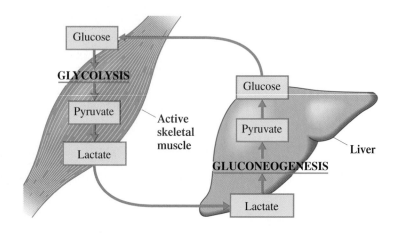

FIGURE 24.13 The relationships among four common metabolic pathways that involve glucose.

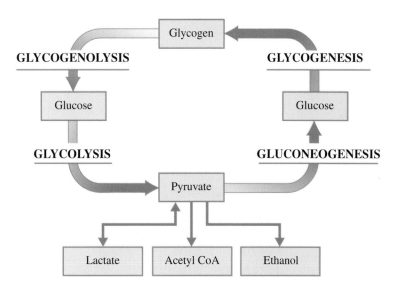

(left side of Figure 24.13) have names ending in -*lysis,* which means "breakdown." The pathways associated with glycogen synthesis (right side of Figure 24.13) have names ending in -*genesis,* which means "making."

24.8 The Pentose Phosphate Pathway

Glycolysis is not the only pathway by which glucose may be degraded. Depending on the type of cell, various amounts of glucose are degraded by the pentose phosphate pathway, a pathway whose main focus is *not* subsequent ATP production as is the case for glycolysis. Major functions of this alternative pathway are (1) synthesis of the coenzyme NADPH needed in lipid biosynthesis (Section 25.7), and (2) production of ribose 5-phosphate, a pentose derivative needed for the synthesis of nucleic acids and many coenzymes. The **pentose phosphate pathway** *is the metabolic pathway by which glucose is used to produce NADPH, ribose 5-phosphate (a pentose), and numerous other sugar phosphates.* The operation of the pentose phosphate pathway is significant in cells that produce lipids: fatty tissue, the liver, mammary glands, and the adrenal cortex (an active producer of steroid lipids).

NADPH, the coenzyme produced in the pentose phosphate pathway, is the reduced form of $NADP^+$ (nicotinamide adenine dinucleotide phosphate). Structurally, $NADP^+/NADPH$ is a phosphorylated version of $NAD^+/NADH$ (see Figure 24.14).

The nonphosphorylated and phosphorylated versions of this coenzyme have significantly different functions. The nonphosphorylated version is involved, mainly in its oxidized form ($NAD^+/NADH$), in the reactions of the common metabolic pathway (Section 23.5). The phosphorylated version is involved, mainly in its reduced form ($NADPH/NADP^+$), in biosynthetic reactions of lipids and nucleic acids.

There are two stages within the pentose phosphate pathway—an oxidative stage and a nonoxidative stage. The oxidative stage, which occurs first, involves three steps through which glucose 6-phosphate is converted to ribulose 5-phosphate and CO_2.

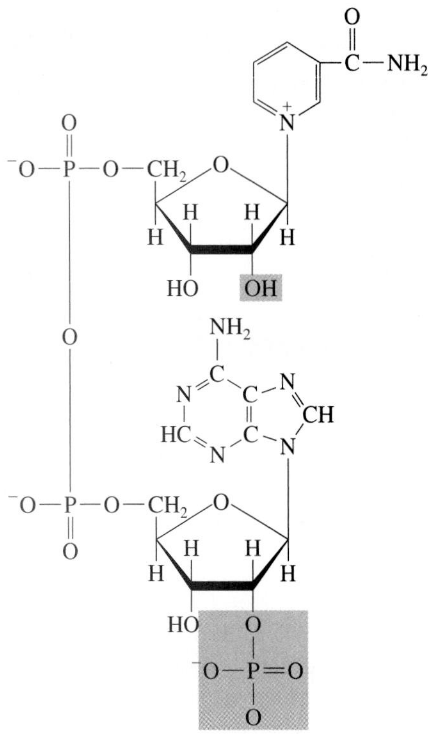

FIGURE 24.14 The Structure of NADPH. The phosphate group shown in color is the structural feature that distinguishes NADPH from NADH.

The net equation for the oxidative stage of the pentose phosphate pathway is

$$\text{Glucose 6-phosphate} + 2NADP^+ + H_2O \longrightarrow$$
$$\text{ribulose 5-phosphate} + CO_2 + 2NADPH + 2H^+$$

Note the production of two NADPH molecules per glucose 6-phosphate processed during this stage.

In the first step of the nonoxidative stage of the pentose phosphate pathway, ribulose 5-phosphate (a ketose) is isomerized to ribose 5-phosphate (an aldose).

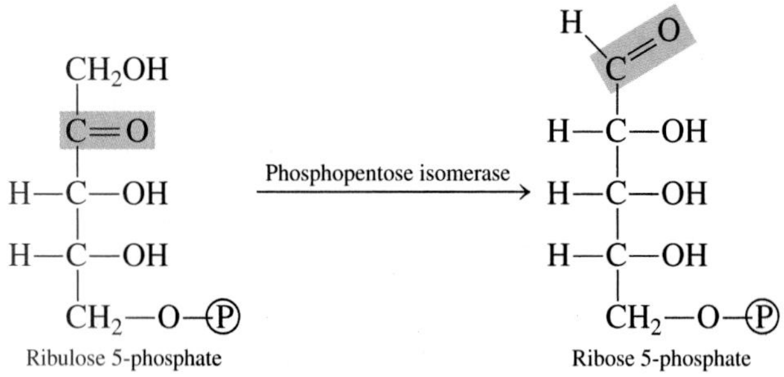

The pentose ribose is a component of ATP, GTP, UTP, CoA, NAD^+/NADH, FAD/FADH$_2$, and RNA. Further steps in the nonoxidative stage contain provision for the conversion of ribose 5-phosphate to numerous other sugar phosphates. Ultimately, glyceraldehyde 3-phosphate and fructose 6-phosphate (both glycolysis intermediates) are formed. The overall net reaction for the pentose phosphate pathway is

$$3 \text{ Glucose 6-phosphate} + 6NADP^+ + 3H_2O \longrightarrow$$
$$2 \text{ fructose 6-phosphate} + 3CO_2 + \text{glyceraldehyde 3-phosphate} + 6NADPH + 6H^+$$

The pentose phosphate pathway, with its many intermediates, helps meet cellular needs in numerous ways.

1. When ATP demand is high, the pathway continues to its end products, which enter glycolysis.

Glucose Metabolism

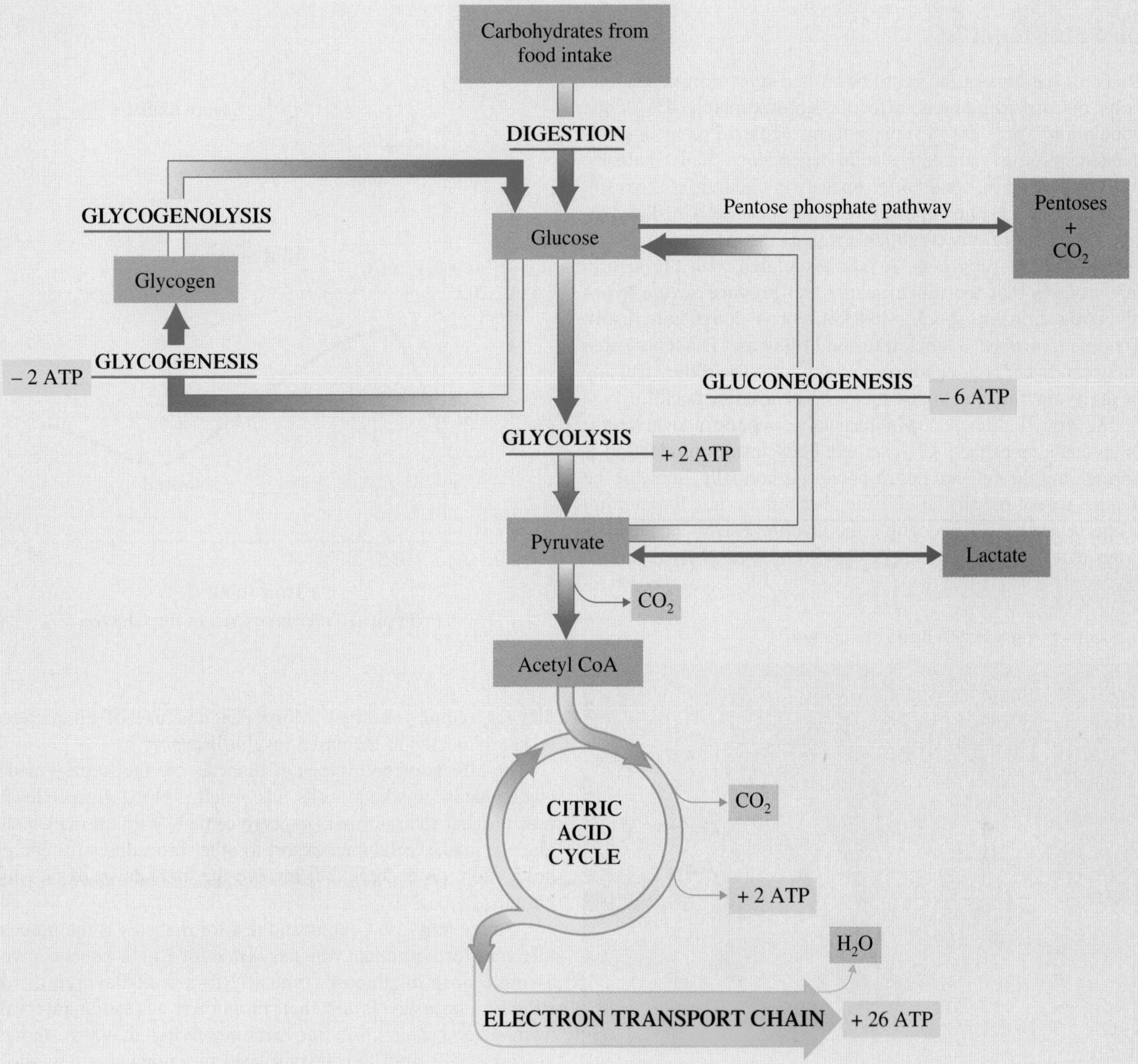

```
                        ┌─────────────────┐
                        │ Carbohydrates from │
                        │   food intake    │
                        └─────────────────┘
                                │
                           DIGESTION
```

GLYCOGENOLYSIS

Glycogen

-2 ATP   GLYCOGENESIS

Glucose

Pentose phosphate pathway → Pentoses + CO_2

GLUCONEOGENESIS -6 ATP

GLYCOLYSIS $+2$ ATP

Pyruvate ←→ Lactate

CO_2

Acetyl CoA

CITRIC ACID CYCLE   → CO_2

$+2$ ATP

H_2O

ELECTRON TRANSPORT CHAIN $+26$ ATP

2. When NADPH demand is high, intermediates are recycled to glucose 6-phosphate (the start of the pathway), and further NADPH is produced.
3. When ribose 5-phosphate demand is high, for nucleic acid and coenzyme production, most of the nonoxidative stage is nonfunctional, leaving ribose 5-phosphate as a major product.

The Chemistry at a Glance feature on this page shows how the pentose phosphate pathway is related to the other major pathways of glucose metabolism that we have considered.

24.9 Hormonal Control of Carbohydrate Metabolism

A second major method for controlling carbohydrate metabolism, besides enzyme inhibition by metabolites (Section 24.2), is hormonal control. Among others, three hormones—insulin, glucagon, and epinephrine—affect carbohydrate metabolism.

Diabetes Mellitus

Diabetes mellitus is the best-known and most prevalent metabolic disease in humans, affecting approximately 4% of the population. There are two major forms of this disease: insulin-dependent (type I) and non–insulin-dependent (type II) diabetes.

Type I diabetes, which often appears in children, is the result of inadequate insulin production by the beta cells of the pancreas. Control of this condition involves insulin injections and special dietary programs. A risk associated with the insulin injections is that too much insulin can produce severe hypoglycemia (insulin shock); blackout or a coma can result. Treatment involves a quick infusion of glucose. Diabetics often carry candy bars (quick glucose sources) for use if they feel any of the symptoms that signal the onset of insulin shock.

In Type II diabetes, which usually occurs in overweight individuals more than 40 years old, body insulin production is normal, but the cells do not respond to it normally. Some of the insulin receptors on the cell membranes are not functioning properly and fail to recognize the insulin. Treatment involves drugs that increase body insulin levels and a carefully regulated

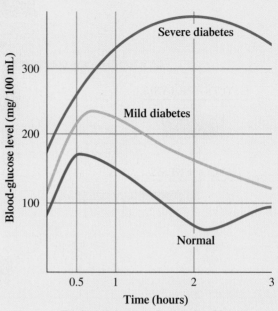

Typical Tolerance Curves for Glucose

A diabetic giving himself a blood glucose test.

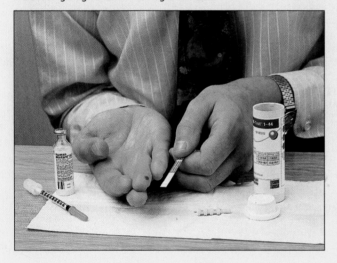

diet (to reduce obesity). More efficient use of undamaged receptors occurs at increased insulin levels.

The effects of both types of diabetes are the same—inadequate glucose uptake by cells. The result is blood-glucose levels much higher than normal (hyperglycemia). With an inadequate glucose intake, cells must resort to other procedures for energy production, procedures that involve the breakdown of fats and protein.

A frequently used diagnostic test for diabetes is the glucose tolerance test. A patient who has fasted for 10–16 hours is given a single dose of glucose, typically in a fruit-flavored drink. Blood-glucose levels are then monitored at regular intervals over several hours. As the accompanying diagram shows, glucose levels drop to a fasting level in a nondiabetic in about 2 hours but remain high in a diabetic.

■ Insulin

Insulin, a 51-amino-acid protein whose structure we considered in Section 20.9, is a hormone produced by the beta cells of the pancreas. Insulin promotes the uptake and utilization of glucose by cells. Thus its function is to lower blood glucose levels. It is also involved in lipid metabolism.

The release of insulin is triggered by *high* blood-glucose levels. The mechanism for insulin action involves insulin binding to protein receptors on the outer surfaces of cells, which facilitates entry of glucose into the cells. Insulin also produces an increase in the rate of glycogen synthesis.

■ Glucagon

Glucagon is a polypeptide hormone (29 amino acids) produced in the pancreas by alpha cells. It is released when blood-glucose levels are *low*. Its principal function is to

FIGURE 24.15 The series of events by which the hormone epinephrine stimulates glucose production.

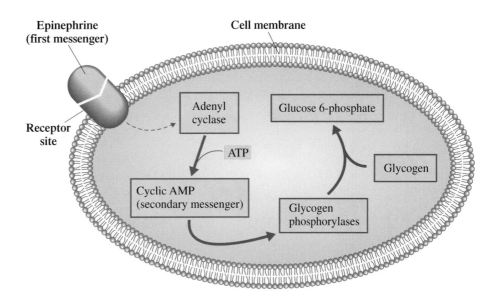

increase blood-glucose concentrations by speeding up the conversion of glycogen to glucose (glycogenolysis) in the liver. Thus glucagon's effects are opposite to those of insulin.

■ Epinephrine

Epinephrine (Section 17.9), also called adrenaline, is released by the adrenal glands in response to anger, fear, or excitement. Its function is similar to that of glucagon—stimulation of glycogenolysis, the release of glucose from glycogen. Its primary target is muscle cells, where energy is needed for quick action. It also functions in lipid metabolism.

Epinephrine acts by binding to a receptor site on the outside of the cell membrane, stimulating the enzyme *adenyl cyclase* to begin production of a *secondary messenger,* cyclic AMP (cAMP) from ATP. The cAMP is released in the cell interior, where, in a series of reactions, it activates *glycogen phosphorylase,* the enzyme that initiates glycogenolysis. The glucose 6-phosphate that is produced from the glycogen breakdown provides a source of quick energy. Figure 24.15 shows the series of events initiated by the release of the hormone epinephrine. Cyclic AMP also inhibits glycogenesis, thus preventing glycogen production at the same time.

CONCEPTS TO REMEMBER

Glycolysis. Glycolysis, a series of ten reactions that occur in the cytosol, is a process in which one glucose molecule is converted into two molecules of pyruvate. A net gain of two molecules of ATP and two molecules of NADH results from the metabolizing of glucose to pyruvate.

Fates of pyruvate. With respect to energy-yielding metabolism, the pyruvate produced by glycolysis can be converted to acetyl CoA under aerobic conditions or to lactate under anaerobic conditions. Some microorganisms convert pyruvate to ethanol, an anaerobic process.

Glycogenesis. Glycogenesis is the process whereby excess glucose is converted into glycogen. The glycogen is stored in the liver and in muscle tissue.

Glycogenolysis. Glycogenolysis is the breakdown of glycogen into glucose. This process occurs when muscles need energy and when the liver is restoring a low blood-sugar level to normal.

Gluconeogenesis. Gluconeogenesis is the formation of glucose from lactate and certain other substances. This process takes place in the liver when glycogen supplies are being depleted and when carbohydrate intake is low.

Cori cycle. The Cori cycle is the cyclic process involving the transport of lactate from muscle tissue to the liver, the resynthesis of glucose by gluconeogenesis, and the return of glucose to muscle tissue.

Pentose phosphate pathway. The pentose phosphate pathway metabolizes glucose to produce ribose (a pentose), NADPH, and other sugars needed for biosynthesis.

Carbohydrate metabolism and hormones. Insulin decreases blood-glucose levels by promoting the uptake of glucose by cells. Glucagon increases blood-glucose levels by promoting the conversion of glycogen to glucose. Epinephrine stimulates the release of glucose from glycogen in muscle cells.

KEY REACTIONS AND EQUATIONS

1. Glycolysis (Section 24.2)

   Glucose + 2P$_i$ + 2ADP + 2NAD$^+$ $\longrightarrow$
   2 pyruvate + 2ATP + 2NADH + 2H$^+$ + 2H$_2$O

2. Oxidation of pyruvate to acetyl CoA (Section 24.3)

   $$CH_3-\overset{\overset{\textstyle O}{\|}}{C}-COO^- + CoA-SH + NAD^+ \xrightarrow[\text{steps}]{\text{Four}}$$

   $$CH_3-\overset{\overset{\textstyle O}{\|}}{C}-S-CoA + NADH + CO_2$$

3. Reduction of pyruvate to lactate (Section 24.3)

   $$CH_3-\overset{\overset{\textstyle O}{\|}}{C}-COO^- + NADH + H^+ \longrightarrow$$

   $$CH_3-\overset{\overset{\textstyle OH}{|}}{CH}-COO^- + NAD^+$$

4. Reduction of pyruvate to ethanol (Section 24.3)

   $$CH_3-\overset{\overset{\textstyle O}{\|}}{C}-COO^- + 2H^+ + NADH \xrightarrow[\text{steps}]{\text{Two}}$$

   $$CH_3-CH_2-OH + NAD^+ + CO_2$$

5. Complete oxidation of glucose (Section 24.4)

   Glucose + 6O$_2$ + 30ADP + 30P$_i$ $\longrightarrow$ 6CO$_2$ + 6H$_2$O + 30ATP

6. Glycogenesis (Section 24.5)

   Glucose $\xrightarrow[\text{steps}]{\text{Three}}$ glycogen

7. Glycogenolysis (Section 24.5)

   Glucose $\xrightarrow[\text{steps}]{\text{Three}}$ glycogen

8. Gluconeogenesis (Section 24.6)

   $\left.\begin{array}{c}\text{Lactate, certain}\\\text{amino acids,}\\\text{citric acid cycle}\\\text{intermediates}\end{array}\right\} \longrightarrow$ pyruvate $\xrightarrow[\text{steps}]{\text{Eleven}}$ glucose

---

KEY TERMS

Cori cycle (24.6)
Digestion (24.1)
Ethanol fermentation (24.3)
Fermentation (24.3)

Gluconeogenesis (24.6)
Glycogenesis (24.5)
Glycogenolysis (24.5)
Glycolysis (24.2)

Lactate fermentation (24.3)
Pentose phosphate pathway (24.8)
Substrate-level phosphorylation (24.2)

EXERCISES AND PROBLEMS

The members of each pair of problems in this section test similar material.

■ Carbohydrate Digestion (Section 24.1)

24.1 Where does starch digestion begin in the body, and what is the name of the enzyme involved in this initial digestive process?

24.2 Very little digestion of starch occurs in the stomach. Why?

24.3 What is the primary site for carbohydrate digestion, and what organ produces the enzymes that are active at this location?

24.4 Where does the final step in carbohydrate digestion take place, and in what form are carbohydrates as they enter this final step?

24.5 Where does the digestion of sucrose begin, and what is the reaction that occurs?

24.6 Where does the digestion of lactose begin, and what is the reaction that occurs?

24.7 Identify the three major monosaccharides produced by digestion of carbohydrates.

24.8 The various stages of carbohydrate digestion all involve the same general type of reaction. What is this reaction type?

■ Glycolysis (Section 24.2)

24.9 What is the starting material for glycolysis?

24.10 What is the end product from glycolysis?

24.11 What coenzyme functions as the oxidizing agent in glycolysis?

24.12 What is meant by the statement that glycolysis is an anaerobic pathway?

24.13 What is the first step of glycolysis, and why is it important in retaining glucose inside the cell?

24.14 Step 3 of glycolysis is the commitment step. Explain.

24.15 What two C$_3$ fragments are formed by the splitting of a fructose 1,6-bisphosphate molecule?

24.16 In one step of the glycolysis pathway, a C$_6$ chain is broken into two C$_3$ fragments, only one of which can be further degraded. What happens to the other C$_3$ fragment?

24.17 How many pyruvate molecules are produced per glucose molecule during glycolysis?

24.18 How many molecules of ATP and NADH are produced per glucose molecule during glycolysis?

24.19 How many steps in the glycolysis pathway produce ATP?

24.20 How many steps in the glycolysis pathway consume ATP?

24.21 Of the 10 steps of glycolysis, which ones involve phosphorylation?

24.22 Of the 10 steps of glycolysis, which ones involve oxidation?

24.23 Where in a cell does glycolysis occur?

24.24 Do the reactions of glycolysis and the citric acid cycle occur at the same location in a cell? Explain.

24.25 Replace the question mark in each of the following word equations with the name of a substance.

a. Glucose + ATP $\xrightarrow{\text{Hexokinase}}$? + ADP + H$^+$

b. ? $\xrightarrow{\text{Enolase}}$ phosphoenolpyruvate + water

c. 3-Phosphoglycerate $\xrightarrow{\;?\;}$ 2-phosphoglycerate

d. 1,3-Bisphosphoglycerate + ? $\xrightarrow[\text{kinase}]{\text{Phosphoglycero-}}$
3-phosphoglycerate + ATP

24.26 Replace the question mark in each of the following word equations with the name of a substance.

a. Glucose 6-phosphate $\xrightarrow[\text{isomerase}]{\text{Phosphogluco-}}$?

b. ? $\xrightarrow{\text{Aldolase}}$ dihydroxyacetone phosphate + glyceraldehyde 3-phosphate

c. Phosphoenolpyruvate + ? + H$^+$ $\xrightarrow[\text{kinase}]{\text{Pyruvate}}$
pyruvate + ATP

d. Dihydroxyacetone phosphate $\xrightarrow{\;?\;}$
glyceraldehyde 3-phosphate

24.27 In which step of glycolysis does each of the following occur?
a. Second substrate-level phosphorylation reaction
b. First ATP-consuming reaction
c. Third isomerization reaction
d. Use of NAD$^+$ as an oxidizing agent

24.28 In which step of glycolysis does each of the following occur?
a. First energy-producing reaction
b. First ATP-producing reaction
c. A dehydration reaction
d. First isomerization reaction

24.29 What is the net ATP production when each of the following molecules is processed through the glycolysis pathway?
a. One glucose molecule
b. One sucrose molecule

24.30 What is the net ATP production when each of the following molecules is processed through the glycolysis pathway?
a. One lactose molecule
b. One maltose molecule

24.31 Draw structural formulas for each of the following pairs of molecules.
a. Pyruvic acid and pyruvate
b. Dihydroxyacetone and dihydroxyacetone phosphate
c. Fructose 6-phosphate and fructose 1,6-bisphosphate
d. Glyceric acid and glyceraldehyde

24.32 Draw structural formulas for each of the following pairs of molecules.
a. Glyceric acid and glycerate
b. Glycerate and pyruvate
c. Glucose 6-phosphate and fructose 6-phosphate
d. Dihydroxyacetone and glyceric acid

24.33 Number the carbon atoms of fructose 1,6-bisphosphate 1 through 6, and show the location of each carbon in the two trioses produced during Step 4 of glycolysis.

24.34 Number the carbon atoms of glucose 1 through 6, and show the location of each carbon in the two molecules of pyruvate produced by glycolysis.

■ Fates of Pyruvate (Section 24.3)

24.35 What are the three common possible fates for pyruvate produced from glycolysis?

24.36 Compare the fates of pyruvate in the body under aerobic and anaerobic conditions.

24.37 What is the overall reaction equation for the conversion of pyruvate to acetyl CoA?

24.38 What is the overall reaction equation for the conversion of pyruvate to lactate?

24.39 Explain how lactate fermentation allows glycolysis to continue under anaerobic conditions.

24.40 How is the ethanol fermentation in yeast similar to lactate fermentation in skeletal muscle?

24.41 In ethanol fermentation, a C$_3$ pyruvate molecule is changed to a C$_2$ ethanol molecule. What is the fate of the third pyruvate carbon?

24.42 What are the structural differences between pyruvate and lactate ions?

24.43 What is the net reaction for the conversion of one glucose molecule to two lactate molecules?

24.44 What is the net reaction for the conversion of one glucose molecule to two ethanol molecules?

■ Complete Oxidation of Glucose (Section 24.4)

24.45 How does the fact that cytosolic NADH/H$^+$ cannot cross the mitochondrial membranes affect ATP production from cytosolic NADH/H$^+$?

24.46 What is the net reaction for the shuttle mechanism involving glycerol 3-phosphate by which NADH electrons are shuttled across the mitochondrial membrane.

24.47 Contrast, in terms of ATP production, the oxidation of glucose to CO$_2$ and H$_2$O with the oxidation of glucose to pyruvate.

24.48 Contrast, in terms of ATP production, the oxidation of glucose to CO$_2$ and H$_2$O with the oxidation of glucose to ethanol.

24.49 How many of the 30 ATP molecules produced from the complete oxidation of 1 glucose molecule are produced during glycolysis?

24.50 How many of the 30 ATP molecules produced from the complete oxidation of 1 glucose molecule are the result of the oxidation of pyruvate to acetyl CoA?

■ Glycogen Metabolism (Section 24.5)

24.51 Compare the meanings of the terms *glycogenesis* and *glycogenolysis.*

24.52 Where is most of the body's glycogen stored?

24.53 Glucose 1-phosphate is the product of the first step of glycogenesis. What is the reactant?

24.54 Glucose 1-phosphate is the product of the first step of glycogenolysis. What are the reactants?

24.55 What is the source of the PP$_i$ produced during the second step of glycogenesis?

24.56 What is the function of the PP$_i$ produced during the second step of glycogenesis?

24.57 How is ATP involved in glycogenesis?

24.58 How many ATP molecules are needed to attach a single glucose molecule to a growing glycogen chain?

24.59 Which step of glycogenolysis is the reverse of Step 1 of glycogenesis?

24.60 What reaction determines whether glucose formed by glycogenolysis can leave a cell?

24.61 What is the difference between glycogenolysis in liver cells and in muscle cells?

24.62 The liver, but not the brain or muscle cells, has the capacity to supply free glucose to the blood. Explain.

24.63 In what form does glycogen enter the glycolysis pathway?

24.64 Explain why one more ATP is produced when glucose is obtained from glycogen than when it is obtained directly from the blood.

■ **Gluconeogenesis (Section 24.6)**

24.65 What organ is primarily responsible for gluconeogenesis?

24.66 What is the physiological function of gluconeogenesis?

24.67 How does gluconeogenesis get around the three irreversible steps of glycolysis?

24.68 Although gluconeogenesis and glycolysis are "reverse" processes, there are 11 steps in gluconeogenesis and only 10 steps in glycolysis. Explain.

24.69 What intermediate in gluconeogenesis is also an intermediate in the citric acid cycle?

24.70 What are the sources of high-energy bonds in gluconeogenesis?

24.71 What is the fate of lactate formed by muscular activity?

24.72 What is the physiological function of the Cori cycle?

■ **The Pentose Phosphate Pathway (Section 24.8)**

24.73 What is the starting material for the pentose phosphate pathway?

24.74 What are two major functions of the pentose phosphate pathway?

24.75 How do the biochemical functions of NADH and NADPH differ?

24.76 How do the structures of NADH and NADPH differ?

24.77 Write a general equation for the oxidative stage of the pentose phosphate pathway.

24.78 Write a general equation for the entire pentose phosphate pathway.

24.79 What compound contains the carbon atom lost from glucose (a hexose) in its conversion to ribose (a pentose)?

24.80 How many molecules of NADPH are produced per glucose 6-phosphate in the pentose phosphate pathway?

■ **Control of Carbohydrate Metabolism (Section 24.9)**

24.81 What effect does insulin have on glycogen metabolism?

24.82 What effect does insulin have on blood-glucose levels?

24.83 What effect does glucagon have on blood-glucose levels?

24.84 What effect does glucagon have on glycogen metabolism?

24.85 What organ is the source of insulin?

24.86 What organ is the source of glucagon?

24.87 The hormone epinephrine generates a "second messenger." Explain.

24.88 What is the relationship between cAMP and the hormone epinephrine?

24.89 Compare the target tissues for glucagon and epinephrine.

24.90 Compare the biological functions of glucagon and epinephrine.

ADDITIONAL PROBLEMS

24.91 Indicate in which of the four processes *glycolysis, glycogenesis, glycogenolysis,* and *gluconeogenesis* each of the following compounds is encountered. There may be more than one correct answer for a given compound.
a. Glucose 6-phosphate
b. Glucose 1-phosphate
c. Dihydroxyacetone phosphate
d. Oxaloacetate

24.92 Indicate in which of the four processes *glycolysis, glycogenesis, glycogenolysis,* and *gluconeogenesis* each of the following situations is encountered. There may be more than one correct answer for a given situation.
a. NAD^+ is consumed.
b. ATP is produced.
c. ATP is consumed.
d. UDP is involved.

24.93 Indicate in which of the four processes *glycolysis, glycogenesis, glycogenolysis,* and *glyconeogenesis* each of the following characterizations applies.
a. Glucose is converted to two pyruvates.
b. Glycogen is synthesized from glucose.
c. Glycogen is broken down into free glucose units.
d. Glucose is synthesized from pyruvate.

24.94 What is the ATP yield *per glucose molecule* in each of the following processes?
a. Glycolysis
b. Glycolysis, acetyl CoA formation, and the common metabolic pathway

c. Glycolysis plus oxidation of pyruvate to acetyl CoA
d. Glycolysis plus reduction of pyruvate to lactate

24.95 Which one of these characterizations, (1) Cori cycle, (2) an anaerobic process, (3) oxidative stage of pentose phosphate pathway, or (4) nonoxidative stage of pentose phosphate pathway, applies to each of the following chemical changes?
a. Pyruvate to lactate
b. Pyruvate to ethanol
c. Glucose 6-phosphate to ribulose 5-phosphate
d. Ribulose 5-phosphate to ribose 5-phosphate

24.96 What condition or conditions determine that pyruvate is involved in each of the following?
a. Gluconeogenesis
b. Converted to lactate
c. Citric acid cycle
d. Converted to ethanol

24.97 In the complete metabolism of 1 mole of sucrose, how many moles of each of the following are produced?
a. CO_2
b. Pyruvate
c. Acetyl CoA
d. ATP

24.98 Under what conditions does glucose 6-phosphate enter each of the following pathways?
a. Glycogenesis
b. Glycolysis
c. Pentose phosphate pathway
d. Hydrolysis to free glucose

25 Lipid Metabolism

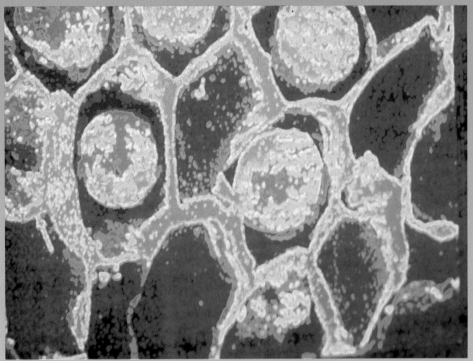

Fat cells.

Certain classes of lipids play an extremely important role in cellular metabolism, because they represent an energy-rich "fuel" that can be stored in large amounts in adipose (fat) tissue. Between one-third and one-half of the calories present in the diet of the average U.S. resident are supplied by lipids. Furthermore, excess energy derived from carbohydrates and proteins beyond normal daily needs is stored in lipid molecules (in adipose tissue), later to be mobilized and used when needed.

25.1 Digestion and Absorption of Lipids

Because 98% of total *dietary* lipids are triacylglycerols (fats and oils; Section 19.4), this chapter focuses on triacylglycerol metabolism. Like all lipids, triacylglycerols (TAGs) are insoluble in water. Hence, water-based salivary enzymes in the mouth have little effect on them. The *major* change that TAGs undergo in the stomach is physical rather than chemical. The churning action of the stomach breaks up triacylglycerol materials into small globules, or droplets, which float as a layer above the other components of swallowed food. The resulting material is called *chyme.*

High-fat foods remain in the stomach longer than low-fat foods. The conversion of high-fat materials into chyme takes longer than the breakup of low-fat materials. This is why a high-fat meal causes a person to feel "full" for a longer period of time.

Lipid digestion also begins in the stomach. Under the action of *gastric lipase* enzymes, hydrolysis of TAGs occurs. Normally, about 10% of TAGs undergo hydrolysis in the stomach, but regular consumption of a high-fat diet can induce the production of higher levels of gastric lipases.

The saliva of infants contains a lipase that can hydrolyze TAGs, so digestion begins in the mouth for nursing infants. Because mother's milk is already a lipid-in-water emulsion, emulsification by stomach churning is a much less important factor in an infant's processing of fat. Mother's milk also contains a lipase that supplements the action of the salivary lipases the infant itself produces. After weaning, infants cease to produce salivary lipases.

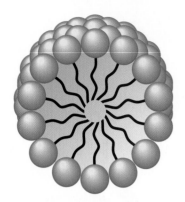

FIGURE 25.1 In a fatty acid micelle, the hydrophobic chains of the fatty acids and monoacylglycerols are in the interior of the micelle.

When freed of the triacylglycerol molecules they "transport" during digestion, bile acids are mostly recycled. Small amounts are excreted.

Chylomicron is pronounced "kye-lo-MY-cron."

FIGURE 25.2 A three-dimensional model of a chylomicron, a type of lipoprotein. Chylomicrons are the form in which TAGs are delivered to the bloodstream via the lymphatic system.

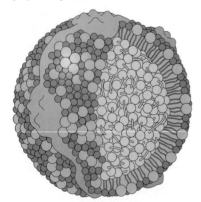

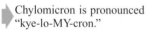 **Triacylglycerols (TAGs)**

■ **Protein**

Membrane lipids

The arrival of chyme from the stomach triggers in the small intestine, through the action of the hormone *cholecystokinin,* the release of bile stored in the gallbladder. The bile (Section 19.11), which contains no enzymes, acts as an emulsifier (Section 19.11). Colloid particle formation (Section 8.6) through bile emulsification "solubilizes" the triacylglycerol globules, and digestion of the TAGs resumes. The major enzymes involved at this point are the *pancreatic lipases,* which hydrolyze ester linkages between the glycerol and fatty acid units of the TAGs. *Complete* hydrolysis does not usually occur; only two of the three fatty acids units are liberated, producing a monoacylglycerol and two free fatty acids. Occasionally, enzymes remove all three fatty acid units, leaving a free glycerol molecule.

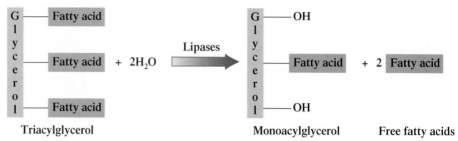

With the help of bile, the free fatty acids and monoacylglycerols produced from hydrolysis are combined into tiny spherical droplets called micelles (Section 19.6). A **fatty acid micelle** *is a micelle in which fatty acids and/or monoacylglycerols and some bile are present.* Fatty acid micelles are very small compared to the original triacylglycerol globules, which contain thousands of triacylglycerol molecules. Figure 25.1 shows a cross section of the three-dimensional structure of a fatty acid micelle.

Micelles, containing free fatty acid and monoacylglycerol components, are small enough to be readily absorbed through the membranes of intestinal cells. Within the intestinal cells, a "repackaging" occurs in which the free fatty acids and monoacylglycerols are reassembled into triacylglycerols. The newly formed triacylglycerols are then combined with membrane lipids (phospholipids and cholesterol) and water-soluble proteins to produce a type of lipoprotein (Section 20.18) called a *chylomicron* (see Figure 25.2). A **chylomicron** *is a lipoprotein that transports triacylglycerols from intestinal cells, via the lymphatic system, to the bloodstream.* Chylomicrons are too large to pass through capillary walls directly into the bloodstream. Consequently, delivery of the chylomicrons to the bloodstream is accomplished through the body's lymphatic system. Chylomicrons enter the lymphatic system through tiny lymphatic vessels in the intestinal lining. They enter the bloodstream through the thoracic duct (a large lymphatic vessel just below the collarbone), where the fluid of the lymphatic system flows into a vein, joining the bloodstream.

Once the chylomicrons reach the bloodstream, the TAGs they carry are again hydrolyzed to produce glycerol and free fatty acids. The products from this hydrolysis are absorbed by the cells of the body and are either broken down to acetyl CoA for energy or stored as lipids (they are again repackaged as TAGs). Figure 25.3 summarizes the events that must occur before triacylglycerols can reach the bloodstream through the digestive process.

Soon after a meal heavily laden with TAGs is ingested, the chylomicron content of both blood and lymph increases dramatically. Chylomicron concentrations usually begin to rise within 2 hours after a meal, reach a peak in 4–6 hours, and then drop rather rapidly to a normal level as they move into adipose cells (Section 25.2) or into the liver.

25.2 Triacylglycerol Storage and Mobilization

Most cells in the body have limited capability for storage of TAGs. However, this activity is the major function of specialized cells called adipocytes, found in adipose tissue. *An* **adipocyte** *is a triacylglycerol-storing cell.* **Adipose tissue** *is tissue that contains large numbers of adipocyte cells.*

Adipose tissue is located primarily directly beneath the skin (subcutaneous), particularly in the abdominal region, and in areas around vital organs. Besides its function as a

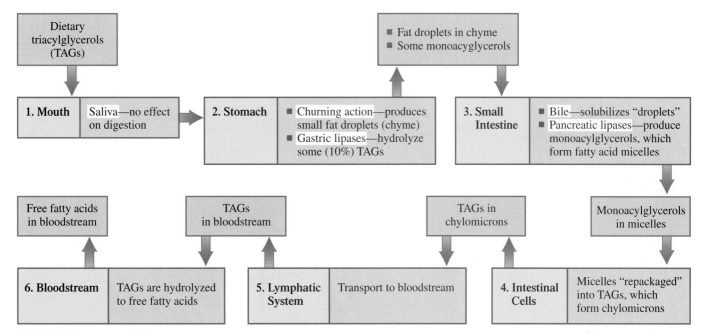

FIGURE 25.3 A summary of the events that must occur before triacylglycerols (TAGs) can reach the bloodstream through the digestive process.

▶ Dietary TAGs deposited in adipose tissue have undergone hydrolysis two times (to form free fatty acids and/or monoacylglycerols) and are repackaged twice (to re-form TAGs) in reaching that state. They undergo hydrolysis for a third time when *triacylglycerol mobilization* occurs.

▶ Adipose tissue is the only tissue in which *free* TAGs occur in appreciable amounts. In other types of cells and in the bloodstream, TAGs are part of lipoprotein particles.

FIGURE 25.4 Structural characteristics of an adipose cell.

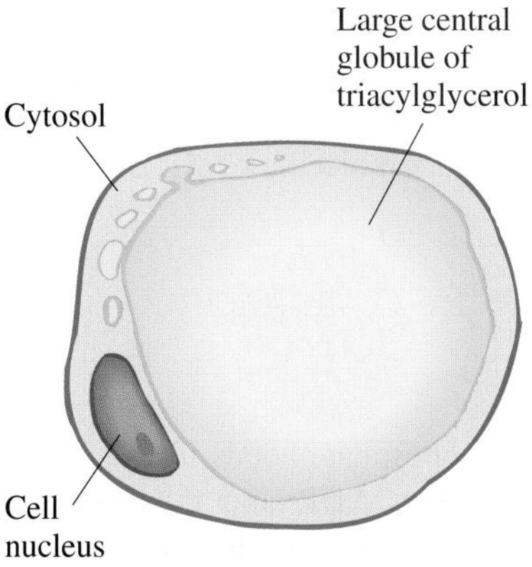

storage location for the chemical energy inherent in TAGs, subcutaneous adipose tissue also serves as an insulator against excessive heat loss to the environment and provides organs with protection against physical shock.

Adipose cells are among the largest cells in the body. They differ from other cells in that most cytoplasm has been replaced with a large triacylglycerol droplet (Figure 25.4). This droplet accounts for nearly the entire volume of the cell. As newly formed TAGs are imported into an adipose cell, they form small droplets at the periphery of the cell that later merge with the large central droplet.

Use of the TAGs stored in adipose tissue for energy production is triggered by several hormones, including epinephrine and glucagon. Hormonal interaction with adipose cell membrane receptors stimulates production of cAMP from ATP inside the adipose cell. In a series of enzymatic reactions, the cAMP activates hormone-sensitive lipase (HSL) through phosphorylation. HSL is the lipase needed for triacylglycerol hydrolysis, a prerequisite for fatty acids to enter the bloodstream from an adipose cell. This cAMP activation process is illustrated in Figure 25.5.

The overall process of tapping the body's triacylglycerol energy reserves (adipose tissue) for energy is called triacylglycerol mobilization. **Triacylglycerol mobilization** *is the hydrolysis of triacylglycerols stored in adipose tissue, followed by release into the bloodstream of the fatty acids and glycerol so produced.* Triacylglycerol mobilization is an ongoing process. On the average, about 10% of the TAGs in adipose tissue are replaced daily by new triacylglycerol molecules.

25.3 Glycerol Metabolism

During triacylglycerol mobilization, one molecule of glycerol is produced for each triacylglycerol completely hydrolyzed. Glycerol metabolism primarily involves processes considered in the previous chapter. After entering the bloodstream, glycerol travels to the liver or kidneys, where it is converted, in a two-step process, to dihydroxyacetone phosphate.

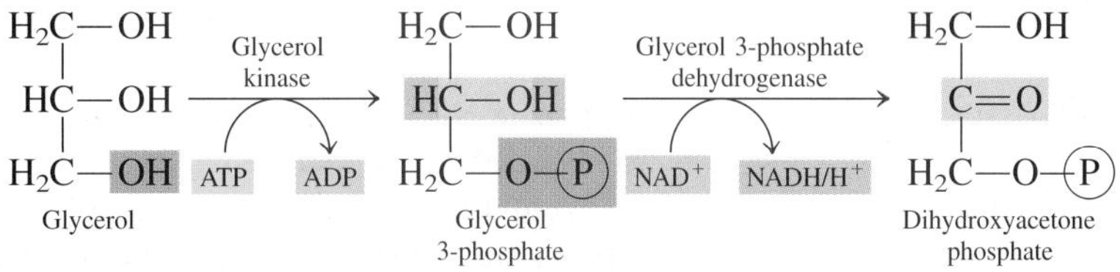

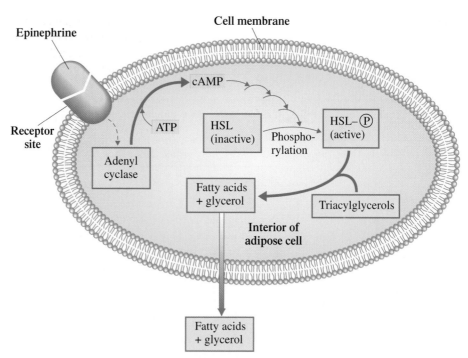

FIGURE 25.5 Hydrolysis of stored triacylglycerols in adipose tissue is triggered by hormones that stimulate cAMP production within adipose cells.

> The use of cAMP in the activation of hormone-sensitive lipase in adipose cells is similar to cAMP's role in the activation of the glycogenolysis process (Section 24.9).

The first step involves phosphorylation of a primary hydroxyl group of the glycerol. In the second step, glycerol's secondary alcohol group (C-2) is oxidized to a ketone.

Dihydroxyacetone phosphate is an intermediate in both glycolysis (Section 24.2) and gluconeogenesis (Section 24.6). It can be converted to pyruvate, then acetyl CoA, and finally carbon dioxide, or it can be used to form glucose. Dihydroxyacetone phosphate formation from glycerol represents the first of several situations we will consider wherein carbohydrate and lipid metabolism are connected.

25.4 Oxidation of Fatty Acids

There are three parts to the process by which fatty acids are broken down to obtain energy.

1. The fatty acid must be *activated* by bonding to coenzyme A.
2. The fatty acid must be *transported* into the mitochondrial matrix by a shuttle mechanism.
3. The fatty acid must be repeatedly *oxidized,* cycling through a series of four reactions, to produce acetyl CoA, $FADH_2$, and NADH.

> The stored TAGs in adipose tissue supply approximately 60% of the body's energy needs when the body is in a resting state.

■ Fatty Acid Activation

The outer mitochondrial membrane is the site of fatty acid *activation,* the first stage of fatty acid oxidation. Here the fatty acid is converted to a high-energy derivative of coenzyme A. Reactants are the fatty acid, coenzyme A, and a molecule of ATP.

> Triacylglycerol reserves would enable the average person to survive starvation for about 30 days, given sufficient water. Glycogen reserves (stored glucose) would be depleted within 1 day.

$$\underset{\text{Free fatty acid}}{R-\overset{\overset{\textstyle O}{\|}}{C}-O^-} + \underset{}{HS-CoA} \xrightarrow[\underset{ATP \quad AMP \; + \; 2P_i}{}]{\text{Acyl CoA}\atop\text{synthetase}} \underset{\text{Acyl CoA}}{R-\overset{\overset{\textstyle O}{\|}}{C}-S-CoA}$$

This reaction requires the expenditure of two high-energy phosphate bonds from a single ATP molecule; the ATP is converted to AMP rather than ADP, and the resulting pyrophosphate (PP_i) is hydrolyzed to $2P_i$.

Acyl is a generic term for

$$R-\overset{\overset{\displaystyle O}{\|}}{C}-$$

which is the species formed when the carboxyl —OH is removed from a carboxylic acid (Section 19.4). The R group can involve a carbon chain of any length.

The activated fatty acid–CoA molecule is called *acyl* CoA. The difference between the designations *acyl* CoA and *acetyl* CoA is that *acyl* refers to a random-length fatty acid carbon chain that is covalently bonded to coenzyme A, whereas *acetyl* refers to a two-carbon chain covalently bonded to coenzyme A.

$$R-\overset{\overset{\displaystyle O}{\|}}{C}-S-CoA \qquad CH_3-\overset{\overset{\displaystyle O}{\|}}{C}-S-CoA$$

Acyl CoA
R = carbon chain of any length

Acetyl CoA
R = CH₃ group

■ Fatty Acid Transport

Acyl CoA is too large to pass through the inner mitochondrial membrane to the mitochondrial matrix, where the enzymes needed for fatty acid oxidation are located. A shuttle mechanism involving the molecule carnitine effects the entry of acyl CoA into the matrix (see Figure 25.6). The acyl group is transferred to a carnitine molecule, which carries it through the membrane. The acyl group is then transferred from the carnitine back to a CoA molecule.

■ The Fatty Acid Spiral

In the mitochondrial matrix, a sequence of four reactions *repeatedly* cleaves two-carbon units from the carboxyl end of a saturated fatty acid. This process is called the *fatty acid spiral,* because of its repetitive nature, or *β oxidation spiral,* because the second, or beta, carbon from the carboxyl end of the chain is oxidized. The **fatty acid spiral** *is the metabolic pathway that degrades fatty acids to acetyl CoA by removing two carbon atoms at a time, with FADH₂ and NADH also being produced.*

For a *saturated* fatty acid, the fatty acid spiral involves the following functional group changes at the β carbon and the following reaction types.

We have encountered an identical set of functional group changes before, in the back side of the citric acid cycle (Section 23.6), Steps 6–8 of this cycle.

$$\text{Alkane} \xrightarrow[\text{(dehydrogenation)}]{\overset{①}{\text{Oxidation}}} \text{alkene} \xrightarrow{\overset{②}{\text{Hydration}}} \begin{array}{c}\text{secondary}\\\text{alcohol}\end{array} \xrightarrow[\text{(dehydrogenation)}]{\overset{③}{\text{Oxidation}}} \text{ketone} \xrightarrow{\overset{④}{\text{Chain}}\\\text{cleavage}} $$

Details about Steps 1–4 of the fatty acid spiral follow.

Step 1: *Oxidation (dehydrogenation).* Hydrogen atoms are removed from the α and β carbons, creating a double bond between these two carbon atoms. FAD is the oxidizing agent, and a FADH₂ molecule is a product.

FIGURE 25.6 Fatty acids are transported across the inner mitochondrial membrane in the form of acyl carnitine.

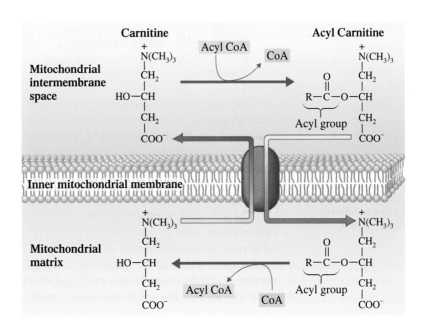

CHEMICAL CONNECTIONS

High-Intensity Versus Low-Intensity Workouts

In a resting state, the human body burns more fat than carbohydrate. The fuel consumed is about one-third carbohydrate and two-thirds fat.

Information about fuel consumption ratios is obtainable from respiratory gas measurements, specifically from the respiratory exchange ratio (RER). The RER is the ratio of carbon dioxide to oxygen inhaled divided by the ratio of carbon dioxide to oxygen exhaled. For 100% fat burning, the RER would be 0.7; for 100% carbohydrate burning, the RER would be 1.0.

When a person at rest begins exercising, his or her body suddenly needs energy at a greater rate—more fuel and more oxygen are needed. It takes 0.7 L of oxygen to burn 1 gram of carbohydrate and 1.0 L of oxygen to burn 1 gram of fat. At the onset of exercise, the body is immediately short of oxygen. Also, there is a time delay in triacylglycerol mobilization. Triacylglycerols have to be broken down to fatty acids, which have

The initial stages of exercise are fueled primarily by glucose; in later stages, triacylglycerols become the primary fuel.

to be attached to protein carriers before they can be carried in the bloodstream to working muscles. At their destination, they must be released from the carriers and then undergo energy-producing reactions. By contrast, glycogen is already present in muscle cells, and it can release glucose 6-phosphate as an instant fuel.

Consequently, the initial stages of exercise are fueled primarily by glucose—it requires less oxygen and can even be burned anaerobically (to lactate). During the first few minutes of exercise, up to 80% of the fuel used comes from glycogen.

With time, increased breathing rates increase oxygen supplies to muscles, and triacylglycerol use increases. Continued activity for three-quarters of an hour achieves a 50–50 balance of triacylglycerol and glucose use. Beyond an hour, triacylglycerol use may be as high as 80%.

Suppose a person is exercising at a moderate rate and decides to speed up. Immediately, body fuel and oxygen needs are increased. The response is increased use of glycogen supplies.

The accompanying table compares exercise on a stationary cycle at 45% and 70% of maximum oxygen uptake sufficient to burn 300 calories.

	Low-intensity exercise	High-intensity exercise
percent of maximum oxygen uptake	45%	70%
time required to burn 300 calories	48 min	30 min
calories obtained from fat	133 cal	65 cal
percent of calories from fat	44%	22%
rate of fat burning per minute	2.8 cal/min	2.1 cal/min

where the body cannot adequately process glucose even though it is present, and (3) *prolonged* fasting conditions, including starvation, where glycogen supplies are exhausted. Under these conditions, the problem of inadequate oxaloacetate arises, which is compounded by the body's using oxaloacetate that is present to produce glucose through gluconeogenesis (Section 24.6).

What happens when oxaloacetate supplies are too low for all acetyl CoA present to be processed through the citric acid cycle? The excess acetyl CoA is diverted to the formation of ketone bodies. A **ketone body** *is one of three substances (acetoacetate, β-hydroxybutyrate, and acetone) produced from acetyl CoA when an excess of acetyl CoA from fatty acid degradation accumulates because of triacylglycerol carbohydrate metabolic imbalances.* The structures for the three ketone bodies are

▶ Ketone bodies are produced when the amount of acetyl CoA is excessive compared with the amount of oxaloacetate available to react with it (Step 1 of the citric acid cycle).

$$\underset{\text{Acetoacetate}}{CH_3-\overset{\overset{\displaystyle O}{\|}}{C}-CH_2-\overset{\overset{\displaystyle O}{\|}}{C}-O^-} \qquad \underset{\beta\text{-Hydroxybutyrate}}{CH_3-\overset{\overset{\displaystyle OH}{|}}{CH}-CH_2-\overset{\overset{\displaystyle O}{\|}}{C}-O^-} \qquad \underset{\text{Acetone}}{CH_3-\overset{\overset{\displaystyle O}{\|}}{C}-CH_3}$$

The structure of β-hydroxybutyrate does not actually include a ketone group, but it is still classified as a ketone body.

For a number of years, ketone bodies were thought of as degradation products that had little physiological significance. It is now known that ketone bodies can serve as sources of energy for various tissues and are very important energy sources in heart muscle and the renal cortex. During starvation, the brain adapts to the use of ketone bodies for energy production.

■ Ketogenesis

> Even when ketogenic conditions are not present in the human body, the liver produces a *small amount* of ketone bodies.

Ketogenesis *is the metabolic pathway by which ketone bodies are synthesized from acetyl CoA.* The primary site for ketogenesis is liver mitochondria. After they are produced, the ketone bodies diffuse from these structures into the bloodstream, where they are transported to peripheral tissues. The reactions that constitute ketogenesis are shown in Figure 25.8.

Step 1: *Condensation.* Ketogenesis begins as two acetyl CoA molecules combine to produce acetoacetyl CoA, a reversal of the last step of the fatty acid spiral (Section 25.4).

$$CH_3-\overset{\overset{O}{\|}}{C}-S-CoA + CH_3-\overset{\overset{O}{\|}}{C}-S-CoA \xrightarrow{\text{Thiolase}}$$
Acetyl CoA   Acetyl CoA

$$CH_3-\overset{\overset{O}{\|}}{C}-CH_2-\overset{\overset{O}{\|}}{C}-S-CoA + CoA-SH$$
Acetoacetyl CoA

Step 2: *Condensation.* Acetoacetyl CoA then reacts with a third acetyl CoA and water to produce 3-hydroxy-3-methylglutaryl CoA (HMG-CoA) and CoA-SH.

$$CH_3-\overset{\overset{O}{\|}}{C}-CH_2-\overset{\overset{O}{\|}}{C}-S-CoA + CH_3-\overset{\overset{O}{\|}}{C}-S-CoA + H_2O \xrightarrow[\text{synthase}]{\text{HMG-CoA}}$$
Acetoacetyl CoA   Acetyl CoA

$$^-OOC-CH_2-\overset{\overset{OH}{|}}{\underset{\underset{CH_3}{|}}{C}}-CH_2-\overset{\overset{O}{\|}}{C}-S-CoA + CoA-SH + H^+$$
HMG-CoA

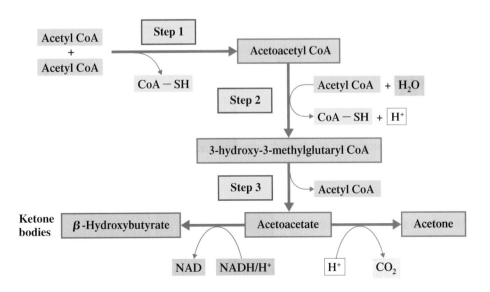

FIGURE 25.8 Ketogenesis involves the production of ketone bodies from acetyl CoA.

FIGURE 25.13 An overview of the biosynthetic pathway for cholesterol synthesis.

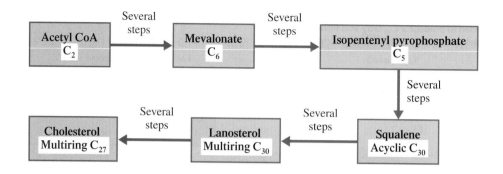

CHEMICAL CONNECTIONS

Statins: Drugs That Lower Plasma Levels of Cholesterol

Over half of all deaths in the United States are directly or indirectly related to heart disease, in particular to atherosclerosis. Atherosclerosis results from the buildup of plaque (fatty acid deposits) on the inner walls of arteries. Cholesterol, obtained from low-density-lipoproteins (LDL) that circulate in blood plasma, is also a major component of plaque.

Because most of the cholesterol in the human body is synthesized in the liver, from acetyl CoA, much research has focused on finding ways to inhibit its biosynthesis. The rate-determining step in cholesterol biosynthesis involves the conversion of 3-hydroxy-3-methylglutaryl CoA (HMG-CoA) to mevalonate, a process catalyzed by the enzyme HMG-CoA reductase.

3-Hydroxy-3-methylglutaryl-CoA
(HMG-CoA)

Mevalonate

In 1976, as the result of screening more than 8000 strains of microorganisms, a compound now called *mevastatin*—a potent inhibitor of HMG-CoA reductase—was isolated from culture broths of a fungus. Soon thereafter, a second, more active compound called *lovastatin* was isolated.

R₁ = R₂ = H, mevastatin
R₁ = H, R₂ = CH₃, lovastatin (Mevacor)
R₁ = R₂ = CH₃, simvastatin (Zocor)

These "statins" are very effective in lowering plasma concentrations of LDL by functioning as competitive inhibitors of HMG-CoA reductase.

After years of testing, the statins are now available as prescription drugs for lowering blood cholesterol levels. Clinical studies indicate that use of these drugs lowers the incidence of heart disease in individuals with mildly elevated blood cholesterol levels. A later-generation statin, with a ring structure distinctly different from that of earlier statins—atorvastatin (Lipitor)—became the most prescribed medication in the United States in the year 2000. Note the structural resemblance between part of the structure of Lipitor and that of mevalonate.

Mevalonate

Atorvastatin (Lipitor)

Recent research studies have unexpectedly shown that the cholesterol-lowering statins have two added benefits.

Laboratory studies with animals indicate that statins prompt growth cells to build new bone, replacing bone that has been leached away by osteoporosis ("brittle-bone disease"). A retrospective study of osteoporosis patients who also took statins shows evidence that their bones became more dense than did bones of osteoporosis patients who did not take the drugs.

Statins have also been shown to function as anti-inflammatory agents that counteract the effects of a common virus, cytomegalovirus, which is now believed to contribute to the development of coronary heart disease. Researchers believe that by age 65, more than 70% of all people have been exposed to this virus. The virus, along with other infecting agents in blood, may actually trigger the inflammation mechanism for heart disease.

Interrelationships Between Carbohydrate and Lipid Metabolism

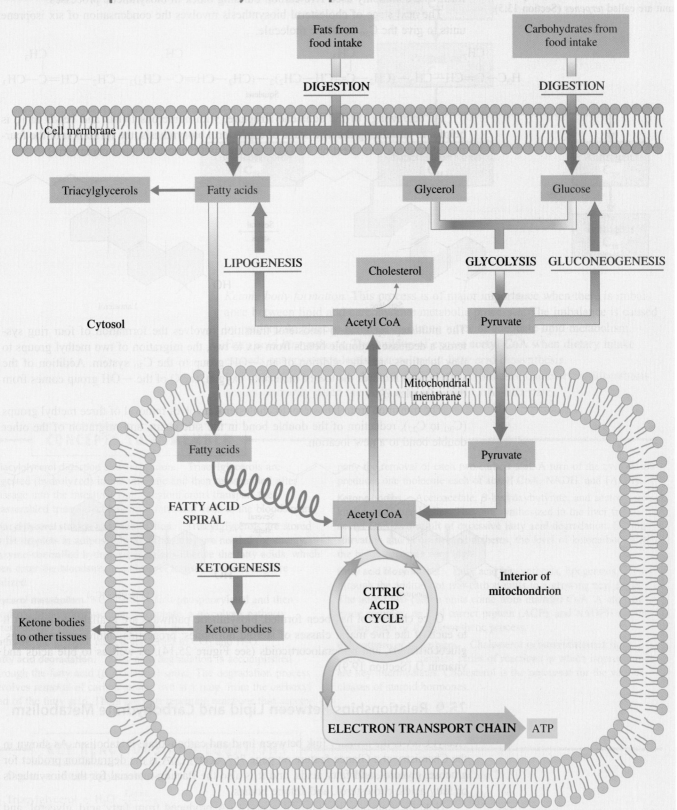

26 Protein Metabolism

False color SEM of bone marrow with developing red blood cells.

From an energy production standpoint, proteins supply only a small portion of the body's needs. With a normal diet, carbohydrates and fats supply 90% of the body's energy, and only 10% comes from proteins. However, despite its minor role in energy production, protein metabolism plays an important role in maintaining good health. The amino acids obtained from proteins are needed for both protein synthesis and synthesis of other nitrogen-containing compounds in the cell. In this chapter, we examine protein digestion, the oxidative degradation of amino acids, and amino acid biosynthesis.

26.1 Protein Digestion and Absorption

Protein digestion begins in the stomach rather than in the mouth, because saliva contains no enzymes that affect proteins. Both protein denaturation (Section 20.16) and protein hydrolysis (Section 20.15) occur in the stomach. The partially digested protein (large polypeptides) passes from the stomach into the small intestine, where digestion is completed (Figure 26.1).

Proteins are denatured in the stomach by the hydrochloric acid present in gastric juice. The acid gives gastric juice a pH of between 1.5 and 2.0. The enzyme pepsin effects the hydrolysis of about 10% of peptide bonds in proteins, producing a variety of polypeptides. In the small intestine, trypsin, chymotrypsin, and carboxypeptidase in pancreatic juice attack peptide bonds. The pH of pancreatic juice is between 7.0 and 8.0, and it neutralizes the acidity of the material from the stomach. Aminopeptidase, secreted by intestinal mucosal cells, also attacks peptide bonds. Pepsin, trypsin, chymotrypsin, carboxypeptidase, and aminopeptidase are all examples of *proteolytic* enzymes (Section

A very small number of people are unable to synthesize enough stomach acid, and these individuals must ingest capsules of dilute hydrochloric acid with every meal.

(c) Al, P, S, O (d) Ca, Mg, C, O **5.37** (a) N, O, Cl, F, Br (b) Li, Na, K, Ca (c) F, O, Cl, N (d) 0.5 units **5.39** (a) $\overset{\delta^+}{B}$—$\overset{\delta^-}{N}$ (b) $\overset{\delta^+}{Cl}$—$\overset{\delta^-}{F}$ (c) $\overset{\delta^-}{N}$—$\overset{\delta^+}{C}$ (d) $\overset{\delta^-}{F}$—$\overset{\delta^+}{O}$ **5.41** (a) H—Br, H—Cl, H—O (b) O—F, P—O, Al—O (c) Br—Br, H—Cl, B—N (d) P—N, S—O, Br—F **5.43** (a) polar covalent (b) ionic (c) nonpolar covalent (d) polar covalent **5.45** (a) nonpolar (b) polar (c) polar (d) polar **5.47** (a) nonpolar (b) polar (c) polar (d) polar **5.49** (a) polar (b) polar (c) nonpolar (d) polar **5.51** (a) sulfur tetrafluoride (b) tetraphosphorus hexoxide (c) chlorine dioxide (d) hydrogen sulfide **5.53** (a) ICl (b) N_2O (c) NCl_3 (d) HBr **5.55** (a) H_2O_2 (b) CH_4 (c) NH_3 (d) PH_3 **5.57** (a) 26 (b) 24 (c) 14 (d) 8 **5.58** (a) same, both single (b) same, both triple (c) different, double and single (d) same, both single **5.59** (a) not enough electron dots (b) not enough electron dots (c) improper placement of a correct number of electron dots (d) too many electron dots **5.60** (a) tetrahedral; tetrahedral (b) tetrahedral; tetrahedral (c) trigonal planar; angular (d) trigonal planar; trigonal planar **5.61** (a) can't classify (b) nonpolar (c) can't classify (d) polar **5.62** (a) BrI (b) SO_2 (c) NF_3 (d) H_3CF **5.63** BA, CA, DB, DA **5.64** (a)

(b) all are tetrahedral (c) nonpolar, polar, polar, polar, nonpolar **5.65** (a) sodium chloride (b) bromine monochloride (c) potassium sulfide (d) dichlorine monoxide

Chapter 6 **6.1** (a) 342.34 amu (b) 100.23 amu (c) 183.20 amu (d) 132.17 amu **6.3** (a) 6.02×10^{23} apples (b) 6.02×10^{23} elephants (c) 6.02×10^{23} Zn atoms (d) 6.02×10^{23} CO_2 molecules **6.5** (a) 9.03×10^{23} atoms Fe (b) 9.03×10^{23} atoms Ni (c) 9.03×10^{23} atoms C (d) 9.03×10^{23} atoms Ne **6.7** (a) 0.200 mole (b) Avogadro's number (c) 1.50 moles (d) 6.50×10^{23} atoms **6.9** (a) 28.0 g (b) 44.0 g (c) 58.4 g (d) 342 g **6.11** (a) 6.7 g (b) 3.7 g (c) 48.0 g (d) 96.0 g **6.13** (a) 0.179 mole (b) 0.114 mole (c) 0.0937 mole (d) 0.0210 mole

6.15 (a) $\dfrac{2\ \text{moles H}}{1\ \text{mole H}_2\text{SO}_4}$ $\dfrac{1\ \text{mole H}_2\text{SO}_4}{2\ \text{moles H}}$ $\dfrac{1\ \text{mole S}}{1\ \text{mole H}_2\text{SO}_4}$

$\dfrac{1\ \text{mole H}_2\text{SO}_4}{1\ \text{mole S}}$ $\dfrac{4\ \text{moles O}}{1\ \text{mole H}_2\text{SO}_4}$ $\dfrac{1\ \text{mole H}_2\text{SO}_4}{4\ \text{moles O}}$

(b) $\dfrac{1\ \text{mole P}}{1\ \text{mole POCl}_3}$ $\dfrac{1\ \text{mole POCl}_3}{1\ \text{mole P}}$ $\dfrac{1\ \text{mole O}}{1\ \text{mole POCl}_3}$

$\dfrac{1\ \text{mole POCl}_3}{1\ \text{mole O}}$ $\dfrac{3\ \text{moles Cl}}{1\ \text{mole POCl}_3}$ $\dfrac{1\ \text{mole POCl}_3}{3\ \text{moles Cl}}$

6.17 (a) 2.00 moles S, 4.00 moles O (b) 2.00 moles S, 6.00 moles O (c) 3.00 moles N, 9.00 moles H (d) 6.00 moles N, 12.0 moles H **6.19** (a) 16.0 moles of atoms (b) 14.0 moles of atoms (c) 45.0 moles of atoms (d) 15.0 moles of atoms **6.21** (a) 5.57×10^{23} atoms (b) 4.81×10^{23} atoms (c) 6.0×10^{22} atoms (d) 3.0×10^{23} atoms **6.23** (a) 63.6 g (b) 31.8 g (c) 5.88×10^{-20} g (d) 1.06×10^{-22} g **6.25** (a) 2.50 moles (b) 0.227 mole (c) 6.6×10^{-14} mole (d) 6.6×10^{-14} mole **6.27** (a) 6.14×10^{22} atoms S (b) 1.50×10^{23} atoms S (c) 3.61×10^{23} atoms S (d) 2.41×10^{24} atoms S **6.29** (a) 32.1 g S (b) 6.39×10^{-22} g S (c) 64.1 g S (d) 1150 g S **6.31** (a) balanced (b) balanced (c) not balanced (d) balanced **6.33** (a) 4 N, 6 O (b) 10 N, 12 H, 6 O (c) 1 P, 3 Cl, 6 H (d) 2 Al, 3 O, 6 H, 6 Cl **6.35** (a) $2Na + 2H_2O \rightarrow 2NaOH + H_2$ (b) $2Na + ZnSO_4 \rightarrow Na_2SO_4 + Zn$ (c) $2NaBr + Cl_2$

$\rightarrow 2NaCl + Br_2$ (d) $2ZnS + 3O_2 \rightarrow 2ZnO + 2SO_2$ **6.37** (a) $CH_4 + 2O_2 \rightarrow CO_2 + 2H_2O$ (b) $2C_6H_6 + 15O_2 \rightarrow 12CO_2 + 6H_2O$ (c) $C_4H_8O_2 + 5O_2 \rightarrow 4CO_2 + 4H_2O$ (d) $C_5H_{10}O + 7O_2 \rightarrow 5CO_2 + 5H_2O$ **6.39** (a) $3PbO + 2NH_3 \rightarrow 3Pb + N_2 + 3H_2O$ (b) $2Fe(OH)_3 + 3H_2SO_4 \rightarrow Fe_2(SO_4)_3 + 6H_2O$ **6.41** $\dfrac{2\ \text{moles Ag}_2\text{CO}_3}{4\ \text{moles Ag}}$

$\dfrac{2\ \text{moles Ag}_2\text{CO}_3}{2\ \text{moles CO}_2}$ $\dfrac{2\ \text{moles Ag}_2\text{CO}_3}{1\ \text{mole O}_2}$ $\dfrac{4\ \text{moles Ag}}{2\ \text{moles CO}_2}$

$\dfrac{4\ \text{moles Ag}}{1\ \text{mole O}_2}$ $\dfrac{2\ \text{moles CO}_2}{1\ \text{mole O}_2}$ The other six are the reciprocals of these six factors. **6.43** (a) 14.0 moles CO_2 (b) 1.00 mole CO_2 (c) 4.00 moles CO_2 (d) 2.00 moles CO_2 **6.45** (a) 24.3 g NH_3 (b) 1.80×10^2 g $(NH_4)_2Cr_2O_7$ (c) 22.9 g N_2H_4 (d) 24.3 g NH_3 **6.47** 5.09 g O_2 **6.49** 14.3 g O_2 **6.51** 5.63 g H_2O **6.53** $y = 8$ **6.54** (a) 1.00 mole S_8 (b) 28.0 g Al (c) 30.0 g Mg (d) 6.02×10^{23} atoms He **6.55** (a) 0.03560 mole SiH_4 (b) 2.139 g SiO_2 (c) 2.144×10^{22} molecules $(CH_3)_3SiCl$ (d) 2.144×10^{22} atoms Si **6.56** 59.0 g Si **6.57** C_4H_6 **6.58** 8.33 g N_2, 21.4 g H_2O, and 45.2 g Cr_2O_3 **6.59** 109 g Ag and 16.2 g S **6.60** 43.4 g Be

Chapter 7 **7.1** (a) Velocity increases with increasing temperature. (b) potential energy (c) Increasing temperature increases disruptive force magnitude. (d) gaseous state **7.3** (a) Increase in vibrational movement is limited because particles are already close together; hence, there is little change in volume. (b) Particles of a gas are widely separated because disruptive forces are greater than cohesive forces. **7.5** (a) 0.967 atm (b) 403 mm Hg (c) 403 torr (d) 0.816 atm **7.7** 7.2 atm **7.9** 2.71 L **7.11** 3.64 L **7.13** 144°C **7.15** (a) $T_1 = \dfrac{P_1V_1T_2}{P_2V_2}$

(b) $P_2 = \dfrac{P_1V_1T_2}{V_2T_1}$ (c) $V_1 = \dfrac{P_2V_2T_1}{P_1T_2}$ **7.17** (a) 5.90 L (b) 2.11 atm (c) −171°C (d) 3.70×10^3 mL **7.19** −209°C **7.21** 1.12 L **7.23** (a) 4.11 L (b) 3.16 atm (c) −98°C (d) 16,300 mL **7.25** 0.42 atm **7.27** 98 mm Hg **7.29** (a) endothermic (b) endothermic (c) exothermic **7.31** (a) no (b) yes (c) yes **7.33** (a) boiling point (b) vapor pressure (c) boiling (d) boiling point **7.35** (a) They differ in the strength of intermolecular forces. (b) Vapor pressure becomes equal to atmospheric pressure at a lower temperature. (c) Evaporation is a cooling process. (d) Low-heat and high-heat boiling water have the same temperature and the same heat content. **7.37** Molecules must be polar. **7.39** Boiling point increases as intermolecular force strength increases. **7.41** (a) London (b) hydrogen bonding (c) dipole–dipole (d) London **7.43** (a) no (b) yes (c) yes (d) no **7.45** four (see Figure 7.21) **7.47** (a) 0.871 atm (b) 298°C (c) 869°C **7.48** (a) 915°C (b) −199°C (c) 24°C (d) 172°C **7.49** (a) Boyle's law, Charles's law and combined gas law (b) Charles's law (c) Boyle's law (d) Boyle's law **7.50** 8.08×10^6 L He **7.51** (a) 4.6 atm (b) 29 atm (c) 0.14 atm (d) 0.90 atm **7.52** 5.37×10^{22} molecules H_2S **7.53** 24.7 L **7.54** 0.22 g N_2 **7.55** He, 9.0 atm; Ne, 5.0 atm; Ar, 15.0 atm **7.56** (a) boils (b) does not boil (c) does not boil (d) does not boil **7.57** (a) PBr_3 (b) PI_3 (c) PI_3 **7.58** (a) Br_2, larger mass (b) H_2O, hydrogen bonding (c) CO, dipole–dipole (d) C_3H_8, larger size

Chapter 8 **8.1** (a) true (b) true (c) true (d) false **8.3** (a) solute: sodium chloride; solvent: water (b) solute: sucrose; solvent: water (c) solute: water; solvent: ethyl alcohol (d) solute: ethyl alcohol; solvent: methyl alcohol **8.5** (a) first solution (b) first solution (c) first solution (d) second solution **8.7** (a) saturated (b) unsaturated (c) unsaturated (d) saturated **8.9** (a) dilute (b) concentrated (c) dilute (d) concentrated

8.11 (a) hydrated ion (b) hydrated ion (c) oxygen atom (d) hydrogen atom **8.13** (a) decrease (b) increase (c) increase (d) increase **8.15** (a) slightly soluble (b) very soluble (c) slightly soluble (d) slightly soluble **8.17** (a) soluble with exceptions (b) soluble (c) insoluble with exceptions (d) soluble **8.19** (a) all are soluble (b) all are soluble (c) CaS, Ca(OH)$_2$, CaCl$_2$ (d) NiSO$_4$ **8.21** (a) 7.10%(m/m) (b) 6.19%(m/m) (c) 9.06%(m/m) (d) 0.27%(m/m) **8.23** (a) 3.62 g (b) 14.5 g (c) 68.8 g (d) 124 g **8.25** 0.6400 g **8.27** 276 g **8.29** (a) 4.21%(v/v) (b) 4.60%(v/v) **8.31** 18%(v/v) **8.33** (a) 2.0%(m/v) (b) 15%(m/v) **8.35** 0.500 g **8.37** 3.75 g **8.39** (a) 6.0 M (b) 0.456 M (c) 0.342 M (d) 0.500 M **8.41** (a) 273 g (b) 0.373 g (c) 136 g (d) 88 g **8.43** (a) 85.6 mL (b) 2.64 mL (c) 9180 mL (d) 0.24 mL **8.45** (a) 0.183 M (b) 0.0733 M (c) 0.0120 M (d) 0.00275 M **8.47** (a) 1450 mL (b) 18.0 mL (c) 85,600 mL (d) 7.5 mL **8.49** (a) 3.0 M (b) 3.0 M (c) 4.5 M (d) 1.5 M **8.51** The presence of solute molecules decreases the ability of solvent molecules to escape. **8.53** It is a more concentrated solution and thus has a lower vapor pressure. **8.55** (a) same as (b) greater than (c) less than (d) greater than **8.57** 2 to 1 **8.59** (a) swell (b) remain the same (c) swell (d) shrink **8.61** (a) hemolyze (b) remain unaffected (c) hemolyze (d) crenate **8.63** (a) hypotonic (b) isotonic (c) hypotonic (d) hypertonic **8.65** (a) K$^+$ and Cl$^-$ leave the bag. (b) K$^+$, Cl$^-$, and glucose leave the bag. **8.67** (a) like (both soluble) (b) unlike (c) unlike (d) like (both insoluble) **8.68** (a) 4.02 g (b) 7.303 g (c) 12.6 g (d) 0.148 g **8.69** 0.0700 qt **8.70** (a) 7.1 L (b) 9.5 L (c) 11 L (d) 14 L **8.71** (a) 0.472 M (b) 0.708 M (c) 1.04 M (d) 1.60 M **8.72** (a) 37.5%(m/v) (b) 2.23 M **8.73** (a) 4.00 M (b) 3.22 M **8.74** (a) NaCl (b) MgCl$_2$

Chapter 9 **9.1** (a) single-replacement (b) decomposition (c) double-replacement (d) combination **9.3** (a) combination, single-replacement, combustion (b) decomposition, single-replacement (c) combination, decomposition, single-replacement, double-replacement, combustion (d) combination, decomposition, single-replacement, double-replacement, combustion **9.5** (a) +2 (b) +6 (c) 0 (d) +5 **9.7** (a) +3 (b) +4 (c) +6 (d) +6 (e) +6 (f) +6 (g) +6 (h) +5 **9.9** (a) +3P, −1F (b) +1Na, −2O, +1H (c) +1Na, +6S, −2O (d) +4C, −2O **9.11** (a) redox (b) nonredox (c) redox (d) redox **9.13** (a) H$_2$ oxidized, N$_2$ reduced (b) KI oxidized, Cl$_2$ reduced (c) Fe oxidized, Sb$_2$O$_3$ reduced (d) H$_2$SO$_3$ oxidized, HNO$_3$ reduced **9.15** (a) N$_2$ oxidizing agent, H$_2$ reducing agent (b) Cl$_2$ oxidizing agent, KI reducing agent (c) Sb$_2$O$_3$ oxidizing agent, Fe reducing agent (d) HNO$_3$ oxidizing agent, H$_2$SO$_3$ reducing agent **9.17** Reactant molecules have greater freedom of movement. **9.19** Molecular collisions are not effective if the activation energy requirement is not met. **9.21** (a) exothermic (b) endothermic (c) endothermic (d) exothermic

9.23

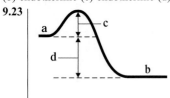

9.25 (a) As temperature increases, so does the number of collisions per second. (b) A catalyst lowers the activation energy. **9.27** The concentration of O$_2$ has increased from 21% to 100%.

9.29

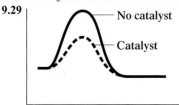

9.31 (a) 1 (b) 3 (c) 4 (d) 3 **9.33** rate of forward reaction = rate of reverse reaction

9.35

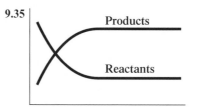

9.37 (a) $K_{eq} = \dfrac{[NO_2]^2}{[N_2O_4]}$ (b) $K_{eq} = \dfrac{[Cl_2][CO]}{[COCl_2]}$ (c) $K_{eq} = \dfrac{[CH_4][H_2S]^2}{[CS_2][H_2]^4}$ (d) $K_{eq} = \dfrac{[SO_3]^2}{[O_2][SO_2]^2}$ **9.39** (a) $K_{eq} = [SO_3]$ (b) $K_{eq} = \dfrac{1}{[Cl_2]}$ (c) $K_{eq} = \dfrac{[NaCl]^2}{[Na_2SO_4][BaCl_2]}$ (d) $K_{eq} = [O_2]$ **9.41** 4.8 × 10^{-5} **9.43** (a) more products than reactants (b) essentially all reactants (c) significant amounts of both reactants and products (d) significant amounts of both reactants and products **9.45** (a) right (b) left (c) left (d) right **9.47** (a) left (b) left (c) left (d) left **9.49** (a) left (b) no effect (c) right (d) no effect **9.51** (a) redox, single replacement (b) redox, combustion (c) redox, decomposition (d) nonredox, double replacement **9.52** (a) redox (b) redox (c) redox (d) can't classify **9.53** (a) gain (b) reduction (c) decrease (d) increase **9.54** (a) decrease (b) increase (c) increase (d) decrease **9.55** (a) no (b) no (c) yes (d) no **9.56** CH$_4$(g) + 2H$_2$S(g) ⟶ CS$_2$(g) + 4H$_2$(g) **9.57** (a) yes (b) yes (c) no (d) yes **9.58** (a) no effect (b) right (c) right (d) right

Chapter 10 **10.1** (a) H$^+$ (b) OH$^-$ **10.3** (a) Arrhenius acid (b) Arrhenius base **10.5** (a) HI $\xrightarrow{H_2O}$ H$^+$ + I$^-$ (b) HClO $\xrightarrow{H_2O}$ H$^+$ + ClO$^-$ (c) LiOH $\xrightarrow{H_2O}$ Li$^+$ + OH$^-$ (d) CsOH $\xrightarrow{H_2O}$ Cs$^+$ + OH$^-$ **10.7** (a) acid (b) base (c) acid (d) acid **10.9** (a) HClO + H$_2$O → H$_3$O$^+$ + ClO$^-$ (b) HClO$_4$ + NH$_3$ → NH$_4^+$ + ClO$_4^-$ (c) H$_3$O$^+$ + OH$^-$ → H$_2$O + H$_2$O (d) H$_3$O$^+$ + NH$_2^-$ → H$_2$O + NH$_3$ **10.11** (a) HSO$_3^-$ (b) HCN (c) C$_2$O$_4^{2-}$ (d) H$_2$PO$_4^-$ **10.13** (a) HS$^-$ + H$_2$O → H$_3$O$^+$ + S^{2-}, HS$^-$ + H$_2$O → H$_2$S + OH$^-$ (b) HPO$_4^{2-}$ + H$_2$O → H$_3$O$^+$ + PO$_4^{3-}$, HPO$_4^{2-}$ + H$_2$O → H$_2$PO$_4^-$ + OH$^-$ (c) NH$_3$ + H$_2$O → H$_3$O$^+$ + NH$_2^-$, NH$_3$ + H$_2$O → NH$_4^+$ + OH$^-$ (d) OH$^-$ + H$_2$O → H$_3$O$^+$ + O^{2-}, OH$^-$ + H$_2$O → OH$^-$ + H$_2$O **10.15** (a) monoprotic (b) diprotic (c) monoprotic (d) diprotic **10.17** H$_3$C$_6$H$_5$O$_7$ + H$_2$O → H$_3$O$^+$ + H$_2$C$_6$H$_5$O$_7^-$, H$_2$C$_6$H$_5$O$_7^-$ + H$_2$O → H$_3$O$^+$ + HC$_6$H$_5$O$_7^{2-}$, HC$_6$H$_5$O$_7^{2-}$ + H$_2$O → H$_3$O$^+$ + C$_6$H$_5$O$_7^{3-}$ **10.19** (a) 1, 0 (b) 2, 4 (c) 1, 7 (d) 0, 4 **10.21** To show that it is a monoprotic acid **10.23** Monoprotic; only one H atom is involved in a polar bond. **10.25** (a) strong (b) weak (c) weak (d) strong **10.27** (a) $K_a = \dfrac{[H^+][F^-]}{[HF]}$ (b) $K_a = \dfrac{[H^+][C_2H_3O_2^-]}{[HC_2H_3O_2]}$ **10.29** (a) $K_b = \dfrac{[NH_4^+][OH^-]}{[NH_3]}$ (b) $K_b = \dfrac{[C_6H_5NH_3^+][OH^-]}{[C_6H_5NH_2]}$ **10.31** (a) H$_3$PO$_4$ (b) HF (c) H$_2$CO$_3$ (d) HNO$_2$ **10.33** 4.9 × 10^{-5} **10.35** (a) acid (b) salt (c) salt (d) base (e) salt (f) base (g) acid (h) acid **10.37** (a) Ba(NO$_3$)$_2$ $\xrightarrow{H_2O}$ Ba^{2+} + 2NO$_3^-$ (b) Na$_2$SO$_4$ $\xrightarrow{H_2O}$ 2Na$^+$ + SO$_4^{2-}$ (c) CaBr$_2$ $\xrightarrow{H_2O}$ Ca^{2+} + 2Br$^-$ (d) K$_2$CO$_3$ $\xrightarrow{H_2O}$ 2K$^+$ + CO$_3^{2-}$ **10.39** (a) no (b) yes (c) yes (d) no **10.41** (a) 1 to 1 (b) 1 to 2 (c) 1 to 1 (d) 2 to 1 **10.43** (a) HCl + NaOH → NaCl + H$_2$O (b) HNO$_3$ + KOH → KNO$_3$ + H$_2$O (c) H$_2$SO$_4$ + 2LiOH → Li$_2$SO$_4$ + 2H$_2$O

(d) $2H_3PO_4 + 3Ba(OH)_2 \rightarrow Ba_3(PO_4)_2 + 6H_2O$ **10.45** (a) $H_2SO_4 +$ $2LiOH \rightarrow Li_2SO_4 + 2H_2O$ (b) $HCl + NaOH \rightarrow NaCl + H_2O$ (c) $HNO_3 + KOH \rightarrow KNO_3 + H_2O$ (d) $2H_3PO_4 + 3Ba(OH)_2 \rightarrow$ $Ba_3(PO_4)_2 + 6H_2O$ **10.47** (a) 3.3×10^{-12} M (b) 1.5×10^{-9} M (c) 1.1×10^{-7} M (d) 8.3×10^{-4} M **10.49** (a) acidic (b) basic (c) basic (d) acidic **10.51** (a) 4.00 (b) 11.00 (c) 11.00 (d) 7.00 **10.53** (a) 7.68 (b) 7.40 (c) 3.85 (d) 11.85 **10.55** (a) 1×10^{-2} (b) 1×10^{-6} (c) 1×10^{-8} (d) 1×10^{-10} **10.57** (a) 2.1×10^{-4} M (b) 8.1×10^{-6} M (c) 4.5×10^{-8} M (d) 3.6×10^{-13} M **10.59** (a) 3.35 (b) 6.37 (c) 7.21 (d) 1.82 **10.61** acid B **10.63** (a) strong acid–strong base salt (b) weak acid–strong base salt (c) strong acid–weak base salt (d) strong acid–strong base salt **10.65** (a) none (b) $C_2H_3O_2^-$ (c) NH_4^+ (d) none **10.67** (a) neutral (b) basic (c) acidic (d) neutral **10.69** (a) no (b) yes (c) no (d) yes **10.71** (a) HCN and CN^- (b) H_3PO_4 and $H_2PO_4^-$ (c) H_2CO_3 and HCO_3^- (d) HCO_3^- and CO_3^{2-} **10.73** (a) $F^- + H_3O^+ \rightarrow HF + H_2O$ (b) $H_2CO_3 + OH^- \rightarrow HCO_3^- + H_2O$ (c) $CO_3^{2-} + H_3O^+ \rightarrow HCO_3^- + H_2O$ (d) $H_3PO_4 + OH^- \rightarrow H_2PO_4^- + H_2O$ **10.75** 7.06 **10.77** 5.17 **10.79** (a) weak (b) strong (c) strong (d) strong **10.81** (a) 0.0500 M (b) 0.800 M (c) 0.952 M (d) 0.120 M **10.83** (a) yes (b) no (c) no (d) yes **10.84** (a) no (b) yes, both strong (c) no (d) yes, both weak **10.85** (a) B (b) A **10.86** 7.0 **10.87** HCl, HCN, KCl, NaOH **10.88** HCN/CN^- **10.89** CN^- ion undergoes hydrolysis to a greater extent than NH_4^+ ion, resulting in a basic solution; NH_4^+ ion and $C_2H_3O_2^-$ ion hydrolyze to an equal extent, resulting in a neutral solution. **10.90** 0.0035 g NaOH

Chapter 11 **11.1** (a) $^{10}_4Be$, Be-10 (b) $^{25}_{11}Na$, Na-25 (c) $^{96}_{41}Nb$, Nb-96 (d) $^{257}_{103}Lr$, Lr-257 **11.3** (a) $^{14}_7N$ (b) $^{197}_{79}Au$ (c) Sn-121 (d) B-10 **11.5** (a) $^4_2\alpha$ (b) $^0_{-1}\beta$ (c) $^0_0\gamma$ **11.7** 2 protons and 2 neutrons **11.9** (a) $^{200}_{84}Po \rightarrow ^4_2\alpha + ^{196}_{82}Pb$ (b) $^{240}_{96}Cm \rightarrow ^4_2\alpha + ^{236}_{94}Pu$ (c) $^{244}_{96}Cm \rightarrow ^4_2\alpha + ^{240}_{94}Pu$ (d) $^{238}_{92}U \rightarrow ^4_2\alpha + ^{234}_{90}Th$ **11.11** (a) $^{10}_4Be \rightarrow ^0_{-1}\beta + ^{10}_5B$ (b) $^{14}_6C \rightarrow ^0_{-1}\beta + ^{14}_7N$ (c) $^{21}_9F \rightarrow ^0_{-1}\beta + ^{21}_{10}Ne$ (d) $^{25}_{11}Na \rightarrow ^0_{-1}\beta + ^{25}_{12}Mg$ **11.13** $A \rightarrow A - 4$, $Z \rightarrow Z - 2$ **11.15** (a) $^0_{-1}\beta$ (b) $^{28}_{12}Mg$ (c) $^4_2\alpha$ (d) $^{200}_{80}Hg$ **11.17** (a) $^4_2\alpha$ (b) $^0_{-1}\beta$ **11.19** (a) $\frac{1}{4}$ (b) $\frac{1}{64}$ (c) $\frac{1}{8}$ (d) $\frac{1}{64}$ **11.21** (a) 1.4 days (b) 0.90 day (c) 0.68 day (d) 0.54 day **11.23** 0.250 g **11.25** 2000 **11.27** 92 **11.29** (a) $^4_2\alpha$ (b) $^{25}_{12}Mg$ (c) $^4_2\alpha$ (d) ^1_1p **11.31** Termination of a decay series requires a stable nuclide. **11.33** $^{232}_{90}Th \rightarrow ^4_2\alpha + ^{228}_{88}Ra$, $^{228}_{88}Ra \rightarrow ^0_{-1}\beta + ^{228}_{89}Ac$, $^{228}_{89}Ac \rightarrow ^0_{-1}\beta + ^{228}_{90}Th$, $^{228}_{90}Th \rightarrow ^4_2\alpha + ^{224}_{88}Ra$ **11.35** the electron and positive ion that are produced during an ionizing interaction between a molecule (or atom) and radiation **11.37** (a) yes (b) no (c) yes (d) no **11.39** It continues on, interacting with other atoms and forming more ion pairs. **11.41** Alpha is stopped; beta and gamma go through. **11.43** alpha, 0.1 the speed of light; beta, up to 0.9 the speed of light; gamma, the speed of light **11.45** (a) no detectable effects (b) nausea, fatigue, lowered blood cell count **11.47** 19% human-made, 81% natural sources **11.49** to monitor the extent of radiation exposure **11.51** so radiation can be detected externally **11.53** (a) bone tumors (b) circulatory problems (c) iron metabolism in blood (d) intercellular space problems **11.55** They are usually α or β emitters. **11.57** (a) fusion (b) fusion (c) both (d) fission **11.59** (a) fusion (b) fission (c) neither (d) neither **11.61** (a) $^{206}_{80}Hg \rightarrow ^0_{-1}\beta + ^{206}_{81}Tl$ (b) $^{109}_{46}Pd \rightarrow ^0_{-1}\beta + ^{109}_{47}Ag$ (c) $^{245}_{96}Cm \rightarrow ^4_2\alpha + ^{241}_{94}Pu$ (d) $^{249}_{100}Fm \rightarrow ^4_2\alpha + ^{245}_{98}Cf$ **11.62** (a) 54 hr (b) 90 hr (c) 108 hr (d) 126 hr **11.63** (a) $^{243}_{94}Pu + ^4_2\alpha \rightarrow ^{246}_{96}Cm + ^1_0n$ (b) $^{246}_{96}Cm + ^{12}_6C \rightarrow ^{254}_{102}No + 4^1_0n$ (c) $^{27}_{13}Al + ^4_2\alpha \rightarrow ^1_0n + ^{30}_{15}P$ (d) $^{23}_{11}Na + ^2_1H \rightarrow ^{21}_{10}Ne + ^4_2\alpha$ **11.64** $^{142}_{60}Nd + ^1_0n \rightarrow ^{143}_{61}Pm + ^0_{-1}\beta$ **11.65** 12 elements **11.66** 4 neutrons **11.67** A = 0 (negligible amount), B = 0 (negligible amount), C = 63 atoms, D = 937 atoms **11.68** ^{228}Ac, ^{228}Th, ^{224}Ac

Chapter 12 **12.1** (a) false (b) false (c) true (d) true **12.3** (a) meets (b) does not meet (c) does not meet (d) does not meet **12.5**

Hydrocarbons contain C and H, and hydrocarbon derivatives contain at least one additional element besides C and H. **12.7** All bonds are single bonds in a saturated hydrocarbon, and at least one carbon–carbon multiple bond is present in an unsaturated hydrocarbon. **12.9** (a) saturated (b) unsaturated (c) unsaturated (d) unsaturated **12.11** (a) 18 (b) 4 (c) 13 (d) 22

12.13 (a) $CH_3-CH_2-CH_2-CH_3$

(b) $CH_3-CH_2-\underset{\underset{CH_3}{|}}{CH}-CH_2-CH_3$

(c) $CH_3-CH_2-\underset{\underset{CH_3}{|}}{CH}-CH_2-\underset{\underset{CH_3}{|}}{CH}-CH_3$

(d) $CH_3-CH_2-\underset{\underset{\underset{\underset{CH_3}{|}}{CH_2}}{|}}{CH}-CH_2-CH_3$

12.15 (a) $CH_3-\underset{\underset{CH_3}{|}}{CH}-CH_2-CH_3$

(b) $CH_3-\underset{\underset{CH_3}{|}}{CH}-\underset{\underset{CH_3}{|}}{CH}-\underset{\underset{CH_3}{|}}{CH}-CH_2-CH_3$

(c) $CH_3-CH_2-CH_2-CH_2-CH_2-CH_3$

(d) $CH_3-\underset{\underset{CH_3}{\overset{\overset{CH_3}{|}}{|}}}{C}-CH_2-CH_3$

12.17 (a)

(b)

(c) $CH_3-(CH_2)_8-CH_3$ (d) C_6H_{14} **12.19** (a) different compounds that are not structural isomers (b) different compounds that are structural isomers (c) different conformations of the same molecule (d) different compounds that are structural isomers **12.21** (a) seven-carbon chain (b) eight-carbon chain (c) eight-carbon chain (d) seven-carbon chain **12.23** (a) 2-methylpentane (b) 2,4,5-trimethylheptane (c) 3-ethyl-2,3-dimethylpentane (d) 3-ethyl-2,4-dimethylhexane (e) decane (f) 4-propylheptane **12.25** horizontal chain, because it has more substituents (two)

12.27 (a) $CH_3-\underset{\underset{CH_3}{|}}{CH}-CH_2-CH_3$

(b) $CH_3-CH_2-\underset{\underset{CH_3}{|}}{CH}-\underset{\underset{CH_3}{|}}{CH}-CH_2-CH_3$

(c) $CH_3-CH_2-\underset{\underset{\underset{\underset{CH_3}{|}}{CH_2}}{|}}{\overset{\overset{CH_3}{|}}{C}}-CH_2-CH_3$

(d) CH₃—CH—CH—CH—CH—CH₂—CH₃
          | | | |
        CH₃ CH₃ CH₃ CH₃

(e) CH₃—CH₂—CH—CH₂—CH—CH₂—CH₂—CH₃
            | |
           CH₂   CH₂
           | |
           CH₃   CH₃

(f) CH₃—CH₂—CH₂—CH—CH₂—CH₂—CH₂—CH₂—CH₃
               |
              CH₂
              |
              CH₂
              |
              CH₃

12.29 (a) 1, 1 (b) 2, 2 (c) 2, 2, (d) 4, 4 (e) 2, 2 (f) 1, 1 **12.31** (a) carbon chain numbered from wrong end; 2-methylpentane (b) not based on longest carbon chain; 2,2-dimethylbutane (c) carbon chain numbered from wrong end; 2,2,3-trimethylbutane (d) not based on longest carbon chain; 3,3-dimethylhexane (e) carbon chain numbered from wrong end and alkyl groups not listed alphabetically; 3-ethyl-4-methylhexane (f) like alkyl groups listed separately; 2,4-dimethylhexane **12.33** (a) 3, 2, 1, 0 (b) 5, 2, 3, 0 (c) 5, 2, 1, 1 (d) 5, 2, 3, 0 (e) 2, 8, 0, 0 (f) 3, 6, 1, 0 **12.35** (a) isopropyl (b) isobutyl (c) isopropyl (d) *sec*-butyl

12.37 (a)
CH₃—CH₂—CH₂—CH₂—CH—CH₂—CH₂—CH₂—CH₂—CH₃
                   |
                  CH—CH₃
                  |
                  CH₂
                  |
                  CH₃

(b)
                  CH₃
                  |
                  CH—CH₃
                  |
CH₃—CH₂—CH₂—C—CH₂—CH₂—CH₂—CH₃
                  |
                  CH—CH₃
                  |
                  CH₃

(c) CH₃—CH—CH—CH₂—CH—CH₂—CH₂—CH₂—CH₃
        | | |
      CH₃ CH₃   CH₂
                  |
                  CH—CH₃
                  |
                  CH₃

(d)
             CH₃
             |
CH₃—CH₂—CH—CH—CH₂—CH₂—CH₃
            | |
           CH₃—C—CH₃
              |
             CH₃

12.39 (a) 16 (b) 6 (c) 5 (d) 15 **12.41** (a) C₆H₁₂ (b) C₆H₁₂ (c) C₄H₈ (d) C₇H₁₄ **12.43** (a) cyclohexane (b) 1,2-dimethylcyclobutane (c) methylcyclopropane (d) 1,2-dimethylcyclopentane **12.45** (a) must locate methyl groups with numbers (b) wrong numbering system for ring (c) no number needed (d) wrong numbering system for ring

12.47 (a) [cyclobutane]—CH₂—CH₂—CH₃ (b) [cyclobutane]—CH—CH₃ with CH₃

(c)
     CH₂—CH₃
     CH₂—CH₃
  [cyclohexane] H
      H

(d)
     CH₂—CH₃
  [cyclopentane] H H
     CH₂—CH₂—CH₃

12.49 (a) not possible

(b) CH₃—CH₂△CH₂—CH₃   CH₃—CH₂△H
         H   H   H CH₂—CH₃
        cis   *trans*

(c) not possible   (d) [cyclohexane with CH₃, CH₃, H, H substituents]
                cis   *trans*   **12.51** boiling point

12.53 (a) octane (b) cyclopentane (c) pentane (d) cyclopentane **12.55** (a) different states (b) same states (c) same states (d) same states **12.57** (a) CO₂ and H₂O (b) CO₂ and H₂O (c) CO₂ and H₂O (d) CO₂ and H₂O **12.59** CH₃Br, CH₂Br₂, CHBr₃, CBr₄

12.61 (a) CH₃—CH₂
           |
           Cl

(b) CH₂—CH₂—CH₂—CH₃   CH₃—CH—CH₂—CH₃
    | |
    Cl   Cl

(c)   Cl   (d) [cyclopentane]—Cl
                    |
CH₂—CH—CH₃   CH₃—C—CH₃
 | | |
Cl CH₃   CH₃

12.63 (a) iodomethane, methyl iodide (b) 1-chloropropane, propyl chloride (c) 2-fluorobutane, sec-butyl fluoride (d) chlorocyclobutane, cyclobutyl chloride

12.65 (a)   Cl   (b)   F F
       | | |
   H—C—Cl   F—C—C—F
       | | |
       Cl   Cl Cl

(c) CH₃—CH—Br   (d)   Br
        | H
      CH₃   [cyclopentane]
                      H
                   Cl

12.67 (a) 16 (b) 6 (c) 5 (d) 22 (e) liquid (f) less dense (g) insoluble (h) flammable **12.68** (a) no (b) yes (c) no (d) no

12.69 (a) F   H   (b) H   CH₃
    [cyclohexane]   [cyclohexane]
    H   F   Cl   H

(c)   CH₃   (d) CH₃—CH—CH₂—I
        | |
CH₃—C—Br   CH₃
       |
     CH₃

12.70 4-tert-butyl-5-ethyl-2,2,6-trimethyloctane **12.71** (a) C₁₈H₃₈ (b) C₇H₁₄ (c) C₇H₁₄F₂ (d) C₆H₁₀Br₂

12.72

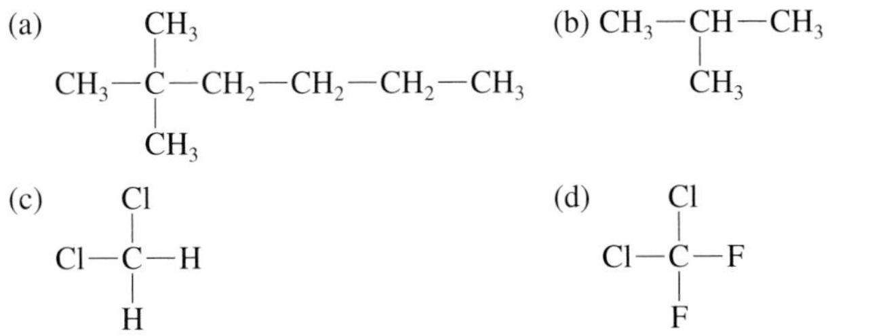

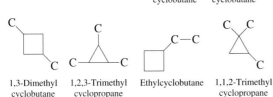

(a)

$$CH_3-\underset{\underset{CH_3}{|}}{\overset{\overset{CH_3}{|}}{C}}-CH_2-CH_2-CH_2-CH_3$$

(b) $CH_3-\underset{\underset{CH_3}{|}}{CH}-CH_3$

(c)

$$Cl-\underset{\underset{H}{|}}{\overset{\overset{Cl}{|}}{C}}-H$$

(d)

$$Cl-\underset{\underset{F}{|}}{\overset{\overset{Cl}{|}}{C}}-F$$

12.73 (a) alkane (b) halogenated cycloalkane (c) halogenated alkane (d) cycloalkane

12.74 (a)

Cyclohexane   Methylcyclopentane   1,1-Dimethyl cyclobutane   1,2-Dimethyl cyclobutane

1,3-Dimethyl cyclobutane   1,2,3-Trimethyl cyclopropane   Ethylcyclobutane   1,1,2-Trimethyl cyclopropane

1-Ethyl-2-methyl cyclopropane   1-Ethyl-1-methyl cyclopropane   Propylcyclopropane

Isopropylcyclopropane

(b)

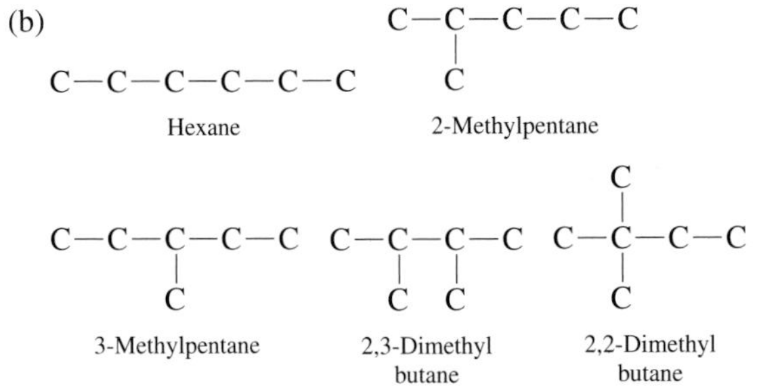

Hexane   2-Methylpentane

3-Methylpentane   2,3-Dimethyl butane   2,2-Dimethyl butane

(c)

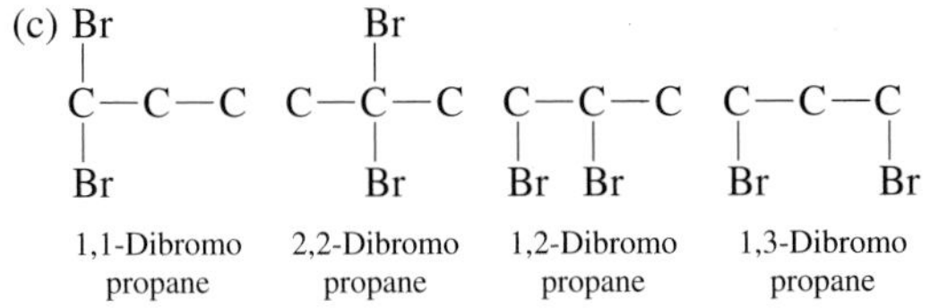

1,1-Dibromo propane   2,2-Dibromo propane   1,2-Dibromo propane   1,3-Dibromo propane

12.75 (a) 1,2-diethylcyclohexane (b) 3-methylhexane (c) 2,3-dimethyl-4-propylnonane (d) 1-isopropyl-3,5-dipropylcyclohexane
12.76 5-(1-ethylpropyl)nonane

Chapter 13 **13.1** (a) unsaturated, alkene with one double bond (b) saturated (c) unsaturated, alkene with one double bond (d) unsaturated, diene (e) unsaturated, triene (f) unsaturated, diene **13.3** (a) C_4H_{10} (b) C_5H_{10} (c) C_5H_8 (d) C_7H_{10} **13.5** (a) C_nH_{2n-2} (b) C_nH_{2n-2} (c) C_nH_{2n-2} (d) C_nH_{2n-6} **13.7** (a) 2-butene (b) 2,4-dimethyl-2-pentene (c) cyclohexene (d) 1,3-cyclopentadiene (e) 2-ethyl-1-pentene (f) 2,4,6-octatriene **13.9** (a) 2-pentene (b) pentane (c) 2,3,3-trimethyl-1-butene (d) 2-methyl-1,4-pentadiene (e) 1,3,5-hexatriene (f) 2,3-pentadiene

13.11 (a) $CH_2=CH-\underset{\underset{CH_3}{|}}{CH}-CH_2-CH_3$

(b)

(c) $CH_2=CH-CH=CH_2$

(d) $CH_2=CH-\underset{\underset{\underset{CH_3}{|}}{\overset{|}{CH_2}}}{CH}-CH=CH_2$

(e) $CH_3-CH=CH-\underset{\underset{\underset{\underset{CH_3}{|}}{\overset{|}{CH_2}}}{\overset{|}{CH_2}}}{CH}-CH_2-CH_2-CH_3$

(f)

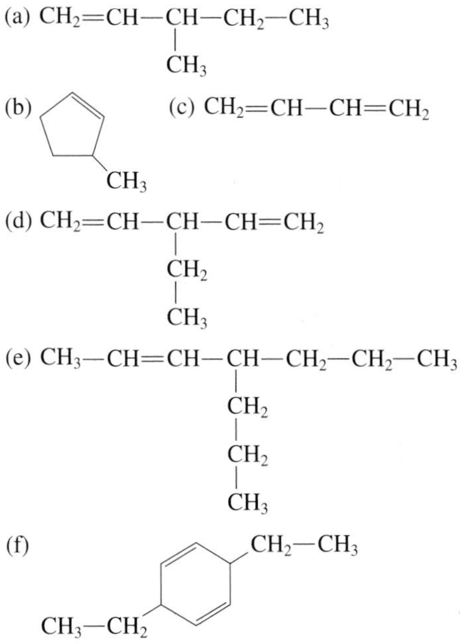

13.13 (a) 3-methyl-3-hexene (b) 2,3-dimethyl-2-hexene (c) 1,3-cyclopentadiene (d) 4,5-dimethylcyclohexene

13.15

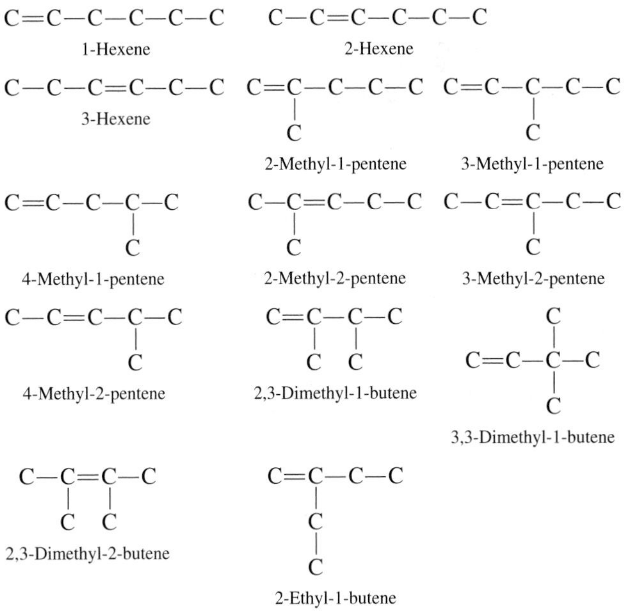

C=C-C-C-C-C   1-Hexene   C-C=C-C-C-C   2-Hexene

C-C-C=C-C-C   3-Hexene

2-Methyl-1-pentene   3-Methyl-1-pentene

4-Methyl-1-pentene   2-Methyl-2-pentene   3-Methyl-2-pentene

4-Methyl-2-pentene   2,3-Dimethyl-1-butene   3,3-Dimethyl-1-butene

2,3-Dimethyl-2-butene   2-Ethyl-1-butene

13.17 (a) no (b) no (c) no

(d)

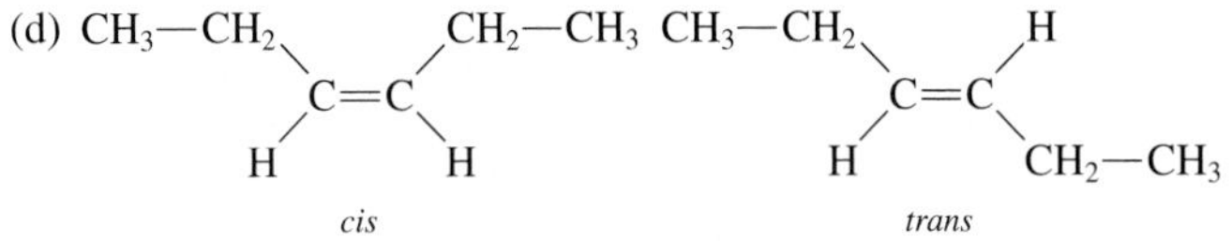

cis   trans

(e)

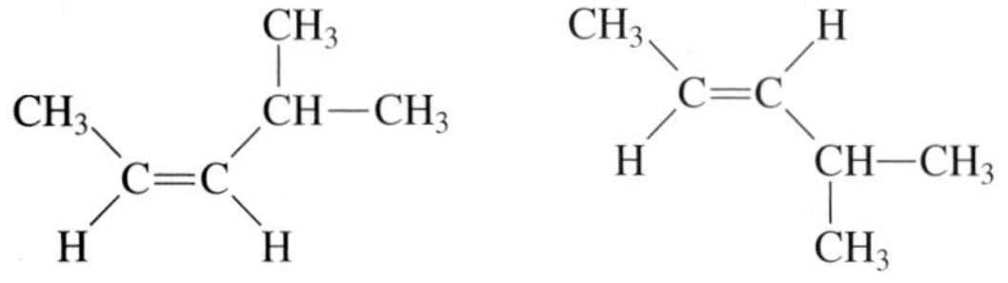

$$CH_3-C=C\begin{matrix} CH_3 \\ | \\ CH-CH_3 \\ | \\ H \end{matrix}$$

cis

$$CH_3-C=C\begin{matrix} H \\ CH-CH_3 \\ | \\ CH_3 \end{matrix}$$

trans

(f)

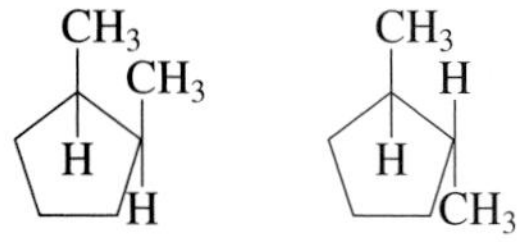

cis   *trans*

13.19 (a) *cis*-2-pentene (b) *trans*-1-bromo-2-iodoethene (c) tetrafluoroethene (d) 2-methyl-2-butene

13.21 (a) CH_3-CH_2 ... $C=C$ ... H, CH_3, CH_2-CH_3

(b) CH_3, $C=C$, CH_2-CH_3, H, H

(c) CH_3, H, $C=C$, H, $CH_2-CH-CH_2-CH_3$, CH_3

(d) $CH_2=CH$, H, $C=C$, H, CH_3

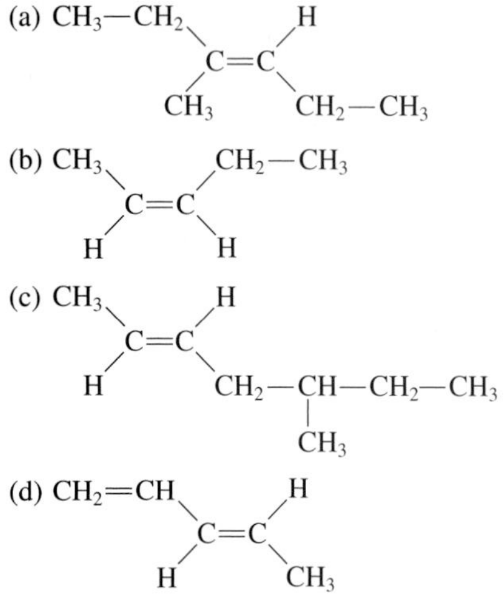

13.23 a compound used by insects (and some animals) to transmit messages to other members of the same species **13.25** Isoprene, the building block for terpenes, contains 5 carbon atoms. **13.27** (a) gas (b) liquid (c) liquid (d) liquid **13.29** (a) yes (b) no (c) yes (d) no

13.31 (a) $CH_2{=}CH_2 + Cl_2 \longrightarrow CH_2-CH_2$ (Cl, Cl)

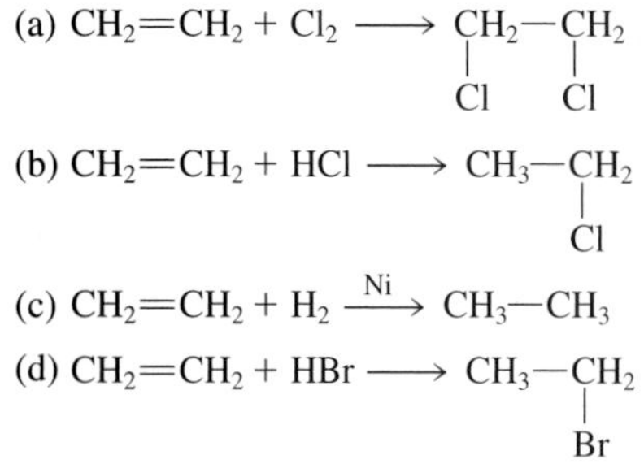

(b) $CH_2{=}CH_2 + HCl \longrightarrow CH_3-CH_2$ (Cl)

(c) $CH_2{=}CH_2 + H_2 \xrightarrow{Ni} CH_3-CH_3$

(d) $CH_2{=}CH_2 + HBr \longrightarrow CH_3-CH_2$ (Br)

13.33 (a) $CH_2{=}CH-CH_3 + Cl_2 \longrightarrow CH_2-CH-CH_3$ (Cl, Cl)

(b) $CH_2{=}CH-CH_3 + HCl \longrightarrow CH_3-CH-CH_3$ (Cl)

(c) $CH_2{=}CH-CH_3 + H_2 \xrightarrow{Ni} CH_3-CH_2-CH_3$

(d) $CH_2{=}CH-CH_3 + HBr \longrightarrow CH_3-CH-CH_3$ (Br)

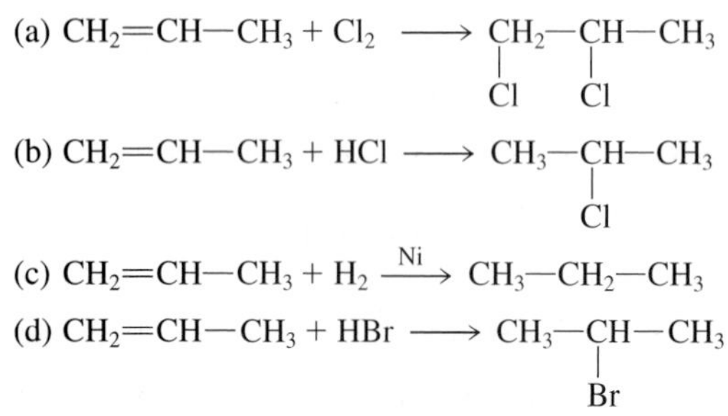

13.35
(a) $CH_3-CH-CH-CH_3$ (Cl, Cl) (b) CH_3-C-CH_3 (Br top, CH_3 bottom)

(c) $CH_3-CH_2-CH-CH_3$ (Cl) (d) [cyclopentane] (e) [cyclopentene] (f) HO—[cyclobutane]

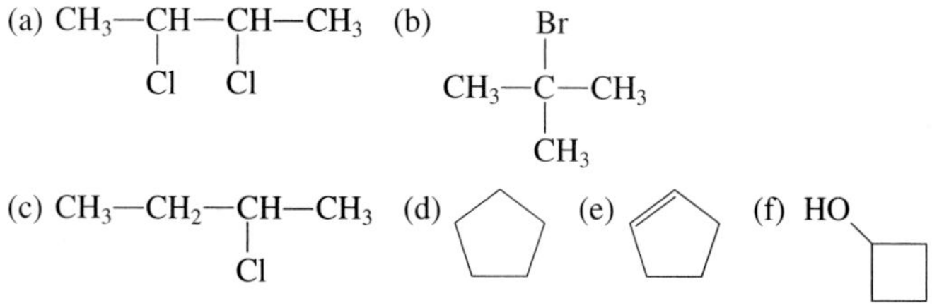

13.37 (a) Br_2 (b) H_2 + Ni catalyst (c) HCl (d) H_2O + H_2SO_4 catalyst

13.39 (a) 2 (b) 2 (c) 2 (d) 3 **13.41** (a) $CF_2{=}CF_2$
(b) $CH_2{=}C-CH{=}CH_2$ (Cl) (c) $CH_2{=}CH$ (Cl) (d) $CH_2{=}CH$ [phenyl]

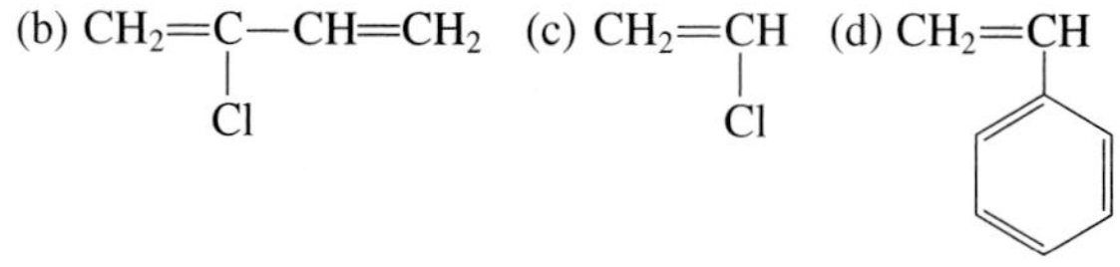

13.43 (a) $-CH_2-CH_2-CH_2-CH_2-CH_2-CH_2-$
(b) $-CH_2-CH-CH_2-CH-CH_2-CH-$ (Cl, Cl, Cl)
(c) $-CH-CH-CH-CH-CH-CH-$ (Cl, Cl, Cl, Cl, Cl, Cl)
(d) $-CH_2-CH-CH_2-CH-CH_2-CH-$ (Cl, Cl, Cl)

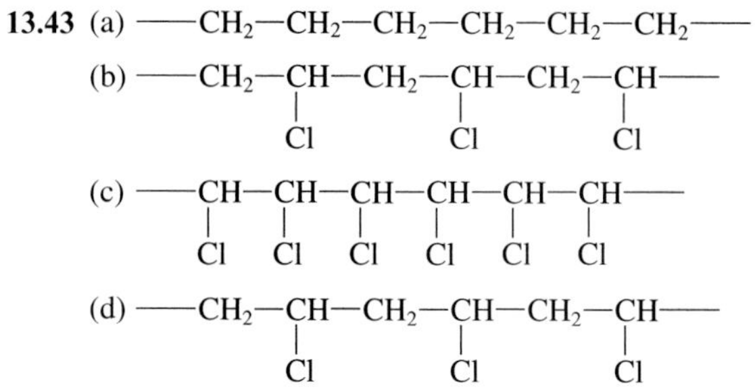

13.45 (a) 1-hexyne (b) 4-methyl-2-pentyne (c) 2,2-dimethyl-3-heptyne (d) 2-butyne (e) 3-methyl-1,4-hexadiyne (f) 4-methyl-2-hexyne

13.47
(a) CH_3-CH_3 (b) CH_3-C-CH (Br Br / Br Br) (c) CH_3-C-CH_3 (Br top, Br bottom)
(d) $CH_2{=}CH$ (Cl) (e) [cyclohexane]—CH_2-CH_3 (f) $CH_3-CH_2-C{=}CH_2$ (Br)

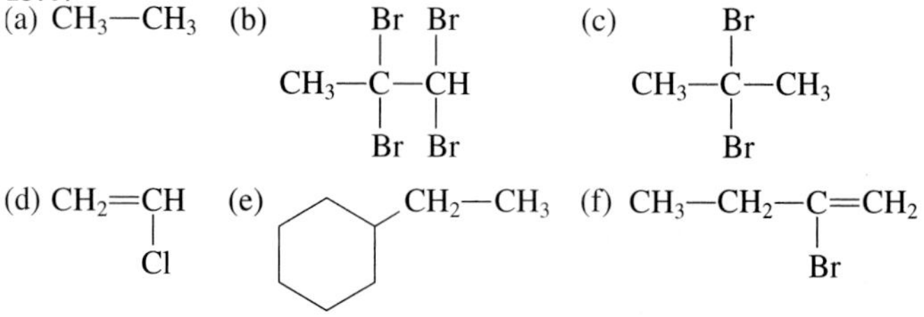

13.49 (a) 1,3-dibromobenzene (b) 1-chloro-2-fluorobenzene (c) 1-chloro-4-fluorobenzene (d) 3-chlorotoluene (e) 1-bromo-2-ethylbenzene (f) 4-bromotoluene **13.51** (a) *m*-dibromobenzene (b) *o*-chlorofluorobenzene (c) *p*-chlorofluorobenzene (d) *m*-chlorotoluene (e) *o*-bromoethylbenzene (f) *p*-bromotoluene **13.53** (a) 2,4-dibromo-1-chlorobenzene (b) 3-bromo-5-chlorotoluene (c) 1-bromo-3-chloro-2-fluorobenzene (d) 1,4-dibromo-2,5-dichlorobenzene **13.55** (a) 2-phenylbutane (b) 3-phenyl-1-butene (c) 3-methyl-1-phenylbutane (d) 2,4-diphenylpentane

13.57
(a) [benzene ring with CH_2-CH_3 and CH_2-CH_3]
(b) [benzene ring with CH_3 and CH_3]
(c) [benzene ring with CH_3 and CH_2-CH_3]
(d) [biphenyl]
(e) [two rings joined by CH_2-CH_2]

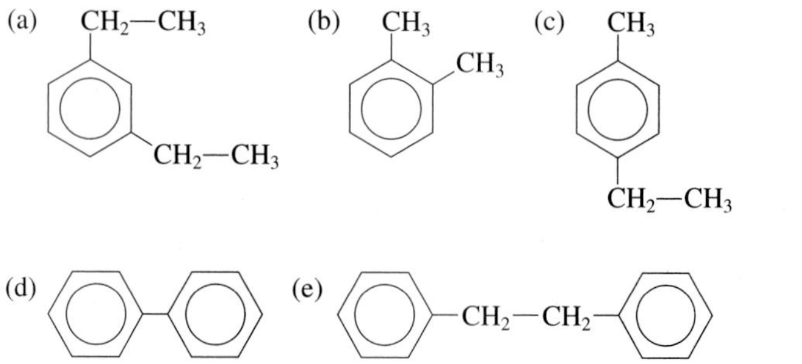

(f)

CH₃
|
CH₃—CH₂—C—CH₂—CH₃
(with phenyl ring below)

13.59 (a) substitution (b) addition (c) substitution (d) addition

13.61 (a) Br₂ (b)

CH₃
|
(phenyl)—CH—CH₃

(c) CH₃—CH₂—Br

13.63 (a) C₂H₄ (b) C₃H₄ (c) C₂H₂ (d) CH₄ **13.64** (a) more (b) more (c) more (d) the same number **13.65** (a) no (b) yes (c) no (d) yes **13.66** (a) All have six carbon atoms. (b) Cyclohexane has 12 H atoms, cyclohexene 10, and benzene 6. (c) Cyclohexane and benzene undergo substitution; cyclohexene undergoes addition. (d) All are liquids.

13.67 (a) CH₃—C≡C—CH₂—CH—CH₃
                                            |
                                           CH₃

(b) CH₂—CH=CH—CH₃
       |
       Cl

(c) CH₃—C≡C—CH₂—CH—CH—CH₃
                          | |
                         CH₃ CH₃

(d) CH₂=CH—CH—CH₂—CH₂—CH₃
               |
               CH—CH₃
               |
               CH₃

(e) CH₂=CH—CH₂—CH₂—CH₂—CH=CH₂

(f) CH≡C—CH—C≡CH
           |
          CH₃

13.68 (a) two (b) one (c) two (d) one (e) two (f) four
13.69

(a) CH₂=CH—(phenyl)

(b) CH₂=CH—CH₂—Cl   (c) CH₃—CH₂—CH₂—C≡CH

(d) CH₃—CH₂—CH₂—C≡C—CH₂—CH₂—CH₃

(e) (phenyl with CH₃ and CH₃)   (f) (biphenyl with CH₃)

13.70 It would require a carbon atom that formed five bonds. **13.71** Each carbon atom in 1,2-dichlorobenzene has only one substituent.
13.72

CH₂=CH—CH₂—CH₂—CH₃   CH₃—CH=CH—CH₂—CH₃
                                    (*cis–trans* forms)

CH₂=C—CH₂—CH₃   CH₂=CH—CH—CH₃
       | |
      CH₃   CH₃

CH₃—C=CH—CH₃   (cyclopentane)   (cyclobutane with CH₃)
       |
      CH₃

(cyclopropane with CH₃ CH₃)   (cyclopropane with CH₃ and CH₃)
                                         (*cis–trans* forms)

13.73 1,2,3-trimethylbenzene; 1,2,4-trimethylbenzene; 1,3,5-trimethylbenzene; 2-ethyltoluene; 3-ethyltoluene; 4-ethyltoluene; propylbenzene; isopropylbenzene **13.74** (a) 3 (b) 3 (c) 11 (d) 3

Chapter 14   14.1 (a) 2 (b) 1 (c) 4 (d) 1 **14.3** R—OH **14.5** R—O—H versus H—O—H **14.7** (a) 2-pentanol (b) ethanol (c) 3-methyl-2-butanol (d) 2-ethyl-1-pentanol (e) 2-butanol (f) 3,3-dimethyl-1-butanol

14.9 (a) CH₃—CH₂—CH—CH₂—CH₃
                                   |
                                  OH

(b)
                OH
                |
CH₃—CH₂—C—CH₂—CH₂—CH₃
                |
               CH₂
                |
               CH₃

(c) CH₂—CH—CH₃
       | |
      OH CH₃

(d) CH₃—CH—CH₂—CH—CH₃
          | |
         OH   CH₃

(e)   OH
              |
CH₃—C—CH₃
              |
          (phenyl)

(f) (cyclobutane with OH and CH₃)

14.11 (a) CH₂—CH₂—CH₂—CH₂—CH₃
              |
             OH
           1-Pentanol

(b) CH₂—CH₂—CH₃
       |
      OH
  1-Propanol

(c) CH₃—CH—CH₂—OH
          |
         CH₃
   2-Methyl-1-Propanol

(d) CH₃—CH₂—CH—OH
                    |
                   CH₃
              2-Butanol

14.13 (a) 1,2-propanediol (b) 1,4-pentanediol (c) 1,3-pentanediol (d) 3-methyl-1,2,4-butanetriol **14.15** (a) cyclohexanol (b) *trans*-3-chlorocyclohexanol (c) *cis*-2-methylcyclohexanol (d) 1-methylcyclobutanol

14.17 (a) CH₃—CH—CH₂—CH=CH₂
                       |
                      OH

(b) CH≡C—CH—CH₂—CH₃
              |
             OH

(c) CH₃—CH—C=CH₂
          | |
         OH CH₃

Glossary

Acetal An organic compound in which a carbon atom is bonded to two alkoxy groups (—OR).

Acetyl group The portion of an acetic acid molecule (CH_3—COOH) that remains after the —OH group is removed from the carboxyl carbon atom.

Achiral molecule A molecule whose mirror images are superimposable.

Acid–base indicator A compound that exhibits different colors depending on the pH of its solution.

Acid–base titration A neutralization reaction in which a measured volume of an acid or a base of known concentration is completely reacted with a measured volume of a base or an acid of unknown concentration.

Acid ionization constant (K_a) The equilibrium constant for the reaction of a weak acid with water.

Acidic solution An aqueous solution in which the concentration of H_3O^+ ion is higher than that of OH^- ion; an aqueous solution whose pH is less than 7.0.

Acidosis A human body condition in which the pH of blood drops from its normal value of 7.4 to 7.1–7.2.

Activation energy The minimum combined kinetic energy that colliding reactant particles must possess in order for their collision to result in a reaction.

Active site The relatively small part of an enzyme's structure that is actually involved in catalysis.

Active transport The transport process in which a substance moves across a cell membrane, with the aid of membrane proteins, against a concentration gradient with the expenditure of cellular energy.

Acyl group The portion of a carboxylic acid that remains after the —OH group is removed from the carboxyl carbon atom.

Addition polymer A polymer in which the monomers simply "add together" with no other products formed besides the polymer.

Addition reaction A reaction in which atoms or groups of atoms are added to each carbon atom of a carbon–carbon multiple bond in a hydrocarbon or hydrocarbon derivative.

Adipocyte A triacylglycerol-storing cell.

Adipose tissue Tissue that contains large numbers of adipocyte cells.

Alcohol An organic compound in which an —OH group is bonded to a saturated carbon atom.

Aldehyde An organic compound in which a carbonyl carbon atom is bonded to at least one hydrogen atom.

Aldose A monosaccharide that contains an aldehyde functional group.

Alkaloid A nitrogen-containing organic compound extracted from plant material.

Alkalosis A human body condition in which the pH of blood increases from its normal value of 7.4 to 7.5.

Alkane A saturated hydrocarbon in which the carbon atom arrangement is acyclic.

Alkene An acyclic unsaturated hydrocarbon that contains one or more carbon–carbon double bonds.

Alkenyl group A noncyclic hydrocarbon substituent in which a carbon–carbon double bond is present.

Alkoxy group An —OR group, an alkyl (or aryl) group attached to an oxygen atom.

Alkyl group The group of atoms that would be obtained by removing a hydrogen atom from an alkane.

Alkyne An acyclic unsaturated hydrocarbon that contains one or more carbon–carbon triple bonds.

Allosteric enzyme An enzyme with two or more protein chains (quaternary structure) and two kinds of binding sites (substrate and regulator).

Alpha-amino acid An amino acid in which the amino group and the carboxyl group are attached to the alpha carbon atom.

Alpha particle A positively charged particle consisting of two protons and two neutrons that is emitted by certain radioactive nuclides.

Alternative splicing A process by which several different proteins that are variations of a basic structural motif can be produced from a single gene.

Amide A carboxylic acid derivative in which the carboxyl —OH group has been replaced with an amino or a substituted amino group.

Amidification reaction The reaction of a carboxylic acid with an amine (or ammonia) to produce an amide.

Amine An organic derivative of ammonia (NH_3) in which one or more alkyl, cycloalkyl, or aryl groups are attached to the nitrogen atom.

Amine salt An ionic compound in which the positive ion is a mono-, di-, or trisubstituted ammonium ion and the negative ion comes from an acid.

Amino acid An organic compound that contains both an amino (—NH_2) group and a carboxyl (—COOH) group.

Amino acid pool The total supply of free amino acids available for use within the human body.

Amino group The —NH_2 functional group.

Amphoteric substance A substance that can either lose or accept a proton and thus can function as either a Brønsted–Lowry acid or a Brønsted–Lowry base.

Anabolism All metabolic reactions in which small biochemical molecules are joined together to form larger ones.

Antibiotic A substance that kills bacteria or inhibits their growth.

Antibody A biochemical molecule that counteracts a specific antigen.

Anticodon A three-nucleotide sequence on a tRNA molecule that is complementary to a codon on a mRNA molecule.

Antigen A foreign substance, such as a bacterium or virus, that invades the human body.

Antioxidant A substance that protects other substances from being oxidized by being oxidized itself in preference to the other substances.

Apoenzyme The protein part of a conjugated enzyme.

Aqueous solution A solution in which water is the solvent.

Aromatic hydrocarbon An unsaturated cyclic hydrocarbon that does not readily undergo addition reactions.

Aromatic ring system   A highly unsaturated carbon ring system in which both localized and delocalized bonds are present.

Arrhenius acid   A hydrogen-containing compound that, in water, produces hydrogen ions (H^+ ions).

Arrhenius base   A hydroxide-containing compound that, in water, produces hydroxide ions (OH^- ions).

Aryl group   An aromatic carbon ring system from which one hydrogen atom has been removed.

Atom   The smallest particle of an element that can exist and still have the properties of the element.

Atomic mass   The calculated average mass for the isotopes of an element expressed on a scale where ^{12}C serves as the reference point.

Atomic number   The number of protons in the nucleus of an atom.

Avogadro's number   The name given to the numerical value 6.02×10^{23}.

Balanced chemical equation   A chemical equation that has the same number of atoms of each element involved in the reaction on both sides of the equation.

Balanced nuclear equation   A nuclear equation in which the sums of the subscripts (atomic numbers or particle charges) on both sides of the equation are equal, and the sums of the superscripts (mass numbers) on both sides of the equation are equal.

Barometer   A device used to measure atmospheric pressure.

Base ionization constant (K_b)   The equilibrium constant for the reaction of a weak base with water.

Basic solution   An aqueous solution in which the concentration of OH^- ion is higher than that of H_3O^+ ion; an aqueous solution whose pH is greater than 7.0.

Beta particle   A negatively charged particle whose charge and mass are identical to those of an electron that is emitted by certain radioactive nuclides.

Bile   A fluid containing emulsifying agents that is secreted by the liver, stored in the gallbladder, and released into the small intestine during digestion.

Bile acid   A cholesterol derivative that functions as a lipid-emulsifying agent in the aqueous environment of the digestive tract.

Bile pigment   A colored tetrapyrrole degradation product present in bile.

Binary compound   A compound in which only two elements are present.

Biochemical substance   A chemical substance found within a living organism.

Biochemistry   The study of the chemical substances found in living organisms and the chemical interactions of these substances with each other.

Biological wax   A lipid that is a monoester of a long-chain fatty acid and a long-chain alcohol.

Boiling   A form of evaporation where conversion from the liquid state to the vapor state occurs within the body of the liquid through bubble formation.

Boiling point   The temperature at which the vapor pressure of the liquid becomes equal to the external (atmospheric) pressure exerted on the liquid.

Bombardment reaction   A nuclear reaction brought about by bombarding stable nuclei with small particles traveling at very high speeds.

Bond polarity   A measure of the degree of inequality in the sharing of electrons between two atoms in a chemical bond.

Bonding electrons   Pairs of valence electrons that are shared between atoms in a covalent bond.

Boyle's law   The volume of a fixed amount of a gas is inversely proportional to the pressure applied to the gas if the temperature is kept constant.

Branched-chain alkane   An alkane in which one or more branches (of carbon atoms) are attached to a continuous chain of carbon atoms.

Brønsted–Lowry acid   A substance that can donate a proton (H^+ ion) to some other substance.

Brønsted–Lowry base   A substance that can accept a proton (H^+ ion) from some other substance.

Buffer   An aqueous solution containing substances that prevent major changes in solution pH when small amounts of acid or base are added to it.

Calorie   The amount of heat energy needed to raise the temperature of 1 gram of water by 1 degree Celsius.

Carbohydrate   A polyhydroxy aldehyde, a polyhydroxy ketone, or a compound that yields polyhydroxy aldehydes or polyhydroxy ketones upon hydrolysis.

Carbonyl group   A carbon atom double bonded to an oxygen atom.

Carboxyl group   A carbonyl group ($C=O$) with a hydroxyl group ($-OH$) bonded to the carbonyl carbon atom.

Carboxylate ion   The negative ion produced when a carboxylic acid loses one or more acidic hydrogen atoms.

Carboxylic acid   An organic compound whose functional group is the carboxyl group.

Carboxylic acid salt   An ionic compound in which the negative ion is a carboxylate ion.

Catabolism   All metabolic reactions in which large biochemical molecules are broken down to smaller ones.

Catalyst   A substance that increases a reaction rate without being consumed in the reaction.

Cell membrane   A lipid-based structure that separates a cell's aqueous-based interior from the aqueous environment surrounding the cell.

Change of state   A process in which a substance is changed from one physical state to another physical state.

Charles's law   The volume of a fixed amount of gas is directly proportional to its Kelvin temperature if the pressure is kept constant.

Chemical bond   The attractive force that holds two atoms together in a more complex unit.

Chemical change   A process in which a substance undergoes a change in chemical composition.

Chemical equation   A written statement that uses chemical symbols and chemical formulas instead of words to describe the changes that occur in a chemical reaction.

Chemical equilibrium   A state in which forward and reverse chemical reactions occur simultaneously at the same rate.

Chemical formula   A notation made up of the symbols of the elements present in a compound and numerical subscripts (located to the right of each symbol) that indicate the number of atoms of each element present in a molecule of the compound.

Chemical property   A characteristic of a substance that describes the way the substance undergoes or resists change to form a new substance.

Chemical reaction   A process in which at least one new substance is produced as a result of chemical change.

Chemical symbol   A one- or two-letter designation for an element derived from the element's name.

Chemiosmotic coupling   An explanation for the coupling of ATP synthesis with electron transport chain reactions that requires a proton gradient across the inner mitochondrial membrane.

Chemistry   The field of study concerned with the characteristics, composition, and transformations of matter.

Chiral center   An atom in a molecule that has four different groups tetrahedrally bonded to it.

Chiral molecule   A molecule whose mirror images are not superimposable.

Cholesterol   A C_{27} steroid molecule that is a component of cell membranes and a precursor for other steroid-based lipids.

Chromosome   An individual DNA molecule bound to a group of proteins.

Chylomicron   A lipoprotein that transports triacylglycerols from intestinal cells, via the lymphatic system, to the bloodstream.

***Cis* isomer**   A disubstituted cycloalkane isomer in which substituents on different carbon atoms are on the same side of the carbon ring; an alkene isomer in which two similar (or identical) double-bond substituents are on the same side of the double bond.

***Cis–trans* isomers**   Isomers that have the same molecular and structural formulas but different arrangements of atoms in space because of restricted rotation about bonds.

Citric acid cycle (CAC)   The series of biochemical reactions in which the acetyl portion of acetyl CoA is oxidized to carbon dioxide and the reduced coenzymes $FADH_2$ and NADH are produced.

Clones   Cells with identical DNA that have descended from a single cell.

Codon   A three-nucleotide sequence in a mRNA molecule that codes for a specific amino acid.

Coenzyme   An organic molecule that serves as a cofactor in a conjugated enzyme.

Cofactor   The nonprotein part of a conjugated enzyme.

Colligative property   A physical property of a solution that depends only on the number (concentration) of solute particles (molecules or ions) present in a given quantity of solvent and not on their chemical identities.

Collision theory   A set of statements that give the conditions necessary for a chemical reaction to occur.

Colloidal dispersion   A mixture that contains dispersed particles that are intermediate in size between those of a true solution and those of an ordinary heterogeneous mixture.

Combination reaction   A reaction in which a single product is produced from two (or more) reactants.

Combined gas law   The product of the pressure and volume of a fixed amount of gas is inversely proportional to its Kelvin temperature.

Combustion reaction   A reaction between a substance and oxygen (usually from air) that proceeds with the evolution of heat and light (usually from a flame).

Common metabolic pathway   The sum total of the biochemical reactions of the citric acid cycle, the electron transport chain, and oxidative phosphorylation.

Competitive enzyme inhibitor   A molecule that sufficiently resembles an enzyme substrate in shape and charge distribution that it can compete with the substrate for occupancy of the enzyme's active site.

Complementary bases   Pairs of bases in a nucleic acid structure that can hydrogen-bond to each other.

Complementary DNA strands   Strands of DNA in a double helix with base pairing such that each base is located opposite its complementary base.

Complete dietary protein   A protein that contains all the essential amino acids in the same relative amounts in which the human body needs them.

Complex carbohydrate   A dietary polysaccharide.

Compound   A pure substance that can be broken down into two or more simpler pure substances by chemical means.

Compressibility   A measure of the change in volume in a sample of matter resulting from a pressure change.

Concentrated solution   A solution that contains a large amount of solute relative to the amount that could dissolve.

Concentration   The amount of solute present in a specified amount of solution.

Condensation polymer   A polymer formed by reacting difunctional monomers to give a polymer and some small molecule (such as water) as a by-product of the process.

Condensation reaction   A reaction in which two molecules combine to form a larger one while liberating a small molecule, usually water.

Condensed structural formula   A structural formula that uses groupings of atoms, in which central atoms and the atoms connected to them are written as a group, to convey molecular structural information.

Conformation   The specific three-dimensional arrangement of atoms in an organic molecule at a given instant that results from rotations about carbon–carbon single bonds.

Conjugate acid   The species formed when a Brønsted–Lowry base accepts a proton (H^+ ion).

Conjugate acid–base pair   Two substances in a Brønsted–Lowry acid–base reaction whose chemical formulas differ by one proton (H^+ ion).

Conjugate base   The species formed when a Brønsted–Lowry acid loses a proton (H^+ ion).

Conjugated enzyme   An enzyme that has a nonprotein part in addition to a protein part.

Conjugated protein   A protein in which one or more other components in addition to amino acids are present.

Continuous-chain alkane   An alkane in which all carbon atoms are connected in a continuous nonbranching chain.

Conversion factor   A ratio that specifies how one unit of measurement is related to another.

Coordinate covalent bond   A covalent bond in which both electrons of a shared pair come from one of the two atoms involved in the bond.

Copolymer   A polymer in which two different monomers are present.

Cori cycle   A cyclic biochemical process in which glucose is converted to lactate in muscle tissue, the lactate is reconverted to glucose in the liver, and the glucose is returned to the muscle tissue.

Coupled reactions   Pairs of biochemical reactions that occur concurrently in which energy released by one reaction is used in the other reaction.

Covalent bond   A chemical bond formed through the sharing of one or more pairs of electrons between two atoms.

Cycloalkane   A saturated hydrocarbon in which carbon atoms connected to one another in a cyclic (ring) arrangement are present.

Cycloalkene   A cyclic unsaturated hydrocarbon that contains one or more carbon–carbon double bonds within the ring system.

Cytochrome A heme-containing protein in which reversible oxidation and reduction of an iron atom occur.

Cytoplasm The water-based material of a eukaryotic cell that lies between the nucleus and the outer membrane of the cell.

Cytosol The water-based fluid part of the cytoplasm of a cell.

Dalton's law of partial pressures The total pressure exerted by a mixture of gases is the sum of the partial pressures of the individual gases present.

Daughter nuclide The nuclide that is produced in a radioactive decay process.

Decomposition reaction A reaction in which a single reactant is converted into two (or more) simpler substances (elements or compounds).

Dehydration reaction A reaction in which the components of water (H and OH) are removed from a single reactant or from two reactants (H from one and OH from the other).

Delocalized bond A covalent bond in which electrons are shared among more than two atoms.

Density The ratio of the mass of an object to the volume occupied by that object.

Deoxyribonucleic acid (DNA) A nucleotide polymer in which each of the monomers contains deoxyribose, phosphate, and one of the heterocyclic bases adenine, cytosine, guanine, or uracil.

Dextrorotatory compound A chiral compound that rotates the plane of polarized light in a clockwise direction.

Dialysis The process in which a semipermeable membrane allows the passage of solvent, dissolved ions, and small molecules but blocks the passage of colloidal-sized particles and large molecules.

Diastereomers Stereoisomers whose molecules are not mirror images of each other.

Diatomic molecule A molecule that contains two atoms.

Dicarboxylic acid A carboxylic acid that contains two carboxyl groups, one at each end of a carbon chain.

Digestion The biochemical process by which food molecules, through hydrolysis, are broken down into simpler chemical units that can be used by cells for their metabolic needs.

Dilute solution A solution that contains a small amount of solute relative to the amount that could dissolve.

Dilution The process in which more solvent is added to a solution in order to lower its concentration.

Dimensional analysis A general problem-solving method in which the units associated with numbers are used as a guide in setting up calculations.

Dipole–dipole interaction An intermolecular force that occurs between polar molecules.

Diprotic acid An acid that supplies two protons (H^+ ions) per molecule in an acid–base reaction.

Disaccharide A carbohydrate that contains two monosaccharide units covalently bonded to each other.

Dissociation The process in which individual positive and negative ions are released from an ionic compound that is dissolved in solution.

Distinguishing electron The last electron added to the electron configuration for an element when electron subshells are filled in order of increasing energy.

DNA replication The biochemical process by which DNA molecules produce exact duplicates of themselves.

DNA sequencing A method by which the base sequence in a DNA molecule (or a portion of it) is determined.

Double covalent bond A covalent bond in which two atoms share two pairs of electrons.

Double-replacement reaction A reaction in which two substances exchange parts with one another and form two different substances.

Eicosanoid An oxygenated C_{20} fatty acid derivative that functions as a messenger lipid.

Electrolyte A substance whose aqueous solution conducts electricity.

Electron A subatomic particle that possesses a negative electrical charge.

Electron configuration A statement of how many electrons an atom has in each of its electron subshells.

Electron orbital A region of space within an electron subshell where an electron with a specific energy is most likely to be found.

Electron shell A region of space about a nucleus that contains electrons that have approximately the same energy and that spend most of their time approximately the same distance from the nucleus.

Electron subshell A region of space within an electron shell that contains electrons that have the same energy.

Electron transport chain (ETC) A series of biochemical reactions in which electrons and hydrogen ions from NADH and $FADH_2$ are passed to intermediate carriers and then ultimately react with molecular oxygen to produce water.

Electronegativity A measure of the relative attraction that an atom has for the shared electrons in a bond.

Electrophoresis The process of separating charged molecules on the basis of their migration toward charged electrodes associated with an electric field.

Electrostatic interaction An attraction or repulsion that occurs between charged particles.

Element A pure substance that cannot be broken down into simpler pure substances by ordinary chemical means such as a reaction, an electric current, heat, or a beam of light; a pure substance in which all atoms present have the same atomic number.

Elimination reaction A reaction in which two groups or two atoms on neighboring carbon atoms are removed, or eliminated, from a molecule, leaving a multiple bond between the two carbon atoms.

Emulsifier A substance that can disperse and stabilize water-insoluble substances as colloidal particles in an aqueous solution.

Enantiomers Stereoisomers whose molecules are nonsuper-imposable mirror images of each other.

Endothermic change of state A change of state in which heat energy is absorbed.

Endothermic reaction A reaction in which a continuous input of energy is needed for the reaction to occur.

Enzyme An organic compound that acts as a catalyst for a biochemical reaction.

Enzyme activity A measure of the rate at which an enzyme converts substrate to products in a biochemical reaction.

Enzyme inhibitor A substance that slows or stops the normal catalytic function of an enzyme by binding to it.

Enzyme–substrate complex   The intermediate reaction species that is formed when a substrate binds to the active site of an enzyme.

Epimers   Diastereomers whose molecules differ only in the configuration at one chiral center.

Equation coefficient   A number that is placed to the left of the chemical formula of a substance in a chemical equation and that changes the amount, but not the identity, of the substance.

Equilibrium   A condition in which two opposite processes take place at the same rate.

Equilibrium constant   A numerical value that characterizes the relationship between the concentrations of reactants and products in a system at chemical equilibrium.

Equilibrium position   A qualitative indication of the relative amounts of reactants and products present at equilibrium for a chemical reaction.

Essential amino acid   An amino acid needed in the human body that must be obtained from dietary sources because it cannot be synthesized within the body, from other substances, in adequate amounts.

Essential fatty acid   A fatty acid needed in the human body that must be obtained from dietary sources because it cannot be synthesized within the body, from other substances, in adequate amounts.

Ester   A carboxylic acid derivative in which the —OH portion of the carboxyl group has been replaced with an —OR group.

Esterification reaction   The reaction of a carboxylic acid with an alcohol (or phenol) to produce an ester.

Ethanol fermentation   The enzymatic anaerobic conversion of pyruvate to ethanol and carbon dioxide.

Ether   An organic molecule in which an oxygen atom is bonded to two carbon atoms by single bonds.

Eukaryotic cell   A cell in which the DNA is found in a membrane-enclosed nucleus.

Evaporation   The process in which molecules escape from the liquid phase to the gas phase.

Exact number   A number whose value has no uncertainty associated with it.

Exon   A gene segment that conveys (codes for) genetic information.

Exothermic change of state   A change of state in which heat energy is given off.

Exothermic reaction   A reaction in which energy is released as the reaction occurs.

Expanded structural formula   A structural formula that shows all atoms in a molecule and all bonds connecting the atoms.

Facilitated transport   The transport process in which a substance moves across a cell membrane, with the aid of membrane proteins, from a region of higher concentration to a region of lower concentration without the expenditure of cellular energy.

Fat   A triacylglycerol mixture that is a solid or semi-solid at room temperature.

Fatty acid   A naturally occurring monocarboxylic acid.

Fatty acid micelle   A micelle in which fatty acids and/or monoacylglycerols and some bile are present.

Fatty acid spiral   The metabolic pathway that degrades fatty acids to acetyl CoA by removing two carbon atoms at a time, with $FADH_2$ and NADH also being produced.

Feedback control   A process in which activation or inhibition of the first reaction in a reaction sequence is controlled by a product of the reaction sequence.

Fermentation   A biochemical process by which NADH is oxidized to NAD^+ without the need for oxygen.

Fibrous protein   A protein in which polypeptide chains are arranged in long strands or sheets.

Fischer projection   A two-dimensional structural notation for showing the spatial arrangement of groups about chiral centers in molecules.

Formula mass   The sum of the atomic masses of all the atoms represented in the chemical formula of a substance.

Formula unit   The smallest whole-number repeating ratio of ions present in an ionic compound that results in charge neutrality.

Free radical   An atom, molecule, or ion that contains an unpaired electron.

Functional group   The part of an organic molecule where most of its chemical reactions occur.

Fused-ring aromatic hydrocarbon   An aromatic hydrocarbon whose structure contains two or more rings fused together.

Gamma ray   A form of high-energy radiation without mass or charge that is emitted by radioactive nuclides.

Gas   The physical state characterized by an indefinite shape and an indefinite volume; the physical state characterized by a complete dominance of kinetic energy (disruptive forces) over potential energy (cohesive forces).

Gas law   A generalization that describes in mathematical terms the relationships among the amount, pressure, temperature, and volume of a gas.

Gene   A segment of a DNA strand that contains the base sequence for the production of a specific hnRNA/mRNA molecule.

Genetic code   The assignment of the 64 mRNA codons to specific amino acids (or stop signals).

Genome   All of the genetic information in the chromosomes of an organism.

Globular protein   A protein is which polypeptide chains are folded into spherical or globular shapes.

Glucogenic amino acid   An amino acid that has a carbon-containing degradation product that can be used to produce glucose via gluconeogenesis.

Gluconeogenesis   The metabolic pathway by which glucose is synthesized from noncarbohydrate materials.

Glycerophospholipid   A lipid that contains two fatty acids and a phosphate group esterified to a glycerol molecule and an alcohol esterified to the phosphate group.

Glycogenesis   The metabolic pathway by which glycogen is synthesized from glucose.

Glycogenolysis   The metabolic pathway by which free glucose units are obtained from glycogen.

Glycol   A diol in which the two —OH groups are on adjacent carbon atoms.

Glycolysis   The metabolic pathway by which glucose (a C_6 molecule) is converted into two molecules of pyruvate (a C_3 molecule).

Glycoprotein   A conjugated protein that contains carbohydrates or carbohydrate derivatives in addition to amino acids.

Glycoside   An acetal formed from a cyclic monosaccharide by replacement of the hemiacetal carbon atom —OH group with an —OR group.

Glycosidic linkage   The bond in a disaccharide resulting from the reaction between the hemiacetal carbon atom —OH group of one monosaccharide and a —OH group on the other monosaccharide.

Gram   The base unit of mass in the metric system.

Group   A vertical column of elements in the periodic table.

Half-life (t$_{1/2}$)   The time required for one-half of a given quantity of a radioactive substance to undergo decay.

Halogenated alkane   An alkane derivative in which one or more halogen atoms are present.

Halogenated cycloalkane   A cycloalkane derivative in which one or more halogen atoms are present.

Halogenation reaction   A substitution or addition reaction between a substance and a halogen (F_2, Cl_2, Br_2, or I_2) in which one or more halogen atoms are incorporated into molecules of an organic compound.

Haworth projection   A two-dimensional structural notation that specifies the three-dimensional structure of a cyclic form of a monosaccharide.

Hemiacetal   An organic compound in which a carbon atom is bonded to both a hydroxyl group (—OH) and an alkoxy group (—OR).

Heteroatomic molecule   A molecule in which two or more kinds of atoms are present.

Heterocyclic amine   An organic compound in which nitrogen atoms of amino groups are part of either an aromatic or a nonaromatic ring system.

Heterocyclic organic compound   A cyclic organic compound in which one or more of the carbon atoms in the ring have been replaced with atoms of other elements.

Heterogeneous mixture   A mixture that contains visibly different phases (parts), each of which has different properties.

Heterogeneous nuclear RNA (hnRNA)   RNA formed directly by DNA transcription and from which messenger RNA is formed.

Heteropolysaccharide   A polysaccharide in which more than one type of monosaccharide monomer (usually two) is present.

High-energy compound   A compound that has a greater free energy of hydrolysis than that of a typical compound.

Holoenzyme   The biochemically active conjugated enzyme produced from an apoenzyme and a cofactor.

Homoatomic molecule   A molecule in which all atoms present are of the same kind.

Homogeneous mixture   A mixture that contains only one visibly distinct phase (part), which has uniform properties throughout.

Homopolysaccharide   A polysaccharide in which only one type of monosaccharide monomer is present.

Hormone   A biochemical substance, produced by a ductless gland, that has a messenger function.

Hydration reaction   An addition reaction in which H_2O is incorporated into molecules of an organic compound.

Hydrocarbon   A compound that contains only carbon atoms and hydrogen atoms.

Hydrocarbon derivative   A compound that contains carbon and hydrogen and one or more additional elements.

Hydrogen bond   An extra strong dipole–dipole interaction between a hydrogen atom covalently bonded to a small, very electronegative element (F, O, or N) and a lone pair of electrons on another small, very electronegative element (F, O, or N).

Hydrogenation reaction   An addition reaction in which H_2 is incorporated into molecules of an organic compound.

Hydrohalogenation reaction   An addition reaction in which a hydrogen halide (HCl, HBr, or HI) is incorporated into molecules of an organic compound.

Hydrolase   An enzyme that catalyzes hydrolysis reactions in which the addition of a water molecule to a bond causes the bond to break.

Hydrolysis reaction   The reaction of a salt with water to produce hydronium ion or hydroxide ion or both; the reaction of a compound with H_2O, in which the compound splits into two or more fragments as the elements of water (H— and —OH) are added to the compound.

Hydroxyl group   The —OH functional group.

Hypertonic solution   A solution with a higher osmotic pressure than that found within cells.

Hypotonic solution   A solution with a lower osmotic pressure than that found within cells.

Ideal gas law   A general mathematical expression relating pressure, temperature, volume, and amount for a gas.

Immunoglobulin   A glycoprotein produced by an organism as a protective response to the invasion of microorganisms or foreign molecules.

Inexact number   A number whose value has a degree of uncertainty associated with it.

Inner transition element   An element located in the f area of the periodic table.

Inorganic chemistry   The study of all substances other than hydrocarbons and their derivatives.

Intermolecular force   An attractive force that acts between a molecule and another molecule.

Intron   A gene segment that does not convey (code for) genetic information.

Ion   An atom (or group of atoms) that is electrically charged as a result of loss or gain of electrons.

Ion pair   The electron and positive ion that are produced during an interaction between an atom or a molecule and ionizing radiation.

Ion product constant for water   The numerical value 1.0×10^{-14}, obtained by multiplying together the molar concentrations of H_3O^+ ion and OH^- ion present in pure water at 24°C.

Ionic bond   A chemical bond formed through the transfer of one or more electrons from one atom or group of atoms to another atom or group of atoms.

Ionic compound   A compound in which ionic bonds are present.

Ionization   The process in which individual positive and negative ions are produced from a molecular compound that is dissolved in solution.

Ionizing radiation   Radiation with sufficient energy to remove an electron from an atom or a molecule.

Irreversible enzyme inhibitor   A molecule that inactivates enzymes by forming a strong covalent bond to an amino acid side-chain group at the enzyme's active site.

Isoelectric point   The pH at which the concentration of the zwitterion is at a maximum in an amino acid solution.

Isoelectronic species   An atom and an ion, or two ions, that have the same number and configuration of electrons.

Isoenzymes   Isomeric forms of the same enzyme with slightly different amino acid sequences.

Isomerase   An enzyme that catalyzes the rearrangement of functional groups within a molecule, converting the molecule into another molecule isomeric with it.

Isotonic solution   A solution with an osmotic pressure that is equal to that within cells.

Isotopes   Atoms of an element that have the same number of protons and same number of electrons but different numbers of neutrons.

Ketogenesis   The metabolic pathway by which ketone bodies are synthesized from acetyl CoA.

Ketogenic amino acid   An amino acid that has a carbon-containing degradation product that can be used to produce ketone bodies.

Ketone   An organic compound in which a carbonyl carbon atom is bonded to two carbon groups.

Ketone body   One of the three substances (acetoacetate, beta-hydroxybutyrate, and acetone) produced from acetyl CoA when an excess of acetyl CoA from fatty acid degradation accumulates because of triacylglycerol–carbohydrate metabolic imbalances.

Ketose   A monosaccharide that contains a ketone functional group.

Kinetic energy   Energy that matter possesses because of particle motion.

Kinetic molecular theory of matter   A set of five statements that are used to explain the physical behavior of the three states of matter (solids, liquids, and gases).

Lactate fermentation   The enzymatic anaerobic reduction of pyruvate to lactate.

Le Châtelier's principle   If a stress (change of conditions) is applied to a system at equilibrium, the system will readjust (change equilibrium position) in the direction that best reduces the stress imposed on the system.

Leukotriene   A messenger lipid that is a C_{20}-fatty-acid derivative that contains three conjugated double bonds.

Levorotatory compound   A chiral compound that rotates the plane of polarized light in a counterclockwise direction.

Lewis structure   A combination of Lewis symbols that represents either the transfer or the sharing of electrons in chemical bonds.

Lewis symbol   The chemical symbol of an element surrounded by dots equal in number to the number of valence electrons present in atoms of the element.

Ligase   An enzyme that catalyzes the bonding together of two molecules into one, with the participation of ATP.

Line-angle drawing   An abbreviated structural formula in which an angle represents a carbon atom and a line represents a bond.

Lipid   An organic compound found in living organisms that is insoluble (or only sparingly soluble) in water but soluble in nonpolar organic solvents.

Lipid bilayer   A two-layer-thick structure of phospholipids and sphingoglycolipids in which the nonpolar tails of the lipids are in the middle of the structure and the polar heads are on the outside surfaces of the structure.

Lipogenesis   The metabolic pathway by which fatty acids are synthesized from acetyl CoA.

Lipoprotein   A conjugated protein that contains lipids in addition to amino acids.

Liquid   The physical state characterized by an indefinite shape and a definite volume; the physical state characterized by potential energy (cohesive forces) and kinetic energy (disruptive forces) of about the same magnitude.

Liter   The base unit of volume in the metric system.

London force   A weak temporary intermolecular force that occurs between an atom or molecule (polar or nonpolar) and another atom or molecule (polar or nonpolar).

Lyase   An enzyme that catalyzes the addition of a group to a double bond or the removal of a group from a double bond in a manner that does not involve hydrolysis or oxidation.

Lysosome   An organelle that contains hydrolytic enzymes needed for cellular rebuilding, repair, and degradation.

Markovnikov's rule   When an unsymmetrical molecule of the form HQ adds to an unsymmetrical alkene, the hydrogen atom from the HQ becomes attached to the unsaturated carbon atom that already has the most hydrogen atoms.

Mass   A measure of the total quantity of matter in an object.

Mass number   The sum of the number of protons and the number of neutrons in the nucleus of an atom.

Mass-volume percent   The mass of solute in a solution (in grams) divided by the total volume of solution (in milliliters), multiplied by 100.

Matter   Anything that has mass and occupies space.

Measurement   The determination of the dimensions, capacity, quantity, or extent of something.

Messenger RNA (mRNA)   RNA that carries instructions for protein synthesis (genetic information) to sites for protein synthesis.

Metabolic pathway   A series of consecutive biochemical reactions used to convert a starting material into an end product.

Metabolism   The sum total of all the biochemical reactions that take place in a living organism.

Metal   An element that has the characteristic properties of luster, thermal conductivity, electrical conductivity, and malleability.

Meter   The base unit of length in the metric system.

Micelle   A spherical cluster of molecules in which the polar portions of the molecules are on the surface and the nonpolar portions are located in the interior.

Mineral wax   A mixture of long-chain alkanes obtained from the processing of petroleum.

Mirror image   The reflection of an object in a mirror.

Mitochondrion   An organelle that is responsible for the generation of most of the energy for a cell.

Mixed triacylglycerol   A triester formed from the esterification of glycerol with more than one kind of fatty acid molecule.

Mixture   A physical combination of two or more pure substances in which each substance retains its own chemical identity.

Molar mass   The mass, in grams, of a substance that is numerically equal to the substance's formula mass.

Molarity   The moles of solute in a solution divided by the liters of solution.

Mole 6.02×10^{23} objects; the amount of a substance that contains as many elementary particles (atoms, molecules, or formula units) as there are atoms in exactly 12 grams of ^{12}C.

Molecular compound   A compound in which atoms are joined via covalent bonds.

Molecular geometry   The three-dimensional arrangement of atoms within a molecule.

Molecular polarity   A measure of the degree of inequality in the attraction of bonding electrons to various locations within a molecule.

Molecule   A group of two or more atoms that functions as a unit because the atoms are tightly bound together.

Monatomic ion   An ion formed from a single atom through loss or gain of electrons.

Monocarboxylic acid A carboxylic acid in which one carboxyl group is present.

Monomer The small molecule that is the structural repeating unit in a polymer.

Monoprotic acid An acid that supplies one proton (H^+ ion) per molecule in an acid–base reaction.

Monosaccharide A carbohydrate that contains a single polyhydroxy aldehyde or polyhydroxy ketone unit.

Monounsaturated fatty acid A fatty acid with a carbon chain in which one carbon–carbon double bond is present.

Mucopolysaccharide A polysaccharide that occurs in connective tissue associated with joints in humans and animals.

Mutagen A substance or agent that causes a change in the structure of a gene.

Mutation An error in base sequence in a gene that is reproduced during DNA replication.

Network polymer A polymer in which monomers are connected in a three-dimensional cross-linked network.

Neurotransmitter A chemical substance that is released at the end of a nerve, travels across the synaptic gap between the nerve and another nerve, and then bonds to a receptor site on the other nerve, triggering a nerve impulse.

Neutral solution An aqueous solution in which the concentrations of H_3O^+ ion and OH^- ion are equal; an aqueous solution whose pH is 7.0.

Neutralization reaction A reaction between an acid and a hydroxide base in which a salt and water are the products.

Neutron A subatomic particle that has no charge associated with it.

Nitrogen balance The state that results when the amount of nitrogen taken into the human body as protein equals the amount of nitrogen excreted from the body in waste products.

Noble-gas element An element located in the far right column of the periodic table.

Nonaqueous solution A solution in which a substance other than water is the solvent.

Nonbonding electrons Pairs of valence electrons on an atom that are not involved in electron sharing.

Noncompetitive enzyme inhibitor A molecule that decreases enzyme activity by binding to a site on an enzyme other than the active site.

Nonelectrolyte A substance whose aqueous solution does not conduct electricity.

Nonionizing radiation Radiation with sufficient energy to excite an electron in an atom or a molecule, but not enough energy to remove the electron from the atom or the molecule.

Nonmetal An element characterized by the absence of the properties of luster, thermal conductivity, electrical conductivity, and malleability.

Nonoxidation–reduction reaction A reaction in which there is no transfer of electrons from one reactant to another reactant.

Nonpolar amino acid An amino acid that contains one amino group, one carboxyl group, and a nonpolar side chain.

Nonpolar covalent bond A covalent bond in which there is equal sharing of electrons between atoms.

Nonpolar molecule A molecule in which there is a symmetrical distribution of electron charge.

Nonsuperimposable mirror images Mirror images that do not coincide at all points when the images are laid upon each other.

Normal boiling point The temperature at which a liquid boils when under a pressure of 760 mm Hg.

Nuclear equation An equation in which the chemical symbols present represent atomic nuclei rather than atoms.

Nuclear fission A nuclear reaction in which a large nucleus (high atomic number) splits into two medium-sized nuclei, with the release of several free neutrons and a large amount of energy.

Nuclear fusion A nuclear reaction in which two small nuclei are put together to make a larger one.

Nuclear reaction A reaction in which the nucleus of an atom undergoes change.

Nucleic acid A polymer in which the monomer units are nucleotides.

Nucleon Any subatomic particle found in the nucleus of an atom.

Nucleotide A three-subunit molecule in which a pentose sugar is bonded to both a phosphate group and a nitrogen-containing heterocylic base.

Nucleus The very small, dense, positively charged center of an atom.

Nuclide An atom of an element with a specific number of protons and neutrons in its nucleus.

Octet rule In compound formation, atoms of elements lose, gain, or share electrons in such a way that their electron configurations contain eight valence electrons.

Oil A triacylglycerol mixture that is a liquid at room temperature.

Oligosaccharide A carbohydrate that contains two to ten monosaccharide units covalently bonded to each other.

Omega-3 fatty acid An unsaturated fatty acid with its endmost double bond three carbon atoms away from its methyl end.

Omega-6 fatty acid An unsaturated fatty acid with its endmost double bond six carbon atoms away from its methyl end.

Optically active compound A compound that rotates the plane of polarized light.

Optimum pH The pH at which an enzyme exhibits maximum activity.

Optimum temperature The temperature at which an enzyme exhibits maximum activity.

Orbital diagram A statement of how many electrons an atom has in each of its electron orbitals.

Organelle A minute structure within the cytoplasm of a cell that carries out a specific cellular function.

Organic chemistry The study of hydrocarbons and their derivatives.

Osmolarity The product of a solution's molarity and the number of particles produced per formula unit if the solute dissociates.

Osmosis The passage of solvent through a semipermeable membrane separating a dilute solution (or pure solvent) from a more concentrated solution.

Osmotic pressure The pressure that must be applied to prevent the net flow of solvent through a semipermeable membrane from a solution of lower concentration to a solution of higher concentration.

Oxidation The process whereby a reactant in a chemical reaction loses one or more electrons.

Oxidation number A number that represents the charge that an atom appears to have when the electrons in each bond it is participating in are assigned to the more electronegative of the two atoms involved in the bond.

Oxidation–reduction reaction A reaction in which there is a transfer of electrons from one reactant to another reactant.

Oxidative deamination reaction   A biochemical reaction in which an alpha-amino acid is converted into an alpha-keto acid with release of an ammonium ion.

Oxidative phosphorylation   The biochemical process by which ATP is synthesized from ADP as a result of the transfer of electrons and hydrogen ions from NADH and $FADH_2$ to O_2 through the electron carriers of the electron transport chain.

Oxidizing agent   A reactant that causes oxidation of another reactant by accepting electrons from it.

Oxidoreductase   An enzyme that catalyzes oxidation–reduction reactions.

Parent nuclide   The nuclide that undergoes decay in a radioactive decay process.

Partial pressure   The pressure that a gas in a mixture of gases would exert if it were present alone under the same conditions.

Passive transport   The transport process in which a substance moves across a cell membrane by diffusion from a region of higher concentration to a region of lower concentration without the expenditure of cellular energy.

Pentose phosphate pathway   The metabolic pathway by which glucose is used to produce NADPH, ribose 5-phosphate (a pentose), and numerous other sugar phosphates.

Peptide   A sequence of amino acids in which the amino acids are joined together through peptide (amide) bonds.

Peptide bond   A covalent bond between the carboxyl group of one amino acid and the amino group of another amino acid.

Percent by mass   The mass of solute in a solution divided by the total mass of solution, multiplied by 100.

Percent by volume   The volume of solute in a solution divided by the total volume of solution, multiplied by 100.

Period   A horizontal row of elements in the periodic table.

Periodic law   When elements are arranged in order of increasing atomic number, elements with similar properties occur at periodic (regularly recurring) intervals.

Periodic table   A graphical display of the elements, arranged in order of increasing atomic number, in which elements with similar properties fall in the same column of the display.

pH   The negative logarithm of an aqueous solution's molar hydronium ion concentration.

pH scale   A scale of small numbers used to specify molar hydrogen ion concentrations in aqueous solution.

Phenol   An organic compound in which an —OH group is attached to a carbon atom that is part of an aromatic ring system.

Pheromone   A compound used by insects (and some animals) to transmit a message to other members of the same species.

Phosphate ester   An organic compound formed from the reaction of an alcohol with phosphoric acid.

Phospholipid   A lipid that contains one or more fatty acids, a phosphate group, a platform molecule to which the fatty acids and phosphate group are attached, and an alcohol that is attached to the phosphate group.

Physical change   A process in which a substance changes its physical appearance but not its chemical composition.

Physical property   A characteristic of a substance that can be observed without changing the basic identity of the substance.

Polar acidic amino acid   An amino acid that contains one amino group and two carboxyl groups, the second carboxyl group being part of the side chain.

Polar basic amino acid   An amino acid that contains two amino groups and one carboxyl group, the second amino group being part of the side chain.

Polar covalent bond   A covalent bond in which there is unequal sharing of electrons between two atoms.

Polar molecule   A molecule in which there is an unsymmetrical distribution of electron charge.

Polar neutral amino acid   An amino acid that contains one amino group, one carboxyl group, and a side chain that is polar but neutral.

Polyamide   A condensation polymer in which the monomers are joined through amide linkages.

Polyatomic ion   An ion formed from a group of atoms (held together by covalent bonds) through loss or gain of electrons.

Polyester   A condensation polymer in which the monomers are joined through ester linkages.

Polymer   A large molecule formed by the repetitive bonding together of many smaller molecules.

Polymerase chain reaction (PCR)   A method for rapidly reproducing multiple copies of a DNA nucleotide sequence.

Polymerization reaction   A reaction in which the repetitious combining of many small molecules (monomers) produces a very large molecule (the polymer).

Polypeptide   A long chain of amino acids, each joined to the next by a peptide bond.

Polyprotic acid   An acid that supplies two or more protons (H^+ ions) per molecule in an acid–base reaction.

Polysaccharide   A polymeric carbohydrate that contains many monosaccharide units covalently bonded to each other.

Polysome   A complex of mRNA and several ribosomes.

Polyunsaturated fatty acid   A fatty acid with a carbon chain in which two or more carbon–carbon double bonds are present.

Potential energy   Stored energy that matter possesses as a result of its position, condition, and/or chemical composition.

Pressure   The force applied per unit area on an object; the force on a surface divided by the area of that surface.

Primary alcohol   An alcohol in which the hydroxyl-bearing carbon atom is bonded to only one other carbon atom.

Primary amide   An amide in which two hydrogen atoms are bonded to the amide nitrogen atom.

Primary amine   An amine in which the nitrogen atom is bonded to one hydrocarbon group and two hydrogen atoms.

Primary carbon atom   A carbon atom in an organic molecule that is bonded to only one other carbon atom.

Primary nucleic acid structure   The order in which nucleotides are linked together in a nucleic acid.

Primary protein structure   The order in which amino acids are linked together in a protein.

Property   A distinguishing characteristic of a substance that is used in its identification and description.

Prostaglandin   A messenger lipid that is a C_{20}-fatty-acid derivative that contains a cyclopentane ring and oxygen-containing functional groups.

Prosthetic group   A non-amino-acid group permanently associated with a protein.

Protein   A biochemical polymer in which the monomer units are amino acids; a polypeptide in which at least 50 amino acid residues are present.

Protein denaturation   The partial or complete disorganization of a protein's characteristic three-dimensional shape as a result of disruption of its secondary, tertiary, and quaternary structural interactions.

Protein turnover   The repetitive process in which proteins are degraded and resynthesized within the human body.

Proteolytic enzyme   An enzyme that catalyzes the breaking of peptide bonds that maintain the primary structure of a protein.

Proton   A subatomic particle that possesses a positive electrical charge.

Pure substance   A single kind of matter that cannot be separated into other kinds of matter by any physical means.

Quaternary ammonium salt   An ammonium salt in which all four groups attached to the nitrogen atom of the ammonium ion are hydrocarbon groups.

Quaternary carbon atom   A carbon atom in an organic molecule that is bonded to four other carbon atoms.

Quaternary protein structure   The organization among the various polypeptide chains in an oligomeric protein.

Radioactive decay   The process whereby a radionuclide is transformed into a nuclide of another element as a result of the emission of radiation from its nucleus.

Radioactive decay series   A series of radioactive decay processes beginning with a long-lived radionuclide and ending with a stable nuclide of lower atomic number.

Radioactive nuclide   A nuclide with an unstable nucleus from which radiation is spontaneously emitted.

Radioactivity   The radiation spontaneously emitted from an unstable nucleus.

Reaction rate   The rate at which reactants are consumed or products produced, in a given time period, in a chemical reaction.

Recombinant DNA   DNA that contains genetic material from two different organisms that has been combined into one DNA molecule.

Reducing agent   A reactant that causes reduction of another reactant by providing electrons for the other reactant to accept.

Reducing sugar   A carbohydrate that gives a positive test with Tollens and Benedict's solutions.

Reduction   The process whereby a reactant in a chemical reaction gains one or more electrons.

Representative element   An element located in the *s* area or the first five columns of the *p* area of the periodic table.

Restriction enzyme   An enzyme that recognizes specific base sequences in DNA and cleaves the DNA in a predictable manner at these sequences.

Reversible reaction   A reaction in which the conversion of reactants to products (the forward reaction) and the conversion of products to reactants (the reverse reaction) occur simultaneously.

Ribonucleic acid (RNA)   A nucleotide polymer in which each of the monomers contains ribose, phosphate, and one of the heterocyclic bases adenine, cytosine, guanine, or thymine.

Ribosomal RNA (rRNA)   RNA that combines with specific proteins to form ribosomes, the physical sites for protein synthesis.

Ribosome   A rRNA-protein complex that serves as the site for the translation phase of protein synthesis.

Rounding off   The process of deleting unwanted (nonsignificant) digits from calculated numbers.

Salt   An ionic compound containing a metal or polyatomic ion as the positive ion and a nonmetal or polyatomic ion (except hydroxide ion) as the negative ion.

Saponification reaction   The hydrolysis of an organic compound, under basic conditions, in which a carboxylic acid salt is one of the products.

Saturated fatty acid   A fatty acid with a carbon chain in which all carbon–carbon bonds are single bonds.

Saturated hydrocarbon   A hydrocarbon in which all carbon–carbon bonds are single bonds.

Saturated solution   A solution that contains the maximum amount of solute that can be dissolved under the conditions at which the solution exists.

Scientific notation   A system in which a decimal number is expressed as the product of a number between 1 and 10 and 10 raised to a power.

Secondary alcohol   An alcohol in which the hydroxyl-bearing carbon atom is bonded to two other carbon atoms.

Secondary amide   An amide in which an alkyl (or aryl) group and a hydrogen atom are bonded to the amide nitrogen atom.

Secondary amine   An amine in which the nitrogen atom is bonded to two hydrocarbon groups and one hydrogen atom.

Secondary carbon atom   A carbon atom in an organic molecule that is bonded to two other carbon atoms.

Secondary protein structure   The arrangement in space adopted by the backbone portion of a protein.

Semipermeable membrane   A membrane that allows certain types of molecules to pass through it but prohibits the passage of other types of molecules.

Significant figures   The digits in a measurement that are known with certainty plus one digit that has uncertainty.

Simple carbohydrate   A dietary monosaccharide or a dietary disaccharide.

Simple enzyme   An enzyme composed only of protein (amino acid residues).

Simple protein   A protein that consists solely of amino acid residues.

Simple triacylglycerol   A triester formed from the esterification of glycerol with three identical fatty acid molecules.

Single covalent bond   A covalent bond in which two atoms share one pair of electrons.

Single-replacement reaction   A reaction in which an atom or molecule replaces an atom or group of atoms in a compound.

Small nuclear ribonucleoprotein particle (snRNP)   A complex formed from an snRNA molecule and several proteins.

Small nuclear RNA (snRNA)   RNA that facilitates the conversion of heterogeneous nuclear RNA to messenger RNA.

Solid   The physical state characterized by a definite shape and a definite volume; the physical state characterized by a dominance of potential energy (cohesive forces) over kinetic energy (disruptive forces).

Solubility   The maximum amount of solute that will dissolve in a given amount of solvent under a particular set of conditions.

Solute   A component of a solution that is present in a lesser amount relative to that of the solvent.

Solution   A homogeneous mixture of two or more substances in which each substance retains its own chemical identity.

Solvent   The component of a solution that is present in the greatest amount.

Specific heat   The quantity of heat energy, in calories, necessary to raise the temperature of 1 gram of a substance by 1 degree Celsius.

Sphingoglycolipid   A lipid that contains both a fatty acid and a carbohydrate component attached to a sphingosine molecule.

Sphingophospholipid   A lipid that contains one fatty acid and one phosphate group attached to a sphingosine molecule and an alcohol attached to the phosphate group.

Spliceosome   A large assembly of snRNA molecules and proteins involved in the conversion of hnRNA molecules to mRNA molecules.

Splicing   The process of joining the exons of an hnRNA molecule together to form an mRNA molecule.

Stable nuclide   An atom whose nucleus does not easily undergo change.

Standard amino acid   One of the 20 alpha-amino acids normally found in proteins.

Stereoisomers   Isomers whose atoms are connected in the same way but that differ in the orientation of these atoms in space.

Steroid   A lipid whose structure is based on a fused-ring system that involves three 6-membered rings and one 5-membered ring.

Steroid hormone   A hormone that is a cholesterol derivative.

Strong acid   An acid that transfers 100%, or very nearly 100%, of its protons (H^+ ions) to water when in an aqueous solution.

Strong electrolyte   A substance that completely (or almost completely) ionizes/dissociates into ions in aqueous solution.

Structural formula   A two-dimensional structural representation that shows how the various atoms in a molecule are bonded to each other.

Structural isomers   Compounds with the same molecular formula but different structural formulas—that is, different bonding arrangements between atoms.

Subatomic particle   A very small particle that is a building block for atoms.

Substituent   An atom or group of atoms attached to a chain (or ring) of carbon atoms.

Substituted ammonium ion   An ammonium ion in which one or more alkyl, cycloalkyl, or aryl groups have been substituted for hydrogen atoms.

Substitution reaction   A reaction in which part of a small reacting molecule replaces an atom or a group of atoms on a hydrocarbon or hydrocarbon derivative.

Substrate   The reactant in an enzyme-catalyzed reaction.

Substrate-level phosphorylation   The biochemical process by which a high-energy phosphate group from an intermediate compound (substrate) is directly transferred to ADP to produce ATP.

Sugar   A general designation for either a monosaccharide or a disaccharide.

Sulfhydryl group   An —SH functional group.

Superimposable mirror images   Mirror images that coincide at all points when the images are laid upon each other.

Supersaturated solution   An unstable solution that temporarily contains more dissolved solute than that present in a saturated solution.

Symmetrical addition reaction   An addition reaction in which identical atoms (or groups of atoms) are added to each carbon of a carbon–carbon multiple bond.

Terpene   An organic compound whose carbon skeleton is composed of two or more 5-carbon isoprene structural units.

Tertiary alcohol   An alcohol in which the hydroxyl-bearing carbon atom is bonded to three other carbon atoms.

Tertiary amide   An amide in which two alkyl (or aryl) groups and no hydrogen atoms are bonded to the amide nitrogen atom.

Tertiary amine   An amine in which the nitrogen atom is bonded to three hydrocarbon groups and no hydrogen atoms.

Tertiary carbon atom   A carbon atom in an organic molecule that is bonded to three other carbon atoms.

Tertiary protein structure   The overall three-dimensional shape of a protein that results from the interactions between amino acid side chains (R groups) that are widely separated from each other within the peptide chain.

Thermal expansion   A measure of the change in volume of a sample of matter resulting from a temperature change.

Thioester   A sulfur-containing analog of an ester in which an —SR group has replaced the —OR group.

Thioether   An organic compound in which a sulfur atom is bonded to two carbon atoms by single bonds.

Thiol   An organic compound in which a sulfhydryl group is present.

Thromboxane   A messenger lipid that is a C_{20}-fatty-acid derivative that contains a cyclic ether ring and oxygen-containing functional groups.

***Trans* isomer**   A disubstituted cycloalkane isomer in which substituents on different carbon atoms are on opposite sides of the carbon ring; an alkene isomer in which two similar (or identical) double-bond substituents are on opposite sides of the double bond.

Transamination reaction   A biochemical reaction that involves the interchange of the amino group of an alpha-amino acid with the keto group of an alpha-keto acid.

Transcription   The process by which DNA directs the synthesis of mRNA molecules that carry the coded information needed for protein synthesis.

Transfer RNA (tRNA)   RNA that delivers amino acids to the sites of protein synthesis.

Transferase   An enzyme that catalyzes the transfer of a functional group from one molecule to another.

Transformation   The process of incorporating recombinant DNA into a host cell.

Transition element   An element located in the *d* area of the periodic table.

Translation   The process by which mRNA codons are deciphered and a particular protein is synthesized.

Translocation   The part of translation in which a ribosome moves down an mRNA molecule three base positions (one codon) so that a new codon can occupy the ribosomal A site.

Transmutation reaction   A nuclear reaction in which a nuclide of one element is changed into a nuclide of another element.

Triacylglycerol   A lipid formed by esterification of three fatty acids to a glycerol molecule.

Triacylglycerol mobilization   The hydrolysis of triacylglycerols stored in adipose tissue, followed by the release into the bloodstream of the fatty acids and glycerol so produced.

Triatomic molecule   A molecule that contains three atoms.

Triple covalent bond   A covalent bond in which two atoms share three pairs of electrons.

Triprotic acid   An acid that supplies three protons (H^+ ions) per molecule in an acid–base reaction.

Turnover number   The number of substrate molecules transformed per minute by one molecule of enzyme under optimum conditions of temperature, pH, and saturation.

Unsaturated hydrocarbon   A hydrocarbon in which one or more carbon–carbon multiple bonds (double bonds, triple bonds, or both) are present.

Unsaturated solution   A solution that contains less than the maximum amount of solute that can be dissolved under the conditions at which the solution exists.

Unstable nuclide   An atom whose nucleus spontaneously undergoes change.

Unsymmetrical addition reaction   An addition reaction in which different atoms (or groups of atoms) are added to the carbon atoms of a carbon–carbon multiple bond.

Urea cycle   The series of biochemical reactions in which urea is produced from ammonium ions and carbon dioxide.

Vaccine   A preparation containing an inactive or weakened form of a virus or bacterium.

Valence electron   An electron in the outermost electron shell of a representative or noble-gas element.

Vapor   A gas that exists at a temperature and pressure at which it would ordinarily be thought of as a liquid or a solid.

Vapor pressure   The pressure exerted by a vapor above a liquid when the liquid and vapor are in equilibrium with each other.

Virus   A small particle that contains DNA or RNA (but not both) surrounded by a coat of protein and that cannot reproduce without the aid of a host cell.

Vitamin   An organic compound that is essential in small amounts for the proper functioning of the human body and that must be obtained from dietary sources because the body cannot synthesize it.

Volatile substance   A substance that readily evaporates at room temperature because of a high vapor pressure.

VSEPR electron group   A collection of valence electrons present in a localized region about the central atom in a molecule.

VSEPR theory   A set of procedures for predicting the three-dimensional geometry of a molecule using the information contained in the molecule's Lewis structure.

Wax   A pliable, water-repelling substance used particularly in protecting surfaces and producing polished surfaces.

Weak acid   An acid that transfers only a small percentage of its protons (H^+ ions) to water in an aqueous solution.

Weak electrolyte   A substance that incompletely ionizes/dissociates into ions in aqueous solution.

Weight   A measure of the force exerted on an object by the pull of gravity.

Zaitsev's rule   The major product in an intramolecular alcohol dehydration reaction is the alkene that has the greatest number of alkyl groups attached to the carbon atoms of the double bond.

Zwitterion   A molecule that has a positive charge on one atom and a negative charge on another atom but that has no net charge.

Zymogen   The inactive precursor of a proteolytic enzyme.

Index

(Entries in bold face refer to pages where terms are defined.)

Photo Credits

Photographs from the Edgar Fahs Smith Collection, Van Pelt-Dietrich Library, University of Pennsylvania: Pages 47, 80, 130 (bottom), 155, 157, 225, 340, 506

Credits for all other photographs are listed below, with page numbers in boldface.

vii © Jeff Hunter/Getty Images; **viii** © Arnuls Husmo/Getty Images; **x** © F.H. Kolwicz/Visuals Unlimited; **1** © Richard Hamilton Smith/CORBIS; **2 (top)** © Phil Degginger/Color-Pic; **3 (top)** © Ed Simpson/Getty Images; **(bottom)** © Andy Levin/Photo Researchers; **4** © Phil Degginger/Color-Pic; **5** © Andy Sacks; **6 (two photos)** James Scherer; **9 (emerald)** National Museum of Natural History © 2002 Smithsonian Institution; **13** Image supplied by Digital Instruments, Santa Barbara, California; **21** © Richard Hamilton Smith/CORBIS OUTLINE; **22** © David Frazier/Photo Researchers; **23 (left)** © E.R. Degginger/Color-Pic; **38** Permission from Jeffrey L. Gage (photographer) and Michael Pollock, Ph.D., Center for Exercise Science, University of Florida; **41 (left)** © David Olsen/Getty Images; **(right)** Grant Heilman/Grant Heilman Photography; **64** © William S. Helsel/Getty Images; **77** M.S. Davidson/Photo Researchers; **87** © E.R. Degginger/Color-Pic; **88 (four photos)** © E.R. Degginger/Color-Pic; **89** © Blair Seitz/Photo Researchers; **101** Kennedy Space Center/NASA; **116** Archive Photos/Getty Images; **127** © Jeff Hunter/Getty Images; **143** Georg Gerster/Photo Researchers; **145** Courtesy of Chrysler Corporation; **149** © Steven Fuller/Peter Arnold, Inc.; **150** Betty Weiser/Photo Researchers; **153** Phil Degginger/Color-Pic; **159** Courtesy Manchester Literary and Philosophical Society; **163 (two photos)** James Scherer; **166** Brian Bailey/Network Aspen; **177** Steve Allen/Peter Arnold, Inc.; **178 (bottom)** © Coco McCoy/Rainbow; **182** David Woodfall/Getty Images; **196** John Mead/Photo Researchers; **197 (left and center)** David M. Phillips/Visuals Unlimited; **(right)** Stanley Flegler/Visuals Unlimited; **205** Science Photo Library/Photo Researchers; **206** James Scherer; **207** James Scherer; **211** James Scherer; **217 (top left)** Vince Streano/Getty Images; **(top right)** Cecile Brunswick/Peter Arnold, Inc.; **(bottom left)** S.C. Fried/Photo Researchers; **(bottom right)** Myrleen Ferguson/PhotoEdit; **219** Mark Gibson; **221** © Tom McHugh/Photo Researchers; **234** Bois/Yvette Tavernier/Peter Arnold, Inc.; **236** Ken O'Donoghue; **238** Ken O'Donoghue; **256 (two photos)** Ken O'Donoghue; **267** © PhotoDisc, Inc.; **268** © Bettmann/CORBIS; **274** © Bettmann/CORBIS; **275** © Myrleen Ferguson/PhotoEdit; **281** © Doug Plummer/Photo Researchers; **282** © Charles D. Winters/Photo Researchers; **284** © Simon Fraser/Science Photo Library/Custom Medical Stock Photo; **285 (top)** © Yoav Levy/Phototake; **(bottom)** © Dr. Mony De Leon/Peter Arnold, Inc.; **287** © Bettmann/CORBIS; **288 (top)** Albert J. Copley/Visuals Unlimited; **(bottom)** NASA; **294** Arnuls Husmo/Getty Images; **298** © Natalie Fobes/Getty Images; **312** Richard Megna/Fundamental Photos; **313** Daryl Solomon/Envision; **314** © Michael Newman/PhotoEdit; **315** Phil Degginger/Color-Pic; **327** Junebug Clark/Photo Researchers; **332** © Alan & Linda Detrick/Photo Researchers; **335** © Donald C. Booth/Color-Pic; **337** Martha Cooper/Peter Arnold, Inc.; **338** © PhotoDisc, Inc.; **343** James Scherer; **345** © Bill Stanton/Rainbow; **361** © Ed Reschke/Peter Arnold, Inc.; **365** © Gary Gold/Getty Images; **368 (top)** © Hank Morgan/Rainbow; **(bottom)** © James Cotier/Getty Images; **377** Bob Daemmrich/Stock Boston/Picturequest; **381** © Henryk T. Kaiser/Envision; **382** © Michael Viard/Peter Arnold, Inc.; **384** © SIU/Peter Arnold, Inc.; **388** Jeff Lapore/Photo Researchers; **397** © PhotoDisc, Inc.; **403** © Harvey Lloyd/Peter Arnold, Inc.; **404 (top)** © Norbert Wu; **(bottom)** © Steven Needham/Envision; **406** © S. McCutcheon/Visuals Unlimited; **417** © Coco McCoy/Rainbow; **426** Hans Pfletschinger/Peter Arnold, Inc.; **430** AP/Wide World Photos; **431** © George Mattei/Envision; **432** © Charles D. Winters/Photo Researchers; **433** © F. Stewart Westmoreland/Photo Researchers; **438** © Sherman Thomson/Visuals Unlimited; **445** Animals Animals © Fred Whitehead; **462** © J.F. Causse/Getty Images; **477** © Nardin/Jacana/Photo Researchers; **477** © Scott Camazine/Photo Researchers; **488** © Peter Skinner/Photo Researchers; **489** © Dan McCoy/Rainbow; **499** © Norris Blake/Visuals Unlimited; **500** © PictureNet/CORBIS; **515** © Tom Raymond/Medichrome; **519** © Hulton-Deutsch Collection/CORBIS; **523** © Larry Mulvehill/Rainbow; **528** © Erica Stone/Peter Arnold, Inc.; **529** © Paul Skelcher/Rainbow; **532** © Steven Needham/Envision; **534** © Dr. Dennis Kunkel/Phototake NYC; **534** © Don Kreuter/Rainbow; **535** © J.M. Barey/Vandystadt/Photo

Common Functional Groups

Name of class	Structural feature
Alkane	$-\overset{\displaystyle |}{\underset{\displaystyle |}{C}}-$
Alkene	$\overset{\diagup}{\underset{\diagdown}{C}}=\overset{\diagdown}{\underset{\diagup}{C}}$
Alkyne	$-C\equiv C-$
Aromatic hydrocarbon	(benzene ring) or (benzene ring)
Alcohol	$-\overset{\displaystyle |}{\underset{\displaystyle |}{C}}-OH$
Phenol	(benzene ring)$-OH$
Ether	$-\overset{\displaystyle |}{\underset{\displaystyle |}{C}}-O-\overset{\displaystyle |}{\underset{\displaystyle |}{C}}-$
Thiol	$-\overset{\displaystyle |}{\underset{\displaystyle |}{C}}-SH$
Aldehyde	$-\overset{\displaystyle O}{\overset{\displaystyle \|}{C}}-H\ (-CHO)$
Ketone	$-\overset{\displaystyle |}{\underset{\displaystyle |}{C}}-\overset{\displaystyle O}{\overset{\displaystyle \|}{C}}-\overset{\displaystyle |}{\underset{\displaystyle |}{C}}-$
Carboxylic acid	$-\overset{\displaystyle O}{\overset{\displaystyle \|}{C}}-OH\ (-COOH\ \text{or}\ -CO_2H)$
Ester	$-\overset{\displaystyle O}{\overset{\displaystyle \|}{C}}-O-\overset{\displaystyle |}{\underset{\displaystyle |}{C}}-\ (-COOR\ \text{or}\ -CO_2R)$
Amine	$-\overset{\displaystyle |}{\underset{\displaystyle |}{C}}-NH_2$
Amide	$-\overset{\displaystyle O}{\overset{\displaystyle \|}{C}}-NH_2$